GW00697115

Numerical Methods for Passive Microwave and Millimeter Wave Structures

OTHER IEEE PRESS BOOKS

Analog MOS Integrated Circuits, II, *Edited by P. R. Gray, B. A. Wooley, and R. W. Brodersen*
Electrostatic Discharge and Electronic Equipment, *By W. Boxleitner*
Instrumentation and Techniques for Radio Astronomy, *Edited by P. F. Goldsmith*
Network Interconnection and Protocol Conversion, *Edited by P. E. Green, Jr.*
VLSI Signal Processing, III, *Edited by R. W. Brodersen and H. S. Moscovitz*
Microcomputer-Based Expert Systems, *Edited by A. Gupta and B. E. Prasad*
Principles of Expert Systems, *Edited by A. Gupta and B. E. Prasad*
High Voltage Integrated Circuits, *Edited by B. J. Baliga*
Microwave Digital Radio, *Edited by L. J. Greenstein and M. Shafi*
Oliver Heaviside: Sage in Solitude, *By P. J. Nahin*
Radar Applications, *Edited by M. I. Skolnik*
Principles of Computerized Tomographic Imaging, *By A. C. Kak and M. Slaney*
Selected Papers on Noise in Circuits and Systems, *Edited by M. S. Gupta*
Spaceborne Radar Remote Sensing: Applications and Techniques, *By C. Elachi*
Engineering Excellence, *Edited by D. Christiansen*
Selected Papers on Logic Synthesis for Integrated Circuit Design, *Edited by A. R. Newton*
Planar Transmission Line Structures, *Edited by T. Itoh*
Introduction to the Theory of Random Signals and Noise, *By W. B. Davenport, Jr. and W. L. Root*
Teaching Engineering, *Edited by M. S. Gupta*
Selected Papers on Computer-Aided Design of Very Large Scale Integrated Circuits, *Edited by A. L. Sangiovanni-Vincentelli*
Robust Control, *Edited by P. Dorato*
Writing Reports to Get Results: Guidelines for the Computer Age, *By R. S. Blicq*
Multi-Microprocessors, *Edited by A. Gupta*
Advanced Microprocessors, II, *Edited by A. Gupta*
Adaptive Signal Processing, *Edited by L. H. Sibul*
Selected Papers on Statistical Design of Integrated Circuits, *Edited by A. J. Strojwas*
System Design for Human Interaction, *Edited by A. P. Sage*
Microcomputer Control of Power Electronics and Drives, *Edited by B. K. Bose*
Selected Papers on Analog Fault Diagnosis, *Edited by R. Liu*
Advances in Local Area Networks, *Edited by K. Kümmerle, J. O. Limb, and F. A. Tobagi*
Load Management, *Edited by S. Talukdar and C. W. Gellings*
Computers and Manufacturing Productivity, *Edited by R. K. Jurgen*
Selected Papers on Computer-Aided Design of Analog Networks, *Edited by J. Vlach and K. Singhal*
Being the Boss, *By L. K. Lineback*
Effective Meetings for Busy People, *By W. T. Carnes*
Selected Papers on Integrated Analog Filters, *Edited by G. C. Temes*
Electrical Engineering: The Second Century Begins, *Edited by H. Freitag*
VLSI Signal Processing, II, *Edited by S. Y. Kung, R. E. Owen, and J. G. Nash*
Modern Acoustical Imaging, *Edited by H. Lee and G. Wade*
Low-Temperature Electronics, *Edited by R. K. Kirschman*
Undersea Lightwave Communications, *Edited by P. K. Runge and P. R. Trischitta*
Multidimensional Digital Signal Processing, *Edited by the IEEE Multidimensional Signal Processing Committee*
Adaptive Methods for Control System Design, *Edited by M. M. Gupta*
Residue Number System Arithmetic, *Edited by M. A. Soderstrand, W. K. Jenkins, G. A. Jullien, and F. J. Taylor*
Singular Perturbations in Systems and Control, *Edited by P. V. Kokotovic and H. K. Khalil*
Getting the Picture, *By S. B. Weinstein*
Space Science and Applications, *Edited by J. H. McElroy*
Medical Applications of Microwave Imaging, *Edited by L. Larsen and J. H. Jacobi*
Modern Spectrum Analysis, II, *Edited by S. B. Kesler*
The Calculus Tutoring Book, *By C. Ash and R. Ash*
Imaging Technology, *Edited by H. Lee and G. Wade*
Phase-Locked Loops, *Edited by W. C. Lindsey and C. M. Chie*
VLSI Circuit Layout: Theory and Design, *Edited by T. C. Hu and E. S. Kuh*
Monolithic Microwave Integrated Circuits, *Edited by R. A. Pucel*

A complete listing of IEEE PRESS books is available upon request.

Numerical Methods for Passive Microwave and Millimeter Wave Structures

Edited by
Roberto Sorrentino
Professor
Tor Vergata University of Rome

A volume in the IEEE PRESS Selected Reprint Series, prepared under the sponsorship of the IEEE Microwave Theory and Techniques Society.

The Institute of Electrical and Electronics Engineers, Inc., New York

PRINTED IN THE UNITED STATES OF AMERICA

IEEE Order Number: PC0240-2

Library of Congress Cataloging-in-Publication Data

Numerical methods for passive microwave and millimeter wave structures / edited by Roberto Sorrentino ; prepared under the sponsorship of the IEEE Microwave Theory and Techniques Society.
p. cm.—(IEEE Press selected reprint series)
Includes indexes.
ISBN 0-87942-249-1
1. Microwave devices—Mathematics. 2. Numerical calculations.
I. Sorrentino, Roberto. II. IEEE Microwave Theory and Techniques Society.
TK7876.N85 1989
621.381'3—dc19 88-32077
CIP

Contents

Preface

A NUMERICAL method is a mathematical technique to convert a *continuum* into a discrete set which is amenable to numerical processing. In this manner, the infinite degrees of freedom of an unknown analytical solution are reduced to a finite number of unknowns which can be evaluated by a digital computer. Numerical methods are the only means by which to attack and solve with the required degree of accuracy complicated boundary value problems associated with modern microwave structures.

In recent years, microwave and millimeter wave integrated circuits have experienced impressive advances. The technology of monolithic circuits has led to the development of numerous circuits for applications in the millimeter wave range. Once they have been fabricated it is virtually impossible to adjust the performances of the monolithic circuits. The only possibility is the analytical adjustment of the parameters of the circuits. The availability of efficient and accurate computer-aided design (CAD) tools is essential to the design of such circuits.

The effectiveness of CAD is largely based on the analysis procedures implemented in the computer programs. The development of an efficient and accurate method of analysis is the key element for the design of microwave and millimeter wave monolithic circuits.

Parallel to the increasing complexity of these circuits, a tremendous evolution of computing resources has occurred. Today, very small and compact personal computers (PCs) provide computing power equivalent to that of the largest computers of a few years ago. On the other hand, new design capabilities have been made possible by the development of supercomputers. The advent of extremely powerful and easily accessible computers has opened new possibilities in the area of millimeter wave circuits and, in general, in applied electromagnetics and has made it possible to attack a variety of problems that were considered to be inaccessible till a few years ago.

This has had a notable impact on computer techniques and numerical methods being implemented in CAD programs. Old methods seldom used because of the excessive computing effort required have become feasible and found new possibilities of application. Others have been improved in efficiency or in ranges of applicability. Methods from other disciplines have been applied in the area of electromagnetic waves.

As an example, the use of integral equations was limited to a great extent to merely theoretical investigations of existence and uniqueness of solutions. The advent of high-speed computers made it possible to readily achieve numerical solutions. A strong interest in numerical methods and CAD techniques is therefore growing again among researchers, working engineers, and students; most have easy and immediate access to computing facilities powerful enough to analyze rather complicated electromagnetic structures, design elaborate microwave circuits, etc.

It has appeared desirable to collect in one book the most significant contributions in the area of numerical methods and computer techniques in order to provide working engineers with background information on the most representative and useful numerical approaches to the solution of microwave and millimeter wave structures.

It is hoped that this book will serve also as a reference for students and beginners, since necessary information is scattered throughout the literature.

Owing to the great variety of numerical methods, this book cannot provide extensive information about each numerical method along with a detailed explanation of its variations, improvements, etc. The aim is mainly to introduce the reader to those numerical techniques which have a broad enough field of application and to provide him with a sufficient understanding to apply the methods at least to the solution of some simple problems. The material provided here is expected to be sufficient for a general understanding of the method and its possible applications and limitations. For further details the reader is referred to the reference lists of the reprint papers as well as to the additional reference lists provided in the introductions to the parts. Overview papers of Part 1 and tutorial papers of the specific parts are an extremely useful starting point for further study. Other excellent sources of information are two recent special issues of the *IEEE Transactions on Microwave Theory and Techniques* devoted to numerical methods (October 1985) and computer-aided design (February 1988).

In editing this book an attempt was made to present the most popular and effective numerical techniques including both historically important articles and the more recent developments.

Even within the relatively limited domain of the microwave circuit applications this book is concerned with, there is an incredible number of good and useful papers on numerical techniques. An extremely small fraction could be included here in order to keep the volume to a manageable size. The Editor is conscious of numerous serious omissions and apologizes for them. To partially compensate for a number of omissions, some additional papers are listed in the introductions to each part.

The book is divided into 11 parts. The first consists of overview papers and serves as an introduction and a general reference to numerical computation in the field of microwave and millimeter wave structures. Each subsequent part is devoted to a different numerical technique. Typically, each contains a tutorial paper or one or two historical papers that have contributed significantly to the development of that specific technique. Subsequent refinements, improvements, or extensions together with examples of applications are then illustrated by additional papers. Emphasis has been placed on problems involving modern planar and quasi-planar structures, but also more conventional structures have been considered. Very recent contributions are included to provide the reader with an up-to-date reference.

Each part starts with a brief introduction. This has the purpose, particularly when no overview paper is included, of setting in a historical perspective not only the reprint papers but also some cited papers which could not be included for reasons of space.

Each method has its own advantages and disadvantages, but it must be emphasized that more than one may be suitable for the solution of a given problem. An initial selection can be made on the basis of the structure or the problem to be solved. Choice must then be made in terms of several aspects, namely, accuracy, computing efficiency, memory requirements, flexibility, and versatility. One of the purposes of the introductory part is to provide critical information about the various numerical approaches, which is crucial for making a well-reasoned choice of the method best suited for the problem at hand.

A number of similarities can be recognized among the various numerical techniques. Actually, it was found in many instances that two apparently different approaches led to the same final result. In an effort to develop in a systematic manner one specific method, it is natural to try to generalize the formulation so as to include other methods in the same general framework. The result of such a generalization process is that it is often difficult to state whether one method includes another or vice versa. Any systematic classification of numerical methods in general is outside the scope of this book. The classification adopted, though perhaps questionable in some instances, reflects, on the one hand, the usual viewpoint adopted among the microwave community, and, on the other, the personal views of the Editor.

A typical feature of a numerical method is the amount of analytical work required for it to be implemented. The less the analytical preprocessing, the easier the implementation of the method. This aspect is somewhat related to other characteristics of the method, such as efficiency and versatility. Roughly speaking, computer efficiency increases with the amount of analytical effort required, but, at the same time, the versatility of the method is generally reduced. Analytical forms in fact can be developed only for a limited number of simple shapes or under some simplifying hypotheses. On the contrary, the most general structures can only be attacked by essentially numerical methods with no restrictions on the type of geometry.

These aspects are summarized in Table I, where the numerical methods treated in this book are compared in terms of storage requirements, computer time, versatility, and analytical preprocessing. Needless to say, these are just rough indications with no absolute validity. They mainly reflect the personal biased experience of the Editor. The method of moments is not quoted in the table; due to the extreme variety of formulations which can be adopted in connection with this general approach, no classification has been attempted. For similar reasons, the planar-circuit approach, which is a mathematical model rather than a numerical method, is classified only by its generality, other aspects being dependent on the specific method adopted within the general framework of the planar approach.

TABLE I

	Storage	CPU	Versatility	Preprocessing
Finite-difference method	L	L	+ +	0
Transmission-line matrix method	L	L	+ +	+
Finite-element method	L	M/L	+ + +	+
Boundary-element method	M	M	+ +	+
Method of lines	M	S/M	+ +	+ +
Mode-matching	M	S/M	+	+ +
Transverse resonance method	M	S/M	+	+ +
Spectral-domain approach	S	S	0	+ + +
Planar-circuit approach			+	

Note: L = large, M = medium, S = small.

Numerical methods are presented, roughly, in the order of increasing analytical complexity, starting with the more numerically oriented methods. Parts 2 to 5 are devoted to the least analytical but most versatile methods (finite-difference, transmission-line matrix, finite-element, and boundary-element methods, respectively). The method of lines, which combines a two-dimensional finite-difference procedure with an analytical formulation in the third direction, is presented in Part 6. The method of moments, which represents a wide variety of numerical techniques and is somewhat related to many other methods, is presented later, in Part 7. Techniques requiring a progressively increasing amount of analytical preprocessing are included in Parts 8–10 (mode-matching and field-matching techniques, transverse resonance techniques, and spectral-domain approach, respectively). Circuit formalisms for the analysis of planar structures and related numerical methods (planar waveguide model, planar-circuit approach, segmentation and desegmentation methods) are included in the final part.

ACKNOWLEDGMENTS

The Editor is highly indebted to Professor T. Itoh for his strong encouragement. The structure of this book has been greatly influenced by some of his work; however, all deficiencies are the responsibility of the Editor. Professor A. A. Oliner is also gratefully acknowledged for very useful comments and suggestions.

Part 1
Tutorial and Overview Papers

THIS part is a general introduction to the numerical methods used to attack boundary value problems inherent to passive microwave and millimeter wave components. The first reprint paper (Paper 1.1) is a classic paper by Wexler. It refers to strictly numerical methods to treat algebraic as well as analytical problems that may be encountered in the solution of both electromagnetic and physical problems, and reviews the state of the art of numerical methods at the end of the sixties. Apart from this historical significance, it is particularly useful for the beginner or the nonspecialist, as it introduces many basic concepts and methodologies that must be familiar to the research and working engineer such as the numerical solution of a system of linear equations, finite-difference differentiation and integration, the variational approach, the Rayleigh-Ritz method, the concept of natural boundary conditions, and the finite-element method.

Not reprinted here, but worth referring to, is the tutorial paper by Silvester and Csendes [1.1] which appeared in 1974. This paper focuses more on the modeling of microwave passive components than on numerical methods themselves. The various transmission-line media and discontinuities are reviewed and numerical techniques for their modeling are discussed. Though analytical details are avoided, this paper provides a clear general picture of the numerical modeling of passive components.

The two other review papers reprinted in this part are recent papers referring to the solutions for integrated circuit structures, which represent the most popular transmission-line configurations nowadays. A detailed treatment of integrated structures on anisotropic substrates is presented in Paper 1.2 by Alexópoulos, but the interest of this paper is clearly not restricted to anisotropic materials. Quasi-static and dynamic approaches, along with the methods to obtain numerical solutions for uniform printed circuit lines, are discussed.

An overview of numerical methods for miniaturized passive components is given in Paper 1.3 by Itoh. The structure of this reprint book essentially reflects the general outline of this paper. It therefore serves as a key to a better understanding of the various parts. This paper also describes other methods which, although significant, did not merit a separate part in this book due to their restricted application.

Space limitation prevents the inclusion of other excellent review papers. Among them are a classic paper by Davies [1.2] which discusses the numerical methods for arbitrarily shaped hollow waveguides as of the beginning of the seventies and a recent review by Saad [1.3] of numerical methods for the analysis of microwave and optical waveguides with arbitrary cross section, including inhomogeneously and anisotropically filled guides.

References

[1.1] P. Silvester and Z. J. Csendes, "Numerical modeling of passive microwave devices," *IEEE Trans. Microwave Theory Tech.*, vol. MTT-22, pp. 190–201, Mar. 1974.

[1.2] J. B. Davies, "Review of methods for numerical solution of the hollow-waveguide problem," *Proc. IEE*, vol. 119, pp. 33–37, Jan. 1972.

[1.3] S. M. Saad, "Review of numerical methods for the analysis of arbitrarily-shaped microwave and optical dielectric waveguides," *IEEE Trans. Microwave Theory Tech.*, vol. MTT-33, pp. 894–899, Oct. 1985.

Computation of Electromagnetic Fields

ALVIN WEXLER, MEMBER, IEEE

Invited Paper

Abstract—**This paper reviews some of the more useful, current and newly developing methods for the solution of electromagnetic fields. It begins with an introduction to numerical methods in general, including specific references to the mathematical tools required for field analysis, e.g., solution of systems of simultaneous linear equations by direct and iterative means, the matrix eigenvalue problem, finite difference differentiation and integration, error estimates, and common types of boundary conditions. This is followed by a description of finite difference solution of boundary and initial value problems. The paper reviews the mathematical principles behind variational methods, from the Hilbert space point of view, for both eigenvalue and deterministic problems. The significance of natural boundary conditions is pointed out. The Rayleigh-Ritz approach for determining the minimizing sequence is explained, followed by a brief description of the finite element method. The paper concludes with an introduction to the techniques and importance of hybrid computation.**

I. Introduction

WHENEVER ONE devises a mathematical expression to solve a field quantity, one must be concerned with numerical analysis. Other than those engineers involved exclusively in measurements or in the proof of general existence theorems or in administration, the remainder of us are numerical analysts to a degree. Whenever we prescribe a sequence of mathematical operations, we are designing an *algorithm*. A theory is to us a means of extrapolating our experiences in order to make predictions. A mathematical theory, of a physical problem, produces numbers and a good theory produces accurate numbers easily.

To obtain these numbers, we must employ processes that produce the required accuracy in a finite number of steps performed in a finite time upon machines having finite word length and store. Accepting these constraints we ask how best to perform under them.

As far as engineers are concerned, there is no principal difference between *analytical* and *numerical* approaches. Both are concerned with numbers. Oftentimes, the so-called "analytical" approach—as though to claim inefficiency as a virtue—is none other than the algorithmically impotent one.

The contrast between efficient and inefficient computation is emphasized whenever anything beyond the most trivial work is performed. As an example, consider the solution of a set of simultaneous, linear equations by Cramer's rule in which each unknown is found as a ratio of two determinants. A moment's reflection will show that calculating these determinants by evaluating all their cofactors is an insurmountable task when the order is not small. Expanding along each row, n cofactors, each of order $n-1$, are involved. Each of these has cofactors of order $n-2$ and so on. Continuing in this way, we find that at least $n!$ multiplications are needed to expand the determinant. Neglecting all other operations, and assuming that the computer does one million multiplications per second, we find that it would take more years than the universe is old to expand a determinant of order 24. Even then, due to roundoff errors, the reliability of the result will be questionable. Cramer's rule then is surely not a useful algorithm except for the smallest systems. It is certainly of the greatest importance in general proofs and existence theorems but is virtually useless for getting numerical answers.

Wilkinson, in a statement referring to the algebraic eigenvalue problem [25, Preface] said "... the problem has a deceptively simple formulation and the background theory has been known for many years; yet the determination of accurate solutions presents a wide variety of challenging problems." He would likely agree that the statement applies equally to most of numerical analysis.

This paper is concerned with the solution of fields by efficient, current and newly developing, computer practices. It is intended as an introduction for those totally unfamiliar with numerical analysis, as well as for those with some experience in the field. For the sake of consistency, many topics had to be deleted including numerical conformal mapping, point matching, mode matching and many others. Proponents of these techniques should not feel slighted as the author has himself been involved in one of them. In addition, it was felt advisable to concentrate on general numerical methods relevant to the microwave engineer, rather than to review particular problems. Companion papers, in this special issue, furnish many specific examples.

The principal methods surveyed are those of finite differences, variational, and hybrid computer techniques. Appropriate mathematical concepts and notations are introduced as an aid to further study of the literature. Some of the references, found to be the most useful for the author, are classified and listed in the bibliography.

II. Systems of Linear Equations

The manipulation and solution of large sets of linear, simultaneous equations is basic to most of the techniques employed in computer solution of field problems. These linear equations are the consequence of certain approximations made to expedite the solution. For example, approximation of the operator (∇^2, often) over a finite set of points

Manuscript received February 27, 1969; revised May 7, 1969. This work was carried out with financial assistance from Atomic Energy of Canada Ltd. and the National Research Council of Canada.

The author is with the Numerical Applications Group, Electrical Engineering Department, University of Manitoba, Winnipeg, Man., Canada.

Reprinted from *IEEE Trans. Microwave Theory Tech.*, vol. MTT-17, no. 8, pp. 416–439, Aug. 1969.

at which the field is to be computed (as in finite differences), or approximation of the field estimate (as in variational methods) causes the problem to be modeled by simultaneous equations. These equations are conveniently represented by matrices. It happens, and we shall see why in succeeding sections, that the finite difference approach creates huge, sparse matrices of order 5000 to 20 000. *Sparseness* means that there are very few nonzero elements, perhaps a tenth of one percent. On the other hand, variational methods tend to produce matrices that are small (typically of order 50), but *dense* by comparison.

The size and density of these matrices is of prime importance in determining how they should be solved. Direct methods requiring storage of all matrix elements are necessarily limited to about 150 unknowns in a reasonably large modern machine. On the other hand, it turns out that iterative techniques often converge swiftly for sparse matrices. In addition, if the values and locations of nonzero elements are easily computed (as occurs with finite differences) there is no point in storing the large sparse matrix in its entirety. In such cases, only the solution vector need be stored along with five nonzero elements (the number produced by many finite difference schemes) at any stage of the iterative process. This maneuver allows systems having up to 20 000 unknowns to be solved.

This section deals with examples of direct and iterative techniques for the solution of systems of simultaneous linear equations. This is followed by a method for solving the eigenvalue problem in which the entire matrix must be stored but all eigenvectors and eigenvalues are found at once. Solution of the eigenvalue problem, by an iterative technique, is reserved for Section IV where it is more appropriate.

A. Solution by Direct Methods

One of the most popular algorithms for the solution of systems of linear equations is known as the method of Gauss. It consists of two parts—*elimination* (or *triangularization*) and *back substitution.* The procedure is, in principle, very simple and is best illustrated by means of an example.

We wish to solve the system

$$Ax = b \tag{1}$$

where A is an $n \times n$ matrix, x and b are n-element vectors (column matrices). b is known and x is to be computed. Consider, for convenience, a small set of equations in three unknowns x_1, x_2, and x_3.

$$\begin{aligned} x_1 + 4x_2 + x_3 &= 7 \\ x_1 + 6x_2 - x_3 &= 13 \\ 2x_1 - x_2 + 2x_3 &= 5. \end{aligned}$$

This example is from [9, pp. 187–188]. Subtract the first equation from the second of the set. Then subtract twice the first from the last. This results in

$$\begin{aligned} x_1 + 4x_2 + x_3 &= 7 \\ 2x_2 - 2x_3 &= 6 \\ -9x_2 + 0x_3 &= -9. \end{aligned}$$

The first equation, which was used to clear away the first column, is known as the *pivot equation.* The coefficient of the equation, which was responsible for this clearing operation, is termed the *pivot.* Now, the first term of the altered second equation is made the pivot. Adding 4.5 times the second equation to the last, we obtain

$$\begin{aligned} x_1 + 4x_2 + x_3 &= 7 \\ 2x_2 - 2x_3 &= 6 \\ -9x_3 &= 18. \end{aligned}$$

Clearly, in the computer there is no need to store the equations as in the preceding sets. In practice, only the numerical values are stored in matrix form with one vector reserved for the as yet unknown variables. When the above procedure is completed, the matrix is said to have been triangularized.

The final step, back substitution, is now initiated. Using the last equation of the last set, it is a simple matter to solve for $x_3 = -2$. Substituting the numerical value of x_3 into the second equation, $x_2 = 1$ is then easily found. Finally, from the first equation, we get $x_1 = 5$. And so on back through all the equations in the set, each unknown is computed, one at a time, whatever the number of equations involved.

It turns out that only $n^3/3$ multiplications are required to solve a system of n real, linear equations. This is a reasonable figure and should be compared with the phenomenal amount of work involved in applying Cramer's rule directly. In addition, the method is easy to program.

This simplified account glosses over certain difficulties that may occasionally occur. If one of the pivots is zero, it is impossible to employ it in clearing out a column. The cure is to rearrange the sequence of equations such that the pivot is nonzero. One can appreciate that even if the pivot is nonzero, but is very small by comparison with other numbers in its column, numerical problems will cause error in the solution. This is due to the fact that most numbers cannot be held exactly in a limited word-length store. The previous example, using integers only, is exceptional. Typically, depending on the make of machine, computers can hold numbers with 7 to 11 significant digits in *single precision.*

To illustrate the sort of error that can occur, consider the example, from [15, p. 34]. Imagine that our computer can hold only three significant digits in *floating point* form. It does this by storing the three significant digits along with an exponent of the base 10. The equations

$$\begin{aligned} 1.00 \times 10^{-4}x_1 + 1.00x_2 &= 1.00 \\ 1.00x_1 + 1.00x_2 &= 2.00 \end{aligned}$$

are to be solved. Triangularizing, as before, while realizing the word-length limitation, we obtain

$$\begin{aligned} 1.00 \times 10^{-4}x_1 + 1.00x_2 &= 1.00 \\ -1.00 \times 10^{4}x_2 &= -1.00 \times 10^{4}. \end{aligned}$$

Solving for x_2 then back-substituting, the computed result is $x_2 = 1.00$ and $x_1 = 0.00$ which is quite incorrect. By the simple expedient of reversing the order of the equations, we find the result $x_2 = 1.00$ and $x_1 = 1.00$ which is correct to three significant digits. The rule then is, at any stage of the computation, to select the largest numerical value in the relevant column as the pivot and to interchange the two equations accordingly. This procedure is known as a *partial pivoting strategy. Complete pivoting strategy* involves the interchange of columns as well as rows in order to use the numerically largest number available in the remaining equations as the pivot. In practice, the results due to this further refinement do not appear to warrant the additional complications and computing time.

The previous example suffered from a pure scaling problem and is easily handled by a pivoting strategy. On the other hand, except for the eigenvalue problem, nothing can be done about solving a system of linear equations whose coefficient matrix is *singular*. In the two dimensional case, singularity (i.e. the vanishing of the coefficient matrix determinant) means that the two lines are parallel. Hence, no solution exists. A similar situation occurs with parallel planes or hyperplanes when the order is greater than two. A *well-conditioned* system describes hyperplanes that intersect at nearly 90°. The intersection point (or solution) is relatively insensitive to roundoff error as a consequence. If hyperplanes intersect at small angles, roundoff error causes appreciable motion of the intersection point with a low degree of trust in the solution. Such systems are termed *ill-conditioned*.

It is unnecessary, and wasteful, to compute the inverse A^{-1} in order to solve (1). Occasionally, however, a matrix must be inverted. This is reasonably easy to accomplish. One of the simplest methods is to triangularize A, as previously described, and then to solve (1) n times using a different right-hand side $\boldsymbol{b}$ each time. In the first case $\boldsymbol{b}$ should have 1 in its first element with zeros in the remainder. The 1 should then appear in the second location for the next back-substitution sequence, and so on. Each time, the solution gives one column of the inverted matrix and n back-substitutions produce the entire inverted matrix. See [12, pp. 32–33]. The procedure is not quite so simple when, as is usually advisable, a pivoting strategy is employed and when one wishes to triangularize A only once for all n back-substitutions. IBM supplies a program called MINV which does this, with a complete pivoting strategy, in an efficient fashion.

Westlake [24, pp. 106–107] reports results of inverting a matrix of order $n=50$ on CDC 6600 and CDC 1604A machines. Computing times were approximately 0.3 and 17 seconds, respectively. These figures are about double that predicted by accounting only for basic arithmetic operations. The increased time is due to other computer operations involved and is some function of how well the program was written. Tests run on the University of Manitoba IBM 360/65 showed that MINV inverts a matrix of order 50 in 3·38 seconds. The matrix elements were random numbers in the range 0 to 10.

Note that the determinant of a triangular matrix is simply the product of the diagonal terms.

The elimination and back-substitution method is easily adapted to the solution of simultaneous, linear, *complex* equations if the computer has a complex arithmetic facility, as most modern machines do. Failing this, the equations can be separated into real and imaginary parts and a program designed for real numbers can then be used. The latter procedure is undesirable as more store and computing time is required.

Other inversion algorithms exist, many of which depend upon particular characteristics of the matrix involved. In particular, when a matrix is symmetric, the symmetric Cholesky method [3, pp. 76–78 and 95–97] appears to be the speediest, perhaps twice as fast as Gauss's method. Under the above condition, we can write

$$A = L\tilde{L} \tag{2}$$

where the tilde denotes transposition. L is a lower triangular matrix, i.e., nonzero elements occur only along and below the diagonal as in

$$L = \begin{bmatrix} l_{11} & 0 & 0 \\ l_{21} & l_{22} & 0 \\ l_{31} & l_{32} & l_{33} \end{bmatrix}. \tag{3}$$

The elements of L may be easily determined by successively employing the following equations:

$$\begin{gathered} a_{11} = l_{11}^2,\ a_{12} = l_{11}l_{21},\ a_{13} = l_{11}l_{31},\ a_{22} = l_{21}^2 + l_{22}^2, \\ a_{23} = l_{21}l_{31} + l_{22}l_{32},\ a_{33} = l_{31}^2 + l_{32}^2 + l_{33}^2. \end{gathered} \tag{4}$$

Thus, the first equation of (4) gives l_{11} explicitly. Using l_{11}, the second equation supplies l_{21}, and so on. This operation is known as *triangular decomposition*.

The inverse of A is

$$A^{-1} = \tilde{L}^{-1}L^{-1} = (\widetilde{L^{-1}})L^{-1} \tag{5}$$

which requires the inverse of L and one matrix multiplication. The inverse of a triangular matrix is particularly easy to obtain by the sequences

$$\begin{gathered} l_{11}x_{11} = 1,\quad l_{21}x_{11} + l_{22}x_{21} = 0, \\ l_{31}x_{11} + l_{32}x_{21} + l_{33}x_{21} = 0,\quad l_{22}x_{22} = 1, \\ l_{32}x_{22} + l_{33}x_{32} = 0,\quad l_{33}x_{33} = 1. \end{gathered} \tag{6}$$

In this way, a symmetric matrix is most economically inverted.

From (4), it is clear that complex arithmetic may have to be performed. However, if in addition to being symmetric A is positive definite as well, we are assured [3, p. 78] that only real arithmetic is required. In addition, the algorithm is extremely stable.

B. Solution by Iterative Methods

Iterative methods offer an alternative to the direct methods of solution previously described. As a typical ith equation of the system (1), we have

$$\sum_{j=1}^{n} a_{ij}x_j = b_i. \tag{7}$$

Rearranging for the ith unknown

$$x_i = \frac{b_i}{a_{ii}} - \sum_{\substack{j=1 \\ j\neq i}}^{n} \frac{a_{ij}}{a_{ii}} x_j. \tag{8}$$

This gives one unknown in terms of the others. The intention is to be able to make a guess at all variables x_i and then successively correct them, one at a time, as indicated above. Different strategies are available. One can, for example, compute revised estimates of all x_i using the previously assumed values. Upon completion of the scan of all equations, the old values are then overwritten by the new ones.

Intuitively, it seems reasonable to use an updated estimate, just as soon as required, rather than to store it until the equation scan is completed. With the understanding that variables are immediately overwritten in computation, we have

$$x_i^{(m+1)} = -\sum_{j=1}^{i-1} \frac{a_{ij}}{a_{ii}} x_j^{(m+1)} - \sum_{j=i+1}^{n} \frac{a_{ij}}{a_{ii}} x_j^{(m)} + \frac{b_i}{a_{ii}} \tag{9}$$

with $1 \leq i \leq n$, $m \geq 0$. m denotes the iteration count. The two-part summation indicates that some variables are newly updated. With this method, only n storage locations need be available for the x_i (in comparison with $2n$ by the previous system), and convergence to the solution is more rapid [22, p. 71].

The difference between any two successive x_i values corresponds to a correction term to be applied in updating the current estimate. In order to speed convergence, it is possible to overcorrect at each stage by a factor ω. ω usually lies between 1 and 2 and is often altered between successive scans of the equation set in an attempt to maximize the convergence rate. The iteration form is

$$\begin{aligned} x_i^{(m+1)} &= x_i^{(m)} + \omega\{x_i^{(m+1)} - x_i^{(m)}\} \\ &= x_i^{(m)} + \omega\Bigg\{-\sum_{j=1}^{i-1} \frac{a_{ij}}{a_{ii}} x_j^{(m+1)} \\ &\qquad - \sum_{j=i+1}^{n} \frac{a_{ij}}{a_{ii}} x_j^{(m)} - x_i^{(m)} + \frac{b_i}{a_{ii}}\Bigg\}. \end{aligned} \tag{10}$$

It turns out that convergence to the solution is guaranteed if matrix A is symmetric and positive definite [34, pp. 237–238] and if $0 \leq \omega \leq 2$. If $\omega < 1$ we have *underrelaxation*, and *overrelaxation* if $\omega > 1$. This procedure is known as successive overrelaxation (SOR), there being no point in underrelaxing.

Surprisingly, (10) can be described as a simple matrix iterative procedure [22, pp. 58–59]. By expansion, it is easy to prove that

$$(D - \omega E)\mathbf{x}^{(m+1)} = \{(1-\omega)D + \omega F\}\mathbf{x}^{(m)} + \omega b \tag{11}$$

where D is a diagonal matrix consisting of the diagonal elements of A, E consists of the negative of all its elements beneath the diagonal with zeros elsewhere, and F is a matrix having the negative of those elements above the diagonal of A. Rearranging (11),

$$\begin{aligned} \mathbf{x}^{(m+1)} &= (D - \omega E)^{-1}\{(1-\omega)D + \omega F\}\mathbf{x}^{(m)} \\ &\quad + \omega(D - \omega E)^{-1}b. \end{aligned} \tag{12}$$

Define the matrix accompanying $\mathbf{x}^{(m)}$ as

$$\mathfrak{L}_\omega = (D - \omega E)^{-1}\{(1-\omega)D + \omega F\}. \tag{13}$$

$\mathfrak{L}_\omega$ is known as the SOR iteration matrix and plays an important role in convergence properties of the method.

The iterative procedure can be rewritten

$$\mathbf{x}^{(m+1)} = \mathfrak{L}_\omega x^{(m)} + \mathbf{c} \tag{14}$$

where $\mathbf{c}$ is the last term of (12). Of course, one does not literally set up $\mathfrak{L}_\omega$ in order to perform the SOR process, but $\mathfrak{L}_\omega$ and (14) express mathematically what happens when the algorithm described by (10) is implemented.

The process must be *stationary* when the solution is attained, i.e., $\mathbf{x}^{(m+1)} = \mathbf{x}^{(m)} = \mathbf{x}$. Therefore, we must have that

$$\mathbf{c} = (I - \mathfrak{L}_\omega)\mathbf{x}. \tag{15}$$

Substitute (15) into (14) and rearrange to obtain

$$\mathbf{x}^{(m+1)} - \mathbf{x} = \mathfrak{L}_\omega(\mathbf{x}^{(m)} - \mathbf{x}). \tag{16}$$

These terms are clearly error vectors as they consist of elements giving the difference between exact and computed values. At the mth iteration, the form is

$$\boldsymbol{\varepsilon}^{(m)} = \mathbf{x}^{(m)} - \mathbf{x}. \tag{17}$$

Therefore (16) becomes

$$\begin{aligned} \boldsymbol{\varepsilon}^{(m+1)} &= \mathfrak{L}_\omega \boldsymbol{\varepsilon}^{(m)} = \mathfrak{L}_\omega{}^2 \boldsymbol{\varepsilon}^{(m-1)} \\ &\cdots = \mathfrak{L}_\omega{}^{m+1} \boldsymbol{\varepsilon}^{(0)} \end{aligned} \tag{18}$$

due to the recursive definition of (16). It is therefore clear that the error tends to vanish when matrix $\mathfrak{L}_\omega$ tends to vanish with exponentiation. It can be proved [22, pp. 13–15] that any matrix A raised to a power r vanishes as $r \rightarrow \infty$ if and only if each eigenvalue of the matrix is of absolute value less than one. It is easy to demonstrate this for the special case of a real, nonsymmetric matrix with distinct eigenvalues. We cannot be sure that $\mathfrak{L}_\omega$ (which is not symmetric in general) has only distinct eigenvalues, but we will assume this for purposes of illustration. In this case all eigenvectors are linearly independent. Therefore, any arbitrary error vector $\boldsymbol{\varepsilon}$ may be expressed as a linear combination of eigenvectors I_i of $\mathfrak{L}_\omega$, i.e.,

$$\boldsymbol{\varepsilon} = a_1 I_1 + a_2 I_2 + \cdots + a_n I_n. \tag{19}$$

If the eigenvectors are known, the unknown a_i may be found by solving a system of simultaneous linear equations. As the I_i are linearly independent, the square matrix consisting of elements of all the I_i has an inverse, and so a solution must exist.

Performing the recursive operations defined by (18), we obtain

$$\boldsymbol{\varepsilon}^{(m)} = a_1\mu_1{}^m I_1 + a_2\mu_2{}^m I_2 + \cdots + a_n\mu_n{}^m I_n \tag{20}$$

where μ_i is the eigenvalue of $\mathfrak{L}_\omega$ corresponding to the eigenvector I_i. It is obvious, from (20), that if all eigenvalues of $\mathfrak{L}_\omega$ are numerically less than unity, SOR will converge to the solution of (1). In the literature, the magnitude of the largest eigenvalue is known as the *spectral radius* $\rho(\mathfrak{L}_\omega)$.

It is also easily shown that the displacement vector $\boldsymbol{\delta}$, which states the difference between two successive $\mathbf{x}$ estimates, can replace $\boldsymbol{\varepsilon}$ in (18)–(20).

A sufficient, although often overly stringent condition, for the convergence of SOR is that the coefficient matrix A display *diagonal dominance*. This occurs if, in each row of A, the sum of the absolute values of all off-diagonal terms is greater than the absolute value of the diagonal term itself.

In practice, we are concerned not only with whether or not convergence will occur, but also with its speed. Equation (20) indicates that the smaller the spectral radius of $\mathcal{L}_\omega$, the more rapidly convergence occurs. As indicated by the subscript, $\mathcal{L}_\omega$ and its eigenvalues are some function of the acceleration factor ω. The relationship between ω and the spectral radius $\rho(\mathcal{L}_\omega)$ is a very complicated one and so only very rarely can the optimum acceleration factor ω_{opt} be predicted in advance. However, some useful schemes exist for successively approximating ω_{opt} as computation proceeds (see [14], [15], [17]–[20]).

To conclude, the main advantage of SOR is that it is unnecessary to store zero elements of the square matrix as is required by many direct methods. This is of prime importance for the solution of difference equations which require only the vector to be stored. Also, the iteration procedure tends to be self-correcting and so roundoff errors are somewhat restricted. There are direct methods that economize on empty matrix elements, but SOR is perhaps the easiest way to accomplish this end.

C. The Matrix Eigenvalue Problem

The general matrix eigenvalue problem is of the form

$$(A - \lambda B)\mathbf{x} = 0 \tag{21}$$

where A and B are square matrices. This is the form that results from a variational solution (see Section V). The problem is to find eigenvalues λ and associated eigenvectors $\mathbf{x}$ such that (21) holds. This represents a system of linear, homogeneous equations and so a solution can exist only if determinant of $(A-\lambda B)$ vanishes. If A and B are known matrices, then a solution can exist only for values of λ which make the determinant vanish. Eigenvectors can be found that correspond to this set of eigenvalues. It is not a practical proposition to find the eigenvalues simply by assuming trial λ values and evaluating the determinant each time until it is found to vanish. Such a procedure is hopelessly inefficient.

An algorithm for matrices that are small enough to be held entirely in the fast store (perhaps $n=100$) is the Jacobi method for real, symmetric matrices. Although it is not a very efficient method, it is fairly easy to program and serves as an example of a class of methods relying upon symmetry and rotations to secure the solution of the eigensystem.

If $B=I$, the unit matrix, (21) becomes

$$(A - \lambda I)\mathbf{x} = 0 \tag{22}$$

which is the form obtained by finite difference methods. If A can be stored in the computer, and if A is symmetric as well, a method using matrix rotations would be used. Equation (21) may not be put into this form, with symmetry preserved, simply by premultiplying by B^{-1}. The product of two symmetric matrices is not in general symmetric. However, if we know that B is symmetric and positive definite, triangular decomposition (as described in Section II-B) may be employed using only real arithmetic. Therefore

$$\begin{aligned} A - \lambda B &= A - \lambda L\tilde{L} \\ &= L(C - \lambda I)\tilde{L} \end{aligned} \tag{23}$$

where

$$C = L^{-1}A\tilde{L}^{-1}. \tag{24}$$

Note that $\tilde{C}$ equals C and so it is symmetric.

Taking the determinant of both sides of (23),

$$\det(A - \lambda B) = \det(L)^2 \det(C - \lambda I). \tag{25}$$

Since det (L) is nonzero, we can see that eigenvalues of A are those of C. Therefore, instead of (21), we can solve the eigenvalue problem

$$(C - \lambda I)y = 0 \tag{26}$$

where

$$y = \tilde{L}\mathbf{x}. \tag{27}$$

We therefore obtain the required eigenvalues by solving (26). Eigenvectors y are easily transformed to the required $\mathbf{x}$ by inverting $\tilde{L}$ in (27).

Orthogonal matrices are basic to Jacobi's method. A matrix T is orthogonal if

$$T\tilde{T} = I. \tag{28}$$

If T consists of real elements, it is orthogonal if the sum of the squares of the elements of each column equals one and if the sum of the products of corresponding elements in two different columns vanishes. One example is the unit matrix and another is

$$T = \begin{bmatrix} \cos\phi & -\sin\phi \\ \sin\phi & \cos\phi \end{bmatrix}. \tag{29}$$

Matrix (29) can be inserted in an otherwise unit matrix such that the cos ϕ terms occur along the diagonal. This is also an orthogonal matrix and we shall denote it T as well.

Since the determinant of a product of several matrices equals the product of the determinants, we can see from (28) that det $(T)=1$ or -1. Therefore

$$\begin{aligned} \det(A - \lambda I) &= \det(T(A - \lambda I)\tilde{T}) \\ &= \det(TA\tilde{T} - \lambda I) \end{aligned} \tag{30}$$

and so the eigenvalues of A and $TA\tilde{T}$ are the same.

It is possible to so position (29) in any larger unit matrix, and to fix a value of ϕ, such that the transformed matrix $TA\tilde{T}$ has zeros in any chosen pair of symmetrically placed elements. Usually one chooses to do this to the pair of elements having the largest absolute values. In doing this, certain other elements are altered as well. The procedure is repeated successively with the effect that all off-diagonal terms gradually vanish leaving a diagonal matrix. The elements of a diagonal matrix are the eigenvalues and so these are all given simultaneously. It is possible to calculate the maximum error of the eigenvalues at any stage of the process and so to terminate the operation when sufficient accuracy is guaranteed. The product of all the transformation matrices is continuously computed. The columns of this matrix are the eigenvectors of A. See [3, pp. 109–112] and [7, pp. 129–131].

Probably the most efficient procedure, in both storage and time, is that due to Householder. It is described by Wilkinson [26] and, in a very readable form, by Walden [23].

A method, employing SOR for large sparse matrices, is described in Section IV.

III. Finite Differences

The importance of finite differences lies in the ease with which many logically complicated operations and functions may be discretized. Operations are then performed not upon continuous functions but rather, approximately, in terms of values over a discrete point set. It is hoped that as the distance between points is made sufficiently small, the approximation becomes increasingly accurate. The great advantage of this approach is that operations, such as differentiation and integration, may be reduced to simple arithmetic forms and can then be conveniently programmed for automatic digital computation. In short, complexity is exchanged for labor.

A. Differentiation

First of all, consider differentiation. The analytic approach often requires much logical subtlety and algebraic innovation. The numerical method, on the other hand, is very direct and simple in principle. However, implementation in many instances presents problems of specific kinds.

Consistent with a frequent finite difference notation, a function f evaluated at any x is often written $f(x)=f_x$. At a point, distance h to the right $f(x+h)=f_{x+h}$. To the left we have f_{x-h}, f_{x-2h}, etc. Alternatively, *nodes* (or *pivotal points*) are numbered i yielding function values f_i, f_{i+1}, f_{i-1}, etc. It is understood that the distance between nodes is h and node i corresponds to a particular x. Refer to Fig. 1. The derivative $f_x' = df/dx$, at any specified x, may be approximated by the *forward difference formula*

$$f_x' = \frac{f_{x+h} - f_x}{h} \tag{31}$$

where the function is evaluated explicitly at these two points. This, for very small h, corresponds to our intuitive notion of a derivative. Equally, the derivative may be expressed by the *backward difference formula*

$$f_x' = \frac{f_x - f_{x-h}}{h}. \tag{32}$$

These are, in general, not equal. The first, in our example, gives a low value and the second a high value. We can therefore expec' the average value

$$f_x' = \frac{f_{x+h} - f_{x-h}}{2h} \tag{33}$$

to give a closer estimate. This expression is the *central difference* formula for the first derivative. In the figure, we see that the forward and backward differences give slopes of chords on alternate sides of the point being considered. The central

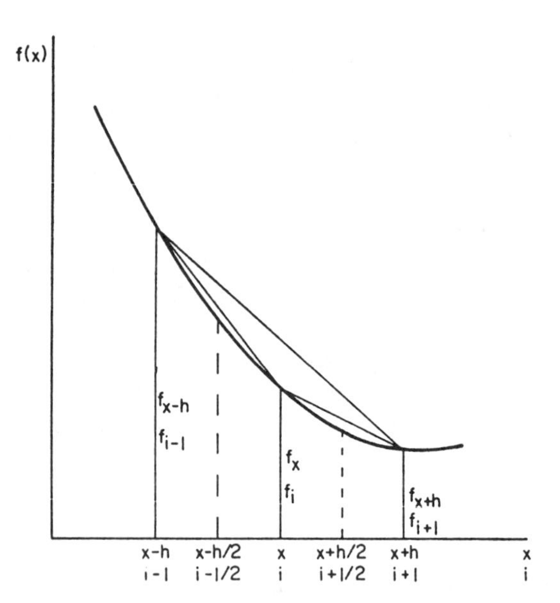

Fig. 1. Forward, backward, and central differences.

difference, given by the slope of the chord passing through f_{x+h} and f_{x-h}, certainly appears to approximate the true derivative most closely. Usually, this is so. Forward and backward difference formulas are required to compute derivatives at extreme ends of series of tabulated data. The central difference formula is preferred and should be used if data are available on both sides of the point being considered.

An idea of the accuracy of (31)–(33) is obtained from a Taylor's expansion at $x+h$. This gives

$$f_{x+h} = f_x + hf_x' + \tfrac{1}{2}h^2 f_x'' + \tfrac{1}{6}h^3 f_x''' + \cdots. \tag{34}$$

Similarly, at $x-h$

$$f_{x-h} = f_x - hf_x' + \tfrac{1}{2}h^2 f_x'' - \tfrac{1}{6}h^3 f_x''' + \cdots. \tag{35}$$

Subtracting f_x from (34), then rearranging, gives

$$\begin{aligned} f_x' &= (f_{x+h} - f_x)/h - \tfrac{1}{2}hf_x'' - \tfrac{1}{6}h^2 f_x''' - \cdots \\ &= (f_{x+h} - f_x)/h + 0(h). \end{aligned} \tag{36}$$

$0(h)$ means that the leading correction term deleted from this forward difference derivative approximation is of order h. Similarly, from (35), the backward difference formula can be seen to be of order h as well.

Subtracting (35) from (34), and rearranging, we obtain

$$\begin{aligned} f_x' &= (f_{x+h} - f_{x-h})/2h - \tfrac{1}{6}h^2 f_x''' + \cdots \\ &= (f_{x+h} - f_{x-h})/2h + 0(h^2) \end{aligned} \tag{37}$$

which is the central difference formula (33) with a leading error term or order h^2. Assuming that derivatives are well behaved, we can consider that the h^2 term is almost the sole source of error. We see that decreasing the interval results in higher accuracy, as would be expected. In addition, we see that the error decreases quadratically for the central difference derivative and only linearly for the forward or backward difference ones.

Only two of the three pivotal points shown in Fig. 1 were needed for evaluating the first derivative. Using the three available pivots, a second derivative may be computed. Consider the derivative at $x+h/2$. Using central differences, with half the previous interval,

$$f'_{x+h/2} = (f_{x+h} - f_x)/h. \tag{38}$$

Similarly,

$$f'_{x-h/2} = (f_x - f_{x-h})/h \tag{39}$$

giving the values of the first derivative at two points distance h apart. The second derivative at x is then

$$\begin{aligned} f''_x &= (f'_{x+h/2} - f'_{x-h/2})/h \\ &= (f_{x+h} - 2f_x + f_{x-h})/h^2 \end{aligned} \tag{40}$$

in which (33) has been applied with (38) and (39) supplying two first derivative values. By adding (34) and (35), (40) may be derived with the additional information that the error is of order h^2.

Another point of view for understanding finite difference differentiation is the following. Assume that a parabola

$$f(x) = ax^2 + bx + c \tag{41}$$

is passed through the three points f_{x-h}, f_x, and f_{x+h} of Fig. 1. For simplicity let $x=0$. Evaluating (41) at $x=-h$, 0, and h, we easily find that

$$a = (f_h - 2f_0 + f_{-h})/2h^2, \tag{42}$$

$$b = (f_h - f_{-h})/2h, \tag{43}$$

and

$$c = f_0. \tag{44}$$

Differentiating (41), then setting $x=0$, we find we get the same forms as (33) and (40) for the first and second derivatives. Thus, differentiation of functions specified by discrete data is performed by fitting a polynomial (either explicitly or implicitly) to the data and then differentiating the polynomial. It is clear then that an nth derivative of a function can be obtained only if at least $n+1$ data points are available. A word of warning—this cannot be pursued to very high orders if the data is not overly accurate. A high-order polynomial, made to fit a large number of approximate data points, may experience severe undulations. Under such circumstances, the derivative will be unreliable due to higher order terms of the Taylor series. Also note that accuracy of a numerical differentiation cannot be indefinitely increased by decreasing h, even if the function can be evaluated at any required point. Differentiation involves differences of numbers that are almost equal over small intervals. Due to this cause, roundoff error may become significant and so the lower limit to h is set largely by the computer word length.

Bearing in mind that high-order polynomials can cause trouble, some increase in accuracy is possible by using more than the minimum number of pivots required for a given derivative. For example, as an alternative to (40), the second derivative can be obtained from

$$f''_x = (-f_{x-2h} + 16f_{x-h} - 30f_x + 16f_{x+h} - f_{x+2h})/12h^2 \tag{45}$$

with an error of $0(h^4)$.

Not always are evenly spaced pivotal points available. Finite difference derivative operators are available for such cases as well. This is obvious as approximating polynomials can be fitted to data not equispaced. For example, if pivots exist at $x-h$, x, and $x+\alpha h$ (where α is a real number), an appropriate derivative expression is

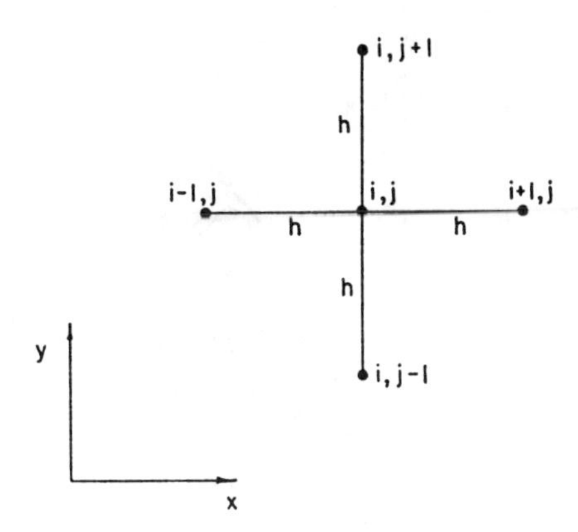

Fig. 2. Five point, finite difference operator.

$$f''_x = \frac{2}{\alpha(\alpha+1)h^2}(\alpha f_{x-h} - (1+\alpha)f_x + f_{x+h}) \tag{46}$$

which is identical to (40) when $\alpha=1$. However, if $\alpha \neq 1$, (46) and other differentiation equations for unsymmetric pivots have reduced accuracy.

The preceding, and many other differentiation expressions, are given in [12, pp. 64–87]. For instance, since numerical differentiation involves the determination and subsequent differentiation of an interpolating polynomial, there is no reason that the derivative need be restricted to pivotal points.

Finally, the most important derivative operator for our purposes is the Laplacian ∇^2. In two dimensions it becomes ∇_t^2. Acting upon a potential $\phi(x, y)$ we have

$$\nabla_t^2\phi = \frac{\partial^2\phi}{\partial x^2} + \frac{\partial^2\phi}{\partial y^2} \tag{47}$$

where x and y are Cartesian coordinates. Using a double subscript index convention, and applying (40) for each coordinate, the finite difference representation of (47) is

$$\nabla_t^2\phi = \frac{\phi_{i,j+1} + \phi_{i-1,j} - 4\phi_{i,j} + \phi_{i+1,j} + \phi_{i,j-1}}{h^2}. \tag{48}$$

Fig. 2 illustrates this five point, finite difference operator which is appropriate for equispaced data. Often it is necessary to space one or more of the nodes irregularly. This is done in the same fashion employed in the derivation of (46) [12, pp. 231–234]. Finite difference Laplacian operators are also available in many coordinate systems other than the Cartesian. For example, see [12, pp. 237–252].

B. Integration

Numerical integration is used whenever a function cannot easily be integrated in closed form or when the function is described by discrete data. The principle behind the usual method is to fit a polynomial to several adjacent points and integrate the polynomial analytically.

Refer to Fig. 3 in which the function $f(x)$ is expressed by data at n equispaced nodes. The most obvious integration

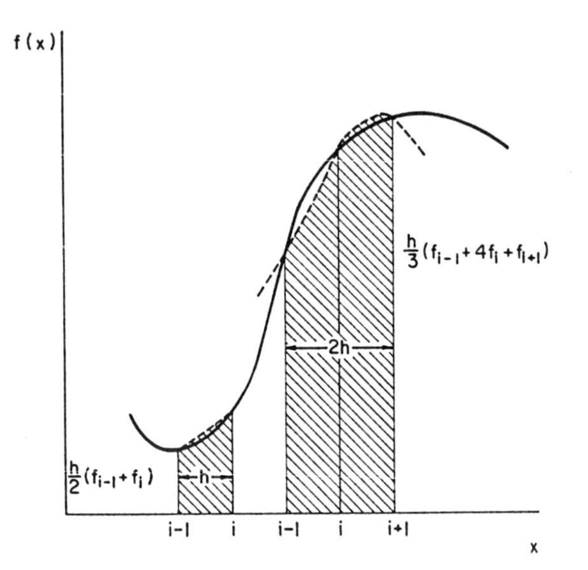

Fig. 3. Integration under the curve $f(x)$. Trapezoidal rule for a single strip and Simpson's 1/3 rule for pairs of strips.

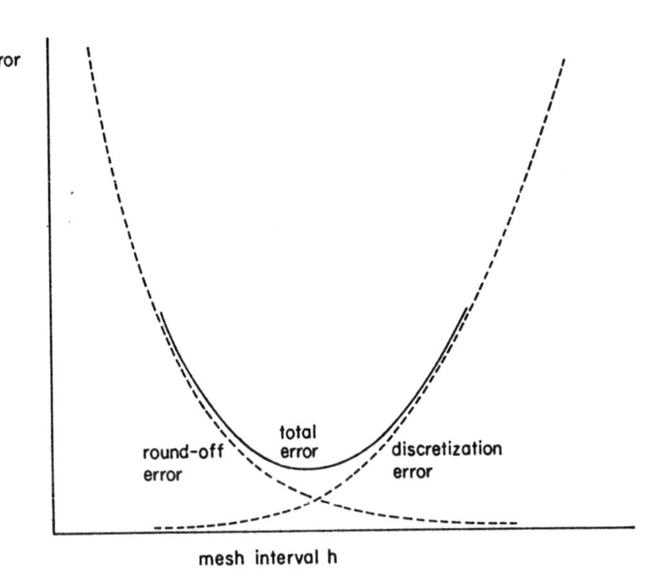

Fig. 4. Error as a function of mesh interval.

formula results by assuming that the function $f(x)$ consists of piecewise-linear sections. Each of these subareas is easily computed through

$$A_i = \int_{x_i-h}^{x_i} f(x)dx \cong \frac{h}{2}(f_{i-1} + f_i) \tag{49}$$

where x_i is the value of x at the ith pivot. The total area is found by summing all A_i. Equation (49) is the trapezoidal rule for approximating the area under one strip with error of order h^3.

The trapezoidal formula (49) may be applied, sequentially, to a large number of strips. In this way we obtain

$$\int_a^b f(x)dx \cong h(\tfrac{1}{2}f_0 + f_1 + f_2 + \cdots + f_{n-1} + \tfrac{1}{2}f_n) \tag{50}$$

which has a remainder of order h^2. The decrease in accuracy, compared with the single-strip case, is due to error accumulated by adding all the constituent subareas.

A more accurate formula (and perhaps the most popular one) is Simpson's $\frac{1}{3}$ rule

$$A = \frac{h}{3}(f_{i+1} + 4f_i + f_{i-1}) \tag{51}$$

which, by using a parabola, approximates the integral over two strips to $0(h^5)$. If the interval (a, b) is divided into an even number of strips, (51) can be applied at each pair in turn. Therefore

$$\int_a^b f(x)dx \cong \frac{h}{3}(f_0 + 4f_1 + 2f_2 + 4f_3 + \cdots + 2f_{n-2} + 4f_{n-1} + f_n) \tag{52}$$

with an error of $0(h^4)$. Another Simpson's formula, known as the $\frac{3}{8}$ rule (because the factor $\frac{3}{8}$ appears in it), integrates groups of three strips with the same accuracy as the $\frac{1}{3}$ rule. Thus, the two Simpson's rules may be used together to cater for an odd number of strips.

By and large, integration is a more reliable process than differentiation, as the error in integrating an approximating polynomial tends to average out over the interval.

It is not possible to indefinitely reduce h with the expectation of increased accuracy. Although smaller intervals reduce the discretization error, the increased arithmetic causes larger roundoff error. A point is reached where minimum total error occurs for any particular algorithm using any given word length. This is indicated in Fig. 4.

Highly accurate numerical integration procedures are provided by Gauss' quadrature formulas [7, pp. 312–367]. Rather than using predetermined pivot positions, this method chooses them in order to minimize the error. As a result, it can only be used when the integrand is an explicit function.

Multiple integration [12, pp. 198–206] is, in theory, a simple extension of one-dimensional integration. In practice, beyond double or triple integration the procedure becomes very time consuming and cumbersome.

To integrate

$$V = \int_a^b \int_c^d f_{x,y}dxdy \tag{53}$$

over the specified limits, the region is subdivided (see Fig. 5(a)). As one would expect, to perform a double integration by the trapezoidal rule, two applications of (49) are required for each elemental region. Integrating along x, over the element shown, we obtain

$$g_j = \frac{h}{2}(f_{i,j} + f_{i+1,j}) \tag{54}$$

$$g_{j+1} = \frac{h}{2}(f_{i,j+1} + f_{i+1,j+1}). \tag{55}$$

It now remains to perform the integration

$$\begin{aligned} V &= \int_{y_j}^{y_{j+1}} g_y dy = \frac{h}{2}(g_j + g_{j+1}) \\ &= \frac{h^2}{4}(f_{i,j} + f_{i+1,j} + f_{i+j,j+1} + f_{i,j+1}) \end{aligned} \tag{56}$$

which results from two applications of the trapezoidal rule.

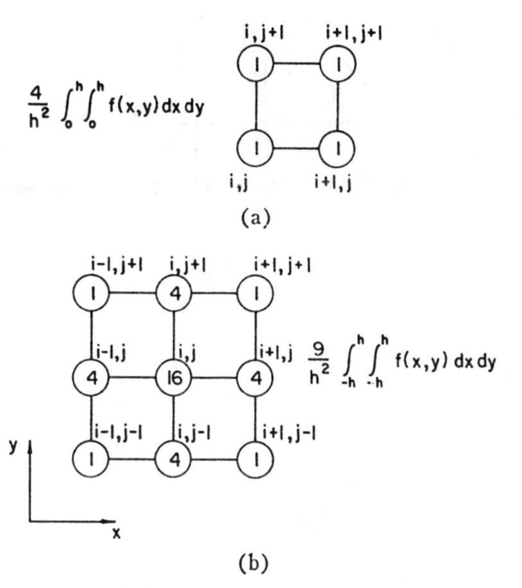

Fig. 5. Double integration molecules. (a) Trapezoidal rule. (b) Simpson's 1/3 rule.

Similarly, three applications of Simpson's $\frac{1}{3}$ rule produce

$$V = \frac{h^2}{9}\{16f_{i,j} + 4(f_{i,j-1} + f_{i+1,j} + f_{i,j+1} + f_{i-1,j}) + f_{i+1,j-1} + f_{i+1,j+1} + f_{i-1,j+1} + f_{i-1,j-1}\} \quad (57)$$

which gives the double integral of $f_{x,y}$ over the elemental two dimensional region of Fig. 5(b). *Molecules* in the figure illustrate the algorithm.

IV. Finite Difference Solution of Partial Differential Equations

The finite difference technique is perhaps the most popular numerical method for the solution of ordinary and partial differential equations. In the first place, the differential equation is transformed into a difference equation by methods described in Section III. The approximate solution to the continuous problem is then found either by solving large systems of simultaneous linear equations for the *deterministic* problem or by solving the algebraic *eigenvalue* problem as outlined in Section II. From the point of view of fields, the resulting solution is then usually a potential function ϕ defined at a finite number of points rather than continuously over the region.

A. Boundary Conditions

The most frequently occurring boundary conditions are of several, very general forms. Consider, first of all, the Dirichlet boundary condition defined by

$$\phi(s) = g(s) \quad (58)$$

which states the values of potential at all points or along any number of segments of the boundary. See Fig. 6 which represents a general two-dimensional region, part of it obeying (58). If we visualize this region as a sheet of resistive material, with surface resistivity r, the Dirichlet border is simply one maintained at a potential $g(s)$. At ground potential, $g(s)=0$ which is the homogeneous Dirichlet boundary condition.

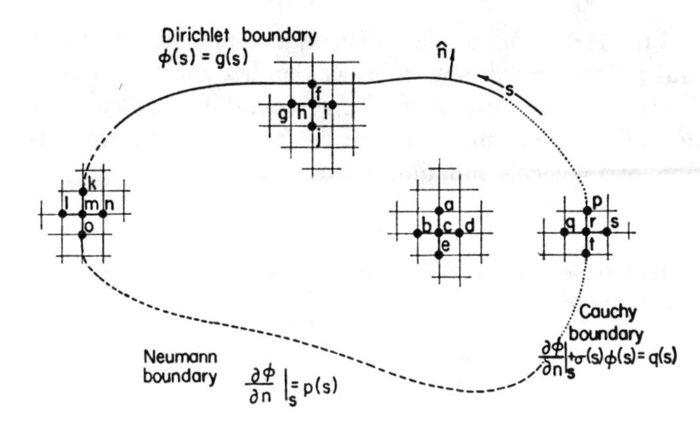

Fig. 6. A mixed boundary value problem.

The Neumann boundary condition

$$\left.\frac{\partial \phi}{\partial n}\right|_s = p(s) \quad (59)$$

is also easy to interpret physically. Imagine current to be forced into the region, from across the boundary, at a rate independent of the potential. A large number of constant current sources, strung along the boundary, would simulate this effect. The normal, linear current flow density is $-i_n(s)$, the negative sign signifying a flow direction in a sense opposite to the unit normal. The surface resistivity r divided into the normal electric field strength at the boundary $\partial\phi/\partial n|_s = -i_n(s)$, i.e., $(1/r)(\partial\phi/\partial n)|_s = -i_n(s)$. Another analogy is the flux emanating from a distributed sheet of charge backed by a conductor. In general then, the Neumann boundary condition is written as in (59). Along an impermeable border, an open circuit in Fig. 6, $p(s)=0$.

The remaining boundary condition, of concern to us, is the Cauchy (or *third*) condition. Imagine that the border is a film offering resistance R to current flowing across it. Such a film occurs in heat transfer between, say, a metal and a fluid. A series of resistors strung out across the boundary would simulate such an effect. Let the potential just outside the conducting region be $\phi_0(s)$, a function of position along the curve. The potential just inside is $\phi(s)$ and so the linear current density transferred across is $i_n(s) = (\phi(s) - \phi_0(s))/R$ where R is the film resistance. Since $(1/r)(\partial\phi/\partial n)|_s = -i_n(s)$, we have by eliminating $i_n(s)$, $(R/r)(\partial\phi/\partial n)|_s + \phi(s) = \phi_0(s)$. In general, this is written

$$\left.\frac{\partial \phi}{\partial n}\right|_s + \sigma(s)\phi(s) = q(s). \quad (60)$$

A harmonic wave function ϕ, propagating into an infinite region in the z direction, obeys $\partial\phi/\partial z = -j\beta\phi$ or $\partial\phi/\partial z + j\beta\phi = 0$ at any plane perpendicular to z. This is a homogeneous case of (60).

A region having two or more types of boundary conditions is considered to constitute a *mixed* problem.

B. Difference Equations of the Elliptic Problem

Second-order partial differential equations are conveniently classified as elliptic, parabolic, and hyperbolic [12, pp. 190–191]. Under the elliptic class fall Laplace's, Poisson's, and the Helmholtz partial differential equations. By way of illustration, we will now consider the first and last ones.

Fig. 6 shows several five-point operators in somewhat typical environments. To be specific, consider the discretization of Laplace's equation

$$\nabla^2\phi = 0 \tag{61}$$

consistent with the applied boundary conditions. From (48), (61) becomes

$$\phi_a + \phi_b - 4\phi_c + \phi_d + \phi_e = 0. \tag{62}$$

A reasonably fine mesh results in a majority of equations having the form of (62). The trouble occurs in writing finite difference expressions near the boundaries. Near the Dirichlet wall we have $\phi_f = g(s_1)$ where s_1 denotes a particular point on s. Making this substitution, the finite difference expression about node h is

$$\phi_g - 4\phi_h + \phi_i + \phi_j = -g_f \tag{63}$$

where s_1 and node f coincide. Along the Neumann boundary, (59) must be satisfied. To simplify matters for illustration, the five-point operator is located along a flat portion of the wall. Using the central difference formula (33), (59) becomes $(\phi_l - \phi_n)/2h = p_m$. This equation can then be used to eliminate node l from the difference equation written about node m. Therefore

$$\phi_k - 4\phi_m + 2\phi_n + \phi_0 = -2hp_m. \tag{64}$$

Finally, at the Cauchy boundary, $(\phi_s - \phi_q)/2h + \sigma_r\phi_r = q_r$ which makes the node r difference equation

$$\phi_p + 2\phi_q - 4(1 + \sigma_r h/2)\phi_r + \phi_t = -2hq_r. \tag{65}$$

SOR is employed to solve for each node potential in terms of all other potentials in any given equation. As there are at most five potentials per equation, each node is therefore written in terms of its immediately adjacent potentials. In other words, it is the central node potential of each operator that is altered in the iterative process. As all points on a Dirichlet boundary must maintain fixed potentials, no operator is centered there. This does not apply to Neumann and Cauchy boundaries, however, and so potentials on these walls must be computed.

It is no mean feat to program the logic for boundaries (particularly non-Dirichlet ones) which do not correspond to mesh lines, and this is one of the major drawbacks of the finite difference approach. The literature gives a number of appropriate schemes. See for example [28], [34, pp. 198–204], [35, pp. 262–266], [37, pp. 36–42].

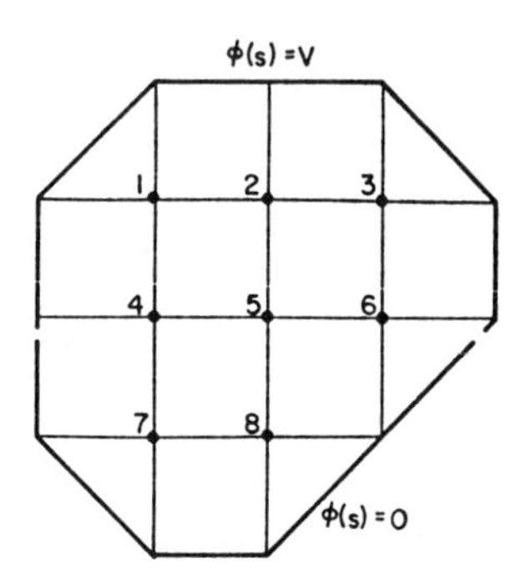

Fig. 7. Finite difference mesh for solution of Laplace's equation under Dirichlet boundary conditions.

In reference [27] a simple, although somewhat inaccurate, method for fitting an arbitrary shape is given. For simplicity, the boundary was permitted to deform, at each horizontal mesh level, to the nearest node point.

A crude-mesh difference scheme, for the solution of Laplace's equation under Dirichlet boundary conditions, is shown in Fig. 7. The appropriate difference equations can be written in matrix form in the following fashion:

$$\begin{bmatrix} -4 & 1 & 0 & 1 & 0 & 0 & 0 & 0 \\ 1 & -4 & 1 & 0 & 1 & 0 & 0 & 0 \\ 0 & 1 & -4 & 0 & 0 & 1 & 0 & 0 \\ 1 & 0 & 0 & -4 & 1 & 0 & 1 & 0 \\ 0 & 1 & 0 & 1 & -4 & 1 & 0 & 1 \\ 0 & 0 & 1 & 0 & 1 & -4 & 0 & 0 \\ 0 & 0 & 0 & 1 & 0 & 0 & -4 & 1 \\ 0 & 0 & 0 & 0 & 1 & 0 & 1 & -4 \end{bmatrix} \begin{bmatrix} \phi_1 \\ \phi_2 \\ \phi_3 \\ \phi_4 \\ \phi_5 \\ \phi_6 \\ \phi_7 \\ \phi_8 \end{bmatrix} = \begin{bmatrix} -2v \\ -v \\ -2v \\ -v \\ 0 \\ -v \\ 0 \\ 0 \end{bmatrix} \tag{66}$$

or

$$A\phi = b. \tag{67}$$

There are several significant features about (66). In the first place, there are a fairly large number of zeros. In a practical case, the mesh interval in Fig. 7 would be much smaller and the square matrix would then become very sparse indeed. In addition, the nonzero elements of any row of the square matrix and any element of the right-hand side vector may be easily generated at will. It is clear therefore that SOR (or any similar iterative scheme) is the obvious choice for solving the system of linear equations. All computer store may then be reserved for the ϕ vector and the program, thus allowing for a very fine mesh with consequent high accuracy. Certainly, a direct solution scheme, such as triangularization and back substitution, could not be considered feasible.

Green [36] discusses many practical aspects and problems associated with the solution of Laplace's equation in TEM transmission lines. Seeger [42] and Green describe the form of the difference equations in multidielectric regions.

Now, let Fig. 7 represent an arbitrarily shaped waveguide. For this purpose, the boundaries must be closed. The technique is to solve for a potential ϕ in finite difference form.

If z is along the axial direction, $\hat{u}_z\phi$ is proportional to E_z or H_z depending on whether TM or TE modes are being considered. For TM modes the boundary condition is the homogeneous Dirichlet one (58), i.e., with $g(s)=0$. Fields are then derived from

$$\begin{aligned} \overline{E}_t &= -\frac{\gamma}{k_c^2}\nabla_t\phi \\ E_z &= \phi \\ \overline{H}_t &= -\frac{j\omega\epsilon}{k_c^2}\hat{u}_z \times \nabla_t\phi \\ H_z &= 0 \end{aligned} \tag{68}$$

if E_z is made equal to ϕ. Likewise, the homogeneous Neumann condition (59) (with $p(s)=0$) applies to TE modes, with fields obtainable from

$$\begin{aligned} \overline{E}_t &= \frac{j\omega\mu}{k_c^2}\hat{u}_z \times \nabla_t\phi \\ E_z &= 0 \\ \overline{H}_t &= -\frac{\gamma}{k_c^2}\nabla_t\phi \\ H_z &= \phi. \end{aligned} \tag{69}$$

k_c is the cutoff wavenumber and γ the propagation constant. Of course, one does not solve for ϕ but rather for $\boldsymbol{\phi}$, a vector of discrete potentials, and so the differentiations indicated by (68) and (69) must be performed by techniques outlined in Section III-A.

Discretization of the Helmholtz equation

$$(\nabla_t^2 + k_c^2)\phi = 0 \tag{70}$$

produces a matrix eigenvalue problem of the form

$$(A - \lambda I)\boldsymbol{\phi} = 0. \tag{71}$$

About a typical internal node such as 5 in Fig. 7, the finite difference form of (70) is

$$-\phi_2 - \phi_4 + (4 - \lambda)\phi_5 - \phi_6 - \phi_8 = 0 \tag{72}$$

where

$$\lambda = (k_c h)^2. \tag{73}$$

Equations such as (72), suitably amended for boundary conditions, make up the set expressed by (71). Signs are changed in (72) to make, as is the frequent convention, the matrix positive semidefinite rather than negative semidefinite.

Successive overrelaxation can be used to solve the matrix eigenvalue problem (71). In the first place a fairly crude mesh, with perhaps 50–100 nodes, is "drawn" over the guide cross section. A guess at the lowest eigenvalue (either an educated one or perhaps the result of a direct method on an even coarser mesh) is taken as a first approximation. The elements of vector $\boldsymbol{\phi}$ are set to some value, perhaps all unity. SOR is then initiated generating only one row at a time of $A-\lambda I$ as required. A nontrivial solution of $(A-\lambda I)\boldsymbol{\phi}$ can exist only if the determinant vanishes, i.e., the guess at λ is a true eigenvalue. In general, the λ estimate will be in error and so $\boldsymbol{\phi}$ cannot be found by SOR alone and an outer iteration employing the Rayleigh quotient (defined later) must be employed.

Application of SOR to a homogeneous set of equations causes (14) to assume the form

$$\boldsymbol{\phi}^{(m+1)} = \mathfrak{L}_{\omega,\lambda}\boldsymbol{\phi}^{(m)}. \tag{74}$$

The subscript λ has been added to the iteration matrix to indicate a further functional dependence. As in Section II-B, for illustration, assume that $\mathfrak{L}_{\omega,\lambda}$ is a real (generally nonsymmetric) matrix with distinct eigenvalues. Then, $\boldsymbol{\phi}^{(m)}$ may be expressed as a linear combination of eigenvectors of $\mathfrak{L}_{\omega,\lambda}$, i.e.,

$$\boldsymbol{\phi}^{(m)} = a_1 l_1 + a_2 l_2 + \cdots + a_n l_n. \tag{75}$$

Iterating through (74) s times, we find that

$$\boldsymbol{\phi}^{(m+s)} = a_1\mu_1^s l_1 + a_2\mu_2^s l_2 + \cdots + a_n\mu_n^s l_n. \tag{76}$$

If μ_1 is the eigenvalue having the greatest absolute value, then if s is large, we have substantially that

$$\boldsymbol{\phi}^{(m+s)} = a_1\mu_1^s l_1. \tag{77}$$

Equation (74) must represent a stationary process when SOR has converged and so $\mu_1 = 1$ at the solution point. It is interesting to note that the eigenvector l_1 of $\mathfrak{L}_{\omega,\lambda}$ is, or is proportional to, the required eigenvector of A when the correct λ is substituted into (71).

Rewrite (71) as

$$B\boldsymbol{\phi} = 0 \tag{78}$$

where

$$B = A - \lambda I \tag{79}$$

with an assumed or computed λ approximation. The convergence theorem [34, p. 240] states that if B is symmetric and positive semidefinite, with all diagonal terms greater than zero (which can always be arranged unless one of them vanishes), and if the correct λ is employed, then the method of successive displacements (SOR with $\omega=1$) converges to the solution whatever $\boldsymbol{\phi}$ is initially. It will also converge [34, pp. 260–262] for $0<\omega<2$ if the elements $b_{ij}=b_{ji}\leq 0$ $(i\neq j)$ and $b_{ii}>0$.

As will usually happen, an eigenvalue estimate will not be correct. If it deviates from the exact eigenvalue by a small amount we can expect μ_1, in (77), to be slightly greater than or less than unity. Therefore $\boldsymbol{\phi}^{(m+s)}$ will grow or diminish slowly as SOR iteration proceeds. It cannot converge to a solution as B is *nonsingular*. (B is termed *singular* if its determinant vanishes.) However, the important point is that whether $\boldsymbol{\phi}$ tends to vanish or grow without limit, its elements tend to assume the correct relative values. In other words, the "shape" of $\boldsymbol{\phi}$ converges to the correct one. After several SOR iterations the approximate $\boldsymbol{\phi}$ is substituted into the Rayleigh quotient [35, pp. 74–75]

$$\lambda^{(r+1)} \cong \frac{\tilde{\boldsymbol{\phi}}^{(r)} A \boldsymbol{\phi}^{(r)}}{\tilde{\boldsymbol{\phi}}^{(r)} \boldsymbol{\phi}^{(r)}}. \tag{80}$$

Equation (80), which depends only upon the "shape" of ϕ, is stationary about the solution point. In other words, *if* ϕ is a reasonable estimate to the eigenvector then (80) produces an improved eigenvalue estimate. The bracketted superscripts give the number of the successive eigenvalue estimate and are therefore used in a different context from that in (74). Using the new eigenvalue approximation, and returning to the SOR process with the most recent field estimate, a second and better estimate to ϕ is found, and so on until sufficient accuracy is obtained. Whether or not convergence has been achieved may be gauged firstly by observing the percentage change in two or more successive eigenvalue estimates. If the change is considered satisfactory, perhaps less than one-tenth of a percent, then the displacement norm as a percentage of the vector norm should be inspected. (The norm of a column matrix is often defined as the square root of the sum of squares of all elements.) When this is within satisfactory limits, the process may be terminated. These requirements must be compatible in that one cannot expect the displacement norm to be very small if the eigenvalue estimate is very inaccurate. What constitutes sufficient stationarity of the process is largely a matter of practical experience with the particular problem at hand and no general rule can be given. This entire computing procedure, including ω optimization, is described in [34, pp. 375–376] and [38, pp. 114–129]. Moler [38] points out that no proof exists guaranteeing convergence with the Rayleigh quotient in the outer loop. However, experience indicates that with a reasonable λ estimate to begin with, and with other conditions satisfied, we can be fairly confident.

Generally, the higher the accuracy required, the smaller the mesh interval and the larger the number of equations to be solved. The number of eigenvalues of $\mathcal{L}_{\omega,\lambda}$ equals the order of the matrix and a large number of eigenvalues means that they are closely packed together. It is therefore clear, from (76), that if the *dominant* and *subdominant* eigenvalues of $\mathcal{L}_{\omega,\lambda}$ (μ_1 and μ_2) are nearly equal, the process (74) will need a great number of iterations before the dominant eigenvector "shape" emerges. In fact, successive overrelaxation corrections could be so small that roundoff errors destroy the entire process and convergence never occurs. The answer is to start off with a crude mesh having, perhaps, one hundred nodes in the guide cross section. Solve that matrix eigenvalue problem, halve the mesh interval, interpolate (quadratically in two dimensions, preferably) for the newly defined node potentials, and then continue the process. The rational behind this approach is that each iterative stage (other than the first) begins with a highly accurate field estimate and so few iterations are required for the fine meshes. For arbitrary boundaries, programming for mesh halving and interpolation can be an onerous chore.

As one additional point, the effect of Neumann boundary conditions is to make B slightly nonsymmetric and so convergence of the SOR process cannot be guaranteed. This occasionally causes iteration to behave erratically, and sometimes fail, for coarse meshes in which the asymmetry is most pronounced. Otherwise, the behaviour of such almost symmetric matrices is similar to that of symmetric ones. The most convenient way to guarantee symmetric matrices is to employ variational methods in deriving difference equations near boundaries. Forsythe and Wasow [34, pp. 182–184] give a good account of this approach.

When the solution for the first mode is obtained, $\det(B)=0$ with the correct λ substituted into (79). If one then wanted to solve for a higher mode, a new and greater λ estimate would be used. If this differs from the first by a, this is equivalent to subtracting a from all diagonal terms of B. Now, if p is any eigenvalue of B, then

$$B\phi = p\phi. \tag{81}$$

Subtracting aI from both sides,

$$(B - aI)\phi = (p - a)\phi \tag{82}$$

we see that all eigenvalues of the new matrix $(B-aI)$ are shifted to the left by a units. Since B had a zero eigenvalue, at least one eigenvalue must now be negative. A symmetric matrix is positive definite if and only if all of its eigenvalues are positive [16, p. 105] and positive semidefinite if all are nonnegative. Therefore, $(B-aI)$ is not positive semidefinite and so the convergence theorem is violated. Consequently, the previous SOR scheme cannot be employed for modes higher than the first.

Davies and Muilwyk [32] published an interesting account of the SOR solution of several arbitrarily shaped hollow waveguides. Typical cutoff wavenumber accuracies were a fraction of one percent. This is an interesting result as reasonable accuracy was obtained even for those geometries containing internal corners. Fields are often singular near such points. The finite difference approximation suffers because Taylor's expansion is invalid at a singularity. If errors due to reentrant corners are excessive, there are several approaches available. The reader is referred to Motz [39] and Whiting [45] in which the field about a singularity is expanded as a truncated series of circular harmonics. Duncan [33] gives results of a series of numerical experiments employing different finite difference operators, mesh intervals, etc.

An algorithm has recently been developed [27] which guarantees convergence by SOR iteration. The principle is to define a new matrix

$$C = \tilde{B}B \tag{83}$$

C is symmetric whether or not B is. Equation (78) becomes

$$C\phi = 0 \tag{84}$$

which is solved by SOR. Note that

$$\det(C) = \det(\tilde{B})\det(B) = (\det(B))^2 \tag{85}$$

and so (84) is satisfied by the same eigenvalues and eigenvectors as (71) and (78).

SOR is guaranteed to be successful on (84) as C is positive semidefinite for any real B. Note that $\tilde{x}x>0$ for any real column matrix x. Substitute the transformation $x=By$ giving $\tilde{y}\tilde{B}By=\tilde{y}Cy>0$ which defines a positive definite matrix C. If $\det(C)=0$, as happens at the solution point, then C is positive semidefinite and so convergence is guaranteed. This much is well known. The usefulness of the algorithm is that it describes a method of deriving the nonzero elements of one row at a time of C, as required by SOR, without recourse to B in its entirety. It is shown that the gth row of C requires only the gth node point potential and those of twelve other nodes in its immediate vicinity. The operations are expressed in the form of a thirteen-point finite difference operator. Thus, because storage requirements are minimal, and C is positive semidefinite, SOR can be employed for higher modes.

This method requires considerably more logical decisions, while generating difference equations near boundaries, than does the usual five-point operator. The process can be speeded up considerably by generating (and storing) these exceptional difference equations only once for each mesh size. In this way, the computer simply selects the appropriate equation for each node as required. In the internal region, difference equations are generated very quickly so that storing them would be wasteful. This is an entirely feasible approach because nodes near boundaries increase in number only as h^{-1} while the internal ones increase as h^{-2} approximately. This boundary-node storage procedure would likely be profitable for the five-point difference operator as well.

The method can be adapted to the deterministic problem (67). Normally, this would not be required, but if one attempts higher order derivative approximations at the boundary, for the Neumann or Cauchy problem, SOR often fails [37, pp. 50–53]. Because it guarantees positive definiteness, a suggested abbreviation is PDSOR.

Recently, Cermak and Silvester [29] demonstrated an approach whereby finite differences can be used in an open region. An arbitrary boundary is drawn about the field of interest. The interior region is solved in the usual way and then the boundary values are altered iteratively, until the effect of the boundary vanishes. Then, the solution in the enclosed space corresponds to a finite part of the infinite region.

Davies and Muilwyk [40] have employed finite differences in the solution of certain waveguide junctions and discontinuities. The method is applicable when the structure has a constant cross section along one coordinate. If this is so, the ports are closed by conducting walls and their finite difference technique for arbitrarily shaped waveguides [32] may be used. A limitation is that the ports must be sufficiently close together so that one seeks only the first mode in the newly defined waveguide. Otherwise, SOR will fail as described previously.

C. Parabolic and Hyperbolic Problems

Prime examples of these classes of differential equations are furnished by the wave equation

$$\nabla^2\phi = \frac{1}{c^2}\frac{\partial^2\phi}{\partial t^2} \tag{86}$$

which is *hyperbolic* and the source-free diffusion equation

$$\nabla^2\phi = \frac{1}{K}\frac{\partial\phi}{\partial t} \tag{87}$$

which is *parabolic*. Note that if ϕ is time harmonic, (86) becomes the Helmholtz equation. If ϕ is constant in time, both becomes Laplace's equation.

The solutions of partial differential equations (86) and (87) are the transient responses of associated physical problems. The solution of (86) gives the space-time response of a scalar wave function. Equation (87) governs the transient diffusion of charge in a semiconductor, heat flow through a thermal conductor, or skin effect in an imperfect electrical conductor [78, pp. 235–236]. K is the diffusion constant. It is a function of temperature, mobility, and electronic charge or thermal conductivity, specific heat, and mass density, depending upon the physical problem. For example, if a quantity of charge (or heat) is suddenly injected into a medium, the electric potential (or temperature) distribution is given by the solution of (87). The result ϕ is a function of space and time.

Such problems are more involved computationally than the elliptic problem is, due partly to the additional independent variable. The function and sufficient time derivatives at $t=0$ must be specified in order to eliminate arbitrary constants produced by integration. It is then theoretically possible to determine ϕ for all t. Problems specified in this way are known as *initial-value* problems. To be really correct, the partial differential equation furnishes us with a boundary-value, initial-value problem.

The finite difference approach is to discretize all variables and to solve a boundary value problem at each time step. For simplicity, consider the one dimensional diffusion equation

$$\frac{\partial^2\phi}{\partial^2 x} = \frac{1}{K}\frac{\partial\phi}{\partial t}. \tag{88}$$

To solve for $\phi(x, t)$, the initial value $\phi(x, 0)$ and boundary conditions, say, $\phi(0, t)=0$ and $(\partial\phi/\partial x)|_{x=1}=0$ are given. Discretization of (88) gives

$$\frac{\phi_{i-1,j} - 2\phi_{i,j} + \phi_{i+1,j}}{h^2} = \frac{\phi_{i,j+1} - \phi_{i,j}}{Kk} \tag{89}$$

where h and k are the space and time intervals respectively. Rather than the central difference formula for second derivatives, forward or backward differences must be used at the boundary points—unless Dirichlet conditions prevail.

The first of each subscript pair in (89) denotes the node number along x and the second denotes the time-step number. Therefore

$$\begin{aligned} x &= ih; \quad & i &= 0, 1, 2, \cdots \\ t &= jk; \quad & j &= 0, 1, 2, \cdots. \end{aligned} \tag{90}$$

Rearranging (89)

$$\phi_{i,j+1} = \phi_{i,j} + r(\phi_{i-1,j} - 2\phi_{i,j} + \phi_{i+1,j}) \tag{91}$$

with

$$r = Kk/h^2. \tag{92}$$

In Fig. 8, the problem is visualized as a two-dimensional region, one dimension t being unbounded. This algorithm presents an *explicit* method of solution as each group of

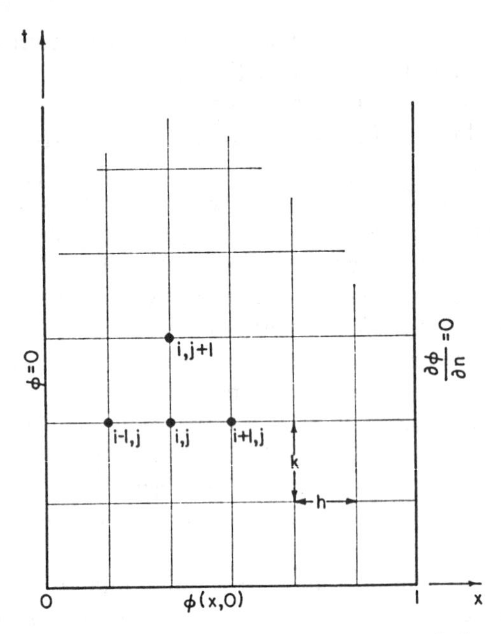

Fig. 8. Finite difference mesh for explicit solution of an initial value problem.

three adjacent pivots can be used to predict one potential at the next time step. In this way, the solution is advanced in time as long as required or until error accumulation becomes unacceptable.

There is a *stability* criterion that must be satisfied. It can be shown that the explicit method with one space coordinate is valid only when $0<r\leq\frac{1}{2}$. This restriction, in conjunction with (92), indicates that the increased amount of computing required for improved accuracy is considerable. If h is halved then k must be quartered. The stability criterion is still more stringent for problems having two space dimensions, requiring that $0<r\leq\frac{1}{4}$. The explicit solution of the wave equation is also subject to a stability constraint.

Another approach, known as the Crank-Nicolson method, requires the solution of all node potentials before advancing the time step. It is unconditionally stable and so does not require terribly fine time intervals. This advantage is partially offset, however, by the fact that all potentials at each time step must be solved as a system of simultaneous, linear equations. Thus it is called an *implicit* method. The final result is that the implicit method is some three or four times faster than the explicit one.

The reader will find very fine introductions to this subject in [12] and [44]. Three books, dealing generally with finite differences and with special sections of interest here, are [31], [34], and [35]. In [30] and [41], initial value problems are discussed. Recently, Yee [46, pp. 302–307] reported some results on transient electromagnetic propagation.

It is disappointing to note that in spite of the great amount of work done on the subject, in practice the solution of many initial value problems exhausts the capabilities of modern digital machines. Problems having two spatial dimensions can easily take many hours to solve with moderate accuracy. Forsythe and Wasow [34, pp. 11–14] have estimated one week for such a problem having 10 000 nodes. Using a modern computer, the time would be reduced to perhaps one-third of that. A three-dimensional problem, solved in fine detail, could easily take 1000 years! There appears to be an answer, however, and that is through *hybrid computation*, the subject of Section VI.

D. Integral Equations

As an alternative to posing a problem in terms of partial differential equations, it may be cast into the form of an integral equation. This approach is particularly useful for certain antenna problems where the Green's function is known in advance. Its efficacy is questionable in arbitrarily shaped closed regions because the numerical solution of the Green's function, for each source point, is as difficult as the solution of the original problem itself. The integral approach is therefore useful in many free-space studies, and when the Green's function may be found analytically without too much trouble.

Insofar as this section is concerned, it is sufficient to point out that the finite difference approach can be used. For a thin, arbitrary antenna, the integral

$$A_z = \mu \int_l I_z \frac{e^{-jkR}}{4\pi R}\, dl \tag{93}$$

gives the z component of vector potential. R is the distance between the source and the observation point, i.e.,

$$R = |r' - r|. \tag{94}$$

If the antenna is excited by a source at a given point, the approximate current distribution can be computed. This is accomplished by assuming I_z to be constant, but unknown, over each of the n subintervals. The integration in (93) is performed with the trapezoidal rule, thus producing an equation in n unknowns. Enforcing the required boundary conditions, n equations are produced and so the unknowns are found by solving a set of simultaneous, linear equations. (Higher order integration schemes may be used if the current distribution along each subinterval is presumed to be described by a polynomial.) With the current distribution known, the potential may be calculated at any point in space.

A good, descriptive introduction is furnished by [53]. In [47], [49]–[51], and [54] the solution of integral equations, in radiation and scattering problems, through matrix methods is described. Fox [35] discusses mathematical and practical aspects of Fredholm (corresponding to the elliptic problem) and Volterra (initial-value problem) integral equations. The quasi-TEM microstrip problem is dealt with in [48] and [52]. The major difficulty is the derivation of the Green's function; the numerical problem is insignificant by comparison.

Variational methods (Section V) offer another approach to the solution of integral equations.

V. Variational Methods

This subject, although not terribly new, is becoming increasingly important for several reasons. In the first place, it is relatively easy to formulate the solution of certain differential and integral equations in variational terms. Secondly, the method is very accurate and gives good results without making excessive demands upon computer store and time. The solution is found by selecting a field which minimizes a certain integral. This integral is often proportional to the energy contained in the system and so the method embodies a close correspondence with the real world.

The literature on variational methods is so scattered that there is good reason to collate and review the principles here. It is hoped that by reviewing these ideas, and relating them to microwave problems, the engineer will be encouraged to make immediate and more general use of them. Otherwise, the initiate could well spend many months accumulating the required information before being able to apply it.

The following theory is concerned almost exclusively with the solution of scalar potentials. Obviously then, static fields are the immediate beneficiaries. In addition, time-varying fields, that may be derived from a single vector potential, are also easily catered for. Although there are some indications of how to proceed, the author has not seen any general computer methods for fields with all six components of electric and magnetic field present. Such fields require both an electric and magnetic vector potential function to generate them. Perhaps it would be just as well to solve the electric and magnetic fields directly rather than through two potential functions.

A. Hilbert Function Spaces

The concept of a Hilbert function space is, in principle, very simple and most useful as well. It consists of a set of functions that obey certain rules. Typically, we will consider those functions belonging to this space as being all those that are possible solutions of any particular field problem we wish to solve. For example, the field within a three-dimensional region bounded by a perfectly conducting surface, having some distribution of charge enclosed, is the solution of the Poisson equation

$$-\nabla^2\phi = \frac{\rho}{\epsilon}\cdot \tag{95}$$

ϕ is some function of position, i.e., $\phi(P)$. We know that the solution must be one of or a combination of functions of the form $u(P)=\sin(l\pi/a)x\cdot\sin(m\pi/b)y\cdot\sin(n\pi/c)z$ in a rectangular region. a, b, and c are the dimensions of the rectangular region and l, m, and n are integers. If the conducting boundary of the box is held at zero potential any one or summation of harmonic functions u will vanish at the walls and will likewise give zero potential there. These components of a Fourier series are akin to vector components of a real space insofar as a summation of particular proportions of functions yields another function whereas vector summation of components defines a point in space. Thus, a summation of harmonic "components" of the above form defines a particular function which, by analogy, we consider to be a point in an abstract function space. For this reason, such functions are often called coordinate functions. The number of dimensions may be finite or perhaps infinite. The Fourier series is an example of a particular function space consisting of orthogonal coordinate functions. In general, however, these functions need not be orthogonal.

The requirements we have placed upon functions belonging to the function space is that they be twice differentiable (at least) and that they satisfy the homogeneous Dirichlet condition $\phi(s)=0$ at the conducting walls. Such functions are considered to belong to a linear set. By this is meant that if any two functions u and v belong to the set, then $u+v$ and av (where a is a constant) likewise belong to it. In other words, the functions $u+v$ and av are also twice differentiable and satisfy the relevant boundary conditions.

An *inner product*

$$\langle u, v\rangle = \int_\Omega uv^* d\Omega \tag{96}$$

is defined which, in a sense, gives the "component" of one function in the "direction" of the other. This appears reasonable when we recall that it is precisely in this way that a Fourier component, v (omitting, for the moment, the complex conjugate*) of an arbitrary function u is found. It is really here that the analogy between vector and function spaces becomes obvious. The reason for including the complex conjugate sign will be shown in a moment. The integration is performed over Ω which may be a one, two, or three-dimensional physical space depending on the problem. In our example, the limits correspond to the walls. If u and v are vector functions of position, we alter (96) slightly to include a dot between them, thus signifying the integral of the vector product $\boldsymbol{u}\cdot\boldsymbol{v}$. In this work, however, we will consider them to be scalars although the generalization of the subsequent derivations should be fairly straightforward.

In addition to linearity, functions that are elements of a Hilbert space must satisfy the following axioms. For each pair of functions u and v belonging to the linear set, a number $\langle u, v\rangle$ is generated that obeys the following axioms:

$$\langle u, v\rangle = \langle v, u\rangle^*; \tag{97}$$

$$\langle a_1u_1 + a_2u_2, v\rangle = a_1\langle u_1, v\rangle + a_2\langle u_2, v\rangle; \tag{98}$$

$$\langle u, u\rangle \geq 0; \tag{99}$$

if

$$\langle u, u\rangle = 0 \quad \text{then} \quad u = 0. \tag{100}$$

Note that the definition of inner product (96) satisfies requirements (97)–(100) for all well-behaved functions. Note also that in a *real* Hilbert space (i.e., one spanned by real functions), $\langle u, v\rangle=\langle v, u\rangle$. From axioms (97) and (98) it is easy to see that

$$\langle u, av\rangle = a^*\langle v, u\rangle^* = a^*\langle u, v\rangle \tag{101}$$

where a is a complex number here.

As a result of these definitions it is clear that an inner product whose factors are sums can be expanded according to the rules for multiplication of polynomials. The essential difference is that the numerical coefficient of the second factor must be replaced by its complex conjugate in carrying it outside of the brackets.

It is clear now why the complex conjugate must be employed in axiom (97). Property (99) states that $\langle u, u\rangle\geq 0$. Therefore, due to the linearity of the function space, new elements au must also satisfy $\langle au, au\rangle\geq 0$ where a is any complex number. If property (100) were of the form $\langle u, v\rangle$

$=\langle v, u\rangle$ then we would have that $\langle au, au\rangle = a^2\langle u, u\rangle$ which, for arbitrary complex a, would be a complex quantity and not positive or equal to zero, perhaps even negative. Thus (99) would be violated.

In the following, we will assume implicitly that the *norm* of each function, defined by

$$\|u\| = \sqrt{\langle u, u\rangle} \tag{102}$$

is finite. The operation beneath the radical is akin to the inner product of a vector with itself and so $\|u\|$ is, by analogy, a measure of the "length" or "magnitude" of the function. Insofar as a field is concerned, it is its rms value.

B. The Extremum Formulation

Among elliptic differential equations, there are two classes we are interested in: the deterministic and the eigenvalue problem. Forsythe and Wasow [58, pp. 163–164] point out that variational approaches are "computationally significant for elliptic problems but not for hyperbolic problems." They leave its application to parabolic problems open. The principle behind the variational method in solving elliptic problems rests on an approach which is an alternative to direct integration of the associated partial differential equation. The latter approach is often attempted by means of a Green's function conversion of a boundary value problem to an integral equation. It frequently becomes overly complicated, if not altogether impossible to handle, because the Green's function itself is difficult to derive. On the other hand, a variational formulation presents an alternative choice—to find the function that minimizes the value of a certain integral. The function that produces this minimal value is the solution of the field problem. On the face of it the alternative seems as unappealing as the original problem. However, due to certain procedures available for determining this minimizing function, the variational formulation has great computational advantages. In addition, convergence can be guaranteed under certain very broad conditions and this is of considerable theoretical and numerical consequence.

Illustrative of the generality of the method is the use of a general *operator* notation L. In practice, a great variety of operations may be denoted by this single letter.

1) *The deterministic problem:* The deterministic problem is written

$$Lu = f \tag{103}$$

where $f=f(P)$ is a function of position. If

$$L = -\nabla^2 \tag{104}$$

and

$$f = \frac{\rho}{\epsilon} \tag{105}$$

is a known charge distribution, we require to find u, the solution of the problem under appropriate boundary conditions. (The minus sign in (104) makes the operator positive definite, as will be shown.) Commonly, these boundary conditions take the forms (58)–(60). In the first instance we will concentrate on certain homogeneous boundary conditions, i.e., (58)–(60) with $g=p=q=0$.

If $f=0$, (103) becomes Laplace's equation. If $L=-(\nabla^2+k^2)$ then (103) represents one vector component of an inhomogeneous Helmholtz equation. u is then one component of the vector potential and f is an impressed current times μ.

To begin the solution, we must consider a set of all functions that satisfy the boundary conditions of the problem and which are sufficiently differentiable. Each such element u of the space belongs to the field of definition of the operator L. Symbolically, $u \in D_L$. We then seek a solution of (103) from this function space.

We consider only *self-adjoint operators.* The self-adjointness of L means that $\langle Lu, v\rangle - \langle u, Lv\rangle$, in which $u, v \in D_L$, is a function of u and v and their derivatives on s only. (s is the boundary of Ω and may be at infinity.) To have a *self-adjoint problem* we must have

$$\langle Lu, v\rangle = \langle u, Lv\rangle. \tag{106}$$

It will be seen that (106) is required in the proof of the minimal functional theorem and therefore is a requirement on those problems treated by variational methods in the fashion described here. Whether or not an operator is self-adjoint depends strongly upon the associated boundary conditions.

In addition, the self-adjoint operator will be required to be *positive definite*. The mathematical meaning of this is that

$$\langle Lu, u\rangle > 0 \tag{107}$$

whenever u is not identically zero and vanishes only when $u \equiv 0$.

The significance of these terms is best illustrated by a simple example. Let $L = -\nabla^2$. Therefore

$$\langle Lu, v\rangle = -\int_\Omega v\nabla^2 u d\Omega. \tag{108}$$

For convenience, take u and v to be real functions. Green's identity

$$\int_s v\frac{\partial u}{\partial n} ds = \int_\Omega \nabla u \cdot \nabla v d\Omega + \int_\Omega v\nabla^2 u d\Omega \tag{109}$$

converts (108) to the form

$$\langle Lu, v\rangle = \int_\Omega \nabla u \cdot \nabla v d\Omega - \int_s v\frac{\partial u}{\partial n} ds \tag{110}$$

in which the last integration is performed over the boundary. n is the outward normal. The one-dimensional analogue of (110) is integration by parts. Similarly

$$\langle u, Lv\rangle = \int_\Omega \nabla u \cdot \nabla v d\Omega - \int_s u\frac{\partial v}{\partial n} ds. \tag{111}$$

Under either the homogeneous Dirichlet or Neumann boundary conditions, the surface integrals in (110) and (111) vanish. Under the homogeneous Cauchy boundary condition, they do not vanish but become equal. At any rate, L is therefore self-adjoint under any one of these boundary conditions or under any number of them holding over various sections of the boundary. Property (106) is akin to matrix symmetry.

Positive definiteness of L is readily observed by making $u=v$ in (110) and substituting any of the previous homogeneous boundary conditions. An additional requirement, for the homogeneous Cauchy condition to satisfy (107), is that $\sigma>0$.

It is a consequence of these properties that we can make the following statement: if the operator L is positive definite then the equation $Lu=f$ cannot have more than one solution. The proof is simple. Suppose the equation to have two solutions u_1 and u_2 such that $Lu_1=f$ and $Lu_2=f$. Let $w=u_1-u_2$. Since the operator is a linear one (a further requirement) we obtain $\langle Lw, w\rangle=0$. Since L is positive definite, we must then have $w=0$ and so $u_1=u_2$, thus proving that no more than one solution can exist. This is simply a general form of the usual proofs for uniqueness of solution of boundary-value problems involving elliptic partial differential equations.

For the solution of a partial differential equation, it was stated earlier that we will attempt to minimize a certain integral. The rule for forming this integral, and subsequently, ascribing a value to it, is a particular example of a *functional*. Whereas a function produces a number as a result of giving values to one or more of independent variables, a functional produces a number that depends on the entire form of one or more functions between prescribed limits. It is, in a sense, some measure of the function. A simple example is the inner product $\langle u, v\rangle$.

The functional we are concerned with, for the solution of the deterministic problem, is

$$F = \langle Lu, u\rangle - 2\langle u, f\rangle \tag{112}$$

in which we assume that u and f are real functions. The more general form, for complex functions is

$$F = \langle Lu, u\rangle - \langle u, f\rangle - \langle f, u\rangle. \tag{113}$$

It is seen that the last two terms of (113) give twice the real part of $\langle u, f\rangle$. Concentrating on (112), we will now show that if L is a positive definite operator, and if $Lu=f$ has a solution, then (112) is minimized by the solution u_0. (The proof of (113) is not much more involved.) Any other function $u\in D_L$ will give a larger value to F. The proof follows.

Take the function u_0 to be the unique solution, i.e.,

$$Lu_0 = f. \tag{114}$$

Substitute (114) into (112) for f. Thus

$$F = \langle Lu, u\rangle - 2\langle u, Lu_0\rangle. \tag{115}$$

Add and subtract $\langle Lu_0, u_0\rangle$ to the right-hand side of (115) and rearrange noting that if L is self-adjoint $\langle Lu_0, u\rangle = \langle Lu, u_0\rangle$ in a real Hilbert space. Finally, we obtain

$$F = \langle L(u-u_0), u-u_0\rangle - \langle Lu_0, u_0\rangle. \tag{116}$$

As L is positive definite, the last term on the right is positive always and the first is ≥ 0. F assumes its least value if and only if $u=u_0$. To summarize, this minimal functional theorem requires that the operator be positive definite and self-adjoint under the stated boundary conditions. It is also required that the trial functions come from the field of definition of the operator L; i.e., they must be sufficiently differentiable and satisfy the boundary conditions. Otherwise, a function, which is not a solution, might give a lesser value to $F(u)$ and delude us into thinking that it is a better approximation to the solution.

From (116), note that the minimal value of the functional is

$$F_{\min} = -\langle Lu_0, u_0\rangle \tag{117}$$

which occurs for the exact solution u_0. Taking $L=-\nabla^2$, we get

$$F_{\min} = \int_\Omega u_0\nabla^2 u_0 d\Omega \tag{118}$$

where the integration is over a volume Ω. Now, say that $\phi(s)=0$. Therefore, using Green's theorem,

$$F_{\min} = -\int_\Omega |\nabla u_0|^2 d\Omega. \tag{119}$$

This integral is proportional to the energy stored in the region. The field arranges itself so as to minimize the contained energy!

The most common approach, used for finding the minimizing function, is the Rayleigh-Ritz method. As it relies upon locating a stationary point, we wish to ensure that once such a point is located, it in fact corresponds to the solution. This is important because a vanishing derivative is a necessary, although not a sufficient condition for a minimum. In other words, if a function $u_0\in D_L$ causes (112) to be stationary, is u_0 then the solution of (103)? It is easy to show that this is so. Let

$$\delta(\epsilon) = F(u_0+\epsilon\eta) - F(u_0) \tag{120}$$

where ϵ is an arbitrary real number. Using (112), substituting the appropriate expressions, and taking L to be self-adjoint, we obtain

$$\delta(\epsilon) = 2\epsilon\langle Lu_0 - f, \eta\rangle + \epsilon^2\langle L\eta, \eta\rangle \tag{121}$$

after some algebraic manipulation. Differentiating

$$\frac{d\delta}{d\epsilon} = 2\langle Lu_0 - f, \eta\rangle + 2\epsilon\langle L\eta, \eta\rangle \tag{122}$$

which must vanish at a stationary point. By hypothesis, this occurs when $\epsilon=0$, therefore

$$\langle Lu_0 - f, \eta\rangle = 0. \tag{123}$$

If this is to hold, for arbitrary η, then we must have that $Lu_0=f$ identically. In other words, the stationary point corresponds to the solution.

2) *The eigenvalue problem:* Functional (115) cannot a priori be used for the eigenvalue problem

$$Lu = \lambda u \tag{124}$$

because the right-hand side of (124) is not a known function as is f in (103). The relevant functional, for the eigenvalue problem, is

$$F = \frac{\langle Lu, u\rangle}{\langle u, u\rangle} \tag{125}$$

where $u \in D_L$. Equation (80) is one particular instance of it. It is often called the Rayleigh quotient. If F_{min} is the lowest bound, attained for some $u_1 \neq 0$, then $F_{min} = \lambda$ is the lowest eigenvalue of operator L and u_1 is the corresponding eigenfunction.

The proof of the preceding statements is quite direct. Let η be an arbitrary function from the field of definition of operator L, i.e., $\eta \in D_L$. Let α be an arbitrary real number. Therefore $u_1 + \alpha\eta \in D_L$. We want to investigate the conditions under which F is stationary about u_1. Substitute

$$u = u_1 + \alpha\eta \tag{126}$$

into (125) giving

$$F = \frac{\langle L(u_1 + \alpha\eta), u_1 + \alpha\eta\rangle}{\langle u_1 + \alpha\eta, u_1 + \alpha\eta\rangle}. \tag{127}$$

As u_1 and η are fixed functions, F is a function only of α. Differentiate (127) with respect to α. Then, by hypothesis, the derivative vanishes when $\alpha = 0$. We therefore get

$$\langle Lu_1, \eta\rangle\langle u_1, u_1\rangle - \langle Lu_1, u_1\rangle\langle u_1, \eta\rangle = 0. \tag{128}$$

With $F = F_{min}$ and $u = u_1$, substitute $\langle Lu_1, u_1\rangle$ from (125) into (128). Rearranging

$$\langle Lu_1 - F_{min}\, u_1, \eta\rangle = 0. \tag{129}$$

Since η is arbitrary,

$$Lu_1 - F_{min}\, u_1 = 0 \tag{130}$$

and so F_{min} is the lowest eigenvalue λ_1 and u_1 the corresponding eigenfunction. It is fairly easy to show [62, pp. 220–221] that if the minimization of (125) is attempted with trial functions u orthogonal to u_1, in the sense

$$\langle u, u_1\rangle = 0, \tag{131}$$

that F_{min} equals the second eigenvalue λ_2. Similarly, defining a Hilbert space orthogonal to u_1 and u_2, the next eigenvalue and eigenvector results, and so on.

As the minimum of (125) corresponds to the lowest eigenvalue λ, we can rewrite the functional in the following form:

$$\lambda \leq \frac{\langle Lu, u\rangle}{\langle u, u\rangle}. \tag{132}$$

The numerator of (132) is positive as L is a positive definite operator. The denominator is positive by (99). Therefore, rearranging the inequality

$$\langle Lu, u\rangle - \lambda\langle u, u\rangle \geq 0. \tag{133}$$

We know that (133) is an equality only for the correct eigenvalue and eigenfunction. Consequently, the left-hand side is otherwise greater than zero. Therefore, as an alternative to minimizing (125), we can seek the solution of (124) by minimizing

$$F = \langle Lu, u\rangle - \lambda\langle u, u\rangle \tag{134}$$

instead. Successive eigenvalues and eigenvectors are found by defining orthogonal spaces as before.

The differential equation, whose solution is a minimizing function, is known as the *Euler's equation*. We are interested in finding functionals whose Euler's equations we wish to solve, e.g., the Helmholtz equation, Laplace's equation, etc.

C. Inhomogeneous Boundary Conditions

In solving $Lu = f$ we have considered homogeneous boundary conditions exclusively. Such a restriction causes the more important problems (e.g., multiconductor lines at various potentials) to be excluded. This happens because self-adjointness cannot be proved. Substitute (58), say, into the surface integrals of (110) and (111) to verify this statement. In addition, the space is nonlinear as well.

Let us express inhomogeneous boundary conditions in the form

$$B_1 u\,|_s = b_1,\; B_2 u\,|_s = b_2, \cdots \tag{135}$$

where the B_i are linear operators and the b_i are given functions of position on the boundary. Equations (58)–(60) are the most common examples. The number of boundary conditions required depends upon whether or not u is a vector and upon the order of the differential equation.

Assume that a function of position w exists which is sufficiently differentiable and satisfies boundary conditions (135). w is not necessarily the solution. As w satisfies the boundary conditions,

$$B_1 w\,|_s = b_1, B_2 w\,|_s = b_2, \cdots. \tag{136}$$

Putting

$$v = u - w, \tag{137}$$

$$B_1 v\,|_s = 0,\quad B_2 v\,|_s = 0 \tag{138}$$

as the B_i are linear operators. We have now achieved homogeneous boundary conditions.

Instead of attempting a solution of

$$Lu = f \tag{139}$$

we examine

$$\begin{aligned} Lv &= L(u - w) \\ &= f - Lw. \end{aligned} \tag{140}$$

Let

$$f_1 = f - Lw \tag{141}$$

and so we can now attempt a solution of

$$Lv = f_1 \tag{142}$$

under homogeneous boundary conditions (138). In any particular case, it still remains to prove self-adjointness for operator L with functions satisfying (138). If this can be accomplished, then we may seek the function that minimizes

$$F = \langle Lv, v\rangle - 2\langle v, f_1\rangle. \tag{143}$$

Substitute (137) and (141) into (143). After expansion,

$$\begin{aligned} F = \langle Lu, u\rangle - 2\langle u, f\rangle + \langle u, Lw\rangle - \langle Lu, w\rangle \\ + 2\langle w, f\rangle - \langle Lw, w\rangle. \end{aligned} \tag{144}$$

f is fixed and w is a particular function selected (which we need not actually know). Therefore, the last two terms are constant and can play no part in minimizing the functional as we have assumed that u is selected from the set of functions that satisfies the required boundary condition. Otherwise w would depend on u. The last two terms may be deleted from (144) because of this.

It now remains to examine $\langle u, Lw\rangle - \langle Lu, w\rangle$ in the hope that u and w may be separated. If this attempt is successful, then an amended version of (144) may be written which excludes the unknown w.

Let us illustrate these principles with a practical example. Solve

$$-\nabla^2 u = f \tag{145}$$

under the boundary condition

$$u(s) = g(s). \tag{146}$$

The symmetrical form of Green's theorem is

$$\int_\Omega (w\nabla^2 u - u\nabla^2 w)d\Omega = \int_s \left(w\frac{\partial u}{\partial n} - u\frac{\partial w}{\partial n}\right) ds \tag{147}$$

where n is the external normal to s. Therefore, the third and fourth terms of (144) are

$$\begin{aligned}\langle u, Lw\rangle - \langle Lu, w\rangle &= \int_\Omega (w\nabla^2 u - u\nabla^2 w)d\Omega \\ &= \int_s \left(w\frac{\partial u}{\partial n} - u\frac{\partial w}{\partial n}\right)ds.\end{aligned} \tag{148}$$

Since

$$u(s) = w(s) = g(s) \tag{149}$$

we have

$$\langle u, Lw\rangle - \langle Lu, w\rangle = \int_s \left(g\frac{\partial u}{\partial n} - g\frac{\partial w}{\partial n}\right)ds. \tag{150}$$

Only the first term on the right-hand side of (150) is a function of u. In addition to neglecting the last two terms of (144), the last term of (150) may be disregarded as well. We are then left with a new functional to be minimized

$$F = -\int_\Omega u\nabla^2 u d\Omega - 2\int_\Omega fu d\Omega + \int_s g\frac{\partial u}{\partial n} ds. \tag{151}$$

Simplify (151) using identity (109)

$$\begin{aligned}F = \int_\Omega |\nabla u|^2 d\Omega - \int_s u\frac{\partial u}{\partial n} ds + \int_s g\frac{\partial u}{\partial n} ds \\ - 2\int_\Omega fu d\Omega.\end{aligned} \tag{152}$$

Because of (146), the second and third integrals cancel and we are left with

$$F = \int_\Omega |\nabla u|^2 d\Omega - 2\int_\Omega fu d\Omega \tag{153}$$

which is to be minimized for the solution of (145) under inhomogeneous boundary conditions (146). It happens, in this case, that (153) has the same form that homogeneous boundary conditions would produce. It also turns out, here, that the existence of a function w was an unnecessary assumption.

Although we shall not demonstrate it here, functionals may be derived for the inhomogeneous Neumann and Cauchy problems. It is also not too difficult to formulate the solution of electrostatic problems involving media that are functions of position.

D. Natural Boundary Conditions

Thus far, we have required that trial functions substituted into (112) should each satisfy the stipulated boundary conditions. Except for the simplest of boundary shapes, such a constraint makes it practically impossible to select an appropriate set of trial functions. It turns out, however, that $\partial u/\partial n|_s = p(s)$ and $\partial u/\partial n|_s + \sigma(s)u(s) = q(s)$ are *natural* boundary conditions for the operator $L = -\nabla^2$. The meaning of this is that we are now permitted to test *any* sufficiently differentiable functions with the certainty that the minimal value attained by (112) will be due to the solution and none other. On the other hand, we cannot entertain this confidence under the Dirichlet boundary condition.

It is easy to show this for the homogeneous Neumann problem. The form of the functional is

$$F = \int_\Omega (|\nabla u|^2 - 2fu)d\Omega. \tag{154}$$

Substitute $u = u_0 + \alpha\eta$ where η need not be in the field of definition of L. Differentiate with respect to α, make $\alpha = 0$, and set the result to zero. Finally, employing Green's formula,

$$\int_\Omega \eta(\nabla^2 u_0 + f)d\Omega - \int_s \eta\frac{\partial u_0}{\partial n} ds = 0. \tag{155}$$

Nowhere has η been required to satisfy boundary conditions. As η is arbitrary (155) can hold only if $-\nabla^2 u_0 = f$ and $\partial u_0/\partial n = 0$. Thus the solution is found with appropriate boundary conditions.

E. Solution by the Rayleigh-Ritz Method

A number of functionals have been derived, the minimization of which produce solutions of differential or integral equations. The remaining question is how to locate the minimizing function. The most popular approach is the Rayleigh-Ritz method.

Assume a finite sequence of n functions

$$u_n = \sum_{j=1}^{n} a_j\phi_j \tag{156}$$

where the a_j are arbitrary numerical coefficients. Substitute (156) into (112). Therefore

$$\begin{aligned}F &= \left\langle \sum_{j=1}^{n} a_j L\phi_j, \sum_{k=1}^{n} a_k\phi_k \right\rangle - 2\left\langle \sum_{j=1}^{n} a_j\phi_j, f\right\rangle \\ &= \sum_{j,k=1}^{n} \langle L\phi_j, \phi_k\rangle a_j a_k - 2\sum_{j=1}^{n} \langle \phi_j, f\rangle a_j.\end{aligned} \tag{157}$$

We now wish to select coefficients a_j so that (157) is a minimum, i.e.,

$$\frac{\partial F}{\partial a_i} = 0; \qquad i = 1, 2, \cdots n. \tag{158}$$

Rearranging (157) into powers of a_i,

$$F = \langle L\phi_i, \phi_i\rangle a_i^2 + \sum_{k\neq i} \langle L\phi_i, \phi_k\rangle a_i a_k + \sum_{j\neq i} \langle L\phi_j, \phi_i\rangle a_j a_i - 2\langle f, \phi_i\rangle a_i \tag{159}$$
$$+ \text{ terms not containing } a_i.$$

Now, write k instead of j in the second summation, and assuming that L is self-adjoint

$$F = \langle L\phi_i, \phi_i\rangle a_i^2 + 2 \sum_{k\neq i} \langle a\phi_i, \phi_k\rangle a_i a_k - 2\langle f, \phi_i\rangle a_i + \cdots. \tag{160}$$

Differentiating (160) with respect to a_i and setting equal to zero

$$\sum_{k=1}^{n} \langle L\phi_i, \phi_k > a_k = \langle f, \phi_i\rangle \tag{161}$$

where $i = 1, 2, \cdots, n$. Writing (161) in matrix form

$$\begin{bmatrix} \langle L\phi_1, \phi_1\rangle & \cdots & \langle L\phi_1, \phi_n\rangle \\ \vdots & & \vdots \\ \langle L\phi_n, \phi_1\rangle & \cdots & \langle L\phi_n, \phi_n\rangle \end{bmatrix} \begin{bmatrix} a_1 \\ \vdots \\ a_n \end{bmatrix} = \begin{bmatrix} \langle f, \phi_1\rangle \\ \vdots \\ \langle f, \phi_n\rangle \end{bmatrix} \tag{162}$$

which may be solved for the coefficients $a_1, a_2, \cdots, a_n$ by the methods of Section II-*A*.

By a very similar approach [62, pp. 226–229], [63, pp. 193–194], the Rayleigh-Ritz method applied to the eigenvalue problem gives

$$\begin{bmatrix} \langle L\phi_1, \phi_1\rangle - \lambda\langle \phi_1, \phi_1\rangle & \cdots & \langle L\phi_1, \phi_n\rangle - \lambda\langle \phi_1, \phi_n\rangle \\ \vdots & & \vdots \\ \langle L\phi_n, \phi_1\rangle - \lambda\langle \phi_n, \phi_1\rangle & \cdots & \langle L\phi_n, \phi_n\rangle - \lambda\langle \phi_n, \phi_n\rangle \end{bmatrix} \cdot \begin{bmatrix} a_1 \\ \vdots \\ a_n \end{bmatrix} = 0 \tag{163}$$

which is a matrix eigenvalue problem of form (21). If the trial functions are orthonormal, the eigenvalue λ occurs only along the diagonal giving

$$\begin{bmatrix} \langle L\phi_1, \phi_1\rangle - \lambda\langle \phi_1, \phi_1\rangle & \cdots & \langle L\phi_1, \phi_n\rangle \\ \vdots & & \vdots \\ \langle L\phi_n, \phi_1\rangle & \cdots & \langle L\phi_n, \phi_n\rangle - \lambda\langle \phi_n, \phi_n\rangle \end{bmatrix} \cdot \begin{bmatrix} a_1 \\ \vdots \\ a_n \end{bmatrix} = 0. \tag{164}$$

Similarly, (162) would have diagonal terms only, giving the solution by the Fourier analysis method.

Because the functions are all real and the operator is self-adjoint, from (97) and (106) we see that both the A and B matrices are symmetric. Also, B is positive definite which permits it to be decomposed, as in (23), using real arithmetic only, thus resulting in an eigenvalue problem of form (22).

It can be shown that (162) and (163) approach the solution of $Lu=f$ and $Lu=\lambda u$ as n approaches infinity. In practice, the matrices need not be very large for a high degree of accuracy to result. Notice also that these matrices are dense. These characteristics determine that direct methods should be employed in their solution.

F. Some Applications

The bibliography lists several useful references for the principles of variational methods. See, for example, [58], [61], and [63]. One of the most detailed treatments available is [62].

Bulley [57] solves the TE modes in an arbitrarily shaped guide by a Rayleigh-Ritz approach. The series of trial functions, representing the axial magnetic field, are each of the form $x^m y^n$ thus constituting a two-dimensional polynomial over the waveguide cross section. Obviously, these trial functions cannot be chosen to satisfy all boundary conditions. However, it turns out that the homogeneous Neumann condition (which H_z must satisfy) is natural and so no such constraint need be placed on the trial functions. On the other hand, the homogeneous Dirichlet condition (which is imposed upon E_z) is not satisfied naturally and so Bulley's method is inapplicable in this case. If the guide boundary is fairly complicated, a single polynomial has difficulty in approximating the potential function everywhere. In such a case, Bulley subdivides the waveguide into two or more fairly regular regions and solves for the polynomial coefficients in each. In doing this, his approach is virtually that of the *finite-element* method. It differs from the usual finite-element method in the way that he defines polynomials that straddle subdivision boundaries while others vanish there.

Thomas [67] solves the TE problem by the use of Lagrange multipliers. In this way, he permits all trial functions while constraining the final result to approximate the homogeneous Dirichlet boundary condition. Although not essential to his method, he employs a polar coordinate system with polynomials in r and trigonometric θ dependence.

Another possible approach, when boundary conditions are not natural, is to alter the functional in order to allow trial functions to be unrestricted [64, pp. 1131–1133].

By a transformation, Yamashita and Mittra [68] reduce the microstrip problem to one dimension. They then solve the fields and line capacitance, of the quasi-TEM mode.

The finite element method is an approach whereby a region is divided into subintervals and appropriate trial functions are defined over each one of them. The most convenient shape is the triangle, for two-dimensional problems, and the tetrahedron in three dimensions. These shapes appear to offer the greatest convenience in fitting them toegether, in approximating complicated boundary shapes, and in satisfying boundary conditions whether or not they are natural.

Silvester [65], [66] has demonstrated the method in waveguide problems. An arbitrary waveguide is divided into triangular subintervals [65]. If the potential is considered to be a surface over the region, it is then approximated by an array of planar triangles much like the facets on a diamond. Higher approximations are obtained by expressing the potential within each triangle by a polynomial [66].

Because of high accuracy, the finite element approach

appears useful for three-dimensional problems without requiring excessive computing [69].

The method was originally expounded for civil engineering applications [70], [71], but has recently seen increasing application in microwaves (e.g., [55], [56] as well as the previous references).

Other implementations of variational methods are described in [59] and [60].

VI. Hybrid Computation

In Section IV-C two methods for the solution of initial value (transient) problems were introduced. Through the explicit approach, one is able to predict the potential of any node at the next time increment as a function of a few adjacent node potentials. The disadvantage is that a stability constraint demands very small time steps. On the other hand, the implicit method does not suffer from instability, and permits larger time steps, but requires simultaneous solution of all node potentials for each step. As a result, both techniques are very time consuming and sometimes impossibly slow.

The hybrid computer offers a significantly different approach to the problem. It consists of two major parts—an *analog* and a *digital* computer. The analog is a model that obeys the same mathematical laws as the problem being considered. So the analog, which is presumably easier to handle, simulates the response of the system being studied. More precisely, the particular form of analog intended here is known as an *electronic differential analyzer*. By connecting electronic units (which perform integration, multiplication, etc.) together, it is possible to solve ordinary differential equations under appropriate initial conditions [74]. The solution is given as a continuous waveform. Whereas the digital computer will solve the problem in a number of discrete time-consuming steps, the analog gets the answer almost immediately. The analog computer is faster; it is a natural ordinary differential equation solver.

This is fairly obvious for an ordinary differential equation, but how is a partial differential equation to be solved? Consider a one-dimensional diffusion equation

$$\frac{\partial^2 \phi}{\partial x^2} = \frac{\partial \phi}{\partial t} \cdot \tag{165}$$

(Many of the following comments apply to the wave equation as well.) Discretize the spatial coordinates at the ith node. We have, using central differences,

$$\frac{d\phi_i}{dt} = \frac{1}{h^2}(\phi_{i-1} - 2\phi_i + \phi_{i+1}). \tag{166}$$

At boundaries, forward or backward differences must be used.

We have therefore reduced the partial differential equation to a system of ordinary differential equations, one at each node point. This is known as the DSCT (discrete-space-continuous-time) analog technique. Other formulations exist as well. The time response of the potential at i, i.e., $\phi_i(t)$, may be found by integrating (166). Other functions of

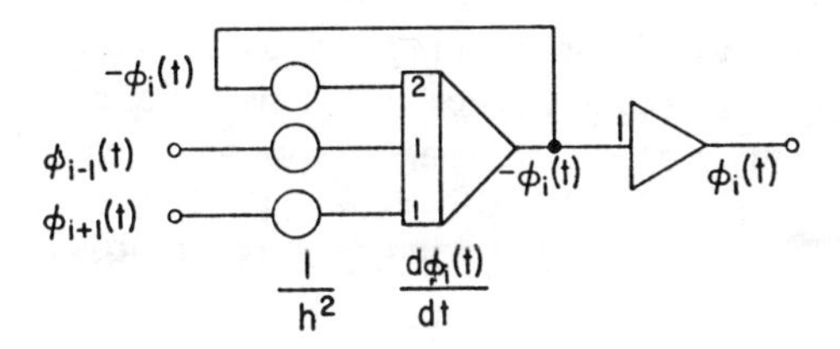

Fig. 9. Single node, DSCT analog of the one-dimensional diffusion equation.

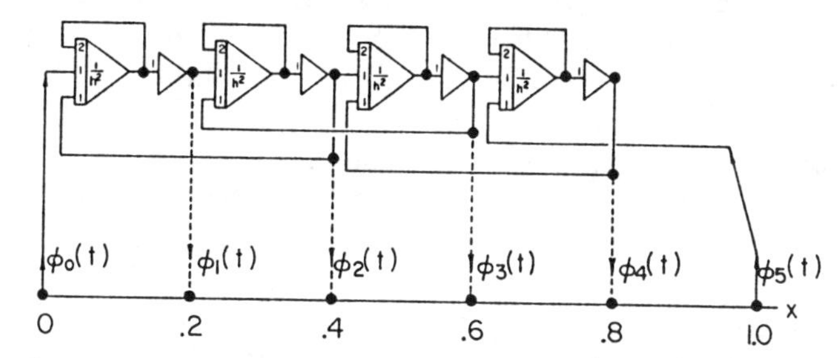

Fig. 10. Simultaneous solution of the one-dimensional diffusion equation, by DSCT analog, with four internal nodes.

time $\phi_{i-1}(t)$ and $\phi_{i+1}(t)$ are forcing functions which are as yet unknown. Assume, for the moment, that we know them and let us see how the analog computer can produce the time response $\phi_i(t)$.

Fig. 9 indicates, symbolically, the operation of an analog computer in solving (166). Circles indicate multipliers, the larger triangle represents an integrator, and the smaller one an inverter. Assume that functions $\phi_{i-1}(t)$ and $\phi_{i+1}(t)$ are known, recorded perhaps, and played back into the integrator with the initial value $\phi_i(0)$. They are all multiplied by $1/h^2$ and fed into the integrator in the ratios indicated. This is then integrated giving

$$\phi_i(t) = \frac{1}{h^2}\int_0^t (\phi_{i-1}(t) - 2\phi_i(t) + \phi_{i+1}(t))dt. \tag{167}$$

In fact, the integrator produces the negative of (167), i.e., $-\phi_i(t)$. This is fed back thus completing the circuit and allowing the process to continue to any time t. An inverter follows the integrator to alter the sign if required.

We do not, of course, know the forcing functions $\phi_{i-1}(t)$ and $\phi_{i+1}(t)$. They are the responses of adjacent nodes, and they in turn depend upon forcing functions defined at other nodes. However, the boundary conditions $\phi_0(t)$ and $\phi_5(t)$ are known in advance as well as the initial conditions $\phi_i(0)$ for all nodes, or $\phi(x, t)$ where $t=0$.

To solve the entire finite difference system at one time requires as many integrators as there are internal nodes available. This is demonstrated in Fig. 10. (The factor $1/h^2$ is incorporated into the integrators for simplicity.) What this scheme does, in fact, is to solve a system of four coupled ordinary differential equations *in parallel*. The analog has two obvious advantages: 1) rapid integration; and 2) parallel processing.

In order to reduce discretization error, one increases the number of nodes. If the intention is to solve all node potentials simultaneously, due to the limited number of integrators available, the number of nodes must be small. One alternative strategy is to attempt an iterative technique reminiscent of the digital relaxation procedure. This is done by making an initial guess at the transient response of each internal node, frequently choosing just constant time responses as shown by the uppermost curve in Fig. 11. Having made this initial guess at each node's time response, each $\phi_i(t)$ is solved sequentially as described previously. Each $\phi_i(t)$, upon being solved, is transmitted via an ADC (analog-digital converter) to the digital computer where it is stored as discrete data in time. Attention is then focused upon node $i+1$ with adjacent potential responses transmitted from store through DAC (digital-analog conversion) equipment. $\theta_{i+1}(t)$ is then computed by the analog with the smoothed $\phi_i(t)$ and $\phi_{i+2}(t)$ acting as forcing functions. The flow of data is indicated by arrowheads in Fig. 11. In sequence then, node transient responses are updated by continually scanning all nodes until convergence is deemed to be adequate. In effect, this procedure involves the solution of coupled ordinary differential equations, one for each node, by iteration.

An additional refinement is to solve, not one node potential at a time but groups of them. The number that can be catered for is, as pointed out before, limited by the amount of analog equipment available. This *parallel processing* speeds the solution of the entire problem and is one of the advantages over the purely digital scheme. Furthermore, because of the continuous time response (i.e., infinitesimal time steps), we have an explicit method without the disadvantage of instability.

Parallel solution of blocks of nodes is the logical approach for two dimensional initial value problems. Fig. 12(a) shows one possible format involving the solution at nine nodes. Forcing functions correspond to solutions at the twelve nodes excluded from the enclosed region. The set of nodes being considered would scan the region with alterations to the block format near boundaries. The making of such logical decisions, as well as storage, is the job of the digital computer.

Hsu and Howe [75] have presented a most interesting feasibility study on the solution of the wave and diffusion equations by hybrid computation. The procedures mentioned above are more fully explained in their paper. Hsu and Howe did not actually have a hybrid computer available at the time of their experiments; their results were obtained through a digital simulation study. Hybrid computation of partial differential equations is still in its early stages and little has been reported on actual computing times. However, some preliminary reports indicate speeds an order of magnitude greater than digital computing for such problems.

An intriguing possibility, for hybrid solution of the elliptic problem, is the method of lines. The mathematical theory is presented in [62, pp. 549–566]. The principle behind it is that discretization is performed along one coordinate only, in a two-dimensional problem, giving us a sequence of strips (Fig. 12(b)). In three dimensions, two coordinates are discretized, producing prisms. We thus obtain a number of coupled difference-differential equations. Each equation is then integrated under boundary conditions at each end of its strip and with forcing functions supplied by adjacent strips.

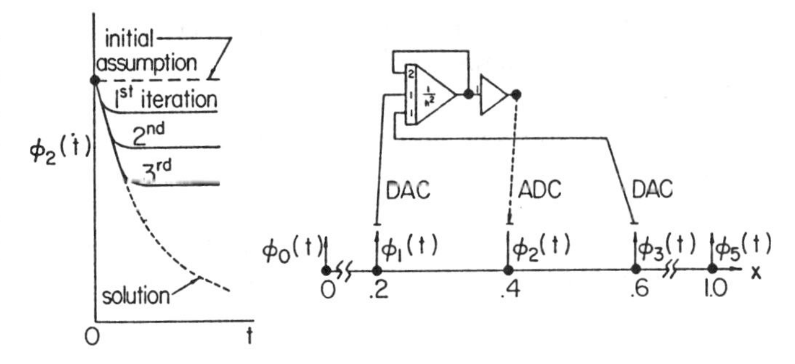

Fig. 11. Iterative DSCT solution of the diffusion equation.

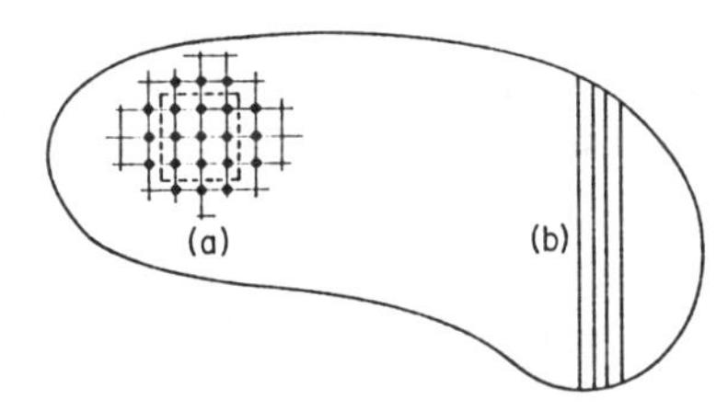

Fig. 12. (a) Iterative, parallel processing scheme for the initial value problem. (b) Method of lines for the elliptic problem.

The analog computer does not as easily solve two-point boundary value problems as it does initial value problems. A common technique is to attempt various initial values at one end of the strip until the far-end boundary condition is satisfied. This can often be very wasteful. A better idea is to solve each two-point boundary value problem as two initial value problems [7, pp. 239–240], [74, pp. 83–85]. This could be done for each strip in turn or perhaps in groups. The entire region must then be scanned repeatedly, in this fashion, until convergence occurs.

For other approaches to the solution of partial differential equations by hybrid computation, see [72], [73], [74], [76], and [77]. Finally, the hybrid system makes the Monte Carlo method [74, pp. 239–242, 360] a more attractive one.

It should be emphasized that hybrid solution of partial differential equations is still in its infancy. This section, in part an optimistic forecast, is intended to show that digital computing has no monopoly and certainly should not be considered a more "respectable" branch of computing. The analog machine integrates rapidly, and the digital machine has the ability to store information and make logical decisions. It therefore stands to reason that working as a pair, each in its own domain, substantial advantages will be gained.

VII. Concluding Remarks

The intention of this paper is to familiarize the reader with the principles behind the numerical analysis of electromagnetic fields and to stress the importance of a clear understanding of the underlying mathematics. Improper numerical technique causes one to run up against machine limitations prematurely.

In the immediate future, emphasis should perhaps be placed upon the development of finite difference and variational techniques for solving fields having all components of electric and magnetic field present everywhere. To date, with a few exceptions (e.g., [59, pp. 172–188]), most methods appear to permit solution only of fields deriveable from a single scalar potential. It is not difficult to formulate and solve field problems in a multidielectric region, but it may not correspond to the actual electromagnetic problem. This is indicated by the continuing discussion of "quasi-TEM" microstrip waves.

Variational methods are being implemented now to a greater extent than ever before. Hybrid computation is likely to assume a significant, or perhaps commanding, role in field computation due to its greater speed and flexibility. Almost certainly, general purpose scientific computers will permit optional analog equipment to be added, the analog elements to be connected and controlled from the program. Beyond this, it is virtually impossible to predict. It is futile wishing for computer technology to keep pace with our problem solving requirements. We have outstripped all machine capabilities already. The greatest hope is in the development of new numerical techniques. Due to his insight into the physical processes and his modeling ability, the engineer is ideally suited to this task.

Acknowledgment

The author wishes to express his appreciation to P. A. Macdonald and G. Oczkowski for discussions on numerical and programming techniques and to Dr. J. W. Bandler for conversations on microwave problems. Views expressed by M. J. Beaubien and B. H. McDonald, on finite differences and hybrid computing respectively, were invaluable. Dr. W. G. Mathers, of Atomic Energy of Canada Ltd., made a number of helpful suggestions. Also to be thanked are librarians, R. Thompson, Mrs. F. Ferguson, and Mrs. D. Globerman, for locating and checking many references, and Miss L. Milkowski for carefully typing the manuscript.

Thanks are due W. J. Getsinger, Guest Editor of this Special Issue, for the invitation to write this review paper.

References

General Numerical Analysis

[1] I. S. Berezin and N. P. Zhidkov, *Computing Methods*, vols. I and II. Oxford: Pergamon, 1965.
[2] D. K. Faddeev and V. N. Faddeeva, *Computational Methods of Linear Algebra*. San Francisco: Freeman, 1963.
[3] C.-E. Fröberg, *Introduction to Numerical Analysis*. Reading, Mass.: Addison-Wesley, 1965.
[4] E. T. Goodwin, *Modern Computing Methods*. London: H. M. Stationery Office, 1961.
[5] R. W. Hamming, *Numerical Methods for Scientists and Engineers*. New York: McGraw-Hill, 1962.
[6] P. Henrici, *Elements of Numerical Analysis*. New York: Wiley, 1964.
[7] F. B. Hildebrand, *Introduction to Numerical Analysis*. New York: McGraw-Hill, 1956.
[8] E. Isaacson and H. B. Keller, *Analysis of Numerical Methods*. New York: Wiley, 1966.
[9] M. L. James, G. M. Smith, and J. C. Wolford, *Applied Numerical Methods for Digital Computation with Fortran*. Scranton, Pa.: International Textbook, 1967.
[10] L. G. Kelly, *Handbook of Numerical Methods and Applications*. Reading, Mass.: Addison-Wesley, 1967.
[11] C. Lanczos, *Applied Analysis*. Englewood Cliffs, N. J.: Prentice-Hall, 1956.
[12] M. G. Salvadori and M. L. Baron, *Numerical Methods in Engineering*. Englewood Cliffs, N. J.: Prentice-Hall, 1964.
[13] J. H. Wilkinson, *Rounding Errors in Algebraic Processes*. London: H. M. Stationery Office, 1963.

Matrices and Linear Equations

[14] B. A. Carré, "The determination of the optimum accelerating factor for successive over-relaxation," *Computer J.*, vol. 4, pp. 73–78, 1961.
[15] G. E. Forsythe and C. B. Moler, *Computer Solution of Linear Algebraic Systems*. Englewood Cliffs, N. J.: Prentice-Hall, 1967.
[16] J. N. Franklin, *Matrix Theory*. Englewood Cliffs, N. J.: Prentice-Hall, 1968.
[17] L. A. Hageman and R. B. Kellogg, "Estimating optimum overrelaxation parameters," *Math. Computation*, vol. 22, pp. 60–68, January 1968.
[18] T. Lloyd and M. McCallion, "Bounds for the optimum overrelaxation factor for the S.O.R. solution of Laplace type equations over irregular regions," *Computer J.*, vol. 11, pp. 329–331, November 1968.
[19] T. J. Randall, "A note on the estimation of the optimum successive overrelaxation parameter for Laplace's equation," *Computer J.*, vol. 10, pp. 400–401, February 1968.
[20] J. K. Reid, "A method for finding the optimum successive overrelaxation parameter," *Computer J.*, vol. 9, pp. 201–204, August 1966.
[21] J. F. Traub, *Iterative Methods for the Solution of Equations*. Englewood Cliffs, N. J.: Prentice-Hall, 1964.
[22] R. S. Varga, *Matrix Iterative Analysis*. Englewood Cliffs, N. J.: Prentice-Hall, 1962.
[23] D. C. Walden, "The Givens-Householder method for finding eigenvalues and eigenvectors of real symmetric matrices," M.I.T. Lincoln Lab., Cambridge, Mass., Tech. Note 1967-51, October 26, 1967.
[24] J. R. Westlake, *A Handbook of Numerical Matrix Inversion and Solution of Linear Equations*. New York: Wiley, 1968.
[25] J. H. Wilkinson, *The Algebraic Eigenvalue Problem*. New York: Oxford, 1965.
[26] J. H. Wilkinson, "Householder's method for the solution of the algebraic eigenproblem," *Computer J.*, pp. 23–27, April 1960.

Finite Difference Solution of Partial Differential Equations

[27] M. J. Beaubien and A. Wexler, "An accurate finite-difference method for higher order waveguide modes," *IEEE Trans. Microwave Theory and Techniques*, vol. MTT-16, pp. 1007–1017, December 1968.
[28] C. T. Carson, "The numerical solution of TEM mode transmission lines with curved boundaries," *IEEE Trans. Microwave Theory and Techniques* (Correspondence), vol. MTT-15, pp. 269–270, April 1967.
[29] I. A. Cermak and P. Silvester, "Solution of 2-dimensional field problems by boundary relaxation," *Proc. IEE* (London), vol. 115, pp. 1341–1348, September 1968.
[30] F. Ceschino, J. Kuntzmann, and D. Boyanovich, *Numerical Solution of Initial Value Problems*. Englewood Cliffs, N. J.: Prentice-Hall, 1966.
[31] L. Collatz, *The Numerical Treatment of Differential Equations*. Berlin: Springer, 1960.
[32] J. B. Davies and C. A. Muilwyk, "Numerical solution of uniform hollow waveguides with boundaries of arbitrary shape," *Proc. IEE* (London), vol. 113, pp. 277–284, February 1966.
[33] J. W. Duncan, "The accuracy of finite-difference solutions of Laplace's equation," *IEEE Trans. Microwave Theory and Techniques*, vol. MTT-15, pp. 575–582, October 1967.
[34] G. F. Forsythe and W. R. Wasow, *Finite-Difference Methods for Partial Differential Equations*. New York: Wiley, 1960.
[35] L. Fox, *Numerical Solution of Ordinary and Partial Differential Equations*. Oxford: Pergamon, 1962.
[36] H. E. Green, "The numerical solution of some important transmission-line problems," *IEEE Trans. Microwave Theory and Techniques*, vol. MTT-13, pp. 676–692, September 1965.
[37] D. Greenspan, *Introductory Numerical Analysis of Elliptic Boundary Value Problems*. New York: Harper and Row, 1965.
[38] C. B. Moler, "Finite difference methods for the eigenvalues of Laplace's operator," Stanford University Computer Science Dept., Stanford, Calif., Rept. CS32, 1965.
[39] H. Motz, "The treatment of singularities of partial differential equations by relaxation methods," *Quart. Appl. Math.*, vol. 4, pp. 371–377, 1946.
[40] C. A. Muilwyk and J. B. Davies, "The numerical solution of rectangular waveguide junctions and discontinuities of arbitrary

cross section," *IEEE Trans. Microwave Theory and Techniques*, vol. MTT-15, pp. 450–455, August 1967.

[41] R. D. Richtmyer and K. W. Morton, *Difference Methods for Initial-Value Problems*. New York: Interscience, 1967.

[42] J. A. Seeger, "Solution of Laplace's equation in a multidielectric region," *Proc. IEEE* (Letters) vol. 56, pp. 1393–1394, August 1968.

[43] D. H. Sinnott, "Applications of the numerical solution to Laplace's equation in three dimensions," *IEEE Trans. Microwave Theory and Techniques* (Correspondence), vol. MTT-16, pp. 135–136, February 1968.

[44] G. D. Smith, *Numerical Solutions of Partial Differential Equations*. London: Oxford, 1965.

[45] K. B. Whiting, "A treatment for boundary singularities in finite difference solutions of Laplace's equation," *IEEE Trans. Microwave Theory and Techniques* (Correspondence), vol. MTT-16, pp. 889–891, October 1968.

[46] K. S. Yee, "Numerical solution of initial boundary value problems involving Maxwell's equations in isotropic media," *IEEE Trans. Antennas and Propagation*, vol. AP-14, pp. 302–307, May 1966.

Finite Difference Solution of Integral Equations

[47] M. G. Andreasen, "Scattering from cylinders with arbitrary surface impedance," *Proc. IEEE*, vol. 53, pp. 812–817, August 1965.

[48] T. G. Bryant and J. A. Weiss, "Parameters of microstrip transmission lines and of coupled pairs of microstrip lines," *IEEE Trans. Microwave Theory and Techniques*, vol. MTT-16, pp. 1021–1027, December 1968.

[49] R. F. Harrington [59, pp. 62–81] and [60].

[50] K. K. Mei, "On the integral equations of thin wire antennas," *IEEE Trans. Antennas and Propagation*, vol. AP-13, pp. 374–378, May 1965.

[51] J. H. Richmond "Digital computer solutions of the rigorous equations for scattering problems," *Proc. IEEE*, vol. 53, pp. 796–804, August 1965.

[52] P. Silvester, "TEM wave properties of microstrip transmission lines," *Proc. IEE* (London), vol. 115, pp. 43–48, January 1968.

[53] R. L. Tanner and M. G. Andreasen, "Numerical solution of electromagnetic problems," *IEEE Spectrum*, pp. 53–61, September 1967.

[54] P. C. Waterman, "Matrix formulation of electromagnetic scattering," *Proc. IEEE*, vol. 53, pp. 805–812, August 1965.

Variational Methods

[55] S. Ahmed, "Finite element method for waveguide problems," *Electron. Letts.*, vol. 4, pp. 387–389, September 6, 1968.

[56] P. F. Arlett, A. K. Bahrani, and O. C. Zienkiewicz, "Application of finite elements to the solution of Helmholtz's equation," *Proc. IEE* (London), vol. 115, pp. 1762–1766, December 1968.

[57] R. M. Bulley, "Computation of approximate polynomial solutions to the Helmholtz equation using the Rayleigh-Ritz method," Ph.D. dissertation, University of Sheffield, England, July 1968.

[58] G. F. Forsythe and W. R. Wasow [34, pp. 159–175].

[59] R. F. Harrington, *Field Computation by Moment Methods*. New York: Macmillan, 1968.

[60] ——, "Matrix methods for field problems," *Proc. IEEE*, vol. 55, pp. 136–149, February 1967.

[61] L. V. Kantorovich and V. I. Krylov, *Approximate Methods of Higher Analysis*. Groningen, Netherlands: Interscience, 1958.

[62] S. G. Mikhlin, *Variational Methods in Mathematical Physics*. New York: Macmillan, 1964.

[63] S. G. Mikhlin and K. L. Smolitskiy, *Approximate Methods for Solution of Differential and Integral Equations*. New York: Elsevier, 1967.

[64] P. M. Morse and H. Feshbach, *Methods of Theoretical Physics*, Part II. New York: McGraw-Hill, 1953, pp. 1106–1172.

[65] P. Silvester, "Finite-element solution of homogeneous waveguide problems," 1968 URSI Symp. on Electromagnetic Waves, paper 115 (to appear in *Alta Frequenza*).

[66] ——, "A general high-order finite-element waveguide analysis program," *IEEE Trans. Microwave Theory and Techniques*, vol. MTT-17, pp. 204–210, April 1969.

[67] D. T. Thomas, "Functional approximations for solving boundary value problems by computer," *IEEE Trans. Microwave Theory and Techniques*, this issue, pp. 447–454.

[68] E. Yamashita and R. Mitrra, "Variational method for the analysis of microstrip lines," *IEEE Trans. Microwave Theory and Techniques*, vol. MTT-16, pp. 251–256, April 1968.

[69] O. C. Zienkiewicz, A. K. Bahrani, and P. L. Arlett, "Numerical solution of 3-dimensional field problems," *Proc. IEE* (London), vol. 115, pp. 367–369, February 1968.

[70] O. C. Zienkiewicz and Y. K. Cheung, *The Finite Element Method in Structural and Continuum Mechanics*. New York: McGraw-Hill, 1967.

[71] ——, "Finite elements in the solution of field problems," *The Engineer*, pp. 507–510, September 24, 1965.

Hybrid Computing

[72] J. R. Ashley, "Iterative integration of Laplace's equation within symmetric boundaries," *Simulation*, pp. 60–69, August 1967.

[73] J. R. Ashley and T. E. Bullock, "Hybrid computer integration of partial differential equations by use of an assumed sum separation of variables," *1968 Spring Joint Computer Conf., AFIPS Proc.*, vol. 33. Washington, D.C.: Thompson, pp. 585–591.

[74] G. A. Bekey and W. J. Karplus, *Hybrid Computation*. New York: Wiley, 1968.

[75] S. K. T. Hsu and R. M. Howe, "Preliminary investigation of a hybrid method for solving partial differential equations. *1968 Spring Joint Computer Conf., AFIPS Proc.*, vol. 33. Washington, D. C.: Thompson, pp. 601–609.

[76] R. Tomovic and W. J. Karplus, *High Speed Analog Computers*. New York: Wiley, 1962.

[77] R. Vichnevetsky, "A new stable computing method for the serial hybrid computer integration of partial differential equations," *1968 Spring Joint Computer Conf., AFIPS Proc.*, vol. 32. Washington, D. C.: Thompson, pp. 143–150.

Miscellaneous

[78] S. Ramo, J. R. Whinnery, and T. VanDuzer, *Fields and Waves in Communication Electronics*. New York: Wiley, 1965.

Integrated-Circuit Structures on Anisotropic Substrates

NICÓLAOS G. ALEXÓPOULOS, SENIOR MEMBER, IEEE

(Invited Paper)

Abstract —**This paper addresses the problem of anisotropy in substrate materials for microwave integrated-circuit applications. It is shown that in modeling the circuit characteristics, a serious error is incurred which becomes larger with increasing frequency when the substrate anisotropy is neglected. Quasi-static, dynamic, and empirical methods employed to obtain the propagation characteristics of microstrip, coplanar waveguides, and slotlines on anisotropic substrates are presented. Numerical solutions such as the method of moments and the transmission-line matrix technique are outlined. The modified Wiener–Hopf, the Fourier series techniques, and the method of lines are also discussed. A critique of the aforementioned methods and suggestions for future research directions are presented. The paper includes new results as well as a review of established methods.**

I. Introduction

MANY MATERIALS used as substrates for integrated microwave circuits or printed-circuit antennas exhibit dielectric anisotropy which either occurs naturally in the material or is introduced during the manufacturing process. The development of accurate methods and optimization techniques for the design of integrated microwave circuits requires a precise knowledge of the substrate material dielectric constant. It is well recognized that variations in the value of the substrate material relative dielectric constant, as well as possible variations in the value of ϵ for different material batches, introduce errors in integrated-circuit design and reduce integrated-circuit repeatability. For these reasons and because in certain applications anisotropy serves to improve circuit performance, it must be fully and accurately accounted for.

The plurality of substrate materials used for microwave integrated circuits belong to the alumina family. Permittivity variations occurring from batch to batch necessitate repeated measurements for the accurate determination of the dielectric constant [1]; in addition, these materials are slightly anisotropic [2]. Teflon-type substrates are usually ceramic-impregnated, which introduces anisotropic behavior. It is known, e.g., that the E-10 ceramic-impregnated teflon (commonly known as Epsilam 10) is anisotropic with a relative dielectric constant $\epsilon_{yy} = 10.3$ perpendicular and $\epsilon_{xx} = \epsilon_{zz} = 13.0$ parallel to the substrate plane. Similar anisotropies are exhibited by a variety of other teflon substrates such as the TFE/glass cloth and loaded TFE/glass cloth [3].

Manuscript received May 21, 1985; revised June 3, 1985. This work was supported in part by the National Science Foundation, under Research Grant ESC 82-15408 and in part by the U.S. Army, under Research Contract DAAG 29-83-K-0067.

The author is with the Electrical Engineering Department, University of California, Los Angeles, CA 90024.

Among the crystalline substrates, single-crystal sapphire ($\epsilon_{xx} = \epsilon_{zz} = 9.4$, $\epsilon_{yy} = 11.6$) has attracted considerable attention [4], [5]. Sapphire exhibits several very desirable properties in that it is optically transparent, it is compatible with high-resistivity silicon, its electrical properties are reproducible from batch to batch, and it exhibits a 30 percent higher thermal conductivity than alumina [2]. On the other hand, it is produced in rather small area samples (about 22 mm square) and it is quite expensive. Pyrolitic boron nitride is another anisotropic material suggested for potential use as a substrate for microwave applications [6], [7]. Boron nitride exhibits anisotropy with $\epsilon_{xx} = \epsilon_{zz} = 5.12$ and $\epsilon_{yy} = 3.4$.

There are applications where magnetic anisotropy is employed (as in nonreciprocal devices). For such applications, magnetized ferrite materials are used whose magnetic properties are depicted by a second-rank tensor permeability $\bar{\bar{\mu}}$. The elements of $\bar{\bar{\mu}}$ are related to the externally applied dc magnetic field, microwave frequency, as well as the inherent physical properties of the ferrite material [8]. Recently, microstrip [9], [10] and finline [11] have been analyzed on ferrite substrate layers.

The basic interaction of electromagnetic waves with anisotropic materials is well understood. Extensive results exist in the literature for plane-wave propagation through anisotropic materials as well as for guided waves in waveguides loaded with gyrotropic slabs [15]–[26]. As far as the determination of the characteristics of integrated microwave circuits on anisotropic substrates is concerned, however, the existing publications relate mostly to microstrip structures, with a few publications on the analysis of coupled slots and slotlines.

The intent of this paper is to present existing empirical, quasi-static, and dynamic solution methods for the derivation of the propagation characteristics for a variety of structures such as microstrip, coplanar waveguides, and slotlines. Among the quasistatic approaches, the finite differences method [4], [5], the method of moments [27]–[33], and the variational principle [34]–[43] are emphasized. The transmission-line matrix method [44]–[49], the Fourier spectrum approach [51]–[56], and the method of lines [10] constitute the dynamic solution techniques presented in

Reprinted from *IEEE Trans. Microwave Theory Tech.*, vol. MTT-33, no. 10, pp. 847–881, Oct. 1985.

this paper. Empirical methods are discussed, a critique of the accuracy and applicability of each approach is given, and finally, future research directions are suggested.

II. General Background on Anisotropic Substrates

Dielectric substrate materials are either naturally (e.g., crystalline materials) or artificially anisotropic (as a result of the manner under which they are manufactured). In either case, the permittivity of these materials is a second-rank tensor or dyadic, and it is expressed as

$$\bar{\bar{\epsilon}} = [\epsilon_{ij}] = \begin{pmatrix} \epsilon_{11} & \epsilon_{12} & \epsilon_{13} \\ \epsilon_{21} & \epsilon_{22} & \epsilon_{23} \\ \epsilon_{31} & \epsilon_{32} & \epsilon_{33} \end{pmatrix}. \tag{1}$$

For lossless crystals, $\bar{\bar{\epsilon}}$ is symmetric (i.e., $\epsilon_{ij} = \epsilon_{ji}$). For this case, $\bar{\bar{\epsilon}}$ can always be transformed into a diagonalized form

$$\bar{\bar{\epsilon}} = \begin{pmatrix} \epsilon_1 & 0 & 0 \\ 0 & \epsilon_2 & 0 \\ 0 & 0 & \epsilon_3 \end{pmatrix} \tag{2}$$

where the diagonal elements $\epsilon_1, \epsilon_2, \epsilon_3$ are the eigenvalues of $\bar{\bar{\epsilon}}$ and their directions constitute the principal dielectric axes of the crystal. Furthermore, $\bar{\bar{\epsilon}}$ is positive definite, and guarantees that the inverse $\bar{\bar{\epsilon}}^{-1}$ exists. In general, the values of $\epsilon_1, \epsilon_2, \epsilon_3$ are distinct, in which case the crystal is called biaxial [57]. Most of the crystalline substrates considered in this paper are characterized by a single axis of symmetry (optic axis) or equivalently by a diagonal tensor with two equal elements. These crystals are defined as uniaxial. With reference to the geometries shown in Fig. 1, the most general dyadic form of $\bar{\bar{\epsilon}}$ considered in this paper will be

$$\bar{\bar{\epsilon}} = \begin{pmatrix} \epsilon_{xx} & \epsilon_{xy} & 0 \\ \epsilon_{yx} & \epsilon_{yy} & 0 \\ 0 & 0 & \epsilon_{zz} \end{pmatrix}. \tag{3}$$

The dyadic elements ϵ_{xy} and ϵ_{yx} may represent misalignment of the substrate coordinate system with respect to that of the integrated circuit.

As far as magnetic substrates are concerned, the permeability tensor may take the form

$$\bar{\bar{\mu}} = \begin{pmatrix} \mu_{xx} & \mu_{xy} & 0 \\ \mu_{yx} & \mu_{yy} & 0 \\ 0 & 0 & \mu_{zz} \end{pmatrix} \tag{4}$$

when an external dc magnetic field is applied in the $\hat{x}$-direction, or the form

$$\bar{\bar{\mu}} = \begin{pmatrix} \mu_{xx} & 0 & \mu_{xz} \\ 0 & \mu_{yy} & 0 \\ \mu_{zx} & 0 & \mu_{zz} \end{pmatrix} \tag{5}$$

when the external dc magnetic field is applied in the $\hat{y}$-direction. It is the latter case which will be referred to in this paper. In the analytical development which follows, a

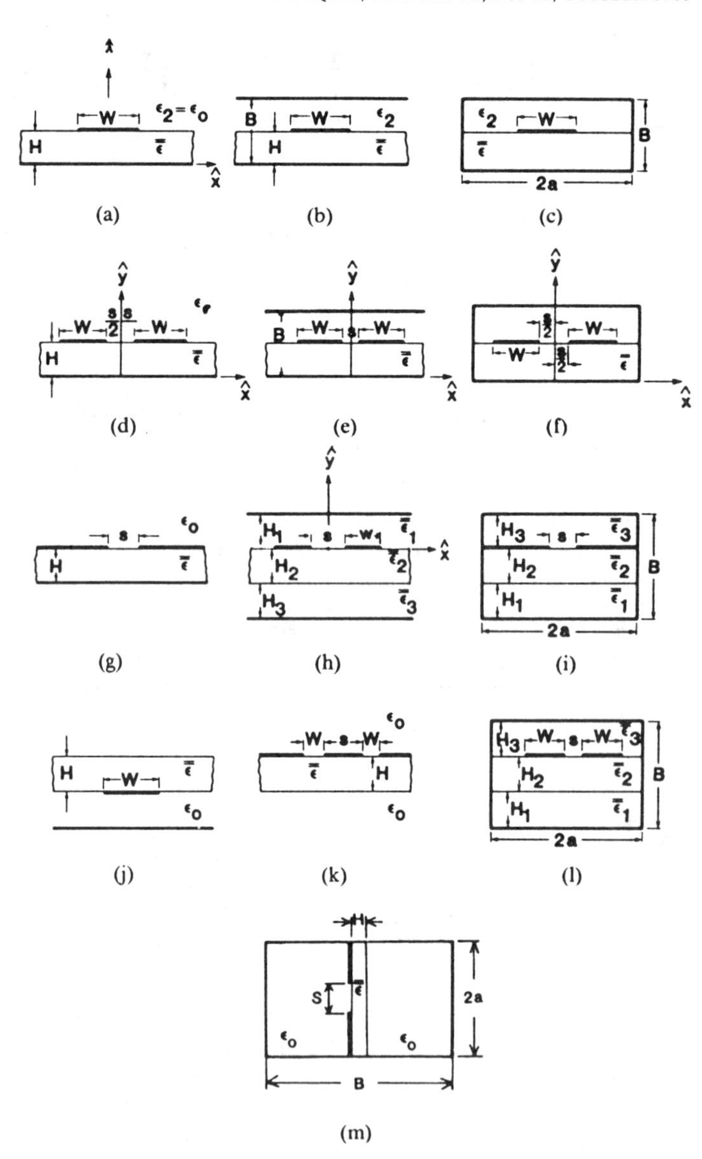

Fig. 1. A variety of integrated-circuit structures on anisotropic substrates.

tensor conductivity will also be allowed, that is,

$$\bar{\bar{\sigma}} = \begin{pmatrix} \sigma_{xx} & 0 & 0 \\ 0 & \sigma_{yy} & 0 \\ 0 & 0 & \sigma_{zz} \end{pmatrix}. \tag{6}$$

Maxwell's equations will be considered in their generalized form; thus, the system of equations to be solved is

$$\nabla \times \boldsymbol{E}(\boldsymbol{r}, t) = -\frac{\partial \boldsymbol{B}}{\partial t}(\boldsymbol{r}, t) \tag{7}$$

$$\nabla \times \boldsymbol{H}(\boldsymbol{r}, t) = \boldsymbol{J} + \frac{\partial \boldsymbol{D}}{\partial t}(\boldsymbol{r}, t) \tag{8}$$

$$\nabla \cdot \boldsymbol{B} = 0 \tag{9}$$

$$\nabla \cdot \boldsymbol{D} = \rho \tag{10}$$

with the constitutive equations

$$\boldsymbol{D}(\boldsymbol{r}, t) = \epsilon_0 \bar{\bar{\epsilon}} \cdot \boldsymbol{E}(\boldsymbol{r}, t) \tag{11}$$

$$\boldsymbol{B}(\boldsymbol{r}, t) = \mu_0 \bar{\bar{\mu}} \cdot \boldsymbol{H}(\boldsymbol{r}, t) \tag{12}$$

and

$$J(r,t)=\bar{\bar{\sigma}}\cdot E(r,t). \tag{13}$$

With the aid of these relations, Maxwell's equations can be written in the rectangular coordinate system as

$$\frac{\partial E_z}{\partial y}-\frac{\partial E_y}{\partial z}=-\mu_0\left(\mu_{xx}\frac{\partial H_x}{\partial t}+\mu_{xz}\frac{\partial H_z}{\partial t}\right) \tag{14}$$

$$\frac{\partial E_x}{\partial z}-\frac{\partial E_z}{\partial x}=-\mu_0\mu_{yy}\frac{\partial H_y}{\partial t} \tag{15}$$

$$\frac{\partial E_y}{\partial x}-\frac{\partial E_x}{\partial y}=-\mu_0\left(\mu_{zx}\frac{\partial H_x}{\partial t}+\mu_{zz}\frac{\partial H_z}{\partial t}\right) \tag{16}$$

and

$$\frac{\partial H_z}{\partial y}-\frac{\partial H_y}{\partial z}=\sigma_{xx}E_x+\epsilon_0\left(\epsilon_{xx}\frac{\partial E_x}{\partial t}+\epsilon_{xy}\frac{\partial E_y}{\partial t}\right) \tag{17}$$

$$\frac{\partial H_x}{\partial z}-\frac{\partial H_z}{\partial x}=\sigma_{yy}E_y+\epsilon_0\left(\epsilon_{yx}\frac{\partial E_x}{\partial t}+\epsilon_{yy}\frac{\partial E_y}{\partial t}\right) \tag{18}$$

$$\frac{\partial H_y}{\partial x}-\frac{\partial H_x}{\partial y}=\sigma_{zz}E_z+\epsilon_0\epsilon_{zz}\frac{\partial E_z}{\partial t}. \tag{19}$$

This system of equations encompasses all the cases to be treated in this paper, and solutions to this system will be provided for particular quasi-static as well as time- and frequency-domain cases. Propagation will be assumed in the z-direction. In the frequency domain, a harmonic time dependence will be considered of the form $e^{+j\omega t}$. This implies that $\partial/\partial z \rightarrow -\gamma$, where γ is the propagation constant ($\gamma = j\beta$ for lossless materials) and $\partial/\partial t \rightarrow j\omega$. For the time-harmonic solutions, the field vectors will be denoted by capital letters as, e.g., $A(r,t)=A(r)e^{j\omega t}$.

III. Anisotropic Materials in Integrated-Circuit Applications

The development of sophisticated analytical methods for the design of microwave integrated circuits on substrates with anisotropy is meaningful only to the extent that the physical parameters describing the anisotropy ($\bar{\bar{\epsilon}}$, $\bar{\bar{\mu}}$, or $\bar{\bar{\sigma}}$) can be accurately determined. For uniaxial crystals, $\epsilon_{\parallel}=\epsilon_{yy}$ is defined as the relative permittivity parallel and $\epsilon_{\perp}=\epsilon_{xx}=\epsilon_{zz}$ as the component perpendicular to the crystal optic axis.

Among the crystalline substrate materials, sapphire has been measured at low [58], microwave [59], [60], infrared [61], and optical frequencies [62]. At 1 KHz, the relative permittivity values were determined as $\epsilon_{\perp}=9.395\pm0.005$ and $\epsilon_{\parallel}=11.589\pm0.005$, while at 3 GHz as $\epsilon_{\perp}=9.39$ and $\epsilon_{\parallel}=11.584$ [59]. More recent results on sapphire in the microwave frequency range of 2–12 GHz indicate $\epsilon_{\perp}=9.34$ and $\epsilon_{\parallel}=11.49$ with ±0.5-percent error [60]. In this case, the measurements were performed on completely and partially metallized sapphire substrates cut with the optic axis either parallel or perpendicular to the substrate surface. The formula

$$\epsilon_{yy}=\left(\frac{c}{2lf_{n,m}}\right)^2(n^2+m^2) \tag{20}$$

is used for the computation of ϵ normal to the broad walls of the cavity, where l is the length of each side of a square substrate sample, and $f_{n,m}$ the measured resonant frequency of the n,mth node. It is estimated that, with this type of procedure, the measured resonant frequency is lower than the actual one by the fraction $\Delta f/f=1/2Q$, where Q is the loaded quality factor of the resonator [12]. This indicates that due to radiation loss at the open ends of the cavity, the method predicts a permittivity ϵ_{yy} higher than the actual value by the factor $\Delta\epsilon/\epsilon_{yy}=2(\Delta f/f)=1/Q$. The Q measurements for the $(n,0)$ mode yield $Q>200$, and therefore the correction to the measured permittivity, due to radiation losses from the cavity, is much smaller than 1 percent. When a completely metallized cavity is used, the measurements produce (due to coupling errors) higher $\bar{\bar{\epsilon}}$ than the actual value. Correct estimates of these types of error are not available, but it is suggested that $\epsilon_{\parallel}=9.40\pm0.01$ and $\epsilon_{\parallel}=11.6\pm0.01$ should be considered as the typical sapphire substrate relative permittivity values for this frequency range [2]. The ±0.01 error range is recommended by both the low-frequency measurements [58] as well as those in the infrared [61] (the latter have shown less than 0.1-percent bulk material dispersion below 300 GHz).

Single-crystal α-quartz is also a useful substrate for both microwave and millimeter-wave applications, with the permittivity tensor elements having been measured as $\epsilon_{\parallel}=4.6368\pm0.001$ and $\epsilon_{\perp}=4.5208\pm0.001$ at 1 KHz [58]. Data extrapolated to zero frequency from measurements in the far-infrared yield $\epsilon_{\parallel}=4.635\pm0.004$ [61], $\epsilon_{\parallel}=4.693\pm0.004$ [62], and $\epsilon_{\parallel}=4.635\pm0.01$ [63]. On the other hand, $\epsilon_{\perp}$ has been measured as $\epsilon_{\perp}=4.436\pm0.004$ [61], $\epsilon_{\perp}=4.46\pm0.004$ [62], and $\epsilon_{\perp}=4.418\pm0.01$ [63], indicating a discrepancy of the latter two measurements from the data obtained in [61] and [63]. Typically, quoted values for $\epsilon_{\parallel}$ and $\epsilon_{\perp}$ are 4.6 and 4.5, respectively [58].

With the exception of crystalline substrates such as sapphire and quartz, the bulk of materials used as substrates for microwave integrated-circuit applications and which exhibit varying degrees of anisotropy are the soft, high-permittivity substrates such as 3M's Epsilam 10® (E-10), Roger's RT/Duroid® 6010, and Keene Corporation's Dieclad® 810. As an example, consideration is given to Epsilam-10, which is a ceramic-impregnated teflon material (low-loss PTFE- (Polytetrafluoroethylene) based substrate). As in all cases where impregnant (fill) materials are introduced so as to obtain substrate dimensional stability, a varying degree of dielectric anisotropy is generated. The permittivity tensor elements of E-10 have the values $\epsilon_{yy}=10.2$ and $\epsilon_{xx}=\epsilon_{zz}=13.0$. The larger permittivity occurs in the xz-plane due to the shear introduced in that plane during processing. The anisotropy of impregnated PTFE materials can be measured by the plated disk test [64] to determine ϵ_{yy} and by the TE_{111} cavity test [3] (estimated accuracy of this method is 0.1–0.2 percent) for the ϵ_{xx} and ϵ_{zz} elements. A list of data from such measurements is shown in Table I for PTFE materials.

Substrate materials such as woven glass PTFE laminates consist of glass fibers oriented along planes parallel to the

TABLE I
DIELECTRIC ANISOTROPY FOR PTFE SUBSTRATES [3]

Material	Description	Sample Thickness (cm)	ϵ_{yy}	$\epsilon_{xx}, \epsilon_{zz}$
Unfilled PTFE	2 Discs	0.522	2.08	2.09
PTFE Glass	CuClad 217	0.051	2.15	2.34
PTFE Cloth	Old CuClad 2.45	0.153	2.45	2.89/2.95
PTFE Cloth	New CuClad 2.45	0.153	2.43	2.88
Filled PTFE (Glass Cloth)	GL 606	0.153	6.24	6.64/5.56

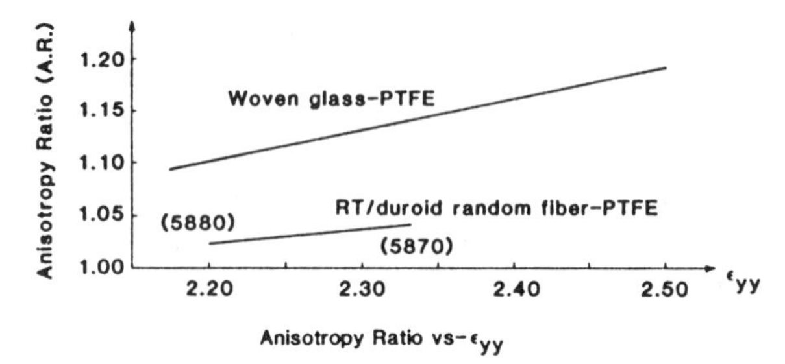

Fig. 2. Anisotropy Ratio versus ϵ_{yy}. Reprinted by permission from Rogers Co., TR 2692, July 1981.

xz-plane. These glass fiber planes are interspersed in the y-direction with the polymer matrix. From an equivalent network point of view, the substrate appears as a three-dimensional capacitance network with series connections in the $\hat{y}$-direction and parallel connections in the x, y-direction. This equivalent representation indicates that $\epsilon_{yy} < \epsilon_{xx}$, $\epsilon_{yy} < \epsilon_{zz}$. If it is desired to minimize anisotropy in composite substrate materials, the obvious solution is to (again considering the woven glass PTFE as an example) orient the glass fibers randomly. The effectiveness of this approach is shown in Fig. 2 where the measured anisotropy ratio $\mathrm{AR} = \epsilon_{xx}/\epsilon_{yy} = \epsilon_{zz}/\epsilon_{yy}$ is graphed as a function of ϵ_{yy} for the woven glass PTFE and RT/Duroid random fiber PTFE substrates [65]. Clearly there is considerable reduction in anisotropy when a random rather than an ordered orientation of the impregnant glass fiber is enforced in the PTFE base material. Where anisotropy is not accounted for analytically, low ϵ composite substrates with randomly oriented filling to reduce anisotropy should result in more successful designs for microwave printed-circuit antennas. On the other hand, high ϵ soft substrate materials which are useful for microwave integrated circuits may exhibit considerable anisotropy even for random orientation of the filling substance. This is readily observed if a linear, albeit arbitrary, extrapolation of the random fiber-filled PTFE substrate curve is constructed. Such an extrapolation indicates, with $\epsilon_{xx} = \epsilon_{zz} \simeq 16$ and $\epsilon_{yy} = 10$, an anisotropy ratio of 1.6, which is perhaps high. The example shows, however, that for high ϵ soft substrates which are impregnated with another material matrix, anisotropy is not negligible and should be accounted for in the development of high-accuracy design procedures.

IV. QUASI-STATIC METHODS

A full-wave analysis of integrated-circuit structures (such as microstrip) on anisotropic substrates involves the development of guided waves in terms of hybrid modes. A much simpler approach is to consider that the structure supports a dominant TEM mode, an argument especially valid at low frequencies. Under the assumption of a dominant TEM mode, a simplified design procedure evolves in a rather straightforward manner, since the guided-wave field components can be derived from the solution to Laplace's equation.

The design parameters of the microstrip structures of Fig. 1 are the characteristic impedance Z_0 and effective dielectric constant ϵ_{eff}. These parameters are defined for nonmagnetic substrates by

$$Z_0 = \frac{1}{c\sqrt{C_a C_s}} \quad \text{and} \quad \epsilon_{\text{eff}} = C_s/C_a \tag{21}$$

where c is the speed of light in vacuum, C_a and C_s denote the capacitance of the strip conductor in the absence and presence of the substrate, respectively, and ϵ_{eff} is the effective dielectric constant of the structure. The computation of C_a or C_s is obtained from the definition $C_r = Q_r/V_0$ ($r = a$ or $r = s$), where V_0 is the potential of the strip with respect to ground. The total charge Q is given as

$$Q_r = \int_{-w/2}^{w/2} \rho_r(x')\, dx' \tag{22}$$

where $\rho_r(x')$ is the unknown charge density on the strip. It is clear now that the central problem of a quasi-static method is the determination of $\rho_r(x)$. If $\rho_r(x)$ is known, then the potential at any point (x, y) is given by

$$\phi_r(x, y) = \int_{-w/2}^{w/2} \rho_r(x') G_r(x - x', y - H)\, dx' \tag{23}$$

where $G_r(x - x', y - y')$ is the Green's function pertinent to the boundary-value problem. On the conductor strip ($y = H$, $|x| \leq w/2$, $|x'| \leq w/2$), $\phi(x, H) = V_0$ and therefore

$$V_{0_r} = \int_{-w/2}^{w/2} \rho_r(x') G_r(x - x')\, dx' \qquad (|x| \leq w/2,\ |x'| \leq w/2). \tag{24}$$

This is a Fredholm integral equation of the first kind to be solved for $\rho_r(x')$. The Green's function $G_r(x - x', y - y')$ is obtained by considering Laplace's equation for the given boundary-value problem. In the anisotropic medium, Laplace's equation is obtained from $\nabla \cdot \boldsymbol{D} = 0$, $\boldsymbol{D} = \bar{\bar{\epsilon}} \cdot \boldsymbol{E}$, and $\boldsymbol{E} = -\nabla\phi$, in the form

$$\nabla \cdot [\bar{\bar{\epsilon}} \cdot \nabla\phi(x, y)] = 0. \tag{25}$$

Equation (25) can be solved by the finite differences technique [4], variable substitution [66]–[71], or by Fourier transform methods [27], [72], [73]. The method is easily extendable to the characterization of coupled microstrip lines, as those shown in Fig. 1. This is readily achieved by composing the solution in terms of even- $(+V_0, V_0)$ and odd- $(+V_0, -V_0)$ mode excitation of the coupled lines (see Fig. 1(f)). Under this scheme, the even- and odd-mode impedances, and the effective dielectric constants are defined as

$$Z_0^{\left\{ e \atop o \right\}} = \frac{1}{c\sqrt{C_a^{\left\{ e \atop o \right\}} C_s^{\left\{ e \atop o \right\}}}} \qquad \epsilon_{\text{eff}} = \frac{C_s^{\left\{ e \atop o \right\}}}{C_a^{\left\{ e \atop o \right\}}} \tag{26}$$

where

$$C_r^{\left\{ e \atop o \right\}} = Q_r^{\left\{ e \atop o \right\}} / V_0 \qquad Q_r^{\left\{ e \atop o \right\}} = \int_{s/2}^{s/2+w} \rho_r^{\left\{ e \atop o \right\}}(x')\, dx'. \tag{27}$$

In addition, $\rho_r^{\left\{ e \atop o \right\}}(x')$ is obtained by solving the integral equations

$$1 = \int_{s/2}^{s/2+w} \rho_r^{\left\{ e \atop o \right\}}(x') G_r^{\left\{ e \atop o \right\}}(x-x')\, dx' \tag{28}$$

where

$$G_r^{(e)}(x-x') = G_r(x-x') + G_r(x+x'+s) \tag{29}$$

and

$$G_r^{(0)}(x-x') = G_r(x-x') - G_r(x+x'+s). \tag{30}$$

The even- and odd-mode excitations are equivalent to erecting magnetic and electric walls, respectively, on the $x=0$ plane. The single-strip case can be obtained from the even-mode excitation as $s/H \rightarrow 0$.

Various quasi-static design procedures have evolved for the determination of Z_0 and ϵ_{eff}. These procedures are 1) finite differences, 2) empirical, 3) method of moments, 4) coordinate transformation (variable substitution), and 5) variational methods. Each of these procedures will now be presented and their advantages as well as limitations discussed.

A. *Finite Differences*

The finite differences technique has been employed to obtain design parameters for microstrip without cover on a sapphire substrate [4], [5]. The $\bar{\bar{\epsilon}}$ is assumed diagonal, and therefore Laplace's equation becomes

$$\epsilon_{ixx} \frac{\partial^2 \phi_i(x,y)}{\partial x^2} + \epsilon_{iyy} \frac{\partial^2 \phi_i(x,y)}{\partial y^2} = 0 \tag{31}$$

where $i=1,2$ denotes the region of validity. At the interface between the two layers, the tangential electric field and its gradient must be continuous and

$$\frac{\partial^2 \phi_1(x,H)}{\partial x^2} = \frac{\partial^2 \phi_2(x,H)}{\partial x^2} \tag{32}$$

is obtained. In addition, the normal $\boldsymbol{D}$ component is continuous, i.e.,

$$\epsilon_{1yy} \frac{\partial \phi_1(x,H)}{\partial y} = \epsilon_{2yy} \frac{\partial \phi_2(x,H)}{\partial y}. \tag{33}$$

If the grid shown in Fig. 3 is considered, then by the relaxation method, the potential at A, ϕ_{iA} can be obtained in terms of the potentials ϕ_{iB}, ϕ_{iC}, ϕ_{iD}, and ϕ_{iE}. Now the finite difference equation can be derived in its general form

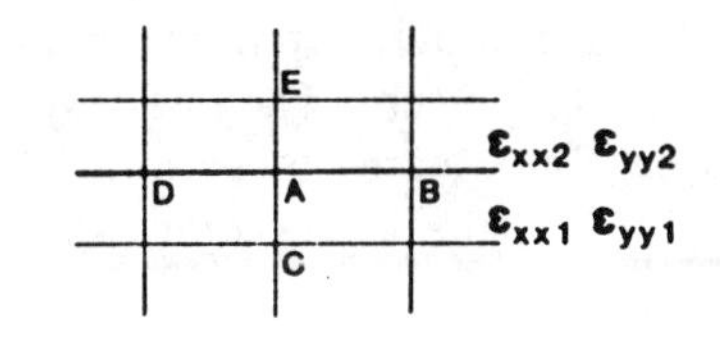

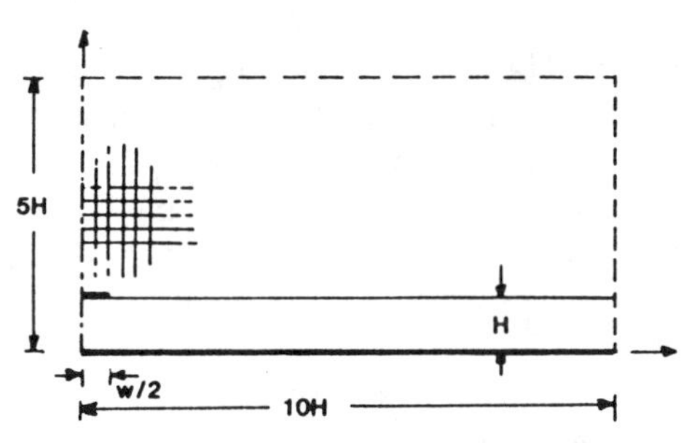

Fig. 3. Rectangular grid for application of the method of finite differences.

by eliminating derivatives in the previous equations, i.e.,

$$\frac{1}{2}(\epsilon_{2xx} + \epsilon_{1xx})(\phi_{iB} + \phi_{iD}) + \epsilon_{2yy}\phi_{iE} + \epsilon_{1yy}\phi_{iC} - (\epsilon_{2xx} + \epsilon_{1xx} + \epsilon_{2yy} + \epsilon_{1yy})\phi_{iA} = 0. \tag{34}$$

There are three regions where this generalized finite difference equation applies and where it takes distinct forms.

1) *Anisotropic Region 1 ($0<y<H$):* In this region, $\epsilon_{2yy} = \epsilon_{1yy}$, $\epsilon_{2xx} = \epsilon_{1xx}$, and (34) becomes

$$\epsilon_{xx}(\phi_{1B} + \phi_{1D}) + \epsilon_{yy}(\phi_{1E} + \phi_{1C}) - 2(\epsilon_{xx} + \epsilon_{yy})\phi_{1A} = 0. \tag{35}$$

2) *Interface ($y=H$):* Here the top points of the grid are in air, and therefore $\epsilon_{2xx} = \epsilon_{2yy} = 1$. In addition, by writing $\epsilon_{1xx} = \epsilon_{xx}$, $\epsilon_{1yy} = \epsilon_{yy}$, $\phi_{1A} = \phi_{2A} = \phi_A$, $\phi_{1B} = \phi_{2B} = \phi_B$, $\phi_{1D} = \phi_{2D} = \phi_D$, the equation reduces to

$$\frac{1}{2}(1+\epsilon_{xx})(\phi_B + \phi_D) + \phi_{2E} + \epsilon_{yy}\phi_{1C} - (2 + \epsilon_{xx} + \epsilon_{yy})\phi_A = 0. \tag{36}$$

3) *$y>H$:* The entire grid is in the second region, which is assumed to be a vacuum, and therefore $\epsilon_{1xx} = \epsilon_{1yy} = \epsilon_{2xx} = \epsilon_{2yy} = 1$. The difference equation is simplified now into

$$\phi_{2B} + \phi_{2C} + \phi_{2D} + \phi_{2E} - 4\phi_{2A} = 0. \tag{37}$$

The microstrip capacitance may be computed using the definition

$$C_s = \epsilon_0 \int\!\!\int_S \epsilon_n E_n\, ds \tag{38}$$

where the subscript n denotes the direction normal to the strip. Thus, under the strip $\epsilon_n = \epsilon_{yy}$, and in the air region $\epsilon_n = 1$. The choice of ϵ_n at the strip edges (substrate–air interface) is dictated by the coefficient of the term $\phi_B + \phi_D$ in (36), i.e., $\epsilon_n = (1+\epsilon_{xx})/2$. In applying the relaxation method, an overrelaxation factor of $\alpha \simeq 1.8$ has been assumed [4]. The grid is established by choosing the substrate thickness $H = 4N$, where N is the number of grid points. A zero potential boundary is assumed at $x = \pm 10H(\pm 40N)$ and at $y=0$ and $y=5H(20N)$. Using this scheme, the capacitance is computed iteratively until the value changes by less than 0.1 percent. The final asymptotic value of $C_{s\infty}$ is subsequently obtained by extrapolation [2], [4].

For $w/H \geqslant 1$, the asymptotic capacitance value $C_{s\infty}$ is reached quickly, i.e., the finite differences method converges rapidly. When $w/H < 1$, however, the method converges very slowly and it becomes increasingly costly to determine $C_{s\infty}$. The evaluation of $C_{a\infty}$ proves more difficult, even over the range of $w/H > 1$. An accuracy of 0.5 percent is estimated [4] in computing $C_{s\infty}$ when $w/H = 0.125$ with $N = 12$. With $w/H = 9$, the method converges rapidly but the result is sensitive to box size (the estimated error for the range is stated to be of the order of 0.5 percent [4]).

This method suffers from sensitive convergence problems depending on the w/H range and box size. Even though the method is theoretically extendable to treat coupled lines, the computer cost to obtain desirable accuracy would be prohibitive. In fact, the method was not found accurate enough to obtain Z_0 and ϵ_{eff} [4] by direct computation of C_a and C_s. Consequently, an important equivalent permittivity ϵ_{req} was defined as that for isotropic substrate permittivity, which yields the same Z_0 and ϵ_{eff} [4] as the anisotropic layer. The parameter ϵ_{req} is computed by using the method of finite differences from the definition

$$\epsilon_{req} = \epsilon_b + (\epsilon_c - \epsilon_b)\left(\frac{C_{s\infty} - C_{b\infty}}{C_{c\infty} - C_{b\infty}}\right) \tag{39}$$

where ϵ_b and ϵ_c are the isotropic permittivities above and below the anticipated ϵ_{req} ($C_{b\infty}, C_{c\infty}$ are the asymptotic capacitance values for the cases corresponding to ϵ_b, ϵ_c). The behavior of ϵ_{req} with respect to the linewidth ratio w/H as obtained by the method of finite differences through the use of (39) is shown in Fig. 4(a) for a sapphire substrate [4].

B. *Empirical Methods*

In order to obtain a design method for microstrip on a sapphire substrate, it is possible to utilize the results obtained for ϵ_{req} with the method of finite differences and develop an empirical design approach. To this end, the empirical formula

$$\epsilon_{req} = 12.0 - \frac{1.21}{1 + 0.39\left[\log\left(\frac{10w}{H}\right)\right]^2} \tag{40}$$

has been developed [4]. The accuracy of this formula has been estimated to be ± 0.5 percent in the range $0.1 \leqslant w/H \leqslant 10.0$, and it may be used with existing methods [74], [75] for microstrip on isotropic substrates to yield design graphs for ϵ_{req} and Z_0 as shown in Fig. 4(b) [4]. The accuracy of Z_0 obtained in this manner is reported to be 4 percent for $w/H = 0.1$, and it is claimed that it improves to 0.5 percent when $w/H = 1.0$ [4]. These accuracy estimates have been found actually to be on the conservative side. A quasi-static method of moments solution indicates that (40) yields a better than 0.4-percent accuracy for $0.1 \leqslant w/H \leqslant 10.0$, while the error in Z_0 is less than 1 percent for the same range in w/H. The empirical formula given for ϵ_{req} is simple and useful but it suffers from lack of generality, as it applies only to sapphire substrates.

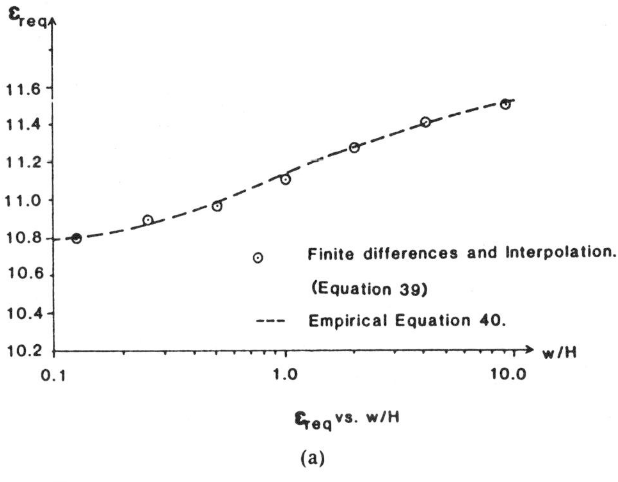

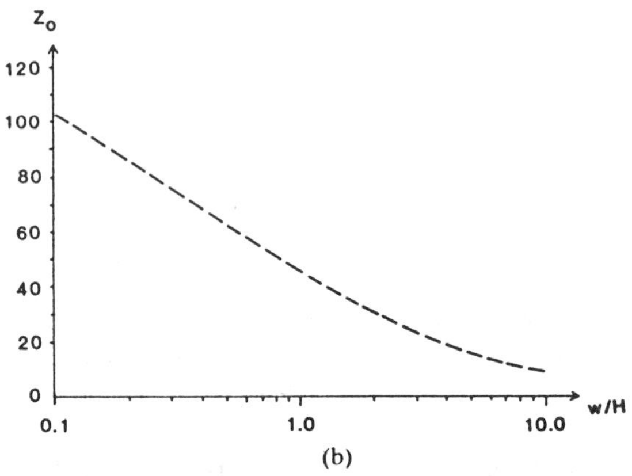

Fig. 4. (a) ϵ_{req} versus w/H. (b) Characteristic impedance Z_o versus w/H.

C. *The Method of Moments*

The method of moments [76] has proven to be a very useful numerical technique in solving a variety of engineering problems in electromagnetics. It will be adopted here to obtain the total charge per unit length for a single line, and the even- and odd-mode total charges for coupled microstrip lines. The unknown charge density $\rho_f^{\{e \atop o\}}(x)$ is expanded into a series of the form

$$\rho^{\{e \atop o\}}(x) = \sum_{n=1}^{N} \alpha_n^{\{e \atop o\}} f_n(x) \tag{41}$$

where $\{f_n(x)\}$, $n = 1, 2, 3, \cdots, N$, is a set of known basis (expansion) functions, and the α_n are unknown coefficients. Substitution of this expansion into (28) yields the following pair of integral equations for the even- and odd-mode unknown coefficients $\alpha_n^{\{e \atop o\}}$:

$$1 = \sum_{n=1}^{N} \alpha_n^{\{e \atop o\}} \int_{s/2}^{s/2+w} G^{\{e \atop o\}}(x - x') f_n(x')\, dx'. \tag{42}$$

At this point, another set of known functions $\{w_m(x)\}$,

$m = 1, 2, 3, \cdots, N$, (the testing functions) is selected and the inner product is formed with both sides of (42), i.e.,

$$\langle 1, w_m(x)\rangle = \sum_{n=1}^{N} \alpha_n^{\left\{ e \atop o \right\}} \cdot \left\langle w_m(x), \int_{s/2}^{s/2+w} G^{\left\{ e \atop o \right\}}(x - x') f_n(x')\, dx' \right\rangle \tag{43}$$

where the inner product is defined by

$$\langle f(x), w(x)\rangle = \int_{s/2}^{s/2+w} f(x) w(x)\, dx. \tag{44}$$

Equation (43) can now be written as

$$\beta_m^{\left\{ e \atop o \right\}} = \sum_{n=1}^{N} \alpha_n^{\left\{ e \atop o \right\}} g_{mn}^{\left\{ e \atop o \right\}} \tag{45}$$

where

$$\beta_m^{\left\{ e \atop o \right\}} = \langle 1, w_m(x)\rangle \tag{46}$$

and

$$g_{mn}^{\left\{ e \atop o \right\}} = \left\langle w_m(x), \int_{s/2}^{s/2+w} G^{\left\{ e \atop o \right\}}(x - x') f_n(x')\, dx' \right\rangle \tag{47}$$

or in matrix form

$$[\beta_m] = [g_{mn}][\alpha_n]. \tag{48}$$

Upon inversion, the coefficients α_n are

$$[\alpha_n] = [g_{mn}]^{-1}[\beta_m]. \tag{49}$$

The total line charge per unit length finally takes the form

$$Q^{\left\{ e \atop o \right\}} = \sum_{N=1}^{N} \alpha^{\left\{ e \atop o \right\}} \int_{s/2}^{s/2+w} f_n(x)\, dx \tag{50}$$

and therefore the even- and odd-mode capacitance is $C^{\left\{ e \atop o \right\}} = Q^{\left\{ e \atop o \right\}}$, since $V^{\left\{ e \atop o \right\}} = 1$. These capacitance values are subsequently employed to determine $Z^{\left\{ e \atop o \right\}}_0$, $\epsilon^{\left\{ e \atop o \right\}}_{\text{eff}}$, and $v^{\left\{ e \atop o \right\}}_p$. The single-line results can be readily obtained by allowing $s/2 \to 0$ in the formulation for the even mode. Recognize that C_r for a single strip $= 2C_r$ for coupled lines, Z_0 for a single strip $= \frac{1}{2}Z_0^e$ for coupled lines, and w/H of a single strip $= \frac{1}{2}w/H$ for coupled lines. A significant feature of the method is the choice of expansion and testing functions. These functions are critical both in the complexity of analysis as well as rate of convergence. Often the choice $\{w_n(x)\} = \{f_n(x)\}$ is made; this selection of testing and expansion functions is known as the Galerkin method [76]. For the problem at hand, $\{f_n(x)\} = \{p_n(x)\}$ is chosen where $\{p_n(x)\}$ is a set of pulse functions defined by

$$p_n(x) = P_n(x - x_n) = \begin{cases} 1, & x_n - \Delta/2 \leqslant x \leqslant x_n + \Delta/2 \\ 0, & x < x_n - \Delta/2,\ x > x_n + \Delta/2 \end{cases} \tag{51}$$

and $\Delta = (w/H)/N$. Furthermore, the testing functions $\{w_m(x)\}$ are chosen as $w_m(x) = \delta(x - x_m)$ the point-matching method, with $x_m = \Delta(m - 1/2)$. N denotes here the number of subsections the metallic strip is divided into. Finally, the $g^{\left\{ e \atop o \right\}}_{mn}$ matrix elements reduce to

$$g_{mn}^{\left\{ e \atop o \right\}} = \int_{x_n - \Delta/2}^{x_n + \Delta/2} G^{\left\{ e \atop o \right\}}(x_m - x')\, dx'. \tag{52}$$

This approach applies to a variety of cases.

1) Microstrip on Anisotropic Substrate with Optic Axis Misalignment: It is assumed that the principal axes (ξ, η) of the anisotropic substrate form an angle θ with respect to the microstrip coordinate system (x, y) [77] (see Fig. 7). The elements of the relative permittivity tensor are given in the microstrip coordinate system by

$$\begin{aligned} \epsilon_{xx} &= \epsilon_{\xi\xi}\cos^2\theta + \epsilon_{\eta\eta}\sin^2\theta \\ \epsilon_{yy} &= \epsilon_{\xi\xi}\sin^2\theta + \epsilon_{\eta\eta}\sin^2\theta \\ \epsilon_{xy} &= \epsilon_{yx} = (\epsilon_{\xi\xi} - \epsilon_{\eta\eta})\sin\theta\cos\theta \end{aligned} \tag{53}$$

and

$$\epsilon_{zz} = \epsilon_{zz}.$$

To proceed with the method of moments, the Green's function is obtained [27] by solving the boundary-value problem for the potential functions $\phi_1(x, y)$ and $\phi_2(x, y)$. The potentials must satisfy Laplace's equation and the pertinent boundary conditions.

In Region 1:

$$\epsilon_{xx}\frac{\partial^2\phi_1}{\partial x^2} + 2\epsilon_{xy}\frac{\partial^2\phi_1}{\partial x\,\partial y} + \epsilon_{yy}\frac{\partial^2\phi_1}{\partial y^2} = 0. \tag{54}$$

In Region 2:

$$\frac{\partial^2\phi_2}{\partial x^2} + \frac{\partial^2\phi_2}{\partial y^2} = 0. \tag{55}$$

The boundary conditions are

$$\phi_1(x, 0) = 0 \tag{56}$$

$$\phi_2(x, B) = 0 \tag{57}$$

$$\phi_1(x, H) = \phi_2(x, H) \tag{58}$$

and

$$\left[\epsilon_{yx}\frac{\partial\phi_1(x, y)}{\partial x} + \epsilon_{yy}\frac{\partial\phi_1(x, y)}{\partial y} - \frac{\partial\phi_2(x, y)}{\partial y}\right]_{y=H} = \frac{\rho_l}{\epsilon_0}\delta(x - x') \tag{59}$$

where ρ_l is the line of charge at $x = x'$, $y = H$, generating the potentials $\phi_1(x, y)$ and $\phi_2(x, y)$. This problem can be solved by using a Fourier transform in x and integrating the resulting ordinary differential equation in y. The transform potential in region 1 is found to be

$$\tilde{\phi}_1(\zeta, y) = \frac{\rho_l}{2\pi\epsilon_0}\frac{\exp[-j\zeta(y - H)]}{\zeta[\alpha\delta\coth(\zeta\delta H) + \coth(\zeta H\nu)]} \cdot \frac{\sinh(\zeta\delta y)}{\sinh(\zeta\delta H)} \tag{60}$$

with

$$\delta = \left[\frac{\epsilon_{xx}}{\epsilon_{yy}} - \left(\frac{\epsilon_{xy}}{\epsilon_{yy}}\right)^2\right]^{1/2} \quad \alpha = \epsilon_{yy} \text{ and } \nu = \frac{B}{H} - 1. \tag{61}$$

After deriving the inverse transforms of $\tilde{\phi}_1$ and $\tilde{\phi}_2$, the Green's function pertinent to the characterization of single and coupled microstrip is obtained from either $\phi_1(x, y)$ or $\phi_2(x, y)$ by letting $\rho_l = 1$ and $y = H$. The Green's function is

$$G(x-x') = \frac{1}{2\pi\epsilon_0}\int_{-\infty}^{\infty} d\zeta \frac{\cos[\zeta|x-x'|]}{\zeta[\alpha\delta\coth(\zeta\delta H)+\coth(\zeta H\nu)]} \quad (62)$$

from which $G^{\{e \atop o\}}(x-x')$ is obtained by considering (29) and (30). The elements of the $[g_{mn}^{\{e \atop o\}}]$ matrix are derived in the form

$$g_{mn}^{\{e \atop o\}} = \frac{4}{\pi\epsilon_0}\int_0^{\infty} \frac{\sin\left(\frac{\alpha\zeta}{2}\right)\left[{\cos \atop \sin}\left(\zeta\left[x_m+\frac{s}{2}\right]\right)\right]}{\zeta^2[\alpha\delta\coth(\delta H\zeta)+\coth(\nu H\zeta)]} \cdot \left[{\cos \atop \sin}\left(\zeta\left[x_n+\frac{s}{2}\right]\right)\right] d\zeta. \quad (63)$$

The upper limit of integration in (63) is chosen as $\zeta_m = \max(A/\delta, A/\nu)$, where A is determined from $\tanh(A) \simeq 1.0$ (for $\tanh(A) \simeq 0.999$, $A = 5$). This value of A results in the negligible error of 0.009 percent. The Green's function can also be written in the following series form by considering analytic continuation and the Cauchy residue theorem [28]:

$$G(x-x') = \frac{1}{\epsilon_0 H}\sum_{l=1}^{\infty} \frac{\exp[-v_l|x-x'|]}{v_l[\alpha\delta^2\csc^2(v_l\delta H)+\nu\csc^2(v_l H\nu)]} \quad (64)$$

where v_l is the lth root of the transcendental equation

$$\sin[v_l(\delta+\nu)H] + M\sin[v_l(\nu-\delta)H] = 0 \quad (65)$$

with

$$M = \frac{\alpha\delta-1}{\alpha\delta+1}.$$

For this representation, the matrix elements are given by

$$g_{mn}^{\{e \atop o\}} = \frac{2}{\epsilon_0 H}\sum_{l=1}^{\infty}\frac{1}{v_l^2[\alpha\delta^2\csc^2(v_l\delta H)+\nu\csc^2(v_l H\nu)]} \times \begin{cases} 1-\exp\left[-v_l\frac{\Delta}{2}\right] \pm \sinh\left(\frac{v_l\Delta}{2}\right)\exp[-v_l|x_m+x_n+s|] \\ \sinh\left(\frac{v_l\Delta}{2}\right)\exp[-v_l|x_m-x_n|] \pm \exp[-v_l|x_m+x_n+s|] \end{cases} \quad (66)$$

where the upper form is valid for $x_m = x_n$, while the lower is for $x_m \neq x_n$.

Computations have been carried out for single and coupled microstrip lines with a convergence accuracy better than 0.5 percent (any desired convergence accuracy is obtainable by increasing the number of subsections N). The method of moments has been compared with other techniques for microstrip on isotropic substrates. The Bryant and Weiss approach [74] agrees well with the method of moments (to within 1 percent) except for very small linewidths where for coupled lines a discrepancy on the order of 3.5 percent is observed for the odd-mode impedance ($w/H = 0.1$, $s/H = 0.1$). It appears that when $w/H < 0.2$, the accuracy of the Bryant and Weiss results for coupled lines is somewhat questionable due to a coarse subdivision of the lines [74]. In addition, the increase in error for small linewidths may be due to the sensitivity of the finite differences method to mesh size.

The Finite Differences–Capacitance Interpolation (FD–CI) procedure (which includes the incorporation of the Bryant and Weiss algorithm) has been compared against the method of moments for a single microstrip line on a sapphire substrate. The method of moments is used in two computations whereby in one case the tensor permittivity $\bar{\bar{\epsilon}}$ is involved (MMA), while in the subsequent case ϵ_{req} is employed (MMEI). Table II summarizes the results for this comparison. For the method of moments, the geometry of Fig. 1(b) is considered with $B/H = 5.0$ and $B/H = 20.0$. The case of $B/H = 5.0$ is chosen since the empirical formula for ϵ_{req} is derived in [4] for the equivalent box size of $B/H = 5.0$. Table II indicates that when $B/H = 5.0$, the percent error is considerable when the FD–CI procedure is compared with the method of moments. On the other hand, when $B/H = 20.0$ (essentially an open structure), the agreement is very good. This may be due to the possible use in [4] of the Bryant and Weiss algorithm for an open structure. The method of moments quasi-static results shown in Table II have also been verified with excellent agreement by considering the low-frequency limit of a dynamic solution [56] for the geometry of Fig. 1(c) with the proper dimensions.

An investigation has also been carried out to determine the error introduced when the anisotropic nature of a given substrate is neglected and in addition to clarify the effect of the anisotropy ratio (AR) on line characteristics.

Table III provides a comparison of results for Z_0 and ϵ_{eff} versus w/H for an Epsilam-10 substrate with $\epsilon_{xx} = \epsilon_{zz} = 13$, and $\epsilon_{yy} = 10.3$. As Table III indicates, the error increases for narrow linewidths. This is due to the fact that the fringing field is not taken into account correctly when anisotropy is neglected, an ommission which leads to erroneous calculation of the guided wavelength, resonant length, and subsequently inaccurate equivalent-circuit representations. The method of moments yields a faster convergence in computing ϵ_{eff} than Z_0. For 0.5-percent convergence accuracy, the largest number of subsections needed for a single line was $N = 16$.

Table IV provides an understanding of the rate of convergence on the UCLA IBM 3033 computer. The results

TABLE II
COMPARISON OF FINITE DIFFERENCES (FD–CI) AND THE EMPIRICAL TECHNIQUE (MMET) WITH THE METHOD OF MOMENTS (MMA)

w/H	Z_0	Z_0	% Error	Z_0	% Error
0.125	97.095	97.76	0.68	96.982	0.80
	97.858		0.1	97.752	0.01
1.0	45.845	46.60	1.62	45.765	1.79
	46.569		0.07	46.439	0.35
9.0	9.651	10.17	5.11	9.634	5.27
	10.143		0.27	10.126	0.43
w/H	ϵ_{eff}	ϵ_{eff}	% Error	ϵ_{eff}	% Error
0.125	6.4048	6.522	1.80	6.4197	1.57
	6.4924		0.45	6.5065	0.24
1.0	7.1699	7.391	2.99	7.1949	2.65
	7.3633		0.37	7.3875	0.05
9.0	8.8135	9.718	9.31	8.8439	8.99
	9.5671		1.55	9.5977	1.24

Upper data: $B/H = 5.0$; lower data: $B/H = 20.0$.

TABLE III
ERROR IN NEGLECTING ANISOTROPY—SINGLE LINE

w/H	Z_0 $\epsilon=10.3$ isotropic	Z_0 $\epsilon_{xx}=\epsilon_{zz}=13$ $\epsilon_{yy}=10.3$	%Error in Z_0	ϵ_{eff} $\epsilon=10.3$ isotropic	ϵ_{eff} $\epsilon_{xx}=\epsilon_{zz}=13$ $\epsilon_{yy}=10.3$	% Error in ϵ_{eff}
0.1	105.3763	101.2740	3.89	6.1854	6.6967	8.27
1.0	48.0923	46.9167	2.44	6.8345	7.1813	5.07
2.0	32.7080	32.1597	1.68	7.2877	7.5384	3.44
3.0	25.0139	24.6982	1.26	7.6128	7.8086	2.57
4.0	20.3157	20.1101	1.01	7.8537	8.0151	2.06
5.0	17.1296	16.9846	0.85	8.0379	8.1758	1.72
7.0	13.0643	12.9792	0.65	8.2980	8.4073	1.32
9.0	10.5852	10.5243	0.58	8.4645	8.5628	1.16

All results within 0.5-percent convergence accuracy.

for a case of coupled lines ($s/H = 0.1$, $B/H = 10.0$) are also summarized in Table V. The subscripts i and a refer to Epsilam-10 with anisotropy omitted or taken into account, respectively. It is observed that the error is larger for the odd-mode characteristic impedance and ϵ_{eff} than for the even mode. The discrepancy in ϵ_{eff} is also shown in Fig. 5. Further computations demonstrate that the error increases as s/H decreases (computations for $s/H = 1.0$ yield a 6.77-percent error in ϵ_{eff}^e, which is lower than the corresponding cases when $s/H = 0.1$). This is due to higher intensity fringing fields between the lines for the anisotropic than for the isotropic substrate. The coupled-line algorithm required $N = 32$ for the desired convergence accuracy (5.39-s CPU) with the odd-mode converging more slowly than the even mode.

Equalization of even–odd-mode phase velocities is a goal for improving integrated-circuit performance such as the directivity D of directional couplers [77], [78]. In theory, anisotropic substrates can equalize the even- and odd-mode phase velocities for coupled microstrip without a cover, but the required AR is not realizable with known substrate materials. On the other hand, if a cover is used, the requirement $v_p^e = v_p^o$ is possible for practical isotropic as well as anisotropic substrates. Fig. 6 illustrates the behavior of ϵ_{eff} and Z_0 for coupled lines with AR > 1 (AR = 1.26 for Epsilam-10), AR = 1 (isotropic substrate with $\epsilon = 10.3$), and AR < 1 (AR = 0.89 for sapphire) versus B/H. Equalization of phase velocities is achieved in all three cases. Note, however, that the smaller B/H is, the more sensitive the coupler design is to tolerance errors. Substrates with AR > 1 should be utilized where phase velocity equalization and lower sensitivity to tolerance errors are desired. Table VI summarizes eight different directional coupler

TABLE IV
METHOD OF MOMENTS CPU REQUIREMENT

w/H	N	Z_0	ϵ_{eff}	IBM 3033 CPU (seconds)
0.1	8	101.6595	6.6951	0.390
1.0	12	47.1137	7.1778	0.585
5.0	16	17.0495	8.1724	0.885
7.0	16	13.0352	8.4043	0.880
9.0	16	10.5705	8.5604	0.885

$B/H = 10.0$, Fig. 1(b). For 0.5-percent convergence accuracy. $\epsilon_{xx} = \epsilon_{zz} = 13$, $\epsilon_{yy} = 10.3$.

TABLE V
COMPARISON OF COUPLED LINE CHARACTERISTICS OF EPSILAM-10 ISOTROPIC VERSUS EPSILAM-10 ANISOTROPIC

w/H	Z^e_{oi}	Z^e_{oa}	% Error	$\epsilon^e_{\mathrm{eff}_i}$	$\epsilon^e_{\mathrm{eff}_a}$	% Error
0.1	158.3048	152.8358	3.455	6.3705	6.8346	7.285
1.0	63.4321	62.3986	1.629	7.3236	7.5682	3.340
3.0	29.2129	29.0010	0.725	8.1945	8.3148	1.468
5.0	19.1763	19.0638	0.587	8.5260	8.6270	1.185
7.0	14.5671	14.4020	1.13	8.5867	8.7847	2.306
w/H	Z^o_{oi}	Z^o_{oa}	% Error	$\epsilon^o_{\mathrm{eff}_i}$	$\epsilon^o_{\mathrm{eff}_a}$	% Error
0.1	50.7637	48.1442	5.160	5.6566	6.2889	10.054
1.0	26.6548	25.4680	4.452	5.8253	6.3809	8.707
3.0	17.4429	16.9438	2.861	6.4355	6.8208	5.987
5.0	13.2197	12.9606	1.960	6.9656	7.2468	3.880
7.0	10.5131	10.4271	0.724	7.4683	7.5920	1.656

$s/H = 0.1$, $B/H = 0.5$-percent convergence accuracy.

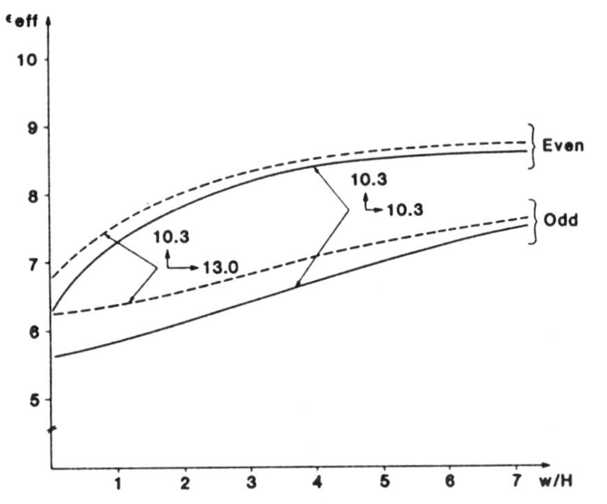

Fig. 5. Error in ϵ_{eff} versus w/H when anisotropy is ignored. Microstrip with cover: $s/H = 0.1$, $B/H = 10$.

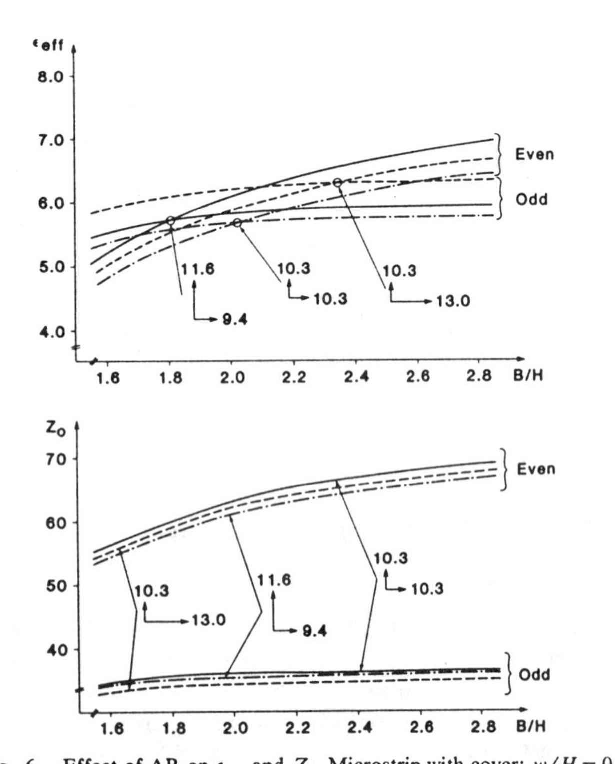

Fig. 6. Effect of AR on ϵ_{eff} and Z_o. Microstrip with cover: $w/H = 0.7$, $s/H = 0.26$.

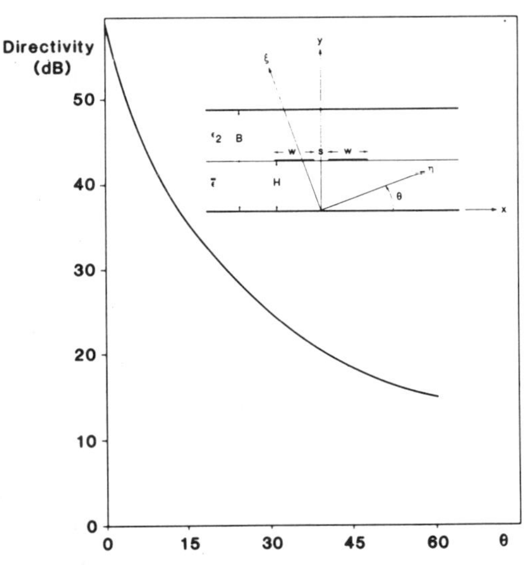

Fig. 7. 10-dB coupler boron-nitride substrate directivity versus rotation θ. $B/H = 2.8$, $s/H = 0.095$, $w/H = 1.6$.

configurations on Epsilam-10, PBN, and sapphire substrates. For each case shown, the parameters s/H and B/H were varied to obtain $Z^e_0 \simeq 69.4$ and $Z^o_0 = 36\ \Omega$ for a 10-dB coupler matched to a 50-Ω line. Whenever $v^e_p \simeq v^o_p$, the coupler directivity tends to infinity ($D \to \infty$); as shown in Table VI, $D \simeq 58$ dB when $B/H \simeq 2.80$ for a PBN substrate ($v^e_p = 1.876 \times 10^8$ m/s, $v^o_p = 1.875 \times 10^8$ m/s). An indication of the tight tolerances needed is clearly observed when the substrate optic axis is misaligned with respect to the microstrip coordinate system. Fig. 7 depicts the variation of D as a function of misalignment angle θ, where it is observed that even for small θ there is a significant reduction in coupler directivity [77].

2) Microstrip Couplers on an Anisotropic Substrate with an Isotropic Overlay: It has been established that coupler directivity improvement results on isotropic substrates when an isotropic overlay is used [79]–[84]. An overlay will also improve coupler directivity on anisotropic substrates by relaxing the tight tolerance requirements on B/H. This is particularly true for materials with AR < 1 (e.g., sapphire). For this design, phase velocities have been nearly equalized but, more importantly, both impedance and phase velocity curves vary quite slowly with increasing d/h (decreasing B/H) as shown in Fig. 8. Table VII indicates the usefulness of the overlay in realizing coupler designs with commercially available materials such as a 0.025-mil-thick sapphire substrate with a 0.050-mil alumina overlay. For this two-layer structure (isotropic overlay on an anisotropic substrate), the Green's function is given by [78]

$$G(x - x') = \frac{1}{2\pi\epsilon_0} \int_{-\infty}^{\infty} \frac{\cos[\zeta|x - x'|]}{\zeta} \frac{N(\zeta)}{D(\zeta)} \, d\zeta \quad (67)$$

where

$$N(\zeta) = \epsilon_2 + \epsilon_3 \coth(\zeta H \nu) \tan(\zeta t H) \quad (68)$$

TABLE VI
COUPLER DESIGNS—SUMMARY

Coupler	w/H	B/H	s/H	Z_o^e	Z_o^o	(10^8 m/s) v_p^e	(10^8 m/s) v_p^o	Directivity (dB)	VSWR	Center Freq. Coupling
Epsilam—Shielded	0.700	2.55	0.260	69.0	35.9	1.207	1.210	43	<1.01	10.02
Epsilam—Unshielded	0.800	>6	0.280	69.4	36.0	1.138	1.204	18	1.03	10.00
Alumina Unshielded	0.875	>6	0.260	69.2	35.9	1.150	1.286	12	1.06	10.04
Boron Nitride—Unshielded	1.850	>6	0.120	70.0	35.9	1.772	1.860	19.5	1.03	9.83
Boron Nitride—Shielded	1.60	2.80	0.095	69.3	36.0	1.876	1.875	58	<1.01	10.00
Quartz—Unshielded	1.830	>6	0.110	69.2	36.2	1.708	1.886	13	1.05	10.13
Sapphire—Shielded, 90° Offset ($\epsilon_{xx}=11.6$, $\epsilon_{yy}=9.4$)	0.690	2.20	0.225	69.2	35.9	1.256	1.257	49	1.01	9.98
Sapphire—Unshielded ($\epsilon_{xx}=9.4$, $\epsilon_{yy}=11.6$)	0.730	>6	0.260	69.4	36.2	1.086	1.227	11	1.06	10.12

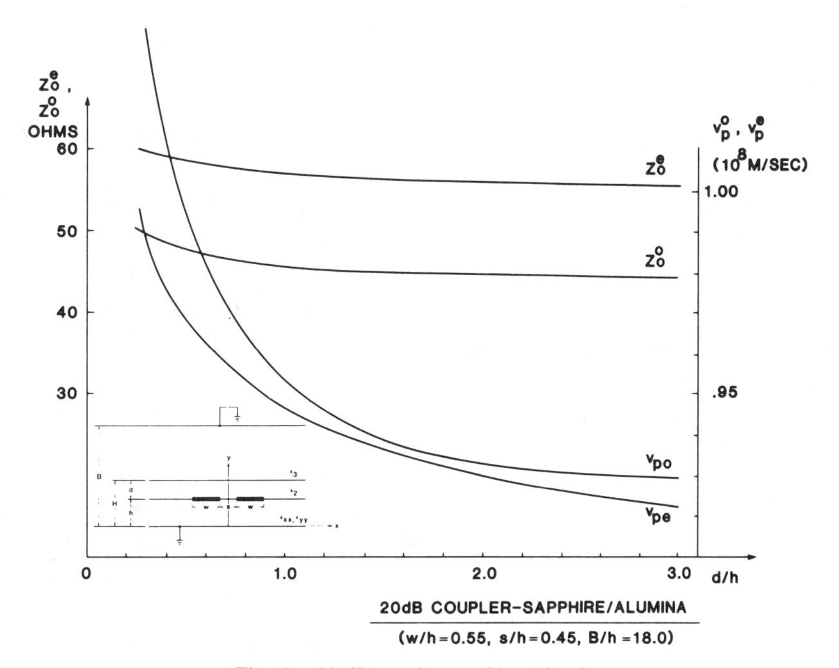

Fig. 8. 20-dB coupler-sapphire/alumina....

TABLE VII
COUPLER DESIGNS WITH AN OVERLAY

No.	Type	B/H	B/h	d/h	s/h	w/h	ϵ_2	Z_o^e	$\times 10^8$ v_p^e	Z_o^o	$\times 10^8$ v_p^e	C (dB)	Isol (dB)	Dir (dB)	VSWR
1.	10 dB (unshielded)	8.6	12.0	0.40	0.45	0.55	9.9	69.4	.9905	36.3	.9885	10.1	44.7	34.6	1.01
2.	20 dB (shielded)	3.4	6.2	0.80	1.60	0.58	9.9	55.5	.9629	45.1	.9646	19.7	57.3	37.6	1.002
3.	20 dB (shielded)	4.0	12.0	2.0	1.60	0.55	9.9	56.0	.9323	44.6	.9333	18.9	61.3	42.4	1.002
4.	20 dB (uncompensated)	12.0	12.0	0.0	1.20	0.86	—	55.5	1.068	45.3	1.180	20.0	22.4	2.4	1.02
5.	20 dB	6.0	18.0	2.0	1.60	0.55	9.9	56.1	.9306	44.6	.9333	18.9	53.0	34.1	1.00

Sapphire $\epsilon_{xx}=9.4$, $\epsilon_{yy}=11.6$. Alumina $\epsilon_2=9.9$, $\epsilon_3=1.0$.

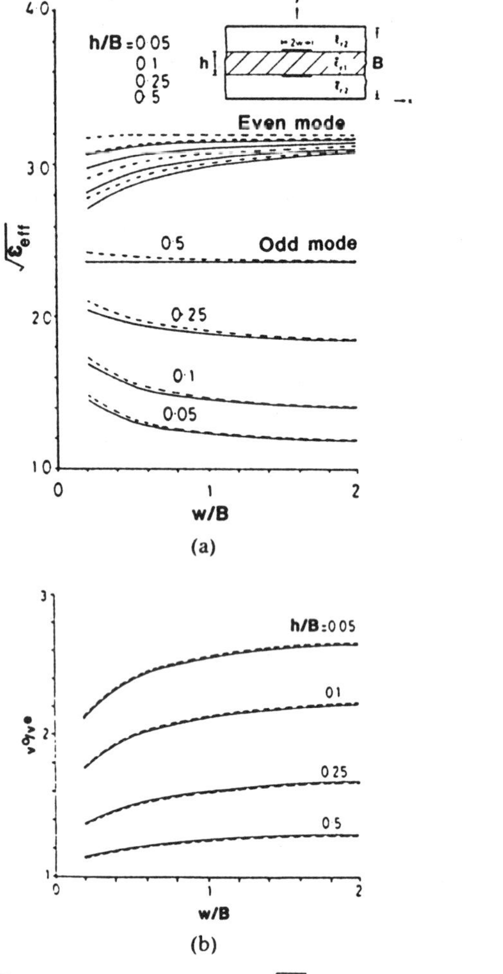

Fig. 9. (a) Even- and odd-mode factor $\sqrt{\epsilon_{\text{eff}}}$. (b) Phase-velocity ratio, v^o/v^e. Reprinted by permission from IEE *Electronics Letters* [29].

$$D(\zeta) = \tan(\zeta t H) \cdot \left[\epsilon_2^2 + \epsilon_3(\epsilon_{xx}\epsilon_{yy})^{1/2}\coth(\zeta H \nu) \cdot \coth\left(\left(\frac{\epsilon_{xx}}{\epsilon_{yy}}\right)^{1/2}\zeta h\right)\right] + \epsilon_2\left[\epsilon_3\coth(\zeta H\nu) + (\epsilon_{xx}\epsilon_{yy})^{1/2} \cdot \coth\left(\left(\frac{\epsilon_{xx}}{\epsilon_{yy}}\right)^{1/2}\zeta h\right)\right]. \quad (69)$$

The even and odd Green's function components $G^{\left\{ e \atop o \right\}}(x - x')$ are obtained by using (29) and (30)

3) Broadside-Coupled Microstrip Lines: Broadside-coupled microstrip lines are considered as another application of the method of moments. The even-odd-mode Green's function for the structure shown in Fig. 9 is developed into the series form [29]

$$G^{\left\{ e \atop o \right\}}(x - x') = \frac{1}{\epsilon_0}\sum_{l=1}^{\infty}\left\{\frac{2\exp(-v_l|x-x'|)}{v_l h} \cdot \left[\epsilon_{xx_1}\left\{ \begin{matrix} \sec^2\left[v_l \dfrac{h}{2}\left(\dfrac{\epsilon_{xx_1}}{\epsilon_{yy_1}}\right)^{1/2}\right] \\ \cos^2\left[v_l \dfrac{h}{2}\left(\dfrac{\epsilon_{xx_1}}{\epsilon_{yy_1}}\right)^{1/2}\right] \end{matrix} \right\} + \left\{\epsilon_{xx_2}\nu \csc^2\left[v_l \frac{h}{2}\nu\left(\frac{\epsilon_{xx_2}}{\epsilon_{yy_2}}\right)^{1/2}\right]\right\}\right]\right\} \quad (70)$$

where $\nu = B/h - 1$, v_l is the lth zero of the transcendental equation

$$\frac{\tan\left[\dfrac{v_l h}{2}\left(\dfrac{\epsilon_{xx_1}}{\epsilon_{yy_1}}\right)^{1/2}\right]^{-p}}{\cot\left[v_l \dfrac{h}{2}\nu\left(\dfrac{\epsilon_{xx_2}}{\epsilon_{yy_2}}\right)^{1/2}\right]} = (-p)\left(\frac{\epsilon_{xx_2}\epsilon_{yy_2}}{\epsilon_{xx_1}\epsilon_{yy_1}}\right)^{1/2} \quad (71)$$

and $p = \pm 1$ for even and odd modes, respectively. Incorporation of the Green's function in the algorithm yields the results [29] shown in Fig. 9(a) and (b) for broadside-coupled lines on anisotropic Epsilam-10 substrate. The curves shown indicate very small differences in $Z^{\left\{ e \atop o \right\}}_0, v^{\left\{ e \atop o \right\}}_p$ values between Epsilam-10 ($\epsilon_{xx} = \epsilon_{zz} = 13$, $\epsilon_{yy} = 10.2$) and AlSiMag 838 ($\epsilon = 10.2$) substrates. For small linewidths, however, the error increases when anisotropy is neglected since the fringing fields between the broadside-coupled lines is not accounted for correctly.

The method of moments proves to be a powerful tool in the effort to obtain the quasi-static characteristics of single and coupled microstrip lines on anisotropic substrates. Multiple material layers, either anisotropic or combinations of isotropic and anisotropic layers, can be incorporated into the algorithm easily by considering the appropriate Green's function for the structure. For the assumed geometries in this section (no sidewalls), the Green's function can be obtained either in integral or series form. For each structure, the series form as obtained by the Cauchy residue theorem, although it requires approximately 1/3 the computer time, is not as accurate as the integral form. This is due to the error accumulation incurred during the location of the roots when (65) or (71) are solved numerically. The results obtained for the characteristic impedance by incorporating the series form in the method of moments algorithm are consistently 1-2-percent low. The integral form should be used where very high accuracy is required.

C. *Coordinate Transformation Method*

A simple approach to analyze the properties of microstrip on anisotropic substrates (without cover) may be developed if the Green's function given by (62) is examined closely. The anisotropy and structural dimensions are involved explicitly in the denominator term given by

$$\zeta\left[\left(\epsilon_{xx}\epsilon_{yy}-\epsilon_{xy}^2\right)^{1/2}\cdot\coth\left\{\zeta\left(\frac{\epsilon_{xx}}{\epsilon_{yy}}-\left(\frac{\epsilon_{xy}}{\epsilon_{yy}}\right)^2\right)^{1/2}H\right\}+\epsilon_2\coth(\zeta H\nu)\right].$$

If there is no top cover, $H\nu\to\infty$ as $B\to\infty$, and therefore $\coth(\zeta H\nu)\to 1$. The denominator may then be rewritten as $\zeta[\epsilon_{eq}\coth(\zeta H_{eq})+\epsilon_2]$ where [68]–[71]

$$\epsilon_{eq}=\left(\epsilon_{xx}\epsilon_{yy}-\epsilon_{xy}^2\right)^{1/2} \tag{72}$$

and

$$H_{eq}=\left[\frac{\epsilon_{xx}}{\epsilon_{yy}}-\left(\frac{\epsilon_{xy}}{\epsilon_{yy}}\right)^2\right]^{1/2}H. \tag{73}$$

This procedure reveals that the anisotropic substrate may be replaced with an "equivalent isotropic" layer whose permittivity and thickness are defined by (72) and (73). This equivalent microstrip problem can then be solved with an appropriate algorithm for isotropic substrates [74], [75]. A more rigorous justification of the "equivalent isotropic" problem has been derived by considering the simple case of a diagonalized tensor $\bar{\bar{\epsilon}}$ ($\epsilon_{xy}=\epsilon_{yx}=0$) [67], [68]. The coordinate transformation

$$\tau=x \qquad \upsilon=y\left(\frac{\epsilon_{xx}}{\epsilon_{yy}}\right)^{1/2}$$

yields Laplace's or Poisson's equation in the τ,υ coordinate system for a substrate characterized by

$$\epsilon_{eq}=(\epsilon_{xx}\epsilon_{yy})^{1/2}\ \text{and}\ H_{eq}=\left(\frac{\epsilon_{xx}}{\epsilon_{yy}}\right)^{1/2}H.$$

Furthermore, the relationship between $\phi(x,y)$ and $\Phi(\tau,\upsilon)$ is readily established as $\phi_P(x,y)=\Phi_Q(\tau,\upsilon)$ and $\partial\phi_P(x,y)/\partial y=(\partial\phi_Q(\tau,\upsilon)/\partial\upsilon)\frac{\partial\upsilon}{\partial y}$ where $Q(\zeta,\upsilon)$ is the point $P(x,y)$ transformed into the (τ,υ) coordinate system. Under this transformation, the boundary conditions at $y=H$ are invariant [68], i.e.,

$$E_x=-\frac{\partial\Phi}{\partial\tau}=E_\tau\ \text{and}\ D_y=-\epsilon_0\epsilon_{eq}\frac{\partial\Phi}{\partial\upsilon}=D_\upsilon. \tag{74}$$

This method has been used in conjunction with the Bryant and Weiss algorithm [74] but the results obtained [68] are in error ranging up to 20 percent, due possibly to erroneous adaptation of the algorithm.

This approach should be used with caution when microstrip with a cover is considered, since the term $\coth(\zeta H\nu)$ cannot be equated to unity. For this case, the denominator of the Green's function takes the form $\zeta[\epsilon_{eq}\coth(\zeta H_{eq})+\epsilon_2\coth(\zeta H\nu)]$, and, strictly speaking, the concept of an equivalent isotropic substrate is no longer valid. Nevertheless, an algorithm for isotropic substrates can still be used with the extra care of proper entry for ϵ_{eq}, H_{eq}, and H.

D. *The Image-Coefficient Method*

An alternate approach, which also leads to the conclusions of the previous section, is to obtain the Green's function by considering the method of images for anisotropic media [30]. As a first step, Poisson's equation is considered with a unit strength per unit length charged line source at x', y'. For the moment, the entire space is anisotropic and is characterized by a diagonalized tensor $\bar{\bar{\epsilon}}$. Poisson's equation takes the form [67]

$$\epsilon_{xx}\frac{\partial^2 G}{\partial x^2}+\epsilon_{yy}\frac{\partial^2 G}{\partial y^2}=-\frac{\delta(x-x',y-y')}{\epsilon_0}. \tag{75}$$

Transforming coordinates with $\tau=x/\sqrt{\epsilon_{xx}}$, $\upsilon=y/\sqrt{\epsilon_{yy}}$, and using the delta function property $\delta(ax)=\delta(x)/|a|$, the equation is written as

$$\frac{\partial^2 G}{\partial\tau^2}+\frac{\partial^2 G}{\partial\upsilon^2}=-\frac{1}{\epsilon_0\epsilon_{xx}\epsilon_{yy}}\delta(\tau-\tau',\upsilon-\upsilon'). \tag{76}$$

This transformed Poisson equation easily yields the solution

$$G(x-x',y-y')=\frac{1}{2\pi\epsilon_0\epsilon_{eq}}\cdot\ln\left[\frac{1}{\sqrt{(x-x')^2+(y-y')^2\left(\frac{\epsilon_{yy}}{\epsilon_{xx}}\right)}}\right] \tag{77}$$

where, as before, $\epsilon_{eq}=\sqrt{\epsilon_{xx}\epsilon_{yy}}$. If the line source is now placed at x', y' above an anisotropic half-space (see Fig. 10), then consideration of the boundary conditions at the interface ($y=0$) and the reciprocity theorem [30] yield for this configuration the solution

$$G(x-x',y-y')=\begin{cases}\dfrac{\rho_1}{2\pi\epsilon_0\epsilon_{eq}}\ln\left[\dfrac{1}{\sqrt{(x-x')^2+(y-y_0')^2\left(\frac{\epsilon_{yy}}{\epsilon_{xx}}\right)}}\right], & y\leqslant 0 \quad (78)\\[2ex] \dfrac{1}{2\pi\epsilon_0}\ln\left[\dfrac{1}{\sqrt{(x-x')^2+(y-y')^2}}\right]+\dfrac{\rho_2}{2\pi\epsilon_0}\cdot\ln\left[\dfrac{1}{\sqrt{(x-x')^2+(y-y')^2}}\right], & y\geqslant 0 \quad (79)\end{cases}$$

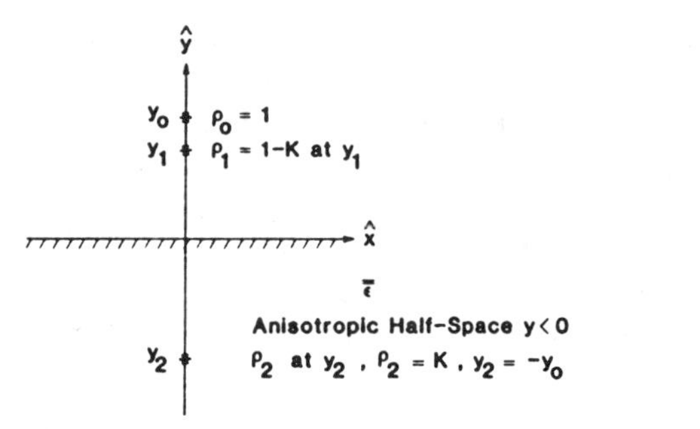

Fig. 10. Imaging a line charge source ($\rho_0 = 1$) over an anisotropic half-space.

where

$$y_0' = \sqrt{\frac{\epsilon_{xx}}{\epsilon_{yy}}}\, y' \tag{80}$$

$$y_1' = -y' \tag{81}$$

$$\rho_1 = 1 - K \tag{82}$$

$$\rho_2 = K \tag{83}$$

and

$$K = \frac{1-\epsilon_{eq}}{1+\epsilon_{eq}}. \tag{84}$$

For the microstrip geometry (without cover), multiple images result, just as for an isotropic substrate. Using image theory for this structure, the Green's function is written in the form

$$G(x-x') = \frac{1}{2\pi\epsilon_0(1+\epsilon_{eq})} \sum_{n=1}^{\infty} K^{n-1} \cdot \ln\left\{\frac{\left[4n^2 + \left(\frac{x-x'}{H_{eq}}\right)^2\right]\left[\left(4n^2 + \frac{x+x'}{H_{eq}}\right)^2\right]}{\left[4(n-1)^2 + \left(\frac{x-x'}{H_{eq}}\right)^2\right]\left[4(n-1)^2 + \left(\frac{x+x'}{H_{eq}}\right)^2\right]}\right\} \tag{85}$$

where

$$H_{eq} = H\sqrt{\frac{\epsilon_{xx}}{\epsilon_{yy}}}.$$

This expression is identical to the Green's function by the method of images for microstrip on an isotropic substrate [86] provided the isotropic layer is characterized by the relative dielectric constant ϵ_{eq} and substrate thickness H_{eq}. This series representation of the Green's function converges rapidly and it yields results for the microstrip capacitance per unit length with excellent accuracy when adopted with an appropriate numerical method [30].

The theory of images is easily extended to obtain the Green's function for an electrooptic modulator structure, i.e., a metallic strip conductor on an anisotropic substrate of thickness H without a ground plane. This Green's function is given by [30]

$$G(x-x') = \frac{1}{2\pi\epsilon_0(1+\epsilon_{eq})} \cdot \sum_{n=1}^{\infty} K^{n-1} \ln\left\{\frac{4(n-1)^2 + \left(\frac{x+x'}{H_{eq}}\right)^2}{4(n-1)^2 + \left(\frac{x-x'}{H_{eq}}\right)^2}\right\}. \tag{86}$$

These series representations, for both the microstrip line and electrooptic modulator structures, converge quite rapidly and they have been adopted in the discretized integral equation [30]

$$V = \sum_{j=1}^{M} \int_{y_j}^{y_{j+1}} \rho(x') G(x-x')\, dx' \tag{87}$$

where

$$\rho(x') = \rho_j + (\rho_{j+1} - \rho_j)\left(\frac{x' - x_j}{x_{j+1} - x_j}\right), \qquad 0 \leqslant x' \leqslant w/2 \tag{88}$$

with

$$x_j = \frac{w}{2}\left\{1 - \left(1 - \frac{j-1}{m}\right)^{\gamma}\right\}, \qquad j = 1,2,3,\cdots,m+1 \quad (\gamma = 1, 2, \text{ or } 3). \tag{89}$$

The expression given for $\rho(x')$ gives an excellent piecewise approximation of the true charge density distribution for $0 \leqslant x' \leqslant w/2$. The total charge on the half-strip is

$$Q = \sum_{j=1}^{m} \int_{y_j}^{y_{j+1}} \rho(x')\, dx'. \tag{90}$$

When the conductor is charged to 1 V, the lineal capacitance is $C = 2Q$.

The accuracy of this approach is remarkable. The error (again comparing to the value obtained with conformal mapping) for microstrip on isotropic substrates is quoted as less than 0.0024 percent when $w/H = 0.01$ and less than 0.001 percent for $w/H > 0.01$ [30]. Similar accuracy is observed for the electrooptic modulator case [30].

E. Application of the Variational Principle

The variational principle is a powerful tool in that it yields results with very good accuracy for a variety of integrated-circuit structures. Furthermore, it provides upper- and lower-bound numerical results for the capacitance of single and coupled printed strip conductors with arbitrary substrate parameters and conductor geometry. In this section, the variational expression for capacitance is presented in the Fourier transform or space domain for a variety of microstrip geometries, as well as for unshielded suspended stripline, coupled slotlines, and coplanar waveguide structures on anisotropic substrates.

1) Fourier Transform Representation: The variational expression for the lower-bound computation of capacitance can be derived in the form [87], [88]

$$\frac{1}{C}=\frac{1}{2\pi Q^2}\int_{-\infty}^{\infty}\tilde{\rho}(\zeta)\tilde{G}(\zeta,H)\,d\zeta \tag{91}$$

where $\tilde{\rho}(\zeta)$ and $\tilde{G}(\zeta)$ are the Fourier transforms of the line charge density and Green's function, respectively, while Q is the total charge per unit length. This variational expression can be rewritten as

$$\frac{1}{C}=\frac{1}{\pi Q^2}\int_{0}^{\infty}[\tilde{\rho}(\zeta)]^2\tilde{g}(\zeta,H)\,d\zeta \tag{92}$$

where $\tilde{g}(\zeta, H)$ is given by the following expressions depending on the structure under consideration.

Microstrip with cover:

$$\tilde{g}(\zeta,H)=\frac{1}{\epsilon_0\zeta\left[\epsilon_{eq}\coth(\zeta H_{eq})+\epsilon_2\coth(\zeta H v)\right]} \tag{93}$$

(usually $\epsilon_2 = 1$).

Microstrip without cover:

$$\tilde{g}(\zeta,H)=\frac{1}{\epsilon_0\zeta\left[\epsilon_{eq}\coth(\zeta H_{eq})+\epsilon_2\right]}. \tag{94}$$

Unshielded suspended stripline or coplanar striplines:

$$\tilde{g}(\zeta,H)=\frac{\epsilon_{eq}\coth(\zeta H_{eq})+1}{\epsilon_0\zeta\left[(1+\epsilon_{eq})+2\epsilon_{eq}\coth(\zeta H_{eq})\right]}. \tag{95}$$

The trial functions which are typically used to minimize the error in the computation of C are

$$\rho(x)=\begin{cases}|x|, & |x|\leqslant w/2\\ 0, & |x|>w/2\end{cases} \tag{96}$$

or

$$\rho(x)=\begin{cases}1+|2x/w|^3, & |x|\leqslant w/2\\ 0, & |x|>w/2\end{cases}. \tag{97}$$

The corresponding Fourier transforms are

$$\frac{\tilde{\rho}(\zeta)}{Q}=\frac{2\sin\left(\frac{\zeta w}{2}\right)}{\left(\frac{\zeta w}{2}\right)}-\left[\frac{\sin\left(\frac{\zeta w}{4}\right)}{\left(\frac{\zeta w}{4}\right)}\right]^2 \tag{98}$$

and

$$\frac{\tilde{\rho}(\zeta)}{Q}=\frac{8}{5}\left\{\frac{\sin(\zeta w/2)}{(\zeta w/2)}\right\}+\frac{12}{5\left(\frac{\zeta w}{2}\right)^2}\cdot\left[\cos\left(\frac{\zeta w}{2}\right)-\frac{2\sin(\zeta w/2)}{\zeta w/2}+\frac{\sin^2(\zeta w/4)}{(\zeta w/4)^2}\right]. \tag{99}$$

The method is easily extended to include the even- and odd-mode capacitance computation by considering the following representation for the Fourier transform of the charge density:

$$\tilde{\rho}^{\left\{\substack{e\\o}\right\}}(\zeta)=2\int_{s/2}^{s/2+w}\rho^{\left\{\substack{e\\o}\right\}}(x)\begin{matrix}\cos\\ \sin\end{matrix}\{\zeta x\}\,dx. \tag{100}$$

The Rayleigh–Ritz procedure has been employed with this method to determine the unknown constants, the a_i, in the expansions [36]–[38]

$$\rho^{(e)}(x)=\sum_{i=1}^{M+1}a_i x^{i-1} \tag{101}$$

and

$$\rho^{(0)}(x)=\sum_{i=1}^{M+1}a_i(w-x)^{i-1}. \tag{102}$$

For microstrip on an isotropic substrate without a cover layer, with $w/H>0.5$ and $s/H>0.5$, the reported error is less than 1 percent for the upper-bound even-mode characteristic impedance and 2 percent for the odd mode. However, for small s/H (less than 0.1), with $w/H>1.0$ or $w/H<0.1$, the error exceeds 6 percent [37]. Large errors result with this method for the case of microstrip with a cover [38]

The trial functions given by (98) and (99) have been considered to analyze the unshielded stripline on a Lithium–Niobate substrate (Li-Nb-O_3, $\epsilon_{xx}=28$, $\epsilon_{yy}=43$) [88]. The results obtained using the variational principle have been found to be in error by 4–10 percent when compared to the approach which uses image theory with the numerical technique discussed previously [30]. The corresponding formulation for even- and odd-mode upper-bound capacitance computation is given by [36]–[38]

$$C^{\left\{\substack{e\\o}\right\}}=\frac{\epsilon_0}{2\pi}\int_0^{\infty}\tilde{\phi}^{\left\{\substack{e\\o}\right\}}(\zeta)\tilde{g}(\zeta)\,d\zeta \tag{103}$$

where the trial functions for the potential may be chosen as

$$\phi^{\left\{\substack{e\\o}\right\}}(x)=\begin{cases}\sum\limits_{i=L}^{L}a_i(d-x)^{-i}, & 0\leqslant x\leqslant s/2\\ 1, & s/2\leqslant x\leqslant s/2+w\\ \sum\limits_{j=1}^{N+1}b_j(x-d)^{-j}, & x\geqslant s/2+w\end{cases} \tag{104}$$

with $L=M+1$ for even modes, $L=M+2$ for odd modes, and $d=(1/2)(s+w)$. The unknown constants a_i and b_j may be determined with the procedure used for the lower-bound computations. For isotropic substrates, the incurred error is found to be 2 percent for the even-mode impedance and 4 percent for the odd mode (by comparison with the results reported in [74] for $s/H\geqslant 0.5$ and $w/H\geqslant 0.5$). For smaller s/H and/or w/H, the error is substantially larger.

2) Space-Domain Formulation: A variational representation for the capacitance per unit length can also be given directly in terms of spatial coordinates, i.e., [12]

$$C=\frac{\int_{-w/2}^{w/2}\rho(x)\,dx}{\int_{-w/2}^{w/2}\int_{-w/2}^{w/2}G(x-x',y-y')\rho(x)\rho(x')\,dx\,dx'}. \tag{105}$$

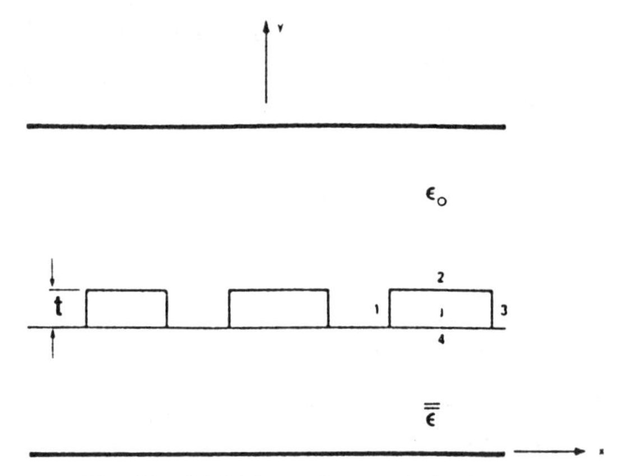

Fig. 11. Multiple conductor microstrip system.

It converges to the true value from the lower side. This variational representation has been implemented to analyze the properties of microstrip, inverted microstrip, and coplanar lines with multiple anisotropic layers in a rectangularly shaped shield [39]–[42].

The variational method can readily be extended to consider the characteristics of the system of N microstrip lines shown in Fig. 11. The conductor thickness may also be taken into account [34], [35]. The potential of the ith conductor is given for this configuration by

$$V_i = \sum_{j=1}^{N} \sum_{p=1}^{M} \int_{\mathscr{L}_{jp}} ds_p' G(s_r, s_p') q_{jp}(s_p') \tag{106}$$

with $i = 1,2,3,\cdots, N$. $\mathscr{L}_{jp}$ denotes integration on the pth side of the jth conductor, M the number of conductor sides ($M = 4$ for the rectangularly shaped strips considered here), and s_p' the integration variable along the pth side of the jth conductor. $s_r(s_p')$ denotes $x(x')$ or $y(y')$ depending on whether the pth side is parallel to the x- or y-axis. Lastly, $q_{jp}(s_p')$ is the unknown charge distribution of the pth side of the jth conductor and $G(s_r, s_p')$ is the Green's function satisfying the Poisson equation. The total charge on the jth conductor is

$$Q_j = \sum_{p=1}^{4} \int_{\mathscr{L}_{jp}} q_{jp}(s_p')\, ds_p' \tag{107}$$

which may be rewritten as

$$Q_j = \sum_j \sum_p \langle 1, q_{jp}(s_p) \rangle_{j,p}. \tag{108}$$

The inner product is defined here by

$$\langle V(s_p), W(s_p) \rangle_{j,p} = \int_{\mathscr{L}_{jp}} ds_p V(s_p) W(s_p). \tag{109}$$

The unknown charge density on the pth side may be written in terms of the following expansion:

$$q_{jp}(s_p) = \sum_{k=-M(j,p)}^{M(j,p)} c_k(j,p) \exp\{M_k(j,p) s_p\} \tag{110}$$

where $c_k(j,p)$ are unknown coefficients and $M_k(j,p)$ are known parameters on the pth side of the jth conductor. Utilizing the inner product definition and substituting (110) into (106), the V_i can be expressed as

$$V_i = \sum_{j=1}^{N} \sum_{p=1}^{4} \cdot \sum_{k=-M}^{M} c_k(j,p) \langle G(s_r, s_p'), \exp\{M_k(j,p) s_p'\} \rangle_{jp}. \tag{111}$$

Subsequently, the inner product of V_i with $\exp\{M_m(i,r)s_r\}$ yields

$$\langle \exp\{M_m(i,r)s_r\}, V_i \rangle_{ir} = \sum_{j=1}^{N} \sum_{p=1}^{4} \sum_{k=-M}^{M} c_k(j,p) \cdot \langle \exp\{M_m(i,r)s_r\}, \langle G(s_r, s_p'), \exp\{M_k(j,p)s_p'\} \rangle\rangle_{ir,jp} \tag{112}$$

where

$$\langle\ \rangle_{ir,jp} = \int_{\mathscr{L}_{ir}} \int_{\mathscr{L}_{jp}} ds_r\, ds_p' \exp\{M_m(i,r)s_r\} \cdot G(s_r, s_p') \exp\{M_k(j,p)s_p'\}. \tag{113}$$

In order to simplify notation, the following matrices are defined, namely the column vector

$$[Y(m,i,r)] = [\langle \exp\{M_m(i,r)s_r\}, V_i \rangle_{ir}] \tag{114}$$

and the square matrix

$$[T(m,i,r,k,j,p)] = [\langle\ \rangle ir, jp]. \tag{115}$$

These definitions enable (112) to be rewritten in the matrix form

$$[T(m,i,r;k,j,p)][c_k(j,p)] = [Y(m,i,r)] \tag{116}$$

which, upon inversion, yields the solution for the unknown coefficients

$$[c_k(j,p)] = [T(m,i,r;k,j,p)]^{-1}[Y(m,i,r)]. \tag{117}$$

The total charge on the jth conductor is finally obtained as

$$Q_j = \sum_{k=-M}^{M} \sum_{p=-1}^{4} c_k(j,p) \langle 1, \exp\{u_k(j,p)s_p\} \rangle_{jp}. \tag{118}$$

In view of the assumption that $V_j = 1$ and $V_1 = \cdots = V_{j-1} = \cdots = \cdots = V_N = 0$, the variational expression (105) is rewritten for $C_{ij} = Q_i$ as

$$Q = \frac{\left(\sum_{r=1,\, r\in\mathscr{L}_{ir}}^{4} \int_{\mathscr{L}_{ir}} ds_r q_r(s_r) \right)^2}{\sum_{\substack{r=1 \\ r\in\mathscr{L}_{ir}}}^{4} \int_{\mathscr{L}_{ir}} ds_r q_{ir}(s_r) \sum_{j=1}^{N} \sum_{\substack{p=1 \\ p\in\mathscr{L}_{jp}}}^{4} \int_{\mathscr{L}_{jp}} ds_p' G(s_r, s_p') q_{jp}(s_p')}. \tag{119}$$

TABLE VIII
CONVERGENCE PATTERN FOR CAPACITANCE PER METER FOR MICROSTRIP ON SAPPHIRE

$u_k(2)=u_k(4)$				$u_k(1)=u_k(3)$	q_1+q_3 (nC)	q_2+q_4 (nC)	C(nF/m)
0.0	0.6			±0.83,	0.01847	0.1832	0.2017
0.0	0.6	1.2	1.8	±0.55, ±1.66	0.02109	0.1789	0.2000
0.0	1.0			±0.083	0.02093	0.1790	0.1999
0.0	1.0			±0.055, ±0.17	0.10977	0.1812	0.2010

$t/H=0.3$, $w/H=1.0$, $B/H=6.0$.

This expression is useful in obtaining the effect of the conductor thickness on the propagation characteristics of microstrip conductors [34], [35]. The Green's function pertinent for this purpose is [34]

$$G(x-x',y-y')=\frac{1}{\pi\epsilon_0}\int_0^\infty \cos\left[\frac{\zeta|x-x'|}{n_x}\right]\sinh\left(\frac{\zeta(B-y_>)}{n_x}\right)\cdot\frac{n_xn_y\sinh\left(\frac{\zeta(y_>-H)}{n_x}\right)\cosh\left(\frac{\zeta H}{n_y}\right)+\sinh\left(\frac{\zeta H}{n_y}\right)\cosh\left(\frac{\zeta(y_<-H)}{n_x}\right)}{n_xn_y\sinh\left(\frac{\zeta(B-H)}{n_x}\right)\cosh\left(\frac{\zeta H}{n_y}\right)+\sinh\left(\frac{\zeta H}{n_y}\right)\cosh\left(\frac{\zeta(B-H)}{n_x}\right)}\frac{d\zeta}{\zeta} \quad (120)$$

where $y_<=\min(y,y')$, $y_>=\max(y,y')$, $n_x=\sqrt{\epsilon_{xx}}$, and $n_y=\sqrt{\epsilon_{yy}}$.

A single conductor of thickness t is considered presently as an application of the just-outlined approach. On the vertical sides ($p=1,3$), the charge distribution is expressed as

$$q_p(y)=\sum_{k=-M}^{M} c_k(p)\exp\{u_k(p)y\}, \qquad p=1,3$$

while on the horizontal sides ($p=2,4$), q is given by

$$q_p(x)=\sum_{k=-M}^{M} c_k(p)\exp\{u_k(p)x\}.$$

Table VIII demonstrates the capacitance convergence pattern of this technique. Even with the choice of $u_k=0.0$, the value of Z_0 is found to be within 5 percent of its convergence value. An example of the dependence of Z_0 and v_p on t/H is shown in Figs. 12 and 13, for $w/H=1.0$ and $B/H=6.0$. Clearly, for $w/H=1.0$, the variation of Z_0 and v_p with t/H is a second-order effect. Convergence is obtained, using this technique, with two or three u_k points. In addition, it has been determined that for increasing values of w/H, faster convergence is achieved if the parameters u_k are chosen as $u_k=H/w$ [34], [35].

3) Extension to Coplanar Waveguides: The variational principle can be extended to yield an expression for the capacitance of the coplanar waveguide (CP) in the form of [89]

$$C=\frac{\int_{s/2}^{s/2+w}\int_{s/2}^{s/2+w}\int_0^\infty e_x(x)G(\zeta;x-x')e_x(x')\,dx\,dx'}{\left\{\int_{s/2}^{s/2+w}e_x(x)\,dx\right\}^2} \quad (123)$$

where $e_x(x)$ is the unknown electric-field distribution across the slot aperture and

$$G(\zeta;x-x')=\frac{4\epsilon_0}{\pi}\left\{1+\frac{1+\epsilon_{eq}\tanh(\zeta H_{eq})}{1+\frac{1}{\epsilon_{eq}}\tanh(\zeta H_{eq})}\right\}\frac{\sin\zeta x\sin\zeta x'}{\zeta}. \quad (124)$$

The aperture electric field is written as the expansion

$$e_x(x)=e_0(x)+\sum_{k=1}^{N}a_ke_k(x) \quad (125)$$

where

$$e_k(x)=\frac{T_k\left\{\frac{2(x-s)}{w}\right\}}{\left[1-\left\{\frac{2(x-s)}{w}\right\}^2\right]^{1/2}}. \quad (126)$$

The Chebyshev polynomials $T_k\{x\}$ of the first kind are used, and the parameters are calculated using the Rayleigh–Ritz method. This technique yields results which are identical to those obtained by conformal mapping when $N\geqslant 2$ in the absence of the substrate, and it is considered to be highly accurate when the anisotropic substrate is included [89]. The variation of ϵ_{eff} and Z_0 versus θ, as obtained by this method, are shown in Fig. 14 for a sapphire substrate (θ is defined here as in (53)).

V. MODELING DISPERSION

The quasi-static methods described previously provide solutions of limited validity since they do not account for dispersive effects. Simple frequency-dependent formulas based on empirical observation and curve fitting have been derived, but they too are of limited value. They either apply exclusively to a sapphire substrate [2], [4] or they are not accurate enough for electrically thick substrates. Although they may lack general applicability, these methods and formulas offer the convenience of closed-form alge-

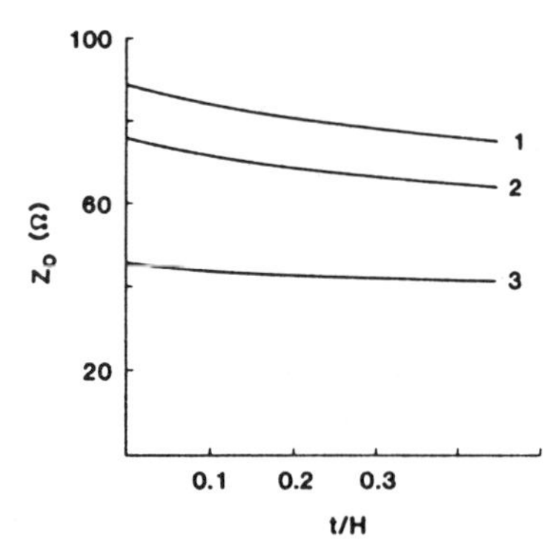

Fig. 12. Z_o versus (t/H) for 1: Polystyrene ($\epsilon_{xx} = \epsilon_{yy} = 2.54$); 2: quartz ($\epsilon_{xx} = \epsilon_{yy} = 3.78$); 3: sapphire ($\epsilon_{xx} = 9.4$, $\epsilon_{yy} = 11.6$). Reprinted by permission from the *Journal of the Franklin Institute* [34].

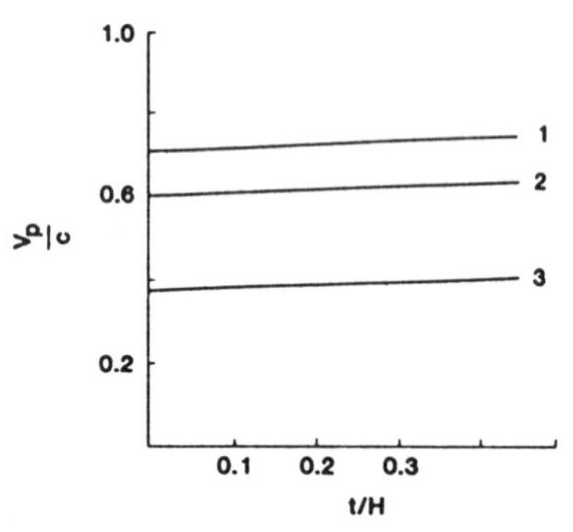

Fig. 13. $v_{p/c}$ versus (t/H) for the microstrip of Fig. 12. Reprinted by permission from the *Journal of the Franklin Institute* [34].

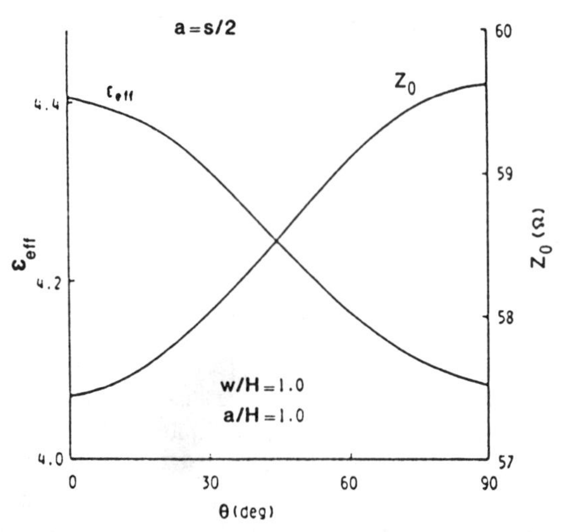

Fig. 14. Effective dielectric constant ϵ_{eff} and characteristic impedance Z_o versus θ.

braic expressions. For reference, then, some of these are presented.

An empirical formula for the frequency-dependent effective dielectric constant for a sapphire substrate is given as [5]

$$\epsilon_{eff} = \epsilon_{req} - \frac{\epsilon_{req} - \epsilon_{e0}}{1 + \left(\frac{H}{Z_0}\right)^{1.33} \left[0.43f^2 - 0.009f^3\right]} \tag{127}$$

where ϵ_{req} is defined by (40) and ϵ_{e0} is the static effective dielectric constant. H is in millimeters and f in gigahertz. This formula is reportedly valid with $a \pm 0.8$-percent error over the frequency range $2 \leqslant f \leqslant 18$ GHz and for a characteristic impedance value of $10 < Z_0 < 100\ \Omega$ [2], [5]. A dynamic solution obtained by the method of moments to within 0.5-percent accuracy [56] indicates that the error estimate in using (127) is within $\leqslant 3.3$ percent for $w/H = 0.5$, $\leqslant 1.5$ percent for $w/H = 1.0$, and $\leqslant 2.5$ percent for $w/H = 5.0$.

An approximate empirical formula which applies for arbitrary substrate anisotropy and thickness has been derived by combining two different dispersive models for isotropic substrates. The effective dielectric constant is defined as [90]

$$\epsilon_{eff} = \begin{cases} \dfrac{4}{[I_1 + I_2]^2}, & \text{if } I_1 > I_2 \\ \dfrac{1}{I_2^2}, & \text{if } I_1 < I_2 \end{cases} \tag{128}$$

where I_1 and I_2 are dispersive models for isotropic substrates [91], [92]. I_1 is expressed in the form

$$I_1 = \frac{\dfrac{1}{\sqrt{\epsilon_{yy}}}\left(\dfrac{f}{f_k}\right)^2 + \dfrac{1}{\sqrt{\epsilon_{e0}}}}{\left(\dfrac{f}{f_k}\right)^2 + 1} \tag{129}$$

with

$$f_k = \frac{v_0 \tan^{-1}\left[\epsilon_{req}\left(\dfrac{\epsilon_{e0} - 1}{\epsilon_{req} - \epsilon_{e0}}\right)^{1/2}\right]}{2\pi H\left(1 + \dfrac{w}{H}\right)[\epsilon_{req} - \epsilon_{e0}]^{1/2}}. \tag{130}$$

For I_2, the expression is

$$I_2 = \frac{\left(\dfrac{f}{f_y}\right)^{3/2} + 4}{\left(\dfrac{f}{f_y}\right)^{3/2}\sqrt{\epsilon_{yy}} + 4\sqrt{\epsilon_{e0}}} \tag{131}$$

with

$$f_y = \frac{v_0}{4H(\epsilon_{req} - 1)^{1/2}\left[\dfrac{1}{2} + \left(1 + 2\log\left(1 + \dfrac{w}{H}\right)\right)^2\right]}. \tag{132}$$

In these definitions, ϵ_{req} is the equivalent relative dielectric constant at zero frequency for an isotropic substrate on which the microstrip line (w, H being identical to the original line) has the same effective dielectric constant ϵ_{e0} as the latter line at zero frequency. Also, v_0 is the speed of light in vacuum.

The accuracy of the ϵ_{eff} formula given by (128) is very good for large w/H and arbitrary H/λ_0. When $H/\lambda_0 > 0.03$ and $w/H < 1.0$, however, the error for a sapphire is of the order of 4 percent. Clearly, for cases of arbitrary anisotropic substrates, a more precise accounting of dispersion is required. Rigorous solutions to Maxwell's equations addressing that need will now be presented.

A. *The Transmission-Line Matrix (TLM) Technique*

The TLM method, as it applies to anisotropic substrates [44] provides a solution to Maxwell's equations in the time domain by determining the impulse response of an equivalent distributed transmission-line network that models the given waveguiding structure [44]–[49]. The equivalence is obtained in terms of ideal two-wire transmission lines of length Δl connected in a three-dimensional lattice arrangement. At the transmission-line crossings, shunt or series nodes are formed which enable accurate characterization of the propagating medium with the incorporation of open- or short-circuited transmission-line stubs at each node. These transmission-line stubs are most instrumental in that they model the relative permittivity, conductivity, and relative permeability of the substrate. In the analysis, Kirchoff's voltage and current laws are applied to the equivalent three-dimensional network to yield a set of equations identifiable as an analog to Maxwell's equations (as they apply to the guiding structure).

The equivalent three-dimensional circuit is a periodic structure and it therefore exhibits the inherent passband and stopband frequency response characteristic of periodic networks. The upper frequency cutoff f_2 of the TLM model is the highest frequency of the lowest passband and it is determined by the mesh size Δl. It is possible to increase f_2 by choosing a smaller mesh size ($f_2 \to \infty$ as $\Delta l \to 0$). Moreover, for a given frequency having a finer mesh or smaller, Δl increases the model accuracy but at the expense associated with rapidly increasing computer run times and storage requirements. Distinct advantages, however, such as simplicity, versatility, and direct modeling of the physical waveguiding processes make this method a very useful engineering tool.

The TLM technique will be adopted herein to solve Maxwell's equations in the general form given by (7)–(19) for the microstrip structure shown in Fig. 1(c). A generalized node is shown in Fig. 15(a). It consists of three shunt and three series nodes $\Delta l/2$ apart from one another. Permeability is modeled by a short-circuited stub attached to each series node, while an open-circuited stub attached to each shunt node models permittivity. In addition, conductivity may be modeled with an infinite or matched line connected to each shunt node, and referred to as a loss stub. The coordinate orientation of the stubs denotes the particular component of the diagonalized tensor modeled by the stub, while the dashed lines in Fig. 15(a) are guide lines (and not equivalent transmission lines or stubs) [44]. The transmission-line representation of this generalized node is shown in Fig. 15(b). It illustrates the three-dimensional formation of shunt and series nodes (for clarity, stubs are not included in this figure). The voltages at the three shunt nodes represent the $\boldsymbol{E}$-field components, while the currents at the three series nodes are associated with the $\boldsymbol{H}$-field components in the three coordinate directions as shown in Fig. 15(b). Guide discontinuities and substrate material properties can be modeled with the appropriate choice of the stub electrical parameters (admittance or impedance). A better understanding of this may be obtained by considering a series and a shunt node individually. The series node shown in Fig. 16(a) is analyzed by considering the equivalent lumped network schematic of Fig. 16(c). The short-circuited stub of length $\Delta l/2$ is in the $\hat{x}$-direction and its input impedance is $Z_{\text{in}} = jZ_{xx}(L/C)^{1/2}\tan(\omega\Delta l/2c)$, where Z_{xx} is the line characteristic impedance. If $\omega\Delta l/2c \ll 1$, then $Z_{\text{in}} = j\omega L'$, where $L' = (Z_{xx}\Delta l/2)L$. Kirchoff's voltage law then yields for this series node

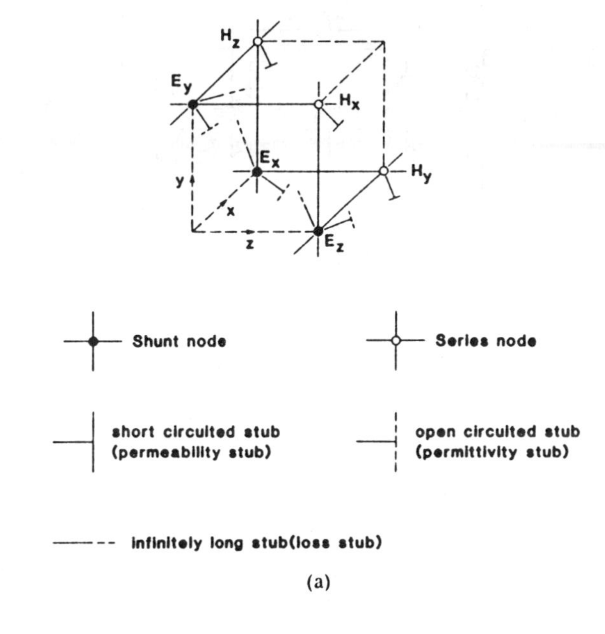

(a)

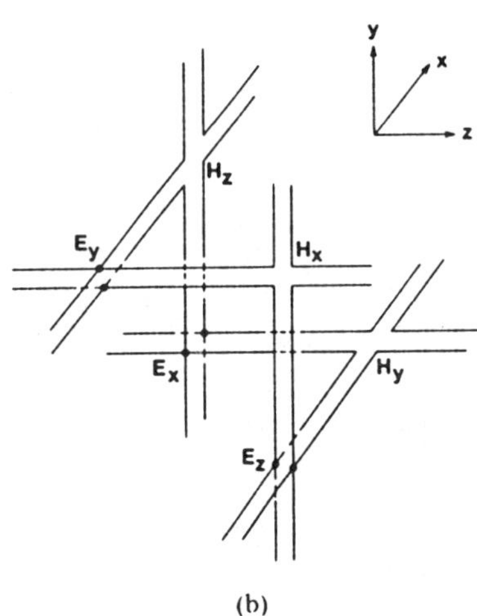

(b)

Fig. 15. (a) A generalized three-dimensional node. (b) Three-dimensional node.

$$\frac{\partial v_z}{\partial y} - \frac{\partial v_y}{\partial z} = -2L\left(1 + \frac{Z_{xx}}{4}\right)\frac{\partial i_x}{\partial t}. \tag{133}$$

This network equation is an analog to Maxwell's equations. Upon identification of v_z with $\mathscr{E}_z$, v_y with $\mathscr{E}_y$, and i_x with $\mathscr{H}_x$, it follows that (133) and (14) are equivalent, provided [44], [45]

$$\mu_0 = 2L \tag{134}$$

and

$$\mu_{xx} = \frac{4 + Z_{xx}}{4}. \tag{135}$$

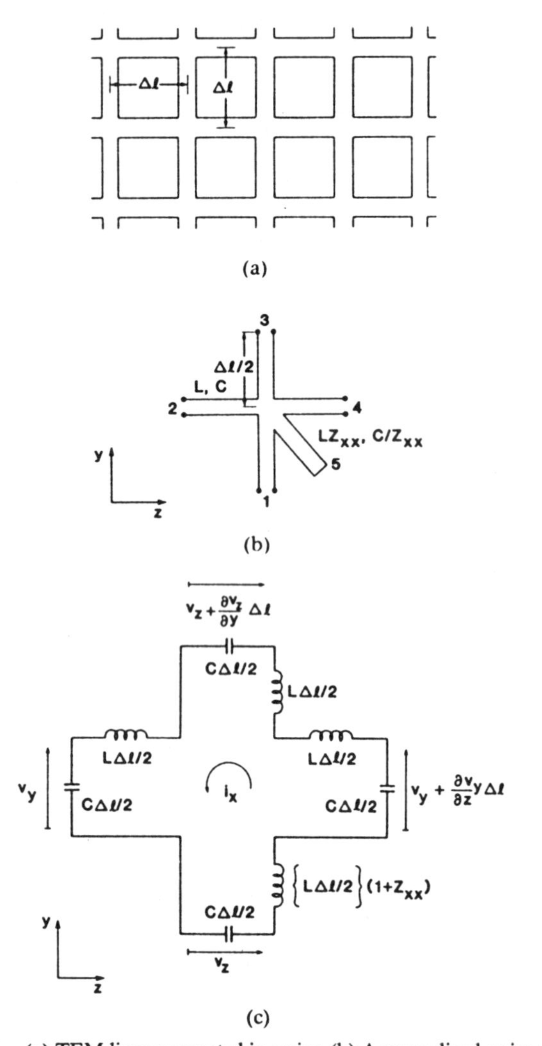

Fig. 16. (a) TEM lines connected in series. (b) A generalized series node. (c) Series node lumped network representation.

A similar analysis for series nodes in the xy- and xz-planes yields upon application of Kirchoff's voltage law

$$\frac{\partial v_y}{\partial x} - \frac{\partial v_x}{\partial y} = -2L\left(1 + \frac{Z_{zz}}{4}\right)\frac{\partial i_z}{\partial t} \tag{136}$$

and

$$\frac{\partial v_x}{\partial z} - \frac{\partial v_z}{\partial x} = -2L\left(1 + \frac{Z_{yy}}{4}\right)\frac{\partial i_y}{\partial t}. \tag{137}$$

Comparison with Maxwell's equations indicates the equivalences $v_y \equiv \mathscr{E}_y$, $v_x \equiv \mathscr{E}_x$, $i_z \equiv \mathscr{H}_z$, and $i_y \equiv \mathscr{H}_y$ hold if the following identifications are made:

$$\mu_{zz} = \frac{4 + Z_{zz}}{4} \tag{138}$$

and

$$\mu_{yy} = \frac{4 + Z_{yy}}{4}. \tag{139}$$

Continuing, the yz-plane shunt node shown in Fig. 17(b) is considered. In this case, the open-circuited stub (for permittivity) and the loss stubs (for conductivity) are in the $\hat{x}$-direction. The input admittance of the open-circuited stub is given by $Y_{\text{in}} = j\omega Y_{xx}C\Delta l/2(\omega \Delta l/2c \ll 1)$, so that

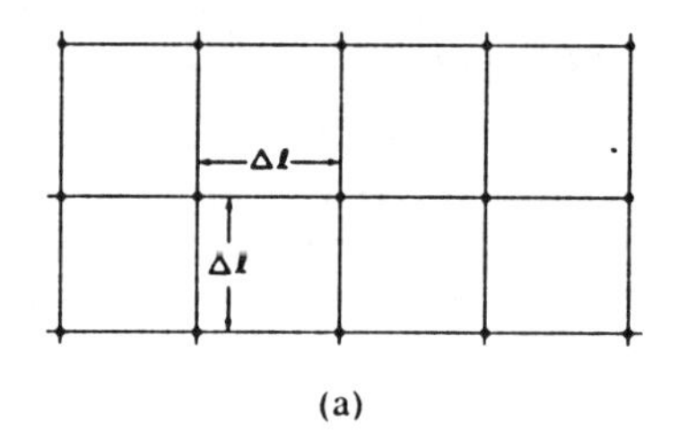

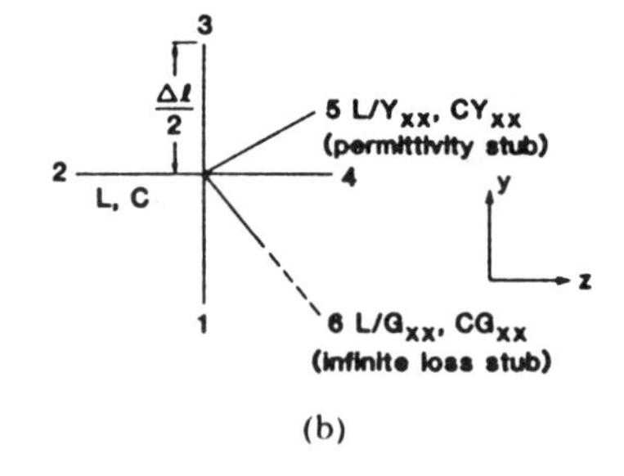

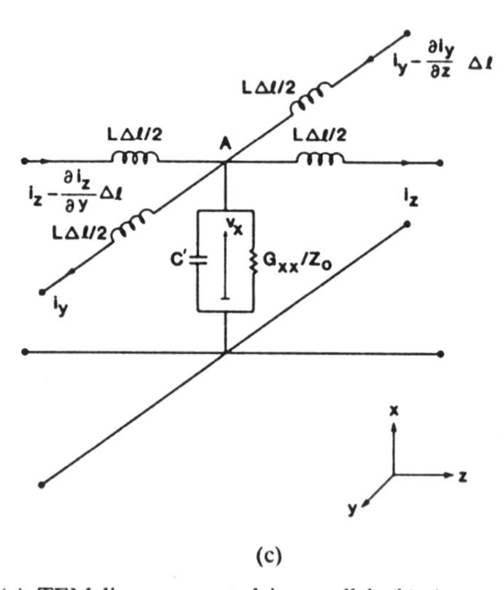

Fig. 17. (a) TEM lines connected in parallel. (b) A generalized shunt node. (c) Shunt node lumped network representation.

the equivalent capacitance is $C' = Y_{xx}C\Delta l/2$, while the total node capacitance is $2C(1 + Y_{xx}/4)\Delta l$. Application of Kirchoff's, current law at node A yields

$$\frac{\partial i_z}{\partial y} - \frac{\partial i_y}{\partial z} = \left(\frac{G_{xx}}{Z_0 \Delta l} + 2C\left[1 + \frac{Y_{xx}}{4}\right]\frac{\partial}{\partial t}\right)v_x. \tag{140}$$

A similar procedure in the other two planes produces the equations

$$\frac{\partial i_x}{\partial z} - \frac{\partial i_z}{\partial x} = \left(\frac{G_{yy}}{Z_0 \Delta l} + 2C\left[1 + \frac{Y_{yy}}{4}\right]\frac{\partial}{\partial t}\right)v_y \tag{141}$$

and

$$\frac{\partial i_y}{\partial x} - \frac{\partial i_x}{\partial y} = \left(\frac{G_{zz}}{Z_0 \Delta l} + 2C\left[1 + \frac{Y_{zz}}{4}\right]\frac{\partial}{\partial t}\right)v_z. \tag{142}$$

Comparing (140)–(142) with Maxwell's equations suggests the following parameter equivalences: $\mathscr{E}_x \equiv v_x$, $\mathscr{H}_z \equiv i_z$, $\mathscr{H}_y \equiv i_y$, $\sigma_{xx} \equiv G_{xx}/Z_0 \Delta l$, $\epsilon_0 = 2C$, and $\epsilon_{xx} = (4 + Y_{xx})/4$. Similarly, $\mathscr{H}_x \equiv i_x$, $\mathscr{E}_y \equiv v_y$, $\sigma_{yy} \equiv G_{yy}/Z_0 \Delta l$, $\epsilon_{yy} \equiv (4 + Y_{yy})/4$, and $\mathscr{E}_z \equiv v_z$, $\sigma_{zz} \equiv G_{zz}/Z_0 \Delta l$, and $\epsilon_{zz} \equiv (4 + Y_{zz})$.

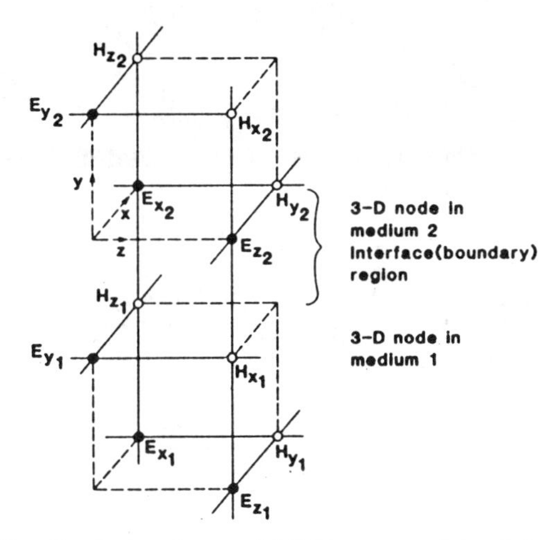

Fig. 18. Continuity of tangential fields across a dielectric boundary.

The previous discussion dealt with the derivation of the equivalent distributed circuit which models Maxwell's equations in the point form. Thus, the equivalent circuit, which is an analog to the differential form of Maxwell's equations, is obtained in terms of a generalized node with three shunt and three series nodes. A permittivity and a conductivity stub are connected to the shunt node, while a permeability stub is connected to the series node to complete the model. With this, a propagating medium can be represented accurately. By connecting appropriately characterized generalized nodes in a three-dimensional mesh, the individual homogeneous regions of the actual waveguiding structure can be modeled. To complete the model, boundary conditions need to be incorporated. A short circuit (electric wall) is obtained by shorting out the shunt nodes in the plane of interest, while an open circuit (magnetic wall) is achieved by open circuiting the appropriate series nodes in the plane of interest. Dielectric interfaces are dealt with in terms of the continuity of tangential field components. An example [44] is the xz-plane boundary between two dielectric materials. In this case, elementary transmission-line sections connect a generalized node in one medium to a generalized node in the second medium as Fig. 18 indicates. If the tangential field components for E_x, E_z, H_x, and H_z are considered on either side of the boundary, then the following equations are obtained:

$$E_{z_2} = E_{z_1} + \frac{\partial E_{z_1}}{\partial y}\Delta l \tag{143}$$

$$E_{x_2} = E_{x_1} + \frac{\partial E_{x_1}}{\partial y}\Delta l \tag{144}$$

$$H_{x_2} = H_{z_1} + \frac{\partial H_{x_1}}{\partial y}\Delta l \tag{145}$$

$$H_{z_2} = H_{z_1} + \frac{\partial H_{z_1}}{\partial y}\Delta l. \tag{146}$$

These relations are obtained from the correspondence of the electric field at a shunt node to voltage, and of the magnetic field at a series node to current.

The TLM method predicts, as stated previously, the impulse response of a given network. According to this technique, an impulse excitation takes place at some circuit location. It propagates throughout the transmission-line sections scattering at the shunt and series node locations. The manner according to which the impulse is scattered at a given node is prescribed by the scattering matrix pertinent to that node. The scattering matrix of a shunt node is

$$[S]_{\text{shunt}} = \frac{2}{Y}\begin{bmatrix} 1 & 1 & 1 & 1 & Y_{ll} \\ 1 & 1 & 1 & 1 & Y_{ll} \\ 1 & 1 & 1 & 1 & Y_{ll} \\ 1 & 1 & 1 & 1 & Y_{ll} \\ 1 & 1 & 1 & 1 & Y_{ll} \end{bmatrix} - [I] \tag{147}$$

where $Y = 4 + Y_{ll} + G_{ll}$, ll denotes $\hat{x}\hat{x}$, $\hat{y}\hat{y}$, or $\hat{z}\hat{z}$, and $[I]$ is the unitary matrix. For a series node

$$[S]_{\text{series}} = \frac{2}{Z}\begin{bmatrix} -1 & 1 & 1 & -1 & -1 \\ 1 & -1 & -1 & 1 & 1 \\ 1 & -1 & -1 & 1 & 1 \\ -1 & 1 & 1 & -1 & -1 \\ -Z_{ll} & Z_{ll} & Z_{ll} & -Z_{ll} & -Z_{ll} \end{bmatrix} + [I] \tag{148}$$

where $Z = 4 + Z_{ll}$.

The voltage–current analog of any electromagnetic field of interest can be excited by imposing the properly weighted voltage and current impulses at the node points of the equivalent network. These impulse fields can be followed as they travel and scatter through the network, and allow a determination of the field value at any point of the guiding structure by way of the analog and the corresponding network point. The response is obtained at the point of observation as the collection of the impulse amplitudes incident at that point. Fourier transformation of this result yields easily the Fourier domain response.

The TLM method just described has been applied to determine the dispersion characteristics of single and coupled microstrip lines, as well as of microstrip discontinuities on an anisotropic substrate as defined by a diagonalized permittivity tensor [44]. The geometry under consideration is illustrated in Fig. 1(c). Due to the even symmetry in x, a magnetic wall is placed at $x = 0$, and as such the input data involves boundary conditions which take the form $E_x = E_z = 0$ at $y = 0$, B, $E_y = E_z = 0$ at $x = a$, $H_y = H_z = 0$ at $x = 0$, and $E_x = E_z = 0$ at $y = H$, $0 \leqslant x \leqslant w/2$. Shorting planes are placed $2L_r$ units apart along the length of the microstrip transmission line to form a resonator. At the lowest resonant frequency of this cavity, the quantity $2L_r$ corresponds to half the guided wavelength of the fundamental propagating mode on the microstrip line, thereby yielding the dispersion characteristics of the line (i.e., at resonance $\beta = \pi/2L_r$).

Results for the particular case where $w/H = 3$, $\Delta l = H$, and $B/H = 6$ are shown in Fig. 19. For these computations, one thousand iterations were used (only a 0.01-percent change in resonant frequency is observed if more iterations are used [44]). The difference in the values as

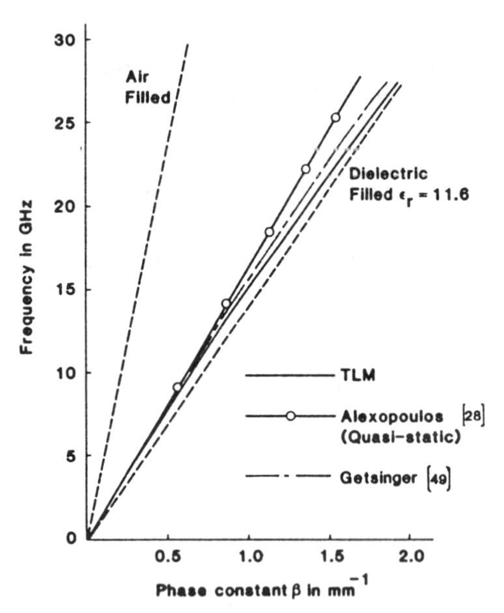

Fig. 19. Dispersion diagram for single microstrip on sapphire substrate $w/H = 3.0$, $H = \Delta 1$.

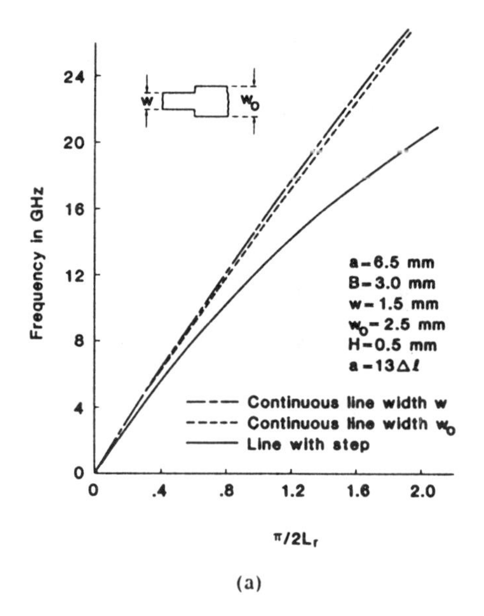

(a)

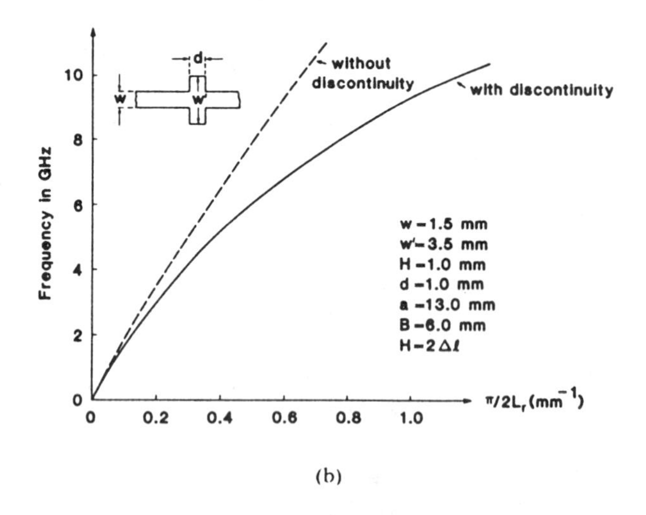

(b)

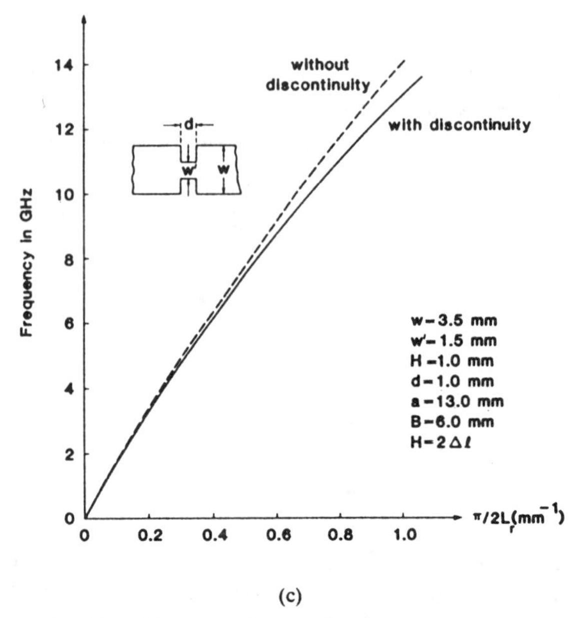

(c)

Fig. 20. Dispersion diagrams for the shown structures.

compared to an isotropic substrate case reported in the literature [50] are 7 percent for $H = \Delta l$, improved to 2 percent for $H = 2\Delta l$, and finally 0.5 percent if $H = 3\Delta l$ is chosen. A crucial point in convergence enhancement is, in this case, in *a priori* knowledge of the fundamental-mode field distribution. It has been found that a more accurate representation of the fundamental mode results if E_y is excited at all nodes lying directly below the strip conductor and E_x is excited along the edge of the strip.

The TLM method has also been used to predict the dispersion properties of microstrip discontinuities [44] such as those shown in Fig. 20. An example of dispersion for coupled microstrip on sapphire is shown in Fig. 21. There the dimensions are: $a = 17\Delta l$, $H = 3\Delta l$, $B = 6\Delta l$, $s = 3\Delta l$, $w = 3\Delta l$, and $\Delta l = 0.5$ mm.

The TLM technique as described is a very simple and versatile method which is easily adopted to obtain the dispersion characteristics of single or coupled microstrip lines, as well as of microstrip discontinuities on anisotropic substrates. An important disadvantage of the technique is the need for *a priori* knowledge or very good initial guess of the dominant-mode field distribution to enhance convergence. In addition, the accuracy is dependent on the number of iterations used to ensure convergence for the selected mesh size. Obviously, the finer the chosen mesh size, the more accurate the solution, at the expense of computer run time and memory storage requirements.

B. Fourier-Domain Methods

The frequency-dependent characteristics of integrated-circuit structures on anisotropic substrates can be analyzed, in addition to the transmission-line matrix method, by solving Maxwell's equations with Fourier-spectrum techniques. The electromagnetic-field components may be expressed either in terms of a continuous or a discrete Fourier spectrum depending on whether the waveguiding

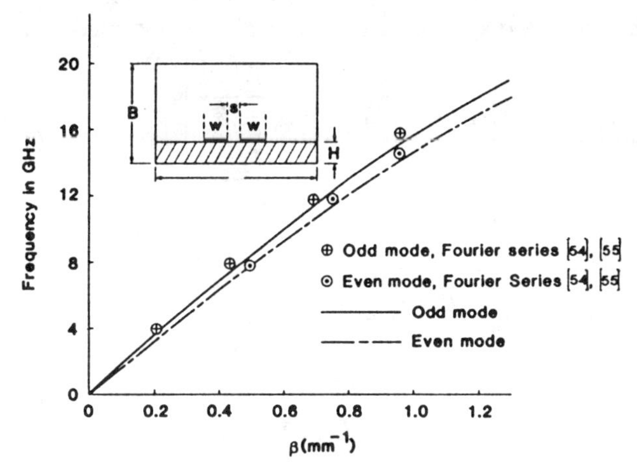

Fig. 21. Dispersion diagram for even and odd fundamental modes of edge-coupled microstrip ($a = 17\Delta 1$, $H = 3\Delta 1$, $B = 9\Delta 1$, $s = 3\Delta 1$, $w = 3\Delta 1$, $\Delta 1 = 0.5$ mm, sapphire substrate).

circuit under consideration is an open or a closed structure. Due to the inhomogeneity in the $\hat{y}$-dimension of the structures under consideration, the complete field solution is obtained by the superposition of LSE and LSM modes. In this section, methods of solution for microstrip, microstrip with cover, microstrip in a rectangularly shaped shield, and the corresponding cases of coplanar waveguide and slotline are investigated. The substrate is characterized by a diagonalized tensor $\bar{\bar{\epsilon}}$, while the material magnetic and conductive properties are assumed isotropic ($\mu_r = 1$, $\sigma = 0$).

1) Continuous Fourier Spectrum—Microstrip and Coplanar Slots: The dispersive properties of microstrip with cover may be obtained by adopting the continuous Fourier spectrum in conjunction with the Wiener–Hopf method [51]. Structures such as microstrip with or without cover, inverted microstrip, coupled microstrip, coplanar lines, and coplanar slots may be analyzed by combining the continuous Fourier-spectrum field representation with the equivalent network method of solution of Maxwell's equations [52]–[54].

a) Modified Wiener–Hopf method; Microstrip with cover: For time-harmonic fields and propagation in the $+\hat{z}$-direction, Maxwell's equations are simplified by allowing $\partial/\partial t \to j\omega$ and $\partial/\partial t \to -j\beta$. Further simplification is obtained if the electromagnetic-field quantities E and H are written as the inverse Fourier transforms

$$\mathscr{A}(x, y) = \frac{1}{2\pi}\int_{-\infty}^{\infty} \tilde{\mathscr{A}}(\zeta, y)\, e^{j\zeta x}\, d\zeta \tag{149}$$

where $\mathscr{A}$ may represent E or H. With this substitution, the LSE and LSM modes are expressed in the spectral domain as follows.

LSE Modes (TE_y):

$$\tilde{E}_x^{TE_y} = \frac{\omega\mu_0\beta}{\zeta^2+\beta^2}\tilde{H}_y^{TE_y} \tag{150}$$

$$\tilde{E}_z^{TE_y} = \frac{\omega\mu_0\zeta}{\zeta^2+\beta^2}\tilde{H}_y^{TE_y} \tag{151}$$

$$\tilde{H}_x^{TE_y} = \frac{j\zeta}{\zeta^2+\beta^2}\frac{\partial \tilde{H}^{TE_y}}{\partial y} \tag{152}$$

and

$$\tilde{H}_z^{TE_y} = \frac{-j\beta}{\zeta^2+\beta^2}\frac{\partial \tilde{H}_y^{TE_y}}{\partial y} \tag{153}$$

where $\tilde{H}_y^{TE_y}$ satisfies the wave equation

$$\frac{\partial^2 \tilde{H}_y^{TE_y}}{\partial y^2} - \left[\zeta^2+\beta^2-\epsilon_t k_0^2\right]\tilde{H}_y^{TE_y} = 0. \tag{154}$$

Similarly, the following can be expressed.

LSM Modes (TM_y):

$$\tilde{E}_x^{TM_y} = \frac{j\zeta(\epsilon_{yy}/\epsilon_t)}{\zeta^2+\beta^2}\frac{\partial \tilde{E}_y^{TM_y}}{\partial y} \tag{155}$$

$$\tilde{E}_z^{TM_y} = -\frac{j\beta(\epsilon_{yy}/\epsilon_t)}{\zeta^2+\beta^2}\frac{\partial \tilde{E}_y^{TM_y}}{\partial y} \tag{156}$$

$$\tilde{H}_x^{TM_y} = -\frac{\omega\epsilon_0\epsilon_{yy}\beta}{\zeta^2+\beta^2}\tilde{E}_y^{TM_y} \tag{157}$$

and

$$\tilde{H}_z^{TM_y} = -\frac{\zeta\omega\epsilon_0\epsilon_{yy}}{\zeta^2+\beta^2}\tilde{E}_y^{TM_y} \tag{158}$$

where, for this case, $\tilde{E}_y^{TM_y}$ satisfies the wave equation

$$\frac{\partial^2 \tilde{E}_y^{TM_y}}{\partial y^2} - \left(\frac{\epsilon_t}{\epsilon_{yy}}\right)\left[\zeta^2+\beta^2-\epsilon_{yy}k_o^2\right]\tilde{E}_y^{TM_y} = 0 \tag{159}$$

and $\epsilon_t = \epsilon_{xx} = \epsilon_{zz}$. The LSE and LSM modes are superposed to yield the following system of equations:

$$\omega\mu_0\tilde{H}_y^{TE_y} = \beta\tilde{E}_x + \zeta\tilde{E}_z \tag{160}$$

$$j\left(\frac{\epsilon_{yy}}{\epsilon_t}\right)\frac{\partial \tilde{E}_y^{TM_y}}{\partial y} = \zeta\tilde{E}_x - \beta\tilde{E}_z \tag{161}$$

$$-\omega\epsilon_0\epsilon_{yy}\tilde{E}^{TM_y} = \beta\tilde{H}_x + \zeta\tilde{H}_z \tag{162}$$

and

$$j\frac{\partial \tilde{H}_y^{TE_y}}{\partial y} = \zeta\tilde{H}_x - \beta\tilde{H}_z \tag{163}$$

with $\tilde{E}_x$, $\tilde{E}_z$, $\tilde{H}_x$, $\tilde{H}_z$ representing the transform of the total field components. These relations are needed so as to determine the boundary conditions which $\tilde{E}_y^{TM_y}$ and $\tilde{H}_y^{TE_y}$ must satisfy. The wave equations to be solved in regions 1 and 2 of the geometry shown in Fig. 1(b) are given by

$$\frac{d^2\tilde{E}_{y_1}^{TM_y}}{dy^2} - \left(\frac{\epsilon_t}{\epsilon_{yy}}\right)R_y^2\tilde{E}_{y_1}^{TM_y} = 0 \tag{164}$$

$$\frac{d^2\tilde{H}_{y_1}^{TE_y}}{dy^2} - R_t^2\tilde{H}_{y_1}^{TE_y} = 0 \tag{165}$$

where $R_t^2 = \zeta^2+\beta^2-\epsilon_t k_0^2$, $R_y^2 = \zeta^2+\beta^2-\epsilon_{yy}k_0^2$ and

$$\frac{d^2\tilde{E}_{y_2}^{TM_y}}{dy^2} - R_0^2\tilde{E}_{y_2}^{TM_y} = 0 \tag{166}$$

$$\frac{d^2\tilde{H}_{y_2}^{TM_y}}{dy^2} - R_0^2\tilde{H}_{y_2}^{TE_y} = 0 \tag{167}$$

with $R_0^2 = \zeta^2 + \beta^2 - k_0^2$. The boundary conditions impose the following requirements.

At $y = 0, B$:

$$\tilde{H}_y^{\mathrm{TE}_y} = \frac{\partial \tilde{E}_y^{\mathrm{TM}_y}}{\partial y} = 0 \text{ for conductive wall.}$$

At $y = H$ and $|x| > w/2$:

$$\left(\frac{\epsilon_{yy}}{\epsilon_t}\right)\frac{\partial \tilde{E}_y^{\mathrm{TM}_y}}{\partial y},\ \tilde{H}_y^{\mathrm{TE}_y},\ \frac{\partial \tilde{H}_y^{\mathrm{TE}_y}}{\partial y},\ \text{and } \epsilon_{yy}\tilde{E}_y^{\mathrm{TM}_y}$$

must be continuous.

At $y = H$, $|x| \leqslant w/2$:

$$\tilde{H}_y^{\mathrm{TE}_y}, \left(\frac{\epsilon_{yy}}{\epsilon_t}\right)\frac{\partial \tilde{E}_y^{\mathrm{TM}_y}}{\partial y}$$

are continuous and $\tilde{J}_x = \tilde{H}_z$, $\tilde{J}_z = -\tilde{H}_x$, where $\tilde{J}_x$ and $\tilde{J}_z$ are the transforms of the current density distribution on the microstrip. Solution to the aforementioned boundary-value problem yields the following relations [51]:

$$U_1(\zeta) = j\omega\epsilon_0\left[\frac{\coth(R_0 H\nu)}{R_0} + \frac{\sqrt{\epsilon_t\epsilon_{yy}}}{R_y}\coth(R_y H_{\mathrm{eq}})\right]F_1(\zeta) \tag{168}$$

and

$$j\omega\mu_0 U_2(\zeta) = \left[R_0\coth(R_0 H\nu) + R_t\coth(R_t H)\right]F_2(\zeta) \tag{169}$$

where the quantities $U_i(\zeta)$, $F_i(\zeta)$, $i = 1,2$, are defined by ($\nu = B/H - 1$)

$$U_1(\zeta) = \omega\epsilon_0\left(\tilde{E}_{y_2}^{\mathrm{TM}_y} - \epsilon_{yy}\tilde{E}_{y_1}^{\mathrm{TM}_y}\right)_{y=H} = -\zeta\tilde{J}_x + \beta\tilde{J}_z \tag{170}$$

$$U_2(\zeta) = -j\left(\frac{\partial \tilde{H}_{y_2}^{\mathrm{TE}_y}}{\partial y} - \frac{\partial \tilde{H}_{y_1}^{\mathrm{TE}_y}}{\partial y}\right)_{y=H} = \beta\tilde{J}_x + \zeta\tilde{J}_z \tag{171}$$

$$F_1(\zeta) = j\left[\frac{\partial \tilde{E}_{y_2}^{\mathrm{TM}_y}}{\partial y} - \left(\frac{\epsilon_{yy}}{\epsilon_t}\right)\frac{\partial \tilde{E}_{y_1}^{\mathrm{TM}_y}}{\partial y}\right]_{y=H} = +\zeta\tilde{E}_x - \beta\tilde{E}_z \tag{172}$$

and

$$F_2(\zeta) = -\omega\mu_0\left(\tilde{H}_{y_2}^{\mathrm{TE}_y} - \tilde{H}_{y_1}^{\mathrm{TE}_y}\right)_{y=H} = -\beta\tilde{E}_x - \zeta\tilde{E}_z. \tag{173}$$

A modified Wiener–Hopf method has been applied to solve this system of equations for the dispersion properties of a single microstrip conductor on sapphire [51]. The equation

$$F_1^+(\mp j\beta) \mp jF_2^+(\mp j\beta) = 0 \tag{174}$$

where

$$F_1(\zeta) = F_1^+(\zeta) + e^{-j\zeta w}F_1^+(-\zeta) \tag{175}$$

and

$$F_2(\zeta) = F_2^+(\zeta) - e^{-j\zeta w}F_1^+(-\zeta) \tag{176}$$

leads to the dispersion equation, which is obtained from

$$\begin{vmatrix} \dfrac{T_1}{\chi_1^+(-j\beta)} & \dfrac{jT_2}{\chi_2^+(-j\beta)} \\ \dfrac{T_3}{\chi_1^+(j\beta)} & \dfrac{jT_4}{\chi_2^+(j\beta)} \end{vmatrix} = 0 \tag{177}$$

where

$$T_1 = 1 + \sum_{n=0}^{\infty}\frac{s_n}{j\beta + \zeta_{n1}}A_n \tag{178}$$

$$T_2 = 1 + \sum_{n=1}^{\infty}\frac{t_n}{j\beta + \zeta_{n2}}B_n \tag{179}$$

$$T_3 = 1 - \sum_{n=0}^{\infty}\frac{s_n}{j\beta - \zeta_{n1}}A_n \tag{180}$$

and

$$T_4 = 1 - \sum_{n=1}^{\infty}\frac{t_n}{j\beta - \zeta_{n2}}B_n. \tag{181}$$

The F_i^+ are analytic on the upper half-plane $\mathrm{Im}\,\zeta > 0$, and (175) and (176) are valid for a symmetric current distribution on the microstrip. The solutions for F_i^+ are obtained as

$$F_1^+(\zeta) = \frac{P}{\chi_1^+}\left[1 - \sum_{n=0}^{\infty}\frac{s_n}{\zeta - \zeta_{n1}}A_n\right] \tag{182}$$

and

$$F_2^+(\zeta) = \frac{Q}{\chi_2^+}\left[1 - \sum_{n=1}^{\infty}\frac{t_n}{\zeta - \zeta_{n2}}B_n\right] \tag{183}$$

where P and Q are constants and χ_i^+ are defined as the plus functions of $\chi_i(\zeta)$ where

$$\chi_1(\zeta) = \frac{\coth(R_0 H\nu)}{R_0} + \frac{\sqrt{\epsilon_t\epsilon_{yy}}}{R_y}\coth(R_y H_{\mathrm{eq}}) \tag{184}$$

and

$$\chi_2(\zeta) = R_0\coth(R_0 H\nu) + R_t\coth(R_t H) \tag{185}$$

and ζ_n are the poles of χ_1 and χ_2 in the lower half-plane. In addition, A_n and B_n are the solutions to the system of equations

$$A_n = 1 + \sum_{m=0}^{\infty}\frac{s_m}{\zeta_{n1} + \zeta_{m1}}A_m, \qquad n = 0,1,2,\cdots \tag{186}$$

and

$$B_n = 1 + \sum_{m=1}^{\infty}\frac{t_m}{\zeta_{n2} + \zeta_{m2}}B_m, \qquad n = 1,2,\cdots \tag{187}$$

with

$$s_n = \frac{\mathrm{Res}\left[\chi_1^+(\zeta_{n1})\right]}{\chi_1^+(-\zeta_{n1})}e^{-j\zeta_n w} \tag{188}$$

and

$$t_n = -\frac{\mathrm{Res}\left[\chi_2^+(\zeta_{n2})\right]}{\chi_2^+(-\zeta_{n2})}e^{-j\zeta_n w}. \tag{189}$$

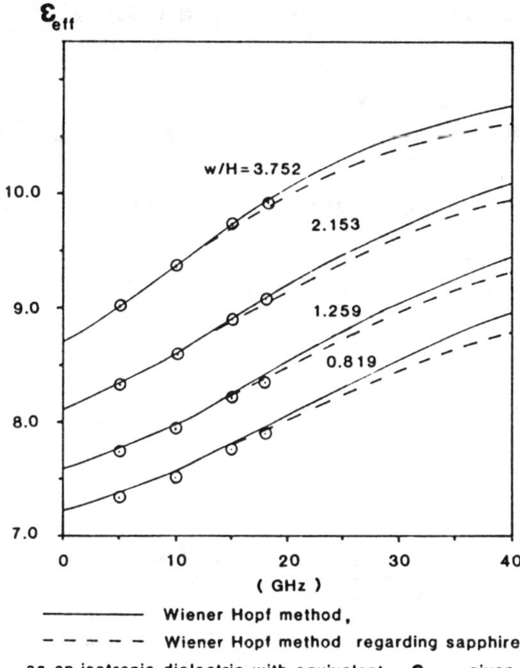

Fig. 22. Dispersion characteristics of microstrip with cover.

The characteristic impedance and the effective dielectric constant have been computed by this method. An expression for Z_0, based on the definition of wave impedance as the ratio of the quasi-static voltage at the center of the strip to the total longitudinal current, has been derived as

$$Z_0 = -\frac{j}{2\zeta_{01}}\frac{\beta F_1^+(-\zeta_{01})}{\omega\epsilon_0 F_1^+(0)\chi_1(0)}e^{-j\zeta_0 w/2} \tag{190}$$

where ζ_{01} is the lowest order pole of $\chi_1(\zeta)$ given by $\zeta_{01} = (k_0^2\epsilon_{yy} - \beta^2)^{1/2}$.

The effective dielectric constant is computed by using the definition $\epsilon_{\text{eff}} = (\beta/k_0)^2$ and it is shown in Fig. 22 as a function of frequency up to 40 GHz. Computations using the equivalent isotropic permittivities $\epsilon_{\text{eq}} = \sqrt{\epsilon_{xx}\epsilon_{yy}}$ and $\epsilon_{r\text{eq}}$ (as given by (40)) are also superimposed for comparison. When ϵ_{eq} is used, an error of 4–10 percent or greater occurs for $f > 5$ GHz, while the linewidth-corrected empirical expression for $\epsilon_{r\text{eq}}$ yields excellent agreement up to about 25 GHz, while at 40 GHz it introduces an error of about 2 percent [51].

b) Equivalent network method: An approach which is straightforward and more general in that it can be readily modified to yield solutions to a variety of integrated-circuit structures on anisotropic substrates with or without cover is the equivalent network technique [52]–[54]. The structure of Fig. 1(h) is considered for which the electric- and magnetic-field components transverse to the y-direction are expressed in the case of microstrip lines as [54]

$$\left.\begin{matrix} \boldsymbol{E}_t^{(i)} \\ \boldsymbol{H}_t^{(i)} \end{matrix}\right\} = \sum_{l=1}^{2}\int_{-\infty}^{\infty}\left\{\begin{matrix} V_l^{(i)}(\zeta,y)\, \boldsymbol{f}_l(\zeta,x) \\ I_l^{(i)}(\zeta,y)\, \hat{y}\times \boldsymbol{f}_l(\zeta,x)\end{matrix}\right\} e^{-j\beta z}\, d\zeta \tag{191}$$

where $i = 1,2,3$ refers to the ith region of the multiple-

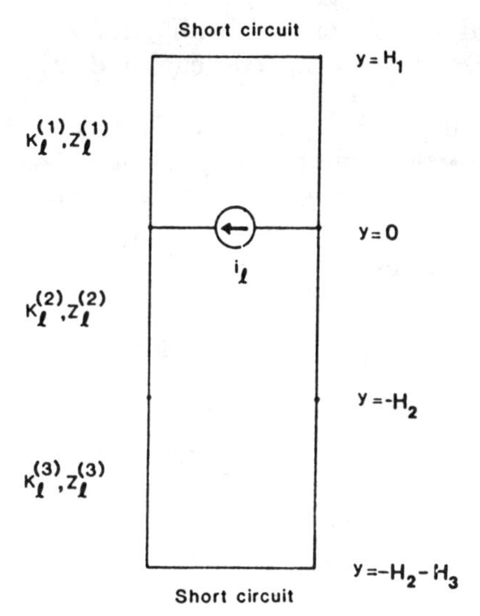

Fig. 23. Equivalent transmission-line circuits for transverse section of coupled strips.

layered anisotropic structure. The index $l = 1$ corresponds to LSM and $l = 2$ to LSE modes. Furthermore

$$\boldsymbol{f}_1(\zeta,x) = \frac{j}{\sqrt{2\pi}}\boldsymbol{K}_0 e^{-j\zeta x} \quad \boldsymbol{f}_2(\zeta,x) = -\hat{y}\times \boldsymbol{f}_1(\zeta,x) \tag{192}$$

and $\boldsymbol{K}_0 = \boldsymbol{K}/|\boldsymbol{K}|$ with $\boldsymbol{K} = \zeta\hat{x} + \beta\hat{z}$.

Substitution of $\boldsymbol{E}_t^{(i)}$ and $\boldsymbol{H}_t^{(i)}$ into Maxwell's equations yields the equivalent transmission-line relations

$$\frac{dV_l^{(i)}}{dy} = -j\kappa_l^{(i)} z_l^{(i)} I_l^{(i)} \tag{193}$$

and

$$\frac{dI_l^{(i)}}{dy} = -j\kappa_l^{(i)} y_l^{(i)} V_l^{(i)} \tag{194}$$

where

$$\kappa_1^{(i)} = \left[\epsilon_{i\perp}k_0^2 - \frac{\epsilon_{i\perp}}{\epsilon_{i_\parallel}}K^2\right]^{1/2} \quad \kappa_2^{(i)} = \left[\epsilon_{i\perp}k_0^2 - K^2\right]^{1/2}$$

$$z_1^{(i)} = \frac{\kappa_1^{(i)}}{\omega\epsilon_0\epsilon_{i\perp}} \quad z_2^{(i)} = \frac{\omega\mu_0}{\kappa_2^{(i)}} \text{ and } y_l^{(i)} = \frac{1}{z_l^{(i)}}.$$

In this development, $\kappa_1^{(i)}$ and $\kappa_2^{(i)}$ are the propagation constants in the $\hat{y}$-direction for TM_y and TE_y waves, while $z_{1,2}^{(i)}$ are the corresponding characteristic wave impedances for these waves (see Fig. 23 for equivalent transmission-line circuit). The (source) current density on the microstrip conductor at $y = 0$ is given by

$$\boldsymbol{j}_s(x,y,z) = \boldsymbol{j}_s(x)\delta(y)e^{-j\beta z} \tag{195}$$

where $\boldsymbol{j}_s(x)$ may be written as

$$\boldsymbol{j}_s(x) = \frac{1}{2\pi}\int_{-\infty}^{\infty} \boldsymbol{i}(\zeta)e^{-j\zeta x}\, d\zeta. \tag{196}$$

It is also possible to define $\boldsymbol{i}(\zeta)e^{-j\zeta x} = -2\pi[i_1(\zeta)\boldsymbol{f}_1(\zeta,x) + i_2(\zeta)\boldsymbol{f}_2(\zeta,x)]$ so that $\boldsymbol{j}_s(x)$ can be formulated as

$$\boldsymbol{j}_s(x) = -\int_{-\infty}^{\infty}\left[i_1(\zeta)\boldsymbol{f}_1(\zeta,x) + i_2(\zeta)\boldsymbol{f}_2(\zeta,x)\right] d\zeta. \tag{197}$$

Vector multiplication of (197) with $f_l^*(\zeta, x)$ and integration over the spectrum yields the current density transform

$$i_l(\zeta) = -\int_{-\infty}^{\infty} f_l^*(\zeta, x')\cdot j_s(x')\, dx'. \tag{198}$$

Proceeding with the derivation of the dispersion equation for the propagation constant β, the boundary conditions at the layer interfaces and the grounded conductors may be expressed in terms of voltages and currents, i.e.,

$$V_l^{(1)}(\zeta, H_1) = 0 \tag{199}$$

$$V_l^{(2)}(\zeta, -H_2) = V_l^{(2)}(\zeta, -H_2) \tag{200}$$

$$V_l^{(1)}(\zeta, 0) = V_l^{(2)}(\zeta, 0) \tag{201}$$

$$V_l^{(3)}(\zeta, -H_2 - H_3) = 0 \tag{202}$$

$$I_l^{(1)}(y = 0^+) - I_l^{(2)}(y = 0^-) = i_l \tag{203}$$

and

$$I_l^{(2)}(y = -H_2) = I_l^{(3)}(y = -H_2). \tag{204}$$

Decoupling of (193) and (194) results in the solutions

$$V_l^{(i)}(\zeta, y) = Z_l^{(i)}(\zeta, y)\, i_l(\zeta) \tag{205}$$

and

$$I_l^{(i)}(\zeta, y) = T_l^{(i)}(\zeta, y)\, i_l(\zeta) \tag{206}$$

where

$$Z_l^{(1)}(\zeta, y) = -\frac{j \sin\left[\kappa_l^{(1)}(y - H_1)\right]}{D_0} \tag{207}$$

$$Z_l^{(2)}(\zeta, y) = \frac{j \sin\left(\kappa_l^{(1)} H_1\right)}{D_0}\left[D_1 \sin\left(\kappa_l^{(2)} y\right) + \cos\left(\kappa_l^{(2)}(y)\right)\right] \tag{208}$$

$$Z_l^{(3)}(\zeta, y) = -\frac{j \sin\left[\kappa_l^{(2)}(y + H_2 + H_3)\right]}{D_0}\, \frac{\sin\left(\kappa_l^{(1)} H_1\right)}{\sin\left(\kappa_l^{(3)} H_3\right)} \times \left[D_1 \sin\left(\kappa_l^{(2)} H_2\right) - \cos\left(\kappa_l^{(2)} H_2\right)\right] \tag{209}$$

$$T_l^{(1)}(\zeta, y) = \frac{y_l^{(1)} \cos\left[\kappa_l^{(1)}(y - H_1)\right]}{D_0} \tag{210}$$

$$T_l^{(2)}(\zeta, y) = \frac{y_l^{(2)} \sin\left(\kappa_l^{(1)} H_1\right)}{D_0}\left[\sin\left(\kappa_l^{(2)} y\right) - D_1 \cos\left(\kappa_l^{(2)} y\right)\right] \tag{211}$$

$$T_l^{(3)}(\zeta, y) = -\frac{y_l^{(3)} \sin\left(\kappa_l^{(1)} H_1\right)}{D_0}\, \frac{\cos\left[\kappa_l^{(3)}(y + H_2 + H_3)\right]}{\sin\left(\kappa_l^{(3)} H_3\right)} \cdot \left[\cos\left(\kappa_l^{(2)} H_2\right) - D_1 \sin\left(\kappa_l^{(2)} H_2\right)\right] \tag{212}$$

with

$$D_0 = y_l^{(1)} \cos\left(\kappa_l^{(1)} H_1\right) + y_l^{(2)} \sin\left(\kappa_l^{(1)} H_1\right) D_1$$

and

$$D_1 = \frac{\dfrac{\cos\left(\kappa_l^{(2)} H_2\right)}{\sin\left(\kappa_l^{(3)} H_3\right)} - \dfrac{y_l^{(2)} \sin\left(\kappa_l^{(2)} H_2\right)}{y_l^{(3)} \cos\left(\kappa_l^{(3)} H_3\right)}}{\dfrac{\sin\left(\kappa_l^{(2)} H_2\right)}{\sin\left(\kappa_l^{(3)} H_3\right)} + \dfrac{y_l^{(2)} \cos\left(\kappa_l^{(2)} H_2\right)}{y_l^{(3)} \cos\left(\kappa_l^{(3)} H_3\right)}}.$$

With the aid of these relations and (198), $E_t^{(i)}$ can be derived in each region. The electric-field components in region 1 can be written as

$$E_x^{(1)} = -\frac{e^{-j\beta z}}{2\pi}\int_{-\infty}^{\infty}\frac{d\zeta}{\zeta^2 + \beta^2} \cdot \int_{-\infty}^{\infty}\left[\left[\zeta^2 Z_1^{(1)}(\zeta, y) + \beta^2 Z_2^{(1)}(\zeta, y)\right] j_x(x') + \zeta\beta\left[Z_1^{(1)}(\zeta, y) - Z_2^{(1)}(\zeta, y)\right] j_z(x')\right] \cdot e^{-j\zeta(x - x')}\, dx' \tag{213}$$

and

$$E_z^{(1)} = -\frac{e^{-j\beta z}}{2\pi}\int_{-\infty}^{\infty}\frac{d\zeta}{\zeta^2 + \beta^2} \cdot \int_{-\infty}^{\infty}\left[\zeta\beta\left[Z_1^{(1)}(\zeta, y) - Z_2^{(1)}(\zeta, y)\right] \times j_x(x') + \left[\beta^2 Z_1^{(1)}(\zeta, y) + \zeta^2 Z_2^{(1)}(\zeta, y)\right] j_z(x')\right] \cdot e^{-j\zeta(x - x')}\, dx'. \tag{214}$$

In order to derive the dispersion relations for the coupled microstrip lines shown in Fig. 1(h), a solution is obtained in terms of even- and odd-mode analysis.

i) Even modes (magnetic wall at $x = 0$): For this particular case, even-mode symmetry implies $j_{xe}(x') = -j_{xe}(-x')$ and $j_{ze}(x') = j_{ze}(-x')$, and therefore (213) and (214) can be rewritten as

$$E_{xe}^{(1)} = -\frac{2e^{-j\beta z}}{\pi}\int_{0}^{\infty}\frac{d\zeta}{\zeta^2 + \beta^2} \cdot \int_{s/2}^{s/2 + w}\left[\left[\zeta^2 Z_1^{(1)}(\zeta, y) + \beta^2 Z_1^{(2)}(\zeta, y)\right] j_{xe}(x') \cdot \sin(\zeta x)\sin(\zeta x') - j\left[\zeta\beta\left(Z_1^{(1)}(\zeta, y) - Z_2^{(1)}(\zeta, y)\right]\right. \cdot j_{ze}(x')\sin(\zeta x)\cos(\zeta x')\, dx' \tag{215}$$

and

$$E_{ze}^{(1)} = -\frac{2e^{-j\beta z}}{\pi}\int_{0}^{\infty}\frac{d\zeta}{\zeta^2 + \beta^2} \cdot \int_{s/2}^{s/2 + w}\left[j(\zeta\beta)\left(Z_1^{(1)}(\zeta, y) - Z_2^{(1)}(\zeta, y)\right] \cdot j_{xe}(x')\cos(\zeta x)\sin(\zeta x') + \left[\beta^2 Z_1^{(1)}(\zeta, y) + \zeta^2 Z_2^{(1)}(\zeta, y)\right] \cdot j_{ze}(x')\cos(\zeta x)\cos(\zeta x')\, dx'. \tag{216}$$

On the strip, i.e., for $s/2 < x < s/2 + w$, $E_{xe}^{(1)} = E_{ze}^{(1)} = 0$, and the Galerkin method is invoked to obtain solutions of (215) and (216) for $j_{xe}(x)$ and $j_{ze}(x)$. These current densities are expanded into the forms [54]

$$j_{ze}(x') = \sum_{n=0}^{N} a_{nze}\frac{T_{2n}\left(\dfrac{2(x' - s)}{w}\right)^2}{\sqrt{1 - \left(\dfrac{2(x' - s)}{w}\right)}} \tag{217}$$

and

$$j_{xe}(x') = -j\sum_{n=0}^{N} a_{nxe} U_{2n}\left(\frac{2(x - s)}{w}\right) \tag{218}$$

where $T_n(x)$ and $U_n(x)$ represent the Chebyshev polynomials of the first and second kind, respectively. The particular choice of this representation for $j_{ze}(x')$ and $j_{xe}(x')$ is dictated by the edge conditions for the current density components at $x=s/2$ and $x=s/2+w$. Consideration of the fact that $E_{xe}^{(1)}=E_{ze}^{(1)}=0$ on the microstrip and substitution of (217) and (218) into (215) and (216) yields the following system of equations to be solved for the propagation constant β, namely:

$$\sum_{n=0}^{N} \alpha_{mne}a_{nze}+\sum_{n+1}^{N}\beta_{mne}a_{nxe}=0, \qquad m=0,1,2,3,\cdots,N \tag{219}$$

and

$$\sum_{n=0}^{N} \gamma_{mne}a_{nze}+\sum_{n=1}^{N}\delta_{mne}a_{nxe}=0, \qquad m=1,2,3,\cdots,N \tag{220}$$

where

$$\alpha_{mne}=(-1)^{m+n}\left(\frac{w}{2}\right)^2\left(\frac{2}{\pi}\right)\cdot\int_0^\infty\frac{d\zeta}{\zeta^2+\beta^2}\left[\beta^2Z_1^{(1)}(\zeta,0)+\zeta^2Z_2^{(1)}(\zeta,0)\right]\cdot J_{2n}\left(\frac{\zeta w}{2}\right)\cdot J_{2m}\left(\frac{\zeta w}{2}\right) \tag{221}$$

$$\beta_{mne}=(-1)^{m+n}nw(2/\pi)\cdot\int_0^\infty\frac{d\zeta}{\zeta^2+\beta^2}\left[\beta\left(Z_1^{(1)}(\zeta,0)-Z_2^{(1)}(\zeta,0)\right)\right]\cdot J_{2n}\left(\frac{\zeta w}{2}\right)J_{2m}\left(\frac{\zeta w}{2}\right) \tag{222}$$

and

$$\gamma_{mne}=\frac{m}{n}\beta_{mne} \tag{223}$$

$$\delta_{mne}=(-1)^{m+n}4mn(2/\pi)\cdot\int_0^\infty\frac{d\zeta}{\zeta^2+\beta^2}\left[Z_1^{(1)}(\zeta,0)+\frac{\beta^2}{\zeta^2}Z_2^{(1)}(\zeta,0)\right]\cdot J_{2n}\left(\frac{\zeta w}{2}\right)J_{2m}\left(\frac{\zeta w}{2}\right)d\zeta. \tag{224}$$

Setting the determinant of the system of equations (219) and (220) equal to zero yields the dispersion equation for the even-mode propagation constant.

ii) Odd modes (electric wall at $x=0$): For odd modes, $j_{xo}(x')=j_{xo}(-x')$, while $j_{zo}(x')=-j_{zo}(-x')$ and the electric-field components in region 1 are now given by

$$E_{xo}^{(1)}=-\frac{2e^{-j\beta z}}{\pi}\int_0^\infty\frac{d\zeta}{\zeta^2+\beta^2}\cdot\int_{s/2}^{s/2+w}\left[\left[\zeta^2Z_1^{(1)}(\zeta,y)+\beta^2Z_2^{(1)}(\zeta,y)\right]\right.$$
$$j_x(x')\cos(\zeta x)\cos(\zeta x')+j\zeta\beta\left[Z_1^{(1)}(\zeta,y)-Z_2^{(1)}(\zeta,y)\right]$$
$$\left.\cdot j_z(x')\cos(\zeta x)\sin(\zeta x')\right]dx' \tag{225}$$

and

$$E_{zo}^{(1)}=-\frac{2e^{-j\beta z}}{\pi}\int_0^\infty\frac{d\zeta}{\zeta^2+\beta^2}\cdot\int_{s/2}^{s/2+w}\left[\left[\beta^2Z_1^{(1)}(\zeta,y)+\zeta^2Z_2^{(1)}(\zeta,y)\right]\right.$$
$$j_z(x')\sin(\zeta x)\sin(\zeta x')$$
$$-j\zeta\beta\left[Z_1^{(1)}(\zeta,y)-Z_2^{(1)}(\zeta,y)\right]$$
$$\left.\cdot j_x(x')\sin(\zeta x)\cos(\zeta x')\right]dx'. \tag{226}$$

In applying the Galerkin procedure for odd modes, the following expansions have been adopted for the current density components $j_{xo}(x)$ and $j_{zo}(x)$, namely [54]:

$$j_{zo}(x)=\sum_{n=1}^{N}a_{nzo}\frac{T_{2n-1}\left(\frac{2(x'-s)}{w}\right)}{\sqrt{1-\left(\frac{2(x-s)}{w}\right)^2}}$$

and

$$-jj_{xo}(x)=\sum_{n=1}^{N}a_{nxo}U_{2n-1}\left(\frac{2(x'-s)}{w}\right).$$

Adaptation of the above current distributions and the fact that $E_{xo}^{(1)}=E_{zo}^{(1)}=0$ on the strip yields, in this case, the system of equations

$$\sum_{n=1}^{N}\alpha_{mno}a_{nzo}+\sum_{n=1}^{N}\beta_{mno}a_{nxo}=0, \qquad m=1,2,3,\cdots,N \tag{227}$$

and

$$\sum_{n=1}^{N}\gamma_{mno}a_{nzo}+\sum_{n=1}^{N}\delta_{mno}a_{nxo}=0, \qquad m=1,2,3,\cdots,N \tag{228}$$

whose solution yields the dispersive properties of the propagation constant for the odd mode. For this particular case

$$\alpha_{mno}=(-1)^{m+n}\left(\frac{w}{2}\right)^2\frac{2}{\pi}\int_0^\infty\frac{d\zeta}{\zeta^2+\beta^2}\cdot\left[\beta^2Z_1^{(1)}(\zeta,0)+\zeta^2Z_2^{(1)}(\zeta,0)\right]\cdot J_{2n-1}\left(\frac{\zeta w}{2}\right)J_{2m-1}\left(\frac{\zeta w}{2}\right) \tag{229}$$

$$\beta_{mno}=(-1)^{m+n}\left(\frac{w}{2}\right)(2n-1)\left(\frac{2}{\pi}\right)\int_0^\infty\frac{d\zeta}{\zeta^2+\beta^2}\cdot\left[\beta\left(Z_0^{(1)}(\zeta,0)-Z_2^{(1)}(\zeta,0)\right)\right]\cdot J_{2n-1}\left(\frac{\zeta w}{2}\right)J_{2m-1}\left(\frac{\zeta w}{2}\right) \tag{230}$$

$$\gamma_{mno}=\frac{(2m-1)}{(2n-1)}\beta_{mno}$$

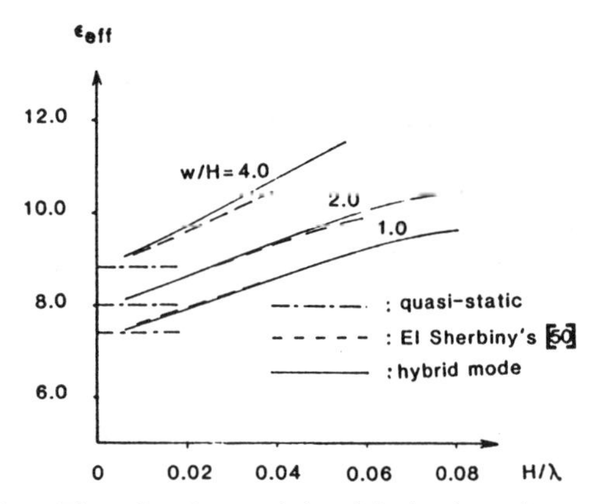

Fig. 24. Dispersion characteristics of single microstrip on sapphire.

and

$$\delta_{mno} = (-1)^{m+n}(2n-1)(2m-1)\left(\frac{2}{\pi}\right)\int_0^\infty \frac{d\zeta}{\zeta^2+\beta^2} \cdot \left[Z_1^{(1)}(\zeta,0)+\frac{\beta^2}{\zeta^2}Z_2^{(1)}(\zeta,0)\right] \cdot J_{2n-1}\left(\frac{\zeta w}{2}\right)J_{2m-1}\left(\frac{\zeta w}{2}\right). \tag{231}$$

Numerical computations based on this analysis have been performed for various microstrip geometries. The results show that even for $N=2$, sufficient convergence accuracy is obtained [54]. Comparison of this technique with the Wiener–Hopf method indicates excellent agreement for the single microstrip with cover case in the computations for ϵ_{eff} when $w/H \leq 4.0$. The discrepancy between the two techniques becomes larger for increasing w/H, as Fig. 24 indicates. Furthermore, this disagreement, as Fig. 25 shows, is even more prominent when the dispersive behavior of Z_0 is compared between the two methods. This discrepancy is also due to the fact that the characteristic impedance in this case is defined in terms of the ratio P_{ave}/I^2, where P_{ave} is the average power flowing in the $\hat{z}$-direction along the microstrip as computed by the Poynting vector, and I is the total current on the microstrip. This definition of Z_0 accounts for dispersion more accurately than the definition used in [52], which is based on the ratio of the quasi-static voltage at the center of the strip to the total longitudinal strip current.

2) Discrete Fourier Spectrum—Structures with a Rectangular Shield: When a waveguiding structure is enclosed entirely within a rectangular shield, the discrete Fourier spectrum may be used to determine the dispersion properties of the distributed circuit [55], [56]. For coupled lines, the method requires the following Fourier transform definitions.

Even Modes (Magnetic Wall at $x=0$):

$$\tilde{E}_y^{\text{TM}_y} = \int_0^a E_y^{\text{TM}_y}\cos(k_n x)\,dx \tag{232}$$

$$\tilde{H}_y^{\text{TE}_y} = \int_0^a H_y^{\text{TE}_y}\sin(k_n x)\,dx \tag{233}$$

where $k_n = (2n-1)/2a\pi$.

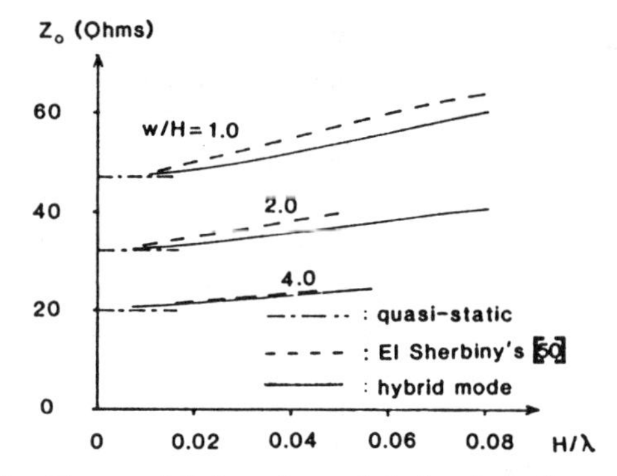

Fig. 25. Characteristic impedance of single microstrip on sapphire.

Odd Modes (Electric Wall at $x=0$):

$$\tilde{E}_y^{\text{TM}_y} = \int_0^a E_y^{\text{TM}_y}\sin(k_n x)\,dx \tag{234}$$

$$\tilde{H}_y^{\text{TE}_y} = \int_0^a H_y^{\text{TE}_y}\cos(k_n x)\,dx \tag{235}$$

where $k_n = n\pi/a$.

These representations yield for the LSE and LSM modes the following equations.

LSE Modes:

$$\tilde{E}_x^{\text{TE}_y} = \frac{\omega\mu_0\beta}{\beta^2+k_n^2}\tilde{H}_y^{\text{TE}_y} \tag{236}$$

$$\tilde{E}_z^{\text{TE}_y} = \pm\frac{j\omega\mu_0 k_n}{\beta^2+k_n^2}\tilde{H}_y^{\text{TE}_y} \tag{237}$$

$$\tilde{H}_x^{\text{TE}_y} = \mp\frac{k_n}{\beta^2+k_n^2}\frac{\partial\tilde{H}_y^{\text{TE}_y}}{\partial y} \tag{238}$$

and

$$\tilde{H}_z^{\text{TE}_y} = -\frac{j\beta}{\beta^2+k_n^2}\frac{\partial\tilde{H}_y^{\text{TE}_y}}{\partial y} \tag{239}$$

where $H_y^{\text{TE}_y}$ satisfies the wave equation with $\zeta^2 = k_n^2$.

LSM Modes:

$$\tilde{E}_x^{\text{TM}_y} = \pm\frac{k_n\left(\frac{\epsilon_{yy}}{\epsilon_t}\right)}{\beta^2+k_n^2}\frac{\partial\tilde{E}_y^{\text{TM}_y}}{\partial y} \tag{240}$$

$$\tilde{E}_z^{\text{TM}_y} = -\frac{j\beta\left(\frac{\epsilon_{yy}}{\epsilon_t}\right)}{\beta^2+k_n^2}\frac{\partial\tilde{E}_y^{\text{TM}_y}}{\partial y} \tag{241}$$

$$\tilde{H}_x^{\text{TM}_y} = -\frac{\omega\epsilon_0\epsilon_{yy}\beta}{\beta^2+k_n^2}\tilde{E}_y^{\text{TM}_y} \tag{242}$$

$$\tilde{H}_z^{\text{TM}_y} = \pm\frac{j\omega\epsilon_0\epsilon_{yy}k_n}{\beta^2+k_n^2}\tilde{E}_y^{\text{TM}_y} \tag{243}$$

and $E_y^{\text{TM}_y}$ satisfies the wave equation with $\zeta^2 = k_n^2$. The lower (upper) signs refer to even (odd) modes, respectively. The current distribution Fourier components $\tilde{J}_x$ and $\tilde{J}_z$ can be obtained by inverting the system of equations [56]

$$\begin{bmatrix} j\tilde{E}_z \\ \tilde{E}_x \end{bmatrix} = \begin{bmatrix} \tilde{G}_{11} & \tilde{G}_{12} \\ \tilde{G}_{21} & \tilde{G}_{22} \end{bmatrix}\begin{bmatrix} j\tilde{J}_x \\ -\tilde{J}_z \end{bmatrix} \tag{244}$$

which is valid on the plane of the strips. The matrix components G_{lm} are elements of the Green's dyadic function for the multiple-layered geometry shown in Fig. 1(l). The boundary-value problem solution yields for G_{lm} the following result:

$$\tilde{G}_{11}=\tilde{G}_{22}=\frac{\pm k_0 k_n \beta}{\gamma_n^2 \omega \epsilon_0}\left[F_1(k_n,\beta)+F_2(k_n,\beta)\right] \tag{245}$$

$$\tilde{G}_{12}=\frac{k_0}{\gamma_n^2 \omega \epsilon_0}\left[k_n^2 F_1(k_n,\beta)-\beta^2 F_2(k_n,\beta)\right] \tag{246}$$

$$\tilde{G}_{21}=\frac{k_0}{\gamma_n^2 \omega \epsilon_0}\left[\beta^2 F_1(k_n,\beta)-k_n^2 F_2(k_n,\beta)\right] \tag{247}$$

where

$$F_1(k_n,\beta)=\left[1/f_{t1}+\left(\alpha_t^{(2)^2} f_{t2} f_{t3}+1\right)/\left(f_{t2}+f_{t3}\right)\right]^{-1} \tag{248}$$

$$F_2(k_n,\beta)=\left[1/g_{y1}+\left(g_{y2}g_{y3}+\alpha_y^{(2)^2}\right)/\left\{\alpha_y^{(2)^2}\left(g_{y2}+g_{y3}\right)\right\}\right]^{-1} \tag{249}$$

$$\gamma_n^2=\beta^2+k_n^2 \tag{250}$$

$$\alpha_y^{(i)}=\sqrt{\left\{(\gamma_n/k_0)^2-\epsilon_y^{(i)}\right\}/\left(\epsilon_y^{(i)}\epsilon_t^{(i)}\right)} \tag{251}$$

$$\alpha_t^{(i)}=\sqrt{(\gamma_n/k_0)^2-\epsilon_t^{(i)}} \tag{252}$$

$$f_{ti}=\tanh\left(k_0\alpha_t^{(i)}h_i\right)/\alpha_t^{(i)} \tag{253}$$

and

$$g_{yi}=\alpha_y^{(i)}\tanh\left(\epsilon_t^{(i)}k_0\alpha_y^{(i)}h_i\right). \tag{254}$$

The subscript/superscript i refers to the ith anisotropic layer in the structure.

For the slotline or coplanar waveguide problem, duality may be invoked to show that the conductor currents are related to the slot-field components through the relation [56]

$$\begin{bmatrix} j\tilde{J}_x \\ -\tilde{J}_z \end{bmatrix}=\begin{bmatrix} \tilde{Q}_{11} & \tilde{Q}_{12} \\ \tilde{Q}_{21} & \tilde{Q}_{22} \end{bmatrix}\begin{bmatrix} j\tilde{E}_z \\ \tilde{E}_x \end{bmatrix} \tag{255}$$

where

$$\tilde{Q}_{11}=\tilde{G}_{22}/\Delta \quad \tilde{Q}_{12}=-\tilde{G}_{12}/\Delta \quad \tilde{Q}_{21}=-\tilde{G}_{21}/\Delta \tag{256}$$

and

$$\Delta=\tilde{G}_{11}\tilde{G}_{22}-\tilde{G}_{12}\tilde{G}_{21}. \tag{257}$$

The microstrip current density or slot-field distribution may be expanded in terms of a set of known basis functions $\{f_k(x)\}$ and $\{g_k(x)\}$ in the form

$$\left.\begin{matrix} -J_z(x) \\ E_x(x) \end{matrix}\right\}=\sum_{k=1}^{N} c_k f_k(x) \tag{258}$$

and

$$\left.\begin{matrix} jJ_x(x) \\ jE_z(x) \end{matrix}\right\}=\sum_{k=1}^{M} d_k g_k(x) \tag{259}$$

where $f_k(x)$ and $g_k(x)$ are defined only on the microstrip line or on the slot. The basis functions are chosen so that the edge effect is properly included, i.e., $f_k(x)=[(x-\frac{1}{2}(s+w))/w/2]^{k-1}$ and $g_k(x)=\sin[k\pi(x-s/2)/w]$, [56]. The microstrip current density or slot-field distribution as represented by (258) and (259) are Fourier transformed and the result is substituted in (244) or (255). The Galerkin procedure is applied to yield a system of $(N+M)\times(N+M)$ eigenequations for the unknown constants c_k and d_k. On the microstrip or the slot, this is a homogeneous system of equations whose determinant is set equal to zero to yield the dispersive behavior of the propagation constant β. The elements of the determinantal equation are given by

$$\begin{pmatrix} X_{rp}^{(1,1)} \\ X_{rq}^{(1,2)} \\ X_{sp}^{(2,1)} \\ X_{sq}^{(2,2)} \end{pmatrix}=\sum_{l=1}^{\infty}\begin{pmatrix} \tilde{J}_{zr}(k_l)\tilde{G}_{11}(\beta,k_l)\tilde{J}_{xp}(k_l) \\ \tilde{J}_{zr}(k_l)\tilde{G}_{12}(\beta,k_l)\tilde{J}_{zq}(k_l) \\ \tilde{J}_{xs}(k_l)\tilde{G}_{21}(\beta,k_l)\tilde{J}_{xp}(k_l) \\ \tilde{J}_{xs}(k_l)\tilde{G}_{22}(\beta,k_l)\tilde{J}_{yq}(k_l) \end{pmatrix} \tag{260}$$

where $p=s=1$ to M and $r=q=1$ to N.

This technique has been tested against already discussed quasi-static as well as dynamic solutions. In the quasi-static case, the results of the microstrip couplers with a superstrate layer shown in Table VII have been checked. For each case, a difference of less than 0.03 percent was found for $N=2$, $M=10$, $a=20$, $k_0=10^{-4}$, and $l=1000$. The dispersion curves for $\epsilon_{\text{eff}}^{\{e,o\}}$ as determined by the TLM method have also been verified. In using the discrete Fourier technique, a convergence accuracy better than 0.5 percent has been enforccd. This convergence requirement is satisfied when $N=M=4$ and $l=300$ for the results shown in Fig. 26, and it has been determined that for the mesh size chosen the TLM computations are consistently lower by 3 percent for $\epsilon_{\text{eff}}^{(o)}$ and 1.5 percent for $\epsilon_{\text{eff}}^{(e)}$. A particular case of interest is shown in Fig. 27 where as observed equalization of even- and odd-mode phase velocities is obtained at those normalized frequencies where $\epsilon_{\text{eff}}^{(e)}=\epsilon_{\text{eff}}^{(o)}$. In order to emphasize the versatility of this technique, the dispersive properties of coupled inverted microstrip lines are demonstrated in Fig. 28, while Fig. 29 shows the variation of ϵ_{eff} for a shielded slotline.

The dispersion curves of Fig. 28 highlight the frequency dependence of the error incurred when anisotropy is not included in the computation. The error becomes larger with increasing frequency and it is of the order of 17 percent when, e.g., normalized $k_0=0.70$. The results calculated by this method have been found to be in excellent agreement with those obtained by the equivalent network approach for microstrip with cover and for coupled slots without cover. In addition to the excellent accuracy, this approach provides a generalized algorithm which can resolve all the waveguiding structures shown in Fig. 1 [56]. For this reason, it is perhaps the most useful of the tools presented in this paper for the analysis of the dispersive properties of a variety of integrated-circuit waveguiding structures.

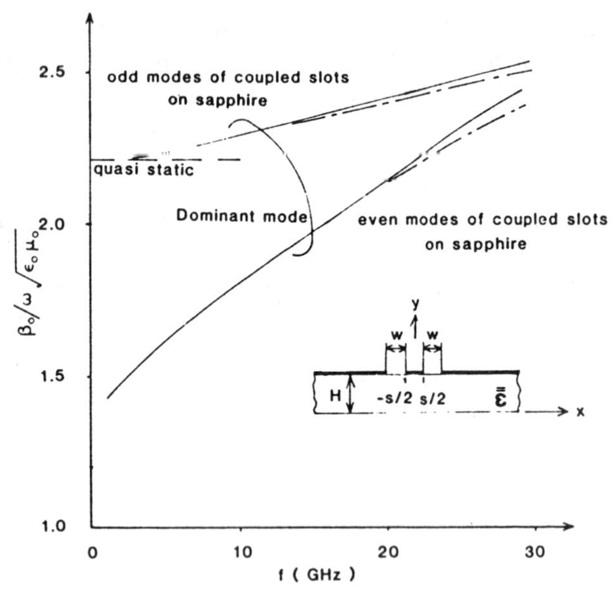

Fig. 26. Normalized propagation constant.

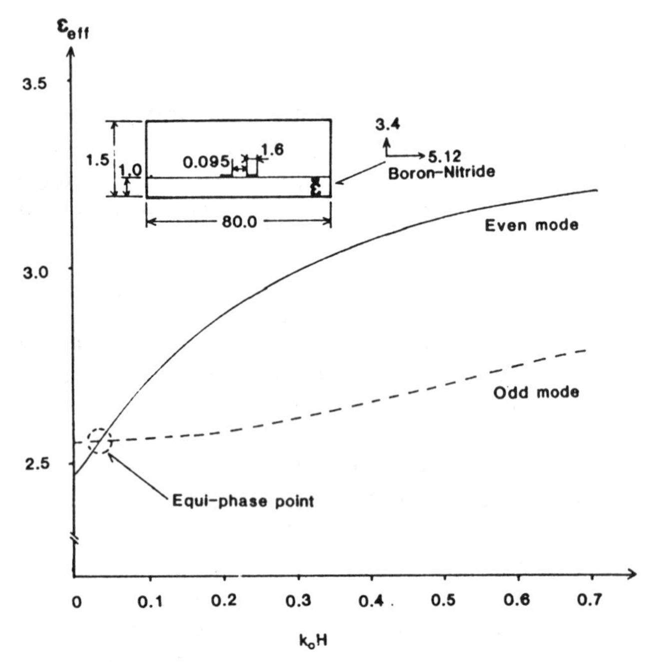

Fig. 27. Dispersive behavior of ϵ for coupled microstrip in a rectangular shield.

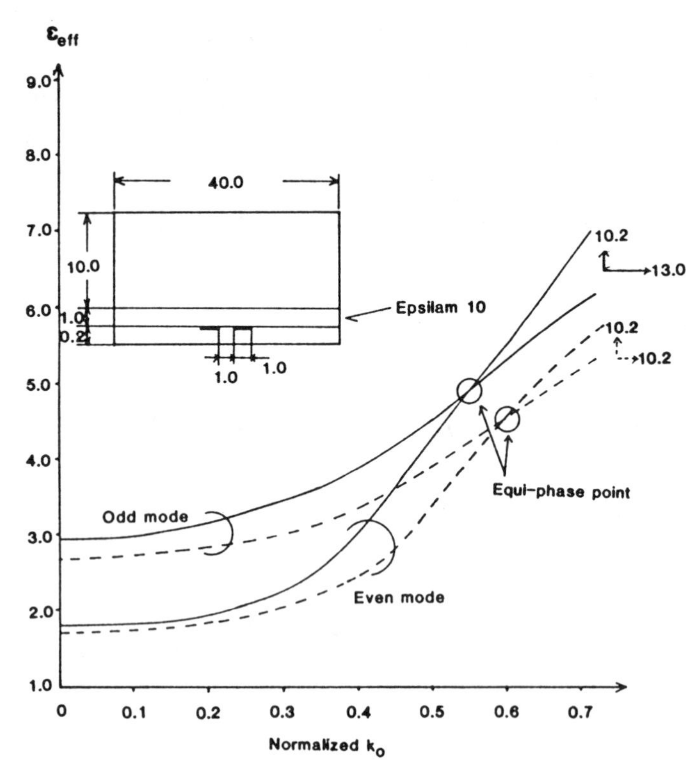

Fig. 28. Dispersion behavior of inverted coupled microstrip lines.

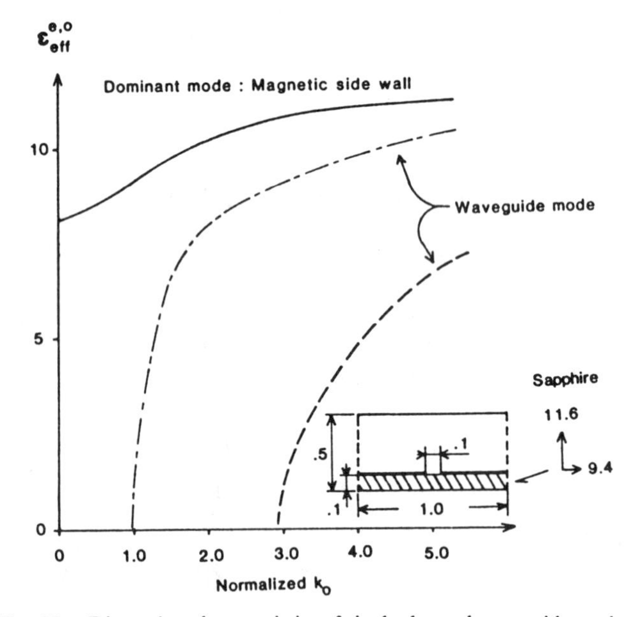

Fig. 29. Dispersion characteristics of single slot and waveguide modes.

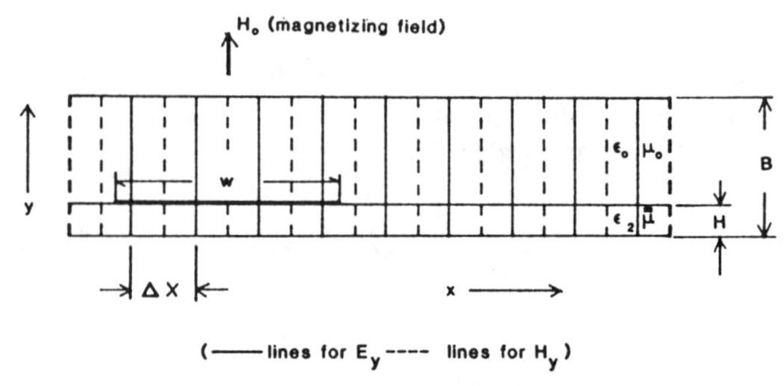

Fig. 30. Cross section of planar microwave structure on magnetized ferrite substrate.

C. *The Method of Lines*

This method solves directly the wave equation and associated boundary conditions for waveguiding structures in a rectangular shield. It is essentially a simplified version of the finite differences method, it is more accurate, and requires less computation time. The system of partial differential equations which describes the nature of the propagating modes is discretized in all directions, except the direction which is transverse to the electrical inhomogeneities of the structure under consideration. The procedure requires, e.g., that the $\hat{x}$-dimension of the circuit of Fig. 1(c) be divided into N subsections by defining $x_n = x_o + n\,\Delta x$ with $n = 1, 2, \cdots, N$ (see Fig. 30). This discretization forces the replacement of derivatives in the $\hat{x}$-direction by finite differences, and it yields a system of N-coupled ordinary differential equations. Upon the introduction of the proper transformed potentials, this system is reduced to N-uncoupled ordinary second-order differential equations which can be solved easily. The steps leading to this latter system of differential equations may be chosen carefully so as to account correctly for the boundary conditions at the side walls of the enclosure, as well as for the edge condition at the waveguiding circuit edges [93]–[95].

This technique has been adopted to obtain the dispersive properties of a microstrip transmission line on a gyrotropic substrate [10]. For the case of a dc magnetizing field $\boldsymbol{H} = \hat{y}H_0$, the permeability tensor given by (5) is used with $\mu_{xx} = \mu_0\mu_1$, $\mu_{xz} = j\mu_0\mu_2$, $\mu_{yy} = \mu_0$, $\mu_{zx} = \mu_{xz}^*$, and $\mu_{zz} = \mu_0\mu_1 = \mu_{xx}$ where

$$\mu_1 = 1 + \frac{\gamma^2 M_s H_0}{(\gamma H_0)^2 - \omega^2} \tag{261}$$

and

$$\mu_2 = \frac{\omega\gamma M_s}{(\gamma H_0)^2 - \omega^2}. \tag{262}$$

In these equations, γ is the gyromagnetic ratio and M_s represents the saturation magnetization [8]. Maxwell's equations are solved in this gyrotropic medium in terms of the electric- and magnetic-field components in the direction of H_0, i.e., in terms of E_y and H_y [10]. A coupled system of second-order partial differential equations results for this case in the form

$$\frac{\partial^2 E_y}{\partial x^2} + \frac{\partial^2 E_y}{\partial y^2} - \beta^2 E_y + \mu_e k_e^2 E_y = k_e \eta \left(\frac{\mu_2}{\mu_1}\right)\frac{\partial H_y}{\partial y} \tag{263}$$

and

$$\frac{\partial^2 H_y}{\partial x^2} + \frac{1}{\mu_0}\frac{\partial^2 H_y}{\partial y^2} - \beta^2 H_y + k_e^2 H_y = -\frac{k_e}{\eta}\left(\frac{\mu_2}{\mu_1}\right)\frac{\partial E_y}{\partial y} \tag{264}$$

where

$$\eta = (\mu_0/\epsilon_2)^{1/2} \quad \mu_e = \mu_1 - \frac{\mu_2^2}{\mu_1} \text{ and } k_e^2 = \omega^2\epsilon_2\mu_0.$$

A discretization procedure is adopted in the $\hat{x}$-direction [10] as suggested in [93]–[95] and [10] which reduces this pair of second-order coupled partial differential equations to a system of second-order ordinary differential equations in the form

$$\left(\left(\frac{d^2}{dy^2} + \mu_e k_e^2 - \beta^2\right) - \frac{1}{(\Delta x)^2}[\lambda^2]\right)\underline{E}_y = \eta k_e \frac{\mu_2}{\mu_1}\frac{\partial}{\partial y}\underline{H}_y \tag{265}$$

and

$$\left(\left(\frac{1}{\mu_1}\frac{d^2}{dy^2} + k_e^2 - \beta^2\right) - \frac{1}{(\Delta x)^2}[\lambda^2]\right)\underline{H}_y = -\frac{1}{\eta}k_e\left(\frac{\mu_2}{\mu_1}\right)\frac{\partial}{\partial y}\underline{E}_y \tag{266}$$

where $\underline{E}_y$ and $\underline{H}_y$ are column vectors with elements E_{yn}, H_{yn} $(n = 1, 2, 3, \cdots, M)$. In order to arrive at this system of coupled ordinary second-order differential equations, the following boundary conditions have been invoked.

On Electric Wall:

$$E_y = E_z = 0 \quad \text{and} \quad \frac{\partial H_y}{\partial x} = -j\frac{\mu_2}{\mu_1}\frac{\partial H_z}{\partial y}. \tag{267}$$

On Magnetic Wall:

$$H_y = H_z = 0 \quad \text{and} \quad \frac{\partial E_y}{\partial x} = -\eta\mu_2 k_e H_x. \tag{268}$$

In (265) and (266), $[\lambda]$ is a diagonal matrix and its elements represent the eigenvalues of the discretization matrix [10]. Equations (265) and (266) may be decoupled in the spectral domain to yield a system of fourth-order ordinary differential equations in the form

$$\left[\left(\frac{1}{\mu_1}\frac{d^2}{dy^2} + [\xi^h]\right)\left(\frac{d^2}{dy^2} + [\xi^e]\right) + k_e^2\frac{\mu_2^2}{\mu_1^2}\frac{d^2}{dy^2}\right]\underline{G}_y = 0 \tag{269}$$

where $\underline{G}_y$ represents either $\underline{E}_y$ or $\underline{H}_y$. In addition, $\xi_n^h = k_e^2 - \beta^2 - \frac{1}{(\Delta x)^2}\lambda_n^2$ and $\xi_n^e = \mu_e k_e^2 - \beta^2 - \frac{1}{(\Delta x)^2}\lambda_n^2$.

This fourth-order differential equation may be solved easily in terms of hyperbolic sines and cosines to yield the solutions

$$E_{yn} = A_{1n}\cosh k_{y1n}y + A_{2n}\cosh k_{y2n}y \tag{270}$$

and

$$H_{yn} = B_{1n}\sinh k_{y1n}y + B_{2n}\sinh k_{y2n}y \tag{271}$$

with

$$k_{y1n,2n}^2 = -\mu_1\left(x_n \pm \sqrt{x_n^2 - 4\frac{\xi_n^e\xi_n^h}{\mu_1}}\right) \tag{272}$$

and

$$x_n = \xi_n^h + \frac{1}{\mu_1}\xi_n^e + k_e^2\frac{\mu_2^2}{\mu_1^2}. \tag{273}$$

The coefficients B_{1n} and B_{2n} are obtained in terms of A_{1n} and A_{2n} by substituting (270) and (271). A similar approach is followed in the air region, and subsequently the boundary conditions are applied at the interface to yield after some manipulations the dispersion equation in β/k_0. This procedure as adopted in [10] yields the dispersion diagrams shown in Fig. 31 for $H_0/M_s = 2.0$ and 8.1 Comparison with the results obtained by the mode-matching technique [9] indicates excellent agreement for $H_0/M_s = 8.1$, but a serious discrepancy exists between the two methods for increasing frequency when $H_0/M_s = 2.0$. This disagreement has not been clarified as yet, but previous results on isotropic substrates have in general verified the accuracy of the method of lines. The discussion in this section simply indicates that this is a useful technique which can be extended to analyze the properties of integrated-circuit structures on anisotropic substrates.

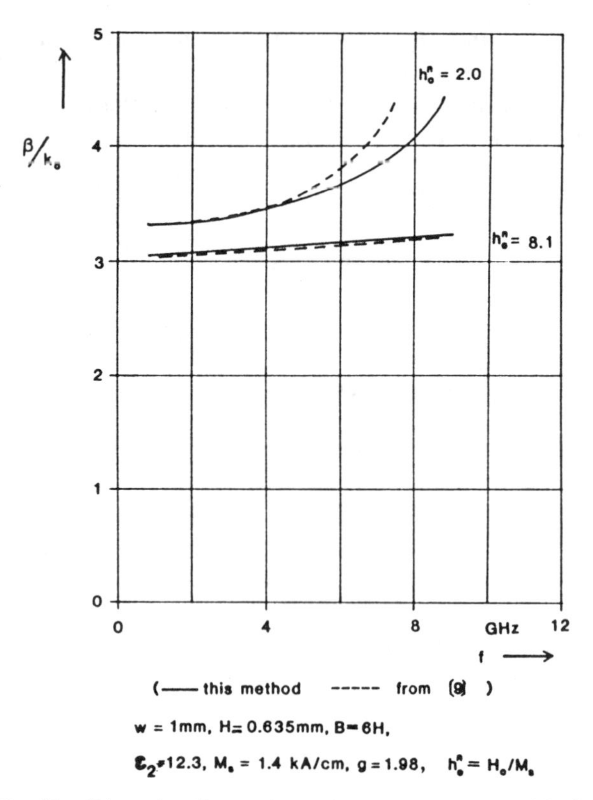

Fig. 31. Dispersion diagram for a microstrip on magnetized ferrite.

IV. CONCLUSIONS

A collection of new results has been presented in this paper aiming at the clarification of anisotropic substrate effects on the propagation properties of various integrated-circuit structures. In addition, the bulk of this paper has been devoted to the presentation of analytical and numerical methods which are useful in the accurate modeling of the effective dielectric constant and characteristic impedance of various structures such as microstrip, slotline, and coplanar waveguide on anisotropic substrates.

The various quasi-static and dynamic analytical methods summarized in this paper have shown that when anisotropy is not accounted for in the computation of the waveguiding structure properties, an error is incurred which increases with decreasing linewidth and/or increasing frequency. The concept of anisotropy ratio (AR) has been introduced, which may be used as an indication of dimensional tolerance sensitivity for coupled lines with cover, where the cover is used for achieving equalization of even- and odd-mode phase velocities. It has been found that, when $AR > 1$, the equalization of phase velocities is less sensitive to small variations in the cover height to substrate thickness ratio compared to when $AR \leqslant 1$. Also, it has been seen that by introducing an equivalent relative dielectric constant and an equivalent substrate thickness, the anisotropic layer may be replaced by an equivalent isotropic substrate for microstrip without cover. In this case, computations are simplified since existing design methods for microstrip without cover may be used provided ϵ_r and H are replaced by ϵ_{eq} and H_{eq}, respectively.

The quasi-static methods summarized in this paper prove to be of practical use when the largest (and transverse to the direction of propagation) characteristic dimension of the circuit structure under consideration is small by comparison to the source wavelength. Among the various techniques presented, the method of moments in conjunction with the pertinent Green's function provides a straightforward solution to the integral equation for the charge density of single or coupled lines. This solution may be obtained to within desired convergence accuracy even for open structures since the boundary condition at infinity is included in the Green's function representation. On the other hand, the finite differences technique suffers from convergence sensitivity problems, especially for open structures, while the variational method provides results only to within an upper or lower bound from the true answer. The method of moments is considered to be the superior of all the other quasi-static techniques discussed due to its versatility and excellent accuracy.

The dispersive properties of various integrated-circuit structures have also been addressed in this paper. It has been found that the Fourier series method is the most generalized dispersion modeling procedure since it yields solutions to essentially all the structures of practical interest. It provides, in addition, results to within desired convergence accuracy.

Similarly, the equivalent network method also offers a generalized approach since it can deal with most of the integrated-circuit geometries of practical interest with excellent accuracy. The modified Wiener–Hopf procedure, on the other hand, is a mathematically elegant technique, but it has been applied only to microstrip with cover on an anisotropic substrate and it has not been adopted to tackle the question of coupled microstrip lines and that of more general integrated-circuit structures.

The aforementioned dispersive models suffer from a major disadvantage in that they fail to characterize the dispersive properties of circuit discontinuities and structure transitions. The transmission-line matrix and the method of lines procedures, on the other hand, are adoptable to modeling effectively the dispersive properties of a waveguiding circuit in a rectangular waveguide as well as certain circuit discontinuities. As stated previously, the TLM technique has been applied to the geometry of single and coupled microstrip on sapphire in a rectangular waveguide for wide microstrip lines. The graphs shown in Fig. 21 for $\epsilon_{eff}^{(e,o)}$ have been found to be in error by 1.5 and 3 percent, respectively, when compared to the corresponding cases computed by the Fourier series method when the latter is applied with a convergence accuracy of 0.5 percent. This error observation, coupled with the fact that the computations in [44] refer only to wide lines ($w/H \geqslant 1$), suggests that there is no sufficient evidence for the degree of accuracy provided by this method, especially for lines with $w/H < 1$ on anisotropic substrates. In summary, the inherent disadvantages of this method are: a) the need of *a priori* knowledge or a very good initial guess of the dominant field distribution; b) the dependence of the method on the chosen mesh size and large number of iterations necessary to achieve desired convergence accuracy, factors which may lead to excessive computer run time and memory storage requirements; c) the method is effective for modeling dispersion in closed structures only; d) the method does not yield readily an equivalent circuit for the discontinuities of interest; e) there is no theoretical or experimental verification for the dispersive models derived by this method for the step and gap discontinuities presented in this paper. The method of lines is also an effective method to study the dispersive properties of distributed circuits in a rectangular waveguide with very good accuracy. Its major advantage is the simplicity of the resulting computer algorithm which allows efficient circuit parameter computation on a personal computer. The method has also been used to model discontinuities such as a periodic meander microstrip line and a periodically slotted microstrip. In addition, it has been used effectively to model a slotline short circuit. The versatility of this method in resolving the dispersive properties of step, gap, and other nonperiodic types of useful discontinuities has yet to be demonstrated.

Neither the TLM technique nor the method of lines are adequately general to provide solutions for the dispersion properties for the majority of the structures in Fig. 1. In fact, these techniques are ineffective as far as open structures are concerned and, in particular, in the modeling of discontinuities associated with open structures.

A novel approach was developed recently which resolves the dispersion properties of microstrip transmission lines, and it provides very accurate frequency-dependent equivalent circuits for microstrip discontinuities such as microstrip gap, open-circuited microstrip, etc., on isotropic substrates [15]–[99]. The method accounts for line and discontinuity radiation loss, conductor thickness, as well as all substrate effects, including the excitation of substrate surface waves. The model involves derivation of the standing-wave pattern for the current density along the circuit from which the line dispersion properties and discontinuity frequency-dependent equivalent circuits are derived [96]. The current density standing-wave pattern is obtained by solving Pocklington's integral equation by the method of moments. The radiation aspects of the problem, as well as the substrate effects, are taken into account by the Green's function which is obtained by solving the boundary-value problem of radiation by an infinitesimally short electric dipole printed on a substrate [96]. This approach can be extended to all the microstrip geometries of Fig. 1, where the Green's function must be derived for anisotropic substrates. A dual direction may be followed for the geometries which involve slotlines.

Acknowledgment

The author is indebted to his students J. Castaneda, D. Jackson, and A. Nakatani for their critical review and helpful comments. Thanks are also due to S. Spurrier, P. Parris, and M. Schoneberg for typing the manuscript and to K. Abolhassani and A. Nakatani for drawing the figures. This paper was written at Phraxos.

References

[1] K. C. Gupta, R. Garg, and I. J. Bahl, *Microstrip Lines and Slotlines.* Dedham, MA: Artech House, 1979.
[2] T. C. Edwards, *Foundations for Microstrip Circuit Design.* New York: Wiley, 1981.
[3] M. Olyphant, Jr., "Measuring anisotropy in microwave substrates," in *IEEE MTT-S 1979 Int. Microwave Symp. Dig.*, Apr. 30–May 2, 1979, pp. 91–94.
[4] R. P. Owens, J. E. Aitken, and T. C. Edwards, "Quasi-static characteristics of microstrip on an anisotropic sapphire substrate," *IEEE Trans. Microwave Theory Tech.*, vol. MTT-24, pp. 499–505, Aug. 1976.
[5] T. C. Edwards and R. P. Owens, "2–18-GHz dispersion measurements on 10–100-Ω microstrip lines on sapphire," *IEEE Trans. Microwave Theory Tech*, vol. MTT-24, pp. 506–513, Aug. 1976.
[6] C. M. Krowne, "Microstrip transmission lines on pyrolytic boron nitride," *Electron. Lett.*, vol. 12, no. 24, pp. 642–643, Nov. 25, 1976.
[7] C. M. Krowne, "Pyrolytic boron nitride as a microstrip substrate material," in *Electrical Electronics Insulation Conf. Proc.*, Sept. 1977, pp. 35–38.
[8] B. Lax and K. J. Button, *Microwave Ferrites and Ferrimagnetics.* New York: McGraw-Hill, 1962
[9] N. Krause, "Ein Verfahren zur Berechnung der Dispersion einer Mikrostreifenleitung auf gyrotropem Substrat," *Arch. Elek. Übertragung.*, vol. 31, pp. 205–211, 1977.
[10] R. Pregla and S. B. Worm, "The method of lines for the analysis of planar waveguides with magnetized ferrite substrate," in *IEEE MTT-S 1984 Int. Microwave Symp. Dig.*, May 28–June 1, 1984, pp. 348–350.
[11] A. Beyer and I. Wolff, "Power density distribution analysis of ferrite loaded finlines for the development of integrated nonreciprocal millimeter wave elements," in *IEEE MTT-S Int. Symp. Dig.*, May 28–June 1, 1984, pp. 342–344.
[12] R. E. Collin, *Field Theory of Guided Waves.* New York: McGraw-Hill, 1960.
[13] J. A. Kong, *Theory of Electromagnetic Waves.* New York: Wiley, 1975.
[14] H. C. Chen, *Theory of Electromagnetic Waves.* New York: McGraw-Hill, 1983.
[15] A. A. Th. M. van Trier, "Guided electromagnetic waves in anisotropic media," *Appl. Sci. Res.*, vol. B3, pp. 305–308, 1953.
[16] M. L. Kales, "Modes in waveguides that contain ferrites," *J. Appl. Phys.*, vol. 24, pp. 604–608, May 1953.
[17] R. F. Soohoo, *Theory and Application of Ferrites.* Englewood Cliffs, NJ: Prentice Hall, 1960.
[18] J. B. Davies, "An analysis of the *m*-port symmetrical *H*-plane waveguide junction with central ferrite post," *IRE Trans. Microwave Theory Tech.*, vol. MTT-10, pp. 596–604, Nov. 1962.
[19] J. B. Davies and P. Cohen, "Theoretical design of symmetrical junction stripline circulator," *IEEE Trans. Microwave Theory Tech.*, vol. MTT-11, pp. 506–512, Nov. 1963.
[20] H. Bosma, "On stripline *Y*-circulation at UHF," *IEEE Trans. Microwave Theory Tech.*, MTT-12, pp. 61–72, Jan. 1964.
[21] W. J. Ince and E. Stern, "Nonreciprocal remanence phase shifters in rectangular waveguide," *IEEE Trans. Microwave Theory Tech.*, vol. MTT-15, pp. 87–95, Feb. 1967.
[22] J. B. Castillo, Jr., and L. E. Davis, "Computer-aided three-port waveguide junction circulator," *IEEE Trans. Microwave Theory Tech.*, vol. MTT-18, pp. 25–34, Jan. 1970.
[23] J. Helszajn, *Principles of Microwave Ferrite Engineering.* New York: Wiley, 1970.
[24] J. Helszajn, *Nonreciprocal Microwave Junctions and Circulators.* New York: Wiley, 1975.
[25] R. F. Soohoo, *Microwave Magnetics.* New York: Harper and Row, 1985.
[26] K. J. Button, "Microwave ferrite devices: The first ten years," *IEEE Trans. Microwave Theory Tech.*, vol. MTT-32, pp. 1088–1096, Sept. 1984.

[27] N. G. Alexopoulos, S. R. Kerner, and C. M. Krowne, "Dispersionless coupled microstrip over fused silica-like anisotropic substrates," *Electron. Lett.*, vol. 12, no. 22, pp. 579–580, Oct. 28, 1976.

[28] N. G. Alexopoulos and C. M. Krowne, "Characteristics of single and coupled microstrips on anisotropic substrates," *IEEE Trans. Microwave Theory Tech.*, vol. MTT-26, pp. 387–393, June 1978.

[29] A. G. D'Assunção, A. J. Giarola, and D. A. Rogers, "Characteristics of broadside coupled microstrip lines with iso/anisotropic substrates," *Electron. Lett.*, vol. 17, no. 7, pp. 264–265, Apr. 2, 1981.

[30] M. Kobayashi, "Analysis of the microstrip and the electrooptic light modulator," *IEEE Trans. Microwave Theory Tech.*, vol. MTT-26, pp. 119–126, Feb. 1978.

[31] ——, "Green's function technique for solving anisotropic electrostatic field problems," *IEEE Trans. Microwave Theory Tech.*, vol. MTT-26, pp. 510–512, July 1978.

[32] H. Shibata, S. Minakawa, and R. Terakado, "Analysis of the shielded-strip transmission line with an anisotropic medium," *IEEE Trans. Microwave Theory Tech.*, vol. MTT-30, pp. 1264–1267, Aug. 1982.

[33] E. Yamashita, K. Atsuki, and T. Mori, "Application of MIC formulas to a class of integrated-optics modulator analyses: A simple transformation," *IEEE Trans. Microwave Theory Tech.*, vol. MTT-25, pp. 146–150, Feb. 1977.

[34] N. G. Alexopoulos and N. K. Uzunoglu, "An efficient computation of thick microstrip properties on anisotropic substrates," *J. Franklin Inst.*, vol. 306, no. 1, pp. 9–22, July 1978.

[35] N. G. Alexopoulos and N. K. Uzunoglu, "A simple analysis of thick microstrip anisotropic substrates," *IEEE Trans. Microwave Theory Tech.*, vol. MTT-26, pp. 455–456, June 1978.

[36] M. Horno, "Quasistatic characteristics of microstrip on arbitrary anisotropic substrates," *Proc. IEEE*, vol. 67, pp. 1033–1034, Aug. 1980.

[37] M. Horno, "Upper and lower bounds on capacitances of coupled microstrip lines with anisotropic substrates," *Proc. Inst. Elec. Eng.*, vol. 129, pt. H, no. 3, pp. 89–93, June 1982.

[38] M. Horno and R. Marqués, "Coupled microstrips on double anisotropic layers," *IEEE Trans. Microwave Theory Tech.*, vol. MTT-32, pp. 467–470, Apr. 1984.

[39] S. K. Koul and B. Bhat, "Inverted microstrip and suspended microstrip with anisotropic substrates," *Proc. IEEE*, vol. 70, pp. 1230–1231, Oct. 1982.

[40] S. K. Koul and B. Bhat, "Shielded edge-coupled microstrip mixture with anisotropic substrates," *Arch. Elek. Übertragung*, vol. 37, pp. 269–274, July/Aug. 1983.

[41] S. K. Koul and B. Bhat, "Transverse transmission line method for the analysis of broadside-coupled microstrip lines with anisotropic substrates," *Arch. Elek. Übertragung.*, vol. 37, pp. 59–64, Jan./Feb. 1983.

[42] S. K. Koul and B. Bhat, "Generalized analysis of microstrip-like transmission lines and coplanar strips with anisotropic substrates for MIC, electrooptic modulator, and SAW applications," *IEEE Trans. Microwave Theory Tech.*, vol. MTT-31, pp. 1051–1058, Dec. 1983.

[43] T. Kitazawa and Y. Hayashi, "Quasi-static characteristics of coplanar waveguide on a sapphire substrate with its optical axis inclined," *IEEE Trans. Microwave Theory Tech.*, vol. MTT-30, pp. 920–922, June 1982.

[44] G. E. Mariki, "Analysis of microstrip lines on inhomogeneous anisotropic substrates," Ph.D. dissertation, Univ. of California, Los Angeles, June 1978.

[45] G. E. Mariki and C. Yeh, "Dynamic three-dimensional TLM analysis of microstrip lines on anisotropic substrates," *IEEE Trans. Microwave Theory Tech.*, vol. MTT-33, pp. 789–799, Sept. 1985.

[46] P. B. Johns and R. L. Beurle, "Numerical solution of two dimensional scattering problems using a transmission line matrix," *Proc. Inst. Elec. Eng.*, vol. 118, pp. 1203–1208, Sept. 1971.

[47] P. B. Johns, "Application of the transmission line matrix method to homogeneous waveguides of arbitrary cross section," *Proc. Inst. Elec. Eng.*, vol. 119, pp. 1086–1091, Aug. 1972.

[48] S. Akhtarzad and P. B. Johns, "The solution of Maxwell's equations in three space dimensions and time by the TLM method," *Proc. Inst. Elec. Eng.*, vol. 122, 1975.

[49] S. Akhtarzad and P. B. Johns, "Generalised elements for the TLM method of numerical analysis," *Proc. Inst. Elec. Eng.*, vol. 122, 1975.

[50] W. Getsinger, "Microstrip dispersion model," *IEEE Trans. Microwave Theory Tech.*, vol. MTT-21, pp. 34–39, Jan. 1973.

[51] A-M. A. El-Sherbiny, "Hybrid mode analysis of microstrip lines on anisotropic substrates," *IEEE Trans. Microwave Theory Tech.*, vol. MTT-29, pp. 1261–1265, Dec. 1981.

[52] Y. Hayashi and T. Kitazawa, "Analysis of microstrip transmission line on a sapphire substrate," *J. Inst. Electron. Commun. Eng. Jap.*, vol. 62-B, pp. 596–602, June 1979.

[53] T. Kitazawa and Y. Hayashi, "Coupled slots on an anisotropic sapphire substrate," *IEEE Trans. Microwave Theory Tech.*, vol. MTT-29, pp. 1035–1040, Oct. 1981.

[54] T. Kitazawa and Y. Hayashi, "Propagation characteristics of striplines with multilayered anisotropic media," *IEEE Trans. Microwave Theory Tech.*, vol. MTT-31, pp. 429–433, June 1983.

[55] K. Shibata and K. Hatori, "Dispersion characteristics of coupled microstrip with overlay on anisotropic dielectric substrate," *Electron. Lett.*, Jan. 1984.

[56] A. Nakatani and N. G. Alexopoulos, "A generalized algorithm for structures on anisotropic substrates," presented at IEEE MTT-S Int. Microwave Symp., St. Louis, June 1985.

[57] N. J. Damaskos, R. B. Mack, A. L. Maffett, W. Parmon, and P. L. E. Uslenghi, "The inverse problem for biaxial materials," *IEEE Trans. Microwave Theory Tech.*, vol. MTT-32, pp. 400–405, Apr. 1984.

[58] J. Fontanella, C. Andeen, and D. Schuele, "Low-frequency dielectric constants of α-quartz, sapphire, MgFz and MgO," *J. Appl. Phys.*, vol. 45, no. 7, pp. 2852–2854, July 1974.

[59] W. B. Westphal and A. Sils, "Dielectric constant and ion data," Tech. Rep. AFML-TR-72-39, Apr. 1972.

[60] P. H. Ladbrooke, M. H. N. Potok, and E. H. England, "Coupling errors in cavity-resonance measurements on MIC dielectrics," *IEEE Trans. Microwave Theory Tech.*, vol. MTT-21, pp. 560–562, Aug. 1973.

[61] E. V. Loewenstein, D. R. Smith, and R. L. Morgan, "Optical constants for infra-red materials, 2: Crystalline solids," *Appl. Optics*, vol. 12, pp. 398–406, Feb. 1973.

[62] E. E. Russell and E. E. Bell, "Optical constants of sapphire in the far infrared," *J. Opt. Soc. Am.*, vol. 57, pp. 543–544, Apr. 1967.

[63] S. Roberts and D. D. Coon, "Far-infrared properties of quartz and sapphire," *J. Opt. Soc. Am.*, vol. 52, pp. 1023–1029, Sept. 1962.

[64] M. Olyphant, Jr., "Microwave permittivity measurements using disk cavity specimens," *IEEE Trans. Instrum. Meas.*, vol. 20, no. 4, pp. 342–344, Nov. 1971.

[65] RT/Duroid®, Rogers Corporation, TR 2692, July 1981.

[66] W. R. Smythe, *Static on Dynamic Electricity*. New York: McGraw-Hill, 1950.

[67] L. D. Landau and E. M. Lifshitz, *Electrodynamics of Continuous Media*. Addison-Wesley, 1960.

[68] B. T. Szentkuti, "Simple analysis of anisotropic microstrip lines by a transform methods," *Electron. Lett.*, vol. 25, no. 12, pp. 672–673, Dec. 9, 1976.

[69] M. Kobayashi and R. Terakado, "New view on an anisotropic medium and its application to transformation from anisotropic to isotropic problems," *IEEE Trans. Microwave Theory Tech.*, vol. MTT-27, pp. 769–775, Sept. 1979.

[70] ——, "Accurately approximate formula of effective filling fraction for microstrip line with isotropic substrate and its application to the case with anisotropic substrate," *IEEE Trans. Microwave Theory Tech.*, vol. MTT-27, pp. 776–778, Sept. 1979.

[71] ——, "Method for equalizing phase velocities of coupled microstrip lines by using anisotropic substrate," *IEEE Trans. Microwave Theory Tech.*, vol. MTT-28, pp. 719–722, July 1980.

[72] E. Yamashita, K. Atsuki, and T. Akamatsu, "Application of microstrip analysis to the design of a broad-band electrooptical modulator," *IEEE Trans. Microwave Theory Tech.*, vol. 22, pp. 462–464, Apr. 1974.

[73] E. Yamashita and K. Atsuki, "Distributed capacitance of a thin-film electrooptic light modulator," *IEEE Trans. Microwave Theory Tech.*, vol. 23, pp. 177–178, Jan. 1975.

[74] T. G. Bryant and J. A. Weiss, "Parameters of microstrip transmission lines and of coupled pairs of microstrip lines," *IEEE Trans. Microwave Theory Tech.*, vol. MTT-16, pp. 1021–1027, Dec. 1968.

[75] A. Farrar and A. T. Admas, "Characteristic impedance of microstrip by the method of moments," *IEEE Trans. Microwave Theory Tech.*, vol. 18, pp. 65–66, Jan. 1980.

[76] R. F. Harrington, *Field Computation by Moment Methods*. New York: Macmillan Co., 1968.

[77] N. G. Alexopoulos and S. A. Maas, "Characteristics of microstrip directional couplers on anisotropic substrates," *IEEE Trans. Microwave Theory Tech.*, vol. MTT-30, pp. 1267–1270, Aug. 1982.

[78] N. G. Alexopoulos and S. A. Maas, "Performance of microstrip couplers on an anisotropic substrate with an isotropic superstate," *IEEE Trans. Microwave Theory Tech.*, vol. MTT-31, pp. 671–673, Aug. 1983.
[79] B. Sheleg and B. E. Spielman, "Broad-band directional couplers with dielectric overlays," *IEEE Trans. Microwave Theory Tech.*, vol. MTT-22, pp. 1216–1220, Dec. 1974.
[80] D. Paolino, "MIC overlay coupler design using spectral domain techniques," *IEEE Trans. Microwave Theory Tech.*, vol. MTT-26, pp. 646–649, Sept. 1978.
[81] G. Haupt and H. Delfs, "High directivity microstrip couplers," *Electron. Lett.*, vol. 10, no. 9, pp. 142–143, May 2, 1974.
[82] R. N. Karekar and M. K. Pande, "MIC coupler with improved directivity using thin film BiO_3 overlay," *IEEE Trans. Microwave Theory Tech.*, vol. MTT-25, pp. 74–75, Jan. 1977.
[83] S. Rhenmark, "High directivity CTL couplers and a new technique for measurement of CTL coupler parameters," *IEEE Trans. Microwave Theory Tech.*, vol. MTT-25, p. 1116, Dec. 1977.
[84] J. E. Dalley, "A stripline directional coupler utilizing a non-homogeneous dielectric medium," *IEEE Trans. Microwave Theory Tech.*, vol. MTT-17, pp. 706–712, Sept. 1969.
[85] K. Atsuki and E. Yamashita, "Three methods for equalizing the even- and odd-mode phase velocity of coupled strip lines with an inhomogeneous medium," *Trans. IECE (Japan)*, vol. 55-B, pp. 424–426, July 1972 (in Japanese).
[86] W. T. Weeks, "Calculation of coefficients of capacitance of multiconductor transmission lines in the presence of a dielectric interface," *IEEE Trans. Microwave Theory Tech.*, vol. MTT-19, pp. 35–43, Jan. 1970.
[87] E. Yamashita, K. Atsuki, and T. Akamatsu, "Application of microstrip analysis to the design of a broad-band electrooptical modulator," *IEEE Trans. Microwave Theory Tech.*, vol. MTT-22, pp. 462–464, Apr. 1974.
[88] E. Yamashita and K. Atsuki, "Distributed capacitance of a thin-film electrooptic light modulator," *IEEE Trans. Microwave Theory Tech.*, vol. MTT-23, pp. 177–178, Jan. 1975.
[89] T. Kitazawa and Y. Hayashi, "Quasi-static characteristics of coplanar waveguide on a sapphire substrate with its optical axis inclined," *IEEE Trans. Microwave Theory Tech.*, vol. MTT-30, pp. 920–922, June 1982.
[90] M. Kobayashi, "Frequency dependent characteristics of microstrips on anisotropic substrates," *IEEE Trans. Microwave Theory Tech.*, vol. MTT-30, pp. 2054–2057, Nov. 1982.
[91] M. V. Schneider, "Microstrip lines for microwave integrated circuits," *Bell Syst. Tech. J.*, vol. 48, pp. 1421–1444, May/June 1969.
[92] E. Yamashita, K. Atsuki, and T. Ueda, "An approximate dispersion formula of microstrip lines for computer-aided design of microwave integrated circuits," *IEEE Trans. Microwave Theory Tech.*, vol. MTT-27, pp. 1036–1038, Dec. 1979.
[93] U. Schulz and R. Pregla, "A new technique for the analysis of the dispersion characteristics of planar waveguides," *Arch. Elek. Übertragung.*, vol. 34, pp. 169–173, 1980.
[94] U. Schulz and R. Pregla, "A new technique for the analysis of the dispersion characteristics of planar waveguides and its application to microstrips with tuning septums," *Radio Sci.*, vol. 16, pp. 1173–1178, Nov.–Dec. 1981.
[95] S. B. Worm and R. Pregla, "Hybrid mode analysis of arbitrarily shaped planar microwave structures by the method of lines," *IEEE Trans. Microwave Theory Tech.*, vol. MTT-32, pp. 191–196, Feb. 1984.
[96] P. B. Katehi and N. G. Alexopoulos, "On the modeling of electromagnetically coupled microstrip antennas—The printed strip dipole," *IEEE Trans. Antennas Propagat.*, vol. 32, pp. 1179–1186, Nov. 1984.
[97] P. B. Katehi and N. G. Alexopoulos, "Microstrip discontinuity modelling for millimetric integrated circuits," in *1985 IEEE MTT-S Int. Microwave Symp. Proc.*, pp. 571–573.
[98] R. W. Jackson and D. M. Pozar, "Surface wave losses at discontinuities in millimeter wave integrated transmission lines," in *1985 IEEE MTT-S Int. Microwave Symp. Proc.*, pp. 563–565.
[99] P. B. Katehi and N. G. Alexopoulos, "Frequency dependent characteristics of microstrip discontinuities in millimeter wave integrated circuits," *IEEE Trans. Microwave Theory Tech.*, this issue, pp. 1029–1035.

An overview on numerical techniques for modeling miniaturized passive components*

Tatsuo ITOH **

Abstract

This paper presents a number of numerical techniques used for characterization and modeling of passive components for microwave and millimeter-wave circuits. Brief descriptions are provided for each technique with discussions on advantages, limitation, typical structures amenable to the method, and cautions to be exercised where appropriate. The text is intended to serve as a tutorial. Representative publications on each topic are included in the reference list.

Key words : Numerical method, Modelization, Electronic component, Passive component, Microwave circuit, Review, Finite difference method, Finite element method, Three dimensional model, Moment method, Transmission line, Planar waveguide, Microstrip line, Discontinuity.

VUE D'ENSEMBLE SUR LES TECHNIQUES NUMÉRIQUES DE MODÉLISATION DE COMPOSANTS PASSIFS MINIATURISÉS

Analyse

Cet article présente un certain nombre de techniques numériques utilisées pour la caractérisation et la modélisation de composants passifs pour les circuits hyperfréquences ou en ondes millimétriques. Chaque technique est décrite brièvement avec discussion des avantages, des limitations, des structures typiques justifiant la méthode et des précautions à prendre. Le texte est présenté de manière didactique. Les publications représentatives de chaque thème sont incluses dans la liste des références.

Mots clés : Méthode numérique, Modélisation, Composant électronique, Composant passif, Circuit hyperfréquence, Article synthèse, Méthode différence finie, Méthode élément fini, Modèle tridimensionnel, Méthode moment, Ligne transmission, Guide onde plan, Ligne microruban, Discontinuité.

Contents

1. INTRODUCTION

Although numerical, and possibly analytical, characterizations and modelings of miniaturized passive components have been exercised by a large number of researchers and engineers for some time, the necessity of such activities are increasingly important in recent years. This is in line with increased research and development in the millimeter-wave circuits and the monolithic integrated circuits. It is no longer economical or, in many cases not even feasible, to tune the circuits once they are built. Therefore, extremely accurate characterization methods are needed to model the structures.

Because most structures used in today's printed and planar integrated circuits are not amenable to closed form analytical expressions, the numerical methods needed for characterizations are in fact a necessary evil. The circuit designers would like to use CAD packages which in most cases consist of curve-fitting or empirical formulas. However, these formulas must be backed up by accurate characterizations. In addition, any numerical methods for characterizations need to be as efficient as possible both in CPU time and temporary storage requirement, although recent rapid advances in computers impose less severe restrictions on the efficiency of the method. Another effort in the development of numerical methods has been the versatility of the method.

* This work has been supported by Army Research Office contract DAAG29-84-K-0076, Office of Naval Research contract N00014-79-C-0553 and Joint Services Electronics Program contract F-49620-C-0033.
** Department of Electrical & Computer Engineering, The University of Texas, Austin, Texas 78712, USA.

Reprinted with permission from *Annales des Télécommunications,* vol. 41, pp. 449–462, Sept.–Oct. 1986.

In reality, however, choice of the numerical methods is the product of trade-off between accuracy, speed, storage requirement, versatility, etc. and is quite structure dependent. Since the advent of microwave integrated circuits, a number of methods have been invented in addition to the somewhat more classical methods refined for these modern structures.

2. SURVEY OF NUMERICAL METHODS

When a specific structure is analyzed, one has to make a choice of the method best suited for the structure. Obviously, the choice is not unique. Therefore, the user must make a critical assessment for each candidate method. In this chapter, we list a number of numerical methods and present generally accepted appraisals for these methods. It is not possible to make an exhaustive list of all available methods. Rather, most representative ones are reviewed. Although the assessment will be aimed at the characterization of 3-dimensional passive components, many methods are also effective for 2-dimensional problems. Further, 2-dimensional methods can be used in an integral part of the composite characterization program for 3-dimensional structures.

2.1. Finite difference method [1].

The method is best illustrated by means of a problem characterized by the two-dimensional Laplace equation :

$$\frac{\partial^2 \varphi}{\partial x^2} + \frac{\partial^2 \varphi}{\partial y^2} = 0. \tag{1}$$

An extension to a three-dimensional problem is more complicated but is straightforward. The region of interest is divided into mesh points separated by the distance h. Instead of solving (1) directly, the method entails the solution of its discretized version. Let us take the coordinate origin at the point A in Figure 1. The potentials φ_B, φ_C, φ_D, and φ_E at the points B, C, D, and E can be expressed in terms of Taylor expansions :

$$\varphi_B = \varphi_A + h\left(\frac{\partial \varphi}{\partial x}\right)_A + \frac{h^2}{2!}\left(\frac{\partial^2 \varphi}{\partial x^2}\right)_A + \frac{h^3}{3!}\left(\frac{\partial^3 \varphi}{\partial x^3}\right)_A + O(h^4),$$

$$\varphi_D = \varphi_A - h\left(\frac{\partial \varphi}{\partial x}\right)_A + \frac{h^2}{2!}\left(\frac{\partial^2 \varphi}{\partial x^2}\right)_A - \frac{h^3}{3!}\left(\frac{\partial^3 \varphi}{\partial x^3}\right)_A + O(h^4),$$

$$\varphi_E = \varphi_A + h\left(\frac{\partial \varphi}{\partial y}\right)_A + \frac{h^2}{2!}\left(\frac{\partial^2 \varphi}{\partial y^2}\right)_A + \frac{h^3}{3!}\left(\frac{\partial^3 \varphi}{\partial y^3}\right)_A + O(h^4),$$

$$\varphi_C = \varphi_A - h\left(\frac{\partial \varphi}{\partial y}\right)_A + \frac{h^2}{2!}\left(\frac{\partial^2 \varphi}{\partial y^2}\right)_A - \frac{h^3}{3!}\left(\frac{\partial^3 \varphi}{\partial y^3}\right)_A + O(h^4),$$

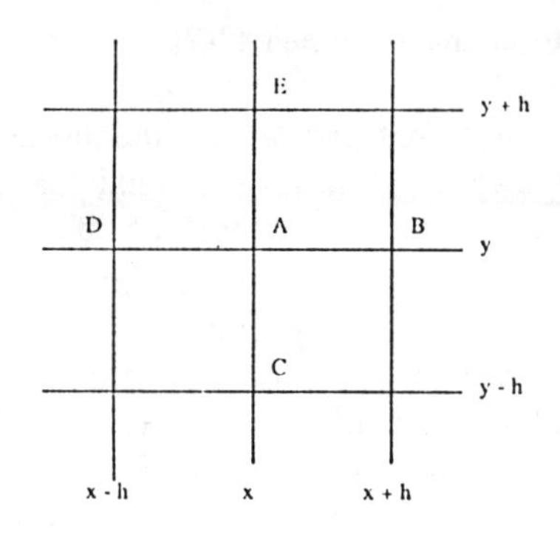

FIG. 1. — A typical mesh for a two-dimensional finite difference method.

Maillage typique pour la méthode des différences finies à deux dimensions.

where the subscript A indicates the quantities evaluated at A and $O(h^4)$ is a quantity of the order of h^4. Adding these equations, one obtains :

$$\varphi_B + \varphi_C + \varphi_D + \varphi_E = 4\varphi_A + h^2\left[\left(\frac{\partial^2 \varphi}{\partial x^2}\right) + \left(\frac{\partial^2 \varphi}{\partial y^2}\right)\right]_A + O(h^4).$$

The second term on the right hand side vanishes as φ is required to satisfy (1) everywhere. Hence :

$$\frac{1}{4}(\varphi_B + \varphi_C + \varphi_D + \varphi_E) = \varphi_A, \tag{2}$$

is a good approximation of (1) as long as h is small enough to neglect $O(h^4)$ terms. Somewhat different equations are used if the point A is located on the boundary between two media [2]. At the boundary point, φ itself, its derivative in the form of a finite difference or a combination thereof is specified [2].

All of these procedures are repeated at each mesh point. The result is a matrix equation :

$$M\boldsymbol{\varphi} = \boldsymbol{B}. \tag{3}$$

The right hand side vector $\boldsymbol{B}$ contains information given by the boundary points. It is readily seen from (2) that the coefficient matrix M contains a large number of zero elements and only the diagonal and nearby elements are filled. For this reason, in most cases, (3) is solved not by matrix inversion but by an interactive method. A certain scheme called the successive over relaxation method is employed to accelerate convergence of the solution [2].

This method is well known to be the least analytical. The mathematical preprocessing is minimal and the method can be applied to a wide range of structures including those with odd shapes. A price one has to pay is numerical inefficiency. Certain precautions have to be taken into account when the method is used for an open region problem in which the region is truncated to a finite size. Also, the method requires that mesh points lie on the boundary.

2.2. Finite element method [3-7].

This is somewhat similar to the finite difference method. However, it has variational features in the algorithm and contains several flexible features.

In this method, instead of the partial differential equations with boundary conditions, corresponding functionals are set up and variational expressions are applied to each of small areas or volumes subdividing the region of interest. Usually, these small segments are polygons such as triangles and rectangles for two-dimensional problems and tetrahedral elements for three-dimensional problems. Because of this type of discretization, hardly any restrictions can be imposed on the shape of the structure.

The essence of this method is illustrated below for a problem of Laplace equation (1) in a two-dimensional region in Figure 2. The solution of (1)

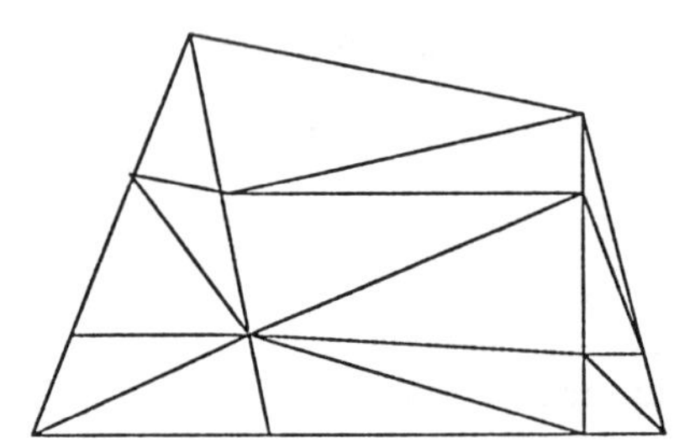

FIG. 2. — Typical subdivision of a cross section in a two-dimensional finite element analysis.

Subdivision typique d'une section transversale pour l'analyse par la méthode des éléments finis.

subject to the boundary condition is equivalent to minimizing the following functional :

$$(4)\quad I(\varphi) = < \varphi, \nabla^2\varphi >,$$

$$= \iint_s \varphi \left(\frac{\partial^2\varphi}{\partial x^2} + \frac{\partial^2\varphi}{\partial y^2}\right) dx\, dy,$$

$$= -\iint_s \left[\left(\frac{\partial\varphi}{\partial x}\right)^2 + \left(\frac{\partial\varphi}{\partial y}\right)^2\right] dx\, dy.$$

This integral is carried out as a collection of the contributions from all of the small polygonal (triangular in this example) areas. In each polygon φ is approximated as a polynomial of the variables x and y :

$$\varphi = a + a_x x + a_y y.$$

The coefficients a, a_x and a_y can be expressed in terms of the values of φ at each vertex of the triangle :

$$\varphi_p = a + a_x x_p + a_y y_p, \quad p = i, j, k,$$

where the subscript $p = i, j, k$ identifies three vertices. Since only a_x and a_y are required for evaluation of (4), they are written as :

$$\begin{bmatrix} a_x \\ a_y \end{bmatrix} = A \begin{bmatrix} \varphi_i \\ \varphi_j \\ \varphi_k \end{bmatrix}.$$

The value of $I(\varphi)$ for one polygon is :

$$(5)\quad I_{ijk}(\varphi) = [\varphi_i, \varphi_j, \varphi_k]\, A^t A \begin{bmatrix} \varphi_i \\ \varphi_j \\ \varphi_k \end{bmatrix} |\Delta s|,$$

where t indicates transpose and $|\Delta s|$ is the area of this polygon given by :

$$\Delta s = \frac{1}{2} \begin{bmatrix} 1 & x_i & y_i \\ 1 & x_j & y_j \\ 1 & x_k & y_k \end{bmatrix}.$$

To minimize $I_{ijk}(\varphi)$, the Rayleigh-Ritz technique is used :

$$(6)\quad \frac{\partial I_{ijk}}{\partial \varphi_i} = \frac{\partial I_{ijk}}{\partial \varphi_j} = \frac{\partial I_{ijk}}{\partial \varphi_k} = 0.$$

Use of (5) in (6) results in :

$$A^t A \begin{bmatrix} \varphi_i \\ \varphi_j \\ \varphi_k \end{bmatrix} = 0,$$

when this process is applied to all of the polygons in s, one obtains :

$$(7)\quad Z \begin{bmatrix} \varphi_1 \\ \varphi_2 \\ \vdots \\ \varphi_N \end{bmatrix} = 0.$$

Since some of φ_i's located on the boundaries are known, (7) can be solved for potentials of all interior points. Algorithms for wave equations in two- and three-dimensions have been worked out extensively [8]. One of the problems of the finite element methods is existence of the so-called spurious zeros. Such zeros correspond to unphysical field structures. The exact cause of this phenomenon is not yet resolved. Several schemes are available to reduce or eliminate these zeros. Typically, they are based on the variational expression which contains an additional constraint $\nabla \cdot H = 0$ [9].

A certain precaution needs to be exercised when the finite element method is applied to an open region problem such as a dielectric waveguide circuit. In many cases, the region to which the method is applied is truncated at a finite extent. In some situations, for instance, the waveguide near cutoff such truncations are not straightforward as the field decays very slowly [6].

Recently, the boundary-element method has been proposed [10, 11]. This is a combination of the boundary integral equation and a discretization technique similar to the finite-element algorithm as applied to the boundary. Essentially, the wave equation for the volume is converted to the surface integral equation by way of the Green's identity. The surface integrals are discretized into N segments (elements) and their evaluation in each element is performed after the field quantities are approximated by polynomials.

One of the advantages of this method is reduction of storage locations and CPU time resulted from the reduction of one of the dimensions.

2.3. TLM method [12-13].

In this method, the field problem is converted to a three dimensional equivalent network problem. This method is essentially for simulation of the wave propagation phenomena in the time domain rather than characterization of the structure. As such, the method is very versatile. In the generic form of the 3-dimensional TLM method, the space is discretized into a three dimensional lattice with a period Δl. Six field components are represented by a hybrid TLM cell shown in Figure 3. Boundaries corresponding

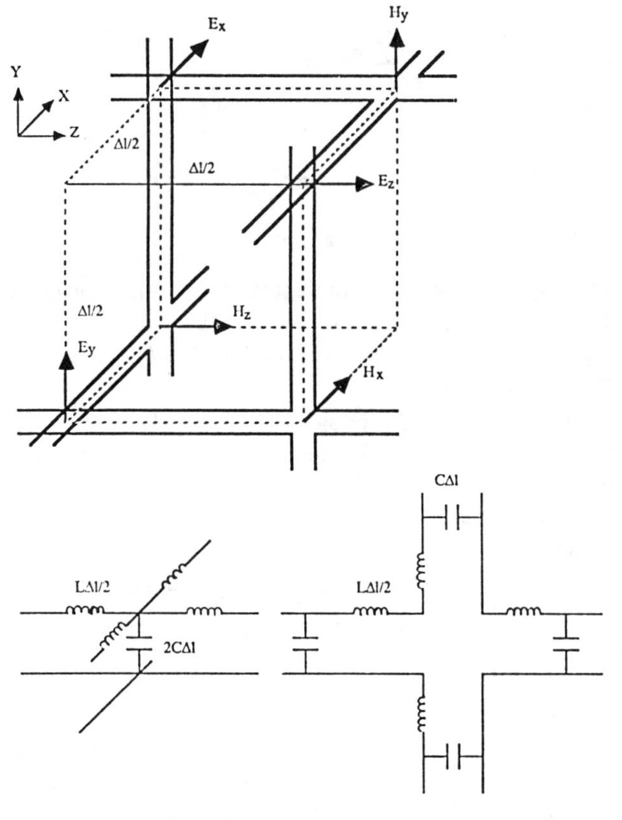

FIG. 3. — A hybrid TLM cell.

Cellule TLM hybride.

to the electric wall and magnetic wall are represented by short-circuiting shunt nodes and open-circuiting shunt nodes on the boundary. Magnetic and dielectric materials can be introduced by adding short-circuited series stubs of length $\Delta l/2$ at the series nodes (magnetic field components) and open-circuited $\Delta l/2$ stubs at the shunt nodes (electric field components). The losses can be represented by resistively loading the shunt nodes. After the time domain response is obtained, the frequency response is found by the Fourier transform.

There are several precautions to be exercised. Due to the introduction of periodic lattice structures, a typical passband-stopband phenomenon appears in the frequency domain data. The frequency range must be below the upper bound of the lowest passband and is determined by the mesh size Δl. There are a number of sources of error. Several remedial procedures have also been reported. In a recent article Hoefer reviews the TLM method extensively [14].

The structures that can be analyzed by the TLM method are quite varied. A typical problem is a shielded microstrip cavity containing a step discontinuity [13].

2.4. Integral equation method [15].

The field in a 3-dimensional structure can be found from the unknown quantities over a certain boundary that are solved by this method. Moderate to extensive analytical preprocessings are often required.

A typical integral equation for a 3-dimensional passive component may be derived formally in the following manner. Consider a microstrip resonator problem in Figure 4. From the superposition principle

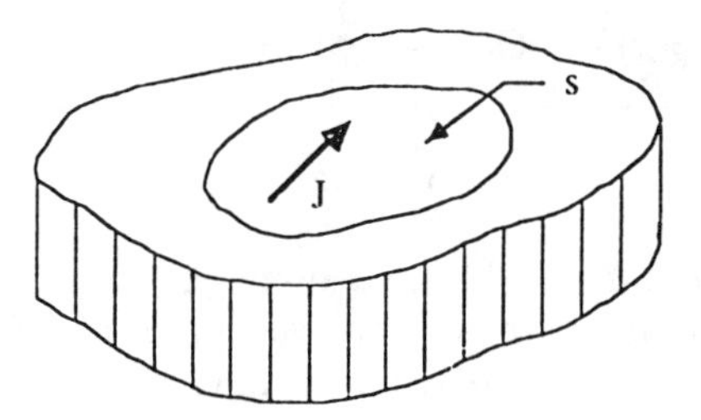

FIG. 4. — Microstrip resonator.

Résonateur microruban.

of a linear system, the total electric field tangential to the surface of the substrate is given by :

$$\mathbf{E}^{\mathrm{i}}(r) + \int \mathrm{Z}(r, r')\, \mathbf{J}(r')\, \mathrm{d}s = \mathbf{E}_{\mathrm{t}}(r), \tag{8}$$

where $\boldsymbol{J}$ is the vector current density on the microstrip surface S and Z is the dyadic Green's function. The incident field $\boldsymbol{E}^{\mathrm{i}}$ vanishes if the formulation is for an eigenvalue problem. The integral equation is derived from the recognition that the total electric field $\boldsymbol{E}_{\mathrm{t}}$ must be zero on the strip S. Hence :

$$\int_S \mathrm{Z}(r, r')\, \mathbf{J}(r')\, \mathrm{d}s = -\, \mathbf{E}^{\mathrm{i}}(r), \quad r \in S. \tag{9}$$

Obviously, a homogeneous equation is found for an eigenvalue problem for which $\boldsymbol{E}^{\mathrm{i}} = 0$.

Derivation of $\mathrm{Z}(r, r')$ is obviously an important and often difficult task. One possible way is to first introduce a two-dimensional Fourier transform with respect to two directions parallel to the substance surface, transform the Helmholtz equation to a one-dimensional ordinary differential equation with respect to the vertical direction and to find the solution. The Green's function is then found from the two-dimensional inverse Fourier transform.

The integral equation itself is transformed to a set of linear simultaneous equations for numerical inversion. The transformation is done by one of several methods such as the moment method [16]. In some cases, a variational expression derived from the integral equation is sufficient for the solution [17]. For instance, in his now-historical paper, Yamashita solved a quasi-TEM microstrip line problem by a variational expression in the spectral (Fourier transform) domain. Under the quasi-TEM approximation, the capacitance per unit length of the microstrip line need to be computed from the charge distribution on the strip of width 2 W. The integral equation for the unknown charge distribution is :

$$\int_{-w}^{w} G(x-x' ; y=y'=d)\,\rho(x')\,dx' = V, \quad |x| < w. \tag{10}$$

Instead of solving this equation, the variational expression is used for finding the line capacitance by :

$$\frac{1}{C} = \frac{\int_{-w}^{w} \rho(x) \int_{-w}^{w} G(x-x' ; y=y'=d)\,\rho(x')\,dx'\,dx}{\left[\int_{-w}^{w} \rho(x)\,dx\right]^2}. \tag{11}$$

Although this can be used directly, Yamashita evaluated a Fourier transformed version of (11) for more express processing of numerical calculations.

In the variational method, the first order error in the choice of an approximate $\rho(x)$ results in a second order (quadratic) error in C. It should be noted however that the above statement is not a guarantee for an accurate solution for C. The magnitude of the error in C is directly related to the choice of an approximate $\rho(x)$. It is important to select $\rho(x)$ as closely as possible to the true but unknown charge distribution.

2.5. Moment methods and Galerkin's method.

These are popular means for discretizing a continuous operator equation such as an integral equation [16]. In the narrowest sense, the moment method employs step functions as the basis functions and delta functions as the testing functions. However, choices for basis and testing functions can be much more flexible. The basis and testing functions are identical in the Galerkin's method and the resulting solutions are known to be variational [18].

The formal procedure of the moment method is now presented. Assume that the integral equation given is :

$$\int_{D} G(r, r')\, f(r')\, dr' = p(r), \quad r \in D, \tag{12}$$

where G is the Green's function and p is the known *excitation* term. The first step in the moment method is to expand the unknown function f in terms of a linear combination of known basis functions $\varphi_n(r)$ with $n = 1, 2, ..., N$:

$$f(r) = \sum_{n=1}^{N} c_n \varphi_n(r), \tag{13}$$

where c_n is the unknown coefficient to be determined. When (13) is substituted into (12), one obtains :

$$\sum_{n=1}^{N} c_n \int_{D} G(r, r')\, \varphi_n(r')\, dr' = p(r). \tag{14}$$

The second step is to take inner products of (14) with testing functions $\chi_m(r)$, $m = 1, 2, ..., N$. The results are :

$$\sum_{n=1}^{N} K_{mn} c_n = b_m, \tag{15}$$

where :

$$K_{mn} = < \chi_m(r), \int_{D} G(r, r')\, \varphi_m(r')\, dr' >, \tag{16}$$

$$b_m = < \chi_m(r), p(r) >. \tag{17}$$

The symbol $< \, >$ indicates the inner product and is typically an integral with respect to r over the region D. It is clear that (15) is a set of linear equations of size $N \times N$.

There are several choices available for the basis functions $\varphi_n(r)$ and the testing functions $\chi_n(r)$. One of the simplest is the choice in the so-called point matching method. In this method, the following selection is made :

$$\varphi_n(r) = U(r_n) = \begin{cases} 1, & r \in [r_n - \Delta/2, r_n + \Delta/2], \\ 0, & \text{otherwise}, \end{cases}$$

$$\chi_n(r) = \delta(r - r_n),$$

where U is the unit pulse function which is zero outside the narrow range of Δ around a discretized point r_n in the domain and δ is the Dirac delta function. It is clear now that, if $|\Delta|$ is small enough :

$$K_{mn} = G(r_m, r_n)\, |\Delta|, \tag{18}$$

$$b_m = p(r_m). \tag{19}$$

Due to the choice of the functions, no integral operations are needed. Hence, the analytical preprocessing is extremely simple. The price one has to pay for this simplicity is the large matrix size N for accurate solutions. The method is quite structure independent and can be applied to a large class of odd shaped geometry.

There are several improved versions of point matching method. For instance, higher order functions or piecewise sinusoidal functions can be used for basis functions [19].

Another popular method is the Galerkin's method which essentially results in the same procedures for the Rayleigh-Ritz method. In the Galerkin's method, the basis functions and the testing functions are identical and are defined over the entire range :

$$\chi_n(r) = \varphi_n(r) \quad r \in D.$$

In this case, the matrix element K_{mn} and the vector element b_m become :

$$(20) \qquad K_{mn} = \int_D dr\, \varphi_m(r) \int_D dr'\, G(r, r')\, \varphi_n(r'),$$

$$(21) \qquad b_m = \int_D dr\, \varphi_m(r)\, p(r).$$

It is known that the *results* from the Galerkin's method with a real operator are variational. The *results* could be the eigenvalues for $p(r) = 0$ and could be some scalar product quantities such as $< p(r)\, f(r) >$. A problem of this method is that double integrals have to be evaluated for each matrix element. However, the size N of the matrix can be substantially smaller than the one required in the point matching method. In many cases, $N = 1$ can result in a reasonably accurate solution as long as a good choice is made for the basis function. The Galerkin's method is more flexible than the straightforward variational method in the derivation of a variational quantity. It is possible to improve the accuracy of the approximation of the unknown function $f(r)$ simply by increasing the matrix size N. However, proper choice of the basis functions is still important. If they are substantially different from the correct solution, convergence of the solution is poor and a large matrix is required.

2.6. Mode matching method.

This method is typically applied to the problem of scattering into waveguiding structures on both sides of the discontinuity such as the one in Figure 5. The fields on both sides of the discontinuity are expanded in terms of the modes in the respective regions with unknown coefficients [20]. To illustrate the method, let us choose a simple step discontinuity

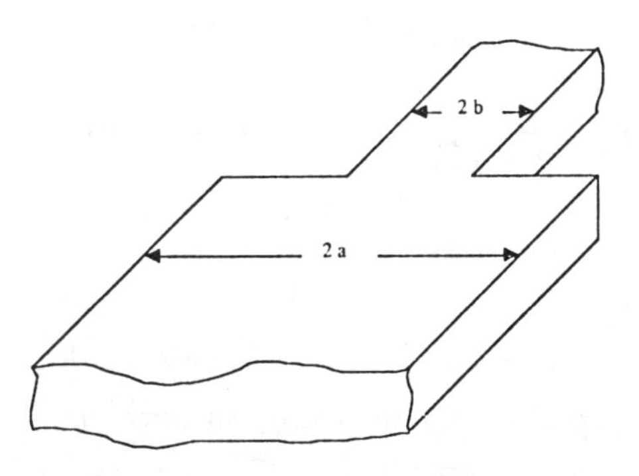

FIG. 5. — Waveguide discontinuity.
Discontinuité de guide d'onde.

with a TE_{n0} excitation. The first step is to expand the E_y and H_x in terms of modal functions $\varphi_{an}(x)$ and $\varphi_{bn}(x)$, $n = 1, 2, \ldots$ Next, continuity of E_y and H_x are applied at the discontinuity $z = 0$:

$$(22) \quad E_y: \sum_{n=1}^{\infty} (A_n^+ + A_n^-)\, \varphi_{an} = \begin{cases} \sum_{n=1}^{\infty} (B_n^+ + B_n^-)\, \varphi_{bn} & 0 < |x| < b, \\ 0 & b < |x| < a, \end{cases}$$

$$(23) \quad H_x: \sum_{n=1}^{\infty} (A_n^+ - A_n^-)\, Y_{an}\varphi_{an} = \sum_{n=1}^{\infty} (B_n^+ - B_n^-)\, Y_{bn}\varphi_{bn} \quad 0 < |x| \leqslant b,$$

where the superscripts + and — indicate the amplitude of the modal wave propagating in the positive and negative directions of z. Hence, A_n^+ and B_n^- are the incident terms and usually only one of them with a designated n is nonzero for a single mode excitation. The next step is to eliminate the x dependence in (22) and (23). If we use the orthogonality of $\varphi_{bn}(x)$ in the region $0 < |x| < b$, we find :

$$(24) \qquad \sum_{n=1}^{\infty} H_{nm}(A_n^+ + A_n^-) = B_m^+ + B_m^-,$$

$$(25) \quad \sum_{n=1}^{\infty} H_{nm} Y_{an}(A_n^+ - A_n^-) = Y_{bm}(B_m^+ - B_m^-).$$

On the other hand, the orthogonality of $\varphi_{an}(x)$ for $0 < |x| < a$ can only be used for (22) because (23) is defined only for $0 < |x| < b$:

$$(26) \qquad A_m^+ + A_m^- = \sum_{n=1}^{\infty} H_{mn}(B_n^+ + B_n^-),$$

where :

$$H_{mn} = \int_0^b \varphi_{am}(x)\, \varphi_{bn}(x)\, dx.$$

From this point, several approaches are possible. One way is to eliminate the unknown B_m^+ from (24) and (25).

Then :

$$(27) \quad \sum_{n=1}^{\infty} (Y_{an} + Y_{bm})\, H_{nm} A_n^- = \sum_{n=1}^{\infty} (Y_{an} - Y_{bm})\, H_{nm} A_n^+ + 2\, Y_{nm} B_n^-, \; m = 1, 2, \ldots$$

This is a set of linear simultaneous equations for the unknown A_n^- and is called a formulation of the first kind. In the solution process, the matrix size must be truncated to a finite size so that $n, m = 1, 2, \ldots, N$.

There are several alternative formulations in addition to (27). All of them are theoretically equivalent. They may be different numerically, however [21].

The mode matching is often applied to find the guided mode in a waveguide with a complicated cross sectional structure. Strictly speaking, this application, however, should be called the field matching method. Let us consider finding the guided mode in a shielded microstrip line with a thick center conductor shown

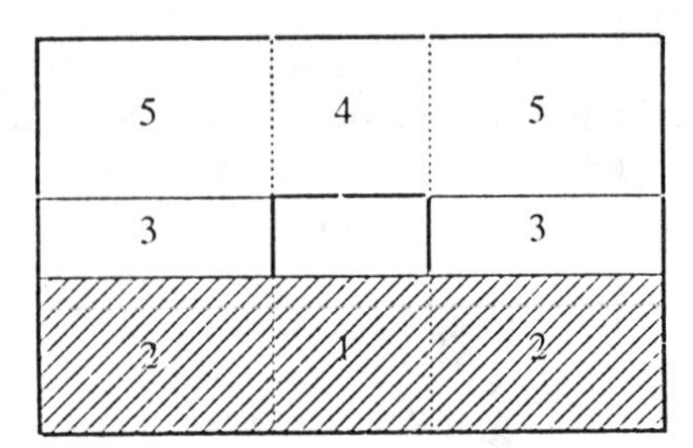

FIG. 6. — Cross section of a shielded microstrip line with a finitely thick strips.

Section transversale d'une ligne microruban blindée à rubans d'épaisseur finie.

in Figure 6. In this method, the fields in the subdivided regions in the cross section are expanded in terms of appropriate orthogonal sets with a common but unknown propagation constant. Some of the boundary conditions are satisfied by individual terms in the expansions. For instance, by expanding the fields into sinusoidal series, the boundary conditions on the metal conductors can be satisfied. The continuity conditions of the tangential electric and magnetic fields are now imposed along each interface. After the orthogonality of the expansion functions is used, we obtain a linear simultaneous homogeneous equation for unknown expansion coefficients in each region. We look for a value of the propagation constant that makes the determinant of this system of equations zero [22].

2.7. Transverse resonance technique.

This technique is somewhat similar to mode matching method and is suited for characterization of the discontinuity in a planar waveguide structure. The method is illustrated by way of a finline discontinuity shown in Figure 7.

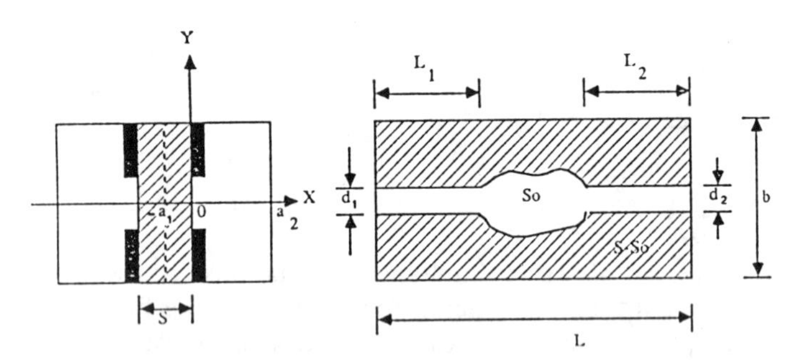

FIG. 7. — Finline discontinuity.

Discontinuité d'une ligne à ailette.

First, two shorting end plates are placed in the waveguide case at such distances away from the discontinuity that all the higher order modes excited at the discontinuity are negligible. Only the dominant modes can propagate in the two finline sections. The objective of the analysis is to find the resonant frequency from which one can extract information on the discontinuity [23].

Due to the symmetry of the bilateral finline configuration, only one half of the cross section $-a_1 < x < a_2$ in Figure 8 is considered.

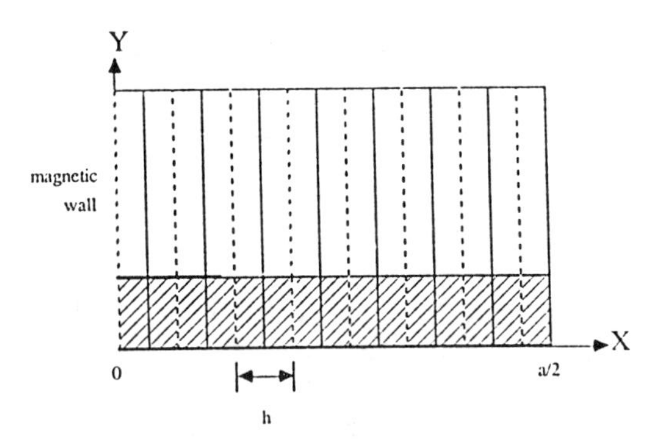

FIG. 8. — One half of the cross section of a microstrip line for the method of lines procedure.

Demi-section transversale d'une ligne microruban pour la procédure utilisant la méthode des lignes.

The electromagnetic field in the dielectric region (region 1 : $-a_1 \leqslant x \leqslant 0$) and in the air region (region 2 : $0 \leqslant x \leqslant a_2$) can be expanded in terms of TE and TM modes of a rectangular waveguide with inner dimensions L and b. We obtain the following expressions for the transverse E- and H-field components in the two regions :

Dielectric region : $-a_1 \leqslant x \leqslant 0$

$$E_{t_1} = \sum_{mn} A'_{mn} \cos k'_{mn}(\hat{x} + a_1)\, x \times \nabla_t \Psi'_{mn} + \frac{1}{j\omega\varepsilon_0\varepsilon_r} \sum_{nm} B'_{mn}\, k'_{mn} \cos k'_{mn}(x + a_1)\, \nabla_t \varphi_{mn}\,,$$

$$(28)\quad H_{t_1} = \frac{-1}{j\omega\mu_0} \sum A'_{mn}\, k'_{mn} \sin k'_{mn}(x + a_1)\, \nabla_t \Psi'_{mn} + \sum B'_{mn} \sin k'_{mn}(x + a_1)\, \nabla_t \varphi_{mn} \times \hat{x}.$$

Air region : $0 \leqslant x \leqslant a_2$

$$E_{t_2} = \sum_{mn} A_{mn} \sin k_{mn}(x - a_2)\, \hat{x} \times \nabla_t \Psi'_{mn} - \frac{1}{j\omega\varepsilon_0} \sum_{mn} B_{mn}\, k_{mn} \sin k_{mn}(x - a_2)\, \nabla_t \varphi_{mn}\,,$$

$$(29)\quad H_{t_2} = \frac{1}{j\omega\mu_0} \sum_{mn} A_{mn}\, k_{mn} \cos k_{mn}(x - a_2)\, \nabla_t \Psi'_{mn} + \sum_{mn} B_{mn} \cos k_{mn}(x - a_2)\, \nabla_t \varphi_{mn} \times \hat{x},$$

where :

$$\Psi_{mn} = P_{mn} \cos(m\pi z/l) \cos(n\pi y/b),$$

$$\varphi_{mn} = P_{mn} \sin(m\pi z/l) \sin(n\pi y/b),$$

$$P_{mn} = \sqrt{\frac{\delta_m \delta_n}{lb}}\, \frac{1}{\gamma_{mn}} \qquad \delta_i = \begin{cases} 1, & i = 0 \\ 2, & i \neq 0 \end{cases},$$

$$Y_{mn}^2 = (m\pi/l)^2 + (n\pi/b)^2,$$

$$k_{mn}^2 = k_0^2 - Y_{mn}^2 \qquad k_{mn}'^2 = k_0^2\,\varepsilon_r - Y_{mn}^2,$$

$$k_0^2 = \omega^2\mu_0\varepsilon_0,$$

when Ψ_{mn} and φ_{mn} are the TE and TM scalar potentials. Notice that (28) and (29) already satisfy the boundary conditions at $x = -a_1$ and a_2. The boundary conditions at $x = 0$ are :

$$(30) \qquad E_{t_1} = E_{t_2} = \begin{cases} E_{t_0}, & \text{on } S_0, \\ 0, & \text{on } S - S_0, \end{cases}$$

$$(31) \qquad H_{t_1} = H_{t_2} = H_{t_0}, \qquad \text{on } S_0,$$

where E_{t_0} and H_{t_0} are unknown functions of z, y.

From this point, we could proceed in a manner similar to the one in the mode matching method in Section II.6. However, we will take a different approach. E_{t_0} and H_{t_0} are expanded in terms of a set of orthonormal vector functions e_ν and h_μ defined on the aperture S_0.

$$(32) \qquad E_{t_0} = \Sigma V_\nu e_\nu,$$

$$(33) \qquad H_{t_0} = \Sigma I_\mu h_\mu.$$

Substituting (28), (29), (32) and (33) into (30) and (31) and using the orthogonal properties of Ψ_{mn}, φ_{mn}, e_ν and h_μ, we obtain homogeneous equation. We eliminate A_{mn}, A'_{mn}, B_{mn}, B'_{mn} and I_μ and obtain a homogeneous equation :

$$(34) \qquad \Sigma k_{\mu\nu} V_\nu = 0,$$

when $K_{\mu\nu}$ contains summations over m and n. The nontrivial solutions of (34) results in the resonant frequency of the structure.

As it has been demonstrated above, this technique is useful when the discontinuity is located only over a plane including the guide axis, that is the discontinuity not involving the change in height.

2.8. Method of lines.

In this method, two of the three dimensions are discretized for numerical processing while the analytical expressions are sought in the remaining dimension. The essential feature of this method is first explained by way of a simple two dimensional problem of finding the propagation constant of a microstrip line in Figure 9 [24]. First, the x direction is discretized by a family of N straight lines parallel to the y axis separated by h. When the partial derivative with respect to the x coordinate is replaced with the difference formula, the two scalar potentials Ψ^e and Ψ^h necessary for describing the hybrid field satisfy :

$$(35) \qquad \frac{d^2\Psi_i}{dy^2} + \frac{1}{h^2}[\Psi_{i-1}(y) - 2\Psi_i(y) + \Psi_{i+1}(y)] + (k^2 - \beta^2)\Psi_i(y) = 0, \qquad i = 1, 2, \ldots, N,$$

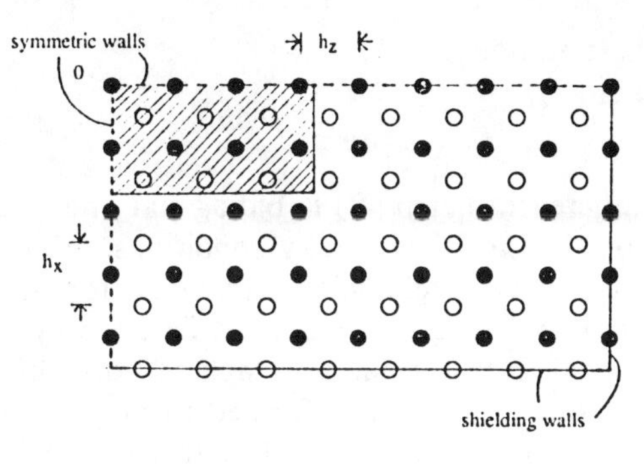

FIG. 9. — One quarter of the top view of a microstrip resonator for the method of lines procedure.

Quart de la vue de dessus d'un résonateur microruban pour la procédure utilisant la méthode des lignes.

or in a matrix form :

$$(36) \qquad h^2\frac{d\psi}{dy^2} - [P - h^2(k^2 - \beta^2)\,I]\,\Psi = 0,$$

where I is the identity matrix and P is a tri-diagonal matrix determined by the lateral boundary conditions at $x = 0$ and $a/2$. The discretization lines for Ψ^e and Ψ^h are shifted by half the discretization distance $h/2$ so that the lateral boundary conditions are easily implemented. The essential feature of the method lies in the diagonalization of (36) so that the equation for the potential can be solved independently for each discretization i. This is accomplished by the transformation :

$$(37) \qquad T^t\psi = U,$$

where t denotes transpose of T which is an orthogonal matrix and is determined by the lateral boundary conditions. The uncoupled equations take the form of :

$$(38) \qquad h^2\frac{d^2U_i}{dy^2} - [\lambda_i - h^2(k^2 - \beta^2)]\,U_i = 0, \qquad i = 1, 2, \ldots, N,$$

when λ_i is the eigenvalue of P. The equations of this form for the two scalar potentials are solved for each homogeneous regions. Then we impose the boundary conditions at the substrate-air interface.

Finally, the condition that the tangential electric fields on the strip are zero is imposed in the original domain, and the following matrix equations are derived :

$$(39) \qquad R\begin{bmatrix} J_x \\ J_z \end{bmatrix} = \begin{bmatrix} 0 \\ 0 \end{bmatrix},$$

where J_x and J_z are the current components and are vectors with the elements consisting of the values at each discretized point.

The method can be extended to three dimensional problems such as the microstrip resonator [25]. Instead of the central difference formula, the forward difference formula is used for the first derivative of the potentials Ψ with respect to the x variable. In the matrix notation :

$$h_x \frac{\partial \Psi}{\partial x} \to D_x \psi .$$

The difference matrix D_x is bidiagonal and is dependent on the lateral boundary conditions. Once again, the discretized Helmholtz equations for the two potentials Ψ^e and Ψ^h are treated. By way of the orthogonal transformation matrices, the difference matrix equations are transformed to diagonal forms. From the applications of the interface conditions, the matrix relation between the electric field and the current is obtained in the transformed domain. This relation is transformed back to the original domain and the final boundary condition on the strip is imposed. From nontriviality of the solution, the eigenvalue equation of the resonant frequency is obtained.

The method of lines has been applied to a number of practical but analytically complex structures. Examples include a triangular microstrip resonator and a periodic microstrip structure.

2.9. Generalized S matrix method.

Although this has been developed for analyzing complicated discontinuity problems, it can be used for characterization of cascaded discontinuities often seen in a passive components such as the E-plane filter [26]. The generalized S matrix combines the mutual interaction of two discontinuities via the dominant and higher order modes. This method need to be used with other techniques such as the mode matching method that characterizes a single discontinuity.

Let us illustrate the method by means of a cascaded discontinuity in Figure 10 [27]. The first step is to characterize all the discontinuities involved in the microwave circuit. This characterization is expressed in terms of the generalized scattering matrix which is closely related to the scattering matrix used in microwave network theory but differs in that the higher order modes are included in addition to the dominant mode. Hence, the generalized scattering matrix is in general of infinite order. Consider that the junction 1 is excited with the p-th mode with unit amplitude from the left. If the complex amplitude of the n-th mode of the reflected wave to the left is A_n, the (n, p) entry of the generalized scattering matrix $S^{11}(n, p)$ is A_n. Similarly, if the amplitude of the m-th mode transmitted to the right is B_m, $S^{21}(m, p)$ is B_m. The generalized scattering matrix S_1 of the junction 1 is :

$$S_1 = \begin{bmatrix} S^{11} & S^{12} \\ S^{21} & S^{22} \end{bmatrix} . \quad (40)$$

Similarly, for the junction 2 is :

$$S_2 = \begin{bmatrix} S^{33} & S^{34} \\ S^{43} & S^{44} \end{bmatrix} . \quad (41)$$

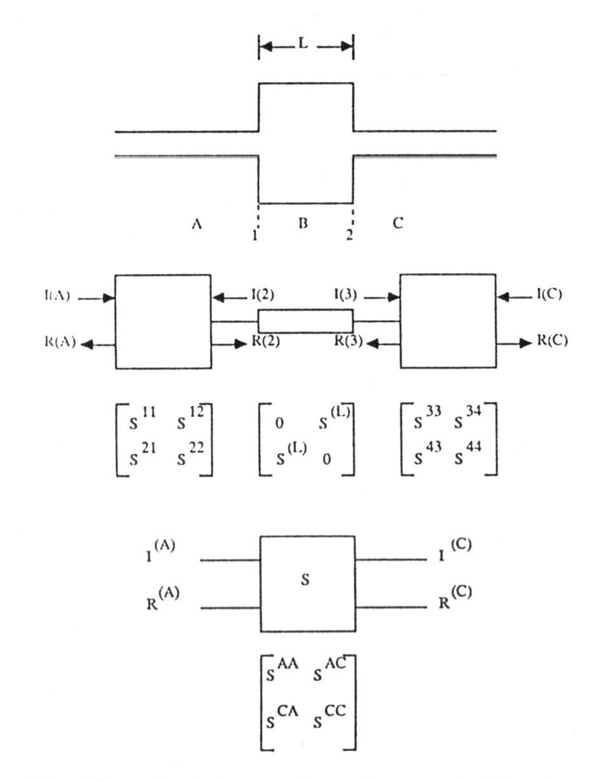

FIG. 10. — Analysis procedure of a cascaded junction by the generalized scattering matrix analysis.

Analyse d'une jonction en cascade utilisant la matrice de répartition généralisée.

Obviously, we need to find all the scattering matrix elements by some means such as the mode matching technique before proceeding further.

The next step is to combine S_1 and S_2 to find the composite matrix :

$$S = \begin{bmatrix} S^{AA} & S^{AC} \\ S^{CA} & S^{CC} \end{bmatrix} , \quad (42)$$

of the cascaded junctions. It turns out that :

$$S^{AA} = S^{11} + S^{12} S^{(L)} U_2 S^{33} S^{(L)} S^{21} , \quad (43a)$$

$$S^{AC} = S^{12} S^{(L)} U_2 S^{34} , \quad (43b)$$

$$S^{CA} = S^{43} S^{(L)} U_1 S^{21} , \quad (43c)$$

$$S^{CC} = S^{44} + S^{43} S^{(L)} U_1 S^{22} S^{(L)} S^{34} , \quad (43d)$$

where :

$$U_1 = (I - S^{22} S^{(L)} S^{33} S^{(L)})^{-1} ,$$

$$U^2 = (I - S^{33} S^{(L)} S^{22} S^{(L)})^{-1} ,$$

and $S^{(L)}$ is the transmission matrix for the waveguide between the two junctions :

$$S^{(L)} = \begin{bmatrix} e^{-\gamma_1 L} & & 0 \\ & e^{-\gamma_2 L} & \\ 0 & & \ddots \end{bmatrix} . \quad (44)$$

I is the identity matrix and γ_n is the propagation constant of the n-th mode.

The method can be applied to a complicated discontinuity by decomposing it to several of less complicated geometry for which the solution is available. This application is illustrated by the offset discontinuity shown in Figure 11 *a* [27]. To this end, an auxiliary structure in Figure 11 *b* is introduced. Notice that the original offset discontinuity can be recovered by letting δ to zero after all the formulations are carried out. Therefore, the generalized scattering matrices for the junctions J_1 and J_2 are first obtained. They can be combined to find the composite matrix by way of (43) except that $S^{(L)} = I$ when $\delta \to 0$.

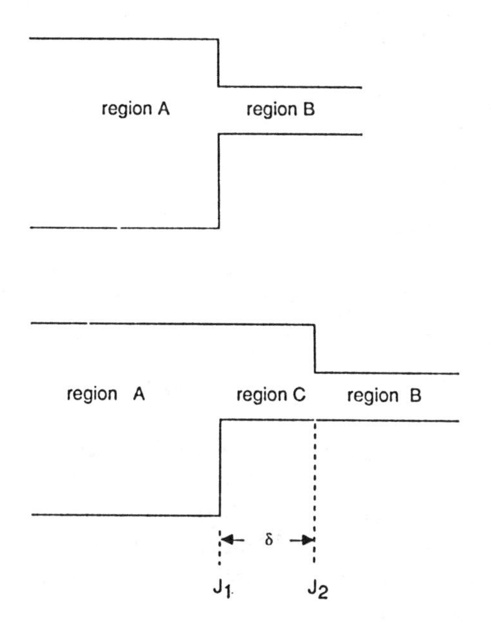

FIG. 11. — Offset microstrip step discontinuity and an auxiliary structure for analysis.

Discontinuité en échelon décalée et structure d'analyse auxiliaire.

Notice that, in the above formulations, all the interactions via higher order modes are included in addition to the dominant mode contributions. All the matrices are of infinite order. In practice, however, they must be truncated to a finite size. It turns out that small matrices such as 2×2 and 3×3 provide excellent results even when $\delta = 0$ [27].

2.10. Spectral domain method.

This is a Fourier transformed version of the integral equation method applied to microstrip or other printed line structures. It is one of the most preferred methods in recent years. The method is known to be efficient, but is restricted in general to well-shaped structures that involves infinitely thin conductors. The method is illustrated by means of a microstrip resonator in Figure 12 [28].

It is known that the hybrid fields in the structure can be found from two scalar potentials φ and Ψ associated with the E_y and H_y fields in both the substrate and air regions. When all the field components are Fourier transformed in both the x and z directions with the transform variables α and β,

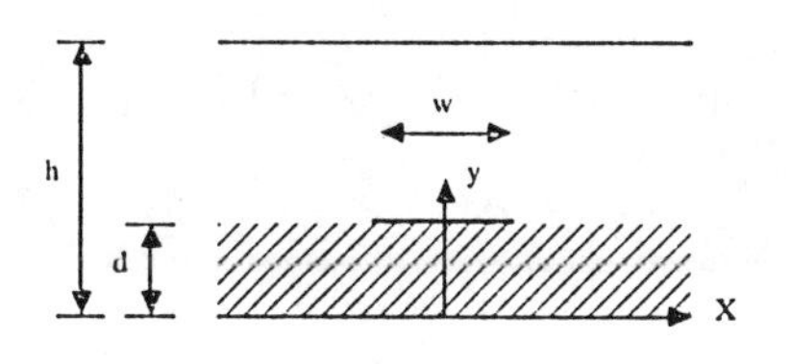

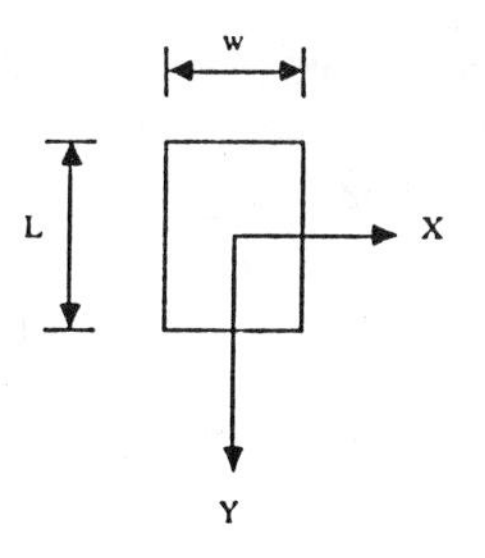

FIG. 12. — Shielded rectangular microstrip resonator.

Résonateur microruban rectangulaire blindé.

the Helmholtz equations to be satisfied with φ, Ψ and all the field components are now reduced to one-dimensional ordinary differential equations for y only. Appropriate solutions to these equations are now found so that the boundary conditions at the bottom and top conducting planes are satisfied. This implies that $E_x = E_z = 0$ there and hence $\tilde{E}_x = \tilde{E}_z = 0$ where the latters with $\sim$ are Fourier transformed quantities.

Next, the boundary conditions at the interface $y = d$ are applied in the Fourier transform domain (spectral domain). Notice that in the space domain they are :

$$(45) \quad E_{x_1} = E_{x_2}, \qquad E_{z_1} = E_{z_2}$$

$$(46) \quad E_{x_1} = \begin{cases} 0 \\ f(x) \end{cases} \qquad E_{z_1} = \begin{cases} 0 & \text{on strip} \\ g(x) & \text{outside} \end{cases}$$

$$(47) \quad H_{x_1} - H_{x_2} = \begin{cases} J_z \\ 0 \end{cases}, \quad H_{z_1} - H_{z_2} = \begin{cases} J_x & \text{on strip} \\ 0 & \text{outside} \end{cases}$$

These conditions are Fourier transformed with respect to x and z directions. All the field components expressed in the spectral domain are substituted into the boundary conditions in the spectral domain. When all the unknown coefficients of the field expressions are eliminated, the following coupled algebraic equations are obtained :

$$(48a) \quad \tilde{E}_x(\alpha, \beta) = \tilde{G}_{xx}(\alpha, \beta, k)\, \tilde{J}_x(\alpha, \beta) + \tilde{G}_{xz}(\alpha, \beta, k)\, \tilde{J}_z(\alpha, \beta),$$

$$(48b) \quad \tilde{E}_z(\alpha, \beta) = \tilde{G}_{zx}(\alpha, \beta, k)\, \tilde{J}_x(\alpha, \beta) + \tilde{G}_{zz}(\alpha, \beta, k)\, \tilde{J}_z(\alpha, \beta).$$

The above equations correspond to the coupled homogeneous integral equations obtainable in the space domain :

$$(49a) \quad 0 = \int_{\text{strip}} G_{xx}(x, x', z, z', k)\, J_x(x', z')\, dx'\, dz' + \int_{\text{strip}} G_{xz}(x, x', z, z', k)\, J_z(x', z')\, dx'\, dz',$$

(49b) $0 = \int_{strip} G_{zx}(x, x', z, z', k)\, J_x(x', z')\, dx'\, dz' +$

$\int_{strip} G_{zz}(x, x', z, z', k)\, J_z(x', z')\, dx'\, dz' \quad (x, z) \in \text{strip}.$

Also, notice that (48) contain four unknowns $\tilde{E}_x$, $\tilde{E}_z$, $\tilde{J}_x$ and $\tilde{J}_z$. In the solution process, however, $\tilde{E}_x$ and $\tilde{E}_z$ are eliminated and (48) can be solved only for $\tilde{J}_x$ and $\tilde{J}_z$.

The solution of (48) is undertaken by means of the Galerkin's procedure. To this end, $\tilde{J}_x$ and $\tilde{J}_z$ are first expanded in terms of known basis functions :

(50a) $$\tilde{J}_x(\alpha, \beta) = \sum_{m=1}^{M} a_m \tilde{J}_{xm}(\alpha, \beta),$$

(50b) $$\tilde{J}_z(\alpha, \beta) = \sum_{n=1}^{N} b_n \tilde{J}_{zn}(\alpha, \beta).$$

It is important to select $\tilde{J}_{xm}$ and $\tilde{J}_{zn}$ such that their inverse transforms $J_{xm}(x, z)$ and $J_{zn}(x, z)$ are nonzero only on the strip. Further, they should incorporate appropriate edge conditions for faster convergence of the solution [29].

The expressions (50) are substituted into (48) and inner products with each of $\tilde{J}_{xm}$ and $\tilde{J}_{zn}$ are formed. The results are the following linear simultaneous equations :

(51a) $$\sum_{m=1}^{M} K_{pm}^{xx}(k)\, a_m + \sum_{n=1}^{N} K_{pn}^{xz}(k)\, b_n = 0 \qquad p = 1, 2, \ldots, M,$$

(51b) $$\sum_{m=1}^{M} K_{qm}^{zx}(k)\, a_m + \sum_{n=1}^{N} K_{qn}^{zz}(k)\, b_n = 0 \qquad q = 1, 2, \ldots, N.$$

The right hand sides become zero by virtue of Parseval's relation. The above equation is solved for the unknowns a_m and b_n. To have meaningful solution, the determinant of the coefficients matrix must be zero. From this requirement the resonant frequency ($k = \omega\sqrt{\varepsilon_0 \mu_0}$) is obtained. All the field components are obtained from a_m and b_n.

The method has been applied to two-dimensional waveguide problems [30] and has been extended to discontinuity problems [31].

It is pointed out that the derivation of (48) is often involved although straightforward. This is particularly true if one deals with multilayered structures or structures with conductors at several interfaces. The process can be significantly simplified by means of the immittance approach based on the coordinate transformation and the equivalent transmission lines [32]. The result for the structure in Figure 12 is :

(52a) $$G_{xx} = N_x^2 Z_{11}^e + N_z^2 Z_{11}^h,$$

(52b) $$G_{zx} = G_{xz} = N_x N_z(-Z_{11}^e + Z_{11}^h),$$

(52c) $$G_{zz} = N_z^2 Z_{11}^e + N_x^2 Z_{11}^h,$$

(53) $$N_x = \frac{\alpha}{\sqrt{\alpha^2 + \beta^2}}, \quad N_z = \frac{\beta}{\sqrt{\alpha^2 + \beta^2}},$$

(54) $$Z_{11} = \frac{1}{Y_u^e + Y_d^e}, \quad Z_{11} = \frac{1}{Y_u^h + Y_d^h},$$

Y_u^e and Y_d^e are input admittance looking upward and downward at 1 — 1 for the TM transmission line in Figure 13. Y_u^h and Y_d^h are similarly defined. It is clear that G_{xx}, etc. are now written down almost by inspection of the structure.

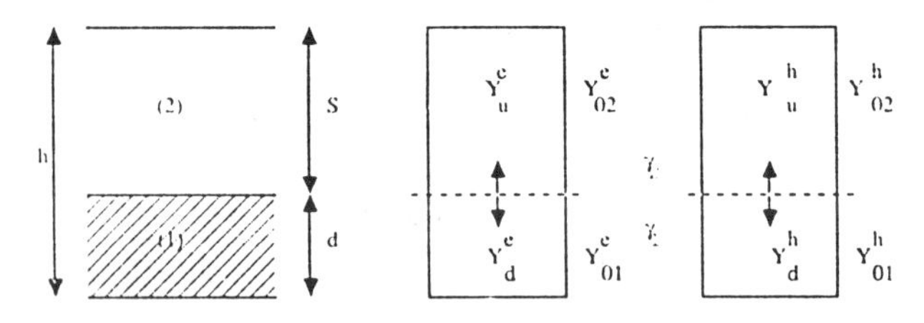

FIG. 13. — Equivalent transmission lines for derivations of Green's functions in the spectral domain.

Lignes de transmission équivalentes pour l'obtention des fonctions de Green du domaine spectral.

2.11. Equivalent waveguide model.

This is not the numerical method but a formalism used for analysis of microstrip discontinuity problems. After the microstrip problem is converted to the equivalent waveguide model, one of the suitable numerical methods is used for characterizing the discontinuities.

The technique has been introduced by the research group headed by I. Wolff [33, 34]. Let us illustrate this technique by means of a microstrip step discontinuity shown in Figure 14. The essential feature

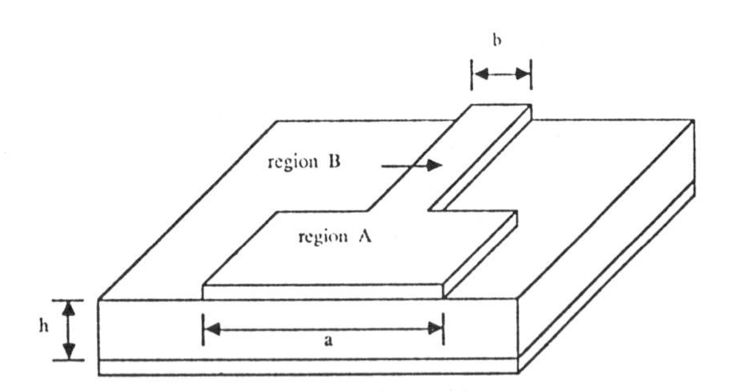

FIG. 14. — Microstrip step discontinuity.

Discontinuité en échelon de ligne microruban.

of this technique is identification of a hypothetical waveguide representing the microstrip line. For instance, the microstrip discontinuity is replaced with the equivalent waveguide structure. The equivalent waveguide has the same height as the substrate thickness, the two perfectly conducting top and bottom walls and the two magnetic sidewalls. It is filled with a hypothetical medium with the effective dielectric constant and the width of the waveguide is equal to the effective width. The effective dielectric constants of regions A and B are given by :

(55*a*) $\varepsilon_A = (\beta_A/k)^2,$

(55*b*) $\varepsilon_B = (\beta_B/k)^2,$

and the effective widths are :

(56*a*) $a = (120\,\pi/\sqrt{\varepsilon_A})\,h/Z_{01}\,,$

(56*b*) $b = (120\,\pi/\sqrt{\varepsilon_B})\,h/Z_{02}\,.$

In the above β_A and β_B are the phase constants and Z_{01} and Z_{02} are the characteristic impedances of the dominant microstrip mode in the respective regions. These four quantities must be calculated by a standard technique such as the spectral domain method. Naturally, they are functions of frequency.

Once the equivalent waveguide structure is obtained for each microstrip section, the discontinuity problem is transformed to that of the closed waveguide configuration. A number of techniques are available including the mode matching method to characterize such a discontinuity.

The method is inherently limited to the case where the surface wave excitation and radiation phenomena at the discontinuity are negligible. Reasonably accurate data have been obtained as long as the frequency is relatively low.

2.12. Planar circuit model.

This is also a formalism to analyze planar passive components. The eigenmode expansion and the integral equation are often used for this model. In addition, the so-called segmentation and desegmentation techniques are powerful supplements for the planar circuit approach.

The concept of planar circuits has been introduced by Okoshi and Miyoshi [35]. A planar circuit is defined as a microwave structure in which one of the three dimensions, say z, is much smaller than the wavelength whereas the remaining two are comparable to the wavelength. It is hence possible to assume that the field is invariant in the z direction ($\partial/\partial z = 0$). One needs then to deal with two dimensional Helmholtz equation. When the magnetic side wall is assumed in Figure 15 except the i-th port where a transmission line of width W_i is connected, the equation

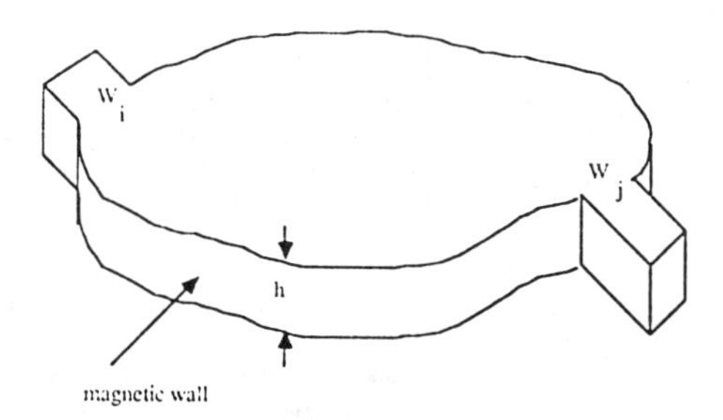

FIG. 15. — Planar circuit model.

Modèle de circuit plan.

for E_z and the boundary conditions are :

(57) $$\nabla_t^2 E_z + k^2 E_z = 0,$$

(58) $$\hat{n}\,.\,\nabla E_z = \begin{cases} 0 & \text{magnetic wall,} \\ -\mathrm{j}\omega\mu J_S\,.\,\hat{n} & \text{on } w_i\,, \end{cases}$$

where J_S is the surface current density. The solution to this problem can be found once the Green's function is available :

(59) $$\nabla^2 G + k^2 G = -\,\delta(\boldsymbol{r} - \boldsymbol{r}').$$

One of the methods is the direct solution of integral equation derived from the Green's theorem. The Green's function is :

(60) $$G = -\frac{\mathrm{j}}{4}\,\mathrm{H}_0^{(2)}(kr), \quad r = |\boldsymbol{r} - \boldsymbol{r}'|,$$

where H_0 is the zeroth order Hankel function of the second kind. Then :

(61) $$\mathrm{v}(\boldsymbol{r}) = \frac{1}{4\,\mathrm{j}}\int_C k\cos\theta\, H_1^{(2)}(kr)\,\mathrm{v}(\boldsymbol{r}')\,\mathrm{d}\boldsymbol{r} - \frac{\omega\mu}{4}\sum\int_{w_i} \mathrm{H}_0^{(2)}(kr)\,\hat{n}\,.\,J_S(\boldsymbol{r}')\,\mathrm{d}\boldsymbol{r},$$

where $v = hE_z$ and θ is the angle between the normals at $\boldsymbol{r}'$ and $\boldsymbol{r}' - \boldsymbol{r}$. Discretization of (61) provides an impedance relation between the terminal voltage and the current at each port.

The above method is applicable to a structure with an arbitrary shape. However, when the shape of the circuit is more regular such as rectangular or circular, another method is more convenient and informative. In this second method, the Green's function with the boundary condition $\hat{n}\,.\,\nabla G = 0$ is expanded in terms of its eigenfunctions :

(62) $$\mathrm{G}(\boldsymbol{r}, \boldsymbol{r}') = \mathrm{j}\omega\mu \sum_{\nu=0}^{\infty} \frac{\varphi_\nu(\boldsymbol{r})\,\varphi_\nu(\boldsymbol{r}')}{k_\nu^2 - k^2}\,.$$

From the expression, the impedance relationship between the terminal voltage and current can be found.

In general, the voltage v and the injected current j at the i-th port can be written in terms of Fourier expansions [36] :

(63) $$v = \sum_{n=0}^{\infty} V_i^{(m)}\,\sqrt{\delta_m}\cos\frac{m\pi l}{w_i}\,,$$

(64) $$j = \sum_{m=0}^{\infty} I_i^{(n)}\,\delta_n\cos\frac{n\pi l}{w_i}\,,$$

$$\delta_m = \begin{cases} 1 & m = 0, \\ 2 & m \neq 0, \end{cases}$$

where l is the coordinate along the i-th port ($0 \leqslant l \leqslant w_i$). The generalized impedance matrix is defined as :

(65) $$\mathrm{V}_i^{(m)} = \sum_{n=0}^{\infty}\sum_{j=1}^{N} Z_{ij}^{(mn)}\,I_j^{(n)},$$

(66) $$Z_{ij}^{(mn)} = \frac{\mathrm{j}\omega\mu h\sqrt{\delta_m\delta_n}}{w_i w_j}\sum_{\nu=0}^{\infty}\frac{g\nu_i^{(m)}\,g\nu_j^{(n)}}{k_\nu^2 - k^2}\,,$$

$$g_{V_i}^{(m)} = \int_{w_i} \varphi_v \cos \frac{m\pi l}{w_i} \, dl, \tag{67}$$

$Z_{ij}^{(mn)}$ gives the m-th order voltage in the i-th port when a unit n-th order current is injected at the j-th port with all other currents zero.

The planar circuit approach can be extended by introduction of segmentation of planar elements. It is recognized that the Green's function and eigenfunctions are known only for a limited number of structural shapes. However, a more complicated shape can be segmented into elementary shapes for which the impedance matrix can be calculated by the method described above. The common ports are connected to form the original circuit (Fig. 16).

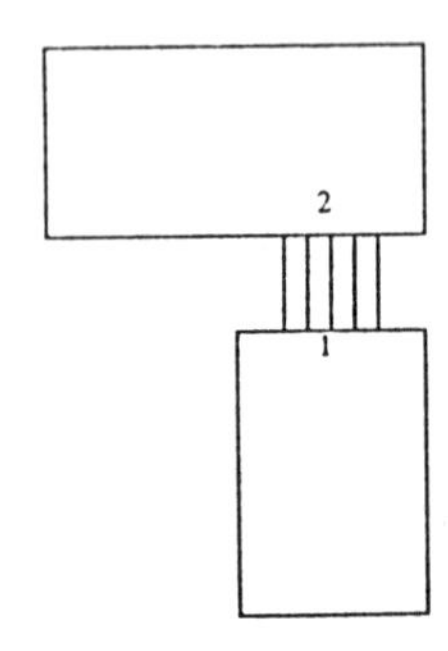

FIG. 16. — Segmentation procedure.
Procédure de segmentation.

The same physical port 1 (only in Fig. 16) can be decomposed to infinite electrical ports corresponding to the Fourier expansion in (63) and (64) which are truncated to a finite size.

In the so-called segmentation method [37] the interconnection is discretized into a finite number of physical ports after the voltage and current along the interconnection are approximated by step functions.

Another extension of the planar circuit approach is the desegmentation method [38]. This technique is applicable to the geometry which, if a simple element is added, can be analyzed easily by either the original planar circuit method or the segmentation method.

3. DISCUSSION AND CONCLUSIONS

In the paper, a number of numerical methods are described that are applicable to the analysis of passive components appearing in miniature microwave networks.

In concluding, the author would like to include some personal, and most likely biased, feelings toward the use of computers and CAD in the future. It seems that accessibility to increasingly more powerful computers is being improved every day. Certainly, the availability of large main frames and even super computers will help solve many complicated problems accurately. We can make use of these machines for CAD where the modeling methods described above may be used directly instead of curve fitting procedures. On the other hand, availability of personal computers to design engineers is rapidly improving. Besides, new generation PC's are quite powerful. Some simple characterizations can be done on these machines using BASIC or even FORTRAN. Another approach for PC based CAD is to use the PC's as the look-up tables and data base or interpolation mechanism of the stored data base. These data bases can be generated by a larger machine.

Manuscrit reçu le 16 *décembre* 1985,
accepté le 20 *mars* 1986.

REFERENCES

[1] MUR (G.). Finite difference method for the solution of electromagnetic waveguide discontinuity problem. *IEEE Trans. MTT*, USA (Jan. 1974), **22**, n° 1, pp. 54-57.
[2] GREEN (H. E.). The numerical solution of some important transmission-line problems. *IEEE Trans. MTT*, USA (sept. 1965), n° 5, pp. 676-692.
[3] SILVESTER (P.). Finite elements for electrical engineers. *Cambridge University Press*, New York (1983).
[4] DALY (P.). Hybrid-mode analysis of microstrip by finite element method. *IEEE Trans. MTT*, USA (Jan. 1971), **19**, n° 1, pp. 19-25.
[5] THOMSON (A. F.), GOPINATH (A.). Calculation of microstrip discontinuity inductances. *IEEE Trans. MTT*, USA (Aug. 1975), **23**, n° 8, pp. 648-655.
[6] RAHMAN (B. M. A.), DAVIES (J. B.). Finite element analysis of optical and microwave waveguide problems. *IEEE Trans. MTT*, USA (Jan. 1984), **32**, n° 1, pp. 20-28.
[7] SILVESTER (P.). Finite element analysis of planar microwave networks. *IEEE Trans. MTT*, USA (Feb. 1973), **21**, n° 2, pp. 104-108.
[8] FERRARI (R. L.). Finite element analysis of three-dimensional electromagnetic devices. *15th European Microwave Conf. Digest*, Paris (Sept. 1985), pp. 1064-1069.
[9] RAHMAN (B. M. A.), DAVIES (J. B.). Penalty function improvement of waveguide solution by finite elements. *IEEE Trans. MTT*, USA (Aug. 1984), **32**, n° 8, pp. 922-928.
[10] BREBBIA (C. A.). The boundary element method for engineers. *Pentech Press*, London (1978).
[11] KAGAMI (S.), FUKAI (I.). Application of boundary element method to electromagnetic field problems. *IEEE Trans. MTT*, USA (April 1984), **32**, n° 4, pp. 455-461.
[12] HOEFER (W. J. R.), ROS (A.). Fin line parameters calculated with the TLM-method. *IEEE MTT-S, International Microwave Symposium*, Orlando, Florida (April 30 - May 2, 1979), pp. 341-343.
[13] AKHTARZAD (S.), JOHNS (P. B.). Three-dimensional transmission-line matrix computer analysis of microstrip resonators. *IEEE Trans. MTT*, USA (Dec. 1975), **23**, n° 12, pp. 990-997.

[14] HOEFER (W. J. R.). The transmission-line matrix method - theory and applications. *IEEE Trans. MTT*, USA (oct. 1985), **33**, nº 10, pp. 882-893.

[15] CHEW (W. C.), KONG (J. A.). Resonance of the axial-symmetric modes in microstrip disk resonators. *J. Math Phys.*, USA (March 1980), **21**, nº 3, pp. 582-591.

[16] HARRINGTON (R. F.). Field computation by moment methods. *Macmillan*, New-York (1968).

[17] YAMASHITA (E.), MITTRA (R.). Variational method for the analysis of microstriplines. *IEEE Trans. MTT*, USA (August 1968), **16**, nº 4, pp. 251-256.

[18] JONES (D. S.). The theory of electromagnetism. *Pergamon*, New York (1964).

[19] JACKSON (R. W.), POZER (D. M.). Full-wave analysis of microstrip open-end and gap discontinuities. *IEEE Trans. MTT*, USA (Oct. 1985), **33**, nº 10, pp. 1036-1042.

[20] SHIH (Y. C.), GRAY (K. G.). Convergence of numerical solutions of step-type waveguide discontinuity problems by modal analysis. *IEEE/MTT-S Intl. Microwave Symp. Dig.* (May 1983), pp. 233-235, Boston, Ma.

[21] CHU (T. S.), ITOH (T.), SHIH (Y.-C.). Comparative study of mode-matching formulations for microstrip discontinuity problems. *IEEE Trans. MTT*, USA (Oct. 1985), **33**, nº 10, pp. 1018-1023.

[22] KOWALSKI (G.), PREGLA (R.). Dispersion characteristics of shielded microstrips with finite thickness. *Arch Elektron Ubertragungstech.*, Dtsch (April 1971), **25**, nº 4, pp. 193-196.

[23] SORRENTINO (R.), ITOH (T.). Transverse resonance analysis of finline discontinuities. *IEEE Trans. MTT*, USA (Dec. 1984), **32**, nº 12, pp. 1633-1638.

[24] SCHULZ (U.), PREGLA (R.). A new technique for the analysis of the dispersion characteristics of planar waveguides and its application to microstrips with tuning septums. *Radio Sci.*, USA (Nov.-Dec. 1981), **16**, nº 6, pp. 1173-1178.

[25] WORM (S. B.), PREGLA (R.). Hybrid-mode analysis of arbitrarily shaped planar microwave structures by the method of lines. *IEEE Trans. MTT*, USA (Feb. 1984), **32**, nº 2, pp. 191-196.

[26] SHIH (Y.-C.), ITOH (T.), BUI (L. Q.). Computer-aided design of millimeter-wave E-plane filters. *IEEE Trans. MTT*, USA (Feb. 1983), **31**, nº 2, pp. 135-142.

[27] CHU (T. S.), ITOH (T.). Analysis of cascaded and offset microstrip step discontinuities by the generalized scattering matrix technique. *IEEE Trans. MTT*, USA (Feb. 1986), **34**, nº 2.

[28] ITOH (T.). Analysis of microstrip resonators. *IEEE Trans. MTT*, USA (Nov. 1974), **22**, nº 11, pp. 946-952.

[29] JANSEN (R. H.). Unified user-oriented computation of shielded, covered and open planar microwave and millimeter-wave transmission-line characteristics. *IEE J. Microwaves, Optics and Acoustics*, USA (Jan. 1979), **3**, nº 1, pp. 14-22.

[30] SCHMIDT (L. P.), ITOH (T.), HOFMANN (H.). Characteristics of unilateral fin-line structures with arbitrarily located slots. *IEEE Trans. MTT*, USA (April 1981), **29**, nº 4, pp. 352-355.

[31] BOUKAMP (J.), JANSEN (R. H.). The high-frequency behaviour of microstrip open ends in microwave lintegrated circuits including energy leakage. *14th European Microwave Conf. Digest*, Liège, Belgium (Sept. 1984), pp. 142-147.

[32] ITOH (T.). Spectral domain immittance approach for dispersion characteristics of generalized printed transmission lines. *IEEE Trans. MTT*, USA (July 1980), **28**, nº 7, pp. 733-736.

[33] WOLFF (I.), KNOPPIK (N.). Rectangular and circular microstrip disk capacitors and resonators. *IEEE Trans. MTT*, USA (Oct. 1974), **22**, nº 10, pp. 857-864.

[34] KOMPA (G.). Frequency dependent behaviour of microstrip offset junction. *Electron. Lett.*, UK (Oct. 1975), **11**, nº 22, pp. 537-538.

[35] OKOSHI (T.), MIYOSHI (T.). The planar circuit - an approach to microwave integrated circuitry. *IEEE Trans. MTT*, USA (April 1972), **20**, nº 4, pp. 245-252.

[36] SORRENTINO (R.). Planar circuits, waveguide models, and segmentation method. *IEEE Trans. MTT*, USA (Oct. 1985), **33**, nº 10, pp. 1057-1066.

[37] OKOSHI (T.), UEHARA (Y.), TAKEUCHI (T.). The segmentation method - an approach to the analysis of microwave planar circuits. *IEEE Trans. MTT*, USA (Oct. 1976), **24**, nº 10, pp. 662-668.

[38] SHARMA (P. C.), GUPTA (K. C.). Desegmentation method for analysis of two-dimensional microwave circuits. *IEEE Trans. MTT*, USA (Oct. 1981), **29**, nº 10, pp. 1094-1098.

Part 2
Finite-Difference Method

THE finite-difference method (FDM) is the oldest and least analytical method to transform a differential equation into a system of algebraic equations. Derivatives are simply replaced by finite differences. The region of interest is divided into nodes located on a two- or three-dimensional grid. Because of the very simple algorithm required, the method has great versatility. It requires, however, a large number of mesh points, thus a large memory storage, and numerical efficiency is rather low. Another shortcoming is the difficulty of fitting curved boundaries with a rectangular mesh.

Brief descriptions of this method have been given in the introductory papers of Part 1. One of the classic applications of this method for the solution of arbitrarily shaped hollow waveguides is detailed in a paper by Davies and Muilwyk [2.1]. The finite-difference equations are implemented iteratively in an overrelaxation procedure in conjunction with the use of a variational Rayleigh quotient to improve the solution for the eigenvalue. The method was limited to the computation of the dominant mode. It was extended by Beaubien and Wexler to the computation of higher order modes (Paper 2.1). Later, the same authors further improved the algorithm by introducing unequal arm operators [2.2]. It can easily be recognized by virtue of the transverse resonance concept that the analysis of waveguide discontinuities can be reduced to the analysis of a waveguide cross section, provided that the discontinuity is uniform in some transverse direction. This idea was used by Muilwyk and Davies to extend the applicability of their finite-difference (FD) technique [2.1] to the analysis of such a class of discontinuities [2.3]. A more general approach to the FD analysis of waveguide discontinuities using the so-called boundary relaxation was presented a few years later by Mur [2.4].

The FDM can be formulated also in conjunction with a variational [1.2] or, in general, an integral formulation of the problem. A paper by Albani and Bernardi (Paper 2.2) shows the possibility of solving inhomogeneously loaded guiding and resonant structures by direct discretization of Maxwell's equations in integral form.

Next, a recent paper by Bierwirth *et al.* (Paper 2.3) is included. This paper presents an application of the FDM to the analysis of rectangular dielectric waveguide structures. A formulation in terms of the transverse H-field components is developed, which avoids one of the troublesome phenomena associated with FD application to inhomogeneous structures, i.e., the existence of spurious nonphysical solutions. The existence of complex modes in these structures is also pointed out.

Pioneered by Yee [2.5] in 1966, the time-domain formulation of the FDM (FD-TD method) has been attracting increasing attention from researchers. This is not only due to the possibility of analyzing transients, but also because it is possible to perform wide-band analyses with a single computation process.

This method has the problem of treating unbounded structures, as it is necessary to simulate open space with proper absorbing boundary conditions [2.6]. Microstrip dispersion has been evaluated recently by Zhang *et al.* [2.7], using the FD-TD method, by solving the problem twice, once with electric wall and once with magnetic wall terminations. No such problem arises in closed electromagnetic structures. Choi and Hoefer recently showed the applicability of the FD-TD method to two- and three-dimensional eigenvalue problems (Paper 2.4). Gwarek has developed a two-dimensional version of the FD-TD method for the analysis of arbitrarily shaped planar circuits [2.8] and in his latest paper (Paper 2.5) a comparison with other analysis techniques for planar circuits is given. It is to be noted that the FD-TD method has some close similarities with the transmission-line matrix method of Part 3 as discussed in Papers 2.4 and 2.5 and in [2.9], [2.10].

References

[2.1] J. B. Davies and C. A. Muilwyk, "Numerical solution of uniform hollow waveguides with boundaries of arbitrary shapes," *Proc. IEE*, vol. 113, pp. 277–284, Feb. 1966.

[2.2] M. J. Beaubien and A. Wexler, "Unequal-arm finite-difference operators in the positive-definite successive overrelaxation (PDSOR) algorithm," *IEEE Trans. Microwave Theory Tech.*, vol. MTT-18, pp. 1132–1149, Dec. 1970.

[2.3] C. A. Muilwyk and J. B. Davies, "The numerical solution of rectangular waveguide junctions and discontinuities of arbitrary cross section," *IEEE Trans. Microwave Theory Tech.*, vol. MTT-15, pp. 450–455, Aug. 1967.

[2.4] G. Mur, "A finite difference method for the solution of electromagnetic waveguide discontinuity problems," *IEEE Trans. Microwave Theory Tech.*, vol. MTT-22, pp. 54–57, Jan. 1974.

[2.5] K. S. Yee, "Numerical solution of initial boundary value problems involving Maxwell's equations in isotropic media," *IEEE Trans. Antennas Propagat.*, vol. AP-14, pp. 302–307, May 1966.

[2.6] G. Mur, "Absorbing boundary conditions for the finite-difference approximation of the time-domain electromagnetic-field equations," *IEEE Trans. Electromag. Compat.*, vol. EMC-23, pp. 377–382, Nov. 1981.

[2.7] X. Zhang, J. Fang, and K. K. Mei, "Calculations of the dispersive characteristics of microstrips by the time-domain finite difference method," *IEEE Trans. Microwave Theory Tech.*, vol. MTT-36, pp. 261–267, Feb. 1988.

[2.8] W. K. Gwarek, "Analysis of an arbitrarily-shaped planar circuit—A time-domain approach," *IEEE Trans. Microwave Theory Tech.*, vol. MTT-33, pp. 1067–1072, Oct. 85.

[2.9] P. B. Johns, "On the relationship between TLM and finite-difference methods for Maxwell's equations," *IEEE Trans. Microwave Theory Tech.*, vol. MTT-35, pp. 60–61, Jan. 1987.

[2.10] W. K. Gwarek and P. B. Johns, "Comments on 'On the relationship between TLM and finite-difference methods for Maxwell's equations'," *IEEE Trans. Microwave Theory Tech.*, vol. MTT-35, pp. 872–873, Sept. 1987.

An Accurate Finite-Difference Method for Higher Order Waveguide Modes

M. J. BEAUBIEN AND A. WEXLER, MEMBER, IEEE

Abstract—**The study of new waveguide shapes requires an accurate knowledge of their higher order modes for bandwidth consideration, waveguide discontinuity analysis, and multimode launching and propagation studies.**

The finite-difference method solved by successive overrelaxation is a very accurate and general technique for dominant mode solution. The method makes minimum demand on computer store, the number of storage locations required being equal to the order of the matrix defining the system of linear equations used. Convergence criteria require that this matrix be positive semidefinite. For modes higher than the dominant, the method fails to converge as this condition is violated.

Solution is obtained by redefining the problem such that the matrix is positive semidefinite for all modes. An algorithm is described which produces one row at a time of the new matrix, as required for successive overrelaxation, and thus reserves almost all computer store for the eigenvector. The eigenvector elements give field potentials at discrete points in the guide cross section. Iteration for higher order modes is successfully applied with all the computational benefits realized by previous finite-difference schemes for the dominant mode. Examples studied are the rectangular, circular, asymmetric ridge, and lunar guides. Results for higher order cutoff wave numbers are usually accurate to within 0.1 percent.

I. INTRODUCTION

TO DESIGN new waveguide shapes and to assess their performance in larger systems, propagation constants and field patterns must be found. Very often, solution of the dominant mode alone is not sufficient for the designer. The following examples are typical of studies requiring knowledge of higher order modes: bandwidth considerations—the upper limit is set by the inception of higher order modes; waveguide discontinuity analysis—a set of modes is required to solve scattering problems [1]; and multimode launching and propagation studies—applications include prediction of undesirable linear accelerator resonances, multimode techniques in aerial improvement [2], etc. For the same reasons that they are needed in studies of systems involving standard guides (e.g., rectangular, circular, etc.), higher order modes in arbitrarily shaped guides must be known.

One of the most general techniques for the numerical solution of the Helmholtz equation is the finite-difference method. In a very enlightening paper, Davies and Muilwyk [3] illustrated the power and usefulness of the method. Rather than attempt to solve for the field by classical means, the potential function ϕ was solved only at a finite number of discrete points in the waveguide cross section, the more points employed the higher the accuracy. This formulation resulted in a matrix eigenvalue problem. Using successive overrelaxation [4], with up to 20 000 discrete points, they computed fields and cutoff wavenumbers for guide shapes ranging from familiar ones to eccentric shapes such as the club-shaped guide. Typical error in cutoff wavenumber was less than 0.1 percent. The disadvantage of their technique is that it fails to converge to higher eigenvalues. Therefore it is applicable to the dominant mode only and, in cases of certain guide symmetries, to one or two higher order modes as well.

Because of computer storage economies available to successive overrelaxation, very large numbers of discrete points are feasible. On the other hand, direct methods [5], which have the advantage of giving higher order modes, had to be rejected because high computer store requirements limit the number of discrete node points to less than 150 in a fairly large computer. The resulting accuracy was considered unsatisfactory in comparison to that attainable using several thousand points.

There is a remaining possibility for large systems. By keeping the current solution iterate orthogonal to all previous lower order solutions [5, pp. 112–113], matrix powering [5, pp. 103–105] can be employed for higher order modes. This has two major disadvantages. In the first place, as each mode solution depends upon all previously determined solutions, inaccuracies accumulate and results become increasingly unreliable as mode order increases. Secondly, potential values at each node point, for each mode, must be stored with a consequent reduction in maximum system size and resulting accuracy. Otherwise, a large number of magnetic disk transfers are required, resulting in excessive computing time. For these reasons, this approach was rejected as well.

This paper describes a method employing successive overrelaxation that allows the determination of higher order modes with all the computational advantages exhibited by Davies and Muilwyk for the dominant mode.

A. Potentials and Fields

Consider a waveguide, uniform along the z direction, with an arbitrarily shaped cross section located in the x-y plane. Let the waveguide be loaded homogeneously with isotropic lossless material. Assume that modes vary as $e^{j\omega t-\gamma z}$ where γ is the propagation constant.

All TE and TM mode fields may be derived from the scalar potential function ϕ which satisfies the two-dimensional Helmholtz equation

$$(\nabla_t^2 + k_c^2)\phi = 0 \tag{1}$$

Manuscript received June 10, 1968; revised September 3, 1968. Results of this paper were presented in part at the 1968 International Microwave Symposium, Detroit, Mich. This research was supported by the National Research Council of Canada and Atomic Energy of Canada, Ltd.

The authors are with the Numerical Applications Group, Department of Electrical Engineering, University of Manitoba, Winnipeg, Manitoba, Canada.

Reprinted from *IEEE Trans. Microwave Theory Tech.*, vol. MTT-16, no. 12, pp. 1007–1017, Dec. 1968.

where k_c is the cutoff wave number. For any given mode these constants are related through

$$k_c^2 - \gamma^2 = k^2 \tag{2}$$

where

$$k = \omega\sqrt{\mu\epsilon}. \tag{3}$$

μ and ϵ are, respectively, the permeability and permittivity of the isotropic medium filling the guide.

The potential function is subject to certain boundary contions. For TM modes,

$$\phi = 0 \tag{4}$$

at a perfectly conducting surface. This is a Dirichlet condition. For TE modes the Neumann condition is that

$$\frac{\partial\phi}{\partial n} = 0 \tag{5}$$

along such a surface.

Once solutions of (1) are obtained subject to boundary conditions (4) and (5), the electric and magnetic fields may then be derived from the following equations [6]:

$$\text{TM modes:}\quad \overline{E}_t = -\frac{\gamma}{k_c^2}\nabla_t\phi,\quad E_z = \phi,\quad \overline{H}_t = -\frac{j\omega\epsilon}{k_c^2}\bar{u}_z \times \nabla_t\phi,\quad H_z = 0 \tag{6}$$

$$\text{TE modes:}\quad \overline{E}_t = \frac{j\omega\mu}{k_c^2}\bar{u}_z \times \nabla_t\phi,\quad E_z = 0,\quad \overline{H}_t = -\frac{\gamma}{k_c^2}\nabla_t\phi,\quad H_z = \phi. \tag{7}$$

B. Finite-Difference Equations

An arbitrary cross section may be specified to the computer as a sequence of straight lines, circular arcs, polynomials or numerical data. The computer then interprets the boundary in a discrete fashion and not as a continuous smooth curve. Computations proceed with several mesh sizes and so the program discretizes the boundary in the first instance with the largest mesh size, e.g., Fig. 1. The computer superimposes a mesh over the guide cross section and at each node level it marks the left- and right-hand boundaries of the guide and other internal conducting surfaces as well. Since these marks are made at the node point nearest the actual boundary, the mesh is fitted neither internally nor externally, but in a somewhat random fashion about the true boundary. It was felt that these random perturbations to the true boundary would tend to cancel out their effects on the eigenvalue computation.

It is now easy to see the equivalent waveguide shape that

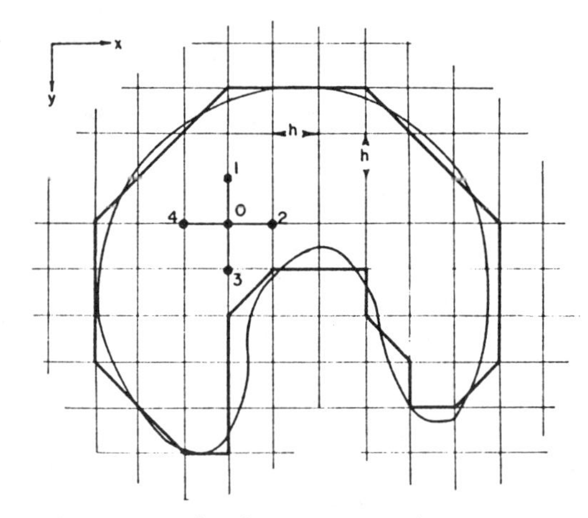

Fig. 1. Polygon approximation to a waveguide boundary for a large mesh size and a typical five-point, finite-difference operator.

the computer analyzes by noting the limits beyond which no nodes exist. Because of local perturbations to the actual waveguide, the boundary of the equivalent guide is defined by horizontal or vertical links between nodes or 45° diagonals across meshes. As a result, the boundary is defined by an appropriate polygon through boundary nodes. This leads to considerable simplification in application of boundary conditions.

A square mesh, of side length h, is presumed drawn over the guide cross section (e.g., Fig. 1). Equation (1) is written in its five-point finite-difference form [4, pp. 3–5]. Thus

$$-\phi_1 - \phi_2 - \phi_3 - \phi_4 + (4 - \lambda)\phi_0 = 0 \tag{8}$$

where the eigenvalue λ is

$$\lambda = (k_c h)^2. \tag{9}$$

Equation (8) gives ϕ_0, a typical node potential of the mesh in terms of the immediately adjacent node potentials ϕ_1 to ϕ_4. The error involved in this approximation is usually of order h^2; near reentrant corners, errors are higher due to field singularities [7], [8]. Now, instead of defining the potential $\phi(x, y)$ at all points (x, y), the discrete potential function ϕ_i is defined only at points i. Equation (8) is evaluated at each node point with appropriate modification to include boundary conditions. Because we are left with a set of simultaneous linear homogeneous equations, a nontrivial solution can occur only when the determinant of the matrix vanishes. This occurs only for particular values of the eigenvalue λ with corresponding eigenvector

$$\Phi = \begin{bmatrix} \phi_1 \\ \phi_2 \\ \cdot \\ \cdot \\ \phi_N \end{bmatrix} \tag{10}$$

whose elements give field potentials at the N discrete points in the guide cross section. The square N-order matrix of coefficients is denoted A.

The Dirichlet condition $\phi = 0$ is particularly easy to apply. If, say, node 1 (Fig. 1) of the five-point operator is on a conducting boundary, the ϕ_1 term in (8) vanishes.

For Neumann boundary conditions $\partial\phi/\partial n = 0$ the polygon boundary nodes appear as variables in the matrix equation. To approximate Neumann boundary conditions, a simplified approach is employed. If a node is outside the boundary, on an arm of an operator with central node on the boundary, its potential is set equal to the potential on the opposite arm distance $2h$ away. Because of the polygon fit employed, and exclusion of channels of width less than h as well, all possible cases are catered for in this way.

One advantage in approximating the boundary as we have done is that interpolation between boundary points and adjacent nodes is unnecessary. Computing economies are considerable, and this is of great importance. However, this method of defining the boundary is not an integral part of the theory in Part II and so other more accurate and complicated methods, such as the use of unequal arm operators [7, pp. 262–266], may be employed.

C. Matrix Formulation

With the finite-difference equation defined for interior points (8) and with rules for imposing Dirichlet or Neumann boundary conditions, a finite-difference equation may be written at each node point. As an example, refer to Fig. 2 which gives a waveguide of the shape shown or the perturbed equivalent of some slightly different shape. For the Dirichlet problem, the solution of which gives a TM mode set, we wish to solve for the internal nodes numbered 1 to 8.

Writing the finite-difference equation at each node point (the computer does this automatically), with due regard to the condition $\phi=0$ on the boundary, we get the following matrix eigenvalue problem:

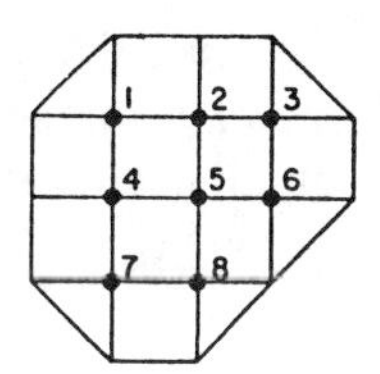

Fig. 2. A TM problem example with Dirichlet boundary conditions.

$$\begin{bmatrix} 4-\lambda & -1 & 0 & -1 & 0 & 0 & 0 & 0 \\ -1 & 4-\lambda & -1 & 0 & -1 & 0 & 0 & 0 \\ 0 & -1 & 4-\lambda & 0 & 0 & -1 & 0 & 0 \\ -1 & 0 & 0 & 4-\lambda & -1 & 0 & -1 & 0 \\ 0 & -1 & 0 & -1 & 4-\lambda & -1 & 0 & -1 \\ 0 & 0 & -1 & 0 & -1 & 4-\lambda & 0 & 0 \\ 0 & 0 & 0 & -1 & 0 & 0 & 4-\lambda & -1 \\ 0 & 0 & 0 & 0 & -1 & 0 & -1 & 4-\lambda \end{bmatrix} \begin{bmatrix} \phi_1 \\ \phi_2 \\ \phi_3 \\ \phi_4 \\ \phi_5 \\ \phi_6 \\ \phi_7 \\ \phi_8 \end{bmatrix} = 0 \tag{11}$$

or

$$(A - \lambda I)\Phi = 0 \tag{12}$$

where I is the unit matrix. It is required to solve for one or more λ and corresponding Φ values.

The Neumann problem for TE modes requires the solution of node potentials on the boundary and so the matrix order increases to 19 in this particular case.

There are two significant features of this system. First of all, the square matrix is sparse; i.e., there is a high proportion of zero elements. As the mesh size decreases the ratio of zero to nonzero elements becomes increasingly larger. This has significance relative to the computer and the technique employed for solution of the eigenvalues and eigenvectors of A. Because the matrix is sparse, there is no sense in storing anything but the nonzero elements. An even more economical method, in both storage and time, is to use successive overrelaxation. This iterative method of solution requires only one equation (matrix row) at a time and successively corrects each element of Φ in terms of all other elements in an attempt to satisfy each equation. This procedure is repeated over all equations a number of times until the solution is deemed to be sufficiently accurate. The computer therefore generates the matrix row by row as required. As no row contains more than five nonzero elements, nearly all computer store can then be saved for the vector Φ. For a moderately large computer this makes it possible to locate somewhere around 20 000 node points over the waveguide cross section, representing a matrix of order 20 000! This results in highly accurate boundary and field definition with a consequent accurate determination of cutoff wavenumber and fields that are unmatched by any direct method.

The second feature of matrix $(A-\lambda I)$ is that it is symmetric, a characteristic affecting convergence of the successive overrelaxation method. For the Neumann problem, symmetry is slightly upset through equations generated by points on the boundary. This occasionally causes iteration to behave erratically, and sometimes fail, for coarse meshes. The lack of symmetry is of decreasing importance as the mesh becomes finer. This is because the number of internal nodes is approximately proportional to h^{-2} while the number of boundary points increases roughly as h^{-1}. The behavior of such almost-symmetric matrices is somewhat similar to symmetric ones.

D. Solution of the Five-Point Operator System

To solve (11) or (12), i.e., to find eigenvalues of A, it is first necessary to assume a value of λ. Let us therefore write

$$B = A - \lambda I \tag{13}$$

where λ is an estimate to an eigenvalue. We therefore wish to solve the matrix equation

$$B\Phi = 0 \tag{14}$$

by successive overrelaxation. A nontrivial solution for Φ

(i.e., $\Phi \neq 0$) can exist only when the determinant of B vanishes—in other words when λ is an eigenvalue. Because λ is not as yet known, the equation cannot be solved. However, there is meaning in attempting a solution of the simultaneous equations (14) through formal application of successive overrelaxation.

We will see in Section II-B that if the λ estimate is slightly less than the correct lowest eigenvalue, all ϕ_i tend to zero, the trivial solution, as iteration proceeds. If the λ estimate is slightly more than the eigenvalue, Φ_1 increases without limit and the solution fails to converge at all. This does not happen quickly but only gradually as Φ is scanned a number of times. However, as Φ is scanned over and over, the elements tend to assume the correct ratios relative to one another, although they may all be growing or diminishing simultaneously. Thus the first approximation $\Phi^{(1)}$ to the shape of the correct field pattern emerges gradually for the initial eigenvalue estimate $\lambda^{(1)}$. Using $\Phi^{(1)}$, an improved eigenvalue $\lambda^{(2)}$ is calculated from the Rayleigh quotient [7, pp. 74–75]:

$$\lambda^{(r+1)} \cong \frac{\Phi^{(r)T} A \Phi^{(r)}}{\Phi^{(r)T} \Phi^{(r)}}, \tag{15}$$

where r is the number of the successive eigenvalue estimate, and the transpose is indicated by T. The advantage of (15) is that it is stationary at the solution point and so approximate potential values ϕ_i yield a more accurate eigenvalue estimate. Using the new eigenvalue, a second and better estimate to Φ is found through (14), and so on until sufficient accuracy is obtained. This sequence of operations is described in [9] and [10].

Notice that at the solution point of the first mode (or eigenvalue of A), $\det(B)=0$. Therefore the lowest eigenvalue of B is zero under this condition. Also note that if one subtracts any number from all diagonal elements of matrix B, all eigenvalues of B decrease by that amount. This is easily seen by letting p be an eigenvalue of B. Therefore

$$B\Phi = p\Phi. \tag{16}$$

Subtract a from diagonal terms of B by subtracting aI from both sides of (16). Thus

$$(B - aI)\Phi = (p - a)\Phi, \tag{17}$$

and it is clear that the new eigenvalue $p-a$ is shifted a units to the left of p.

When higher order eigenvalues are searched for, λ must be increased and so the eigenvalue spectrum of B shifts further to the left. At least one eigenvalue then goes negative. A matrix is positive definite only if all if its eigenvalues are positive [11], and so B is no longer positive definite under these conditions. It is known [9, p. 240] that (14) converges with successive overrelaxation if and only if B is a positive semidefinite matrix. It follows then that convergence cannot be obtained for higher order modes by the method outlined in this section.

II. Positive Definiting

To allow successive overrelaxation to succeed, the problem was reformulated. Define a new matrix

$$C = B^T B. \tag{18}$$

It is easy to see that C is symmetric even when B is not. We therefore have

$$C\Phi = 0 \tag{19}$$

in place of (14). Since

$$\begin{aligned} \det(C) &= \det(B^T)\det(B) \\ &= (\det(B))^2, \end{aligned} \tag{20}$$

(19) is satisfied by the same eigenvalues λ and corresponding eigenvectors that satisfy (12) and (14).

The prime advantage of solving (19) in lieu of (14) is that C is positive semidefinite for correct λ values and positive definite otherwise. To show this let x be any real nonzero vector (column matrix). Therefore

$$x^T x > 0. \tag{21}$$

Define the transformation

$$x = By \tag{22}$$

where B is a real symmetric or unsymmetric matrix. By reversing the transformation, it is clear that y is a real vector when B is nonsingular. Substituting (22) into (21) we have the quadratic form [11, pp. 16–20]

$$y^T B^T B y = y^T C y > 0. \tag{23}$$

Since C is singular at the solution point, it is clear that C is positive definite for all approximations to λ and semidefinite at the exact solution point. Therefore, (19) is always guaranteed to converge by successive overrelaxation—even for higher order modes.

Convergence occurs because, as the first eigenvalue of A is approached from the left, the lowest eigenvalue of C approaches the origin from the right. C is singular at the solution point. As λ is further increased, eigenvalues of C do not go negative (as do those of B), but instead reverse and return along the positive real axis. In other words, the eigenvalue spectrum of C folds back upon itself at the origin, thus always remaining positive.

A. The Positive Definite Operator

In order to solve (19), one must know matrix C. Certainly, direct multiplication of $B^T B$ is impossible for the large systems envisaged. No computer can keep $N^2 = 20\,000^2$ numbers in its immediate-access store. $N^2 = 150^2$ is an approximate upper limit for a reasonably large machine. To use successive overrelaxation, as with the B system, one must do row-by-row matrix generation in order to reserve almost all store for Φ. Rather than deriving B and then producing C, we want an automatic method for finding any one row of C directly without recourse to B. All the advantages of successive overrelaxation (accuracy and speed) exhibited by Davies and Muilwyk for the dominant mode can then be applied to the determination of higher order modes as well. This turns out to be possible for the finite-difference form of the Helmholtz equation, and an algorithm was developed to do this while paying due regard to boundary conditions.

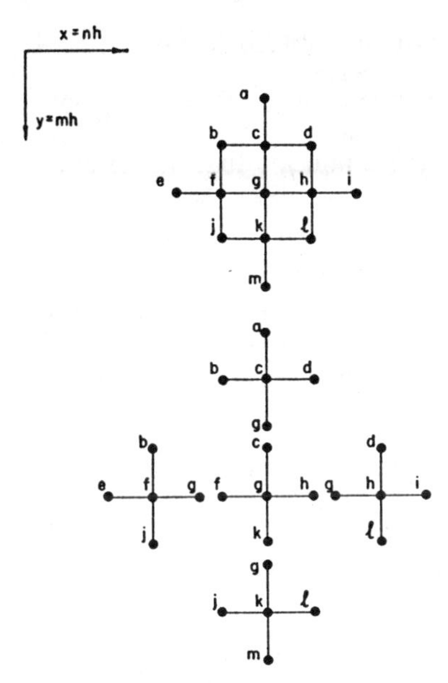

Fig. 3. Five-point positive definite finite-difference operator and its constituent five-point operators.

In the first place (8) is written in a more general and convenient form. For the cross centered at c (Fig. 3), we have

$$b_{ca}\phi_a + b_{cb}\phi_b + b_{cc}\phi_c + b_{cd}\phi_d + b_{cg}\phi_g = 0. \tag{24}$$

Double subscripting of coefficients is necessary. The first subscript gives the operator's central node number, and the second subscript locates any other relevant node number on that operator. Thus, elements of matrix B are denoted b_{ij}. Element $b_{cc}=4-\lambda$ and others are either -1 or are set by boundary condition requirements as previously discussed.

The typical element c_{gs} of C is formed by premultiplying column s of B by the transpose of column g. It is given by

$$c_{gs} = \sum_{p=1}^{N} b_{pg}b_{ps} \tag{25}$$

where N is the order of the matrix. This element c_{gs} may or may not be zero depending upon certain physical considerations.

Nonzero contributions to the sum (25), and hence to the element c_{gs} of the C matrix, are made when the pth elements of column g (b_{pg}) and column s (b_{ps}) are both nonzero. For this to happen it is necessary for the central node of the operator centered at p to be directly connected to nodes g and s. In other words, g and s must be located on arms of operator p—or, as a special case, g and/or s could be at p.

Conversely, for any given g, node p must be located on nodes of the g operator only. These are the only rows having nonzero gth elements. For example, the center cross of Fig. 3 must have node p coinciding with any of c, f, g, h, or k for contributions to c_{gs}.

For b_{pg} and b_{ps} to both exist on the same row p, g and s must be located on the same operator p. Therefore, the only nodes that need to be considered are all those on crosses centered on the nodes of the gth cross.

Fig. 3 illustrates the only nodes in the guide cross section which influence the form of the gth row of matrix C. One need look no further! Having an estimate to λ, for each g, we wish to generate a row of C. Notice that the maximum number of elements in a row of B is five while in C it is thirteen. Since the order of C is the same as that of B, the new matrix must be denser.

It is clear that only thirteen nodes are involved in the generation of any row. We must now see how they are involved; i.e., the rules for use of the positive definite finite-difference operator must be established.

For convenience only, let us assume that nodes are numbered left to right in top to bottom sequence over the guide cross section. Therefore, the node numbers corresponding to the lettered points on the positive definite operator are in such a sequence that

$$a < b < c < \cdots < l < m. \tag{26}$$

Now, to find the gth row of C, the thirteen columns of B corresponding to nodes $a, b, c, \cdots, m$ are premultiplied by the nonzero elements of the transposed gth column of B. Recall that these are the only ones we need be concerned with. Postmultiplying by the vector consisting of potentials ϕ_a to ϕ_m, the gth equation results.

$$[b_{cg} \quad b_{fg} \quad b_{gg} \quad b_{hg} \quad b_{kg}] \begin{bmatrix} b_{ca} & b_{cb} & b_{cc} & b_{cd} & 0 & 0 & b_{cg} & 0 & 0 & 0 & 0 & 0 & 0 \\ 0 & b_{fb} & 0 & 0 & b_{fe} & b_{ff} & b_{fg} & 0 & 0 & b_{fj} & 0 & 0 & 0 \\ 0 & 0 & b_{gc} & 0 & 0 & b_{gf} & b_{gg} & b_{gh} & 0 & 0 & b_{gk} & 0 & 0 \\ 0 & 0 & 0 & b_{hd} & 0 & 0 & b_{hg} & b_{hh} & b_{hi} & 0 & 0 & b_{hl} & 0 \\ 0 & 0 & 0 & 0 & 0 & 0 & b_{kg} & 0 & 0 & b_{kj} & b_{kk} & b_{kl} & b_{km} \end{bmatrix} \begin{bmatrix} \phi_a \\ \phi_b \\ \phi_c \\ \phi_d \\ \phi_e \\ \phi_f \\ \phi_g \\ \phi_h \\ \phi_i \\ \phi_j \\ \phi_k \\ \phi_l \\ \phi_m \end{bmatrix} = 0. \tag{27}$$

Performing the operations indicated in (27) we get

$$c_{ga}\phi_a + c_{gb}\phi_b + c_{gc}\phi_c + \cdots + c_{gm}\phi_m = 0 \quad (28)$$

which is the gth equation of the matrix C. The coefficients are given by the following:

$$\begin{aligned} c_{ga} &= b_{cg}b_{ca} \\ c_{gb} &= b_{cg}b_{cb} + b_{fg}b_{fb} \\ c_{gc} &= b_{cg}b_{cc} + b_{gg}b_{gc} \\ c_{gd} &= b_{cg}b_{cd} + b_{hg}b_{hd} \\ c_{ge} &= b_{fg}b_{fe} \\ c_{gf} &= b_{fg}b_{ff} + b_{gg}b_{gf} \\ c_{gg} &= b_{cg}^2 + b_{fg}^2 + b_{gg}^2 + b_{hg}^2 + b_{kg}^2 \\ c_{gh} &= b_{gg}b_{gh} + b_{hg}b_{hh} \\ c_{gi} &= b_{hg}b_{hi} \\ c_{gj} &= b_{fg}b_{fj} + b_{kg}b_{kj} \\ c_{gk} &= b_{gg}b_{gk} + b_{kg}b_{kk} \\ c_{gl} &= b_{hg}b_{hl} + b_{kg}b_{kl} \\ c_{gm} &= b_{kg}b_{km}. \end{aligned} \quad (29)$$

These elements are not all consecutive along the gth row of C. Referring to Fig. 3, c_{ga} may be followed by any number of zeros before c_{gb}. Also, c_{gb}, c_{gc}, and c_{gd} are consecutive along the gth row. c_{gd} may then be followed by a large number of zeros, and so on.

In summary, the elements of B are determined by boundary conditions. Those relevant matrix elements in the vicinity of node g are evaluated, and then the above elements of the gth row of $C = B^T B$ are found directly through (29). Having the gth equation, ϕ_g is solved by successive overrelaxation, in terms of the other potentials in the equation. This is repeated sequentially for each g corresponding to the potentials $\phi_1, \phi_2, \cdots, \phi_N$ which are solved as described in the following section. We have now gone from a five-point to a thirteen-point operator.

B. Solution by Successive Overrelaxation

Assuming that an eigenvalue λ is known, the problem is the solution of (19). Our main concern is the convergence of the elements of Φ to the correct field shape with any arbitrary accompanying factor.

Successive overrelaxation [4] for (19) takes the form

$$\begin{aligned} \phi_g^{(r+1)} = (1-\omega)\phi_g^{(r)} - \frac{\omega}{c_{gg}} \{ & c_{ga}\phi_a^{(r+1)} \\ & + c_{gb}\phi_b^{(r+1)} + \cdots + c_{gf}\phi_f^{(r+1)} + c_{gh}\phi_h^{(r)} \\ & + c_{gi}\phi_i^{(r)} + \cdots + c_{gm}\phi_m^{(r)} \} \end{aligned} \quad (30)$$

where r is the rth iteration. ω is an acceleration factor with allowable range $0<\omega<2$. An appropriate choice of ω causes rapid convergence of the iteration process.

Equation (30) is applied at each node of the mesh in sequence, the superscript r giving the number of complete applications of it to all internal nodes of the mesh. Equation (30) may be written in matrix form as

$$\Phi^{(r+1)} = \mathcal{L}_{\omega,\lambda}\Phi^{(r)} = \{\mathcal{L}_{\omega,\lambda}\}^r \Phi^{(0)}. \quad (31)$$

Subscripts ω and λ indicate that the iteration matrix $\mathcal{L}$ has a functional dependence upon these parameters. $\mathcal{L}$ is derived directly from C [4, p. 59]. The convergence theorem [9, p. 240], [4, pp. 77–78] states, in effect, that the modulus of every eigenvalue μ_i of $\mathcal{L}$ is less than unity if and only if a symmetric matrix C is positive definite and $0<\omega<2$. Expressing $\Phi^{(0)}$ as a sum of eigenvectors $a_i\mathbf{x}_i$ of $\mathcal{L}$, then substituting into (31), we obtain

$$\Phi^{(r+1)} = a_1\mu_1{}^r\mathbf{x}_1 + a_2\mu_2{}^r\mathbf{x}_2 + \cdots + a_N\mu_N{}^r\mathbf{x}_N. \quad (32)$$

Assuming that μ_1 is the greatest eigenvalue in modulus (but $|\mu_1|<1$), then as iteration proceeds the first term of (32) predominates, i.e.,

$$\Phi^{(r+1)} = \lim_{r\to\infty} a_1\mu_1{}^r\mathbf{x}_1. \quad (33)$$

For the Φ iterate to stabilize, it is clear that at a solution point

$$\Phi = a_1\mathbf{x}_1 \quad (34)$$

with $\mu_1 = 1$. This can only happen when the eigenvalue λ of A is correctly chosen; i.e., a solution cannot exist otherwise. Notice that in theory, neglecting effects due to roundoff error, (30) is stationary for correct ϕ_a to ϕ_m for *all* real ω.

We can now see why iteration of (14) causes Φ to converge to the correct shape while continuously diminishing or growing depending on whether the assumed eigenvalue is slightly less than or greater than the lowest eigenvalue λ_1. From the convergence theorem, if λ is slightly less than λ_1, then μ_1 should be slightly less than 1; thus successive iterates of Φ must tend to zero while converging to the correct shape. The opposite is true for an assumed λ slightly greater than λ_1. This argument is based on slight perturbation to the correct eigenvalue. If $\lambda \gg \lambda_1$ or $\lambda \ll \lambda_1$, we cannot be sure what will happen.

The remaining problem is to determine an acceleration parameter ω that will cause maximum rate of convergence of the initial approximation $\Phi^{(0)}$ to the solution Φ. We have used a search technique that seems reasonably successful. The displacement column vector

$$\boldsymbol{\delta}^{(r)} = \Phi^{(r)} - \Phi^{(r-1)} \quad (35)$$

may be written in the form

$$\begin{aligned} \boldsymbol{\delta}^{(r)} = \mathcal{L}_{\omega,\lambda}\boldsymbol{\delta}^{(r-1)} &= d_1\mu_1{}^{r-1}\mathbf{x}_1 + d_2\mu_2{}^{r-1}\mathbf{x}_2 + \cdots \\ &+ d_N\mu_N{}^{r-1}\mathbf{x}_N \end{aligned} \quad (36)$$

and

$$\begin{aligned} \boldsymbol{\delta}^{(r-1)} = \mathcal{L}_{\omega,\lambda}\boldsymbol{\delta}^{(r-2)} &= d_1\mu_1{}^{r-2}\mathbf{x}_1 + d_2\mu_2{}^{r-2}\mathbf{x}_2 + \cdots \\ &+ d_N\mu_N{}^{r-2}\mathbf{x}_N. \end{aligned} \quad (37)$$

The norm of a column matrix can be defined in several different ways. We have taken the displacement norm $\|\boldsymbol{\delta}^{(r)}\|$ as the sum of the absolute values of the elements of $\boldsymbol{\delta}^{(r)}$.

For reasonably large r, the ratio of two successive displacement norms is

$$\frac{\|\delta^{(r)}\|}{\|\delta^{(r-1)}\|} \cong \frac{\|d_1\mu_1^{r-1}\mathbf{x}_1 + d_2\mu_2^{r-1}\mathbf{x}_2\|}{\|d_1\mu_1^{r-2}\mathbf{x}_1 + d_2\mu_2^{r-2}\mathbf{x}_2\|}. \tag{38}$$

If λ approaches the exact value, μ_1 approaches unity, and $d_1\mathbf{x}_1$ ($\mathbf{x}_1$ is proportional to Φ) approaches zero since there can be no displacement of the exact solution component of the iterates $\Phi^{(r)}, \Phi^{(r-1)}, \cdots, \Phi^{(0)}$. Thus $d_1=0$ for exact λ. In practice, however, λ is not exact and d_1 is a very small quantity near the solution. Therefore, minimization of (38) requires the magnitude of the subdominant eigenvalue μ_2, and others next to it, be minimum. ω affects the relative values of eigenvalues μ_i to some extent and so ω may be adjusted to attain approximate minimization of (38). For a specific λ, (38) is evaluated for ω, $\omega-\Delta\omega$, and $\omega+\Delta\omega$. The result that yields a minimum indicates the direction that ω should be perturbed to increase convergence rate.

There is a possible danger in the above approach that should be borne in mind. Because $\mathfrak{L}_{\omega,\lambda}$ is usually unsymmetric, its eigenvalues μ_i are in general complex. δ is always real, implying that some d_i terms are likely to be complex. δ may therefore not decrease monotonically, and the process given for improving the estimate of the optimum ω may be unreliable. Some work has been reported on the problem of optimum ω estimation when a certain matrix characteristic known as Property A is lacking [9, pp. 260–266]. However, more remains to be done. In spite of possible defects, the procedure worked to our entire satisfaction.

C. Sequence of Computations

To solve (27) an initial approximation $\Phi^{(0)}$ to Φ is chosen. For the TM case, all components of $\Phi^{(0)}$, or equivalently all internal node potentials, are set to unity. In the TE case, all node potentials are set to plus and minus one alternately. This initial alternating approximation is used since the uniform field is the first TE mode solution and represents error in the determination of higher order modes. Note from (32) that, since all $|\mu_i| < 1$ for the positive definite system, the vector diminishes. It must be rescaled after a number of iterations.

As each node is relaxed its displacement is stored and added to the previous displacement. A starting ω of 1.9 is used. As computation progresses ω is optimized as previously described.

To determine a number of starting eigenvalues, a direct method was used on the five-point difference equations of a coarse mesh consisting of about 60 nodes. Using any one of these eigenvalue approximations, successive overrelaxation was applied to (19). After a number of iterations the vector was substituted into the Rayleigh quotient (15) and the residual

$$R = \sqrt{\frac{(B\Phi)^T B\Phi}{\Phi^T\Phi}} \tag{39}$$

was calculated. B was generated only one row at a time. The residual R is a measure of how well the system of linear equations is satisfied by the potential column matrix. If R was very small (<0.001), and the Rayleigh quotient agreed with the eigenvalue approximation to some specified tolerance (within 0.001), relaxation on the coarse mesh was terminated. Experience has shown that these tolerances yield adequate results for reasonable computing times.

All desired coarse mesh eigenvectors, consisting of about sixty elements each, were generated in like manner and held in storage. Any one of these vectors, with the corresponding eigenvalue, was then chosen. The mesh size was halved and a Taylor series interpolation used to initialize potential values for the newly created nodes. Iteration was then performed on the finer mesh and an improved eigenvalue estimate was calculated using the Rayleigh quotient. The mesh size was halved again and the process repeated. This was done several times until the final mesh was 1/8 or 1/16 of its original size, depending on the accuracy required.

D. Reduction of Computation

The computer program superimposes a rectangular grid over the waveguide cross section and reserves one storage location for each node point. As the outer boundary of the grid is a rectangle, there is a small wastage of store near the borders consisting of nodes excluded from the waveguide. On the other hand, a considerable simplification is gained in a double-subscript notation which is convenient for the computer.

Let the location of nodes be given by coordinates x and y. Thus, they may also be located by stating coordinates in numbers of mesh lengths from the origin. In this way

$$m = y/h \tag{40}$$

and

$$n = x/h. \tag{41}$$

Thus, the potential at any node point p can be denoted $\phi_{m,n}$ rather than ϕ_p.

Rewriting (30) we get the following for a central node potential $\phi_{m,n}$:

$$\begin{aligned}\phi_{m,n} = {} & (1-\omega)\phi_{m,n} \\ & -\{b_{hg}[b_{hh}\phi_{m,n+1}+b_{hd}\phi_{m-1,n+1}+b_{hl}\phi_{m+1,n+1}+b_{hi}\phi_{m,n+2}] \\ & +b_{kg}[b_{kk}\phi_{m+1,n}+b_{kl}\phi_{m+1,n+1}+b_{kj}\phi_{m+1,n-1}+b_{km}\phi_{m+2,n}] \\ & +b_{fg}[b_{ff}\phi_{m,n-1}+b_{fj}\phi_{m+1,n-1}+b_{fb}\phi_{m-1,n-1}+b_{fe}\phi_{m,n-2}] \\ & +b_{cg}[b_{cc}\phi_{m-1,n}+b_{cd}\phi_{m-1,n+1}+b_{cb}\phi_{m-1,n-1}+b_{ca}\phi_{m-2,n}] \\ & +b_{gg}[b_{gh}\phi_{m,n+1}+b_{gk}\phi_{m+1,n}+b_{gf}\phi_{m,n-1}+b_{gc}\phi_{m-1,n}]\} \\ & \cdot\omega/(b_{cg}^2+b_{fg}^2+b_{gg}^2+b_{hg}^2+b_{kg}^2). \end{aligned} \tag{42}$$

The coefficients of the potential terms are left with their previous notation for logical simplicity. Note that terms of the denominator and numerator of (42) each consist of individual contributions from each constituent operator in Fig. 3. Each of these five-point operators has its own coefficient logic associated with it, depending upon its location and boundary conditions. It is now a simple matter to program the proper logic to evaluate (42) at all nodes of the mesh that are to be relaxed.

If $\phi_{m,n}$ is an internal node potential, located $>2h$ for the TM case and $\geq 2h$ for the TE case, from the boundary on all sides, (42) simplifies to

$$\begin{aligned}\phi_{m,n} = & (1-\omega)\phi_{m,n} \\ & -[-2(4-\lambda)(\phi_{m,n-1}+\phi_{m,n+1}+\phi_{m-1,n}+\phi_{m+1,n}) \\ & +2(\phi_{m-1,n-1}+\phi_{m-1,n+1}+\phi_{m+1,n-1}+\phi_{m+1,n+1}) \\ & +\phi_{m-2,n}+\phi_{m+2,n}+\phi_{m,n-2}+\phi_{m,n+2}]\omega/[(4-\lambda)^2+4]\end{aligned} \tag{43}$$

since $b_{gg}=b_{ff}=b_{cc}=b_{kk}=b_{hh}=4-\lambda$, and the other coefficients are -1. Only four multiplications are really necessary since other quantities may be defined before evaluating (43) at these internal nodes. This is a significant computational saving due to the large number of internal nodes compared with the number near the boundary.

Let us now consider $\phi_{m,n}$ located $\leq 2h$ from the boundary. For the TM case (42) simplifies to

$$\begin{aligned}\phi_{m,n} = & (1-\omega)\phi_{m,n} \\ & -\{b_{hg}[(4-\lambda)\phi_{m,n+1}-\phi_{m-1,n+1}-\phi_{m+1,n+1}-\phi_{m,n+2}] \\ & +b_{kg}[(4-\lambda)\phi_{m+1,n}-\phi_{m+1,n+1}-\phi_{m+1,n-1}-\phi_{m+2,n}] \\ & +b_{fg}[(4-\lambda)\phi_{m,n-1}-\phi_{m+1,n-1}-\phi_{m-1,n-1}-\phi_{m,n-2}] \\ & +b_{cg}[(4-\lambda)\phi_{m-1,n}-\phi_{m-1,n+1}-\phi_{m-1,n-1}-\phi_{m-2,n}] \\ & +(4-\lambda)[-\phi_{m,n+1}-\phi_{m+1,n}-\phi_{m,n-1}-\phi_{m-1,n}]\} \\ & \cdot\omega/[(4-\lambda)^2+b_{hg}^2+b_{kg}^2+b_{fg}^2+b_{cg}^2]\end{aligned} \tag{44}$$

since potentials other than the node potentials $\phi_{m-1,n}$, $\phi_{m+1,n}$, $\phi_{m,n-1}$, and $\phi_{m,n+1}$ of the star centers (Fig. 3) have the following coefficient logic. If a potential is zero, its coefficient in (42) is zero; if nonzero its coefficient is -1. Note that the boundary and exterior node potentials act as logical triggers, since they are at all times equal to zero. They were initially set to zero, and remain so since they are never relaxed. Thus the final form of (44) is completely controlled by the zero or nonzero property of the four potentials $\phi_{m-1,n}$, $\phi_{m+1,n}$, $\phi_{m,n-1}$, and $\phi_{m,n+1}$, which determine whether b_{cg}, b_{kg}, b_{fg}, and b_{hg}, respectively, are -1 or 0. Therefore there are only $2^4=16$ forms of the relaxation equation. Note that the case of $\phi_{m-1,n}=\phi_{m+1,n}=\phi_{m,n-1}=\phi_{m,n+1}=0$ represents an isolated node, and thus 15 cases are really present.

No more than effectively four multiplications is necessary for any one case. Also there is actually a reduction in labor in the vicinity of the boundary. The computer program chooses the correct relaxation equation by sensing the "ON-OFF" properties of the four controlling node potentials.

For the TE cases a similar reduction in computation for nodes in the vicinity of the boundary has not been effected, due to more complicated boundary conditions. This results in increased computing time for TE problems.

A considerable reduction in computing time could be effected by storing coefficients of (42) in the vicinity of the boundary, thus performing coefficient logic only once for a given mesh size. (Nodes remote from the boundary present no problem.) Storage requirements are moderate as the number of nodes near the boundary are a small fraction of the total for a fine mesh.

E. Results

Figs. 4 through 7 illustrate higher order mode field patterns in shapes ranging from the rectangle to the lunar guide. Constant E_z contours from TM modes and constant H_z contours for TE modes were plotted by the computer. The cases where there is some asymmetry in the fields could have been improved with more iterations. Also there are regions where nodal lines should join, but these regions are very flat, and a slight error in field distribution causes these nodal lines to shift significantly.

Tables I through IV illustrate the convergence trends of the various modes. Since modes in arbitrary shapes cannot have names corresponding to the number of field variations along coordinates, we have denoted them as TM_1, TM_2, $\cdots$, and TE_1, TE_2, $\cdots$, where the subscripts 1, 2, $\cdots$ refer to the increasing magnitude of the cutoff propagation constants. Table I lists the computed iterative solution of the eigenvalues of the finite-difference system of linear equations, the exact analytic eigenvalue solution [12] of these equations:

$$k_c a = \frac{2a}{h}\left[\sin^2\left(\frac{i\pi}{2m+2}\right)+\sin^2\left(\frac{k\pi}{2n+2}\right)\right]^{1/2} \tag{45}$$
$$(i = 1, 2 \cdots m;\ k = 1, 2, \cdots n)$$

and the continuous solution in a rectangle. In (45), m is the number of internal nodes in a row and n is the number of internal nodes in a column of the rectangle (Fig. 4). h corresponds to an exact mesh fit, and a, to the guide characteristic dimension. There is less than 0.015 percent disagreement between computed and analytic eigenvalues up to $h/8$ (4785 equations), and then the gap widens to 0.0915 percent at $h/16$ (19 425) equations. Thus successive overrelaxation used to solve the equations generated by the positive definite operator yields very accurate results up to 5000 equations, and then the accuracy deteriorates slightly with 20 000 equations due to roundoff errors. Therefore between 4000 and 10 000 equations could be considered optimum. These results are very encouraging, considering the fact that only a seven-digit word length was used, whereas Davies and Muilwyk used eleven. Note the convergence of the computed and analytic eigenvalues to the continuous answer as the mesh size becomes finer. Even for the eleventh mode, TM_{15}, there is very small deterioration in accuracy. These are the most accurate results expected since the mesh fitting is exact.

Degradation of accuracy with mode order must be expected due to the increased number of oscillations over the cross section relative to a given mesh size. This increased discretization error occurs because the mesh appears coarser for higher order modes. As an example of how far this method can be pressed, the thirtieth mode of the Dirichlet problem (TM_{37}) was investigated analytically. It is known that the continuous normalized cutoff wavenumber is $k_c a=26.5133$. Calculating $k_c a$ from (45) for 1161, 4785, and 19 425 internal nodes, the discretized $k_c a$ was found low by only 0.86, 0.22, and 0.05 percent, respectively.

The method was then tested on a classical curved model—the circle, Fig. 5. For TM modes, errors of less than 0.013

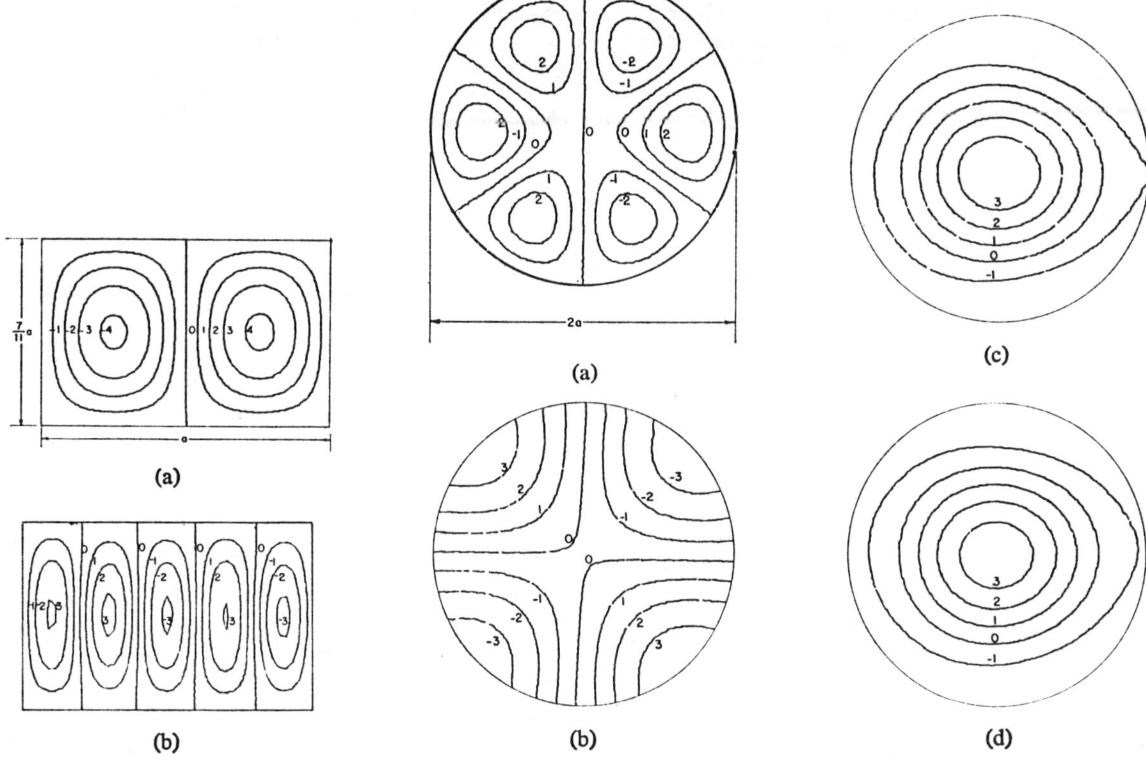

Fig. 4. TM modes in a rectangular waveguide. (a) TM_{11}. (b) TM_{15}.

Fig. 5. Modes in circular waveguide. (a) TM_{31}. (b) TE_{21}. (c) TE_{01}. (d) Improved TE_{01}.

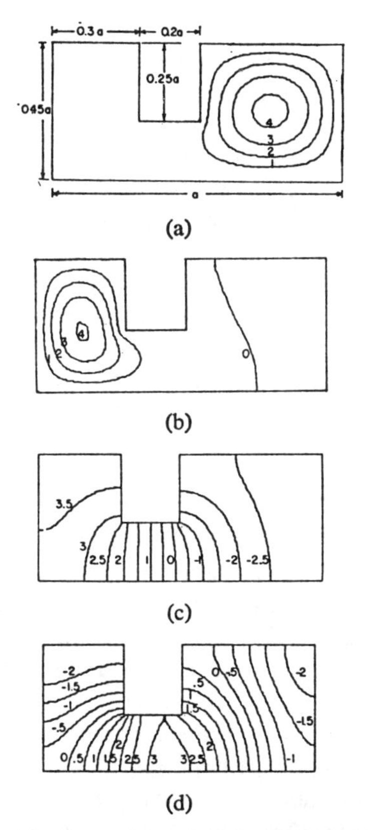

Fig. 6. Asymmetric ridge guide. (a) TM_1. (b) TM_2. (c) TE_1. (d) TE_2.

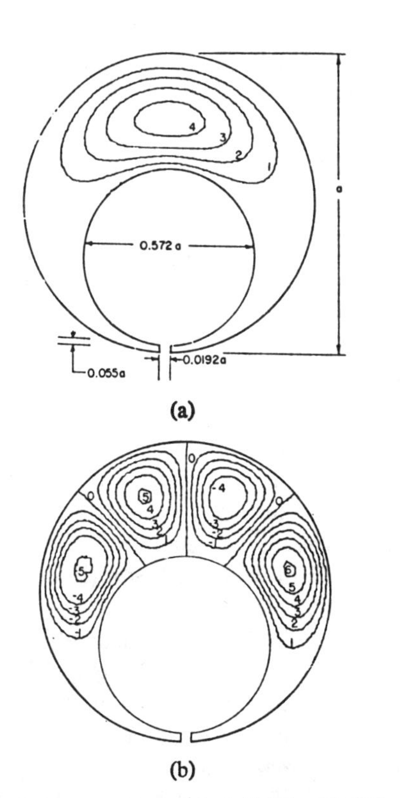

Fig. 7. Lunar waveguide. (a) TM_1. (b) TM_4.

TABLE I

CONVERGENCE OF TM MODES IN A RECTANGULAR GUIDE $h=\pi/7$, $a=11/7\pi$

		Mesh Size				
		h	$h/2$	$h/4$	$h/8$	$h/16$
TM_{11} iterative k_ca		5.8111	5.8412	5.8495	5.8513	5.8567
Fig. 6(a) analytic k_ca		5.8110	5.8415	5.8491	5.8510	5.8515
continuous k_ca	5.8516					
TM_{12} iterative k_ca		7.8982	7.9676	7.9854	7.9904	7.9976
Fig. 6(b) analytic k_ca		7.8982	7.9675	7.9848	7.9892	7.9903
continuous k_ca	7.9906					
TM_{23} iterative k_ca		13.215	13.540	13.623	13.644	13.660
Fig. 6(c) analytic k_ca		13.215	13.540	13.622	13.643	13.648
continuous k_ca	13.650					
TM_{15} iterative k_ca		15.216	16.146	16.385	16.449	16.465
Fig. 6(d) analytic k_ca		15.216	16.146	16.385	16.445	16.460
continuous k_ca	16.465					

TABLE II

k_ca FOR VARIOUS MODES AND MESH SIZES FOR CIRCULAR WAVEGUIDE $h=0.6$, $a=3.0$

Mesh Size	TM_{01}	TM_{11}	TM_{31} (Fig. 5(a))	TE_{11}	TE_{21} (Fig. 5(b))	TE_{01} (Fig. 5(d))
h	2.3439	3.6935	5.9385	1.7460	3.0582	3.7727
$h/2$	2.3898	3.7971	6.2685	1.7933	3.0354	3.8685
$h/4$	2.4013	3.8322	6.3670	1.8337	3.0116	3.8530
$h/8$	2.4050	3.8351	6.3809	1.8383	3.0027	3.8416
$h/16$	2.4079	3.8375	6.3870			
Exact	2.4048	3.8317	6.3801	1.8411	3.054	3.832

TABLE III

k_ca FOR VARIOUS MODES AND MESH SIZES FOR AN ASYMMETRICAL RIDGE GUIDE $h=0.1$, $a=1.0$

Mesh Size	TM_1 (Fig. 6(a))	TM_2 (Fig. 6(b))	TM_5	TE_1 (Fig. 6(c))	TE_2 (Fig. 6(d))	TE_4
h	8.6969	11.589	14.644	2.1595	5.0486	7.1761
$h/2$	9.2603	12.103	16.048	2.2502	5.6959	7.3621
$h/4$	9.2785	12.128	16.249	2.2376	5.8494	7.2741
$h/8$	9.2820	12.125	16.294	2.2402	5.9213	7.2244
$h/16$	9.2865	12.126	16.312	2.2412	5.9690	7.1984

percent in cutoff frequency are illustrated in Table II. Error increased to 0.62 percent for the TE case. Notice the considerable TE_{01} contour error in Fig. 5(c). The potential function is fairly flat near the wall, and the contour is therefore very error sensitive. A significant improvement was obtained by increasing the number of iterations by approximately 50 percent for each mesh size (Fig. 5(d)). The eigenvalue was not significantly altered.

Fig. 6 illustrates the field patterns, and Table III lists the cutoff propagation constants associated with an asymmetrical ridge guide. Guide regions without contours have fields too small to be plotted. As TM mode order increases, propagation first occurs mainly in the larger section, then in the smaller, and finally for the fifth mode the connecting arm was found to propagate substantially. Note the field singularities at the internal corners. The TE propagation constants obtained oscillated with mesh size h (possibly due to these singularities), and it was necessary to go to $h/16$ to see if significant digits had stabilized. The lunar waveguide (Fig. 7) was the most complicated shape attempted. Note the migration of the TM fields to the lower region as mode order increases. Here again, field symmetry could be improved with more iterations. Convergence data are given in Table IV.

TABLE IV

k_ca FOR VARIOUS MODES AND MESH SIZES FOR A LUNAR GUIDE $h=2.6$, $a=26.0$

Mesh Size	TM_1 (Fig. 7(a))	TM_2	TM_4 (Fig. 7(b))
h	8.4443	9.9565	12.625
$h/2$	8.7917	10.671	13.782
$h/4$	9.0573	10.896	13.983
$h/8$	9.1594	10.896	14.058
$h/16$	9.1723	10.880	14.045

Computing times for TM mode computation are roughly one minute per mode. For TE modes the times were five minutes per mode, due to the more complicated logic that was required. It is expected to reduce both times with improved programming. Computation was done on an IBM 360/65 computer.

CONCLUSIONS

In conclusion we have shown that higher order waveguide modes may be accurately computed for a large class of shapes. Restrictions are that the guide must be filled homogeneously with an isotropic medium and must be continuous in the direction of propagation. We have not found it necessary to make use of waveguide symmetries that occur, although this would increase the accuracy with less computing time and fewer node equations required. TE modes are less accurate since Neumann boundary conditions are more difficult to satisfy accurately than Dirichlet.

Acknowledgment

The authors wish to acknowledge their appreciation for the expert programming and unstinting help of G. Oczkowski. The University of Manitoba Institute for Computer Studies is gratefully thanked for its cooperation.

References

[1] A. Wexler, "Solution of waveguide discontinuities by modal analysis," *IEEE Trans. Microwave Theory and Techniques*, vol. MTT-15, pp. 508–517, September 1967.
[2] S. W. Drabowitch, "Multimode antennas," *Microwave J.*, vol. 9, pp. 41–51, January 1966.
[3] J. B. Davies and C. A. Muilwyk, "Numerical solution of uniform hollow waveguides with boundaries of arbitrary shape," *Proc. IEE* (London), vol. 113, pp. 277–284, February 1966.
[4] R. S. Varga, *Matrix Iterative Analysis.* Englewood Cliffs, N. J.: Prentice-Hall, 1965, pp. 56–78.
[5] C.-E. Fröberg, *Introduction to Numerical Analysis.* Reading, Mass.: Addison-Wesley, 1965, pp. 109–123.
[6] R. F. Harrington, *Time-Harmonic Electromagnetic Fields.* New York: McGraw-Hill, 1961, p. 130.
[7] L. Fox, Ed., *Numerical Solution of Ordinary and Partial Differential Equations.* Oxford: Pergamon, 1962, pp. 301–308.
[8] J. W. Duncan, "The accuracy of finite-difference solutions of Laplace's equation," *IEEE Trans. Microwave Theory and Techniques*, vol. MTT-15, pp. 575–582, October 1967.
[9] G. F. Forsythe and W. R. Wasow, *Finite-Difference Methods for Partial Differential Equations.* New York: Wiley, 1967, pp. 375–376.
[10] C. Moler, "Finite difference methods for the eigenvalues of Laplace's operator," Stanford University, Stanford, Calif., Computer Sci. Rept. CS22, 1965, pp. 114–129.
[11] J. Heading, *Matrix Theory for Physicists.* London: Longmans, 1958, pp. 66–67.
[12] G. N. Polozhii, *The Method of Summary Representation for Numerical Solution of Problems of Mathematical Physics.* Oxford, England: Pergamon, 1965, p. 78.

Paper 2.2

A Numerical Method Based on the Discretization of Maxwell Equations in Integral Form

M. ALBANI AND P. BERNARDI, SENIOR MEMBER, IEEE

Abstract—A method is described for the solution of the electromagnetic field inside resonant cavities and waveguides of arbitrary shape, whether homogeneously or inhomogeneously filled. The method, suitably programmed for use with a digital computer, is based on the direct discretization of the Maxwell equations in integral form. Since the method works with the components of the electromagnetic field, the numerical solution directly gives the distributions of the field in the structure, in addition to the resonant frequencies of cavities or the propagation constants of waveguides. Some numerical applications of the method are given.

I. INTRODUCTION

Numerous satisfactory numerical methods are available today for determining the electromagnetic field, both in structures in which the field can be derived from a single scalar potential, as in the case of empty guides of arbitrary shape [1]–[3], and in more general structures in which the field has all the components differing from 0, such as waveguides loaded with axial dielectrics [4]–[9] or resonant cavities of arbitrary shape, whether empty or loaded with dielectric regions [8]. Comparative discussions of these methods [2], [3], [7] show that "no single solution method has proved to be best for all requirements that might be imposed."

In this short paper, a method based on the direct discretization of the Maxwell equations in integral form is presented. The method does not require the introduction of auxiliary potential functions or the use of particular analytical procedures to formulate the problem in a computationally convenient form, and it therefore represents a very direct approach for the solution of a large class of structures. Moreover, the method presented allows the solution, with unified treatment, of both two- and three-dimensional structures.

II. DISCRETIZATION OF MAXWELL'S EQUATIONS IN INTEGRAL FORM

Considering a source-free region and assuming $\exp[j\omega t]$ as time dependence, Maxwell's equations in integral form may be written

$$\oint_s t\cdot E\,ds = -\int_S n\cdot H\,dS \tag{1a}$$

$$\oint_s t\cdot H\,ds = -\int_S \epsilon_r(P)n\cdot E\,dS \tag{1b}$$

where $\epsilon_r(P)$ is the permittivity of the medium in the structure. In (1) the lengths and the electric field are normalized to $1/\omega(\mu_0\epsilon_0)^{1/2}$ and $j(\mu_0/\epsilon_0)^{1/2}$, respectively; that is,

$$s = \omega(\mu_0\epsilon_0)^{1/2}\bar{s} \tag{2a}$$

$$E = -j\bar{E}/(\mu_0/\epsilon_0)^{1/2} \tag{2b}$$

where $\bar{s}$ and $\bar{E}$ are the effective length and electric field.

A. *Cavities of Arbitrary Shape Inhomogeneously Filled*

A cavity of arbitrary shape, bounded by a perfect conductor, and loaded with an inhomogeneous dielectric medium is considered first. In order to obtain a finite set of algebraic equations, a finite-difference procedure of discretization, the cell method [10], [11], is followed. The method consists of subdividing the cavity into cubic cells of side h, each assumed homogeneously filled, and considering the field as a function defined on the cells. Two types of cell may be considered: internal and boundary cells (Fig. 1). For all the cells of the structure, we assume the following hypotheses on the distribution of the electromagnetic field. 1) Inside each cell the components of the field have constant value. 2) On the interface between two contiguous cells, the components of the field have a value equal to the mean of the values in the two cells considered.

In this way, the continuous electromagnetic field is replaced by a set of discrete values. By applying (1a) and (1b) to each cell of the structure, we obtain a finite system of simultaneous algebraic equations. Assuming a rectangular coordinates set (x, y, z), for the generic internal cell we have

$$2hH_x + E_y(z-h) - E_y(z+h) + E_z(y+h) - E_z(y-h) = 0 \tag{3a}$$

$$2hE_x + \epsilon_r^{-1}[H_y(z-h) - H_y(z+h) + H_z(y+h) - H_z(y-h)] = 0 \tag{3b}$$

where H_x stands for $H_x(x,y,z)$ and $E_y(z-h)$ stands for $E_y(x,y,z-h)$.

The other four equations are obtained with two successive permutations of the coordinate index in (3a) and (3b). For each internal cell, six equations analogous to the preceding ones may be written; the only point to be noted is that the value of ϵ_r must be that of the medium filling the cell. At the boundary cells the electric field is assumed to be 0, while the magnetic field is assumed to be different from 0 because of the surface currents J_s on the boundary, which have not been taken into account in (1). With these hypotheses, from (1a) we obtain for the boundary cell Q of Fig. 1:

$$2hH_{xQ} = 0 \tag{4a}$$

$$2hH_{yQ} - E_{zP} = 0 \tag{4b}$$

$$2hH_{zQ} - E_{yP} = 0. \tag{4c}$$

Applying (3) and (4) to all the cells of the structure, we obtain a homogeneous system of equations that can be expressed as a matrix eigenvalue problem:

$$(A - 2hI)x = 0 \tag{5}$$

where A has not more than four nonzero elements for each row.

Because of the high number of equations necessary to obtain the field distribution with a fair degree of approximation, the eigenvalue problem can be solved numerically only with iterative methods [1]. However, the solution of (5) by iterative methods is not easy in view of the particular structure of the matrix A [10]. In fact, this matrix has all zeros on the main diagonal and, moreover, it has pairs of equal and opposite eigenvalues. It is therefore advisable to reformulate the problem in such a way as to obtain an eigenvalue problem again, but relative to a new matrix having a structure more suitable for the use of iterative methods. The procedure consists in eliminating from (5) the components of the magnetic field, obtaining a system of equations having as unknowns only the components of the electric field in the internal cells of the structure. In particular,

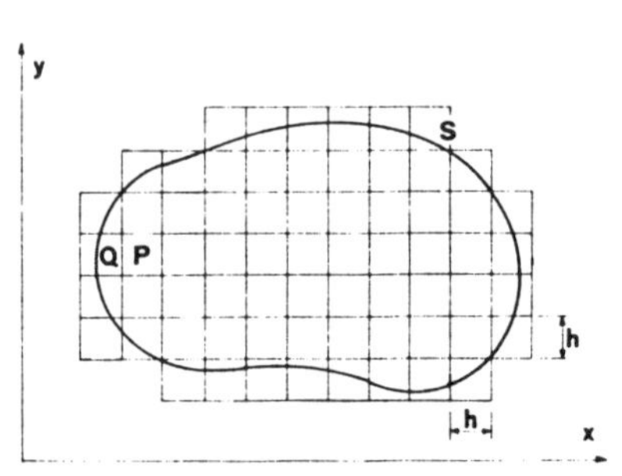

Fig. 1. Cross section of a cavity of arbitrary shape. S is the cavity boundary, P an internal cell, Q a boundary cell.

Manuscript received January 31, 1973; revised October 3, 1973. This work was supported by the Consiglio Nazionale delle Ricerche, Italy.
M. Albani is with Texas Instruments Semiconductors, Rome, Italy.
P. Bernardi is with the Istituto di Elettronica, Università di Roma, Rome, Italy.

Reprinted from *IEEE Trans. Microwave Theory Tech.*, vol. MTT-22, pp. 446–450, Apr. 1974.

for each cell we obtain

$$\begin{aligned}(4h^2 - 4\epsilon_r^{-1})E_x + \epsilon_r^{-1}[E_x(z-2h) - E_z(x-h,z-h)\\ + E_z(x+h,z-h) + E_x(z+2h) + E_z(x-h,z+h)\\ - E_z(x+h,z+h) - E_y(x+h,y+h) + E_y(x-h,y+h)\\ + E_x(y+2h) + E_y(x+h,y-h) - E_y(x-h,y-h)\\ + E_x(y-2h) = 0\end{aligned} \qquad (6)$$

the remaining two equations being obtained with two successive permutations of the coordinate index.

Applying (6) to all internal cells, with due regard to the condition $E = 0$ on the boundary cells, a matrix eigenvalue problem is obtained:

$$(B - 4h^2 I)x = 0. \qquad (7)$$

Matrix B is less sparse than A, since up to 13 elements may be different from 0 in each row, but the dimensions of B are less than half those of A. Moreover, it is to be noted that all the elements of the main diagonal of B are different from 0. Another advantage of (7), compared with (5), is found in the cases in which the eigenvalues of A are all real as, for example, in the case of a cavity homogeneously filled with a lossless dielectric. In such a case, B is positive semidefinite, and it is therefore particularly easy to solve the eigenvalue problem by iterative methods. In conclusion, (7) is adopted as a basis for the solution of inhomogeneously filled cavities of arbitrary shape: the eigenvalues give, through (2a), the resonant frequencies and the eigenvectors give the relative distributions of the electric field.

The simplicity of the method discussed lies in the fact that (6) is directly applicable by assigning the appropriate permittivity to each cell, without taking account of the conditions of continuity of the tangential components at the interface between different media. Also, for the boundary conditions no problem arises, since in all the boundary cells we can put directly $E = 0$, and (6) is not applied.

B. Dielectric Loaded Cylindrical Waveguides

We consider a waveguide section of length h, subdivided into cubic cells, each homogeneously filled. A space dependence exp $(-k_z z)$, where z is the longitudinal axis of the guide, is assumed.

In the transverse xy plane, the same hypotheses on the distribution of the electromagnetic field as in Section I are advanced. Following the procedure described in Section I and taking account of the exponential dependence on z, for the generic internal cell we derive six equations for the components of the field. As in the case of the cavities, it is good to eliminate the magnetic field components, obtaining three equations in the electric field components:

$$\begin{aligned}[4h^2 + \epsilon_r^{-1}(4k_z^2h^2 - 2)]E_x + 2\epsilon_r^{-1}k_zh[E_z(x+h) - E_z(x-h)]\\ + \epsilon_r^{-1}[E_x(y+2h) + E_x(y-2h) - E_y(x+h,y+h)\\ + E_y(x-h,y+h) + E_y(x+h,y-h) - E_y(x-h,y-h)]\\ = 0\end{aligned} \qquad (8a)$$

$$\begin{aligned}[4h^2 + \epsilon_r^{-1}(4k_z^2h^2 - 2)]E_y + 2\epsilon_r^{-1}k_zh[E_z(y+h) - E_z(y-h)]\\ + \epsilon_r^{-1}[E_y(x-2h) + E_y(x+2h) - E_x(x-h,y-h)\\ + E_x(x-h,y+h) + E_x(x+h,y-h) - E_x(x+h,y+h)]\\ = 0\end{aligned} \qquad (8b)$$

$$\begin{aligned}(4h^2 - 4\epsilon_r^{-1})E_z + 2\epsilon_r^{-1}k_zh[E_x(x+h) - E_x(x-h)\\ + E_y(y+h) - E_y(y-h)] + \epsilon_r^{-1}[E_z(y-2h)\\ + E_z(y+2h) + E_z(x+2h) + E_z(x-2h)] = 0.\end{aligned} \qquad (8c)$$

Equations (8a)–(8c) are evaluated at each internal cell, with the condition $E = 0$ at the boundary cells. A set of simultaneous equations is obtained that may be reduced to an eigenvalue problem by putting

$$k_z h = \text{constant}. \qquad (9)$$

With this position, the matrix eigenvalue problem may be written

$$(C - 4h^2 I)x = 0 \qquad (10)$$

where the elements of the matrix C are independent of h.

The eigenvalues of C give the frequencies of the waveguide modes corresponding to the value of the propagation constant given by (9); the relative distributions of E are given directly by the eigenvectors. In this way it is also possible to obtain the dispersion curves of the various modes by solving (10) for various values of k_zh. In particular, by putting $k_zh = 0$, the cutoff eigenvalues h_c can be obtained, as well as the relative field distributions. It may be noted that with the proposed method, the eigenvalue problem (10) is relative to a matrix C that, for homogeneously loaded guides, is always symmetric, for both TE and TM modes. As is known, this does not occur for TE modes when using the finite-difference method, unless a variational formulation is followed. Moreover, the method proposed gives, for any structure, a matrix eigenvalue problem in standard form, while, for instance, the conventional finite-element method [6] leads to a matrix eigenvalue problem in general form.

It is interesting to derive directly from (8) some well-known properties of waveguides as, for instance: in a uniform waveguide the transverse distribution of the field is independent of the propagation constant. Let us refer, for example, to TM modes. Putting $k_z = 0$, the eigenvalue problem is expressed by

$$(D - 4h_c^2 I)x = 0. \qquad (11)$$

For $k_z \neq 0$, on the other hand, we have

$$[D - 4h^2(1 + \epsilon_r^{-1}k_z^2)I]x = 0. \qquad (12)$$

Since in (11) and (12) the matrix D is the same, the eigenvectors (and hence the field distributions) do not vary with k_z. Moreover, from (11) and (12)

$$h_c^2 = h^2(1 + \epsilon_r^{-1}k_z^2) \qquad (13)$$

which represents the well-known dispersion equation for uniform waveguides. On the contrary, for inhomogeneously loaded waveguides, from (10) we obtain a group of eigenvectors that is different for each value of k_z. This means that the distribution of the field for a given mode does not depend only on the geometry of the structure but also on the value of k_z.

III. COMPUTED RESULTS

A. General Remarks

Two general programs have been written to analyze dielectric loaded waveguides and cavities by the proposed method. The programs require as input the number and the coordinates of the cells and the value of the permittivity on each cell for the case of cavities, while for the case of waveguides it is necessary to give also the desired value of the propagation constant. In the numerical computation, account has been taken of the fact that the resulting set of equations consists of independent groups of equations. This means that in order to determine the electromagnetic field in a given structure, it is sufficient to solve an eigenvalue problem for a matrix considerably smaller than the initial one. For example, in [12] it is shown that for a resonator subdivided into 75 internal cells it is sufficient to solve an eigenvalue problem of order 20 instead of the initial one of order 225.

B. Results

1) *Uniform Waveguides*: For TM waves the eigenvalue problem obtained by (8) with $k_z = 0$ is identical to that obtained by applying the finite-difference method to the Helmholtz equation for the scalar potential $\phi = E_z$ [13], provided that the mesh length is $2h$. The results obtained are therefore equal to those in [13] and [14]. It is more interesting to examine the solutions relative to TE waves. In fact, (8) operates on the transverse E-field components instead of on the longitudinal H_z component, as in the finite-difference method. Of course, the relative boundary conditions are also posed in a different manner. Table I shows the results obtained for an empty rectangular guide for the dominant mode, for a TE higher order mode and for a TE_{11} mode. It may be noted that the error varies with h in the same way as in the finite-difference method [13].

2) *Inhomogeneously Loaded Waveguides*: In this case, the eigenvalues can be obtained by (8) with assigned values of k_za ($k_z = \alpha_z + j\beta_z$). As examples, the structures of Figs. 2–4 have been solved.

For the structure of Fig. 2, the results for the dominant mode are shown in Table II. The associated electric field distributions are also shown in Fig. 2. It may be noted that in this case, as h/a decreases, the error decreases about linearly with h. This result may be used, as already done for empty guides [13], for computing extrapolated eigenvalues. For the structure of Fig. 3, in which there are material discontinuities in two dimensions, the results are given in Table III and are compared with those obtained by Schlosser and Unger [15]. The structure of Fig. 4, consisting of a square waveguide loaded with a dielectric rod, has been chosen as an example of the application of the method to a structure in which the material boundary does not consist of parts of straight lines; the results are compared, in Table IV, with those obtained by Bates and Ng [9]. It may be noted that with a mesh of 20 $\times$ 20 cells, corresponding to 261 equations, the results are almost coincident with those given in [9].

In the above examples the numerical computation has been carried out starting from the general equations (8a)–(8c) without taking any advantage of the particular symmetries. For example, in the case of Table IV, since the computation is carried out at cutoff, we could have obtained the TE modes by putting $E_z = 0$ directly in (8), thus considerably reducing the computer time. Therefore, with the same time shown in Table IV, we can calculate points of the dispersion curves for hybrid modes at any frequency. In Table II, where account has been taken of the existence of TE zero-order modes, the computer time is much shorter.

TABLE I
EMPTY RECTANGULAR WAVEGUIDE: TE MODES

Mode	h/a	Computed $k_o a$	Error, %	Error Dependence
TE_{10}	1/18	3.1257	0.51	
	1/36	3.1376	0.13	$\simeq h^2$
true $k_o a$ = 3.1416	1/72	3.1406	0.03	
TE_{30}	1/18	9.0000	4.51	
	1/36	9.3175	1.14	$\simeq h^2$
true $k_o a$ = 9.4248	1/72	9.3979	0.29	
TE_{11} (b/a = 2/3)	1/6	5.1962	8.25	
	1/12	5.5439	2.11	$\simeq h^2$
true $k_o a$ = 5.6637	1/24	5.6334	0.53	

TABLE II
WAVEGUIDE IN FIG. 2: DOMINANT TE MODE

$\beta_z a$	h/a	True $k_o a$ $k_o a = \omega\sqrt{\mu_0 \varepsilon_0}\, a$	Computed $k_o a$	Error, %	Error Dependence	Computer Time
0	1/12	1.0674	1.1828	10.81	$\simeq h$	< 20s
	1/36		1.0239	- 4.08		
	1/108		1.0817	1.34		
	1/324		1.0630	- 0.41		< 60s
2	1/12	1.2513	1.3744	9.84	$\simeq h$	< 20s
	1/36		1.2017	- 3.96		
	1/108		1.2674	1.29		
	1/324		1.2470	- 0.34		< 60s
4	1/12	1.6613	1.7807	7.19	$\simeq h$	< 20s
	1/36		1.5991	- 3.74		
	1/108		1.6810	1.19		
	1/324		1.6560	- 0.32		< 60s

TABLE III
WAVEGUIDE IN FIG. 3: DOMINANT MODE

$\beta_z a$	h/a	Computed $k_o a$	Schlosser and Unger $k_o a$	Error, %	Computer Time
1	1/10	1.98	2.12	6.6	< 20s
	1/20	2.05		3.3	< 60s
2	1/10	2.11	2.25	6.2	< 20s
	1/20	2.18		3.1	< 60s

TABLE IV
WAVEGUIDE IN FIG. 4: DOMINANT TE MODE AT CUTOFF

d/a	Computed $k_c a$	Bates and Ng $k_o a$ Measured	Bates and Ng $k_o a$ Computed	Computer Time
0.0	2.936	2.865	2.958	< 60s
0.206	2.982	2.904	2.982	< 60s

Note: $h/a = 1/20$.

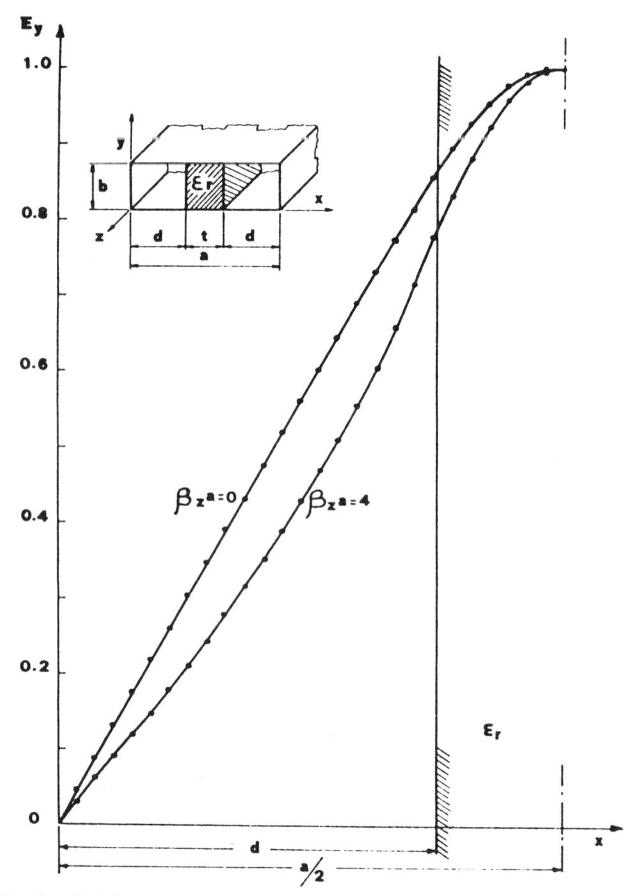

Fig. 2. E-field distribution for the dominant mode in the considered guide. Lines denote theoretical curves, dots denote computed values. $\epsilon_r = 16$, $t/a = 1/4$, $b/a = 4/9$.

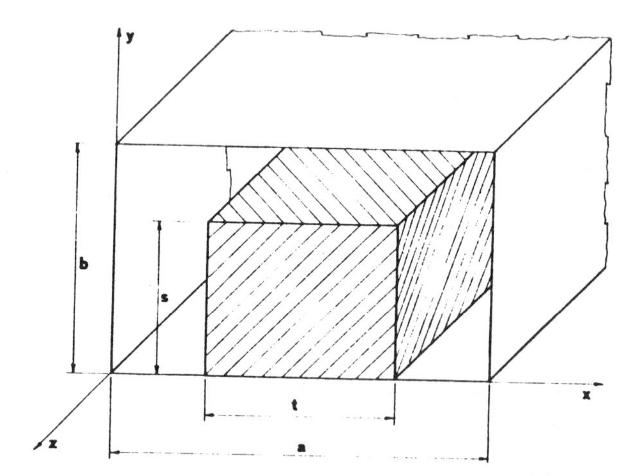

Fig. 3. Rectangular waveguide with insert at center of bottom wall. $\epsilon_r = 6$, $t/a = 1/2$, $b/a = 3/5$, $s/a = 2/5$.

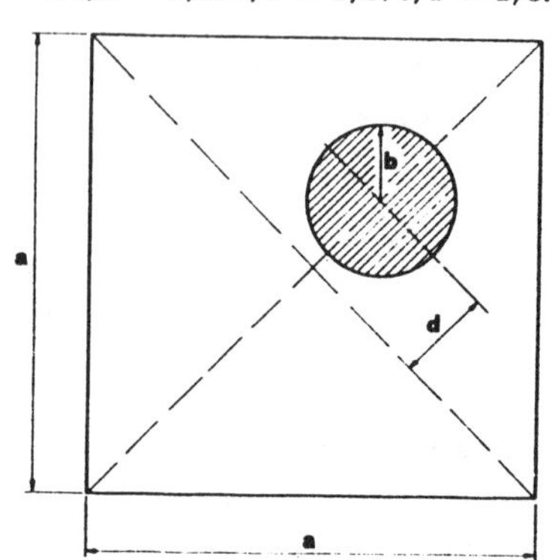

Fig. 4. Square waveguide loaded with dielectric rod. $\epsilon_r = 2.30$, $b/a = 0.161$.

Finally, we considered the structure studied by Franceschetti [16] consisting of a rectangular guide completely filled with an inhomogeneous dielectric whose permittivity is $\epsilon_r(x) = 5 \exp(-1.61x/a)$ (a is the larger dimension of the guide). The results are shown in Table V and Fig. 5.

3) *Homogeneous Cavities*: By applying (6) to the case of a rectangular cavity, the results shown in Table VI are obtained. As may be seen, even with a moderate number of equations the resonant frequencies of the first modes are obtained with errors smaller than 1 percent.

4) *Inhomogeneously Loaded Cavities*: As numerical examples we considered: a) the cavities obtained from the waveguides in Figs. 2 and 3; and b) the cavity of Fig. 6.

The structures in Figs. 2 and 3, because of their cylindrical symmetry, may be solved either by applying (8) of the waveguides after assuming $\beta_z h = m\pi h/c$ $(m = 1, 2, \cdots)$, or by directly applying (6) of the cavities. For these structures both methods have been applied. For cavities of arbitrary shape, on the other hand, like that shown in Fig. 6, only (6) can be applied, in view of the absence of any symmetry.

The results given in Table VII show the following. 1) For cavities with cylindrical symmetry, it is advisable to use (8) instead of (6). 2) The errors obtained via (6) are of the same order of magnitude for structures inhomogeneous in one direction only (Fig. 2), in two directions (Fig. 3), and in all three directions (Fig. 6).

IV. CONCLUSIONS

The method presented may be applied to a wide class of structures, such as cylindrical waveguides and cavities of arbitrary shape, even if inhomogeneously filled. Although the electromagnetic field has been approximated with constant values in the cells, the method gives, with good approximation, both the field distributions and the resonant frequencies or the propagation constants with a moderate number of equations. The complexity of the numerical solution and the computing time depend only on the number of cells into which the structure is subdivided and are not influenced by the presence of one or more dielectrics inside it.

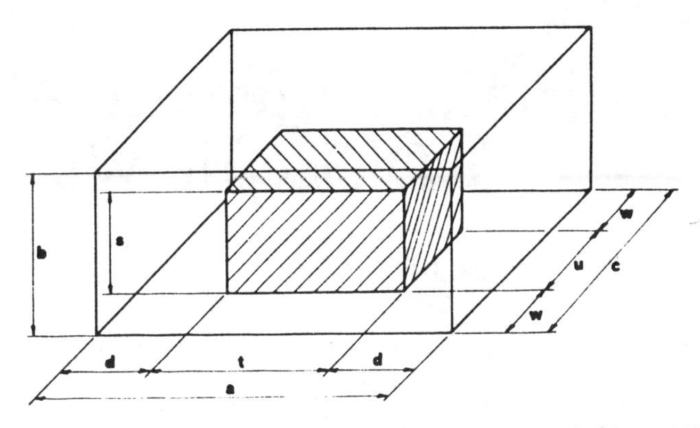

Fig. 6. Rectangular cavity with dielectric insert. $\epsilon_r = 2.05$, $b/a = 4/9$, $c/a = 5/9$, $u/a = 2/9$, $s/a = 5/18$, $t/a = 1/2$.

TABLE VII
INHOMOGENEOUSLY LOADED RESONATORS: DOMINANT MODE

Resonator	Resonant Frequency $k_c a$		Error, %	Computer Time
Fig. 2	True $k_c a$ (TE_{101} mode)	2.583	-	
Resonator length c/a=4/10	$k_c a$ computed via (8)	2.429	6.0	< 10s
h/a = 1/20	$k_c a$ computed via (6)	2.383	7.7	< 60s
Fig. 3	Schlosser & Unger $k_c a$	2.73	-	
Resonator length c/a=4/5	$k_c a$ computed via (8)	2.59	5.1	< 10s
h/a = 1/10	$k_c a$ computed via (6)	2.56	6.2	< 60s
Fig. 6	Measured $k_c a$	5.22	-	
h/a = 1/18	Computed $k_c a$	5.55	6.3	< 60s

TABLE V
RECTANGULAR GUIDE COMPLETELY FILLED WITH INHOMOGENEOUS DIELECTRIC: DOMINANT TE MODE

$\beta_z a$	h/a	Computed $k_c a$	Franceschetti $k_c a$	Computer Time
10.209	1/40	6.276	6.283	< 20s
	1/60	6.280		
	1/100	6.282		

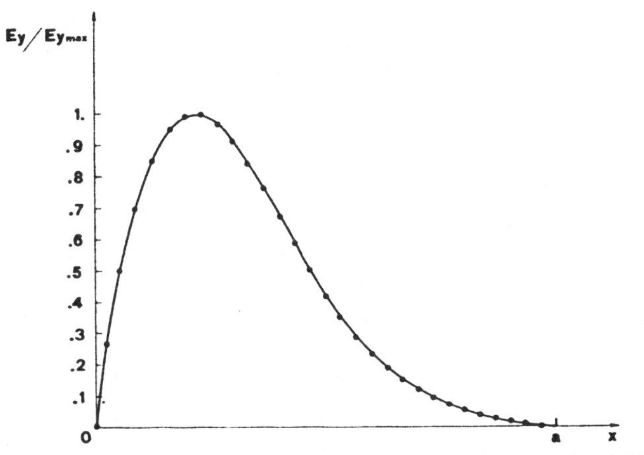

Fig. 5. E-field distribution for dominant mode in waveguide in Table V ($\beta_z a = 10.209$). Line denotes Franceschetti curve, dots denote computed values.

TABLE VI
EMPTY RECTANGULAR RESONATOR (a,b,c)

Mode	h/a	Computed $k_c a$	Error, %	Number of Equations
TM_{110}	1/12	5.5439	2.11	121
True resonant freq. $k_c a$ = 5.6636	1/24	5.6335	0.53	1322
TE_{101}	1/12	6.7560	3.83	121
True $k_c a$ = 7.0249	1/24	6.9571	0.97	1322
TM_{210} , TE_{011}	1/12	7.5558	3.80	121
True $k_c a$ = 7.8540	1/24	7.7784	0.96	1322

Note: $b/a = 2/3$, $c/a = 1/2$.

REFERENCES

[1] A. Wexler, "Computation of electromagnetic fields," *IEEE Trans. Microwave Theory Tech. (Special Issue on Computer-Oriented Microwave Practices)*, vol. MTT-17, pp. 416–439, Aug. 1969.
[2] J. B. Davies, "Numerical solution of the hollow waveguide problem," in *Progress in Radio Science*, vol. 3. Brussels, Belgium: International Scientific Radio Union, 1971, pp. 307–320.
[3] B. E. Spielman and R. F. Harrington, "Waveguides of arbitrary cross section by solution of nonlinear integral eigenvalue equation," *IEEE Trans. Microwave Theory Tech.*, vol. MTT-20, pp. 578–585, Sept. 1972.
[4] W. J. English, "Vector variational solutions of inhomogeneously loaded cylindrical waveguide structures," *IEEE Trans. Microwave Theory Tech.*, vol. MTT-19, pp. 9–18, Jan. 1971.
[5] W. J. English and F. J. Young, "An E vector variational formulation of the Maxwell equations for cylindrical waveguide problems," *IEEE Trans. Microwave Theory Tech.*, vol. MTT-19, pp. 40–46, Jan. 1971.
[6] Z. J. Csendes and P. Silvester, "Numerical solution of dielectric loaded waveguides: I—Finite-element analysis," *IEEE Trans. Microwave Theory Tech. (1970 Symposium Issue)*, vol. MTT-18, pp. 1124–1131, Dec. 1970.
[7] ——, "Numerical solution of dielectric loaded waveguides: II—Modal approximation technique," *IEEE Trans. Microwave Theory Tech.*, vol. MTT-19, pp. 504–509, June 1971.
[8] R. F. Harrington, in *Field Computation by Moment Methods*. New York: Macmillan, 1968.
[9] R. H. T. Bates and F. L. Ng, "Polarisation-source formulation of electromagnetism and dielectric-loaded waveguides," *Proc. Inst. Elec. Eng.*, vol. 119, pp. 1568–1574, Nov. 1972.
[10] G. E. Forsythe and W. R. Wasow, *Finite-Difference Methods for Partial Differential Equations*. New York: Wiley, 1960.
[11] G. Liebmann, "Resistance-network analogues with unequal meshes or subdivided meshes," *Brit. J. Appl. Phys.*, vol. 5, pp. 362–366, Oct. 1954.
[12] M. Albani and P. Bernardi, "A numerical method for six-component electromagnetic fields," Inst. Elettronica, Univ. Roma, Rome, Italy, Internal Rep. 51, Feb. 1973.
[13] J. B. Davies and C. A. Muilwyk, "Numerical solution of uniform hollow waveguides with boundaries of arbitrary shape," *Proc. Inst. Elec. Eng.*, vol. 113, pp. 277–284, Feb. 1966.
[14] M. J. Beaubien and A. Wexler, "An accurate finite-difference method for higher order waveguide modes," *IEEE Trans. Microwave Theory Tech. (1968 Symposium Issue)*, vol. MTT-16, pp. 1007–1017, Dec. 1968.
[15] W. Schlosser and H. G. Unger, "Partially filled waveguides of rectangular cross section," in *Advances in Microwaves*, vol. 1. New York: Academic, 1966.
[16] G. Franceschetti, "Eigenvalues and eigenfunctions of transversely inhomogeneous rectangular waveguides," *Alta Freq.*, vol. 32, pp. 133–141, Feb. 1963.

Finite-Difference Analysis of Rectangular Dielectric Waveguide Structures

KARLHEINZ BIERWIRTH, NORBERT SCHULZ, AND FRITZ ARNDT, SENIOR MEMBER, IEEE

***Abstract*—A class of dielectric waveguide structures using a rectangular dielectric strip in conjunction with one or more layered dielectrics is analyzed with a finite-difference method formulated directly in terms of the wave equation for the transverse components of the magnetic field. This leads to an eigenvalue problem where the nonphysical, spurious modes do not appear. Moreover, the analysis includes hybrid-mode conversion effects, such as complex waves, at frequencies where the modes are not yet completely bound to the core of the highest dielectric constant, as well as at frequencies below cutoff. Dispersion characteristic examples are calculated for structures suitable for millimeter-wave and optical integrated circuits, such as dielectric image lines, shielded dielectric waveguides, insulated image guides, ridge guides, and inverted strip, channel, strip-slab, and indiffused inverted ridge guides. The numerical examples are verified by results available from other methods.**

I. Introduction

DIELECTRIC WAVEGUIDE structures of the class shown in Fig. 1(a) have found increasing interest for integrated circuit applications in the millimeter-wave and optical frequency range [1]–[21]. As this class includes a wide variety of specially shaped dielectric waveguides (Fig. 1(b)–(k)), in the design of integrated circuits utilizing such structures, it is important to find a reliable computer analysis which is sufficiently general and flexible to allow dominant- and higher order mode solutions of all desired cases and which avoids the troublesome problem of non-physical or "spurious" modes [28]–[33], [38].

Various methods of analyzing one or several of the structures in Fig. 1(b)–(k) have been the subject of many papers, e.g. [1]–[33], including, in particular, different kinds of mode-matching techniques [2]–[13], [16], [17], [22]–[27] and the finite-element method [15], [28]–[32]. Although the finite-difference method is a common technique for the solution of boundary value problems [33]–[37], it has only been recently [33] that this method has been applied to one of the structures of Fig. 1, the dielectric waveguide (Fig. 1(d)). The variational formulation in [33], however, brings this method close to a finite-element technique [38], and since an $E_z - H_z$ formulation is utilized, spurious modes occur [33].

Following [37], [38] in judging the appropriateness of a method to solve a dielectric waveguide problem with a particular cross-sectional shape, such as that shown in Fig. 1(a) one has to weigh the following criteria: flexibility to deal with a large number of regions and with the hybrid-mode nature of all interesting modes; accuracy and computational efficiency; and the possibility of modifications to eliminate unwanted nonphysical modes. Among the candidates mentioned above, the finite-difference method is considered to meet all these criteria very well.

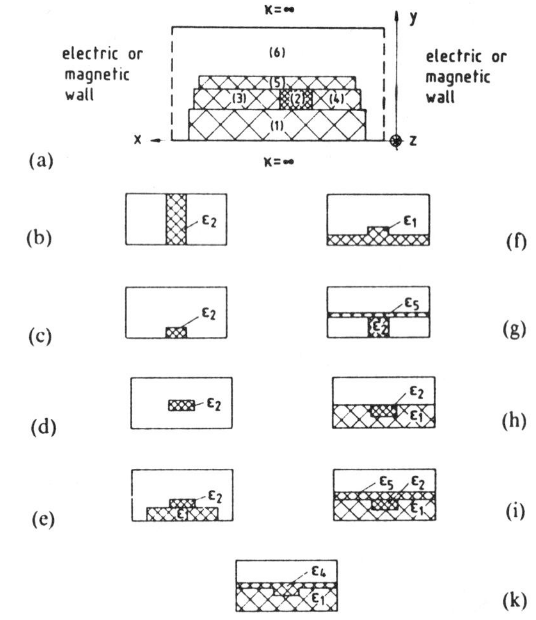

Fig. 1. Layered dielectric waveguide structures for integrated circuits in the millimeter-wave and optical frequency range. (a) General structure. (b) Nonradiative dielectric waveguide (cf. [20]). (c) Dielectric image line ([1]). (d) Dielectric waveguide ([22]). (e) Strip dielectric guide ($\epsilon_2 < \epsilon_1$) or insulated image guide ($\epsilon_2 > \epsilon_1$) ([3]). (f) Ridge guide ([16]). (g) Inverted strip guide ([13]). (h) Channel or embedded strip guide ([2]). (i) Strip-slab guide ([3]). (k) Indiffused inverted ridge guide ([17]).

This paper presents a simple, flexible, versatile finite-difference solution for analyzing the inhomogeneous waveguide structures of Fig. 1 which is free from the problem of spurious modes. Instead of the vector potential formulation of [39], which is solved by searching for the dominant-mode propagation factor [39], a simpler direct wave equation solution formulated in terms of the transverse magnetic field components, H_x and H_y, is utilized which leads advantageously to a conventional eigenvalue problem [40]. A graded mesh permits the investigation of structures with realistic dimensions by making the mesh

Manuscript received March 10, 1986; revised June 24, 1986. This work was supported in part by the Deutsche Forschungsgemeinschaft under Contract No. AR 138/7-1.
The authors are with the Microwave Department of the University of Bremen, Kufsteiner Str., NW-1, D-2800 Bremem 33, West Germany.
IEEE Log Number 8610511.

Reprinted from *IEEE Trans. Microwave Theory Tech.*, vol. MTT-34, no. 11, pp. 1104–1114, Nov. 1986.

finer in regions of particular interest; if necessary, the enclosing box is sufficiently large so that it does not perturb the modes perceptibly. Related coupled structures are implicated by suitable electric or magnetic wall symmetry. Moreover, the finite-difference analysis given allows the investigation of hybrid-mode conversion effects, such as complex waves [26], [43]–[45], at frequencies where the modes are not yet completely bound to the core of the highest permittivity, as well as at frequencies below cutoff. Numerical results compared with available data from other methods verify the theory given.

II. Theory

A finite cross section is defined by enclosing the guide in a rectangular box (Fig. 1(a)) where the side walls may be either electric or magnetic walls in order to include coupled structures. An exponential decay factor may be introduced to approximate the infinite exterior region for related "open" structures [28]. Since the finite-difference analysis given includes mode investigations below cutoff, which makes the decay factor modeling difficult to be applied, it is preferred for these cases to make the box large enough [33], [36], [37] so that the influence on the modes may be neglected. A graded mesh permits the optimum use of the available computer capabilities for these cases as well.

The wave equation describing the propagation in a waveguide with inhomogeneous cross section can be expressed in terms of two field components, which are usually taken to be the longitudinal components E_z and H_z [36], [37]. The formulation in terms of the transverse components H_x and H_y is preferred, however, since it circumvents the spurious-mode problem.

The Helmholtz equations in the homogeneous subregions $\nu = 1,2,3,4$ (Fig. 2(a)) are

$$\nabla_t^2 H_x^{(\nu)} + k_\nu^2 H_x^{(\nu)} = 0$$

$$\nabla_t^2 H_y^{(\nu)} + k_\nu^2 H_y^{(\nu)} = 0$$

$$\nabla_t^2 = \frac{\partial^2}{\partial x^2} + \frac{\partial^2}{\partial y_2} \tag{1}$$

where [42]

$$k_\nu^2 = \omega^2 \mu \epsilon_\nu + \gamma^2$$

$$\gamma = \begin{cases} j\beta \\ \alpha \end{cases}$$

and a z dependence of $\exp(-\gamma z)$ of the wave propagation is understood. At the electric or magnetic walls, the boundary conditions $\partial H/\partial n = 0$ or $H = 0$, respectively, have to be satisfied ($\vec{n}$ = unit vector normal to the walls).

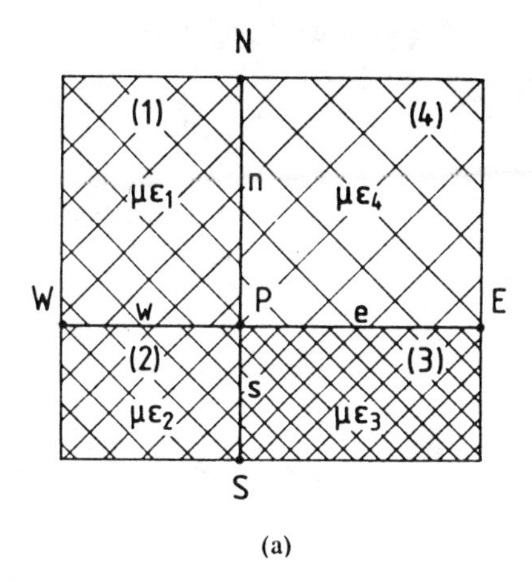

(a)

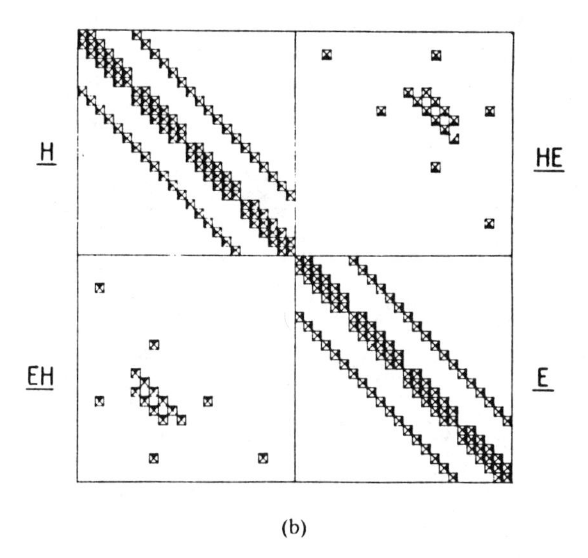

(b)

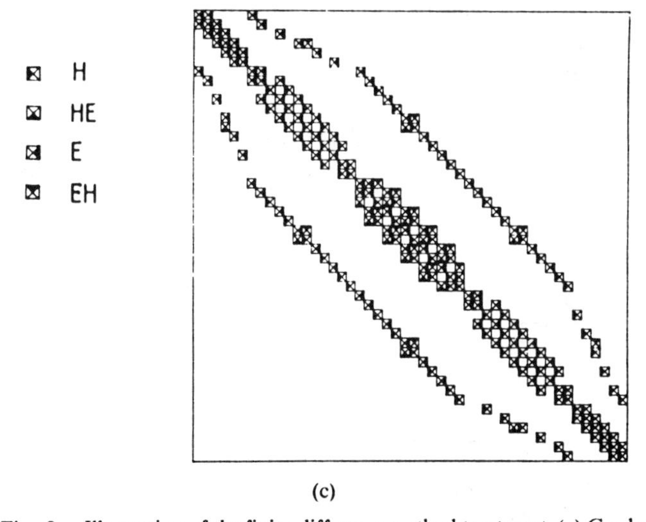

(c)

Fig. 2. Illustration of the finite-difference method treatment. (a) Graded mesh of the five-point representation. (b) Typical original structure of the matrix (A) in (6) and (7). (c) Rearranged structure of (A).

A graded mesh, of side lengths w, n, e, s (Fig. 2(a)), is presumed drawn over the guide cross section. In the general case, there are four subregions with four different dielectric constants. Equations (1) may then be written in its five-point finite-difference form [34], which in the case of inhomogeneous cross sections leads to four coupled equations for each component H_x or H_y, respectively,

$$0=-\frac{n}{w}H_W-\frac{w}{n}H_N+\left(\frac{w}{n}+\frac{n}{w}\right)H_P-\frac{1}{2}wnk_1^2H_P+w\frac{\partial H}{\partial y}\bigg|_1-n\frac{\partial H}{\partial x}\bigg|_1$$
$$0=-\frac{s}{w}H_W-\frac{w}{s}H_S+\left(\frac{w}{s}+\frac{s}{w}\right)H_P-\frac{1}{2}wsk_2^2H_P-w\frac{\partial H}{\partial y}\bigg|_2-s\frac{\partial H}{\partial x}\bigg|_2$$
$$0=-\frac{s}{e}H_E-\frac{e}{s}H_S+\left(\frac{e}{s}+\frac{s}{e}\right)H_P-\frac{1}{2}esk_3^2H_P-e\frac{\partial H}{\partial y}\bigg|_3+s\frac{\partial H}{\partial x}\bigg|_3$$
$$0=-\frac{n}{e}H_E-\frac{e}{n}H_N+\left(\frac{n}{e}+\frac{e}{n}\right)H_P-\frac{1}{2}enk_4^2H_P+e\frac{\partial H}{\partial y}\bigg|_4+n\frac{\partial H}{\partial x}\bigg|_4 \qquad (2)$$

where H denotes H_x or H_y, respectively, and the designations W, N, E, S, etc., are elucidated in Fig. 2(a).

Consideration is next given to the boundary conditions for H_x: $E_{z1}=E_{z2}$, $E_{z3}=E_{z4}$, $H_{z1}=H_{z2}$, $H_{z3}=H_{z4}$, $H_{z1}=H_{z4}$, i.e.

$$-\frac{1}{\epsilon_1}\frac{\partial H_x}{\partial y}\bigg|_1+\frac{1}{\epsilon_2}\frac{\partial H_x}{\partial y}\bigg|_2+\frac{1}{\epsilon_1}\frac{\partial H_y}{\partial x}\bigg|_1-\frac{1}{\epsilon_2}\frac{\partial H_y}{\partial x}\bigg|_2=0$$
$$\frac{1}{\epsilon_3}\frac{\partial H_x}{\partial y}\bigg|_3-\frac{1}{\epsilon_4}\frac{\partial H_x}{\partial y}\bigg|_4-\frac{1}{\epsilon_3}\frac{\partial H_y}{\partial x}\bigg|_3+\frac{1}{\epsilon_4}\frac{\partial H_y}{\partial x}\bigg|_4=0$$
$$\frac{\partial H_x}{\partial x}\bigg|_1-\frac{\partial H_x}{\partial x}\bigg|_2+\frac{\partial H_y}{\partial y}\bigg|_1-\frac{\partial H_y}{\partial y}\bigg|_2=0$$
$$\frac{\partial H_x}{\partial x}\bigg|_3-\frac{\partial H_x}{\partial x}\bigg|_4+\frac{\partial H_y}{\partial y}\bigg|_3-\frac{\partial H_y}{\partial y}\bigg|_4=0$$
$$\frac{\partial H_x}{\partial x}\bigg|_1-\frac{\partial H_x}{\partial x}\bigg|_4+\frac{\partial H_y}{\partial y}\bigg|_1-\frac{\partial H_y}{\partial y}\bigg|_4=0 \qquad (3a)$$

and for H_y: $E_{z1}=E_{z4}$, $E_{z2}=E_{z3}$, $H_{z1}=H_{z4}$, $H_{z2}=H_{z3}$, $H_{z1}=H_{z2}$, i.e.

$$-\frac{1}{\epsilon_1}\frac{\partial H_x}{\partial y}\bigg|_1+\frac{1}{\epsilon_4}\frac{\partial H_x}{\partial y}\bigg|_4+\frac{1}{\epsilon_1}\frac{\partial H_y}{\partial x}\bigg|_1-\frac{1}{\epsilon_4}\frac{\partial H_y}{\partial x}\bigg|_4=0$$
$$\frac{1}{\epsilon_3}\frac{\partial H_x}{\partial y}\bigg|_3-\frac{1}{\epsilon_2}\frac{\partial H_x}{\partial y}\bigg|_2-\frac{1}{\epsilon_3}\frac{\partial H_y}{\partial x}\bigg|_3+\frac{1}{\epsilon_2}\frac{\partial H_y}{\partial x}\bigg|_2=0$$
$$\frac{\partial H_x}{\partial x}\bigg|_1-\frac{\partial H_x}{\partial x}\bigg|_4+\frac{\partial H_y}{\partial y}\bigg|_1-\frac{\partial H_y}{\partial y}\bigg|_4=0$$
$$\frac{\partial H_x}{\partial x}\bigg|_3-\frac{\partial H_x}{\partial x}\bigg|_2+\frac{\partial H_y}{\partial y}\bigg|_3-\frac{\partial H_y}{\partial y}\bigg|_2=0$$
$$\frac{\partial H_x}{\partial x}\bigg|_2-\frac{\partial H_x}{\partial x}\bigg|_1+\frac{\partial H_y}{\partial y}\bigg|_2-\frac{\partial H_y}{\partial y}\bigg|_1=0. \qquad (3b)$$

These conditions are satisfactory to properly continue the wave solution from one subregion to the next such that the whole coupled solution governed by Maxwell's equations is obtained. Utilizing these conditions, the finite-difference equations (2) result in

$$\frac{2}{w(w+e)}H_{xW}+\frac{2}{e(w+e)}H_{xE}+\frac{2}{n(w+e)}\left(\frac{w\epsilon_2}{s\epsilon_1+n\epsilon_2}+\frac{e\epsilon_3}{n\epsilon_3+s\epsilon_4}\right)H_{xN}+\frac{2}{s(w+e)}\left(\frac{w\epsilon_1}{s\epsilon_1+n\epsilon_2}+\frac{e\epsilon_4}{n\epsilon_3+s\epsilon_4}\right)H_{xS}$$
$$-\frac{2}{w+e}\left[\frac{\epsilon_1}{s\epsilon_1+n\epsilon_2}\left(\frac{w}{s}+\frac{s}{w}\right)+\frac{\epsilon_2}{s\epsilon_1+n\epsilon_2}\left(\frac{w}{n}+\frac{n}{w}\right)+\frac{\epsilon_3}{n\epsilon_3+s\epsilon_4}\left(\frac{e}{n}+\frac{n}{e}\right)+\frac{\epsilon_4}{n\epsilon_3+s\epsilon_4}\left(\frac{e}{s}+\frac{s}{s}\right)\right]H_{xP}$$
$$+\omega^2\mu\frac{n+s}{w+e}\left(\frac{w\epsilon_1\epsilon_2}{s\epsilon_1+n\epsilon_2}+\frac{e\epsilon_3\epsilon_4}{n\epsilon_3+s\epsilon_4}\right)H_{xP}+\gamma^2H_{xP}$$
$$-\frac{2}{w+e}\frac{\epsilon_1-\epsilon_2}{s\epsilon_1+n\epsilon_2}H_{yW}-\frac{2}{w+e}\frac{\epsilon_3-\epsilon_4}{n\epsilon_3+s\epsilon_4}H_{yE}-\frac{2}{w+e}\left(\frac{\epsilon_2}{s\epsilon_1+n\epsilon_2}-\frac{\epsilon_3}{n\epsilon_3+s\epsilon_4}\right)H_{yN}$$
$$-\frac{2}{w+e}\left(\frac{\epsilon_4}{n\epsilon_3+s\epsilon_4}-\frac{\epsilon_1}{s\epsilon_1+n\epsilon_2}\right)H_{yS}=0 \qquad (4)$$

and

$$\frac{2}{w(n+s)}\left(\frac{s\epsilon_3}{e\epsilon_2+w\epsilon_3}+\frac{n\epsilon_4}{e\epsilon_1+w\epsilon_4}\right)H_{yW}+\frac{2}{e(n+s)}\left(\frac{n\epsilon_1}{e\epsilon_1+w\epsilon_4}+\frac{s\epsilon_2}{e\epsilon_2+w\epsilon_3}\right)H_{yE}+\frac{2}{n(n+s)}H_{yN}+\frac{2}{s(n+s)}H_{yS}$$
$$-\frac{2}{n+s}\left[\frac{\epsilon_1}{e\epsilon_1+w\epsilon_4}\left(\frac{e}{n}+\frac{n}{e}\right)+\frac{\epsilon_2}{e\epsilon_2+w\epsilon_3}\left(\frac{e}{s}+\frac{s}{e}\right)+\frac{\epsilon_3}{e\epsilon_2+w\epsilon_3}\left(\frac{w}{s}+\frac{s}{w}\right)+\frac{\epsilon_4}{e\epsilon_1+w\epsilon_4}\left(\frac{w}{n}+\frac{n}{w}\right)\right]H_{yP}$$
$$+\omega^2\mu\frac{w+e}{n+s}\left(\frac{n\epsilon_1\epsilon_4}{e\epsilon_1+w\epsilon_4}+\frac{s\epsilon_2\epsilon_3}{e\epsilon_2+w\epsilon_3}\right)H_{yP}+\gamma^2H_{yP}$$
$$+\frac{2}{n+s}\left(\frac{\epsilon_3}{e\epsilon_2+w\epsilon_3}-\frac{\epsilon_4}{e\epsilon_1+w\epsilon_4}\right)H_{xW}+\frac{2}{n+s}\left(\frac{\epsilon_1}{e\epsilon_1+w\epsilon_4}-\frac{\epsilon_2}{e\epsilon_2+w\epsilon_3}\right)H_{xE}$$
$$+\frac{2}{n+s}\frac{\epsilon_4-\epsilon_1}{e\epsilon_1+w\epsilon_4}H_{x_N}+\frac{2}{n+s}\frac{\epsilon_2-\epsilon_3}{e\epsilon_2+w\epsilon_3}H_{xS}=0. \qquad (5)$$

Equations (4) and (5) give H_{xP} and H_{yP}, the magnetic field components at the discrete node point P, in terms of the immediately adjacent node points W, E, N, S (Fig. 2(a)). These equations are evaluated at each node point P with appropriate modification to include the related boundary conditions (electric or magnetic wall) of the structure to be investigated (Fig. 1(a)). In this way, a set of linear homogeneous equations is derived of the form

$$\begin{pmatrix}(H) & (HE)\\(EH) & (E)\end{pmatrix}\begin{pmatrix}(H_x)\\(H_y)\end{pmatrix}=-\gamma^2\begin{pmatrix}(H_x)\\(H_y)\end{pmatrix} \qquad (6)$$

where (H) and (E) denote the submatrices given by the coefficients in (4) and (5) related to H_x or H_y, respectively, whereas (HE) and (EH) denote the coupled terms related to $H_x \leftrightarrow H_y$ and $H_y \leftrightarrow H_x$, respectively. The resulting eigenvalue equation

$$((A)-\lambda(U))(X)=0 \qquad (7)$$

where

$$\lambda=-\gamma^2$$
$$(X)=(H_{x1},H_{x2},\cdots;H_{y1},H_{y2},\cdots)^T$$
$$(A)=\begin{pmatrix}(H) & (HE)\\(EH) & (E)\end{pmatrix}$$
$$(U)=\text{unity matrix}$$

is solved numerically with routines of the well-known EISPACK package [41]. The original form of (A) (a typical structure is illustrated in Fig. 2(b)) is rearranged in the numerically more convenient form of a banded matrix (Fig. 2(c)). The matrix (A) is real, but not symmetric, and the eigenvalue solutions of (7) may include conjugate complex solutions. For appropriate wave-guiding structures, complex solutions are propagation factors of complex waves, e.g., [26], [43]–[45]. The matrix eigensystem (7) solution by the EISPACK package utilizes the QR-procedure [41], i.e., the decomposition into a product of a unitary matrix Q and the upper right triangular matrix R. The eigenvalues and eigenvectors are found by an iterative process [41], including the complex solutions.

Instead of introducing the boundary conditions given by (3), equivalently satisfactory sets of conditions are possible

$$H_x:\ H_{z1}=H_{z4},\ H_{z2}=H_{z3},\ E_{z1}=E_{z4},\ E_{z2}=E_{z3},\ E_{z3}=E_{z4}$$
$$H_y:\ H_{z1}=H_{z2},\ H_{z3}=H_{z4},\ E_{z1}=E_{z2},\ E_{z3}=E_{z4},\ E_{z2}=E_{z3}. \qquad (8)$$

The third possibility, a mixture of (3) and (8), i.e.

$$H_x:\ H_{z1}=H_{z4},\ H_{z2}=H_{z3},\ E_{z1}=E_{z4},\ E_{z2}=E_{z3},\ E_{z3}=E_{z4}$$
$$H_y:\ E_{z1}=E_{z4},\ E_{z2}=E_{z3},\ H_{z1}=H_{z4},\ H_{z2}=H_{z3},\ H_{z1}=H_{z2} \qquad (9a)$$

or

$$H_x:\ E_{z1}=E_{z2},\ E_{z3}=E_{z4},\ H_{z1}=H_{z2},\ H_{z3}=H_{z4},\ H_{z1}=H_{z4}$$
$$H_y:\ H_{z1}=H_{z2},\ H_{z3}=H_{z4},\ E_{z1}=E_{z2},\ E_{z3}=E_{z4},\ E_{z2}=E_{z3} \qquad (9b)$$

has also been investigated. The boundary conditions given by (8) and (9) lead to expressions similar to those of (4) and (5). The convergence behavior for these three equivalent, but for the numerical treatment somewhat different, cases is illustrated in Fig. 3 in the example of the shielded dielectric image line [26] for the first and the fourth mode. The finite-difference results of the boundary condition sets based on (3) (curve 1), the sets based on (8) (curve 2), and the mixture of the two based on (9a) (curve 3) are plotted against the number N of node points in the x and y direction of a uniform mesh. Equation (9b) leads to nearly identical results with (9a) (curve 3) and therefore is not shown. The results are compared with the mode-matching method [26] for a variable number of higher order modes $M=N/2$ considered in the mode-matching process. Good agreement may be stated for $N=20$; the relative errors ΔF correspond to the mode-matching solution for $M=10$. Further, nearly identical convergence behavior for the

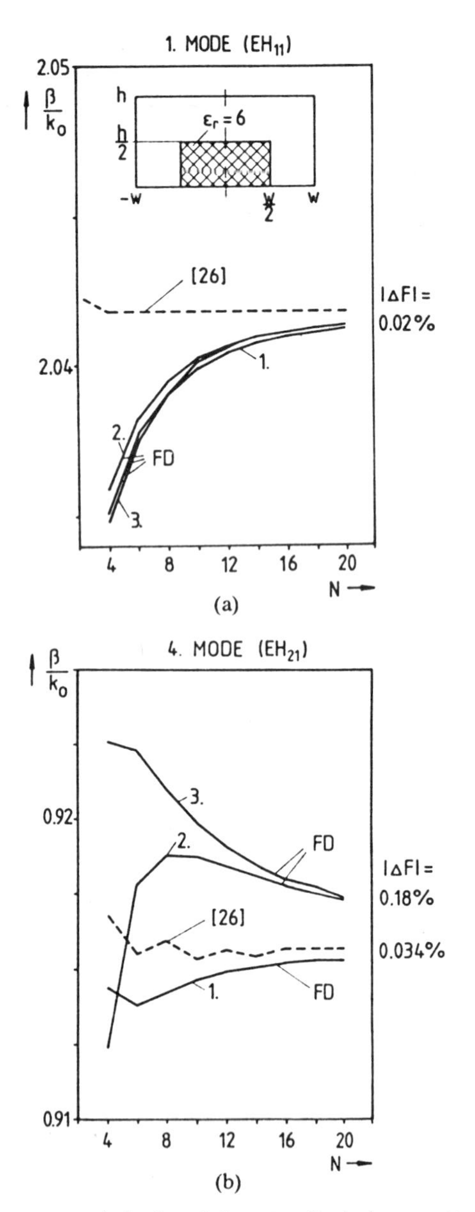

Fig. 3. Convergence behavior of the normalized phase constant β/k_0 (k_0 = free-space wavenumber) versus the number N of node points in x and y direction, respectively. Finite-difference solution for three different sets of boundary conditions (1,2,3), where 1 corresponds to equations (3) and is utilized in this paper. -------- solution of the mode-matching technique of [26] where $M = N/2$ higher order modes are considered. (a) Fundamental mode. (b) Fourth mode.

three boundary sets (curves 1,2,3) may be perceived concerning the fundamental mode (Fig. 3(a)). The same is true for most higher order modes. For some particular modes, however, e.g., the EH_{21} mode, curve 1 (i.e., the boundary conditions of (3)) provides the best convergence; these equations are therefore utilized in this paper.

III. Results

As stated above, the finite-difference formulation in terms of the transverse components H_x and H_y given in (1)–(7) circumvents the spurious-mode problem, in contrast to the formulation in terms of the longitudinal components E_z and H_z. This is demonstrated by Figs. 4 and 5 in the examples of the channel guide (Fig. 4(a) and (b)) and the dielectric waveguide (Fig. 5(a) and (b)). Figs. 4(b) and 5(b), respectively, correspond to the E_z, H_z formulation. Consequently, spurious modes occur, whereas the results of the H_x, H_y formulation (Figs. 4(a) and 5(a)) are free from spurious modes. For the structures under consideration, the mode designations prevailing in the literature have been chosen throughout this paper, i.e., for Figs. 4 and 5 those of [29], [2], and [11].

Good agreement between our results and those of Marcatili [2] may be observed in Fig. 4(a) for the channel guide. This is especially true for higher frequencies, where the approximations of [2] are considered to be more accurate since the field is concentrated increasingly in the regions taken into account in the mode-matching procedure of [2]. To facilitate the comparison, the normalizations B and V_1 [2] for the propagation constant and the frequency, respectively, are used. Concerning the higher order modes, only moderate agreement with [29] is obtained, but like there, an additional H_{41}^y mode may be perceived.

In Fig. 5(a), for the square dielectric waveguide, good agreement between Marcatili's results [2] and our results may be stated, whereas Schweig's results [33], with the exception of the fundamental mode, deviate considerably, especially concerning his "degenerate" modes. The two additionally observed modes between E_{11}^y, E_{21}^y and E_{21}^y,'E_{12}^y are designated with E_{12}^x, E_{21}^x, according to Goell [11]. The comparison with the phase constants of Goell [11] (Fig. 5(c)) shows good agreement. The same is true for a comparison with the available electric wall symmetric results by the mode-matching technique of [23] (Fig. 5(d)). Our calculation of the E_z, H_z formulation for the square dielectric waveguide of high permittivity, Fig. 5(b), indicates that the spurious solutions are found to exist mostly in the range $0.5 < B < 1.0$, as has already been stated in [33] for lower permittivity values. In Fig. 5, the normalized frequency V_2 according to [33] is used. The dispersion curve of the E_{21}^x mode crosses those of the E_{21}^y, E_{12}^x modes (Fig. 5(c)), corresponding to Goell [11]; this effect is increased for higher permittivity values (Fig. 5(a)). The E_{21}^x mode has not been calculated in [2] and [23].

Many examples analyzed by the exact mode-matching technique of [23], [25], and [26] are available for the dielectric image line for many frequency and permittivity ranges. A verification of the finite-difference method by comparison with the previous results for this structure (Fig. 6) is particularly indicated, therefore.

The normalized phase constants β/k_0 (k_0 = free-space wavenumber) versus normalized frequency V_S for a shielded dielectric image line (Fig. 6(a)) agree increasingly well with the related lateral open structure of [23] if the shield dimensions are chosen to be sufficiently large: curve 1 ($d = 2h$, $a = 2w$), curve 2 ($d = 4.8h$, $a = 4.8w$), curve 3 ($d = 4.8h$, $a = 10w$). Deviations are expected near the

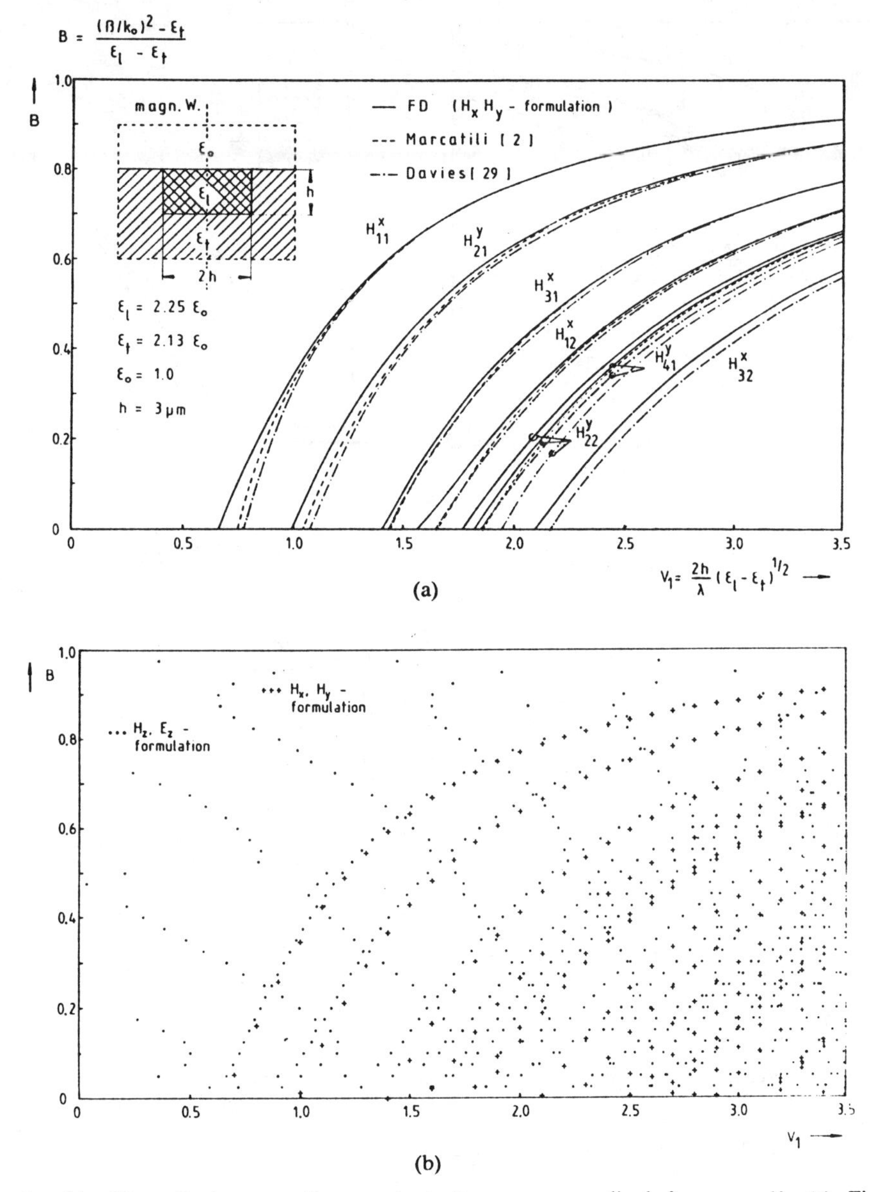

Fig. 4. Channel guide. Normalized propagation constant B versus normalized frequency V_1. (a) Finite-difference formulation in terms of H_x, H_y. (b) Finite-difference formulation in terms of H_z, E_z (own calculations).

cutoff frequency of the fundamental EH_{11} mode because of the influence of the shield, whereas the real open structure [23] exhibits no low-frequency cutoff. The modes are designated according to [23].

Fig. 6(b) shows the propagation constant $\gamma = j\beta$ (or α below cutoff) normalized with k_0 of a dielectric image line shielded with a conventional rectangular *Ku*-band waveguide housing (15.8 mm×7.9 mm). Included is the nonpropagating mode range below the corresponding cutoff frequencies. For simplicity, the corresponding real α values are plotted in the same diagram as in [42], but, for lucidity, in the opposite direction. Between about 13.8 and 16.2 GHz, the eigenvalue solution leads to a complex propagation constant $\gamma_{cw} = \pm \alpha_{cw} \pm j\beta_{cw}$, in spite of the assumption that the shielded image line is lossless. This apparent contradiction is already explained in [26] by complex waves [43], [44], which indicate power transmission with opposite signs: in the forward z direction inside the dielectric region, in the backward direction outside, or vice versa [45]. The affinity to leakage effects stated in [16]–[18] is obvious.

Fig. 6(c), where the normalized propagation constants are plotted against the permittivity ϵ_r, allows the modes to be assigned directly to rectangular waveguide modes ($\epsilon_r = 1$) at finite frequencies (e.g., 14 GHz). A comparison between the finite-difference results and those obtained by the mode-matching method [26] shows good agreement, as indicated in Fig. 6(b) and (c).

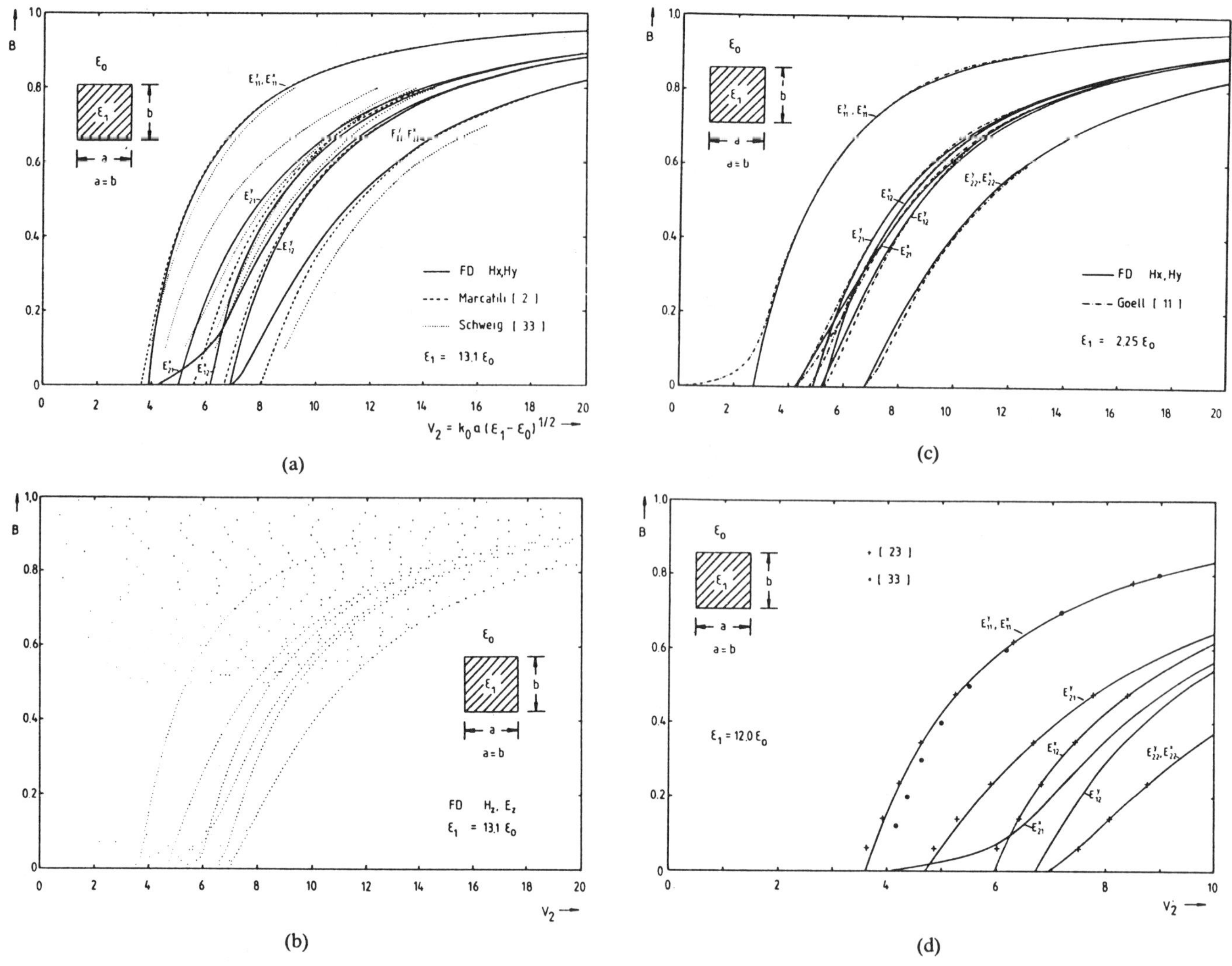

Fig. 5. Dielectric waveguide. Normalized propagation constant B versus normalized frequency V_2, with k_0 the free-space wavenumber. (a) $\epsilon_1 = 13.1\epsilon_0$; comparison with Marcatili [2] and Schweig and Bridges [33]. (b) Finite-difference formulation in terms of H_z, E_z; $\epsilon_1 = 13.1\epsilon_0$ (own calculations). (c) $\epsilon_1 = 2.25\epsilon_0$; comparison with Goell [11]. (d) $\epsilon_1 = 12\epsilon_0$; comparison with Solbach and Wolff [23], and with Schweig and Bridges [33].

Fig. 7 presents normalized propagation constants versus the normalized frequency hk_0 (h = rib height, k_0 = free-space wavenumber) of the ridge guide. The comparison with mode-matching results of [27] shows excellent agreement; this has also been stated for the coupled ridged guide.

For the insulated image guide (Fig. 8), the results of the finite-difference method agree well with the related mode-matching values[1] of [46]. For the inverted strip guide (Fig. 9), excellent agreement with the mode-matching results of [46] is obtained.

Since no spurious modes occur, the finite-difference analysis described here is particularly appropriate for more complicated waveguides, such as the strip-slab guide (Fig. 10).

[1] Note that Fig. 5(a) and (b) in [46] should obviously be interchanged.

IV. Conclusions

A finite-difference analysis for a class of rectangular dielectric waveguide structures is presented which is formulated directly in terms of the simple wave equation for the transverse components of the magnetic fields. This leads to an eigenvalue problem which is free from the troublesome problem of spurious modes. A graded mesh permits the optimum use of the available computer capabilities. The analysis allows the investigation of hybrid-mode conversion effects, such as complex waves, at frequencies where the modes are not yet completely bound to the core of the highest permittivity, as well as at frequen-

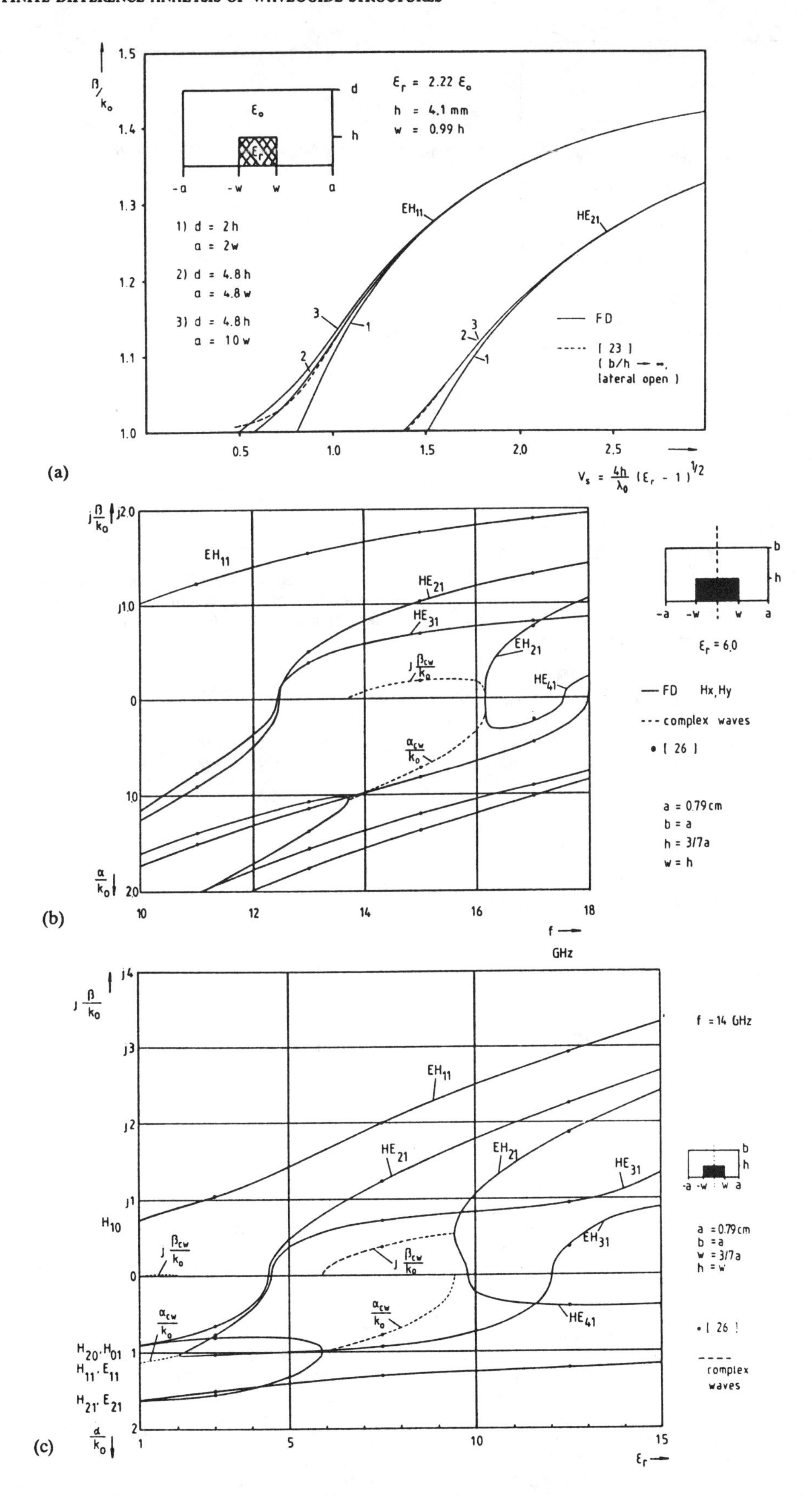

Fig. 6. Dielectric image line. (a) Normalized phase constant β/k_0 (k_0 = free-space wavenumber) versus normalized frequency. Approximation of the lateral open structure of [23] by shield dimensions 1) $d=2h$, $a=2w$; 2) $d=4.8h$, $a=4.8w$; 3) $d=4.8h$, $a=10w$. (b) Propagation constant $\gamma=\{j\beta;\alpha\}$ normalized to k_0 plotted against frequency; $\epsilon_r=6$; shield dimensions $2a=15.799$ mm, $b=a$ (*Ku*-band waveguide housing). ------- complex wave. (c) Propagation constant $\gamma=\{j\beta;\alpha\}$ normalized to k_0 plotted against permittivity ϵ_r; $f=14$ GHz. --------- complex waves.

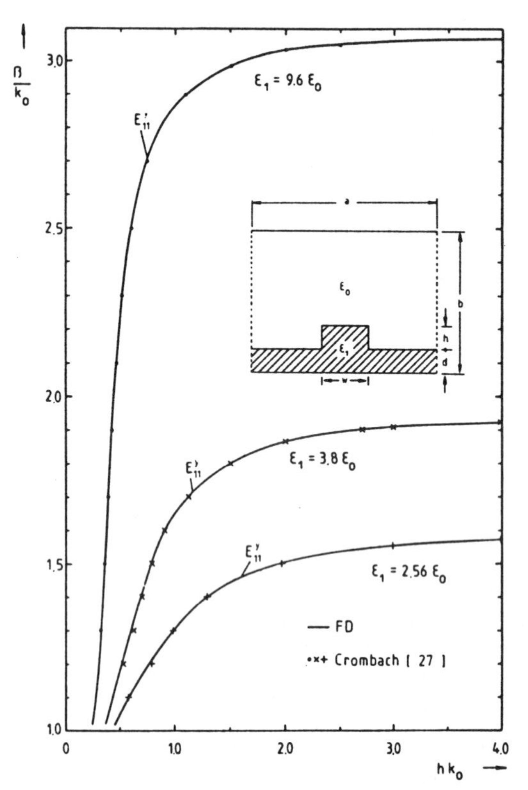

Fig. 7. Ridge guide with $w = 2h$, $d = h$, $a = 100h$, and $b = 6h$. Normalized propagation constant versus normalized frequency (h = rib height, k_0 = free-space wavenumber) for different permittivities.

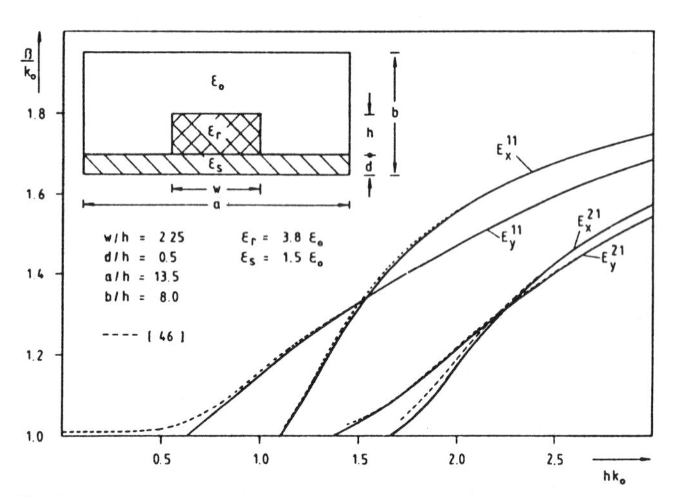

Fig. 8. Insulated image guide with structure according to [46]. Normalized propagation constant versus normalized frequency.

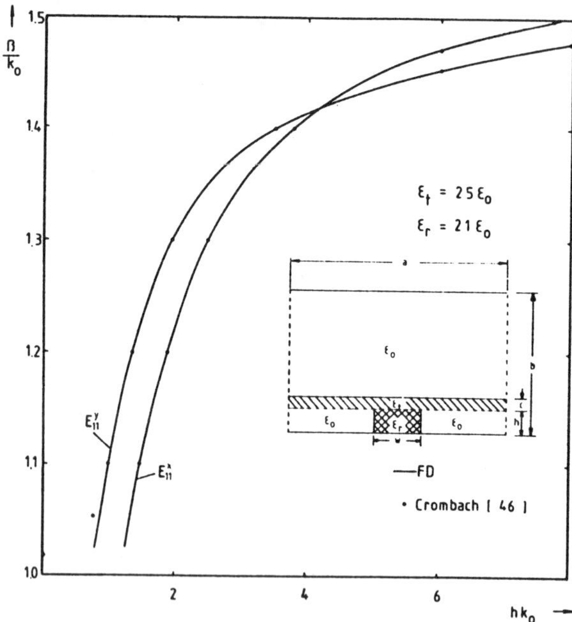

Fig. 9. Inverted strip guide. Normalized propagation constant versus normalized frequency (h = inverted strip height, k_0 = free-space wavenumber): $w/h = 2$, $c/h = 0.5$, $b/h = 8$, $a/b = 20$.

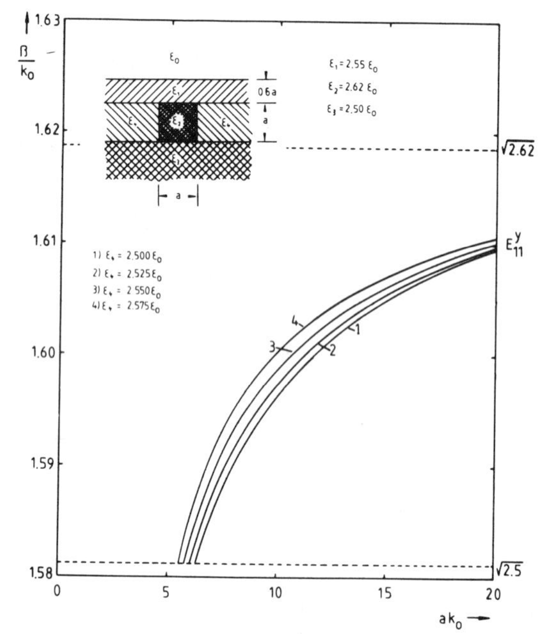

Fig. 10. Strip-slab guide. Normalized propagation constant versus normalized frequency (a = strip height, k_0 = free-space wavenumber); single structure with different permittivity values for ϵ_4.

cies below cutoff. The calculated dispersion curves compare well with results available from other methods.

References

[1] D. D. King, "Circuit components in dielectric images lines," *IRE Trans. Microwave Theory Tech.*, vol. MTT-3, pp. 35–39, Dec. 1955.

[2] E. A. J. Marcatili, "Dielectric rectangular waveguide and directional coupler for integrated optics," *Bell Syst. Tech. J.*, vol. 48, pp. 2071–2102, Sept. 1969.

[3] W. V. McLevidge, T. Itoh, and R. Mittra, "New waveguide structures for millimeter-wave and optical integrated circuits," *IEEE Trans. Microwave Theory Tech.*, vol. MTT-23, pp. 788–794, Oct. 1975.

[4] R. M. Knox, "Dielectric waveguide microwave integrated circuits —An overview," *IEEE Trans. Microwave Theory Tech.*, vol. MTT-24, pp. 806–814, Nov. 1976.

[5] T. Itoh, "Application of gratings in a dielectric waveguide for leaky-wave antennas and band-reject filters," *IEEE Trans. Microwave Theory Tech.*, vol. MTT-25, pp. 1134–1138, Dec. 1977.

[6] J. A. Paul and Y.-W. Chang, "Millimeter wave image-guide integrated passive devices," *IEEE Trans. Microwave Theory Tech.*, vol. MTT-26, pp. 751–754, Oct. 1978.

[7] K. Solbach, "The calculation and the measurements of the coupling properties of dielectric image lines of rectangular cross section," *IEEE Trans. Microwave Theory Tech.*, vol. MTT-27, pp. 54–58, Jan. 1979.

[8] J. A. Paul and P. C. H. Yen, "Millimeter-wave components and

six-port network analyzer in dielectric waveguide," *IEEE Trans. Microwave Theory Tech.*, vol. MTT-29, pp. 948–954, Sept. 1981.

[9] T. Itoh, "Open guiding structures for mmW integrated circuits," *Microwave J.*, pp. 113–126, Sept. 1982.

[10] S. E. Miller, "Integrated optics: An introduction," *Bell Syst. Tech. J.*, vol. 48, pp. 2059–2069, Sept. 1969.

[11] J. E. Goell, "A circular-harmonic computer analysis of rectangular dielectric waveguides," *Bell Syst. Tech. J.*, vol. 48, pp. 2133–2160, Sept. 1969.

[12] R. M. Knox and P. D. Toulios, "Integrated circuits for the millimeter through optical frequency range," in *Proc. of the Symp. Submillimeter Waves* (New York), Mar. 1970, pp. 497–516.

[13] T. Itoh, "Inverted strip dielectric waveguide for millimeter-wave integrated circuits," *IEEE Trans. Microwave Theory Tech.*, vol. MTT-24, pp. 821–827, Nov. 1976.

[14] R. M. Knox, "Dielectric waveguides: A low-cost option for IC's," *Microwaves*, pp. 56–64, Mar. 1976.

[15] M. Ikeuchi, H. Swami, and H. Niki, "Analysis of open-type dielectric waveguides by the finite element iterative method," *IEEE Trans. Microwave Theory Tech.*, vol. MTT-29, pp. 234–239, Mar. 1981.

[16] S.-T. Peng and A. A. Oliner, "Guidance and leakage properties of a class of open dielectric waveguides: Part I—Mathematical formulations," *IEEE Trans. Microwave Theory Tech.*, vol. MTT-29, pp. 843–855, Sept. 1981.

[17] A. A. Oliner, S.-T. Peng, T. I. Hsu, and A. Sanchez, "Guidance and leakage properties of a class of open dielectric waveguides: Part II—New physical effects," *IEEE Trans. Microwave Theory Tech.*, vol. MTT-29, pp. 855–869, Sept. 1981.

[18] G. L. Matthaei, "A note concerning modes in dielectric waveguide gratings for filter applications," *IEEE Trans. Microwave Theory Tech.*, vol. MTT-31, pp. 309–312, Mar. 1983.

[19] G. L. Matthaei, D. Park, Y. M. Kim, and D. L. Johnson, "A study of the filter properties of single and parallel-coupled dielectric-waveguide gratings," *IEEE Trans. Microwave Theory Tech.*, vol. MTT-31, pp. 825–835, Oct. 1983.

[20] T. Yoneya and S. Nishida, "Nonradiative dielectric waveguide for millimeter-wave integrated circuits," *IEEE Trans. Microwave Theory Tech.*, vol. MTT-29, pp. 1188–1192, Nov. 1981.

[21] T. Yoneyama, F. Kuroki, and S. Nishida, "Design of nonradiative dielectric waveguide filters," *IEEE Trans. Microwave Theory Tech.*, vol. MTT-32, pp. 1659–1662, Dec. 1984.

[22] W. O. Schlosser and H. G. Unger, "Partially filled waveguides and surface waveguides of rectangular cross section," in *Advances in Microwaves*, vol. 1. New York: Academic Press, 1966, pp. 319–387.

[23] K. Solbach and I. Wolff, "The electromagnetic fields and the phase constants of dielectric image lines," *IEEE Trans. Microwave Theory Tech.*, vol. MTT-26, pp. 266–274, Apr. 1978.

[24] N. Deo and R. Mittra, "A technique for analyzing planar dielectric waveguides for millimeter wave integrated circuits," *Arch. Elek. Übertragung*, vol. 37, pp. 236–244, July/Aug. 1983.

[25] J. Strube and F. Arndt, "Three-dimensional higher-order mode analysis of transition from waveguide to shielded dielectric image line," *Electron. Lett.*, vol. 19, pp. 306–307, Apr. 1983.

[26] J. Strube and F. Arndt, "Rigorous hybrid-mode analysis of the transition from rectangular waveguide to shielded dielectric image guide," *IEEE Trans. Microwave Theory Tech.*, vol. MTT-33, pp. 391–401, May 1985.

[27] U. Crombach, "Analysis of single and coupled rectangular dielectric waveguides," *IEEE Trans. Microwave Theory Tech.*, vol. MTT-29, pp. 870–874, Sept. 1981.

[28] B. M. A. Rahman and J. B. Davies, "Finite-element analysis of optical and microwave waveguide problems," *IEEE Trans. Microwave Theory Tech.*, vol. MTT-32, pp. 20–28, Jan. 1984.

[29] B. M. A. Rahman and J. B. Davies, "Penalty function improvement of waveguide solution by finite elements," *IEEE Trans. Microwave Theory Tech.*, vol. MTT-32, pp. 922–928, Aug. 1984.

[30] B. M. A. Rahman and J. B. Davies, "Finite-element solution of integrated optical waveguides," *J. Lightwave Technol.*, vol. LT-2, pp. 682–687, Oct. 1984.

[31] M. Koshiba, K. Hayata, and M. Suzuki, "Approximate scalar finite-element analysis of anisotropic optical waveguides with off-diagonal elements in a permittivity tensor," *IEEE Trans. Microwave Theory Tech.*, vol. MTT-32, pp. 587–593, June 1984.

[32] M. Koshiba, K. Hayata, and M. Suzuki, "Improved finite-element formulation in terms of the magnetic field vector for dielectric waveguides," *IEEE Trans. Microwave Theory Tech.*, vol. MTT-33, pp. 227–233, Mar. 1985.

[33] E. Schweig and W. B. Bridges, "Computer analysis of dielectric waveguides: A finite-difference method," *IEEE Trans. Microwave Theory Tech.*, vol. MTT-32, pp. 531–541, May 1984.

[34] J. B. Davies and C. A. Muilwyk, "Numerical solution of uniform hollow waveguides of arbitrary shape," *Proc. Inst. Elec. Eng.*, vol. 113, pp. 277–284, Feb. 1966.

[35] M. J. Baubien and A. Wexler, "An accurate finite-difference method for higher order waveguide modes," *IEEE Trans. Microwave Theory Tech.*, vol. MTT-16, pp. 1007–1017, Dec. 1968.

[36] J. S. Hornsby and A. Gopinath, "Numerical analysis of a dielectric-loaded waveguide with a microstrip line—Finite difference methods," *IEEE Trans. Microwave Theory Tech.*, vol. MTT-17, pp. 684–690, Sept. 1969.

[37] D. G. Corr and J. B. Davies, "Computer analysis of the fundamental and higher order modes in single and coupled microstrip," *IEEE Trans. Microwave Theory Tech.*, vol. MTT-20, pp. 669–678, Oct. 1972.

[38] S. M. Saad, "Review of numerical methods for the analysis of arbitrarily-shaped microwave and optical dielectric waveguides," *IEEE Trans. Microwave Theory Tech.*, vol. MTT-33, pp. 894–899, Oct. 1985.

[39] J.-D. Decotignie, O. Parriaux, and F. Gardiol, "Birefringence properties of twin-core fibers by finite differences," *J. Opt. Commun.*, vol. 3, no. 1, pp. 8–12, Mar. 1982.

[40] C. G. Williams and G. K. Cambrell, "Numerical solution of surface waveguide modes using transverse field components," *IEEE Trans. Microwave Theory Tech.*, vol. MTT-22, pp. 329–330, Mar. 1974.

[41] B. S. Garbow, J. M. Boyle, J. J. Dougarra, and C. B. Moler, "Matrix eigensystem routines—EISPACK guide extension," in *Lecture Notes in Computer Science*, vol. 51. Heidelberg: Springer-Verlag, 1977.

[42] R. F. Harrington, *Time-Harmonic Electromagnetic Fields*. New York: McGraw-Hill, 1961, ch. 2.7.

[43] P. J. B. Clarricoats and K. R. Slinn, "Complex modes of propagation in dielectric loaded circular waveguide," *Electron. Lett.*, vol. 1, pp. 145–146, 1965.

[44] V. A. Kalmyk, S. B. Rayevskiy, and V. P. Ygvyumov, "An experimental verification of existence of complex waves in a two-layer circular, shielded waveguide," *Radio Eng. Electron. Phys.*, vol. 23, pp. 16–19, 1978.

[45] H. Katzier and F. J. K. Lange, "Grundlegende Eigenschaften komplexer Wellen am Beispiel der geschirmten kreiszylindrischen dielektrischen Leitung," *Arch. Elek. Übertragung*, vol. 37, pp. 1–5, Jan./Feb. 1983.

[46] U. Crombach, "Wellentypen auf einzelnen und gekoppelten dielektrischen Wellenleitern," *Frequenz*, vol. 39, pp. 26–33, Jan./Feb. 1985.

The Finite-Difference–Time-Domain Method and its Application to Eigenvalue Problems

DOK HEE CHOI AND WOLFGANG J. R. HOEFER, SENIOR MEMBER, IEEE

Abstract —This paper describes the application of the finite-difference method in the time domain to the solution of three-dimensional (3-D) eigenvalue problems. Maxwell's equations are discretized in space and time, and steady-state solutions are then obtained via Fourier transform. While achieving the same accuracy and versatility as the TLM method, the finite-difference–time-domain (FD–TD) method requires less than half the CPU time and memory under identical simulation conditions. Other advantages over the TLM method include the absence of dielectric boundary errors in the treatment of 3-D inhomogeneous planar structures, such as microstrip. Some numerical results, including dispersion curves of a microstrip on anisotropic substrate, are presented.

I. Introduction

THE TLM METHOD has been successfully applied to various microwave circuit problems for more than ten years. The special advantages of the TLM technique over other numerical methods are well illustrated by Johns and Beurle [1] in their original paper on the method. Since then, several improvements have been made to this technique by various authors in order to enhance the accuracy of the solution and economize CPU time and memory space [2]–[5]. Mariki [7] has extended the TLM method to analyze anisotropic media, and Saguet and Pic [4] as well as Al-Mukhtar and Sitch [5] have employed a graded mesh to make the algorithm faster and more efficient.

Although the graded mesh algorithm reduces memory space requirements, it demands far more iterations than the original method for equal frequency resolution [5]. This is especially obvious in the case of three-dimensional (3-D) simulations where the grade ratio N requires an additional N^2-1 iterations. These requirements of large computer resources may critically limit the applicability of the TLM technique.

Thus, Saguet [16] has proposed a simplified node which reduces the number of variables to be processed and stored at each node by one third. However, this modification increases the velocity error. A further reduction in computational expenditure has been proposed by the authors [8]; instead of the original vector solution, we obtain a scalar potential solution using a scalar 3-D network. However, the scalar approach is limited to problems which lead to uncoupled modal solutions, i.e., TE and TM or LSE and LSM fields.

The major reason for the large CPU memory demand of this technique resides in the basic 3-D TLM concept. In order to represent each electromagnetic component, each 3-D unit cell requires 26 real memory spaces, 12 for pulse storage and 14 for additional network parameters. Therefore, each operation on each node involves a large number of variables, requiring considerable computer CPU space and time. Furthermore, experience has shown that the number of iterations increases with the complexity of the structure under study. For example, the accurate analysis of a finline requires easily over 1000 iterations. Given these massive requirements, we have searched for an alternative numerical technique that possesses the advantages of the TLM approach but needs fewer computer resources. As a result, a new algorithm is proposed based on both the finite-difference–time-domain (FD–TD) and TLM methods.

The FD–TD method was first formulated by Yee [6], and has been applied extensively to scattering and coupling problems with open boundaries [9]–[15], i.e., to the solution of deterministic problems. We noted the similarity between this method and the TLM method, which has been widely used in the numerical solution of the electromagnetic eigenvalue problems in the time domain. Since the TLM method is based on the computation of the impulse response of a large mesh of transmission lines, much unwanted information is usually generated.

We have therefore developed a novel procedure which increases the numerical efficiency of the time-domain approach without sacrificing its advantages. The method differs from the classical FD–TD method in the assignment of initial field values and the application of the Fourier transform to the time-domain solution. In the following, we will describe this method and its application to some typical microwave problems.

II. Yee's Algorithm

Maxwell's equations have been expressed in finite-difference form by Yee [6] to solve two-dimensional (2-D) wave scattering problems. Subsequently, 3-D scattering problems have been solved by Taflove and other workers

Manuscript received March 22, 1986; revised July 8, 1986. This work was supported in part by the National Science and Engineering Research Council of Canada.

The authors are with the Department of Electrical Engineering, University of Ottawa, Ottawa, Ontario, Canada K1N 6N5.

IEEE Log number 8610560.

Reprinted from *IEEE Trans. Microwave Theory Tech.*, vol. MTT-34, no. 12, pp. 1464–1470, Dec. 1986.

[9]–[15]. We will adopt Yee's original algorithm for the three-dimensional Maxwell's equations. Also, we will extend the concept further to include anisotropic media.

In a rectangular coordinate system, the source-free Maxwell's equations can be written as first-order hyperbolic equations

$$d\bar{E}/dt = |A_x|\,d\bar{H}/dx + |A_y|\,d\bar{H}/dy + |A_z|\,d\bar{H}/dz$$
$$d\bar{H}/dt = |B_x|\,d\bar{E}/dx + |B_y|\,d\bar{E}/dy + |B_z|\,d\bar{E}/dz \tag{1}$$

where $\bar{E} = (E_x, E_y, E_z)'$, $\bar{H} = (H_x, H_y, H_z)'$, and

$$|A_x| = \begin{bmatrix} 0 & 0 & 0 \\ 0 & 0 & -\epsilon_{yy}^{-1} \\ 0 & \epsilon_{zz}^{-1} & 0 \end{bmatrix}, |A_y| = \begin{bmatrix} 0 & 0 & \epsilon_{xx}^{-1} \\ 0 & 0 & 0 \\ -\epsilon_{zz}^{-1} & 0 & 0 \end{bmatrix}$$

$$|A_z| = \begin{bmatrix} 0 & -\epsilon_{xx}^{-1} & 0 \\ \epsilon_{yy}^{-1} & 0 & 0 \\ 0 & 0 & 0 \end{bmatrix}.$$

Here, ϵ_{xx}, ϵ_{yy}, and ϵ_{zz} are the diagonal elements of the permittivity tensor. All B_x, B_y, and B_z are expressed by replacing ϵ_{xx}, ϵ_{yy}, and ϵ_{zz} by $-\mu_{xx}$, $-\mu_{yy}$, and $-\mu_{zz}$ components, which are the diagonal elements of the permeability tensor. Then each of these scalar equations can be expressed in finite-difference form. Following Yee's nomenclature, any function of space and time is discretized

$$F^n(i,j,k) = F(i\Delta x, j\Delta y, k\Delta z, n\Delta t)$$

where $\Delta x = \Delta y = \Delta z = \Delta l$ is the space increment and Δt is the time increment. By positioning the components of $\bar{E}$ and $\bar{H}$ on the mesh as depicted in Fig. 1 and evaluating $\bar{E}$ and $\bar{H}$ at alternate half time steps, we obtain the components of Maxwell's equations.

$$\begin{aligned} H_x^{n+1/2}&(i, j+1/2, k+1/2) \\ &= H_x^{n-1/2}(i, j+1/2, k+1/2) \\ &\quad + s/\mu_{xx}(i, j+1/2, k+1/2)\big[E_y^n(i, j+1/2, k+1) \\ &\quad - E_y^n(i, j+1/2, k) + E_z^n(i, j, k+1/2) \\ &\quad - E_z^n(i, j+1, k+1/2)\big] \end{aligned} \tag{2a}$$

$$\begin{aligned} H_y^{n+1/2}&(i+1/2, j, k+1/2) \\ &= H_y^{n-1/2}(i+1/2, j, k+1/2) \\ &\quad + s/\mu_{yy}(i+1/2, j, k+1/2)\big[E_z^n(i+1, j, k+1/2) \\ &\quad - E_z^n(i, j, k+1/2) + E_x^n(i+1/2, j, k) \\ &\quad - E_x^n(i+1/2, j, k+1)\big] \end{aligned} \tag{2b}$$

$$\begin{aligned} H_z^{n+1/2}&(i+1/2, j+1/2, k) \\ &= H_z^{n-1/2}(i+1/2, j+1/2, k) \\ &\quad + s/\mu_{zz}(i+1/2, j+1/2, k)\big[E_x^n(i+1/2, j+1, k) \\ &\quad - E_x^n(i+1/2, j, k) + E_y^n(i, j+1/2, k) \\ &\quad - E_y^n(i+1, j+1/2, k)\big] \end{aligned} \tag{2c}$$

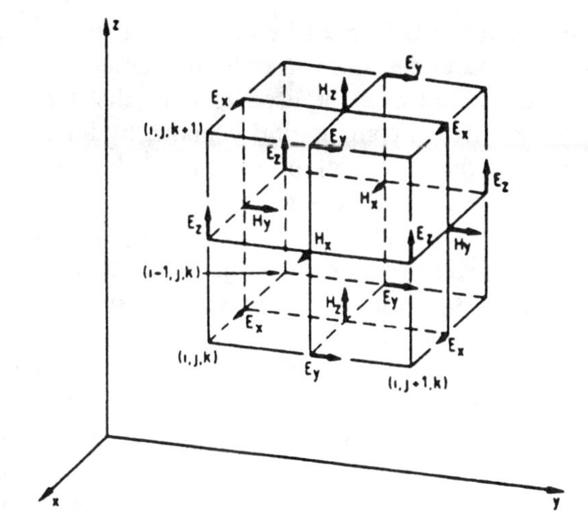

Fig. 1. The position of the field components in Yee's mesh.

$$\begin{aligned} E_x^{n+1}&(i+1/2, j, k) \\ &= E_x^n(i+1/2, j, k) + s/\epsilon_{xx}(i+1/2, j, k) \\ &\quad \cdot\big[H_z^{n+1/2}(i+1/2, j+1/2, k) \\ &\quad - H_z^{n+1/2}(i+1/2, j-1/2, k) \\ &\quad + H_y^{n+1/2}(i+1/2, j, k-1/2) \\ &\quad - H_y^{n+1/2}(i+1/2, j, k+1/2)\big] \end{aligned} \tag{2d}$$

$$\begin{aligned} E_y^{n+1}&(i, j+1/2, k) \\ &= E_y^n(i, j+1/2, k) + s/\epsilon_{yy}(i, j+1/2, k) \\ &\quad \cdot\big[H^{n+1/2}(i, j+1/2, k+1/2) \\ &\quad - H_x^{n+1/2n}(i, j+1/2, k-1/2) \\ &\quad + H_z^{n+1/2}(i-1/2, j+1/2, k) \\ &\quad - H_z^{n+1/2}(i+1/2, j+1/2, k)\big] \end{aligned} \tag{2e}$$

$$\begin{aligned} E_z^{n+1}&(i, j, k+1/2) \\ &= E_z^n(i, j, k+1/2) + s/\epsilon_{zz}(i, j, k+1/2) \\ &\quad \cdot\big[H_y^{n+1/2}(i+1/2, j, k+1/2) \\ &\quad - H_y^{n+1/2}(i-1/2, j, k+1/2) \\ &\quad + H_x^{n+1/2}(i, j-1/2, k+1/2) \\ &\quad - H_x^{n+1/2}(i, j+1/2, k+1/2)\big] \end{aligned} \tag{2f}$$

where the stability factor $s = c\Delta t/\Delta l$, and c is the velocity of light. In these expressions, E and H are normalized such that the characteristic impedance of space is unity. The condition for stability of (1) in free space is [10]

$$s \leqslant 1/\sqrt{3}. \tag{3}$$

III. Boundary Conditions

So far, a space–time mesh has been introduced and Maxwell's equations have been replaced by a system of finite-difference equations. Difficulties arise when the domain in which the field must be computed is unbounded. Since no computer can store an unlimited amount of data, a special technique must be used to limit the domain in which the numerical computation is made, by introducing so-called absorbing or soft boundary conditions. These conditions have been described by Taylor *et al.* [9], who use a simple extrapolation method, and by Taflove and Brodwin [10], who simulate the outgoing waves and use an averaging process in an attempt to account for all possible angles of propagation of the outgoing waves. Kunz and Lee [12] use the radiation condition at a large distance from the center of the scatterer to obtain an absorbing boundary condition. Mur [13] employs a second-order radiation condition to improve the accuracy of the results. Although these schemes have been used in scattering problems in the past, no ideal reflection-free boundary condition has been proposed so far.

However, in the formulation of eigenvalue problems, only "hard boundaries"—usually represented by conducting walls—occur. At these boundaries, the tangential electric and the normal magnetic field components are maintained at zero. For example, on a perfectly conducting wall in the plane $i=1$ (see Fig. 1)

$$\text{for all } n \begin{cases} E_y^n(1, j+1/2, k) = 0 \\ E_z^n(1, j, k+1/2) = 0 \\ H_x^n(1, j+1/2, k+1/2) = 0 \end{cases}$$

(the third condition is implicit in the previous two, but its implementation reduces numerical errors).

IV. Initial Values

In most scattering problems, an impulsive or sinusoidal plane wave is injected at the beginning of the computation. However, in eigenvalue problems, the direction of the propagation vector is usually not known and depends on the space coordinates and on the eigenvalue that is to be found. In these cases, the logical choice is an isotropic pulse that propagates in the radial direction. The spatial pulse envelope should be wide enough with respect to the mesh size not to accumulate numerical errors due to overshoot and ringing as it propagates through the space lattice.

A better way to start the computation is to estimate the field distribution of the desired mode in the structure first and then choose the initial value accordingly. The experienced researcher usually has a good idea of the approximate modal field distributions in a structure and is therefore able to make an educated guess of the steady-state field pattern for a particular eigenmode. This procedure is equivalent to the excitation of a TLM mesh with a weighted impulse distribution, and is somewhat similar to the way in which one chooses appropriate basis functions in the spectral-domain approach.

V. Outline of the Numerical Procedure

The application of the FD–TD method will be discussed using a rectangular resonator as an example. A continuous medium is replaced with a 3-D uniform mesh. To solve the system of equations (2) in this mesh, initial values must be assigned first as described in the previous section. For a rectangular-type resonator, a simple sinusoidal function is an appropriate choice for the dominant mode eigenvalue. As n increases, the discrete time functions for $\bar{E}$ and $\bar{H}$ fields evolve towards the steady state which is characteristic of the desired mode in the geometry. In this way, the evolution of all six field components is obtained simultaneously at discrete time points $n\,\Delta t$. The final steady-state field distribution may be calculated by taking the time average of the time-domain solution at each mesh point. Thus, the steady-state solution is given by

$$F(i_0, j_0, k_0) = \sum_n |F^n(i_0, j_0, k_0)|/N. \tag{4}$$

This simple procedure to obtain the final field distribution is another advantage over the TLM method, which requires two simulations for finding the fields of a given mode.

In eigenvalue problems, the steady-state solution is a time-harmonic function, from which the eigenvalues can

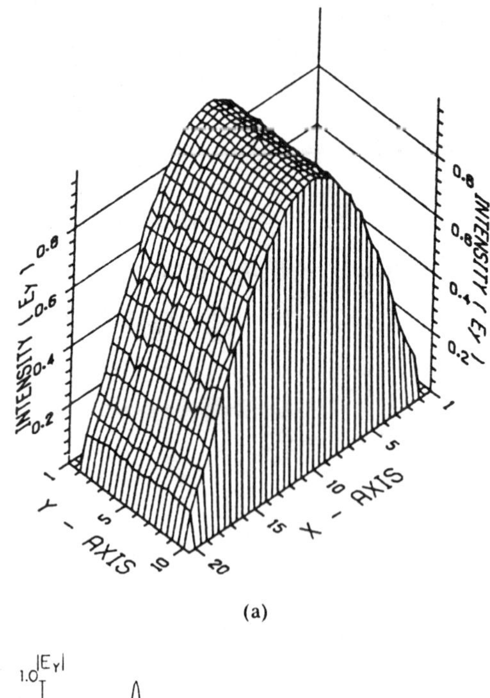

(a)

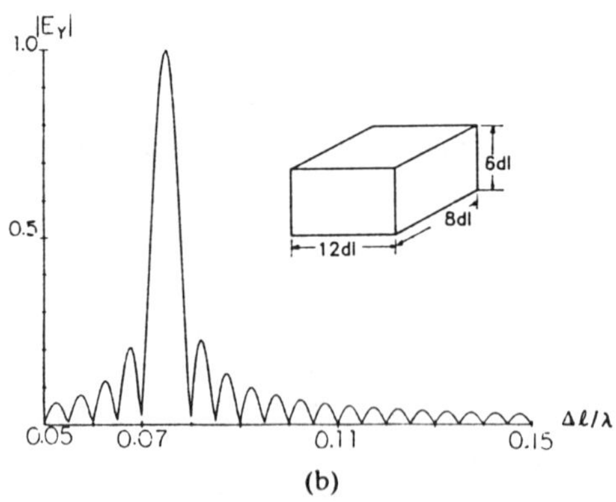

(b)

Fig. 2. (a) Field distribution of an empty rectangular resonator obtained with the FD–TD method. (b) Output spectrum obtained with both the FD–TD and TLM methods under identical conditions.

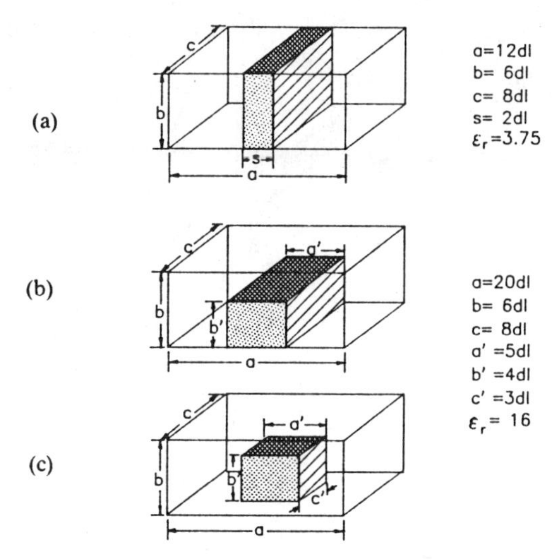

Fig. 3. Three-dimensional inhomogeneous resonators analyzed in this study.

TABLE I
COMPARISON OF RESULTS FOR THE NORMALIZED RESONANT FREQUENCIES $\Delta l/\lambda$ OBTAINED WITH THE TLM AND FD-TD METHODS UNDER IDENTICAL SIMULATION CONDITIONS FOR THE GEOMETRIES IN FIG. 3

Fig.3	Mode	s/a	TRM	TLM (CPU time)	FD-TD (CPU time)
a)	LSE	1/6	0.05220	0.0516 (144)	0.0517 (51)
		1/3	0.0445	0.0440 (145)	0.0442 (51)
b)	Hybrid			0.0278 (357)	0.0278 (117)
c)	Hybrid			0.0405 (357)	0.0405 (117)

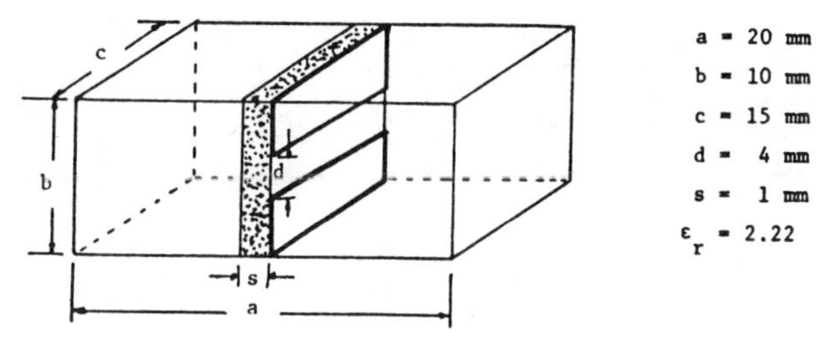

Fig. 4. Finline cavity.

TABLE II
RESONANT FREQUENCIES OF THE UNILATERAL FINLINE CAVITY IN FIG. 4, OBTAINED WITH VARIOUS METHODS

	Saguet [16]		This method	
	S.D.A.	Variable Mesh TLM	TDFD	T.L.M.
Resonant Frequency (GHz)	10.77	10.14	10.74	10.74
Number of Iterations		1000	600	600
CPU Time(s)			170	380

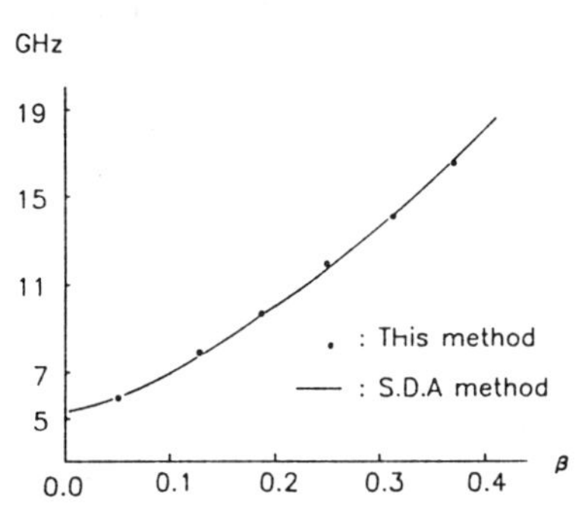

Fig. 5. Dispersion diagram of unilateral finline with the cross-sectional geometry given in Fig. 4.

be extracted by discrete Fourier transform, as in the TLM method

$$S(f)=\sum_{n} F^{n}(i_0, j_0, k_0)\exp(-j2\pi snf). \qquad (5)$$

Both the stability factor s and the number of iterations n strongly affect the spectral response.

In order to test this algorithm for validity, it has been applied to a simple rectangular cavity with sides $12\,\Delta l \times 6\,\Delta l \times 8\,\Delta l$. We have assumed a dominant TE_{101} mode in the initial value assignment. The time-domain solution is given in Fig. 2(a). Discontinuous field figures are due to the numerical error caused by the finite-difference form of (2). Fig. 2(b) compares the frequency responses obtained with the FD–TD and TLM methods under identical conditions. The responses are not distinguishable. Five hundred iterations have been used. The peak of the solution is located at $\Delta l/\lambda = 0.0750$ in both methods. The exact analytical solution is 0.07511. Even though a small number of meshes is used in this algorithm when compared with the scattering problems in [9]–[13], it is noted that the accuracy in the solution of the eigenvalue problems is better than that of the scattering problems by one order of magnitude.

VI. NUMERICAL RESULTS

We have applied this technique to most of the examples described in the TLM literature and obtained practically identical results. The method requires less than one-half of the CPU time spent by the equivalent TLM program under identical conditions, including the initial excitation distribution. Furthermore, while the TLM procedure requires 22 real memory stores per 3-D node in an isotropic dielectric, the FD–TD method requires only seven real memory stores per node. Fig. 3(a), (b), and (c) shows structures for which solutions have been computed with this method. The dominant resonant frequencies of these structures are given and compared with the TLM solution in Table I. The inhomogeneous rectangular cavity of Fig. 3 (b) and (c) illustrates the capability of this algorithm to solve hybrid field problems. The number of nodes chosen in each problem is the same as that employed in the TLM solution.

Furthermore, we have computed the resonant frequency of a finline resonator (Fig. 4.) treated previously by Saguet [16]. Results are compared in Table II, which includes a value obtained with the spectral-domain method by Saguet. Fig. 5 shows the dispersion characteristics of a finline with the same cross section, as obtained with our method. Results calculated with our spectral-domain program are also shown in the same figure. In order to compare convergence of both time-domain methods, solutions obtained after every fifth iteration are drawn in Fig. 6. The results show virtually identical convergence.

To show the versatility of this method, the characteristics of a microstrip resonator on anisotropic substrate are

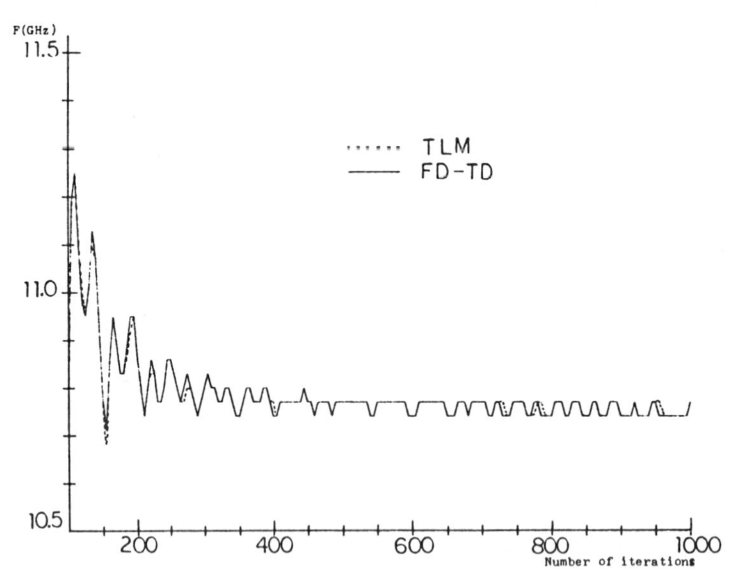

Fig. 6. Stability and convergence of the TLM and FD–TD methods as a function of the number of iterations. The solution of the finline problem in Fig. 4 is represented.

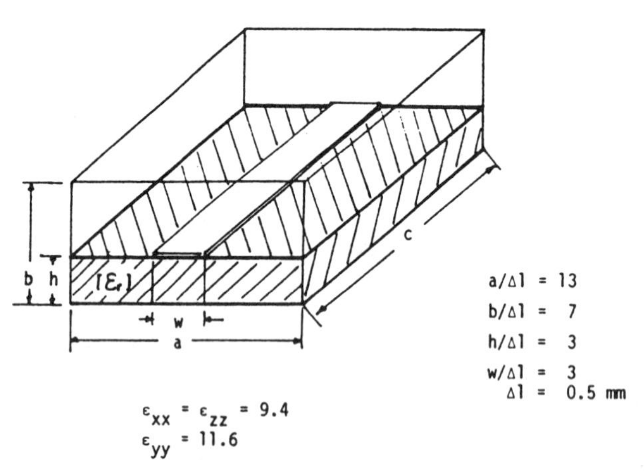

Fig. 7. Microstrip cavity on anisotropic substrate.

TABLE III
DOMINANT RESONANT FREQUENCIES OBTAINED BY BOTH THE TLM AND THE FD–TD METHODS

	h \ c/Δl	6	8	10	15	20
TLM (GHz)	3Δl	15.72	12.24	10.14	7.14	5.46
	4Δl	14.22	11.16	9.18	6.36	4.98
	Average	14.97	11.70	9.66	6.75	5.22
FD - TD (GHz)		14.64	11.52	9.54	6.78	5.28

computed in the last example, shown in Fig. 7. Several different resonant frequencies obtained by changing the length c are tabulated in Table III. It is well known [17] that the TLM simulation of 3-D inhomogeneous planar structures involves dielectric interface ambiguity. The best way to resolve this error is to employ two dielectric substrate thicknesses differing by one Δl. In our case, $3\Delta l$ and $4\Delta l$ are used. The final result is obtained by taking the average of the solutions obtained for these two values. In order to illustrate this process, frequency spectra obtained with the TLM method for the two cases where c is equal to $10\Delta l$ are shown in Fig. 8. The solution obtained with the FD–TD method is also drawn in the same figure. As expected, the latter solution is located exactly between the two values obtained with the TLM method. This clearly illustrates the accuracy and convenience of the FD–TD method in such situations. Fig. 9 shows the dispersion characteristics of the microstrip which has the same cross section as that in Fig. 7. Again, both methods give very similar results except at higher frequencies, where the discretization errors associated with both methods become more pronounced, and their differences are more visible.

VII. CONCLUSIONS

The proposed new application of the FD–TD method to 3-D eigenvalue problems gives practically the same results

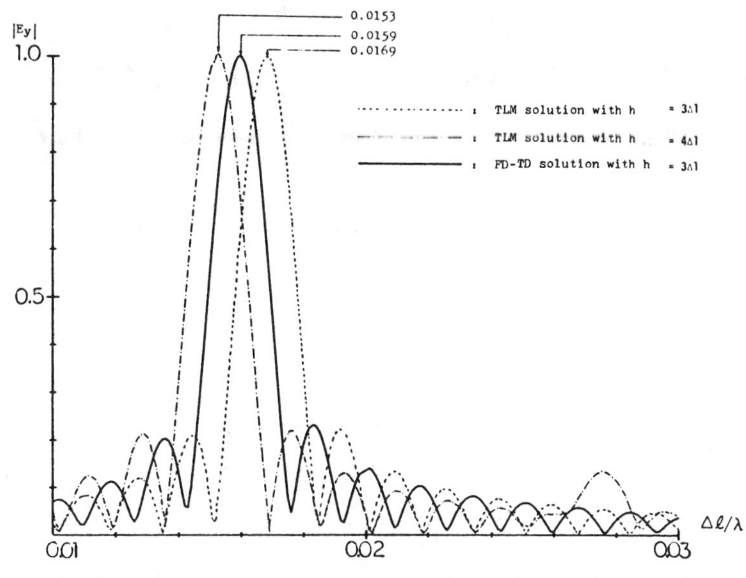

Fig. 8. Frequency spectra for the microstrip problem in Fig. 7, showing the effect of dielectric interface error in the TLM solution.

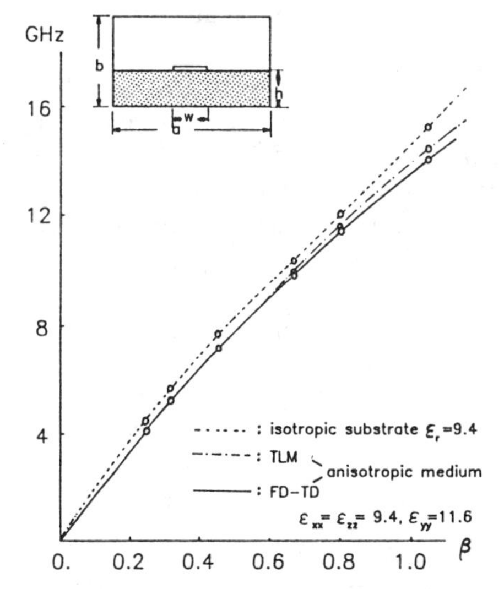

Fig. 9. Dispersion diagram of the anisotropic microstrip in Fig. 7.

as the TLM method under identical simulation conditions. However, the overall CPU time and storage requirements are typically less than half those needed in the TLM solution. Other advantages reside in the ease with which field distributions can be computed, and in the elimination of dielectric boundary error in planar structures. Further efforts are being made to implement losses, variable mesh size, and nonlinarity in the FD–TD procedure.

References

[1] P. B. Johns and R. L. Beurle, "Numerical solution of two-dimensional scattering problems using a transmission line matrix," *Proc. Inst. Elec. Eng.*, vol. 118, no. 9, pp. 1203–1208, Sept. 1971.

[2] S. Akhtarzad and P. B. Johns, "Solution of Maxwell's equations in three space dimensions and time by the TLM method of numerical analysis," *Proc. Inst. Elec. Eng.*, vol. 122, no. 12, pp. 1344–1348, Dec. 1975.

[3] Y. C. Shih and W. J. R. Hoefer, "The accuracy of TLM analysis of finned rectangular waveguides," *IEEE Trans. Microwave Theory Tech.*, vol. MTT-28, pp. 743–746, July 1980.

[4] P. Saguet and E. Pic, "Le maillage rectangulaire et le changement de maille dans la methode TLM en deux dimensions," *Electron. Lett.*, vol. 17, no. 7, pp. 277–279, Apr. 1981.

[5] D. A. Al-Mukhtar and J. E. Sitch, "Transmission line matrix method with irregularly graded space," *Proc. Inst. Elec. Eng.*, vol. 128, no. 6, pp. 299–305, Dec. 1981.

[6] K. S. Yee, "Numerical solution of initial boundary value problems involving Maxwell's equations in isotropic media," *IEEE Trans. Antennas Propagat.*, vol. AP-14, pp. 302–307, May 1966.

[7] G. E. Mariki, "Analysis of microstrip lines on inhomogeneous anisotropic substrates by the TLM numerical technique," Ph.D. thesis, Univ. California, Los Angeles, CA, June 1978.

[8] D. Choi and W. J. R. Hoefer, "The simulation of three dimensional wave propagation by a scalar TLM model," *IEEE MTT-S Symp. Dig.*, 1984, pp. 70–71.

[9] C. D. Taylor, D. H. Lam, and T. H. Shumpert, "Electromagnetic pulse scattering in time varying inhomogeneous media," *IEEE Trans. Antennas Propagat.*, vol. AP-17, pp. 586–589, Sept. 1969.

[10] A. Taflove and M. E. Brodwin, "Numerical solution of steady state electromagnetic scattering problems using the time dependent Maxwell's equations," *IEEE Trans. Microwave Theory Tech.*, vol. MTT-23, pp. 623–630, Aug. 1975.

[11] R. Holland, "Threde: A free-field EMP coupling and scattering code," *IEEE Trans. Nucl. Sci.*, vol. NS-24, pp. 2416–2421, Dec. 1977.

[12] K. S. Kunz and K. M. Lee, "A three-dimensional finite-difference solution of the external response of an aircraft to a complex transient EM environment: Part 1—The method and its implementation," *IEEE Trans. Electromagn. Compat.*, vol. EMC-20, pp. 328–33, May 1978.

[13] G. Mur, "Absorbing conditions for the finite-difference approximation of the time-domain electromagnetic-field equations," *IEEE Trans. Electromagn. Compat.*, vol. EMC-23, Nov. 1981.

[14] A. Taflove and K. R. Umashankar, "A hybrid moment method/finite difference time domain approach to electromagnetic coupling and aperture penetration into complex geometries," *IEEE Trans. Antennas Propagat.*, vol. AP-30, pp. 617–627, July 1982.

[15] K. R. Umashankar and A. Taflove, "A novel method to analyze electromagnetic scattering of complex objects," *IEEE Trans. Electromagn. Compat.*, vol. EMC-24, pp. 397–405, Nov. 1982.

[16] P. Saguet, "Le maillage parallelépipédique et le changement de maille dans la méthode TLM en trois dimensions," *Electron. Lett.* vol. 20, no. 5, pp. 222–24, Mar. 15, 1984.

[17] W. J. R. Hoefer, "The transmission-line matrix method—Theory and applications," *IEEE Trans. Microwave Theory Tech.*, vol. MTT–33, no. 10, pp. 882–893, Oct. 1985.

Analysis of Arbitrarily Shaped Two-Dimensional Microwave Circuits by Finite-Difference Time-Domain Method

WOJCIECH K. GWAREK

Abstract —The paper presents a version of the finite-difference time-domain method adapted to the needs of S matrix calculations of microwave two-dimensional circuits. The analysis is conducted by simulating the wave propagation in the circuit terminated by matched loads and excited by a matched pulse source. Various aspects of the method's accuracy are investigated. Practical computer implementation of the method is discussed and an example of its application to an arbitrarily shaped microstrip circuit is presented. It is shown that the method in the proposed form is an effective tool of circuit analysis in engineering applications. The method is compared to two other methods used for a similar purpose, namely the contour integral method and the transmission-line matrix method.

I. Introduction

A TWO-DIMENSIONAL circuit as understood here is a circuit in which the fields may be characterized by a scalar function $V(x,y)$ which obeys a two-dimensional wave equation

$$\nabla_{xy}^2 V(x,y,t) - \beta^2 \frac{\partial^2 V(x,y,t)}{\partial t^2} = 0 \tag{1}$$

with proper boundary conditions.

There are many microwave circuits characterized accurately or approximately by (1). As examples we suggest stripline or microstrip junctions or resonators, and also some types of waveguide discontinuities. That is why solving this equation was investigated by many researchers. If the boundary is of complicated shape, only numerical methods can be used. These methods can be divided into two groups. The methods of the first group require a certain amount of analytical preprocessing before the particular problem may be solved by a computer. As an example we propose the Green's function method supported by segmentation techniques [1]. Another group of methods assumes that the analytical preprocessing should be practically nonexistent and that it is up to the computer to do the entire job. In this group we find the contour integral method [2], the transmission-line matrix method [3], and the finite-difference method, which will be the subject of this paper.

The finite-difference time-domain method (FD-TD) was first introduced by Yee [4] and since then has been applied by many authors. These applications, however, were concentrated in the domain of scattering [5], wave absorption [17], and accelerator physics [6]. Application of the FD-TD method to microwave circuit analysis has so far attracted little attention.

The aim of this paper is to show that the FD-TD method adapted to the needs of 2-D circuit analysis can be a strong competitor to the other mentioned methods. It can lead to a universal and effective computer program capable of solving a wide range of practical problems. The paper extends the ideas presented in [7] and [8]. It presents an FD-TD method of 2-D circuit analysis based on pulse excitation by a matched source. It discusses various aspects of the method's accuracy and presents the author's experience with its implementation on an IBM PC AT computer.

II. Outline of the FD-TD Method

In the FD-TD method instead of solving the second-order equation (1) a pair of first-order equations is solved:

$$\nabla V(x,y,t) = -L_s \frac{\partial \boldsymbol{J}(x,y,t)}{\partial t} \tag{2}$$

$$\nabla \cdot \boldsymbol{J}(x,y,t) = -C_s \frac{\partial V(x,y,t)}{\partial t}. \tag{3}$$

In microwave planar circuits, the variables and constants in (2) and (3) have the following interpretations: V—voltage, $\boldsymbol{J}$—surface current density, C_s—capacitance of a unitary square of the circuit, L_s—inductance of an arbitrary square of the circuit. The xy plane is divided into a set of meshes which are basically square but may have their shapes modified to match the boundary line. The coordinates of the middle of a mesh in the kth row and the lth column are denoted by x_l and y_k. Replacing the differentials in (2) and (3) by finite differences Δt and a

Manuscript received June 29, 1987; revised November 16, 1987.
The author is with the Institute of Radioelectronics, Warsaw University of Technology, 00-665 Warsaw, Nowowiejska 15/19, Poland.
IEEE Log Number 8719199.

Reprinted from *IEEE Trans. Microwave Theory Tech.*, vol. 36, no. 4, pp. 738–744, Apr. 1988.

yields

$$J_x\left(x_l+\frac{a}{2}, y_k, t_0+\frac{\Delta t}{2}\right) = J_x\left(x_l+\frac{a}{2}, y_k, t_0-\frac{\Delta t}{2}\right) - (V(x_l+a, y_k, t_0) - V(x_l, y_k, t_0))\frac{\Delta t}{L_s a f_1(l,k)} \quad (4)$$

$$J_y\left(x_l, y_k+\frac{a}{2}, t_0+\frac{\Delta t}{2}\right) = J_y\left(x_l, y_k+\frac{a}{2}, t_0-\frac{\Delta t}{2}\right) - (V(x_l, y_k+a, t_0) - V(x_l, y_k, t_0))\frac{\Delta t}{L_s a f_2(l,k)} \quad (5)$$

$$V(x_l, y_k, t_0+\Delta t) = V(x_l, y_k, t_0) - \left(J_x\left(x_l+\frac{a}{2}, y_k, t_0+\frac{\Delta t}{2}\right) - J_x\left(x_l-\frac{a}{2}, y_k, t_0+\frac{\Delta t}{2}\right) + J_y\left(x_l, y_k+\frac{a}{2}, t_0+\frac{\Delta t}{2}\right) - J_y\left(x_l, y_k-\frac{a}{2}, t_0+\frac{\Delta t}{2}\right)\right)\frac{\Delta t}{C_s a f_3(l,k)} \quad (6)$$

where $f_1(l,k)$, $f_2(l,k)$, and $f_3(l,k)$ are mesh shape functions which are equal to unity for all square meshes (that is, those inside the circuit) but adopt different values (calculated by a boundary matching procedure like that of [7]) for the meshes modified to match boundary lines.

Consecutive calculations of (4), (5), and (6) simulate the process of the wave propagation in the circuit.

III. Problems of FD-TD Analysis of Two-Dimensional Circuits

A. Modeling of Matched Loads and Sources

The most convenient way of describing a microwave linear circuit is by its S matrix. Since the FD-TD method models the energy flow in the circuit it may be used to compute the S matrix directly, provided that the matched loads of the output lines are properly modeled in the algorithm. Matching the source, although not absolutely necessary, is highly desirable, since any reflections from the input would prolong the transient response of the circuit, thus prolonging the computing time.

The absorbing boundary conditions have been investigated in [14] and [15], but the procedures developed there are not useful in our case. They cannot handle the situation of matched source and also, being designed for general absorbing conditions, they are unnecessarily complicated when applied to the case of normal incidence. That is why a different approach will be used here.

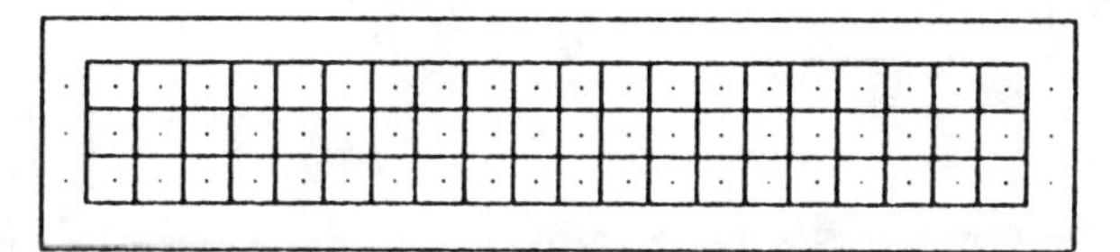

Fig. 1. A uniform transmission line as a grid of meshes.

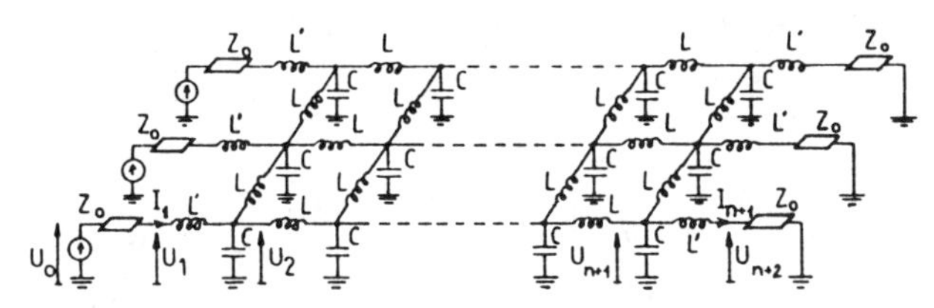

Fig. 2. Lumped circuit model of the line of Fig. 1.

Let us consider a model of a uniform transmission line of length $n = 20a$ and width $w = 3a$, where a is the mesh size (Fig. 1). The lumped circuit model of this line corresponding to the FD approximation is presented in Fig. 2. Propagation inside the circuit is described by (4), (5), and (6). The input and output matching is obtained by introducing in each of the rows of meshes at the input and output the following operations:

$$I_1\left(t_0+\frac{\Delta t}{2}\right) = I_1\left(t_0-\frac{\Delta t}{2}\right) - (V_2(t_0) - V_1(t_0))\frac{\Delta t}{L'} \quad (7)$$

$$V_1(t_0+\Delta t) = V_0(t_0+\Delta t) - I_1\left(t_0+\frac{\Delta t}{2}\right)Z_0 \quad (8)$$

$$I_{n+1}\left(t_0+\frac{\Delta t}{2}\right) = I_{n+1}\left(t_0-\frac{\Delta t}{2}\right) + (V_{n+1}(t_0) - V_{n+2}(t_0))\frac{\Delta t}{L'} \quad (9)$$

$$V_{n+2}(t_0+\Delta t) = I_{n+1}\left(t_0+\frac{\Delta t}{2}\right)Z_0 \quad (10)$$

where

$$Z_0 = \sqrt{\frac{L}{C}} \qquad L' = \frac{L}{2} + L'' = \frac{L}{2} + \frac{\Delta t\omega}{2a\beta}L \qquad C = C_s a^2 \qquad L = L_s.$$

The meaning of the voltages and the currents used in (7)–(10) is explained in Fig. 2. The additional inductance L'' was introduced as a correction element to lower the matching errors caused by the fact that in the FD-TD algorithm the voltage is defined at points of time and space which are different from the points where the current is defined. Because of this it is also necessary to introduce some correction terms in the formulas for the S matrix elements. Let us assume that $V_0(\omega), V_1(\omega), \cdots$ are the complex amplitudes calculated by the Fourier transformation of $V_0(t), V_1(t), \cdots$ and $I_1(\omega), I_2(\omega), \cdots$ are the complex amplitudes calculated by the Fourier transforma-

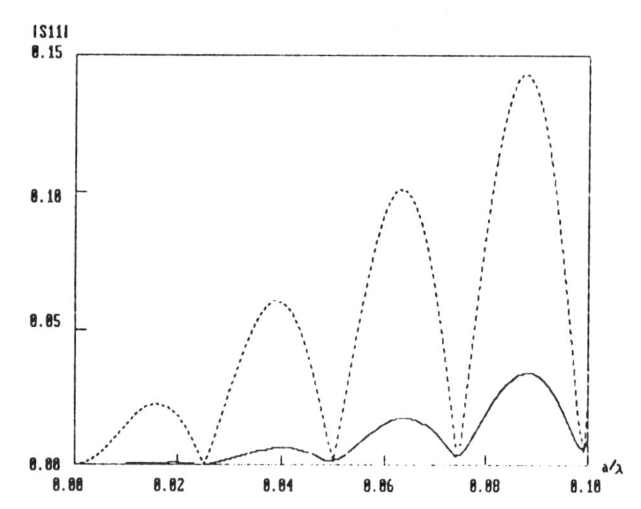

Fig. 3. Results of calculations of $|S_{11}|$ versus a/λ for the line of Fig. 1 with the special correction (continuous line) and without it (dashed line).

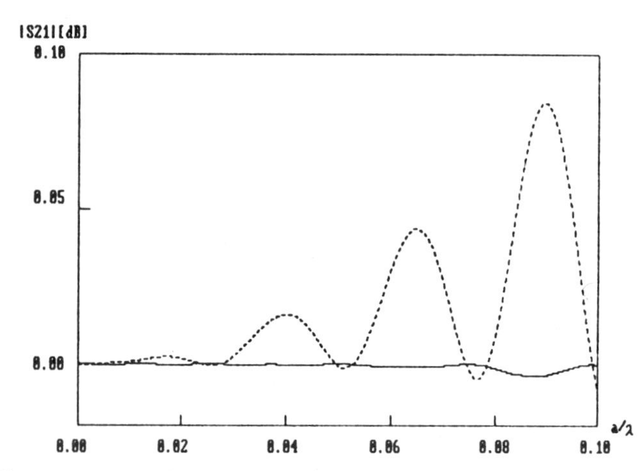

Fig. 4. Results of calculations of $|S_{21}|$ versus a/λ for the circuit of Fig. 1 with the special correction (continuous line) and without it (dashed line).

tion of $I_1(t-\Delta t/2), I_2(t-\Delta t/2), \cdots$. Resolving the circuit equations at the input and output we obtain

$$S_{21} = \frac{2V_{n+2}(\omega)\cos(\phi)}{V_0(\omega)} e^{j\phi} \tag{11}$$

$$S_{11} = \frac{Z_{in} - Z_o}{Z_{in} + Z_o} \qquad \text{with } Z_{in} = \frac{V_1(\omega)}{I_1(\omega)} e^{-j\phi} - j\omega L'' \tag{12}$$

where $\phi = \omega \Delta t/2$.

Figs. 3 and 4 present the results of calculations of $|S_{11}|$ and $|S_{21}|$ versus frequency for the line of Fig. 1 with the correction terms containing ϕ and L'' (continuous line) and without them (dashed line). The frequency is characterized by the mesh size to wavelength ratio a/λ. It is seen that when the corrections are applied, the errors of matching drop to negligible values even for relatively high a/λ ratios.

Modeling of a matched source and a matched load was introduced for a uniform transmission line. It can be applied to an arbitrarily shaped circuit provided that at the input and output the circuit includes segments of uniform transmission lines long enough to ensure effective attenuation of all the width modes except the dominant one.

B. Pulse Excitation

In the approach used in [7] the circuit with zero initial values of $V(x, y, t)$ and $J(x, y, t)$ was excited by a sinusoidal source of frequency ω. The circuit's parameters at that frequency were obtained from the steady state achieved in the circuit after a sufficiently long time. This caused the time for computing a wide-band frequency response to be very long.

Equations (4), (5), and (6) show that all the operations in the FD-TD algorithm are linear. Thus it is possible to use Fourier analysis to obtain the circuit's frequency response from the transient response. However, we need a proof that this approach will not sacrifice the accuracy of the method.

Let us consider a relation between a voltage at the input, V_0, and a voltage at the output, V_l. We may write

$$V_l(t) = \boldsymbol{L}(V_0(t)) = \boldsymbol{L}'(V_0(t)) + \Delta\boldsymbol{L}(V_0(t)) \tag{13}$$

where $\boldsymbol{L}$ is a linear operator of the FD-TD algorithm. $\boldsymbol{L}'$ is an operator of the transformation between input and output of the original circuit, which is linear due to the linearity of the Maxwell's equations; and $\Delta\boldsymbol{L}$ is an operator describing the error of the FD-TD method, which has to be linear due to the linearity of the two former operators.

Let us write the Fourier transform of (13):

$$V_l(\omega) = T(\omega)V_0(\omega) = (T'(\omega) + \Delta T(\omega))V_0(\omega) \tag{14}$$

where $V_0(\omega), V_l(\omega)$ are the Fourier transforms of $V_0(t)$, $V_l(t)$ and $T(\omega), T'(\omega)$, and $\Delta T(\omega)$ describe the operators $\boldsymbol{L}$, $\boldsymbol{L}'$, and $\Delta\boldsymbol{L}$ in the frequency domain. This yields

$$\frac{V_l(\omega)}{V_0(\omega)} = T(\omega) = T'(\omega) + \Delta T(\omega). \tag{15}$$

Equation (15) suggests that for any particular frequency ω, the error of the FD-TD method does not depend on the shape of the input signal and that the function $V_0(t)$ may be chosen to be the most convenient for use in the computer algorithm. However, there is one aspect of the pulse excitation which needs additional checking. The Fourier transforms in (15) are assumed to be calculated in an infinite period of time. The limited period of time assumed in any computer calculation causes additional error. Thus we must determine how much this error depends on the shape of the source.

Let us consider two types of pulses. The first is a δ pulse which in the algorithm adopts the form

$$V_0(t) = \begin{cases} 0 & \text{for } t = 0 \\ 1 & \text{for } t = n\Delta t;\ n = \pm 1, \pm 2, \cdots. \end{cases} \tag{16}$$

The second is a pulse of limited spectrum approximating a δ pulse after passing through a bandpass filter of cutoff

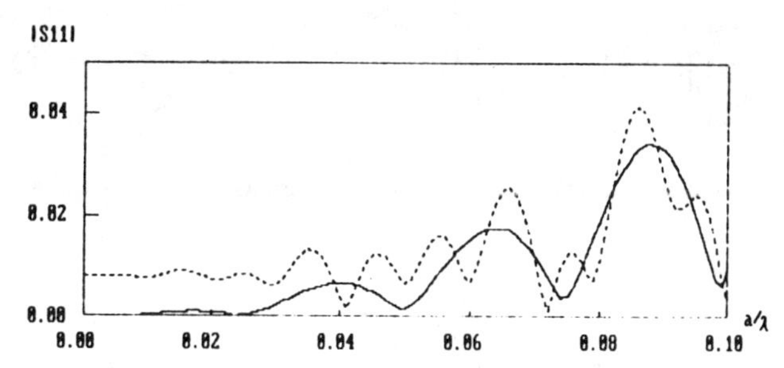

Fig. 5. $|S_{11}|$ of the circuit of Fig. 1 calculated after 200 iterations; with δ-type excitation (dashed line) and with excitation of limited spectrum (continuous line).

frequency ω_c:

$$V_0(t) = \begin{cases} 2\omega_c \Delta t/\pi & \text{for } t = 0 \\ \dfrac{2\,\Delta t \sin(\omega_c t)}{\pi t} & \text{for } t = m\,\Delta t;\ m = \pm 1, \pm 2, \pm 3, \cdots, \pm M \\ 0 & \text{for } t = m\,\Delta t;\ |m| > M \end{cases} \tag{17}$$

Fig. 5 shows the results of calculations of $|S_{11}|$ of the line of Fig. 1 after 200 iterations ($t = 200\Delta t$) with the two types of excitation. The dashed line was obtained with δ type excitation, while the continuous line was obtained with the excitation of limited spectrum (with ω_c corresponding to $a/\lambda = 0.2$ and $M = 50$). When the computation is prolonged, the results obtained with the second type of pulse change very little while the ripples on the curve obtained with the δ type of excitation decrease slowly, and after about 1500 iterations the shape of the curve approaches the result obtained before with the second type of pulse.

The reason for this effect is that the FD approximation produces some resonances above the investigated frequency band. A δ pulse excites the circuit at these frequencies and produces a ringing-type response. Cutting off the high frequencies from the exciting pulse eliminates the effect.

However, it should be noted that when the resonances of the investigated circuit lie within the band of interest their effect cannot be eliminated by changing the exciting pulse spectrum. An example of such a circuit will be shown later in this paper.

We may conclude that since using a δ type of pulse simplifies the algorithm and speeds it up (because the Fourier transform of the source does not have to be computed), it is a reasonable choice in many cases. However if we consider a relatively resonance free band and the resonances are grouped outside that band, the time of computing can be brought substantially down by applying a pulse of the spectrum limited in such a way that the unwanted resonances are not excited.

C. Microstrip Circuits Analysis

As was already shown in [7], the FD-TD method is effective in arbitrarily shaped stripline circuit analysis. The fringing fields were included in the calculations by assuming that the circuit is bounded by a magnetic wall shifted by some distance from the real edge of the circuit. For a microstrip circuit a more complicated model is needed to describe the complicated nature of the fringing fields. Here is a proposal of such a model.

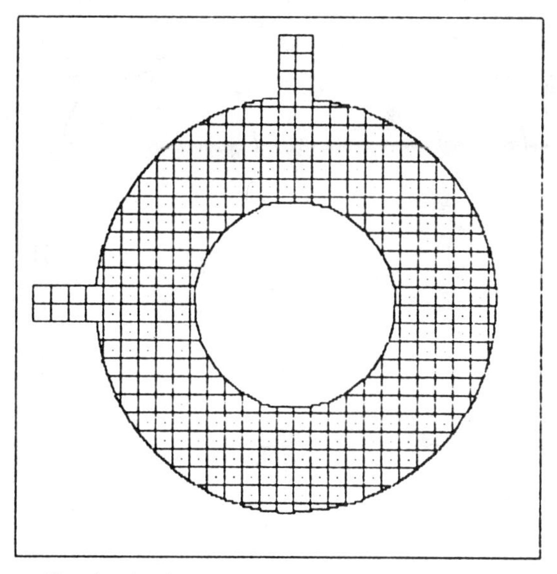

Fig. 6. A microstrip ring circuit as a grid of meshes.

Let us consider a microstrip line of width w. This line is characterized by its unit capacitance $C(w)$, unit inductance $L(w)$, and effective permittivity $\epsilon_{\text{eff}}(w)$ (with permeability $\mu = \mu_0$). Let us now imagine a line filled with a uniform dielectric characterized by $\epsilon' = \text{const}$ and $\mu' = \text{const}$ and having such properties that its unit capacitance C' and unit inductance L' obey the relations

$$C'(w + \Delta w) = C(w) \tag{18}$$

$$L'(w + \Delta w + \Delta w') = L(w). \tag{19}$$

In many practical cases it is possible to find such values of Δw and $\Delta w'$ that (18) and (19) are obeyed with good accuracy over a wide range of w. For example, for a duroid substrate of $\epsilon = 10\epsilon_0$ and height $h = 0.635$ mm, drawing the functions $C(w)$ and $L(w)$ from the closed-form expressions for Z_o and ϵ_{eff} after [9] we find that (18) and (19) are obeyed with 1 percent accuracy in the range $2h < w < 10h$, assuming that $\Delta w = 1.05h$, $\Delta w' = h$, $\epsilon' = 10\epsilon_0$, and $\mu' = 0.93\mu_0$. For wider strips the error of this approximation rises only slightly. For very narrow strips it is bigger but can be corrected by adopting for the edges of these strips different values of Δw and $\Delta w'$.

We will now apply the FD-TD method to the analysis of a ring circuit already analyzed by another method by D'Inzeo *et al.* [10]. The circuit was built on the duroid substrate, had the dimensions of $r_{\text{out}} = 7$ mm, $r_{\text{in}} = 4$ mm, and was connected to two 50 Ω lines making a 90° angle. For FD-TD calculations we assumed the circuit as presented in Fig. 6. Its dimensions were obtained by shifting the circuit's boundary by $\Delta w/2 = 0.525h$ and it was assumed that $\epsilon' = 10\epsilon_0$ and $\mu' = 0.93\mu_0$. We applied the boundary matching procedure described in [7] but took into account additional inductance distributed along the

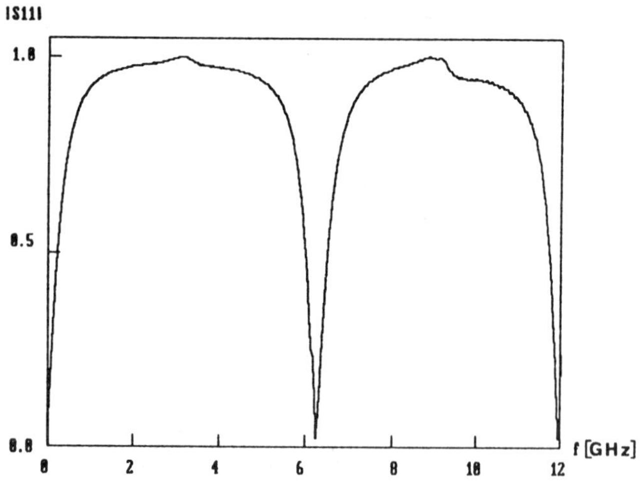

Fig. 7. $|S_{11}|$ versus frequency of the circuit of Fig. 6.

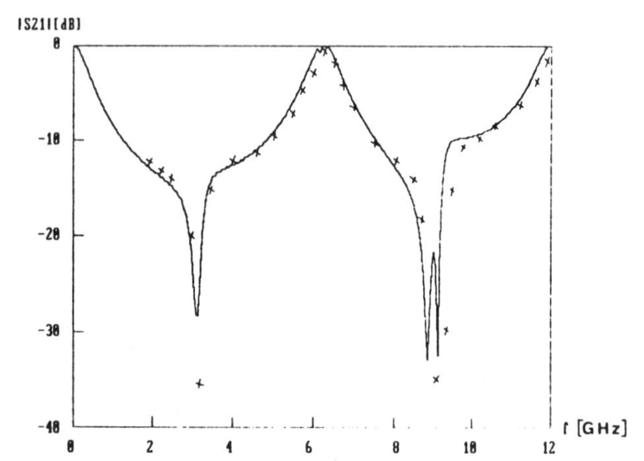

Fig. 8. $|S_{21}|$ versus frequency of the circuit of Fig. 6. —— calculations, ××××× measurements after [10].

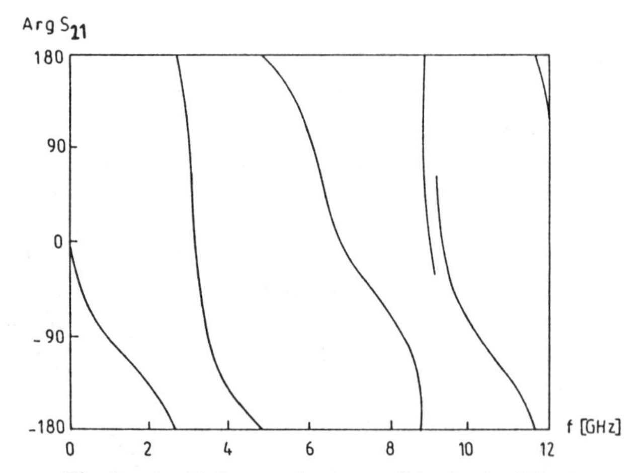

Fig. 9. Arg(S_{21}) versus frequency of the circuit of Fig. 6.

border. This corresponded to the inductance of a line of width $\Delta w'/2 = 0.5h$. The results of calculations of $|S_{11}|$, $|S_{21}|$, and Arg(S_{21}) are presented in Figs. 7, 8, and 9. In the case of $|S_{21}|$ they are compared with the measurements after [10]. The agreement is quite good. It has to be stressed that the method presented here does not assume any regularities of the circuit's shape. Thus it is a method for truly arbitrarily shaped circuits.

We may conclude that the presented example of the FD-TD method application to an arbitrarily shaped microstrip circuit is encouraging, but more work has to be done to check its value for various circuits, especially in the higher frequency band, when dispersion becomes important.

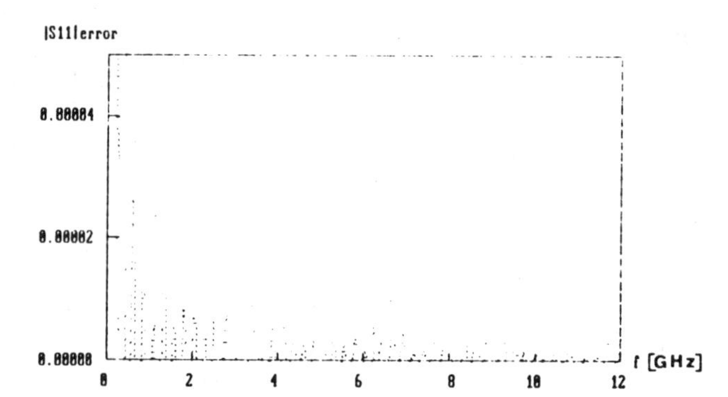

Fig. 10. Error in calculations of $|S_{11}|$ in Fig. 7 caused by a simulated low accuracy of the computer arithmetics equal to 10^{-6}.

D. Roundoff Errors

In many programming languages there is a choice of the precision of the floating point arithmetic and it is important to know which precision to choose to keep the computer roundoff errors negligible while not boosting the memory requirements and the computing time. To check the level of the roundoff errors the calculations of the circuit of Fig. 6 were repeated with a simulated low computer precision of 10^{-6}. The difference between the results of $|S_{11}|$ obtained with full and with limited precision with δ-type excitation are presented in Fig. 10. It is seen that computer precision even as low as 10^{-6} introduces negligible errors into FD-TD calculations. This and other numerical experiments have shown that the error level increases with decreasing a/λ ratio (as seen in Fig. 10) and that it is slightly smaller for other than δ-type excitations but these dependencies have negligible effect due to generally low level of the roundoff errors. We may draw the conclusion that the FD-TD method is very resistant to roundoff errors and in most cases can be applied even with the lowest (4 bytes) precision of the floating point arithmetics used in personal computers.

E. Algorithm Implementation

The described problems of the FD-TD analysis were checked with a Pascal program prepared for an IBM PC AT computer. The program was written in a "user friendly" manner under the assumption that it will serve a variety of users.

As an example, the time needed in computing the S matrix of the circuit of Fig. 6 was about 20 min. for 1000 iterations. The results presented in Figs. 7–9 were obtained after 3000 iterations, but reasonably accurate results can be obtained even after 1000 iterations. In this case the

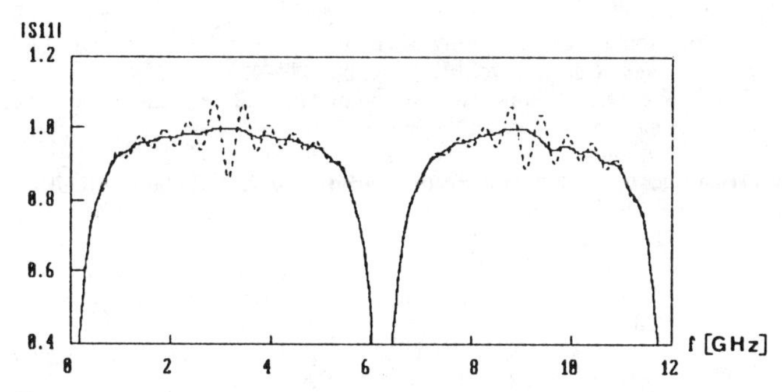

Fig. 11. $|S_{21}|$ of the circuit of Fig. 6 after 1000 iterations (dashed line) and the same result after smoothing procedure using the criterion of power conservation (continuous line).

functions obtained have some ripples (see, for example, the dotted line in Fig. 11) but these ripples can be eliminated using the criterion $|S_{11}|^2 + |S_{21}|^2 = 1$, valid for lossless circuits (see the continuous line in Fig. 11, which is practically the same as that of Fig. 7).

F. Comparison to Other Methods

As was mentioned in the introduction, the methods which can be assigned to the same class as possible competitors to the FD-TD method are the contour integral (CI) method [2], [13] and the transmission-line matrix (TLM) method [3], [11], [12].

In the contour integral method [2], the circuit's boundary is divided into a set of N elements of finite width. Using the properties of cylindrical waves for a particular frequency, we can obtain the relations between the voltages at the elements. This leads to a set of N linear equations to be solved.

When comparing the FD-TD and the CI methods we have to note first that their domains of application are not the same. The FD-TD method can be applied to circuits filled with nonuniform media and also to circuits with nonlinear elements (when restricted to sinusoidal excitation). These two classes of circuits cannot be treated by the CI method.

It is difficult to say which of the two methods needs less computer time. There are three reasons for this.

1) It depends on the type of circuit. Many factors contribute, for example,

- the computer time needed by the CI method depends on the length of the boundary line while the time of the FD-TD calculations depends on the surface of the circuit;
- for some circuits greater advantage may be taken from the flexibility of the CI method in applying varying boundary element length.
- the time of computing by the FD-TD method depends on the presence of high-Q resonances in the circuit.

2) Specialized hardware and software can bring a dramatic drop in computing time for either of the two methods, but the gain obtained with particular computer resources may be quite different for each of them.

3) In the CI method calculations are conducted for a particular frequency and have to be repeated from the very beginning when the circuit's parameters are to be calculated for another frequency. On the contrary, in the FD-TD method calculating the circuit's parameters for several hundred frequencies brings only a fractional increase in the computing time with respect to one-frequency calculations.

As an example, the computing time by both methods was compared for the circuit of Fig. 6 (but with no additional inductance along the border, a situation which cannot be handled by the CI method) on an IBM PC AT computer using standard library routines. In the time needed for calculating the wide-band frequency characteristics by the FD-TD method only a few (about five) frequency points could be calculated by the CI method.

An important advantage of the FD-TD method is that all algorithm operations have clear physical interpretations. That is why if a computing error occurs its cause can be quite easily spotted. This is in contrast to calculations by the CI method involving large-scale matrix operations which are very difficult to control.

The FD-TD method and the transmission-line matrix (TLM) method [3], [11] are similar in many respects. They both use a net of square meshes and both consider a pulse propagation in the net (although the algorithm of simulating the pulse propagation is different). Comparison of the FD-TD and the TLM methods has been discussed in recent publications [12], [16]. The general conclusion is that those methods are developed in a parallel way. Progress in one of them triggers development of the other. For example, to the author's knowledge the TLM method was not yet applied to S matrix calculations by modeling the circuit with matched input and output, but taking into account the discussion of this paper such an approach seems straightforward. Although the author found that in the considered application his version of the FD-TD method is faster and more convenient than the versions of the TLM known to him, he thinks of his work as a step forward in developing the entire group of wave-simulating methods, which includes the FD-TD and the TLM.

III. CONCLUSIONS

The paper has presented a version of the finite-difference time-domain method adapted to the needs of the S matrix calculations of two-dimensional microwave circuits. To allow direct S matrix calculations, matched loads and matched sources were modeled in the algorithm. This modeling gave good results over a wide frequency range (covering mesh size to wavelength ratios $0 < a/\lambda < 0.1$).

Pulse excitation was introduced in the method for fast calculations of frequency-dependent circuit characteristics. It was shown that the pulse excitation does not degrade the FD-TD accuracy. Different types of pulses were studied. In some cases a δ-type pulse is the most practical choice but in the cases when the circuit's main resonances are grouped outside the band of interest a pulse of limited spectrum can bring a substantial drop in the time of computing.

It was shown that the FD-TD method is not sensitive to the computer roundoff errors. Low precision of the floating point numbers may be chosen in computer programs to keep down the computer time and memory.

An analysis of an arbitrarily shaped microstrip circuit was presented as an example. Good agreement between the results of calculations and measurements was obtained but more work has to be done to check the applied fringing fields models for various circuits.

Implementation of the described method on an IBM PC AT computer shows that it can become a practical tool in engineering applications. In many cases it performs better than other widely used methods such as the contour integral method and the transmission-line matrix method.

References

[1] R. Sorrentino, "Planar circuits, waveguide models, and segmentation method," *IEEE Trans. Microwave Theory Tech.*, vol. MTT-33, pp. 1057–1066, Oct. 1985.

[2] T. Okoshi and T. Miyoshi, "The planar circuit—An approach to microwave integrated circuitry," *IEEE Trans. Microwave Theory Tech.*, vol. MTT-20, pp. 245–252, Apr. 1972.

[3] P. B. Johns and R. L. Beurle, "Numerical solution of 2-dimensional scattering problems using a transmission-line matrix," *Proc. Inst. Elec. Eng.*, vol. 118, no. 9, pp. 1203–1208, Sept. 1971.

[4] K. S. Yee, "Numerical solution of initial boundary value problems involving Maxwell's equations in isotropic media," *IEEE Trans. Antennas Propagat.*, vol. AP-14, pp. 302–307, May 1966.

[5] K. Umashankar and A. Taflove, "A novel method to analyze electromagnetic scattering of complex objects," *IEEE Trans. Electromagn. Compat.*, vol. EMC-24, pp. 397–405, Nov. 1982.

[6] T. Weiland, "Numerical solution of Maxwell's equations for static, resonant and transient problems," in *Proc. U.R.S.I. Int. Symp. Electromagn. Theory* (Budapest), 1986, pp. 537–542.

[7] W. K. Gwarek, "Analysis of an arbitrarily shaped planar circuit A time domain approach," *IEEE Trans. Microwave Theory Tech.*, vol. MTT-33, pp. 1067–1072, Oct. 1985.

[8] W. K. Gwarek, "Two-dimensional circuit analysis by wave simulation," in *Proc. U.R.S.I. Int. Symp. Electromagn. Theory* (Budapest), 1986, pp. 580–582.

[9] K. C. Gupta, R. Garg, and R. Chadha, *Computer-Aided Design of Microwave Circuits.* Dedham, MA: Artech House, 1981, ch. 3.

[10] G. D'Inzeo *et al.*, "Wide-band equivalent circuits of microwave planar networks," *IEEE Trans. Microwave Theory Tech.*, vol MTT-28, pp. 1107–1113, Oct. 1980.

[11] W. J. R. Hoefer, "The transmission-line matrix method—Theory and applications," *IEEE Trans. Microwave Theory Tech.*, vol. MTT-33, pp. 882–893, Oct. 1985.

[12] P. B. Johns, "On the relationship between TLM and finite-difference methods for Maxwell's equations," *IEEE Trans. Microwave Theory Tech.*, vol. MTT-35, pp. 60–61, Jan. 1987.

[13] T. Okoshi, *Planar Circuits for Microwaves and Lightwaves.* Berlin; Heidelberg: Springer-Verlag, 1985.

[14] B. Engquist and A. Majda, "Absorbing boundary conditions for the numerical simulation of waves," *Math. Comput.*, vol. 31, pp. 629–651, July 1977.

[15] G. Mur, "Absorbing boundary conditions for the finite-difference approximation of the time-domain electromagnetic-field equations," *IEEE Trans. Electromagn. Compat.*, vol. EMC-23, pp. 377–382, Nov. 1981.

[16] W. K. Gwarek and P. B. Johns, "Comments on 'On the relationship between TLM and finite difference methods for Maxwell's equations'," *IEEE Trans. Microwave Theory Tech.*, vol. MTT-35, pp. 872–873, Sept. 1987.

[17] D. T. Borup, D. M. Sullivan, and O. P. Gandhi, "Comparison of the FFT conjugate gradient method and the finite-difference time-domain method for the 2-D absorbtion problem," *IEEE Trans. Microwave Theory Tech.*, vol. MTT-35, pp. 383–395, Apr. 1987.

Part 3
Transmission-Line Matrix Method

THE transmission-line matrix (TLM) method simulates the wave propagation in the time domain by discretizing the space into a two- or three-dimensional transmission-line matrix. It therefore has a number of similarities with the finite-difference time-domain (FD-TD) method discussed in Part 2 (see Papers 2.4 and 2.5) and in [2.9], [2.10]. The method is founded on the modeling of the spatial electromagnetic field in terms of a distributed transmission-line network. Electric and magnetic fields are made equivalent to voltage and current on the network. The numerical calculation starts by exciting the matrix at specific points by voltage or current pulses. Propagation of pulses is then evaluated at discrete time intervals. Time synchronism is required so that all pulses reach nodes at the same time. Simplicity of formulation and programming, and the calculation of transients are the main attractions of the method.

A very enlightening overview of the TLM method is given in the first paper (Paper 3.1) by Hoefer. This paper serves as a thorough introduction to the method. It describes its theoretical foundations, the basic algorithm of implementation as well as the various improvements and extensions, the sources of errors, and limitations. A vast bibliography on the TLM method is also provided. The interested reader is referred to it.

This collection then includes the pioneer work by Johns and Beurle (Paper 3.2), where this numerical technique was first proposed for solving two-dimensional waveguide scattering problems. A number of papers were then published extending the method to three-dimensional problems and including losses and dielectric loading [3.1]–[3.4]. The restriction associated with the use of squares or cubes for space discretization was removed by Al-Mukhtar and Sitch (Paper 3.3) and, independently, by Saguet and Pic [3.5]. Recent modifications of the method that led to the modified TLM method are described in the paper by Saguet and Hoefer (Paper 3.4).

A numerical method called Bergeron's method has been under development in Japan by Yoshida and Fukai since 1979, but the method was published in the western literature only in 1984 in a paper dealing with the transient analysis of the field propagation through a bend in a stripline (Paper 3.5). The method is very similar and substantially equivalent to the TLM method, but the formulation differs in some details such as the introduction of magnetic currents so that it seemed worth including this paper in the present collection. It also makes reference to papers previously published in Japan. Another application of the method appeared recently in [3.6].

References

[3.1] P. B. Johns, "The solution of inhomogeneous waveguide problems using a transmission-line matrix," *IEEE Trans. Microwave Theory Tech.*, vol. MTT-22, pp. 209–215, Mar. 1974.

[3.2] S. Akhtarzad and P. B. Johns, "Three-dimensional transmission-line matrix computer analysis of microstrip resonators," *IEEE Trans. Microwave Theory Tech.*, vol. MTT-23, pp. 990–997, Dec. 1975.

[3.3] S. Akhtarzad and P. B. Johns, "Solution of Maxwell's equations in three space and time by the t.l.m. method of numerical analysis," *Proc. IEE*, vol. 122, pp. 1344–1348, Dec. 1975.

[3.4] S. Akhtarzad and P. B. Johns, "Generalised elements for t.l.m. method of numerical analysis," *Proc. IEE*, vol. 122, pp. 1349–1352, Dec. 1975.

[3.5] P. Saguet and E. Pic, "Le maillage rectangulaire et le changement de la maille dans la méthode TLM en deux dimensions," *Electron. Lett.*, vol. 17, no. 7, pp. 277–279, Apr. 1981.

[3.6] S. Koike, N. Yoshida, and I. Fukai, "Transient analysis of a directional coupler using a coupled microstrip slot line in three-dimensional space," *IEEE Trans. Microwave Theory Tech.*, vol. MTT-34, pp. 353–357, Mar. 1986.

The Transmission-Line Matrix Method—Theory and Applications

WOLFGANG J. R. HOEFER, SENIOR MEMBER, IEEE

(Invited Paper)

Abstract—This paper presents an overview of the transmission-line matrix (TLM) method of analysis, describing its historical background from Huygens's principle to modern computer formulations. The basic algorithm for simulating wave propagation in two- and three-dimensional transmission-line networks is derived. The introduction of boundaries, dielectric and magnetic materials, losses, and anisotropy are discussed in detail. Furthermore, the various sources of error and the limitations of the method are given, and methods for error correction or reduction, as well as improvements of numerical efficiency, are discussed. Finally, some typical applications to microwave problems are presented.

I. Introduction

Before the advent of digital computers, complicated electromagnetic problems which defied analytical treatment could only be solved by simulation techniques. In particular, the similarity between the behavior of electromagnetic fields, and of voltages and currents in electrical networks, was used extensively during the first half of the twentieth century to solve high-frequency field problems [2]–[4].

When modern computers became available, powerful numerical techniques emerged to predict directly the behavior of the field quantities. The great majority of these methods yield harmonic solutions of Maxwell's equations in the space or spectral domain. A notable exception is the transmission-line matrix (TLM) method of analysis which represents a true computer simulation of wave propagation in the time domain.

In this paper, the theoretical foundations of the TLM method are reviewed, its basic algorithm for simulating the propagation of waves in unbounded and bounded space is derived, and it is shown how the eigenfrequencies and field configurations of resonant structures can be determined with the Fourier transform. Sources and types of errors are discussed, and possible pitfalls are pointed out. Then, various methods of error correction are presented, and the most significant improvements to the conventional TLM approach are described. A referenced list of typical applications of the method is included as well. In the conclusion, the advantages and disadvantages of the method are summarized, and it is indicated under what circumstances it is appropriate to select the TLM method rather than other numerical techniques for solving a particular problem.

Manuscript received February 22, 1985; revised May 31, 1985.

The author is with the Department of Electrical Engineering, University of Ottawa, Ottawa, Ontario, Canada K1N 6N5.

II. Historical Background

Two distinct models describing the phenomenon of light were developed in the seventeenth century: the corpuscular model by Isaak Newton and the wave model by Christian Huygens. At the time of their conception, these models were considered incompatible. However, modern quantum physics has demonstrated that light in particular, and electromagnetic radiation in general, possess both granular (photons) and wave properties. These aspects are complementary, and one or the other usually dominates, depending on the phenomenon under study.

At microwave frequencies, the granular nature of electromagnetic radiation is not very evident, manifesting itself only in certain interactions with matter, while the wave aspect predominates in all situations involving propagation and scattering. This suggests that the model proposed by Huygens, and later refined by Fresnel, could form the basis for a general method of treating microwave propagation and scattering problems.

Indeed, Johns and Beurle [5] described in 1971 a novel numerical technique for solving two-dimensional scattering problems, which was based on Huygens's model of wave propagation. Inspired by earlier network simulation techniques [2]–[4], this method employed a Cartesian mesh of open two-wire transmission lines to simulate two-dimensional propagation of delta function impulses. Subsequent papers by Johns and Akhtarzad [6]–[16] extended the method to three dimensions and included the effect of dielectric loading and losses. Building upon the groundwork laid by these original authors, other researchers [17]–[34] added various features and improvements such as variable mesh size, simplified nodes, error correction techniques, and extension to anisotropic media.

The following section describes briefly the discretized version of Huygens's wave model which is suitable for implementation on a digital computer and forms the algorithm of the TLM method. A detailed description of this model can be found in a very interesting paper by P. B. Johns [9].

III. Huygen's Principle and Its Discretization

According to Huygens [1], a wavefront consists of a number of secondary radiators which give rise to spherical wavelets. The envelope of these wavelets forms a new

Reprinted from *IEEE Trans. Microwave Theory Tech.*, vol. MTT-33, no. 10, pp. 882–893, Oct. 1985.

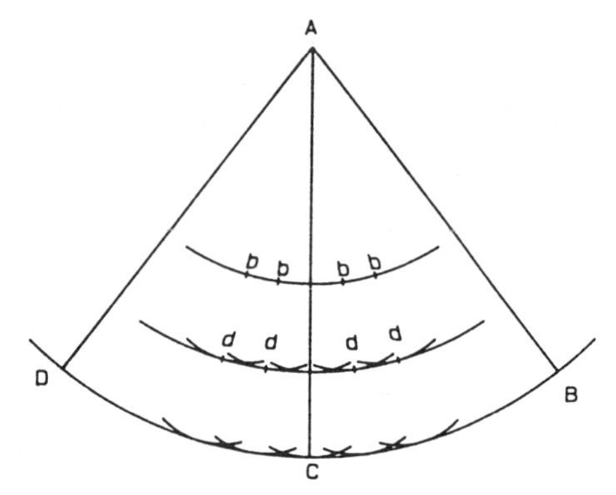

Fig. 1. Huygens's principle and formation of a wavefront by secondary wavelets.

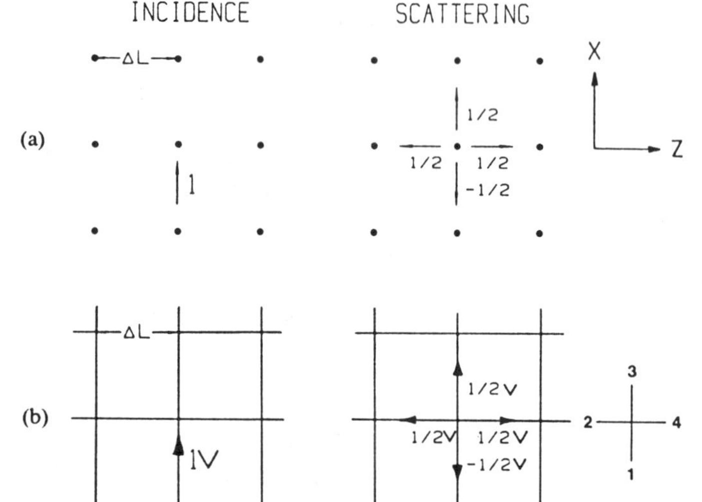

Fig. 2. The discretized Huygens's wave model (a) in two-dimensional space and (b) in an equivalent Cartesian mesh of transmission lines (after Johns [9]).

wavefront which, in turn, gives rise to a new generation of spherical wavelets, and so on (Fig. 1). In spite of certain difficulties in the mathematical formulation of this mechanism, its application nevertheless leads to an accurate description of wave propagation and scattering, as will be shown below.

In order to implement Huygens's model on a digital computer, one must formulate it in discretized form. To this end, both space and time are represented in terms of finite, elementary units Δl and Δt, which are related by the velocity of light such that

$$\Delta t = \Delta l / c. \tag{1}$$

Accordingly, two-dimensional space is modeled by a Cartesian matrix of points or nodes, separated by the mesh parameter Δl (see Fig. 2(a)). The unit time Δt is then the time required for an electromagnetic pulse to travel from one node to the next.

Assume that a delta function impulse is incident upon one of the nodes from the negative x-direction. The energy in the pulse is unity. In accordance with Huygen's principle, this energy is scattered isotropically in all four directions, each radiated pulse carrying one fourth of the incident energy. The corresponding field quantities must then be 1/2 in magnitude. Furthermore, the reflection coefficient "seen" by the incident pulse must be negative in order to satisfy the requirement of field continuity at the node.

This model has a network analog in the form of a mesh of orthogonal transmission lines, or transmission-line matrix (Fig. 2(b)), forming a Cartesian array of shunt nodes which have the same scattering properties as the nodes in Fig. 2(a). It can be shown that there is a direct equivalence between the voltages and currents on the line mesh and the electric and magnetic fields of Maxwell's equations [5].

Consider the incidence of a unit Dirac voltage-impulse on a node in the TLM mesh of Fig. 2(b). Since all four branches have the same characteristic impedance, the reflection coefficient "seen" by the incident impulse is indeed $-1/2$, resulting in a reflected impulse of -0.5 V and three transmitted impulses of $+0.5$ V.

The more general case of four impulses being incident on the four branches of a node can be obtained by superposition from the previous case. Hence, if at time $t = k\,\Delta t$, voltage impulses ${}_kV_1^i$, ${}_kV_2^i$, ${}_kV_3^i$, and ${}_kV_4^i$ are incident on lines 1–4, respectively, on any junction node, then the total voltage impulse reflected along line n at time $(k+1)\,\Delta t$ will be

$${}_{k+1}V_n^r = \frac{1}{2}\left[\sum_{m=1}^{4} {}_kV_m^i\right] - {}_kV_n^i. \tag{2}$$

This situation is conveniently described by a scattering matrix equation [7] relating the reflected voltages at time $(k+1)\,\Delta t$ to the incident voltages at the previous time step $k\,\Delta t$

$${}_{k+1}\begin{pmatrix} V_1 \\ V_2 \\ V_3 \\ V_4 \end{pmatrix}^r = 1/2 \begin{pmatrix} -1 & 1 & 1 & 1 \\ 1 & -1 & 1 & 1 \\ 1 & 1 & -1 & 1 \\ 1 & 1 & 1 & -1 \end{pmatrix} \times {}_k\begin{pmatrix} V_1 \\ V_2 \\ V_3 \\ V_4 \end{pmatrix}^i. \tag{3}$$

Furthermore, any impulse emerging from a node at position (z, x) in the mesh (reflected impulse) becomes automatically an incident impulse on the neighboring node. Hence

$$\begin{aligned} {}_{k+1}V_1^i(z,x) &= {}_{k+1}V_3^r(z, x-1) \\ {}_{k+1}V_2^i(z,x) &= {}_{k+1}V_4^r(z-1, x) \\ {}_{k+1}V_3^i(z,x) &= {}_{k+1}V_1^r(z, x+1) \\ {}_{k+1}V_4^i(z,x) &= {}_{k+1}V_2^r(z=1, x). \end{aligned} \tag{4}$$

Consequently, if the magnitudes, positions, and directions of all impulses are known at time $k\,\Delta t$, the corresponding values at time $(k+1)\,\Delta t$ can be obtained by operating (3) and (4) on each node in the network. The impulse response of the network is then found by initially fixing the magni-

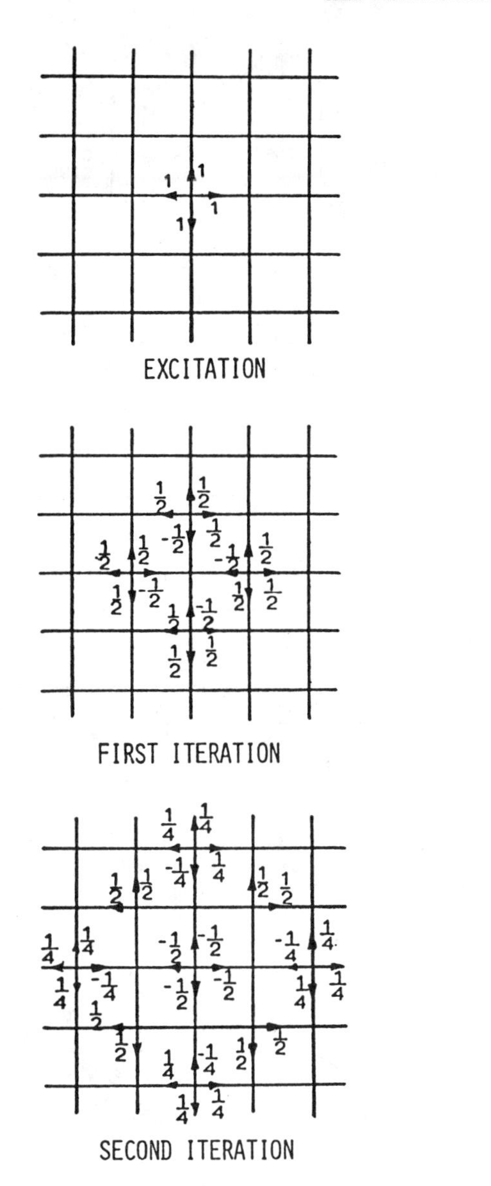

Fig. 3. Three consecutive scatterings in a two-dimensional TLM network excited by a Dirac impulse.

tudes, directions, and positions of all impulses at $t = 0$ and then calculating the state of the network at successive time intervals.

The scattering process described above forms the basic algorithm of the TLM method. Three consecutive scatterings are shown in Fig. 3, visualizing the spreading of the injected energy across the two-dimensional network.

This sequence of events closely resembles the disturbance of a pond due to a falling drop of water. However, there is one obvious difference, namely the discrete nature of the TLM mesh which causes dispersion of the velocity of the wavefront. In other words, the velocity of a signal component in the mesh depends on its direction of propagation as well as on its frequency.

In order to appreciate the importance of this dispersion, note that the process in Fig. 3 depicts a short episode of the response of the TLM network to a single impulse which contains all frequencies. Thus, harmonic solutions to a problem are obtained from the impulse response via the

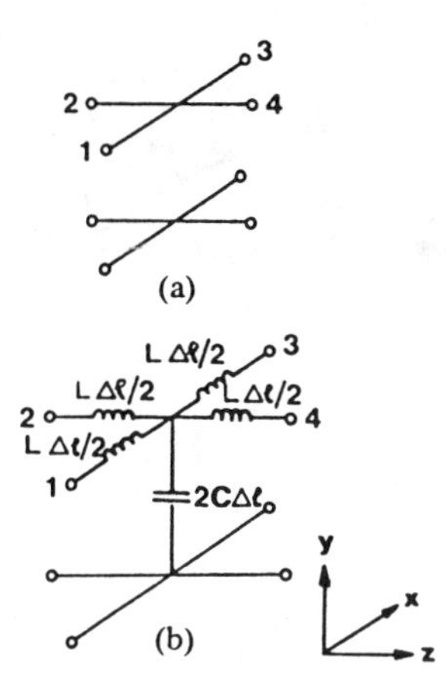

Fig. 4. The building block of the two-dimensional TLM network. (a) Shunt node. (b) Equivalent lumped-element model.

Fourier transform. Accurate solutions will be obtained only at frequencies for which the dispersion effect can be neglected. This aspect will be discussed in Section IV.

The TLM mesh can be extended to three dimensions, leading to a rather complex network containing series as well as shunt nodes. Each of the six field components is simulated by a voltage or a current in that mesh. Three-dimensional TLM networks will be discussed in Section V.

IV. The Two-Dimensional TLM Method

A. *Wave Properties of the TLM Network*

The basic building block of a two-dimensional TLM network is a shunt node with four sections of transmission lines of length $\Delta l/2$ (see Fig. 4(a)). Such a configuration can be approximated by the lumped-element model shown in Fig. 4(b). Comparing the relations between voltages and currents in the equivalent circuit with the relations between the H_z-, H_x-, and E_y-components of a TE_{m0} wave in a rectangular waveguide, the following equivalences can be established [5]:

$$E_y \equiv V_y \qquad -H_z \equiv (I_{x3} - I_{x1})$$
$$-H_x \equiv (I_{z2} - I_{z4}) \quad \mu \equiv L \quad \epsilon \equiv 2C. \tag{5}$$

For elementary transmission lines in the TLM network, and for $\mu_r = \epsilon_r = 1$, the inductance and capacitance per unit length are related by

$$1/\sqrt{LC} = 1/\sqrt{\epsilon_0 \mu_0} = c \tag{6}$$

where $c = 3 \times 10^8$ m/s.

Hence, if voltage and current waves on each transmission-line component travel at the speed of light, the complete network of intersecting transmission lines represents a medium of relative permittivity twice that of free space. The means that as long as the equivalent circuit in Fig. 4 is valid, the propagation velocity in the TLM mesh is $1/\sqrt{2}$ the velocity of light.

Note that the dual nature of electric and magnetic fields also allows us to simulate, for example, the longitudinal

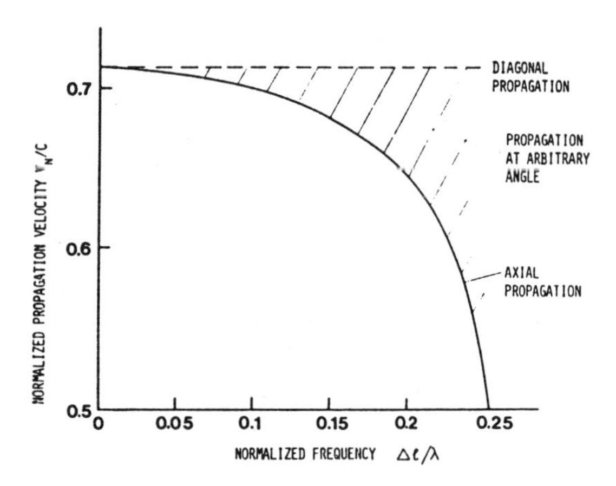

Fig. 5. Dispersion of the velocity of waves in a two-dimensional TLM network (after Johns and Beurle [5]).

magnetic field of TE modes by the network voltage, while the network currents simulate the transverse electric-field components. Whatever the relationship between field and network variables, the wave properties of the mesh, which will be discussed next, remain the same. Considering the mesh as a periodic structure, Johns and Beurle [5] calculate the following dispersion relation for propagation along the main mesh axes:

$$\sin(\beta_n \Delta l/2) = \sqrt{2}\sin[\omega \Delta l/(2c)] \tag{7}$$

where β_n is the propagation constant in the network. The resulting ratio of velocities on the matrix and in free space, $v_n/c = \omega/(\beta_n c)$, is shown in Fig. 5. It appears that a first cutoff occurs for $\Delta l/\lambda = 1/4$ (λ is the free-space wavelength). However, no cutoff occurs in the diagonal direction, where the velocity is frequency-independent, while in intermediate directions, the velocity ratio lies somewhere between the two curves shown in Fig. 5.

In conclusion, the TLM network simulates an isotropic propagating medium only as long as all frequencies are well below the network cutoff frequency, in which case the network propagation velocity may be considered constant and equal to $c/\sqrt{2}$.

B. Representation of Lossless and Lossy Boundaries

Electric and magnetic walls are represented by short and open circuits, respectively, at the appropriate positions in the TLM mesh. To ensure synchronism, they must be placed halfway between two nodes. In practice, this is achieved by making the mesh parameter Δl an integer fraction of the structure dimensions. Curved walls are represented by piecewise straight boundaries as shown in Fig. 6.

In the computation, the reflection of an impulse at a magnetic or electric wall is achieved by returning it, after one unit time step Δt, with equal or opposite sign to its boundary node of origin.

Lossy boundaries can be represented in the same way as lossless boundaries, with the difference that the reflection coefficient in each boundary branch is now

$$\rho = (R-1)/(R+1) \tag{8}$$

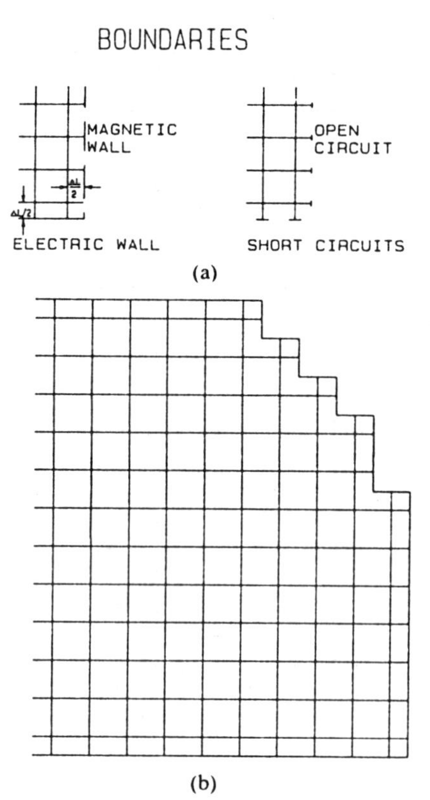

Fig. 6. Representation of boundaries in the TLM mesh. (a) Electric and magnetic walls. (b) Curved wall represented by a piecewise straight boundary.

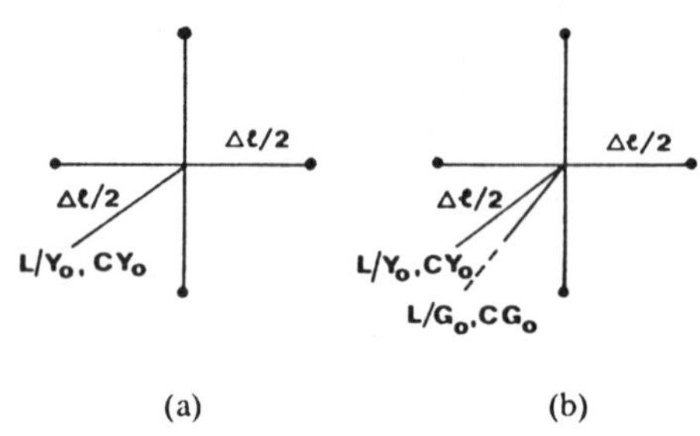

Fig. 7. Simulation of permittivity and losses. (a) Permittivity stub. (b) Permittivity stub and loss stub.

instead of unity. R is the normalized surface resistance of the boundary.

For a good but imperfect conductor of conductivity σ, the reflection coefficient ρ is approximately

$$\rho \simeq -1 + 2[\epsilon_0\omega/(2\sigma)]^{1/2}. \tag{9}$$

Note that since ρ depends on the frequency ω, the loss calculations are accurate only for that frequency which has been selected in determining ρ.

C. Representation of Dielectric and Magnetic Materials

The presence of dielectric or magnetic material (for example, in partial dielectric or magnetic loading of a waveguide) can be taken into account by loading inside nodes with reactive stubs of appropriate characteristic impedance and a length equal to half the mesh spacing [7], as shown in Fig. 7(a). For example, if the network voltage

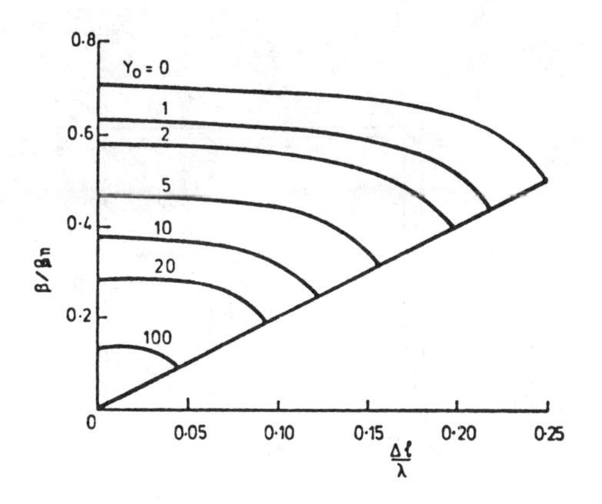

Fig. 8. Dispersion of the velocity of waves in a two-dimensional stub-loaded TLM network with the characteristic admittance of the reactive stubs as a parameter (after Johns [7]).

simulates an electric field, an open-circuited shunt stub of length $\Delta l/2$ will produce the effect of additional capacity at the nodes. This reduces the phase velocity in the structure and, at the same time, satisfies the boundary conditions at the air–dielectric interface [7]. At low frequencies, the velocity of waves in the stub-loaded TLM mesh is given by

$$v_n^2 = 0.5c^2/(1+Y_0/4) \tag{10}$$

where c is the free-space velocity, and Y_0 is the characteristic admittance of the stubs, normalized to the admittance of the main network lines. Note that the velocity in the network is now made variable by altering the single constant Y_0. The relationship between ϵ_r of the simulated space and Y_0 is

$$\epsilon_r = 2(1+Y_0/4). \tag{11}$$

The velocity characteristic along the main axes of the stub-loaded network is shown in Fig. 8 for various values of Y_0. Again, for relatively low frequencies, the mesh velocity is practically the same in all directions.

In cases where the voltage on the TLM mesh represents a magnetic field, the open shunt stubs describe a permeability. The velocity of waves in a magnetically loaded medium will be simulated correctly by such a mesh. However, the interface conditions are not satisfied, and a correction must be introduced in the form of local reflection and transmission coefficients at the interface between the media, as described in [7].

D. *Description of Dielectric Losses*

Losses in a dielectric can be accounted for in two different ways. One can either consider the TLM mesh to consist of lossy transmission lines, or one can load the nodes of a lossless mesh with so-called loss-stubs (Fig. 7(b)).

In the first case, the magnitude of each pulse is reduced by an appropriate amount while traveling from one node to the next, and the ensuing change in velocity is accounted for by increasing the time required to reach the next node [8]. This method is particularly suited for homogeneous structures.

In the second case, each node is resistively loaded with a matched transmission line of appropriate characteristic admittance G_0, extracting energy from each node at every iteration [10]. This technique is more suitable for inhomogeneous structures since it describes the interface conditions as well as the loss mechanism.

The normalized admittance of the loss-stub is related to the conductivity σ of the lossy medium by

$$G_0 = \sigma \Delta l (\mu_0/\epsilon_0)^{1/2}. \tag{12}$$

E. *Computation of the Frequency Response of a Structure*

The previous sections have described how the wave properties of two-dimensional unbounded and bounded space can be simulated by a two-dimensional mesh of transmission lines, and how the impulse response of such a mesh can be computed by iteration of (3) and (4). Any node (or several nodes) may be selected as input and/or output points. The output function is an infinite series of discrete impulses of varying magnitude, representing the response of the system to an impulsive excitation (see Fig. 12). The output corresponding to any other input may be obtained by convolving it with this impulse response.

Of particular interest is the response to a sinusoidal excitation which is obtained by taking the Fourier transform of the impulse response. Since the latter is a series of delta functions, the Fourier integral becomes a summation, and the real and imaginary parts of the output spectrum are

$$\mathrm{Re}[F(\Delta l/\lambda)] = \sum_{k=1}^{N} {}_kI \cos(2\pi k \Delta l/\lambda) \tag{13}$$

$$\mathrm{Im}[F(\Delta l/\lambda)] = \sum_{k=1}^{N} {}_kI \sin(2\pi k \Delta l/\lambda) \tag{14}$$

where $F(\Delta l/\lambda)$ is the frequency response, ${}_kI$ is the value of the output response at time $t = k\Delta l/c$, and N is the total number of time intervals for which the calculation has been made, henceforth called the "number of iterations."

In the case of a closed structure, this frequency response represents its mode spectrum. A typical example is Fig. 9(a), which shows the cutoff frequencies of the modes in a WR-90 waveguide.

Note that, as in a real measurement, the position of input and output points as well as the nature of the field component under study will affect the magnitudes of the spectral lines. For example, if input and output nodes are situated close to a minimum of a particular mode field, the corresponding eigenfrequency will not appear in the frequency response. This feature can be used either to suppress or enhance certain modes.

F. *Computation of Fields and Impedances*

Since the network voltages and currents are directly proportional to field quantities in the simulated structure, the TLM method also yields the field distribution. In order to obtain the configuration of a particular mode, its eigen-

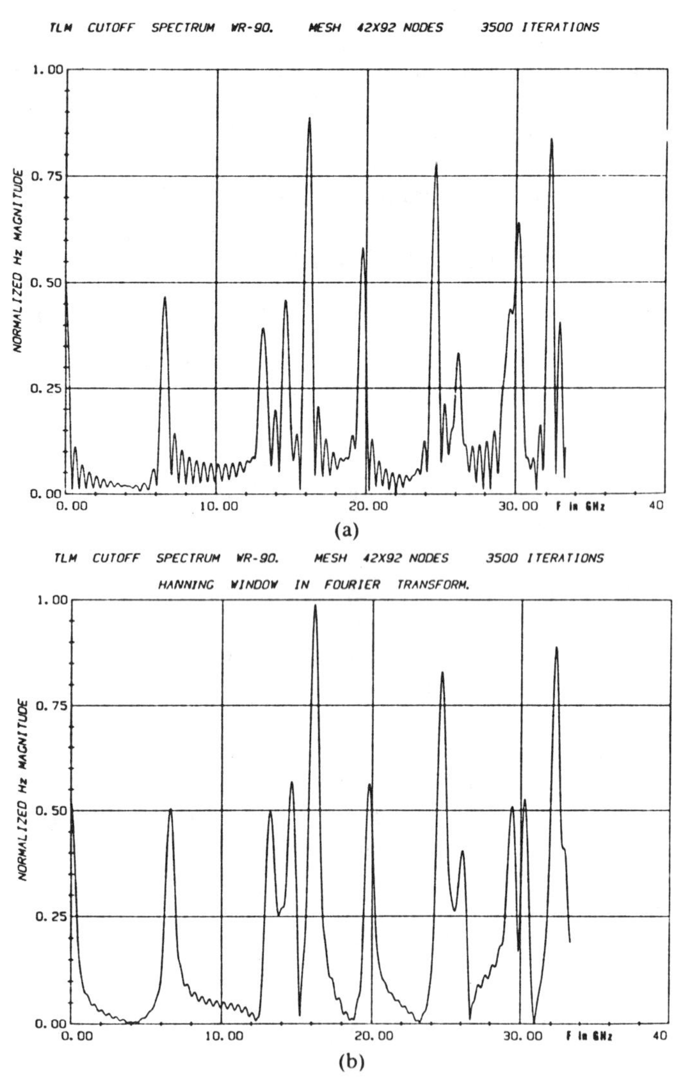

Fig. 9. Typical output from a two-dimensional TLM program. (a) Cutoff spectrum of a WR-90 waveguide. (b) The same spectrum after convolution of the output impulse function with a Hanning window (TLM) mesh: 42 × 92 nodes, 3500 iterations).

frequency must be computed first. Then the Fourier transform of the network variable representing the desired field component is computed at each node during a second run. In this process, (13) and (14) are computed for each node, with $\Delta l/\lambda$ corresponding to the eigenfrequency of the mode. The field between nodes can be obtained by interpolation techniques.

Impedances can, in turn, be obtained from the field quantities. Local field impedances can be found directly as the ratio of voltages and currents at a node, while impedances defined on the basis of particular field integrals (such as the voltage–power impedance in a waveguide) are computed by stepwise integration of the discrete field values. This procedure is identical to that used in finite-element and finite-difference methods of analysis.

V. The Three-Dimensional TLM Method

The two-dimensional method described above can be extended to three dimensions at the expense of increased complexity [10] to [15]. In order to simultaneously describe

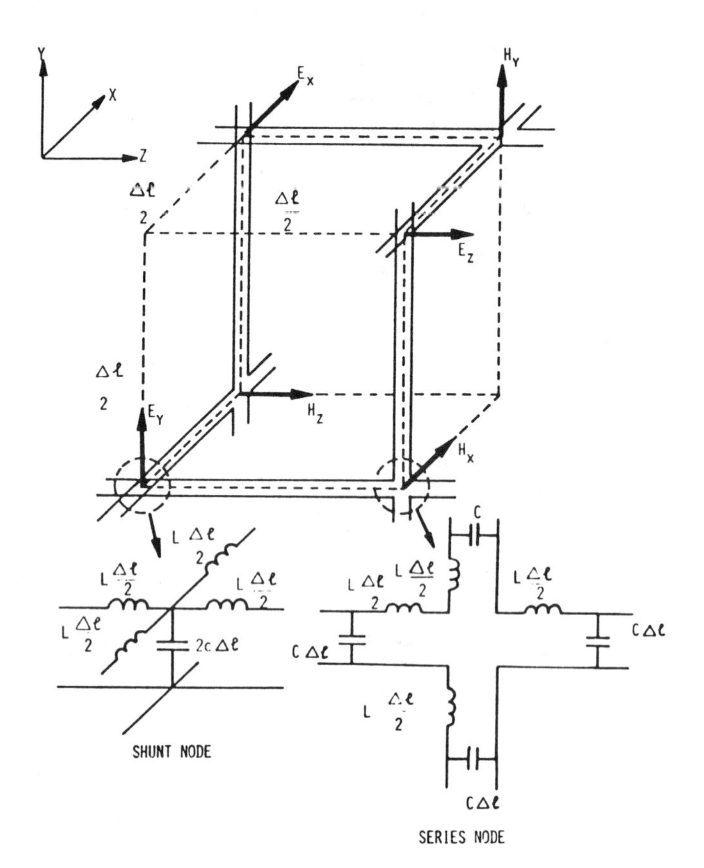

Fig. 10. The three-dimensional TLM cell featuring three series and three shunt nodes.

all six field components in three-dimensional space, the basic shunt node must be replaced by a hybrid TLM cell consisting of three shunt and three series nodes as shown in Fig. 10. The side of the cell is $\Delta l/2$. The voltages at the three series nodes represent the three electric-field components, while the currents at the series nodes represent the magnetic-field components.

The wave properties of the three-dimensional mesh are similar to that of its two-dimensional counterpart with the difference that the low-frequency velocity is now $c/2$ instead of $c/\sqrt{2}$ [15].

Boundaries are simulated by short-circuiting shunt nodes (electric wall) or open-circuiting shunt nodes (magnetic wall) situated on a boundary. Wall losses are included by introducing imperfect reflection coefficients.

Magnetic and dielectric materials may be introduced by adding short-circuited $\Delta l/2$ series stubs at the series nodes and open-circuited $\Delta l/2$ shunt stubs at the shunt nodes, respectively. Furthermore, losses are taken into account by resistively loading the shunt nodes in the network (see Fig. 11). Even anisotropic materials may be simulated by introducing at each of the three series or shunt nodes of a cell a stub with a different characteristic admittance [17]. Finally, losses as well as permittivities and permeabilities can be varied in space and in time by controlling the admittances of the dissipative and reactive stubs. The relationships between material parameters and stub admittances are the same as in the two-dimensional case.

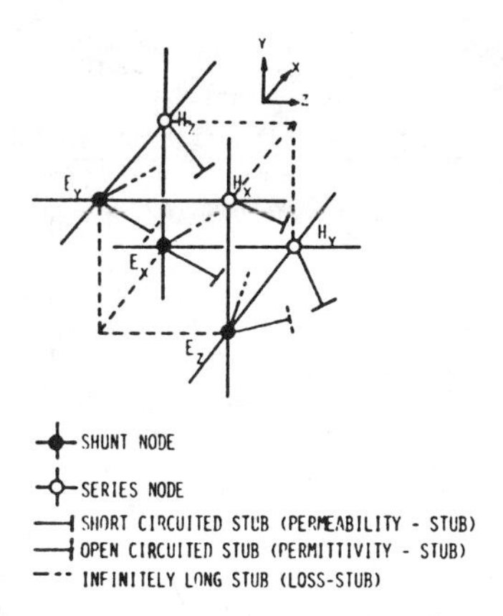

Fig. 11. Simulation of permittivity, permeability, and losses in a three-dimensional TLM network (after Akhtarzad [13]).

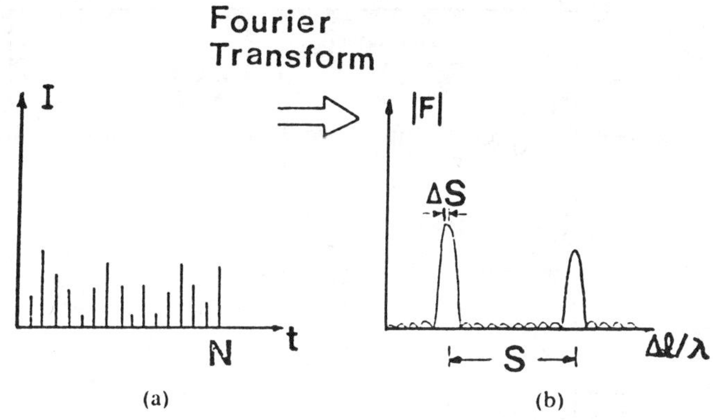

(a) (b)

Fig. 12. (a) Truncated output impulse response and (b) resulting truncation error in the frequency domain.

The impulse response of a three-dimensional network is found in the same way as in the two-dimensional case, and everything that has been said about the computation of eigenfrequencies, fields, and impedances, applies here as well.

VI. Errors and Their Correction

Like all other numerical techniques, the TLM method is subject to various sources of error and must be applied with caution in order to yield reliable and accurate results. The main sources of error are due to the following circumstances:

a) The impulse response must be truncated in time.
b) The propagation velocity in the TLM mesh depends on the direction of propagation and on the frequency.
c) The spatial resolution is limited by the finite mesh size.
d) Boundaries and dielectric interfaces cannot be aligned in the 3-D TLM model.

The resulting errors will be discussed below, and ways of eliminating or, at least, significantly reducing these errors will be described.

A. Truncation Error

The need to truncate the output impulse function leads to the so-called truncation error: Due to the finite duration of the impulse response, its Fourier transform is not a line spectrum but rather a superposition of $\sin x/x$ functions (Gibbs's phenomenon) which may interfere with each another such that their maxima are slightly shifted. The resulting error in the eigenfrequency, or truncation error, is given by

$$E_T \leqq \Delta S/(\Delta l/\lambda_c) = 3\lambda_c/(SN^2\pi^2\Delta l) \quad (15)$$

where N is the number of iterations and S is the distance in the frequency domain between two neighboring spectral peaks (see Fig. 12).

This expression shows that the truncation error decreases with increasing separation S and increasing number of iterations N. It is thus desirable to suppress all unwanted modes close to the desired mode by choosing appropriate input and output points in the TLM network. Another technique, proposed by Saguet and Pic [20], is to use a Hanning window in the Fourier transform, resulting in a considerable attenuation of the sidelobes.

In this process, the output impulse response is first convolved with the Hanning profile

$$f_k(k) = 0.5(1+\cos \pi k/N), \qquad k = 1,2,3,\cdots,N \quad (16)$$

where k is the iteration variable or counter. The filtered impulse response is then Fourier transformed. The resulting improvement can be appreciated by comparing Figs. 9(a) and 9(b).

Finally, the number of iterations may be made very large, but this leads to increased CPU time. It is recommended that the number of iterations be chosen such that the truncation error given by (16) is reduced to a fraction of a percent and can be neglected.

B. Velocity Error

If the wavelength in the TLM network is large compared with the network parameter Δl, it can be assumed that the fields propagate with the same velocity in all directions. However, when the wavelength decreases, the velocity depends on the direction of propagation (see Fig. 5). At first glance, the resulting velocity error can be reduced only by choosing a very dense mesh, unless propagation occurs essentially in an axial direction (e.g., rectangular waveguide), in which case the error can be corrected directly using the dispersion relation (7). Fortunately, the velocity error responds to the same remedial measures as the coarseness error (which will be described next), and it therefore does not need to be corrected separately.

C. Coarseness Error

The coarseness error occurs when the TLM mesh is too coarse to resolve highly nonuniform fields as can be found at corners and wedges. This error is particularly cumbersome when analyzing planar structures which contain such regions. A possible but impractical measure would be to choose a very fine mesh. However, this would lead to large memory requirements, particularly for three-dimensional

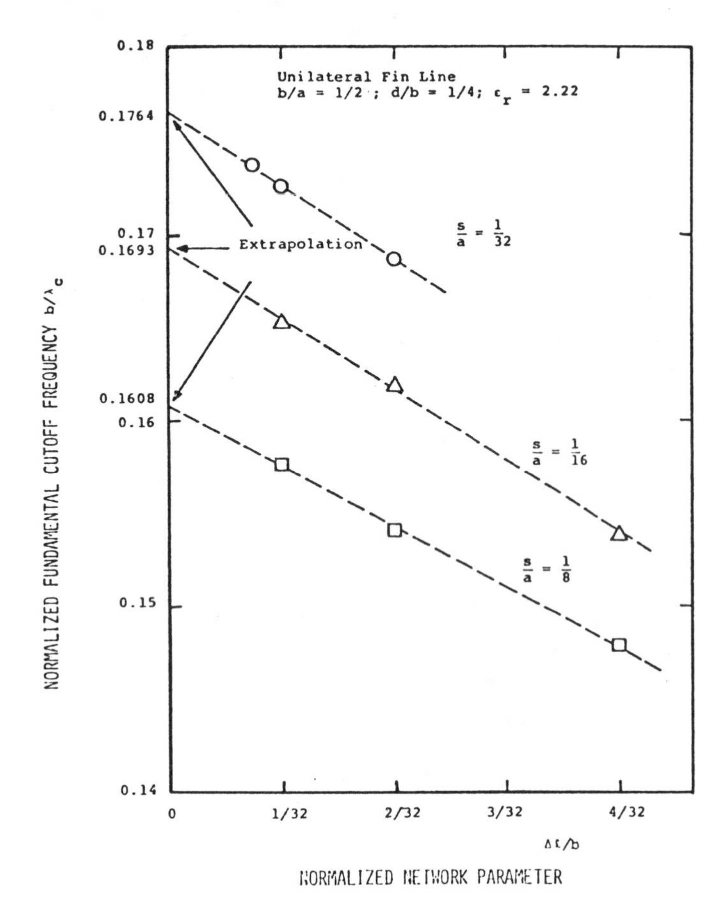

Fig. 13. Elimination of coarseness error by linear extrapolation of results obtained with TLM meshes of different parameter $\Delta l/b$ (after Shih and Hoefer [25]).

problems. A better response is to introduce a network of variable mesh size to provide higher resolution in the nonuniform field region [27]–[29]. This approach is described in the next section; however, it requires more complicated programming. Yet another approach, proposed by Shih and Hoefer [25], is to compute the structure response several times using coarse meshes of different mesh parameter Δl, and then to extrapolate the obtained results to $\Delta l = 0$ as shown in Fig. 13.

Both measures effectively reduce the error by one order of magnitude and simultaneously correct the velocity error.

D. Misalignment of Dielectric Interfaces and Boundaries in Three-Dimensional Inhomogeneous Structures

Due to the particular way in which boundaries are simulated in a three-dimensional TLM network, dielectric interfaces appear halfway between nodes, while electric and magnetic boundaries appear across such nodes. This can be a problem when simulating planar structures such as microstrip or finline. In the TLM model, the dielectric either protrudes or is undercut by $\Delta l/2$, as shown in Fig. 14. Unless the resulting error is acceptable, one must make two computations, one with recessed and one with protruding dielectric, and take the average of the results. The problem does not occur in a variation of the three-dimensional TLM method involving an alternative node configuration proposed by Saguet and Pic [31], and described in the next section.

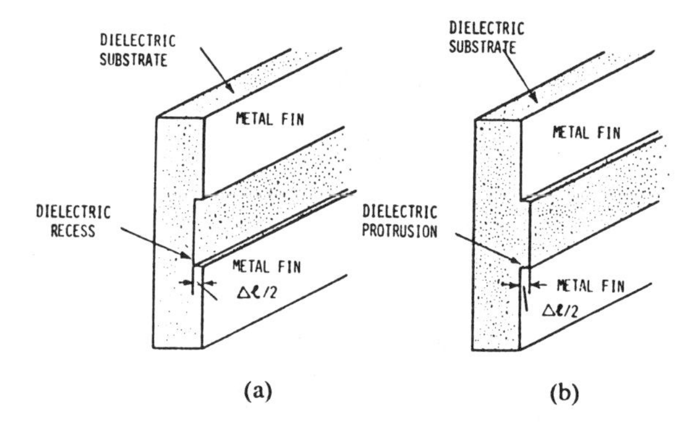

Fig. 14. Misalignment of conducting boundaries and dielectric interfaces in the three-dimensional TLM simulation of planar structures (after Shih [26]).

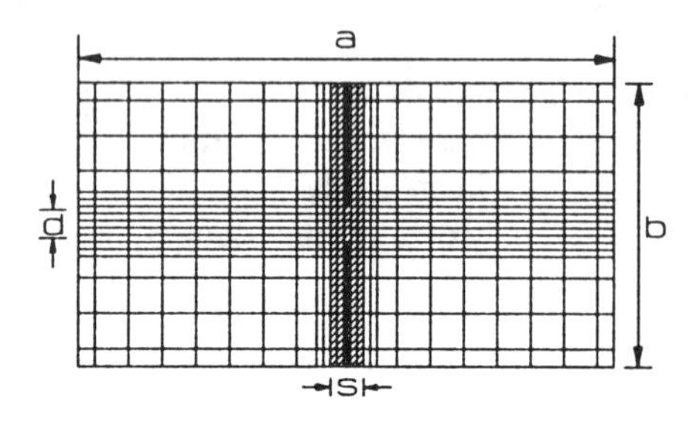

Fig. 15. Two-dimensional TLM network with variable mesh size for the computation of cutoff frequencies of finlines (after Saguet and Pic [28]).

VII. Variations of the TLM Method

A number of modifications of the conventional TLM method have been proposed over the last few years with the aim of reducing errors, memory requirements, and CPU time. Some of them have already been mentioned, such as the introduction of a Hanning window [20], and extrapolation from coarse mesh calculations [21], [25]. Some effort has also been directed towards improving the efficiency of programing techniques [24].

In the following, three other interesting and significant innovations will be discussed briefly.

A. TLM Networks with Nonuniform Mesh

In order to ensure synchronism, the conventional TLM network uses a uniform mesh parameter throughout. This can lead to considerable numerical expenditure if the structure contains sharp corners or fins producing highly nonuniform fields and thus demands a high density mesh. Saguet and Pic [28] and Al-Mukhtar and Sitch [29] have independently proposed ways to implement irregularly graded TLM meshes which, as in the finite-element method, allow the network to adapt its density to the local nonuniformity of the fields. Fig. 15 shows such a network as proposed by Saguet and Pic [28] for the computation of cutoff frequencies in a finline. Note, however, that the size of the mesh cells is not arbitrary as in the case of finite elements; the length of each side is an odd integer multiple P of the smallest cell length in the network. To keep the

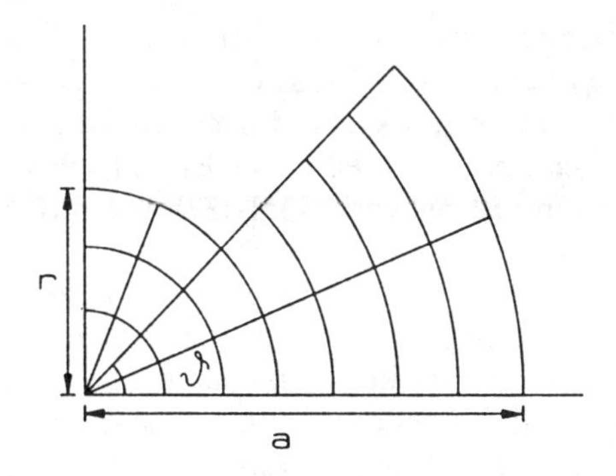

Fig. 16. A radial TLM mesh for the treatment of circular ridged waveguides (after Al-Mukhtar and Sitch [29]).

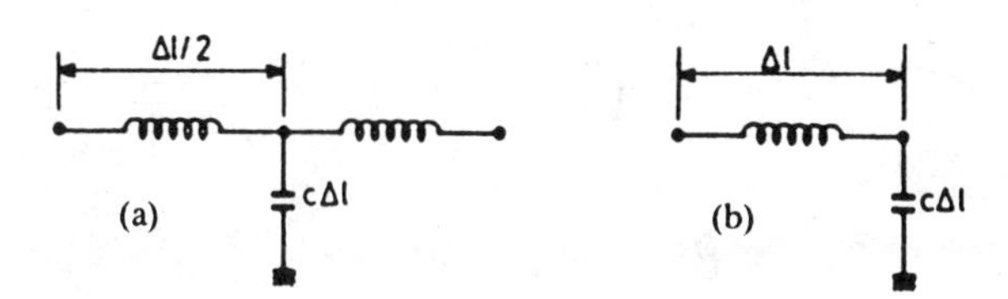

Fig. 17. Alternative lumped-element network for the two-dimensional shunt node. (a) Classical representation. (b) Representation proposed by Saguet and Pic [31].

velocity of traveling impulses the same in all branches, the inductivity per unit length of the longer mesh lines is increased by a factor P while their capacity per unit length is reduced by $1/P$. This, in turn, increases their characteristic impedance by a factor P, and the scattering matrix of nodes connecting cells of different size must be modified accordingly.

To preserve synchronism, impulses traveling on longer branches are kept in store for P iterations before being reinjected at the next node.

For the configuration shown in Fig. 15, Saguet and Pic found that computing time was reduced between 3.5 and 5 times over a uniform mesh, depending on the relative size of the larger cells.

A different approach has been proposed by Al-Mukhtar and Sitch [27], [29]. They describe two possible ways to modify the characteristics of mesh elements in order to ensure synchronism, one involving the insertion of series stubs between nodes and loading of nodes by shunt stubs, the other involving modification of inductivity and capacity per unit length in such a way that propagation velocity in a branch becomes proportional to its length. The work by Al-Mukhtar and Sitch also covers the representation of radial meshes (see Fig. 16) as well as three-dimensional inhomogeneous structures. They report an economy of 45 percent in computer expenditure for a two-dimensional ridged waveguide problem, and a 40-percent reduction in storage and an 80-percent reduction in run time for a three-dimensional finline problem thanks to mesh grading.

B. A Punctual Node for Three-Dimensional TLM Networks

Conventional three-dimensional TLM networks require three shunt and three series nodes for the representation of one single cell (see Fig. 10). Saguet and Pic [31] have proposed an alternative method of interconnection. Representation of the short transmission-line sections by two rather than three lumped elements (see Fig. 17) makes it possible to realize both shunt and series connections in one point, resulting in a punctual node with 12 branches. This node is equivalent to a cell, such as that in Fig. 10, in which the inner connections have been eliminated. Losses and dielectric or magnetic loading can be simulated with stubs in the same way as discussed earlier.

This new node representation reduces, according to Saguet and Pic [31], the computation time by about 30 percent. By employing both the punctual node and variable mesh size in a three-dimensional program, Saguet [32] has computed the resonant frequencies of a finline cavity 35 times faster than with a program based on the traditional TLM method.

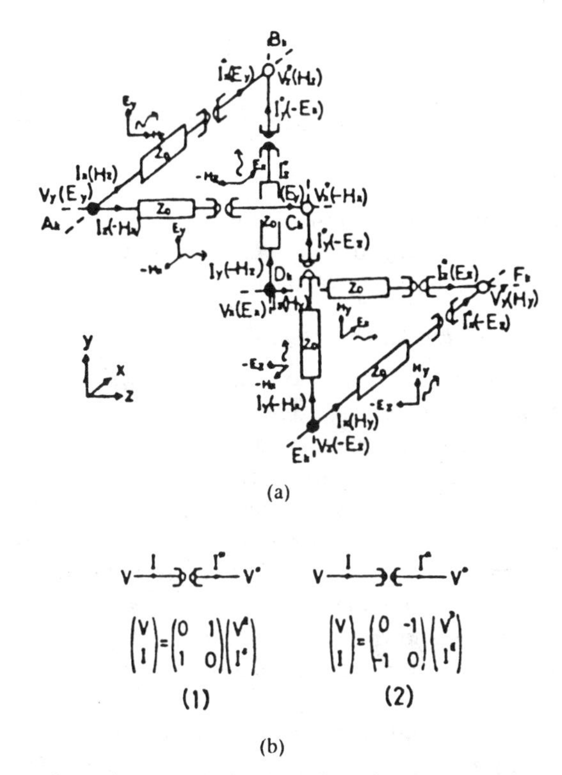

Fig. 18. Alternative network for three-dimensional TLM analysis proposed by Yoshida *et al.* [33]. (a) Equivalent circuit of an alternative three-dimensional TLM cell. (b) Definition of gyrators in (a): 1) positive gyrator, 2) negative gyrator.

C. Alternative Network Simulating Maxwell's Equations

Yoshida, Fukai, and Fukuoka [19], [23], [30], [33] have described a network similar to the TLM mesh, differing only in the way the basic cell element has been modeled. Instead of series and shunt nodes, this network contains so-called electric and magnetic nodes which are both "shunt-type nodes": while at the electric node, the voltage variable represents an electric field, it symbolizes a magnetic field at the magnetic node. The resulting ambivalence in the nature of the network voltage and current must be removed by inserting gyrators between the two types of nodes, as shown in Fig. 18. The wave properties of this network are identical with that of the conventional TLM

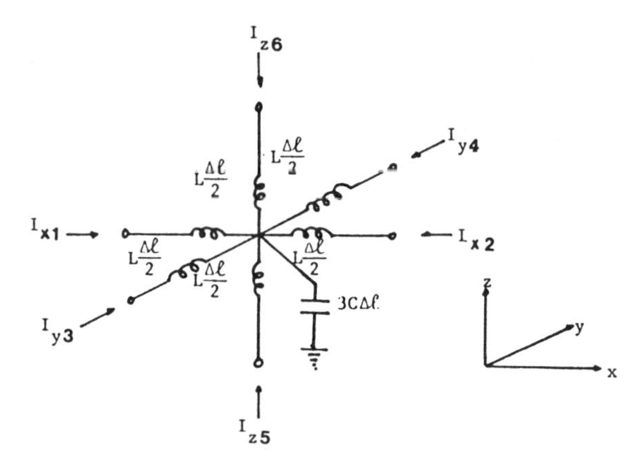

Fig. 19. Basic node of a three-dimensional scalar TLM network (after Choi and Hoefer [34]).

mesh. Errors and limitations are the same, and so are the possibilities of introducing losses and isotropic as well as anisotropic dielectric and magnetic materials.

D. *The Scalar TLM Method*

In those cases where electromagnetic fields can be decomposed into TE and TM modes (or LSE and LSM modes), it is only necessary to solve the scalar wave equation. Choi and Hoefer [34] have described a scalar TLM network to simulate a single field component or a Hertzian potential in three-dimensional space. The scalar TLM mesh can be thought of as a two-dimensional network to which additional transmission lines are connected orthogonally at each node as shown in Fig. 19. Such a structure could be realized in the form of a three-dimensional grid of coaxial lines.

The voltage impulses traveling across such a network represent the scalar variable to be simulated. Boundary reflection coefficients depend on both the nature of the boundary and that of the quantity to be simulated. For example, impulses will be subject to a reflection coefficient of -1 at a lossless electric wall if they represent either a tangential electric- or a normal magnetic-field component. A normal electric or a tangential magnetic field will be reflected with a coefficient of $+1$ in the same circumstances.

The slow-wave velocity in the three-dimensional scalar mesh is $c/\sqrt{3}$ as opposed to $c/2$ in the conventional TLM network. Dielectric or magnetic material as well as losses may be simulated using reactive and dissipative stubs.

The scalar method requires only 1/4 of the memory space and is seven times faster than the conventional method for a commensurate problem. However, its application is severely restricted, as it can be applied to scalar wave problems only.

VIII. Applications of the TLM Method

In the previous sections, the flexibility, versatility, and generality of the transmission-line matrix method has been demonstrated. In the following, an overview of potential applications of the method is given, and references describing specific applications are indicated. This list is not exhaustive, and many more applications can be found, not only in electromagnetism, but also in other fields dealing with wave phenomena, such as optics and acoustics.

For completeness, it should be mentioned that the TLM procedure can also be used to model and solve linear and nonlinear lumped networks [35]–[37] and diffusion problems [38]. Readers with a special interest in these applications should consult these references for more details.

Wave problems can be simulated in unbounded and bounded space, either in the time domain or—via Fourier analysis—in the frequency domain. Arbitrary homogeneous or inhomogeneous structures with anisotropic, space- and time-dependent electrical properties, including losses, can be simulated in two and three dimensions.

Below are some typical application examples.

A. *Two-Dimensional Scattering Problems in Rectangular Waveguides (Field Distribution of Propagating and Evanescent Modes, Wave Impedance, Scattering Parameters)*

— Open-circuited rectangular waveguide (TE_{n0}) [5].
— Bifurcation in rectangular waveguide (TE_{n0}) [5].
— Scattering by arbitrarily-shaped two-dimensional discontinuities in rectangular waveguide (TE_{n0}) including losses.

B. *Two-Dimensional Eigenvalue Problems (Eigenfrequencies, Mode Fields)*

— Cutoff frequencies and mode fields in homogeneous waveguides of arbitrary cross section, such as ridged waveguides [6], [8], [13], [26], [27].
— Cutoff frequencies and mode fields in inhomogeneous waveguides of arbitrary cross section, such as dielectric loaded waveguides, finlines, image lines [7], [13], [16], [18], [21], [22], [25], [26], [28].

C. *Three-Dimensional Eigenvalue and Hybrid Field Problems (Dispersion Characteristics, Wave Impedances, Losses, Eigenfrequencies, Mode Fields, Q-Factors)*

— Characteristics of dielectric loaded cavities [10], [13], [14], [15], [18], [26], [27], [29], [32], [34].
— Dispersion characteristics and scattering in inhomogeneous planar transmission-line structures, including anisotropic substrate [11], [12], [13], [14], [17], [26], [27], [29], [30], [32].
— Transient analysis of transmission-line structures [19], [23], [33].

General-purpose two-dimensional and three-dimensional TLM programs can be found in Akhtarzad's Ph.D. thesis [13]. They can be adapted to most of the applications described above. If the various improvements and modifications described in Section VII are implemented in these programs, versatile and powerful numerical tools for the solution of complicated field problems are indeed obtained.

IX. Discussion and Conclusion

This paper has described the physical principle, the formulation, and the implementation of the transmission-line matrix method of analysis. Numerous features and applications of the method have been discussed, in particular the principal sources of error and their correction, the inclusion of losses, inhomogeneous and anisotropic properties of materials, and the capability to analyze transient as well as steady-state wave phenomena.

A general-purpose two-dimensional TLM program can be written in about 80 lines of FORTRAN, while a three-dimensional program is about 110 lines long. [13].

The method is limited only by the amount of memory storage required, which depends on the complexity of the structure and the nonuniformity of fields set up in it. In general, the smallest feature in the structure should at least contain three nodes for good resolution. The total storage requirement for a given computation can be determined by considering that each two-dimensional node requires five real number storage places, and an additional number equal to the number of iterations is needed to store the output impulse function. A basic three-dimensional node requires twelve number locations; if it is completely equipped with permittivity, permeability, and loss stubs, the required number of stores goes up to 26. Again, one real number must be stored per output function and per iteration. The number of iterations required varies between several hundred and several thousand, depending on the size and complexity of the TLM mesh.

As far as computational expenditure is concerned, the TLM method compares favorably with finite-element and finite-difference methods. Its accuracy is even slightly better by virtue of the Fourier transform, which ensures that the field function between nodes is automatically circular rather than linear as in the two other methods.

The main advantage of the TLM method, however, is the ease with which even the most complicated structures can be analyzed. The great flexibility and versatility of the method reside in the fact that the TLM network incorporates the properties of the electromagnetic fields and their interaction with the boundaries and materials. Hence, the electromagnetic problem need not to be reformulated for every new structure; its parameters are simply entered into a general-purpose program in the form of codes for boundaries, losses, permeability and permittivity, and excitation of the fields. Furthermore, by solving the problem through simulation of wave propagation in the time domain, the solution of large numbers of simultaneous equations is avoided. There are no problems with convergence, stability, or spurious solutions.

Another advantage of the TLM method resides in the large amount of information generated in one single computation. Not only is the impulse response of a structure obtained, yielding, in turn, its response to any excitation, but also the characteristics of the dominant and higher order modes are accessible in the frequency domain through the Fourier transform.

In order to increase the numerical efficiency and reduce the various errors associated with the method, more programing effort must be invested. Such an effort may be worthwhile when faced with the problem of scattering by a three-dimensional discontinuity in an inhomogeneous transmission medium, or when studying the overall electromagnetic properties of a monolithic circuit.

Finally, the TLM method may be adapted to problems in other areas such as thermodynamics, optics, and acoustics. Not only is it a very powerful and versatile numerical tool, but because of its affinity with the mechanism of wave propagation, it can provide new insights into the physical nature and the behavior of electromagnetic waves.

References

[1] C. Huygens, "Traité de la Lumière" (Leiden, 1690).

[2] J. R. Whinnery and S. Ramo, "A new approach to the solution of high-frequency field problems," *Proc. IRE*, vol. 32, pp. 284–288, May 1944.

[3] G. Kron, "Equivalent circuit of the field equations of Maxwell—I," *Proc. IRE*, vol. 32, pp. 289–299, May 1944.

[4] J. R. Whinnery, C. Concordia, W. Ridgway, and G. Kron, "Network analyzer studies of electromagnetic cavity resonators," *Proc. IRE*, vol. 32, pp. 360–367, June 1944.

[5] P. B. Johns and R. L. Beurle, "Numerical solution of 2-dimensional scattering problems using a transmission-line matrix," *Proc. Inst. Elec. Eng.*, vol. 118, no. 9, pp. 1203–1208, Sept. 1971.

[6] P. B. Johns, "Application of the transmission-line matrix method to homogeneous waveguides of arbitrary cross-section," *Proc. Inst. Elec. Eng.*, vol. 119, no. 8, pp. 1086–1091, Aug. 1972.

[7] P. B. Johns, "The solution of inhomogeneous waveguide problems using a transmission-line matrix," *IEEE Trans. Microwave Theory Tech.*, vol. MTT-22, pp. 209–215, Mar. 1974.

[8] S. Akhtarzad and P. B. Johns, "Numerical solution of lossy waveguides: T.L.M. computer program," *Electron. Lett.*, vol. 10, no. 15, pp. 309–311, July 25, 1974.

[9] P. B. Johns, "A new mathematical model to describe the physics of propagation," *Radio Electron. Eng.*, vol. 44, no. 12, pp. 657–666, Dec. 1974.

[10] S. Akhtarzad and P. B. Johns, "Solution of 6-component electromagnetic fields in three space dimensions and time by the T.L.M. method," *Electron. Lett.*, vol. 10, no. 25/26, pp. 535–537, Dec. 12, 1974.

[11] S. Akhtarzad and P. B. Johns, "T.L.M. analysis of the dispersion characteristics of microstrip lines on magnetic substrates using 3-dimensional resonators," *Electron. Lett.*, vol. 11, no. 6, pp. 130–131, Mar. 20, 1975.

[12] S. Akhtarzad and P. B. Johns, "Dispersion characteristic of a microstrip line with a step discontinuity," *Electron. Lett.*, vol. 11, no. 14, pp. 310–311, July 10, 1975.

[13] S. Akhtarzad, "Analysis of lossy microwave structures and microstrip resonators by the TLM method," Ph.D dissertation, Univ. of Nottingham, England, July 1975.

[14] S. Akhtarzad and P.B. Johns, "Three-dimensional transmission-line matrix computer analysis of microstrip resonators," *IEEE Trans. Microwave Theory Tech.*, vol. MTT-23, pp. 990–997, Dec. 1975.

[15] S. Akhtarzad and P. B. Johns, "Solution of Maxwell's equations in three space dimensions and time by the T.L.M. method of analysis," *Proc. Inst. Elec. Eng.*, vol. 122, no. 12, pp. 1344–1348, Dec. 1975.

[16] S. Akhtarzad and P. B. Johns, "Generalized elements for T.L.M. method of numerical analysis," *Proc. Inst. Elec. Eng.*, vol. 122, no. 12, pp. 1349–1352, Dec. 1975.

[17] G. E. Mariki, "Analysis of microstrip lines on inhomogeneous anisotropic substrates by the TLM numerical technique," Ph.D. thesis, Univ. of California, Los Angeles, June 1978.

[18] W. J. R. Hoefer and A. Ros, "Fin line parameters calculated with the TLM method," in *IEEE MTT Int. Microwave Symp. Dig.* (Orlando, FL), Apr. 28–May 2, 1979.

[19] N. Yoshida, I. Fukai, and J. Fukuoka, "Transient analysis of two-dimensional Maxwell's equations by Bergeron's method," *Trans. IECE Japan*, vol. J62B, pp. 511–518, June 1979.

[20] P. Saguet and E. Pic, "An improvement for the TLM method," *Electron. Lett.*, vol. 16, no. 7, pp. 247–248, Mar. 27, 1980.

[21] Y.-C. Shih, W. J. R. Hoefer, and A. Ros, "Cutoff frequencies in fin lines calculated with a two-dimensional TLM-program," in *IEEE MTT Int. Microwave Symp. Dig.* (Washington, DC), June 1980, pp. 261–263.

[22] W. J. R. Hoefer and Y.-C. Shih, "Field configuration of fundamental and higher order modes in fin lines obtained with the TLM method," presented at URSI and Int. IEEE-AP Symp., Quebec, Canada, June 2–6, 1980.

[23] N. Yoshida, I. Fukai, and J. Fukuoka, "Transient analysis of three-dimensional electromagnetic fields by nodal equations," *Trans. IECE Japan*, vol. J63B, pp. 876–883, Sept. 1980.
[24] A. Ros, Y.-C. Shih, and W. J. R. Hoefer, "Application of an accelerated TLM method to microwave systems," in *10th Eur. Microwave Conf. Dig.* (Warszawa, Poland), Sept. 8 11, 1980, pp. 382–388.
[25] Y.-C. Shih and W. J. R. Hoefer, "Dominant and second-order mode cutoff frequencies in fin lines calculated with a two-dimensional TLM program," *IEEE Trans. Microwave Theory Tech.*, vol. MTT-28, pp. 1443–1448, Dec. 1980.
[26] Y.-C. Shih, "The analysis of fin lines using transmission line matrix and transverse resonance methods," M.A.Sc. thesis, Univ. of Ottawa, Canada, 1980.
[27] D. Al-Mukhtar, "A transmission line matrix with irregularly graded space," Ph.D. thesis, Univ. of Sheffield, England, Aug. 1980.
[28] P. Saguet and E. Pic, "Le maillage rectangulaire et le changement de maille dans la méthode TLM en deux dimensions," *Electron. Lett.*, vol. 17, no. 7, pp. 277–278, Apr. 2, 1981.
[29] D.A. Al-Mukhtar and J. E. Sitch, "Transmission-line matrix method with irregularly graded space," *Proc. Inst. Elec. Eng.*, vol. 128, pt. H, no. 6, pp. 299–305, Dec. 1981.
[30] N. Yoshida, I. Fukai, and J. Fukuoka, "Application of Bergeron's method to anisotropic media," *Trans. IECE Japan*, vol J64B, pp. 1242–1249, Nov. 1981.
[31] P. Saguet and E. Pic, "Utilisation d'un nouveau type de noeud dans la méthode TLM en 3 dimensions," *Electron. Lett.*, vol. 18, no. 11, pp. 478–480, May 1982.
[32] P. Saguet, "Le maillage parallelepipédique et le changement de maille dans la méthode TLM en trois dimensions," *Electron. Lett.*, vol. 20, no. 5, pp. 222–224, Mar. 15, 1984.
[33] N. Yoshida and I. Fukai, "Transient analysis of a stripline having a corner in three-dimensional space," *IEEE Trans. Microwave Theory Tech.*, vol. MTT-32, pp. 491–498, May 1984.
[34] D. H. Choi and W. J. R. Hoefer, "The simulation of three-dimensional wave propagation by a scalar TLM model," in *IEEE MTT Int. Microwave Symp. Dig.* (San Francisco), May 1984.
[35] P. B. Johns, "Numerical modelling by the TLM method," in *Large Engineering Systems*, A. Wexler, Ed. Oxford: Pergamon Press, 1977.
[36] J. W. Bandler, P. B. Johns, and M. R. M. Rizk, "Transmission-line modeling and sensitivity evaluation for lumped network simulation and design in the time domain," *J. Franklin Inst.*, vol. 304, no. 1, pp. 15–23, 1977.
[37] P. B. Johns and M. O'Brien, "Use of the transmission-line modelling (T.L.M.) method to solve non-linear lumped networks," *Radio Electron. Eng.*, vol. 50, no. 1/2, pp. 59–70, Jan./Feb. 1980.
[38] P. B. Johns, "A simple explicit and unconditionally stable numerical routine for the solution of the diffusion equation," *Int. J. Num. Meth. Eng.*, vol. 11, pp. 1307–1328, 1977.

Numerical solution of 2-dimensional scattering problems using a transmission-line matrix

P. B. Johns, B.Sc.(Eng.), M.Sc., C.Eng., M.I.E.E., and
R. L. Beurle, B.Sc.(Eng.), Ph.D., A.C.G.I., F.Inst.P., C.Eng., M.I.E.E.

Indexing terms: Guided-electromagnetic-wave propagation, Electromagnetic-wave scattering, Transmission-line theory

Abstract

A numerical method using impulse analysis of a transmission-line matrix is introduced and used to obtain wave-impedance values in a waveguide. The method is demonstrated by applying it to the scattering caused by a waveguide bifurcation.

1 Introduction

The use of electrical networks for the solution of electromagnetic-field problems is well established,[1-3] and network analysers have been built and used to solve 2-dimensional waveguide scattering problems.[4,5] The principles involved in these analogue methods are used to formulate a numerical procedure for the solution of 2-dimensional scattering problems. The method is tested by calculating the wave impedance in a rectangular waveguide under various load conditions, and comparing the results with established analytical results. The examples include the scattering effects of a bifurcation in a waveguide, both for the case where the separate guides (after bifurcation) support only evanescent modes, and also for the case where one guide supports a propagating mode.

One of the important features of this method is that it is a transient analysis which gives results at the solution point in terms of the response to a delta-function impulse at the input. This means that one calculation contains all the necessary information about the frequency response for frequencies from zero to infinity. In practice, the network itself has a filtering effect, the frequency response having an upper limit which depends on the coarseness of the mesh. In contrast, for example, the method of ray optics,[6,7] which complements this method in many ways, has solutions in terms of the frequency, and separate calculations are made for each frequency.

More conventional methods[8,9] of solving waveguide discontinuity problems usually have harmonic solutions also, and often apply over a limited frequency range. As indicated in Reference 13, many of these methods are geared to producing solutions to particular problems which can be calculated by hand, often entailing considerable mathematical agility on the part of the user. With the advent of high-speed digital computers, a need for straightforward techniques suited to computers arises, and this method, like that of Reference 13, attempts to meet these requirements. In most of the numerical methods available for the solution of this type of problem,[10-13] the analysis leads to a set of a large number of linear simultaneous equations, and the solution of these may require some specialised knowledge. In the method proposed, however, this problem does not arise, and, in fact, the programming is of a very elementary nature indeed.

In common with other techniques of numerical solution of Helmholtz's equation,[10-12] the method proposed here is limited by the size of the matrix that can be stored in a computer. For the simple boundary conditions of the examples in the paper, however, the matrix sizes are small enough to allow use of a computer having a total store of only 8 kbyte. Since boundary dimensions must be an integral number of mesh points long, problems with more complicated boundaries than the ones illustrated here require correspondingly more storage.

Paper 6501 E, first received 19th April and in revised form 14th June 1971

Mr. Johns and Prof. Beurle are with the Department of Electrical & Electronic Engineering, University of Nottingham, University Park, Nottingham NG7 2RD, Notts., England

2 Representation of waves by a transmission-line matrix

Calculations are carried out on a 2-dimensional Cartesian mesh of open 2-wire transmission lines which are parallel to the z and x axes. Each node in the mesh corresponds to a junction between a pair of transmission lines, shown in Fig. 1A, and this forms an impedance discontinuity in each line.

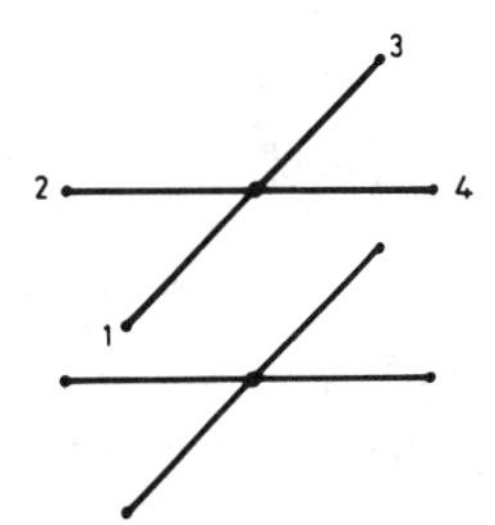

Fig. 1A
Junction between transmission lines

In order to show how Maxwell's equations may be represented by the transmission-line matrix, the elementary length of the transmission line between two nodes of the mesh is represented by lumped inductors and capacitors in the first instance. If the inductance and capacitance, per unit length for an individual line, are L and C, respectively, the junction between a pair of lines at a mesh node point can be represented by the basic elementary network of Fig. 1B. The complete

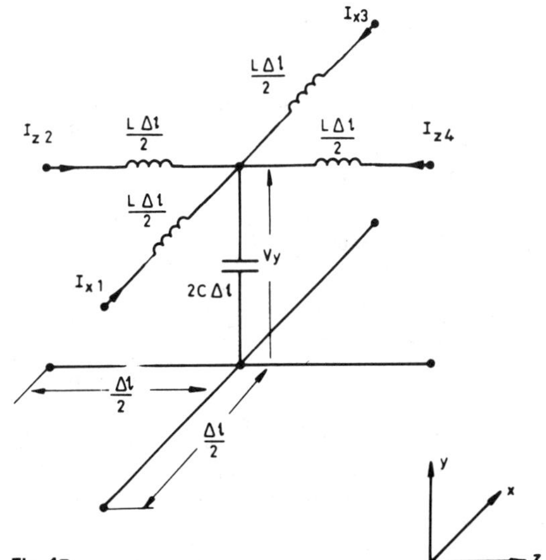

Fig. 1B
Equivalent network of a transmission-line junction

network is made up of a large number of such building blocks, connected as a 2-dimensional array.

For a distance Δl between node points, the following approximate equations apply:

$$-\frac{\partial}{\partial x}(I_{x1} - I_{x3}) - \frac{\partial}{\partial z}(I_{z2} - I_{z4}) - 2C\frac{\partial V_y}{\partial t} \quad . \quad . \quad (1)$$

$$-\frac{\partial V_y}{\partial x} = L\frac{\partial}{\partial t}(I_{x1} - I_{x3}) \quad . \quad . \quad . \quad . \quad . \quad . \quad . \quad . \quad (2)$$

$$-\frac{\partial V_y}{\partial z} = L\frac{\partial}{\partial t}(I_{z2} - I_{z4}) \quad . \quad . \quad . \quad . \quad . \quad . \quad . \quad . \quad (3)$$

These equations may be combined to give the wave equation

$$\frac{\partial^2 V_y}{\partial x^2} + \frac{\partial^2 V_y}{\partial z^2} = 2LC\frac{\partial^2 V_y}{\partial t^2} \quad . \quad . \quad . \quad . \quad . \quad . \quad (4)$$

For an H_{m0} mode with field components H_z, H_x and E_y, Maxwell's curl equations may be written as

$$-\frac{\partial H_x}{\partial z} - \frac{\partial H_z}{\partial x} = \epsilon\frac{\partial E_y}{\partial t} \quad . \quad . \quad . \quad . \quad . \quad . \quad . \quad . \quad . \quad (5)$$

$$\frac{\partial E_y}{\partial x} = -\mu\frac{\partial H_z}{\partial t} \quad . \quad . \quad . \quad . \quad . \quad . \quad . \quad . \quad . \quad . \quad (6)$$

$$\frac{\partial E_y}{\partial z} = \mu\frac{\partial H_x}{\partial t} \quad . \quad . \quad . \quad . \quad . \quad . \quad . \quad . \quad . \quad . \quad (7)$$

These combine to give the wave equation

$$\frac{\partial^2 E_y}{\partial x^2} + \frac{\partial^2 E_y}{\partial z^2} = \mu\epsilon\frac{\partial^2 E_y}{\partial t^2} \quad . \quad . \quad . \quad . \quad . \quad . \quad (8)$$

A direct comparison between eqns. 1–4 and eqns. 5–8 can be made with the following equivalences between parameters:

$$E_y \equiv V_y \qquad -H_z \equiv (I_{x3} - I_{x1})$$

$$-H_x \equiv (I_{z2} - I_{z4}) \qquad \mu \equiv L \qquad \epsilon \equiv 2C$$

Now, for the elementary transmission lines in the transmission line matrix, and for $\mu_r = \epsilon_r = 1$, the inductance and capacitance per unit length are related by

$$\frac{1}{\sqrt{LC}} = \frac{1}{\sqrt{\mu_0\epsilon_0}} = c$$

where $c = 3 \times 10^8$ m/s.

Thus, for the approximations made in eqns. 1, 2 and 3, if voltage and current waves on each transmission-line component propagate at the speed of light c, the complete network of intersecting transmission lines represents a medium of relative permittivity twice that of free space.

3 Wave propagation on a transmission-line matrix

For a given length of transmission line Δl between nodes, it is necessary to estimate the range of frequencies for which the approximate analysis of the previous Section is valid. This may be done by examining propagation of the matrix shown in Fig. 2, and treating it as a slow-wave structure.

For the purpose of the analysis, it is assumed that waves are travelling in the positive $z$direction and are invariant in the x and $y$directions. Under these conditions, any wave travelling transversely along a component line from A to B (for example) will, by symmetry, be accompanied by a similar wave travelling from B to A. It follows that a wave travelling over the matrix may be represented by the passage of a wave down a transmission line having open-circuited stubs, of length $\Delta l/2$, as shown in Fig. 3.

The propagation properties of a periodic structure of this type may be analysed by dividing the line into individual cells, and expressing the voltage and current at the output of one cell in terms of the voltage and current at the input. If the ratio of the length of the transmission-line element to the free-space wavelength of the propagating wave is $\theta/2\pi = \Delta l/\lambda$, then

$$\begin{bmatrix} V_i \\ I_i \end{bmatrix} = \begin{bmatrix} \cos\theta/2 & j\sin\theta/2 \\ j\sin\theta/2 & \cos\theta/2 \end{bmatrix} \begin{bmatrix} 1 & 0 \\ 2j\tan\theta/2 & 1 \end{bmatrix}$$

$$\begin{bmatrix} \cos\theta/2 & j\sin\theta/2 \\ j\sin\theta/2 & \cos\theta/2 \end{bmatrix} \begin{bmatrix} V_{i+1} \\ I_{i+1} \end{bmatrix} \quad . \quad . \quad . \quad (8)$$

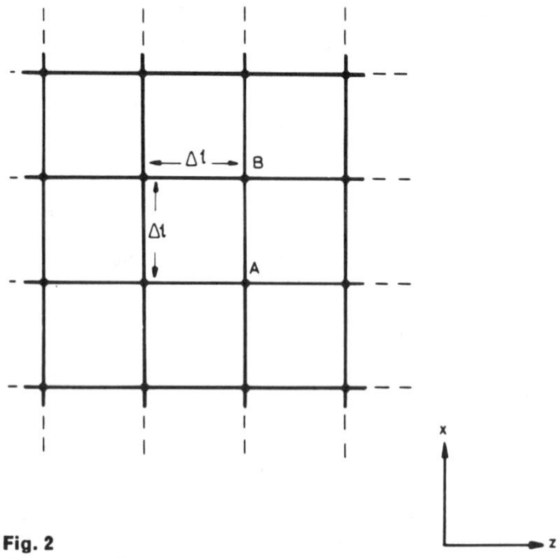

Fig. 2
Transmission-line matrix in xz plane

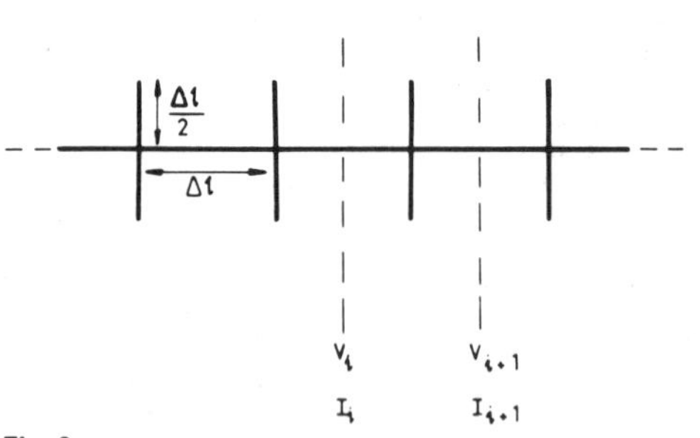

Fig. 3
Equivalent network for waves propagating in the z direction only

If the waves on the periodic structure have a propagation constant $\gamma_n = \alpha_n + j\beta_n$, then

$$\begin{bmatrix} V_i \\ I_i \end{bmatrix} = \begin{bmatrix} e^{\gamma_n\Delta l} & 0 \\ 0 & e^{\gamma_n\Delta l} \end{bmatrix} \begin{bmatrix} V_{i+1} \\ I_{i+1} \end{bmatrix} \quad . \quad . \quad . \quad . \quad (10)$$

Solution of eqns. 9 and 10 gives

$$\cosh\gamma_n\Delta l = \cos\theta - \tan\theta/2\sin\theta$$

This equation describes the manner in which waves propagate on the transmission-line matrix. There are ranges of frequencies over which propagation can take place (passbands), corresponding to $|\cos\theta - \tan(\theta/2)\sin\theta| < 1$, and ranges of frequencies over which propagation cannot occur (stopbands), corresponding to $|\cos\theta - \tan(\theta/2)\sin\theta| > 1$.

For the lowest-frequency propagation region, $\gamma_n = j\beta_n$, and

$$\cos(\beta_n\Delta l) = \left(\cos\frac{\omega\Delta l}{c}\right) - \tan\left(\frac{\omega\Delta l}{2c}\right)\sin\left(\frac{\omega\Delta l}{c}\right)$$

which reduces to

$$\sin\left(\frac{\beta_n\Delta l}{2}\right) = \sqrt{2}.\sin\left(\frac{\omega\Delta l}{2c}\right) \quad . \quad . \quad . \quad . \quad . \quad (11)$$

The ratio of the velocity v_n of the waves on the matrix network to the free-space velocity c is given by

$$v_n/c = \frac{\omega}{\beta_n c}$$

and is plotted in Fig. 4 as a function of the ratio of the distance between network nodes to the free-space wavelength $\Delta l/\lambda$.

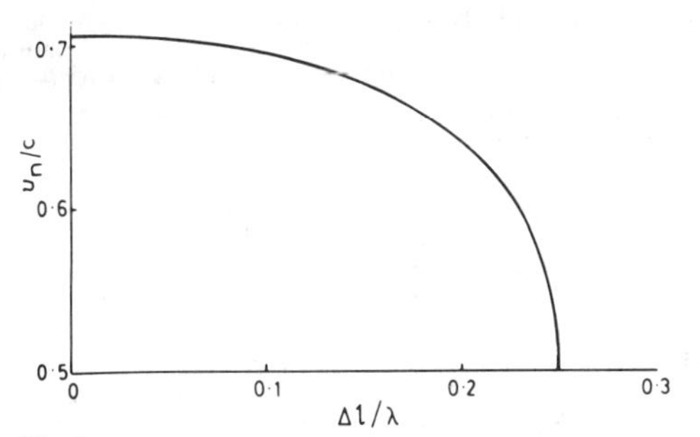

Fig. 4
Velocity characteristic of waves propagating on a transmission-line matrix

Thus, for waves propagating in the z direction, the transmission-line matrix can only represent Maxwell's equations over specific frequency ranges, and, in particular, the range of interest here is from zero frequency to the first network cutoff frequency, which occurs at $\omega\Delta l/c = \pi/2$, i.e. $\Delta l/\lambda = 1/4$ (λ is the free-space wavelength). Over this range of frequencies, the velocity of the waves behaves according to the characteristic of Fig. 4, and this may be taken into account when applying the network to the solution of propagation problems in one dimension. For frequencies well below the network cutoff frequency, the propagation velocity approximates to $1/\sqrt{2}$ of the free-space velocity.

For the case of 2-dimensional propagation, where waves may travel in any direction in the x–z plane, the velocity characteristic differs for different directions. For example, for propagation at 45° to the x or z axis, there is no cutoff, since identical pulses on two lines approaching a matrix node will 'see' a match. The velocity of waves remains constant, in this case, at $1/\sqrt{2}$ of the free-space velocity for all frequencies. For 2-dimensional problems, therefore, the characteristic of Fig. 4 serves only as a guide to the accuracy obtained, and solutions must be taken for frequencies well below the network cutoff frequency where the ratio of the network velocity to free-space velocity may be taken as constant at $1/\sqrt{2}$.

4 Numerical solution for TEM waves on a transmission-line matrix

Consider a voltage-impulse delta function of unit magnitude launched into terminal 1 of the basic line element of Fig. 5*a*.

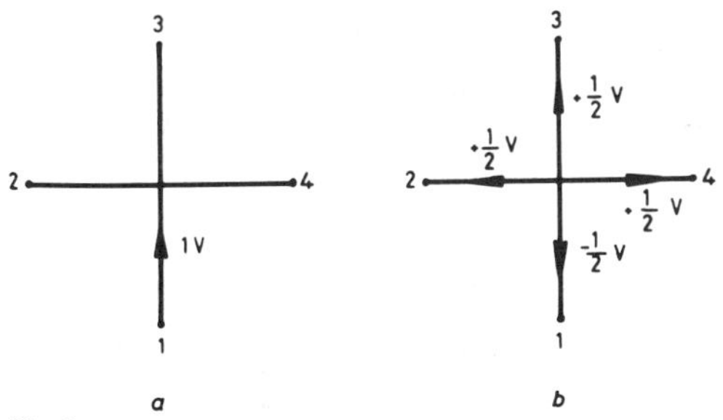

Fig. 5
Impulse response of basic line element

If the characteristic impedance of the line is normalised, a unit-magnitude delta function of voltage and current will travel towards the junction. Since three other lines meet at the junction, the effective terminal impedance of line 1 at the junction is 1/3, and the magnitude of the impulse voltage at the junction is 1/2. Thus, voltage impulses of 1/2 will be launched into each of lines 2, 3 and 4, while a reflected pulse of $-1/2$ will appear in line 1 (Fig. 5*b*).

Further, if at time $t = k\Delta l/c$, voltage impulses ${}_kV_1^i$, ${}_kV_2^i$, ${}_kV_3^i$ and ${}_kV_4^i$ on lines 1–4, respectively, are incident on any junction node in the transmission-line matrix, then the combined voltage reflected along line 1 at time $(k+1)\Delta l/c$ will be

$$ {}_{k+1}V_1^r = \tfrac{1}{2}({}_kV_2^i + {}_kV_3^i + {}_kV_4^i - {}_kV_1^i) \quad . \quad . \quad . \quad (12) $$

If this pulse is reflected from a node at position (z, x) in the matrix, it becomes an incident pulse on node $(z, x-1)$, i.e.

$$ {}_{k+1}V_1^r(z,x) = {}_{k+1}V_3^i(z,x-1) \quad . \quad . \quad . \quad . \quad (13) $$

In general,

$$ \left. \begin{aligned} {}_{k+1}V_n^r &= \tfrac{1}{2}\left[\sum_{m=1}^{4} {}_kV_m^i\right] - {}_kV_n \\ \text{and} \quad {}_{k+1}V_1^i(z,x) &= {}_{k+1}V_3^r(z,x-1) \\ {}_{k+1}V_2^i(z,x) &= {}_{k+1}V_4^r(z-1,x) \\ {}_{k+1}V_3^i(z,x) &= {}_{k+1}V_1^r(z,x+1) \\ {}_{k+1}V_4^i(z,x) &= {}_{k+1}V_2^r(z+1,x) \end{aligned} \right\} \quad . \quad . \quad . \quad (14) $$

Thus, if the magnitudes, positions and directions of travel of impulses on the matrix at time k are known, the corresponding parameters at time $(k+1)$ may be calculated by operating eqn. 14 for each node in the matrix. The impulse response may, therefore, be found by initially fixing the magnitude, position and direction of travel of impulse voltages at time $t = 0$, and examining the output impulse voltage at successive time intervals of $\Delta l/c$.

5 Boundary representation

5.1 TEM-wave boundary condition for the analysis of 1-dimensional waves

A plane TEM wave of infinite extent and no variation in the x direction (or the y direction) may be simulated in a computer on a rectangular matrix of the form shown in Fig. 6.

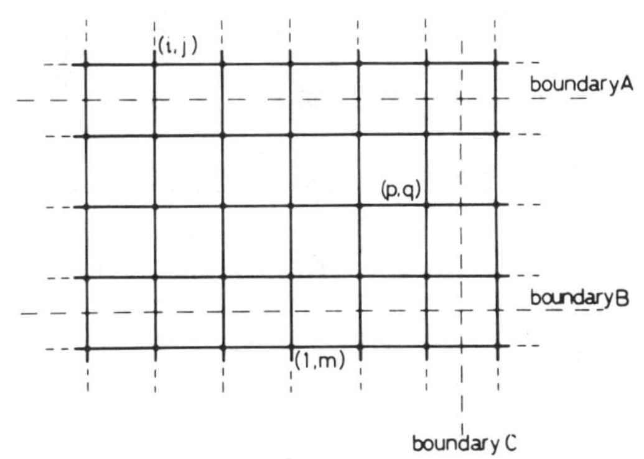

Fig. 6
Transmission-line matrix and boundaries

For the point $(z = i, x = j)$ immediately outside the upper boundary A, the computation

$$ {}_kV_1^r(i,j) = {}_kV_1^r(i,j-1) \quad . \quad . \quad . \quad . \quad . \quad . \quad . \quad (15) $$

is made.

This ensures that pulses transmitted in the x direction at any part of the matrix on the line $z = i$ are also reproduced at the boundary.

A similar arrangement,

$$ {}_kV_3^r(l,m) = {}_kV_3^r(l,m+1) \quad . \quad . \quad . \quad . \quad . \quad . \quad (16) $$

is made for the lower boundary B.

5.2 General boundaries

Boundary C in Fig. 6 represents a short circuit if pulses proceeding from points like (p, q) to the boundary are returned equal in magnitude but reversed in sign, i.e.

$$ {}_kV_4^i(p,q) = {}_kV_2^r(p+1,q) = -{}_kV_4^r(p,q) \quad . \quad . \quad (17) $$

Any other resistive load at C may be simulated by introducing a reflection coefficient ρ

$$_kV_4^i(p, q) = {_kV_2^r}(p+1, q) = \rho\{_kV_4^r(p, q)\} \quad . \quad . \quad (18)$$

where $$\rho = \frac{R-1}{R+1} \quad . \quad . \quad . \quad . \quad . \quad . \quad . \quad . \quad . \quad (19)$$

and R is the required resistive load.

6 Numerical solutions for 1-dimensional TEM waves

From eqns. 1–7, the values of E_y and H_x for the point $(z = m, x = n)$ correspond to the node voltage at this point and the net current entering the point in the $z$direction, i.e.

$$_kE_y(m,n) = \tfrac{1}{2}\{_kV_1^i(m,n) + {_kV_2^i}(m,n) + {_kV_3^i}(m,n) + {_kV_4^i}(m,n)\} \quad . \quad . \quad . \quad (20)$$

$$-{_kH_x}(m,n) = {_kV_2^i}(m,n) - {_kV_4^i}(m,n) \quad . \quad . \quad . \quad (21)$$

Note that, from the symmetry of the waves automatically set up by the TEM boundary conditions,

$$_kH_z(m,n) = {_kV_1^i}(m,n) - {_kV_3^i}(m,n) = 0$$

which is required by definition of a TEM wave.

$x = 1{\cdot}5$ and $x = 10{\cdot}5$. The initial impulse excitation was on all points along the line $z = 4$, and at all subsequent time intervals the field along this line was set to zero. In this way, interference from boundaries to the left of the excitation line was avoided. Calculations in the $z$direction were terminated at $z = 24$, so that no reflections were received from points at $z = 25$ in the matrix, and the boundary C, situated at $z = 24{\cdot}5$, was therefore matched to free space. The output-impulse response for E_y and H_x was taken at the point $(z = 14, x = 6)$, which is 10·5 mesh points away from the boundary C, for 100, 150 and 200 iterations of eqn. 14. Fig. 7 shows the results for $|E_y|$ and $|H_x|$ plotted against $\Delta l/\lambda$ for 200 iterations.

The sinusoidal-like variation of $|E_y|$ and $|H_x|$ with $\Delta l/\lambda$ is due to the mismatch presented by the boundary at C. Since the velocity of waves on the matrix is less than that in free space by a factor v_n/c (see Fig. 4), the effective intrinsic impedance presented by the network matrix is less by the same factor. The magnitude of the wave impedance on the matrix, nomalised to the intrinsic impedance of free space, is given by $|Z| = |E_y|/|H_x|$, and is tabulated in Table 1, together with Arg (Z), for the various numbers of iterations of eqn. 14 made. A comparison is made with the exact impedance values obtained by simple analysis of standing waves for the same mismatch.

Table 1

FREE-SPACE DISCONTINUITY FOR A TEM WAVE

Number of iterations	$\lvert Z\rvert$ 100	150	200	Exact	Arg (Z) 100	150	200	Exact
$\Delta l/\lambda$					rad	rad	rad	rad
0·002	0·9789	0·9730	0·9782	0·9747	−0·1368	−0·1396	−0·1253	−0·1282
0·004	0·9028	0·8980	0·9072	0·9077	−0·2432	−0·2322	−0·2400	−0·2356
0·006	0·8114	0·8229	0·8170	0·8185	−0·3068	−0·2979	−0·3046	−0·3081
0·008	0·7238	0·7328	0·7287	0·7256	−0·3307	−0·3457	−0·3404	−0·3390
0·010	0·6455	0·6367	0·6396	0·6414	−0·3201	−0·3350	−0·3281	−0·3263
0·012	0·5783	0·5694	0·5742	0·5731	−0·2730	−0·2619	−0·2680	−0·2707
0·014	0·5272	0·5313	0·5266	0·5255	−0·1850	−0·1712	−0·1797	−0·1765
0·016	0·4993	0·5043	p·5009	0·5018	−0·0609	−0·0657	−0·0538	−0·0545
0·018	0·5502	0·4987	0·5057	0·5057	0·0790	0·0748	0·0785	0·0768

The impedance Z of a TEM wave at a solution point 105 mesh points from a free-space discontinuity, calculated for various numbers of iterations of the transmission-line matrix. The exact results are obtained for a medium of intrinsic impedance given by the velocity characteristic of Fig. 4 (Δl = mesh size, λ = free-space wavelength)

A series of discrete delta-function magnitudes for E_y and H_x corresponding to time intervals $\Delta l/c$ are thus obtained by iteration of eqn. 14. This is the output-impulse function for the point, and may be used to obtain the output waveform corresponding to any input excitation. For example, the output waveform corresponding to a pulse input may be obtained by convolving the output-impulse function with the shape of the input pulse. In order to demonstrate the method, however, only the response to sinusoidal inputs is considered, and this may be most conveniently obtained by taking the Fourier transform of the output-impulse function. Since the function is a series of delta functions, the Fourier-transform integral may be replaced by a summation, and the real and imaginary parts of the output spectrum are given by

$$\mathrm{Re}\,[F(\Delta l/\lambda)] = \sum_{k=1}^{N} {_kI} \cos\left(\frac{2\pi k \Delta l}{\lambda}\right)$$

$$\mathrm{Im}\,[F(\Delta l/\lambda)] = \sum_{k=1}^{N} {_kI} \sin\left(\frac{2\pi k \Delta l}{\lambda}\right)$$

where $F(\Delta l/\lambda)$ is the frequency response, $_kI$ is the output-impulse response value for E_y or H_x at time $t = k\Delta l/c$ and N is the total number of time intervals for which the calculation is made.

Numerical calculations for TEM waves were carried out on a 25×11 rectangular matrix. TEM field-continuation boundaries (eqns. 15 and 16) were fixed along the $x = 2$ and $x = 10$ lines, producing boundaries, in effect, along the lines

For low frequencies, the intrinsic impedance of the network matrix is $1/\sqrt{2}$ times the intrinsic impedance of free space, which corresponds (from eqn. 19) to $\rho = -0{\cdot}1716$. Table 2 shows the results obtained for this reflection coefficient at the boundary C.

Table 2

MATCHED TEM WAVE

$\Delta l/\lambda$	$\lvert Z\rvert$	Arg (Z)
		rad
0·002	0·7088	−0·0023
0·004	0·7071	0·0033
0·006	0·7057	−0·0024
0·008	0·7097	0·0006
0·010	0·7055	0·0025
0·012	0·7076	−0·0028
0·014	0·7086	0·0032
0·016	0·7048	−0·0004
0·018	0·7092	−0·0022

The impedance Z of a TEM wave at a solution point 10·5 mesh points from a medium of intrinsic impedance $1/\sqrt{2}$ (0·7071). The results are calculated for 200 iterations of the transmission-line matrix (Δl = mesh size, λ = free-space wavelength)

7 Numerical solution of 2-dimensional H_{m0} modes in waveguide

7.1 Simple load

A 25×11 matrix was used for the numerical analysis of waveguide modes. Short-circuit boundaries were placed at

$x = 2$ and $x = 10$, the boundaries being effectively at $x = 1 \cdot 5$ and $10 \cdot 5$, the width between the waveguide walls thus being nine mesh points. The system was excited at all points along the line $z = 2$, and the impulse function of the output was

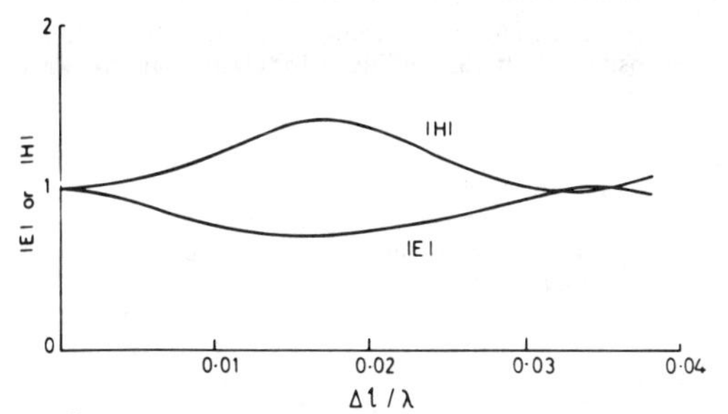

Fig. 7
Magnitudes of a TEM wave 10·5mesh points from a 'freespace discontinuity

taken from the point ($x = 6$, $z = 12$), which is 12·5mesh points from the boundary C at $z = 24$ (see Fig. 8). The boundary C again represented an abrupt change to the intrinsic impedance of free space.

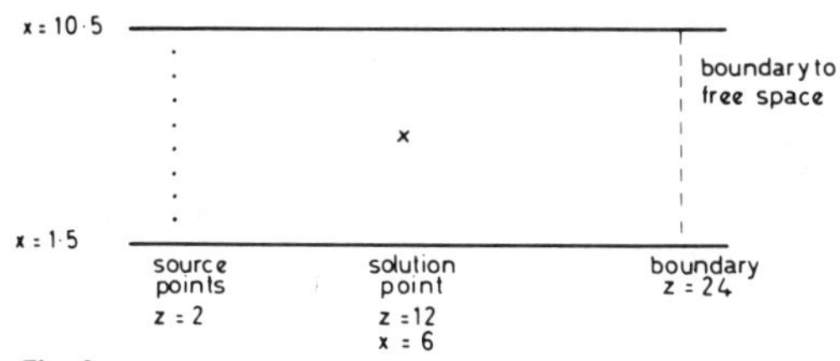

Fig. 8
Waveguide geometry for 'free-space' discontinuity

The cutoff frequency for the waveguide occurs at $\Delta l/\lambda_n = 1/18$ (λ_n is the network-matrix wavelength), and this corresponds to $\Delta l/\lambda = 0 \cdot 0394$. Fig. 9 shows the variation of

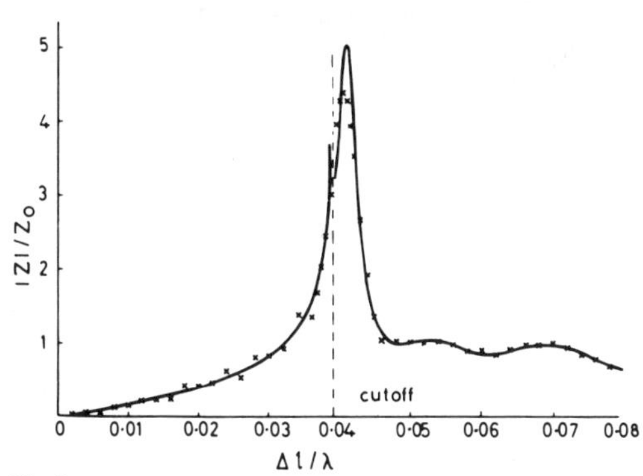

Fig. 9
Impedance (normalised to intrinsic impedance of free space) in a waveguide with a 'free-space' discontinuity

Evanescent mode
——— exact for infinitely long guide
× × × numerical method
Propagating mode
——— exact
× × × numerical method

the magnitude of the guide impedance with $\Delta l/\lambda$, for frequencies above and below the cutoff frequency. The results are compared with the exact impedance obtained from simple waveguide theory for the same mismatched termination, showing that the numerical method is not only accurate for propagating waves, but also for evanescent modes. The discontinuity of the exact results at the guide cutoff frequency arises because the calculations above the cutoff frequency were made for the termination mismatch as described, but an infinitely long guide was assumed for the evanescent-mode calculations.

Increased accuracy for problems which have an axis of symmetry along the z direction can be obtained by making one of the boundaries of the waveguide an open circuit. This is achieved by setting $\rho = 1$ in eqn. 18, and in this way a waveguide of twice the width (and therefore twice the number of mesh points) is simulated on the same matrix. A comparison between the results for the guide impedance using this method is made with the exact results in Table 3. The guide

Table 3

FREE-SPACE DISCONTINUITY IN WAVEGUIDE

$\Delta l/\lambda$	$\|Z\|$		Arg (Z)	
	Numerical method	Exact	Numerical method	Exact
			rad	rad
0·020	1·9391	1·9325	0·8936	0·9131
0·021	2·0594	2·0964	0·6175	0·6415
0·022	1·9697	2·0250	0·3553	0·3603
0·023	1·7556	1·7800	0·1530	0·1438
0·024	1·5173	1·5132	0·0189	0·0163
0·025	1·3036	1·2989	−0·0518	−0·0388
0·026	1·1370	1·1471	−0·0648	−0·0457
0·027	1·0297	1·0482	−0·0350	−0·0249
0·028	0·9776	0·9900	0·0088	0·0075
0·029	0·9620	0·9622	0·0416	0·0396
0·030	0·9623	0·9556	0·0554	0·0632

The impedance Z in a waveguide of width 18mesh points at a solution point 12·5mesh points from a free-space discontinuity. The exact results are calculated for a medium of intrinsic impedance $1/\sqrt{2}$ within the guide (Δl = mesh size, λ = free-space wavelength)

is again terminated in the mismatch caused by an abrupt change in the free-space intrinsic impedance at the boundary C.

7.2 Scattering due to a waveguide bifurcation

The scattering due to a waveguide bifurcation provides a convenient example to test the numerical method, since exact results are available.[9] Two examples are given, the first being for the symmetrically placed bifurcation of Fig. 10 and the results being presented in Fig. 11.

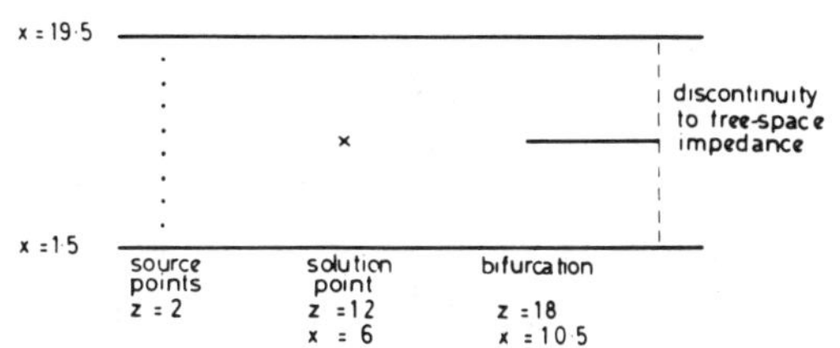

Fig. 10
Waveguide geometry for a symmetrical bifurcation

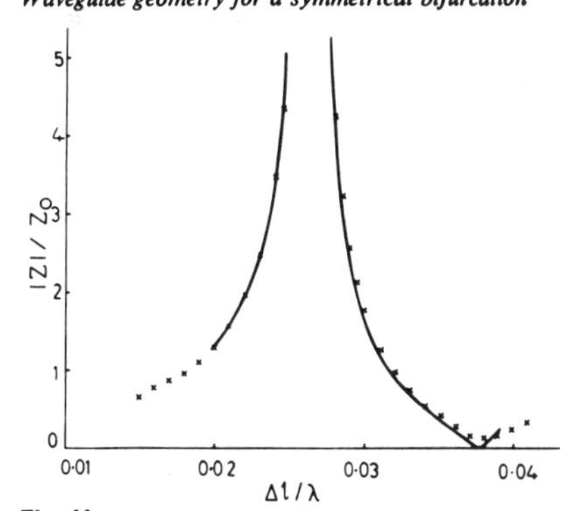

Fig. 11
Impedance (normalised to intrinsic impedance of free space) in a waveguide with a symmetrical bifurcation

——— exact
× × × numerical method

The results are given for the frequency range where only the H_{10} mode propagates, and in this region the bifurcation presents a short circuit. The exact results (available in Reference 9) for the position of the plane of the short circuit enable the exact impedance at the solution point to be obtained. The numerical method utilises the symmetry along the line $x = 10{\cdot}5$ to double the number of mesh points for the calculation.

The second example is for the asymmetrical bifurcation shown in Fig. 12, and the results are given in Fig. 13. The

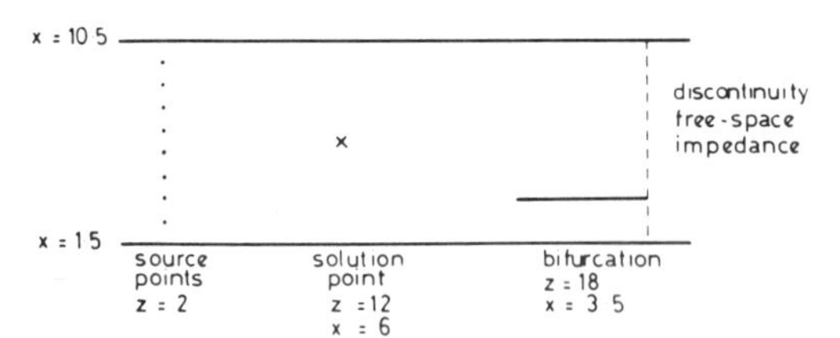

Fig. 12
Waveguide geometry for an asymmetrical bifurcation

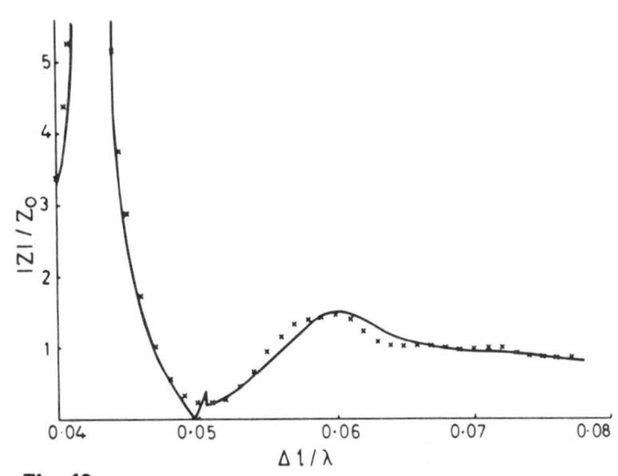

Fig. 13
Impedance (normalised to intrinsic impedance of free space) in a waveguide with an asymmetrical bifurcation

—— exact
× × × numerical method

frequency range is again that for which only the H_{10} mode is propagating. In this case, as the frequency rises, the bifurcation initially presents a short circuit, as before, but at higher frequencies the waveguide A becomes propagating, and the mismatch presented by the boundary C also controls the impedance at the solution point. Again, using Reference 9, the exact results for this impedance have been calculated and compared with those obtained from the numerical method.

It should be noted that there is no line of symmetry in this problem, and the number of mesh points is severely limited. There are only seven mesh points between the walls of guide A, and only two between the walls of guide B, and so good accuracy cannot be expected. Nevertheless, the results of Fig. 13 show that the numerical method is still useful under these circumstances.

8 Discussion and conclusions

A new numerical technique for the calculation of 2-dimensional scattering problems for electromagnetic waves has been introduced. Simple waveguide-impedance problems have been evaluated on a computer having a modest core store (8kbyte) and no backing store, but the results demonstrate the possibility of applying the method to problems with much more complicated boundaries and which therefore require a larger store. The technique is accurate at the cutoff frequency of the dominant waveguide mode, as well as above it and below it, and the transient nature of the analysis means that information at all frequencies within the working range is obtained from a single calculation. There is no limit to the lowest frequency for which the approximations made are accurate, but the upper frequency is limited to that corresponding to a mesh coarseness of about $\Delta l/\lambda = 0{\cdot}1$ (depending on the degree of discrepancy that can be tolerated).

The type of problem solved by this method is similar to those exemplified by the method of ray optics,[6,7] where the known free-space scattering properties of a discontinuity are used to generate rays from which the various modes produced in a waveguide may be obtained. The method of ray optics is not a numerical method, and the geometry of the problem is not divided up into a mesh. The computer-storage requirement for the computation, therefore, is not likely to be affected by the complication of the boundaries or the frequency at which the results are required in the same way as the numerical method proposed.

An advantage of the transmission-line matrix method, however, is that it is not necessary to know the free-space scattering properties of a discontinuity.

9 References

1 KRON, G.: 'Equivalent circuit of the field equations of Maxwell—Pt. I', *Proc. Inst. Radio Engrs.*, 1944, **32**, pp. 289–299

2 WHINNERY, J. R., and RAMO, S.: 'A new approach to the solution of high-frequency field problems', *ibid.*, 1944, **32**, pp. 284–288

3 VINE, J.: 'Impedance networks' *in* VITKOVITCH, D. (Ed.): 'Field analysis: experimental and computational methods' (Van Nostrand, 1966), chap. 7

4 WHINNERY, J. R., CONCORDIA, C., RIDGWAY, W., and KRON, G.: 'Network analyzer studies of electromagnetic cavity resonators', *Proc. Inst. Radio Engrs.*, 1944, **32**, pp. 360–367

5 SPANGENBERG, K., WALTERS, G., and SCHOTT, F.: 'Electrical network analyzers for the solution of electromagnetic field problems', *ibid.*, 1949, **37**, pp. 724–729

6 HAMID, M. A. K., KASHYAP, S. C., MOHSEN, A., BOERNER, W. M., and BOULANGER, R. J.: 'A ray-optical approach to the analysis of microwave filters', *IEE Conf. Publ.* 58, 1969, pp. 189–193

7 YEE, H. Y., and FELSEN, L. B.: 'Ray-optics—a novel approach to scattering by discontinuities in a waveguide', *IEEE Trans.*, 1967, **MTT-17**, pp. 73–85

8 COLLIN, R. E.: 'Field theory of guided waves' (McGraw-Hill, 1960)

9 MARCUVITZ, N.: 'Waveguide handbook' (McGraw-Hill, 1951)

10 ARLETT, P. L., BAHRANI, A. K., and ZIENKIEWICZ, O. C.: 'Application of finite elements to the solution of Helmholtz's equation', *Proc. IEE*, 1968, **115**, (12), pp. 1762–1766

11 HORNSBY, J. S., and GOPINATH, A.: 'Numerical analysis of a dielectric-loaded waveguide with a microstrip line–finite-difference methods', *IEEE Trans.*, 1969, **MTT-17**, pp. 684–690

12 AHMED, S., and DALY, P.: 'Finite-element methods for inhomogeneous waveguides', *Proc. IEE*, 1969, **116**, (10), pp. 1661–1664

13 MASTERMAN, P. H., and CLARRICOATS, P. J. B.: 'Computer field-matching solution of waveguide transverse discontinuities', *ibid.*, 1971, **118**, (1), pp. 51–63

Transmission-line matrix method with irregularly graded space

D.A. Al-Mukhtar, B.Sc., M.Tech., and J.E. Sitch, B.A., M.Eng., Ph.D.

Indexing terms: *Transmission-line theory, Numerical analysis, Microwave systems, Wave propagation, Waveguides*

Abstract: The paper describes a proposed modification to the transmission-line-matrix method of numerical analysis (TLM). The conventional formulation of this technique uses squares and cubes for space quantisation in two and three dimensions, respectively. This restriction, which is due to the necessity of time synchronism, severely affects the efficiency of the method. The paper introduces a technique in which the propagation space can be irregularly graded according to the nature of the problem under investigation. A new formulation for TLM in the $r\theta$-plane of polar coordinates is also introduced; here the space elements are non-uniform by nature.

1 Introduction

The transmission-line-matrix (TLM) method of numerical analysis can be used to provide time-domain solutions for wave-propagation problems [1–5]. The main feature of this method is the simplicity of formulation and programming for a wide range of applications; in waveguides for example it can provide solutions for imhomogeneous structures with arbitrary cross-sections and lossy fillings [2, 3], it is also used in the analysis of a wide range of microwave integrated circuits (MIC) [5]. The basic foundation of this method is the well known concept of modelling field space by lumped networks [8], where these networks have been built to solve the wave equation. Using a distributed parameter transmission-line network model, the propagation space is then represented by a mesh of TEM transmission lines. Electric and magnetic fields are made equivalent to voltages and currents on the network. Two-dimensional propagation space can be modelled by either a shunt or a series matrix [4]. Three-dimensional space is modelled by a matrix composed of alternate series and shunt node connections [5]. Any inhomogeneity in the form of dielectric or magnetic material can be accounted for by introducing further lengths of transmission lines as series or shunt stubs. The numerical calculation usually starts by exciting the matrix at specific points by voltage or current impulses and follows the propagation of these impulses over the matrix as they are scattered by the nodes and bounce at boundaries. Clearly such a process cannot be carried out on a digital computer unless all the pulses reach the nodes or boundaries at the same time; the technique is therefore limited to the use of square mesh elements in two dimensions and cubes in the three-dimensional model, to ensure a constant time delay for pulses travelling between the nodes. Boundaries are usually taken halfway between the nodes for the same reason. This time synchronism allows the calculations to be repeated on an iterative basis. The output, which is taken from a chosen point, consists of a series of impulses separated by constant time intervals. The Fourier transform of this output function reveals the matrix response to a sinusoidal input [1]. The transient nature of the calculations ensures that the whole frequency spectrum (up to the matrix cutoff) exists in the output, and this is one of the attractive features of TLM.

The brief outline given above shows that there is a restriction on the shape and size of the mesh elements. This undoubtedly affects the efficiency of the method, as for example in the application to the problem of ridged waveguide which necessitates fine mesh deployment around the ridge corners [2]. Also the application to the problem of slot line represents a considerable burden on the computer as far as storage and run time are concerned, as the narrow slot requires fine mesh to represent the fields accurately and at the same time the dimensions of the structure must be large compared to the slot width for realistic representation of a slot line. To overcome these problems, this paper proposes modifications to 'conventional TLM' which enable the mesh to be irregularly graded in the two- and three-dimensional models. Thus fine mesh can be used only where it is necessary, for example at field concentration regions, such as edges, corners etc. The technique is also extended to produce a TLM which fits the co-ordinates of the $r\theta$-plane in cylindrical co-ordinates. In such a matrix the elements are not uniform by nature.

2 The physical parameters of a TLM space

In drawing equivalences between the fields and network parameters in conventional TLM the dimensions of the space elements have always been normalised to unity. The capacitances and inductances per unit length for these lines are therefore kept constant, and time synchronism is achieved. However, in order to grade the space irregularly the physical dimensions of the elements must be incorporated in the equations describing the currents and voltages of the network; the total capacitances and inductances of the lines then become functions of the element dimensions. These relationships are derived below.

Consider a two-dimensional space element of dimensions u and v, Fig. 2. This element is to be represented by a pair of transmission lines forming a shunt junction at its centre. If C is the total capacitance formed by this junction and L_x, L_y are the total inductances for the x-directed and y-directed lines, respectively, then the voltage/current difference equations for this junction represent the following differential equations:

$$\begin{aligned} \frac{\partial V_z}{\partial x} &= -\frac{L_x}{u}\frac{\partial I_x}{\partial t} \\ \frac{\partial V_z}{\partial y} &= -\frac{L_y}{v}\frac{\partial I_y}{\partial t} \\ -u\frac{\partial I_x}{\partial x} - v\frac{\partial I_y}{\partial y} &= C\frac{\partial V_z}{\partial t} \end{aligned} \quad (1)$$

The appropriate expansion of Maxwell's equations with $\partial/\partial z = 0$ gives

$$\begin{aligned} \frac{\partial E_z}{\partial x} &= \mu\frac{\partial H_y}{\partial t} \\ \frac{\partial E_z}{\partial y} &= -\mu\frac{\partial H_x}{\partial t} \\ \frac{\partial H_y}{\partial x} - \frac{\partial H_x}{\partial y} &= \epsilon\frac{\partial E_z}{\partial t} \end{aligned} \quad (2)$$

Paper 1659H, received 17th November 1980

The authors are with the Department of Electronic & Electrical Engineering, University of Sheffield, Mappin Street, Sheffield S1 3JD, England

Reprinted with permission from *IEE Proc.*, vol. 128, pt. H, no. 6, pp. 299–305, Dec. 1981.

where μ is the permeability of the space and ϵ its permittivity. Using the equivalences

$$E_z \equiv \frac{V_z}{w}, \quad I_x \equiv -vH_y, \quad I_y \equiv uH_x$$

where w is an arbitrary distance put in to retain correct dimensionality. A direct comparison between eqns. 1 and 2 can be made, from which

$$C = \epsilon \frac{uv}{w} \tag{3a}$$

$$L_y = \mu \frac{wv}{u} \tag{3b}$$

$$L_x = \mu \frac{wu}{v} \tag{3c}$$

Eqns. 3 give the fundamental physical parameters in terms of element dimensions.

It is appropriate at this stage to state that for a general three-dimensional element, Fig. 2, the total capacitances at the shunt nodes, and inductances at the series nodes, can be derived in the same way, and these are

$$C_x = \epsilon \frac{vw}{u} \tag{4a}$$

$$C_y = \epsilon \frac{uw}{v} \tag{4b}$$

$$C_z = \epsilon \frac{uv}{w} \tag{4c}$$

$$L_{px} = \mu \frac{vw}{u} \tag{5a}$$

$$L_{py} = \mu \frac{uw}{v} \tag{5b}$$

$$L_{pz} = \mu \frac{uv}{w} \tag{5c}$$

where L_{px} is the inductance seen by current circulating in the yz-plane, i.e. perpendicular to the x-direction. Finally, for an element in the matrix of the $r\theta$-plane of cylindrical coordinates, Fig. 3, it can be shown that

$$C = \epsilon \frac{r\alpha u}{w} \tag{6a}$$

$$L_r = \mu \frac{uw}{r\alpha} \tag{6b}$$

$$L_\theta = \mu \frac{r\alpha w}{u} \tag{6c}$$

where L_r and L_θ are the total inductances for the radial and circular lines, respectively.

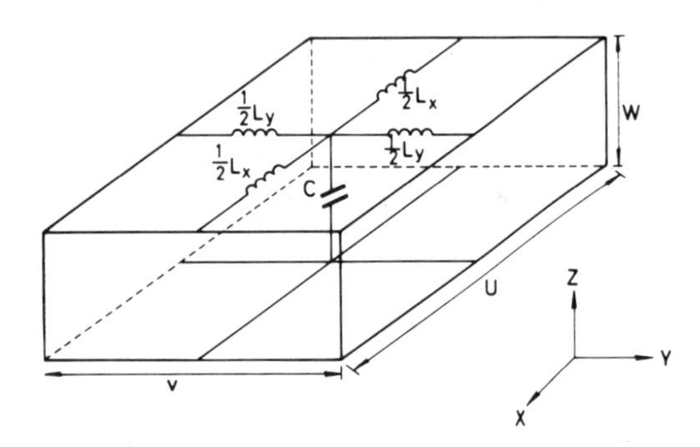

Fig. 1 *Basic element and its equivalent circuit*

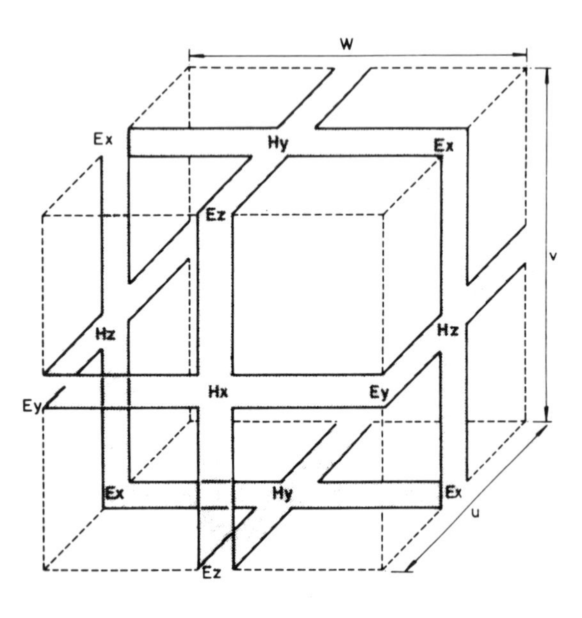

Fig. 2 *Basic 3D element and its nodal structure*

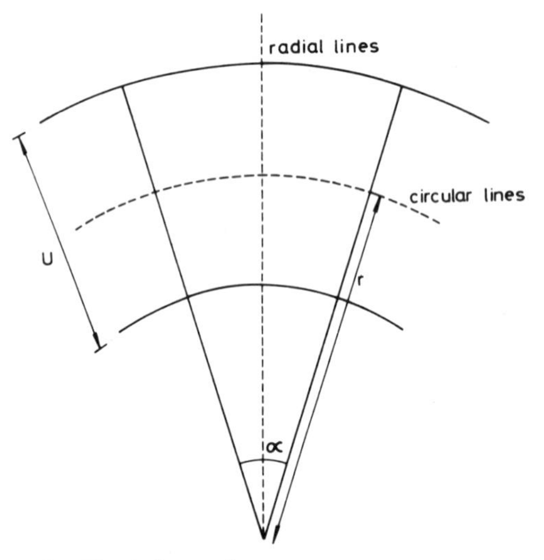

Fig. 3 *The rθ-plane mesh*

Eqns. 3–6 show that the inductance and capacitance per unit length of the TEM lines would be no longer constant in an irregularly graded matrix. Therefore, unless the matrix configuration is modified, time synchronism will be lost. Two possible methods of modification are discussed in the following Sections.

3 Two-dimensional graded TLM in cartesian coordinates

The most important condition to fulfil in building a graded TLM is the time synchronism. The pulses must reach the nodes or boundaries at the same instant of time, regardless of the distance they cover between any two concecutive nodes in either direction. Any solution must assume that propagation velocity is proportional to line length. This can be achieved by assuming a constant total inductance and capacitance for all the link lines, thus fixing the time step and link line characteristic impedance, and then model any additional inductances and capacitances needed due to mesh grading as short and open circuit stubs distributed over the matrix wherever necessary. The other alternative is to vary the total inductance and capacitance of the link lines, while keeping the delay constant. Stubs are still needed but fewer than in the first case. The first option results in a matrix loaded with two kinds of stubs (stub loaded matrix), the second gives a matrix with lines of different characteristic impedance and at least one kind of stub (hybrid matrix). Each of these two types is now considered separately.

3.1 *Stub-loaded matrix*

The realisation of this type of matrix is summarised in the following steps:

(i) In the mesh under consideration the shortest piece of line that links two consecutive nodes is taken to be of unit length. This will be called the link line.

(ii) Each link will be assumed to have ϵ_0/h (or ϵ_0) of capacitance and μ_0 (or μ_0/h) of inductance. This means that a pulse that travels on this line would have speed such as to cover it in a time step τ, given by

$$\tau = \sqrt{\frac{\mu_0 \epsilon_0}{h}} \text{ seconds} \tag{7}$$

the significance of h will appear shortly.

(iii) The next step is to imagine that the whole matrix is built from this elementary unit line, i.e. the nodes are to be linked with lines that allow the same time step.

(iv) From eqns. 3, it is clear that if u or v is greater than 1, then extra capacitances and inductances need to be supplied to the matrix in the regions where this condition appears.

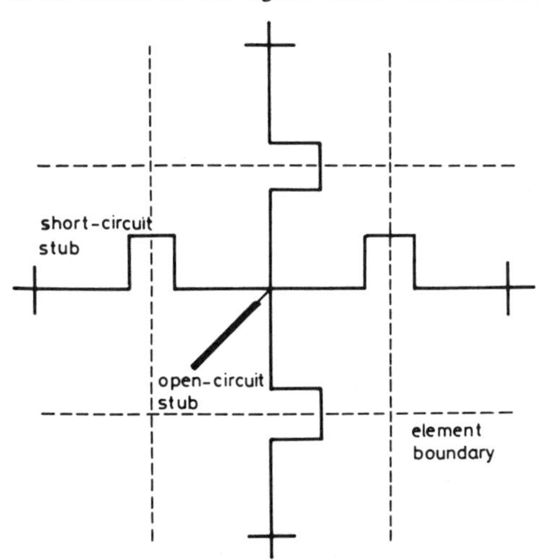

Fig. 4 *Element in stub-loaded matrix*

These extra components will be accounted for by adding open-circuited stubs shunted at the nodes for the extra capacitances and short-circuited stubs in series with the links at their mid-points for extra inductances, as shown in Fig. 4. Accordingly, these stubs have characteristic impedances that depend on their location in the matrix. This arrangement makes it also possible to represent inhomogeneous propagation space by introducing further changes to the impedances of the stubs in the form of weighting parameters. According to the relative permittivity and permeability of the space, the impedances for these stubs are derived as follows.

The stubs' lengths must be chosen such that the pulses cover them in half a timestep ($\tau/2$). For short-circuit stubs using the low-frequency lumped representation

$$L = Z_0 \frac{\tau}{2} \tag{8}$$

where L is the inductance modelled by the stub.

Using eqn. 8 in eqns. 3*b* or 3*c*, and assuming that the link has μ_0/h for its inductance, the impedance for the stubs in the y-direction lines is

$$Z_y = 2\left[\frac{wvh}{u}\mu_r - 1\right] \tag{9}$$

and for the x-directed line

$$Z_x = 2\left[\frac{wuh}{v}\mu_r - 1\right] \tag{10}$$

where Z_x and Z_y are normalised to $\sqrt{\mu_0/\epsilon_0 h}$.

The low-frequency lumped approximation for open-circuit stubs gives

$$C = \frac{\tau Y_0}{2} \tag{11}$$

where Y_0 is the characteristic admittance for this stub. From eqns. 11 and 3*c*, assuming that the links have ϵ_0 of capacitance, the characteristic admittance for the open-circuit stub is

$$Y = 2\left[\frac{uv}{w}\epsilon_r - 2\right] \tag{12}$$

this is also normalised to $\sqrt{\mu_0/\epsilon_0 h}$.

The significance of h can be seen from eqns. 9 and 10, where sometimes because of the degree of grading the factor $\mu_r\ wu/v$ or $\mu_r\ wv/u$ becomes less than 1, in which case the corresponding stub will have negative impedance. This is incorrect modelling for passive media. Therefore the value of h is chosen to offset the above inequality so that all the matrix stubs have positive or zero impedance.

For time-synchronism purposes the short-circuit stubs are located midway between the nodes, and as a result of this the stub shares two elements: hence the average dimension is used in calculating the corresponding stub impedance. In view of this arrangement, the scattering process should be modified as there are now two stages of scattering: first, at the nodes where the scattering matrix is the same as given in Reference 3. However pulses travelling from node to node hit the short-circuit stubs at the middle of the interval. If v_1 and v_2 are the voltage pulses moving toward the stub from both sides, and v_3 is the voltage pulse reflected back from the end of the stub, the scattering matrix is

$$\begin{bmatrix} V_1 \\ V_2 \\ V_2 \end{bmatrix}^r = \frac{1}{2+Z}\begin{bmatrix} Z & 2 & -2 \\ 2 & Z & 2 \\ -2Z & 2Z & 2-Z \end{bmatrix}\begin{bmatrix} V_1 \\ V_2 \\ V_3 \end{bmatrix}^i \tag{13}$$

The location of the stubs and the proper choice of their length make time synchronism possible.

Considering the bounderies it is clear that by taking them midway between the nodes they cut the short-circuit stubs in half, Fig. 5; therefore, the scattering at the boundary is described by the following matrix:

$$\begin{bmatrix} V_L \\ V_S \end{bmatrix}^r = \frac{1}{Y_{1/2}+1} \begin{bmatrix} (1-Y_{1/2}) & 2Y_{1/2} \\ 2 & -(1-Y_{1/2}) \end{bmatrix} \begin{bmatrix} V_L \\ V_S \end{bmatrix}^i \tag{14}$$

where $Y_{1/2} = 1/.5Z$. The above matrix is for short-circuit boundaries; however, close inspection of this equation shows that when the boundary is represented as an open circuit the link voltage pulse v_L reflects unchanged.

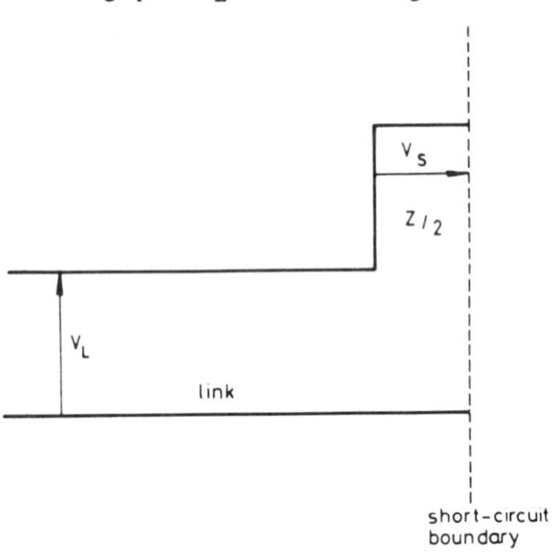

Fig. 5 *Short-circuit stub at boundary*

The transmission-line matrix is a slow-wave structure [1], and further reduction of the wave velocity is introduced by the action of the stubs. This of course limits the degree of grading that could be used due to the velocity error; however, a wide range of practical examples has been tested with large grading ratios, and yet the results have good accuracy. A quantitative picture of the velocity reduction due to the stubs' action can be obtained by using TEM wave analysis similar to that given in Reference 3; this type of analysis gives the following characteristics for wave propagation in one direction:

$$\frac{\tau_0}{\tau} = \frac{\pi f \tau_0}{\sin^{-1}\sqrt{2(1+Y/4)(1+Z/2)}\,\sin\dfrac{\omega\tau}{2}} \tag{15}$$

where τ is the time step on the matrix and τ_0 is the equivalent parameter in free space.

The first propagation band cutoff occurs at

$$(f\tau_0)_{cutoff} = \frac{1}{\pi}\sin^{-1}\frac{1}{\sqrt{2(1+Y/4)(1+Z/2)}} \tag{16}$$

Eqn. 16 is plotted in Fig. 6 for different values of Y and Z, clearly showing their effects on matrix cutoff.

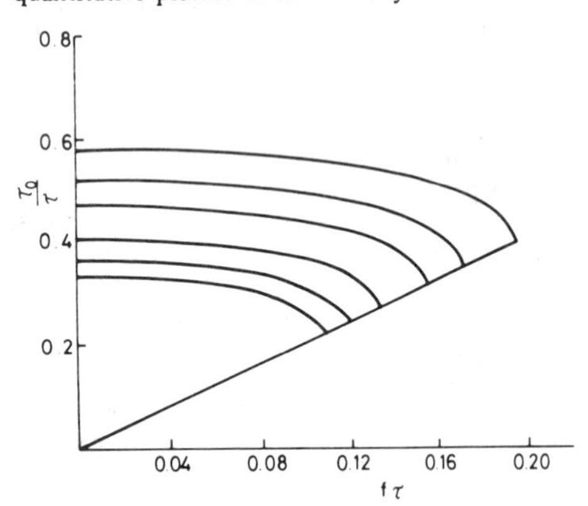

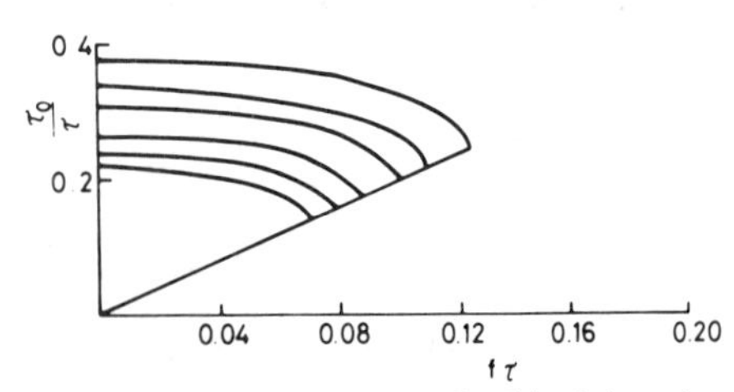

Fig. 6 *Velocity characteristics of stub-loaded matrix*

a $z = 1$,
b $z = 5.0$

3.2 The hybrid matrix

The hybrid matrix is an alternative approach to the stub-loaded matrix, described in the preceding Section. The emphasis here is on assuming different velocities for the pulses on the lines, i.e fast pulse on long lines, slow pulses on the short and a unique time step for all. The velocity of the pulses on a line of length l is

$$v_n = \frac{1}{\sqrt{L_0 C_0}} = \frac{l}{\tau} \tag{17}$$

where L_0 and C_0 are the inductance and capacitance per unit length for this line. In fixing the time step τ (this is taken as eqn. 7) one is left with two degrees of freedom, either fixing the inductance L_0 and calculating C_0 from eqn. 17 or vice versa. Here the first option is used which leaves only the open-circuit stub. The inductance per unit length for the y and x lines can be calculated from eqns. 3*a* and 3*b*, and, substituting them into eqn. 17, the capacitances per unit length for these lines are

$$C_{y0} = \frac{u}{v^2 h w \mu_r}\epsilon_0 \tag{18a}$$

$$C_{x0} = \frac{v}{u^2 h w \mu_r}\epsilon_0 \tag{18b}$$

Using eqns. 18*a* and 3*a*, the characteristic impedance for the y-directed line is

$$Z_y = \frac{wv}{u}\mu_r \tag{19a}$$

and, using eqns. 18*b* and 3*b*, the characteristic impedance for the x-line is

$$Z_x = \frac{wu}{v}\mu_r \tag{19b}$$

both are normalised to $\sqrt{\mu_0 h/\epsilon_0}$.

Due to the grading conditions the total capacitances assumed for the x and y lines are sometimes less than the total junction capacitance given by eqn. 3*c*. The residue is therefore modelled by open-circuit stubs at the nodes. This residual capacitance is given by

$$C_s = \frac{\epsilon_0}{w}\left(uv\epsilon_r - \frac{u^2+v^2}{huv\mu_r}\right) \tag{20}$$

from eqns. 18 and 3*c*.

The normalised admittance of these stubs is

$$Y = 2h\frac{C_s}{\epsilon_0} \tag{21}$$

As in the stub-loaded matrix the scattering process here is done in two stages; at the nodes and on the element boundaries. The nodal scattering matrix is similar to that given in Reference 3, with slight modifications because the lines have different characteristic impedances. The second scattering experienced by the pulses is at the element boundaries where an abrupt change in the line impedance occurs due to the variation in element sizes. The scattering takes the form of simple reflection and transmission as only two lines are involved. If a pulse propagates on a line of impedance Z_i which meets a line of impedance Z_j, the reflection coefficient at the junction is

$$\rho_{ij} = \frac{Z_j - Z_i}{Z_j + Z_i}$$

The transmission coefficient is related to this as

$$T_{ij} = 1 - \rho_{ij}$$

In general there are two pulses, one on line i the other on j, propagating in opposite directions; the scattering at the junction is summarised as

$$\begin{bmatrix} V_i \\ V_j \end{bmatrix}^r = \begin{bmatrix} \rho_{ij} & T_{ji} \\ T_{ij} & -\rho_{ij} \end{bmatrix} \begin{bmatrix} V_i \\ V_j \end{bmatrix}^i \qquad (22)$$

The boundaries are also taken halfway between the nodes to achieve time synchronism.

4 TLM model for two-dimensional polar coordinates

Space quantisation for the $r\theta$-plane of cylindrical co-ordinates, shown in Fig.3, produces elements which are nonuniform, i.e. they get bigger away from the centre. The principle of graded mesh can be extended to produce a matrix which fits such co-ordinates.

The stub-loaded matrix is found to be not very useful for such an application; because it would require to have very large stub impedances near the centre and at the outer circumference, this would increase the velocity error and reduce the workable frequency range. The hybrid matrix, however, is a convenient type of matrix to use and can be easily formulated. Such a matrix is composed of two groups of lines, circular and radial. The characteristic impedances can be found as in Section 3.2, and these are

$$Z_\theta = \frac{r\alpha w}{u}\mu_r \qquad (23a)$$

$$Z_r = \frac{uw}{r\alpha}\mu_r \qquad (23b)$$

The normalised admittance for the open-circuit stub is

$$Y = \frac{2hMr\theta}{w} \qquad (23c)$$

where

$$M = u\,\epsilon_r - \frac{1 + \left(\dfrac{u}{r\theta}\right)^2}{uh\mu_r}$$

5 Three-dimensional model of graded TLM

The graded matrix in three dimensions can be built using the conventional three-dimension model. In this model the three-dimensional node consists of three shunt nodes and three series nodes. At each shunt node there is an open-circuit stub, and at each series node there is a short-circuit stub. These stubs can therefore be used to account for any extra capacitances and inductances needed owing to extra lengths of lines. This means that only a small amount of work is needed to transform the conventional TLM to a graded one. Extra calculations needed are those necessary to obtain the admittances and impedances of the open- and short-circuit stubs. These are given by

$$Y_x = 2\left(\frac{vw\sqrt{h}}{u}\epsilon_r - 2\right) \qquad (24a)$$

$$Y_y = 2\left(\frac{uw\sqrt{h}}{v}\epsilon_r - 2\right) \qquad (24b)$$

$$Y_z = 2\left(\frac{vu\sqrt{h}}{w}\epsilon_r - 2\right) \qquad (24c)$$

$$Z_x = 2\left(\frac{vw\sqrt{h}}{u}\mu_r - 2\right) \qquad (25a)$$

$$Z_y = 2\left(\frac{uw\sqrt{h}}{v}\mu_r - 2\right) \qquad (25b)$$

$$Z_z = 2\left(\frac{vu\sqrt{h}}{w}\mu_r - 2\right) \qquad (25c)$$

these are normalised to $\sqrt{\mu_0/\epsilon_0}$.

The form in which the above equations are given should be modified before using them in a general computer program, since, in general, in an irregularly graded matrix representing inhomogeneous space, the volumes of the elements and their relative permittivities or permeabilities will be different. For a shunt open-circuit stub at one edge of an element the dimensions u, v, w and ϵ_r should be averaged over the four elements which meet at that edge, see Fig. 2. Similarly, for a short-circuit stub at the centre of a face between two elements, the dimensions u, v, w and μ_r should be averaged over these two elements [6]. This averaging procedure solves the problem of unrealistic representation of the interface between two media, which is present in the conventional formulation [5]. This is because the stubs in an element representing the properties in different directions are in different places, as can be seen from Fig. 2. So the process of working out the stubs' values individually, which is essential for the graded mesh, also yields an improvement in the representation of the problem.

6 Some applications of graded TLM

6.1 Ridged waveguide

This example is chosen to demonstrate the usability of two-dimensional graded TLM. The interesting point in this

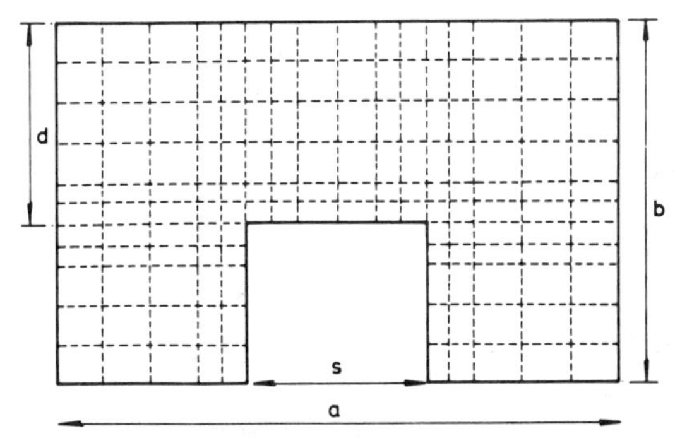

Fig. 7 *Single-ridge waveguide and its graded mesh representation*

Table 1: Effect of mesh grading on accuracy of single-ridge-waveguide dominant mode

$\frac{a}{\Delta l}$	Grading ratio	$K_0 a$	Error, %
20	1·1	1.73	1, 6
40	1:2	1.750	.05
40	1:3	1.78	1. 2

$s/a = 0.2, d/b = 0.25, b/a = 0.4$, exact 1.758

$K_0 = \frac{2\pi}{\lambda}$, λ cutoff wavelength

particular type of waveguide is the way in which the field of the dominant mode depend on the scattering caused by the ridge corners [2]. The deviation of this mode cutoff in the ridge waveguide from that with no ridge is about 42% for the geometry chosen here. In applying conventional TLM to this problem, the error in the numerical calculations decreases from 2% to 0.3% upon doubling the number of nodes used to represent a given geometry. However, as the field is affected more by the corners than anything else it seems much more efficient, especially when the ridge width is small, to use fine mesh around the corners and coarser mesh elsewhere, Fig. 7. Using this approach, the cutoff for the dominant mode in single-ridge waveguide is calculated for different grading conditions, Table 1. In these calculations the saving in the computer storage and run time was more than 45% and yet very good accuracy was obtained.

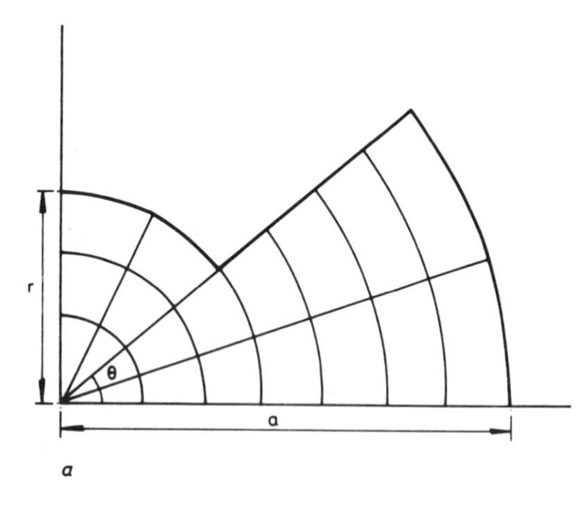

a

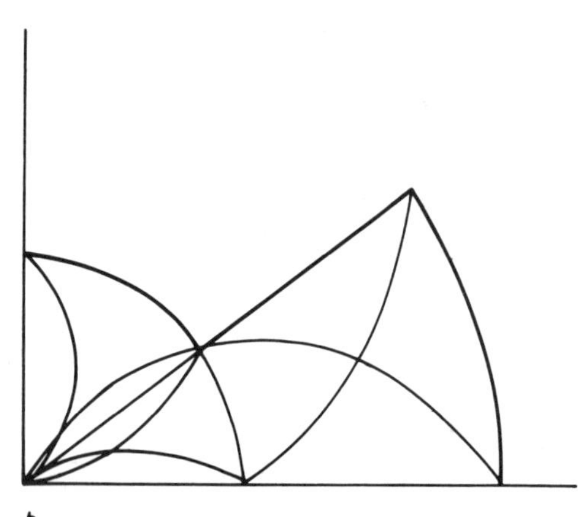

b

Fig. 8 *Quadrant of double-ridged circular waveguide*

a **TLM representation,**
b **finite-element representation**

6.2 Double-ridged circular waveguide

This class of waveguides represents an interesting application of the $r\theta$-plane TLM, as its boundaries can be exactly represented by such a matrix, Fig. 8. This type of waveguide has been considered by Daly [9] as an example for the finite-element method in polar geometry. In the finite elements the centre of the waveguide needs special attention because elements joining at the centre have only two vertices instead of three. These elements should have special matrix representation. In the present TLM the radial lines are terminated at the centre of the waveguide, at a circle of infinitesimal circumference. This circle is taken as a short-circuit or open-circuit boundary, depending on the mode choice. The cutoff frequency for the dominant TE_{11} mode (which is a perturbed version of the TE_{11} mode in circular waveguide) is computed for a range of values of ridge dimensions and compared with those given by Daly, Fig. 9. The results obtained with TLM technique are always lower than Daly's. This could be due to the fact that in finite elements all eigenvalues of the matrix equation are the upper bounds to the exact cutoff wavelengths, as reported by Daly. It should be noted, however, than in both techniques accurate estimation of the absolute error does not exist.

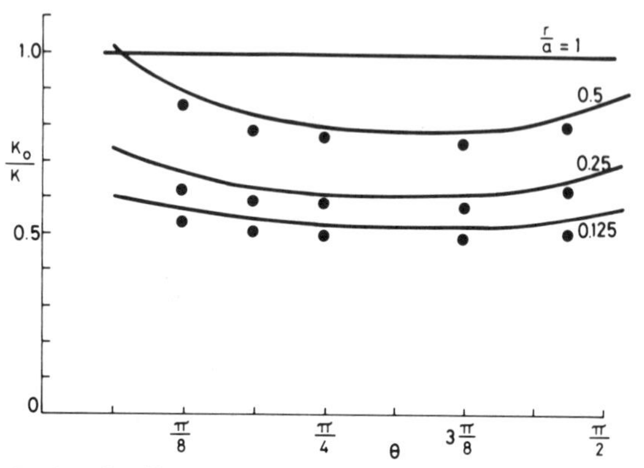

Fig. 9 *Cutoff wave number of double-ridged waveguide as function of ridge parameters*

—— Daly; • TLM

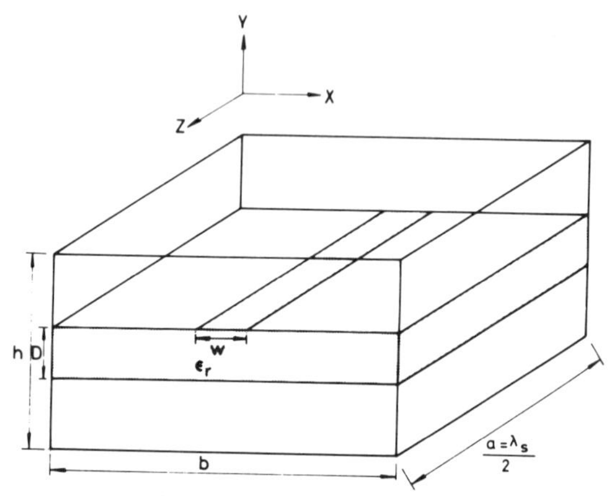

Fig. 10 *Enclosed slot-line structure*

6.3 Slot line

A slot-line cavity, shown in Fig. 10, is considered as an example for the three-dimensional graded TLM model. We start the analysis of such a line by investigating the effect of the enclosure on its dispersion characteristics. A fixed length cavity is chosen and the corresponding free space wavelength is found for different cavity dimensions. This analysis shows that for $b/\lambda_0 > 0.43$ and $d/\lambda_0 > 0.4$ (λ_0 is the free space wavelength) the enclosure walls would have no effects. Using these conditions the dispersion characteristics for typical slot lines are given in Fig. 11, and comparison is made with results obtained by Cohn [10]. Good agreement is noted. It should be stated here that for the geometry used and the condition imposed on the boundaries the problem becomes a real burden on the computer, and could only be handled when graded mesh is used. The grading pattern used in this particular problem consists of a fine mesh inside the dielectric substrate and around the slot, and the element size is gradually increased above and below the dielectric substrate and on both sides of the slot. The maximum grading ratio used is 1:4, and the cavity dimensions are taken as real-number multiples of the shortest element dimension in the mesh.

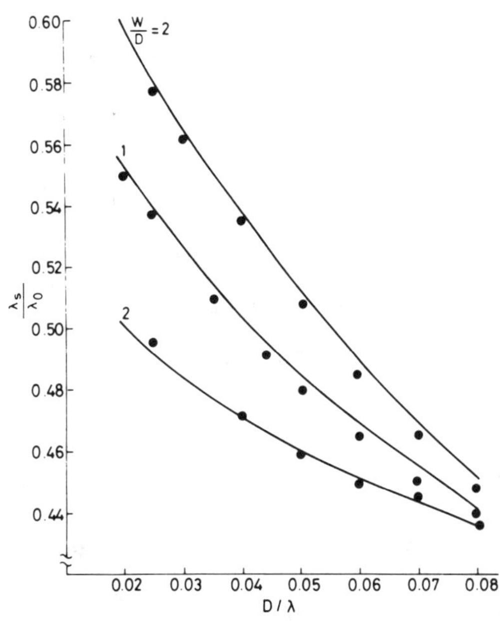

Fig. 11 *Dispersion characteristics for slot line*

$\epsilon_r = 9.6$, $b = 20$ units, $h = 15$ units, grading ratio 1:2:4
• TLM
—— Cohn

7 Discussion and conclusion

A modification to the traditional routine of the transmission line matrix is presented. Use of this modification breaks the limitation of square and cube elements required for time synchronism. The immediate advantage of this is the employment of more efficient space quantisation without introducing unacceptable levels of error. The saving in computer run time and storage in two-dimensional models is more than 45% of the original requirements. This is quite useful especially when the geometry of the problem is large and with dimensions that are not small integer multiples of a common length. An interesting extension to the technique is the formulation of TLM in the cases of irregular space elements such as the $r\theta$-plane. The only drawback of this technique is the need to use the factor h, which prevents the impedances of some stubs from being negative. The effect of h can be seen on examining the spectrum equations representing the output, these are

$$\mathrm{Re}\,[I(f\tau)] = \sum_{n=1}^{N} I_n \cos\, 2\pi n f \frac{\tau_0}{\sqrt{h}}$$

$$\mathrm{Im}\,[I(f\tau)] = \sum_{n=1}^{N} I_n \sin\, 2\pi n f \frac{\tau_0}{\sqrt{h}}$$

where Re and Im are real and imaginary parts of the spectrum. A close examination of the above equations shows that an increase of $\sqrt{h}$ in the number of iterations is needed for the same frequency resolution as in the uniform-mesh case. The value of h, however, can be kept to a minimum in practical applications. For example, in the analysis of inhomogeneous media the mesh elements inside the dielectric can be taken as small as required and coarser mesh used outside the dielectric; this of course reduces the velocity error. The use of h itself introduces a reduction in the velocity error since it decreases the values of $f\tau$, see Fig. 6. The use of h in the $r\theta$-plane matrix is essential as a result of the matrix nature.

Three-dimensional TLM has the advantage of retaining the simplicity inherent in the traditional model, since the added impedance and admittance calculations are straightforward. Furthermore, this technique has been tested for the inhomogeneous cavity problem used before to test conventional TLM [7]. Calculations showed that up to 40% reduction in storage and 80% in run time is possible without introducing unacceptable error [6]. The computer effort required to calculate and store the stub impedances is found to be only a small fraction of the total requirement. These advantages make it possible for TLM to handle structures of large geometries such as slot lines, coupled microstrips and many others [6].

8 References

1 JOHNS, P.B., and BEURLE, R.L.: 'Numerical solution of 2-dimensional scattering problems using a transmission-line matrix', *Proc. IEE*, 1971, **118**, (9), pp. 1203–1208

2 JOHNS, P.B.: 'Application of the transmission-line-matrix method to homogeneous waveguides of arbitrary cross-section' *Proc. IEE*, 1972, **199**, (8), pp. 1086–1091

3 JOHNS, P.B.: 'The solution of inhomogeneous waveguide problems using a transmission line matrix', *IEEE Trans.*, 1974, **MTT-22**, pp. 209–215

4 AKHTARZAD, S., and JOHNS, P.B.: 'Three-dimensional transmission-line matrix computer analysis of microstrip resonators', *IEEE Trans.*, 1975, **MTT-23**, (12), pp. 990–997

5 AKHTARZAD, S.: 'Analysis of lossy microwave structures and microstrip resonators by the TLM method'. Ph.D dissertation, University of Nottingham, May 1975

6 AL-MUKHTAR, D.: 'A transmission-line matrix with irregularly graded space'. Ph.D thesis, University of Sheffield, Aug. 1980

7 AKHTARZAD, S.: 'TLMRES-TLM computer program manual for the three-dimensional space'. Department of Electronic & Electrical Engineering, University of Nottingham, July 1975

8 WHINNERY, J.R., and RAMO, S.: 'A new approach to the solution of high-frequency field problems', *Proc. IRE*, 1944, **32**, pp. 284–288

9 DALY, P.: 'Polar geometry waveguides by finite element methods', *IEEE Trans.*, 1974, **MTT-22**, pp. 202–209

10 COHN, S.B.: 'Slot lines on dielectric substrate', *IEEE Trans.*, 1969, **MTT-17**, pp. 768–778

THE MODELLING OF MULTIAXIAL DISCONTINUITIES IN QUASI-PLANAR STRUCTURES WITH THE MODIFIED TLM METHOD

PIERRE SAGUET

Laboratoire d'Electromagnétisme, Institut National Polytechnique de Grenoble, ENSERG, 23 Avenue des Martyrs, 38031 Grenoble Cedex, France

AND

WOLFGANG J. R. HOEFER

Laboratory for Electromagnetics and Microwaves, Department of Electrical Engineering, University of Ottawa, Ottawa, Ontario, Canada K1N 6N5

SUMMARY

The principal features of the modified TLM method such as the asymmetrical condensed node, graded mesh and dispersion characteristics, are briefly presented. Its general application to multiaxial discontinuities in quasi-planar structures is described, and the modelling of bilateral finline T-junctions is demonstrated in particular.

INTRODUCTION

The modelling of multiaxial discontinuities in planar waveguides is one of the most challenging problems in microwave circuit analysis and design. Difficulties are accentuated by the presence of inhomogeneous dielectric loading and sharp metallization edges. The highly non-uniform, hybrid electromagnetic field in a multiaxial junction cannot be expressed simply as a sum of propagating and evanescent modes in a generic waveguide. Only the most general field solution techniques which impose no restrictions on the shape of the boundaries offer the flexibility required to model such a problem. Furthermore, the fine discretization needed to resolve highly concentrated fields results in large computer expenditures which must be minimized by careful modelling and programming.

In these situations, field modelling techniques such as the transmission line matrix (TLM) method can be used to their full potential. The TLM method was formulated in 1971 by Johns and Beurle,[1] and was inspired by earlier network simulation techniques[2–4]. It simulates the wave properties of space by a three-dimensional mesh of interconnected transmission lines, and employs a discretized form of Huygens's model of wave propagation[5] to compute the impulse time response of a transmission line network. This network, or mesh, embodies all electromagnetic properties of the original structure, including boundary reflections, losses, permeability and permittivity of materials, and refraction at dielectric interfaces. The frequency characteristics of the TLM mesh can be extracted from its impulse response by Fourier transformation. Since it properly simulates both propagating and evanescent fields, it is appropriate for the solution of discontinuity problems.

The TLM algorithm is very simple — it describes the scattering of Dirac impulses at series and shunt nodes of ordinary transmission lines. The subsequent Fourier transform routine is a straightforward multiply-and-add procedure. The method converges unconditionally.

However, the TLM method has its difficulties and pitfalls. Coarseness, velocity and truncation errors must be eliminated, or at least minimized. As in many other numerical procedures, high accuracy requires large amounts of CPU time and memory, so much effort has been directed

Received 12 October 1987
Revised 3 December 1987

Reprinted with permission from *Int. J. Numerical Modelling: Electronic Networks, Devices and Fields,* vol. 1, no. 1, pp. 7–17, Mar. 1988.

towards increasing the efficiency of the TLM method. Some innovations concern the excitation of the TLM network,[6] the treatment of the output impulse response[6–9] and the TLM network itself.[9–13]

The purpose of this paper is to give a brief description of the modifications leading from John's original TLM concept to the procedure used in this study, and to present the application of the modified approach to the modelling of multiaxial discontinuities, choosing a bilateral finline T-junction as an example.

THE MODIFIED TLM METHOD

The original three-dimensional TLM method of Akhtarzad and Johns[14, 15] employs a regularly spaced mesh of transmission lines. The voltages and currents on the mesh obey the same differential equations as the electric and magnetic field components (Maxwell's equations), provided the wavelength is large compared with the mesh parameter. The mesh lines are interconnected at two-dimensional shunt and series nodes which are half a mesh parameter apart (Figure 1(a)). For this reason the classical TLM network is termed an expanded-node network.

This network has been used successfully in many different applications.[16] Its topology is quite complicated, and the separation in space of the six field components can introduce errors in the description of boundaries and dielectric interfaces. This inconvenience has stimulated the development of condensed-node schemes by Saguet,[9] Saguet and Pic,[10] Amer[17] and Johns.[13] While References 9, 10 and 17 describe an asymmetrical node, Reference 13 introduces a symmetrical node, a new concept which breaks away from the traditional lumped-element representation of the unit cell by deriving the scattering matrix of the node directly from the field equations.

All condensed-node schemes lead to considerable savings in computer resources, particularly when they are combined with a graded mesh technique in which the density of the mesh can be adapted to the degree of field nonuniformity. In this paper, such a modified version of the TLM method (henceforth called the 'MTLM' method) is presented, and its application to the modelling of multiaxial discontinuities is demonstrated. The MTLM method, developed by Saguet and others,[9–12] employs a graded mesh scheme with asymmetrical condensed nodes.

Properties of the condensed asymmetrical node

When an elementary transmission line cell is represented by a half-T instead of a full T, the series and the shunt nodes can be connected at one point, A, resulting in a single three-dimensional node with 12 branches, as shown schematically in Figure 1(b).

As in the distributed-node mesh, a dielectric can be simulated by adding open-circuited $\lambda l/2$ stubs at shunt nodes. The normalized characteristic stub admittance is $Y_0 = 4(\epsilon_r - 1)$. Shortcircuited $\Delta l/2$ series stubs at the series nodes simulate magnetic permeability. Their normalized characteristic impedance is $Z_0 = 4(\mu_r - 1)$. Finally, loss stubs of normalized admittance $G_0 = \sigma Z_{air} \Delta l$ at the shunt nodes simulate finite conductivity.

A completely equipped cubic MTLM node has 21 branches (six less than the distributed node). Its equivalent circuit is shown in Figure 2. Each branch is numbered and represented by its immittance. An impulse incident on any given branch 'sees' the driving immittance of the network presented to that branch.

At first glance the computation of the scattering matrix of this network seems to be quite arduous. However, as many of its elements are identical it is rather easy to program and requires 30 per cent less computing time than the regular TLM procedure. The wave propagation characteristics of the condensed-node mesh are slightly different from those of the regular mesh, as the comparison in Figure 3 reveals. However, the resulting velocity error is virtually the same in both methods and can be corrected, as described below.

The graded mesh technique

The use of a graded mesh is essential if structures containing highly concentrated nonuniform field regions are to be modelled efficiently. The theory of the graded MTLM mesh has been

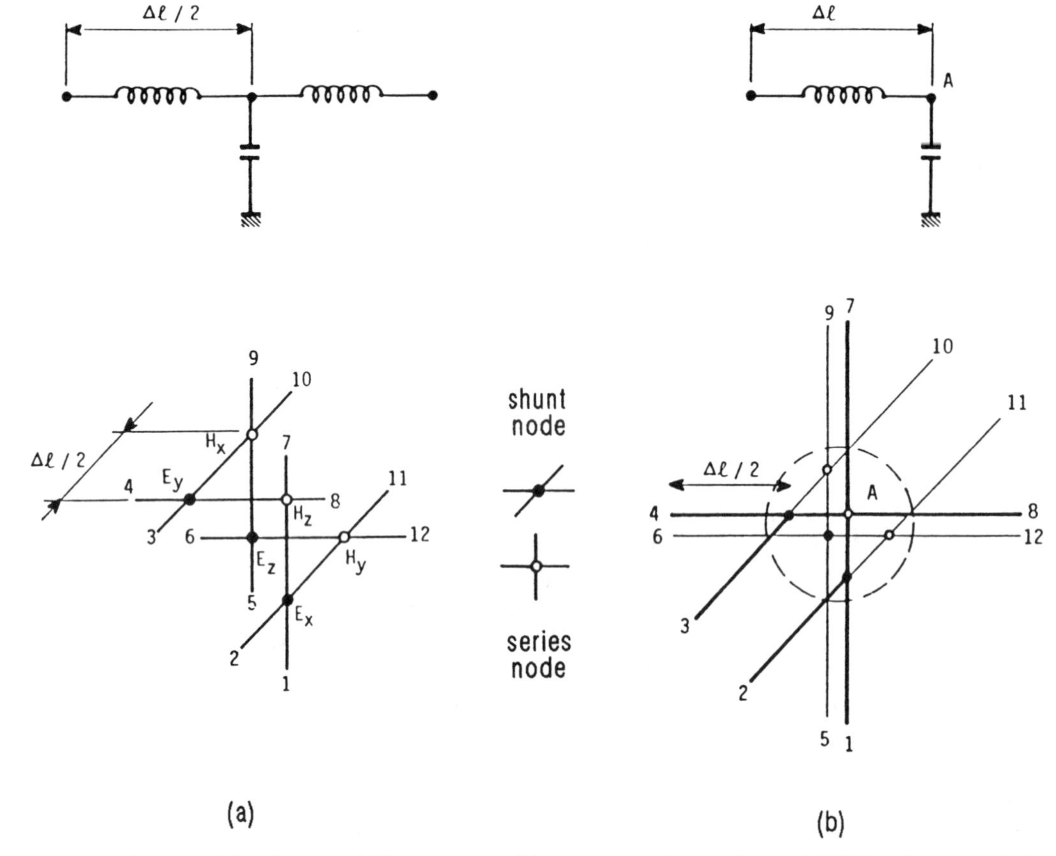

Figure 1. Topologies of (a) a distributed TLM node and (b) an asymmetrical condensed MTLM node; the equivalent L/C circuits show the structure of the generic shunt nodes

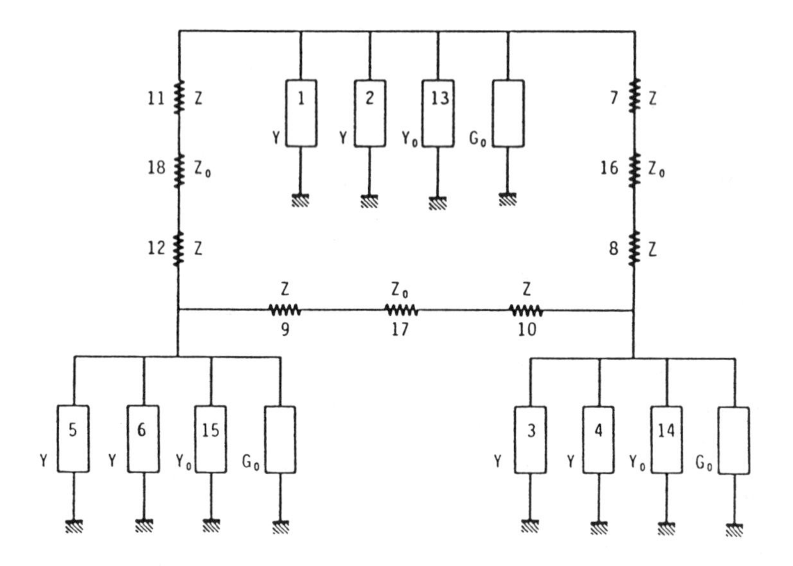

Figure 2. Equivalent circuit of a three-dimensional condensed MTLM node equipped with reactive and lossy stubs

described in References 9 and 12, so only its most important features are summarized here.

Consider the elementary condensed series and shunt nodes shown in Figure 4. The parameters $N_x\Delta l$, $N_y\Delta l$ and $N_z\Delta l$ represent the length of their branches in x-, y- and z-directions, respectively; Δl is the basic mesh parameter, and N_x, N_y and N_z are integers. The node equations can be related to Maxwell's field equations by the following equivalences: for the shunt node (Figure 4(a))

$$N_z I_z \equiv -H_x, \qquad N_x I_x \equiv H_z, \qquad V_y \equiv E_y, \qquad \epsilon \equiv 2C \tag{1}$$

and for the series node (Figure 4(b))

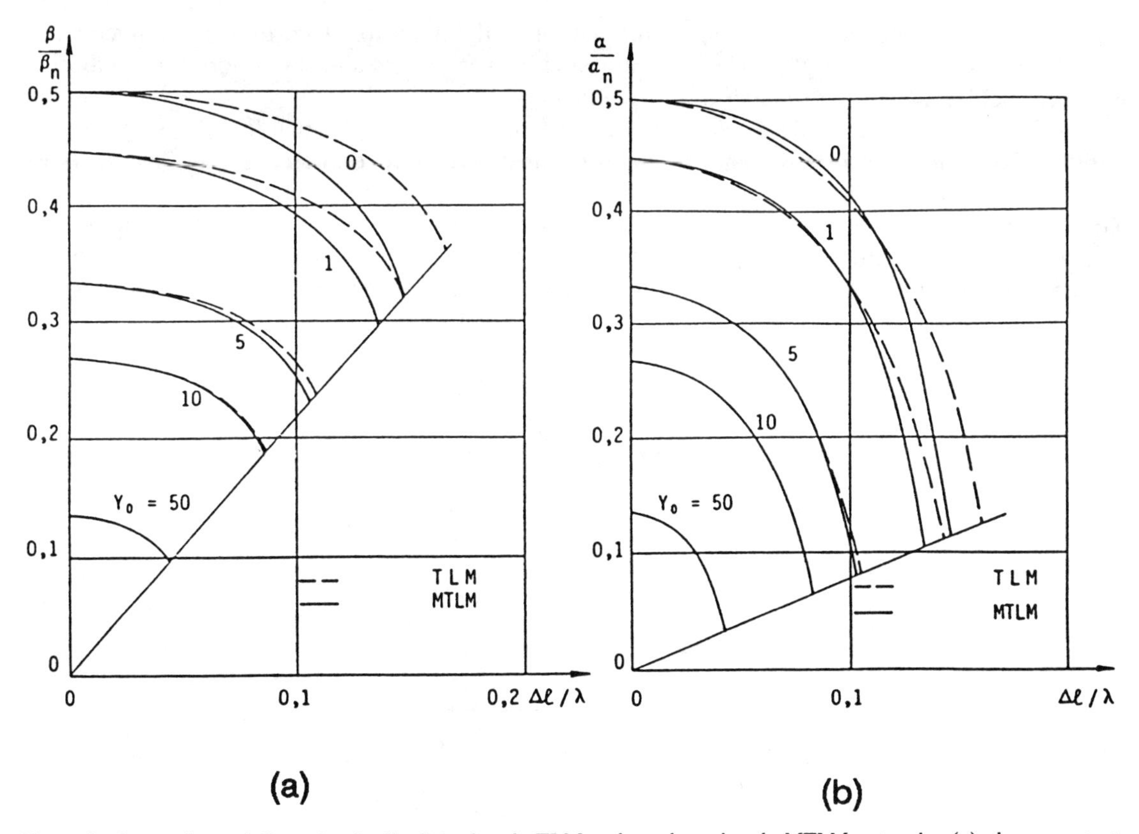

Figure 3. Comparison of dispersion in distributed-node TLM and condensed-node MTLM networks: (a) phase constant; (b) attenuation constant

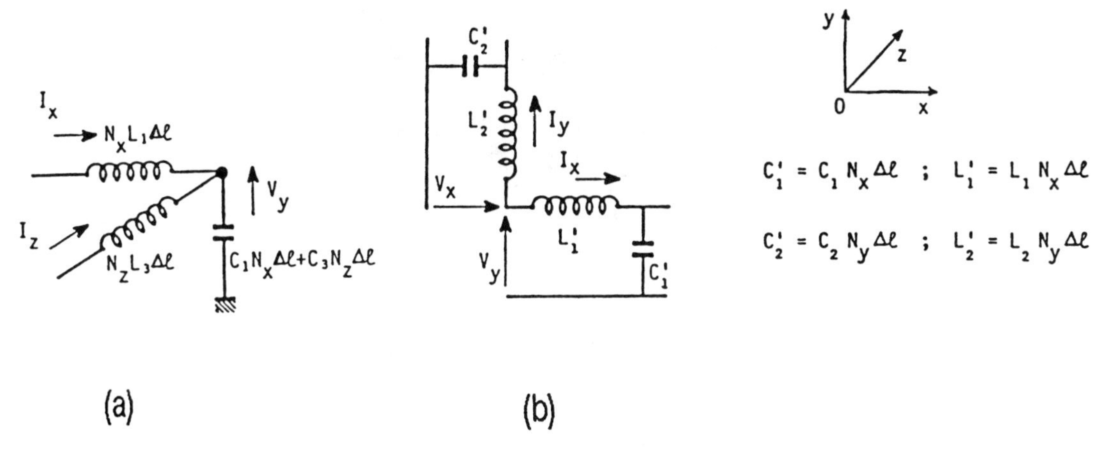

Figure 4. Structure of basic nodes in a graded MTLM mesh: (a) elementary shunt node; (b) elementary series node

$$I \equiv H_z, \qquad N_x V_y \equiv E_y, \qquad N_y V_x \equiv E_x, \qquad \mu \equiv 2L \tag{2}$$

The characteristic admittances of the mesh lines in the three coordinate directions are for the shunt node

$$\begin{aligned} Y_x &= \sqrt{(C_1/L_1)} = (1/N_x)\sqrt{(C/L)} \\ Y_y &= \sqrt{(C_2/L_2)} = (1/N_y)\sqrt{(C/L)} \\ Y_z &= \sqrt{(C_3/L_3)} = (1/N_z)\sqrt{(C/L)} \end{aligned} \tag{3}$$

and for the series node

$$\begin{aligned} Z_x &= \sqrt{(L_1/C_1)} = (1/N_x)\sqrt{(L/C)} \\ Z_y &= \sqrt{(L_2/C_2)} = (1/N_y)\sqrt{(L/C)} \\ Z_z &= \sqrt{(L_3/C_3)} = (1/N_z)\sqrt{(L/C)} \end{aligned} \tag{4}$$

For convenience, the reference admittance $Y = \sqrt{(C/L)}$ is set equal to unity in all computations. Note that for a given direction, the characteristic impedance ratio of the mesh lines is N^2

(N_x^2, N_y^2 and N_z^2, respectively) according to the nature of the node to which they are connected. Thus it becomes necessary to place ideal transformers between series and shunt nodes to establish the correct field values and to match the line impedances.

The equivalent circuit of a 3D condensed node with three different branch lengths is shown in Figure 5. For simplicity, reactive and dissipative stubs have been omitted; Z_i and Y_j are the characteristic branch immittances of the series and shunt nodes, respectively.

For field modelling, the structure under study is divided into subregions within which the mesh size as well as the permittivity and permeability are constant, so that all nodes in one subregion have the same impulse scattering matrix. This matrix is computed only once for each subregion.

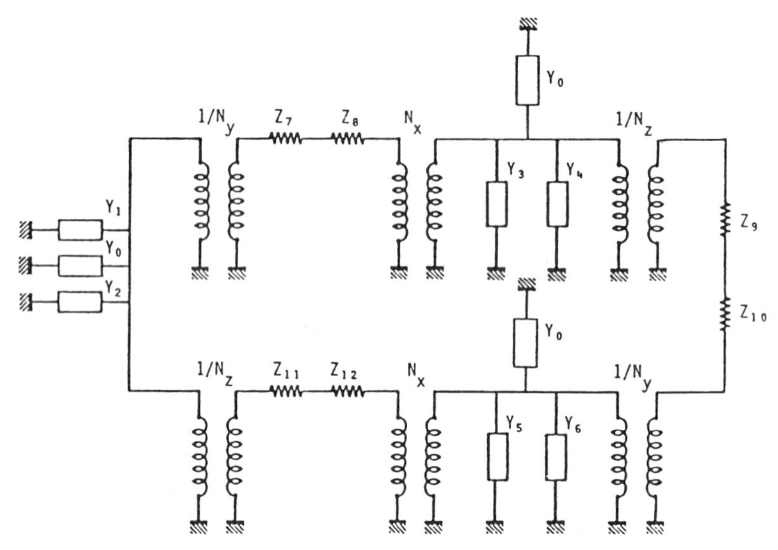

Figure 5. Equivalent circuit of a three-dimensional condensed node in a graded MTLM mesh; stubs are not included

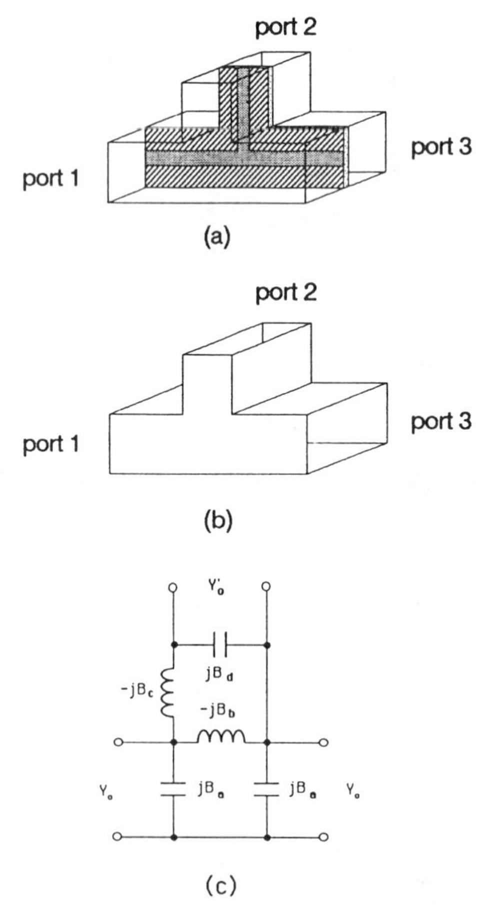

Figure 6. (a) Bilateral finline T-junction; (b) commensurate waveguide E-plane T-junction; (c) equivalent network recommended by El Hennawy *et al.*[18]

During the iteration process impulses travelling on branches longer than the elementary length Δl are stored during a number of iterations corresponding to N_x, N_y or N_z, as the case may be, before being re-injected into the neighbouring node. To ensure synchronism of scattering events, N_x, N_y and N_z must be integers. This requirement is in no way restrictive; it simplifies programming and accelerates the computation of the scattering matrix of the nodes. All further processing is the same as in the conventional TLM procedure.

MTLM MODELLING OF MULTIAXIAL DISCONTINUITIES

The TLM and MTLM methods simulate in a very physical way the propagation of electromagnetic waves in the time domain. The S-parameters of multiaxial discontinuities (represented as multiports with waveguide arms) can thus be computed by emulating the traditional measurement: one port is excited, while all other ports are matched. The only difference is in the Dirac excitation of the TLM model. The impulse time response is computed and stored for a string of nodes along the axis of each arm. Subsequent Fourier transformation yields the amplitude and phase of the fields at these nodes for any desired frequency within the first passband of the TLM mesh. Finally, complex S-parameters are derived from the voltage standing wave ratio (VSWR) and the position of minima in the input arm, and from the transmitted field amplitude in each matched output arm.

Along with the dominant mode, evanescent higher-order modes are also excited by the Dirac injection. Thus, the input arm must be long enough to ensure that only the dominant mode reaches the discontinuity. Furthermore, the scattering on the discontinuity generates higher-order modes as well. Hence the 'measurement' of amplitudes and phases must be done sufficiently far away from the discontinuity area if correct values are to be obtained for the incident, reflected and transmitted fields.

The propagation constant for the dominant mode is obtained by observing the wavelength of the standing wave in the input arm. For those frequencies at which the velocity error is significant, it must be corrected as described below.

CORRECTION OF SYSTEMATIC ERRORS

Correction of the velocity error

In order to keep CPU-time and memory-size requirements within reasonable limits, a graded mesh must be used. As the arms leading to the multiaxial discontinuitiy must be rather long to avoid higher-order mode interference, the mesh size in propagation direction becomes large enough to introduce a sizeable velocity error. Hence, if we compute the electric field along the slot at a particular frequency f_p by convolving the impulse response at nodes in the slot with a sine function of that frequency, we obtain a guided wavelength λ_{TLM} which is too short because of the velocity error. However, the accurate guided wavelength in the finline can be obtained with a fast spectral domain program to within, typically, 0·1 per cent. For the frequency f_p, that wavelength λ_{SDM} is somewhat longer. If all arms have the same geometry, (as in the example below) the velocity error can be treated as a frequency error which is corrected by assuming that the structure has been excited not by a frequency f_p but by a frequency f'_p, such that

$$f'_p = f_p\,\lambda_{SDM}/\lambda_{TLM} \tag{5}$$

Correction of the mismatch error

A second error arises from imperfect matching of the output arms of the multiport. In the TLM simulation, a matching boundary can be simulated by terminating each mesh line with the wave impedance of the structure. Therefore, impulses incident on the matching boundary see the following local reflection coefficient:

$$\rho = (1 - \sqrt{\epsilon_{eff}})/(1 + \sqrt{\epsilon_{eff}}) \tag{6}$$

where ϵ_{eff} is the effective dielectric constant of the arm. This quantity is defined as

$$\epsilon_{eff} = (\lambda/\lambda_g)^2 \tag{7}$$

where λ and λ_g are the free-space and the guided wavelength, respectively. The latter is obtained from the standing wave pattern in the input arm, as explained above. Since for non-TEM structures the effective dielectric constant is dispersive, the match is perfect only at the frequency for which it has been computed. However, as long as the dispersion is sufficiently small, the residual reflection coefficient Γ is small over the frequency range of interest. It can easily be taken into account when computing *S*-parameters since the TLM method provides the magnitude and phase of Γ in the output arms, as well as S_{11} in the input arm. Hence, a set of complex relations between incident and transmitted fields is obtained, and the values of the correct *S*-parameters can easily be deduced, for example by an iterative procedure.

MODELLING EXAMPLE: A BILATERAL FINLINE T-JUNCTION

The MTLM modelling procedure is demonstrated here for bilateral finline T-junctions (Figure 6(a)). Data on this type of junction are scarce, in spite of its importance in circuit applications. El Hennawy *et al.*[18] have proposed that Marcuvitz's model[19] be adopted for the equivalent rectangular waveguide T (Figure 6(b) and (c)), and to modify the given expressions for the equivalent circuit elements by replacing the guided wavelength in its arms by the finline wavelength. However, such a model is unable to fit MTLM data over an entire waveguide band, particularly for the phase angles of *S*-parameters. However, we have found that another model given in Reference 19 (Figure 7) can reproduce the behaviour of the finline T-junction, both in magnitude

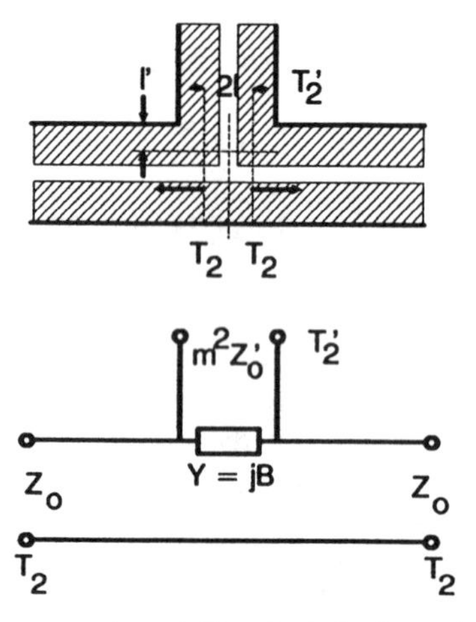

Figure 7. Equivalent network modelling the behaviour of the finline T-junction

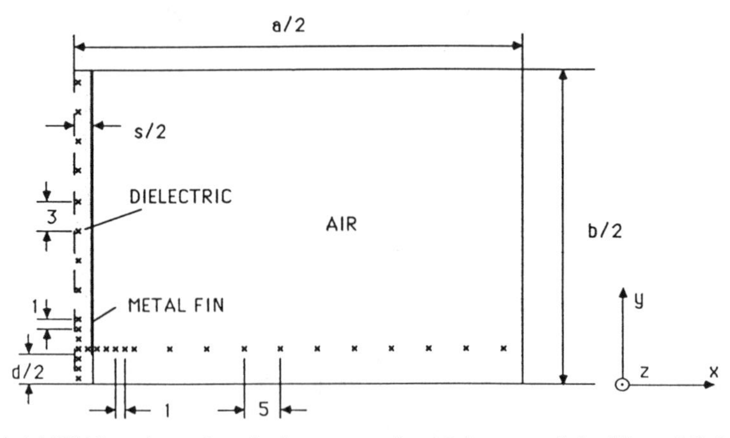

Figure 8. Graded MTLM mesh topology in the cross-section of the arms of the bilateral finlineT-junction

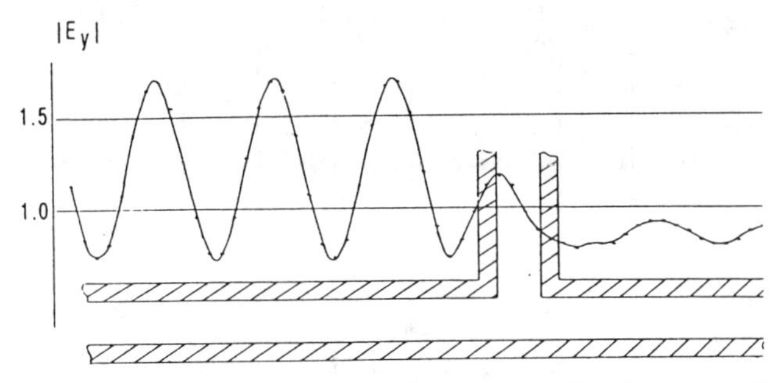

Figure 9. Electric slot field pattern in ports 1 and 3, obtained with the MTLM program after Fourier transformation (WR–28, d = 2 mm)

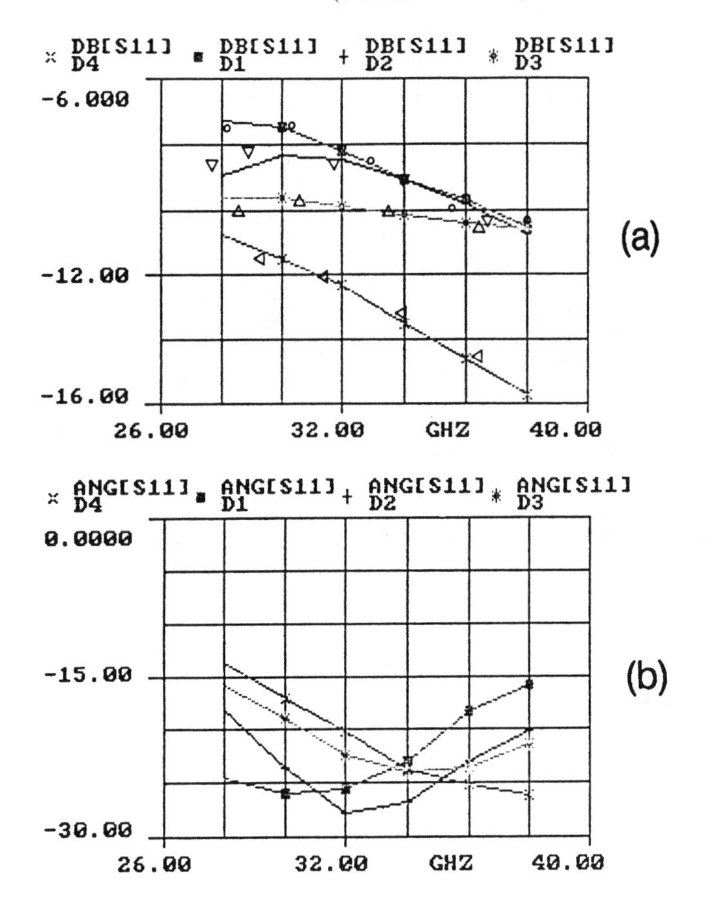

Figure 10. (a) Magnitude and (b) phase of S_{11} at terminals T_2. Continuous lines were produced by the model in Figure 7; triangles and circles were obtained with the MTLM program

d = 1·0 mm
d = 1·5 mm
d = 2·0 mm
d = b = 3·556 mm

a = 7·112 mm; b = 3·556 mm
s = 0·254 mm; ϵ_r = 2·22

and phase, when the shunt susceptance, the transformer ratio and the position of the reference planes T and T′ are adjusted according to the frequency and the structural parameters of the finline T-junction. This model has therefore been adopted in this study. The results given below have been obtained for a bilateral finline enclosed in a WR–28 waveguide housing. The junction is symmetrical and the same slotwidth is used in the three arms. Thus, the S-parameters can be derived by considering only the transverse electric field in the centre of the finline slots.

Figure 8 demonstrates a typical mesh topology in the cross-section of the arms (only one quarter is shown for reasons of symmetry). It also defines the spatial parameters of the T-junction. The mesh is densest in the substrate and the adjacent air-space close to the slot. These nodes are

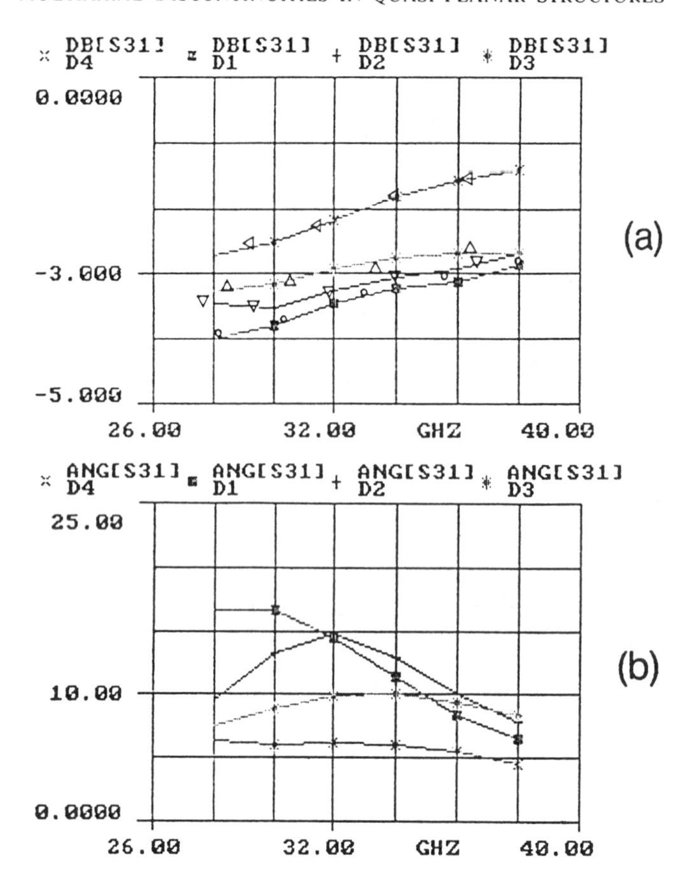

Figure 11. (a) Magnitude and (b) phase of S_{31} at terminals T_2 and T_2'. Continuous lines were produced by the model in Figure 7; triangles and circles were obtained with the MTLM program; dimensions as in Figure 10

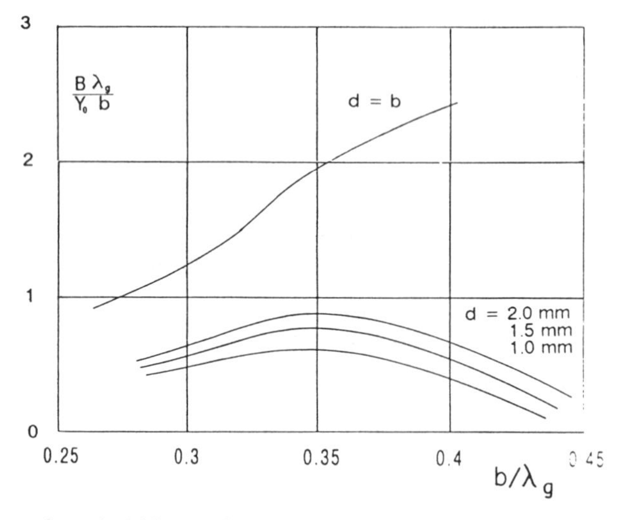

Figure 12. Shunt susceptance of matched bilateral finline T-junction at terminals T_2 and T_2'; dimensions as in Figure 10

separated by the basic mesh parameter Δl. The spacing of the remaining mesh is $5\Delta l$ in the x-direction, $3\Delta l$ in the y-direction, and $8\Delta l$ in the longitudinal (z) direction.

Figure 9 shows a typical plot of the transverse E-field amplitude in the slots of arms 1 and 3, obtained after 4000 iterations. As expected, standing waves arise not only in the input arm, but also in the output arms. Nevertheless, the VSWR in the output ports never exceeds 1·25 (Γ = 0·111). Effects of higher-order modes generated by the junction can be observed at the beginning of the output arm.

Figures 10 and 11 show the S_{11} and S_{13} parameters of the T-junction for four different gapwidths including $d = b$ (no metallization on the dielectric substrate). Continuous lines have been drawn

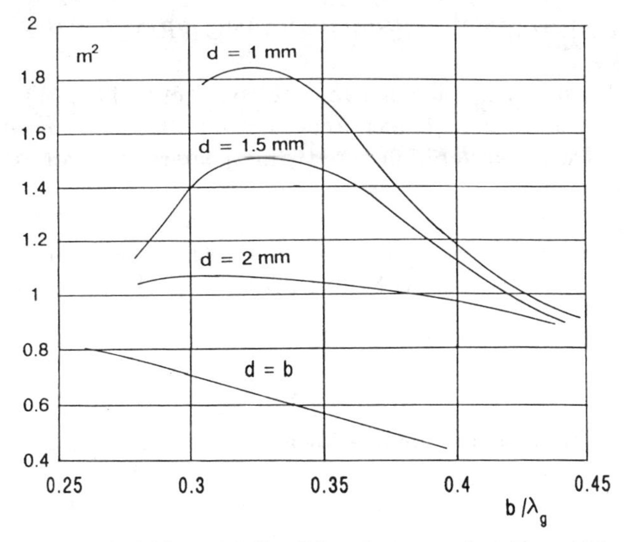

Figure 13. Transformer ratio of matched bilateral finline T-junction at terminals T_2 and T_2'; dimensions as in Figure 10

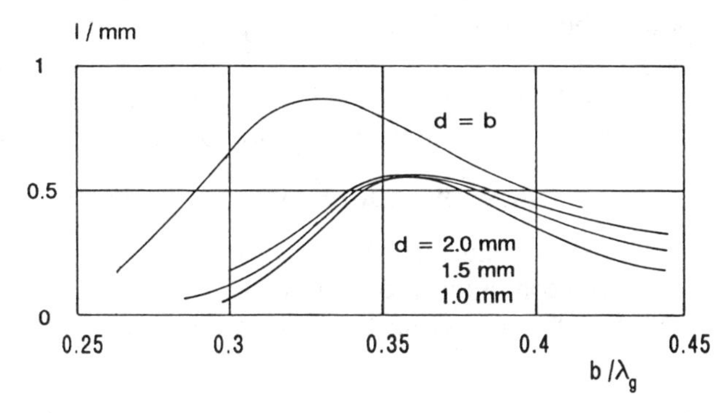

Figure 14. Position of terminal T_2 for matched bilateral finline T-junction; dimensions as in Figure 10

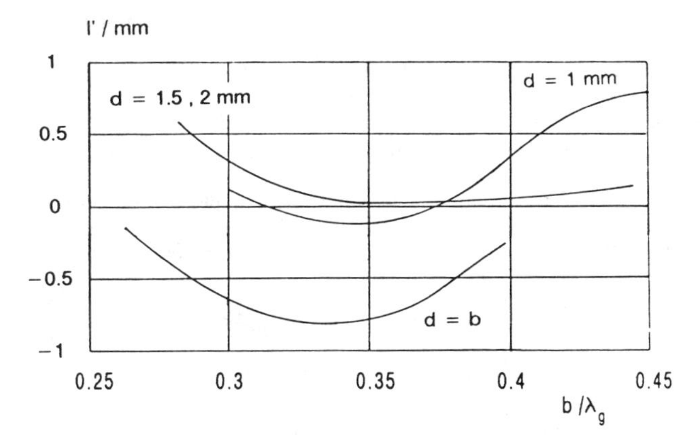

Figure 15. Position of terminal T_2' for matched bilateral finline T-junction; dimensions as in Figure 10

with Touchstone; they represent the characteristics of the model. Discrete points in the magnitude plots are the MTLM data to which the model has been fitted.

The corresponding values of the susceptance and the transformer ratio are shown in Figures 12 and 13. The positions of the terminals T and T′ are shown in Figures 14 and 15.

Regarding the *S*-parameters of the junction, it appears that two distinct behaviours occur. Beyond a certain frequency corresponding to an effective dielectric constant of unity, the structure behaves like a slotline T-junction with *S*-parameters which are not very sensitive to the slot width. At lower frequencies, it acts more like a waveguide T. When $d = b$, ϵ_{eff} never reaches unity, and the junction behaves like a waveguide T over the whole waveguide band.

ACCURACY OF THE MODELLING PROCEDURE

The velocity error is corrected as mentioned above. The asymmetry of the MTLM nodes introduces an error which is at least one order of magnitude smaller than the coarseness error. As the truncation error is negligible (4000 iterations used), the coarseness error is the only remaining source of inaccuracies. The small mesh size used in the highly nonuniform field regions keeps the estimated error for S_{11} to less than 1 dB, and that for S_{13} to less than 0·5 dB. The error in phase angle is difficult to evaluate but is likely to be less than 10°. The equivalent circuit model emulates numerical data within 0·5 dB for S_{11} and within 0·2 dB for S_{13}. The phase angles are reproduced to within 5°.

CONCLUSIONS

The modified TLM (MTLM) method is suitable for modelling multiaxial discontinuities in planar structures. This has been demonstrated by the example of the bilateral finline T-junction. The condensed node and the graded mesh employed in the MTLM method keep the computational expenditure within acceptable limits. The modelling procedure is really a computer simulation of an S-parameter measurement. For the purpose of computer-aided design, the behaviour of the junction can either by emulated by an equivalent circuit, as in this paper, or it could be described by a multidimensional look-up table. In the latter case, appropriate interpolation quickly yields the S-parameters for arbitrary frequencies and geometries. Velocity and truncation errors can be reduced to negligible values. Thus the accuracy of the simulation is determined essentially by the ability of the mesh to resolve the highly nonuniform fields around the metallic edges.

ACKNOWELDGEMENT

The authors gratefully acknowledge the financial support received from the following institutions: Centre National d'Etudes des Télécommunications (CNET), the French Ministry of External Affairs and the Natural Science and Engineering Research Council of Canada (NSERCC).

REFERENCES

1. P. B. Johns and R. L. Beurle, 'Numerical solution of 2-dimensional scattering problems using a transmission-line matrix', *Proc. IEE,* **118** (9), 1203–1208 (1971).
2. J. R. Whinnery and S. Ramo, 'A new approach to the solution of high-frequency field problems', *Proc. IRE,* **32**, 284–288 (1944).
3. G. Kron, 'Equivalent circuit of the field equations of Maxwell — I', *Proc. IRE,* **32**, 289–299 (1944).
4. J. R. Whinnery, C. Concordia, W. Ridgway and G. Kron, 'Network anlyzer studies of electromagnetic cavity resonators', *Proc. IRE,* **32**, 360–367 (1944).
5. P. B. Johns, 'A new mathematical model to describe the physics of propagation', *Radio Electron. Eng.,* **44**, (12), 657–666 (1974).
6. D. Pompei and E. Rivier, 'Étude des micro-circuits en régime dynamique', *Ann. Télécommun.,* **37**, (1/2), 63–73 (1982).
7. P. Saguet and E. Pic, 'Un traitement du signal simple pour améliorer la méthode TLM', (An improvement for the TLM method), *Electron. Lett.,* **16** (7), 247–248 (1980).
8. M. Leroy, 'Original improvements of TLM method', *Electron. Lett.,* **17** (19), 684–685 (1981).
9. P. Saguet, 'Analyse des milieux guides — La méthode MTLM', Doctoral Thesis, Institut National Polytechnique de Grenoble, 1985.
10. P. Saguet and E. Pic, 'La maillage rectangulaire et le changement de maille dans la méthode TLM en deux dimensions', *Electron. Lett.,* **17** (7), 277–278 (1981).
11. D. A. Al-Mukhtar and J. E. Sitch, 'Transmission-line matrix method with irregularly graded space', *Proc. IEE,* Part H, **128**(6), 299–305 (1981).
12. P. Saguet and S. Tedjini, 'Méthode des lignes de transmission en trois dimensions: Modification du processus de simulation', *Ann. Télécommun.,* **40** (3/4), 145–152 (1985).
13. P. B. Johns, 'A symmetrical condensed node for the TLM method', *IEEE Trans. Microw. Theory Techniques,* **MTT-35** (4), 370–377 (1987).
14. S. Akhtarzad and P. B. Johns, 'Solution of Maxwell's equations in three space dimensions and time by the TLM method of analysis', *Proc. IEE,* **122** (12), 1344–1348 (1975).
15. S. Akhtarzad and P. B. Johns, 'Generalised elements for TLM method of numerical analysis', *Proc. IEE,* **122** (12), 1349–1352 (1975).
16. W. J. R. Hoefer, 'The transmission line matrix method — Theory and applications', *IEEE Trans. Microw. Theory Techniques,* **MTT-33** (10), 882–893 (1985).
17. A. Amer, 'The condensed node TLM method and its application to transmission in power systems', Ph.D. Thesis, Nottingham University, 1980.
18. H. El Hennawy, R. Knoechel and K. Schuenemann, 'Wideband branchline couplers in finline technology', AEU, **37**, 1/2, pp. 4046 (1983).
19. N. Marcuvitz, *Waveguide Handbook*, McGraw-Hill, New York, 1951.

Transient Analysis of a Stripline Having a Corner in Three-Dimensional Space

NORINOBU YOSHIDA AND ICHIRO FUKAI

Abstract —The transient analysis of electromagnetic fields has shown its utility not only in clarifying the variation of the fields in time but also in gaining information on mechanisms by which the distributions of an electromagnetic field at the stationary state are brought about. We have recently proposed a new numerical method for the transient analysis in three-dimensional space by formulating the equivalent circuit based on Maxwell's equation by Bergeron's method. The resultant nodal equation is uniquely formulated in the equivalent circuit for both the electric field and the magnetic field. In this paper, we deal with the stripline which should be analyzed essentially in three-dimensional space because of its structure. The time variation of the electric and magnetic field of the stripline having a corner is analyzed and the remarkable changing of distribution of the field is presented as a parameter of time and of conditions imposed by the corner structure.

I. Introduction

THE TRANSIENT ANALYSIS of electromagnetic fields not only clarifies the variation of the fields in time but also provides information on mechanisms by which the distributions of electromagnetic fields at the stationary state are brought about. We have recently proposed a new numerical method for the transient analysis in three-dimensional space [1], [2]. The method was based on the equations obtained by Bergeron [3]. The equations show the character of the propagation of electromagnetic waves in the equivalent circuit based on Maxwell's equation [4]. This method has two important advantages for the analysis. One is the formulation of the electromagnetic fields in terms of the variables in the equivalent circuits. This treatment enables us to see that the nodal equation is uniquely formulated in the equivalent circuit for both the electric field and the magnetic field because of the duality of both field components. The other advantage is the formulation by Bergeron's method with its many merits, such as the representation of the medium by the lumped elements at each node and its reactive characteristics which are represented by the trapezoidal rule of the differential equation in the time domain. This treatment is based on an iterative computation in time using only the values obtained after the previous step. Consequently, the savings in memory storage space and computer time is remarkable. The formulation of this method is fundamentally equivalent to that of the Transmission-Line Matrix (TLM), because both methods are based on the property of the traveling wave, that is formulated as the general solution of one-dimensional wave equation by d'Alembert. But the Bergeron's formulation [5], in terms of the voltage variable and current variable, presents the direct handling of the electromagnetic field variables and the characteristics of the medium instead of the division of each variable into the incident and reflective components and the composition of those in TLM method [6].

Manuscript received July 20, 1983; revised November 28, 1983.
The authors are with the Department of Electrical Engineering, Faculty of Engineering, Hokkaido University, Sapporo, 060 Japan.

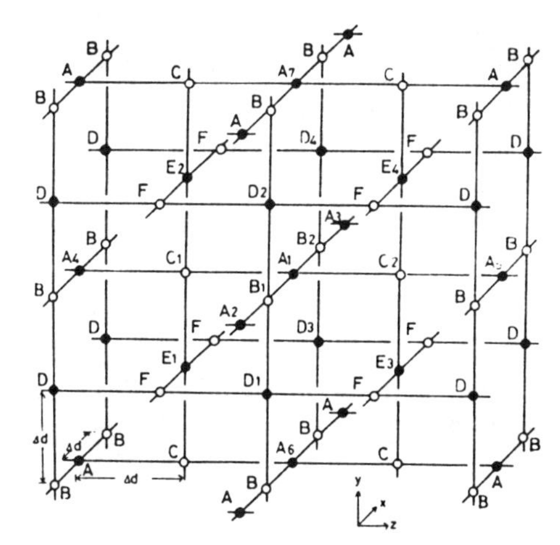

Fig. 1. Three-dimensional lattice network model of Maxwell's equations.

In this paper, we deal with the stripline which should be analyzed essentially in three-dimensional space because of its structure. The stripline is widely used as the transmission medium in MIC design, and its small size compared to the wavelength is the main reason for the good performance of microwave components and usually permits the treatment of circuits as a system composed of lumped elements in the analysis. But in high-frequency application, such as millimeter-wave devices, especially when using the pulse-wave technique that has progressed remarkably with the digital technique, the exact treatment of the higher components in the spectrum of the waves are indispensable, so the distributed formulation of the devices in three-dimensional space is essential.

In the following sections, the fundamental formulations for the stripline by our method are briefly described, and derived parameters, such as the characteristic impedance and wavelength, are examined and compared with those obtained analytically. Lastly, the time variation of the field distribution in the stripline with a corner is shown for several conditions of the structure.

II. Three-Dimensional Nodal Formulation of Maxwell's Equation

A. Three-Dimensional Equivalent Circuit

We now consider the formulation of nodal equations for three-dimensional analysis of the electromagnetic field in

Reprinted from *IEEE Trans. Microwave Theory Tech.*, vol. MTT-32, no. 5, pp. 491–498, May 1984.

TABLE I
CORRESPONDENCES BETWEEN THE FIELD VARIABLES IN MAXWELL'S EQUATION AND THE EQUIVALENT CIRCUIT AT EACH KIND OF NODE IN THE EQUIVALENT CIRCUIT

	Electric node			Magnetic node	
	Maxwell's Equ.	Variables		Maxwell's Equ.	Variables
An	$\frac{\partial H_x}{\partial z}-\frac{\partial H_z}{\partial x}=\varepsilon_0\frac{\partial E_y}{\partial t}$ $-\frac{\partial E_y}{\partial z}=-\mu_0\frac{\partial H_x}{\partial t}$ $\frac{\partial E_y}{\partial x}=-\mu_0\frac{\partial H_z}{\partial t}$	$V_y=E_y$ $I_z=-H_x$ $I_x=H_z$	Fn	$\frac{\partial E_x}{\partial z}-\frac{\partial E_z}{\partial x}=-\mu_0\frac{\partial H_y}{\partial t}$ $-\frac{\partial H_y}{\partial z}=\varepsilon_0\frac{\partial E_x}{\partial t}$ $\frac{\partial H_y}{\partial x}=\varepsilon_0\frac{\partial E_z}{\partial t}$	$V_y^*=H_y$ $I_z^*=E_x$ $I_x^*=-E_z$
Dn	$\frac{\partial H_z}{\partial y}-\frac{\partial H_y}{\partial z}=\varepsilon_0\frac{\partial E_x}{\partial t}$ $\frac{\partial E_x}{\partial z}=-\mu_0\frac{\partial H_y}{\partial t}$ $-\frac{\partial E_x}{\partial y}=-\mu_0\frac{\partial H_z}{\partial t}$	$V_x=E_x$ $I_z=H_y$ $I_y=-H_z$	Bn	$\frac{\partial E_y}{\partial x}-\frac{\partial E_x}{\partial y}=-\mu_0\frac{\partial H_z}{\partial t}$ $\frac{\partial H_z}{\partial y}=\varepsilon_0\frac{\partial E_x}{\partial t}$ $-\frac{\partial H_z}{\partial x}=\varepsilon_0\frac{\partial E_y}{\partial t}$	$V_z^*=H_z$ $I_y^*=-E_x$ $I_x^*=E_y$
En	$\frac{\partial H_y}{\partial x}-\frac{\partial H_x}{\partial y}=\varepsilon_0\frac{\partial E_z}{\partial t}$ $\frac{\partial E_z}{\partial y}=-\mu_0\frac{\partial H_x}{\partial t}$ $-\frac{\partial E_z}{\partial x}=-\mu_0\frac{\partial H_y}{\partial t}$	$V_z=-E_z$ $I_y=-H_x$ $I_x=H_y$	Cn	$\frac{\partial E_z}{\partial y}-\frac{\partial E_y}{\partial z}=-\mu_0\frac{\partial H_x}{\partial t}$ $\frac{\partial H_x}{\partial z}=\varepsilon_0\frac{\partial E_y}{\partial t}$ $-\frac{\partial H_x}{\partial y}=\varepsilon_0\frac{\partial E_z}{\partial t}$	$V_x^*=-H_x$ $I_z^*=E_y$ $I_y^*=-E_z$
dielectric const.	$C_0=\varepsilon_0/2$		dielectric const.	$L_0^*=\varepsilon_0/2$	
permeability	$L_0=\mu_0/2$		permeability	$C_0^*=\mu_0/2$	
poralization	$\Delta C=\varepsilon_0\chi_e/2\cdot\Delta d$		magnetization	$\Delta C^*=\mu_0\chi_m/2\cdot\Delta d$	
conductivity	$G=\sigma/2\cdot\Delta d$		magnetic current loss	$G^*=\sigma^*/2\cdot\Delta d$	

the time domain. In Fig. 1, the three-dimensional network model is shown. It is well known that this network gives a fundamental connection between the field variables in Maxwell's equation. This model is used in other methods, such as the "TLM" by P. B. Johns. In this network, each set of two-dimensional equations for the propagation of waves in each plane is related to a node and the connected lines. We interpret this network as the equivalent circuit, in which the line between nodes is a one-dimensional transmission line and the node is the point where the continuity of currents occurs. In Table I, the correspondence between the equivalent circuit variables and field quantities are shown at every kind of node of the network. The nodes are classified into two types. One is the electric node at which an electric field component is treated as a voltage variable and the other is a magnetic node at which a magnetic field component is treated as a voltage variable. The electric node corresponds to the shunt node and the magnetic node correspond to the series node in the "TLM". However, in our method the introduction of the magnetic current in the magnetic nodes results in the existence of the shunt node only in a sense of "TLM", where the continuity of current is postulated. In this paper, all variables at the magnetic nodes are characterized by the symbol " * " because of the duality of their physical meaning, as compared with their interpretation at the electric node. In Fig. 2, the fundamental connection between the nodes in the network is expressed. The correspondence of the variables is also illustrated in each node and each transmission line. The direction of the Poynting vector, which is decided by the set of an electric and a magnetic field component supposed in each one-dimensional transmission line, is also shown. Each of the supposed directions of the Poynting vector coincides with that of the currents in both nodes of the transmission line, so the currents are defined as the usual conduction currents in the electrical circuit. The gyrator is

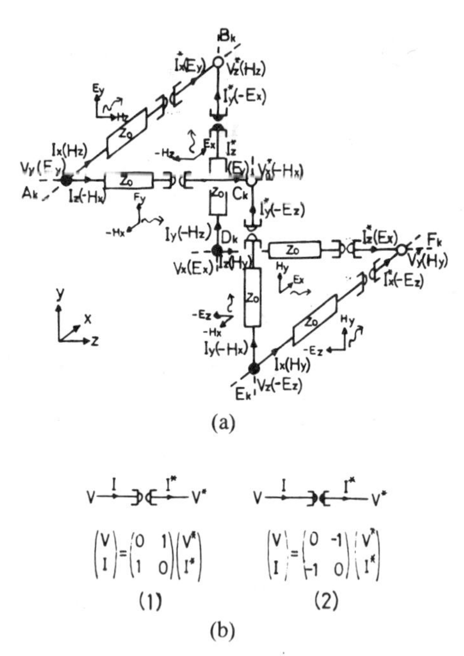

Fig. 2. (a) Fundamental connection of the node in the network and detailed expression of the variables in the equivalent circuit. The direction of the Poynting vector is shown by the symbol " ⤳ " at each transmission line. (b) Definition of gyrator in (a), (1) positive gyrator, and (2) negative gyrator.

inserted in series with each magnetic node to show the duality of the physical meaning of the circuit variables of both nodes of each transmission line. At D_n nodes, negative gyrators are inserted, since the corresponding circuit variables and electromagnetic variables have a polarity opposite to that at the nodes B_n. We interpret this negative gyrator as a circuit representation of the self-consistence of the Maxwell's equations, and the node to be inserted is determined by the correspondence of circuit variables and electromagnetic variables.

B. Bergeron's Method

Next, we formulate the propagation characteristics of a one-dimensional transmission line by Bergeron's method. In Fig. 3(b), showing a section of lossless line, the propagation characteristics of waves in the time domain are given by the one-dimensional wave-equation

$$v(k,t)+z\cdot i(k,t)=v(k-1,t-\Delta t)+z\cdot i(k-1,t-\Delta t) \tag{1a}$$

$$v(k-1,t)-z\cdot i(k-1,t)=v(k,t-\Delta t)-z\cdot i(k,t-\Delta t) \tag{1b}$$

where the parameter k denotes the node numbers and z is the characteristic impedance of the line, t is time and Δt is the transit time between two adjacent nodes, which also becomes the fundamental time step in the numerical computation. Each lumped element to be connected with the line at nodes is characterized as follows: the conductance G is expressed in terms of its branch voltage v_g and current i_g by

$$v_g(k,t)=G(k)i_g(k,t). \tag{2}$$

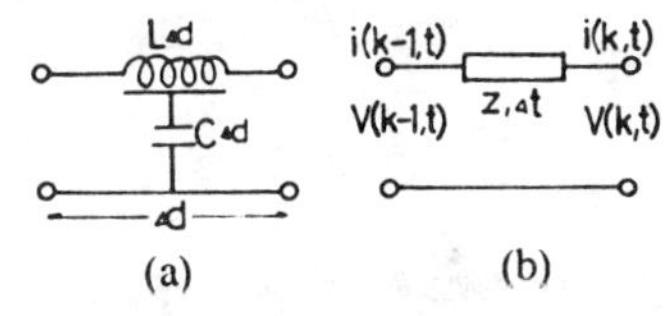

Fig. 3. Typical equivalent circuit (a) of the one-dimensional transmission line and its description (b) by means of Bergeron's method.

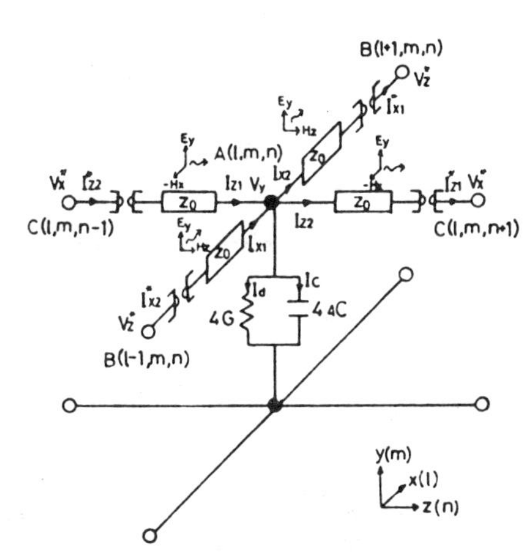

Fig. 4. Equivalent circuit at each A_n node in the dielectric medium. Dielectric loss is expressed by the parallel conductance.

In this equation, the conductance G is a function of the parameter k at each node. The capacitance $C(k)$ is defined through the trapezoidal rule and is given by

$$v_c(k,t) - R_c i_c(k,t) = v_c(k,t-\Delta t) + R_c i_c(k,t-\Delta t) \tag{3a}$$

where

$$R_c = \frac{\Delta t}{2C(k)}. \tag{3b}$$

In this equation, (v_c, i_c) is a pair of branch voltages and currents of the capacitance. A comparison of both sides in (1a), (1b), and (3a) shows that all values calculated at time t are only a function of ones at the previous time $t-\Delta t$. The nodal equation for each node at time t is independent of the values of the adjacent node at time t. The time responses are thus iteratively computed at each time, t from values of the circuit values at every node obtained at the previous steps.

C. Three-Dimensional Nodal Equations

For the three-dimensional network, the characteristics of the transmission line are formulated by Bergeron's method. In Fig. 4, the example of the equivalent circuit is shown at the node A_n where the electric field E_y is supposed to be a voltage variable and the magnetic fields $-H_x$ and H_z are supposed to be the current components in the directions z and x, respectively. Application of (1a) and (1b) to each line connected to the node yields the following equation:

$$V_y(l,m,n,t) + z_0 I_{z1}(l,m,n,t) = I_{z2}^*(l,m,n-1,t-\Delta t) + z_0 V_x^*(l,m,n-1,t-\Delta t) \tag{4a}$$

$$V_y(l,m,n,t) - z_0 I_{z2}(l,m,n,t) = I_{z1}^*(l,m,n+1,t-\Delta t) - z_0 V_x^*(l,m,n+1,t-\Delta t) \tag{4b}$$

$$V_y(l,m,n,t) + z_0 I_{x1}(l,m,n,t) = I_{x2}^*(l-1,m,n,t-\Delta t) + z_0 V_z^*(l-1,m,n,t-\Delta t) \tag{4c}$$

$$V_y(l,m,n,t) - z_0 I_{x2}(l,m,n,t) = I_{x1}^*(l+1,m,n,t-\Delta t) - z_0 V_z^*(l+1,m,n,t-\Delta t). \tag{4d}$$

The parameters l, m, and n denote the described position numbers of x, y, and z directions, respectively. Then (2) is written with the notation of this case as follows:

$$V_y(l,m,n,t) = 4G(l,m,n) I_d(l,m,n,t). \tag{5}$$

Equations (3a) and (3b) are again written as follows:

$$V_y(l,m,n,t) - R_c I_c(l,m,n,t) = V_y(l,m,n,t-\Delta t) + R_c I_c(l,m,n,t-\Delta t) \tag{6a}$$

where

$$R_c = \frac{\Delta t}{8\Delta C(l,m,n)}. \tag{6b}$$

The conductance G and capacitance ΔC are listed in Table I and are shown to correspond to the conductive loss and the electrical displacement of the dielectric medium, respectively. The conductance G also corresponds to the equivalent dielectric loss in the medium. These quantities are considered to be a function of the position variables l, m, and n. The continuity of the current at node $A(l,m,n)$ is given by

$$I_{z1} - I_{z2} + I_{x1} - I_{x2} - I_d - I_c = 0. \tag{7}$$

Substituting (4a)–(4d), (5), and (6a) into (7), the unified nodal equation in a dielectric medium with conductive or equivalent dielectric loss is given by

$$V_y(1,m,n,t) = \frac{R_c \cdot (\Psi_1^* + \Psi_2^* + \Psi_3^* + \Psi_4^*) + z_0 \Psi_c}{z_0 + R_c \cdot (4 + z_0 \cdot 4G(l,m,n))} \tag{8}$$

where Ψ_1^*, Ψ_2^*, Ψ_3^*, and Ψ_4^* correspond to the right-hand sides of (4a)–(4d), respectively, and Ψ_c is equal to the right side of (6a). Equation (8) is iteratively evaluated at every A_n node, and the time response of the field in the overall region is analyzed by the same procedure at other kinds of nodes. Each component of the currents at the time t is evaluated by substituting $V_y(t)$ in (8) by $V_y(t)$ in the left sides of (4a)–(4d), (5), and (6a). The other variables at other nodes also are obtained in the same manner. For example, the voltage variable $V_y^*(t)$ in the magnetic nodes F_n is given as follows:

$$V_y^*(l',m',n',t) = \frac{R_c^* \cdot (\Psi_1 + \Psi_2 + \Psi_3 + \Psi_4) + z_0^* \Psi_c^*}{z_0^* + R_c^* \cdot (4 + z_0^* \cdot 4G^*(l',m',n'))} \tag{9a}$$

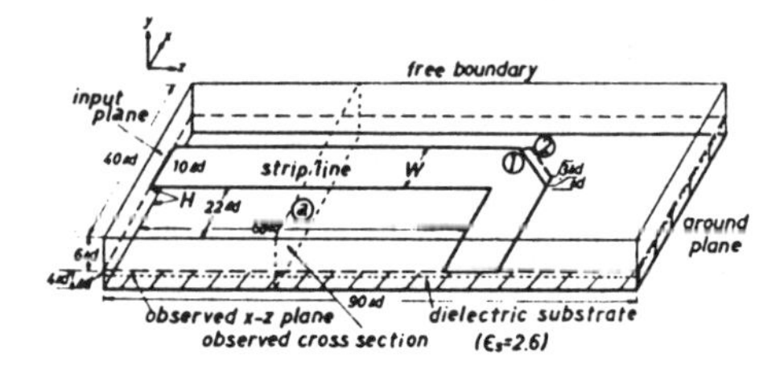

Fig. 5. Geometry of a stripline with a 90° bend.

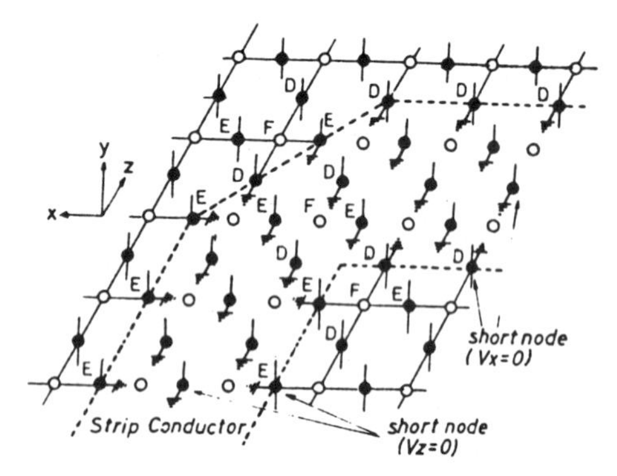

Fig. 6. Equivalent circuit of the surface of the stripline with infinite conductivities.

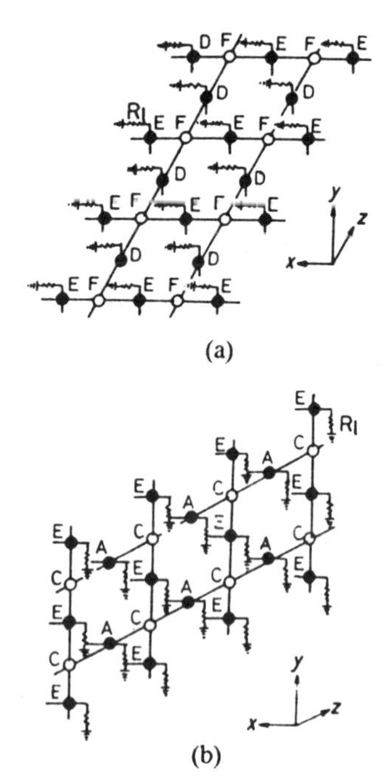

(a)

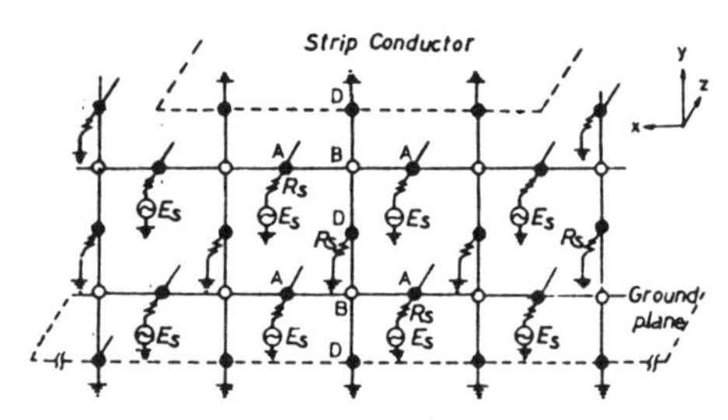

(b)

Fig. 7. Equivalent circuit of the free boundary. (a) The xz-plane of the top of the analyzed region. (b) The yz-plane of the side of the region, where R_1 is the characteristic impedance of the free space.

where

$$R_c^* = \frac{\Delta t}{8\Delta C^*(l', m', n')}. \tag{9b}$$

In such magnetic nodes, the conductance G^* and the capacitance ΔC^* have duality with those in electric nodes. These then correspond to losses of the magnetic current and the magnetization. But (8) and (9) have the same form and calculations are the same as for the electric nodes. These characteristics occur as a consequence of the application of the duality between the electric field and the magnetic field in Maxwell's equations. Thus, the introduction of magnetic currents is an important concept in this method and constitutes the difference between this method and the "TLM" [7]. Both methods are fundamentally based on the d'Alembert's general solution for one-dimensional wave equation, but the use of both voltage and current variables in this method enables us to express the characteristic equation of the medium by lumped circuit element instead of the artificial stub in "TLM". This formulation is extended to more complex characteristics such as dispersive, resonance, and anisotropic media [8].

III. Numerical Results and Discussion

The transient analysis for the stripline with a corner has been performed by the method described in the preceding section. In Fig. 5, the model of the stripline with the corner is shown. In this figure, Δd is the interval between adjacent nodes in the equivalent circuit. In order to describe this model by the "Nodal Equation", three different conditions are introduced, namely, the boundary condition at the strip conductor, the boundary condition at the free boundary, which is supposed to be the surface of the analyzed region in air and dielectric medium, and the condition of the dielectric. Firstly, the boundary condition of the conductor is described. The conductor is supposed to have infinite conductivity, so the tangential component of electric fields on the surface of the conductor should be zero. This

Fig. 8. Equivalent circuit of the input condition, where R_s is the characteristic impedance of the stripline, and E_s is the voltage source, in this analysis, of the sinusoidal wave expression as $E_s = E_o \sin(2\pi/T)n\Delta t$, ($E_o$: Amplitude, T: Period of the sinusoidal wave, n: Number of iteration).

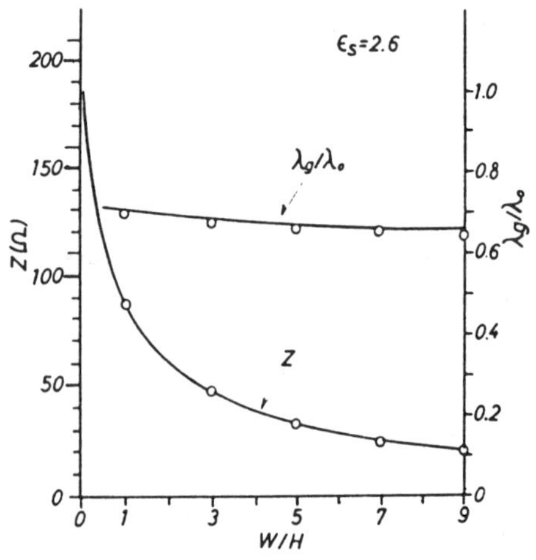

Fig. 9. Characteristic impedance and wavelength as a function W/H of a straight stripline. ∘ --- result computed by our method, ——--- analytical results by E. Yamashita and R. Mittra.

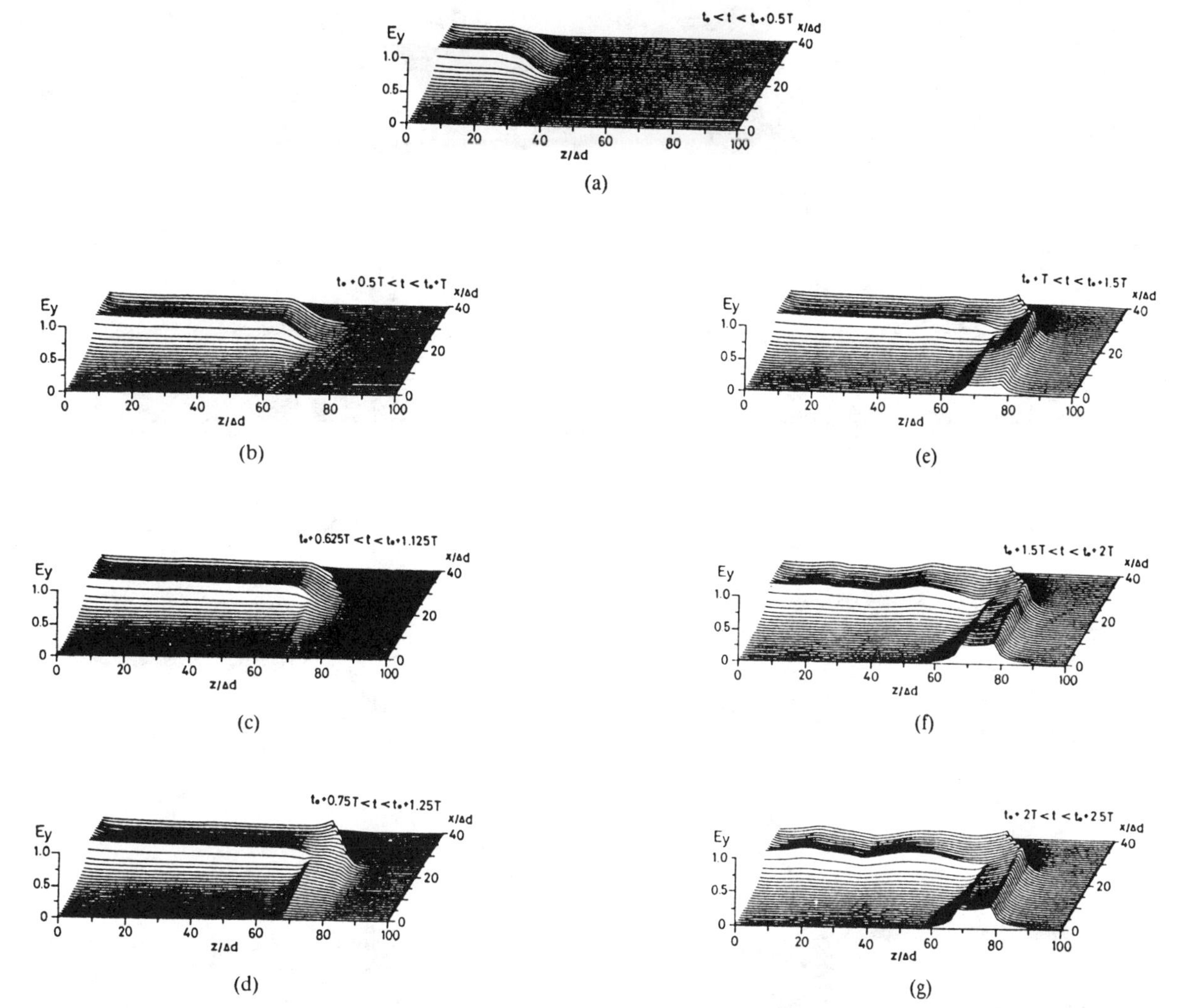

Fig. 10. Time variation of the distribution of the electrical field E_y on the observed xz-plane shown in Fig. 5 and in the case of the corner pattern ①, where t_o is the initial time at which the input wave is applied to the input plane. T is the period of the applied sinusoidal wave.

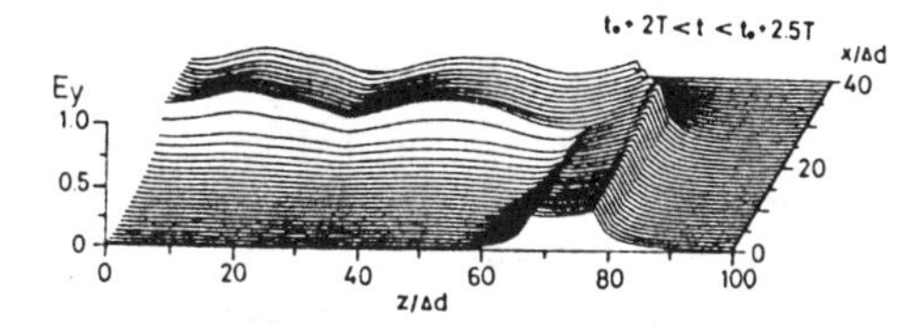

Fig. 11. Time variation of the distribution of the electrical field E_y in the case of the corner pattern ② in Fig. 5.

condition is realized by short-circuiting the appropriate electric node, in which the tangential electric field at the surface is the voltage variable, and by open-circuiting the appropriate magnetic node, in which the tangential component of the electric field is the current variable, that is, it is considered as a magnetic current. These situations are shown in Fig. 6. In the equivalent circuit, the plane in which the strip conductor is positioned is arbitrarily defined. In this analysis, the plane is situated at the plane containing D_n, F_n, and E_n nodes. Thus, the D_n and E_n nodes are short-circuited, because the electric fields E_x and E_z are to be zero at both surfaces of the conductor, and F_n nodes are neglected because all the field components E_x, E_z, and H_y are equal to zero on the surface. Next, the free boundary condition is expressed as a nonreflective termination, at which the load resistance, equal to the characteristic impedance of the free space, approximates the matching condition. The equivalent circuit of this condition is shown in Fig. 7, in which (a) shows the upper plane of the analyzed region and (b) shows the side plane. Finally, the characteristics of the dielectric are expressed in terms of the equivalent parallel circuit composed of the capacitance and the conductance at the electric node in the dielectric medium, as shown in Fig. 4. The physical meaning of the lumped element is shown in Table I and the formulation of this elements is expressed in (5) and (6). At the node situated on the dielectric–air interface, the value of the capacitance is assigned to be one half of that in the inner node. In this analysis, losses in the dielectric medium have been neglected.

Using this model of the stripline, the transient analysis of the stripline with a corner has been performed. The input condition is assumed as follows in the equivalent circuit: A sinusoidal voltage wave is applied through the source resistance at the A_n nodes under the stripline on the

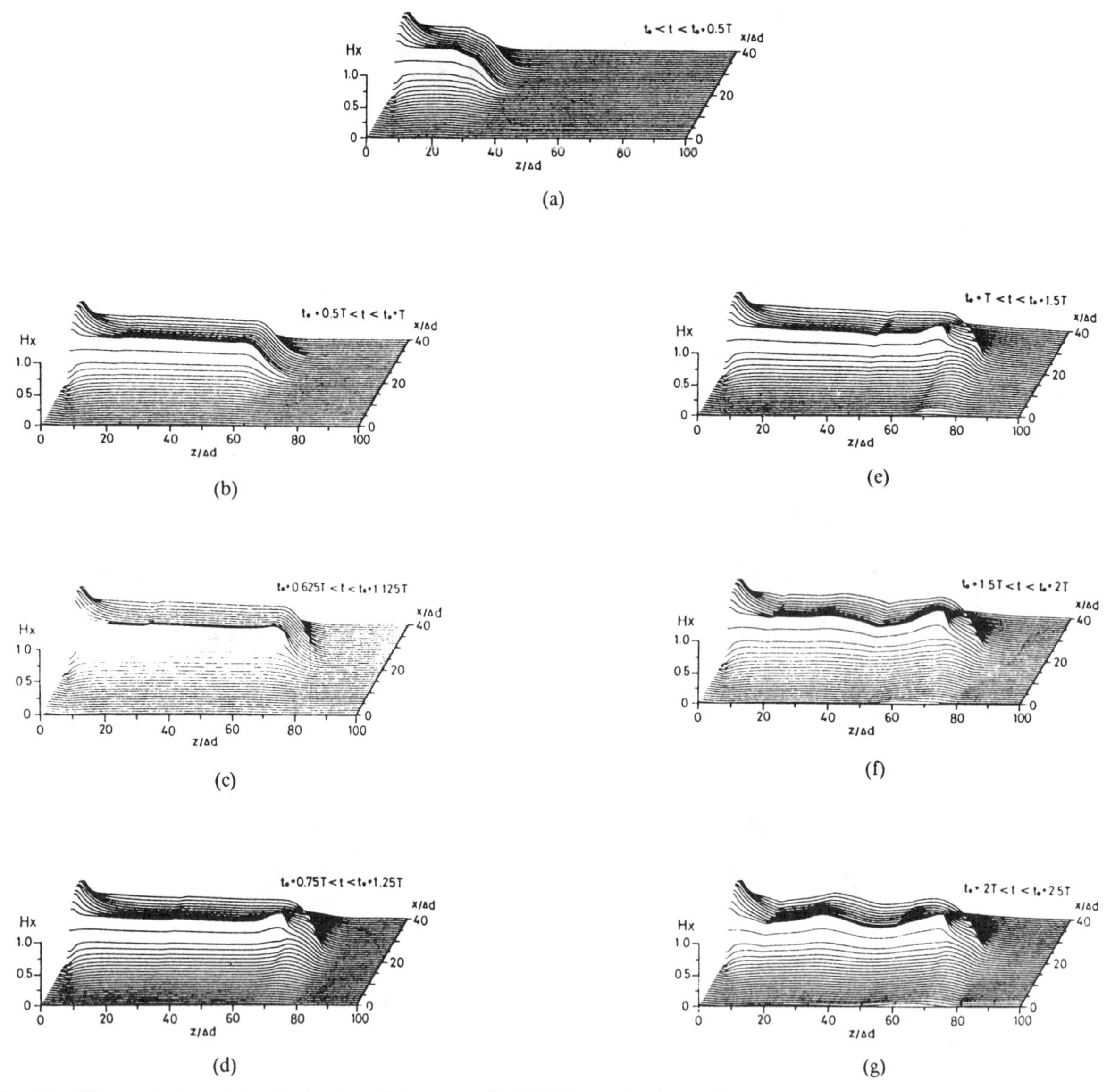

Fig. 12. Time variation of the distribution of the magnetic field H_x on the observed xz-plane in Fig. 5 for the same case as Fig. 10.

input plane. The source impedance is equal to the characteristic impedance of the dominant mode in the dielectric medium. This input condition approximates the excitation of the TEM-wave. The other A_n nodes and D_n nodes in the input plane are terminated by a matching impedance. The configuration of the input plane is shown in Fig. 8. The shape of the input wave is expressed as a pulse train with a spacing Δt in the time domain. In this numerical evaluation, the interval Δd between adjacent nodes in the equivalent is circuit is chosen, for example, to be 0.005 cm. Then, the time interval Δt becomes 8.333×10^{-5} ns. The period of the applied sinusoidal wave is $213\Delta t$ in this analysis, so its frequency is about 56 GHz. These values of Δd and Δt are sufficiently small so that the resolution of the spatial and time function is satisfactory. In the numerical computation, all parameters in space and time are normalized to Δd and Δt, respectively.

In Fig. 9, the numerical results for the characteristic impedance and wavelength are plotted as a function of W/H, and compared with analytically obtained curves. This figure shows that the numerical results are in close agreement with the analytical ones. In the following figures, the spatial distribution of the field at each time is obtained by taking the maximum values in the half period of the applied wave because of computations on the time axis. The xz-plane on which we observe the field is that of Δd beneath the upper strip conductor as shown in Fig. 5 as "observed plane". The initial point of the time axis is assumed to be the point at which the incident wave is applied at the input plane.

Fig. 10 shows the time variations of the electrical field E_y in the case of the corner cut pattern ① given in Fig. 5. It is observed that the propagating wave curves the corner smoothly and the VSWR at the incident side is small. This result clearly shows that the cutting pattern of the corner is suitable. However, Fig. 11 shows a comparatively large VSWR for the other cutting pattern ② shown in Fig. 5. These results show that the cutting pattern of the corner influences the propagation characteristics considerably. Figs. 12 and 13 present the time variations of the magnetic

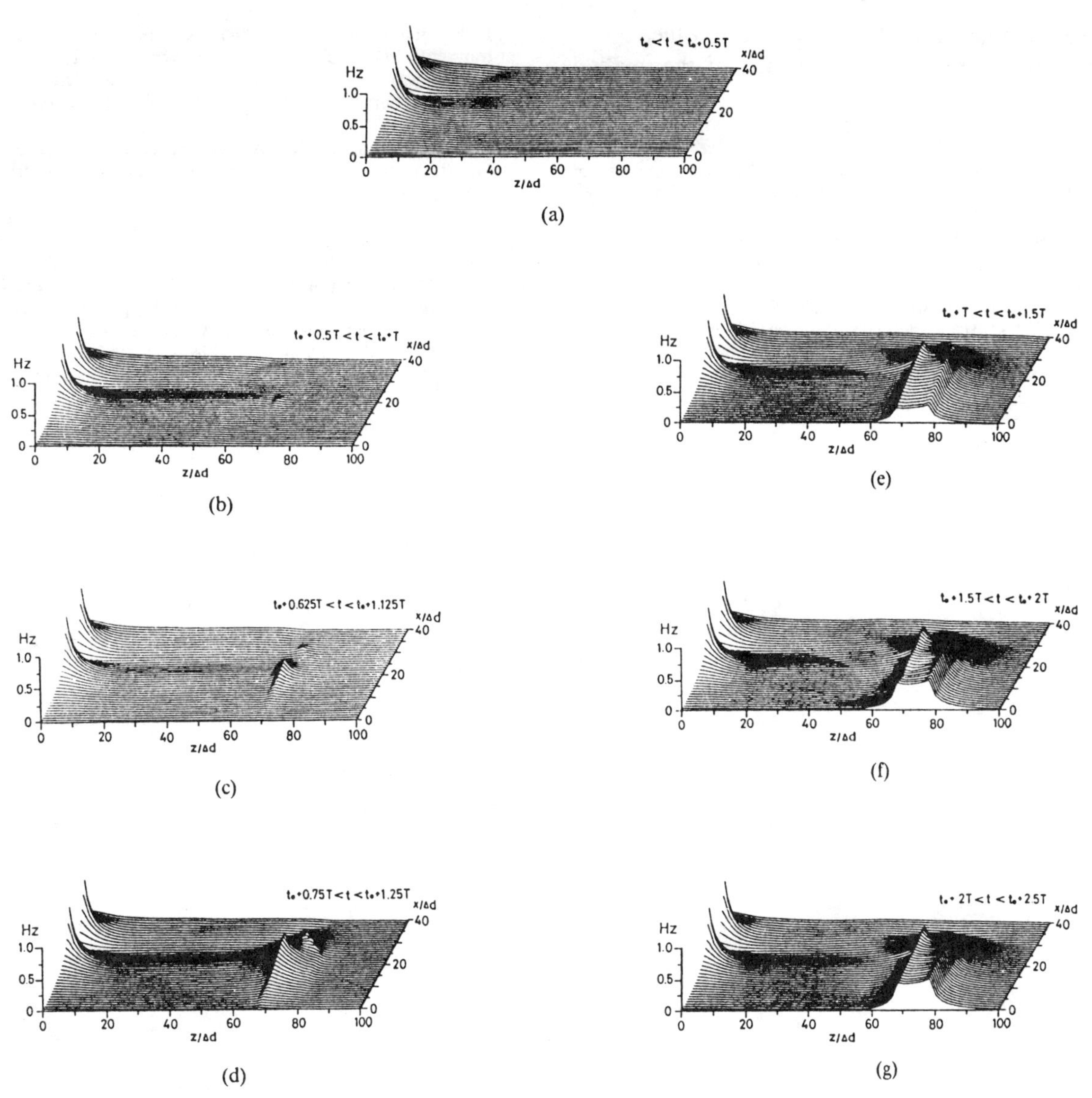

Fig. 13. Time variation of the distribution of the magnetic field H_z in the same case as Fig. 10.

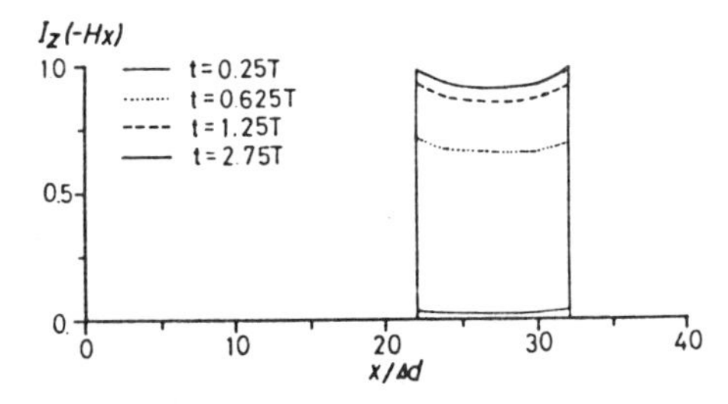

Fig. 14. Time variation of the magnetic field H_x at the cross section ⓐ in Fig. 5. This magnetic component corresponds to the current in the z-direction.

field H_x and H_z, respectively, for the case of Fig. 10. Each figure clearly shows that the conversion of the magnetic component H_x to the component H_z occurs at the corner. These magnetic components are the dominant terms of the Poynting vector in each longitudinal section of the stripline. Both figures show that near the input port, both magnetic components H_x and H_z are generated by diffraction of the input wave because of the plane-wave approximation of the input conditions. Fig. 14 shows the time variation of the magnetic field H_x at the cross section ⓐ shown in Fig. 5, where the steady-state distribution is established and the edge effect is evient.

The size of the program used is about 4.5 MB, and the computed time for the transient analysis from $t = 0$ to $t = 2.5\mathrm{T}$ is about 80 s.

IV. Conclusion

The present study verifies that our method is appropriate for the time-domain analysis of the stripline in three-dimensional space. The obtained results demonstrate the propagation of the wave through a 90° bend and show how the direction of the magnetic field is changed by the corner contour. We are now studying in more detail the time variations of fields in striplines as a function of many other parameters and examine the relation between the propagation characteristics for transient distributions and stationary continuous waves and the pulsed wave. These results will be reported in later papers.

The present method can also be applied to other three-dimensional problems by using all merits of the method [9]–[11].

REFERENCES

[1] N. Yoshida, I. Fukai, and J. Fukuoka, "Transient analysis of two-dimensional Maxwell's equations by Bergeron's method," *Trans. IECE Japan*, vol. J62-B, pp. 511–518, Jun. 1979.

[2] ____, "Transient analysis of three-dimensional electromagnetic fields by nodal equations," *Trans. IECE Japan*, vol. J63-B, pp. 876–883, Sept. 1980.

[3] G. Metzger and J-P. Vabre, *Transmission Lines with Pulse Excitation*. New York: Academic Press, 1969, pp. 65–96.

[4] S. Akhtarzad and P. B. Johns, "Solution of Maxwell's equations in three space dimensions and time by the t.l.m. method of numerical analysis," *Proc. IEE*, vol. 122, pp. 1344–1348, Dec. 1975.

[5] H. W. Dommel and W. S. Meyer, "Computation of electromagnetic transients," *Proc. IEEE*, vol. 62, pp. 983–993, July 1974.

[6] S. Akhtarzad and P. B. Johns, "Generalized elements for TLM method of numerical analysis," *Proc. IEE*, vol. 122, pp. 1349–1352, Dec. 1975.

[7] P. B. Johns and M. O'Brien, "Use of the transmission-line modelling (t.l.m.) method to solve non-linear lumped networks," *Radio Electron. Eng.*, vol. 50, pp. 59–70, Jan./Feb. 1980.

[8] N. Yoshida, I. Fukai, and J. Fukuoka, "Application of Bergeron's method to anisotropic media," *Trans. IECE Japan*, vol. J64-B, pp. 1242–1249, Nov. 1981.

[9] ____, "Adaptation of Bergeron's method to complicated boundary problems," *Trans. IECE Japan*, vol E64, pp. 455–462, July 1981.

[10] ____, "Transient analysis of waveguide having *H*-corner," *Trans. IECE Japan*, vol. E65, pp. 125–126, Feb. 1982.

[11] ____, "Transient analysis of waveguide having tuning window," *Trans. IECE Japan*, vol. E66, pp. 161–162, Feb. 1983.

Part 4
Finite-Element Method

THE finite-element method (FEM) is probably the most popular strictly numerical method that can be applied to the solution of an extremely wide variety of analytical problems concerning not only electromagnetics but also the general domain of engineering and physics. The number of publications dealing with this technique and its applications is enormous. The method itself can be approached and formulated according to many different viewpoints and in connection with other methods in various manners related to it.

The FEM is related to other general methods such as the method of weighted residuals, or the method of moments. There are different ways to categorize the various numerical methods according to some general scheme, but we will not attempt to do so here. For a general treatment of the FEM the reader is referred to many excellent books such as, for instance, [4.1]. Here we limit ourselves to some applications and specializations of this method to microwave structures.

It should be emphasized that even within the more limited domain of the microwave circuit applications this book is concerned with, a reasonably comprehensive collection of papers on the FEM would exceed the space limitations of a manageable volume, so that, unavoidably, many more significant papers have been left out here than could be included.

The FEM is implemented on an integral formulation of the boundary value problem. In the conventional and more usual applications, this is a variational expression in the form of a functional to be minimized. The region of interest is subdivided into surface or volume elements, depending on whether a two- or three-dimensional structure is being examined. The unknown function, which may be a scalar potential or a vectorial field component, is approximated within each element by a polynomial function. Application of the Rayleigh-Ritz procedure transforms the functional minimization into a linear system of equations. It could easily be shown that such a procedure is equivalent to Galerkin's form of the weighted residual, or moment, method.

The FEM has the advantage over the FDM of using triangular elements which can fit curved boundaries with less difficulty. In addition, some boundary conditions, e.g. the Neumann condition in the solution of Helmholtz's equation, are included in the functional itself, so they do not need to be imposed.

One of the first applications of the FEM to guided electromagnetic (EM) wave propagation by solving scalar Helmholtz equations is reported in Paper 4.1 due to Zienkiewicz (one of the most distinguished contributors to the development of the FEM) and collaborators. In this paper the basic formulation of the FEM is illustrated in conjunction with linear (first order) polynomial approximation within each element. The use of higher order polynomials was introduced by Silvester to improve the accuracy and efficiency of the method (Paper 4.2). The analytical derivations were detailed in a parallel paper (Paper 4.3).

A variational expression for dielectric loaded waveguides using the longitudinal components of the electric and magnetic fields was then developed by Csendes and Silvester [4.2]. In this paper the existence of spurious (nonphysical) solutions was recognized, one of the problems of the FEM in the analysis of inhomogeneous structures. At the same time, a similar approach was applied by Daly to perform a hybrid mode numerical analysis of the quasi-TEM mode of the closed microstrip [4.3], but discontinuities in the computed eigenvalues were observed. Interest in microwave planar-circuit analysis began to develop at the beginning of the seventies. Silvester applied a finite-element (FE) technique to the analysis of two-dimensional problems [4.4]. This paper is related to the development of the planar-circuit approach, discussed here in Part 11.

A unified variational formulation in terms of either **H** or **E** valid for lossless anisotropic media was presented by Konrad in 1976 [4.5]. This vector formulation was shown not to produce singularities unlike the formulation in terms of the longitudinal components and allowed the occurrence of spurious solutions to be predicted.

Three-dimensional problems involving inhomogeneous media, including scattering by dielectric objects in a waveguide, have been attacked using a three-component vector variational formulation of the FEM by Webb *et al.* (Paper 4.4).

In its generalized formulation the FEM can be applied without resorting to a variational principle, by simply applying the weighted residual concept (see Part 7 of this book). One such example is given in Paper 4.5 by Tzuang and Itoh, where FEM formulation using longitudinal E_z-H_z components is used to analyze monolithic microwave integrated circuit (MMIC) slow-wave transmission lines.

One difficulty with the FEM is in the analysis of open structures, like dielectric waveguides. The most elementary, but clearly rather inefficient, way to deal with open boundaries is just truncation, assuming far enough artificial boundaries. The use of infinite elements, i.e., elements extending to the infinite where proper shape functions are defined, has been proposed by Rahman and Davies [4.6]. Infinite elements and singular elements have been adopted by Pantic and Mittra [4.7] to enhance the accuracy and efficiency of the numerical analysis of open structures having edge field singularities, such as the microstrip line. Such an analysis, however, is limited to the quasi-TEM case. The boundary integral equation technique, treated in Part 5, provides a general approach to circumvent such a difficulty.

Another recurrent difficulty, as already mentioned, is the occurrence of spurious solutions in the analysis of inhomogeneous waveguides by the FEM. Recently, novel formulations have been proposed which apparently eliminate spurious modes. In a paper by Hayata *et al.* [4.8] spurious modes are eliminated by including in the FE technique the condition of zero divergence of the magnetic field. In the new variational formulation proposed by Angkaew *et al.* (Paper 4.6) spurious-mode solutions are not real, while physical solutions are, so they can easily be discriminated.

It is finally worth considering that, as in the case of finite difference, the FEM can be associated with a time-domain computation. It is believed that finite-element time-domain methods will be the subject of a number of investigations in the future. Though not strictly related to the topic of microwave passive components, a recent tutorial paper on this topic by Cangellaris *et al.* [4.9] may be quite usefully consulted by the microwave engineer.

References

[4.1] O. C. Zienkiewicz, *The Finite Element Method*, 3rd ed. London: McGraw-Hill, 1977.

[4.2] Z. J. Csendes and P. Silvester, "Numerical solution of dielectric loaded waveguides: I—Finite-element analysis," *IEEE Trans. Microwave Theory Tech.*, vol. MTT-18, pp. 1124–1131, Dec. 1970.

[4.3] P. Daly, "Hybrid-mode analysis of microstrip by finite-element methods," *IEEE Trans. Microwave Theory Tech.*, vol. MTT-19, pp. 19–25, Jan. 1971.

[4.4] P. Silvester, "Finite element analysis of planar microwave networks," *IEEE Trans. Microwave Theory Tech.*, vol. MTT 21, pp. 104–108, Feb. 1973.

[4.5] A. Konrad, "Vector variational formulation of electromagnetic fields in anisotropic media," *IEEE Trans. Microwave Theory Tech.*, vol. MTT-24, pp. 553–559, Sept. 1976.

[4.6] M. M. A. Rahman and J. B. Davies, "Finite element analysis of optical and microwave waveguide problems," *IEEE Trans. Microwave Theory Tech.*, vol. MTT-32, pp. 20–28, Jan. 1984.

[4.7] Z. Pantic and R. Mittra, "Quasi-TEM analysis of microwave transmission lines by the finite-element method," *IEEE Trans. Microwave Theory Tech.*, vol. MTT-34, pp. 1096–1103, Nov. 86.

[4.8] K. Hayata, M. Koshiba, M. Eguchi, and M. Suzuki, "Vectorial finite-element method without any spurious solutions for dielectric waveguiding problems using transverse magnetic-field component," *IEEE Trans. Microwave Theory Tech.*, vol. MTT-34, pp. 1120–1124, Nov. 1986.

[4.9] A. C. Cangellaris, C.-C. Lin, and K. K. Mei, "Point-matched time domain finite element methods for electromagnetic radiation and scattering," *IEEE Trans. Antennas Propagat.*, vol. AP-35, pp. 1160–1173, Oct. 1987.

Application of finite elements to the solution of Helmholtz's equation

P. L. Arlett, B.Sc.(Eng.), Mem.I.E.E.E., C.Eng., M.I.E.E., A. K. Bahrani, M.Sc., Ph.D., and O. C. Zienkiewicz, B.Sc., Ph.D., D.Sc., Dipl.Eng., C.Eng., F.I.C.E.

Synopsis

A novel method, that of finite elements, for the solution of Helmholtz's equation is suggested. Various 2- and 3-dimensional problems are solved using this method, and the results are compared with more conventional techniques, particularly the finite-difference method, which it may be regarded to supersede. The ease with which various boundary conditions may be handled is discussed and illustrated. Nonhomogeneous configurations present no difficulty, nor do they require any special formulation.

There is considerable scope for the further development of the technique, which has, until now, been applied mainly to the solution of Laplace or Poisson equations.

1 Introduction

The use of Helmholtz's equation for the solution of waveguide problems is well known. Various methods exist for the solution of this equation. Exact methods, using the separation of variables, are generally only applicable to the solution of problems associated with waveguides having simple cross-sections. It is usual to use approximate methods when the cross-section of the waveguide becomes more complicated, though conformal-transformation methods may also be used to advantage.[1,2] The use of the latter entails the transformation of complicated cross-sections into simpler ones, enabling the boundary conditions to be more easily satisfied. Application of such techniques are, however, limited. The Raleigh–Ritz method may also be used to obtain an approximate solution for the cutoff wavelength of waveguides with particular cross-sections.[3] The perturbation process has also been successfully used for such cross-sections as the T septate lunar line.[4] When the contour of the cross-section of a guide forms a closed curve, point matching[5] has proved to be useful. Even impedance networks have been employed in attempts to solve Helmholtz's equation. Vine[6] explains the use of such an analogue.

Most of the methods stated above differ considerably in the basic techniques employed. In general, though, it is not possible to achieve a high degree of accuracy for a wide range of problems with any one of them. The finite-difference approximations often form a more attractive method for the solution of general problems. In its range of applications, it often compares favourably with the previous methods for the solution of Helmholtz's equation when applied to waveguides. Much has been published on such applications,[7] though there are still disadvantages which are generally accepted. Amongst these are the requirements of a regular 'grid' (or a change of the basic co-ordinates), approximations at the boundaries etc.

We would suggest that the use of finite-element approximations possesses all the advantages of the finite-difference method and a few of its disadvantages. The method has been applied to the solution of Laplace's and Poisson's equations in the electrical-engineering field, with considerable success.[8] Here we hope to show that its application to the solution of Helmholtz's equation is not only feasible, but that it may be used to advantage in the solution of general problems associated with waveguides which involve complicated boundaries.

2 Mathematical statement

The basic steady-state equation requiring solution for the general waveguide problem is

$$\left\{\frac{\partial}{\partial x}\left(\frac{1}{\epsilon_d}\frac{\partial}{\partial x}\right)+\frac{\partial}{\partial y}\left(\frac{1}{\epsilon_d}\frac{\partial}{\partial y}\right)+\frac{\partial}{\partial z}\left(\frac{1}{\epsilon_d}\frac{\partial}{\partial z}\right)+K^2\right\}\phi=0 \quad (1)$$

Paper 5659 E, first received 7th May and in revised form 17th July 1968
The authors are with the School of Engineering, University College of Swansea, Swansea, Wales

where $K^2 = \omega^2\mu_0\epsilon_0$

ϵ_d = dielectric permittivity

ϵ_0 = permittivity of free space

μ_0 = permeability of free space

This Helmholtz's equation may be used in the 2- or 3-dimensional form for the solution of waveguide problems. For H waves, the equation must satisfy the Neumann boundary condition that $\frac{\partial\phi}{\partial n} = 0$, and for E waves the Dirichlet boundary condition that $\phi = 0$. K^2 is the eigenvalue which represents the cutoff frequency.

If use is made of Euler's equations associated with the calculus of variations, it may be shown that eqn. 1 may be restated identically as the determination of function ϕ, which minimises the functional X defined by[9]

$$X=\int\frac{1}{2}\frac{1}{\epsilon_d}\left\{\left(\frac{\partial\phi}{\partial x}\right)^2+\left(\frac{\partial\phi}{\partial y}\right)^2+\left(\frac{\partial\phi}{\partial z}\right)^2\right\}d(\text{vol}) - \int\frac{1}{2}K^2\phi^2\,d(\text{vol}) \quad (2)$$

for 3-dimensional analysis, or

$$=\int\frac{1}{2}\frac{1}{\epsilon_d}\left\{\left(\frac{\partial\phi}{\partial x}\right)^2+\left(\frac{\partial\phi}{\partial y}\right)^2\right\}d(\text{area}) - \int\frac{1}{2}K^2\phi^2\,d(\text{area}) \quad (3)$$

for 2-dimensional analysis.

Now only the Dirichlet boundary condition has to be satisfied, as the Neumann condition is included in the functional.

Each of these equations may be written in the following form, for convenience of solution:

$$X = X_f + X_s \quad (4)$$

The solution is outlined using the 3-dimensional analysis, though examples are given later of both forms of solution.

3 The finite-element approximation

In the finite-element process, the region under analysis is divided into a number of subregions, or elements. For a

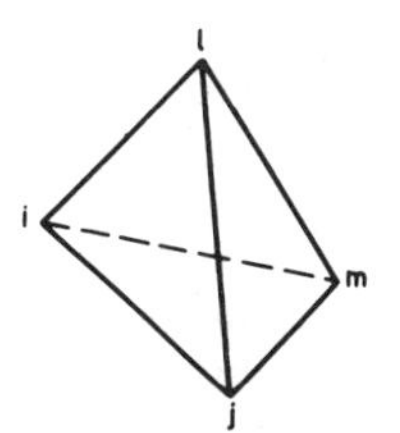

Fig. 1
Tetrahedron lijm as an element

typical 3-dimensional situation, a normal volume element is associated with a number of nodes defining its shape (as in Fig. 1). It is at these nodes that the value of ϕ is to be determined. The value ϕ at each of these nodes under consideration in the whole region must be defined as a vector:

$$\{\phi\} = \begin{bmatrix} \phi_1 \\ \phi_2 \\ \cdot \\ \cdot \\ \phi_n \end{bmatrix} \quad . \quad . \quad . \quad (5)$$

It is stipulated that, within each element, the variation of ϕ is prescribed by the values of ϕ associated with the nodes of the element; thus for a 4-noded element:

$$\{\phi\}^e = \begin{bmatrix} \phi_i \\ \phi_j \\ \phi_l \\ \phi_m \end{bmatrix} \quad . \quad . \quad . \quad (6)$$

where i, j, l and m are nodes of an element and

$$\phi(x, y, z) = N\{\phi\}^e = [N_i, N_j, N_l, N_m]\{\phi\}^e \quad . \quad . \quad (7)$$

in which the matrix N involves suitable functions of the co-ordinates. For any set of values of $\{\phi\}$, X may be determined and minimisation accomplished approximately by writing a set of equations

$$\left.\begin{aligned} \frac{\partial X_f}{\partial \phi_i} + \frac{\partial X_s}{\partial \phi_i} &= 0 \\ \frac{\partial X_f}{\partial \phi_j} + \frac{\partial X_s}{\partial \phi_j} &= 0 \\ \frac{\partial X_f}{\partial \phi_l} + \frac{\partial X_s}{\partial \phi_l} &= 0 \\ \frac{\partial X_f}{\partial \phi_m} + \frac{\partial X_s}{\partial \phi_m} &= 0 \end{aligned}\right\} \quad . \quad . \quad . \quad (8)$$

If X^e is the contribution of an element e to the total integration,

$$X_f = \Sigma X_f^e \text{ and } X_s = \Sigma X_s^e \quad . \quad . \quad . \quad (9)$$

provided none of the terms in X becomes infinite at the interfaces. This establishes the first necessary condition which has to be satisfied by the 'shape function' N. Since only the first derivation of ϕ enters the definition of X, it is necessary that these should be finite at the interfaces; thus no discontinuity of ϕ should arise between adjacent elements.

If the definition of X_f (eqns. 3 and 4) is now examined, it will be seen that it is a quadratic function of the nodal values, and thus, for each element, there is a linear relationship:

$$\left\{\begin{matrix} \frac{\partial X_f}{\partial \phi_i} \\ \frac{\partial X_f}{\partial \phi_j} \\ \cdot \\ \cdot \end{matrix}\right\} = \frac{\partial X_f^e}{\partial \{\phi\}^e} = S^e\{\phi\}^e \quad . \quad . \quad . \quad (10)$$

In this, the matrix S will be dependent on only the shape function $N = [N_i, N_j \ldots]$ associated with the element nodes and the coefficient ϵ_d.

The matrix S^e is symmetric, and a typical term may be given as

$$S_{rs}^e = \left(\int \frac{\partial N_r}{\partial x}\frac{\partial N_s}{\partial x} d(\text{vol}) + \int \frac{\partial N_r}{\partial y}\frac{\partial N_s}{\partial y} d(\text{vol}) + \int \frac{\partial N_r}{\partial z}\frac{\partial N_s}{\partial z} d\,\text{vol} \right) \frac{1}{\epsilon_d} \quad . \quad (11)$$

The term X_s, defined by

$$X_s = -\int \tfrac{1}{2} K^2 \phi^2 d(\text{vol}) \quad . \quad . \quad . \quad (12)$$

is also minimised by differentiating with respect to the nodal values

$$\frac{\partial X_s^e}{\partial \phi_i} = -\int K^2 \phi \frac{\partial \phi}{\partial i} d(\text{vol}) \quad . \quad . \quad . \quad (13)$$

and, in a similar way, this may be redefined as

$$\frac{\partial X_s^e}{\partial \phi_i} = -\int K^2 N_i [N_i, N_j, N_m, N_e]\, d(\text{vol}) \{\phi^e\} \quad . \quad (14)$$

with equivalent equation for the other nodes of any one element. The total contribution for a 4-noded element can be stated in matrix form as

$$\left\{\begin{matrix} \frac{\partial X_s}{\partial \phi_i} \\ \frac{\partial X_s}{\partial \phi_i} \\ \frac{\partial X_s}{\partial \phi_m} \\ \frac{\partial X_s}{\partial \phi_e} \end{matrix}\right\}^e = -\int K^2 \begin{bmatrix} N_i^2 & N_iN_j & N_iN_m & N_iN_e \\ N_jN_i & N_j^2 & N_jN_m & N_jN_e \\ N_mN_i & N_mN_j & N_m^2 & N_mN_e \\ N_eN_i & N_eN_j & N_eN_m & N_e^2 \end{bmatrix} d(\text{vol}) \left\{\begin{matrix} \phi_i \\ \phi_j \\ \phi_m \\ \phi_e \end{matrix}\right\}^e \quad . \quad . \quad . \quad (15)$$

or

$$\frac{\partial X_s^e}{\partial \{\phi^e\}} = -K^2 M^e \{\phi\}^e \quad . \quad . \quad . \quad (16)$$

The two portions of X, X_f and X_s may now be combined to give

$$\frac{\partial X^e}{\partial \{\phi^e\}} = \frac{\partial X_f^e}{\partial \{\phi^e\}} + \frac{\partial X_s^e}{\partial \{\phi^e\}} \quad . \quad . \quad . \quad (17)$$

If summation is now taken over all elements,

$$\frac{\partial X}{\partial \phi} = \Sigma \frac{\partial X^e}{\partial \{\phi^e\}} = 0 \quad . \quad . \quad . \quad (18)$$

Thus, for the complete problem, from eqns. 8, 10 and 16,

$$S\{\phi\} = K^2 M\{\phi\} \quad . \quad . \quad . \quad (19)$$

The solution of this equation gives the eigenvalues K^2, which represent the cutoff frequencies.

The relationships are valid for 2- or 3-dimensional spaces and, by suitable extension, for any number of associated nodes.

4 Basic elements

4.1 The tetrahedral element

The volume element shown in Fig. 1 represents the simplest division of a volumetric region (in 2-dimensional analysis, this basic element would be a triangular element). The shape function, previously referred to, may be that giving ϕ in terms of a linear polynomial with four constants:

$$\phi = \alpha_1 + \alpha_2 x + \alpha_3 y + \alpha_4 z \quad . \quad . \quad . \quad (20)$$

Since there are four nodes (i, j, l and m) involved for the simplest volumetric element, α may be found in terms of the nodal values; thus:

$$\left.\begin{aligned} \phi_i &= \alpha_1 + \alpha_2 x_i + \alpha_3 y_i + \alpha_4 z_i \\ \phi_j &= \alpha_1 + \alpha_2 x_j + \alpha_3 y_j + \alpha_4 z_j \\ \phi_l &= \alpha_1 + \alpha_2 x_l + \alpha_3 y_l + \alpha_4 z_l \\ \phi_m &= \alpha_1 + \alpha_2 x_m + \alpha_3 y_m + \alpha_4 z_m \end{aligned}\right\} \quad . \quad . \quad . \quad (21)$$

which, on solution, gives

$$\phi = [N_i, N_j, N_l, N_m]\{\phi\}^e \quad . \quad . \quad . \quad . \quad . \quad . \quad . \quad . \quad (22)$$

where $N_i = (a_i + b_i x + c_i y + d_i z)/6V$ etc.

with

$$6V = \begin{vmatrix} 1 & x_i & y_i & z_i \\ 1 & x_j & v_j & z_j \\ 1 & x_l & y_l & z_l \\ 1 & x_m & y_m & z_m \end{vmatrix}$$

where V is the volume of the tetrahedron and

$$a_i = \begin{vmatrix} x_j & y_j & z_j \\ x_l & y_l & z_l \\ x_m & y_m & z_m \end{vmatrix}$$

$$b_i = - \begin{vmatrix} 1 & y_j & z_j \\ 1 & y_l & z_l \\ 1 & y_m & z_m \end{vmatrix}$$

$$c_i = \begin{vmatrix} 1 & x_j & z_j \\ 1 & x_l & z_l \\ 1 & x_m & z_m \end{vmatrix}$$

$$d_i = - \begin{vmatrix} 1 & x_j & y_i \\ 1 & x_l & y_l \\ 1 & x_m & y_m \end{vmatrix}$$

etc.

This chosen-shape function will satisfy the criteria necessary for convergence. Its linear form indicates that ϕ along any common face between adjacent tetradedra is uniquely defined, and that a constant slope is attainable.

A typical term of the matrix S now becomes

$$S_{rs} = (b_r b_s + c_r c_s + d_r d_s)/36V \, . \, \epsilon_d$$

where again V is the volume of the tetrahedron being dealt with.

4.2 8-noded element assembled from a number of tetrahedrons

Fig. 2 indicates a possible division of an 8-noded element into five tetrahedrons. This former is a much more

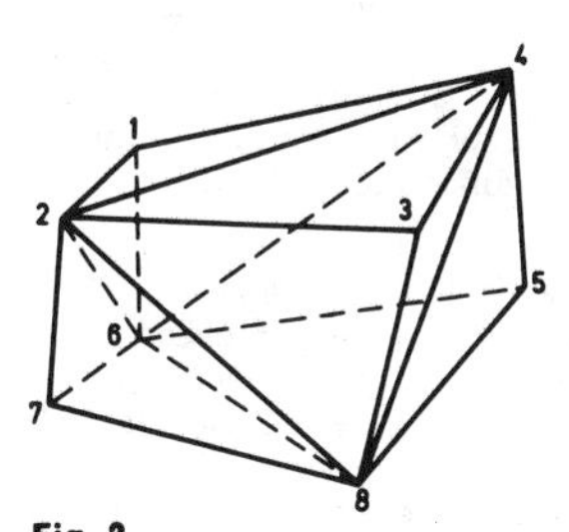

Fig. 2
8-corner element built up from 5 tetrahedra

convenient 'building block' to handle than the less easily visualised tetrahedrons. The data preparation is eased considerably, and no nodes are neglected. Use is made of the computer to divide the 8-noded element into five tetrahedrons, on which the actual calculations are performed.

No limitation is placed on the relative size of the elements used (in any unsymmetrical 8-noded element, the size of each tetrahedron would differ). Since the only criterion is that all nodes must be common to adjacent elements, it is possible to vary the size and shape of the elements, as shown in Fig. 3. The optimum condition for a given number of available nodes is approached when most nodes are in a region of rapidly varying ϕ with fewer nodes in regions of lesser interest. This is similar to having a variable-sized grid when using the finite difference method, without the need to change the basic form of solution formulas.

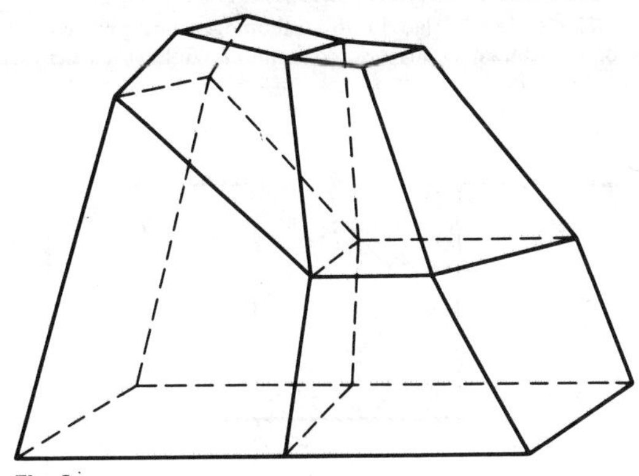

Fig. 3
Combination of '8-cornered elements'

5 Examples

5.1 Simple rectangular waveguide

This example is given because the results of the finite-difference method and the exact results are available. The number of nodes used is the same for both the calculated methods, and the example merely indicates that, for such conditions, similar results are obtained. However, by judicious use of variable elements, the accuracy of the finite-element method may be increased at no extra cost in computer time and storage, or in data preparation. Use is made of symmetry, and the boundary conditions thus imposed are as in Fig. 4. Table 1 gives a comparison of the results of the finite-difference method,[10] the exact method [calculated from the equation $(KA)^2 = \pi$; $\phi = \sin \pi X/A$)] and the finite-element method.

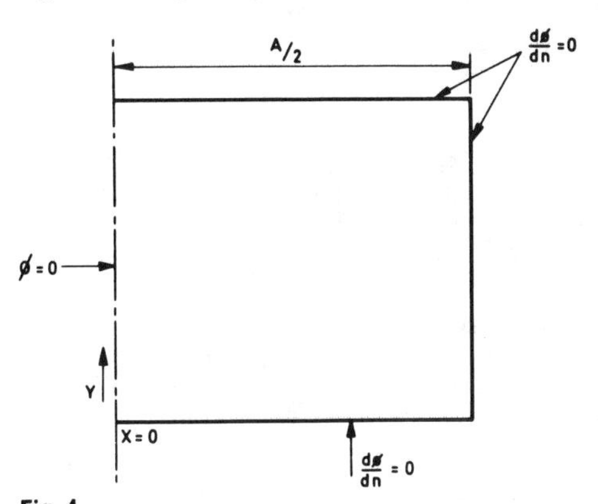

Fig. 4
Rectangular-waveguide half cross-section

Table 1

RECTANGULAR WAVEGUIDE H MODE

	Finite difference	Finite element	Exact	X/A
$(KA)^2$	9·79	9·9	9·87	
Field vector $[\phi]$	1·00000	1·00000	1·00000	0·5
	0·95105	0·95105	0·95105	0·4
	0·80902	0·80902	0·809017	0·3
	0·58778	0·58778	0·58778	0·2
	0·30902	0·30902	0·30902	0·1
	0·0000	0·00000	0·0000	0

The results given are for the fundamental mode of the H wave. Similar results are obtained for the higher modes.

5.2 Dielectric-slab loaded waveguide

The dielectric slab in this inhomogeneous problem may be distributed in any desired fashion. We have chosen the configuration shown in Fig. 5, because results of such a problem obtained by Vartiarian[11] are available for comparison. The ratio of width of the slab to that of the waveguide is 1/3 in this case. Various values of permittivity are used, and Table 2 gives the cutoff wavelengths for each case.

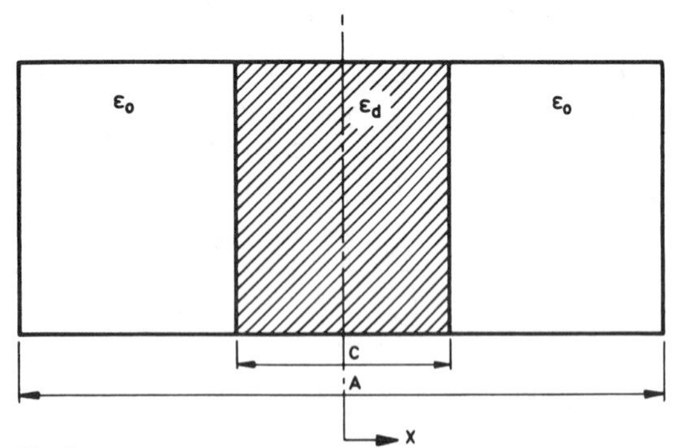

Fig. 5
Cross-section of dielectric-loaded waveguide

Table 2
RESULTS FOR THE CUTOFF WAVELENGTH OF FIG. 5 WHEN $\frac{C}{A} = 0{\cdot}333$ (TE_{10})

	Finite element	Reference (10)
ε_d	$\lambda C/A$	$\lambda C/A$
1	2·04	2·0
2	2·56	2·54
4	3·43	3·4
6	4·08	4·08
9	5·0	4·94
12	5·7	5·7

5.3 Eccentric waveguides

The results of such an example using the point-matching method are available,[5] and are compared with those obtained using the finite-element method, in Table 3. A rather coarse grid was used, and the results could have been improved upon, using more nodes.

Table 3
ECCENTRIC WAVEGUIDE H MODE

Method	Ka	h/d	L/a	b/a
Finite-element	0·68	$\frac{1}{10}$ to $\frac{1}{20}$ approx.	0·8	2
	0·7	$\frac{1}{10}$ to $\frac{1}{20}$ approx.	0·56	2
Point-matching	0·66		0·8	2
	0·68		0·56	2

a = radius of inner circle
b = radius of outer circle
$d = 2b$
L = distance between the two centres

5.4 Waveguide with more complicated cross-sections

The finite element is most suitable for the solution of irregular cross-sections, because of the degree of freedom it allows. The lunar waveguide is chosen here as an example. A typical cross-section with the field pattern is in Fig. 6.

Table 4 gives the computed results, using the finite-element and finite-difference methods,[7] and compares these with experimental results.[2] It is worthy of note, in these and in the results given, that a similar order of accuracy may be obtained using far fewer nodes with the finite-element method than with the finite-difference method.

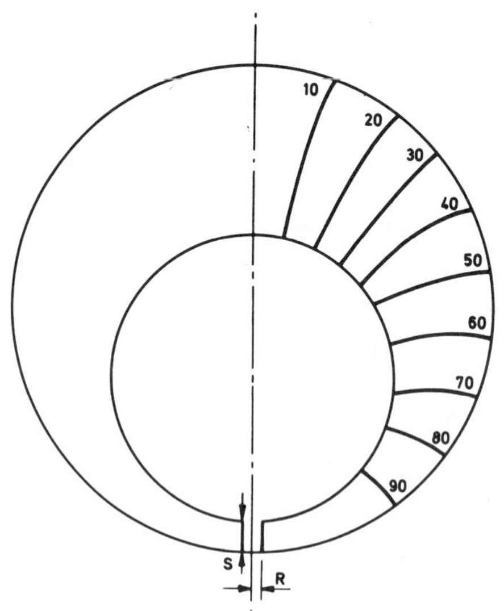

Fig. 6
Lunar waveguide, H mode
Outer diameter = d
Inner diameter = $0{\cdot}57d$
$S = 0{\cdot}055d$
$R = 0{\cdot}02d$

Table 4
LUNAR WAVEGUIDE H MODE

Method	Kd	h/d
Finite-element	1·96	$\frac{1}{10}$ to $\frac{1}{20}$ approx.
Finite-difference	2·01	$\frac{1}{50}$
Finite-difference	1·954	$\frac{1}{208}$
Experimental	1·925	

5.5 The T septate lunar waveguide

The considerably more complicated modification of the lunar waveguide may typically have a cross-section such as that shown in Fig. 7. Here the irregular subdivision and the field distribution are indicated.

Table 5 lists the finite-element results, with those obtained

Table 5
T SEPTATE LUNAR WAVEGUIDE H MODE

Method	Kd	h/d
Finite-element	1·052	$\frac{1}{10}$ to $\frac{1}{20}$ approx.
Finite difference	1·039	$\frac{1}{52}$
Finite difference	1·036	$\frac{1}{208}$
Experimental	1·042	

using a regular grid by the finite-difference method,[7] and experimental results.[7]

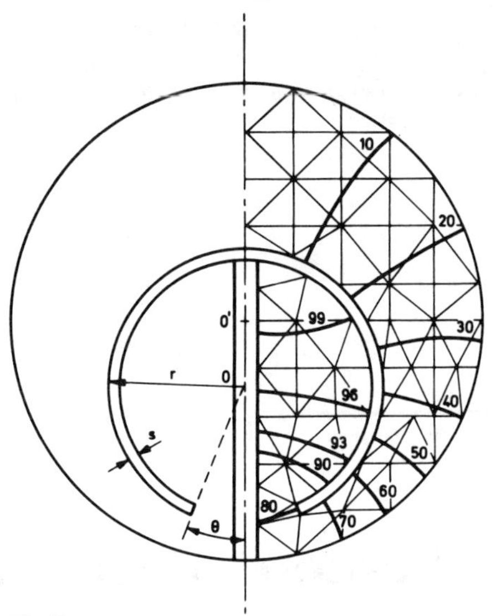

Fig. 7
T septate lunar waveguide, H mode
Outer diameter = d
$00' = 1\cdot 3d$
$r = 0\cdot 29d$
$s = 0\cdot 055d$
$\theta = 22°$

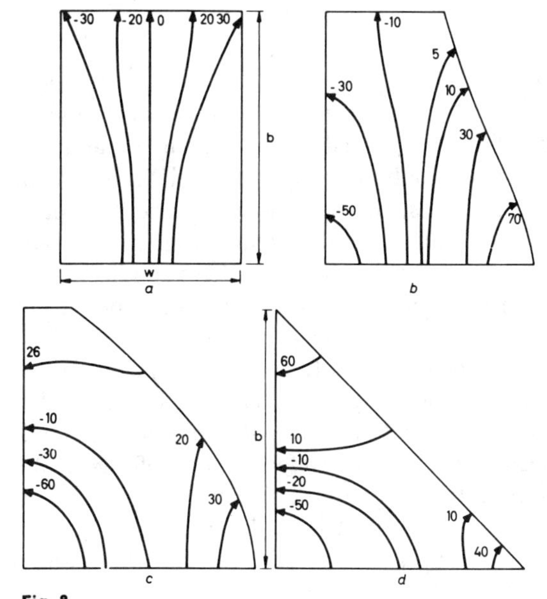

Fig. 8
Mode transducer

5.6 The mode transducer

Here the 3-dimensional solution is used. The problem was concerned with an attempt to transform an H_{10} rectangular mode to an H_{11} right-angled triangular mode. The dimensions of the waveguides at either end of the transducer were modified, so that each had a similar cutoff frequency.

The shape of the transducer used was approximately that given by Solymar,[12] and various sections are shown in Fig. 8. The resultant wave cutoff number for the 3-dimensional transducer was 9·13, while that of the rectangular section and the triangular section were 9·0 and 9·08, respectively, these latter being obtained by 2-dimensional analysis. The transducer did not provide perfect matching; hence the non-ideal field pattern at both ends. Available computer-storage limitations did not allow more than four planes through the transducer to be examined. It is in such a study as this that the benefits of the 8-noded element as the basic building block is best shown. It would have been extremely difficult to attempt to fit tetrahedrons into such a volume.

6 Conclusions

The finite-element method, by approaching the problem variationally, permits a suitably graded subdivision of the region to be obtained—thus increasing the accuracy for a given computational time, as compared with standard finite-difference approaches. Further, by including the Neumann-boundary condition in the variational function, no use of special meshes etc. on such boundaries is necessary. Non-homogeneous configuration presents again no computational difficulty or special formulation.

In this paper, only the simplest 2- and 3-dimensional elements are used. Further improvements of accuracy can be obtained by the use of more elaborate 2- and 3-dimensional elements with a larger number of associated nodes.[9]

7 References

1 CHI, M., and LAURA, P. A.: 'Approximate method of determining the cutoff frequencies of waveguides of arbitrary cross-section', *IEEE Trans.*, 1965, **MTT-13**, pp. 248–257

2 MEINKE, H. H., LANGE, K. P., and RUGER, J. F.: 'TE and TM waves in waveguides of very general cross-section', *Proc. Inst. Elect. Electron. Engrs.*, 1963, **51**, pp. 1443–1454

3 VALENZUELA, G. R.: 'The cutoff wavelength of composite waveguides', *IRE Trans.*, 1961, **MTT-0**, pp. 363–368

4 HU WANG, A. Y.: 'Dominant cutoff wavelength of a T septate lunar line', *Proc. IEE*, 1964, **111**, (7), pp. 1262–1270

5 YEE, H. Y., and AUDEH, N. F.: 'Uniform waveguide with arbitrary cross-section considered by the point matching method', *IEEE Trans.*, 1965, **MTT-13**, (6), pp. 847–852

6 VINE, J.: 'Impedance networks' *in* VITKOVITCH, D. (Ed.): 'Field analysis: experimental and computational methods' (Van Nostrand, 1966), chap. 7

7 DAVIES, J. B., and MUILWYK, C. A.: 'Numerical solution of uniform hollow waveguides with boundaries of arbitrary shape', *Proc. IEE*, 1966, **113**, (2), pp. 277–284

8 ZIENKIEWICZ, O. C., BAHRANI, A. K., and ARLETT, P. L.: 'Solution of 3-dimensional field problems by the finite element method', *Engineer*, 27th Oct. 1967, **224**, (5831), pp. 547–550

9 ZIENKIEWICZ, O. C., and CHEUNG, Y. K.: 'The finite-element method in continuous structural mechanics' (McGraw-Hill, 1967)

10 COLLINS, S. H., and DALY, P.: 'Calculations for guided electromagnetic waves using finite difference methods', *J. Electron. Control*, 1963, **14**, pp. 361–364

11 VARTANIAN, P. H., AYRES, W. P., and HELGESSON, A. L.: 'Propagation in dielectric slab loaded rectangular waveguide', *IRE Trans.*, 1958, **MTT-6**, (2), pp. 215–219

12 SOLYMAR, L., and EAGLESFIELD, C. C.: 'Design of mode transducers', *ibid.*, 1960, **MTT-8**, (1), pp. 61–68

Paper 4.2

A General High-Order Finite-Element Waveguide Analysis Program

P. SILVESTER, MEMBER, IEEE

Abstract—A very general computer program for determining sets of propagating modes and cutoff frequencies of arbitrarily shaped waveguides is described. The program uses a new method of analysis based on approximate extremization of a functional whose Euler equation is the scalar Helmholtz equation, subject to homogeneous boundary conditions. Subdividing the guide cross section into triangular regions and assuming the solution to be representable by a polynomial in each region, the variational problem is approximated by a matrix eigenvalue problem, which is solved by Householder tridiagonalization and Sturm sequences. For reasonably simple convex polygonal guide shapes, the dominant eigenfrequencies are obtained to 5–6 significant figures; for nonconvex or complicated shapes, the accuracy may fall to 3 significant figures. Use of the program is illustrated by calculating the propagating modes of a class of degenerate mode guides of current interest, for which experimental data are available. Numerical studies of convergence rate and discretization error are also described. It is believed that the new program produces waveguide analyses of higher accuracy than any general program previously available.

INTRODUCTION

IN THE DESIGN and construction of various microwave devices it is often desirable to employ waveguides of cross-sectional shapes other than rectangular, circular, or elliptical. To analyze the behavior of such guides, it is necessary to solve the eigenvalue problem of Helmholtz's equation

$$(\nabla^2 + \lambda^2)\phi = 0 \tag{1}$$

subject to homogeneous Dirichlet or Neumann boundary conditions [1]. Analytic solution by separation of variables is possible only for the three special shapes mentioned. For other shapes, the Helmholtz equation is not separable, so other solution techniques are required. In recent years, considerable effort has been devoted to finding suitable methods.

In order to know the propagation properties of a given type of guide, it is of paramount importance to be able to determine the cutoff frequency and field pattern of the dominant propagating mode. This can be accomplished by solving the Helmholtz equation problem by finite-difference techniques [2], [3]. Determination of higher modes requires substantial program modification and has only very recently been accomplished [4]. The computational effort required can, at times, become quite considerable with these iterative methods, for if degenerate modes are involved, convergence difficulties arise. Consequently, several workers have attempted to develop explicit rather than iterative methods, usually based on variational principles, to determine complete sets of waveguide modes.

For certain cross-sectional shapes, it is comparatively easy to find conformal transformations that map the given shape into a rectangle, thereby replacing the Helmholtz differential operator, subject to conditions imposed on an inconvenient boundary, by a more complicated elliptic differential operator but a very simple region. The latter problem can be solved by a Rayleigh–Ritz method, using sines and cosines as trial functions. This approach has been extensively explored [5], [6], and it is now possible to solve a number of

Manuscript received August 20, 1968; revised December 6, 1968. This work was supported in part by the National Research Council of Canada.

The author is with the Department of Engineering, McGill University, Montreal 2, P. Q., Canada.

Reprinted from *IEEE Trans. Microwave Theory Tech.*, vol. MTT-17, no. 4, pp. 204–210, Apr. 1969.

problems without recourse to laborious electrolytic tank measurements. Where the guide shape is a union of rectangles, a direct method of variation of parameters may be used [7], [8]; this method has lately gained popularity, and examples abound in the literature.

Abandoning the trigonometric functions, which are inherently bound up with rectangular regions, the family of polynomials readily suggests itself as a reasonable alternative for the Rayleigh–Ritz method. This possibility is as yet little pursued [9] and clearly more work needs to be done to develop it to the fullest. Certain difficulties are still encountered where Dirichlet boundary conditions must be satisfied.

The method used in this paper is related to the variational solution using polynomials, but it restricts the polynomial trial functions to a special class which readily permits matching any natural boundary conditions, and guarantees continuity of solution everywhere. It represents a broad generalization of the simple finite-element method described earlier [10]; in fact, the earlier method is here included as a particularly simple special case. Strictly speaking, the method is usable for any waveguide shape whose boundary is made up of straight-line segments. Although there is, in principle, no limit to the number of segments, 30–40 represents a reasonable upper bound with current computers; within this restriction, curved boundaries may be approximated by linear segments.

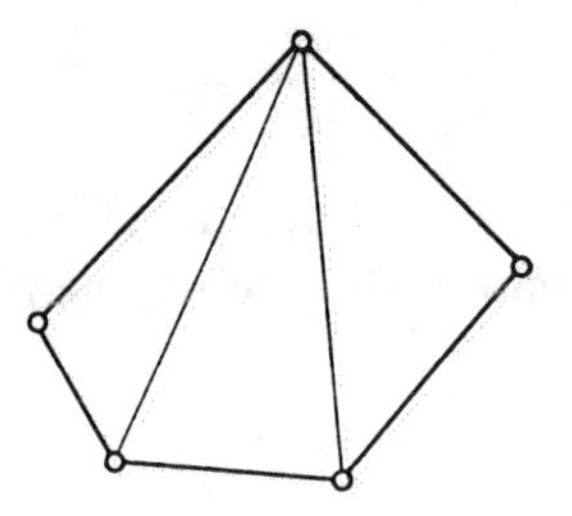

Fig. 1. Subdivision of a polygonal waveguide cross section into triangular elements.

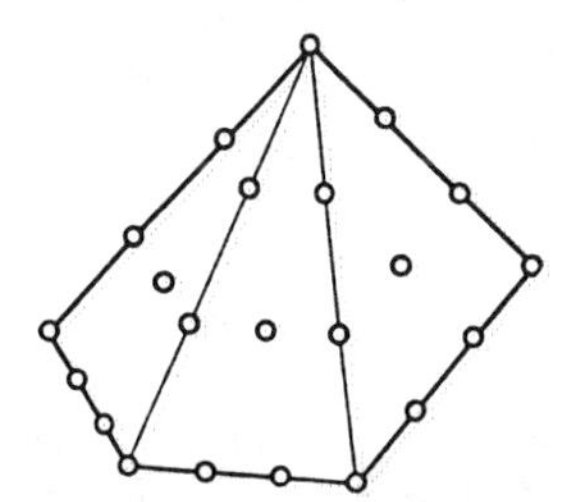

Fig. 2. Point distribution to permit cubic approximation in each triangle of Fig. 1.

High-Order Finite-Element Analysis

In the finite-element method discussed here, the guide cross section is subdivided into a set of triangular subregions, as indicated in Fig. 1. These subregions need not be any more numerous than required to describe the boundary shape; as will be seen below, it is in fact advantageous to use only a few triangles. There is no restriction on the triangle shapes, except that where extremely different side lengths (e.g., 100:1) are encountered in one triangle, large roundoff error and consequent poor accuracy will result. Within each triangular subregion, the solution is assumed to be given by a polynomial expression in the coordinates. It is possible to restrict the polynomials so that the assumed solution surface is always continuous across the subregion boundaries, and therewith continuous everywhere. An approximate solution of the waveguide eigenvalue problem is then obtained by finding the polynomial coefficients in each triangle, using a variational procedure.

In order to guarantee solution isotropy, the polynomial expression applicable to each subregion must be a complete polynomial; that is, if the highest order term in it is x^N or y^N, it must also contain all possible terms $x^m y^n$, $0 \leq m+n \leq N$, but no other terms. Geometrically, this requirement implies that the solution surface is permitted an equal complexity of curvature in any direction. The polynomial will then contain altogether $M=\frac{1}{2}(N+1)(N+2)$ terms. The parameters of the solution, which are to be determined, could be the M coefficients of these terms, or any M linearly independent combinations of coefficients.

It is convenient, and physically useful, to choose the M independent parameters to be the values of the solution function ϕ at M independently chosen points within each triangular subregion. To restrict the solution surface to be continuous, the points are selected to be the three triangle vertices, and $N-1$ other points along each triangle edge; the remaining points (if any) may be placed anywhere in the interior. All points on the triangle edges are shared by adjacent triangles, so that the polynomial expressions for ϕ will match exactly at the interfaces [11]. Fig. 2 illustrates such a distribution of points for the case $N=3$. Here every triangle must have four points along each edge, leaving one point to be placed somewhere inside. In practice, it is of course most convenient to place the points in a regularly spaced triangular arrangement, for such arrangements are easily generated by computer once the three vertices have been specified.

The value of ϕ anywhere within a triangle may be written as

$$\phi(x, y) = \sum_{n=1}^{M} \alpha_n(x, y)\phi_n \tag{2}$$

where the α_n are polynomials of degree N in the coordinates, and the ϕ_n are the values of ϕ at the M specified points of the triangle. The problem may be considered solved once the ϕ_n are known, for it is possible to define suitable polynomials α_n for any triangle [12].

The Helmholtz equation, as is well known, is the Euler equation of the functional

$$F = \frac{1}{2}\iint (|\operatorname{grad} \phi|^2 - \lambda^2\phi^2)dxdy \tag{3}$$

so that instead of solving the Helmholtz problem, one may attempt to extremize F [13]. Since ϕ is everywhere continuous, it is a straightforward matter to evaluate F which becomes an ordinary function of the numerous ϕ_n defined in all the triangles constituting the guide cross section. The functional extremization is then equivalent to requiring that

$$\frac{\partial F}{\partial \phi_n} = 0 \tag{4}$$

for all n. It is shown readily, though not without lengthy algebraic manipulation [10] that (4) may be written as the matrix equation

$$S\Phi = \lambda^2 T\Phi \tag{5}$$

where Φ is the column vector of values of ϕ_n, while S and T are square matrices whose (i, j)th elements are given by

$$S_{ij} = \int\int\left(\frac{\partial \alpha_i}{\partial x}\frac{\partial \alpha_j}{\partial x} + \frac{\partial \alpha_i}{\partial y}\frac{\partial \alpha_j}{\partial y}\right)dxdy \tag{6}$$

and

$$T_{ij} = \int\int \alpha_i\alpha_j dxdy. \tag{7}$$

S and T are obviously symmetric, guaranteeing that the eigenvalues λ^2 are real. Solution of the algebraic eigenvalue problem (5) furnishes eigenvalues and eigenvectors which form good approximations to the eigenvalues and eigenfunctions of the Helmholtz problem, i.e., the cutoff wavelengths and field distribution patterns of the various modes possible in the given guide.

Systematic Formulation of the Matrix Problem

However attractive in principle, the solution method represented by (5) through (7) is not computationally sound unless the differentiations and surface integrations can be carried out in an economic fashion. To show that this is indeed possible, consider the triangular subregions to be initially disjoint, as in Fig. 3. A polynomial expression like (2) may be written for each triangle separately, and matrices S and T formed for each individual triangle. The resulting matrices S^* and T^* for the entire guide cross section are, it is readily appreciated, block diagonal, for no connection between triangles exists. Continuity across triangle interfaces is imposed by requiring ϕ at corresponding points to be equal. In terms of the matrix representation, this amounts to saying that the vector Φ, whose components are the ϕ_n of the connected triangles, is related to the vector Φ^* for unconnected triangles by a connection matrix C:

$$\Phi^* = C\Phi. \tag{8}$$

This implies, as is well known [14], that

$$S = C'S^*C \tag{9}$$

$$T = C'T^*C. \tag{10}$$

These equations permit S and T for any union of triangles to be constructed easily from the corresponding matrices for the disjoint triangles. It suffices, therefore, to construct an algorithm for finding the matrices S_0, T_0 for one triangle. The connection matrix C, it might be added, is not ever needed explicitly; it is not difficult to encode transformations (9) and (10) so that the elements of C, which are either 0 or 1, are generated by the computer as required.

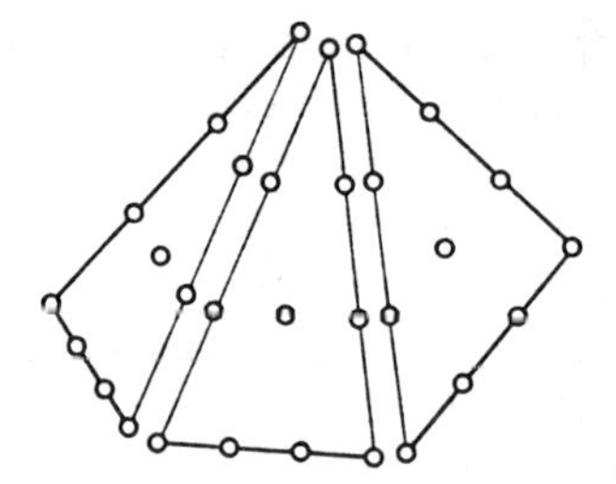

Fig. 3. Disjoint set of triangles corresponding to Fig. 2, used to construct matrices S and T.

It can be shown that for a single triangle of any shape or size, matrices S_0 and T_0 may be written in the form

$$S_0 = \sum_{i=1}^{3} Q_i \cot \theta_i \tag{11}$$

$$T_0 = AT_e \tag{12}$$

where θ_i represents the included angle at the ith triangle vertex, and A is the triangle area; Q_i and T_e are square matrices of order M whose elements are purely numeric and independent of triangle size and shape. The generation of these universally valid coefficient matrices is a lengthy matter, dealt with elsewhere [12]; for present purposes it is only necessary to acknowledge that they exist, and can be found for any N. In other words, all polynomial differentiations and surface integrations can be taken care of once and for all in computing the elementary matrices Q_i and T_e; none are required to assemble the matrix representation of any given problem. Consequently, the eigenvalue problem for any given waveguide shape may be set up by first calculating the vertex angle cotangents and areas of all the individual triangles, and multiplying by the elementary coefficient matrices so as to form the S^* and T^* matrices for the disjoint set of triangles. These in turn are assembled into two large matrices by using (9) and (10) in suitably programmed form.

A General Computer Program

A very general computer program to analyze waveguides by the method described above has been written and used to analyze various guide shapes. The program itself (not including matrices S and T) requires about 8000 words of core storage. Using an IBM 7094 computer, a typical analysis takes about 40 seconds of machine time if high accuracy is desired or the region is of very complicated shape; simpler shapes and lower accuracies are dealt with in a few seconds. Provision has been made for use of polynomials from linear to quartic. Since explicit solution methods are employed throughout, running time is entirely independent of the problem eigenvalues and depends only slightly on the number of modes actually computed.

Since the majority of problems so far analyzed have required description in terms of only a few, say five or ten, triangles, no attempt has been made to generate the subdivision into triangles by machine. Accordingly, the program reads the problem description in terms of triangles from data cards, and generates the (usually quite numerous) added points necessary to provide sufficient parameters in the polynomial representations. Matrices S and T are constructed, using elementary coefficient matrices permanently stored with the program. The eigenvalue problem is solved in fairly standard fashion [10], [15]. T, which can be shown to be positive definite always, is triangularly decomposed so as to be able to recast (5) in the standard form

$$Ax = \lambda^2 x. \tag{13}$$

This equation is solved by a Householder tridiagonalization. The eigenvalues are located by bisection, using Sturm sequences, and the eigenvectors are computed by Wielandt iteration.

In analyzing guides with purely Neumann boundaries, as is the case when TE modes are required and no line of symmetry exists, the solution $\phi=$constant, with corresponding zero eigenvalue, obtrudes here just as in other methods. This trivial solution is a consequence of formulating the problem in terms of a single Helmholtz equation, not of the computational technique. Since eigenvalues are located by the Sturm sequence technique quite independently of each other, this nuisance has been eliminated in the program by first examining the problem to determine whether or not the boundary conditions are all of Neumann type, and accordingly starting the search for eigenvalues either at the lowest or the second lowest.

The present program makes provision for S and T up to order 100, and works only with the immediate-access memory of the computer. Using magnetic tape for backup storage, matrices up to order 175 could be handled at the price of increased computing time. It has not been considered necessary to do so, however. As will be seen below, the accuracy obtained with high-order elements is sufficient to permit comparatively small matrix representations to be used.

Accuracy of Results

An extensive numerical study of the possible accuracy has been made using rectangular and isosceles right triangular guides as model problems. The selection of these shapes was dictated by the fact that analytic solutions are available. Because both shapes are convex, the accuracy of calculated cutoff wavelengths is very high, in most cases inconveniently so; unfortunately, there does not appear to be available any nonconvex shape for which a full set of eigenvalues and eigenfunctions is known. An exact set of results, however, is highly desirable, since the method reported here appears to equal or surpass in accuracy other numerical methods now in use.

Since the question of major interest is asymptotic error behaviour with increasing fineness of subdivision, the rectangular and triangular shapes were solved repeatedly, using various different numbers of triangular regions and different order polynomials. For the lowest order modes, the error showed an almost random scatter when the higher orders of polynomial were used. This scatter was traced to the effect of finite machine word length. With the 36-bit word used by the

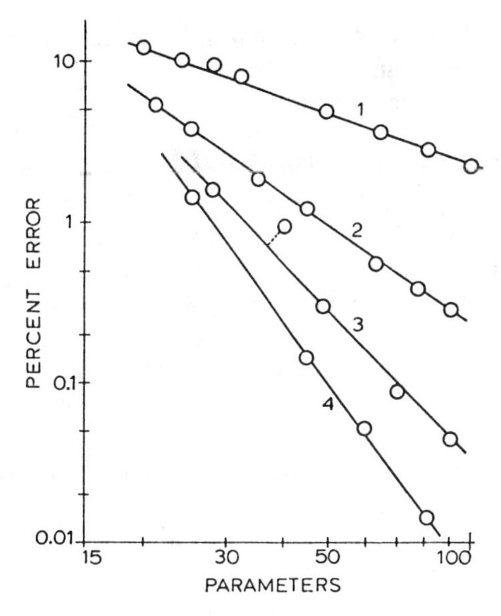

Fig. 4. Error in calculating the cutoff wavelength of the H_{31} mode of a 2.25:1 rectangular guide, as a function of parameter number (matrix size), for polynomials of degree 1 to 4.

7094 computer, about six significant decimal figures appeared to be reliable in the final results, the remaining bits being affected by roundoff in the various arithmetic operations. Consequently, useful conclusions can be drawn only where the error is larger than about 0.005–0.01 percent, but smaller than 5 to 10 percent; in this range the error is known to at least one certain significant figure, yet remains small enough to exhibit asymptotic behaviour. The observed error fell below the minimum useful level in a substantial number of calculations of low-order mode cutoff wavelengths.

Fig. 4 shows a typical set of results of the error study, the error in computed cutoff wavelength of the H_{31} mode of a rectangular waveguide with aspect ratio 2.25:1. In this case, and for a few other modes, the error falls within the useful range for nearly all triangular subdivisions and for every degree of polynomial from linear to quartic. To provide a practically useful basis of comparison, the error is here plotted against the number of independently specified parameters ϕ_n. For TE modes, this also equals the order of matrix problem to be solved. Naturally the number of triangles is dependent on the order of polynomial chosen; given the order of matrix problem, there remains the choice of using few triangles and a high-order polynomial representation, or many triangular subregions with a simpler function.

From the results exhibited in Fig. 4, as well as other results, it appears that the error is asymptotically proportional to p^{-2N}, where p is the number of parameters and N the order of polynomial. This form of variation clearly makes it desirable to use as few triangular subregions as the boundary shape will permit, with as high a degree of polynomial representation as feasible. Low-order polynomials should be reserved for use in problems with complicated boundary shapes, cubics or quartics being used to obtain high accuracy wherever practicable. Keeping in mind that the computing time varies with p but is essentially independent of N, the advantages to be realized in using high-order approximations are obvious.

Analysis of Degenerate H-Mode Guides

Very recently interest has arisen in waveguides capable of propagating two distinct H modes with equal guide wavelength. Such guides are of importance in the design of various microwave devices. One shape, which may be described as a T-septate rectangular guide, has been described by Elliott [16]. Another, the vaned rectangular guide, may be viewed as a limiting case either of the T-septate rectangular guide, or of a ridged guide; both shapes are shown in Fig. 5. These guides are members of a larger class which provides symmetry about the horizontal but not about the vertical axis. All such guides are capable of propagating one H mode with odd, another with even, symmetry about the horizontal center line, and may be made into degenerate H-mode guides by adjusting their dimensions so that the cutoff wavelengths of these dominant modes coincide. In the case of the vaned guide, the even-symmetric mode is obviously the H_{01} mode of the same guide with vane removed. The odd-symmetric dominant mode, on the other hand, is new and must be solved for numerically. Fig. 6 shows the field pattern of the dominant odd-symmetric mode for one particular vaned guide configuration that provides degeneracy of the dominant modes; that is, the cutoff wavelength of the mode illustrated is twice the guide width. A substantial number of vaned guides has now been analyzed, and Fig. 7 summarizes some of the findings. It will be noted that the limiting case of zero vane depth in each case yields simply the H_{10} mode of the empty guide, as would be expected. Which possible combination of guide width and vane depth to select for any particular application will depend on the manner to be used for coupling to the degenerate modes, and is thus a design parameter available to the device designer.

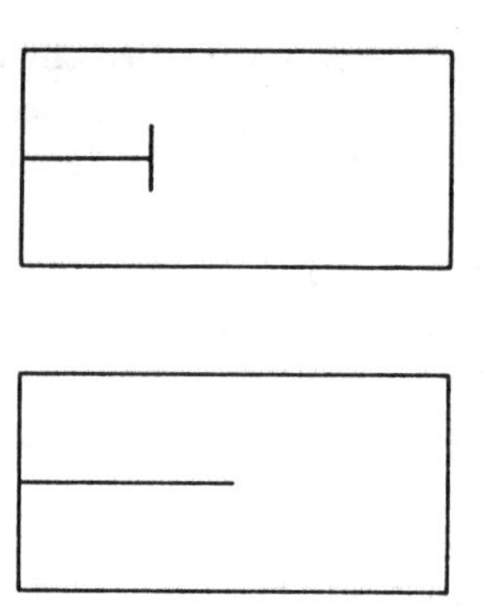

Fig. 5. T-septate rectangular and vaned rectangular guides.

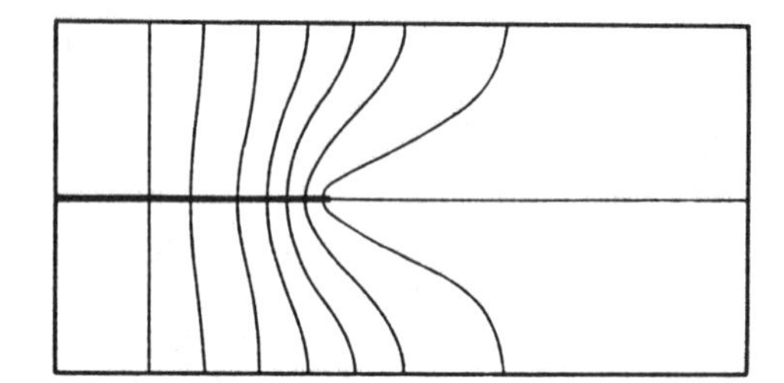

Fig. 6. Dominant odd-symmetric H mode of the degenerate H-mode vaned rectangular guide.

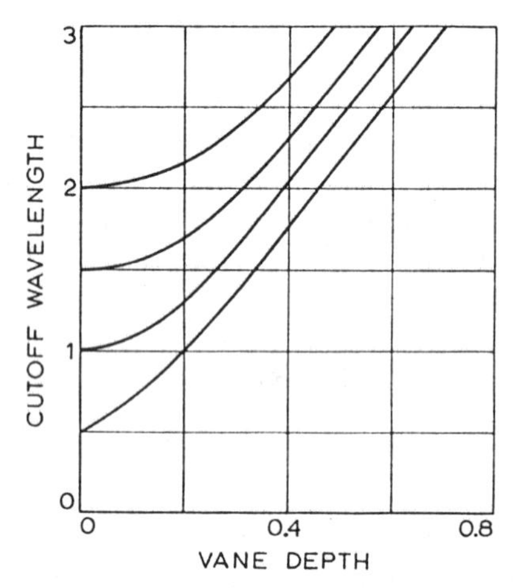

Fig. 7. Variation of cutoff wavelength of the dominant odd-symmetric H mode of vaned rectangular guides, for guide heights of 1 (top curve), 0.75, 0.5, and 0.25 (bottom curve). All dimensions are normalized to guide width.

The T-septate rectangular guide was analyzed by Elliott [16] using the orthonormal block method. The theoretical results did not compare well with experiment; in fact, for mode degeneracy to be obtained, the predicted curves in many cases require a septum depth twice that experimentally determined. It was suggested that the shortcoming in the analysis lay in assuming the T septum to be of infinitesimal thickness. To check on this suggestion and to try to obtain better analyses of T-septate guides, prediction of all the experimental results given by Elliott was attempted, using the program described above. Quartic elements were used, and half the guide cross section modeled by six or seven triangular elements, the number depending on the precise dimensions. From the results it would appear clear that the finite septum thickness has only a small effect. Fig. 8 shows the predicted and measured guide wavelengths at 6 GHz, for a guide in which the crossbar of the T is 0.3 times the guide height. In this analysis, the septum thickness is again assumed infinitesimal, yet the predicted curve agrees with the measured points very well.

For the T-septate guide, both the even and odd modes must be computed numerically. Fig. 9 shows the first two even-symmetric modes for a guide with dimensions chosen to yield degeneracy of the dominant modes. Except for the perturbation of field patterns near the septum tips, they closely resemble the H_{01} and H_{02} modes of the empty guide —a fact reflected in the cutoff wavelengths, which are only about 1.5 percent above those of the empty guide. The corresponding odd-symmetric mode patterns are shown in Fig. 10. These, of course, have no empty-guide counterparts. It is interesting to observe that the subdominant modes are also very nearly degenerate, their cutoff wavelengths differing by less than one percent.

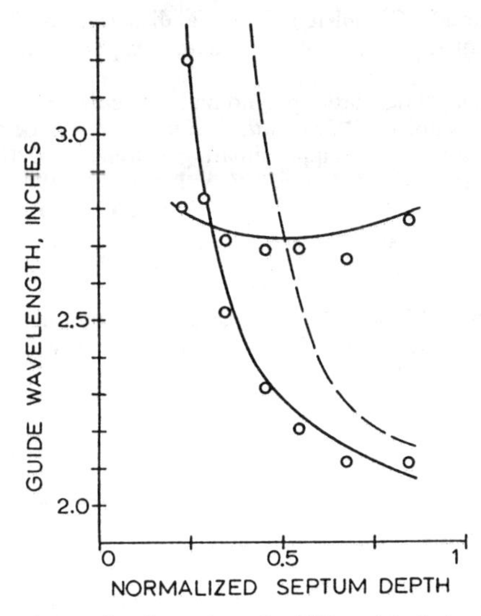

Fig. 8. Experimental values given by Elliott (circles) compared with finite-element analysis (solid curves) and Elliott's theory (dashed curve); for the even-symmetric mode, the two sets of predicted curves nearly match and are not shown separately.

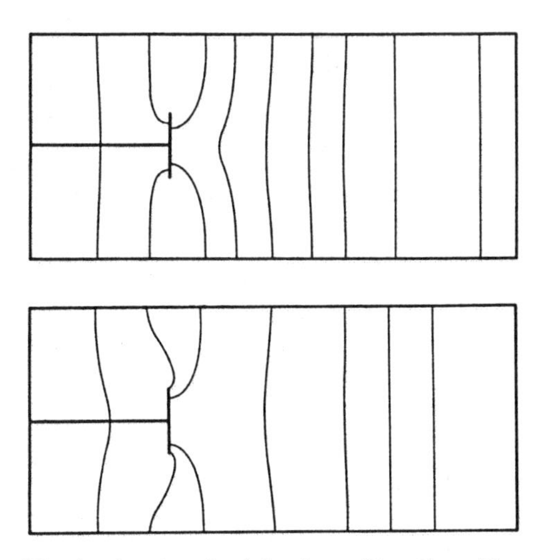

Fig. 9. The dominant and subdominant H modes with even symmetry in a degenerate H-mode T-septate rectangular guide.

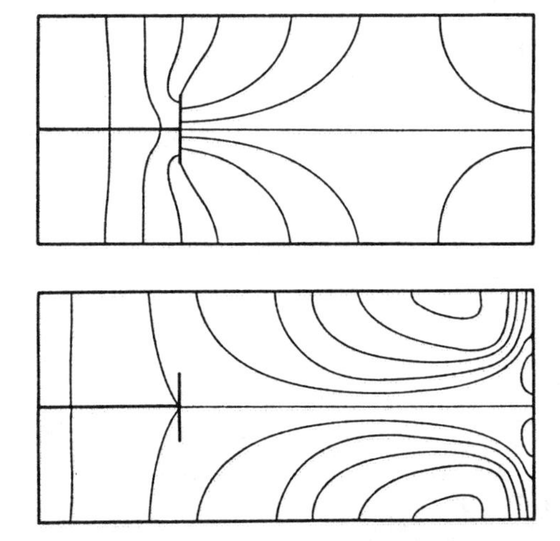

Fig. 10. Dominant and subdominant odd-symmetry H modes in the guide of Fig. 9.

In both the vaned and T-septate rectangular guides, field singularities are to be expected at the septum tips so that solution accuracies are likely to be somewhat lower than for convex shapes. Nevertheless, recomputation with varying fineness of subdivision and different orders of polynomial would seem to indicate that the dominant cutoff wavelengths can probably be relied upon to about 0.1 percent.

Conclusions

The finite-element method of solving homogeneous waveguide problems by approximated extremization of functionals appears to be capable of higher accuracy than numerical methods presently employed, and reliably produces complete sets of propagating modes at little computational cost. Its usefulness for solving problems in both convex and nonconvex regions has been demonstrated by comparing computed results with analytically known solutions in the former case, and with experimental data in the latter. Numerical experiments with the general program described here indicate particularly high accuracies where polynomial elements of relatively high order—cubics and quartics—can be employed. It is therefore believed that the use of this and related programs will in the future obviate the necessity of using those older techniques in which a new program must be written and tested for every new class of guide shapes, such as orthonormal block analysis. As compared to the well-established relaxation methods, the new program is somewhat less adaptable to very complicated boundary shapes, but offers significantly higher accuracy, which may of course be traded for computing time.

The new method is capable of extension to inhomogeneously filled guides and cavity resonators, as well as to other field problems. Some of these possibilities are now under examination.

Acknowledgment

The author wishes to thank Prof. J. Brown, Department of Electrical Engineering, Imperial College of Science and Technology, for the generous provision of research facilities.

References

[1] R. E. Collin, *Field Theory of Guided Waves.* New York: McGraw-Hill, 1960.
[2] H. Motz, "Calculation of the electron field, frequency and circuit parameters of high-frequency resonant cavities," *J. IEE* (London), vol. 93, pt. 3, pp. 335–343, 1946.
[3] J. B. Davies and C. A. Muilwyk, "Numerical solution of uniform hollow waveguides with boundaries of arbitrary shape," *Proc. IEE* (London), vol. 113, pp. 277–284, 1966.
[4] M. J. Beaubien and A. Wexler, "An accurate finite-difference method for higher order waveguide modes," *IEEE Trans. Microwave Theory and Techniques*, vol. MTT-16, pp. 1007–1017, December 1968.
[5] H. H. Meinke, K. P. Lange, and J. F. Ruger, "TE- and TM-waves in waveguides of very general cross section," *Proc. IEEE*, vol. 51, pp. 1436–43, November 1963.
[6] H. H. Meinke and W. Baier, "Die Eigenschaften von Hohlleitern allgemeineren Querschnitts," *Nachrtech. Z.*, pp. 662–70, 1966. (Transl.: "The characteristics of waveguides with general cross-section," *NTZ—Commun. J.*, vol. 7, pp. 1–8, 1968.)

[7] J. J. Skiles and T. J. Higgins, "Determination of the characteristic impedance of UHF coaxial rectangular transmission lines," *1954 Proc. NEC*, vol. 10, pp. 97–108.
[8] M. K. Krage and G. I. Haddad, "The characteristic impedance and coupling coefficient of coupled rectangular strips in a waveguide," *IEEE Trans. Microwave Theory and Techniques*, vol. MTT-16, pp. 302–307, May 1968.
[9] R. M. Bulley, "Computation of approximate polynomial solutions to the Helmholtz equation using the Rayleigh–Ritz method," Ph.D. dissertation, University of Sheffield, Sheffield, England, 1968.
[10] P. Silvester, "Finite-element solution of homogeneous waveguide problems," *1968 URSI Symp. Electromagnetic Waves*, paper 115, (to appear in *Alta Frequenza*).
[11] P. C. Dunne, "Complete polynomial displacement fields for finite element method," *J. Roy. Aeronaut. Soc.*, vol. 72, pp. 245–246, 1968.
[12] P. Silvester, "High-order polynomial triangular finite elements for potential problems," *Internatl. J. Engrg. Sci.* (to be published).
[13] D. Greenspan, "On approximating extremals of functionals—Part II," *Internatl. J. Engrg. Sci.*, vol. 5, p. 571, 1967.
[14] L. V. Bewley, *Tensor Analysis of Electric Circuits and Machines.* New York: Ronald Press, 1961.
[15] J. H. Wilkinson, *The Algebraic Eigenvalue Problem.* New York: Oxford University Press, 1965.
[16] R. S. Elliott, "Two-mode waveguide for equal mode velocities," *IEEE Trans. Microwave Theory and Techniques*, vol. MTT-16, pp. 282–286, May 1968.

HIGH-ORDER POLYNOMIAL TRIANGULAR FINITE ELEMENTS FOR POTENTIAL PROBLEMS

P. SILVESTER

Department of Electrical Engineering, McGill University, Montreal 110, Quebec, Canada

Abstract— An analytic derivation is given for high-accuracy triangular finite elements useful for numerical solution of field problems involving Laplace's, Poisson's, Helmholtz's, or related elliptic partial differential equations in two dimensions. General expressions are developed for complete polynomial fields of arbitrarily high order, and the method for obtaining element describing matrices is shown. These matrices can always be written in terms of trigonometric functions of the vertex angles, and the triangle area, multiplied by certain numerical coefficient matrices which are the same for any triangle. For polynomial fields up to fourth order, the numerical coefficient matrices are given, so that the element matrices for any triangle can be found easily. Use of these new elements is illustrated by a simple vibration problem.

INTRODUCTION

THE FINITE element method of numerical analysis has been successfully applied to a variety of continuum problems in recent years, but especially to those arising in stress analysis. Its application to field problems governed by Poisson's equation was proposed by Zienkiewicz and Cheung[1] some three years ago, and a number of problems involving Poisson's or Laplace's equation, or their anisotropic counterparts, has been solved since. Many of the theoretical questions related to this method have in the meantime been explored by Greenspan[2, 3] and others. However, the use of finite elements in field analysis still cannot be termed widespread. Partly, this may be due to the small variety of available elements. The triangular elements formulated by Zienkiewicz and Cheung, in particular, provide the same degree of precision as the well-known 5-point finite-difference method. In many problems, little is to be gained by using such elements instead of the widely known difference-equation techniques. Of course there is an advantage to be gained in boundary representation with finite elements, but this is often offset by the inconvenience of having to prepare comparatively large volumes of input data for the computer.

The finite element approach becomes considerably more attractive if accuracy can be improved while concurrently reducing the necessary amount of input data. This can be accomplished by the use of triangular elements in which the variation of the desired potential function is given by a polynomial expression of order N, rather than assuming simple linear potential variation within the element. Considerable effort has been devoted to the development of such higher-order elements for use in stress analysis and plate bending problems (e.g. 6-point and 10-point triangles, 10-point tetrahedra[4]). In practice, relatively few high-order elements suffice to model problems with high accuracy. For example, there is not a great deal of computational difference between one quartic triangular element, or sixteen first-order triangles spanning the same area; the resulting matrix problem is of order 15 in either case. However, the higher-order representation provides very substantially improved accuracy. On the other hand, a large number of simple elements is clearly advantageous where a complicated boundary shape needs to be modelled.

This paper presents details of element describing matrices for high-order poly-

Reprinted with permission from *Int. J. Eng. Sci.*, vol. 7, no. 8, pp. 849–861, Aug. 1969.

nomial fields, the element shape being triangular in all cases. The derivation is valid for any arbitrary order of polynomical representation, and may be used to compute element matrices of any order. As will be shown, it is possible to write the element matrices as combinations of numerical coefficient matrices valid for any triangle, with functions of the vertex angles and the element area. These numerical coefficient matrices have universal validity, and hence need only be calculated once and for all.

FORMULATION OF FIELD PROBLEMS

Numerous scalar potential problems arising in engineering may be formulated as boundary-value problems consisting of the inhomogeneous Helmholtz equation

$$\nabla^2\phi = g - \lambda^2\phi \tag{1}$$

and associated boundary conditions; ϕ is the potential to be determined, g represents the corresponding source density function, and λ is a constant. It can be shown [5] that the solution of equation (1) subject to the associated natural boundary conditions is equivalent to minimizing the functional

$$F = \tfrac{1}{2}\int (|\mathrm{grad}\,\phi|^2 + \lambda^2\phi^2)\,\mathrm{d}S - \int \phi g\,\mathrm{d}S \tag{2}$$

the integration being extended over the region within the given boundaries. If other than the natural boundary conditions, i.e. Dirichlet or homogeneous Neumann conditions, must be satisfied, appropriate terms must be added to the functional. Their treatment follows that given below, and will not be discussed separately here.

In principle, the problem may be regarded as that of constructing a solution surface $\phi(x,y)$ over a specified region of the x–y plane, such as to satisfy the boundary conditions as well as the extremum requirement on F. The essence of the finite element method consists of substituting for the true solution surface another surface, made up of elementary sections of a prescribed kind, so as to obtain the best possible approximation to $\phi(x, y)$. Most elements used in the past have been either triangular or rectangular, and characterised by very simple approximating functions for ϕ within each elementary region. The major reason is probably that the equations resulting from any but the simplest approximating functions are usually very complicated. Consequently, many simple elements have generally been preferred to fewer complicated ones.

POLYNOMIAL APPROXIMATIONS TO FIELD VARIABLES

Let the two-dimensional region R within which the boundary-value problem is defined, be subdivided into a set of triangular elements in the manner usual in finite-element analysis [6]. No restrictions need be placed on the triangle areas or shapes. It will now be assumed that ϕ is given within each triangle as a complete polynomial in x and y. As discussed by Dunne [7], complete polynomials are required if the orientation of elements is to be of no importance. This point is of especial significance if the subdivision into elements is to be performed by computer. Since a complete polynomial of order N contains $n = \frac{1}{2}(N+1)(N+2)$ terms, each triangular element must have associated with it n independently specifiable parameters or degrees of freedom. In order for F to be evaluated without additional assumptions, it is necessary for the first derivatives of ϕ to be finite everywhere; that is, ϕ must be continuous everywhere.

This requirement can only be met if ϕ along any triangle edge is given by a polynomial function of order N of distance along that edge. The $N+1$ coefficients of this polynomial must depend only on values of ϕ along the edge, if continuity with the adjoining triangle is to be assured. Consequently it is both necessary and sufficient to specify ϕ at the triangle vertices, and at $N-1$ intermediate points along each edge. In order to define a complete polynomial expression of order $N > 2$, however, more information must be provided. In this paper, the additional specifications are introduced by means of values of ϕ at interior points.

It is convenient to carry out all analysis of triangles in terms of the so-called triangle area coordinates ζ_1, ζ_2, ζ_3 (see, e.g. Zienkiewicz[6]). These specify the position of any point within the triangle by giving the distance measured perpendicularly from each side to the point, distances being expressed as fractions of the triangle altitude. The area coordinates span the range of numerical values from 0 to 1 in any triangle whatever. Lines of constant ζ_n are parallel to side n of the triangle. Although only two of these coordinates can be independent, it is nevertheless useful to retain all three in analysis, for the resulting expressions are often symmetric in the three area coordinates.

A regularly spaced set of points P_{ijk} may be defined in a triangle by the area coordinate values

$$\left(\frac{i}{N}, \frac{j}{N}, \frac{k}{N}\right), \quad 0 \leqslant i, j, k \leqslant N \tag{3}$$

where i, j, k are integers satisfying

$$i+j+k = N. \tag{4}$$

Such a set of points is shown in Fig. 1 for the illustrative case $N = 4$. It is easily verified that the number of points defined in this way is $\frac{1}{2}(N+1)(N+2)$, equal to the number

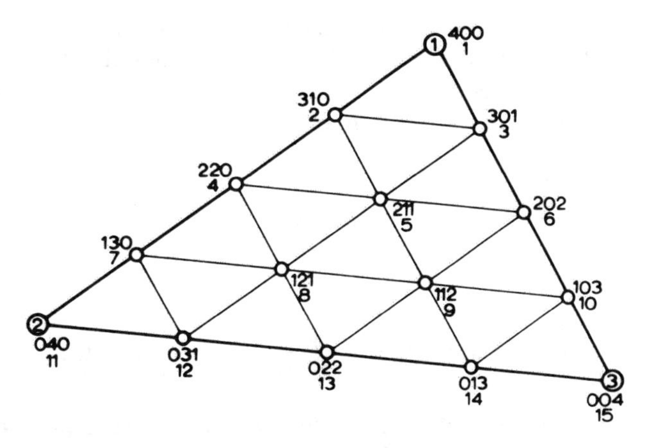

Fig. 1. Point placement in a triangle for $N = 4$, showing the triple and single index numbering schemes, as well as the numbering of triangle vertices.

of independent coefficients in a polynomial expression of order N. There are also N-1 equispaced points along each edge, in addition to the vertices. If a value of ϕ, say

ϕ_{ijk}, is prescribed at each point P_{ijk}, a polynomial expression for ϕ will be uniquely defined throughout the triangle by

$$\phi(\zeta_1, \zeta_2, \zeta_3) = \sum_{i=0}^{N} \sum_{j=0}^{N-i} \alpha_{ijk}(\zeta_1, \zeta_2, \zeta_3)\, \phi_{ijk}. \tag{5}$$

The α_{ijk} are polynomical expressions of order N in $\zeta_1, \zeta_2, \zeta_3$, dependent neither on ϕ nor on the triangle size and shape; suitable polynomials are given below. It is easily seen that (5) specifies a complete polynomial, and satisfies the continuity requirement at the triangle edges. No summation over the index k is shown, since requirement (4) above fixes k for any given i, j. It may be worth emphasizing that a complete polynomial in x, y transforms into a complete polynomial in the area coordinates, and conversely, the number of independent parameters being strictly preserved.

In a similar manner, g will also be expressed in terms of point values, by

$$g(\zeta_1, \zeta_2, \zeta_3) = \sum_{i=0}^{N} \sum_{j=0}^{N-i} \alpha_{ijk}(\zeta_1, \zeta_2, \zeta_3)\, g_{ijk}. \tag{6}$$

Throughout this paper, it will be assumed that the same order of approximation is retained for both functions, i.e. that the polynomials α_{ijk} are the same in equations (5) and (6). However, it will be seen that this assumption is not fundamental, and may be removed easily if desired.

POLYNOMIAL APPROXIMATION TO FUNCTIONAL MINIMISATION

Both the potential and source density functions are now defined and continuous everywhere within the region of interest, so that evaluation of F becomes a straightforward, if lengthy, matter. As is usual in finite–element work, only one element will now be considered; any larger number may be treated by combining element matrices appropriately.

For the following discussion, the triple-subscript notation will be abandoned temporarily in the interest of conciseness. It will be assumed that the points P_{ijk} within the element are numbered sequentially $1, 2, \ldots, n$ in descending order of the triple subscripts, as indicated in Fig. 1. Within any element, the conditions for minimising F may be written

$$\frac{\partial F}{\partial \phi_m} = 0, m = 1, 2, \ldots, n \tag{7}$$

or, substituting (2),

$$\frac{1}{2}\int \frac{\partial}{\partial \phi_m} |\text{grad}\, \phi|^2\, dS = \int \frac{\partial}{\partial \phi_m} (\phi g - \frac{\lambda^2}{2}\phi^2)\, dS. \tag{8}$$

The left-hand term will be evaluated explicitly first. Using equation (5), repeated differentiation in Cartesian coordinates yields

$$\frac{\partial}{\partial \phi_m} |\text{grad}\, \phi|^2 = 2 \sum_{q=1}^{n} \left(\frac{\partial \alpha_m}{\partial x}\frac{\partial \alpha_q}{\partial x} + \frac{\partial \alpha_m}{\partial y}\frac{\partial \alpha_q}{\partial y}\right) \phi_q. \tag{9}$$

To convert to triangle area coordinates, one uses the relationship

$$\zeta_i = \frac{1}{2A}(a_i + b_i x + c_i y) \tag{10}$$

where

$$a_i = \begin{vmatrix} x_{i+1} & y_{i+1} \\ x_{i+2} & y_{i+2} \end{vmatrix} \tag{11}$$

$$b_i = y_{i+2} - y_{i+1} \tag{12}$$

$$c_i = x_{i+1} - x_{i+2} \tag{13}$$

and A represents triangle area. The index i ranges over the triangle vertices (not the points P_{ijk}), and the subscripts in (11–13) must be understood as progressing modulo 3. After a little algebra, equation (9) can be recast in the form

$$\frac{\partial}{\partial \phi_m} |\text{grad}\, \phi|^2 = \frac{1}{2A^2} \sum_{q=1}^{n} \left[\sum_{i=1}^{3} \sum_{j=1}^{3} (b_i b_j + c_i c_j) \frac{\partial \alpha_m}{\partial \zeta_i} \frac{\partial \alpha_q}{\partial \zeta_j} \right] \phi_q. \tag{14}$$

Here the indices i, j range over the triangle vertex numbers, while indices m, q assume values $1, 2, \ldots, n$ corresponding to the points P_{ijk}. It is worth noting that the bracketed expression is symmetric in m and q.

On examining the geometric properties of a triangle, it is easily shown that

$$b_i b_j + c_i c_j = -2A \cot \theta_k \quad i \neq j \tag{15}$$

$$b_i^2 + c_i^2 = 2A(\cot \theta_j + \cot \theta_k) \tag{16}$$

where θ_i denotes the included angle at vertex i. If equations (15) and (16) are substituted into (14), expanded, and terms collected, there results

$$\frac{\partial}{\partial \phi_m} |\text{grad}\, \phi|^2 = \frac{1}{A} \sum_{q=1}^{n} \left[\sum_{i=1}^{3} \left(\frac{\partial \alpha_m}{\partial \zeta_j} - \frac{\partial \alpha_m}{\partial \zeta_k} \right) \left(\frac{\partial \alpha_q}{\partial \zeta_j} - \frac{\partial \alpha_q}{\partial \zeta_k} \right) \cot \theta_i \right] \phi_q. \tag{17}$$

Here i, j, k assume the values 1, 2, 3 cyclically.

A similar but considerably simpler development applies to the right-hand side of equation (8). Substituting equations (5) and (6), there are obtained

$$\frac{\partial}{\partial \phi_m}(\phi g) = \sum_{q=1}^{n} \alpha_q \alpha_m g_q \tag{18}$$

$$\frac{\partial}{\partial \phi_m} = 2 \sum_{q=1}^{n} \alpha_q \alpha_m \phi_q. \tag{19}$$

If matrices S and T are now defined by letting their typical elements be given by

$$S_{mq} = \frac{1}{A} \sum_{i=1}^{3} \cot \theta_i \int \left(\frac{\partial \alpha_m}{\partial \zeta_j} - \frac{\partial \alpha_m}{\partial \zeta_k}\right)\left(\frac{\partial \alpha_q}{\partial \zeta_j} - \frac{\partial \alpha_q}{\partial \zeta_k}\right) \mathrm{d}S \tag{20}$$

and

$$T_{mq} = \int \alpha_m \, \alpha_q \, \mathrm{d}S \tag{21}$$

then (8) may be written in matrix form as

$$S\Phi = T(\mathbf{G} - \lambda^2 \Phi). \tag{22}$$

Here $\mathbf{G}$ and $\mathbf{\Phi}$ represent column vectors of g_q and ϕ_q respectively. A solution to the field problem (1) may now be obtained by solving the matrix equation (22).

The matrix T is essentially similar to the mass matrix encountered in plate vibration problems, and it comes as no surprise that its elements, according to equation (21), do not depend on triangle shape. Matrix S, which corresponds to the stiffness matrix of structural analysis, does depend on shape; but it is readily seen that, from (20), it may be written as

$$S = \sum_{i=1}^{3} Q_i \cot \theta_i \tag{23}$$

where the matrices Q_i do not depend on triangle shape. Furthermore, it is readily verified that the three Q_i are merely row and column permutations of each other, so that the matrix S for any triangular element may be constructed very simply if one coefficient matrix, say Q_1, is known. These numerical coefficient matrices will now be derived.

GENERAL POLYNOMIAL COEFFICIENTS

To evaluate the matrices T and Q_1 above, the polynomials α_{ijk} of equations (5) and (6) are required. For these, it is convenient to define first an auxiliary function $P_m(z)$ by

$$\begin{aligned} P_m(z) &= \prod_{i=1}^{m} \left(\frac{Nz - i + 1}{i}\right), \quad m \geq 1 \\ &= 1 \qquad\qquad\qquad\qquad m = 0 \end{aligned} \tag{24}$$

$P_m(z)$ is clearly a polynomial of order m in z. Let a polynomial α_{ijk} of order N be defined over the triangle by setting

$$\alpha_{ijk} = P_i(\zeta_1) P_j(\zeta_2) P_k(\zeta_3), i + j + k = N. \tag{25}$$

These polynomials satisfy all the conditions necessary to qualify as the polynomials of equations (5) and (6). First, it is immediately clear that there are altogether as many distinct polynomials α_{ijk} as there are terms in a complete polynomial of order N; therefore the use of these expressions makes the polynomials in (5) and (6) complete, as required. Secondly, let the area coordinates assume their values at one of the points given by (3), i.e. at the typical point $(p/N, q/N, r/N)$. From equation (24), one finds that, for i, p integers,

$$P_i\left(\frac{p}{N}\right) = 0, \quad i > p$$
$$= 1, \quad i = p. \tag{26}$$

Therefore, at the typical point all α_{ijk} given by (25) are identically zero, except for α_{pqr}, which assumes the value unity. Equations (5) and (6) therefore become simple identities at the points where ϕ and g are specified. Consequently equation (25) correctly defines the required polynomials.

INTEGRATION OF POLYNOMIAL EXPRESSIONS

It is now possible to develop the integrals given by (20) and (21) in detail. The polynomials of (25) may be differentiated and multiplied in a straightforward manner, yielding simple though lengthy polynomial expressions in the three area coordinates. To perform the surface integrations, it is convenient to reduce the number of independent coordinates to two, using the relationship[6]

$$\sum_{i=1}^{3} \zeta_i = 1. \tag{27}$$

All the required integrations can then be carried out easily if the typical term

$$I_{mn} = \int \zeta_1^m \zeta_2^n \, \mathrm{d}S. \tag{28}$$

Writing the surface element $\mathrm{d}S$ in terms of the area coordinates, some trigonometric manipulation results in

$$I_{mn} = 2A \int_0^1 \int_0^{1-\zeta_2} \zeta_1^m \zeta_2^n \mathrm{d}\zeta_1 \mathrm{d}\zeta_2. \tag{29}$$

Integrating first with respect to ζ_1, expanding the result binomially, then integrating with respect to ζ_2, by parts, there results

$$I_{mn} = 2A \frac{m!\,n!}{(m+n+2)!}. \tag{30}$$

CALCULATION OF THE ELEMENT MATRICES

In principle, calculation of the complete element matrices is now uncomplicated. However, the manipulative labour involved in finding any but the lowest order elements is formidable. In fact, it can be shown that the total number of operations required to construct the element matrices T and Q_1 varies approximately as N^6. Manual solution is easy for $N = 1$ and lengthy though still practial for $N = 2$. Beyond $N = 3$, the volume of work would strain the patience of the most diligent nineteenth-century arithmetician!

Since manual execution of the necessary steps is not practical, appropriate computer programs to perform the algebraic manipulations were written. Individual subprograms were devised to (1) generate arrays of coefficients to represent the polynomial expressions, (2) modify the coefficient arrays according to the rules of differentiation for polynomials, (3) multiply such arrays and expand them in the form required by (28), and (4) to accumulate terms I_{mn} given by (30), themselves generated by yet another subprogram, so as to form the matrix elements. The language used was Fortran IV, and all operations were coded using integer arithmetic. Multiple-length integer arithmetic was unfortunately not available on the computer used, and some difficulties with arithmetic overflows were encountered. However, the final results are exact, since no roundoff error occurs in any calculation. Because the final element matrices are applicable to any triangular element, and need only ever be generated once, it is believed that the necessary extra programming effort and computer time are more than justified.

The element matrices possess several interesting symmetry properties which have proved useful to save computer time and storage. The most important of these are:

(1) Both T and Q_1 are symmetric.

(2) All elements of T must sum to the triangle area. This implies physically that T reapportions the source density g over the triangle; but the total of all sources is invariant.

(3) All row and column sums of Q_1 are zero, so as to guarantee singularity of S. Since in any physical problem the reference point for the potential ϕ may be outside the element, it must be possible to add an arbitrary constant to the ϕ_{ijk}. This is only possible if S has zero row and column sums.

(4) The elements of T must have complete symmetry in the three coordinates. In terms of triple subscripts, all matrix elements are equal for which the two triple subscripts form similar permutations, e.g. $T_{ijk,pqr} = T_{ikj,prq} = T_{kij,rpq}$.

ELEMENT MATRICES FOR TRIANGLES TO ORDER 4

The element describing matrices for the first four orders of triangular elements are given for reference in the Appendix. Only T and Q_1 are shown, Q_2 and Q_3 being derivable by row and column permutations. The sequence of permutations is geometrically obvious, but algebraically obscure if single-subscript notation is used. Employing triple subscripts, the necessary interchanges for forming Q_k are obtained systematically by the following steps. (1) Write the triple subscripts in descending order next to the matrix rows and columns. (2) Permute the three subscript digits so as to have the digit referring to ζ_k appear first, while retaining the order of vertices counter-clockwise. (3) Rearrange the rows and columns so that the new (permuted) triple indices are again in descending order. These steps are readily programmed for a digital computer, and permit S to be formed without actually storing the three permuted matrices Q_k.

Since T and Q_1 are symmetric, only the upper triangle of T (including diagonal elements) and the lower triangle of Q_1 are given in the Appendix. These matrices, whose elements have been computed as integer quotients, have been reduced to the least common denominator, so that it is only necessary to give the matrix of numerators and the denominator. Thus, S is formed by taking the lower triangle of numerators from the Appendix, expanding and dividing to form Q_1, then permuting to find Q_2 and Q_3.

For example, for $N = 1$, this procedure yields

$$S = \tfrac{1}{2}\begin{bmatrix} 0 & 0 & 0 \\ 0 & 1 & -1 \\ 0 & -1 & 1 \end{bmatrix} \cot\theta_1 + \tfrac{1}{2}\begin{bmatrix} 1 & 0 & -1 \\ 0 & 0 & 0 \\ -1 & 0 & 1 \end{bmatrix} \cot\theta_2 + \tfrac{1}{2}\begin{bmatrix} 1 & -1 & 0 \\ -1 & 1 & 0 \\ 0 & 0 & 0 \end{bmatrix} \cot\theta_3. \quad (31)$$

Similarly, T is given by

$$T = \frac{A}{12}\begin{bmatrix} 2 & 1 & 1 \\ 1 & 2 & 1 \\ 1 & 1 & 2 \end{bmatrix}. \quad (32)$$

The higher-order matrices are assembled in precisely the same way.

USE OF HIGH-ORDER ELEMENTS: AN EXAMPLE

To verify the correctness of the element matrices given in the Appendix, as well as to exhibit the quite striking accuracy improvement achieved by the use of high-order matrices, the free vibration modes of an isosceles right triangular flexible membrane were calculated. This problem was selected both because it permits exact modelling of the boundaries, and because its analytic solution is known [10]. For small amplitudes of vibration, this problem may be formulated as the Helmholtz equation within the triangle, subject to homogeneous Dirichlet boundary conditions. Elements from first to fourth order were used, their number being varied so as to keep the total number of degrees of freedom at 55. In other words, the computational problem each time was that of solving an eigenvalue problem of order 55. For a problem involving D degrees of freedom, the number of arithmetic operations required to form the matrices S and T is approx. N^2D, while approx. $\frac{2}{3}D^3$ operations are necessary for solving the eigenvalue problem by direct methods. Since in this case $D = 55$, while $1 \leqslant N \leqslant 4$, the total time of computation is virtually invariant with N. In general, the solution time will naturally depend on the numerical method selected, and on the topology of the matrices S and T, so that no universally valid conclusion can be drawn from the behaviour of the specific problem examined here. Indeed, the object here is to exhibit the different solution accuracies obtained, while keeping as many other aspects of the problem as possible invariant.

The first few frequencies of oscillation are given in Table 1 below, which shows the error incurred for the several orders of element. It is clear that the higher-order elements provide a very significant gain in solution accuracy over the first-order ones, and that accuracy rises with element order.

CONCLUSIONS

The analysis of field problems by finite–element methods has in the past been restricted to large numbers of very simple elements, because the describing matrices for high-order elements have not been known. The method described in this paper has permitted derivation of high-order triangular elements in a very general way, allowing solution of both deterministic and eigenvalue problems to high accuracy. Initial applications of these new field elements have been in the area of waveguide and

Table 1. Percentage error in computed resonant frequencies of triangular membrane

Normalised resonant frequency	$N = 1$ (144 elements)	$N = 2$ (36 elements)	$N = 3$ (16 elements)	$N = 4$ (9 elements)
1·00000	1·46	0·122	0·017	0·0021
1·41421	3·35	0·560	0·125	0·0271
1·61245	3·81	0·725	0·239	0·0448
1·84391	5·38	1·33	0·680	0·160
2·00000	7·25	2·13	0·945	0·628
2·23607	6·84	2·08	1·41	0·418
2·28035	8·04	2·70	1·31	0·700
2·40832	10·2	3·71	2·21	1·04

related microwave device analysis. It appears that several extensions of the method are possible, and some of these are now under investigation [8].

Probably the only serious disadvantage of very high order elements, as compared to a larger number of simpler ones, is the greater difficulty in approximating to complicated boundary shapes. Since several orders of element representation are now available, it is suggested that the element order be chosen in each case to suit the problem at hand.

Acknowledgment—The author wishes to thank Professor J. Brown, Department of Electrical Engineering, Imperical College of Science and Technology, London, for the provision of research facilities.

REFERENCES

[1] O. C. ZIENKIEWICZ and Y. K. CHEUNG, *The Engineer* **222**, 507 (1965).
[2] D. GREENSPAN, *Int. J. Engng Sci.* **5**, 571 (1967).
[3] D. GREENSPAN, *ICC Bull.* **4**, 99 (1965).
[4] J. H. ARGYRIS, *Jl R. aeronaut. Soc.* **69**, 711 (1965).
[5] G. E. FORSYTHE and W. R. WASOW, *Finite-Difference Methods for Partial Differential Equations*, Chap. 3. Wiley (1960).
[6] O. C. ZIENKIEWICZ with Y. K. CHEUNG, *The Finite Element Method in Structural and Continuum Mechanics*, Chap. 10. McGraw-Hill (1967).
[7] P. C. DUNNE, *Jl R. aeronaut. Soc.* **72**, 245 (1968).
[8] P. SILVESTER, *IEEE Trans. Microwave Theory Tech.* **MTT-17**, 247 (1969).
[9] P. SILVESTER, *URSI Symp. electromag. Waves* 115 (1968).
[10] P. M. MORSE and H. FESHBACH, *Methods of Theoretical Physics*, Chap. 6. McGraw-Hill (1953).

(*Received* 4 *March* 1969)

APPENDIX

NUMERICAL TABLE OF ELEMENT MATRICES

The symmetric matrices T and Q are given below, only the upper triangle of T and the lower of Q_1 being shown. To obtain the numerical values of the actual matrix elements, the tabulated numbers (which are really the numerators of integer fractions) must be divided by the denominators given.

$N = 1$
Common denominators: T 12
Q_1 2

2	1	1
0		
	2	1
0	1	
		2
0	−1	1

$N = 2$
Common denominators: T 180
Q_1 6

6	0	0	−1	−4	−1
0					
	32	16	0	16	−4
0	8				
		32	−4	16	0
0	−8	8			
			6	0	−1
0	0	0	3		
				32	0
0	0	0	−4	8	
					6
0	0	0	1	−4	3

$N = 3$
Common denominators: T 6720
Q_1 80

76	18	18	0	36	0	11	27	27	11
0									
	540	270	−189	162	−135	0	−135	−54	27
0	135								
		540	−135	162	−189	27	−54	−135	0
0	−135	135							
			540	162	−54	18	270	−135	27
0	−27	27	135						
				1944	162	36	162	162	36
0	0	0	−162	324					
					540	27	−135	270	18
0	27	−27	27	−162	135				
						76	18	0	11
0	3	−3	3	0	−3	34			
							540	−189	0
0	0	0	0	0	0	−54	135		
								540	18
0	0	0	0	0	0	27	−108	135	
									76
0	−3	3	−3			−7	27	−54	34

$N = 4$
Common denominators: T 56700
Q_1 1890

290	160	160	−80	160	−80	0	−160	−160	0	−27	−112	−12	−112	−27
0														
	2560	1280	−1280	1280	−960	768	256	−256	512	0	512	64	256	−112
0	3968													
		2560	−960	1280	−1280	512	−256	256	768	−112	256	64	512	0
0	−3968	3968												
			3168	384	48	−1280	384	−768	64	−80	−960	48	64	−12
0	−1440	1440	4632											
				10752	384	256	−1536	−1536	256	−160	−256	−768	−256	−160
0	0	0	−5376	10752										
					3168	64	−768	384	−1280	−12	48	−960	−960	−80
0	1440	−1440	744	−5376	4632									
						2560	1280	−256	256	160	1280	−960	512	−112
0	640	−640	−1248	1536	−288	3456								
							10752	−1536	−256	160	1280	384	256	−160
0	0	0	768	−1536	768	−4608	10752							
								10752	1280	−160	256	384	1280	160
0	0	0	768	−1536	768	1536	−7680	10752						
									2560	−112	512	−960	1280	160
0	−640	640	−288	1536	−1248	−384	1536	−4608	3456					
										290	160	−80	0	−27
0	−80	80	80	−160	80	240	−160	−160	80	705				
											2560	−1280	768	0
0	0	0	−128	256	−128	−256	256	256	−256	−1232	3456			
												3168	−1280	−80
0	0	0	96	−192	96	192	−192	−192	192	884	−3680	5592		
													2560	160
0	0	0	−128	256	−128	−256	256	256	−256	−464	1920	−3680	3456	
														290
0	80	−80	80	−160	80	80	−160	−160	240	107	−464	884	−1232	705

Résumé – Dans la présente Etude, on expose un raisonnement analytique relatif à des éléments finis triangulaires de grande précision, utile pour la résolution numérique de problèmes de champ mettant en jeu les équations aux dérivées partielles de Laplace, Poisson, Helmholtz ou elliptiques associées, dans deux dimensions. On établit des expressions générales pour des champs complets représentés par des polynômes de degré arbitrairement élevé et on expose une méthode pour obtenir les matrices décrivant l'élément. Ces matrices peuvent toujours s'écrire sous la forme de fonctions trigonométriques des angles au sommet et de l'aire du triangle, multipliées par certaines matrices à coefficients numériques, qui sont les mêmes pour n'importe quel triangle. Pour des champs représentés par des polynômes jusqu'au quatrième degré, on donne les matrices à coefficients numériques, si bien qu'on peut facilement trouver les matrices de l'élément pour n'importe quel triangle. L'emploi de ces nouveaux éléments est illustré par un problème simple de vibration.

Zusammenfassung – Eine analytische Ableitung wird für sehr genaue dreieckige endliche Elemente gegeben, die für die numerische Lösung von Feldproblemen nützlich sind, die Laplace's, Poisson's, Helmholtz's oder verwandte elliptische partielle Differentialgleichungen in zwei Dimensionen heranziehen. Allgemeine Ausdrücke für vollständige polynomische Felder beliebig hoher Ordnung werden entwickelt und die Methode wird gezeigt, Matrizen, die Elemente beschreiben, zu erhalten. Diese Matrizen können immer als trigonometrische Funktionen der Scheitelwinkel geschrieben werden, und der Dreiecksfläche, multipliziert mit gewissen numerischen Koeffizientenmatrizen, die für alle Dreiecke gleichbleiben. Die numerischen Koeffizientenmatrizen werden für polynomische Felder hinauf bis zur vierten Ordnung gegeben, so dass die Elementmatrizen für jedes Dreieck leicht gefunden werden können. Die Verwendung dieser neuen Elemente wird durch ein einfaches Vibrationsproblem illustriert.

Sommario – Si presenta una derivazione analitica per gli elementi finiti triangolari di grande precisione utile per la soluzione numerica di problemi di campo che richiedano equazioni di Laplace, Poisson, Helmholtz o differenziali parziali ellittiche associate in due dimensioni. Si sviluppano espressioni generali per campi polinomiali completi d'ordine arbitrariamente alto e si dimostra il metodo per ottenere le matrici che descrivano elementi. Tali matrici possono sempre essere scritte alla stregua di funzioni trigonometriche degli angoli di vertice e dell'area del triangolo, moltiplicate per certe matrici di coefficiente numerico che sono identiche per qualsiasi triangolo. Si danno le matrici di coefficiente numerico per campi polinomiali fino alla quarta grandezza, per permettere di scoprire facilmente le matrici di elementi per qualsiasi triangolo. Si illustra l'uso di questi nuovi elementi con un semplice problema di variazione.

Абстракт—Дается аналитическая деривация для конечных трехугольных элементов большой точности, которые могут быть использованы для численного решения полевых задач, связанных с уравнениями Лапласа, Пуассона, Гельмгольца или подобными двухразмерными элиптическими частными дифференциальными уравнениями. Разрабатываются общие выражения для полных многочленных полей любого высокого порядка, причем представлен метод получения матриц описывающих элементы. Эти матрицы могут быть всегда выражены в виде тригонометрических функций вершинных углов и площади треугольника, умноженных на некоторые матрицы численного коэффициента, которые остаются теми же свмыми для любого треугольника. Даются матрицы численного коэжжициента для многочленных полей, вплоть до четвертого порядка, так что легко можно найти матрицы элементов для любого треугольника. Использование этих новых элементов разъясняется на примере по простым колебаниям.

Finite-element solution of three-dimensional electromagnetic problems

J.P. Webb, B.A., Ph.D., G.L. Maile, B.Sc., Ph.D., Mem. I.E.E.E., and
R.L. Ferrari, M.A., D.I.C., C.Eng., M.I.E.E.

Indexing terms: *Electromagnetics, Magnetic fields, Waveguides*

Abstract: A three-dimensional finite-element scheme is proposed for finding directly the electric or magnetic field at a given frequency of excitation inside a closed system containing lossy inhomogeneous materials of arbitrary shape. It is shown that the scheme may be used to find the scattering parameters of general two-port linear waveguide problems. Results are presented for four cases.

1 Introduction

The problem addressed in this paper is that of finding the time-harmonic electric or magnetic field at a given frequency of excitation, in a volume containing arbitrarily shaped, linear, isotropic, sourceless, inhomogeneous materials [1–3]. On a surface surrounding the volume the tangential field components are prescribed. The geometry is taken to be such that a fully three-dimensional analysis is unavoidable.

Problems such as this arise in the scattering of incident energy by waveguide structures, and in the determination of field patterns in microwave cavities and ovens. Despite considerable attention to the two-dimensional version of the problem, which has application in the computation of eigenmodes of waveguides of various cross-sections [4, 5], there are few methods available for the three-dimensional case. Wang [6] has used a volume integral equation technique to deal with arbitrarily shaped lossy dielectric in a rectangular waveguide. More generally, Albani and Bernardi [7] have solved such problems by discretising Maxwell's equations in integral form, using a piecewise-uniform field approximation on a grid of cubical cells; and Akhtarzad and Johns [8] have used the TLM method, which involves what is essentially a finite-difference procedure with a rectangular mesh of nodes.

The method of this paper solves directly for the three field components (electric or magnetic), as do the above techniques. However, it uses finite elements with high-order polynomials, and thus overcomes the restrictions of rectangular meshes and piecewise-uniform or linear trial functions.

2 Variational formulation

The differential equations to be solved in volume V are Maxwell's:

$$\nabla \times \boldsymbol{E} = -j\omega\mu_0\mu \boldsymbol{H}$$
$$\nabla \times \boldsymbol{H} = j\omega\epsilon_0\epsilon \boldsymbol{E} \tag{1}$$

where $\boldsymbol{E}$ and $\boldsymbol{H}$ are the complex phasor electric and magnetic fields at angular frequency ω; the phase factor $\exp(j\omega t)$ is omitted; ϵ_0 and μ_0 are the permittivity and permeability respectively of vacuum; ϵ and μ are the corresponding relative material properties, scalar functions of position and possibly complex to represent loss. In fact in some circumstances the angular frequency ω may also be regarded as complex, representing growth or decay with time, although such a case is not dealt with in this paper.

Eqns. 1 may be solved for $\boldsymbol{E}$ or $\boldsymbol{H}$; the rest of this paper will treat $\boldsymbol{H}$ as the unknown to be determined, but an analogous development exists for the $\boldsymbol{E}$ field case. Indeed, the final computer program solves for $\boldsymbol{E}$ or $\boldsymbol{H}$, depending on the boundary conditions and material properties specified.

On each part of the surface S enclosing V, one of the following boundary conditions is to be satisfied, where $\boldsymbol{n}$ is unit outward normal to the surface:

Homogeneous Neumann (HN)

$$(\nabla \times \boldsymbol{H}) \times \boldsymbol{n} = 0 \tag{2}$$

representing a perfect conductor (short circuit).

Homogeneous Dirichlet (HD) or inhomogeneous Dirichlet (ID)

$$\boldsymbol{H} \times \boldsymbol{n} = 0 \quad \text{or} \quad \boldsymbol{H} \times \boldsymbol{n} = \boldsymbol{P} \tag{3}$$

the homogeneous condition is satisfied on planes of symmetry in the electric field (open circuit); the inhomogeneous condition will be specified on surfaces where the tangential magnetic field is known in advance ($\boldsymbol{P}$ is a prescribed vector function).

It is known [9, 10] that the problem defined by eqns. 1–3 may be formulated variationally. If only lossless materials are present, and $\boldsymbol{H}$ is constrained to satisfy the boundary conditions of eqn. 3, then the stationary point of functional $F(\boldsymbol{H})$

$$F(\boldsymbol{H}) = \int_V \left\{ \frac{1}{\epsilon} (\nabla \times \boldsymbol{H}) \cdot (\nabla \times \boldsymbol{H})^* - k^2 \mu \boldsymbol{H} \cdot \boldsymbol{H}^* \right\} dV \tag{4}$$

with respect to variations in $\boldsymbol{H}$, will be the correct solution $\boldsymbol{H}_0$ to eqns. 1–3. (k is the normalised frequency ω/c, where c is the velocity of light in a vacuum; * denotes complex conjugate.) Eqn. 2 is a natural boundary condition in this variational formulation, and need not be specified in advance.

When lossy material is present, Chen and Lien [11] and Webb [2] have shown that the following complex functional is stationary about the correct solution:

$$F(\boldsymbol{H}) = \int_V \left\{ \frac{1}{\epsilon} (\nabla \times \boldsymbol{H}) \cdot (\nabla \times \boldsymbol{H}) - k^2 \mu \boldsymbol{H} \cdot \boldsymbol{H} \right\} dV \tag{5}$$

Note that, when ϵ, μ and k are purely real, eqns. 4 and 5 are two different valid functionals for the magnetic field. This may be explained as follows. With ϵ, μ and k real put

$$F(\boldsymbol{a}, \boldsymbol{b}) = \int_V \left\{ \frac{1}{\epsilon} (\nabla \times \boldsymbol{a}) \cdot (\nabla \times \boldsymbol{b}) - k^2 \mu \boldsymbol{a} \cdot \boldsymbol{b} \right\} dV \tag{6}$$

Let $\boldsymbol{H}_r$, $\boldsymbol{H}_i$ be the real and imaginary parts of $\boldsymbol{H}$. Then it may

Paper 2358H, first received 17th June and in revised form 20th August 1982

Mr. Ferrari is, and Dr. Webb and Dr. Maile were formerly, with the University Engineering Department, Trumpington Street, Cambridge CB2 1PZ, England. Dr. Webb is now with the Department of Electrical Engineering, McGill University Montreal PQ, Canada H3A 2A7. Dr. Maile is now with Patscentre International, Cambridge Division, Melbourne, Royston, Herts. SG8 6DP, England

Reprinted with permission from *IEE Proc.*, vol. 130, pt. H, no. 2, pp. 153–159, Mar. 1983.

be shown that $F(H_r, H_r)$, $F(H_i, H_i)$ and $F(H_r, H_i)$ are all stationary with respect to their arguments; the stationary points H_{r0} and H_{i0} are the real and imaginary parts of the correct solution H_0 to Maxwell's equations. Thus $\alpha F(H_r, H_r) + \beta F(H_i, H_i) + \gamma F(H_r, H_i)$ has H_0 as a stationary point, for any numbers α, β, γ. The functionals (eqns. 4 and 5) correspond to $(\alpha, \beta, \gamma) = (1, 1, 0)$ and $(1, -1, 2j)$, respectively. In fact with ϵ, μ and k real these two functionals give rise to the same matrix equation on discretisation, and so the discrete solutions from each are identical.

As this is the case, and as also eqn. 5 is more general in its range of application than eqn. 4, the rest of this paper will deal exclusively with eqn. 5. When the material properties and field solutions are known to be purely real, the appropriate arrays in the computational scheme may also be set as real instead of complex, and so there is no loss of efficiency in using the more general functional.

3 Finite-element discretisation

The above three-component vector variational formulation has been used by Konrad [12] as the basis of a two-dimensional finite-element technique for computing the modes of loaded waveguides. Essentially the method involved specifying the field variation to be $\exp(-j\beta z)$ in the z-direction, and modelling it with high-order polynomial triangular finite elements in the x-y plane. The numerical scheme proposed here lifts the restriction on the z-variation, and models the field with tetrahedral finite elements.

The volume V is broken into tetrahedra, in each of which the material properties ϵ and μ are constant, although they may be discontinuous across element boundaries. In each tetrahedron, the magnetic field is approximated by a trial function complete to Mth order in the space co-ordinates; thus

$$H(r) = H^m \alpha^m(\zeta) \tag{7}$$

In expr. 7 $m = 1, \ldots, N$, where $N = (M+1)(M+2)(M+3)/6$, $\alpha^m(\zeta)$ is an interpolation polynomial of Newton-Cotes type [13, 14], and $\zeta = (\zeta_1, \zeta_2, \zeta_3, \zeta_4)$ is the vector of homogeneous (volume) co-ordinates [14] corresponding to the Cartesian coordinates $r = (x, y, z)$. Note that the polynomials are constructed in such a way that the trial function unknown H^m is the magnetic field at the mth node r^m of the tetrahedron (see Fig. 1).

Substitution of expr. 7 into the expression for the functional eqn. 5 allows the value of the functional for the tetrahedron to be written in terms of the unknowns H^m. The final expression is [2]

$$F = \frac{1}{2} H_s^m H_t^n \left(\frac{V}{\epsilon} \hat{Q}^{ij}_{mn} K^{st}_{ij} - V\mu k^2 \delta_{st} T_{mn} \right) \tag{8}$$

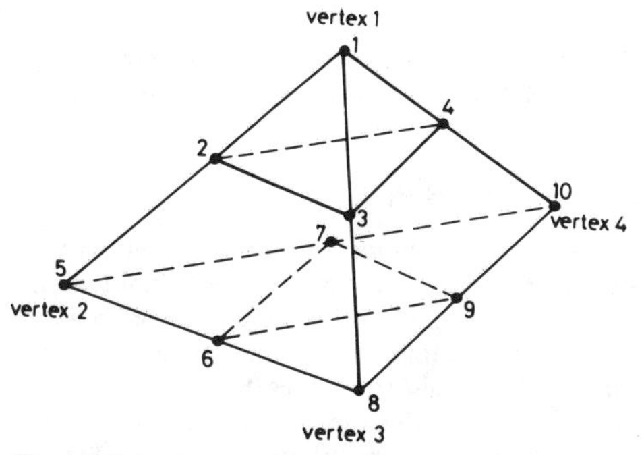

Fig. 1 *Second-order tetrahedron ($M = 2$) showing ten nodes at which magnetic field is unknown vector H^m*

where:

H_s^m is the sth Cartesian component of field at node m

V = volume of the tetrahedron

δ_{st} = the Kronecker delta

$$\hat{Q}^{ij}_{mn} = 6 \int_V \frac{\partial \alpha_m}{\partial \zeta_i} \frac{\partial \alpha_n}{\partial \zeta_j} d\Omega$$

$$T_{mn} = 6 \int_V \alpha_m \alpha_n \, d\Omega$$

$$K^{st}_{ij} = \delta_{st} \frac{\partial \zeta_i}{\partial r_w} \frac{\partial \zeta_j}{\partial r_w} - \frac{\partial \zeta_i}{\partial r_t} \frac{\partial \zeta_j}{\partial r_s}$$

$m, n = 1, \ldots, N$ (tetrahedron node numbering)

$s, t, w = 1, 2, 3$ (Cartesian axis labelling)

$i, j = 1, 2, 3, 4$ (tetrahedron vertex numbering)

$d\Omega = d\zeta_1 d\zeta_2 d\zeta_3 = d\zeta_1 d\zeta_3 d\zeta_4 = \ldots$ as appropriate, and the summation convention for repeated indices is assumed.

Following the example of Silvester [13, 14], F has been written in terms of:

(i) two matrices $\hat{Q}$ and T, which are independent of the tetrahedron geometry, and may be computed once and for all and stored with the program

(ii) a 12×12 matrix K, which must be computed for each element, but which is a simple function of the Cartesian co-ordinates of the tetrahedron vertices, involving no numerical integration.

Notice that the T-matrix is the same as that produced by Silvester in Reference 14. It may be shown that $\sum_{i=1}^{4} K^{st}_{ij} = 0$, and so that $\hat{Q}^{ij}_{mn} K^{st}_{ij}$ can be rewritten

$$\sum_{i=1}^{4} \sum_{\substack{j=1 \\ j \neq i}}^{4} Q^{ij}_{mn} K^{st}_{ij}$$

where

$$Q^{ij}_{mn} = \hat{Q}^{ij}_{mn} - \hat{Q}^{ii}_{mn}$$

As $K^{st}_{ij} \neq K^{st}_{ji}$, this Q-matrix cannot be further reduced to the symmetric Q-matrix of Silvester [14].

We have computed T and $\hat{Q}$ up to order 4, and have them stored. (T-matrices up to order 3 are given in Reference 14.)

Using expr. 8 the functional for the whole volume may be written in terms of a symmetric matrix W and a column vector H_c of all the nodal field components:

$$F = H_c^T W H_c \tag{9}$$

The next step is to apply the Dirichlet boundary conditions. To simplify this, the Dirichlet surface is allowed to consist only of a number of planes, each perpendicular to one of the three Cartesian axes of the problem. Then prescribing $H \times n = P$ on, for example, a plane perpendicular to the z-axis, involves setting $H_x = -P_y$, $H_y = P_x$ and leaving H_z free to vary. The matrix equation (eqn. 9) partitions simply:

$$F = H_F^T W_{FF} H_F + 2 H_F^T W_{FP} H_P + H_P^T W_{PP} H_P \tag{10}$$

where H_F is a column vector of free components and H_P is a column vector of prescribed components.

Finally, to find an approximate solution to the problem, the functional F is made stationary with respect to all variations of vector H_F. This leads to the linear matrix equation

$$W_{FF}H_F = -W_{FP}H_P \quad (11)$$

which may be solved for the unknowns H_F.

4 Application: waveguide problems

An application of the method is in determining scattering by obstacles in waveguide structures. The general case is shown in Fig. 2. Two hollow waveguides with perfectly conducting walls but arbitrary and different cross-sections meet in an irregularly shaped region containing linear, isotropic, sourceless, inhomogeneous materials. The problem is to compute the scattering matrix S of the junction region, with respect to reference planes P_1 and P_2. S is given by

$$\begin{pmatrix} b_1 \\ b_2 \end{pmatrix} = \begin{pmatrix} S_{11} & S_{12} \\ S_{21} & S_{22} \end{pmatrix} \begin{pmatrix} a_1 \\ a_2 \end{pmatrix} \quad (12)$$

where a_1, b_1, a_2, b_2 are the wave amplitudes of forward and reverse dominant-mode waves in the two guides, as shown in Fig. 2. It is assumed that the frequency is such that only one mode can propagate in each guide.

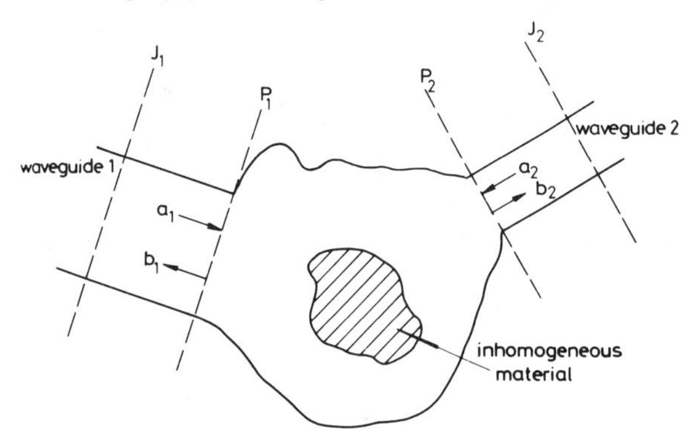

Fig. 2 *Cross-section through typical 2-port waveguide in scattering problem*

The wave amplitudes at reference planes P_1 and P_2 are indicated; J_1 and J_2 are boundary planes.

Notice that this general formulation includes as special cases many commonly occurring waveguide problems: metallic irises, strips and posts in a single guide; junctions between two different guides; bends and changes of cross-section etc. Analytic and semianalytic methods are available for some of these problems [15]. In addition, a number of numerical techniques have been applied to cases where there is variation in only two dimensions [16–18]. However, with the exception of Wang's volume integral technique [6] for rectangular guide, the three-dimensional problem with an arbitrarily shaped obstacle has not been solved.

The present method may be used to find scattering parameters by setting up and solving a series of closed cavity problems. Two planes J_1 and J_2 (Fig. 2) are introduced, sufficiently far from the inhomogeneous region that the field may be taken as purely dominant mode on them. The tangential field H_{ti} is specified on each plane J_i, $i = 1, 2$:

$$H_{ti} = A_i h_i \quad (13)$$

where h_i is the tangential component of the dominant mode field in guide i, and A_i is an arbitrary amplitude. The problem, bounded by J_1, J_2, the guide walls and the walls of the inhomogeneous region, is now of a form which may be modelled by the finite-element method. A field solution is computed, and from it the ratios b_1/a_1 and b_2/a_2 obtained [2]. The procedure is repeated three times, each time with a different pair of dominant-mode amplitudes A_1 and A_2, yielding three pairs of ratios. These may be used with eqn. 12 to find the three parameters S_{11}, S_{12} and S_{22} ($S_{21} = S_{12}$ because only isotropic material is present [23]).

In fact it is convenient to keep one of the amplitudes, say on J_1, equal to zero, so b_1/a_1 is known in advance. Then changing the position of J_1 each time will provide the three cases required. Note that this is very similar to a standard experimental technique for finding scattering parameters [19].

The wave-amplitude ratios (reflection coefficients) may be extracted from the field solution by interpolation of the nodal field values. However, there is a better method, used by Sinnott *et al.* [20] and Lamb [21] with two-dimensional problems. If plane J_2 is the only plane on which an ID boundary condition is prescribed (i.e. on J_1 the amplitude is zero), then it may be shown that

$$F(H_0) = KZ \quad (14)$$

where F is the functional of eqn. 5, H_0 is the correct solution to Maxwell's equations, K is a constant dependent only on the mode concerned, and Z is the normalised impedance at J_2, looking towards P_2. The stationarity of F at H_0 suggests that a good estimate for Z would be Z':

$$Z' = \frac{1}{K}F(H_0') \quad (15)$$

where H_0' is the finite-element solution for the magnetic field. The estimation of $F(H_0')$ from the nodal field values is straightforward because it involves only the Q-, T- and K-matrices used before (eqn. 8). From Z', b_2/a_2 may be readily calculated.

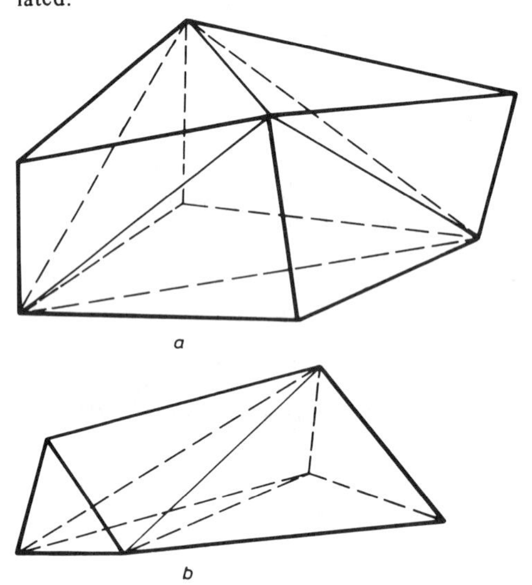

Fig. 3 *Division of hexahedra and pentahedra into tetrahedra*

a Hexahedron = five tetrahedra *b* Pentahedron = three tetrahedra

5 Computer program and results

A general finite-element computer program has been written to implement the numerical method. Element input is via hexahedra (8-noded) bricks) and pentahedra (6-noded bricks), which are broken into tetrahedra by the program (5 per hexahedron and 3 per pentahedron – see Fig. 3). The matrix equation (eqn. 11) is solved by the frontal method of Irons [22]. The program includes a subroutine for evaluating $F(H_0')$ from the field solution (see above).

Results are presented for four problems.

5.1 Slab in rectangular waveguide

A slab of material with complex permeability fits into the end of a piece of short-circuited rectangular waveguide; the geometry and parameters are shown in Fig. 4. TE_{10} dominant mode is incident. This problem is capable of analytic solution, and allows a comparison to be made between the correct magnetic field and the finite-element solution. The boundary conditions are homogeneous Neumann (perfect conductor) everywhere except on the input plane where TE_{10} transverse magnetic field is specified, i.e.

$$H_x = \sin(\pi x/a); \qquad H_y = 0$$

Two third-order hexahedra were used to model the problem.

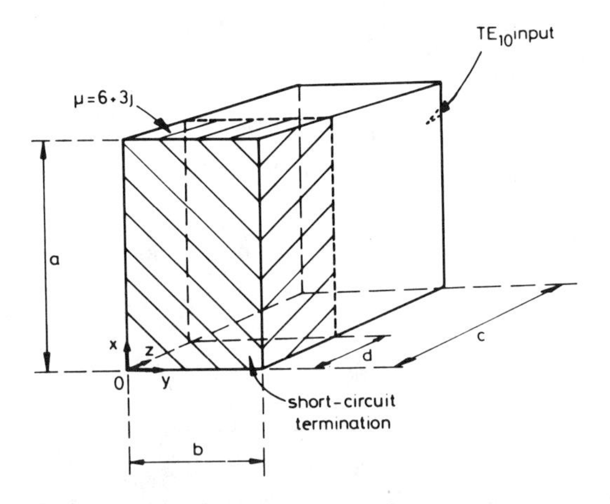

Fig. 4 *Slab with complex permeability in rectangular waveguide*

Dimensions $a = 2$ m, $b = 1$ m, $c = 9$ m, $d = 0.3$ m; normalised frequency $k = 2.22$ rad/m

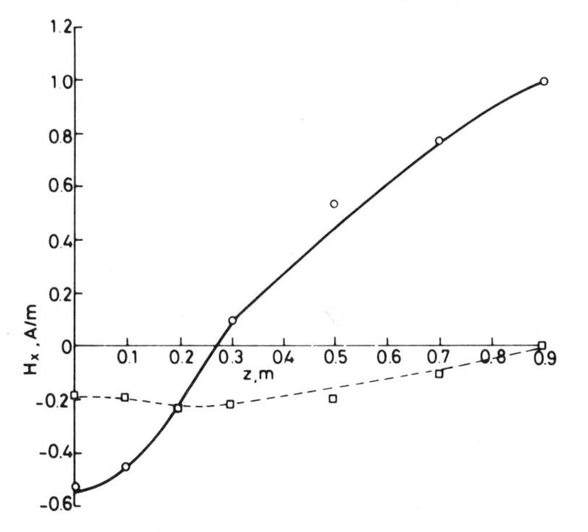

Fig. 5 *$H_x(a/2, y, z)$ against z for all y, in the problem of Fig. 4*

Re H_x: ——— analytical solution ○ computed solution
Im H_x: – – – analytical solution □ computed solution

From the field solution at each plane perpendicular to the z-axis, two complex numbers were extracted by averaging – one representing the value of H_x at $x = a/2$, and the other representing the value of H_z at $x = 0$ (or $x = a$). Figs. 5–7 show plots of these numbers against the z co-ordinate; both analytic and finite-element results are shown.

Note that the finite-element solution for H_x is quite accurate. It is worse for H_z because the latter is discontinuous at $z = d$, but even so the value taken by the solution at the discontinuous point is approximately the average of the two correct values there.

Although this problem is not truly three-dimensional it represents the first instance, to the authors' knowledge of loss being taken into account in a general way in a finite-element electromagnetic-wave analysis.

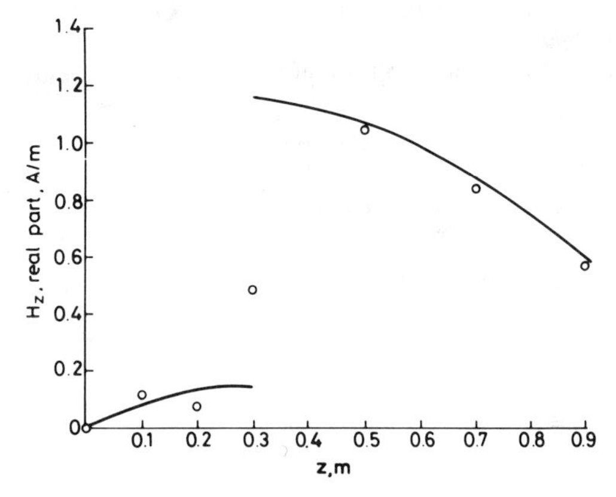

Fig. 6 *Real part of $H_z(0, y, z)$ against z for all y, in the problem of Fig. 4*

——— analytical solution ○ computed solution

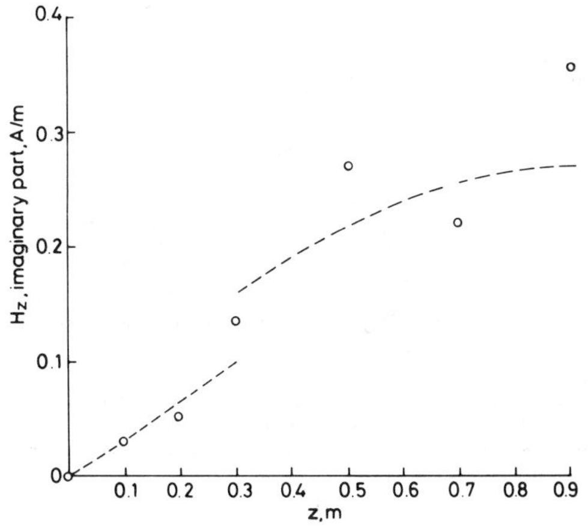

Fig. 7 *Imaginary part of $H_z(0, y, z)$ against z for all y, in the problem of Fig. 4*

——— analytical solution ○ computed solution

5.2 Resonance frequency of a loaded cavity

The rectangular cavity containing a rectangular block of dielectric is shown in Fig. 8. A TE_{10} field distribution was prescribed on plane $z = 0$; all other walls were taken as perfectly conducting. The electric field within the cavity was computed at a given frequency using the finite-element program with 400 first-order tetrahedra. This was repeated for several frequencies until the solution became very large, indicating that the plane $z = 0$ lay close to a field null. In this way, the TE_{101}-limit resonant frequency was estimated, i.e. the frequency of that mode which in the homogeneous limit becomes the empty-cavity TE_{101} mode. Results were obtained for a range of parameters u/c, and are shown in Fig. 9. Also in Fig. 9 are four error bars showing the value and estimated error of other results for the same problem: the two shorter ones were obtained by Akhtarzad and Johns with the TLM method [8] and the two longer ones by Albani and Bernardi [7]. The procedure described here is alternative to that of solving an eigenvalue problem to find the resonant frequency parameter $k_c a$ directly. When using the relatively small number of first-order nodes possible at the time, tests with an unloaded cavity showed confused eigenvalue results. The frequency search method nevertheless gave acceptable agreement with the known solution.

5.3 Scattering parameters of a dielectric cylinder in rectangular guide

The cylinder and guide are shown in Fig. 10. The cylinder has a circular cross-section with its axis parallel to that of the rectangular guide. Dominant mode TE_{10} is incident, and this is the only propagating mode at the frequencies concerned. Using all three planes of symmetry, one eighth of the problem was modelled, with the electric field as the unknown, using

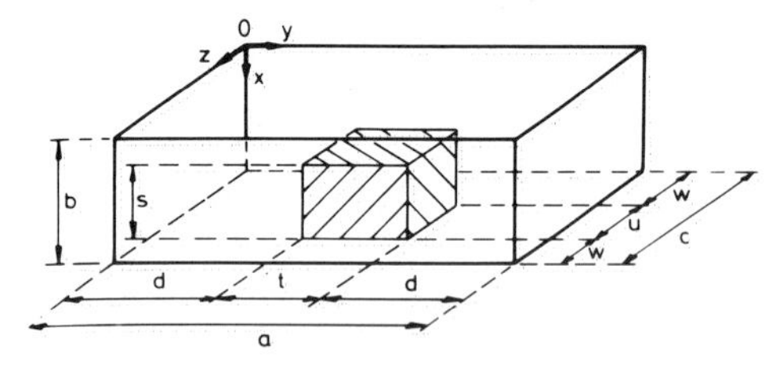

Fig. 8 *Rectangular cavity loaded with dielectric block; $\epsilon = 16$, $t/a = 1/4$, $b/a = 3/10$, $c/a = 4/10$, $s/b = 7/12$*

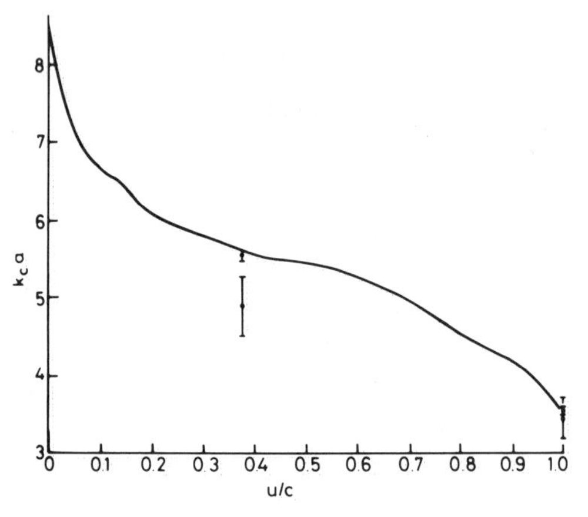

Fig. 9 *Resonant frequency $k_c a$ (TE_{101}-limit mode) against u/c for rectangular cavity loaded with rectangular block*
—— computed solution; long error bars: Albani and Bernadi [7]; short error bars: Akhtarzad and Johns [8]

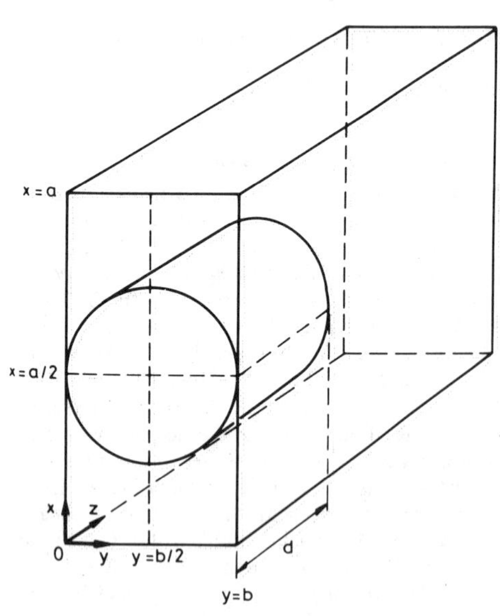

Fig. 10 *Dielectric cylinder in rectangular waveguide; $\epsilon = 2$, $a = 2$ m, $b = 1$ m, $d = 0.5$ m, normalised frequency range $k = 2.0 - 2.7$ rad/m*

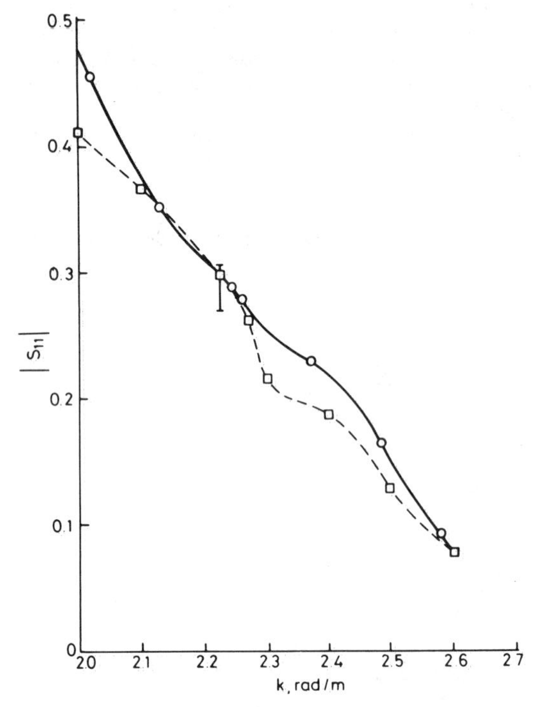

Fig. 11 *$|S_{11}|$ against k for dielectric cylinder of Fig. 10 with an incident TE_{10} waveguide mode*
—o— Lamb's solution [21] --□-- finite-element solution
Error bar gives rigorous limits at $k = 2.22$ rad/m [24]

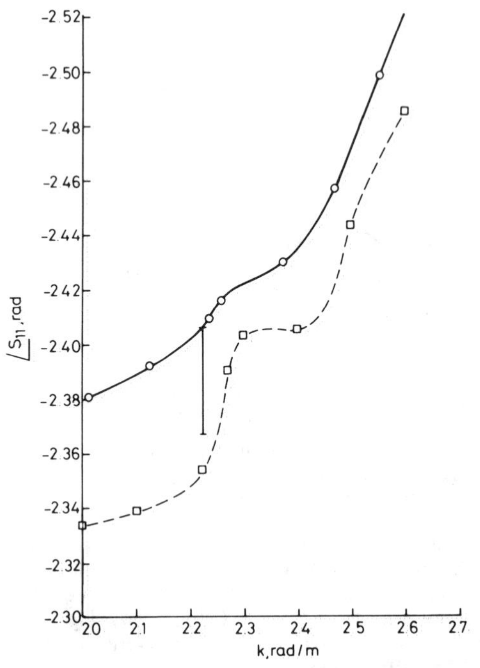

Fig. 12 *$\angle S_{11}$ against k for dielectric cylinder as described in Fig. 11*
—o— Lamb's solution [21] --□-- finite-element solution
Error bar gives rigorous limits at $k = 2.22$ rad/m [24]

20 pentahedra and 15 hexahedra (all second-order). Because the obstacle is lossless and symmetrical, one parameter S_{11} suffices to completely describe the scattering [23]. The magnitude and phase of S_{11} are plotted against the frequency k in Figs. 11 and 12. Also plotted for comparison are Lamb's results [21] for the same problem, obtained variationally with modes of the empty cavity as basis functions and working via the resonant frequency. The error bars are rigorous limits at frequency $k = 2.22$ rad/m, obtained by Bartram and Spruch [24].

5.4 Scattering parameters of a metallic sphere in rectangular guide

The sphere is perfectly conducting, has radius R and is located at the centre of the guide cross-section ($a = 2$ m, $b = 1$ m). TE_{10} mode at frequency $k = 2.22$ rad/m is incident, and again this is the only propagating mode at this frequency. The magnetic field was used as variable (so that the curved surface of the sphere provided a natural boundary condition), and all three symmetry planes were taken into account so that only one eighth of the problem had to be modelled. 19 pentahedra and 20 hexahedra, all second-order, were employed. Plots of magnitude and phase of S_{11} against radius R are given in Figs. 13 and 14, along with results based on the work of Hinken [25], who solved the problem by a perturbation method. The divergence of the finite element and perturbation results for large radii may be explained by the fact that the latter method is only accurate for small obstacles. Nevertheless, there is good agreement for $R < b/4$.

This problem involved 358 nodes (including those generated by the program), required 230 kbytes of immediate access memory and took 28 s of CPU time to run on an IBM 370/165.

6 Conclusions

A numerical scheme has been presented which is capable of solving a very general class of three-dimensional electromagnetic problems. The scheme uses the method of finite elements, allowing arbitrary geometries to be modelled, and employing comparatively sophisticated trial functions for the field.

A simple problem involving a complex material property has been solved, the resulting field being quite close to the analytic solution. A resonant frequency of a loaded cavity has been found accurately. In addition, it has been shown how the scheme can be applied to waveguide scattering problems. Two cases involving lossless obstacles in rectangular guide have been solved with an estimated error of about 0.04 in the modulus and phase (radians) of the scattering parameter S_{11}.

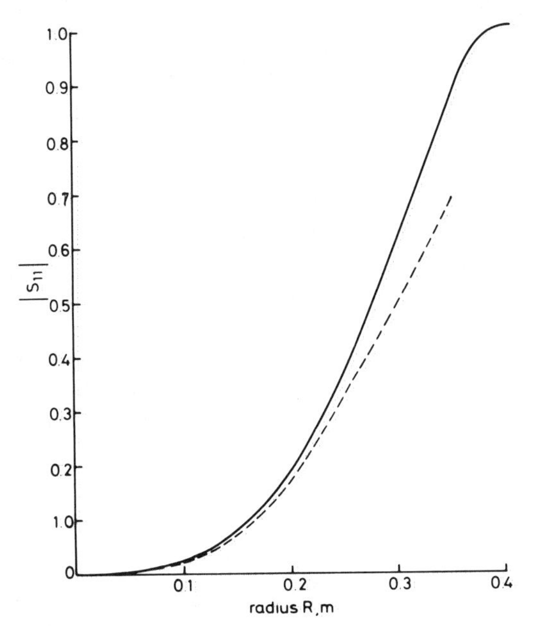

Fig. 13 $|S_{11}|$ *against radius R for a metallic sphere in a rectangular waveguide, TE_{10} mode incident at normalised frequency $k = 2.22$ rad/m*

——— perturbation method [25] – – – finite-element method

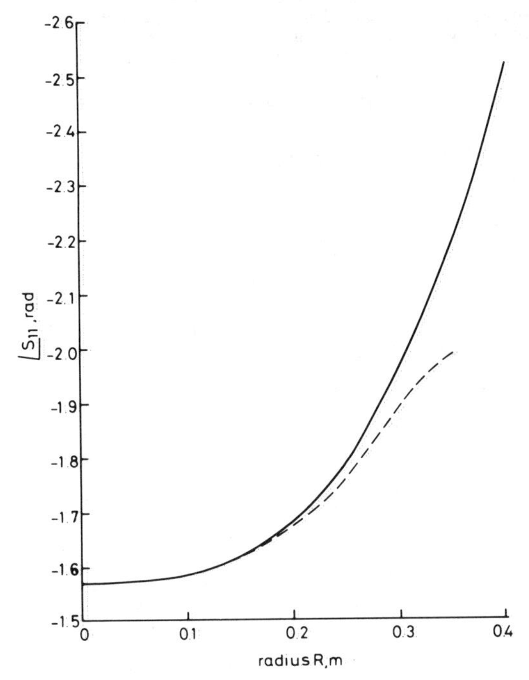

Fig. 14 $/S_{11}$ *against radius R for the metallic sphere in waveguide as described in Fig. 13*

——— perturbation method [25] – – – finite element method

7 Acknowledgments

The authors are grateful to Dr. J.E. Carroll for his helpful remarks and criticisms, particularly with regard to aspects of the variational formulation. We would also like to thank the UK Science & Engineering Research Council for providing financial support for this work.

8 References

1 MAILE, G.L.: 'Three-dimensional analysis of electromagnetic problems by finite element methods'. Ph.D. thesis, University of Cambridge, Dec. 1979
2 WEBB, J.P.: 'Developments in a finite element method for three-dimensional electromagnetic problems'. Ph.D. thesis, University of Cambridge, Sept. 1981
3 FERRARI, R.L., and MAILE, G.L.: 'Three-dimensional finite-element method for solving electromagentic problems', *Electron. Lett.*, 1978, **14**, (15), pp. 467–468
4 CSENDES, Z.J., and SILVESTER, P.: 'Numerical solution of dielectric loaded waveguides: I – finite-element analysis', *IEEE Trans.*, 1970, **MTT-18**, pp. 1124–1131
5 NG, F.L.: 'Tabulation of methods for the numerical solution of the hollow waveguide problem', *ibid.*, 1974, **MTT-22**, pp. 322–329
6 WANG, J.J.H.: 'Analysis of a three-dimensional arbitrarily shaped dielectric or biological body inside a rectangular waveguide', *ibid.*, 1978, **MTT-26**, pp. 457–462
7 ALBANI, M., and BERNARDI, P.: 'A numerical method based on the discretization of Maxwell's equations in integral form', *ibid.*, 1974, **MTT-22**, pp. 446–450

8 AKHTARZAD, S., and JOHNS, P.B.: 'Solution of Maxwell's equations in three space dimensions and time by the t.l.m. method of numerical analysis', *Proc. IEE*, 1975, **122**, (12), pp. 1344–1348

9 MORSE, P.M.: and FESHBACH, H.: 'Methods of theoretical physics. Part 1' (McGraw-Hill, New York, 1953)

10 BERK, A.D.: 'Variational principles for electromagnetic resonators and waveguides', *IRE Trans.*, 1956, **AP-4**, pp. 104–111

11 CHEN, C.H., and LIEN, C.D.: 'The variational principle for non-self-adjoint electromagentic problems', *IEEE Trans.*, 1980, **MTT-28**, pp. 878–886

12 KONRAD, A.: 'Triangular finite elements for vector fields in electromagnetics'. Ph.D. thesis, McGill University, Sept. 1974

13 SILVESTER, P.: 'High-order polynomial triangular finite elements for potential problems', *Int. J. Engng. Sci.*, 1969, **7**, pp. 849–861

14 SILVESTER, P.: 'Tetrahedral polynomial finite elements for the Helmholtz equation', *Int. J. Numer. Meth. Engns.*, 1972, **4**, pp. 405–413

15 MARCUVITZ, N.: 'Waveguide handbook' (McGraw-Hill, New York, 1951)

16 WEXLER, A.: 'Solution of waveguide discontinuities by modal analysis', *IEEE Trans.*, 1967, **MTT-15**, pp. 508–517

17 WU, S., and CHOW, Y.L.: 'An application of the moment method to waveguide scattering problems', *ibid.*, 1972, **MTT-20**, pp. 744–749

18 MUILWYK, C.A., and DAVIES, J.B.: 'The numerical solution of rectangular waveguide junctions and discontinuities of arbitrary cross-section', *ibid.*, 1967, **MTT-15**, pp. 450–455

19 SUCHER, M., and FOX, J.: 'Handbook of microwave measurements. Vol. 1' (Polytechnic Press, New York, 1963)

20 SINNOTT, D.H., CAMPBELL, G.K., CARSON, C.T., and GREEN, H.E.: 'The finite difference solution for wicrowave circuit problems', *IEEE Trans.*, 1969, **MTT-17**, pp. 464–478

21 LAMB, R.I.: 'The numerical solution of discontinuity problems in waveguides'. Ph.D. thesis, University of Cambridge, 1973

22 IRONS, B.M.: 'A frontal solution program for finite element analysis', *Int. J. Numer. Meth. Engng.*, 1970, **2**, pp. 5–32

23 COLLIN, R.E.: 'Foundations of microwave engineering' (McGraw-Hill, 1966)

24 BARTRAM, R., and SPRUCH, L.: 'Bounds on the elements of the equivalent network for scattering in waveguides. II. Application to dielectric obstacles', *J. Appl. Phys.*, 1960, **31**, pp. 913–917

25 HINKEN, J.H.: 'Conducting spheres in rectangular waveguides', *IEEE Trans.*, 1980, **MTT-28**, pp. 711–714

Finite-Element Analysis of Slow-Wave Schottky Contact Printed Lines

CHING-KUANG TZUANG, STUDENT MEMBER, IEEE, AND TATSUO ITOH, FELLOW, IEEE

***Abstract*—Extensive finite-element analyses on MMIC slow-wave structures with both localized and layered models are presented. Good agreement is achieved between the data presented here and other theoretical results and experiments. Higher order elements that improve accuracy are discussed. The comparative studies for Schottky contact microstrip and coplanar waveguide with localized and layered models are presented. Potential applications of the localized models to more general and practical slow-wave circuits are also discussed.**

I. Introduction

THE ADVANCE in monolithic microwave integrated circuits (MMIC's) has led to widespread applications of microstrip and other planar transmission lines such as coplanar waveguide (CPW) and coupled microstrip on semiconductor substrate. In addition to interconnection or transmission-line applications, this class of transmission lines can be employed as circuit elements such as phase shifters, voltage-tunable filters, and voltage-controlled attenuators [1]–[3].

These applications are made possible by the slow-wave propagation resulting from electron–electromagnetic interaction with the lossy semiconductor material. The device was experimentally studied with metal–insulator–semiconductor (MIS) configurations and with Schottky contact microstrip or CPW structures [3]–[6] and was theoretically investigated by a number of techniques, such as spectral-domain analysis (SDA), the mode-matching method, and the finite-element method (FEM) [7]–[10] based on MIS or so-called layered models. However, practical semiconductor devices such as Schottky contact microstrip or CPW are certainly not laminated structures. Instead, they both have localized depletion regions on semiconductors. The effects of these localized depletion regions have never been discussed. It is plausible to contemplate that the layered model commonly used for the analysis may not correctly describe the actual field distributions in the structure with localized depletion regions.

In this paper, the finite-element method based on the $E_z - H_z$ formulation [9], [11] is used for a variety of slow-wave structures which can best be described by the localized model. In conjunction with this study, use of higher order elements, e.g., quadratic isoparametric elements, is discussed and its results are presented.

Manuscript received April 1, 1986; revised June 25, 1986. This work was supported in part by Grant AFOSR-86-0036.

The authors are with the Department of Electrical and Computer Engineering, University of Texas at Austin, Austin, TX 78712.

IEEE Log Number 8610550.

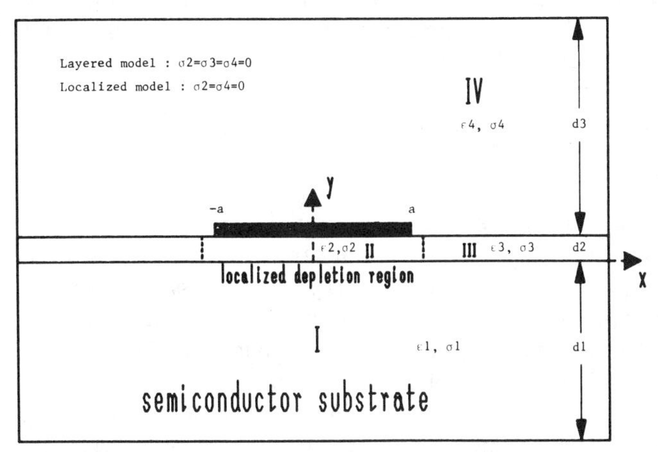

Fig. 1. Localized depletion model for Schottky contact *microstrip*.

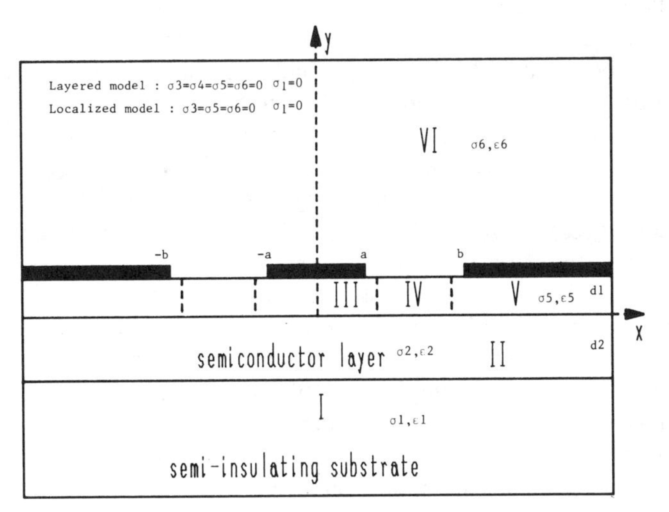

Fig. 2. Localized depletion model for Schottky contact *coplanar waveguide*.

II. Localized Models

The localized depletion models for Schottky contact microstrip and CPW are illustrated in Figs. 1 and 2, respectively. By setting the conductivity in region III of Fig. 1 and that in region IV of Fig. 2 to zero, the localized models will be reduced to the conventional MIS or layered models.

Obviously, the boundaries of the depletion region in actual devices are not straight lines but are curved. Such curved boundaries can be found from solutions of a static Poisson's equation for a given bias condition. The present algorithm can then be applied to such geometry. However,

Reprinted from *IEEE Trans. Microwave Theory Tech.*, vol. MTT-34, no. 12, pp. 1483–1489, Dec. 1986.

the objective of the present paper is to study the effect of localization. The essential feature can be found without solving the structures with curved depletion boundaries. Therefore, the localized depletion region is assumed to be rectangular in shape.

Furthermore, we are also interested in the development of low-loss slow-wave circuit elements. Schottky contacts for both center and ground strips of the CPW are assumed under the same dc bias conditions. It is found that the loss in this CPW structure can be less than the case where only the center conductor is a Schottky contact in the CPW.

III. Derivation of the Matrix Equation

A. Theory

The conventional $E_z - H_z$ formulation that results in a homogeneous coupled symmetric matrix equation is adopted in this paper [9], [11]. In the lossless case, the finite-element method with the $E_z - H_z$ formulation provides a variational solution. The matrix equation can be derived by a functional followed by the Ritz approximation [11] or by weighted residual integration followed by the Galerkin's method [9]. It is possible to obtain a variational statement of the problem even for lossy structure by dividing E_z and H_z components into real and imaginary parts [12]. Alternatively, the three-vector H formulation can be used [13], [14]. This includes mixed boundary conditions, and caution must be exercised in treating the singularity at the corner of conductor edge. The conventional $E_z - H_z$ formulation used here does not provide a variational solution for the lossy system. The method used here forces the residual to be zero by making it orthogonal to each member of a complete set of the trial functions. This method is one manifestation of the method of weighted residuals, which does not require the existence of a variational principle [15], [16]. Not only is the method of weighted residuals simple for implementation; in principle, a systematic improvement in the accuracy of solution can be obtained if enough terms of the trial functions are used. Section IV-B discusses such improvements if one chooses quadratic elements instead of bilinear elements in the lossy waveguide system.

For purposes of clarity, a brief description of the derived $E_z - H_z$ formulation is shown in the Appendix.

B. Application of the Finite-Element Method

The important steps for the actual coding of the final matrix equation obtained can be found in [17]. The hierarchy of the program can accept any isoparametric elements. In the present paper, we use four-node bilinear and quadratic eight-node elements. Their differences in terms of numerical results will be discussed.

By alternating E_z and H_z nodal variables in the column vector X, the matrix A becomes a banded matrix (Appendix). Finally, rows and columns corresponding to the Dirichlet boundary conditions are deleted such that E_z and H_z vanish at electric and magnetic walls, respectively. Therefore, the matrix A has dimension $N \times N$, where N equals $(2M - L)$. M is the total number of nodes and L stands for the sum of the total number of nodes located on electric or magnetic walls. The complex root of the equation, $\det(A) = 0$, is the solution for the propagation constant. The real and imaginary parts of the propagation constant correspond to the attenuation constant and slow-wave factor, respectively.

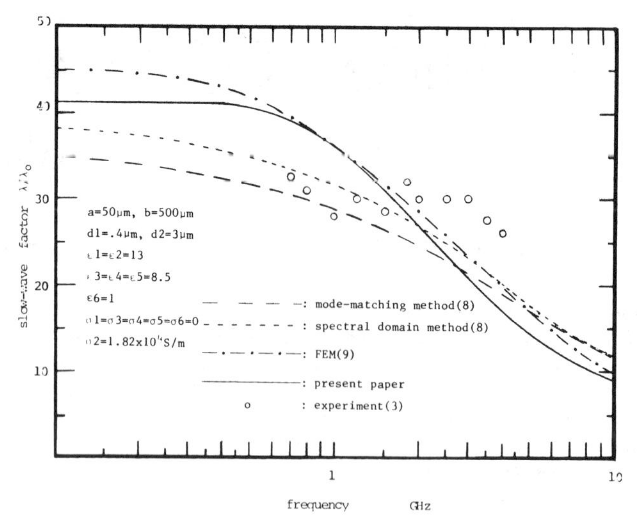

Fig. 3. Plots of *slow-wave factor* as obtained by the present paper and other theories such as mode-matching method [8], SDA [8], and FEM [9] together with experimental data [3] for an MIS CPW.

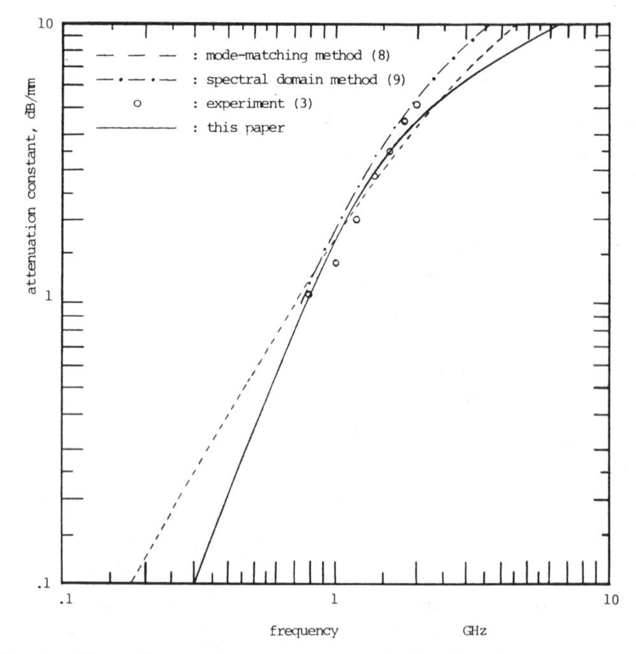

Fig. 4. Plots of *attenuation constant* as obtained by the present paper and other theories such as mode-matching method [8], SDA [9], and FEM [9] together with experimental data [3] for an MIS CPW.

IV. Numerical Results

A. Validity Check

The present FEM code is applied to lossless printed line structures, and excellent agreement has been obtained with the available data, such as [18, fig. 2.7 and fig. 7.11]. For the lossy layered case, Figs. 3 and 4 compare the results

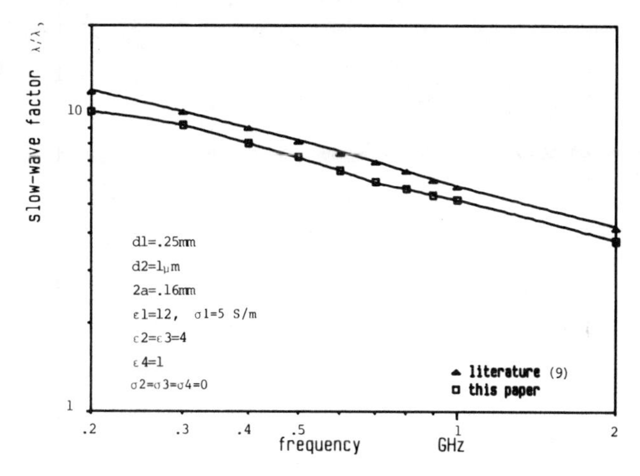

Fig. 5. Validity check of the results obtained by this paper and data in literature [9]; *slow-wave factor* versus frequency.

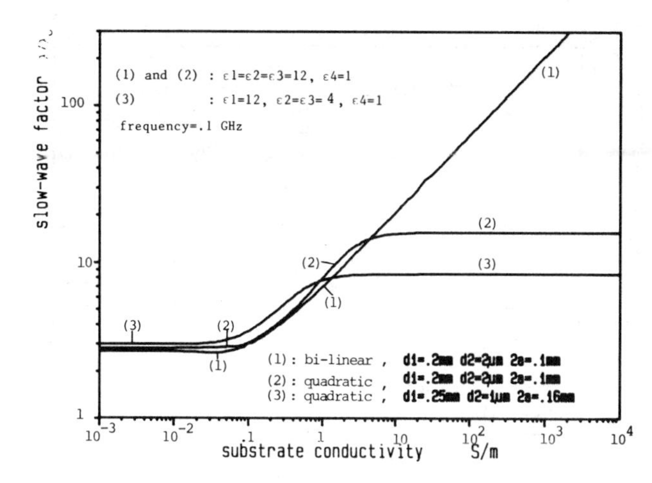

Fig. 7. Comparison of numerical results, *slow-wave factor* versus microstrip substrate conductivity, based on *quadratic and bilinear elements*.

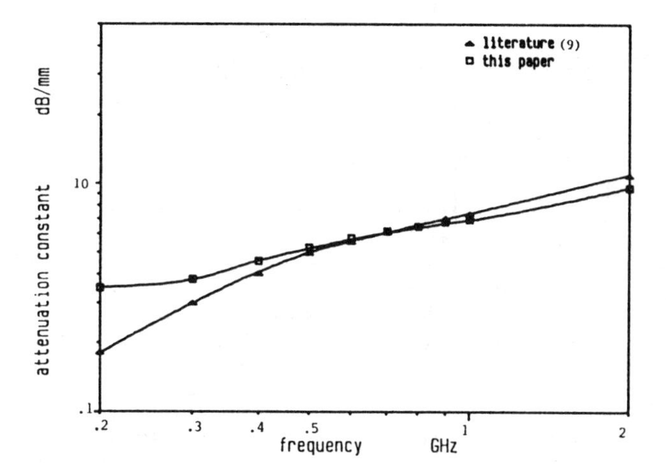

Fig. 6. Validity check of the results obtained by this paper and data in literature [9]; *attenuation constant* versus frequency.

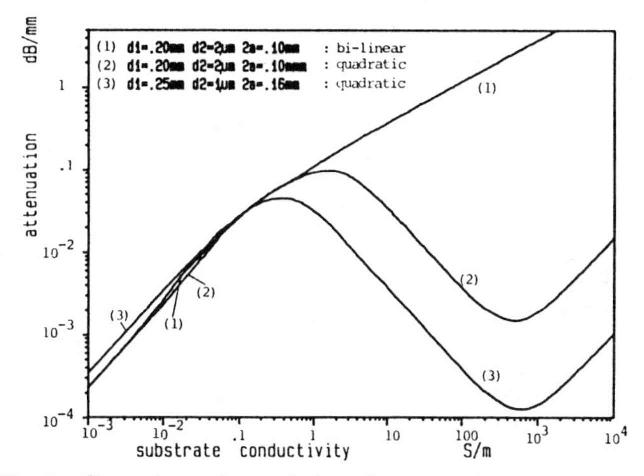

Fig. 8. Comparison of numerical results, *attenuation constant* versus microstrip substrate conductivity, based on *quadratic and bilinear elements*.

obtained by the present method as applied to the layered structure and other existing data for MIS CPW slow-wave propagation both theoretically and experimentally. The small discrepancy among various methods may be attributed to the CPW structure, which has a relatively high aspect ratio (b/a), i.e., 10, and a very thin insulating region. This makes it difficult to find very accurate answers over the frequency span of interest. In particular, it is well known that the slender element may yield poor results in FEM [19]. Owing to the fact that both the depletion and the lossy semiconductor regions are extremely thin, it is inevitable that slender elements exist if we do not use higher order and finer elements. Therefore, we apply quadratic eight-node elements, and a total of 96 nodes are employed for this particular structure.

The case for MIS microstrip slow-wave propagation is shown in Figs. 5 and 6, where the discretization based on bilinear elements and a total of 36 nodes are sufficient to match the data.

B. Bilinear Versus Quadratic Elements

Figs. 7 and 8 compare computational results for slow-wave propagation in an MIS microstrip model based on both bilinear and quadratic elements. Both the slow-wave factor and the attenuation constant agree very well in low substrate conductivity, say, less than 0.1 S/m.

When bilinear elements are used, the slow-wave factor and the attenuation constant approach infinity as substrate conductivity increases. Clearly this is nonphysical because very high conducting substrate can be regarded as a metal layer and the slow-wave factor should be brought down when the substrate turns into metal. In fact, in the case of quadratic elements, the slow-wave factor starts to decline gradually and loss starts to increase again when substrate conductivity is increased from approximately 500 S/m. Similar comparative studies are performed for CPW localized model. We also obtain similar phenomena in that both the slow-wave factor and the attenuation constant approach infinity when we employ bilinear elements and increase substrate conductivity.

Another type of comparative study is performed, i.e., numerical computations of the slow-wave factor and attenuation constant versus frequency by applying both bilinear and quadratic elements on the same CPW structure. The results are illustrated in Figs. 9 and 10. In these figures, curves based on quadratic elements are identical to

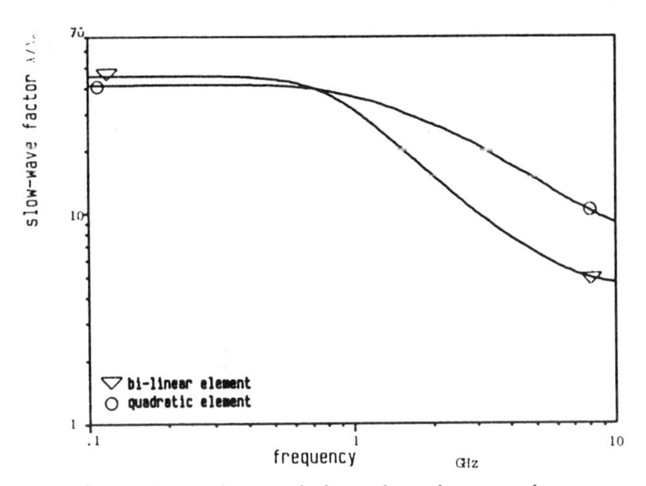

Fig. 9. Comparison of numerical results, *slow-wave factor* versus frequency, on the same MIS CPW structure analyzed in Figs. 3 and 4 based on *quadratic and bilinear elements*.

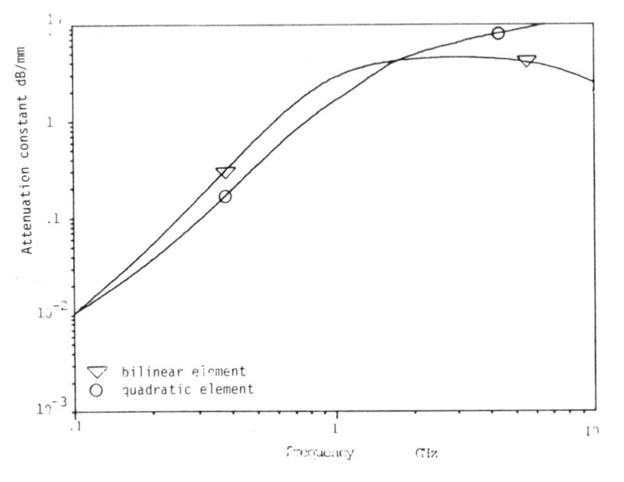

Fig. 10. Comparison of numerical results, *attenuation constant* versus frequency, on the same MIS CPW structure analyzed in Figs. 3 and 4 based on *quadratic and bilinear elements*.

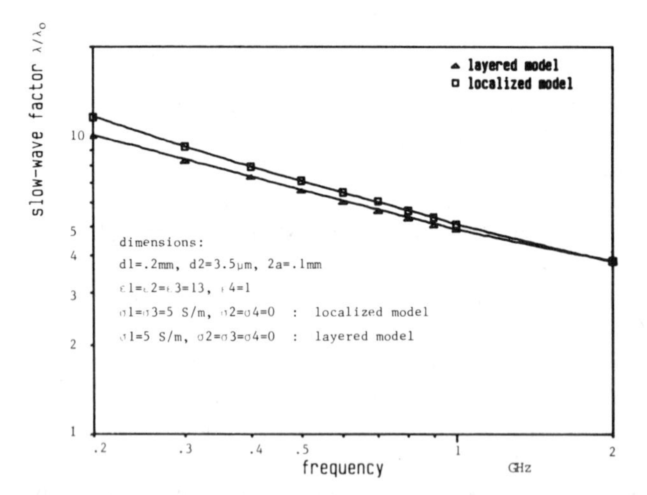

Fig. 11. Comparisons of plots of *slow-wave factor* versus frequency based on *localized and layered* models for Schottky contact microstrips.

those used in Figs. 3 and 4. The solutions obtained by means of quadratic elements are apparently much closer to those obtained by other theoretical methods and experiments. Additional studies on MIS microstrip also draw the same conclusion.

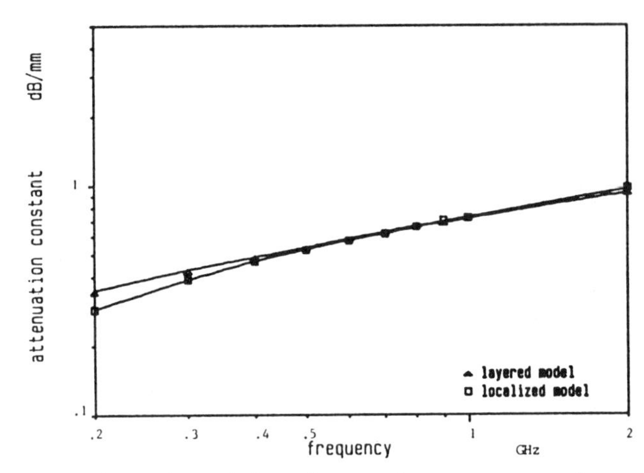

Fig. 12. Comparison of plots of *attenuation constant* versus frequency based on *localized and layered* models for Schottky contact microstrips.

TABLE I
COMPARISON OF SLOW-WAVE PROPAGATION FOR LAYERED AND LOCALIZED MICROSTRIP MODELS

		(slow-wave factor, dB/mm)	
MODEL	σ1(S/m) / σ3(S/m)	5	20×10^3
Layered	0	(8.364661, .0072815)	(8.397601, .0019445)
Localized	5	(8.391247, .0073026)	(8.397601, .0019502)
	20×10^3	(8.398345, .0067541)	(8.399383, .0019613)

Conditions: $a = 80$ μm, d1 = 250 μm, d2 = 1 μm, $\epsilon 1 = 12$, $\epsilon 2 = \epsilon 3 = 4$, $\epsilon 4 = 1$, $\sigma 2 = \sigma 4 = 0$ (microstrip). Frequency = 0.1 GHz.

Since the accuracy of employing the quadratic elements in our particular application is confirmed, the rest of the computations in the paper are based on quadratic elements, except the results shown in Figs. 11 and 12.

C. Effects of Localized Depletion Regions on Slow-Wave Propagation

It is clear in Figs. 11 and 12 that the MIS layered model is a good model for *microstrip* slow-wave propagation. This conclusion is confirmed again by using quadratic elements with improved accuracy. The solutions obtained by layered and localized models are very close and cannot be distinguished by plots. Table I shows how close the solutions are under different combinations of conductivities. For the localized CPW model, the results shown in Figs. 13 and 14 are rather interesting. Structure (2) has about half the gap width of structure (1), and all the rest of the conditions are almost the same. The conducting region IV has a stronger influence on structure (2). A noticeable transition region exists where loss starts to decline and the slow-wave factor starts to change abruptly but settles quickly as σ_4 increases. The skin depth, which is inversely proportional to the square root of the conductivity, is relatively large in region IV, approximately 50 mm when σ_4 equals 0.1 S/m. The electromagnetic field penetrates region IV freely and interacts with it in terms of lossy dielectric material. When σ_4 is increased, say, to 10^5 S/m, the skin depth is reduced to 50 μm. This is very close to

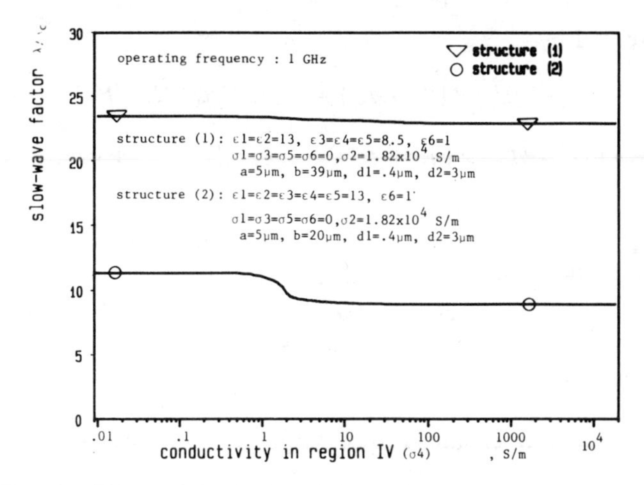

Fig. 13. Effects of *localized* depletion regions on CPW's; slow-wave factor versus conductivity of region IV in Fig. 2.

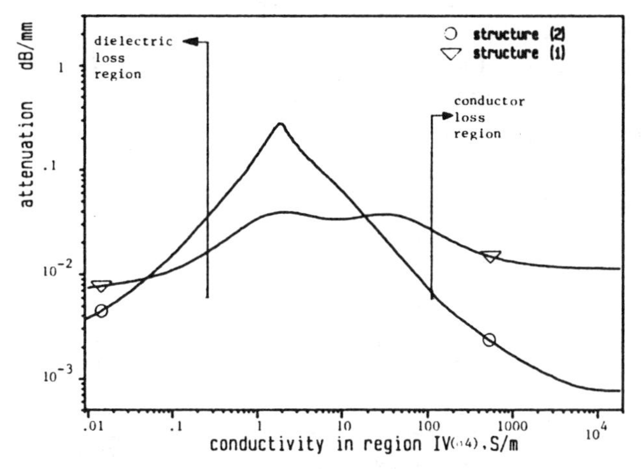

Fig. 14. Effects of *localized* depletion regions on CPW's; attenuation constant versus conductivity of region IV in Fig. 2.

the gap widths in structures (1) and (2). As a result, the highly conducting region IV can in effect be replaced by a piece of thin and imperfect metal inserted into the gap. The loss introduced at higher values of σ_4 is much more closely related to skin-effect ohmic loss. Generally speaking, we may separate the loss mechanism due to region IV in the structure into dielectric loss and conductor loss. The former increases as conductivity increases, and the latter decreases as conductivity increases. Common waveguides and transmission lines also exhibit similar loss behavior [20].

The qualitative discussions can be readily extended to obtain an equivalent transmission-line circuit representation of the localized CPW in Fig. 15. This empirical model happens to be a slight modification of the analytical model of the Schottky contact coplanar line based on semiempirical considerations and is indeed identical if we combine C_1 and C_3 into a single capacitor. Note that in our localized CPW structure, semiconductor Schottky contacts are under the metal strips. L_0 and R_0 represent the inductance and resistance due to ohmic loss and skin loss per unit length, respectively. C_0 represents the capacitance per unit length between center and ground strips in region VI. C_1 represents the capacitance per unit length from the center strip to the edges of region IV and region II. C_3 represents the capacitance per unit length from the ground strip to the other side of region IV and the edge of region II. Parallel R_2 and C_2 represent the circuit contribution of region IV and part of region II. General discussions leading to the determination of these component values can be found in [21] and [22].

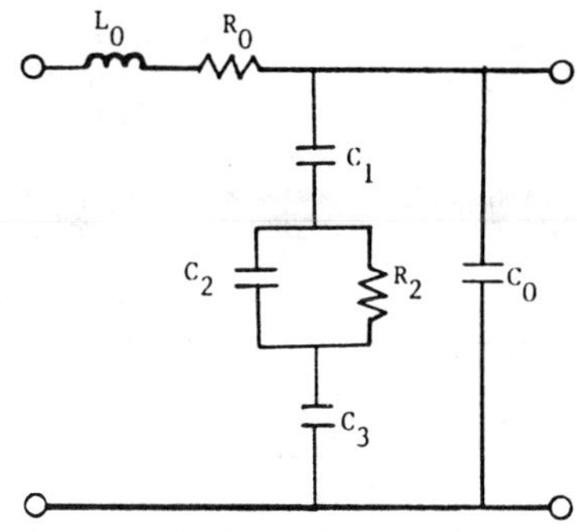

Fig. 15. Approximate transmission-line circuit representation of the Schottky contact CPW with localized depletion regions under metal strips.

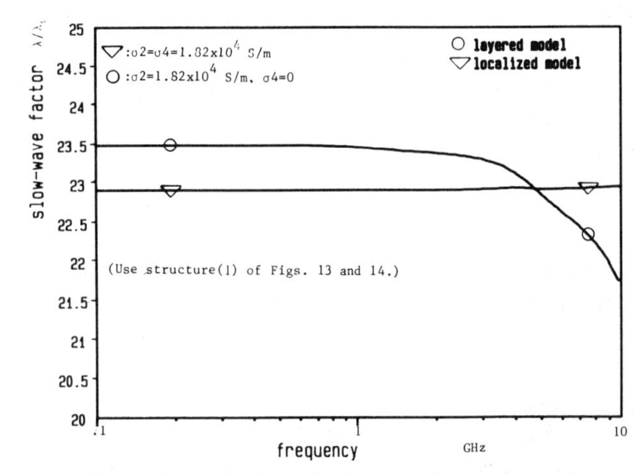

Fig. 16. Comparison of plots of *slow-wave factor* versus frequency based on localized and layered models for CPW's.

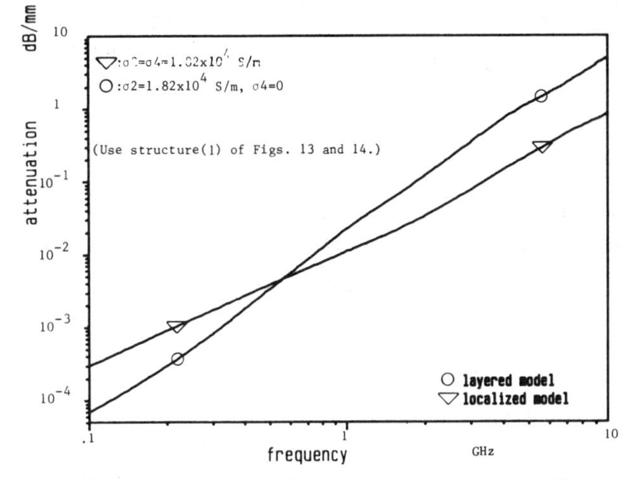

Fig. 17. Comparison of plots of *attenuation constant* versus frequency based on localized and layered models for CPW's.

Figs. 16 and 17 compare numerical results obtained by layered and localized models for slow-wave propagation on CPW's. The localized model tends to show better

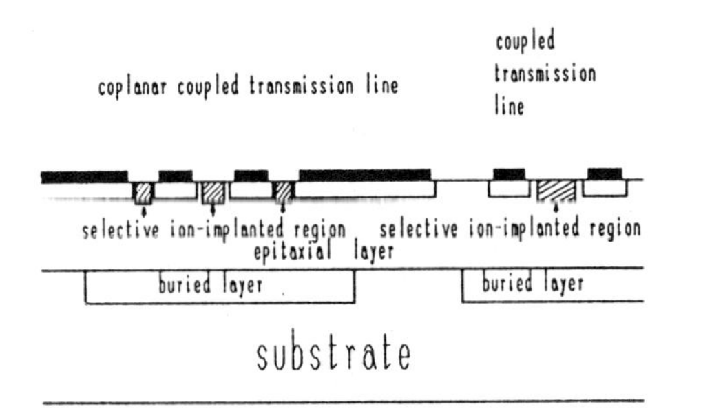

Fig. 18. Potential applications of the extensions of the localized models to MMIC planar waveguide designs, e.g., coupled coplanar lines and coupled lines with selective ion implantations, which add more degrees of freedom to design slow-wave circuits.

performance, with more linear slow-wave characteristics versus frequency and less loss at higher frequency in the structure under analysis. It suggests that the structure with Schottky contacts for all metal strips (Fig. 2) may become a practical slow-wave configuration.

V. Applications

Along with the study of the effects of the localized depletion models on MMIC slow-wave circuits, a general-purpose FEM program is developed for the analysis and design of planar waveguides. Fig. 18 shows several possible applications among a broad class of MMIC waveguides that can be analyzed by extensions of the present code. These structures cannot be modeled accurately by the MIS layered model. The results we obtained from the analyses of CPW and microstrip MMIC's with localized models suggest that we may tune the slow-wave circuit to meet certain design specifications by selective ion implantations on the locations proposed in Fig. 18, with appropriate controls over doping concentration and geometry. A certain mode of propagation may be more subject to the existing selective ion-implanted regions than another mode of propagation, and the additional ion-implanted region can enhance or reduce the loss associated with it depending on what the doping profile is.

VI. Conclusions

Extensive computer simulations employing FEM on CPW and microstrip MMIC slow-wave structures are performed to study the differences in slow-wave propagations between the layered model commonly used in the past and the localized model which is closer to physical situations. The results show that the layered model is a good approximation for the microstrip case. On the other hand, the localized model yields more accurate solutions for Schottky CPW. It is also pointed out that higher order elements improve solutions for the geometry under analysis. The work can be extended to analyze and design more complicated planar waveguides.

Appendix

In the homogeneous, isotropic, linear waveguide with uniform cross section, the Maxwell's equation can be represented by

$$(\nabla_t - \gamma\vec{U}_z)\times(\vec{E}_t + \vec{E}_z) = -j\omega\mu_{re}\mu_0(\vec{H}_t + \vec{H}_z) \quad (1)$$

$$(\nabla_t - \gamma\vec{U}_z)\times(\vec{H}_t + \vec{H}_z) = j\omega\epsilon_{re}\epsilon_0(\vec{E}_t + \vec{E}_z). \quad (2)$$

Note the following.

1) $e^{j\omega t}$ and $e^{-\gamma z}$ factors are assumed for a wave propagating along the positive z direction.

2) $\vec{E} = \vec{E}_t + \vec{E}_z$, $\vec{H} = \vec{H}_t + \vec{H}_z$; $\vec{E}_z = E_z\vec{U}_z$, $\vec{H}_z = H_z\vec{U}_z$. $\vec{U}_x$, $\vec{U}_y$, and $\vec{U}_z$ are unit vectors along the x, y, and z directions, respectively.

3) $\nabla_t = \vec{U}_x\partial/\partial x + \vec{U}_y\partial/\partial y$.

4) μ_0 and ϵ_0 are, respectively, the free-space permeability and permittivity, while μ_{re} and ϵ_{re} are the relative permeability and relative dielectric constant, respectively. Subscript e denotes *element number e* associated with subdomain Ω_e of the entire waveguide cross section region Ω, which contains N_e subdomains.

By defining the inner product as $(\vec{R}, \vec{S}) = \int_{\Omega_e}\vec{R}\cdot\vec{S}\,dx\,dy$ and test functions $\vec{P} = P\vec{U}_z$ and $P = 0$ on the magnetic wall and $\vec{Q} = Q\vec{U}_z$ and $Q = 0$ on the electric wall, one obtains

$$\sum_{e=1}^{N_e}(\vec{E}_t, \nabla_t\times\vec{P}) = \sum_{e=1}^{N_e} -j\omega\mu_{re}\mu_0(\vec{H}_z, \vec{P}) \quad (3)$$

$$\sum_{e=1}^{N_e}(\vec{H}_t, \nabla_t\times\vec{Q}) = \sum_{e=1}^{N_e} +j\omega\epsilon_{re}\epsilon_0(\vec{E}_z, \vec{Q}) \quad (4)$$

satisfying all the necessary boundary conditions.

Now we can derive expressions for $\vec{H}_t$ and $\vec{E}_t$ from (1) and (2) and substitute these into (3) and (4) to eliminate variables $\vec{H}_t$ and $\vec{E}_t$. To obtain an FEM representation of the boundary-value problem, we set

$$U = \sum_{j=1}^{M} u_j W_j^m$$

$$V = \sum_{j=1}^{M} v_j W_j^e$$

where W_j^m and W_j^e are shape functions at the jth node. M is the total number of nodes. Here, we use normalized variables $U = \sqrt{\mu_0}H_z$ and $V = \sqrt{\epsilon_0}E_z$. Note that $u_r = 0$ if the rth node is on the magnetic wall and $v_s = 0$ if the sth node is on the electric wall.

Next, we expand P and Q in terms of shape function W_i^m and W_i^e:

$$P = \sum_{j=1}^{M} p_i W_i^m$$

$$Q = \sum_{j=1}^{M} q_i W_i^e$$

where $W_i^m(x_r, y_r) = 0$ for $i = 1, M$ if the rth node is on the magnetic wall, and $W_i^e(x_s, y_s) = 0$ for $i = 1, M$ if the sth node is on the electric wall

$$A=\begin{bmatrix} K^1_{M\times M} & K^2_{M\times M} \\ K^3_{M\times M} & K^4_{M\times M} \end{bmatrix}$$

$$K^1=\int_\Omega\{\mu_{re}\nabla_t W_i^m\cdot\nabla_t W_j^m/K_e^2-\mu_{re}W_i^m W_j^m\}\,dx\,dy \quad (5)$$

$$K^4=\int_\Omega\{\epsilon_{re}\nabla_t W_i^e\cdot\nabla_t W_j^e/K_e^2-\epsilon_{re}W_i^e W_j^e\}\,dx\,dy \quad (6)$$

$$K^2=\int_\Omega\{+j\gamma C\nabla_t W_i^m\times\nabla_t W_j^e/\omega K_e^2\}\,dx\,dy \quad (7)$$

$$K^3=\int_\Omega\{-j\gamma C\nabla_t W_i^e\times\nabla_t W_j^m/\omega K_e^2\}\,dx\,dy \quad (8)$$

where $K_e^2=k_0^2\mu_{re}\epsilon_{re}+\gamma^2$, $k_0=2\pi/\lambda_0$, and $C=$ the velocity of light.

It can be shown that the matrix A is symmetric. The rows and columns corresponding to the nodes on magnetic walls and electric walls are deleted. Before the final assembly of the matrix A, the node variables are alternated and the column vector X has elements arranged in the following way: $u_1, v_1, u_2, v_2, \cdots$, etc. This results in a banded matrix A for saving computer storage.

References

[1] K. Frike and H. L. Hartnagel, "GaAs MESFET optimization and new device applications based on wave property studies," in *IEEE MTT-S Int. Microwave Symp. Dig.*, June 1985, pp. 192–195.

[2] C. Seguinot, P. Kennis, P. Pribetich, and J. F. Legier, "Performances prediction of an ultra broad-band voltage-controlled attenuator using Schottky contact coplanar line properties," *IEEE Electron Device Lett.*, vol. EDL-7, Feb. 1986.

[3] H. Hasegawa and H. Okizaki, "M.I.S. and Schottky slow-wave coplanar stripline on GaAs substrates," *Electron. Lett.*, vol. 13, pp. 663–664, Oct. 1977.

[4] H. Hasegawa, M. Furukawa, and H. Yanai, "Properties of microstriplines on Si–SiO_2 system," *IEEE Trans. Microwave Theory Tech.*, vol. MTT-19, pp. 869–881, Nov. 1971.

[5] D. Jager, "Slow-wave propagation along variable Schottky-contact microstrip line," *IEEE Trans. Microwave Theory Tech.*, vol. MTT-24, pp. 566–573, Sept. 1976.

[6] D. Jager and W. Rabus, "Bias dependent phase delay of Schottky-contact microstrip line," *Electron Lett.*, vol. 9, no. 9, pp. 201–203, 1973.

[7] Y. C. Shih and T. Itoh, "Analysis of printed transmission lines for monolithic integrated circuits," *Electron. Lett.*, vol. 18, no. 14, pp. 585–586, July 1982.

[8] Y. Fukuoka, Y. C. Shih, and T. Itoh, "Analysis of slow-wave coplanar waveguide for monolithic integrated circuits," *IEEE Trans. Microwave Theory Tech.*, vol. MTT-31, pp. 567–573, July 1983.

[9] M. Aubourg, J. P. Villotte, F. Goon, and Y. Grault, "Analysis of M.I.S. or Schottky contact coplanar lines using the F.E.M. and S.D.A.," in *IEEE MTT-S Int. Microwave Symp. Dig.*, June 1983, pp. 396–398.

[10] R. Sorrentino, G. Leuzzi, and A. Silbermann, "Characteristics of metal–insulator–semiconductor coplanar waveguides for monolithic microwave circuits," *IEEE Trans. Microwave Theory Tech.*, vol. MTT-32, pp. 410–416, Apr. 1984.

[11] P. Daly, "Hybrid-mode analysis of microstrip by finite element methods," *IEEE Trans. Microwave Theory Tech.*, vol. MTT-19, pp. 19–25, Jan. 1971.

[12] A. D. McAulay, "Variational finite-element solution for dissipative waveguides and transportation application," *IEEE Trans. Microwave Theory Tech.*, vol. MTT-25, pp. 382–392, May 1977.

[13] S. R. Cvetkovic and J. B. Davies, "Self-adjoint vector variational formulation for lossy anisotropic dielectric waveguide," *IEEE Trans. Microwave Theory Tech.*, vol. MTT-34, pp. 129–134, Jan. 1986.

[14] C. H. Chen and C. Lien, "The variational principle for non-self-adjoint electromagnetic problems," *IEEE Trans. Microwave Theory Tech.*, vol. MTT-28, pp. 878–886, Aug. 1980.

[15] A. K. Noor, "Multifield (mixed and hybrid) finite element model," in *State-of-the-Art Surveys on Finite Element Technology*, 1983, ch. 5, pp. 127–156.

[16] B. A. Finlayson, *The Method of Weighted Residuals and Variational Principles*. New York: Academic Press, 1972.

[17] E. B. Becker, G. F. Carey, and J. T. Oden, *Finite Elements*, vol. I. Englewood Cliffs, NJ: Prentice-Hall, 1981.

[18] K. C. Gupta, R. Garg, and I. J. Bahl, *Microstrip Lines and Slot Lines*. Dedham, MA: Artech House, 1979.

[19] G. F. Carey and J. T. Oden, *Finite Elements*, vol. II. Englewood Cliffs, NJ: Prentice-Hall, 1983.

[20] S. Ramo, J. R. Whinnery, and T. V. Duzer, *Fields and Waves in Communication Electronics*. New York: Wiley, 1965.

[21] C. Seguinot, P. Kennis, P. Pribetich, and J. F. Legier, "Analytical model of the Schottky contact coplanar line," in *Proc. 14th European Microwave Conf.*, Sept. 1984, pp. 160–165.

[22] H. Hasegawa and S. Seki, "Analysis of interconnection delay on very high-speed LSI/VLSI chips using an MIS microstrip line model," *IEEE Trans. Microwave Theory Tech.*, vol. MTT-32, pp. 1721–1727, Dec. 1984.

Finite-Element Analysis of Waveguide Modes: A Novel Approach That Eliminates Spurious Modes

TUPTIM ANGKAEW, MASANORI MATSUHARA, AND NOBUAKI KUMAGAI, FELLOW, IEEE

Abstract —**An efficient finite-element method for analyzing the propagation characteristics of a wide variety of waveguides is presented. A variational expression suited for the finite-element method is formulated in terms of the transverse electric and magnetic field components. In this approach, all guided-mode solutions are real, while the spurious-mode solutions are not real. Therefore, discrimination of the spurious-mode solutions can be achieved merely by imposing the simple condition that guided-mode solutions be real. Three numerical examples, two for the isotropic case and the other for the magnetic anisotropic case, are carried out.**

I. Introduction

RECENTLY, a method employing finite-element analysis to investigate the propagation characteristics of any arbitrarily shaped waveguide has attracted the attention of many researchers. The finite-element method is a powerful means which enables one to analyze a wide variety of waveguide problems. Several variational formulations for use with the finite-element method have been proposed [1]–[7]. Most of the variational expressions previously used are a functional of frequency and can be classified into the following three types [1]–[6]:

1) variational expressions which are formulated in terms of the longitudinal components of the electric field ($\boldsymbol{e}_z$) and the magnetic field ($\boldsymbol{h}_z$) and can be written as $\omega =$ functional $(\beta, \boldsymbol{e}_z, \boldsymbol{h}_z)$ [1], [2];
2) variational expressions employing the longitudinal component ($\boldsymbol{e}_z$) and the transverse component ($\boldsymbol{e}_t$) of the electric field, which can be written as $\omega =$ functional $(\beta, \boldsymbol{e}_z, \boldsymbol{e}_t)$ [3]; and
3) variational expressions employing all three components of the magnetic field, namely $\boldsymbol{h}_z$ and $\boldsymbol{h}_t$, which can be written as $\omega =$ functional $(\beta, \boldsymbol{h}_z, \boldsymbol{h}_t)$ [4]–[6].

Here, ω is the angular frequency and β is the propagation constant.

The most serious drawback associated with the finite-element method is the appearance of spurious-mode solutions. Up to this time, much effort has been devoted to finding a criterion to eliminate the spurious-mode solutions. A second drawback occurs in the application of variational expressions which are a functional of frequency. Let us consider, for example, the most common requirement in a waveguide problem, namely which mode can exist for a given value of frequency ω and what is the value for the propagation constant β? In order to fulfill this requirement, the calculation of the $\omega-\beta$ diagram must be performed. Further, if the permeability or permittivity of the medium is a function of frequency, the calculation becomes almost impossible.

The approaches proposed by Hano [3] and Koshiba *et al.* [6] are typical methods whereby the mixing of guided-mode solutions with the spurious-mode solutions is prevented. Hano has used a unique method that employs the variational expression of type 2, where the longitudinal and the transverse components of electric field are expressed as quadratic and linear functions of transverse coordinates, respectively. As a consequence of this approach, the values for ω of the spurious-mode solutions do not lie in the region of lower order guided modes. However, an overlap between the existence region of the values for ω^2 of the spurious-mode solutions and guided-mode solutions still occurs. Koshiba *et al.* have modified the variational expression of type 3 given in [4].[1] Besides the continuity condition for $\boldsymbol{i}_z \cdot (\boldsymbol{n} \times \boldsymbol{h}_t)$ and $\boldsymbol{i}_z \cdot \boldsymbol{h}_z$ at the boundary between element and element, a supplementary boundary condition, which requires the continuity of $\boldsymbol{n} \cdot \boldsymbol{h}_t$, has been introduced. Consequently, the values for ω of the spurious-mode solutions do not lie in the existence region of slow guided modes. However, an overlap between the existence region of ω^2 of the spurious-mode solutions and guided-mode solutions cannot be avoided. The supplementary boundary condition also has a constraint which restricts the scope of applications to cases where the permeability of the medium is uniform.

In this paper, a finite-element formulation intended to overcome the two drawbacks previously mentioned is proposed. A novel variational expression is established in terms of the transverse electric and magnetic field components. In our finite-element program, we assign the parameters according to the necessary boundary conditions $\boldsymbol{i}_z \cdot (\boldsymbol{n} \times \boldsymbol{e}_t)$ and $\boldsymbol{i}_z \cdot (\boldsymbol{n} \times \boldsymbol{h}_t)$ required in the variational expres-

Manuscript received July 22, 1986; revised October 10, 1986.
The authors are with the Department of Electrical Communication Engineering, Osaka University, Yamada Oka, Suita, Osaka 565 Japan.
IEEE Log Number 8611939.

[1] Recently, Hayata *et al.* [7] have demonstrated a method more advanced than that outlined in [4].

Reprinted from *IEEE Trans. Microwave Theory Tech.*, vol. MTT-35, no. 2, pp. 117–123, Feb. 1987.

sion. In our approach, the spurious-mode solutions are not real while the guided-mode solutions are real. Thus, the discrimination between spurious-mode solutions and guided-mode solutions can be achieved merely by imposing the simple condition that the latter be real. In addition to the advantage that the spurious-mode solutions can be eliminated, our method can also be applied to the case where the permeability or permittivity of the medium is a function of frequency. Therefore, our finite-element formulation is very useful for the analysis of arbitrarily shaped waveguides.

The application of our finite-element method to isotropic and anisotropic waveguides is discussed. In particular, the case where the permeability of the medium is a function of frequency is considered.

II. Variational Formulation in Terms of the Transverse Electromagnetic Field Components

Consider the anisotropic waveguide with an arbitrary cross section in the $x-y$ plane. The waveguide is assumed to be uniform along its longitudinal z-axis. Then the cross section of the waveguide is subdivided into a finite number of elements according to the finite-element method. The permittivity and permeability tensors of the anisotropic medium are assumed to be Hermite tensors and are defined in matrix form as

$$\epsilon = \begin{bmatrix} \epsilon_{tt} & \epsilon_{tz} \\ \epsilon_{zt} & \epsilon_{zz} \end{bmatrix} \quad \mu = \begin{bmatrix} \mu_{tt} & \mu_{tz} \\ \mu_{zt} & \mu_{zz} \end{bmatrix}$$

where the subscripts tt, tz, zt, and zz refer to 2×2, 2×1, 1×2, and 1×1 submatrices, respectively. From Maxwell's equations, the equations that govern the electromagnetic fields in each element are

$$\begin{aligned} &\omega\epsilon_{tt}\cdot \boldsymbol{e}_t + \omega\epsilon_{tz}\cdot \boldsymbol{e}_z + j\nabla\times \boldsymbol{h}_z + \beta \boldsymbol{i}_z\times \boldsymbol{h}_t = 0 \\ &\omega\mu_{tt}\cdot \boldsymbol{h}_t + \omega\mu_{tz}\cdot \boldsymbol{h}_z - j\nabla\times \boldsymbol{e}_z - \beta \boldsymbol{i}_z\times \boldsymbol{e}_t = 0 \end{aligned} \tag{1}$$

where $\boldsymbol{e}_t$ and $\boldsymbol{h}_t$ are the transverse components of the electric and magnetic fields, respectively. The longitudinal components, $\boldsymbol{e}_z$ and $\boldsymbol{h}_z$, can be expressed in terms of the transverse electromagnetic field components as

$$\begin{aligned} \boldsymbol{e}_z &= \frac{1}{j\omega\epsilon_{zz}}(\nabla\times \boldsymbol{h}_t - j\omega\epsilon_{zt}\cdot \boldsymbol{e}_t) \\ \boldsymbol{h}_z &= -\frac{1}{j\omega\mu_{zz}}(\nabla\times \boldsymbol{e}_t + j\omega\mu_{zt}\cdot \boldsymbol{h}_t). \end{aligned} \tag{2}$$

At the boundary of each element, the boundary conditions for the electromagnetic fields in (1) require the following.

For the boundary between element and element

$$\left.\begin{aligned} \boldsymbol{i}_z\cdot(\boldsymbol{n}\times \boldsymbol{e}_t) &= \text{continuous function} \\ \boldsymbol{i}_z\cdot \boldsymbol{e}_z &= \text{continuous function} \\ \boldsymbol{i}_z\cdot(\boldsymbol{n}\times \boldsymbol{h}_t) &= \text{continuous function} \\ \boldsymbol{i}_z\cdot \boldsymbol{h}_z &= \text{continuous function.} \end{aligned}\right\} \tag{3a}$$

For the boundary between element and electric wall

$$\boldsymbol{i}_z\cdot(\boldsymbol{n}\times \boldsymbol{e}_t) = 0, \quad \boldsymbol{i}_z\cdot \boldsymbol{e}_z = 0. \tag{3b}$$

For the boundary between element and magnetic wall

$$\boldsymbol{i}_z\cdot(\boldsymbol{n}\times \boldsymbol{h}_t) = 0, \quad \boldsymbol{i}_z\cdot \boldsymbol{h}_z = 0. \tag{3c}$$

Here, $\boldsymbol{n}$ is a normal unit vector to the boundary of each element.

Next, the guided-mode solutions that satisfy (1) and the boundary conditions in (3) make the following functional of propagation constant stationary. In other words, we can prove that this functional is a variational expression

$$\beta(\boldsymbol{e}_t, \boldsymbol{h}_t) = \sum A(\boldsymbol{e}_t, \boldsymbol{h}_t) \Big/ \sum B(\boldsymbol{e}_t, \boldsymbol{h}_t) \tag{4}$$

where

$$\begin{aligned} A(\boldsymbol{e}_t, \boldsymbol{h}_t) = \int \Big[& \boldsymbol{e}_t^*\cdot\omega\epsilon_{tt}\cdot \boldsymbol{e}_t + \boldsymbol{h}_t^*\cdot\omega\mu_{tt}\cdot \boldsymbol{h}_t \\ & - \frac{1}{\omega\mu_{zz}}(\nabla\times \boldsymbol{e}_t + j\omega\mu_{zt}\cdot \boldsymbol{h}_t)^* \\ & \cdot(\nabla\times \boldsymbol{e}_t + j\omega\mu_{zt}\cdot \boldsymbol{h}_t) \\ & - \frac{1}{\omega\epsilon_{zz}}(\nabla\times \boldsymbol{h}_t - j\omega\epsilon_{zt}\cdot \boldsymbol{e}_t)^* \\ & \cdot(\nabla\times \boldsymbol{h}_t - j\omega\epsilon_{zt}\cdot \boldsymbol{e}_t) \Big] ds \end{aligned}$$

$$B(\boldsymbol{e}_t, \boldsymbol{h}_t) = \int [\boldsymbol{i}_z\cdot(\boldsymbol{e}_t^*\times \boldsymbol{h}_t + \boldsymbol{e}_t\times \boldsymbol{h}_t^*)]\, ds.$$

Here, Σ denotes a summation taken over all elements and $\int ds$ denotes a surface integral taken in each element.

The trial functions of the transverse electric and magnetic field components used in (4) must necessarily satisfy the following boundary conditions at the boundary of each element.

For the boundary between element and element

$$\left.\begin{aligned} \boldsymbol{i}_z\cdot(\boldsymbol{n}\times \boldsymbol{e}_t) &= \text{continuous function} \\ \boldsymbol{i}_z\cdot(\boldsymbol{n}\times \boldsymbol{h}_t) &= \text{continuous function.} \end{aligned}\right\} \tag{5a}$$

For the boundary between element and electric wall

$$\boldsymbol{i}_z\cdot(\boldsymbol{n}\times \boldsymbol{e}_t) = 0. \tag{5b}$$

For the boundary between element and magnetic wall

$$\boldsymbol{i}_z\cdot(\boldsymbol{n}\times \boldsymbol{h}_t) = 0. \tag{5c}$$

In the following, we shall prove the validity of the variational expression in (4). First, we assume that the trial functions of the transverse electric and magnetic field components used in (4) differ from the true guided-mode solutions satisfying (1) and (3) by small admissible changes $\delta\boldsymbol{e}_t$ and $\delta\boldsymbol{h}_t$, respectively. Let $\delta\beta$ denote a variation of the propagation constant corresponding to $\delta\boldsymbol{e}_t$ and $\delta\boldsymbol{h}_t$. Under the condition that β and $\delta\beta$ be real, substituting the trial functions written in variation notations into (4) results in the following equations:

$$\delta\beta\sum B = \sum(A - \beta B) + \sum(\delta A - \beta\delta B) \tag{6}$$

where

$$A-\beta B=\int[e_t^*\cdot(\omega\epsilon_{tt}\cdot e_t+\omega\epsilon_{tz}\cdot e_z+j\nabla\times h_z+\beta i_z\times h_t)$$
$$+h_t^*\cdot(\omega\mu_{tt}\cdot h_t+\omega\mu_{tz}\cdot h_z$$
$$-j\nabla\times e_z-\beta i_z\times e_t)]\,ds$$
$$+j\oint[h_z\cdot(n_0\times e_t)^*-e_z\cdot(n_0\times h_t)^*]\,dl \quad (7a)$$

and

$$\delta A-\beta\delta B$$
$$=\int[\delta e_t^*\cdot(\omega\epsilon_{tt}\cdot e_t+\omega\epsilon_{tz}\cdot e_z+j\nabla\times h_z+\beta i_z\times h_t)$$
$$+\delta h_t^*\cdot(\omega\mu_{tt}\cdot h_t+\omega\mu_{tz}\cdot h_z-j\nabla\times e_z-\beta i_z\times e_t)$$
$$+\delta e_t\cdot(\omega\epsilon_{tt}\cdot e_t+\omega\epsilon_{tz}\cdot e_z+j\nabla\times h_z+\beta i_z\times h_t)^*$$
$$+\delta h_t\cdot(\omega\mu_{tt}\cdot h_t+\omega\mu_{tz}\cdot h_z-j\nabla\times e_z-\beta i_z\times e_t)^*]\,ds$$
$$+j\oint[h_z\cdot(n_0\times\delta e_t)^*-e_z\cdot(n_0\times\delta h_t)^*$$
$$-h_z^*\cdot(n_0\times\delta e_t)+e_z^*\cdot(n_0\times\delta h_t)]\,dl. \quad (7b)$$

Here $\oint dl$ denotes a line integral taken along the entire boundary line in each element and n_0 denotes an outward normal unit vector at the boundary of each element. The e_t, h_t, e_z, h_z, and β in (7) satisfy (1), (2), and (3). Thus, by substituting (1), (2), and (3) into (7) we get

$$\sum(A-\beta B)=0$$
$$\sum(\delta A-\beta\delta B)=\sum j\oint[h_z\cdot(n_0\times\delta e_t)^*-e_z\cdot(n_0\times\delta h_t)^*$$
$$-h_z^*\cdot(n_0\times\delta e_t)+e_z^*\cdot(n_0\times\delta h_t)]\,dl. \quad (8)$$

Since δe_t and δh_t also satisfy the boundary conditions in (5) according to the trial functions e_t and h_t, the term $\Sigma(\delta A-\beta\delta B)$ in the above equation becomes zero. Hence, (6) can be written as

$$\delta\beta=0. \quad (9)$$

From (9), we can prove that the functional in (4) is the variational expression accounting for (1) and (3). This variational expression has the advantage that for a given value of angular frequency, the values of the propagation constant can be obtained directly. The variational expression is also suited for a finite-element formulation in the sense that all guided-mode solutions are real while the spurious-mode solutions are not real. This is confirmed in the numerical examples described in Section IV.

III. Finite-Element Method

According to the standard finite-element method, the cross section of the waveguide is subdivided into a finite number of triangular elements. Fig. 1 shows an arbitrary triangular element. The node number and the corresponding node coordinate at each vertex are assigned as 1, (x_1, y_1), 2, (x_2, y_2), and 3, (x_3, y_3), respectively. The normal unit vector at each side is assigned as n_{12}, n_{23}, and n_{31} normal to, respectively, the sides (1–2), (2–3), and (3–1). The trial functions of the transverse electric and magnetic field components in each element are expressed by using 12 unknown parameters as

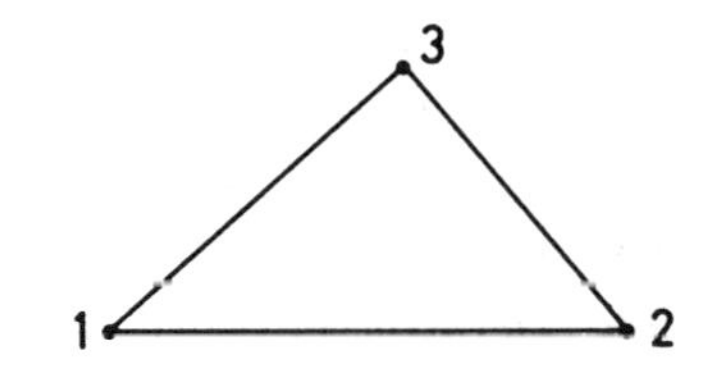

Fig. 1. An arbitrary triangular element.

$$e_t=\sum_{m=1}^{6}N_m(x,y)\phi_m$$
$$h_t=\sum_{m=7}^{12}N_{m-6}(x,y)\phi_m \quad (10)$$

where ϕ_m (for $m=1\sim12$) denotes unknown parameters and $N_m(x,y)$ (for $m=1\sim6$) represents linear vector shape functions. The vector shape functions are determined by the scalar shape functions and the normal unit vectors in the element. These vector shape functions can be expressed as

$$N_{2i-1}=\frac{n_{ij}}{i_z\cdot(n_{ki}\times n_{ij})}N_i$$
$$N_{2i}=\frac{n_{ki}}{i_z\cdot(n_{ij}\times n_{ki})}N_i$$
$$N_i=\frac{(x_jy_k-x_ky_j)+(y_j-y_k)x+(x_k-x_j)y}{x_1y_2+x_2y_3+x_3y_1-x_2y_1-x_3y_2-x_1y_3} \quad (11)$$

where (i, j, k) permutes in a natural order. The equations for the unknown parameters ϕ_m (for $m=1\sim12$) can be recognized from (10) and (11) as follows

$$\phi_{2i-1}=i_z\cdot(n_{ki}\times e_{ti})$$
$$\phi_{2i}=i_z\cdot(n_{ij}\times e_{ti})$$
$$\phi_{2i+5}=i_z\cdot(n_{ki}\times h_{ti})$$
$$\phi_{2i+6}=i_z\cdot(n_{ij}\times h_{ti}) \quad (12)$$

where e_{ti} and h_{ti} (for $i=1\sim3$) denote the transverse electric and magnetic field components at node i. Equations (12) mean that the parameters ϕ_m (for $m=1\sim12$) are $i_z\cdot(n\times e_t)$ and $i_z\cdot(n\times h_t)$ taken at each side of the triangular element. By means of (12), the trial functions can be expressed in terms of parameters ϕ_m instead of explicitly using the transverse electric and magnetic field components. In this way, the trial functions can be forced to satisfy the necessary boundary conditions in (5) in a simple manner.

After substituting (10) into (4) and integrating, the following equations for A and B are obtained:

$$A=\phi^{*t}p\phi$$
$$B=\phi^{*t}q\phi. \quad (13)$$

Here, ϕ denotes a column vector composed of 12 unknown parameters used in the triangular element; p and q are

12×12 matrices composed of the values of permittivity, permeability, and vector shape functions in the element; and t denotes a transposed column vector.

Imposing the boundary conditions in (5) to (13) and summing over all elements, ΣA and ΣB are obtained. Then, substituting ΣA and ΣB into (4), the variational expression can be written as

$$\beta = \frac{\Phi^{*t} P \Phi}{\Phi^{*t} Q \Phi} \tag{14}$$

where Φ is a column vector composed of all unknown parameters used in the waveguide. P is a regular Hermitian matrix[2] and Q is a singular real symmetric matrix. The dimensions of P, Q, and Φ are exactly equal to six times the total number of elements in the case of linear shape functions.

The stationary condition of (4) requires the derivative of (14) with respect to Φ to be equal to zero. From this condition, we can obtain a generalized eigenvalue problem as

$$Q\Phi = \frac{1}{\beta} P \Phi \tag{15}$$

or, in detailed matrix form, as

$$\begin{bmatrix} 0 & Q_2 \\ Q_1 & 0 \end{bmatrix} \begin{bmatrix} \Phi_1 \\ \Phi_2 \end{bmatrix} = \frac{1}{\beta} \begin{bmatrix} P_1 & P_4 \\ P_3 & P_2 \end{bmatrix} \begin{bmatrix} \Phi_1 \\ \Phi_2 \end{bmatrix} \tag{16}$$

where Φ_1 and Φ_2 are subcolumn vectors of Φ composed of the unknown parameters used to express $\boldsymbol{e}_t$ and $\boldsymbol{h}_t$, respectively, in the waveguide. P_1, P_2, P_3, and P_4 are the submatrices of matrix P. Q_1 and Q_2 are the submatrices of matrix Q. For the common case where $\epsilon_{zt} = \mu_{zt} = 0$, the submatrices P_3 and P_4 become equal to zero and the following two equations can be obtained from (16):

$$Q_1 \Phi_1 = \frac{1}{\beta} P_2 \Phi_2 \tag{17}$$

$$Q_2 \Phi_2 = \frac{1}{\beta} P_1 \Phi_1. \tag{18}$$

From (17) and (18), we have

$$\left(P_1^{-1} Q_2 P_2^{-1} Q_1\right) \Phi_1 = (1/\beta)^2 \Phi_1 \tag{19}$$

$$\left(P_2^{-1} Q_1 P_1^{-1} Q_2\right) \Phi_2 = (1/\beta)^2 \Phi_2. \tag{20}$$

It is clear that one of the above two equations can be used to solve for β and that the dimension of both eigenvalue problems is approximately one half the dimension of (15). Thus, the eigenvalue problem obtained by using $\boldsymbol{e}_t$ and $\boldsymbol{h}_t$ can be reduced to an eigenvalue problem in which the unknown parameters are expressed only in terms of $\boldsymbol{e}_t$ or $\boldsymbol{h}_t$. It should also be mentioned that for the case where $\epsilon_{zt} \neq 0$ or $\mu_{zt} \neq 0$, (15) can be transformed into the standard eigenvalue problem: $P^{-1} Q \Phi = 1/\beta \Phi$.

By solving the eigenvalue problem, the eigenvectors Φ and the eigenvalues β are obtained. Substituting the eigenvectors into (10), we obtain the eigenfunctions of the transverse electric and magnetic field components. Thus, by following this method, we can calculate the electromagnetic field distribution in the waveguide.

IV. Numerical Examples and Considerations

A. Dielectric-Loaded Waveguide

As a first numerical example of the method described in Sections II and III, we investigate a dielectric-loaded waveguide. Fig. 2 shows the cross section of the dielectric-loaded waveguide of size $a \times 2a$ bounded by a perfect conductor. Half of the waveguide is filled with dielectric material whose relative permittivity and permeability are equal to 2.25 and 1, respectively. The other half of the waveguide is assumed to be vacuum. The dielectric-loaded waveguide is a test case for the finite-element method, which is widely used in many papers, such as [3] and [6].

Now let us consider our result at, for example, $k_0 a = \omega\sqrt{\epsilon_0 \mu_0}\, a = 3$, where the cross section of the dielectric-loaded waveguide is subdivided into 64 triangular elements, as shown in Fig. 3. Here, k_0, ϵ_0, and μ_0 are the wavenumber, permittivity, and permeability in free space, respectively.

The total number of eigenvalues obtained with our method is equal to six times the total number of elements used in the finite-element mesh. Thus, for the present case, the total number of eigenvalues is equal to $64 \times 6 = 384$. These eigenvalues can be expressed in the form

$$k_0/\beta = u + jv \qquad (u, v \text{ real}).$$

All 384 eigenvalues can be classified into four types, as shown in Table I. From the eigenvalues in Table I, we may conclude that the eigenvalues of types 1, 2, and 3 are definitely nonphysical spurious-mode solutions, because these eigenvalues are not bounded real numbers. All eigenvalues of type 4 are real, as shown in the following:

$$\frac{\beta}{k_0} = \begin{cases} \pm 1.27102 \text{ (the corresponding exact solution is } \pm 1.27576) \\ \pm 0.94546 \text{ (the corresponding exact solution is } \pm 0.97154) \\ \pm 0.67683 \text{ (the corresponding exact solution is } \pm 0.72865) \\ \pm 0.55718 \text{ (the corresponding exact solution is } \pm 0.59390). \end{cases}$$

The eigenvalues of type 4 clearly represent the approximate values of the exact guided-mode solutions, and we also observe the one-to-one correspondence between the values of type 4 and the exact solutions. The positive and negative values of β in type 4 can be regarded as the

[2] In the case of isotropic and some anisotropic waveguides, P can be reduced to a real symmetric matrix.

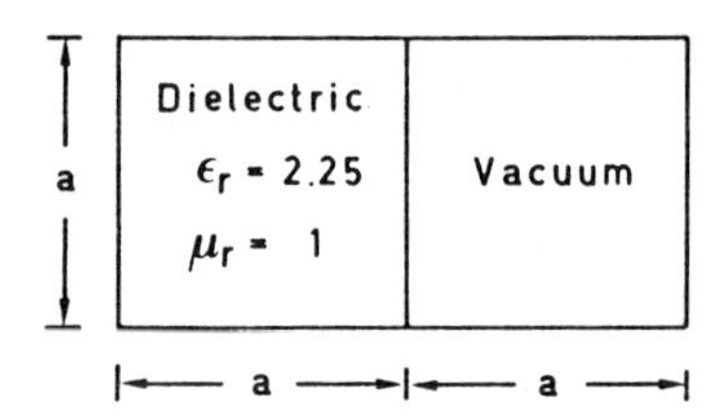

Fig. 2. The cross section of a dielectric-loaded waveguide.

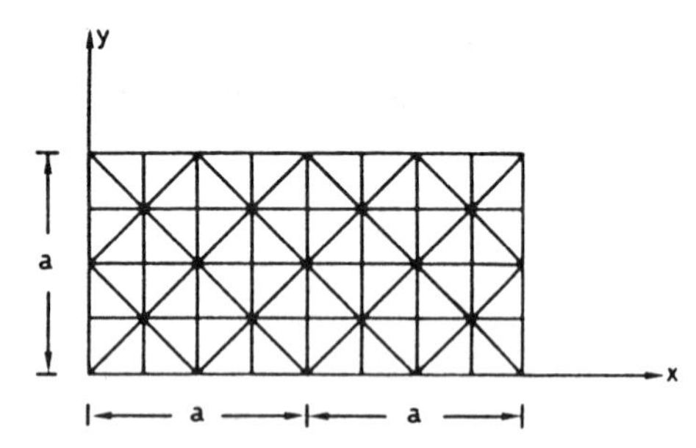

Fig. 3. Illustration of the finite-element division of a dielectric-loaded waveguide.

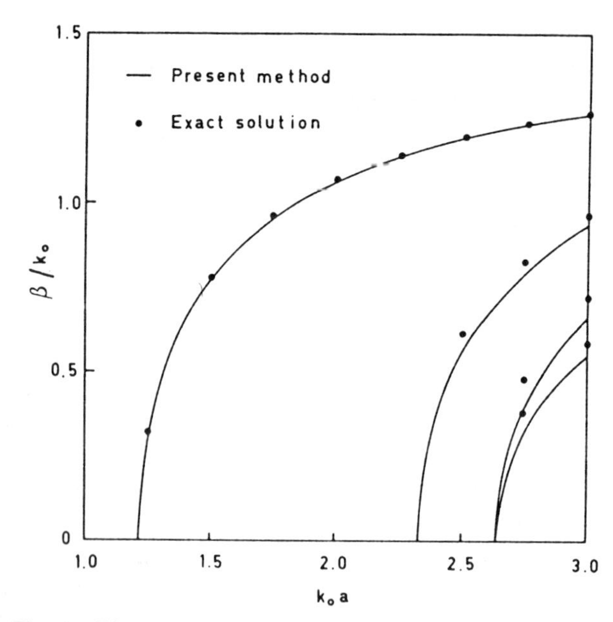

Fig. 4. Dispersion characteristics for a dielectric-loaded waveguide.

TABLE I
CLASSIFICATION OF 384 EIGENVALUES IN THE FINITE-ELEMENT ANALYSIS OF THE DIELECTRIC-LOADED WAVEGUIDE IN FIG. 3

Type 1	$u = 0$	$v = 0$	132 values
Type 2	$u = 0$	$v \neq 0$	240 values
Type 3	$u \simeq 0$	$v \neq 0$	4 values
Type 4	$u \neq 0$	$v = 0$	8 values

values corresponding to the waves propagating in $+z$ and $-z$ directions, respectively. Thus, if the condition that guided-mode solutions be real is imposed, all the spurious-mode solutions can be completely discriminated.

The finite-element analysis for the dielectric-loaded waveguide in Fig. 3 has been carried out by using a mesh with 64 triangular elements. In Fig. 4, the solid lines represent the finite-element analysis, while dots show the exact solutions. In case of the fundamental mode, our finite-element analysis agrees almost exactly with the exact solutions. The higher order modes, obviously, cannot be reproduced so well because of an insufficient number of elements in the finite-element mesh. However the accuracy in the higher order modes can be improved by increasing the number of elements or using quadratic shape functions instead of linear shape functions. Note that the cutoff frequency of each mode in Fig. 4 is the extrapolated value from the values near the cutoff frequency.

B. Ferrite-Loaded Waveguide

As a numerical example for the magnetic anisotropic case, we investigate a ferrite-loaded waveguide, shown in Fig. 5. The ferrite-loaded waveguide is a version of a nonreciprocal microwave components in which a ferrite

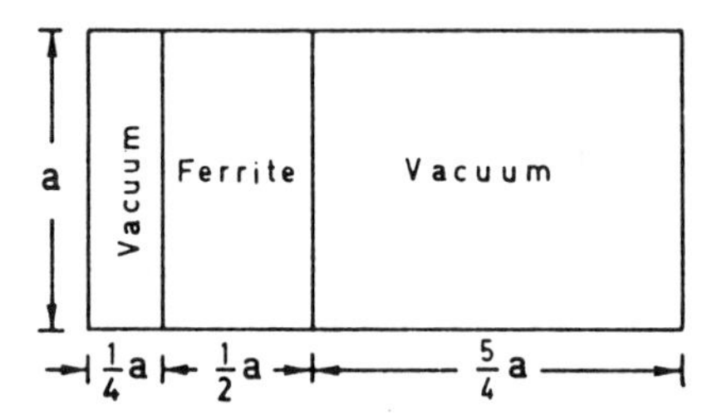

Fig. 5. The cross section of a ferrite-loaded waveguide.

slab is placed asymmetrically into the waveguide, and the static magnetic field is imposed transverse to the direction of propagation. The ferrite slab in the waveguide can be characterized by the tensor permeability $\bar{\mu}$ and the scalar permittivity ϵ as

$$\bar{\mu} = \begin{bmatrix} \mu & 0 & j\kappa \\ 0 & \mu_0 & 0 \\ -j\kappa & 0 & \mu \end{bmatrix}$$

$$\mu = \mu_0\left(1 + \frac{\omega_m \omega_0}{\omega_0^2 - \omega^2}\right)$$

$$\kappa = \mu_0 \frac{\omega_m \omega}{\omega_0^2 - \omega^2}$$

$$\omega_0\sqrt{\epsilon_0 \mu_0}\, a = \omega_m \sqrt{\epsilon_0 \mu_0}\, a = 0.5$$

$$\epsilon = 10\epsilon_0.$$

Here, μ_0 and ϵ_0 are the permeability and permittivity in free space, respectively. The tensor permeability of the ferrite is frequency-dependent. Thus, the ferrite-loaded waveguide can be regarded as a proof for the validity of the finite-element method for the general case where the permeability of the medium is a function of frequency. For the case of the variational expression in (4), the finite-element analysis of the ferrite-loaded waveguide can be

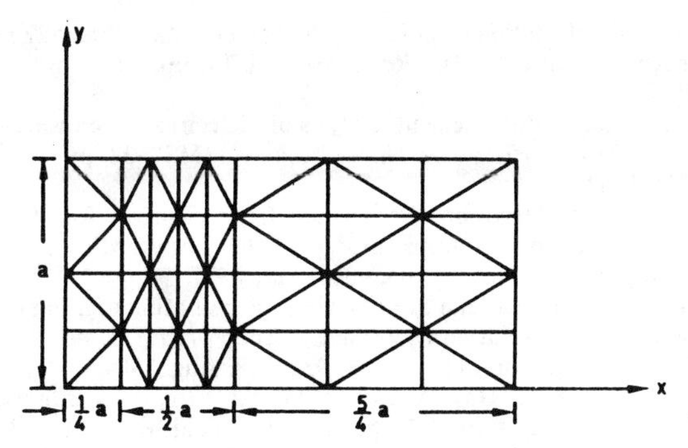

Fig. 6. Illustration of the finite-element division of a ferrite-loaded waveguide.

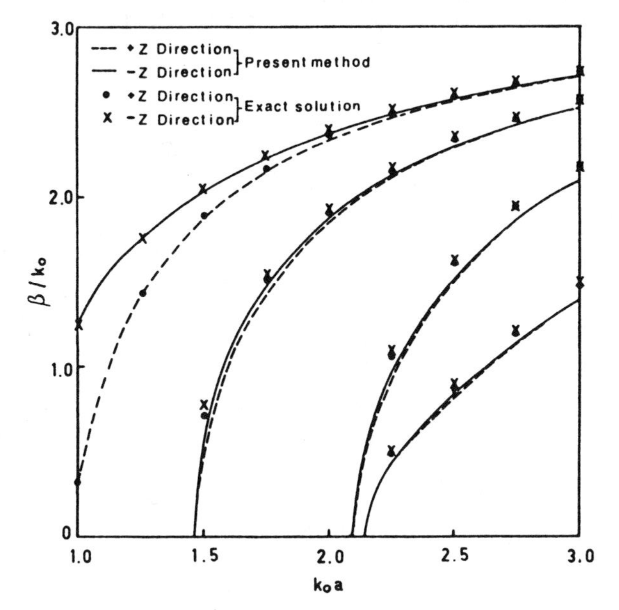

Fig. 7. Dispersion characteristics for a ferrite-loaded waveguide.

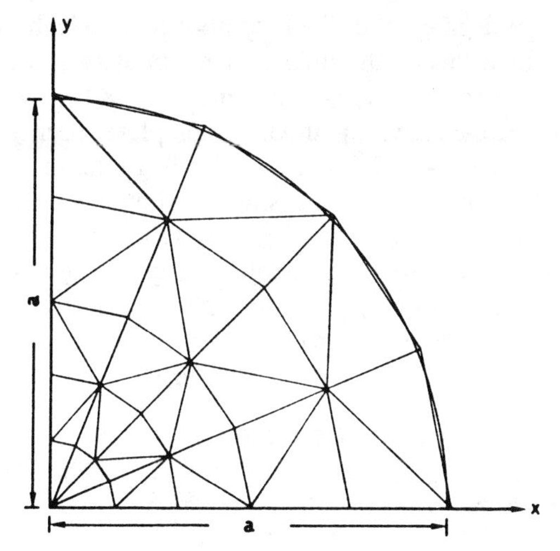

Fig. 8. Illustration of the finite-element division of one quarter of a hollow circular waveguide.

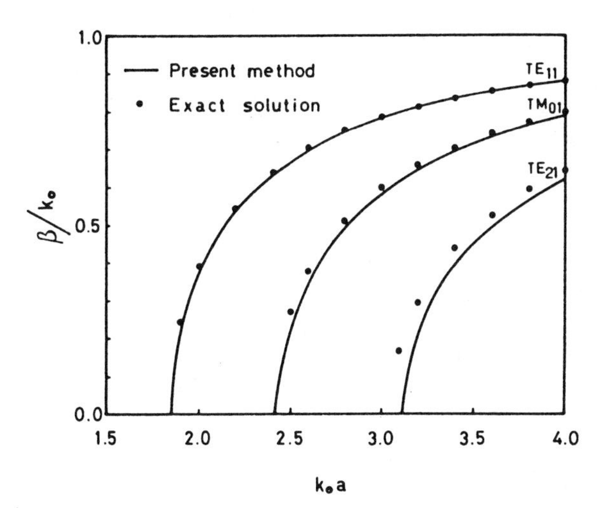

Fig. 9. Dispersion characteristics for a hollow circular waveguide.

accomplished because this variational expression is not a functional of frequency.

In Fig. 6, the cross section of the ferrite-loaded waveguide is subdivided into 64 triangular elements. The calculations are carried out by employing the finite-element mesh shown in Fig. 6. In Fig. 7, the dashed lines and solid lines represent the finite-element analysis corresponding to the waves propagating in the $+z$ and $-z$ directions, respectively, while the exact solutions are represented by dots and crosses. Again, our finite-element analysis is in good agreement with the exact solutions for the fundamental mode, but as in the numerical example of a dielectric-loaded waveguide previously discussed, deviations in the higher order modes still occur. The cause of the deviations in the higher order modes lies in the insufficient number of elements and in the circumstance that the shape functions being used do not yield the correct approximate functions fitted to the electromagnetic field distribution.

Note also that for the case of a ferrite-loaded waveguide, all spurious-mode solutions are not real while all guided-mode solutions are real. Thus, again, all spurious-mode solutions can be discriminated by imposing the condition that the guided-mode solutions be real.

C. *Hollow Circular Waveguide*

An additional numerical example has been carried out with a hollow circular waveguide. This example can be regarded as a test case with curved boundary. The computation is carried out with the mesh in Fig. 8, where a quarter of the waveguide is considered. A one-to-one correspondence between the real eigenvalues in our method and exact solutions has been observed, and no spurious-mode solutions appear to be real numbers. Fig. 9 shows the dispersion characteristics for the hollow circular waveguide. The solid lines and dots show the finite-element solutions and the exact solutions, respectively. Thus, the validity of the present method is also confirmed by this example.

V. Conclusions

A variational expression of the propagation constant, employing transverse electric and magnetic field components has been formulated and used in a finite-element method in order to investigate the propagation characteristics of waveguides. The trial functions of the transverse electric and magnetic field components, which are expressed in terms of the unknown parameters according to the vector products. $\boldsymbol{i}_z \cdot (\boldsymbol{n} \times \boldsymbol{e}_t)$ and $\boldsymbol{i}_z \cdot (\boldsymbol{n} \times \boldsymbol{h}_t)$, were used in this finite-element method. A complete discrimination of the spurious-mode solutions from guided-mode solutions was confirmed by imposing the simple condition that the guided-mode solutions be real. This simple condition does not appear to be available in any of the variational expressions previously proposed. Numerical examples are given for a dielectric-loaded waveguide, ferrite-loaded waveguide, and hollow circular waveguide.

References

[1] C. Yeh, K. Ha, S. B. Dong, and W. P. Brown, "Single-mode optical waveguides," *Appl. Opt.*, vol. 18, pp. 1490–1504, May 1979.

[2] K. Oyamada and T. Okoshi, "Two-dimensional finite-element method calculation of propagation characteristics of axially nonsymmetrical optical fibers," *Radio Sci.*, vol. 17, pp. 109–116, Jan.–Feb. 1982.

[3] M. Hano, "Finite-element analysis of dielectric-loaded waveguides," *IEEE Trans. Microwave Theory Tech.*, vol. MTT-32, pp. 1275–1279, Oct. 1984.

[4] A. Konrad, "High-order triangular finite elements for electromagnetic waves in anisotropic media," *IEEE Trans. Microwave Theory Tech.*, vol. MTT-25, pp. 353–360, May 1977.

[5] B. M. A. Rahman and J. B. Davies, "Penalty function improvement of waveguide solution by finite elements," *IEEE Trans. Microwave Theory Tech.*, vol. MTT-32, pp. 922–928, Aug. 1984.

[6] M. Koshiba, K. Hayata, and M. Suzuki, "Improved finite-element formulation in terms of the magnetic field vector for dielectric waveguides," *IEEE Trans. Microwave Theory Tech.*, vol. MTT-33, pp. 227–233, Mar. 1985.

[7] K. Hayata, M. Koshiba, M. Eguchi, and M. Suzuki, "Novel finite-element formulation without any spurious solutions for dielectric waveguides," *Electron. Lett.*, vol. 22, pp. 295–296, Mar. 1986.

Part 5
Boundary-Element Method

ALTHOUGH the boundary-element method (BEM) is a rather well established numerical technique for the solution of a wide range of problems in continuum mechanics and a viable alternative to other methods such as the FEM, it has to date found a limited number of applications in the analysis of microwave components.

In this method, the wave equation is converted to an integral over the boundary of the region of interest by way of Green's second identity. By a limit process, a boundary integral equation is obtained where the unknown value of the potential over the boundary is expressed in terms of the function itself and its normal derivative. In this manner a reduction of complexity by one dimension is achieved. A discretization procedure is then applied to solve the integral equation.

Besides providing economy in computing time and memory storage by the reduction of the matrix size, compared to the FDM and the FEM, the BEM has the additional advantage of being suited to open structures.

Under the name of contour-integral method, the BEM was applied in 1972 by Okoshi and Miyoshi in the context of planar-circuit analysis. This classic paper is included here in the part devoted to the planar-circuit approach (Part 11, Paper 11.2). More recently, this method has been brought to the attention of the microwave community in western countries by Kagami and Fukai (Paper 5.1). (A paper by Washisu and Fukai had been published in Japanese already in 1981.) An application to planar-circuit analysis including higher order modes on the connecting lines is shown in the letter by Tonye and Baudrand (Paper 5.2). The use of higher order boundary elements has been proposed by Koshiba and Suzuki (Paper 5.3) to increase numerical efficiency in the analysis of waveguide discontinuities. It has been observed that the difficulty in dealing with singularities of Green's function may cause numerical problems. A BEM formulation that eliminates Green's function has recently been proposed by Kishi and Okoshi [5.1]. The suitability of the BEM to open structures has been documented, e.g., by a recent paper by El-Mikati and Davies [5.2].

References

[5.1] N. Kishi and T. Okoshi, "Proposal for a boundary-integral method without using Green's function," *IEEE Trans. Microwave Theory Tech.*, vol. MTT-35, pp. 887–892, Oct. 1987.

[5.2] H. A. El-Mikati and J. B. Davies, "Improved boundary element techniques for two-dimensional scattering problems with circular boundaries," *IEEE Trans. Antennas Propagat.*, vol. AP-35, pp. 539–544, May 1987.

Application of Boundary-Element Method to Electromagnetic Field Problems

SHIN KAGAMI AND ICHIRO FUKAI

Abstract —This paper proposes an application of the boundary-element method to two-dimensional electromagnetic field problems. By this method, calculations can be performed using far fewer nodes than by the finite-element method, and unbounded field problems are easily treated without special additional consideration. In addition, the results obtained have fairly good accuracy. In this paper, analyzing procedures of electromagnetic field problems by the boundary-element method, under special conditions, are proposed and several examples are investigated.

I. Introduction

At present, the finite-element method is widely used in many fields. The main reason may be that, by the finite-element method, it is easy to handle inhomogeneities and complicated structures. However, it requires a large computer memory and long computing time to solve the final matrix equation. In addition, unbounded field problems need some additional techniques [1], [2].

Recently, the boundary-element method has been proposed, which is interpreted as a combination technique of the conventional boundary-integral equation method and a discretization technique, such as the finite-element method, and which has merits of both the above methods, i.e., the required size of the computer memory being small and the obtained results having fairly good accuracy [3], [4]. Namely, the boundary-element method is a boundary method and, therefore, if the region to be analyzed is homogeneous, then it requires nodes, necessary for calculation, on its boundary only. So the problem can be treated with one less dimension. Moreover, it can handle unbounded field problems easily, so that it is suitable for the electromagnetic field analysis which often includes unbounded regions [5], [6].

In this paper, a formulation of two-dimensional electromagnetic field problems by the boundary-element method and its application to several interesting cases, such as the problem of electromagnetic waveguide discontinuities, multi-media problems, and electromagnetic wave scattering problems [6]. In addition, several examples are analyzed and the results obtained with the boundary-element method are compared with rigorous ones, and solutions of the other numerical methods. The propriety of our analyzing procedure of the boundary-element method is verified.

II. General Formulation

A two-dimensional region R enclosed by a boundary B, as illustrated in Fig. 1, is considered. In the region R, Helmholtz's equation

$$(\nabla^2 + k^2)u = 0 \tag{1}$$

holds, where u is the potential used for analysis, we write its outward normal derivative as q, and k denotes the wavenumber

Manuscript received April 15, 1983; revised October 12, 1983.

S. Kagami is with the Department of Electrical Engineering, Asahikawa Technical College, Asahikawa, Japan 070.

I. Fukai is with the Department of Electrical Engineering, Faculty of Engineering, Hokkaido University, Sapporo, Japan 060.

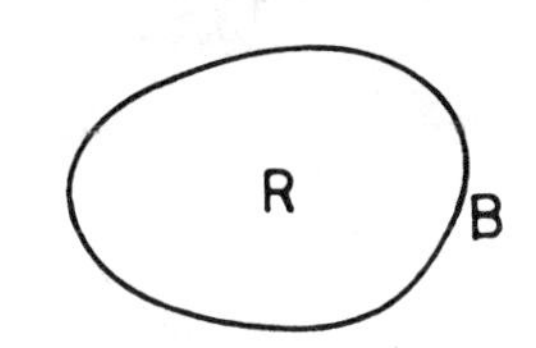

Fig. 1. Two-dimensional region R.

in free space. The boundary condition on B is

$$u = \bar{u} \tag{2}$$

or

$$q = \bar{q} \tag{3}$$

where "$^{-}$" means a known value. Here, Green's function

$$u^* = -\frac{j}{4} H_0^{(2)}(kr) \tag{4}$$

is introduced, where $H_0^{(2)}$ is the Hankel function of the second kind and order zero. By the method of weighted residuals [3], [4] or Green's formula, the following equation is obtained:

$$u_i + \int_B uq^*\, dc = \int_B qu^*\, dc. \tag{5}$$

In (5), the suffix "i" means an arbitrary point in the region R and q^* is the outward normal derivative of u^*

$$q^* = \frac{j}{4} H_1^{(2)}(kr) \frac{\partial (kr)}{\partial n}$$
$$= \frac{j}{4} k H_1^{(2)}(kr) \cos\angle(\boldsymbol{r}; \boldsymbol{n}). \tag{6}$$

In (6), $H_1^{(2)}$ is the Hankel function of the second kind and order one. For the case where the point i is placed on the boundary B, the singular point of Green's function appears and special consideration is necessary. Now, we adopt the integration path going round the node i as shown in Fig. 2. Then, (5) is rewritten as follows:

$$u_i + \lim_{\epsilon \to 0} \int_{B'} uq^*\, dc + \lim_{\epsilon \to 0} \int_{B''} uq^*\, dc$$
$$= \lim_{\epsilon \to 0} \int_{B'} qu^*\, dc + \lim_{\epsilon \to 0} \int_{B''} qu^*\, dc. \tag{7}$$

Here, we estimate the integration over the boundary B'' as follows:

$$\lim_{\epsilon \to 0} \int_{B''} uq^*\, dc = \lim_{\epsilon \to 0} \int_{B''} u \frac{j}{4} k H_1^{(2)}(k\epsilon)\, dc$$
$$= \lim_{\epsilon \to 0} u \frac{j}{4} k H_1^{(2)}(k\epsilon) \epsilon\theta$$
$$= \frac{j}{4} k\theta \lim_{\epsilon \to 0} \left[\epsilon \left\{ \frac{k\epsilon}{2} - j\left(-\frac{2}{\pi}\frac{1}{k\epsilon} \right) \right\} \right]$$
$$= -\frac{\theta}{2\pi} u_i \tag{8}$$

$$\lim_{\epsilon \to 0} \int_{B''} qu^*\, dc = \lim_{\epsilon \to 0} \int_{B''} q \left\{ -\frac{j}{4} H_0^{(2)}(k\epsilon) \right\} dc$$
$$= \lim_{\epsilon \to 0} \left[q\left(-\frac{j}{4} \right) \left\{ 1 - j\frac{2}{\pi}(\ln k\epsilon + \gamma - \ln 2) \right\} \epsilon\theta \right]$$
$$= 0$$
$$\gamma = 0.5772 \cdots \quad \text{(Euler's number)}. \tag{9}$$

Reprinted from *IEEE Trans. Microwave Theory Tech.*, vol. 32, no. 4, pp. 455–461, Apr. 1984.

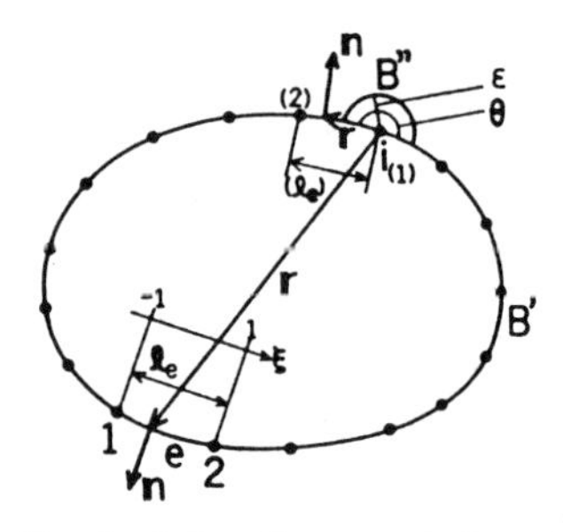

Fig. 2. Integration on each element.

From (5), (8), and (9), we derive the next equation as follows:

$$c_i u_i + \int_B q u^* \, dc = \int_B q u^* \, dc,$$

$$c_i = 1 - \frac{\theta}{2\pi} \quad \int_B = \lim_{\epsilon \to 0} \int_{B'}. \tag{10}$$

In (10), $\int$ denotes the Cauchy's principal value of integration. Next, the boundary B is approximated by the connection of line elements, on each of which u and q are assumed to vary linearly. Then, (10) is discretized as follows:

$$c_i u_i + \sum_{e=1}^{n} [h_1 h_2]_e \begin{Bmatrix} u_1 \\ u_2 \end{Bmatrix}_e = \sum_{e=1}^{n} [g_1 g_2]_e \begin{Bmatrix} q_1 \\ q_2 \end{Bmatrix}_e. \tag{11}$$

In (11), n is the number of elements and u_1, u_2, q_1, and q_2 are the potentials and the potential derivatives on the two nodes of the eth element and h_1, h_2, g_1, and g_2 are contributions of the eth element to integration. When the point i does not belong to the eth element, they are calculated with Gaussian integration as

$$\begin{rcases} h_1 \\ h_2 \end{rcases} = \int_{-1}^{1} \begin{Bmatrix} \phi_1 \\ \phi_2 \end{Bmatrix} \frac{1}{4} jk H_1^{(2)}(kr) \cos\angle(\boldsymbol{r}; \boldsymbol{n}) \, d\xi \cdot \frac{1}{2} l_e \tag{12}$$

$$\begin{rcases} g_1 \\ g_2 \end{rcases} = \int_{-1}^{1} \begin{Bmatrix} \phi_1 \\ \phi_2 \end{Bmatrix} \left\{ -\frac{1}{4} j H_0^{(2)}(kr) \right\} d\xi \cdot \frac{1}{2} l_e \tag{13}$$

where ξ is a normalized coordinate defined on the eth element, l_e is the length of the element, and ϕ_1 and ϕ_2 are the interpolation functions given as follows:

$$\begin{rcases} \Phi_1 \\ \Phi_2 \end{rcases} = \frac{1}{2}(1 \mp \xi). \tag{14}$$

When the point i belongs to the eth element, calculations of g_1, g_2, h_1, and h_2 involve the limitation of $\epsilon \to 0$. In this case, the vector $\boldsymbol{r}$ is at right angles to the outward normal vector $\boldsymbol{n}$ (see Fig. 2), so that $\cos(\boldsymbol{r}; \boldsymbol{n})$ becomes equal to zero and

$$h_1 = h_2 = 0. \tag{15}$$

The remainders are g_1 and g_2. Here, we consider the case where the point i coincides with the first node of the eth element

$$\begin{aligned} g_1 &= \lim_{\epsilon \to 0} \int_{-1+\epsilon}^{1} \frac{1}{2}(1\text{-}\xi) \left\{ -\frac{1}{4} j H_0^{(2)}(kr) \right\} d\xi \cdot \frac{1}{2} l_e \\ &= \frac{jl_e}{16} \lim_{\epsilon \to 0} \int_{-1+\epsilon}^{1} (\xi - 1) H_0^{(2)}\left\{ k(\xi+1)\frac{1}{2} l_e \right\} d\xi \\ &= \frac{jl_e}{16} \lim_{\epsilon \to 0} \int_{\epsilon}^{2} \left\{ \xi H_0^{(2)}\left(k\xi \cdot \frac{1}{2} l_e\right) - 2 H_0^{(2)}\left(k\xi \cdot \frac{1}{2} l_e\right) \right\} d\xi \\ &= \frac{jl_e}{16} \lim_{\epsilon \to 0} \left\{ \left[\frac{2\xi}{kl_e} H_1^{(2)}\left(\frac{1}{2} kl_e \xi\right) \right]_{\epsilon}^{2} - 2 \int_{\epsilon}^{2} H_0^{(2)}\left(\frac{1}{2} kl_e \xi\right) d\xi \right\}. \end{aligned} \tag{16}$$

In the second term of (16), the Hankel function is expanded to infinite series and is integrated by term

$$\begin{aligned} g_1 = \frac{1}{4} jl_e \Bigg[& \frac{1}{kl_e} H_1^{(2)}(kl_e) + j\frac{2}{\pi} \left\{ \ln\left(\frac{1}{2} kl_e\right) + \gamma - 1 - \left(\frac{1}{kl_e}\right)^2 \right\} - 1 \\ & - \sum_{s=1}^{\infty} \frac{(-1)^s (kl_e)^{2s}}{(s!)^2 2^{2s}(2s+1)} \left\{ 1 - j\frac{2}{\pi}\left(\ln\frac{kl_e}{2} + \gamma - h_s - \frac{1}{2s+1} \right) \right\} \Bigg] \end{aligned}$$

$$h_s = 1 + \frac{1}{2} + \frac{1}{3} + \cdots + \frac{1}{s}. \tag{17}$$

From (16) and (17), we obtain the next result for g_2

$$g_2 = -\frac{1}{4} jl_e \left[\frac{1}{kl_e} H_1^{(2)}(kl_e) - j\frac{2}{\pi} \left(\frac{1}{kl_e} \right)^2 \right]. \tag{18}$$

If the point i is the second node of the eth element, then g_1 is estimated with (18) and g_2 with (17). The infinite series in (17) is approximated by the first few terms, since the length of the eth element l_e is chosen as $l_e < \lambda/10$, i.e., $kl_e < 2\pi/10$, so that the series rapidly converges, where λ is the wavelength in free space.

In the above calculations, the estimation of c_i is important for the Dirichlet boundary condition giving the nonzero potential. Because c_i is the coefficient of u_i, any value of c_i is permitted for the boundary condition giving the zero potential; but for the case of the nonzero potential, c_i must be calculated precisely [7].

In the matrix notation, (11) is rewritten as follows:

$$\boldsymbol{Hu} = \boldsymbol{Gq}. \tag{19}$$

On each node, the potential u or its outward normal derivative q must be given as the boundary condition. Then, all the remaining u's and q's can be calculated from (19). The matrices produced by the boundary-element formulation are much smaller in size than the finite-element method.

III. The Case of Waveguides with Discontinuities

For a typical example, a parallel-plane waveguide with discontinuities is considered, and the mode having the z-component of the electric field, which is chosen as the analyzed potential u, is assumed. In this case, a closed region R_w, as shown in Fig. 3, is chosen as the analyzed model, which is enclosed by the boundary of the waveguide wall B_w and pseudo-boundaries at the power supply side and the opposite load side, which we call the input-side boundary B_i and the output-side boundary B_o, respectively.

From (19), the following equation is obtained for the region R_w:

$$[\boldsymbol{H}_i \boldsymbol{H}_o \boldsymbol{H}_w] \begin{Bmatrix} \boldsymbol{u}_i \\ \boldsymbol{u}_o \\ \boldsymbol{u}_w \end{Bmatrix} = [\boldsymbol{G}_i \boldsymbol{G}_o \boldsymbol{G}_w] \begin{Bmatrix} \boldsymbol{q}_i \\ \boldsymbol{q}_o \\ \boldsymbol{q}_w \end{Bmatrix}, \quad \text{in } R_w. \tag{20}$$

In (20), suffixes i, o, and w show the quantities corresponding to the boundaries B_i, B_o, and B_w, respectively. On the waveguide wall, the electric-field component parallel to it vanishes, so that the following boundary condition is taken:

$$\boldsymbol{u}_w = 0, \quad \text{on } B_w. \tag{21}$$

But on the remaining boundaries B_i and B_o, any specified value of u or q cannot be given. If they are given, the phase relation between the field components on the input- and the output-side boundaries are also given, and, therefore, the problem has already been solved. This is a contradiction.

Now, we adopt the following procedure. First, we place the two pseudo-boundaries B_i and B_o at the position where it is consid-

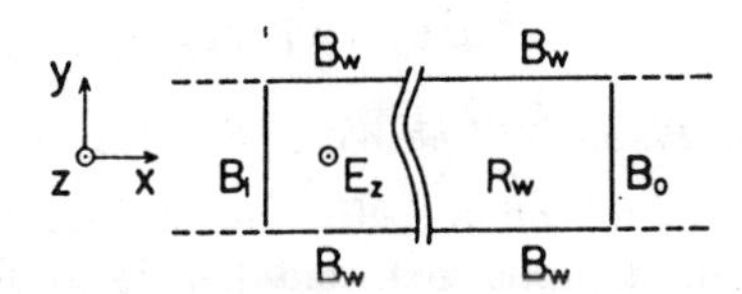

Fig. 3. A parallel-plane waveguide with discontinuities.

ered that the reflecting electromagnetic wave, generated at the discontinuites, attenuates and almost vanishes. The TE_{10}-mode field distribution is assumed on them. Then, on only B_i, the boundary condition, in complex form, is placed

$$u_i = \bar{u}_i, \qquad \text{on } B_i. \tag{22}$$

Next, on B_o, we introduce the TE_{10}-mode propagation constant β, and the electric-field component is written as

$$u_o = E_{z0}\exp(-j\beta y), \qquad \text{on } B_o \tag{23}$$

and its outward normal derivative q_o is also written as

$$q_o = \frac{du_o}{dy} = -j\beta E_{z0}\exp(-j\beta y), \qquad \text{on } B_o. \tag{24}$$

So, on B_o, the next relationship between u_o and q_o is obtained as follows:

$$q_o = -j\beta u_0, \qquad \text{on } B_o. \tag{25}$$

From (20), (21), (22), and (25), the following equation, to be solved finally, is obtained:

$$[G_i - j\beta F_o - H_o G_w]\begin{Bmatrix} q_i \\ u_o \\ q_w \end{Bmatrix} = H_i \bar{u}_i, \qquad \text{in } R_w. \tag{26}$$

In the right-hand side of (26), all quantities are known and q_i, u_o, and q_w are obtained as a solution. Then, q_o is given by (25).

IV. Multi-Media Problems

For multi-media cases, any boundary method requires to make up equations for each homogeneous sub-domain constructed of one media. So it is generally said that boundary methods are weak in multi-media problems and the finite-element method is superior to the boundary-element method in such a case. However, the authors have verified that a bit of effort on the design of the computer program makes this fault of the boundary-element method negligible, and they propose a procedure of programming, for handling multi-media problems, in a slightly different style from those of the references [3], [4].

Consider a two-dimensional region constructed of three different media, as shown in Fig. 4, where R_1, R_2, and R_3 are homogeneous sub-domains, B_1, B_2, and B_3 are the boundaries belonging to only R_1, R_2, and R_3, respectively, and B_{12}, B_{23}, and

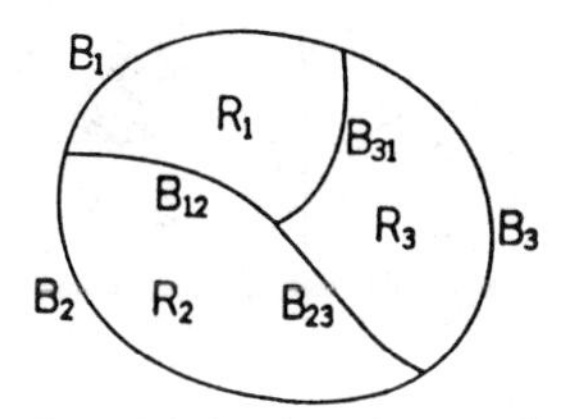

Fig. 4. Two-dimensional region constructed of three media.

B_{31} are the interfaces between two adjacent sub-domains. The ordinary boundary-element technique leads to the following equations for each homogeneous sub-domain:

$$[G_1 G_{12}^{(1)} G_{31}^{(1)}]\begin{Bmatrix} q_1 \\ q_{12}^{(1)} \\ q_{31}^{(1)} \end{Bmatrix} = [H_1 H_{12}^{(1)} H_{31}^{(1)}]\begin{Bmatrix} u_1 \\ u_{12}^{(1)} \\ u_{31}^{(1)} \end{Bmatrix}, \qquad \text{in } R_1 \tag{27}$$

$$[G_2 G_{23}^{(2)} G_{12}^{(2)}]\begin{Bmatrix} q_2 \\ q_{23}^{(2)} \\ q_{12}^{(2)} \end{Bmatrix} = [H_2 H_{23}^{(2)} H_{12}^{(2)}]\begin{Bmatrix} u_2 \\ u_{23}^{(2)} \\ u_{12}^{(2)} \end{Bmatrix}, \qquad \text{in } R_2 \tag{28}$$

$$[G_3 G_{31}^{(3)} G_{23}^{(3)}]\begin{Bmatrix} q_3 \\ q_{31}^{(3)} \\ q_{13}^{(3)} \end{Bmatrix} = [H_3 H_{31}^{(3)} H_{13}^{(3)}]\begin{Bmatrix} u_3 \\ u_{31}^{(3)} \\ u_{13}^{(3)} \end{Bmatrix}, \qquad \text{in } R_3. \tag{29}$$

In (27)–(29), superscript (i) implies the quantity defined in R_i. The boundry conditions on the interfaces are as follows:

$$u_{12}^{(1)} = u_{12}^{(2)} = u_{12}, \quad q_{12}^{(1)} = -q_{12}^{(2)} = q_{12}, \qquad \text{on } B_{12} \tag{30}$$

$$u_{23}^{(2)} = u_{23}^{(3)} = u_{23}, \quad q_{23}^{(2)} = -q_{23}^{(3)} = q_{23}, \qquad \text{on } B_{23} \tag{31}$$

$$u_{31}^{(3)} = u_{31}^{(1)} = u_{31}, \quad q_{31}^{(3)} = -q_{31}^{(1)} = q_{31}, \qquad \text{on } B_{31}. \tag{32}$$

In the above conditions, the minus sign of q originates from the outward normal directions of the adjacent two subregions opposite each other. From (27)–(32), we obtain the next equation to be solved finally.

$$\begin{bmatrix} G_1 & 0 & 0 & G_{12}^{(1)} & -H_{12}^{(1)} & 0 & 0 & -G_{31}^{(1)} & -H_{31}^{(1)} \\ 0 & G_2 & 0 & -G_{12}^{(2)} & -H_{12}^{(2)} & G_{23}^{(2)} & -H_{23}^{(2)} & 0 & 0 \\ 0 & 0 & G_3 & 0 & 0 & -G_{23}^{(3)} & -H_{23}^{(3)} & G_{31}^{(3)} & H_{31}^{(3)} \end{bmatrix}\begin{bmatrix} q_1 \\ q_2 \\ q_3 \\ q_{12} \\ u_{12} \\ q_{23} \\ u_{23} \\ q_{31} \\ u_{31} \end{bmatrix} \tag{33}$$

$$= \begin{bmatrix} H_1 & 0 & 0 \\ 0 & H_2 & 0 \\ 0 & 0 & H_3 \end{bmatrix}\begin{bmatrix} u_1 \\ u_2 \\ u_3 \end{bmatrix}, \qquad \text{in } R_1 + R_2 + R_3.$$

V. Electromagnetic Scattering Problems

In this section, the procedure analyzing the problem of electromagnetic scattering by dielectric bodies is developed. This is considered to be the problem of multi-media and the unbounded field. Therefore, the procedure developed for multi-media problems can be utilized for the scattering problem in the form extended a little.

It is assumed that the incident wave is the E-wave and propa-

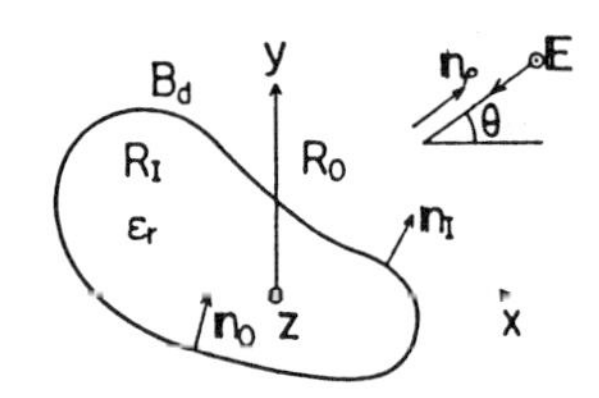

Fig. 5. Electromagnetic scattering by a dielectric cylinder.

gates in the direction parallel to the $x-y$ plane. All quantities are uniform in the z-direction. The analyzed region is constructed of two subregions, as shown in Fig. 5. Subregion R_I is inside the dielectric cylinder and subregion R_O is outside of it. The latter is an unbounded field, but at infinity, the radiation condition can be taken for the scattering wave, so that the boundary-element equations are considered on only the dielectric surface B_d

$$G_I q_I = H_I u_I, \qquad \text{in } R_I$$

$$u_I = u_{I,sc} + u_{I,in} \qquad (34)$$

$$q_I = q_{I,sc} + q_{I,in}$$

$$G_O q_{O,sc} = H_O u_{O,sc}, \qquad \text{in } R_O. \qquad (35)$$

Here, subscripts I and O imply the quantities defined in R_I and R_O, respectively, and sc and in indicate the scattering and incident waves, respectively. In R_I, the ruling equation is defined using the total wave, i.e., the incident wave plus the scattering wave. On the contrary, in R_O, the radiation condition can't be applied for the incident wave, so that the equation is defined by only the scattering wave.

On B_d, the following boundary conditions are taken:

$$u_{I,sc} = u_{O,sc} = u_{sc}, \qquad \text{on } B_d \qquad (36)$$

$$q_{I,sc} = -q_{O,sc} = q_{sc}, \qquad \text{on } B_d. \qquad (37)$$

In addition, the incident wave is described as follows:

$$u_{in} = u_{I,in} = E_{z_0,in} \exp(jk\rho). \qquad (38)$$

Here, the ρ coordinate is chosen in the direction of the incidence, as in Fig. 5. For $q_{I,in}$, we derive the following equation:

$$q_{I,in} = \frac{du_{I,in}}{dn_I} = -jk\mathbf{n}_I \cdot \mathbf{n}_\rho u_{I,in}$$

$$= x\cos\theta + y\sin\theta \qquad (39)$$

where θ is the incident angle. In matrix notation, the above relation is rewritten as

$$q_{I,in} = B_I u_{I,in}. \qquad (40)$$

From (34)–(40), the following equation is obtained for $R_I + R_O$:

$$\begin{bmatrix} -G_I & H_I \\ G_O & -H_O \end{bmatrix} \begin{bmatrix} u_{sc} \\ q_{sc} \end{bmatrix} = \begin{bmatrix} H_I - G_I B_I \\ O \end{bmatrix} u_{I,in}. \qquad (41)$$

By (41), the scattering field is calculated on the boundary B_d provided that the incident wave is given there.

The procedure proposed here is very powerful for the case where the dielectric body has a much larger dimension than the wavelength of the incidence or has a large dielectric constant. The finite-element method is weak because it requires a large computer memory.

VI. Analyzed Results

A. Open-Ended Parallel-Plane Waveguides

We analyzed the reflection coefficients of three kinds of waveguides, i.e., flanged, unflanged, and flared, consisting of two parallel planes. These are unbounded field problems, and closed boundaries extending to infinity should be chosen. But at infinity, the radiation condition can be introduced, and at the point several wavelengths distant from the open-end of the waveguide, the field value becomes negligibly small, so that the contribution to the integration can be neglected. The finite models are chosen as in Fig. 6, which are half-models, because each of them has the symmetry axis parallel to the direction of electromagnetic wave propagation. Fig. 7 denotes the results obtained with the boundary-element method for the above three cases. They have good agreement with Lee's solution by ray theory [8], Vaynshteyn's by the Wiener-Hopf technique [9], and the results by the finite-element method extended to unbounded fields [2]. For the flared waveguide, the boundary-element method results are compared with those of the boundary integral equation method [10]. All of the above boundary-element calculations are performed with 40 nodes, while the corresponding finite-element calculations need at least 400 nodes. The problems analyzed here are unbounded ones, for which the boundary-element method seems to be the most suitable method.

B. A Symmetrical H-Plane Y-Junction

The electric-field distribution in an H-plane Y-junction is analyzed. The electromagnetic wave is assumed to propagate along the y-axis, as in Fig. 8, which also indicates the analyzed model. The right half of the model is considered for the symmetry. On the input- and output-side boundaries, the TE_{10} mode is assumed. Fig. 9 shows the electric-field distribution, represented by the standing-wave on the centerlines. Details of their positions are shown in Fig. 8. From the results in Fig. 8, the reflection coefficient due to this Y-junction is calculated as 0.18. This value leads to the calculated amplitude of the electric field at the output side of the junction, $\sqrt{(1-0.18^2)/2} = 0.6956$, where that of the input side is assumed to be 1.0. The value, 0.6956, agrees well with the boundary-element result of the electric standing-wave amplitude in Fig. 9.

C. A Waveguide with a Dielectric Cylinder Placed at its Open-end

Here, we analyze the electromagnetic-field distribution of the case where the parallel-plane waveguide is open-ended with infinite flanges and a dielectric cylinder is placed at the open-end, as in Fig. 10. Fig. 11 shows the electric-field distributions for three values of dielectric constant, 1.0, 2.0, and 3.0. The case where $\epsilon_r = 1.0$ implies a homogeneous case, i.e., no cylinder is placed. This model brings the reflection coefficient, 0.029, which agrees with that obtained by the computer program for the homogeneous case (cf. Fig. 7(a)) and clarifies the propriety of the multi-media case analyzing procedure. From Fig. 11, it is found that the larger dielectric constant becomes, the denser the electric-field concentrates to the dielectric cylinder.

D. Electromagnetic Scattering of Dielectric Circular Cylinder

As the last example, the plane electromagnetic wave scattering by a dielectric circular cylinder is investigated. The incident wave

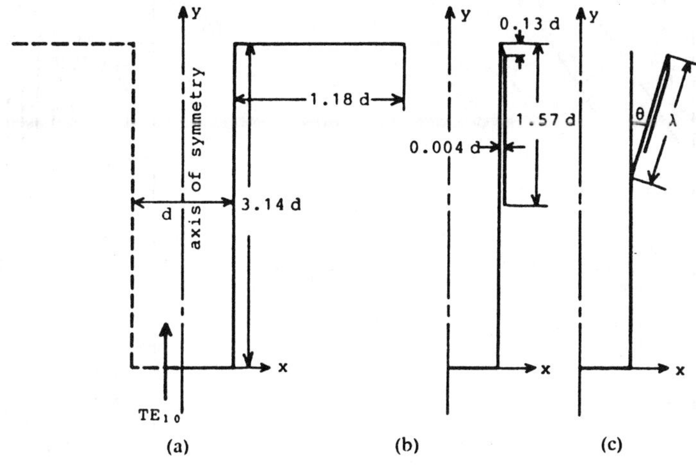

Fig. 6. Analyzed models of open-ended parallel-plane waveguides. (a) Flanged, (b) unflanged, and (c) flared.

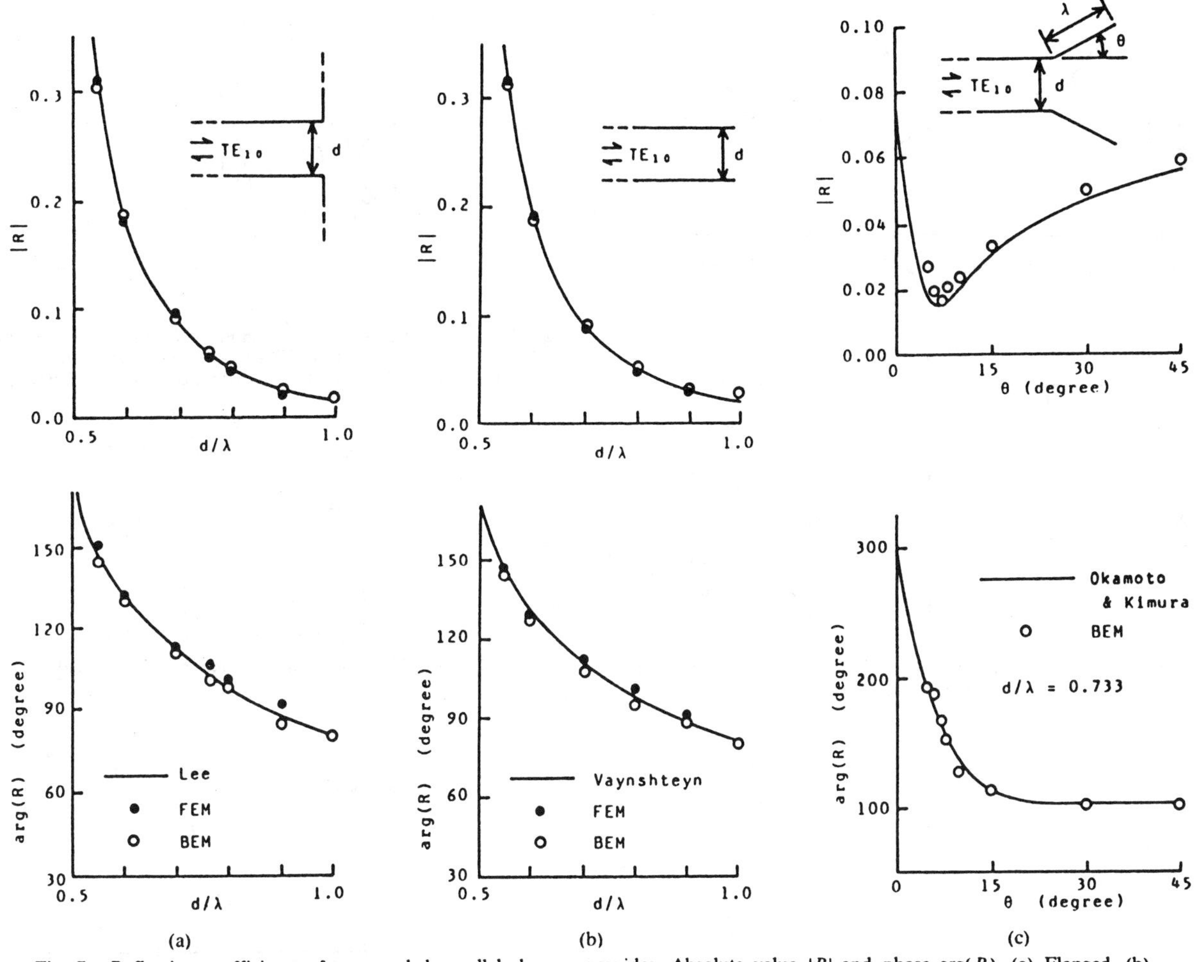

Fig. 7. Reflection coefficients of open-ended parallel-plane waveguides. Absolute value $|R|$ and phase arg(R). (a) Flanged, (b) unflanged, and (c) flared.

is assumed to travel along the x-axis in the negative x-direction. Details are shown in Fig. 12. The dielectric cylinder is assumed to have a dielectric constant 2.0 and a radius 0.408λ, where λ is a wavelength in free space. Fig. 13 is the amplitude and phase of the E-wave scattering far-field pattern in this case. In addition, Fig. 14 shows the electric-field distribution around the dielectric cylinder. In Fig. 13, the results obtained by the boundary-element method are compared with analytical ones and show good accuracy. The calculations are done with only 24 nodes. For smaller cylinders, more accurate results have been obtained. For larger cylinders, no actual analysis has yet been performed, but it seems that the boundary-element method is more suitable than the finite-element method.

E. Computer Implementation

Boundary-element calculations require only a small computer

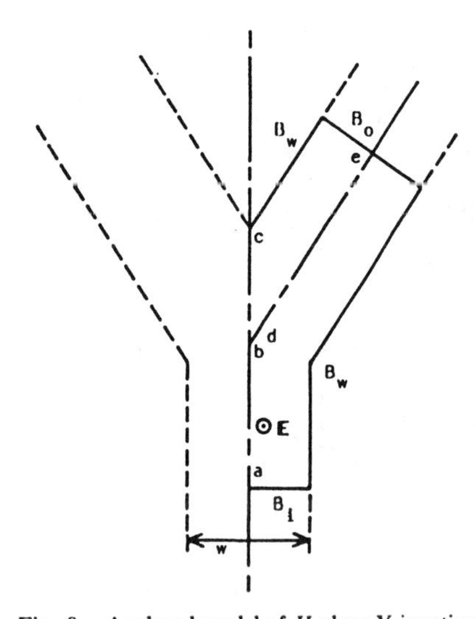

Fig. 8. Analyzed model of H-plane Y-junction.

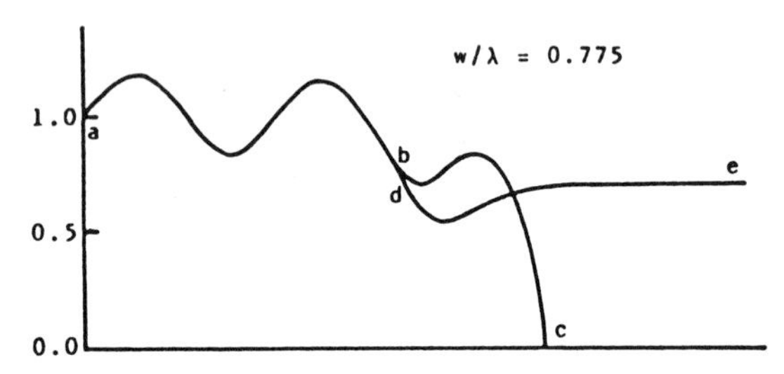

Fig. 9. The electric-field distribution in a H-plane Y-junction.

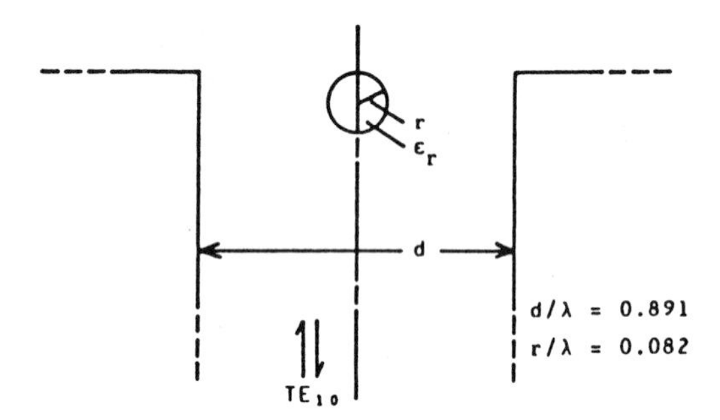

Fig. 10. Flanged open-ended waveguide with a dielectric cylinder placed at its open-end.

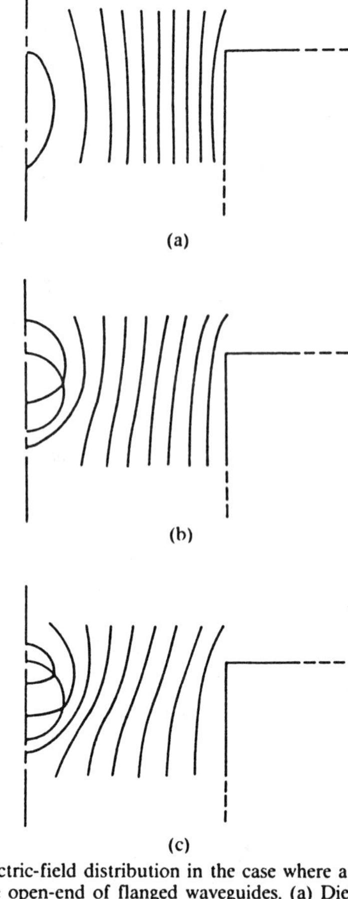

Fig. 11. The electric-field distribution in the case where a dielectric cylinder is placed at the open-end of flanged waveguides. (a) Dielectric constant of the cylinder $\epsilon_r = 1.0$, i.e., no cylinder is placed. (b) $\epsilon_r = 2.0$. (c) $\epsilon_r = 3.0$.

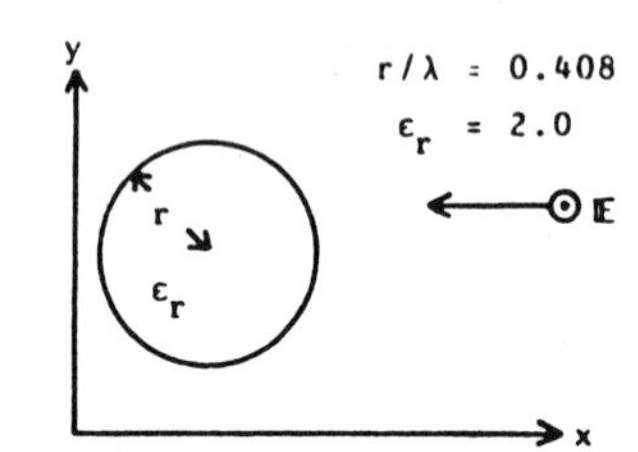

Fig. 12. Dielectric circular cylinder and the incident E-wave.

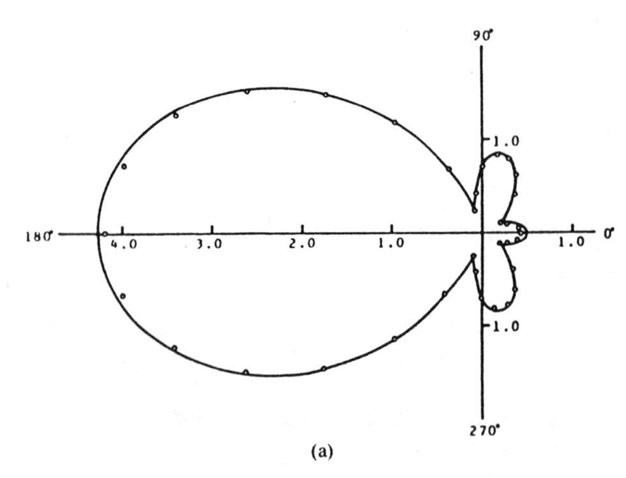

(a)

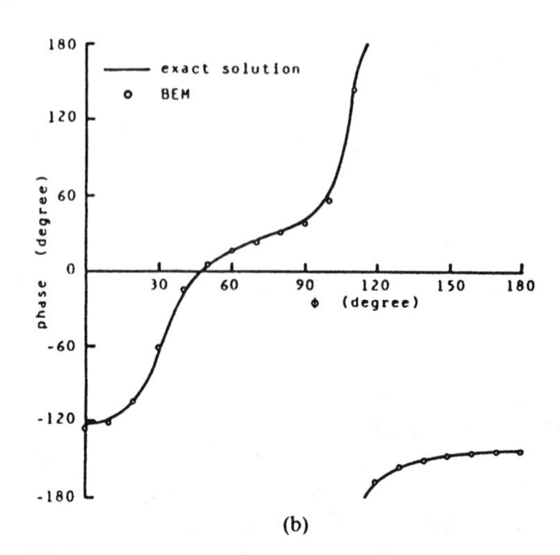

(b)

Fig. 13. E-wave scattering far-field pattern for dielectric circular cylinder in the case where E-wave incidents along x-axis in the negative x-direction. (a) Amplitude. (b) Phase.

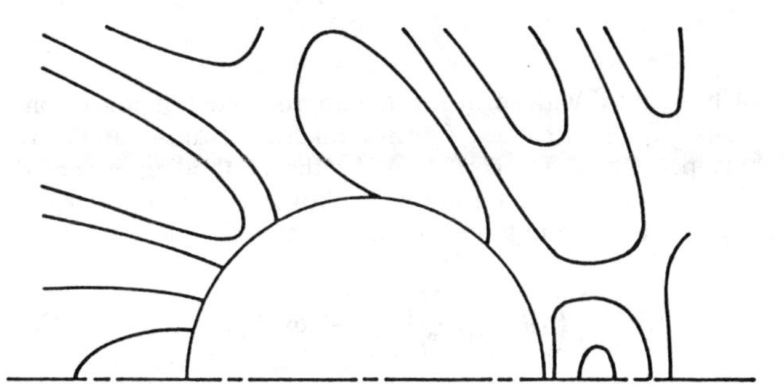

Fig. 14. The electric-field distribution around the dielectric circular cylinder.

memory, so that a microcomputer can handle them. The techniques discussed in this paper have been implemented by a FORTRAN program on a microcomputer, whose main CPU is MC68000 (8 MHz) and whose operating system is the UCSD *p*-system. Typically, the case of a parallel-plane waveguide having 40 nodes took about 20 m of CPU time.

VII. Conclusions

Application of the boundary-element method to electromagnetic-field problems was proposed. Several analyzing procedures for interesting cases were also given. The results obtained show that the boundary-element method is a very powerful numerical method for electromagnetic-field problems. Namely, by using the boundary-element method, far fewer nodes than by the finite-element method bring good accuracy, and unbounded field problems can be treated without any additional technique.

References

[1] B. H. McDonald and A. Wexler, "Finite-element solution of unbounded field problems," *IEEE Trans. Microwave Theory Tech.*, vol. MTT-20, pp. 841–847, Dec. 1972.

[2] S. Washisu, I. Fukai, and M. Suzuki, "Extension of finite-element method to unbounded field problems," *Electron. Lett.*, vol. 15, pp. 772–774, Nov. 1979.

[3] C. A. Brebbia, *The Boundary Element Method for Engineers.* London: Pentech Press, 1978.

[4] C. A. Brebbia and S. Walker, *Boundary Element Techniques in Engineering.* London: Butterworth, 1980.

[5] S. Washisu and I. Fukai, "An analysis of electromagnetic unbounded field problems by boundary element method," *Trans. Inst. Electron. Commun. Eng. Jpn.*, vol. J64-B, pp. 1359–1365, Dec. 1981.

[6] S. Bilgen and A. Wexler, "Spline boundary element solution of dielectric scattering problems," in *12th Eur. Microwave Conf.*, (Helsinki, Finland), Sept. 1982, pp. 372–377.

[7] C. G. Williams and G. K. Cambrell, "Efficient numerical solution of unbounded field problems," *Electron. Lett.*, vol. 8, pp. 247–248, May 1972.

[8] S. W. Lee, "Ray theory of diffraction by open-ended waveguides. Part I," *J. Math. Phys.*, vol. 11, pp. 2830–2850, Sept. 1970.

[9] L. A. Vaynshteyn, *The Theory of Diffraction and the Factorization Method.* Boulder, CO: Golem, 1969.

[10] N. Okamoto and T. Kimura, "Radiation properties of *H*-plane two-dimensional horn antennas of arbitrary shape," *Trans. Inst. Electron. Commun. Eng. Jpn.*, vol. J59-B, pp. 25–32, Jan. 1976.

MULTIMODE *S*-PARAMETERS OF PLANAR MULTIPORT JUNCTIONS BY BOUNDARY ELEMENT METHOD

Indexing terms: Waveguides, Transmission lines

A combination of the boundary element method in the junction, using a Green's function approach and multimode expansions in the outgoing planar transmission lines, has been proposed. The multimode *S*-parameters are calculated with good accuracy from a simple identification in the system of equations given by the boundary element method. In addition to a considerable gain of computing time with this method, it is easy to handle inhomogeneities and complicated structures.

Introduction: Open planar multiport junctions (Fig. 1) are found in practical applications, particularly in the spectral range from microwave to short millimetre-wave frequencies, in the integrated circuits. The frequency-dependent properties of such structures can be investigated using a planar waveguide model.[1–4] Chadha and Gupta[3] have proposed a method using a Green's function and reported the multimode transfer coefficients for a right-angled bend for the case where the signal at the input line is in the lowest-order mode. In their method the reference planes for the external lines have been chosen to be far away from the geometric discontinuities (Fig. 1, Reference 3) so that they do not include evanescent modes in their analysis. In our investigation a boundary element method using Green's function and a planar waveguide model are used. From the system of equations given by the boundary-element method, the identification of the multimode amplitudes in the transmission lines is almost immediate, with the merit of handling evanescent modes.

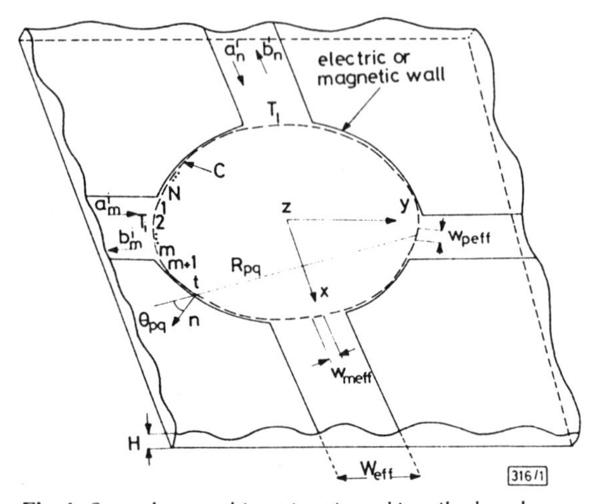

Fig. 1 *Open planar multiport junction arbitrarily shaped*

Theory: In the two-dimensional region D, enclosed by a boundary C (Fig. 1), assuming the thickness of the substrate is much less than the wavelength (in order to ensure that higher-order modes which vary in the *Z*-direction are suppressed), only the (E_z, H_x, H_y) field components exist.[1,2] The E_z field in the planar junction satisfies the bidimensional wave equation

$$(\Delta_x + \Delta_y + k^2)E_z = 0 \qquad k^2 = \omega^2 \varepsilon_{eff} \mu_{eff} \tag{1}$$

which is solved with regard to the appropriate boundary conditions. If we introduce Green's function (whose units are ohms per metre) for eqn. 1, using the method of weighted residual[4,6] or Green's formulas, it follows a Fredholm integral of the second kind in terms of the unknown E_z field:

$$E_z(P) = \int_C \left(G(kR_{pq}) \frac{\partial E_z}{\partial n}(Q) - \cos\theta_{pq} \frac{\partial G}{\partial R} E_z(Q) \right) dt \tag{2a}$$

By use of the discretisation technique shown in Fig. 1, and by assuming that the electric field and its outward normal derivative are constant across each section, the integral equation results in the matrix form

$$\sum_{p=1}^{N} U_{pq} e_p = \sum_{p=1}^{N} T_{pq} \frac{\partial e_p}{\partial n} \qquad q = 1, 2, \ldots, N \tag{2b}$$

or

$$[U][e] = [T][h] \tag{2c}$$

In the case of a homogeneous substrate

$$U_{pq} = \delta_{pq} + k \int_{W_{peff}} \cos\theta_{pq} G'(kR_{pq})\, dt_p$$

$$T_{pq} = \int_{W_{peff}} G(kR_{pq})\, dt_p$$

with W_{peff} the effective width of the *p*th section, it can be straightforward to derive the formulas U_{pq} and I_{pq}. e_p and k are the unknowns in eqns. 2. When the circuit has no coupling port, that is $\partial e_p/\partial n = 0$, from the nontrivial solution of eqns. 2, we have det $U = 0$. This equation gives the resonant frequencies of the circuit, allowing us to calculate k. The unknowns e_p, which belong to the ports, can be calculated by expanding the E_z field in the planar transmission line in the TE_{mo} (or H_{mo}) modes of a waveguide with magnetic walls for smaller sides:[2,4]

$$E_z(t, n) = \sum_{m=1}^{M} (a_m^i \exp \pm k_m n + b_m^i \exp \mp k_m n) \times \cos \frac{(m-1)\pi t}{M W_{m\,eff}} \tag{3}$$

where

$$k_m^2 = \left(\frac{\omega}{C}\right)^2 \varepsilon_{r\,eff} - \left(\frac{(m-1)\pi}{M W_{m\,eff}}\right)^2 \qquad C = 3 \times 10^8 \text{ m/s}$$

a_m^i and b_m^i are amplitudes of the modes of order m of the incident and reflected waves, respectively, M is the number of subports of the *i*th port, and $W_{m\,eff}$ is the effective width of the *m*th subport. We deduce e_l and its outward normal derivative from eqn. 3 as follows:

$$e_l = \frac{1}{W_{l\,eff}} \int_{W_{leff}} E_z(t_l, 0)\, dt_l = \sum_{M-1}^{M} (a_m^i + b_m^i) C_{ml} \tag{4a}$$

Reprinted with permission from *Electron. Lett.*, vol. 20, no. 19, pp. 799–802, Sept. 13, 1984.

$$\frac{\partial e_l}{\partial n} = \int_{W_{leff}} -\frac{\partial E_z}{\partial n}(t_l, 0)\, dt_l$$

$$= \sum_{m=1}^{M} -\frac{k_m}{W_{l\,eff}} (\pm a^i_m \mp b^i_m) C_{ml} \qquad (4b)$$

where

$$C_{ml} = \cos \frac{(m-1)(2l-1)}{2M} \frac{\sin \dfrac{(m-1)}{2M}}{\dfrac{(m-1)}{2M}}$$

$$(m-1)W_{l\,eff} < t_l < mW_{l\,eff}$$

a^i_m, b^i_m, k_m are the unknowns in exprs. 4. Each port has been divided into a number of sections equal to the number of modes propagating and evanescent in the transmission lines. Substituting eqns. 4 into eqns. 2 and finding the identity, the following matrix form is obtained:

$$\begin{bmatrix} B_{pl} & U_{pl} \\ 0 & U_{pl} \end{bmatrix}_{N,N} \begin{bmatrix} b^1_1 \\ \vdots \\ b^i_l \\ \vdots \\ e_1 \\ \vdots \\ e_N \end{bmatrix} = \begin{bmatrix} A_{pl} & 0 \\ 0 & 0 \end{bmatrix}_{N,N} \begin{bmatrix} a^1_1 \\ \vdots \\ a^i_1 \\ \vdots \\ 0 \\ \vdots \\ 0 \end{bmatrix}$$

and then the multimode S-parameters are

$$S^{i,l}_{m,n} = \frac{b^i_m}{a^l_n}$$

$$= \frac{\text{amplitude of the mode of order } m \text{ at the } i\text{th port}}{\text{amplitude of the mode of order } n \text{ at the } l\text{th port}}$$

with

$$B_{pl} = U_{pl} \sum_{m=1}^{M} C_{ml} \mp T_{pl} \sum_{m=1}^{M} k_m C_{ml}$$

$$A_{pl} = -U_{pl} \sum_{m=1}^{M} C_{ml} \mp T_{pl} \sum_{m=1}^{M} k_m C_{ml}$$

Results: This method has been applied to multimode characterisation of asymmetric T and bend junctions considering the frequency dispersion by a hybrid mode investigation. From two computer analyses, a spectral domain approach[7] and the least squares boundary residual, we have calculated the dispersion characteristics (Fig. 2*a*). It follows the accurate calculation of the effective relative dielectric constant $\varepsilon_{r\,eff}$ (Fig. 2*b*) from the dominant mode curve. Next, by use of the cutoff frequency formulas, we deduce the effective width W_{eff} of the waveguide model (Fig. 2*b*).

Figs. 3 and 4 show our S multimode parameter results with a comparison of experimental or theoretical investigation. The symbols used in Fig. 4, $/S^{1;1}_{1;1}/$, $/S^{1;1}_{2;1}/$, denote the magnitude of reflection on the line ① in dominant and first higher modes, respectively. Similarly, $/S^{2;1}_{1;1}/$, $/S^{2;1}_{2;1}/$ denote the magnitude of transmission to the line ② in dominant and first higher modes, respectively, for the case when the signal at the input line ① is in the dominant mode. Results in Fig. 3 have been obtained, when considering only fundamental and evanescent modes in the outgoing planar lines.

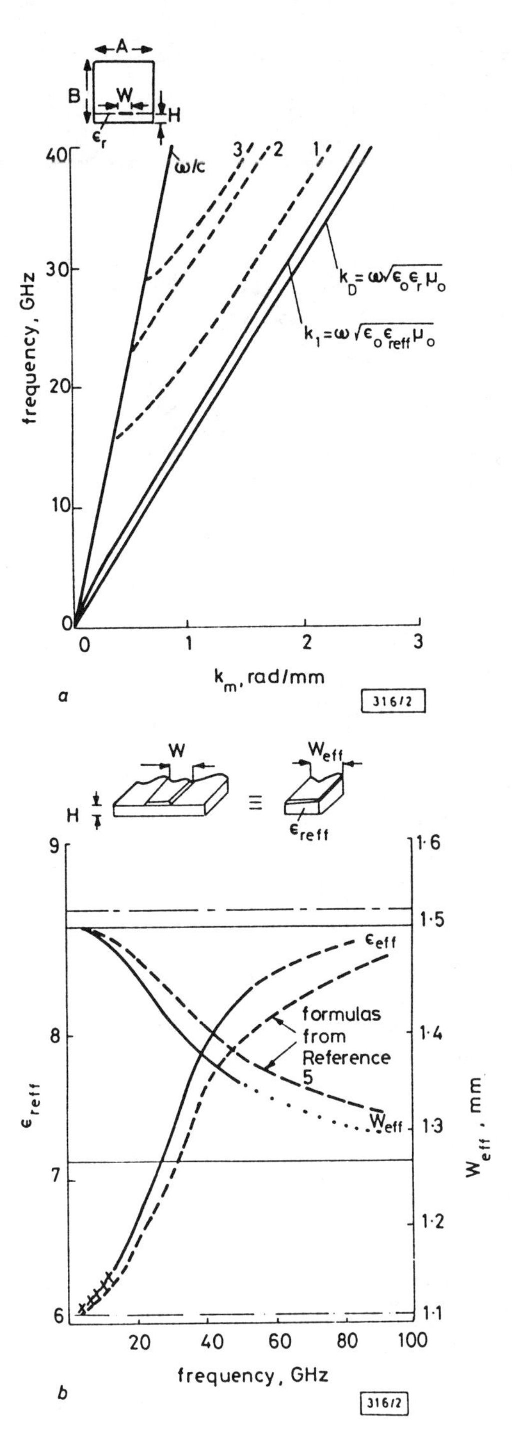

Fig. 2

a Dispersion behaviour of an open microstrip
$A = 12{\cdot}7$ mm $B = 12{\cdot}7$ mm
$H = 1{\cdot}27$ mm $W = 1{\cdot}27$ mm
$\varepsilon_r = 8{\cdot}875$
– – – – higher-order modes

b Effective parameters of a waveguide model
× experimental point from Reference 7

Conclusion: A combination of a planar junction analysis as a resonator by use of the boundary element method and a planar waveguide model from an accurate hybrid-mode investigation, has been proposed. The multimode S-parameter results for simple junctions (T, bend, ...) are in good concordance with other analytical and/or experimental studies. Using this method it is easy to handle complicated structures.

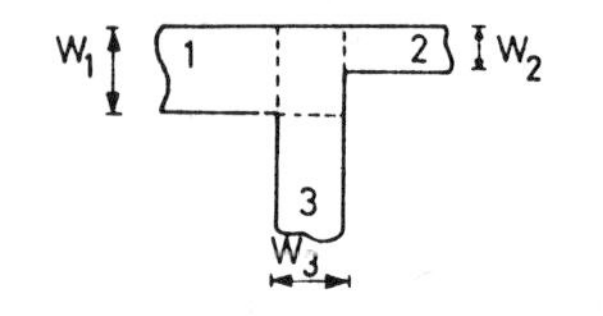

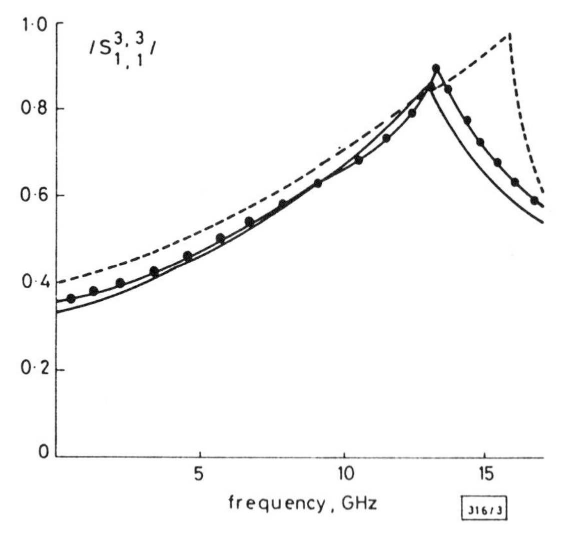

Fig. 3

$W_1 = 0{\cdot}977$ cm
$W_2 = 0{\cdot}780$ cm
$W_3 = 0{\cdot}780$ cm
$H = 0{\cdot}158$ cm
$\varepsilon_r = 2{\cdot}32$
- - - - our method (W_r)
—●— our method (W_{eff})
——— experimental results[1]

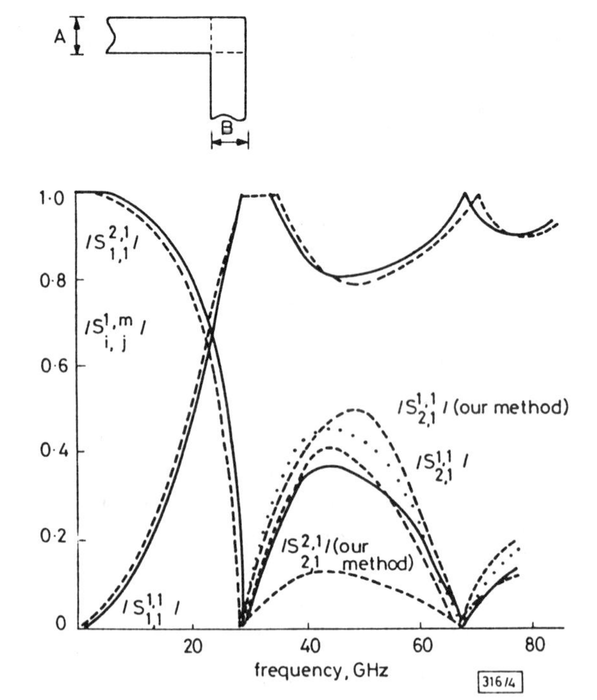

Fig. 4

$A = 0{\cdot}1882$ cm
$B = 0{\cdot}1882$ cm
$H = 0{\cdot}650$ cm
$\varepsilon_r = 6{\cdot}781$
- - - - our method
——— numerical results[3]

Acknowledgment: The authors wish to thank Thomson Bagneux, which supported this project.

E. TONYE
H. BAUDRAND

ENSEEIHT, Laboratoire de Microondes
2 rue C. Camichel, 31071 Toulouse Cedex, France

26th July 1984

References

1 MENZEL, W., and WOLFF, I.: 'A method for calculating the frequency-dependent properties of microstrip discontinuities', *IEEE Trans.*, 1977, **MTT-25**, p. 107

2 MEHRAN, R.: 'Calculation of microstrip bends and Y junctions with arbitrary angle', *ibid.*, 1978, **MTT-26**, p. 400

3 CHADHA, R., and GUPTA, K. C.: 'Green's function approach for obtaining *S*-matrices of multimode planar networks', *ibid.*, 1983, **MTT-31**, p. 224

4 TONYE, E., BAUDRAND, H., and MORISSE, D.: 'Etude de multipoles planaires par less fonctions de Green'. Proc. of the 7th colloquium on Hertian optics and dielectrics, Bordeaux, 26th–28th September 1983, p. 29

5 PRAMANICK, P., and BHARTIA, P.: 'An accurate description of dispersion in microstrip', *Microwave J.*, December 1983, p. 89

6 KAGAMI, S., and FUKAI, I.: 'Application of boundary-element method to electromagnetic field problems', *IEEE Trans.*, 1984, **MTT-32**, p. 455

7 ITOH, T., and MITRA, R.: 'Spectral-domain approach characteristics of microstrip lines', *ibid.*, 1973, **MTT-20**, p. 456

Application of the Boundary-Element Method to Waveguide Discontinuities

MASANORI KOSHIBA, SENIOR MEMBER, IEEE, AND
MICHIO SUZUKI, SENIOR MEMBER, IEEE

Abstract **—A numerical method for the solution of scattering of the *H*- and *E*-plane waveguide junctions is described. The approach is a combination of the boundary-element method and the analytical method. A general computer program has been developed using the quadratic elements (higher order boundary elements). To show the validity and usefulness of this formulation, computed results are given for a right-angle corner bend, a T-junction, an inductive strip-planar circuit mounted in a waveguide, a waveguide-type dielectric filter, and an inhomogeneous waveguide junction, and a linear taper. Comparison of the present results with the results of the finite-element method shows good agreement.**

I. Introduction

Waveguide discontinuities play an important role in designing microwave circuits [1], [2], and theoretical and experimental studies of waveguide discontinuity scattering problems have occupied the attention of numerous researchers for several decades. Recently, a numerical approach based on the finite-element method (FEM) has been developed for the analysis of planar circuits [3], [4], and *H*- and *E*-plane waveguide junctions [5]–[7]. The FEM is very useful for the arbitrarily shaped discontinuities. However, it requires a large computer memory and long computation time to solve the final matrix equation. More recently, the boundary-element method (BEM) [8], [9] has been applied to the *H*-plane junctions [10]–[12] and the planar circuits [13]. The BEM is one of the 'boundary'-type methods based on the integral equation method which has already been successfully applied to open-boundary planar circuits in 1972 [14] and to short-boundary planar circuits in 1975 [15], [16]. It is therefore possible to reduce the matrix dimension and to use computer memory more economically compared with the 'domain'-type method, such as the FEM. However, in [10], [11], [14], and [15], it is assumed that the waveguide propagates a single mode only and the evanescent modes are neglected. Therefore, it seems to be difficult to obtain accurate results over a wide range of frequencies. Furthermore, in [10]–[16], the constant elements [8], [9] or the linear elements [8], [9] are used to divide the boundary of the two-dimensional region. Generally, it is difficult to reduce the energy error with these boundary elements. In [12], the linear elements are used and the condition of power conservation is satisfied to an accuracy of about ± 4 percent. In order to obtain more accurate results, fairly many elements are necessary, and, thus, the merits of the BEM are lost. In the FEM analysis using the quadratic triangular elements (higher order finite elements), on the other hand, the energy error is less than 0.1 percent [5]–[7].

In this paper, the combined method of the BEM with the quadratic line elements (higher order boundary elements) and the analytical method is described for the analysis of scattering by the *H*- or *E*-plane waveguide junctions. To show the validity and usefulness of this formulation, computed results are given for various *H*- and *E*-plane waveguide discontinuities. Comparison of the results of the BEM with those of the FEM [5]–[7] shows good agreement. In the present BEM analysis, the power condition is satisfied to an accuracy of $\pm 10^{-4}$ to 10^{-3}.

II. Basic Equations

In order to minimize the detail, we consider the waveguide junction as shown in Fig. 1, where the boundary Γ_i connects the discontinuities to the rectangular waveguide i ($i = 1, 2$), d_i is the width a_i or the height b_i of the waveguide i for the *H*- or *E*-plane junction, respectively; the region Ω surrounded by Γ_1, Γ_2, and the short-circuit boundary Γ_0 completely encloses the waveguide discontinuities, and the waveguide i is assumed to be filled with dielectric of relative permittivity ϵ_{ri}.

Considering the excitation by the dominant TE_{10} mode, we have the following basic equation:

$$\frac{\partial^2 \phi}{\partial x^2} + \frac{\partial^2 \phi}{\partial y^2} + \hat{k}^2 \phi = 0 \tag{1}$$

$$\hat{k}^2 = k_0^2 \hat{\epsilon}_r \tag{2}$$

$$k_0^2 = \omega^2 \epsilon_0 \mu_0 \tag{3}$$

$$\phi = \begin{cases} E_z, & \text{for } H\text{-plane junction} \\ H_z, & \text{for } E\text{-plane junction} \end{cases} \tag{4}$$

$$\hat{\epsilon}_r = \begin{cases} \epsilon_r, & \text{for } H\text{-plane junction} \\ \epsilon_r - (\pi / k_0 a)^2, & \text{for } E\text{-plane junction} \end{cases} \tag{5}$$

where ω is the angular frequency, E_z and H_z are the electric and magnetic fields, respectively, and ϵ_0 and μ_0 are the permittivity and permeability of free space, respectively.

III. Mathematical Formulation

A. Boundary-Element Approach[1]

Considering the region surrounded by the boundary Γ as shown in Fig. 2, and using the fundamental solution ϕ^* [8], [9] and Green's formula, from (1) we obtain the following equation [10]–[13]:

$$\phi_p + \int_\Gamma \psi^* \phi \, d\Gamma = \int_\Gamma \phi^* \psi \, d\Gamma \tag{6}$$

where

$$\phi^* = \frac{1}{4j} H_0^{(2)}(\hat{k} r) \tag{7}$$

$$\psi^* = \frac{j}{4} \hat{k} H_1^{(2)}(\hat{k} r) \cos \alpha . \tag{8}$$

Manuscript received May 14, 1985; September 27, 1985.
The authors are with the Department of Electronic Engineering, Hokkaido University, Sapporo, 060, Japan.
IEEE Log Number 8406478.

[1]Since a general formulation of the BEM with linear elements for analyzing two-dimensional electromagnetic fields is given in [11], only the outline of the BEM with quadratic elements will be described here.

Reprinted from *IEEE Trans. Microwave Theory Tech.*, vol. MTT-34, no. 2, pp. 301–307, Feb. 1986.

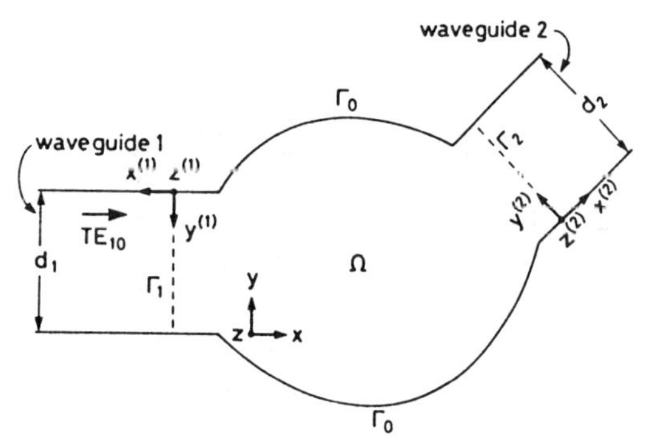

Fig. 1. Geometry of problem.

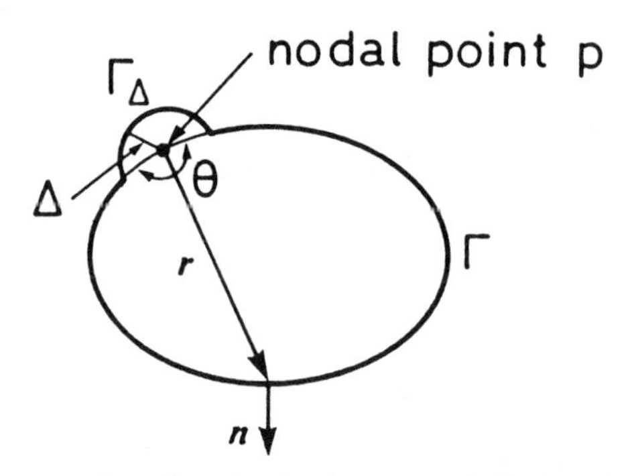

Fig. 2. Two-dimensional region surrounded by boundary Γ.

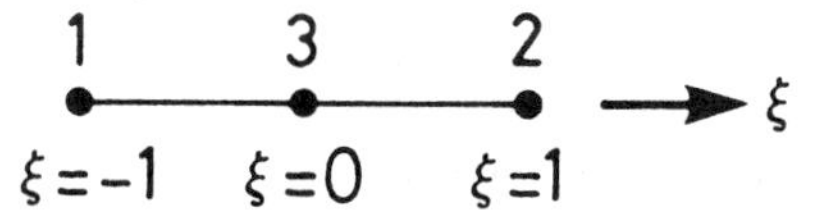

Fig. 3. Quadratic line element.

Here ϕ_p is the value of ϕ at the nodal point p, ψ, and ψ^* are the outward normal derivatives of ϕ and ϕ^*, respectively, $H_0^{(2)}$ and $H_1^{(2)}$ are the zeroth- and first-order Hankel functions of the second kind, respectively, and α is the angle between the vector $\boldsymbol{r}$ and the outward unit normal vector $\boldsymbol{n}$.

Noting that the nodal point p is placed on the boundary Γ and considering the integration path Γ_Δ going around the nodal point p as shown in Fig. 2, we obtain for (6)

$$\frac{\theta}{2\pi}\phi_p + \int_\Gamma \psi^*\phi \, d\Gamma = \int_\Gamma \phi^*\psi \, d\Gamma \tag{9}$$

where $\int$ denotes the Cauchy's principal value of integration, namely $\int_\Gamma = \lim_{\Delta\to 0}\int_{\Gamma-\Gamma_\Delta}$. Dividing the boundary Γ into quadratic line elements as shown in Fig. 3, ϕ and ψ within each element are defined in terms of ϕ_q and ψ_q at the nodal points q ($q=1, 2, 3$), respectively, as follows:

$$\phi = \{N\}^T\{\phi\}_e \tag{10}$$

$$\psi = \{N\}^T\{\psi\}_e \tag{11}$$

where

$$\{\phi\}_e = [\phi_1\phi_2\phi_3]^T \tag{12}$$

$$\{\psi\}_e = [\psi_1\psi_2\psi_3]^T \tag{13}$$

$$\{N\} = [N_1N_2N_3]^T. \tag{14}$$

Here T, $\{\cdot\}$, and $\{\cdot\}^T$ denote a transpose, a column vector, and a row vector, respectively, and the shape function N_q is given by

$$N_q = A_q\xi^2 + B_q\xi + C_q \tag{15}$$

$$A_1 = 1/2, \quad A_2 = 1/2, \quad A_3 = -1 \tag{16a}$$

$$B_1 = -1/2, \quad B_2 = 1/2, \quad B_3 = 0 \tag{16b}$$

$$C_1 = 0, \quad C_2 = 0, \quad C_3 = 1 \tag{16c}$$

with the normalized coordinate ξ defined on the eth element.

Substituting (10) and (11) into (9), we obtain

$$\frac{\theta}{2\pi}\phi_p + \sum_e \{h\}_e^T\{\phi\}_e = \sum_e \{g\}_e^T\{\psi\}_e \tag{17}$$

where

$$\{h\}_e = [h_1h_2h_3]^T \tag{18}$$

$$\{g\}_e = [g_1g_2g_3]^T. \tag{19}$$

Here Σ_e extends over all different elements. When the nodal point p does not belong to the eth element, h_q and g_q are calculated with Gaussian integration as

$$h_q = \frac{L}{2}\int_{-1}^{1} N_q \frac{j}{4}\hat{k}H_1^{(2)}(\hat{k}r)\cos\alpha \, d\xi \tag{20}$$

$$g_q = \frac{L}{2}\int_{-1}^{1} N_q \frac{1}{4j}H_0^{(2)}(\hat{k}r)\, d\xi \tag{21}$$

where L is the length of the element. When the nodal point p belongs to the eth element, calculations of h_q and g_q involve the limitation of $\Delta \to 0$. In this case, $\cos\alpha = 0$, so that

$$h_q = 0. \tag{22}$$

For the case where the nodal point p coincides with the nodal point $q = 1$, 2, or 3 of the eth element in Fig. 2, g_q is given by

$$g_q = (L/2)\left[A_qI_2(2) - (2A_q - B_q)\{I_1(2) - 2/(\pi\hat{k}^2L^2)\} + (A_q - B_q + C_q)I_0(2)\right] \tag{23a}$$

$$g_q = (L/2)\left[A_qI_2(2) - (2A_q + B_q)\{I_1(2) - 2/(\pi\hat{k}^2L^2)\} + (A_q + B_q + C_q)I_0(2)\right] \tag{23b}$$

or

$$g_q = L\left[A_qI_2(1) + C_qI_0(1)\right] \tag{23c}$$

respectively. Here I_0, I_1, and I_2 are calculated as follows:

$$I_0(\eta) = \int \frac{1}{4j}H_0^{(2)}\left(\frac{\hat{k}L}{2}\eta\right)d\eta = -\frac{\eta}{4}\sum_{v=0}^{\infty}\frac{(-1)^v}{(2v+1)(v!)^2}\left(\frac{\hat{k}L}{4}\eta\right)^{2v} \cdot\left[\frac{2}{\pi}\left\{\gamma + \ln\left(\frac{\hat{k}L}{4}\eta\right) - \frac{1}{2v+1} - \sum_{s=1}^{v}\frac{1}{s}\right\} + j\right] \tag{24a}$$

$$I_1(\eta) = \int \frac{\eta}{4j}H_0^{(2)}\left(\frac{\hat{k}L}{2}\eta\right)d\eta = \frac{1}{4j}\frac{2\eta}{\hat{k}L}H_1^{(2)}\left(\frac{\hat{k}L}{2}\eta\right) \tag{24b}$$

$$I_2(\eta) = \int \frac{\eta^2}{4j} H_0^{(2)}\left(\frac{\hat{k}L}{2}\eta\right) d\eta$$
$$= \left(\frac{2}{\hat{k}L}\right)^2 \left[\frac{\hat{k}L}{2}\frac{\eta^2}{4j} H_1^{(2)}\left(\frac{\hat{k}L}{2}\eta\right) + \frac{\eta}{4j} H_0^{(2)}\left(\frac{\hat{k}L}{2}\eta\right) - I_0(\eta)\right] \quad (24c)$$

where γ is the Euler's number.

In the matrix notation, (17) is rewritten as follows [8]–[13]:

$$[H]\{\phi\} = [G]\{\psi\}. \quad (25)$$

From (25), the following equation is obtained for the waveguide junction in Fig. 1:

$$[[H]_0 \quad [H]_1 \quad [H]_2] \begin{bmatrix} \{\phi\}_0 \\ \{\phi\}_1 \\ \{\phi\}_2 \end{bmatrix} = [[G]_0 \quad [G]_1 \quad [G]_2] \begin{bmatrix} \{\psi\}_0 \\ \{\psi\}_1 \\ \{\psi\}_2 \end{bmatrix} \quad (26)$$

where the subscripts 0, 1, and 2 denote the quantities corresponding to the boundaries Γ_0, Γ_1, and Γ_2 in Fig. 1, respectively.

B. Analytical Approach

Assuming that the dominant TE_{10} mode of unit amplitude is incident from the waveguide j ($j = 1, 2$) in Fig. 1, ϕ on Γ_i ($i = 1, 2$) may be expressed analytically as

$$\phi(x^{(i)} = 0, y^{(i)}) = 2\delta_{ij} f_{j1}(y^{(j)}) - \sum_m \frac{1}{j\beta_{im}} \int_0^{d_i} f_{im}(y_0^{(i)}) f_{im}(y_0^{(i)}) \cdot \psi(x^{(i)} = 0, y^{(i)})\, dy_0^{(i)} \quad (27)$$

where

$$f_{im}(y^{(i)}) = \sqrt{2/a_i}\, \sin m\pi y^{(i)}/a_i, \qquad m = 1, 2, 3, \cdots \quad (28)$$

$$\beta_{im} = \sqrt{k_0^2 \epsilon_{ri} - (m\pi/a_i)^2}, \qquad m = 1, 2, 3, \cdots \quad (29)$$

for the H-plane junction

$$f_{im}(y^{(i)}) = \sqrt{\sigma_n/b_i}\, \cos n\pi y^{(i)}/b_i, \qquad n = 0, 1, 2, \cdots \quad (30)$$

$$\beta_{im} = \sqrt{k_0^2 \epsilon_{ri} - (\pi/a)^2 - (n\pi/b_i)^2}, \qquad n = 0, 1, 2, \cdots \quad (31)$$

$$\sigma_n = \begin{cases} 1, & n = 0 \\ 2, & n \neq 0 \end{cases} \quad (32)$$

for the E-plane junction, and δ_{ij} is the Kronecker δ.

Using (10) and (11), (27) can be discretized as follows:

$$\{\phi\}_i = \delta_{ij}\{f\}_j + [Z]_i\{\psi\}_i \quad (33)$$

where

$$\{f\}_j = 2\{f_1\}_j \quad (34)$$

$$[Z]_i = -\sum_m (1/j\beta_{im})\{f_m\}_i \sum_{\dot{e}} \int_{\dot{e}} f_{im}(y_0^{(i)}) \cdot \{N(x^{(i)} = 0, y_0^{(i)})\}\, dy_0^{(i)}. \quad (35)$$

Here the components of the $\{f_m\}_i$ vector are the values of $f_{im}(y^{(i)})$ at the nodal points on Γ_i and $\Sigma_{\dot{e}}$ extends over the elements related to Γ_i.

C. Combination of Boundary-Element and Analytical Relations

Using (33), from (26) we obtain the following final matrix equation:

$$\begin{bmatrix} [H]_0 & [H]_1 & [H]_2 & -[G]_0 & -[G]_1 & -[G]_2 \\ \hline [0] & [1] & [0] & [0] & -[Z]_1 & [0] \\ [0] & [0] & [1] & [0] & [0] & -[Z]_2 \end{bmatrix} \cdot \begin{bmatrix} \{\phi\}_0 \\ \{\phi\}_1 \\ \{\phi\}_2 \\ \{\psi\}_0 \\ \{\psi\}_1 \\ \{\psi\}_2 \end{bmatrix} = \begin{bmatrix} \{0\} \\ \{0\} \\ \{0\} \\ \{0\} \\ \hline \delta_{1j}\{f\}_j \\ \delta_{2j}\{f\}_j \end{bmatrix} \quad (36)$$

where [1] is a unit matrix, [0] is a null matrix, and {0} is a null vector. In (36), $\{\phi\}_0 = \{0\}$ and $\{\psi\}_0 = \{0\}$ should be considered for the H- and E-plane junctions, respectively.

The values of ϕ at nodal points on Γ_i, namely $\{\phi\}_i$, are computed from (36), and then $\phi(x^{(i)} = 0, y^{(i)})$ on Γ_i can be calculated from (10). The solutions on Γ_i allow the determination of the scattering parameters S_{ij} of the TE_{10} mode as follows:

$$S_{jj} = \int_0^{d_j} \phi(x^{(j)} = 0, y^{(j)}) f_{j1}(y^{(j)})\, dy^{(j)} - 1 \quad (37)$$

$$S_{ij} = \sqrt{\beta_{i1}\hat{\epsilon}_{rj}/\beta_{j1}\hat{\epsilon}_{ri}} \cdot \int_0^{d_i} \phi(x^{(i)} = 0, y^{(i)}) f_{i1}(y^{(i)})\, dy^{(i)}, \qquad i \neq j. \quad (38)$$

In (38), for the H-Plane junction, both $\hat{\epsilon}_{ri}$ and $\hat{\epsilon}_{rj}$ should be replaced by 1.

IV. Computed Results

In this section, we present the computed results for various H- and E-plane waveguide discontinuities. Convergence of the solution is checked by increasing m in (35) and the number of the elements. Although the convergence is obtained by using the first three or four evanescent higher modes, in this analysis, the first six evanescent higher modes are used in (35). The results of the BEM agree well with those of the FEM [5]–[7] and agree well with the other theoretical results [15]–[20] and the experimental results [18], [19], [21]. For the H-plane waveguide discontinuities, the experimental results [18], [19] and the results of the integral equation method [15], [16], the normal-mode method [17]–[19], and the moment method [20] are not shown in this paper (these results are cited in [5] and [6]).

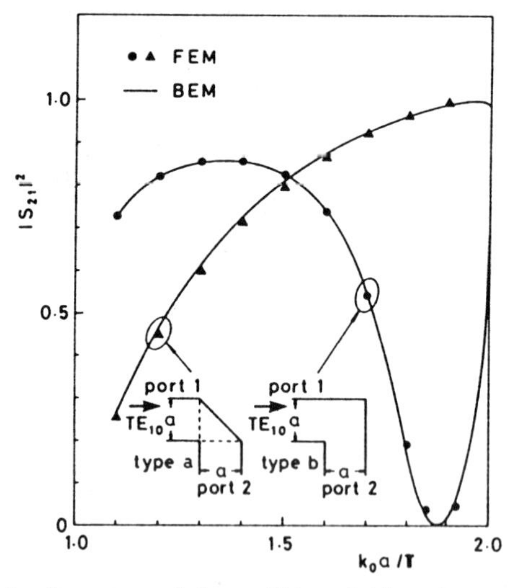

Fig. 4. Power transmission coefficient of right-angle corner bend.

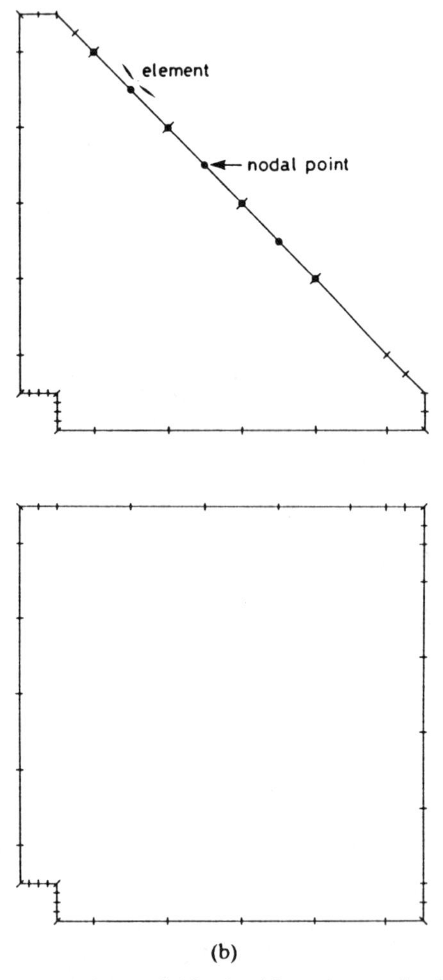

(b)

Fig. 5. Element division for right-angle corner bend.

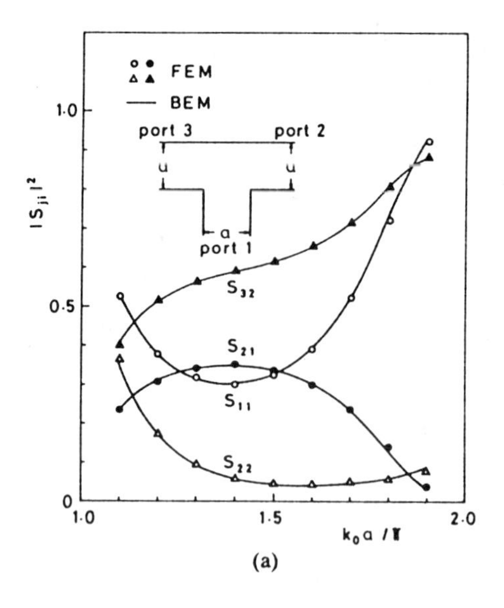

(a)

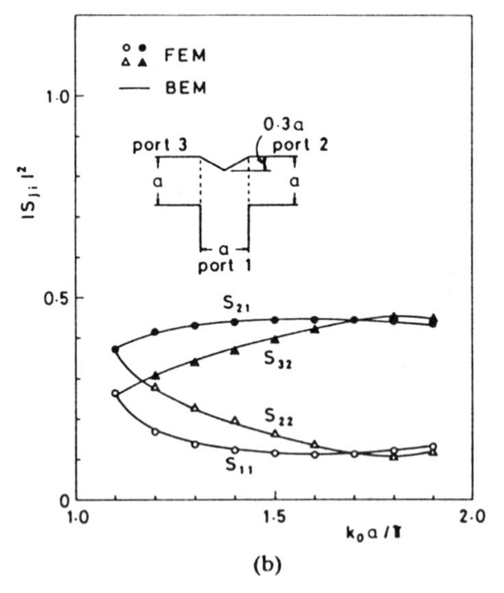

(b)

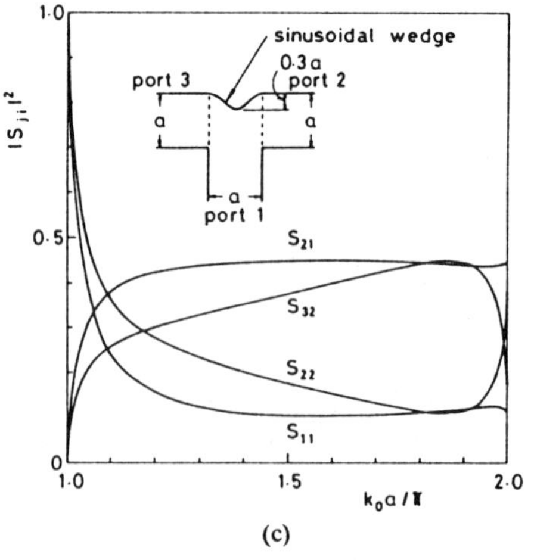

(c)

Fig. 6. Power reflection and transmission coefficients of T-junction.

A. H-Plane Junction

Fig. 4 shows the power transmission coefficient ($|S_{21}|^2$) of a right-angle corner bend. Fig. 5(a) and (b) shows the element divisions for the type *a* and the type *b* in Fig. 4, respectively.

The present approach can be applied easily to the analysis of multi-port junctions. Fig. 6 shows the power reflection coefficients ($|S_{11}|^2$ and $|S_{22}|^2$) and the power transmission coefficients ($|S_{21}|^2$ and $|S_{32}|^2$) of a T-junction. Fig. 7(a) and (b) shows the element divisions for the T-junction in Fig. 6(a) and the T-junction with wedge in Fig. 6(b), respectively. From Fig. 6(a)–(c), it is found that over a wide range of frequencies, the reflection at port 1 is reduced with a linear wedge (Fig. 6(b)) and that this reflec-

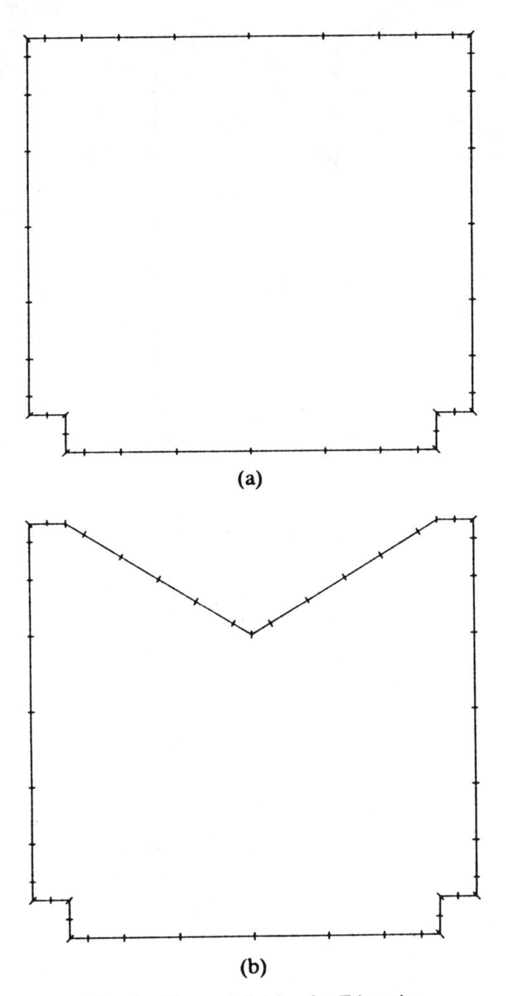

Fig. 7. Element division for T-junction.

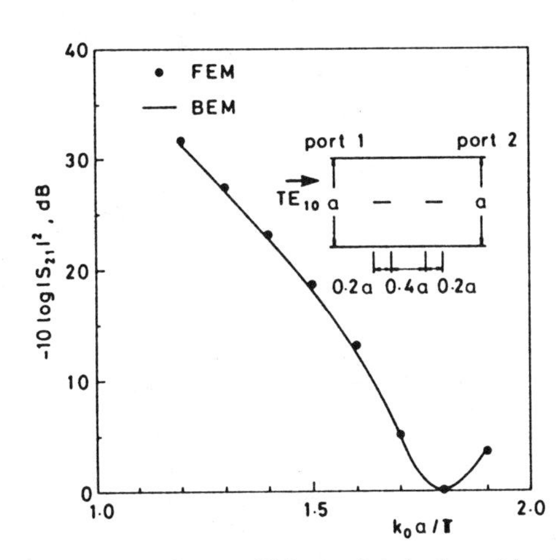

Fig. 8. Power transmission coefficient of inductive strip-planar circuit mounted in a waveguide.

tion at port 1 may be further reduced with a sinusoidal wedge (Fig. 6(c)).

Fig. 8 shows the power transmission coefficient of an inductive strip-planar circuit mounted in a waveguide. Fig. 9 shows the element division for this circuit. In this case, the boundary condition $\phi = 0$ should be considered on strip conductors.

The present approach can also be applied to the analysis of multi-media problems. A procedure of programming for handling multi-media problems is given in [11]. Fig. 10 shows the power transmission coefficient of a waveguide-type dielectric filter. Fig.

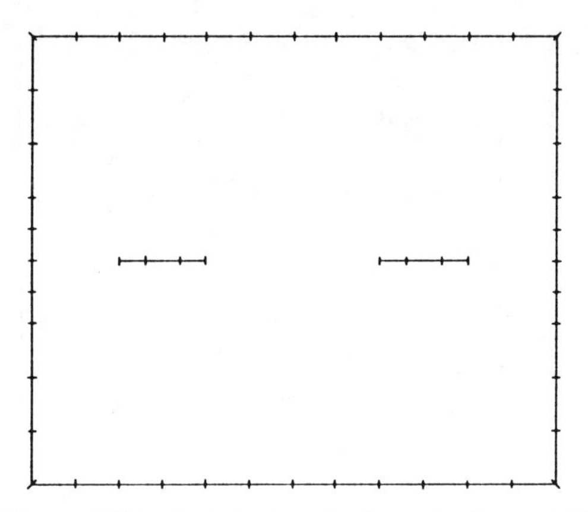

Fig. 9. Element division for inductive strip-planar circuit mounted in a waveguide.

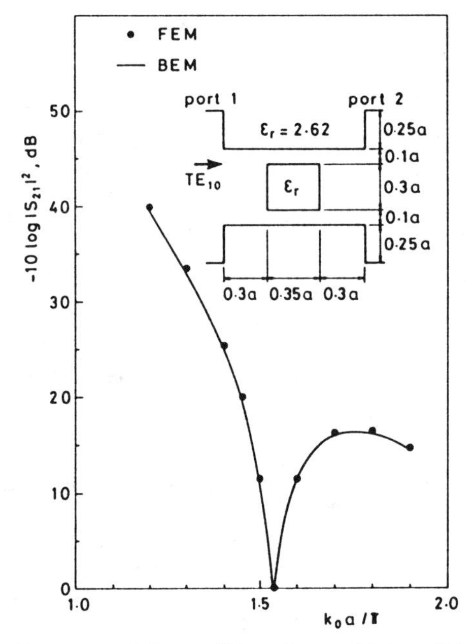

Fig. 10. Power transmission coefficient of waveguide-type dielectric filter.

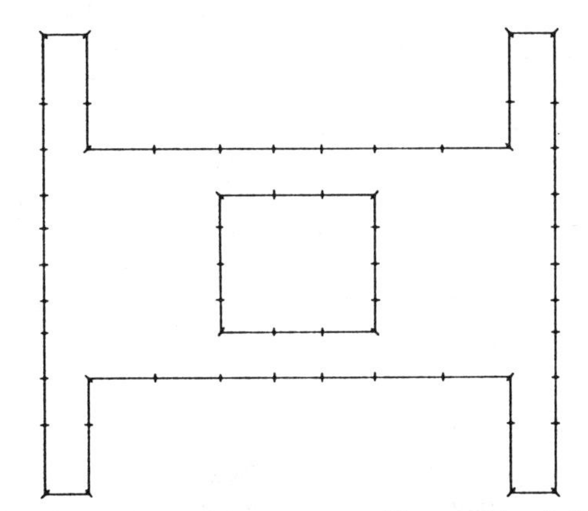

Fig. 11. Element division for waveguide-type dielectric filter.

11 shows the element division for this filter. In this case, the boundary conditions $\phi_{\mathrm{air}} = \phi_{\mathrm{dielectric}}$ and $\psi_{\mathrm{air}} = -\psi_{\mathrm{dielectric}}$ should be considered on the interface between air and dielectric.

The present approach is applicable to the frequency range in which waveguide propagates multi-modes. Fig. 12(a) and (b) shows the magnitudes of reflection and transmission coefficients

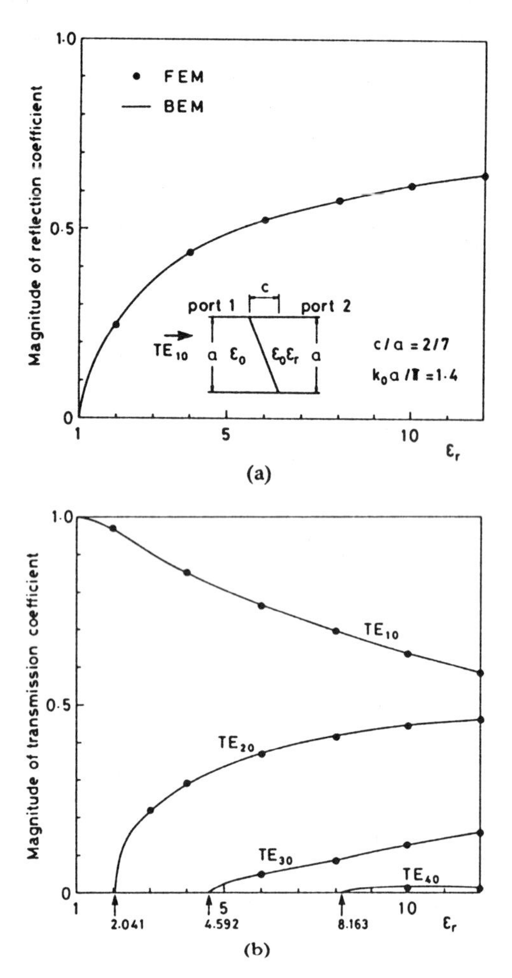

Fig. 12. Magnitudes of reflection and transmission coefficients of inhomogeneous waveguide junction.

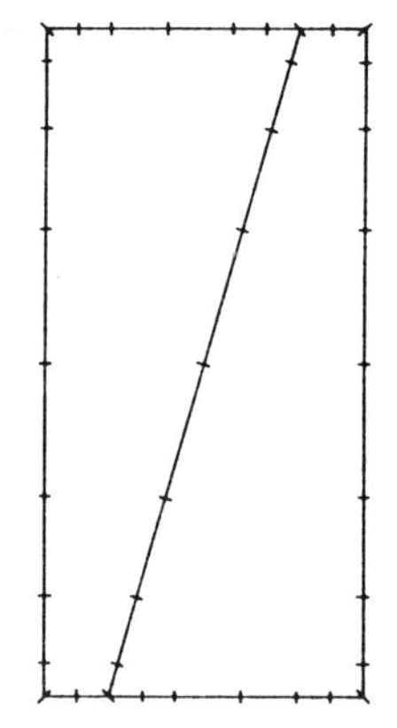

Fig. 13. Element division for inhomogeneous waveguide junction.

TABLE I
NUMBER OF NODAL POINTS USED IN THE BEM AND FEM ANALYSES

H-plane junction	BEM	FEM
Fig.4 (type a)	62	299
Fig.4 (type b)	76	377
Fig.6 (a)	84	385
Fig.6 (b)	96	399
Fig.8	102	609
Fig.10	120	379
Fig.12	79	589

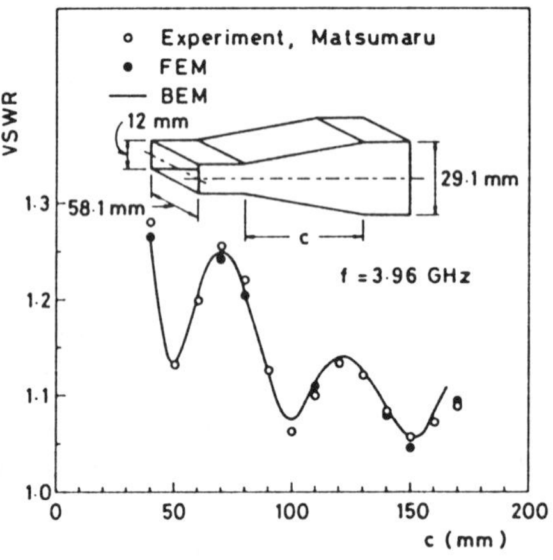

Fig. 14. VSWR characteristics of linear E-plane taper.

of an inhomogeneous waveguide junction, respectively. Fig. 13 shows the element division for this junction. For both reflection and transmission coefficients, the results of the BEM agree well with those of the FEM [6]. The results of the moment method [20] for the transmission coefficient are different from those of the BEM and the FEM. In the moment method, the transmission coefficients of the higher order modes are not zero at the cutoff values of ϵ_r.

Table I shows the number of the nodal points used in the BEM and FEM analyses of H-plane junctions. Here, in both BEM and FEM analyses, the symmetry of a circuit to reduce the dimensions of the matrices is not used. The accuracies of the present boundary-element calculations are almost identical to those of the earlier finite-element calculations [5], [6], and yet the BEM allows the matrix dimension to be reduced by a factor of about 7 to 3.

B. E-Plane Junction

A comparison of the results obtained applying the BEM to the linear E-plane tapers of various lengths with the experimental results [21] and the results of the FEM [7] is given in Fig. 14 and very good agreement is obtained.

V. CONCLUSION

A method of analysis, based on the boundary-element approach and the analytical approach, was developed for the solution of the H- and E-plane junctions. The validity of the method was confirmed by comparing numerical results for various H- and E-plane waveguide discontinuities with the results of the finite-element method.

This approach can be applied easily to the planar circuits [3], [4], [13]. The problem of how to deal with waveguide junctions with lossy media or anisotropic media hereafter still remains.

ACKNOWLEDGMENT

The authors wish to thank T. Miki for his assistance in numerical computations.

REFERENCES

[1] N. Marcuvitz, *Waveguide Handbook*. New York: McGraw-Hill, 1951.
[2] R. E. Collin, *Field Theory of Guided Waves*. New York: McGraw-Hill, 1960.

[3] P. Silvester, "Finite element analysis of planar microwave networks," *IEEE Trans. Microwave Theory Tech.*, vol. MTT-21, pp. 104–108, Feb. 1973.

[4] T. Miyoshi, "The expansion of electromagnetic field in planar circuit," *Trans. Inst. Electron. Commun. Eng. Japan*, vol. J58-B, pp. 84–91, Feb. 1975 (in Japanese).

[5] M. Koshiba, M. Sato, and M. Suzuki, "Application of finite-element method to *H*-plane waveguide discontinuities," *Electron. Lett.*, vol. 18, pp. 364–365, Apr. 1982.

[6] M. Koshiba, M. Sato, and M. Suzuki, "Finite-element analysis of arbitrarily shaped *H*-plane waveguide discontinuities," *Trans. Inst. Electron. Commun. Eng. Japan*, vol. E66, pp. 82–87, Feb. 1983.

[7] M. Koshiba, M. Sato, and M. Suzuki, "Application of finite-element method to *E*-plane waveguide discontinuities," *Trans. Inst. Electron. Commun. Eng. Japan*, vol. E66, pp. 457–458, July 1983.

[8] C. A. Brebbia, *The Boundary Element Method for Engineers*. London: Pentech Press, 1978.

[9] C. A. Brebbia and S. Walker, *Boundary Element Techniques in Engineering*. London: Butterworth, 1980.

[10] S. Washisu and I. Fukai, "An analysis of electromagnetic unbounded field problems by boundary element method," *Trans. Inst. Electron. Commun. Eng. Japan*, vol. J64-B, pp. 1359–1365, Dec. 1981 (in Japanese).

[11] S. Kagami and I. Fukai, "Application of boundary-element method to electromagnetic field problems," *IEEE Trans. Microwave Theory Tech.*, vol. MTT-32, pp. 455–461, Apr. 1984.

[12] K. Kanao and S. Kurazono, "Waveguide-type dielectric filter," *Trans. Inst. Electron. Commun. Eng. Japan*, vol. J67-B, pp. 1177–1178, Oct. 1984 (in Japanese).

[13] E. Tonye and H. Baudrand, "Multimode *S*-parameters of planar multiport junctions by boundary element method," *Electron. Lett.*, vol. 20, pp. 799–802, Sept. 1984.

[14] T. Okoshi and T. Miyoshi, "The planar circuit—An approach to microwave integrated circuitry," *IEEE Trans. Microwave Theory Tech.*, vol. MTT-20, pp. 245–252, Apr. 1972.

[15] T. Okoshi and S. Kitazawa, "Computer analysis of short-boundary planar circuit," *IEEE Trans. Microwave Theory Tech.*, vol. MTT-23, pp. 299–306, Mar. 1975.

[16] T. Okoshi and S. Kitazawa, "Computer analysis of short-boundary planar circuit," *Tech. Res. Rep. Inst. Electron. Commun. Eng. Japan*, MW75-75, Oct. 1975 (in Japanese).

[17] T. Anada and J. P. Hsu, "Analysis of planar circuit with short circuit boundary by normal mode method through impedance matrix," *Trans. Inst. Electron. Commun. Eng. Japan*, vol. J61-B, pp. 646–653, July 1978 (in Japanese).

[18] T. Anada, M. Hatayama, and J. P. Hsu, "Wide-band frequency characteristics of rectangular waveguide *H*-plane T-junctions," *Tech. Res. Rep. Inst. Electron. Commun. Eng. Japan*, MW79-68, Sept. 1979 (in Japanese).

[19] R. Monzen, N. Ogawa, T. Anada, and J. P. Hsu, "Derivation of normal mode for dielectric planar circuit—Waveguide-type dielectric filter," *Tech. Res. Rep. Inst. Electron. Commun. Eng. Japan*, MW79-69, Sept. 1979 (in Japanese).

[20] Y. L. Chow and S. C. Wu, "A moment method with mixed basis functions for scattering by waveguide junctions," *IEEE Trans. Microwave Theory Tech.*, vol. MTT-21, pp. 333–340, May 1973.

[21] K. Matsumaru, "Reflection coefficient of *E*-plane tapered waveguides," *IRE Trans. Microwave Theory Tech.*, vol. MTT-6, pp. 143–149, Apr. 1958.

Part 6
Method of Lines

PRINTED circuit transmission lines, such as striplines, microstrips, coplanar lines, slotlines, finlines, etc., have a layered structure, i.e., all discontinuities (dielectric interfaces and metal strips) in the cross section are perpendicular to the same direction. Such a geometrical property can be used to simplify their analysis. Actually, the most efficient numerical methods for the analysis of printed transmission lines are designed to properly take account of such a characteristic. A typical example is the method of lines (MOL), but the spectral-domain method, the modal analysis, and the transverse resonance technique take similar advantage of the layered characteristic of printed lines to reduce by one the dimensional complexity of the problem. Certain similarities among these methods actually do exist.

Developed by mathematicians to solve partial differential equations, the MOL has recently been introduced into the microwave community by Pregla and his group. The basic idea of this method is to reduce a system of partial differential equations into ordinary differential equations by discretizing all but one of the independent variables.

In the MOL as applied to printed circuit configurations, a finite-difference discretization technique is used in the plane of the substrate, while the analytical formulation is retained in the normal direction. As a result of this procedure, the structure appears as consisting of a number of *lines* orthogonal to the substrate, each line corresponding to a discretization point on the circuit itself. After suitable manipulations and an orthogonal transformation to diagonalize the resulting matrix equation, the boundary conditions lead to the characteristic equation of the structure. The basic idea of the MOL to retain the analytical field solution in the direction normal to the substrate is just that used in the spectral-domain approach (SDA), except the discretization in the substrate's plane replaces the Fourier transform. As a consequence, the MOL has greater versatility and flexibility than the SDA, though its numerical efficiency is lower. The details of the formulation of the method are given in the reprinted papers.

The MOL was proposed for the analysis of planar waveguides by Schulz and Pregla in 1980 (Paper 6.1). The extension to arbitrarily shaped planar resonators and periodic structures appeared a few years later in a paper authored by Worm and Pregla (Paper 6.2), while Diestel and Worm have developed a nonuniform discretization procedure (Paper 6.3). A number of improvements and extensions have very recently been developed. The applicability of the MOL to planar structures on anisotropic substrates has been demonstrated by Pregla in [6.1]. Pascher and Pregla have introduced in Paper 6.4 the use of the Kronecker product of matrices for two-dimensional discretization and a fast algorithm for the solution of the characteristic equation for periodic structures. Unlike the SDA, the MOL is capable of accounting for finite metallization thickness, which may be very important in monolithic microwave integrated circuits. This result has been shown by Pregla and Pascher in [6.2], which is also a very comprehensive report on the MOL. A discussion on the convergence and accuracy of the method and its connection with the mode-matching technique can be found in [6.3].

Finally, it is worth mentioning that work is being done to extend the applicability of the MOL and other numerical techniques, like the FDM and the FEM, to time-domain analysis [6.4].

References

[6.1] R. Pregla, "Analysis of planar microwave structures on magnetized ferrite substrate," *Arch. Elek. Ubertragung.*, vol. 40, no. 5, pp. 270–273, 1986.

[6.2] R. Pregla and W. Pascher, "The method of lines," in *Numerical Techniques for Microwave and Millimeter Wave Passive Structures*, T. Itoh, Ed. New York, NY: John Wiley, in press.

[6.3] R. Pregla, "About the nature of the method of lines," *Arch. Elek. Ubertragung.*, vol. 41, no. 6, pp. 368–370, 1987.

[6.4] S. Nam, S. El-Ghazaly, H. Ling, and T. Itoh, "Time-domain method of lines," in *1988 Int. MTT-S Microwave Symp. Dig.*, pp. 627–630.

A New Technique for the Analysis of the Dispersion Characteristics of Planar Waveguides

by Uwe Schulz* and Reinhold Pregla *

Dedicated to Professor Dr. rer. nat. Hans K. F. Severin on the occasion of his 60th birthday.

An efficient method for calculating the dispersion characteristics of planar waveguide structures is presented, of which the principle is known as the "method of lines" in mathematical literature. The wave equation is discretized in one direction, and the resulting differential-difference-equation can be solved analytically. As an example of application and as a test for the computed results single and coupled microstrip lines were calculated. The obtained results show good agreement with other available data.

Ein neues Verfahren zur Berechnung des Dispersionsverhaltens von planaren Wellenleitern

Ein leistungsfähiges Verfahren zur Berechnung der dispersiven Eigenschaften planarer Wellenleiterstrukturen wird beschrieben, dessen Grundprinzip in der mathematischen Literatur unter dem Namen „Methode der Geraden" bekannt ist. Die Wellengleichung wird in einer Richtung diskretisiert und die daraus entstehende Differential-Differenzengleichung analytisch gelöst. Als Beispiel wurde die einfache und verkoppelte Microstripleitung berechnet. Die erhaltenen Ergebnisse zeigen gute Übereinstimmung mit anderen vergleichbaren Daten.

1. Introduction

For hybrid mode analysis of planar waveguide structures, e.g. microstrip, slotline, coplanar-line, finline etc., several numerical methods exist. In spite of its simplicity the finite difference method is as yet not commonly applied to the more complicated structures, as it leads to a system of equations of very large size. Even improvements, like e.g. such of Corr and Davies [4], who apply a graduell mesh, could not help principally.

In this paper a method is presented, which is basicly a finite difference method too, but essentially more effective with respect to accuracy and computation time than the usual finite difference method, when applied to structures like those in Fig. 1.

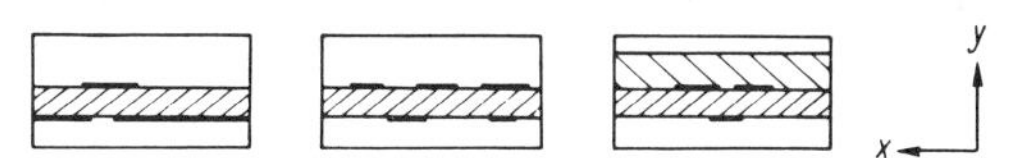

Fig. 1. Some planar structures.

The basic principle of this method is known as the "method of lines" in the mathematical literature, e.g. [1], [2]. To the best of the authors' knowledge this method has not been applied to the calculation of planar microwave structures up to the present. Merely Lennartsson's "network analogue method" [3], which has a certain similarity with the "method of lines", was applied to compute the quasi-TEM-parameters of microstrip structures.

* Dipl.-Ing. U. Schulz, Prof. Dr. R. Pregla, Fernuniversität, Lehrstuhl für Allgemeine und Theoretische Elektrotechnik, Frauenstuhlweg 31, D-5860 Iserlohn.

The "method of lines" is simple in concept — for a given system of partial differential equations all but one of the independent variables are discretized to obtain a system of ordinary differential equations. This semidiscrete procedure is apparently a very useful one in order to calculate planar waveguide structures like those in Fig. 1, since these consist of regions, which are homogeneous in one direction. Moreover, this method has no problem with the so called "relative convergence" as have e.g. the mode-matching- and Galerkin's method. This is an important advantage, especially when structures with more than one metallic strip are considered.

2. Method of Analysis

The electromagnetic field can be described by two scalar potentials $\psi^{(e)}$ and $\psi^{(h)}$, which have to satisfy Helmholtz' equation

$$\frac{\partial^2 \psi^{(e,h)}}{\partial x^2} + \frac{\partial^2 \psi^{(e,h)}}{\partial y^2} + (k^2 - \beta^2)\,\psi^{(e,h)} = 0 \qquad (1)$$

and some boundary conditions within the considered region, where the dependence $e^{j(\omega t - \beta z)}$ has been assumed. The first step is to discretize the x-variable in the partial differential equation (1). For that purpose a family of N straight lines parallel to the y-axis is laid into the cross section. The distance between adjacent lines shall be constant and equal to h. Now, the potential ψ in eq. (1) can be replaced by a set $(\psi_1, \psi_2, \ldots, \psi_N)$ at the lines

$$x_i = x_0 + ih, \quad i = 1, 2, \ldots, N \qquad (2)$$

and the derivatives with respect to x are replaced by finite differences. This procedure yields a system

Reprinted with permission from *Arch. Elek. Ubertragung.*, vol. 34, pp. 169–173, Apr. 1980.

of N *coupled* ordinary differential equations

$$\frac{\partial^2 \psi_i}{\partial y^2} + \frac{1}{h^2}\left[\psi_{i-1}(y) - 2\psi_i(y) + \psi_{i+1}(y)\right] + (k^2 - \beta^2)\psi_i(y) = 0, \quad i = 1, 2, \ldots, N. \tag{3}$$

When a column vector

$$\vec{\psi} = [\psi_1(y), \psi_2(y), \ldots, \psi_N(y)]^t \tag{4}$$

and the matrix

$$\boldsymbol{P} = \begin{pmatrix} p_1 & -1 & & & \\ -1 & 2 & -1 & & 0 \\ & \ddots & \ddots & \ddots & \\ 0 & & -1 & 2 & -1 \\ & & & -1 & p_2 \end{pmatrix} \tag{5}$$

are introduced. Eq. (3) can be expressed in matrix notation as

$$h^2 \frac{\partial^2 \vec{\psi}}{\partial y^2} - [\boldsymbol{P} - h^2(k^2 - \beta^2)\boldsymbol{I}]\vec{\psi} = \vec{0} \tag{6}$$

where $\boldsymbol{I}$ is the identity matrix. The lateral boundary conditions are already included in the matrix $\boldsymbol{P}$, i.e., in the numbers p_1 and p_2. The system of differential equations (6) can not be solved in the present form, because the equations are coupled. But it will be shown, that eq. (6) can be transformed into a system of uncoupled equations. As $\boldsymbol{P}$ is a real symmetric matrix, there exists an orthogonal matrix $\boldsymbol{T}$ such that

$$\boldsymbol{T}^t \boldsymbol{P} \boldsymbol{T} = \boldsymbol{\lambda} \tag{7}$$

is diagonal, where t denotes the transpose. The elements λ_i of the diagonal matrix $\boldsymbol{\lambda}$ are the eigenvalues of $\boldsymbol{P}$.

A transformed potential vector $\vec{U}$ is now introduced,

$$\boldsymbol{T}^t \vec{\psi} = \vec{U} \tag{8}$$

so that instead of eq. (6) one can write a system of N ordinary differential equations, which are now *uncoupled*:

$$h^2 \frac{\partial^2 U_i}{\partial y^2} - [\lambda_i - h^2(k^2 - \beta^2)]\, U_i = 0, \quad i = 1, 2, \ldots, N. \tag{9}$$

It is obvious, that eq. (9) can be solved analytically for each homogeneous region of the structure. The solution has a form similar to the line equations:

$$\begin{pmatrix} U_i(y_1) \\ h\dfrac{\partial U_i}{\partial y}(y_1) \end{pmatrix} = \begin{pmatrix} \cosh \dfrac{\varkappa_i(y_1 - y_2)}{h} & \dfrac{1}{\varkappa_i}\sinh \dfrac{\varkappa_i(y_1 - y_2)}{h} \\ \varkappa_i \sinh \dfrac{\varkappa_i(y_1 - y_2)}{h} & \cosh \dfrac{\varkappa_i(y_1 - y_2)}{h} \end{pmatrix} \begin{pmatrix} U_i(y_2) \\ h\dfrac{\partial U_i}{\partial y}(y_2) \end{pmatrix} \tag{10}$$

with

$$\varkappa_i = [\lambda_i - h^2(k^2 - \beta^2)]^{1/2}, \quad i = 1, 2, \ldots, N. \tag{11}$$

With the aid of eq. (10) the potential vector $\vec{\psi}$ can be transformed from one interface into another. So it is possible to replace the boundary value problem (1) by a linear algebraic system, the unknowns of which are only those potentials at the interfaces with metallic strips. Further, it is possible to reduce the order of the resulting matrix associated with the eigenvalue equation to the number of points on the strips or in the slots between the strips. Thus, the eigenvalue equation has a very low order.

3. The Lateral Boundary Conditions

It has been mentioned before, that the matrix $\boldsymbol{P}$ includes the lateral boundary conditions. The left boundary determines p_1, the right boundary p_2 (see eq. (5)). In Fig. 2 only the left boundary is considered. The consideration of the right boundary is equivalent. The Dirichlet condition (Fig. 2a) requires $\psi_0 = 0$. Inserting this in eq. (3) with $i = 1$ yields $p_1 = 2$. On the other hand, the Neumann-condition (Fig. 2b) requires $\psi_0 = \psi_1$. This condition can be satisfied with $p_1 = 1$.

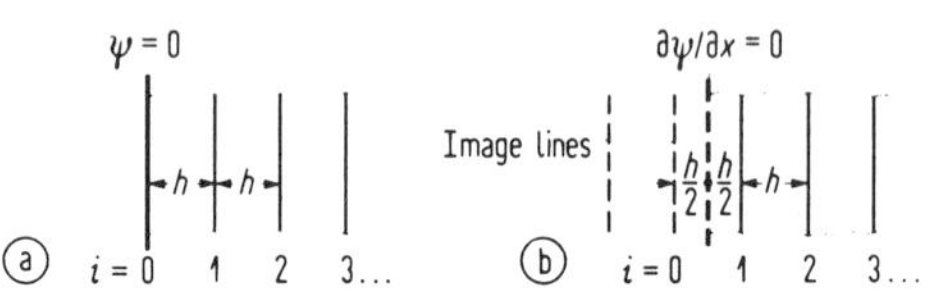

Fig. 2. Boundary conditions,
(a) Dirichlet-boundary $\psi = 0$,
(b) Neumann-boundary $\partial\psi/\partial x = 0$.

It should be mentioned, that this is not the only way to satisfy the boundary conditions. E.g., the Neumann-boundary can be laid at $i = 0$ too, but then the requirement $\psi_{-1} = \psi_1$ leads to a nonsymmetrical matrix $\boldsymbol{P}$, and the transformation matrix $\boldsymbol{T}$ is no longer orthogonal.

4. Discretisation of Hybrid Modes

It is well known that the hybrid field components can be expressed in terms of a superposition of the TE and TM fields, which are in turn derivable from the scalar potentials $\psi^{(e)} \sim E_z$ and $\psi^{(h)} \sim H_z$. Because of the complementary properties of the boundary conditions relative to $\psi^{(e)}$ and $\psi^{(h)}$ it is expedient to relatively shift the lines for $\psi^{(e)}$ and $\psi^{(h)}$ by half the discretisation distance. This is the only way to achieve that the matrix $\boldsymbol{P}$ will be symmetric for both $\psi^{(e)}$ and $\psi^{(h)}$. For instance, in Fig. 3 the left wall is a Neumann-boundary for $\psi^{(e)}$ and

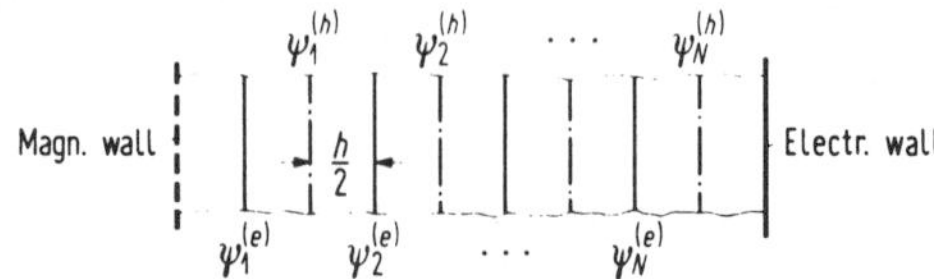

Fig. 3. Shifting of lines for $\psi^{(e)}$ and $\psi^{(h)}$.

a Dirichlet-boundary for $\psi^{(h)}$. At the right wall it is just the other way. The shifting of lines now guaranties a simple fitting of all the boundary conditions.

At dielectric interfaces the two potentials $\psi^{(e)}$ and $\psi^{(h)}$ are coupled by the continuity conditions of the tangential field components. This coupling is always such, that the derivative $\partial\psi_i^{(e)}/\partial y$ is related to the difference of the adjacent potentials $\psi_i^{(h)} - \psi_{i-1}^{(h)}$ and $\partial\psi_i^{(h)}/\partial y$ to the difference $\psi_{i+1}^{(e)} - \psi_i^{(e)}$ (see Fig. 3).

5. Edge Condition

Most of the considered waveguide structures contain metallic strips, which are assumed to have zero thickness. In [7] the discretisation error caused by the singularity due to the edge of such a strip is investigated. It was found there that this kind of discretisation error is negligibly small, if the field at the discrete lines satisfies the edge condition [5], [6]. This is approximately achieved, when the end of the strip exceeds the last $\psi^{(e)}$-line on the strip by $h/4$ and the last $\psi^{(h)}$-line by $\frac{3}{4}h$ (see Fig. 4). It is a further advantage of the shifting of the $\psi^{(e)}$- and $\psi^{(h)}$-lines, that these requirements can be satisfied simultaneously, so that the discretisation error caused by an edge can be compensated very well.

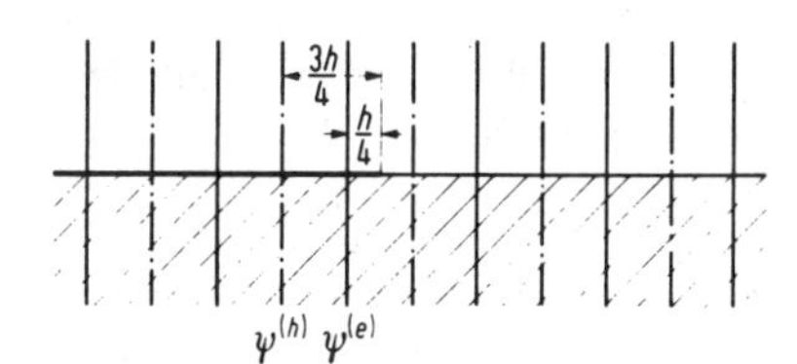

Fig. 4. To the edge condition.

Because of this, the number of lines in the considered structure and thus the order of the characteristic equation need not be very large, so that accurate results are obtained with rather low computational effort.

6. Example

For convenience, the method is demonstrated on a simple structure, the shielded microstrip line (Fig. 5). Because of symmetry only the half cross

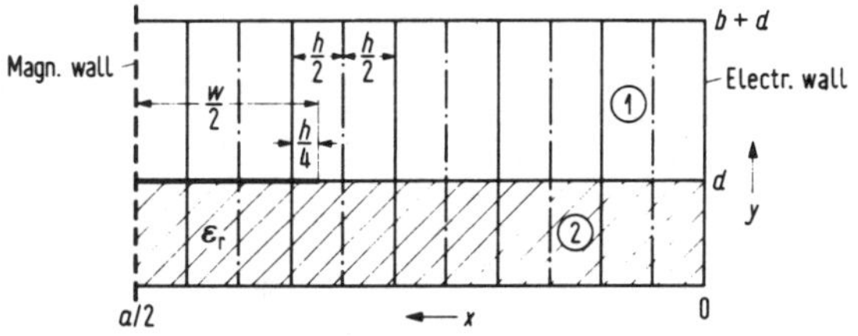

Fig. 5. Cross section of a shielded microstrip line.

section is considered. At the interface $y = d$ both potentials are related by the continuity conditions of the tangential field components. With

$$\psi^{(e)} = \frac{j\omega\varepsilon}{k^2 - \beta^2} E_z \tag{12}$$

and

$$\psi^{(h)} = \frac{j\omega\mu}{k^2 - \beta^2} H_z \tag{13}$$

the continuity conditions are:

$$\frac{\beta}{\omega\varepsilon_0}\frac{\partial}{\partial x}\left(\psi_{\mathrm{I}}^{(e)} - \frac{1}{\varepsilon_r}\psi_{\mathrm{II}}^{(e)}\right) = \frac{\partial\psi_{\mathrm{II}}^{(h)}}{\partial y} - \frac{\partial\psi_{\mathrm{I}}^{(h)}}{\partial y}, \tag{14}$$

$$(k_0^2 - \beta^2)\,\psi_{\mathrm{I}}^{(e)} = \frac{1}{\varepsilon_r}(\varepsilon_r k_0^2 - \beta^2)\,\psi_{\mathrm{II}}^{(e)}, \tag{15}$$

$$\frac{\partial\psi_{\mathrm{I}}^{(e)}}{\partial y} - \frac{\partial\psi_{\mathrm{II}}^{(e)}}{\partial y} = \frac{\beta}{\omega\mu}\frac{\partial}{\partial x}(\psi_{\mathrm{I}}^{(h)} - \psi_{\mathrm{II}}^{(h)}) - J_z, \tag{16}$$

$$(k_0^2 - \beta^2)\,\psi_{\mathrm{I}}^{(h)} = (\varepsilon_r k_0^2 - \beta^2)\,\psi_{\mathrm{II}}^{(h)} - j\omega\mu J_x. \tag{17}$$

The subscripts indicate the subregions I and II. J_x and J_z are the current density distributions at $y = d$.

Now the operator $\partial/\partial x$ is replaced by the difference operator $\boldsymbol{D}$

$$\boldsymbol{D} = \begin{pmatrix} 1 & -1 & & \bigcirc \\ & \ddots & \ddots & \\ \bigcirc & & \ddots & \ddots \end{pmatrix} \tag{18}$$

with

$$\frac{\partial\psi^{(e)}}{\partial x} \to \frac{1}{h}\boldsymbol{D}\vec{\psi}^{(e)}$$

and

$$\frac{\partial\psi^{(h)}}{\partial x} \to -\frac{1}{h}\boldsymbol{D}^{t}\vec{\psi}^{(h)}. \tag{19}$$

The normal derivatives of $\psi^{(e,h)}$ at the interface are replaced by the following matrix expressions with the operators $\boldsymbol{G}^{(e,h)}$:

$$h\frac{\partial\psi_k^{(e)}}{\partial n} \to h\frac{\partial\vec{\psi}^{(e)}}{\partial n} = \boldsymbol{G}_k^{(e)}\vec{\psi}_k^{(e)}, \tag{20}$$

$$h\frac{\partial\psi_k^{(h)}}{\partial n} \to h\frac{\partial\vec{\psi}^{(h)}}{\partial n} = \boldsymbol{G}_k^{(h)}\vec{\psi}_k^{(h)}, \tag{21}$$

$$k = \mathrm{I}, \mathrm{II}$$

where $\partial n = -\partial y$ for $k = \mathrm{I}$ and $\partial n = \partial y$ for $k = \mathrm{II}$.

Eqs. (20) and (21) can be transformed into the diagonal form:

$$h\frac{\partial\vec{U}_k^{(e)}}{\partial n} = \boldsymbol{\gamma}_k^{(e)}\vec{U}_k^{(e)}, \tag{22}$$

$$h\frac{\partial\vec{U}_k^{(h)}}{\partial n} = \boldsymbol{\gamma}_k^{(h)}\vec{U}_k^{(h)}, \quad k = \mathrm{I}, \mathrm{II}. \tag{23}$$

The diagonal matrices $\boldsymbol{\gamma}_k^{(e,h)}$ are calculated analytically with the aid of eq. (10) and the known boundary conditions at $y = 0$ and $y = b$ (see Appendix B).

Now, eqs. (14) to (17) can be written in discretized form. After elimination of $\psi_{\mathrm{II}}^{(e)}$ and $\psi_{\mathrm{II}}^{(h)}$ with eqs. (15) and (17) and transformation with $\boldsymbol{T}^{(e)}$ and $\boldsymbol{T}^{(h)}$ respectively, remain

$$\frac{\beta}{\omega\varepsilon_0}(1-\tau)\underbrace{\boldsymbol{T}^{(h)t}\boldsymbol{D}\,\boldsymbol{T}^{(e)}}_{\boldsymbol{\delta}}\cdot\vec{U}_{\mathrm{I}}^{(e)} = (\boldsymbol{\gamma}_{\mathrm{I}}^{(h)} + \tau\boldsymbol{\gamma}_{\mathrm{II}}^{(h)})\cdot\vec{U}_{\mathrm{I}}^{(h)}, \tag{24}$$

$$-(\boldsymbol{\gamma}_{\rm I}^{(e)} + \varepsilon_{\rm r}\,\tau\,\boldsymbol{\gamma}_{\rm II}^{(e)}) \cdot \vec{U}_{\rm I}^{(e)} = \tag{25}$$
$$= \frac{-\beta}{\omega\mu}(1-\tau)\underbrace{\boldsymbol{T}^{(e)\rm t}\boldsymbol{D}^{\rm t}\boldsymbol{T}^{(h)}}_{\boldsymbol{\delta}^{\rm t}} \cdot \vec{U}_{\rm I}^{(h)} - \boldsymbol{T}^{(e)\rm t}\vec{J}_z$$

with
$$\tau = \frac{1-\varepsilon_{\rm eff}}{\varepsilon_{\rm r} - \varepsilon_{\rm eff}}\,. \tag{26}$$

Note, that the matrix product $\boldsymbol{T}^{(h)\rm t}\boldsymbol{D}\,\boldsymbol{T}^{(e)}$, denoted by $\boldsymbol{\delta}$, is a diagonal matrix, too. The elements δ_i are given by simple analytical expressions (see Appendix B).

Since the transverse current density is negligibly small, only the longitudinal current density distribution is considered in this example. Nevertheless, it is also possible to proceed without this restriction. From eqs. (24) and (25) results

$$\vec{U}_{\rm I}^{(e)} = \boldsymbol{\rho}\,\boldsymbol{T}^{(e)\rm t} \cdot \vec{J}_z \tag{27}$$

with the diagonal matrix

$$\boldsymbol{\rho} = [\boldsymbol{\gamma}_{\rm I}^{(e)} + \varepsilon_{\rm r}\,\tau\,\boldsymbol{\gamma}_{\rm II}^{(e)} - \varepsilon_{\rm eff}(1-\tau)^2 \cdot$$
$$\cdot\,\boldsymbol{\delta}^{\rm t}(\boldsymbol{\gamma}_{\rm I}^{(h)} + \tau\,\boldsymbol{\gamma}_{\rm II}^{(h)})^{-1}\boldsymbol{\delta}]^{-1}\,. \tag{28}$$

After inverse transformation of eq. (27) (see eq. (8)), for the potential vector $\vec{\psi}_{\rm I}^{(e)}$ is found

$$\vec{\psi}_{\rm I}^{(e)} = \boldsymbol{T}^{(e)}\,\boldsymbol{\rho}\,\boldsymbol{T}^{(e)\rm t} \cdot \vec{J}_z\,. \tag{29}$$

At last, the final boundary condition on the strip, which requires

$$\vec{\psi}_{\rm I}^{(e)} = 0 \quad \text{on the strip} \tag{30}$$

leads, with the additional relation

$$\vec{J}_z = \begin{cases} \vec{J}_{z\,\rm red} & \text{on the strip} \\ 0 & \text{elsewhere} \end{cases} \tag{31}$$

to the characteristic equation

$$(\boldsymbol{T}^{(e)}\,\boldsymbol{\delta}\,\boldsymbol{T}^{(e)\rm t})_{\rm red}\,\vec{J}_{z\,\rm red} = \vec{0} \tag{32}$$

where only the number of points on the strip determines the size of the matrix. The eigenvector $\vec{J}_{z\,\rm red}$ is the reduced part of the discretized current density vector.

Finally it should be emphasized, that except for the reduced characteristic equation (32) all operations can be carried out with diagonal matrices. This is an important property of the described method, and permits fast computation.

7. Numerical Results

The method described is so simple, that all numerical computations could be carried out on a HP 9825A desk calculator. Figs. 6 and 7 show the results for a single and a pair of coupled microstrip lines. A comparison with other results available [8], [9] shows good agreement. For the single microstrip line the total number of points at the interface is $N = 18$ and the number of points on the strip is 6. For the coupled lines 32 points were chosen at the interface, three points on each strip and two points in the slot. Of cause, these are the

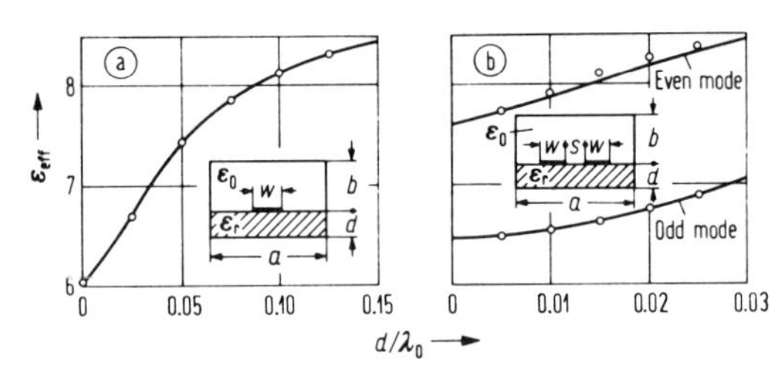

Fig. 6. (a) Effective dielectric constant of a single microstrip; $\varepsilon_{\rm r} = 9$, $w/d = 2$, $a/d = 7$, $b/d = 3$; ○ Kowalski, Pregla [8].
(b) Effective dielectric constant of a pair of coupled microstrips; $\varepsilon_{\rm r} = 10.2$, $w/d = 1.5$, $s/d = 1.5$, $a/d = 20$, $b/d = 19$; ○ Sharma, Bhat [9].

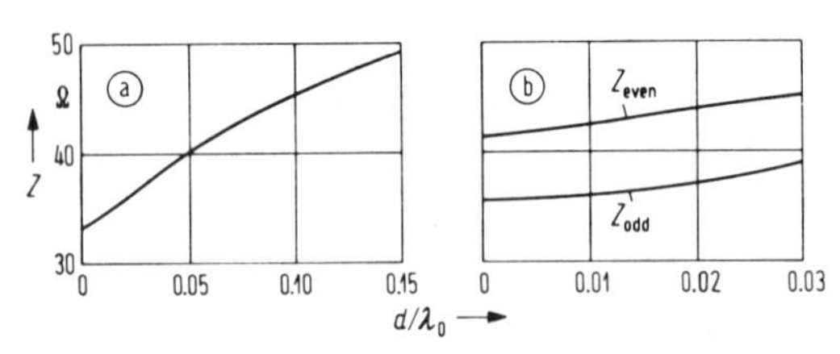

Fig. 7. (a) Characteristic impedance of a single microstrip; parameters as in Fig. 6a.
(b) Characteristic impedance of a pair of coupled microstrips; parameters as in Fig. 6b.

numbers for only one potential, say $\vec{\psi}^{(e)}$. Because of symmetry only one half of each structure has to be considered. Thus, in both cases the size of the matrix associated with the final characteristic equation has the order of only 3×3. Computing $\varepsilon_{\rm eff}$ and Z for one frequency on a HP 9825A takes between 0.5 and 1 minute. The accuracy of the results obtained (see Fig. 8) is better than 0.5%

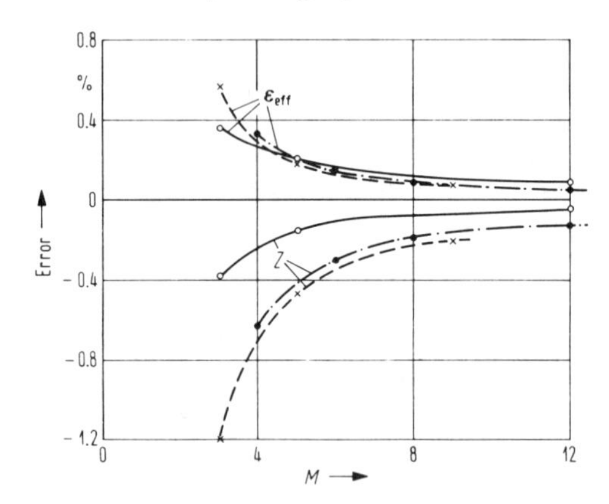

Fig. 8. Relative errors of $\varepsilon_{\rm eff}$ and Z of a single and a pair of coupled microstrip lines; parameters as in Fig. 6; —○— single line, --×-- even mode, —·—●—·— odd mode.

for $\varepsilon_{\rm eff}$ and about 1% for Z. This has been achieved with only three points on a strip! This is a consequence of the described line-shifting making it possible to compensate very well the discretisation error due to the edge of the strip.

(Received August 3rd, 1979.)

Appendix A

The matrix $\boldsymbol{P}$ and consequently the transformation matrix $\boldsymbol{T}$ (eq. (7)) depend on the lateral boundary conditions. The following table gives the matrix $\boldsymbol{T}$ and the eigenvalues λ_i for the various combinations of boundaries.

left boundary	right boundary	T_{ij}	λ_i
Dirichlet	Dirichlet	$\sqrt{\frac{2}{N+1}}\sin\frac{ij\pi}{N+1}$	$4\sin^2\frac{i\pi}{2(N+1)}$
Dirichlet	Neumann	$\sqrt{\frac{2}{N+1/2}}\sin\frac{i(j-1/2)\pi}{N+1/2}$	$4\sin^2\frac{(i-1/2)\pi}{2N+1}$
Neumann	Dirichlet	$\sqrt{\frac{2}{N+1/2}}\cos\frac{(i-1/2)(j-1/2)\pi}{N+1/2}$	$4\sin^2\frac{(i-1/2)\pi}{2N+1}$
Neumann	Neumann	$1/\sqrt{N},\ j=1$ $\sqrt{\frac{2}{N}}\cos\frac{(i-1/2)(j-1)\pi}{N},\ j>1$	$4\sin^2\frac{(i-1)\pi}{2N}$

Appendix B

The following formulas may be seen as a supplement to the considered example in section 6. The two transformation matrices $\boldsymbol{T}^{(e,h)}$ and the related eigenvalues λ_i are not given here, as they can be taken from Appendix A:

$$\boldsymbol{\gamma}_{\mathrm{I}}^{(e)} = \operatorname{diag}[\varkappa_i \coth(\varkappa_i b/h)], \quad (33)$$

$$\boldsymbol{\gamma}_{\mathrm{I}}^{(h)} = \operatorname{diag}[\varkappa_i \tanh(\varkappa_i b/h)], \quad (34)$$

$$\boldsymbol{\gamma}_{\mathrm{II}}^{(e)} = \operatorname{diag}[\eta_i \coth(\eta_i d/h)], \quad (35)$$

$$\boldsymbol{\gamma}_{\mathrm{II}}^{(h)} = \operatorname{diag}[\eta_i \tanh(\eta_i d/h)] \quad (36)$$

with

$$\varkappa_i = \left[4\sin^2\left(\frac{i-0.5}{2N+1}\pi\right) - h^2(k_0^2-\beta^2)\right]^{1/2} \quad (37)$$

and

$$\eta_i = \left[4\sin^2\left(\frac{i-0.5}{2N+1}\pi\right) - h^2(\varepsilon_r k_0^2-\beta^2)\right]^{1/2}, \quad (38)$$

$$\boldsymbol{\delta} = \operatorname{diag}\left[2\sin\left(\frac{i-0.5}{2N+1}\pi\right)\right]. \quad (39)$$

References

[1] Michlin, S. G. and Smolizki, Ch. L., Näherungsmethoden zur Lösung von Differential- und Integralgleichungen. B. G. Teubner-Verlag, Leipzig 1969, pp. 238–243.
[2] Ames, W. F., Numerical methods for partial differential equations. Academic Press, New York 1977, pp. 302–304.
[3] Lennartsson, B. L., A network analogue method for computing the TEM-characteristics of planar transmission lines. Transact. IEEE MTT-**20** [1972], 586–591.
[4] Corr, D. G. and Davies, J. B., Computer analysis of the fundamental and higher order modes in single and coupled microstrip. Transact. IEEE MTT-**20** [1972], 669–678.
[5] Meixner, J., Die Kantenbedingung in der Theorie der Beugung elektromagnetischer Wellen an vollkommen leitenden ebenen Schirmen. Ann. Phys. (Leipzig) **6** [1949], 2–9.
[6] Collin, R. E., Field theory of guided waves. McGraw-Hill Book Co., New York 1960, pp. 18–20.
[7] Schulz, U., On the edge condition with the method of lines in planar waveguides. AEÜ **34** [1980], 176–178.
[8] Kowalski, G. and Pregla, R., Dispersion characteristics of shielded microstrips with finite thickness. AEÜ **25** [1971], 193–196.
[9] Sharma, A. K. and Bhat, B., Dispersion characteristics of shielded coupled microstrip lines. AEÜ **32** [1978], 503–504.

Hybrid-Mode Analysis of Arbitrarily Shaped Planar Microwave Structures by the Method of Lines

STEPHAN B. WORM AND REINHOLD PREGLA, SENIOR MEMBER, IEEE

Abstract —This paper presents a method for analyzing arbitrarily shaped planar microwave structures, which is based on the method of lines and applies to both resonant and periodic structures in microstrip, slotline, and finline circuits. Numerical results are presented for some selected structures.

I. Introduction

THE CHARACTERIZATION of planar structures (e.g., microstrip, slotline, and finline) is important in microwave integrated circuit design. The considered structures consist of two or more homogeneous layers, i.e., dielectric substrates or air regions, with various types of metallization located on the interfaces between the layers. Some typical cross sections are shown in Fig. 1

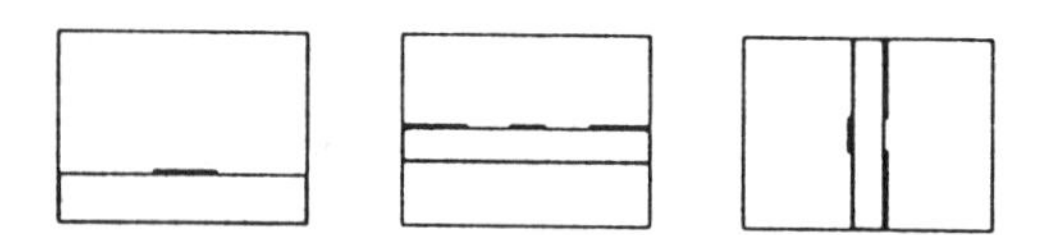

Fig. 1. Cross sections of some planar microwave structures.

In the following, it will be assumed that the metallization has vanishing thickness and that the structures are shielded with perfectly conducting walls. Also, in cases of symmetry, magnetic walls may occur.

Recently, an efficient method for calculating the dispersion characteristics of these types of planar transmission lines was published [1], [2], which is based on the method of lines. With this method, the cross sections are discretized in one direction, whereas the other directions are treated analytically. The object of this paper is to show how the method of lines may be extended to two-dimensional discretization for the analysis of resonant and periodic planar microwave structures.

In principle, arbitrarily shaped resonators may be analyzed, as is demonstrated for a triangular microstrip resonator. The method is accurate enough to derive the characteristic properties of discontinuities from the calculated resonant frequencies. As an example, the end effect of a shorted slotline is investigated.

Another important class of planar microwave structures is formed by periodic structures (e.g., meander-type delay lines). Periodic structures have also been proposed for the phase-equalization of odd and even modes in order to improve the properties of microstrip couplers [3]. Here, the method of lines may also be applied successfully.

Manuscript received June 21, 1983; revised September 26, 1983. This work was supported in part by the Deutsche Forschungs gemeinschaft.

The authors are with the Department of Electrical Engineering, Fernuniversität, P.O. Box 940, D-5800 Hagen, Federal Republic of Germany.

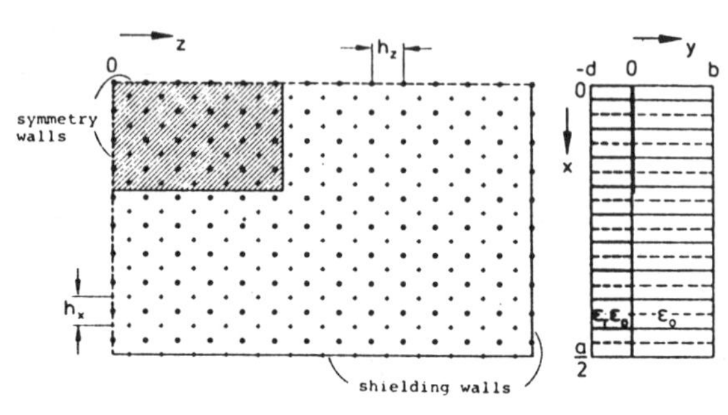

Fig. 2. Position of the discretization lines for a rectangular microstrip resonator (+ for Ψ^e and ∘ for Ψ^h lines).

Advantages of the presented method are its easy formulation and its simple convergence behavior. There is no need to specify specially suited expansion functions, which is particularly advantageous to the analysis of odd-shaped resonators or if the expansion functions would be complex quantities, as is the case with periodic structures. With conventional finite difference methods, large systems of equations are solved directly. With this method a problem-oriented discrete Fourier transform is applied, so the main part of the problem is solved in a transformed domain, where only diagonal matrices occur. The final equations in the original domain are solved with matrices of considerably reduced size.

II. The Analysis of Planar Resonators

The principles of the method of analysis will be demonstrated for a simple rectangular microstrip resonator (Fig. 2). The extension to multilayered structures like slotlines and suspended striplines is straightforward.

The electromagnetic field in each homogeneous region is described by two scalar potential functions Ψ^e and Ψ^h that satisfy the Helmholtz equation and the boundary conditions at the shielding and symmetry walls.

The field components are then found from

$$\vec{E} = \nabla \times \nabla \times (\Psi^e \vec{u}_z)/j\omega\epsilon - \nabla \times (\Psi^h \vec{u}_z)$$
$$\vec{H} = \nabla \times (\Psi^e \vec{u}_z) + \nabla \times \nabla \times (\Psi^h \vec{u}_z)/j\omega\mu_0 \qquad (1)$$

Reprinted from *IEEE Trans. Microwave Theory Tech.*, vol. MTT-32, no. 2, pp. 191–196, Feb. 1984.

where the time-dependence according to $\exp(j\omega t)$ is assumed. At the air–dielectric interface, the continuity conditions for the tangential electric and magnetic field components have to be satisfied as follows:

$$E_{x\mathrm{I}} - E_{x\mathrm{II}} = \frac{1}{j\omega\epsilon_r\epsilon_0}\frac{\partial^2\Psi_{\mathrm{I}}^e}{\partial x\partial z} - \frac{\partial\Psi_{\mathrm{I}}^h}{\partial y} - \frac{1}{j\omega\epsilon_0}\frac{\partial^2\Psi_{\mathrm{II}}^e}{\partial x\partial z} + \frac{\partial\Psi_{\mathrm{II}}^h}{\partial y} = 0$$

$$E_{z\mathrm{I}} - E_{z\mathrm{II}} = \frac{1}{j\omega\epsilon_r\epsilon_0}\left(\frac{\partial^2\Psi_{\mathrm{I}}^e}{\partial z^2} + \epsilon_r k_0^2\Psi_{\mathrm{I}}^e\right) - \frac{1}{j\omega\epsilon_0}\left(\frac{\partial^2\Psi_{\mathrm{II}}^e}{\partial z^2} + k_0^2\Psi_{\mathrm{II}}^e\right) = 0$$

$$H_{x\mathrm{I}} - H_{x\mathrm{II}} = \frac{\partial\Psi_{\mathrm{I}}^e}{\partial_y} + \frac{1}{j\omega\mu_0}\frac{\partial^2\Psi_{\mathrm{I}}^h}{\partial x\partial z} - \frac{\partial\Psi_{\mathrm{II}}^e}{\partial y} - \frac{1}{j\omega\mu_0}\frac{\partial^2\Psi_{\mathrm{II}}^h}{\partial x\partial z} = -J_z$$

$$H_{z\mathrm{I}} - H_{z\mathrm{II}} = \frac{1}{j\omega\mu_0}\left(\frac{\partial^2\Psi_{\mathrm{I}}^h}{\partial z^2} + \epsilon_r k_0^2\Psi_{\mathrm{I}}^h - \frac{\partial^2\Psi_{\mathrm{II}}^h}{\partial z^2} - k_0^2\Psi_{\mathrm{II}}^h\right) = J_x \quad (2)$$

with

$$k_0 = \omega\sqrt{\epsilon_0\mu_0}.$$

The final boundary condition states, that the current density distributions J_z and J_x be nonzero or that the tangential electric field components E_z and E_x be zero on the metallized parts of the interface.

In order to solve this hybrid field problem numerically, the considered region is discretized in the x- and z-direction with meshwidths h_x and h_z, respectively. The discretization lines for Ψ^h are shifted by $h_x/2$ and $h_z/2$ with respect to the lines for Ψ^e. In this way, the lateral boundary conditions at the shielding and symmetry walls can be fitted easily and discretization error is reduced considerably, as was also found for the case of one-dimensional discretization [1], [2].

For the example of Fig. 2, the potential function Ψ^e at $x=(i-0.5)h_x$ and $z=(k-0.5)h_z$ $(i=1,\cdots,N_x;\ k=1,\cdots,N_z)$ will be denoted by Ψ_{ik}^e and interpreted as an element of the matrix $[\Psi^e]$, in which the elements are arranged in the same pattern as in the discretized structure. It should be noted, however, that $[\Psi^e]$ really has a vector character, which will become important later on.

For the first derivative of Ψ^e with respect to the x-direction, one obtains

$$\left.\frac{\partial\Psi^e}{\partial x}\right|_{\substack{x=ih_x\\ z=(k-0.5)h_z}} = \frac{\Psi_{i+1,k}^e - \Psi_{i,k}^e}{h_x} + 0(h_x^2) \quad (3)$$

or, in matrix notation

$$h_x\frac{\partial\Psi^e}{\partial x} \rightarrow \begin{bmatrix} -1 & 1 & & \\ & \ddots & \ddots & \\ & & & 1\\ & & & -1\end{bmatrix}\begin{bmatrix}\Psi_{1,1}^e & \cdots & \Psi_{1,N_z}^e\\ \vdots & & \vdots\\ \Psi_{N_x,1}^e & \cdots & \Psi_{N_x,N_z}^e\end{bmatrix} = [D_x][\Psi^e]. \quad (4)$$

The difference matrix $[D_x]$ depends on the lateral boundary conditions for Ψ^e (see Table I). It is the same operator matrix as used in the case of one-dimensional discretization. Here, it forms the difference between two successive rows of the matrix $[\Psi^e]$.

Because Ψ^e and Ψ^h have dual boundary conditions, the finite difference expression for the first derivative of Ψ^h becomes

$$h_x\frac{\partial\Psi^h}{\partial x} \rightarrow -[D_x]^t[\Psi^h]. \quad (5)$$

Combining (4) and (5), one obtains for the second-order derivatives

$$h_x^2\frac{\partial^2\Psi^e}{\partial x^2} \rightarrow -[D_x]^t[D_x][\Psi^e] = [D_{xx}^e][\Psi^e] \quad (6)$$

$$h_x^2\frac{\partial^2\Psi^h}{\partial x^2} \rightarrow -[D_x][D_x]^t[\Psi^h] = [D_{xx}^h][\Psi^h]. \quad (7)$$

Analogously, the difference operator for the first derivative of Ψ^e with respect to the z-direction should form the difference between two successive columns of the matrix $[\Psi^e]$. Thus, the difference matrix $[D_z]$, as taken from Table I, will operate on the transpose of the matrix $[\Psi^e]$

$$h_z\frac{\partial\Psi^e}{\partial z} \rightarrow [D_z][\Psi^e]^t \quad (8)$$

or rather

$$\rightarrow [\Psi^e][D_z]^t. \quad (9)$$

In a similar way, the finite difference expression for the second-order derivative of Ψ^e with respect to z is written as

$$h_z^2\frac{\partial^2\Psi^e}{\partial z^2} \rightarrow -[\Psi^e][D_z]^t[D_z] = [\Psi^e][D_{zz}^e]^t. \quad (10)$$

This notation provides simple compatibility with the difference operators for the x-direction, e.g.,

$$h_x h_z\frac{\partial^2\Psi^e}{\partial x\partial z} \rightarrow [D_x][\Psi^e][D_z]^t. \quad (11)$$

Working out this expression for the above example yields

$$h_x h_z\left.\frac{\partial^2\Psi^e}{\partial x\partial z}\right|_{\substack{x=ih_x\\ z=kh_z}} \simeq \Psi_{i,k}^e - \Psi_{i,k+1}^e + \Psi_{i+1,k+1}^e - \Psi_{i+1,k}^e. \quad (12)$$

It is evaluated at the discretization line for Ψ_{ik}^h from the function values at the four adjacent Ψ^e lines, so it fits well into the continuity equations (2) using only small discretization distances.

Because of the tri-diagonal structure of the difference matrices $[D_{xx}^e]$ and $[D_{zz}^e]$, the discretized Helmholtz equation

$$\frac{d^2[\Psi^e]}{dy^2} + \frac{[D_{xx}^e][\Psi^e]}{h_x^2} + \frac{[\Psi^e][D_{zz}^e]^t}{h_z^2} + \epsilon_r k_0^2[\Psi^e] = 0 \tag{13}$$

represents a system of $N_x N_z$ differential equations, that are all coupled with each other.

By means of the orthogonal transformation matrices $[T_x^e]$ and $[T_z^e]$ (see Appendix) the difference matrices are transformed into diagonal matrices

$$[T_x^e]^t[D_{xx}^e][T_x^e] = \mathrm{diag}[d_{xx}^e] \tag{14}$$

$$[T_z^e]^t[D_{zz}^e][T_z^e] = \mathrm{diag}[d_{zz}^e]. \tag{15}$$

So, for the elements of the transformed potential matrix

$$[U] = [T_x^e]^t[\Psi^e][T_z^e] \tag{16}$$

one obtains the uncoupled differential equations

$$\frac{d^2[U]_{ik}}{dy^2} - \kappa_{ik}^2[U]_{ik} = 0 \tag{17}$$

with

$$\kappa_{ik}^2 = -\left(\frac{[d_{xx}^e]_{ii}}{h_x^2} + \frac{[d_{zz}^e]_{kk}}{h_z^2} + \epsilon_r k_0^2\right). \tag{18}$$

The general solution to (17) may be written as a relation between $[U]_{ik}$ and its normal derivative in the planes $y = y_1$ and $y = y_2$

$$\begin{pmatrix} U(y_1) \\ \left.\frac{d[U]}{dy}\right|_{y_1} \end{pmatrix}_{ik} = \begin{pmatrix} \cosh\kappa_{ik}(y_1 - y_2) & \frac{\sinh\kappa_{ik}(y_1 - y_2)}{\kappa_{ik}} \\ \kappa_{ik}\sinh\kappa_{ik}(y_1 - y_2) & \cosh\kappa_{ik}(y_1 - y_2) \end{pmatrix} \begin{pmatrix} U(y_2) \\ \left.\frac{d[U]}{dy}\right|_{y_2} \end{pmatrix}_{ik}. \tag{19}$$

By means of this relation, the boundary conditions at the top and bottom shielding can be transformed into the interface plane $y = 0$, e.g., for the substrate region one obtains with $[U(y = -d)]_{ik} = 0$

$$\left.\frac{d[U]_{ik}}{dy}\right|_{y=0} = \frac{\kappa_{ik}}{\tanh\kappa_{ik}d}[U(y=0)]_{ik}. \tag{20}$$

The other potential function Ψ^h is transformed by means of the orthogonal matrices $[T_x^h]$ and $[T_z^h]$ (see Appendix) in a similar way, so that the continuity equations (2) may be solved entirely in the transformed domain. This yields an equation of the following form:

$$\begin{pmatrix} \tilde{E}_z \\ \tilde{E}_x \end{pmatrix} = \begin{bmatrix} [\tilde{Z}_{11}] & [\tilde{Z}_{12}] \\ [\tilde{Z}_{21}] & [\tilde{Z}_{22}] \end{bmatrix} \begin{pmatrix} \tilde{J}_z \\ \tilde{J}_x \end{pmatrix} \tag{21}$$

in which $[\tilde{Z}_{nm}]$ $(n, m = 1,2)$ are diagonal matrices if the transformed quantities $\tilde{E}_z$, $\tilde{E}_x$, $\tilde{J}_z$, and $\tilde{J}_x$ are written in vector notation.

Because the final boundary condition cannot be applied in the transformed domain, (21) has to be transformed back into the original domain [4]. For the example of Fig. 2, the metallic strip makes up the smaller part of the interface, so in this case the reverse transformation is performed only for the (reduced) number of lines that pass through the strip. The resulting equation

$$\begin{pmatrix} E_z \\ E_x \end{pmatrix}_{\text{red}} = \begin{bmatrix} [Z_{11}] & [Z_{12}] \\ [Z_{21}] & [Z_{22}] \end{bmatrix}_{\text{red}} \begin{pmatrix} J_z \\ J_x \end{pmatrix}_{\text{red}} = 0 \tag{22}$$

in which $[Z_{nm}]$ $(n, m = 1,2)$ are now full matrices, will have nontrivial solutions for the resonant frequencies of the structure only, which are found from the determinantal equation

$$\det[Z(f)]_{\text{red}} = 0. \tag{23}$$

At resonance, all field components can be derived from the current density distribution, which occurs as an eigenvector in (22).

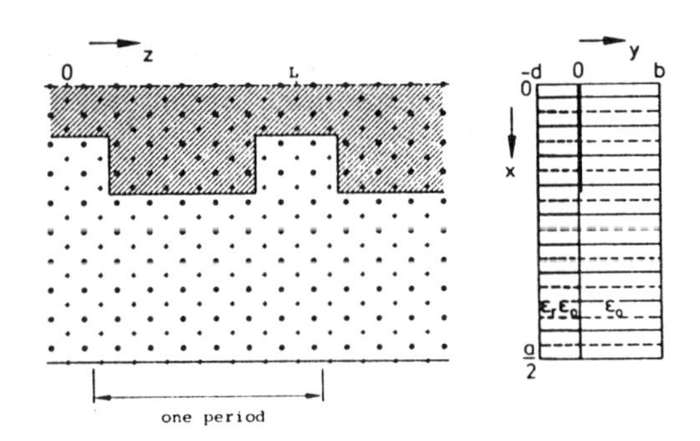

Fig. 3. Discretization lines for a periodic structure.

III. Periodic Structures

The analysis of planar periodic structures proceeds similarly to the analysis of resonant structures described above. For the periodic structure in Fig. 3, the potential functions and all electromagnetic field components must satisfy Floquet's theorem

$$\Psi^{e,h}(x, y, z + L) = e^{-j\beta L}\Psi^{e,h}(x, y, z) \tag{24}$$

where β is the propagation constant in the z-direction and L is the period length. For the x-direction homogeneous boundary conditions apply.

One period of the structure is discretized with Ψ^e lines located at $x = (i - 0.5)h_x$, $z = kh_z$, and Ψ^h lines at $x = ih_x$, $z = (k + 0.5)h_z$ $(i = 1,\cdots,N_x;\ k = 1,\cdots,N_z)$. The finite difference expression for the first derivative of Ψ^e with respect to z is then given by

$$h_z\frac{\partial\Psi^e}{\partial z} \to [\Psi^e][D_z]^t \tag{25}$$

with

$$[D_z] = \begin{bmatrix} -1 & 1 & & \\ & \ddots & \ddots & \\ & & & 1 \\ s & & & -1 \end{bmatrix}, \qquad s = e^{-j\beta L}. \tag{26}$$

The difference operator for Ψ^h and for $\partial\Psi^e/\partial z$ is $-[D_z]^{*t}$, which yields for the second-order derivative of Ψ^e

$$h_z^2\frac{\partial^2\Psi^e}{\partial z^2}\rightarrow-[\Psi^e][D_z]^t[D_z]^*=[\Psi^e][D_{zz}^e]^t. \quad (27)$$

The Hermitian matrix $[D_{zz}^e]$ is transformed into a diagonal matrix $[d_{zz}^e]$ by means of the unitary transformation matrix $[T_z^e]$

$$[T_z^e]^{*t}[D_{zz}^e][T_z^e]=\operatorname{diag}[d_{zz}^e] \quad (28)$$

where

$$[d_{zz}^e]_{kk}=-4\sin^2(\varphi_k/2) \quad (29)$$

and

$$[T_z^e]_{ik}=\sqrt{1/N}\,e^{ji\varphi k} \quad (30)$$

with

$$\varphi_k=\frac{2\pi(k-1)-\beta L}{N_z}. \quad (31)$$

On account of the periodic boundary conditions, one obtains a difference matrix $[D_{zz}^h]$ for Ψ^h equal to $[D_{zz}^e]$, so the same transformation matrix could be used. However, if the elements of $[T_z^h]$ are defined as

$$[T_z^h]_{ik}=\frac{j}{\sqrt{N}}e^{j(i+0.5)\varphi_k} \quad (32)$$

in which the shifting of lines is apparent, the expressions for the first derivatives in the transformed domain remain real, e.g.,

$$[T_z^h]^{*t}[D_z][T_z^e]=\operatorname{diag}[d_z] \quad (33)$$

with

$$[d_z]_{kk}=2\sin(\varphi_k/2). \quad (34)$$

The continuity conditions for the tangential field components are again solved in the transformed domain. After the reverse transformation, the final boundary condition leads to a Hermitian matrix $[Z]_{\text{red}}$, of which the determinant should vanish

$$\det[Z(\omega,\beta)]_{\text{red}}=0. \quad (35)$$

The solutions to this equation are typically represented as the dispersion curves in the $\omega-\beta$ diagram.

IV. Numerical Results

As a first example, a triangular microstrip resonator is considered. It should demonstrate that the analysis is not restricted to rectangular structures. The shielding dimensions are taken large enough to approximate unshielded structures. Fig. 4 shows that the resonant frequencies obtained with this method agree fairly well with the results from the transverse-resonance method and even better with the measurements, both from [5].

The accuracy of about 1 percent will be sufficient for most applications. If the resonant frequencies are used for calculating discontinuities, however, a much higher accuracy is required, as will be illustrated for the end effect of a shorted slotline.

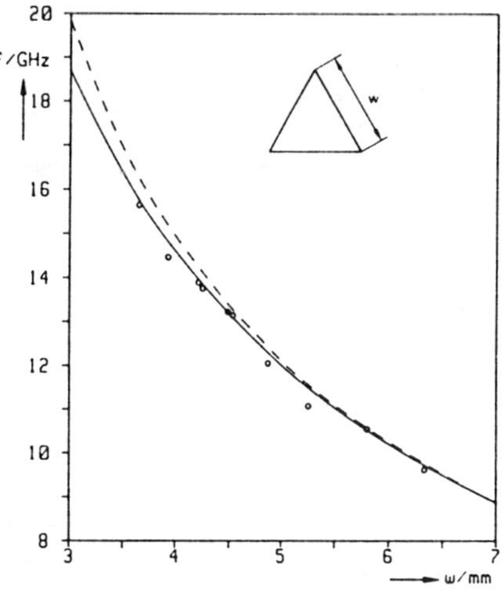

Fig. 4. Resonant frequency for a triangular microstrip resonator ($\epsilon_r=9.7$, $d=0.635$ mm, $b=10d$). ——: this method, - - -: transverse-resonance method [5], ∘ ∘ ∘ ∘: experiment [5].

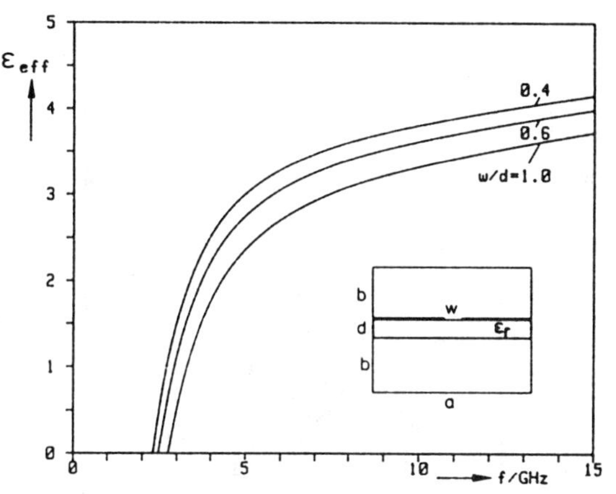

Fig. 5. Dispersion characteristics for shielded slotlines. ($\epsilon_r=9.7$, $d=0.635$ mm, $b=10d$, $a=w+20d$).

A slotline resonator with a cross-section and dispersion characteristic as depicted in Fig. 5 will have resonant frequencies given by

$$f=\frac{n}{2}\frac{c_0}{(L+2\Delta l)\sqrt{\epsilon_{\text{eff}}}} \qquad (n=1,2,\cdots) \quad (36)$$

where L is the resonator length.

The effective dielectric constant ϵ_{eff} is calculated by the method of lines from the cross-sectional problem with negligible error, so the resonant frequency has to be calculated with a relative error, which is about a factor $2\Delta l/L$ smaller than the error tolerated for the end effect Δl.

Fig. 6 shows the convergence behavior of the resonant frequencies in dependence on the discretization distance h_z. The position of the edges is fixed with respect to the discretization lines by means of the edge parameters p_x and p_z. This results in smooth convergence curves, so the discretization error may be represented in good approximation by a quadratic or cubic function, from which the extrapolated values for $h_z\rightarrow 0$ are easily derived. The final result is independent of the actual value of p_z.

In Fig. 6, the resonant frequency calculated with $p_z=0.25$ and $h_z=0.5$ mm has an error of about 1 percent. Taking

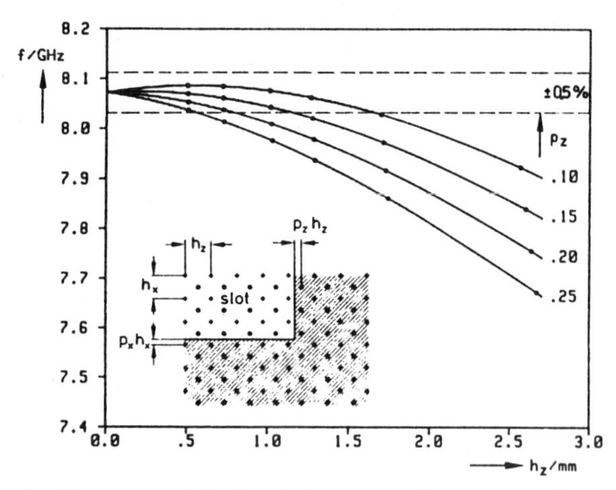

Fig. 6. Convergence behavior of the resonant frequency as a function of the discretization distance h_z and the edge parameter p_z. ($w = d$, $L = 20$ mm).

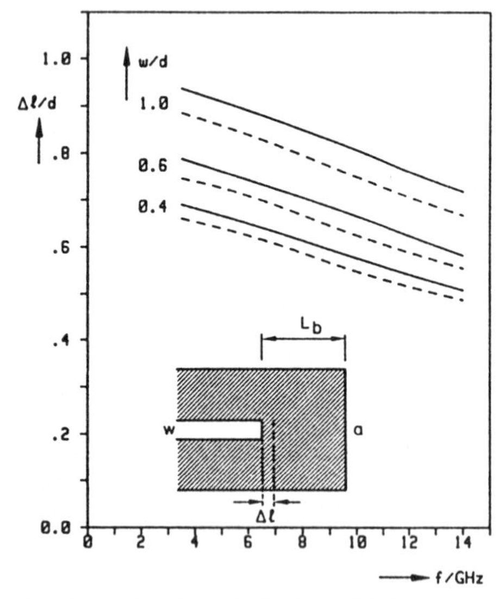

Fig. 7. End effect of a shorted slotline. (Dimensions as in Fig. 5; $L_b > 12$ mm, variable in order to avoid box resonances.) —— this method, ---- Jansen [6].

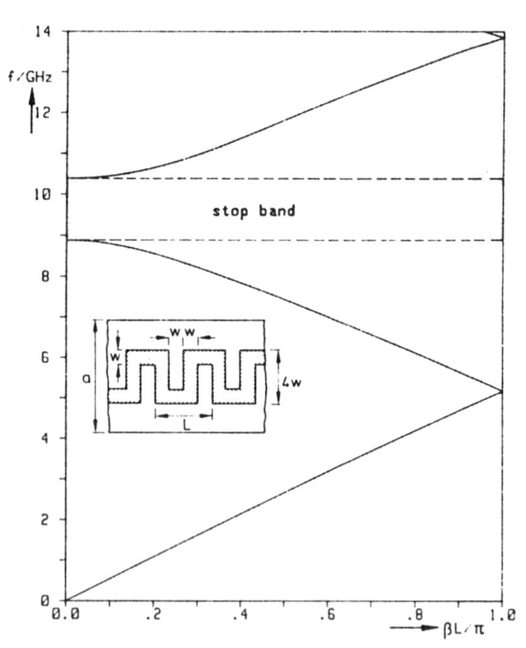

Fig. 8. Dispersion diagram of a microstrip meander line ($\epsilon_r = 2.3$, $w = 2.37$ mm, $d = 0.79$ mm, $b = 10d$, $a = 12w$, $L = 4w$).

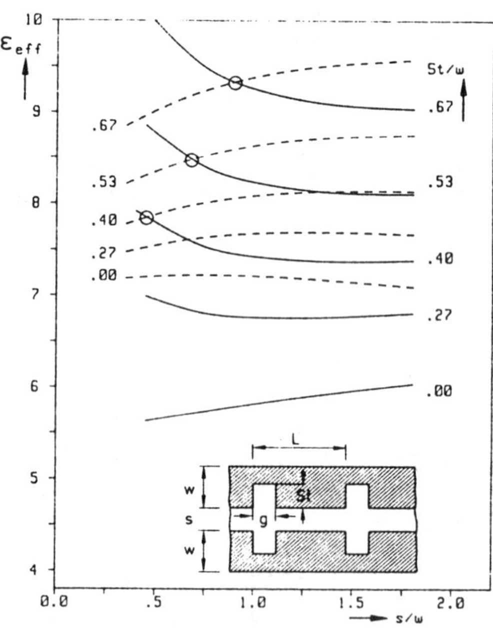

Fig. 9. Effect of periodic slotting on the phase velocities of coupled microstrip lines ($\epsilon_r = 9.6$, $f = 6$ GHz, $w = d = 0.635$ mm, $b = 10d$, $L = 4g = 0.15$ mm). —— odd mode, ---- even mode.

into account the results obtained at some larger discretization distances and applying a least squares method for quadratic or cubic extrapolation increases the accuracy by more than one order of magnitude. In this case, the error for the resonant frequency is estimated to be less than 0.1 percent, including the error from the discretization in the x-direction (about 0.02 percent with $p_x = 0.25$ and $h_x = 0.115$ mm). Thus, the end effect has an accuracy of about 2 percent.

In Fig. 7, the calculated end effect of a shorted slotline is shown as a function of frequency and compared with results from the literature [6]. For $\lambda/2$ resonators, the end effect was found to be a few percent larger, whereas resonators with one full wavelength or more gave identical results.

As a first example for the analysis of periodic structures a microstrip meander line is considered. Fig. 8 shows the calculated dispersion diagram. Up to about 9 GHz, this diagram may be approximated to some extent by considering the meander as a straight line and taking into account the propagation constants β_{even} and β_{odd} of the multiple coupled lines at $\beta L = 0, 2\pi, \cdots$, and $\beta L = \pi, 3\pi, \cdots$, respectively. The stopband, however, would not be obtained by such a simple method.

The final example demonstrates the use of periodic structures in coupled microstrip lines. The difference between the phase velocities of the even and odd mode of these lines leads to bad isolation in microstrip proximity couplers [3]. To eliminate this effect, transverse periodic slotting at the inner edges may be applied. Fig. 9 shows that the slotting has more influence on the odd mode than

TABLE I
FIRST-ORDER DIFFERENCE MATRICES IN THE ORIGINAL AND THE TRANSFORMED DOMAIN FOR THE VARIOUS (N FOR NEUMANN, D FOR DIRICHLET) BOUNDARY CONDITIONS FOR Ψ^e

boundary conditions left - right	[D]	[d]	elements of [d] ≠ 0
N - D	$\begin{bmatrix}-1 & 1 & & \\ & \ddots & \ddots & \\ & & & 1 \\ & & & -1\end{bmatrix}$	$\begin{bmatrix}\ddots & 0 \\ 0 & \ddots\end{bmatrix}$	$[d]_{i,i} = -2 \sin \frac{(i-.5)\pi}{2N+1}$ $(i=1,\ldots,N)$
D - N	$\begin{bmatrix}1 & & & \\ -1 & \ddots & & \\ & \ddots & \ddots & \\ & & -1 & 1\end{bmatrix}$	$\begin{bmatrix}\ddots & 0 \\ 0 & \ddots\end{bmatrix}$	$[d]_{i,i} = 2 \sin \frac{(i-.5)\pi}{2N+1}$ $(i=1,\ldots,N)$
D - D	$\begin{bmatrix}1 & & \\ -1 & \ddots & \\ & \ddots & 1 \\ & & -1\end{bmatrix}$	$\begin{bmatrix}0 & \cdots & 0 \\ \ddots & & 0 \\ 0 & & \ddots\end{bmatrix}$	$[d]_{i+1,i} = 2 \sin \frac{i\pi}{2N+2}$ $(i=1,\ldots,N)$
N - N	$\begin{bmatrix}-1 & 1 & & \\ & \ddots & \ddots & \\ & & -1 & 1\end{bmatrix}$	$\begin{bmatrix}0 & \ddots & 0 \\ \vdots & & \\ 0 & 0 & \ddots\end{bmatrix}$	$[d]_{i-1,i} = -2 \sin \frac{(i-1)\pi}{2N}$ $(i=2,\ldots,N)$

TABLE II
ELEMENTS OF THE TRANSFORMATION MATRICES

boundary conditions left - right	$[T^e]_{ik}$	$[T^h]_{ik}$
N - D	$\sqrt{\frac{2}{N+.5}} \cos \frac{(i-.5)(k-.5)\pi}{N+.5}$ $(i,k=1,\ldots,N)$	$\sqrt{\frac{2}{N+.5}} \sin \frac{i(k-.5)\pi}{N+.5}$ $(i,k=1,\ldots,N)$
D - N	$\sqrt{\frac{2}{N+.5}} \sin \frac{i(k-.5)\pi}{N+.5}$ $(i,k=1,\ldots,N)$	$\sqrt{\frac{2}{N+.5}} \cos \frac{(i-.5)(k-.5)\pi}{N+.5}$ $(i,k=1,\ldots,N)$
D - D	$\sqrt{\frac{2}{N+1}} \sin \frac{ik\pi}{N+1}$ $(i,k=1,\ldots,N)$	$\sqrt{\frac{2}{N+1}} \cos \frac{(i+.5)k\pi}{N+1}$; $k>0$ $\sqrt{\frac{1}{N+1}}$; $k=0$ $(i,k=0,1,\ldots,N)$
N - N	$\sqrt{\frac{2}{N}} \cos \frac{(i-.5)(k-1)\pi}{N}$; $k>1$ $\sqrt{\frac{1}{N}}$; $k=1$ $(i,k=1,\ldots,N)$	$\sqrt{\frac{2}{N}} \sin \frac{(i-1)(k-1)\pi}{N}$ $(i,k=2,\ldots,N)$

on the even mode, so it is possible to achieve a phase-equalization.

APPENDIX

Because the discretizations in both directions may be treated fully independently, the difference and transformation matrices are summarized for one-dimensional discretization only. The difference matrices depend on the lateral boundary conditions. In Fig. 2, for example, the Neumann condition ($\partial\Psi^e/\partial x = 0$) at $x = 0$ and the Dirichlet condition ($\Psi^e = 0$) at $x = a/2$ are taken into account by putting $\Psi^e_{0,k} = \Psi^e_{1,k}$ and $\Psi^e_{N_x+1,k} = 0$ in (3).

The difference and transformation matrices for the various combinations of the lateral boundary conditions of Ψ^e are shown in Tables I and II. The number of lines for Ψ^e is denoted by N, which should be replaced by N_x and N_z for the respective directions.

If one of the potential functions Ψ^e and Ψ^h has Neumann conditions on both side walls, it will have one discretization line and one spectral component (a dc component) more than the other one, which results in rectangular matrices $[D]$ and $[d]$.

The transformations are all derived from the elementary relation $[D][T^e] = [T^h][d]$, e.g.,

$$\begin{aligned}[T^e_x]^t[D^e_{xx}][T^e] &= -[T^e_x]^t[D_x]^t[T^h_x][T^h_x]^t[D_x][T^e_x] \\ &= -[d_x]^t[d_x] \\ &= \operatorname{diag}[d^e_{xx}].\end{aligned}$$

REFERENCES

[1] U. Schulz and R. Pregla, "A new technique for the analysis of the dispersion characteristics of planar waveguides," *Arch. Elec. Übertragung.*, vol. 34, pp. 169–173, 1980.

[2] U. Schulz and R. Pregla, "A new technique for the analysis of planar waveguides and its application to microstrips with tuning septums," *Radio Sci.*, vol. 16, pp. 1173–1178, 1981.

[3] F. C. de Ronde, "Wide-band high directivity in MIC proximity couplers by planar means," in *Proc. MTT Int. Symp.*, (Washington, D.C.), 1980.

[4] S. B. Worm, "Analysis of planar microwave structures with arbitrary contour (in German)," Ph.D. Thesis, Fernuniversitaet Hagen, 1983.

[5] W. T. Nisbet and J. Helszajn, "Mode charts for microstrip resonators on dielectric and magnetic substrates using a transverse-resonance method," *Microwaves Opt. Acoust.*, vol. 3, pp. 69–77, 1979.

[6] R. H. Jansen, "Hybrid mode analysis of end effects of planar microwave and millimeterwave transmission lines," *Inst. Elec. Eng. Proc.*, part H, vol. 128, pp. 77–86, 1981.

Analysis of Hybrid Field Problems by the Method of Lines with Nonequidistant Discretization

HEINRICH DIESTEL AND STEPHAN B. WORM

Abstract —The method of lines, which has been proved to be very efficient for calculating the characteristics of one-dimensional and two-dimensional planar microwave structures, is extended to nonequidistant discretization. By means of an intermediate transformation it is possible to maintain all essential transformation properties that are given in the case of equidistant discretization. The flexibility of the method of lines is increased substantially. As a consequence, the accuracy is improved with reduced computational effort.

I. Introduction

A SUCCESSFUL DESIGN of planar microwave circuits presupposes accurate knowledge of the characteristics of the elementary components.

In principle, an exact determination of the characteristics of passive components like transmission lines, resonators, and filters is possible by means of complete Fourier series expansions. For numerical evaluation, only a finite number of terms can be taken into account. Hence, this method is characterized by the fact that the exactly formulated problem is solved approximately.

Manuscript received November 3, 1983; revised February 6, 1984. This work was supported by Deutsche Forschungsgemeinschaft.

The authors are with the Department of Electrical Engineering, Fernuniversitaet, Hagen, Federal Republic of Germany.

A completely different way is taken by the grid-point method and the method of lines [1], where the approximately formulated problem is solved exactly.

The semi-analytical method of lines has been applied to various problems of physics [2]. An essential extension of this method is given in [3] for the one-dimensional and in [4] for the two-dimensional hybrid problem of planar waveguides. It has been shown that this class of waveguides can be solved accurately and in a simple manner.

In the limiting case of an infinite number of lines, exactly the same solution is obtained as in the limiting case of an infinite number of terms in the Fourier series expansions.

The relative convergence phenomenon, which is a consequence of the Fourier series truncations, does not occur with the method of lines. Optimum convergence is always assured, if the simple condition is satisfied that the strip-edges are located at definite positions with respect to the adjacent ψ^e- and ψ^h-lines [5]. It should be noted, however, that the convergence of the propagation constant, the characteristic impedance or the resonant frequency does not critically depend on the edge parameters, so that the problem of convergence on the whole is not critical.

Reprinted from *IEEE Trans. Microwave Theory Tech.*, vol. MTT-32, no. 6, pp. 633–638, June 1984.

This is the main advantage of the method of lines for planar structures.

In order to satisfy correctly the edge condition for each edge of a given waveguide and to satisfy in addition the lateral boundary conditions, an appropriate number of lines has to be determined. It is obvious that this problem becomes more difficult with an increasing number of conductors. A further deficiency of the method is given by the fact that, in case of extreme differences in the widths of the conductors and the spacings between them, the total number of lines increases considerably.

The reason for these drawbacks lies in the inflexibility of the equidistant discretization.

In the present paper, it will be shown that the nonequidistant discretization, which has been applied successfully in the grid-point method, can also be introduced in the method of lines without changing its special transformation properties. An outline of the method will be given for the one-dimensional nonequidistant discretization. The extension of this method to two-dimensional problems does not cause any difficulties: the procedure is similar to that given in [4].

Numerical results are presented for two selected examples: the coplanar waveguide (one-dim. discretization) and the hair-pin resonator (two-dim. discretization). The convergence behavior is discussed and comparisons are made with the limiting case of equidistant discretization.

II. Formulation

The cross-section of the structure is subdivided into several partial areas, as indicated in Fig. 1. Within each area, constant permittivity is assumed. Conducting strips of vanishing thickness are located at the interfaces between the areas.

The electromagnetic field components $\vec{E}$ and $\vec{H}$ are derived from two independent vector potential functions, which in each case exhibit only one component in z-direction

$$\vec{E} = \nabla \times \nabla \times (\Psi^e \vec{e}_z)/j\omega\epsilon - \nabla \times (\Psi^h \vec{e}_z) \quad (1)$$

$$\vec{H} = \nabla \times (\Psi^e \vec{e}_z) + \nabla \times \nabla \times (\Psi^h \vec{e}_z)/j\omega\mu_0. \quad (2)$$

The harmonic time dependence $\exp(j\omega t)$ has been omitted for brevity.

For waveguides uniform in the direction of propagation (z-direction), the two scalar functions of the vector potentials can be expressed as

$$\Psi^{e,h} = \psi^{e,h}(x, y)\exp(-j\beta z) \quad (3)$$

where β is the propagation constant.

Substituting (3) in the corresponding Helmholtz equations for the scalar potential functions yields

$$\frac{\partial^2 \psi^{e,h}}{\partial y^2} + \frac{\partial^2 \psi^{e,h}}{\partial x^2} + (k^2 - \beta^2)\psi^{e,h} = 0 \quad (4)$$

with $k^2 = \omega^2 \mu_0 \epsilon_0 \epsilon$.

The potential functions are submitted to homogeneous Dirichlet or Neumann conditions on the shielding (and symmetry) walls. Continuity conditions have to be satisfied at the boundaries between the different areas.

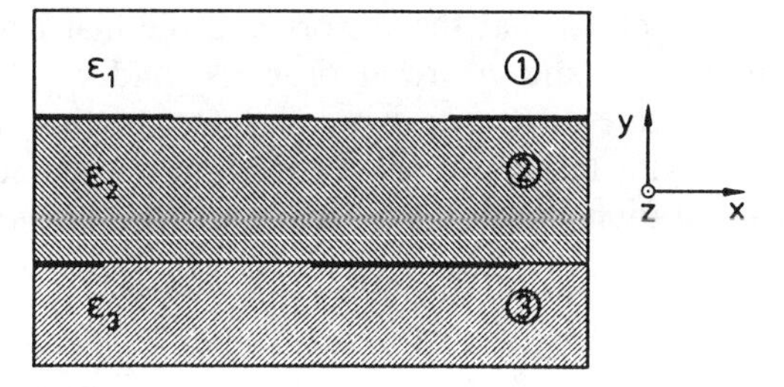

Fig. 1. Cross section of a planar microwave structure.

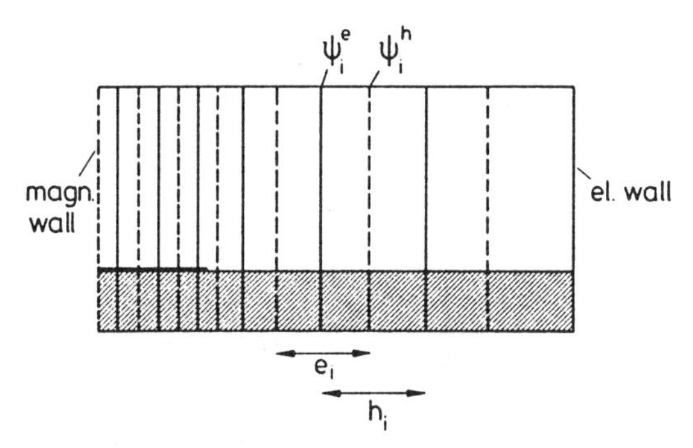

Fig. 2. Position of the discretization lines for the scalar potentials ψ^e and ψ^h; e_i and h_i designate interval sizes.

Because of strip-conductor edges, the electromagnetic fields exhibit singularities. Hence, a discrete representation is chosen along the interfaces (x-direction), whereas in the vertical direction the fields are expressed analytically. This means that the potential functions ψ^e and ψ^h are considered on lines, as illustrated in Fig. 2.

The shifting of the two sets of lines with respect to each other is a necessary condition for the compatibility of the operators applied in the following. As a consequence of the shifting, both the lateral boundary conditions and the edge condition fit in harmoniously.

The sizes of the intervals intersected by the discretization lines for ψ_i^e and ψ_i^h are denoted by e_i $(i = 1, \cdots, N_e)$ and h_i $(i = 1, \cdots, N_h)$, respectively.

In order to obtain symmetric second-order operators, normalized potential functions are introduced next

$$\phi_i^e = \sqrt{e_i/h}\,\psi_i^e \quad (5a)$$

and

$$\phi_i^h = \sqrt{h_i/h}\,\psi_i^h \quad (5b)$$

where h represents the interval size of the limiting case of equidistant discretization.

In matrix notation, (5a) and (5b) lead to the following equations:

$$\vec{\psi}^e = [r_e]\vec{\phi}^e \quad (6a)$$

and

$$\vec{\psi}^h = [r_h]\vec{\phi}^h \quad (6b)$$

with

$$[r_e] = \mathrm{diag}\left(\sqrt{h/e_i}\right), \qquad [r_h] = \mathrm{diag}\left(\sqrt{h/h_i}\right). \quad (7)$$

It should be noted that the vectors and the matrices with the subscripts e and h are of order N_e and N_h, respectively. The finite-difference expression for the first derivative of ψ^e with respect to the x-direction is evaluated on the discretization line for ψ^h. Hence, on the line for ψ_i^h, marked in Fig. 2, the first derivative of ψ^e is approximated as follows:

$$\left.\frac{\partial \psi^e}{\partial x}\right|_i \simeq \frac{\psi_{i+1}^e - \psi_i^e}{h_i}. \tag{8}$$

After normalization, this becomes

$$\sqrt{h_i/h}\left(h\left.\frac{\partial \psi^e}{\partial x}\right|_i\right) \simeq \sqrt{h/h_i}\,(\psi_{i+1}^e - \psi_i^e) \tag{9}$$

or, in matrix notation

$$\begin{aligned}[r_h]^{-1}\left(h\frac{\partial \vec{\psi}^e}{\partial x}\right) &\rightarrow [r_h][D]\vec{\psi}^e \\ &= [r_h][D][r_e]\vec{\phi}^e \\ &= [D_x]\vec{\phi}^e.\end{aligned} \tag{10}$$

In the case of equidistant discretization, characterized by the relation $h_i = e_i = h$ for all i, the bidiagonal matrix $[D_x]$ is identical to the difference operator $[D]$, which is given in [4] for the various combinations of lateral boundary conditions. For the combination magnetic/electric wall of Fig. 2, one obtains the following square matrix:

$$[D] = \begin{bmatrix} -1 & 1 & & \\ & \ddots & \ddots & \\ & & \ddots & 1 \\ & & & -1 \end{bmatrix}. \tag{11}$$

On account of the dual lateral boundary conditions and the shifting of lines, the finite-difference translation for the first derivative of ψ^h can be given immediately

$$[r_e]^{-1}\left(h\frac{\partial \vec{\psi}^h}{\partial x}\right) \rightarrow -[r_e][D]^t[r_h]\vec{\phi}^h = -[D_x]^t\vec{\phi}^h. \tag{12}$$

Combining the first-order operators, one obtains for the second-order derivatives

$$[r_e]^{-1}\left(h^2\frac{\partial^2 \vec{\psi}^e}{\partial x^2}\right) = h^2\frac{\partial^2 \vec{\phi}^e}{\partial x^2} \rightarrow -[D_x]^t[D_x]\vec{\phi}^e \tag{13a}$$

$$[r_h]^{-1}\left(h^2\frac{\partial^2 \vec{\psi}^h}{\partial x^2}\right) = h^2\frac{\partial^2 \vec{\phi}^h}{\partial x^2} \rightarrow -[D_x][D_x]^t\vec{\phi}^h. \tag{13b}$$

The second-order operators

$$\begin{aligned}[D_{xx}^e] &= -[D_x]^t[D_x] \\ [D_{xx}^h] &= -[D_x][D_x]^t\end{aligned} \tag{14}$$

are real-symmetric tridiagonal matrices. Thus, they can be transformed by orthogonal transformation into the diagonal form of their real and distinct eigenvalues

$$[T_e]^t[D_{xx}^e][T_e] = [\lambda^e]$$

and

$$[T_h]^t[D_{xx}^h][T_h] = [\lambda^h] \tag{15}$$

where $[T_e]$ and $[T_h]$ are the matrices of the eigenvectors. It can be proved that the bidiagonal first-order operator $[D_x]$ is transferred to quasi-diagonal form by the following transformation [6]:

$$[T_h]^t[D_x][T_e] = [\delta]. \tag{16}$$

From (14) to (16), the following relations are derived:

$$[\lambda^e] = -[\delta]^t[\delta]$$

and

$$[\lambda^h] = -[\delta][\delta]^t. \tag{17}$$

In case of different lateral boundary conditions (magn. wall/el. wall, and vice versa) $[\delta]$ is a square diagonal matrix and (17) is reduced to

$$[\lambda^e] = [\lambda^h] = -[\delta]^2. \tag{18}$$

The eigenvalues and the matrices of the eigenvectors in (15) are determined by means of the 'Implicit QL-method' [7], an accurate and numerically stable method.

Only in the limiting case of equidistant discretization, these quantities are given in analytical form.

The partial differential equations (4) can now be transferred to the following systems of ordinary differential equations:

$$\frac{d^2\vec{V}^{e,h}}{dy^2} + \left([\lambda^{e,h}]/h^2 + (k^2 - \beta^2)\right)\vec{V}^{e,h} = 0 \tag{19}$$

with $\vec{V}^{e,h} = [T_{e,h}]^t\vec{\phi}^{e,h}$. The solutions V_j^e and V_j^h, respectively, of these one-dimensional Helmholtz equations correspond to the simple transmission line equations.

The boundary conditions at the top and the bottom shielding, as well as the matching of the fields at the interfaces, can be carried out using only diagonal matrices. An inhomogeneous matrix equation is obtained:

$$[\tilde{Z}]\begin{bmatrix}\vec{\tilde{J}}_z \\ \vec{\tilde{J}}_x\end{bmatrix} = \begin{bmatrix}\vec{\tilde{E}}_z \\ \vec{\tilde{E}}_x\end{bmatrix} \tag{20}$$

where $(\vec{\tilde{J}}_z, \vec{\tilde{J}}_x)$ represents the transformed current distribution and $(\vec{\tilde{E}}_z, \vec{\tilde{E}}_x)$ the transformed tangential electric field at the interfaces.

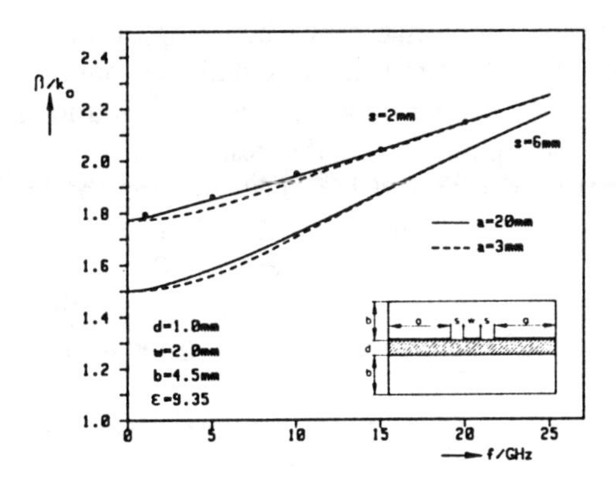

Fig. 3. Normalized phase constant β/k_0 versus frequency for the fundamental mode of the coplanar waveguide; ●... [8].

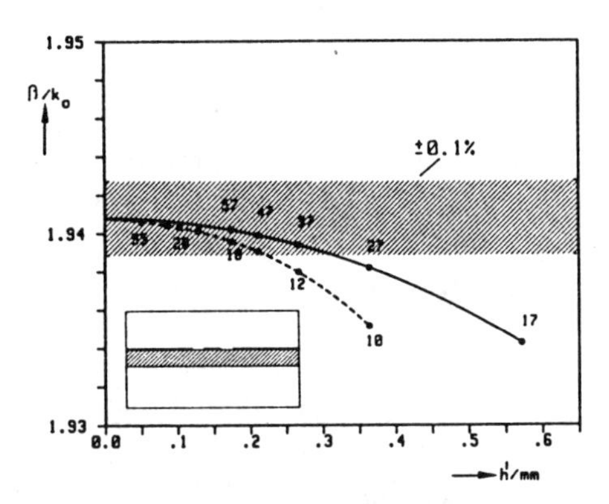

Fig. 4. Convergence behavior of the propagation constant as a function of the interval size h' for the center conductor; structural data as in Fig. 3, $f = 10$ GHz, $a = 7$ mm, $s = 2$ mm.

From (20), a reduced matrix equation in the original domain is derived:

$$[Z(\beta)]\begin{bmatrix}\vec{J}_z\\ \vec{J}_x\end{bmatrix}_{\text{strip}} = \begin{bmatrix}\vec{E}_z\\ \vec{E}_x\end{bmatrix}_{\text{strip}} = 0. \tag{21}$$

The propagation constants are found from the corresponding determinantal equation. By the Gaussian algorithm, one obtains the current distributions, which represent the current densities per interval. A simple multiplication yields the current density per unit length

$$J'_{zi} = e_i J_{zi}$$
$$J'_{xi} = h_i J_{xi}. \tag{22}$$

III. Results and Discussion

The method presented has been applied to the coplanar waveguide. In Fig. 3, the dispersion characteristics are given for different slot-widths and distances of the lateral shielding. As can be seen, the propagation constant is mainly determined by the slot-width. However, in the lower frequency range and for decreasing slot-width, the influence of the lateral shielding cannot be neglected.

The convergence behavior of the propagation constant as a function of the smallest interval size is illustrated in Fig. 4. Equidistant and nonequidistant discretization have the same curve of convergence (drawn curve), as long as the discretization between the symmetry wall and the outer slot-edge is equidistant. The discretization of the remaining distance 'a' has only little effect. However, if in the above-named region of high field concentration a nonequidistant discretization is chosen, the curve of convergence changes; for the dashed curve, the interval size at the outer edge is twice that near the symmetry wall. At the marked points, the total number of ψ^e-lines, needed for half the waveguide, is indicated. The numbers at the drawn curve refer to the equidistant case.

Finally, the distribution of the electric field in the slot and of the surface current on the strips is depicted in Fig. 5. The discretization corresponds to that of the dashed curve in Fig. 4. Near the strip-edges the fields vary rapidly, so that a fine discretization is chosen there. Exterior to these regions of high energy concentration, the functions are smooth. Hence, a coarse discretization is adequate. The

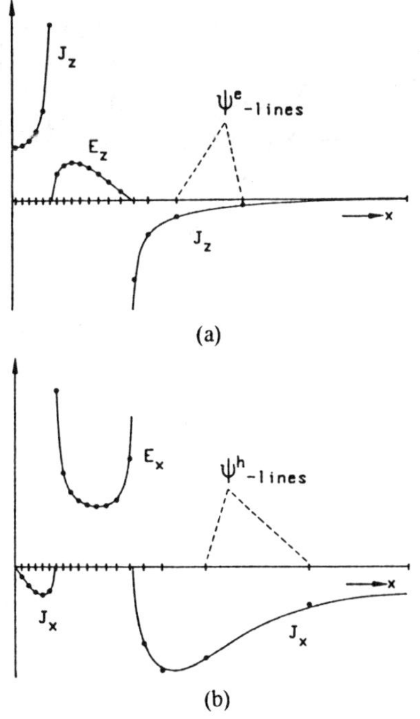

Fig. 5. Distribution of the electric field in the slot and the surface current on the strips at $f = 10$ GHz for the coplanar waveguide. The values from the nonequidistant discretization are given by dots ($N_e = 18$), those from the equidistant discretization are located on the drawn curves ($N_e = 57$); $J_z/J_x \approx 14$, $E_x/E_z \approx 14$.

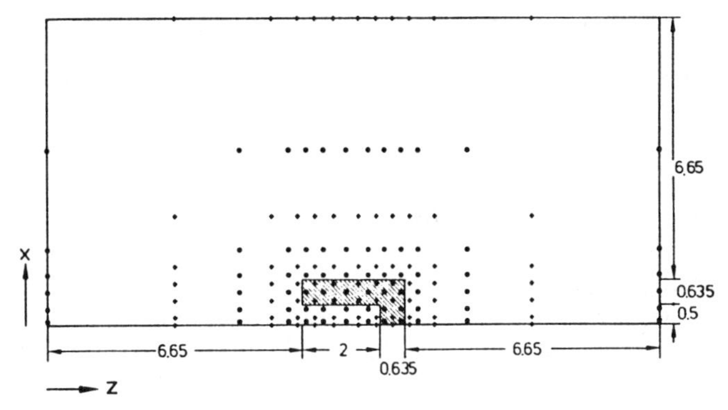

Fig. 6. Discretization lines for the hair-pin resonator (top view): + for ψ^e-lines, ● for ψ^h-lines; $h' \approx 0.42$ mm (cf. Fig. 8).

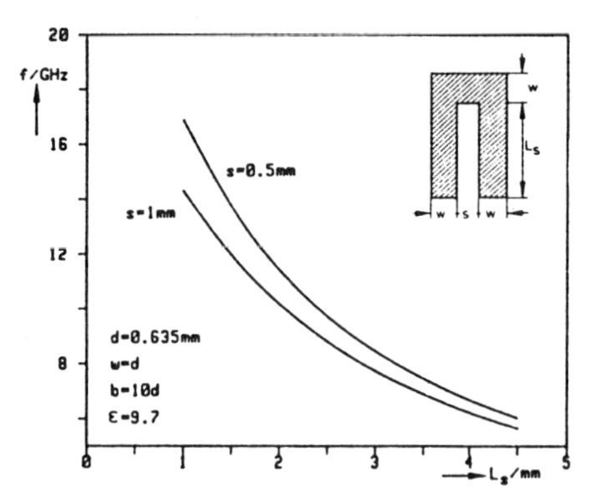

Fig. 7. Resonant frequency of a hair-pin resonator versus the stub length L_s for different spacings s.

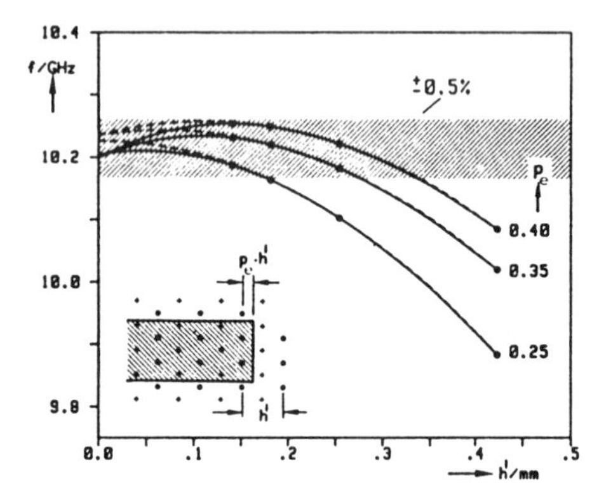

Fig. 8. Convergence behavior of the resonant frequency as a function of the interval size h' at the open end of the resonator; --- parabolic interpolation, ——— cubic interpolation; structural data as in Fig. 7, $L_s = 2$ mm, $s = 1$ mm.

tick-marks on the x-axis indicate the positions of the ψ^e-lines (Fig. 5(a)) and of the ψ^h-lines (Fig. 5(b)), respectively, which are shifted to each other. The corresponding values for the field components (dots) deviate only slightly from the more accurate values of the drawn curves, which are calculated by the equidistant discretization.

The advantages of nonequidistant discretization become particularly evident for two-dimensional problems. In Fig. 6, the pattern of lines for a hair-pin resonator is shown. As the results depend mainly on the accurate incorporation of the edges, fine and equidistant discretization is chosen in the vicinity of the contour. Exterior to the conductor, the fields are evanescent and smooth, so that a coarse discretization is advantageous. Inside the bordered region, 55 ψ^e-lines are located. The equidistant discretization, $h = h'$, would need 629 ψ^e-lines without giving more accurate results.

The computation time depends on the total number of lines, in particular on the number of lines that pass through the conducting structure and that determines the order of the reduced matrix (cf. (21)). By nonequidistant discretization, the computational effort can be reduced by a factor $4 \cdots 10$, compared with the effort in case of equidistant discretization according to [4].

For different spacings 's' the resonant frequency as a function of the stub length is shown in Fig. 7.

Fig. 8 gives the convergence behavior of the resonant frequency. As illustrated in the detail, the parameter 'p_e' determines the spacing between the ψ^h-lines and the conductor at the open end of the resonator. The (marked) results that correspond to $h' \simeq 0.42$ mm are obtained with the pattern of lines from Fig. 6. For h' less than 0.17, all results are within the margins of error of 0.5 percent. It has to be emphasized that an equidistant discretization would, in this range of interval sizes, overstress even large computers.

The two sets of curves represent parabolic (dashed) and cubic (drawn) curves of interpolation, respectively. In case of parabolic interpolation, for each 'p_e' only three values of the interval size h' are taken into account. For verification, a fourth value is calculated with finer discretization. As can be recognized, these points are located fairly well on the corresponding curves of interpolation. The extrapolated results of the dashed curves are in good agreement.

In case of cubic interpolation, all marked points of each 'p_e'-value are included. The extrapolated results, which are also in good agreement, differ less than 0.5 percent from those of the parabolic curves.

IV. Hints for Computation

As has been mentioned above, the discretization should follow the change of the field components. In the vicinity of conductor-edges, equidistant and fine discretization is adequate. Exterior to these regions, the interval sizes may increase. In order to obtain a uniform error distribution for the two scalar potentials, successive interval sizes should not differ too much.

This is easily achieved, if at first the sub-intervals between the ψ^e-lines and the ψ^h-lines are determined, e.g., as a geometrical series where the quotient of successive sub-intervals is a constant 'q', and after that the ψ^e-lines and the ψ^h-lines are assigned alternately. Each interval size, e_i or h_i, is composed of two sub-intervals.

The advantages of the geometrical series are easily shown: if the conductor-width of the coplanar waveguide (detail of Fig. 3) is designated by 'a' and the size of the sub-interval at the strip-edge by 'h_e', the following relation holds:

$$a = h_e \frac{q^M - 1}{q - 1}$$

where 'M' is the sum of the ψ^e- and the ψ^h-lines between the edge and the lateral shielding.

The quotient 'q' should not exceed the range $1 < q < 1.5$.

V. Conclusions

The method of lines as presented in this paper is highly adapted to planar waveguide problems. As the accuracy of the solutions mainly depends on the incorporation of the fields near the strip-conductor edges, the discretization should be fine in these regions. Exterior to the contours the fields are smooth, so that a coarse discretization is adequate. A development of the method to the inhomogeneous (source-type) waveguide problem, as presented in [9] for equidistant discretization, is possible. The difficulty of positioning the sources in a sufficiently large distance from discontinuities is reduced considerably.

In principle, the method presented includes the possibility for calculating planar stripline structures, where the permittivity of the substrate is given by $\epsilon = \epsilon(x)$. In that case, the partial differential equations for the scalar potentials are of the Sturm–Liouville type [6].

REFERENCES

[1] B. P. Demidowitsch, *et al.*, *Numerical Methods of Analysis*, (in German). Berlin: VEB-Verlag, 1968, ch. 5.

[2] O. A. Liskovets, "The method of lines (Review)," *Differentsial'nye Uravneniya*, vol. 1, no. 12, pp. 1662–1678, 1965.

[3] U. Schulz and R. Pregla, "A new technique for the analysis of the dispersion characteristics of planar waveguides and its application to microstrips with tuning septums," *Radio Sci.*, vol. 16, no. 6, pp. 1173–1178, 1981.

[4] S. B. Worm and R. Pregla, "Hybrid mode analysis of arbitrarily shaped planar microwave structures by the method of lines," *IEEE Trans. Microwave Theory Tech.*, vol. MTT-32, pp. 191–196, Feb. 1984.

[5] U. Schulz, "On the edge condition with the method of lines in planar waveguides," *Arch. Elek. Übertragung*, vol. 34, pp. 176–178, 1980.

[6] H. Diestel, "A method for calculating inhomogeneous planar dielectric waveguides" (in German), Ph.D. thesis, Fernuniversitaet Hagen, 1984.

[7] R. S. Martin and J. H. Wilkinson, "The implicit *QL*-algorithm," *Numer. Math.*, vol. 12, pp. 377–383, 1968.

[8] E. Yamashita and K. Atsuki, "Analysis of microstrip-like transmission lines by nonuniform discretization of Integral equations," *IEEE Trans. Microwave Theory Tech.*, vol. MTT-24, pp. 195–200, Apr. 1976.

[9] S. B. Worm, "Analysis of planar microwave structures with arbitrary contour" (in German), Ph.D. thesis, Fernuniversitaet Hagen, 1983.

Full wave analysis of complex planar microwave structures

Wilfrid Pascher and Reinhold Pregla

Fachgebiet Allgemeine und Theoretische Elektrotechnik, FernUniversität Hagen, Federal Republic of Germany

(Received December 8, 1986; revised February 10, 1987; accepted February 24, 1987.)

For the efficient analysis of discontinuities in planar circuits a fast computer algorithm is needed based on an uncomplicated theory. In the past the method of lines was successfully applied to planar waveguiding structures and simple two-dimensional cases. In this paper we adapt the method of lines to structures requiring a high number of discretization lines. The two-dimensional discretization and transformation of the Helmholtz equation into the spectral domain are reformulated in an elegant way using the Kronecker product of two matrices. A fast algorithm for the solution of the characteristic equation is developed for periodic structures employing the inversion of block Toeplitz matrices. Any microstrip, finline, or slotline circuit or discontinuity which is composed of several rectangular patches of metallization can be treated in this way. The current distribution of a periodic microstrip step discontinuity is given.

1. INTRODUCTION

For the design of microwave integrated circuits a characterization of discontinuities is necessary. These two-dimensional components require an approach, which is analytically uncomplicated and numerically efficient. The method of lines takes advantage of the planar structure by discretizing the wave equation in all directions except the direction which is transverse to the substrate. After a transformation to spectral domain an analytical solution is obtained for the last direction. Consequently, the method of lines is more accurate than the finite difference method and requires less time to compute, e.g., the characteristic constants. On the other hand, no prior experiences with expansion functions as in the Galerkin method are necessary.

The method of lines was first applied to planar waveguiding structures [Schulz and Pregla, 1980, 1981]. With a two-dimensional discretization [Worm, 1983], resonators, periodic structures [Worm and Pregla, 1984] and discontinuities were treated. A nonequidistant discretization was introduced [Diestel, 1984] and applied to various planar structures [Diestel and Worm, 1984] to increase the flexibility of the method, also resulting in greater efficiency.

In this paper another improvement of the method of lines is presented. The special properties of periodic two-dimensional structures are elaborated in order to optimize the computation of the eigenvalues. For this purpose we first consider difference operators and transformation matrices for a general two-dimensional structure and present a new, clear mathemathical formulation based on the Kronecker product. For the equidistant periodic case it is shown that the characteristic matrix consists of block Toeplitz submatrices. This matrix structure is used to introduce a fast inversion algorithm. As an example the current distribution on a periodic microstrip step discontinuity (Figure 1) is computed.

Paper number 7S0523.

2. FOUNDATIONS OF THE METHOD OF LINES

Apart from its numerical efficiency, one of the advantages of the method of lines is its easy formulation. We demonstrate it in five basic steps for a symmetrical microstrip line. A more detailed treatment is given by Schulz and Pregla [1981].

First step: Discretization

The electromagnetic field is described by the scalar potentials ψ^e and ψ^h, which fulfill the Helmholtz equation

$$\frac{\delta^2\psi^{e,h}}{\delta x^2} + \frac{\delta^2\psi^{e,h}}{\delta y^2} + (k^2 - \beta^2)\psi^{e,h} = 0 \qquad (1)$$

where the dependence $\exp(j(\omega t - \beta z))$ has been assumed. We discretize the x direction and get for the first derivative

$$\left.\frac{\delta\psi^e}{\delta x}\right|_i \approx \frac{\psi^e_{i+1} - \psi^e_i}{h} \longrightarrow \frac{1}{h} D_x \vec{\psi}^e \qquad (2)$$

where the difference operator D_x is given in Table 1. With the second derivative

$$h^2\frac{\delta^2\psi^e}{\delta x^2} \longrightarrow -D_x^t D_x \vec{\psi}^e = D_{xx}\vec{\psi}^e \qquad (3)$$

Reprinted with permission from *Radio Sci.*, vol. 22, no. 6, pp. 999–1002, Nov. 1987.

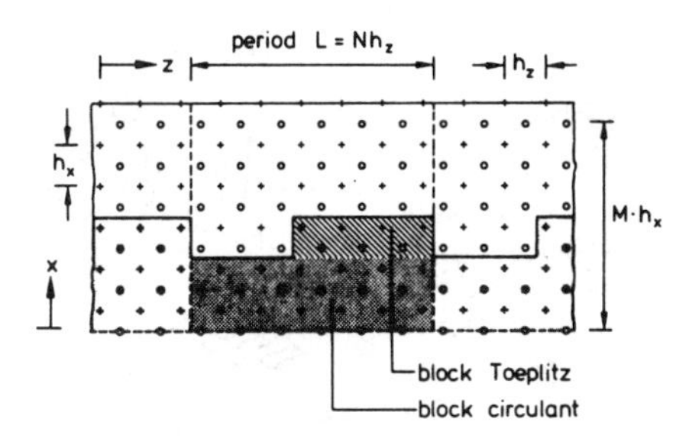

Fig. 1. Position of the discretization lines for a typical periodic structure (pluses are ψ^e and circles are ψ^h lines).

we replace (1) by its discretized form, arriving at a set of ordinary differential equations in y with a tridiagonal coefficient matrix. For the second potential ψ^h, D_x is replaced by $-D_x^*$, where the asterisk stands for complex conjugate transposed.

Second step: Transformation

To solve the system we diagonalize D_{xx}:

$$T_x\, D_{xx}\, T_x^* \;=\; \Lambda_x^2 \tag{4}$$

where the transformation matrix T_x (Table 1) is different for ψ^e and ψ^h. The potentials and the field components are transformed like

$$T_x\, \vec{\psi} \;=\; \tilde{\psi} \tag{5}$$

Third step: Field matching on the interfaces

In the continuity conditions we discretize and transform E_z, H_x, and J_z as ψ^e (using T^e) and E_x, H_z, and J_x as ψ^h (using T^h). After some algebraic manipulations we get

$$j\eta_0\bar{h}_x \begin{bmatrix} \tilde{E}_z \\ \tilde{E}_x \end{bmatrix} = \begin{bmatrix} \tilde{Z}_{11} & \tilde{Z}_{12} \\ \tilde{Z}_{12}^* & \tilde{Z}_{22} \end{bmatrix} \begin{bmatrix} \tilde{J}_z \\ \tilde{J}_x \end{bmatrix} \tag{6}$$

where $\tilde{Z}_{ik}$ are diagonal matrices and $\tilde{E}_z$, $\tilde{E}_x$, $\tilde{J}_z$, $\tilde{J}_x$ are written in vector form; $\bar{h}_x \;=\; k_0 h_x$.

Fourth step: The inverse transform

The tangential electric fields must vanish on the metallized part (reduced part) of the interface. This condition can be applied in the spatial domain only, so we compute the inverse transform for the reduced part:

$$\begin{bmatrix} Z_{11} & Z_{12} \\ Z_{12}^* & Z_{22} \end{bmatrix}_{red} \begin{bmatrix} \vec{J}_z \\ \vec{J}_x \end{bmatrix}_{red} = j\eta_0\bar{h}_x \begin{bmatrix} \vec{E}_z \\ \vec{E}_x \end{bmatrix}_{red} = \begin{bmatrix} \vec{0} \\ \vec{0} \end{bmatrix} \tag{7}$$

where, e.g.,

$$Z_{11} \;=\; T_x^{e*}\, \tilde{Z}_{11} T_x^e \tag{8}$$

Fifth step: The characteristic equation

For nontrivial solutions we have

$$det(Z_{red}(\omega, \beta)) \;=\; 0 \tag{9}$$

The wave impedance Z_0 is determined using the eigenvectors $\vec{J}_z$ and $\vec{J}_x$.

3. TWO-DIMENSIONAL STRUCTURES

To use the above five steps for two-dimensional structures, we replace D_x and T_x by two-dimensional operators in (2) to (8), namely, by means of the Kronecker product. If A and B are $N \times N$ and $M \times M$ matrices, respectively, the Kronecker product is a $NM \times NM$ matrix defined by

$$A \otimes B \;=\; \begin{bmatrix} a_{11}B & \dots & a_{1N}B \\ \vdots & & \vdots \\ a_{N1}B & \cdots & a_{NN}B \end{bmatrix} \tag{10}$$

First step: Discretization

The considered region is discretized as shown in Figure 1. The following formulas are equally valid for resonators, periodic structures and nonequidistant discretization, but with different $D_{x,z}$ and $T_{x,z}$. This approach allows any combination of these cases. The ψ^e-lines are linearly numbered by

$$n \;=\; i + (k-1)M \qquad 1 \le i \le M \qquad 1 \le k \le N \tag{11}$$

TABLE 1. Difference Operators and Transformation Matrices for the x Direction (With Magnetic/Electric Walls) and z Direction (With Periodic Boundaries) as in Figure 1

	x	z
D	$\begin{bmatrix} -1 & 1 & & \\ & \ddots & \ddots & \\ & & \ddots & 1 \\ & & & -1 \end{bmatrix}$	$\begin{bmatrix} -s & s^* & & \\ & \ddots & \ddots & \\ & & \ddots & s^* \\ s^* & & & -s \end{bmatrix}$
T_{ik}^e	$\sqrt{\frac{2}{M+1/2}}\cos(i-\frac{1}{2})\alpha_k$	$\frac{1}{\sqrt{N}}e^{ji\alpha_k}$
T_{ik}^h	$\sqrt{\frac{2}{M+1/2}}\sin i\alpha_k$	$\frac{1}{\sqrt{N}}e^{j(i+1/2)\alpha_k}$
α_k	$\frac{(k-1/2)\pi}{M+1/2}$	$\frac{(k-1)2\pi}{N}$

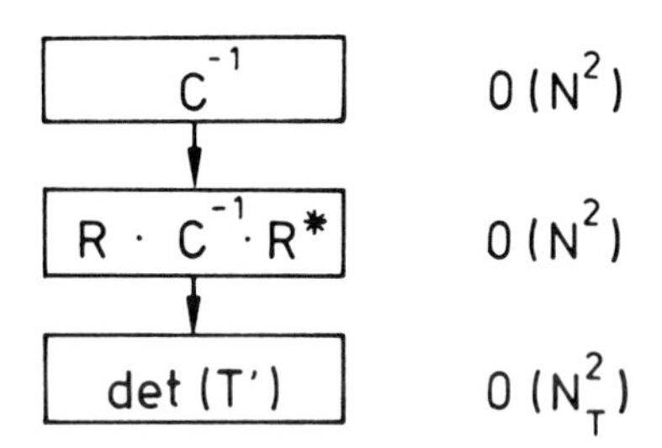

Fig. 2. Flow chart for the solution of the characteristic equation with number of necessary multiplications.

Thus $\vec{\psi}$ has the dimension MN. The two-dimensional difference operators are

$$D_x \longrightarrow \hat{D}_x = I_N \otimes D_x \qquad (12)$$

$$D_z \longrightarrow \hat{D}_z = D_z \otimes I_M \qquad (13)$$

where D_z (Table 1) is derived like in (1) and I_M and I_N are identity matrices of a dimension of M and N, respectively. $\hat{D}_{xx}$ and $\hat{D}_{zz}$ are constructed in the same way as $\hat{D}_x$ and $\hat{D}_z$.

Second step: Transformation

$$T_x \longrightarrow \hat{T} = T_z \otimes T_x \qquad (14)$$

Third step: Field matching

Now the $\tilde{Z}_{ik}$ are block diagonal, e.g.,

$$\tilde{Z}_{11} = diag\,(\tilde{A}_1, \ldots, \tilde{A}_N) \qquad (15)$$

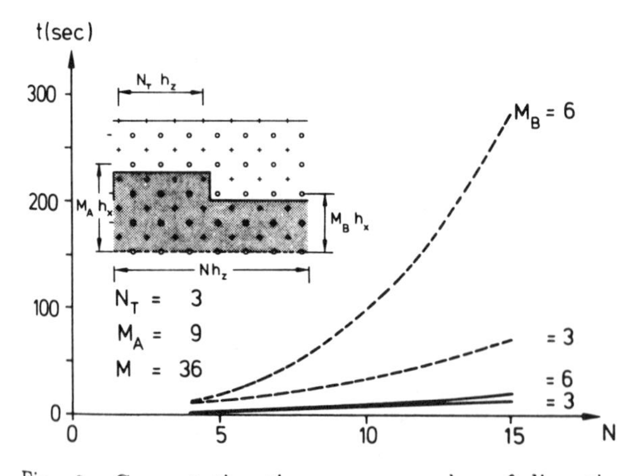

Fig. 3. Computation time versus number of discretization lines in longitudinal (z) direction. Dashed line, old algorithm (Hermitian inversion); solid line, new algorithm (block circulant inversion).

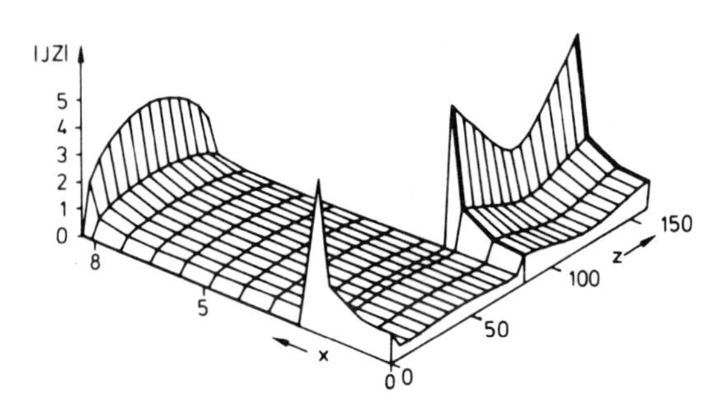

Fig. 4. Current distribution for a periodic microstrip step discontinuity as in Figure 3 ($f = 1GHz$, $L = 160mm$, $w_A = M_A h_x = 8.7mm$; $N = 27$, $N_T = 14$, $M = 14$, $M_A = 11$. $M_B = 3$; substrate thickness $d = 1.58mm$, height of air $h = 8.58mm$, dielectric constant $\epsilon_r = 2.32$).

Fourth step: The inverse transform

Using the properties of block diagonal matrices and of the Kronecker product, we get Z_{11} as result of two successive inverse transformations:

$$Z_{11} = \hat{T}_z^{*}\; diag\,(T_x^{*} \tilde{A}_1 T_x \ldots T_x^{*} \tilde{A}_N T_x)\; \hat{T}_z \qquad (16)$$

where

$$\hat{T}_z = T_z \otimes I_x \qquad (17)$$

Fifth step: The characteristic equation

For the case of resonators, *det* only depends on ω.

4. PERIODIC STRUCTURES: THE STEP DISCONTINUITY

To get periodic potentials $\vec{\psi}^e$ and $\vec{\psi}^h$ and field quantities, they are phase-normalized by S^e and sS^e, respectively.

$$s_{ii}^e = s^{2i} \qquad (18)$$

$$s = e^{j\beta h_z/2} \qquad (19)$$

This results in a more convenient matrix structure. D_z (Table 1) is circulant, which means that the elements of each row of D_z are identical to those in the previous row, but are moved one position to the right and wrapped around [Davies, 1979]. Hence T_z is a Fourier matrix and Z_{11} is block circulant. The same relations apply for the other submatrices Z_{ik}.

The idea of the fast algorithm is demonstrated for Z_{11} in the case of a step discontinuity (Figure 1). Discard of rows and columns gives $Z_{11,red}$:

$$Z_{11,red} = \begin{pmatrix} C & R^{*} \\ R & T \end{pmatrix} \qquad (20)$$

C is a block circulant, R is rectangular block circulant,

and T is a block Toeplitz matrix. In a Toeplitz matrix every diagonal parallel to the principal diagonal consists of identical elements. The circulant corresponds to the long rectangle in Figure 1, the Toeplitz matrix to the short one, and the rectangular block circulant to the interaction between both of them. We compute

$$det\ (Z_{11,r,d}) = det\ (C) \cdot det\ (T') \tag{21}$$

where

$$T' = T - R \cdot C^{-1} \cdot R' \tag{22}$$

T' has the same structure as T, because products and inverses of block circulants are block circulants again [Davies, 1979]. The time consumption of the whole algorithm (Figure 2) is $O(N^2)$ multiplications [Akaike, 1973; Smith, 1977], whereas the well known determinant of a Hermitian matrix H needs $O(N^3)$. The difference can be seen in Figure 3. For a broad microstrip line ($M_B = 6$) it reaches a factor of 13 for $N = 15$.

The current distribution J_z is computed [Worm, 1983] and plotted in Figure 4. Linear interpolation on the transversal edges and an extrapolation near the longitudinal edges according to the edge condition [Schulz, 1980b] were used to get additional data points.

5. SUMMARY

The method of lines has been improved for two-dimensional planar microwave structures. For this purpose the two-dimensional difference operators and transformation matrices have been represented as Kronecker products of the one-dimensional operators and identity matrices. For the case of periodic structures a new fast algorithm for the solution of the characteristic equation has been demonstrated for the microstrip step discontinuity. The computation time was reduced from $O(N^3)$ to $O(N^2)$ for nearly homogeneous structures and for those requiring a high number of discretization lines in longitudinal direction. The current distribution was given.

Acknowledgements. The work reported in this paper was supported by Deutsche Forschungsgemeinschaft.

REFERENCES

Akaike, H., Block Toeplitz matrix inversion, *SIAM J. Appl. Math.*, *24*(2), 234-241, 1973.

Davies, P. J., *Circulant Matrices*, 250 pp., John Wiley, New York, 1979.

Diestel, H., A method for calculating inhomogeneous planar dielectric waveguides (in German), Ph.D. thesis, FernUniv., Hagen, Federal Republic of Germany, 1984.

Diestel, H., and S. B. Worm, Analysis of hybrid field problems by the method of lines with nonequidistant discretization. *IEEE Trans. Microwave Theory Techniques*, *MTT-32*(6), 633-638, 1984.

Schulz, U., On the edge condition with the method of lines in planar waveguides. *Arch. Elektron. Uebertragungstech.*, *34*, 176-178, 1980.

Schulz, U., and R. Pregla, A new technique for the analysis of the dispersion characteristics of planar wavequides, *Arch. Elektron. Uebertragungstech.*, *34*, 169-173, 1980.

Schulz, U., and R. Pregla, A new technique for the analysis of the dispersion characteristics of planar waveguides and its application to microstrips with tuning septums, *Radio Sci.*, *16*(6), 1173-1178, 1981.

Smith, R. L., Moore-Penrose inverses of block circulant and block k-circulant matrices, *Lin. Alg. Appl.*, *16*, 237-245, 1977.

Worm, S. B., Analysis of planar microwave structures with arbitrary contour (in German), Ph.D. thesis, FernUniv., Hagen, Federal Republic of Germany, 1983.

Worm, S. B., and R. Pregla, Hybrid mode analysis of arbitrarily shaped planar microwave structures by the method of lines, *IEEE Trans. Microwave Theory Tech.*, *MTT-32*(2), 191-196, 1984.

W. Pascher and R. Pregla, Fachgebiet Allgemeine und Theoretische Elektrotechnik, FernUniversitaet Hagen, Postfach 9 40, D 5800 Hagen, Federal Republic of Germany.

Part 7
Method of Moments

THE method of moments (MOM) is a general procedure to convert the analytical formulation of a field problem into a numerical formulation in the form of a matrix equation. The procedure is very similar to that of the so-called method of weighted residuals, although the basic viewpoint may be slightly different.

Many numerical methods may be considered as special forms of the MOM, as explained in the two tutorial and overview papers reprinted here (Papers 7.1 and 7.2) which serve as an introduction to the method. While the general procedure to produce a linear system of equations from the field formulation of the problem is common to various methods, each specific application has its particular features, so that the actual details of the solutions vary substantially.

The first paper (Paper 7.1) is a classic one by Harrington. It states the fundamentals of the MOM and gives a number of examples of applications of the method, from electrostatic problems to wire scatterers. It is followed by a paper by Ney (Paper 7.2) which appeared in the October 1985 Special Issue on Numerical Methods of the *IEEE Transactions on Microwave Theory and Techniques*. The different specializations of the MOM are reviewed. It is pointed out, in particular, that appropriate techniques must be applied to solve the matrix problem, once it has been formulated. Two important methods for solving large systems of equations, the conjugate gradient method (CGM) and the pseudo-inverse technique, are briefly described. An overview on such numerical aspects can be found in [7.1], while more detailed treatments on the CGM are given in [7.2].

The method of moments is generally applied to the solution of an integral equation. It has been used in many instances for the solution of discontinuity problems in microstrip lines, generally under the quasi-static approximation. Silvester and Benedek [7.3] and Gopinath and Easter [7.4] used the MOM to compute the capacitances and inductances of microstrip discontinuities, respectively. The same approach is used in the reprinted paper by Anders and Arndt (Paper 7.3) where both capacitances and inductances of double steps, mitered bends, and right-angle bends are computed.

The method of moments is very popular for the solution of open field problems. It is presently recognized as the most powerful approach to the analysis of printed antenna configurations, and, with regard to the objective of this book, for characterizing radiation and coupling phenomena in printed circuit discontinuities. An enormous number of applications of the MOM can be found in the literature. Just two examples on the modeling of discontinuities are included here: A paper by Jackson and Pozar deals with microstrip open-end and gap discontinuities (Paper 7.4), and a very recent paper by Yang and Alexópoulos (Paper 7.5) is concerned with the characterization of a rather intricate discontinuity problem, the microstrip-to-slotline transition.

References

[7.1] T. K. Sarkar, K. R. Siarkowicz, and R. F. Stratton, "Survey of numerical methods for solution of large systems of linear equations for electromagnetic field problems," *IEEE Trans. Antennas Propagat.*, vol. AP-29, pp. 847–856, Nov. 1981.

[7.2] T. Sarkar, "The conjugate-gradient technique as applied to electromagnetic field problems," *IEEE AP-S Newsletter*, vol. 28, pp. 5–14, 1986.

[7.3] P. Silvester and P. Benedek, "Microstrip discontinuity capacitances for right-angle bends, T junctions, and crossings," *IEEE Trans. Microwave Theory Tech.*, vol. MTT-21, pp. 341–346, May 1973.

[7.4] A. Gopinath and B. Easter, "Moment method of calculating discontinuity inductance of microstrip right-angled bends," *IEEE Trans. Microwave Theory Tech.*, vol. MTT-22, pp. 880–883, Oct. 1974.

Matrix Methods for Field Problems

ROGER F. HARRINGTON, SENIOR MEMBER, IEEE

Abstract—**A unified treatment of matrix methods useful for field problems is given. The basic mathematical concept is the method of moments, by which the functional equations of field theory are reduced to matrix equations. Several examples of engineering interest are included to illustrate the procedure. The problem of radiation and scattering by wire objects of arbitrary shape is treated in detail, and illustrative computations are given for linear wires. The wire object is represented by an admittance matrix, and excitation of the object by a voltage matrix. The current on the wire object is given by the product of the admittance matrix with the voltage matrix. Computation of a field quantity corresponds to multiplication of the current matrix by a measurement matrix. These concepts can be generalized to apply to objects of arbitrary geometry and arbitrary material.**

I. Introduction

THE USE of high-speed digital computers not only allows one to make more computations than ever before, it makes practicable methods too repetitious for hand computation. In the past much effort was expended to analytically manipulate solutions into a form which minimized the computational effort. It is now often more convenient to use computer time to reduce the analytical effort. Almost any linear problem of analysis can be solved to some degree of approximation, depending upon the ingenuity and effort expended. In other words, the methods are known, but much work remains to be done on the details.

It is the purpose of this paper to give a brief discussion of a general procedure for solving linear field problems, and to apply it to some examples of engineering interest. The procedure is called a *matrix method* because it reduces the original functional equation to a matrix equation. The name *method of moments* has been given to the mathematical procedure for obtaining the matrix equations. Sometimes the procedure is called an approximation technique, but this is a misnomer when the solution converges in the limit. It is only the computational time for a given accuracy which differs from other solutions, as, for example, an infinite power series. Of course, the method can also be used for truly approximate solutions, that is, ones which do not converge in the limit.

The mathematical concepts are conveniently discussed in the language of linear spaces and operators. However, an attempt has been made to minimize the use of this language, so that readers unfamiliar with it may better follow the discussion. Those concepts which are used are defined as they are introduced. Detailed expositions of linear spaces and operators may be found in many textbooks [1]–[3].

In this paper, only equations of the inhomogeneous type

$$L(f) = g \qquad (1)$$

will be considered. Here L is a *linear operator*, g is the *excitation* or *source* (known function), and f is the *field* or *response* (unknown to be determined). The problem is said to be *deterministic* if the solution is unique, that is, if only one f is associated with each g. The problem of *analysis* involves determining f when L and g are given. The problem of *synthesis* involves determining L when f and g are specified. This paper deals only with analysis.

The method of moments gives a general procedure for treating field problems, but the details of solution vary widely with the particular problem. The examples of this paper have been chosen not only because they illustrate these details, but also because they are problems of engineering interest. It is hoped that these examples will allow the reader to solve similar problems, and also will suggest extensions and modifications suitable for other types of problems. While the examples are all taken from electromagnetic theory, the procedures apply to field problems of all kinds.

II. Formulation of Problems

Given a deterministic problem of the form (1), it is desired to identify the operator L, its domain (the functions f on which it operates), and its range (the functions g resulting from the operation). Furthermore, one usually needs an *inner product* $\langle f, g\rangle$, which is a scalar defined to satisfy[1]

$$\langle f, g\rangle = \langle g, f\rangle \qquad (2)$$

$$\langle \alpha f + \beta g, h\rangle = \alpha\langle f, h\rangle + \beta\langle g, h\rangle \qquad (3)$$

$$\begin{aligned}\langle f^*, f\rangle &> 0, \quad \text{if } f \neq 0\\ &= 0, \quad \text{if } f = 0\end{aligned} \qquad (4)$$

where α and β are scalars, and * denotes complex conjugate. The *norm* of a function is denoted $\|f\|$ and defined by

$$\|f\| = \sqrt{\langle f, f^*\rangle}. \qquad (5)$$

It corresponds to the Euclidean vector concept of length. The *metric* d of two functions is

$$d(f, g) = \|f - g\| \qquad (6)$$

Manuscript received September 15, 1966. This invited paper is one of a series planned on topics of general interest.—*The Editor*. This work was supported partly by Contract AF 30(602)-3724 from the Rome Air Development Center, Griffiss Air Force Base, N. Y., and partly by the National Science Foundation under Grant GK-704.

The author is with the Department of Electrical Engineering, Syracuse University, Syracuse, N. Y.

[1] The usual definition of inner product in Hilbert space corresponds to $\langle f^*, g\rangle$ in our notation. For this paper it is more convenient to show the conjugate operation explicitly wherever it occurs, and to define the adjoint operator without conjugation.

Reprinted from *Proc. IEEE,* vol. 55, no. 2, pp. 136–149, Feb. 1967.

and corresponds to the Euclidean vector concept of distance between two points. It is important for discussing the convergence of solutions.

Properties of the solution of (1) depend on properties of the operator L. The *adjoint operator* L^a and its domain are defined by

$$\langle Lf, g\rangle = \langle f, L^a g\rangle \tag{7}$$

for all f in the domain of L. An operator is *self adjoint* if $L^a = L$ and the domain of L^a is that of L. An operator is *real* if Lf is real whenever f is real. An operator is *positive definite* if

$$\langle f^*, Lf\rangle > 0 \tag{8}$$

for all $f \neq 0$ in its domain. It is *positive semidefinite* if $>$ is replaced by $\geq$ in (8), *negative definite* if $>$ is replaced by $<$ in (8), etc. Other properties of operators will be identified as they are needed.

If the solution to $L(f) = g$ exists and is unique for all g, then the *inverse operator* L^{-1} exists such that

$$f = L^{-1}(g). \tag{9}$$

If g is known, then (9) represents the solution to the original problem. However, (9) is itself an inhomogeneous equation for g if f is known, and its solution is $L(f) = g$. Hence, L and L^{-1} form a pair of operators, each of which is the inverse of the other.

Facility in formulating problems using the concepts of linear spaces comes only with practice, which will be provided by the examples in later sections. For the present, a simple abstract example will be considered, so that mathematical concepts may be illustrated without bringing physical concepts into the picture.

Example: Given $g(x)$, find $f(x)$ in the interval $0 \leq x \leq 1$ satisfying

$$-\frac{d^2 f}{dx^2} = g(x) \tag{10}$$

and

$$f(0) = f(1) = 0. \tag{11}$$

This is a boundary value problem for which

$$L = -\frac{d^2}{dx^2}. \tag{12}$$

The range of L is the space of all functions g in the interval $0 \leq x \leq 1$ which are being considered. The domain of L is the space of those functions f in the interval $0 \leq x \leq 1$, satisfying the boundary conditions (11), and having second derivatives in the range of L. The solution to (10) is not unique unless appropriate boundary conditions are included. In other words, both the differential operator and its domain are required to define the operator.

A suitable inner product for this problem is

$$\langle f, g\rangle = \int_0^1 f(x)g(x)\,dx. \tag{13}$$

It is easily shown that (13) satisfies the postulates (2) to (4), as required. Note that the definition (13) is not unique. For example,

$$\int_0^1 w(x)f(x)g(x)\,dx \tag{14}$$

where $w(x) > 0$ is an arbitrary weighting function, is also an acceptable inner product. However, the adjoint operator depends on the inner product, and it can often be chosen to make the operator self adjoint.

To find the adjoint of a differential operator, form the left-hand side of (7), and integrate by parts to obtain the right-hand side. For the present problem

$$\begin{aligned}\langle Lf, g\rangle &= \int_0^1 \left(-\frac{d^2 f}{dx^2}\right) g\,dx \\ &= \int_0^1 \frac{df}{dx}\frac{dg}{dx}\,dx - \left[\frac{df}{dx}g\right]_0^1 \\ &= \int_0^1 f\left(-\frac{d^2 g}{dx^2}\right)dx + \left[f\frac{dg}{dx} - g\frac{df}{dx}\right]_0^1.\end{aligned} \tag{15}$$

The last terms are boundary terms, and the domain of L^a may be chosen so that these vanish. The first boundary terms vanish by (11), and the second vanish if

$$g(0) = g(1) = 0. \tag{16}$$

It is then evident that the adjoint operator to (12) for the inner product (13) is

$$L^a = L = -\frac{d^2}{dx^2}. \tag{17}$$

Since $L^a = L$ and the domain of L^a is the same as that of L, the operator is self adjoint.

It is also evident that L is a real operator, since Lf is real when f is real. That L is a positive definite operator is shown from (8) as follows:

$$\begin{aligned}\langle f^*, Lf\rangle &= \int_0^1 f^*\left(-\frac{d^2 f}{dx^2}\right)dx \\ &= \int_0^1 \frac{df^*}{dx}\frac{df}{dx}\,dx - \left[f^*\frac{df}{dx}\right]_0^1 \\ &= \int_0^1 \left|\frac{df}{dx}\right|^2 dx.\end{aligned} \tag{18}$$

Note that L is a positive definite operator even if f is complex.

The inverse operator to L can be obtained by standard Green's function techniques.[2] It is

$$L^{-1}(g) = \int_0^1 G(x, x')g(x')\,dx' \tag{19}$$

where G is the Green's function

$$G(x, x') = \begin{cases} x(1 - x'), & x < x' \\ (1 - x)x', & x > x' \end{cases}. \tag{20}$$

[2] See, for example, Friedman [2], ch. 3.

One can verify that (19) is the inverse operator by forming $f=L^{-1}(g)$, differentiating twice, and obtaining (10). Note that no boundary conditions are needed on the domain of L^{-1}, which is characteristic of most integral operators. That L^{-1} is self adjoint follows from the proof that L is self adjoint, since

$$\langle Lf_1, f_2\rangle = \langle g_1, L^{-1}g_2\rangle. \tag{21}$$

Of course, the self-adjointness of L^{-1} can also be proved directly. It similarly follows that L^{-1} is positive definite whenever L is positive definite, and vice versa.

III. Method of Moments

A general procedure for solving linear equations is the *method of moments* [4]. Consider the deterministic equation

$$L(f) = g \tag{22}$$

where L is a linear operator, g is known, and f is to be determined. Let f be expanded in a series of functions $f_1, f_2, f_3, \cdots$ in the domain of L, as

$$f = \sum_n \alpha_n f_n \tag{23}$$

where the α_n are constants. The f_n are called *expansion functions* or *basis functions*. For exact solutions, (23) is usually an infinite summation and the f_n form a complete set of basis functions. For approximate solutions, (23) is usually a finite summation. Substituting (23) into (22), and using the linearity of L, one has

$$\sum_n \alpha_n L(f_n) = g. \tag{24}$$

It is assumed that a suitable inner product $\langle f, g\rangle$ has been determined for the problem. Now define a set of *weighting functions*, or *testing functions*, $w_1, w_2, w_3, \cdots$ in the range of L, and take the inner product of (24) with each w_m. The result is

$$\sum_n \alpha_n \langle w_m, Lf_n\rangle = \langle w_m, g\rangle \tag{25}$$

$m=1, 2, 3, \cdots$. This set of equations can be written in matrix form as

$$[l_{mn}][\alpha_n] = [g_m] \tag{26}$$

where

$$[l_{mn}] = \begin{bmatrix} \langle w_1, Lf_1\rangle & \langle w_1, Lf_2\rangle & \cdots \\ \langle w_2, Lf_1\rangle & \langle w_2, Lf_2\rangle & \cdots \\ \cdots & \cdots & \cdots \end{bmatrix} \tag{27}$$

$$[\alpha_n] = \begin{bmatrix} \alpha_1 \\ \alpha_2 \\ \vdots \end{bmatrix} \qquad [g_m] = \begin{bmatrix} \langle w_1, g\rangle \\ \langle w_2, g\rangle \\ \vdots \end{bmatrix}. \tag{28}$$

If the matrix $[l]$ is nonsingular its inverse $[l^{-1}]$ exists. The α_n are then given by

$$[\alpha_n] = [l_{nm}^{-1}][g_m] \tag{29}$$

and the solution for f is given by (23). For concise expression of this result, define the matrix of functions

$$[\tilde{f}] = [f_1 \;\; f_2 \;\; f_3 \;\; \cdots] \tag{30}$$

and write

$$f = [\tilde{f}_n][\alpha_n] = [\tilde{f}_n][l_{nm}^{-1}][g_m]. \tag{31}$$

This solution may be exact or approximate, depending upon the choice of the f_n and w_n. The particular choice $w_n=f_n$ is known as *Galerkin's method* [5], [6].

If the matrix $[l]$ is of infinite order, it can be inverted only in special cases, for example, if it is diagonal. The classical eigenfunction method leads to a diagonal matrix, and can be thought of as a special case of the method of moments. If the sets f_n and w_n are finite, the matrix is of finite order, and can be inverted by known computational algorithms.

One of the main tasks in any particular problem is the choice of the f_n and w_n. The f_n should be linearly independent and chosen so that some superposition (23) can approximate f reasonably well. The w_n should also be linearly independent and chosen so that the products $\langle w_n, g\rangle$ depend on relatively independent properties of g. Some additional factors which affect the choice of f_n and w_n are a) the accuracy of solution desired, b) the ease of evaluation of the matrix elements, c) the size of the matrix that can be inverted, and d) the realization of a well-conditioned matrix $[l]$.

Example: Consider again the problem stated by (10) and (11). For a power-series solution, choose

$$f_n = x^{n+1} - x \tag{32}$$

$n=1, 2, 3, \cdots, N$, so that the series (23) is

$$f = \sum_{n=1}^{N} \alpha_n(x^{n+1} - x). \tag{33}$$

Note that the term $-x$ is needed in (34), else the f_n will not be in the domain of L, that is, the boundary conditions will not be satisfied. For testing functions, choose

$$w_n = f_n = x^{n+1} - x \tag{34}$$

in which case the method is that of Galerkin. In Section V it is shown that the w_n should be in the domain of the adjoint operator. Since L is self adjoint for this problem, the w_n should be in the domain of L, as are those of (34). Evaluation of the matrix (27) for the inner product (13) and L given by (12) is straightforward. The resultant elements are

$$l_{mn} = \langle w_m, Lf_n\rangle = \frac{mn}{m+n+1}. \tag{35}$$

A knowledge of the matrix elements (35) is fully equivalent to the original differential equation. Hence, a matrix formulation for the problem has been obtained. For any particular excitation g, the matrix excitation $[g_m]$ has elements given by

$$g_m = \langle w_m, g\rangle = \int_0^1 (x^{m+1} - x)g(x)\,dx \tag{36}$$

and a solution to the boundary value problem is given by (31). This solution is a power series, exact if f can be expressed as a power series. In general, it is an infinite power-series solution, in which case a finite number of terms gives an approximate solution. The nature of the approximation is discussed in Section V.

IV. Special Techniques

As long as the operator equation is simple, application of the method of moments gives solutions in a straightforward manner. However, most field problems of engineering interest are not so simple. The physical problem may be represented by many different operator equations, and a suitable one must be chosen. Even then the form of L may be very complicated. There are an infinite number of sets of expansion functions f_n and testing functions w_n that may be chosen. Finally, there are mathematical approximations that can be made in the evaluation of the matrix elements of l_{mn} and g_m. In this section a number of special techniques, helpful for overcoming some of these difficulties, will be discussed in general terms. Some of these concepts will be used in the electromagnetic field problems considered later.

Point-Matching: The integration involved in evaluating the $l_{mn}=\langle w_m, Lf_n\rangle$ of (27) is often difficult to perform in problems of practical interest. A simple way to obtain approximate solutions is to require that (24) be satisfied at discrete points in the region of interest. This procedure is called a *point-matching method.* In terms of the method of moments, it is equivalent to using Dirac delta functions as testing functions.

Subsectional Bases: Another approximation useful for practical problems is the *method of subsections.* This involves the use of basis functions f_n each of which exists only over subsections of the domain of f. Then each α_n of the expansion (23) affects the approximation of f only over a subsection of the region of interest. This procedure often simplifies the evaluation and/or the form of the matrix $[l]$. Sometimes it is convenient to use the point-matching method of the preceding section in conjunction with the subsection method.

Extended Operators: As noted earlier, an operator is defined by an operation (for example, $L=-d^2/dx^2$) plus a domain (space of functions to which the operation may be applied). We can *extend the domain* of an operator by redefining the operation to apply to new functions (not in the original domain) so long as this extended operation does not change the original operation in its domain. If the original operator is self adjoint, it is desirable to make the extended operator also self adjoint. By this procedure we can use a wider class of functions for solution by the method of moments. This becomes particularly important in multivariable problems (fields in multidimensional space) where it is not always easy to find simple functions in the domain of the original operator.

Approximate Operators: In complex problems it is sometimes convenient to approximate the operator to obtain solutions. For differential operators, the finite difference approximation has been widely used [7]. For integral operators, an approximate operator can be obtained by approximating the kernel of the integral operator [5]. Any method whereby a functional equation is reduced to a matrix equation can be interpreted in terms of the method of moments. Hence, for any matrix solution using approximation of the operator there will be a corresponding moment solution using approximation of the function.

Perturbation Solutions: Sometimes the problem under consideration is only slightly different (perturbed) from a problem which can be solved exactly (the unperturbed problem). A first-order solution to the perturbed problem can then be obtained by using the solution to the unperturbed problem as a basis for the method of moments. This procedure is called a *perturbation method.* Higher-order perturbation solutions can be obtained by using the unperturbed solution plus correction terms in the method of moments. Sometimes this is done as successive approximations by including one correction term at a time, but for machine computations it is usually easier to include all correction terms at once.

V. Variational Interpretation

It is known that Galerkin's method ($w_n=f_n$) is equivalent to the Rayleigh-Ritz variational method [5], [6]. The method of moments is also equivalent to the variational method, the proof being essentially the same as that for Galerkin's method. The application of these techniques to electromagnetic field problems is known as the reaction concept [8], [9].

An interpretation of the method of moments in terms of linear spaces will first be given. Let $\mathcal{S}(Lf)$ denote the range of L, $\mathcal{S}(Lf_n)$ denote the space spanned by the Lf_n, and $\mathcal{S}(w_n)$ denote the space spanned by the w_n. The method of moments (25) then equates the projection of Lf onto $\mathcal{S}(w_n)$ to the projection of the approximate Lf onto $\mathcal{S}(w_n)$. In other words, both the approximate Lf and the exact Lf have equal components in $\mathcal{S}(w_n)$. The difference between the approximate Lf and the exact Lf is the error, which is orthogonal to $\mathcal{S}(w_n)$. Because of this orthogonality, a first-order change in the projection produces only a second-order change in the error. In Galerkin's method, $\mathcal{S}(w_n)=\mathcal{S}(f_n)$, and the distance from the approximate Lf to the exact Lf is minimized. In general, the method of moments does not minimize the distance from the approximate f to the exact f, although it may in some special cases.

The variational approach to the same problem is as follows. Given an operator equation $Lf=g$, it is desired to determine a functional of f (number depending on f)

$$\rho(f)=\langle f, h\rangle \tag{37}$$

where h is a given function. If h is a continuous function, then $\rho(f)$ is a *continuous linear functional.* Now let L^a be the adjoint operator to L and define an adjoint function f^a (adjoint field) by

$$L^a f^a = h. \tag{38}$$

By the calculus of variations, it can then be shown that [6]

$$\rho = \frac{\langle f, h\rangle\langle f^a, g\rangle}{\langle Lf, f^a\rangle} \tag{39}$$

is a variational formula for ρ with stationary point (37) when f is the solution of $Lf=g$ and f^a the solution to (38). For an approximate evaluation of ρ, let

$$f = \sum_n \alpha_n f_n \qquad f^a = \sum_m \beta_m w_m. \tag{40}$$

Substitute these into (39), and apply the Rayleigh-Ritz conditions $\partial\rho/\partial\alpha_i = \partial\rho/\partial\beta_i = 0$ for all i. The result is that the necessary and sufficient conditions for ρ to be a stationary point are (25), [6]. Hence, the method of moments is identical to the Rayleigh-Ritz variational method. Sometimes the method of moments is called a *direct method*, in contrast to variational approaches which are often rather circuitous.

The above variational interpretation can be used to give additional insight in how to choose the testing functions. It is evident from (38) and (40) that the w_n should be chosen so that some linear combination of them can closely represent the adjoint field f^a. When we calculate f itself by the method of moments, h of (37) is a Dirac delta function, ρ of (37) is no longer a continuous linear functional, and f^a of (38) is a Green's function. This implies that some combination of the w_n must be able to approximate the Green's function. Since a Green's function is usually poorly behaved, one should expect computation of a field by the method of moments to converge less slowly than computation of a continuous linear functional. This is found to be the case.

VI. Electrostatics

This section is a general discussion of electrostatic problems according to the operational formulation. The static electric intensity $\boldsymbol{E}$ is conveniently found from an electrostatic potential ϕ according to

$$\boldsymbol{E} = -\nabla\phi \tag{41}$$

where ∇ is the gradient operator. In a region of constant permittivity ε and volume change density ρ, the electrostatic potential satisfies the *Poisson equation*

$$-\varepsilon\nabla^2\phi = \rho \tag{42}$$

where ∇^2 is the Laplacian operator. For unique solutions, boundary conditions on ϕ are needed. In other words, the domain of the operator must be specified.

For now, consider fields from charges in unbounded space, in which case

$$r\phi \rightarrow \text{constant as } r \rightarrow \infty \tag{43}$$

for every ρ of finite extent, where r is the distance from the coordinate origin. The differential operator formulation is therefore

$$L\phi = \rho \tag{44}$$

where

$$L = -\varepsilon\nabla^2 \tag{45}$$

and the domain of L is those functions ϕ whose Laplacian exists and which have $r\phi$ bounded at infinity according to (43). The well-known solution to this problem is

$$\phi(x, y, z) = \iiint \frac{\rho(x', y', z')}{4\pi\varepsilon R}\, dx'dy'dz' \tag{46}$$

where $R = \sqrt{(x-x')^2+(y-y')^2+(z-z')^2}$ is the distance from a source point (x', y', z') to a field point (x, y, z). Hence, the inverse operator to L is

$$L^{-1} = \iiint dx'dy'dz' \frac{1}{4\pi\varepsilon R}. \tag{47}$$

It is important to keep in mind that (47) is inverse to (45) only for the boundary conditions (43). If the boundary conditions are changed, L^{-1} changes. Also, the designation of (45) as L and (47) as L^{-1} is arbitrary, and the notation could be reversed if desired.

A suitable inner product for electrostatic problems is

$$\langle\phi, \psi\rangle = \iiint \phi(x, y, z)\psi(x, y, z)\, dxdydz \tag{48}$$

where the integration is over all space. That (48) satisfies the required postulates (2), (3), and (4) is easily verified. It will now be shown that L is self adjoint for this inner product. From the left-hand side of (7)

$$\langle L\phi, \psi\rangle = \iiint (-\varepsilon\nabla^2\phi)\psi\, d\tau \tag{49}$$

where $d\tau = dxdydz$. Green's identity is

$$\iiint_V (\psi\nabla^2\phi - \phi\nabla^2\psi)\, d\tau = \oiint_S \left(\psi\frac{\partial\phi}{\partial n} - \phi\frac{\partial\psi}{\partial n}\right) ds \tag{50}$$

where S is the surface bounding the volume V and n is the outward direction normal to S. Let S be a sphere of radius r, so that in the limit $r\rightarrow\infty$ the volume V includes all space. For ϕ and ψ satisfying boundary conditions (43), $\psi\rightarrow C_1/r$, and $\partial\phi/\partial n\rightarrow C_2/r^2$ as $r\rightarrow\infty$. Hence, $\psi\partial\phi/\partial n\rightarrow C/r^3$ as $r\rightarrow\infty$, and similarly for $\phi\partial\psi/\partial n$. Since $ds = r^2 \sin\theta\, d\theta d\phi$ increases only as r^2, the right-hand side of (50) vanishes as $r\rightarrow\infty$. Equation (50) then reduces to

$$\iiint \psi\nabla^2\phi\, d\tau = \iiint \phi\nabla^2\psi\, d\tau \tag{51}$$

from which it is evident that the adjoint operator L^a is

$$L^a = L = -\varepsilon\nabla^2. \tag{52}$$

Since the domain of L^a is that of L, the operator L is self adjoint. The concept of self adjointness in this case is related to the physical concept of reciprocity.

It is evident from (45) and (47) that L and L^{-1} are real operators. It will now be shown that they are also positive definite, that is, they satisfy (8). As discussed in Section II, this need be shown only for L or L^{-1}. For L, form

$$\langle \phi^*, L\phi \rangle = \iiint \phi^*(-\varepsilon\nabla^2\phi)\,d\tau \tag{53}$$

and use the vector identity $\phi\nabla^2\phi = \nabla\cdot(\phi\nabla\phi) - \nabla\phi\cdot\nabla\phi$ plus the divergence theorem. The result is

$$\langle \phi^*, L\phi \rangle = \iiint_V \varepsilon\nabla\phi^*\cdot\nabla\phi\,d\tau - \oiint_S \varepsilon\phi^*\nabla\phi\cdot ds \tag{54}$$

where S bounds V. Again take S a sphere of radius r. For ϕ satisfying (43), the last term of (54) vanishes as $r\to\infty$ for the same reasons as in (50). Then

$$\langle \phi^*, L\phi \rangle = \iiint \varepsilon|\nabla\phi|^2\,d\tau \tag{55}$$

and, for ε real and $\varepsilon > 0$, L is positive definite. In this case positive definiteness of L is related to the concept of electrostatic energy.

VII. Charged Conducting Plate

Consider a square conducting plate 2a meters on a side and lying on the $z=0$ plane with center at the origin, as shown in Fig. 1. Let $\sigma(x, y)$ represent the surface charge density on the plate, assumed to have zero thickness. The electrostatic potential at any point in space is

$$\phi(x, y, z) = \int_{-a}^{a} dx' \int_{-a}^{a} dy' \frac{\sigma(x', y')}{4\pi\varepsilon R} \tag{56}$$

where $R = \sqrt{(x-x')^2 + (y-y')^2 + z^2}$. The boundary condition is $\phi = V$ (constant) on the plate. The integral equation for the problem is therefore

$$V = \int_{-a}^{a} dx' \int_{-a}^{a} dy' \frac{\sigma(x', y')}{4\pi\varepsilon\sqrt{(x-x')^2 + (y-y')^2}} \tag{57}$$

$|x| < a$, $|y| < a$. The unknown to be determined is the charge density $\sigma(x, y)$. A parameter of interest is the capacitance of the plate

$$C = \frac{q}{V} = \frac{1}{V}\int_{-a}^{a} dx \int_{-a}^{a} dy\,\sigma(x, y) \tag{58}$$

which is continuous linear functional of σ.

A straightforward development of a subsection and point-matching solution [10] will first be given, and later it will be interpreted in terms of more general concepts. Consider the plate divided into N square subsections, as shown in Fig. 1. Define functions

$$f_n = \begin{cases} 1 & \text{on } \Delta s_n \\ 0 & \text{on all other } \Delta s_m \end{cases} \tag{59}$$

and let the charge density be represented by

$$\sigma(x, y) \approx \sum_{n=1}^{N} \alpha_n f_n. \tag{60}$$

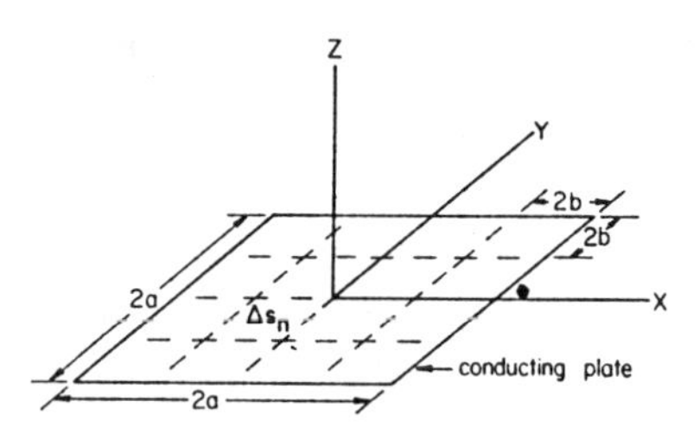

Fig. 1. A square conducting plate.

Substituting (60) into (57) and satisfying the resultant equation at the midpoint (x_m, y_m) of each Δs_m, one obtains the set of equations

$$V = \sum_{n=1}^{N} l_{mn}\alpha_n, \qquad m = 1, 2, \cdots, N \tag{61}$$

where

$$l_{mn} = \int_{\Delta x_n} dx' \int_{\Delta y_n} dy' \frac{1}{4\pi\varepsilon\sqrt{(x_m - x')^2 + (y_m - y')^2}}. \tag{62}$$

Note that l_{mn} is the potential at the center of Δs_m due to a uniform charge density of unit amplitude over Δs_n. A solution to the set (61) gives the α_m, in terms of which the charge density is approximated by (60). The corresponding capacitance of the plate, approximating (58), is

$$C \approx \frac{1}{V}\sum_{n=1}^{N} \alpha_n \Delta s_n = \sum_{mn} l_{nm}^{-1}\Delta s_n. \tag{63}$$

This result can be interpreted as stating that the capacitance of an object is the sum of the capacitances of all its subsections plus the mutual capacitances between every pair of subsections.

To translate the above results into the language of linear spaces and the method of moments, let

$$f(x, y) = \sigma(x, y) \tag{64}$$

$$g(x, y) = V, \quad |x| < a, \quad |y| < a \tag{65}$$

$$L(f) = \int_{-a}^{a} dx' \int_{-a}^{a} dy' \frac{f(x', y')}{4\pi\varepsilon\sqrt{(x-x')^2 + (y-y')^2}}. \tag{66}$$

Then $L(f) = g$ is equivalent to (57). A suitable inner product, satisfying (2) to (4), for which L is self adjoint, is

$$\langle f, g \rangle = \int_{-a}^{a} dx \int_{-a}^{a} dy\, f(x, y)\, g(x, y). \tag{67}$$

To apply the method of moments, use the function (59) as a subsectional basis, and define testing functions as

$$w_m = \delta(x - x_m)\delta(y - y_m) \tag{68}$$

which is the two-dimensional Dirac delta function. Now the elements of the $[l]$ matrix (27) are those of (62) and the $[g]$ matrix of (28) is

$$[g_m] = \begin{bmatrix} V \\ V \\ \vdots \\ V \end{bmatrix}. \tag{69}$$

The matrix equation (26) is, of course, identical to the set of equations (61). In terms of the inner product (67), the capacitance (58) can be written as

$$C = \frac{\langle \sigma, \phi \rangle}{V^2} \tag{70}$$

since $\phi = V$ on the plate. Equation (70) is the conventional stationary formula for the capacitance of a conducting body [11].

For numerical results, the l_{mn} of (62) must be evaluated. Let $2b = 2a/\sqrt{N}$ denote the side length of each Δs_n. The potential at the center of Δs_n due to unit charge density over its own surface is

$$\begin{aligned} l_{nn} &= \int_{-b}^{b} dx \int_{-b}^{b} dy \frac{1}{4\pi\varepsilon\sqrt{x^2 + y^2}} \\ &= \frac{2b}{\pi\varepsilon} \ln (1 + \sqrt{2}) = \frac{2b}{\pi\varepsilon} (0.8814). \end{aligned} \tag{71}$$

This derivation uses Dwight 200.01 and 731.2 [12]. The potential at the center of Δs_m due to unit charge over Δs_n can be similarly evaluated, but the formula is complicated. For most purposes it is sufficiently accurate to treat the charge on Δs_n as if it were a point charge, and use

$$l_{mn} \approx \frac{\Delta s_n}{4\pi\varepsilon R_{mn}} = \frac{b^2}{\pi\varepsilon\sqrt{(x_m - x_n)^2 + (y_m - y_n)^2}} \qquad m \neq n. \tag{72}$$

This approximation is 3.8 percent in error for adjacent subsections, and has less error for nonadjacent ones. Table I shows capacitance, calculated by (63) using the α's obtained from the solution of (61), for various numbers of subareas. The second column of Table I uses the approximation (72), the third column uses an exact evaluation of the l_{mn}. A good estimate of the true capacitance is 40 picofarads. Figure 2 shows a plot of the approximate charge density along the subareas nearest the center line of the plate, for the case $N = 100$ subareas. Note that σ exhibits the well-known square root singularity at the edges of the plate.

TABLE I

CAPACITANCE OF A SQUARE PLATE (PICOFARADS PER METER)

No. of subareas	$C/2a$ approx. l_{mn}	$C/2a$ exact l_{mn}
1	31.5	31.5
9	37.3	36.8
16	38.2	37.7
36	39.2	38.7
100		39.5

VIII. ELECTROMAGNETIC FIELDS

The operator formulation of electromagnetic fields is analogous to that of electrostatic fields, but considerably more complicated. For the time-harmonic case, $e^{j\omega t}$ variation, the Maxwell equations are[3]

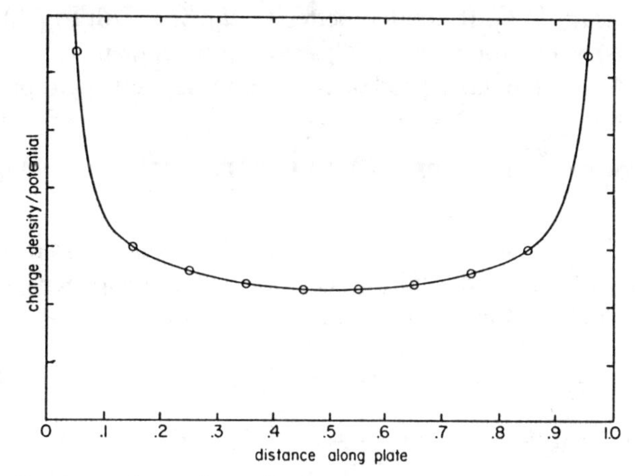

Fig. 2. Approximate charge density on subareas closest to the centerline of a square plate.

$$\begin{aligned} \nabla \times \boldsymbol{E} &= -j\omega\mu \boldsymbol{H} \\ \nabla \times \boldsymbol{H} &= j\omega\varepsilon \boldsymbol{E} + \boldsymbol{J} \end{aligned} \tag{73}$$

where $\boldsymbol{E}$ is the electric field, $\boldsymbol{H}$ the magnetic field, and $\boldsymbol{J}$ the electric current density. Equations (73) can be combined into a single equation for $\boldsymbol{E}$ as

$$\frac{-1}{j\omega} \nabla \times (\mu^{-1} \nabla \times E) - j\omega\varepsilon E = \boldsymbol{J}. \tag{74}$$

This is of the form

$$L(E) = \boldsymbol{J} \tag{75}$$

where the operator L is evident from (74). For a specific case, let the permittivity and permeability be that of free space, that is, $\varepsilon = \varepsilon_0$ and $\mu = \mu_0$. The domain of L must be restricted by suitable differentiability conditions on $\boldsymbol{E}$, and boundary conditions on $\boldsymbol{E}$ must be given. To be specific, let these boundary conditions be the radiation condition, that is, the field must represent outward traveling waves at infinity.

The inverse operator is the well-known potential integral solution to (74), which is

$$\boldsymbol{E} = L^{-1}(\boldsymbol{J}) = -j\omega \boldsymbol{A} - \nabla\Phi \tag{76}$$

where

$$\boldsymbol{A}(\boldsymbol{r}) = \mu \iiint \boldsymbol{J}(\boldsymbol{r}') \frac{e^{-jk|\boldsymbol{r}-\boldsymbol{r}'|}}{4\pi|\boldsymbol{r} - \boldsymbol{r}'|} d\tau' \tag{77}$$

$$\Phi(\boldsymbol{r}) = \frac{1}{\varepsilon} \iiint \rho(\boldsymbol{r}') \frac{e^{-jk|\boldsymbol{r}-\boldsymbol{r}'|}}{4\pi|\boldsymbol{r} - \boldsymbol{r}'|} d\tau' \tag{78}$$

$$\rho = \frac{-1}{j\omega} \nabla \cdot \boldsymbol{J}. \tag{79}$$

These equations can be combined into a single equation

$$\boldsymbol{E} = L^{-1}(\boldsymbol{J}) = \iiint \Gamma(\boldsymbol{r}, \boldsymbol{r}') \cdot \boldsymbol{J}(\boldsymbol{r}') d\tau' \tag{80}$$

where Γ is the dyadic Green's function. However, the derivation of (80) involves an interchange of integration and differentiation which restricts the domain of L^{-1} more than

[3] Only the case of electric sources is considered in this paper. The more general case of electric and magnetic sources is treated by the reaction concept [8], [9].

necessary [13]. It is often better to consider (76) to (78) as the basic equations, with (80) as symbolic of them.

A suitable inner product for electromagnetic field problems is

$$\langle \mathrm{E}, \mathrm{J}\rangle = \iiint \boldsymbol{E}\cdot\boldsymbol{J}\, d\tau \tag{81}$$

which is the quantity called *reaction*. Note that (81) satisfies postulates (2), (3), and (4). The concept of reciprocity is a statement of the self-adjointness of L^{-1}, that is,

$$\langle L^{-1}J_1, J_2\rangle = \langle J_1, L^{-1}J_2\rangle. \tag{82}$$

The operator L is also self adjoint, since (82) can be written as

$$\langle E_1, LE_2\rangle = \langle LE_1, E_2\rangle. \tag{83}$$

Other properties of L can be determined as the need arises.

IX. Wires of Arbitrary Shape

An important engineering problem is the electromagnetic behavior of thin wire objects. A general analysis of such objects according to the method of moments is presented in this section. The impressed field is considered arbitrary, and hence both the antenna and scatterer problems are included in the solution. The distinction between antennas and scatterers is primarily that of the location of the source. If the source is at the object it is viewed as an antenna; if the source is distant from the object it is viewed as a scatterer.

So that the development of the solution may be easily followed, it is given with few references to the general theory. Basically, it involves a) an approximation of the exact equation for conducting bodies by an approximate equation valid for thin wires, b) replacement of the derivatives by finite difference approximations, yielding an approximate operator, c) use of pulse functions for expansion functions, to give a step approximation to the current and charge, and d) the use of point-matching for testing.

A particularly descriptive exposition of the solution can be made in terms of network parameters. To effect a solution, the wire is considered as N short segments connected together. The end points of each segment define a pair of terminals in space. These N pairs of terminals can be thought of as forming an N port network, and the wire object is obtained by short-circuiting all ports of the network. One can determine the impedance matrix for the N port network by applying a current source to each port in turn, and calculating the open circuit voltages at all ports. This procedure involves only current elements in empty space. The admittance matrix is the inverse of the impedance matrix. Once the admittance matrix is known, the port currents (current distribution on the wire) are found for any particular voltage excitation (applied field) by matrix multiplication.

An integral equation for the charge density σ_s and current $\boldsymbol{J}_s$ on a conducting body S in a known impressed field $\boldsymbol{E}^i$ is obtained as follows. The scattered field $\boldsymbol{E}^s$, produced by σ_s and $\boldsymbol{J}_s$, is expressed in terms of retarded potential in-

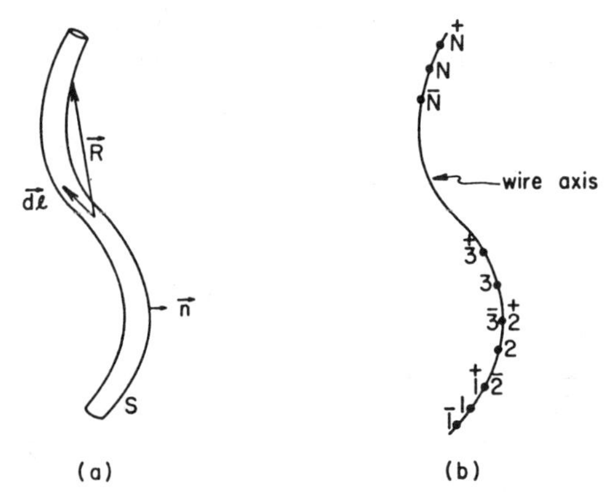

Fig. 3. (a) A wire scatterer. (b) The wire axis divided into N segments.

tegrals, and the boundary condition $\boldsymbol{n}\times(\boldsymbol{E}^i+\boldsymbol{E}^s)=0$ on S is applied. This is summarized by

$$\boldsymbol{E}^s = -j\omega\boldsymbol{A} - \nabla\phi \tag{84}$$

$$\boldsymbol{A} = \mu \oiint_S \boldsymbol{J}_s \frac{e^{-jkR}}{4\pi R}\, dS \tag{85}$$

$$\phi = \frac{1}{\varepsilon}\oiint_S \sigma_s \frac{e^{-jkR}}{4\pi R}\, dS \tag{86}$$

$$\sigma_s = \frac{-1}{j\omega}\nabla_s\cdot\boldsymbol{J}_s \tag{87}$$

$$\boldsymbol{n}\times\boldsymbol{E}^s = -\boldsymbol{n}\times\boldsymbol{E}^i \quad \text{on } S. \tag{88}$$

Figure 3(a) represents an arbitrary thin-wire scatterer, for which the following approximations are made. a) The current is assumed to flow only in the direction of the wire axis. b) The current and charge densities are approximated by filaments of current I and charge σ on the wire axis. c) The boundary condition (88) is applied only to the axial component of $\boldsymbol{E}$ at the wire surface. To this approximation, (84) to (88) become

$$-E_l^i = -j\omega A_l - \frac{\partial\phi}{\partial l} \quad \text{on } S \tag{89}$$

$$\boldsymbol{A} = \mu\int_{\text{axis}} \boldsymbol{I}(l)\frac{e^{-jkR}}{4\pi R}\, dl \tag{90}$$

$$\phi = \frac{1}{\varepsilon}\int_{\text{axis}} \sigma(l)\frac{e^{-jkR}}{4\pi R}\, dl \tag{91}$$

$$\sigma = \frac{-1}{j\omega}\frac{dI}{dl} \tag{92}$$

where l is the length variable along the wire axis, and R is measured from a source point on the axis to a field point on the wire surface.

A solution to the above equations is obtained as follows. Integrals are approximated by the sum of intergrals over N small segments, obtained by treating I and q as constant over each segment. Derivatives are approximated by finite differences over the same intervals used for integration. Figure 3(b) illustrates the division of the wire axis into N segments, and defines the notation. If a wire terminates, the boundary condition $I=0$ is taken into account by starting the first segment 1/2 interval in from the end of the wire. This is suggested in Fig. 3(b) by the extra 1/2 interval shown at each end. The nth segment is identified by its starting point $\bar{n}$, its midpoint n, and its termination $\overset{+}{n}$. An increment Δl_n denotes that between $\bar{n}$ and $\overset{+}{n}$, $\Delta l_{\bar{n}}$ and $\Delta l_{\overset{+}{n}}$ denote increments shifted 1/2 segment minus or plus along l. The desired approximations for (89) to (92) are then

$$-E_l^i(m) \approx -j\omega A_l(m) - \frac{\phi(\overset{+}{m}) - \phi(\bar{m})}{\Delta l_m} \tag{93}$$

$$\boldsymbol{A}(m) = \mu \sum_n \boldsymbol{I}(n) \int_{\Delta l_n} \frac{e^{-jkR}}{4\pi R} dl \tag{94}$$

$$\phi(\overset{+}{m}) \approx \frac{1}{\varepsilon} \sum_n \sigma(\overset{+}{n}) \int_{\Delta l_{\overset{+}{n}}} \frac{e^{-jkR}}{4\pi R} dl \tag{95}$$

$$\sigma(\overset{+}{n}) \approx \frac{-1}{j\omega} \left[\frac{I(n+1) - I(n)}{\Delta l_{\overset{+}{n}}} \right] \tag{96}$$

with equations similar to (95) and (96) for $\phi(\bar{m})$ and $\sigma(\bar{n})$.

The σ's are given in terms of the I's by (96), and hence (93) can be written in terms of the $I(n)$ only. One can view the N equations represented by (93) as the equations for an N port network with terminal pairs $(\overset{+}{n}, \bar{n})$. The voltages applied to each port are approximately $\boldsymbol{E}^i \cdot \Delta \boldsymbol{l}_n$. Hence, by defining

$$[I] = \begin{bmatrix} I(1) \\ I(2) \\ \vdots \\ I(N) \end{bmatrix} \qquad [V] = \begin{bmatrix} \boldsymbol{E}^i(1) \cdot \boldsymbol{l}_1 \\ \boldsymbol{E}^i(2) \cdot \boldsymbol{l}_2 \\ \vdots \\ \boldsymbol{E}^i(N) \cdot \boldsymbol{l}_N \end{bmatrix} \tag{97}$$

one can rewrite (93) in matrix form as

$$[V] = [Z][I]. \tag{98}$$

This corresponds to the method of moment representation (26), with $[Z]$ corresponding to $[l]$, $[V]$ to $[g]$, and $[I]$ to $[\alpha]$. The elements of the matrix $[Z]$ can be obtained by substituting (94) through (96) into (93) and rearranging into the form of (98). Alternatively, one can apply (93) through (96) to two isolated elements and obtain the impedance elements directly. This latter procedure will be used because it is somewhat easier to follow.

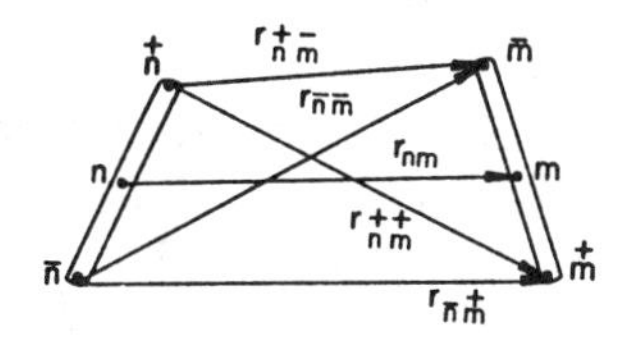

Fig. 4. Two segments of a wire scatterer.

Consider two representative elements of the wire scatterer, as shown in Fig. 4. The integrals in (94) and (95) are of the same form, and are denoted by

$$\psi(n, m) = \frac{1}{\Delta l_n} \int_{\Delta l_n} \frac{e^{-jkR_{mn}}}{4\pi R_{mn}} dl_n. \tag{99}$$

Symbols + and – are used over m and n when appropriate. Evaluation of the ψ in general is considered in the Appendix. Let element n of Fig. 4 consist of a current filament $I(n)$, and two charge filaments of net charge

$$q(\overset{+}{n}) = \frac{1}{j\omega} I(n) \qquad q(\bar{n}) = \frac{-1}{j\omega} I(n) \tag{100}$$

where $q = \sigma \Delta l$. The vector potential at m due to $I(n)$ is, by (94),

$$\boldsymbol{A} = \mu \boldsymbol{I}(n) \Delta \boldsymbol{l}_n \psi(n, m). \tag{101}$$

The scalar potentials at $\overset{+}{m}$ and $\bar{m}$ due to the charges (100) are, by (95)

$$\begin{aligned} \phi(\overset{+}{m}) &= \frac{1}{j\omega\varepsilon} [I(n)\psi(\overset{+}{n}, \overset{+}{m}) - I(n)\psi(\bar{n}, \overset{+}{m})] \\ \phi(\bar{m}) &= \frac{1}{j\omega\varepsilon} [I(n)\psi(\overset{+}{n}, \bar{m}) - I(n)\psi(\bar{n}, \bar{m})]. \end{aligned} \tag{102}$$

Substituting from (101) and (102) into (93), and forming $Z_{mn} = \boldsymbol{E}^i(m) \cdot \Delta \boldsymbol{l}_m / I(n)$, one obtains

$$Z_{mn} = j\omega\mu \Delta \boldsymbol{l}_n \cdot \Delta \boldsymbol{l}_m \psi(n, m) + \frac{1}{j\omega\varepsilon} [\psi(\overset{+}{n}, \overset{+}{m}) - \psi(\bar{n}, \overset{+}{m}) - \psi(\overset{+}{n}, \bar{m}) + \psi(\bar{n}, \bar{m})]. \tag{103}$$

This result applies for self impedances ($m = n$) as well as for mutual impedances. When the two current elements are widely separated, a simpler formula based on the radiation field from a current element can be used.

The wire object is completely characterized by its impedance matrix, subject, of course, to the approximations involved. The object is defined by $2N$ points on the wire axis, plus the wire radius. The impedance elements are calculated by (103), and the voltage matrix is determined by the impressed field, according to (97). The current at N points on the scatterer is then given by the current matrix, obtained from the inversion of (98) as

$$[I] = [Y][V] \qquad [Y] = [Z]^{-1}. \tag{104}$$

Once the current distribution is known, parameters of interest such as field patterns, input impedances, echo areas, etc., can be calculated by numerically evaluating the appropriate formulas.

X. Wire Antennas

A wire antenna is obtained when the wire is excited by a voltage source at one or more points along its length. Hence, for an antenna excited in the nth interval, the applied voltage matrix (97) is

$$[V^s] = \begin{bmatrix} 0 \\ \vdots \\ V_n \\ \vdots \\ 0 \end{bmatrix} \tag{105}$$

i.e., all elements zero except the nth, which is equal to the source voltage. The current distribution is given by (104), which for the $[V]$ of (105) becomes

$$[I] = V_n \begin{bmatrix} Y_{1n} \\ Y_{2n} \\ \vdots \\ Y_{Nn} \end{bmatrix}. \tag{106}$$

Hence, the nth column of the admittance matrix is the current distribution for a unit voltage source applied to the nth interval. Inversion of the impedance matrix therefore gives simultaneously the current distributions when the antenna is excited in any arbitrary interval along its length. The diagonal elements Y_{nn} of the admittance matrix are the input admittances of the wire object fed in the nth interval, and the Y_{mn} are the transfer admittances between a port in the mth interval and one in the nth interval.

The radiation pattern of a wire antenna is obtained by treating the antenna as an array of N current elements $I(n)\Delta l_n$. By standard formulas, the far-zone vector potential is given by

$$A = \frac{\mu e^{-jkr_0}}{4\pi r_0} \sum_n I(n)\Delta l_n e^{jkr_n \cos \xi_n} \tag{107}$$

where r_0 and r_n are the radius vectors to the distant field point and to the source points, respectively, and ξ_n are the angles between r_0 and r_n. The far-zone field components are

$$E_\theta = -j\omega A_\theta \qquad E_\phi = -j\omega A_\phi \tag{108}$$

where θ and ϕ are the conventional spherical coordinate angles.

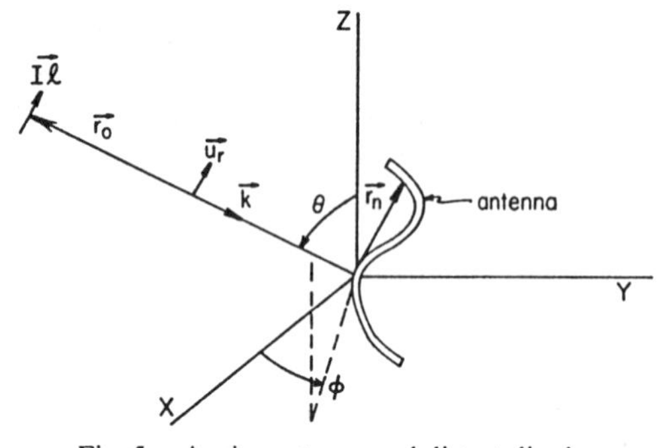

Fig. 5. A wire antenna and distant dipole.

An alternative derivation of the radiation pattern can be obtained by reciprocity. Figure 5 represents a distant current element Il_r (subscripts r denote "receiver"), adjusted to produce the unit plane wave

$$E^r = u_r e^{-jk_r \cdot r_n} \tag{109}$$

in the vicinity of the antenna. Here u_r is a unit vector specifying the polarization of the wave, k_r is a wave number vector pointing in the direction of travel of the wave, and r_n is the radius vector to a point n on the antenna. By reciprocity,

$$E_r = \frac{1}{Il} \int_{\text{antenna}} E^r \cdot I \, dl \tag{110}$$

where E_r is the u_r component of E from the antenna, and I is the current on the antenna. The constant $1/Il$ is that needed to produce a plane wave of unit amplitude at the origin, which is

$$\frac{1}{Il} = \frac{\omega\mu e^{-jkr_0}}{j4\pi r_0}. \tag{111}$$

A numerical approximation to (110) is obtained by defining a voltage matrix

$$[V^r] = \begin{bmatrix} E^r(1)\cdot \Delta l_1 \\ E^r(2)\cdot \Delta l_2 \\ \vdots \\ E^r(N)\cdot \Delta l_N \end{bmatrix} \tag{112}$$

where E^r is given by (109), and expressing (110) as the matrix product

$$E_r = \frac{\omega\mu e^{-jkr_0}}{j4\pi r_0} [\tilde{V}^r][I] = \frac{\omega\mu e^{-jkr_0}}{j4\pi r_0} [\tilde{V}^r][Y][V^s] \tag{113}$$

where $[\tilde{V}]$ denotes the transpose of $[V]$. Note that $[V^r]$ is the same matrix for plane-wave excitation of the wire. Equation (113) remains valid for an arbitrary excitation $[V^s]$; it is not restricted to the single source excitation (105).

The power gain pattern for the u_r component of the radiation field is given by

$$g(\theta, \phi) = \frac{4\pi r_0^2}{\eta} \frac{|E_r(\theta, \phi)|^2}{P_{\text{in}}} \tag{114}$$

where $\eta = \sqrt{\mu/\varepsilon}$ is the intrinsic impedance of space, and P_{in} is the power input to the antenna (* denotes conjugate)

$$P_{\text{in}} = \text{Re}\,\{[\tilde{V}^s][I^*]\} = \text{Re}\,\{[\tilde{V}^s][Y^*][V^{s*}]\}. \tag{115}$$

For the special case of a single source, (105), P_{in} becomes simply Re $(|V_n|^2 Y_{nn})$. Using (113) and (115) in (114), one has

$$g(\theta, \phi) = \frac{\eta k^2}{4\pi} \frac{|[\tilde{V}^r(\theta, \phi)][Y][V^s]|^2}{\text{Re}\,\{[\tilde{V}^s][Y^*][V^{s*}]\}} \tag{116}$$

where $[V^r(\theta, \phi)]$ is given by (112) for various angles of incidence θ, ϕ. Equation (116) gives the gain pattern for only a single polarization of the radiation field. If the total power gain pattern is desired, the g's for two orthogonal polarizations may be added together.

Computations for linear wire antennas have been made using the formulas of this section, and good results obtained. For far-field quantities, such as radiation patterns, as few as 10 segments per wavelength give accurate results. (Radiation patterns are continuous linear functionals, that is, they depend on the weighted integral of the antenna current.) For the current itself, convergence was slower. A typical result for a half-wave antenna was about four percent change in going from 20 to 40 segments, less for other lengths. Faster convergence can be obtained by going from a step approximation to a piecewise-linear approximation to the current. This modification was used for most of the computations, of which Fig. 6 is typical. It shows the input admittance to a center-fed linear antenna with length-to-diameter ratio 74.2 ($\Omega = 2 \log L/a = 10$) using 32 segments.[4]

[4] Because of the extra 1/2 interval at each wire end, this corresponds to an $N = 31$ solution.

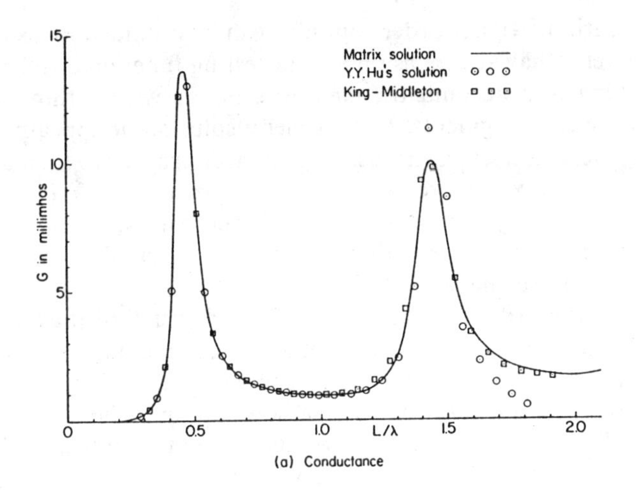

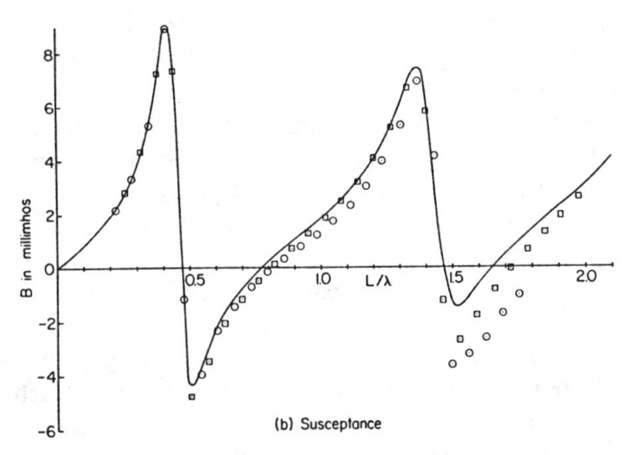

Fig. 6. Input admittance for a center-fed linear antenna of length L and diameter $L/72.4$.

For the points tested, it was almost identical to the 64 segment solution using (103). It is compared to the second-order variational solution of Y. Y. Hu [14], and to the second iteration of Hallén's equation by King and Middleton [15]. The conductances are in close agreement except for Hu's solution $L > 1.3\lambda$, in which case her trial functions are inadequate. The input susceptances are in poorer agreement, which is to be expected because each solution treats the gap differently. The matrix solution of this paper treats it as if it were one segment in length. Hu's solution contains no trial function which can support a singularity in current at the gap, hence gives a low gap capacitance. The King-Middleton method is an iterative procedure, and hence B depends on the number of iterations. Many more computations, as well as a description of the piecewise-linear modification for the current, can be found in the original report [16].

XI. Wire Scatterers

Consider now the field scattered by a wire object in a plane wave incident field. Figure 7 represents a scatterer and two distant current elements, Il_t at the transmitting point, r_t, and Il_r at the receiving point r_r. The Il_t is adjusted to produce a unit plane wave at the scatterer

$$E^t = u_t e^{-jk_t \cdot r_n} \tag{117}$$

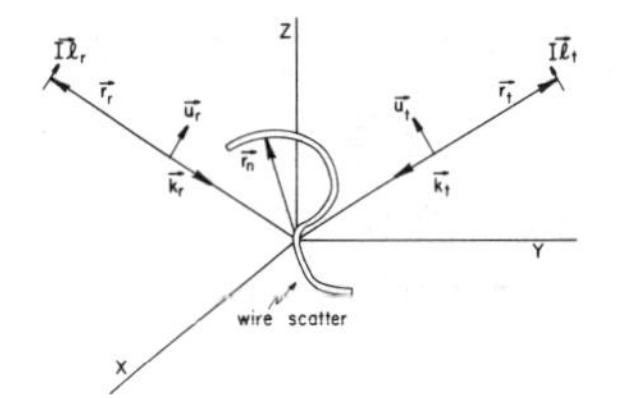

Fig. 7. Definitions for plane-wave scattering.

where the notation is analogous to that of (109). The voltage excitation matrix (97) is then

$$[V^t] = \begin{bmatrix} E^t(1)\cdot \Delta l_1 \\ E^t(2)\cdot \Delta l_2 \\ \vdots \\ E^t(N)\cdot \Delta l_N \end{bmatrix} \tag{118}$$

and the current $[I]$ is given by (104) with $[V]=[V^t]$. The field produced by $[I]$ can then be found by conventional techniques.

The distant scattered field can also be evaluated by reciprocity, the same in the antenna case. A dipole Il_r at the receiving point is adjusted to produce the unit plane wave (109) at the scatterer. The scattered field is then given by (113) with $[V^s]$ replaced by $[V^t]$, that is,

$$E_r = \frac{\omega\mu e^{-jkr_r}}{j4\pi r_r}[\tilde{V}^r][Y][V^t]. \tag{119}$$

A parameter of interest is the bistatic scattering cross section σ, defined as that area for which the incident wave contains sufficient power to produce the field E_r by omnidirectional radiation. In equation form, this is

$$\begin{aligned}\sigma &= 4\pi r_r^2 |E_r|^2 \\ &= \frac{\eta^2 k^2}{4\pi}|[\tilde{V}^r][Y][V^t]|^2.\end{aligned} \tag{120}$$

For the monostatic cross section, set $[V^r]=[V^t]$ in (120). The cross section depends on the polarization of the incident wave and of the receiver. A better description of the scatterer can be made in terms of a scattering matrix.

Another parameter of interest is the total scattering cross section σ_t, defined as the ratio of the total scattered power to the power density of the incident wave. The total power radiated by $[I]$ is given by (115) for any excitation; therefore the scattered power is given by (115) with $[V^s]$ replaced by $[V^t]$. The incident power density is $1/\eta$, hence

$$\sigma_t = \eta \operatorname{Re}[\tilde{V}^t][Y^*][V^{t*}]. \tag{121}$$

Note that σ_t is dependent on the polarization of the incident wave.

Computations for linear wire scatterers have been made using the same $[Y]$ matrix as for antennas. Again far-field quantities, such as echo areas, converged rapidly, with good results obtained with as few as 10 segments per wavelength. Computation of the current converged less rapidly than far-field quantities, but more rapidly than did computation of the current on antennas. This is because the impressed field

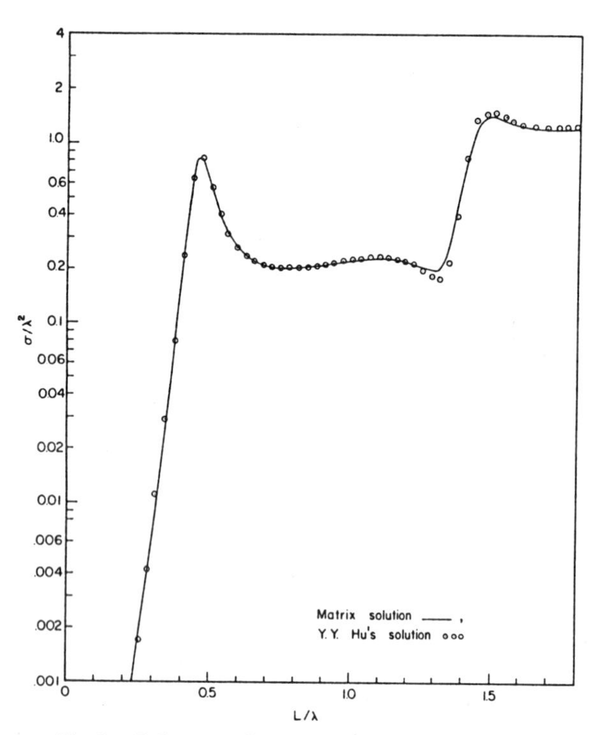

Fig. 8. Echo area of a wire scatterer of length L and diameter $L/72.4$, broadside incidence.

$\boldsymbol{E}^i$ for plane-wave scattering is a well-behaved function, compared with the impulsive impressed field of the antenna problem. Figure 8 shows the echo area for the same wire as was used for an antenna in Fig. 6. Hence, length-to-diameter ratio is 74.2, and a 32 segment piecewise-linear current approximation was used. It is compared to Hu's second-order variational solution [14]. Again good agreement is obtained in the range $L<1.3\lambda$, for which Hu's trial functions are adequate, and a slight discrepancy shows up for $L>1.3\lambda$. Additional computations for linear wire scatterers are given in the original report [16].

XII. Discussion

The method of reducing a functional equation to a matrix equation, and then inverting the matrix for a solution, is particularly well suited to machine computation. Furthermore, the inverted matrix is a representation of the system for arbitrary excitation, hence all responses are solved for at once. As demonstrated by the treatment of wire objects of arbitrary shape, one can also obtain solutions for classes of systems.

In electromagnetic theory, the interpretation of the solution in terms of generalized network parameters is quite general, and applies to bodies of arbitrary shape and arbitrary material. This generalization has been discussed in another paper [17]. The network representation is also useful for the treatment of loaded bodies, both with lumped loads [18] and with continuous loading. Examples of continuously loaded bodies are dielectric coated conductors, magnetic coated conductors, and imperfect conductors.

The solution for wires of arbitrary shape, Section IX, is a first-order solution to the appropriate integrodifferential equation. Higher-order solutions can be obtained by using better-behaved expansion and/or testing functions, and by taking into account the curvature of the wire within the elementary segments. For a general solution, it appears to be more convenient to use a numerical procedure than an analytical procedure. This numerical procedure can be implemented by further subdividing each wire segment, and summing the contributions from the finer subdivisions to obtain the elements of $[Z]$.

As the order of solution is increased, much of the complication comes from the treatment of singularities. The derivative of the current (i.e., charge) is discontinuous at wire ends and at any voltage source along the wire. In the first-order solution this problem has not been accurately treated, and computations appear to justify that this procedure is permissible. For example, at the end of a wire the solution (103) treats the charge as an equivalent line segment extending 1/2 interval beyond the current. The actual charge is singular (or almost so), and could be treated by a special subroutine. While this modification is simple, a similar modification for voltage sources along the wire is not practicable for a general program. This is because the impedance matrix would then depend on the location of the source instead of being a characteristic of the wire object alone. On the basis of experience, it appears that a first-order solution with no special treatment of singularities is adequate for most engineering purposes. This is particularly true for far-zone quantities, such as radiation patterns and echo areas, which are relatively insensitive to small errors in the current distribution.

A number of other electromagnetic field problems have been treated in the literature by procedures basically the same as the method of moments with point matching. Some of these problems are scattering by conducting cylinders [19], [20], scattering by dielectric cylinders [21], [22], and scattering by bodies of revolution [23]. Also available in the literature is an alternative treatment of linear wire scatterers, using sinusoidal expansion functions [24], and an alternative treatment of wire antennas of arbitrary shape, using an equation of the Hallén type [25].

Appendix—Evaluation of ψ

An accurate evaluation of the scalar ψ function of (99) is desired. Let the coordinate origin be located at the point n, and the path of integration lie along the z axis. Then

$$\psi(m, n) = \frac{1}{8\pi\alpha}\int_{-\alpha}^{\alpha} \frac{e^{-jkR_{mn}}}{R_{mn}}\, dz' \tag{122}$$

where

$$2\alpha = \Delta l_n \tag{123}$$

$$R_{mn} = \begin{cases} \sqrt{\rho^2 + (z - z')^2} & m \neq n \\ \sqrt{a^2 + (z')^2} & m = n \end{cases} \tag{124}$$

and a = wire radius. The geometry for these formulas is given in Fig. 9.

One approximation to the ψ's can be obtained by expanding the exponential in a Maclaurin series, giving

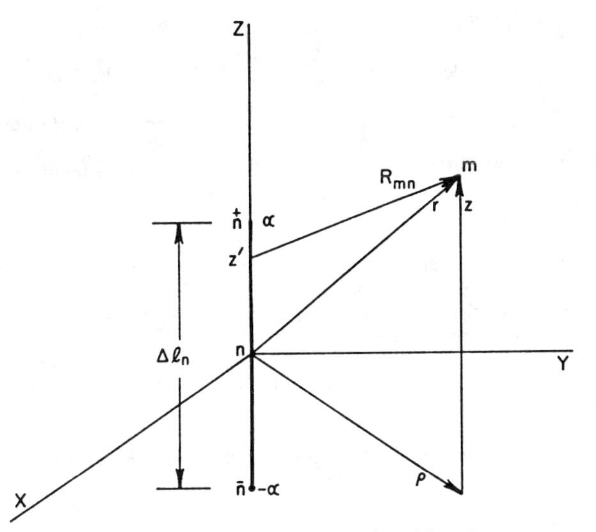

Fig. 9. Geometry for evaluating $\psi(m, n)$.

$$\psi = \frac{1}{8\pi\alpha}\int_{-\alpha}^{\alpha}\left(\frac{1}{R_{mn}} - jk - \frac{k^2}{2}R_{mn} + \cdots\right)dz'. \quad (125)$$

The first term is identical with the static potential of a filament of charge. The second term is independent of R_{mn}. Hence, a two-term approximation of (122) is

$$\psi(m, n) \approx \frac{1}{8\pi\alpha}\log\left[\frac{z + \alpha + \sqrt{\rho^2 + (z + \alpha)^2}}{z - \alpha + \sqrt{\rho^2 + (z - \alpha)^2}}\right] - \frac{jk}{4\pi}. \quad (126)$$

If $r = \sqrt{\rho^2 + z^2}$ is large and $\alpha \ll \lambda$, then

$$\psi(m, n) \approx \frac{e^{-jkr}}{4\pi r}. \quad (127)$$

For a first-order solution, one can take (126) as applying for small r, say $r \leq 2\alpha$, and (127) for large r, say $r > 2\alpha$.

For higher-order approximations, more rapid convergence can be obtained by taking a phase term e^{-jkr} out of the integrand. Then

$$\psi = \frac{e^{-jkr}}{8\pi\alpha}\int_{-\alpha}^{\alpha}\frac{e^{-jk(R_{mn}-r)}}{R_{mn}}dz'$$
$$= \frac{e^{-jkr}}{8\pi\alpha}\int_{-\alpha}^{\alpha}\left(\frac{1}{R_{mn}} - \frac{jk(R_{mn} - r)}{R_{mn}} - \frac{k^2(R_{mn} - r)^2}{2R_{mn}} + \cdots\right)dz'. \quad (128)$$

Term by term integration gives

$$\psi(m, n) = \frac{e^{-jkr}}{8\pi\alpha}\left[I_1 - jk(I_2 - rI_1) - \frac{k^2}{2}(I_3 - 2rI_2 + r^2 I_1) + j\frac{k^3}{6}(I_4 - 3rI_3 + 3r^2 I_2 - r^3 I_1) + \cdots\right] \quad (129)$$

where

$$I_1 = \log\left[\frac{z + \alpha + \sqrt{\rho^2 + (z + \alpha)^2}}{z - \alpha + \sqrt{\rho^2 + (z - \alpha)^2}}\right] \quad (130)$$

$$I_2 = 2\alpha \quad (131)$$

$$I_3 = \frac{\alpha + z}{2}\sqrt{\rho^2 + (\alpha + z)^2} + \frac{\alpha - z}{2}\sqrt{\rho^2 + (z - \alpha)^2} + \frac{\rho^2}{2}I_1 \quad (132)$$

$$I_4 = 2\alpha\rho^2 + \frac{2\alpha^3 + 6\alpha z^2}{3}. \quad (133)$$

An expansion of the type (129) is theoretically valid for all r, but it fails numerically for large r because it involves subtractions of almost equal numbers. For $\rho < a$, one should set $\rho = a$ in the expansion.

An expression suitable for large r is obtained by expanding (122) in a Maclaurin series in z' as

$$\psi = \frac{1}{8\pi\alpha}\int_{-\alpha}^{\alpha}\left[f(0) + f'(0)z' + \frac{1}{2!}f''(0)(z')^2 + \cdots\right]dz' \quad (134)$$

where

$$f(z') = \frac{e^{-jk\sqrt{\rho^2 + (z - z')^2}}}{\sqrt{\rho^2 + (z - z')^2}}.$$

When a five-term expansion of (134) is integrated term by term, there results

$$\psi = \frac{e^{-jkr}}{4\pi r}[A_0 + jk\alpha A_1 + (k\alpha)^2 A_2 + j(k\alpha)^3 A_3 + (k\alpha)^4 A_4] \quad (135)$$

where

$$A_0 = 1 + \frac{1}{6}\left(\frac{\alpha}{r}\right)^2\left[-1 + 3\left(\frac{z}{r}\right)^2\right] + \frac{1}{40}\left(\frac{\alpha}{r}\right)^4\left[3 - 30\left(\frac{z}{r}\right)^2 + 35\left(\frac{z}{r}\right)^4\right]$$
$$A_1 = \frac{1}{6}\left(\frac{\alpha}{r}\right)\left[-1 + 3\left(\frac{z}{r}\right)^2\right] + \frac{1}{40}\left(\frac{\alpha}{r}\right)^3\left[3 - 30\left(\frac{z}{r}\right)^2 + 35\left(\frac{z}{r}\right)^4\right]$$
$$A_2 = -\frac{1}{6}\left(\frac{z}{r}\right)^2 - \frac{1}{40}\left(\frac{\alpha}{r}\right)^2\left[1 - 12\left(\frac{z}{r}\right)^2 + 15\left(\frac{z}{r}\right)^4\right]$$
$$A_3 = \frac{1}{60}\left(\frac{\alpha}{r}\right)\left[3\left(\frac{z}{r}\right)^2 - 5\left(\frac{z}{r}\right)^4\right]$$
$$A_4 = \frac{1}{120}\left(\frac{z}{r}\right)^4. \quad (136)$$

For accuracy of better than one percent, one can use (129) for $r < 10\alpha$ and (135) for $r \geq 10\alpha$.

An alternative derivation of the type of (135) can be obtained as follows. For $r > z'$, one has the expansion

$$\frac{e^{-jkR_{mn}}}{-jkR_{mn}} = \sum_{n=0}^{\infty}(2n + 1)j_n(kz')h_n^{(2)}(kr)P_n\left(\frac{z}{r}\right) \quad (137)$$

where j_n are the spherical Bessel functions of the first kind, $h_n^{(2)}(kr)$ are the spherical Hankel functions of the second kind, and $P_n(z/r)$ are the Legendre polynomials. If (137) is substituted into (122) and integrated term by term, there results

$$\psi(m, n) = \frac{1}{4\pi j} \sum_{n=0}^{\infty} b_n h_n^{(2)}(kr) P_n\left(\frac{z}{r}\right) \tag{138}$$

where

$$b_n = \frac{2n+1}{2\alpha} \int_{-k\alpha}^{k\alpha} j_n(x)\, dx. \tag{139}$$

Equation (138) can be rearranged into the form of (135), although the recurrence formulas for $h_n^{(2)}$ and P_n make computation directly from (138) almost as easy.

Acknowledgment

The computer programming for the numerical results was done by J. Mautz.

References

[1] B. Z. Vulikh, *Introduction to Functional Analysis for Scientists and Technologists*, I. N. Sneddon, trans. Oxford: Pergamon, 1963.

[2] B. Friedman, *Principles and Techniques of Applied Mathematics*. New York: Wiley, 1956.

[3] J. W. Dettman, *Mathematical Methods in Physics and Engineering*. New York: McGraw-Hill, 1962.

[4] L. Kantorovich and G. Akilov, *Functional Analysis in Normed Spaces*, D. E. Brown, trans. Oxford: Pergamon, 1964, pp. 586–587.

[5] L. Kantorovich and V. Krylov, *Approximate Methods of Higher Analysis*, C. D. Benster, trans. New York: Wiley, 1964, ch. 4.

[6] D. S. Jones, "A critique of the variational method in scattering antennas," *IRE Trans. on Antennas and Propagation*, vol. AP-4, pp. 297–301, July 1956.

[7] Forsythe and Wasov, *Finite Difference Methods for Partial Differential Equations*. New York: Wiley, 1960.

[8] V. H. Rumsey, "The reaction concept in electromagnetic theory," *Phys. Rev.*, ser. 2, vol. 94, pp. 1483–1491, June 15, 1954.

[9] R. F. Harrington, *Time-Harmonic Electromagnetic Fields*. New York: McGraw-Hill, 1961, pp. 340–345.

[10] D. K. Reitan and T. J. Higgins, "Accurate determination of the capacitance of a thin rectangular plate," *Trans. AIEE (Communication and Electronics)*, pt. I, vol. 75, pp. 761–766, 1956 (Jan. 1957 section).

[11] J. Van Bladel, *Electromagnetic Fields*. New York: McGraw-Hill, 1964, p. 96.

[12] H. B. Dwight, *Tables of Integrals and Other Mathematical Data*. New York: Macmillan, 1947.

[13] J. Van Bladel, "Some remarks on Green's dyadic for infinite space," *IRE Trans. on Antennas and Propagation*, vol. AP-9, pp. 563–566, November 1961.

[14] Yueh-Ying Hu, "Back-scattering cross section of a center-loaded cylindrical antenna," *IRE Trans. on Antennas and Propagation*, vol. AP-6, pp. 140–148, January 1958.

[15] R. W. P. King, *The Theory of Linear Antennas*. Cambridge, Mass.: Harvard University Press, 1956, p. 172.

[16] R. F. Harrington et al., "Matrix methods for solving field problems," Rome Air Development Center, Griffiss AFB, N. Y., final rept. under Contract AF 30(602)-3724, March 1966.

[17] R. F. Harrington, "Generalized network parameters in field theory," *Proc. Symposium on Generalized Networks*, MRIS series, vol. 16. Brooklyn, N. Y.: Polytechnic Press, 1966.

[18] ——, "Theory of loaded scatterers," *Proc. IEE (London)*, vol. 111, pp. 617–623, April 1964.

[19] K. K. Mei and J. G. Van Bladel, "Scattering by perfectly-conducting rectangular cylinders," *IEEE Trans. on Antennas and Propagation*, vol. AP-11, pp. 185–192, March 1963.

[20] M. G. Andreasen, "Scattering from parallel metallic cylinders with arbitrary cross sections," *IEEE Trans. on Antennas and Propagation*, vol. AP-12, pp. 746–754, November 1964.

[21] J. H. Richmond, "Scattering by a dielectric cylinder of arbitrary cross section shape," *IEEE Trans. on Antennas and Propagation*, vol. AP-13, pp. 334–341, May 1965.

[22] ——, "TE-wave scattering by a dielectric cylinder of arbitrary cross-section shape," *IEEE Trans. on Antennas and Propagation*, vol. AP-14, pp. 460–464, July 1966.

[23] M. G. Andreasen, "Scattering from bodies of revolution," *IEEE Trans. on Antennas and Propagation*, vol. AP-13, pp. 303–310, March 1965.

[24] J. H. Richmond, "Digital computer solutions of the rigorous equations for scattering problems," *Proc. IEEE*, vol. 53, pp. 796–804, August 1965.

[25] K. K. Mei, "On the integral equations of thin wire antennas," *IEEE Trans. on Antennas and Propagation*, vol. AP-13, pp. 374–378, May 1965.

Method of Moments as Applied to Electromagnetic Problems

MICHEL M. NEY, MEMBER, IEEE

(*Invited Paper*)

Abstract—This paper reviews one of the most important general methods for solving electromagnetic-field problems, namely, the moment method. It begins with a brief mathematical foundation of the general method. Then, the various specializations are described, accompanied with relevant references to illustrate the pitfalls and shortcomings, as well as the advantages, as compared to other methods. Deterministic and eigenvalue problems are both discussed separately. Finally, two advanced techniques which have been found to be among the most efficient ones for solving matrix equations resulting from the moment method, namely, the conjugate gradient and the pseudo-inverse, are described. A version of their algorithm which is easily programmable on computer is also presented.

I. Introduction

WITH THE EVER-INCREASING complexity of communication systems, there has been a need for engineers to predict the behavior of such systems by means of computer simulations. As a result, the equations involved in the mathematical description of electromagnetic quantities, which are of interest in most cases, have become more complex. Consequently, more sophisticated numerical methods have been developed to solve electromagnetic problems. These methods are gaining greater and greater success with the constant development of new powerful digital computers.

A theory is meant to extrapolate observations in order to make some prediction. As far as engineers are concerned, a theory is relevant if it can produce numbers in a finite number of steps performed in a reasonable period of time and with sufficient accuracy, taking into account the fact that computers have finite word length. Very elegant theories have been known for decades but had been useless for engineers owing to the lack of appropriate numerical algorithms to produce accurate numbers.

In recent years, most of those theories have received renewed interest with the developments of powerful high-speed computers which have made their numerical solutions within reach. Simultaneously, more sophisticated algorithms have been developed to obtain more accurate numbers while decreasing the number of operations.

Generally, before producing numbers, two steps are necessary. First, the original functional equations must be transformed into structures (such as systems of equations) that can be handled by computers. Then, appropriate techniques must be used to solve numerically the new form of equations hence produced. The degrees of difficulty that may be encountered in both steps are somehow interdependent. Indeed, if an efficient algorithm that transforms the functional equations is found, the size of the systems to be solved may be considerably decreased. However, efficient algorithms may require more computer time to numerically determine the various coefficients of the new form of equations.

This paper is a critical inspection of a general method used extensively in electromagnetic-field problems, namely, the method of moments. Shortcomings and pitfalls of the method will be discussed and some possible improvements proposed. New methods which are gaining interest to solve linear systems of equations that follow from the method of moments will be outlined, as solutions to such systems are an important step towards the production of numbers which are, probably, of most interest for engineers.

This paper is intended as an introduction for those totally unfamiliar with numerical methods, as well as for those with some experience in the field. It was felt by the author that it was more advisable to discuss the different specializations of the general method of moments, rather than to survey specific examples. Basic mathematical concepts are introduced to provide some help for further reading of the relevant literature. References found to be most useful for the specific examples are listed in the bibliography and the reader is urged to consult the papers relevant to his or her interest.

II. Method of Moments

Most of the solutions to linear functional equations can be interpreted in terms of projections onto subspaces of functional spaces. For computational reasons, these subspaces must be finite dimensional. For theoretical work, they may be infinite dimensional. The idea of transforming linear functional equations to linear matrix equations is rather old. Galerkin, a Russian engineer, developed the method, which carries his name, around 1920. It was a specialization of the more general method of moments which was presented later by R. F. Harrington in 1967 [1].

Most of the electromagnetic problems can be expressed under the form of a linear functional equation. One gener-

Manuscript received February 22, 1985; revised June 3, 1985.
The author is with the Department of Electrical Engineering, University of Ottawa, Ottawa, Ontario, K1N 6N5, Canada.

Reprinted from *IEEE Trans. Microwave Theory Tech.*, vol. MTT-33, no. 10, pp. 972–980, Oct. 1985.

ally classifies electromagnetic problems in two categories, namely, *deterministic* and *eigenvalue* problems. In the first category, the linear functional equation enables one to determine the electromagnetic quantity directly. In the second category, parameters for which nontrivial solutions exist are found first. Then, the corresponding solutions called eigensolutions are determined. Both categories of problems can be handled by the method of moments.

A. *Deterministic Problems*

First, consider a deterministic problem for which the corresponding functional is given by

$$L \cdot f = y \tag{1}$$

where L is any linear operator, f is the unknown function to be determined, and y is the input also called the excitation. The space spanned by all functions resulting from the operation L is called the *range* of L. The set of all functions on which L can operate define the *domain* of L.

One needs an inner product $\langle\ \rangle$ associated with the problem, which must satisfy

$$\begin{gathered}\langle u, v\rangle = \langle v, u\rangle^* \\ \langle \alpha u + \beta v, f\rangle = \alpha\langle u, f\rangle + \beta\langle v, f\rangle \\ \langle f, f\rangle \geqslant 0 \\ \text{if } \langle f, f\rangle = 0 \text{ then } f = 0\end{gathered} \tag{2}$$

where α and β are scalars, f, u, v any functions, and $*$ denotes the complex conjugate. For instance, a suitable inner product for function spaces can be given by the functional

$$\langle u, v\rangle = \int_\Omega uv^* \, d\Omega \tag{3}$$

The above integral is performed over any N-dimensional subspace, depending on the application. Equation (3) is called an unweighted or standard inner product and it indicates a "projection" of u in the "direction" of v, from which the similarity between vector and function space becomes obvious. In some situations, one needs the *adjoint operator* of L and its domain defined by

$$\langle Lf, y\rangle = \langle f, L^a y\rangle \tag{4}$$

for all f in the domain of L. An operator is *self-adjoint* if $L^a = L$ and the domain of L^a is that of L. Self-adjointness depends strongly on the associated boundary conditions and also on the selection of an appropriate inner product. For instance, self-adjointness is largely determined by the boundary conditions in the case of differential operators [2]. However, for integral operators, self-adjointness is assured if the kernel of the integral possesses some symmetric properties [3].

The operator L is said to be positive/negative definite if

$$\langle Lf, f\rangle \gtrless 0 \tag{5}$$

for any $f \neq 0$ in the domain of L. The properties of the solution of (1) depend strongly on the properties of the operator L. For instance, if the operator is positive definite, the solution of (1) is unique. Indeed, suppose that u and v are two solutions of (1) such that $L \cdot u = y$ and $L \cdot v = y$. Then, by virtue of the linearity of L, $w = u - v$ is also a solution. Therefore, $Lw = 0$, and, since L is positive definite, w must be zero, yielding $u = v$.

Equations such as (1) can be analytically solved in a very few cases. Most of the time, they require methods that transform the original equation in the form of linear equation systems. The most well known are variational, finite difference, and moment methods. The first two methods are important, and, in certain cases, may have some advantage. However, for the sake of consistency, they will not be discussed here.

First of all, let us express the unknown function in terms of *basis* or *expansion* functions f_j in the domain of L

$$f = \sum_N \alpha_j f_j. \tag{6}$$

The set of basis functions can be finite or infinite. In the latter case, since in practical problems the summation must be truncated, the solution will be an approximation of the true solution. This is the case for orthogonal developments such as Fourier series. Using the property of linearity of L, (1) can now be written as

$$\sum_N \alpha_j L f_j = y. \tag{7}$$

If a set of *weighting* or *testing* functions w_i is chosen in *the range of the operator* L and the inner product of both sides of (7) is taken for each w_i, the original functional equation becomes a set of linear equations that can be written in the matrix form

$$[L]\vec{\alpha} = \vec{y}$$

where

$$\begin{gathered}L_{ij} = \langle w_i, Lf_j\rangle \\ y_i = \langle w_i, y\rangle\end{gathered} \tag{8}$$

and

$$\vec{\alpha} = [\alpha_1, \alpha_2, \cdots, \alpha_j \cdots]^T$$

in which T indicates transposition. If the matrix $[L]$ is regular, $[L]^{-1}$ exists, and the α_i's are given by

$$\vec{\alpha} = [L]^{-1}\vec{y} \tag{9}$$

and the solution is found using (6). The moment method can be interpreted as an error-minimization procedure with the concept of linear spaces. Let $R(L)$ be the range of the operator L. The right-hand side of (8) is the orthogonal projection of the subspace of $R(L)$ spanned by the operation of L on the exact solution f, i.e., the y, onto the subspace W spanned by the w_i's. The left-hand side of (8) is the projection of the subspace spanned by the operations Lf_j onto W. The moment method equates these two projections (see Fig. 1). Since the error (also called weighted residual) is orthogonal to the projection, it is of the second order and, consequently, the method is an error-minimization procedure.

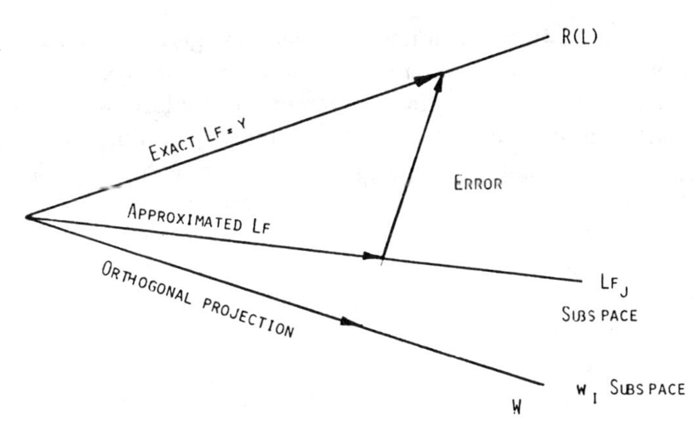

Fig. 1. Illustration of the method of moments in the function space.

There are infinitely many possible sets of basis and weighting functions. The most important task of the engineer for any particular problem is the selection of an appropriate set of f_j's and w_i's. Although the choice of these functions is specific to each problem, one can state rules that can be applied generally to optimize the chance of success by obtaining accurate results in a minimum time and computer memory storage. First of all, they should form a set of linearly independent functions. Second, using (6), the f_j's should approximate the (expected) function f reasonably well. Finally, the w_i's should be in $R(L)$ and so chosen that the inner products $\langle w_i, y\rangle$ depend on relatively independent properties of y. Some additional factors may influence the selection, such as

i) the desired accuracy of the solution,
ii) the size of the matrix $[L]$ to be inverted,
iii) the realization of a well-behaved matrix $[L]$,
iv) the ease of evaluating of the inner products.

The various selections of the w_i's lead to the different specializations of the moment method.

1) Galerkin's Method: In cases where the domain of L is identical to the domain of L^a, one can select $w_i = f_j$, which leads to the well-known Galerkin's method. For self-adjoint operators, the condition is automatically met and they are best suited for this method because, according to (4), the resulting matrix $[L]$ is symmetrical. This may have some numerical advantage for solving the corresponding linear system of equations. However, the elements of the matrix $[L]$ can be more difficult to evaluate than in other methods. This may outweigh the advantage of having a symmetric matrix $[L]$.

The Galerkin's method has been used extensively in electromagnetic problems. Numerous examples of application are given in [4]. This method has been found to yield accurate results with rapid convergence, as compared to others, in the case of low-order solutions, i.e., when few expansion functions are required.

Galerkin's method is also involved in a new method which is gaining interest in transmission-line problems, namely, the spectral-domain method [5]. In this approach, the coupled integral equations, relating field and current, which typically appear in the space domain, are expressed in the spectral domain via Fourier transform. As a result, the original equations are transformed into algebraic equations and convolutions into simple products. The application of the boundary conditions yields a system of linear algebraic equations relating the Fourier transform of both unknowns, respectively, the electric field at dielectric interface, and the current densities in the conducting strips. Finally, the application of the Galerkin's method in the spectral domain produces an homogeneous eigenvalue matrix equation. Equating the determinant of the matrix to zero leads to the solution of the propagation constants for the dominant and higher order modes [6].

2) Subsection and Point-Matching Method: For higher order solutions, i.e., when a large number of expansion functions is required to approximate the unknown function, there may be a certain advantage in using weighting functions that render the inner products, involved in the moment method, easy to determine. This is achieved by choosing the w_i's equal to Dirac's functions. Indeed, by virtue of the property of this function, the inner product which involves an integral in the function spaces, becomes trivial. This specialization of the moment method is called the *point matching* or *collocation* method. The elements of the matrix $[L]$ and the vector $\vec{y}$, hence, become

$$\begin{aligned} l_{ij} &= Lf_j|_{\vec{r}=\vec{r}_i} \\ y_i &= f_j(\vec{r}_i). \end{aligned} \tag{10}$$

This is equivalent to enforce (7) at various points of interest, generally where boundary conditions must be met. The main advantage of this method resides in the ease with which the matrix elements are computed as compared to other specializations of the moment method. The major disadvantage is that for low-order solutions, the accuracy and the convergence of the solution generally depend on the location of the points at which (7) is matched (see, for example, [7]). For higher order solutions, Galerkin's method has been found to give better results and faster convergence in the majority of cases. However, equidistant points, in this case, give satisfactory results for the point-matching method.

Another important aspect is that the point-matching method has been proved inaccurate when the operations Lf_j yield symbolic functions such as Dirac's functions. This can be explained by the fact that the inner product of two distributions such as Dirac's functions is not defined [8]. On the other hand, integral operators do not produce such functions for practical problems. Consequently, they are better suited to be used with the point-matching method. A typical illustration is found in [9] in which the method is used for finding scattering fields produced by infinitely long dielectric cylinders with transverse magnetic irradiation. The operator is integral and the method yields excellent results. On the other hand, for transverse electric irradiation, the operator involved is integro-differential and, since pulse functions as basis are used, the results are less accurate [10].

For problems lacking symmetry, it is difficult to find basis functions that are defined over the entire domain of the solution, and they would imply rather involved calcula-

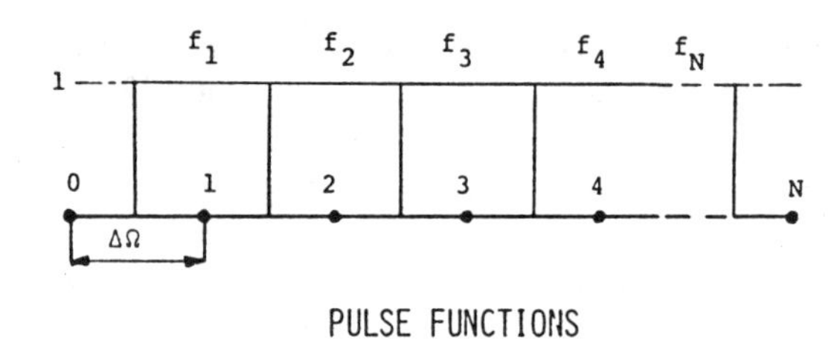

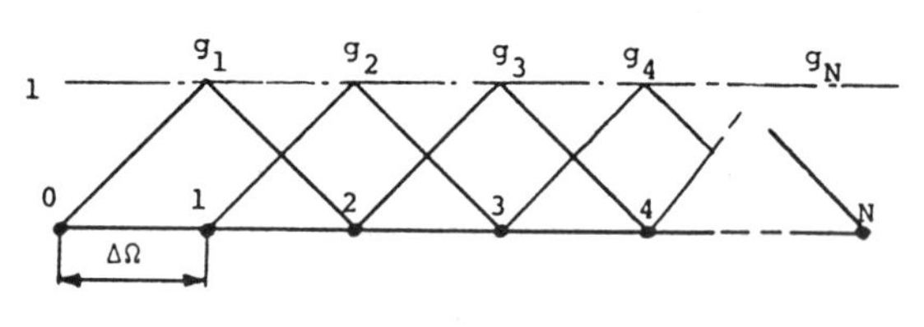

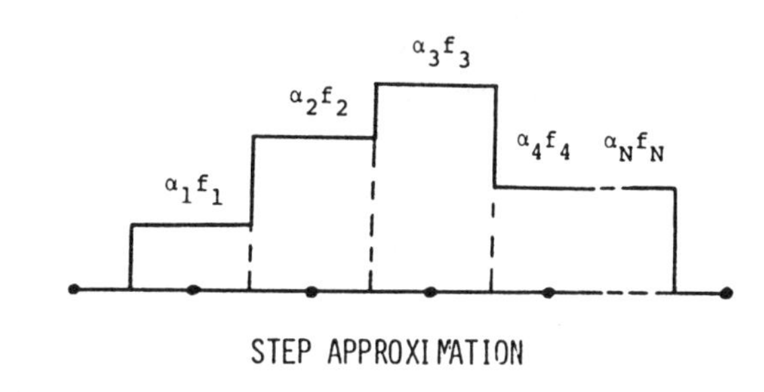

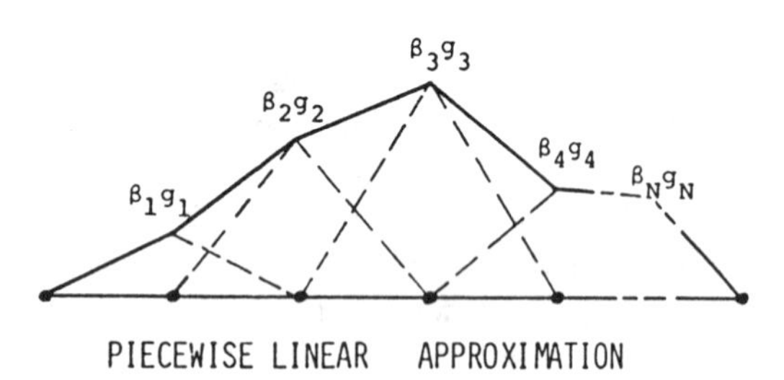

Fig. 2. Examples of subsectional basis and corresponding approximated function.

tions of the inner products. Consequently, it is more judicious to approximate the solution by basis functions which are defined only over subsections of the domain of f. The point-matching method is often used in conjunction with the subsectional basis. The main advantage is a certain facility to compute the elements of the matrix $[L]$ provided that the basis function be simple. For instance, the unknown function can be approximated using step or triangular functions (see Fig. 2). Care must be taken regarding the subsectional basis. The basis functions y should be in the domain of the operator L. In addition, the point-matching procedure should not be used if the operation of L on the f_j's yields symbolic functions, for reasons stated before. For instance, step functions should not be used at all when second-order differential operators are involved. Also, they should not be used for first-order differential operators if point-matching is intended.

The method would be very limited if such simple basis functions could not be used for this type of operators, which are often encountered in electromagnetic problems. Fortunately, the differential operators can always be approximated using a finite-difference procedure which is discussed in the next section. Another possibility is to extend the original domain of the operator such that it can operate on a wider class of functions without changing the operation in the original domain [4]. For problems involving integral operators, the method yields reasonably good results. However, the advantage inherent to the method described above can be outweighed by the size of the system. Indeed, in order to approximate the solution reasonably well, a rather large number of subsections is necessary. Consequently, the method is difficult to be applied in cases for which the wavelength is relatively short as compared to the dimension of the problem.

Many applications that use point-matching in conjunction with the subsection method can be found in the literature [4], [11]–[14]. They most frequently deal with scattering problems. For instance, an interesting comparison between Galerkin's and the point-matching methods for different choices of basis functions can be found in [11] for microstrip antenna problems. It is found that triangular expansions do not give significant improvement as compared to pulse functions. Three-dimensional problems can also be approached with the point-matching and subsection methods [12]. The electric field is expended by 3-D pulses (block model) inside lossy dielectric bodies and point-matching is applied within each subvolume. The method is limited to relatively low frequencies. Surface patch models of conducting objects are proposed in [13] for determining scattering fields of metallic objects with arbitrary shape. The use of triangular patches circumvents the existence of a fictitious line or point of charges at the edges of the subsections [14]. Indeed, by virtue of the continuity equation, line or point charges may appear at the limit of subsections if rectangular pulses are used for current expansion. Potential and field are not defined at those locations, and anomalies or inconsistencies usually appear in the solution. Point-matching used with appropriate basis functions can significantly decrease the order of the system to be solved. Hagman *et al.* [15] proposed a plane-wave correction for scattering problems. The order of the matrix is reduced by one order of magnitude as compared with simple pulse functions presented in [9]. However, the computations of the matrix elements are much more involved.

There are various techniques which have been proposed to improve the point-matching and subsection methods. For instance, in inverse scattering problems, it has been found that the solution is highly sensitive to the points of match [16]. In addition, the problem was found to be ill-posed. One possibility to rendeı the solutions more accurate and less sensitive to the location of the matching points is to use least-square techniques [17]. Some constraints are enforced on the solution, and fairly good solutions are obtained even in the presence of realistic levels of noise in the input function of the integral equation. Another possibility is to match the equation at a larger number of points than required by the number of unknowns and to use least-squares optimization techniques to solve the matrix equation [18]. A theoretical treatment of the so-called overdetermined collocation can be found in [19].

3) Approximate Operators: It is sometimes convenient to approximate the operator L. For instance, differential operators can be replaced by finite-difference operators [2], [4], [20]. This procedure is very useful for point-matching since it allows one to use discontinuous basis functions, such as pulses, for problems involving differential operators. The major inconvenience with the finite-difference approximation is that, for a given approximation of the operator, the solution does not converge when the distance between the mesh nodes decreases. In fact, the corresponding matrix becomes singular [21]. One possibility to obtain a faster convergence and better accuracy without increasing the number of meshes is to retain higher order terms in the finite-difference approximation. This automatically implies more computation for the matrix elements. In practical problems, it is convenient to make the nodes coincide with the boundaries on which the potential and its derivative are known. To achieve this, mesh refinement is required to a level at which the approximation of the boundaries does not affect the solution. Another way is to use irregularly spaced nodes [21], [22].

For integral operators, it is sometimes more convenient to approximate the Kernel of the integral operator [23], [24] so that analytical integration can take place. If this cannot be done, since the elements of $[L]$ generally involve the integral of Green's functions which behave poorly at the origin, numerical methods of integrations such as singularity extraction or statistical Riemann method of integration are necessary to evaluate the diagonal matrix elements. These methods are time consuming if a significant number of points are required, and eventually will not converge. They will not be discussed here.

4) Other Specializations: There are other selections of w_i's and f_j's which have been used. For instance, step functions for testing and triangular functions for expansion functions were used for microstrip antenna problems [11]. Results showed that no significant improvement can be observed as compared to Galerkin's method in which pulses were used for testing and expansion. Therefore, it is not always judicious to use more elaborate basis functions because the computations of the matrix elements may introduce more errors and require more computer time.

Among other possibilities, choosing $w_i = Lf_j$ yields the method of the minimum residual called more commonly the least-squares method. It can be readily seen from Fig. 1 that the error vector (residual) has a minimum norm when it is orthogonal to the space spanned by the Lf_j's.

B. *Eigenvalue Problems*

An eigenvalue equation is an homogeneous equation which can be written in the general linear case as

$$Lf = \lambda Mf \tag{11}$$

where L and M are linear operators. A solution of (11) exists only for particular values of λ called eigenvalues, associated with the corresponding solutions called eigenfunctions, eigenvectors, or eigensolutions. The method of moments transforms (11) to a matrix eigenvalue equation that can be solved by appropriate methods. Eigenvalue problems are important in electromagnetics. Indeed, the eigenvalues correspond to physical quantities which are of major importance for engineers, such as cutoff or resonance frequencies of a system. However, the numerical solution of the matrix eigenvalue equations is, in general, more complicated than that for deterministic matrix equations. Iterative schemes such as Jacobi method are most commonly used [25].

Using the same procedures as described for deterministic equations, (11) can be written as

$$\sum_N \alpha_i Lf_i = \lambda \sum_N \alpha_i Mf_i \tag{12}$$

where the f_j's are in the domain of the operators L and M. In a similar manner, a set of weighting functions w_i is chosen in the range of L and M and the inner product of (12) is taken for each w_i yielding the matrix system

$$[L]\vec{\alpha} = \lambda [M]\vec{\alpha} \tag{13}$$

where

$$m_{ij} = \langle w_i, Mf_j \rangle.$$

The above system can have a nontrivial solution only if

$$\det |[L] - \lambda [M]| = 0. \tag{14}$$

The determinant (14) is a polynomial in which roots $\lambda_1, \lambda_2, \cdots$, correspond to the eigenvalues of the matrix equation (13). They approximate the eigenvalues of the original functional equation (11). The corresponding vectors with coefficients $\alpha_{1n}, \alpha_{2n}, \cdots$, are the eigenvectors of the matrix equation (13) and

$$f_n = \sum_N \alpha_{in} f_i \tag{15}$$

approximates each eigenvector of the original functional equation (11).

If M possesses an inverse, (11) can be written in the canonical form

$$M^{-1}Lf = \lambda f. \tag{16}$$

Thus, the matrix $[L]$ must be multiplied by $[M]^{-1}$ before applying the method of solutions for eigenvalue matrix equations. It is worth noticing that if M and M^{-1} are the identity operators in (11) and (16), $[M]^{-1}$ is not the identity matrix in (13) and (14). Indeed, the elements of $[M]$ involve the scalar product of the weighting functions and the basis functions.

A judicious choice of w_i's is to select $w_i = f_j$ (Galerkin's procedure). It has been found that Galerkin's solutions give eigenvalues higher than the exact values for second-order differential operators, while they give smaller eigenvalues for first-order differential operators [4]. Like deterministic problems, it is sometimes convenient to extend the operator. However, when this is applied to eigenvalue problems, extraneous eigenvalues appear if the basis functions associated with the extended operator violate the boundary conditions of the problem. Fortunately, there are several factors which make extraneous eigenvalues easily recognizable. First of all, they do not converge like the other eigenvalues. Then, even if the original operator is positive definite, they may have negative values because the extended operator is not necessarily positive definite and, finally, the corresponding eigenvectors tend to be irregular and do not generate eigenfunctions which respect the boundary conditions of the problem.

III. New Methods for Solving Matrix Equations

The generation of a linear matrix equation by using any specialization of the moment method is only a step towards the production of numbers which are of most interest for engineers. The last, but not necessarily the least, task is to solve numerically the matrix equation. For small-order well-conditioned systems, classical techniques such as Gauss, diagonal decomposition, and linear iterative techniques are efficient in the majority of cases. They are discussed in great detail in the literature [26], [27] and will not be surveyed here. For large systems, the classical schemes may not yield fast convergence or sufficient accuracy. In addition, if the matrix is ill-conditioned (as, for instance, in the point-matching method), more appropriate techniques must be applied. Finally, if the system is overdetermined because of the application of redundant data technique, the least-squares techniques must be used.

There are two methods for solving linear equation systems that result from the application of the moment method which are becoming increasingly popular among researchers in electromagnetics, namely, the conjugate gradient method and the pseudo-inverse technique. The reasons are the facility with which they can be implemented on a computer and their capability of handling ill-posed problems.

A. *The Conjugate Gradient Method*

Consider the following matrix equation which may result from the application of the moment method:

$$[L]\vec{\alpha}=\vec{y} \tag{17}$$

where the above quantities were defined before. It can be shown [28] that an iterative method, called conjugate gradient method, can produce the desired solution usually in a number of steps less than the order of the matrix $[L]$. The conjugate gradient method is similar to the steepest descent which involves the search for the minimum of a functional in a direction suggested by its negative gradient. A rigorous mathematical treatment of the method can be found in [29]. The conjugate gradient is a nonlinear iterative method, i.e., the new estimate is not a linear function of the past estimate.

The method starts with an initial guess that generates the first residual vector given by

$$\vec{r}_o=[L]\vec{\alpha}_0-\vec{y} \tag{18}$$

and the direction vector

$$\vec{d}_1=-[\tilde{L}]^*\vec{r}_0 \tag{19}$$

where $[\tilde{L}]^*$ is the transposed complex conjugate of $[L]$. Then, the successive iterative steps are given by

$$\vec{\alpha}_{n+1}=\vec{\alpha}_n+t_n\vec{d}_n \tag{20}$$

where

$$t_n=\frac{\|[\tilde{L}]^*\vec{r}_n\|^2}{\|[L]\,\vec{d}_n\|^2}$$

$$\vec{r}_{n+1}=\vec{r}_n+t_n\cdot[L]\,\vec{d}_n \tag{21}$$

$$\vec{d}_{n+1}=-[\tilde{L}]^*\vec{r}_{n+1}+q_n\vec{d}_n \tag{22}$$

where

$$q_n=\frac{\|[\tilde{L}]^*\vec{r}_{n+1}\|^2}{\|[\tilde{L}]^*\vec{r}_n\|^2}$$

in which $\|\cdot\|$ indicates the norm.

The conjugate gradient requires more memory storage as compared to linear iteration schemes [30]. However, it has the great advantage of having a rate of convergence practically insensitive to the initial guess. However, a good initial guess reduces considerably the number of iterations to obtain sufficient accuracy. In addition, as in iterative schemes, the round-off errors are confined in the final step of the solution, regardless of the condition number[1] of the matrix $[L]$. Consequently, the method is also well suited for ill-conditioned matrices. The round-off error can eventually be reduced if the ith residual is computed by $\vec{y}-[L]\vec{\alpha}$ rather than (21). However, more computer time is required in this case.

The application of the conjugate gradient method for electromagnetic problems was originated by Sarkar *et al.* for wire antenna scattering [31]. Solutions exhibit fast convergence. More recently, the problem of induced fields in lossy dielectric cylinders was investigated [32]. The matrix system resulting from a point-matching procedure was iteratively solved by the conjugate gradient method. Successive increases of the number of subsections were used to achieve a faster convergence. The order of the system that could be solved was considerably increased as compared with the direct method of solving matrix systems.

B. *The Pseudoinverse*

It was previously pointed out that if one wants to use the redundant data technique, the resulting system is over determined and, consequently, the least-squares techniques are necessary. These are generally classified into two categories, namely, unconstrained and constrained techniques. This section deals with a constrained least-squares method which has been proved to successfully handle ill-condi-

[1] The condition number of a matrix is defined as the ratio of its highest and its lowest eigenvalue.

tioned problems, in various disciplines such as optics [33], image processing [34], and inverse scattering [18], namely, the pseudoinverse.

Consider the matrix equation (17) in which $[L]$ is the matrix of a bounded operator. Rather than solving directly (17), an estimate $\vec{\alpha}_s$ is generated such that the norm of the residual produced by this estimate is minimum. Simultaneously, the norm of the estimate is constrained to be minimum. This can be written as follows:

$$\text{Among all } \vec{\alpha}_e \text{ which minimize } \| \vec{y} - [L]\vec{\alpha} \| \tag{23}$$

find the particular one which possesses the smallest norm.

The pseudoinverse generates a unique solution for (17) with the constraint (23) even if $[L]$ is singular [35].

A vector space approach is utilized to give a description of the pseudoinverse operation. Let L^a be the adjoint operator of L. Now, consider the nullspace of L generated by the solutions of the homogeneous equation $[L]\vec{\alpha} = \vec{0}$. If L is not onto nor one-to-one, $\vec{y}$ does not belong to the range of L. The orthogonal projection $\vec{y}_p$ of $\vec{y}$ onto the range of L yields a minimum norm of the residual error $\vec{y} - \vec{y}_p$. The set of vector $\vec{\alpha}_e$, which satisfies min $(\|\vec{y} - \vec{y}_p\|)$, can be found by solving the system

$$[L]\vec{\alpha}_e = \vec{y}_p \tag{24}$$

in which $[L]$ is the matrix associated with the operator L. There are, generally, an infinite number of vectors which satisfy (24). The constraint of minimum norm determines the unique pseudoinverse solution. Using the adjoint operator of L, it can be shown that the minimum norm vector is found by the orthogonal projection of the $\vec{\alpha}_e$ onto the range of L^a [35]. This is a consequence of the orthogonality between the nullspace of L and the range of L^a.

Different techniques have been proposed for determining the pseudoinverse of a matrix, among them an iterative technique for sparse matrices [36]. For general applications, the mathematical description of a projection method is given in [37]. Here, a modified version of the projection method is proposed. It comprises a Gram–Schmidt orthogonalization procedure with pivoting, in order to minimize round-off error propagation. Indeed, when two nearly equal vectors are substracted, the error is likely to be significant in both magnitude and direction. Thus, the Gram–Schmidt procedure applied to nearly dependent vectors invariably results in substractions of nearly equal vectors. Error propagation can be avoided if the vectors with relatively small norm are not used until the end of the procedure.

Let $\{\vec{l}_1, \vec{l}_2 \cdots \vec{l}_N\}$ be the set of vectors corresponding to the column of $[L]$ associated with the operator L. The following procedure is recommended for orthogonalizing.

i) Begin with the vector of the largest norm, say, $\vec{l}_k$ (pivot vector).

ii) Make all other vectors orthogonal to it using

$$\underline{l}_i = \vec{l}_i - \left(\langle \vec{l}_i, \vec{l}_k \rangle \| \vec{l}_k \|^2 \right) \underline{l}_k. \tag{25}$$

iii) From the now modified vectors, choose the vector of the largest norm, say, $\underline{l}_{i+q}$ (the second pivot vector).

iv) Make all other vectors (excluding $\vec{l}_k$) orthogonal to it using (25).

v) Repeat iii) and iv) until the search for a new $\underline{l}_i$ finds no vectors whose norm is above a certain threshold.

In order to keep a record of the operations on the column of $[L]$, (25) is applied at each step on the column vectors of a identity matrix $[I]$.

At this stage, the matrices $[L]$ and $[I]$ have been transformed to new matrices $[B]$ and $[Q]$, respectively

$$\frac{[L]}{[I]} \rightarrow \frac{[\underline{l}_1, \underline{l}_2 \cdots \underline{l}_m \;|\; \vec{0}, \vec{0} \cdots \vec{0}]}{[\vec{q}_1, \vec{q}_2 \cdots \vec{q}_m \;|\; \vec{q}_{m+1} \cdots \vec{q}_N]} = \frac{[B]}{[Q]}.$$

Since $[Q]$ records the operations performed on $[L]$, one has

$$[L][Q] = [B]. \tag{26}$$

Consequently, $\{\vec{q}_{m+1} \cdots \vec{q}_N\}$ is a basis for the nullspace of $[L]$. It is, then, orthogonalized with the same procedure described before. In addition, the set $\{\vec{l}_1, \vec{l}_2 \cdots \underline{l}_m\}$ forms an orthogonal basis for the range of L. Consequently, the projection of $\vec{y}$ onto the range of L can be written as

$$\vec{y}_p = [B]\vec{a} = [L][Q]\vec{a} \tag{27}$$

with

$$a_i = \langle \vec{y}, \underline{l}_i \rangle / \|\underline{l}_i\|^2, \qquad i \leqslant m$$
$$a_i = 0, \qquad i > m. \tag{28}$$

By virtue of (27), the solution of $[L]\vec{\alpha}_e = \vec{y}_p$ is clearly

$$\vec{\alpha}_e = [Q]\vec{a}. \tag{29}$$

Only the orthogonal projection of the solutions of (29) onto the range of L^a remains to be carried out. The set of vectors $\{\vec{q}_{m+1}, \vec{q}_{m+2} \cdots \vec{q}_N\}$ constitutes a basis for the orthogonal complement of the range of L^a by virtue of the decomposition theorem [35]. Consequently, the pseudoinverse solution is found by the vector orthogonal to the nullspace of L, which is given by

$$\vec{\alpha}_s = \vec{\alpha}_e - \sum_{i=m+1}^{N} \left(\langle \vec{\alpha}_e, \vec{q}_i \rangle / \|\vec{q}_i\|^2 \right) \vec{q}_i. \tag{30}$$

It can be noticed that the second term of the right-hand side of (30) represents the orthogonal projection of the vector given by (24) onto the nullspace of L.

The procedure described above is easily automated. The sets $\{\vec{l}_i\}$ and $\{\vec{q}_i\}$ need to be orthogonalized only once for a given problem. This is an advantage since most of the computer time involved in the whole procedure takes place during the Gram–Schmidt orthogonalization. Note that the adjoint operator need not be determined in this approach. If the pseudoinverse is to be calculated, the input vector is replaced by the standard basis vector for the N-dimensional space and the procedure is repeated for the N standard basis vectors yielding each time one column vector of the pseudoinverse of $[L]$. The Gram–Schmidt procedure has to be performed only once.

In conclusion, the method just described is, by its simplicity, very attractive. It has the disadvantage of requiring N more memories than the conjugate gradient method. If the matrix is $N \times N$ and has a low condition number, the pseudoinverse is identical to the standard inverse. The condition (23) prevails for ill-conditioned, over- or under-determined systems. To prevent meaningless results, the orthogonalization procedure must be interrupted when the relative norm of the remaining column vectors of the matrix $[L]$ is too small. This is left to the judgment of the user. Decreasing or increasing the number of q_i's may severely affect the solution. Typical examples are illustrated in [18].

IV. Conclusion

The goal of this paper was to familiarize the reader with the principle of numerical analysis of electromagnetic-field problems. It was stressed that the success of obtaining accurate numbers with the method of moments lies mostly in the choice of basis functions. There is always a compromise to be made between the difficulty of computing the inner scalar products involved in the procedure and the size of the corresponding matrix equation. This is where the skill of the person who wishes to use the method can be a major factor yielding successful results.

It is felt that with the rapid development of computers, numerical techniques are becoming increasingly popular among engineers. The numerical methods can be applied for design applications and simulations for antenna, scattering, and transmission-line problems. The reader should be aware that there exist other numerical methods that are widely used in electromagnetic problems. They can be more appropriate in certain situations. One refers to finite-element and variational methods.

Two advanced numerical methods for solving equation systems, namely, the conjugate gradient and pseudoinverse method, were presented. They have the merit of being able to handle ill-posed problems which can easily occur in electromagnetic problems. Again, they are not the only methods that have this feature, but they have an advantage of being easily programmable. In addition, the pseudo-inverse minimizes the norm of the solution which can be of practical interest in certain situations, such as for inverse scattering problems.

It is unrealistic to believe that computer technology will keep pace with problem-solving requirements. One has reached the machine capabilities and the only hope is the development of new algorithms.

References

[1] R. F. Harrington, "Matrix methods for fields problems," *Proc. IEEE*, vol. 55, pp. 136–149, Feb. 1967.

[2] A. Wexler, "Computation of electromagnetic fields," *IEEE Trans. Microwave Theory Tech.*, vol. MTT-17, pp. 416–439, Aug. 1969.

[3] B. H. McDonald, M. Friedman, and A. Wexler, "Variational solution of integral equations," *IEEE Trans. Microwave Theory Tech.*, vol. MTT-22, pp. 237–248, Mar. 1974.

[4] R. F. Harrington, *Field Computation by Moment Method.* New York: Macmillan, 1968.

[5] R. Mittra and T. Itoh, "A new technique for the analysis of the dispersion characteristics of microstrip lines," *IEEE Trans. Microwave Theory Tech.*, vol. MTT-19, pp. 47–56, Jan. 1971.

[6] L.-P. Schmidt and T. Itoh, "Spectral domain analysis of the dominant and higher order modes in fin-lines," *IEEE Trans. Microwave Theory Tech.*, vol. MTT-28, pp. 981–985, Sept. 1980.

[7] M. Gex-Fabry, J. R. Mosis, and F. E. Gardiol, "Reflection and radiation of an open-ended circular waveguide: Application to nondestructive measurement on materials," *Arch. Elek. Übertragung*, vol. 33, no. 12, pp. 473–478, 1979.

[8] M. Lighthill, *Introduction to Fourier Analysis and Generalized Functions.* Cambridge: University Press, 1962.

[9] J. H. Richmond, "Scattering by a dielectric cylinder of arbitrary cross section shape," *IEEE Trans. Antennas Propagat.*, vol. AP-13, pp. 334–341, Mar. 1965.

[10] J. H. Richmond, "TE wave scattering by a dielectric cylinder of arbitrary cross section shape," *IEEE Trans. Antennas Propagat.*, vol. AP-14, pp. 460–464, Apr. 1966.

[11] J. R. Mosis and R. E. Gardiol, "A dynamical radiation model for microstrip structures," *Advances in Electronics and Electron Physics*, vol. 59, pp. 139–235, 1982.

[12] D. E. Livesay and K. M. Chen, "Electromagnetic fields induced inside arbitrary shaped biological bodies," *IEEE Trans. Microwave Theory Tech.*, vol. MTT-22, pp. 1273–1280, 1974.

[13] N. N. Wang, J. H. Richmond, and M. C. Gilreath, "Sinusodial reaction formulation for radiation and scattering from conducting surfaces," *IEEE Trans. Antennas Propagat.*, vol. AP-23, pp. 376–382, Mar. 1975.

[14] A. W. Glisson and D. R. Wilton, "Simple and efficient numerical methods for problems of electromagnetic radiation and scattering from surfaces," *IEEE Trans. Antennas Propagat.*, vol. AP-28, pp. 593–603, May 1980.

[15] M. J. Hagmann, O. P. Gandhi, and C. H. Durney, "Procedures for improving convergence of moment-method solutions in electromagnetics," *IEEE Trans. Antennas Propagat.*, vol. AP-26, pp. 743–748, May 1978.

[16] D. K. Ghodgaonkar, O. P. Gandhi, and M. J. Hagmann, "Estimation of complex permittivities of three-dimensional inhomogeneous biological bodies," *IEEE Trans. Microwave Theory Tech.*, vol. MTT-31, pp. 442–446, June 1983.

[17] M. M. Ney, S. S. Stuchly, A. M. Smith, and M. Goldberg, "Electromagnetic imaging using moment methods," in *URSI Int. Symp. Dig. on Electromagnetic Theory* (Santiago de Compostela, Spain), 1983, pp. 104–107.

[18] M. M. Ney, A. M. Smith, and S. S. Stuchly, "A solution of electromagnetic imaging using pseudoinverse transformation," *IEEE Trans. Medical Imaging*, vol. MI-3, no. 4, pp. 155–162, Dec. 1984.

[19] E. D. Eason, "A review of least-squares methods for solving partial differential equations," *Int. Numerical Methods in Eng.*, vol. 10, pp. 1021–1046, 1976.

[20] G. E. Forsythe and W. R. Wasow, *Finite-Difference Methods for Partial Differential Equations.* New York: Wiley, 1960.

[21] Y. V. Vorbyev, *Method of Moments in Applied Mathematics.* New York: Gordon & Breach, 1965.

[22] M. G. Salvadori and M. L. Baron, *Numerical Methods in Engineering.* Englewood Cliffs, NJ: Prentice-Hall, 1964.

[23] E. H. Newman and P. Tulyathan, "Analysis of microstrip antennas using moment methods," *IEEE Trans. Antennas Propagat.*, vol. AP-29, pp. 47–53, 1981.

[24] J. R. Mosig and F. E. Gardiol, "Analytical and numerical techniques in the Green's function treatment of microwave antennas and scatterers," *Proc. Inst. Elec. Eng.*, pt. H (MOA), vol. 130, pp. 175–182.

[25] P. J. Eberlein, "A Jacobi-like method for the automatic computation of eigenvalues and eigenvectors of an arbitrary matrix," *J. SIAM*, vol. 10, no. 1, pp. 74–88, 1962.

[26] R. W. Hamming, *Numerical Methods for Scientists and Engineers.* New York: McGraw-Hill, 1962.

[27] J. R. Westlake, "A handbook of numerical matrix inversion and solution of linear equations. *Englewood-Cliffs, NJ: Prentice-Hall*, 1968.

[28] M. R. Hestenes and E. Stiefel, "Methods of conjugate gradients for solving linear systems," *J. Res. Nat. Bur. Stand.*, vol. 49, pp. 409–436, 1952.

[29] M. Hestenes, *Conjugate Direction Methods in Optimization.* New York: Springer-Verlag, 1980.

[30] T. K. Sarkar, K. R. Siarkiewicz, and R. F. Stratton, "Survey of numerical methods for solutions of large systems of linear equations

for electromagnetic field problems," *IEEE Trans. Antennas Propagat.* vol. AP-29, pp. 847–856, June 1981.

[31] T. K. Sarkar and S. M. Rao, "The application of the conjugate gradient method for the solution of electromagnetic scattering from arbitrary oriented wire antennas," *URSI Int. Symp. Dig. Electromagnetic Theory* (Santiago de Compostela, Spain), 1983, pp. 93–96.

[32] M. F. Sultan and R. Mittra, "An iterative moment method for analyzing the electromagnetic field distribution inside inhomogeneous lossy dielectric objects," *IEEE Trans. Microwave Theory Tech.*, vol. MTT-33, pp. 163–168, Feb. 1985.

[33] R. Barakat and G. Newsam, "Numerically stable iterative method for the inversion of wave-front aberrations from measured point-spread-function data," *J. Opt. Soc. Amer.*, vol. 70, no. 10, pp. 1255–1263, 1980.

[34] W. K. Pratt and F. Davarian, "Fast computational techniques for pseudoinverse and Wiener image restoration," *IEEE Trans. Comput.*, vol. C-26, no. 6, pp. 571–580, 1977.

[35] C. N. Dorny, *A Vector Space Approach to Models and Optimization.* Hunington, New York: R. E. Krieger, 1980.

[36] A. Bjorck and T. Elfving, "Accelerated projection methods for computing pseudoinverse solutions of systems of linear equations," *J. BIT*, vol. 19, 10, pp. 145–163, 1979.

[37] K. Tanabe, "Projection method for solving a singular system of linear equations and its applications," *Numer. Math.*, vol. 17, pp. 203–214, 1971.

Paper 7.3

Microstrip Discontinuity Capacitances and Inductances for Double Steps, Mitered Bends with Arbitrary Angle, and Asymmetric Right-Angle Bends

PETER ANDERS AND FRITZ ARNDT

Abstract—**The equivalent capacitances and inductances for microstrip double steps, mitered bends with arbitrary angle, and asymmetric right-angle bends are calculated by the moment method. The data for the double step include the coupling effect between the two single steps. The geometry of the mitered bend with arbitrary angle is determined for minimized bend VSWR over a wide range of parameters. The equivalent circuit data of the asymmetric right-angle bend are compared with results of the frequency dependent planar waveguide model.**

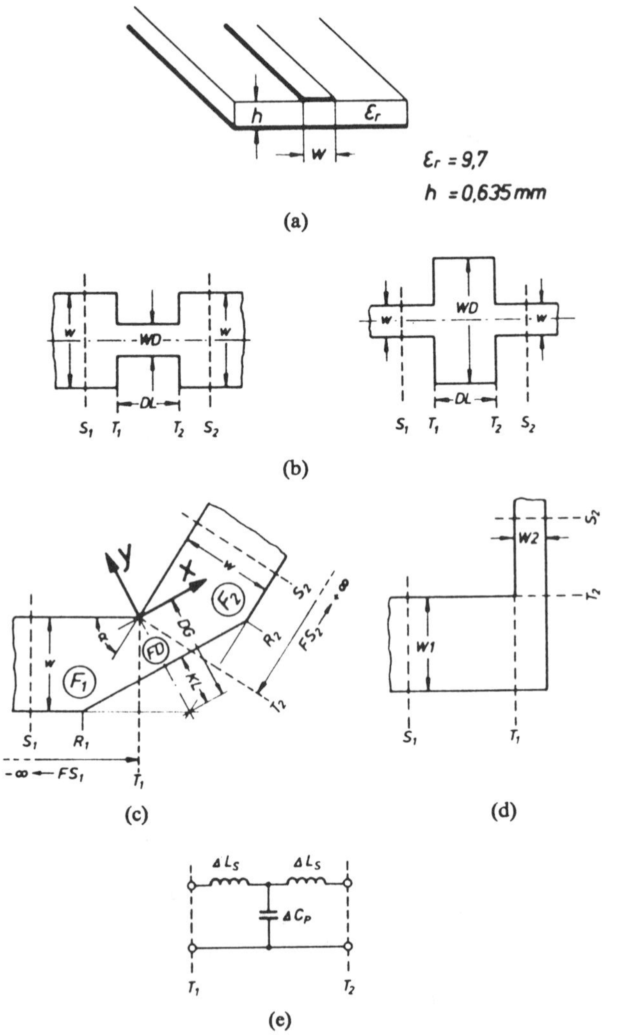

I. Introduction

FOR COMPUTER-AIDED design of microwave circuits microstrip discontinuities are often represented by equivalent circuit parameters [1]–[11]. Benedek and Silvester calculated the equivalent capacitance of symmetrical right-angle bends [6] and of steps in the microstrip line [7]. This was also done by Farrar and Adams using a somewhat different approach [8]. Gopinath, Easter, Thomson, and Stephenson [9]–[11] extended the analysis to the equivalent inductance. Recently, equivalent circuit data have been calculated for microstrip bends with arbitrary angle and with curved transition [12]. The calculation of the microstrip bend in [12] has been compared with results of a waveguide model theory [13], [14] which confirms the validity of the quasi-static approach up to about 9 GHz for usual bend geometries.

In this paper the equivalent capacitances and inductances of some other interesting microstrip discontinuities shown in Fig. 1 are calculated by the moment method. The double step (Fig. 1(b)) is often used in microwave circuits as a quasi-lumped inductance or capacitance [1]–[5]. It is shown that a substantial error exists if, as in the past, the coupling effect between the two single steps is neglected. The mitered bend with arbitrary angle (Fig. 1(c)) is of considerable interest for designing matched microstrip circuits. The miter geometry for minimized bend VSWR is determined over a wide range of practical parameters. The results are compared with some available experimental data. The third example, the asymmetrical right-angle bend (Fig. 1(d)), is found for instance in microstrip phase shifter structures, e.g., [16]. It combines the effects of sudden change of width and of a sharp bend.

$\varepsilon_r = 9.7$
$h = 0.635$ mm

(a)

(b)

(c)

(d)

(e)

Fig. 1. Investigated microstrip discontinuities. (a) Microstrip line. (b) Double step. (c) Mitered bend with arbitrary angle. (d) Asymmetric right-angle bend. (e) Equivalent circuit for all discontinuities.

Manuscript received March 3, 1979; revised July 10, 1980.

P. Anders was with the Department of Microwaves, University of Bremen, Kufsteiner Str. D-28, Bremen 33, West Germany. He is now with the Deutsche ITT Industries, Hans-Bunte-Str. 19, D-7800 Freiburg, West Germany.

F. Arndt is with the Department of Microwaves, University of Bremen, Kufsteiner Str., D-2800 Bremen 33, West Germany.

Reprinted from *IEEE Trans. Microwave Theory Tech.*, vol. MTT-28, no. 11, pp. 1213–1217, Nov. 1980.

II. Method

The computational methods used are similar to those used before [6]–[12], [20] and thus the theoretical formulation can be abbreviated. As in [9] the reference planes S_1, S_2 (Fig. 1) are chosen so as to include a satisfactorily large proportion of the fringing field of the discontinuity in the region of T_1, T_2.

By plotting the amplitude of the surface current versus the coordinates x and y it was seen that the disturbance by the step is practically negligible beyond a distance approximately equal to the actual line width w_1 or w_2, respectively, (cf. also [7], [21]). Similar results are found for the double step. For the microstrip bend (Fig. 1(c)), however, the fringing field is more intense. It was taken into account within the region $\overline{S_1R_1}=\overline{S_2R_2}\approx 1.5w$, $\approx 3w$, $\approx 6w$ for a 45°, 90°, 120° bend, respectively. For the asymmetric bend (Fig. 1(d)) the corresponding values are $3w_1$, and $3w_2$, respectively.

For determination of the equivalent capacitance ΔC_p the microstrip discontinuity is subdivided into suitable rectangular subsections with the dimensions $\Delta l_n, \Delta b_n$ where the charge density σ is assumed to be constant (pulse expansion). The ground plane is assumed to be at zero potential and the conducting strip to be at a potential of $\Phi = 1$ V. The equivalent capacitance ΔC_p is

$$\Delta C_p = \frac{1}{\Phi}\sum_{m=1}^{M}(\sigma_{um}\cdot\Delta l_m\cdot\Delta b_m)+C_{\text{inf}} \tag{1}$$

with

$$C_{\text{inf}} = \frac{1}{\Phi}\sum_{m=1}^{ND}\sigma_{km}\cdot\Delta l_m\cdot\Delta b_m \tag{2}$$

- ND number of the known substrip charge densities σ_k within the region T_1, T_2 due to the infinite strip;
- M total number of finite subsections within S_1, S_2;
- σ_{km} known value of the mth substrip or subsection charge density due to the infinite strip.

The unknown charge densities σ_{um} are obtained by matrix inversion

$$(\sigma_{um}) = (D_{VF})^{-1}(\Phi_{\text{dif}}) \tag{3}$$

with

$$\Phi_{\text{dif}} = \Phi(P_m) - \left[\sum_{n=1}^{N1} D_{VS}(P_m, P_n')\sigma_{kn}(P_n') + \sum_{n=1}^{ND} D_{VF}(P_m, P_n')\sigma_{kn}(P_n') + \sum_{n=1}^{N2} D_{VS}(P_m, P_n')\sigma_{kn}(P_n')\right] \tag{4}$$

where

- $N1, N2$ number of semi-infinite substrips ending at T_1, T_2, respectively;
- P_m field point (x_m, y_m, z_m);
- P_n' source point (x_n', y_n', z_n'); and

$$D_{VF} = \int_{\Delta x}\int_{\Delta y} G(x, y|x', y')\,dy'\,dx' \tag{5}$$

(source function of the finite subsection with the dimension $\Delta x, \Delta y$)

$$D_{VS} = \int_{\Delta y}\lim_{\Delta x\to\infty}\int_{\Delta x} G(x, y|x', y')\,dx'\,dy' \tag{6}$$

(source function of the semi-infinite substrip ($\Delta x\to\infty$)); where G = Green's function [19].

The equivalent inductances ΔL_S are determined using the current loops method [9]. To include all discontinuity geometries investigated the current loops are not required to be rectangular.

The equivalent inductance is

$$2\Delta L_s = \frac{1}{I_t^2}\int_F \vec{A}\vec{K}\,dF \tag{7}$$

where I_t is the total current flowing into or out of the discontinuity. The vector potential A at the field point P can be separated into the parts

$$A(P) = \int_{FS_1} K_{k1}G_c\,dFS_1 + \int_{FS_2} K_{k2}G_c\,dFS_2 + \int_{F_1} K_{u1}G_c\,dF_1 + \int_{F_2} K_{u2}G_c\,dF_2 + \left(\int_{FD} K_{kD}G_c FD + \int_{FD} K_{uD}G_c\,dFD\right) \tag{8}$$

where

- K_{k1}, K_{k2}, K_{kD} known surface currents of the substrips outside of T_1, T_2, and within T_1, T_2, respectively, due to the infinite strip

$$G_c = \frac{\mu_0}{4\pi}\left[\frac{1}{\sqrt{(x-x')^2+(y-y')^2}} - \frac{1}{\sqrt{(x-x')^2+(y-y')^2+2h^2}}\right] \tag{9}$$

FS_1, FS_2, F_1, F_2, FD = areas (cf., Fig. 1(c)).

The unknown surface currents K_u are obtained by requiring the magnetic induction B_{normal} penetrating into the strip to zero

$$B_{\text{normal}} = 0 = \sum_{i=1}^{N_1} I_{k1i}Q_{BS_{ij}} + \sum_{i=1}^{N_2} I_{k2i}Q_{BS_{ij}} + \sum_{i=1}^{ND} I_{kD_i}Q_{BF_{ij}} + \sum_{i=1}^{M}\sum_{n=1}^{4} I_{ui}Q_{BF_{ijn}} \tag{10}$$

with

- N_1, N_2, ND number of the known substrip currents I_k outside of T_1, T_2, and within T_1, T_2, respectively;

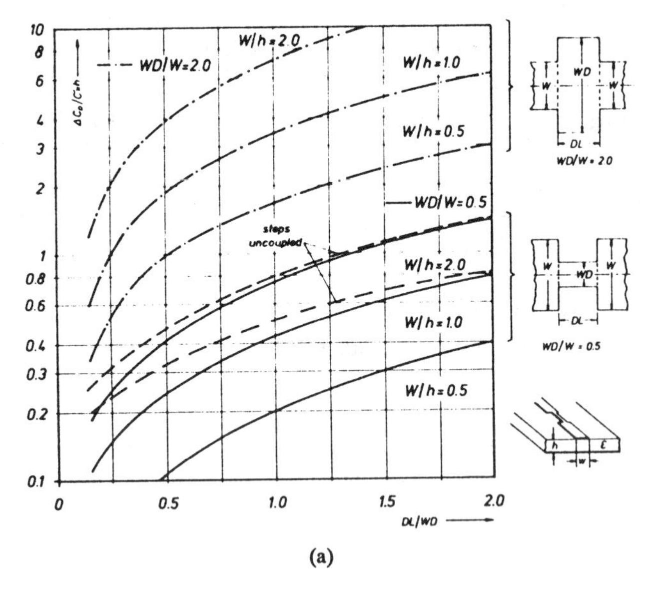

(a)

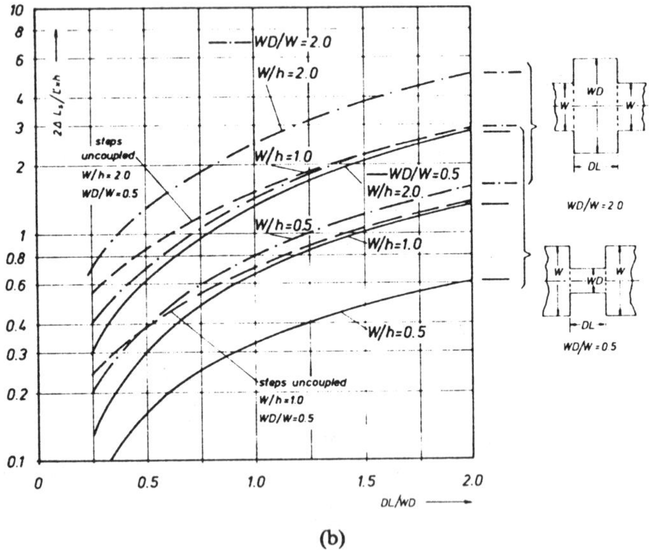

(b)

Fig. 2. Normalized equivalent values of the double step including the coupling effect of the two steps. The dashed lines (---) represent values according to [7], [10], and [11] not including the coupling effect. (a) Equivalent capacitance ΔC_P for $WD/W=2$ (—·—) and $WD/W=0.5$ (——). (b) Equivalent inductance ΔL_S for $WD/W=2$ (—·—) and $WD/W=0.5$ (——). The equivalent values are normalized to the values C', L' of the infinite uniform strip and the height h of the dielectric.

M total number of the unknown loop currents I_u in the four-cornered loop

$$Q_{BF}=-\frac{\partial}{\partial y}\left[\int_{\Delta x} G_c(x,y|x',y')\,dx'\right] \tag{11}$$

$$Q_{BS}=-\frac{\partial}{\partial y}\left[\lim_{\Delta x\to\pm\infty}\int_{\Delta x} G_c(x,y|x',y')\,dx'\right] \tag{12}$$

(source functions). A computer program was developed including (1)–(4), and (7)–(10) for calculating the discontinuity capacitances and inductances.

III. Results

Fig. 2 show the normalized equivalent capacitance ΔC_p and inductances ΔL_S of the entire double step (Fig. 1b) including so the mutual coupling effect of the two steps (solid and dotted lines). The dashed lines represent the values obtained if no coupling effect is considered according to [7], [10], and [11]. It is obvious that, particularly for short relative strip lengths DL/WD, the mutual coupling effect must be taken into account, since the known single step model is based on the assumption of infinitely long uniform strips on both sides and thus leads to wrong results.

In Fig. 3(a)–(c) the normalized equivalent capacitance $\Delta C_p^N=\Delta C_p/(C'\cdot h)$ (dashed line) and inductance $2\Delta L_S^N(L'\cdot h)$ (solid line) of the mitered bend are shown as a function of the miter percentage

$$M=\frac{KL}{DG}\cdot 100 \text{ percent} \tag{13}$$

for various values of the bend angle α and for $\omega/h=0.5$, 1, and 2.

The optimum miter percentage M for minimized bend VSWR can be found approximately if the characteristic impedance Z_W of the equivalent circuit (Fig. 1(e)) is equal to the characteristic impedance Z_0 of the microstrip line which for simplicity is assumed to be quasistatic

$$Z_W=\sqrt{\frac{2\Delta L_S}{\Delta C_p}-\omega^2\Delta L_S^2}\approx\sqrt{\frac{2\Delta L_S}{\Delta C_p}}=Z_0\approx\sqrt{\frac{L'}{C'}}. \tag{14}$$

This means that the points of intersection for $\Delta C_p^N=2\Delta L_S^N$ in Fig. 3 approximately indicate the optimum miter percentage.

The optimum miter is shown to be dependent on ω/h (Fig. 4). For the 90° and the 45° bend the results are compared with Douville and James' empirical values [15] which are considered there to be accurate within ± 4 percent. The two results agree within about 5 percent if the $\pm$4-percent limit is not taken into account. For the 90° bend of the 50-Ω line the optimum miter percentage given by Kelly *et al.* [17] is also indicated. Our result for the 50-Ω line 90° bend lies between the values given by [15] and [17]. For bend angles of 70° and 110°, $\omega/h=1$ ($h=0.635$ mm), some copper metallized mitered bends have been constructed on Al_2O_3 substrate material ($\epsilon_r\approx 9.7$). To reduce the influence by the coax-to-microstrip transitions to the measurement of the bend VSWR a time domain reflectometer measurement was used to locate the impedance discontinuities within the bend region. For constructed bends with miter percentages of $M_{opt}\approx 50$ percent (70° bend) and $M_{opt}\approx 80$ percent (110° bend) which are indicated in Fig. 4 the maximum VSWR measured within the bend region was <1.005 and <1.01, respectively.

Fig. 5(a), (b) show the normalized equivalent circuit parameters of the unsymmetrical right-angle bend. To compare the quasi-static approach (moment method) used here with the frequency dependent planar waveguide model for the right-angle bend [18] the reflection coefficient $/S_{11}/$ and the transmission coefficient $/S_{21}/$ have been calculated. For low frequencies the two models agree well (cf. Fig. 6(a), (b)). For high frequencies the quasi-static

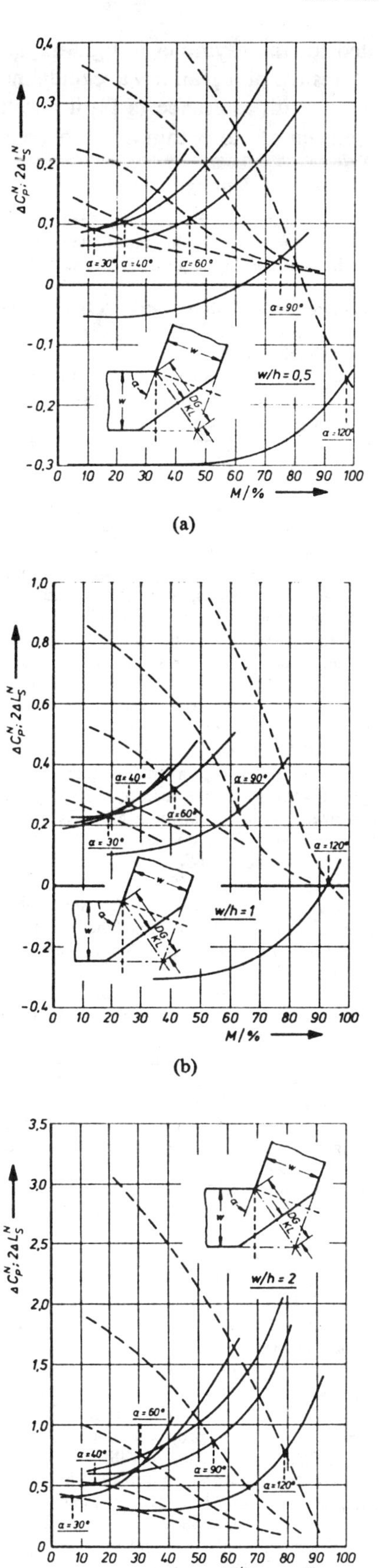

Fig. 3. Normalized equivalent capacitance $\Delta C_P^N = \Delta C_P/(C' \cdot h)$ (dashed line) and inductance $2\Delta L_S^N/(L' \cdot h)$ (solid line) of the mitered bend as a function of the miter percentage $M = KL/I \cdots 00$ percent. (a) $w/h = 0.5$. (b) $w/h = 1$. (c) $w/h = 2$.

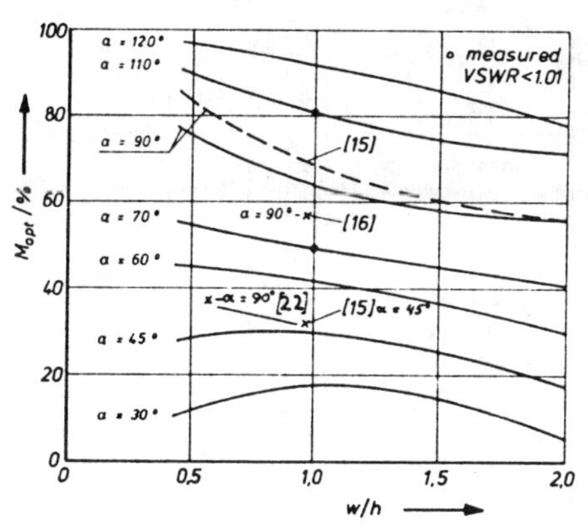

Fig. 4. Optimum miter M_{opt} in percent for minimized bend VSWR of a mitered bend as a function of w/h for bend angles $\alpha = 30° \cdots 120°$.

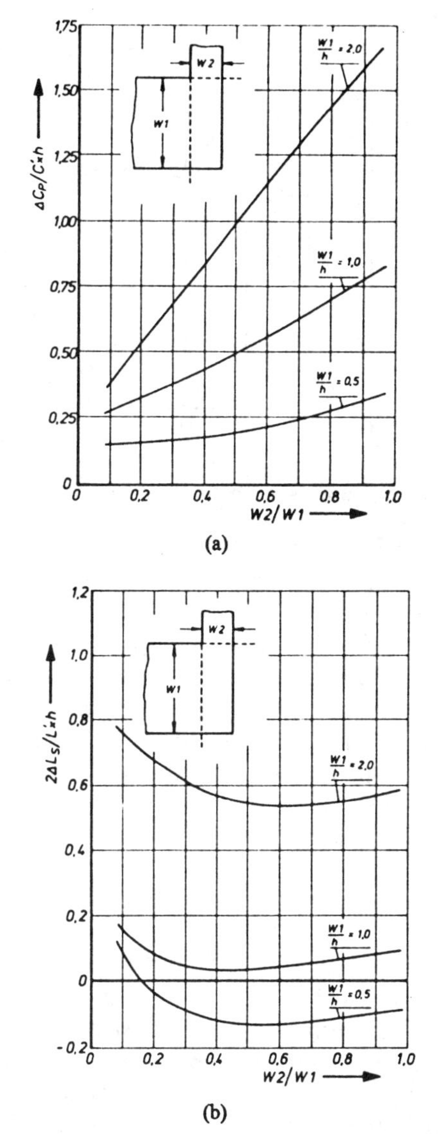

Fig. 5. Asymmetric right-angle bend. (a) Normalized equivalent capacitance. (b) Normalized equivalent inductance.

approach leads to wrong results because the higher order modes cannot be taken into account.

The number of subsections and current loops was chosen to be about 150 for all computations. The dimension

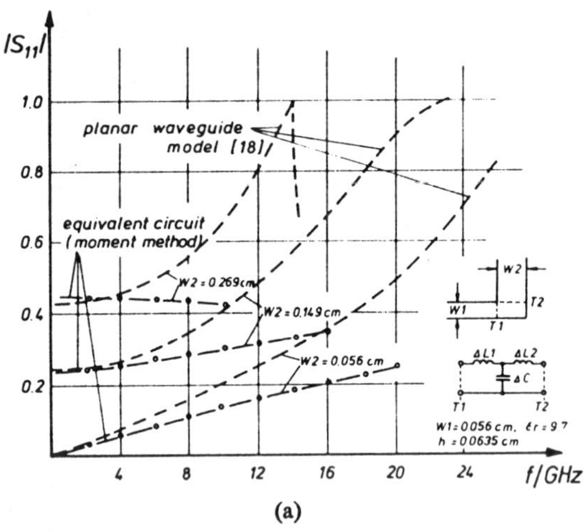

(a)

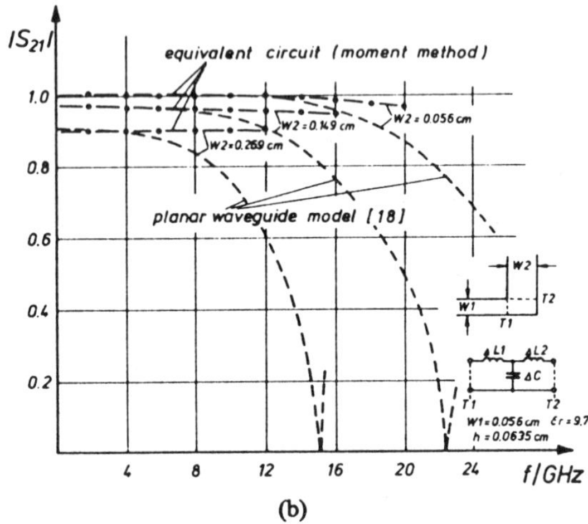

(b)

Fig. 6. Asymmetric right-angle bend. (a) Reflection coefficient $/S_{11}/$. (b) Transmission coefficient $/S_{21}/$. These are as a function of frequency, compared with results of a planar waveguide model [18].

of the subsections was varied, with much bigger subsections near that regions where the charge density is nearly uniform. The total computation time for the capacitance data was about 3 min for each data point on the double steps, about 5 min for each mitered bend capacitance computation, and about 10 min for each data point on the asymmetric bend. The corresponding values for the inductances data were about 4 min, 7 min, 15 min, respectively (IRIS 80 computer was used in all computation). The maximum storage required was 120K.

IV. Conclusions

The known matrix methods were applied to often used microstrip discontinuities: double steps, mitered bends with arbitrary angle and the asymmetrical right-angle bend. The equivalent parameters calculated for the double steps include the mutual coupling effect. The data for the mitered bends were used to calculate an optimum miter performance concerning the minimized bend VSWR. These results agree well with experiment and other data reported in the literature [15], [17]. The equivalent circuit data computed for the asymmetric right-angle bend agree very well with results of a planar waveguide model [18] for low frequencies. With reference to [8] it is estimated that the results are accurate to within a few percent.

References

[1] H.-N. Toussaint and R.-K. Hoffmann, "Gegenwärtiger stand und entwicklungstendenzen der rechnergestützten entwicklung von integrierten mikrowellenschaltungen," *Frequenz*, vol. 28, pp. 148–154, 1974.

[2] E. Sanchez-Sinencio and T. N. Trick, "Computer-aided design of microwave integrated circuits," *IEEE Trans. Microwave Theory Tech.*, vol. MTT-22, pp. 309–316, 1974.

[3] J. W. Bandler, P. C. Lin, and H. Tromp, "Integrated approach to microwave design," *IEEE Trans. Microwave Theory Tech.*, vol. MTT-24, pp. 584–591, 1976.

[4] N. M. Hosseini, V. H. Shurmer, and R. A. Soares, "OPTIMAL, a program for optimizing microstrip networks," *Electron. Lett.*, vol. 12, pp. 190–192, 1976.

[5] R. H. Jansen, "Computer-aided design of transistorized microstrip broadband amplifiers on the base of physical circuit structures," *Arch. Elek. Übertragung.*, vol. AEÜ-32, pp. 145–152, 1978.

[6] P. Silvester and P. Benedek, "Microstrip discontinuity capacitances for right-angle bends, T-junctions and crossings," *IEEE Trans. Microwave Theory Tech.*, vol. MTT-21, pp. 341–346, 1973.

[7] P. Benedek and P. Silvester, "Equivalent capacitances for microstrip gaps and steps," *IEEE Trans. Microwave Theory Tech.*, vol. MTT-20, pp. 729–733, 1972.

[8] A. Farrar and A. T. Adams, "Matrix-methods for microstrip three-dimensional problems," *IEEE Trans. Microwave Theory Tech.*, vol. MTT-20, pp. 497–504, 1972.

[9] A. Gopinath and B. Easter, "Moment method of calculating discontinuity inductance of microstrip right-angle bends," *IEEE Trans. Microwave Theory Tech.*, vol. MTT-22, pp. 880–883, 1974.

[10] A. F. Thomson and A. Gopinath, "Calculation of microstrip discontinuity inductances," *IEEE Trans. Microwave Theory Tech.*, vol. MTT-23, pp. 648–655, 1975.

[11] A. Gopinath. A. F. Thomson, and I. M. Stephenson, "Equivalent circuit parameters of microstrip step chance in width and cross sections," *IEEE Trans. Microwave Theory Tech.*, vol. MTT-24, pp. 142–144, Mar. 1976.

[12] P. Anders and F. Arndt, "Beliebig abgeknickte microstrip-leitungen mit bogenförmigem übergang," *Arch. Elek. Übertragung.*, vol. AEÜ-33, pp. 93–99, 1979.

[13] I. Wolff and W. Menzel, "A universal method to calculate the dynamical properties of microstrip discontinuities," in *Proc. 5th European Microwave Conf.*, (Hamburg, Germany), Sept. 1975.

[14] R. Mehran, "Calculation of microstrip bends and Y-junction with arbitrary angle," *IEEE Trans. Microwave Theory Tech.*, vol. MTT-26, pp. 400–405, 1978.

[15] R. J. P. Douville and D. S. James, "Experimental study of symmetric microstrip bends and their compensation," *IEEE Trans. Microwave Theory Tech.*, vol. MTT-26, pp. 175–182, 1978.

[16] Y. Ho Chen, "Microstrip phase shifter provides improved match," *Microwaves*, vol. 18, pp. 44–48, Oct. 1979.

[17] D. Kelly, A. G. Kramer, and F. C. Willwerth, "Microstrip filters and couplers," *IEEE Trans. Microwave Theory Tech.*, vol. MTT-16, pp. 560–562, Aug. 1968.

[18] R. Mehran, "The frequency dependent scattering matrix of microstrip right-angle bends, T-junctions and crossings," *Arch. Elek. Übertragung.*, vol. AEÜ-29, pp. 454–460, 1975.

[19] P. Silvester, "TEM-wave properties of microstrip-transmission lines," *Proc. IEEE*, vol. 115, pp. 43–48, 1968.

[20] J. R. Mosig and F. E. Gardiol, "Equivalent inductance and capacitance of a microstrip slot," in *Proc. 7th European Microwave Conf.*, (Copenhagen, Denmark), Sept. 1977.

[21] E. O. Hammerstad and F. Bekkadal, *Microstrip Handbook*, SINTEF Report STF 44, A 74169, Feb. 1975.

[22] V. Biontempo and F. Pucci, Experimental characterization of right-angled bends with varying degree of chamfer," *Alta Freq.*, vol. XLIV, pp. 713–717, Nov. 1975.

Full-Wave Analysis of Microstrip Open-End and Gap Discontinuities

ROBERT W. JACKSON, MEMBER, IEEE, AND DAVID M. POZAR, MEMBER, IEEE

Abstract — **A solution is presented for the characteristics of microstrip open-end and gap discontinuities on an infinite dielectric substrate. The exact Green's function of the grounded dielectric slab is used in a moment method procedure, so surface waves as well as space-wave radiation are included. The electric currents on the line are expanded in terms of longitudinal subsectional piecewise sinusoidal modes near the discontinuity, with entire domain traveling-wave modes used to represent incident, reflected, and, for the gap, transmitted waves away from the discontinuity. Results are given for the end admittance of an open-ended line, and the end conductance is compared with measurements. Results are also given for the reflection coefficient magnitude and surface-wave power generation of an open-ended line on substrates with various dielectric constants. Loss to surface and space waves is calculated for a representative gap discontinuity.**

I. INTRODUCTION

THIS PAPER DESCRIBES a "full-wave" solution of the open-end and symmetric gap discontinuities in microstrip line. The solution is rigorous in that space-wave radiation and surface-wave generation from discontinuities is explicitly included through the use of the exact Green's function for a grounded dielectric slab. A moment method procedure is used whereby the electric surface current density on the microstrip line is expanded in terms of four different types of expansion modes: one mode represents a traveling wave incident on the discontinuity, another mode represents a traveling wave reflected from the discontinuity, a third represents a traveling wave transmitted through the discontinuity (gap case only), and a number of subsectional (piecewise sinusoidal) modes are used in the vicinity of the discontinuity to model the nonuniform current in that region. The result is a physically meaningful solution in terms of incident, reflected, and transmitted-wave amplitudes, with only a small number of unknown coefficients to solve for (typically four to five for the open-end case and twice that for the gap case). For the open-end case, the complex reflection coefficient can then be determined, as well as an "end admittance," referred to the end of the microstrip line. For the gap, scattering parameters can be determined. In addition, the amount of real power delivered to radiation and surface waves can be calculated. It is assumed that only the fundamental microstrip mode is propagating on the line away from the open end, although higher order mode fields are accounted for in the vicinity of the discontinuity.

Manuscript received January 11, 1985; revised June 3, 1985. This work was supported in part by the Rome Air Development Center, Hanscom Air Force Base, MA, under U.S. Air Force Contracts F19628-84-K-0022 and F49620-82-C-0035.

The authors are with the Department of Electrical and Computer Engineering, University of Massachusetts, Amherst, MA 01003.

Much of the previous work on the open-circuited microstrip line has used quasi-static approximations, with the results of Hammerstad and Bekkadal [1], [2] being widely referenced. Jansen [3] has calculated length extensions for an enclosed microstrip using a spectral-domain method. Lewin [4] used an assumed current distribution to calculate the radiated power from an open line. James and Henderson [5], [6] developed an improved analysis using a variational technique, including surface-wave effects, and compared their results favorably with measurements of the end conductance of an open-ended line on a thin, low dielectric constant substrate. Compared with the present solution, the results of James and Henderson appear to be quite good for such substrates, and their relatively simple expressions are an advantage computationally.

Likewise, the gap has also been analyzed by predominately quasistatic methods [7]–[10], the results of which are used extensively in computer-aided design routines. Fully electromagnetic solutions have been calculated by Jansen and Koster [11] using a spectral-domain method, but their gap is surrounded on four sides by perfect conductors. None of the aforementioned approaches include surface waves and radiation losses.

There were two motivations for the present work. First, the increasing interest in monolithic and millimeter-wave integrated circuits requires rigorous analyses to characterize such microstrip discontinuities on electrically thick, high dielectric constant substrates (such as GaAs, with $\epsilon_r = 12.8$). Quantities such as radiation and surface waves are more important with such substrates than with thin, low dielectric constant substrates. Second, the present work is an ancillary result from the solution to the problem of a microstrip patch antenna on an electrically thick substrate fed by a microstrip line. This problem is also of interest in terms of MMIC design, and may be addressed in the future.

Section II presents the theory of the solution, which is based on the moment method/Green's function solutions for printed dipole and microstrip patch antennas [12], [13]. The propagation constant for the fundamental mode of an infinite microstrip line is also developed in terms of a "full-wave" solution in this section, and the opportunity is taken to dispel a few myths about propagation on microstrip lines. Section III presents results for terminal conductance and the Δl length extension for $\epsilon_r = 2.32$ and 12.8 substrates, and is compared with measurements and calculations from [1] and [6]. Reflection coefficient magnitudes

Reprinted from *IEEE Trans. Microwave Theory Tech.*, vol. MTT-33, no. 10, pp. 1036–1042, Oct. 1985.

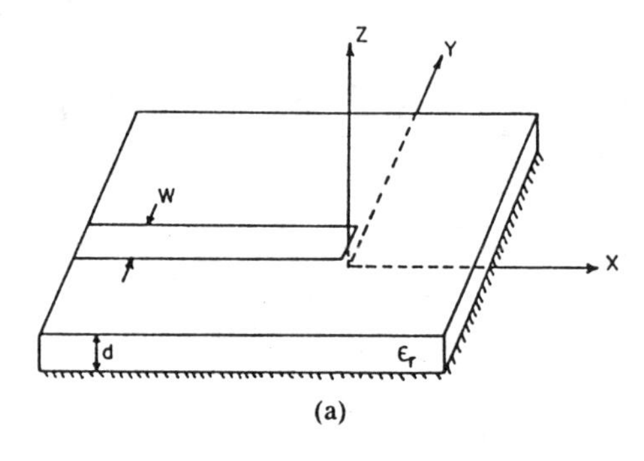

(a)

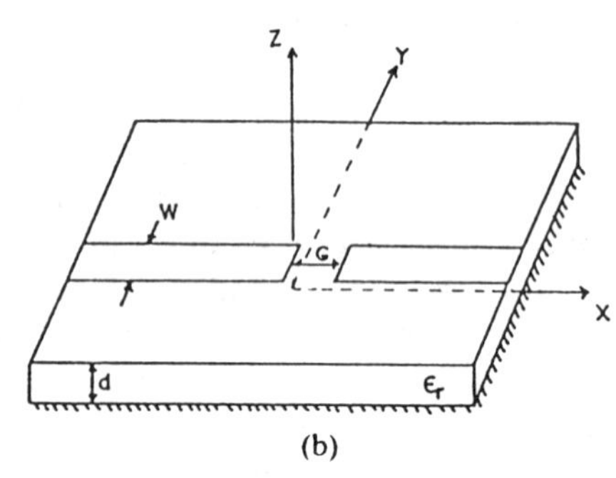

(b)

Fig. 1. Geometry of microstrip open-end and gap discontinuities.

and a radiation efficiency e, defined as the ratio of radiated power to radiated plus surface-wave power [13], are plotted versus substrate thickness for $\epsilon_r = 2.55$ and $\epsilon_r = 12.8$. The fraction of incident power launched into surface waves is seen to increase sharply with increasing substrate thickness and/or dielectric constant. Results are also presented for radiation loss at a representative gap discontinuity on $\epsilon_r = 12.8$.

II. Theory

Fig. 1 shows the geometry of the open-end and gap discontinuities in width W. The substrate is assumed infinitely wide in the x- and y-directions, and of thickness d, and relative permittivity ϵ_r. Only $\hat{x}$-directed electric surface currents are assumed to flow on the microstrip line, which, as was found in [14] and other references, is a good approximation when thin lines (with respect to wavelength) are used on substrates of any thickness.

A. Green's Function for the Grounded Dielectric Slab

The canonical building block for the present solution is the plane-wave spectral representation of the grounded dielectric slab Green's function, representing the $\hat{x}$-directed electric field at (x, y, d) due to an $\hat{x}$-directed infinitesimal dipole of unit strength at (x_0, y_0, d). This field can be written as [12]

$$E_{xx}(x, y|x_0, y_0) = -\iint_{-\infty}^{\infty} Q(k_x, k_y) e^{jk_x(x-x_0)} \cdot e^{jk_y(y-y_0)} \, dk_x \, dk_y \quad (1)$$

where

$$Q(k_x, k_y) = \frac{jZ_0}{4\pi^2 k_0} \frac{(\epsilon_r k_0^2 - k_x^2) k_2 \cos k_1 d + jk_1 (k_0^2 - k_x^2) \sin k_1 d}{T_e T_m} \sin k_1 d \quad (2)$$

$$\begin{aligned} T_e &= k_1 \cos k_1 d + jk_2 \sin k_1 d \\ T_m &= \epsilon_r k_2 \cos k_1 d + jk_1 \sin k_1 d \\ k_1^2 &= \epsilon_r k_0^2 - \beta^2, \qquad \operatorname{Im} k_1 < 0 \\ k_2^2 &= k_0^2 - \beta^2, \qquad \operatorname{Im} k_2 < 0 \\ \beta^2 &= k_x^2 + k_y^2 \\ k_0 &= \omega\sqrt{\mu_0 \epsilon_0} = 2\pi/\lambda_0 \\ Z_0 &= \sqrt{\mu_0/\epsilon_0}. \end{aligned} \quad (3)$$

As discussed in [12], the zeros of the T_e, T_m functions constitute surface-wave poles. During the integration in (1), which is done numerically, special care must be given to these pole contributions, and a method for doing this is presented in [12]. The integration in (1) is further facilitated by a conversion to polar coordinates, as described in [12].

B. Propagation Constant of an Infinite Microstrip line

The solution for the open-circuited line requires the propagation constant of an infinitely long microstrip line. It is assumed that the electrical thickness of the substrate is such that only the fundamental microstrip mode propagates. A quasi-static value [2] could be used with reasonable results, but the more rigorous "full-wave" solution involves only a small fraction of the total effort for the open-circuit problem, and so the propagation constant was computed in this manner. The method is very similar to [14].

Consider an infinitely long microstrip line of width W with a traveling-wave current of the form $e^{-jk_e x_0}$, where k_e is the effective propagation constant to be determined. Substituting this current into (1) and integrating over x_0, y_0 yields the electric field at (x, y, d) due to this line source

$$E_{xx}^l = -2\pi \iint_{-\infty}^{\infty} Q(k_x, k_y)\delta(k_x + k_e) \cdot e^{jk_x x} e^{jk_y y} F_y(k_y) \, dk_x \, dk_y \quad (4)$$

where F_y is the Fourier transform of the distribution of current in the y-direction, which is, for now, assumed uniform. Thus

$$F_y(k_y) = \frac{2\sin(k_y W/2)}{k_y}. \quad (5)$$

Now, the above electric field must vanish at all points on the microstrip line, since it is assumed to be a perfect conductor. This boundary condition is enforced across the width of the strip by integrating on y over the width. After carrying out the k_x integration, the following characteristic equation for k_e results:

$$\int_{-\infty}^{\infty} Q(k_e, k_y) F_y^2(k_y) \, dk_y = 0. \quad (6)$$

This equation can be solved relatively quickly for k_e using a simple search technique, such as the interval halving method. The characteristic impedance of the uniform line (used later) can also be derived from this solution by computing the voltage between the strip and the ground plane [12]. In the interest of brevity, this derivation is not presented.

Two points of interest regarding propagation on uniform microstrip lines can be inferred from the above solution. First, there exists in the literature (for example, [15] and [16]) the idea that surface-wave modes can be excited by the fundamental mode of the uniform microstrip line. This is false, as can be seen by noting that the fundamental propagation constant k_e (as determined numerically) is always greater than any surface-wave pole β_{sw}. Thus, the integration path of (6) never crosses a surface-wave pole, with the result that no surface-wave power is generated by the uniform line. Discontinuities in the line can, of course, excite surface waves, as can higher order propagating modes.

Second, it is sometimes stated that a uniform microstrip line does not radiate *any* power into space waves. Again, this is false, as a stationary phase evaluation of the field above the substrate due to a uniform line will show far-zone radiated power is generated. As a practical matter, however, this loss to radiation is much less than either conductor loss or dielectric loss.

C. Current Expansion Modes

The method of solution for both the open-circuited microstrip line and the gap basically involves expanding the electric surface current density on the line and formulating an integral equation which can be solved by the method of moments for the unknown expansion currents. The choice of basis functions affects the computational efficiency quite significantly, so a judicious choice is important. We first describe in detail the basis functions for the open end and then describe the modifications needed to compute the gap.

In this formulation, only $\hat{x}$-directed currents are assumed, which should be adequate for lines that are not too wide [14]. Sinusoids, several cycles in length, are used to represent incident and reflected traveling waves of the fundamental microstrip mode, and subsectional (piecewise sinusoidal) modes are used near the open end, to represent currents that do not conform to the fundamental mode. This approach thus differs from a recent solution to the microstrip dipole antenna proximity fed by a microstrip line [17], where subsectional expansion modes were used. We also note that, in contrast to [5], the exciting wave is not assumed to be TEM.

Thus, define an incident electric current of unit amplitude as

$$I^{\text{inc}} = e^{-jk_e x} \tag{7}$$

and a reflected current as

$$I^{\text{ref}} = -Re^{jk_e x} \tag{8}$$

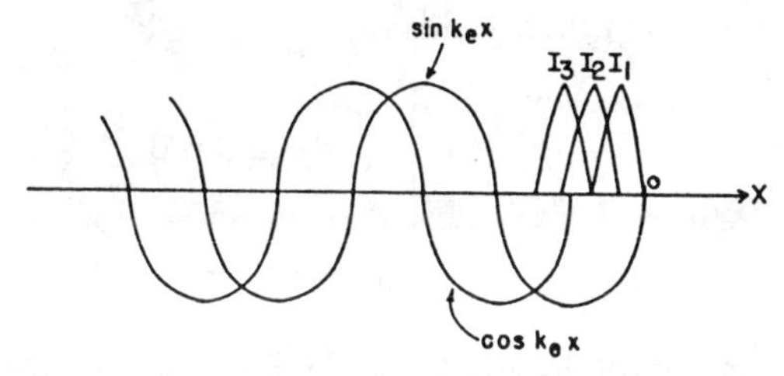

Fig. 2. Layout of expansion modes on the open-ended microstrip line.

where R is the reflection coefficient referenced to the end of the line ($x = 0$). Now, because of the method of numerically integrating (1) [12], it is useful to deal with only real expansion modes; thus, a simple transformation from exponential form to sine and cosine form is made

$$I^{\text{inc}} + I^{\text{ref}} = (1-R)\cos k_e x - j(1+R)\sin k_e x, \qquad x<0. \tag{9}$$

At $x = 0$, the total electric current must be zero. The sine term of the above current satisfies this condition, but not the cosine term, so the cosine term is truncated at $x = -\pi/2k_e$. Also, both terms are truncated after several cycles. The incident and reflected current components can then be written as

$$I^{\text{inc}} + I^{\text{ref}} = (1-R)f_s(k_e x + \pi/2) - j(1+R)f_s(k_e x) \tag{10}$$

where

$$f_s(u) = \begin{cases} \sin u, & 0 > u > -m\pi \\ 0, & \text{otherwise} \end{cases}.$$

It has been found that choosing the length of the sinusoids to be an integer number of half wavelengths speeds the convergence of the integrals shown in the next section. (Physically, this means that no end charges exist on the lines, as would be the case if the end currents were nonzero.) Typically, the solutions are insensitive to sinusoid length for lengths greater than three or four wavelengths.

Piecewise sinusoidal (PWS) modes are defined starting at the end and working left. These modes can be defined as

$$I_n J_n(x,y) = I_n \frac{\sin k_e(h - |x - x_n|)}{\sin k_e h},$$
$$\text{for } |x - x_h| < h, |y| < W/2 \tag{11}$$

where I_n is the unknown expansion coefficient, h is the half-length of the mode, and x_n is the terminal location, which is chosen as $x_n = -nh$, for $n = 1,2,3,\cdots$. The current is assumed uniform across the strip width. Fig. 2 shows how the various modes are arranged on the microstrip line. Typically convergence is achieved with three or four PWS modes.

D. Integral Equation/Moment Method Solution

An integral equation for the discontinuity is written by enforcing the boundary condition that the total $\hat{x}$ electric field due to all the currents on the line must be zero on the line. Equation (1) then yields

$$\int_{x_0}\int_{y_0}\left[I^{\text{inc}}+I^{\text{ref}}+\sum_{n=1}^{N} I_n f_n\right] E_{xx}\, dx_0\, dy_0 = 0,\quad \text{for } x < -\infty, |y| < W/2 \quad (12)$$

where N is the total number of PWS modes. This equation is enforced by multiplying by $N+1$ weighting or test functions (since there are $N+1$ unknowns), taken here as PWS modes as defined in (11) for $n=1$ to $N+1$, and integrating over x and y. Impedance matrix elements can then be defined as

$$Z_{mn} = \iint_{-\infty}^{\infty} Q(k_x, k_y) F_y^2(k_y) F_{xm}(k_x) F_{xn}^*(k_x)\, dk_x\, dk_y \quad (13)$$

$$Z_{mc} = \iint_{-\infty}^{\infty} Q(k_x, k_y) F_y^2(k_y) F_{xm}(k_x) F_{xc}^*(k_x)\, dk_x\, dk_y \quad (14)$$

$$Z_{ms} = \iint_{-\infty}^{\infty} Q(k_x, k_y) F_y^2(k_y) F_{xm}(k_x) F_{xs}^*(k_x)\, dk_x\, dk_y \quad (15)$$

where F_y is the Fourier transform defined in (5), and F_{xn}, F_{xc}, F_{xs} are Fourier transforms of the mode currents

$$F_{xn} = \int_{x_n-h}^{x_n+h} f_n(x) e^{jk_x x}\, dx \quad (16)$$

$$F_{xs} = \int_{-m\pi/k}^{0} \sin k_e x e^{jk_x x}\, dx \quad (17)$$

$$F_{xc} = \exp\left[-jk_x \pi/(2k_e)\right] F_{xs}. \quad (18)$$

This results in a matrix equation for the unknown coefficients $R, I_1, I_2, \cdots, I_N$. For example, with two PWS modes, $N=2$, and a 3×3 matrix equation results

$$\begin{bmatrix} Z_{11} & Z_{12} & -(Z_{1c}+jZ_{1s}) \\ Z_{21} & Z_{22} & -(Z_{2c}+jZ_{2s}) \\ Z_{31} & Z_{32} & -(Z_{3c}+jZ_{3s}) \end{bmatrix} \cdot \begin{bmatrix} I_1 \\ I_2 \\ R \end{bmatrix} = \begin{bmatrix} -Z_{1c}+jZ_{1s} \\ -Z_{2c}+jZ_{2s} \\ -Z_{3c}+jZ_{3s} \end{bmatrix}. \quad (19)$$

Note that the above testing procedure only enforces (12) near the open end, where the testing modes are located. Farther away from the end (but still much greater than $-m\pi/k_e$), (12) is automatically satisfied since then the line looks locally as if it were infinitely long, and (6) implies that the E_x field is near zero. In other words, Z_{nc} and Z_{ns} quickly approach zero as n increases.

E. Surface-Wave Power

With the above formulation, the reflection coefficient of the open-circuited microstrip line can be found. Since the termination is not an *ideal* electrical open circuit, the reflection coefficient magnitude is always less than unity, implying that some incident power is lost to radiation of space and surface waves. In addition, an end admittance Y can be defined using the characteristic impedance Z_0 of the line. Thus

$$Y = \frac{1-R}{Z_0(1+R)}. \quad (20)$$

In order to quantify the separation of the total power loss into space-wave radiation and surface-wave excitation, an efficiency is defined as in [13]

$$e = \frac{P_{\text{rad}}}{P_{\text{rad}} + P_{\text{sw}}} \quad (21)$$

where P_{rad} is the power lost to space waves and P_{sw} is the power lost to surface waves. This efficiency was originally defined for printed antennas [13] and is used here in the interest of consistency. The powers in (21) are found as follows:

$$P_{\text{rad}} + P_{\text{sw}} = \text{Re}\left\{\sum_{i=1}^{N+2}\sum_{j=1}^{N+2} I_i Z_{ij} I_j^*\right\} \quad (22)$$

$$P_{\text{sw}} = \text{Re}\left\{\sum_{i=1}^{N+2}\sum_{j=1}^{N+2} I_i Z_{ij}^{sw} I_j^*\right\} \quad (23)$$

where

$$I_i = \begin{cases} I_i, & \text{for } 1 \leq i \leq N \\ (1-R), & \text{for } i = N+1 \\ -j(1+R), & \text{for } i = N+2 \end{cases} \quad (24)$$

and Z_{ij} is an impedance matrix element with indices i, j for $i, j \leq N$, $i, j = c$ for $i, j = N+1$, and $i, j = s$ for $i, j = N+2$. The Z_{ij}^{sw} elements represent only the surface-wave contribution of the Z_{ij} elements, as computed from the residues of the surface-wave poles [12].

F. Gap Formulation

It is not difficult to modify the formulation for the open end in order to compute gap-discontinuity parameters. The configuration is shown in Fig. 1(b) with a gap G between the input and output microstrip lines. To analyze this discontinuity, three entire domain modes are used to represent incident, reflected, and transmitted currents. In addition to I^{inc} and I^{ref}, we therefore add

$$I^{\text{tr}} = Te^{-jk_e(x-G)} \quad (25)$$

which is then modified in a manner similar to (9) and (10) to eliminate current discontinuities and to impose a finite length. This results in

$$I^{\text{tr}} = T\left[-g\left(k_e[x-G]-\frac{\pi}{2}\right) - jg\left(k_e[x-G]\right)\right] \quad (26)$$

where

$$g(u) = \begin{cases} \sin u, & 0 < u < m\pi \\ 0, & \text{otherwise} \end{cases}$$

which is then placed, with I^{inc} and I^{ref}, in (12) along with additional piecewise sinusoidal modes (eq. (11)). Piecewise modes will now exist at $x_n = -nh$ and at $x_n = nh + G$ for

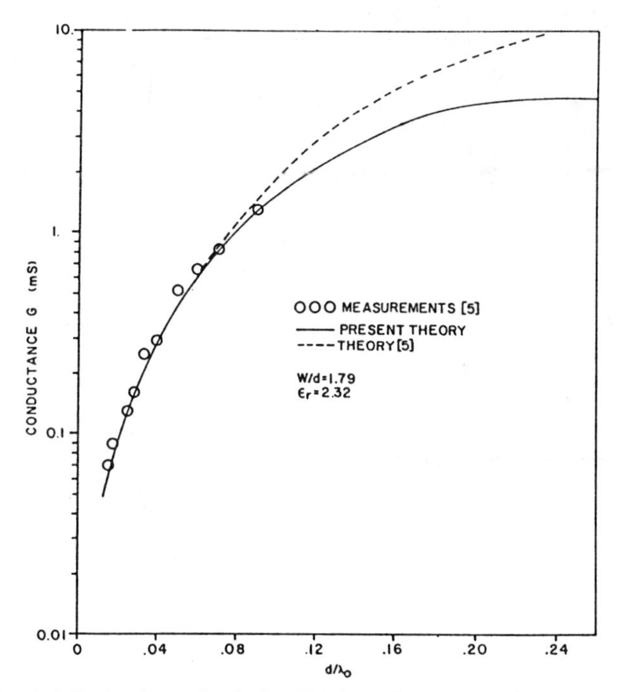

Fig. 3. Comparison of calculated end conductance of an open-circuit microstrip line compared with measurements and calculations of [5].

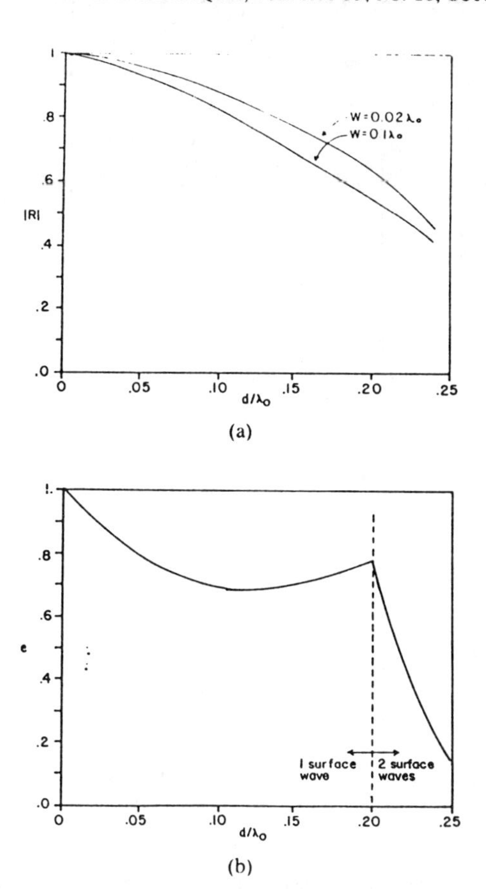

Fig. 4. (a) Reflection coefficient magnitude and (b) efficiency for an open-circuit microstrip line on an $\epsilon_r = 2.55$ substrate.

$n = 1, 2, \cdots, N$. This gives a total of $2N$ piecewise modes. Equation (12) has thus been modified such that R, T, and I_n are $2N+2$ unknowns for which one can solve once (12) has been tested with $2N+2$ piecewise testing functions. The remaining formulation is analogous to (13)–(19). We note that in computing the impedance matrix elements there are several redundancies due to reciprocity and due to the physical symmetry of the gap configuration. Making use of these redundancies considerably reduces computation time.

III. Results

Fig. 3 shows the terminal conductance of an open-circuited microstrip line as computed by this theory and compared with the measurements of [5] and calculations of [5] and [6]. The agreement of both theories and the measured data is good for substrate thickness up to about $0.1\lambda_0$, while the theories depart slightly above this value. In contrast to this theory, James and Henderson assume a TEM parallel-plate mode as an excitation. For thicker, higher dielectric constant substrates, their assumption is questionable and the two theories may diverge more readily. Note the trend that, as the substrate thickness increases, the termination looks less like an ideal open circuit.

Fig. 4(a) and (b) shows the reflection coefficient magnitude and efficiency e versus substrate thickness for $\epsilon_r = 2.55$, and various microstrip line widths. The efficiency e is practically independent of widths. Observe from Fig. 4(a) that the reflection coefficient magnitude drops well below unity for substrate thicknesses greater than a few hundreths of a wavelength, and is smaller for wider strips, as would be expected. The efficiency data of Fig. 4(b) shows that very little of the total power loss is caused by surface-wave excitation when the substrate is thin, but that the surface-wave power increases (e decreases) for thicker substrates, with a cusp in the data at the cutoff point of the TE_1 surface-wave mode. This curve is very similar to that obtained for printed antennas [12], [18]. Fig. 5(a) and (b) shows corresponding data for a substrate with $\epsilon_r = 12.8$. It can be seen that the reflection coefficient magnitude drops off more rapidly with increased permittivity, and that significantly more power is launched into surface waves, for a given substrate thickness.

The data given in Figs. 3 and 4 allow one to determine the amount of power loss and the amount of surface-wave power generation of an open-circuit line. For example, assume that $1w$ of power is incident on an open microstrip line of width $0.1\lambda_0$ on a GaAs ($\epsilon_r = 12.8$) substrate $0.04\lambda_0$ thick. Then, from Fig. 5(a) and (b), $|R| = 0.96$ and $e = 0.53$, so there is $0.922w$ reflected on the line, $0.0416w$ delivered to space-wave radiation, and $0.0368w$ delivered to surface waves.

For MIC design, an open-circuit microstrip line is often modeled as having a reflection coefficient with unit magnitude and a phase accounted for by a length extension $\Delta l/d$. When radiation loss occurs, a conductance in parallel with a length extension is necessary, yielding

$$YZ_0 = GZ_0 + j\tan(k_e \Delta l) \qquad (27)$$

where Y is given in (20). Fig. 6 gives normalized end conductance for common microstrip parameters on a Gal-

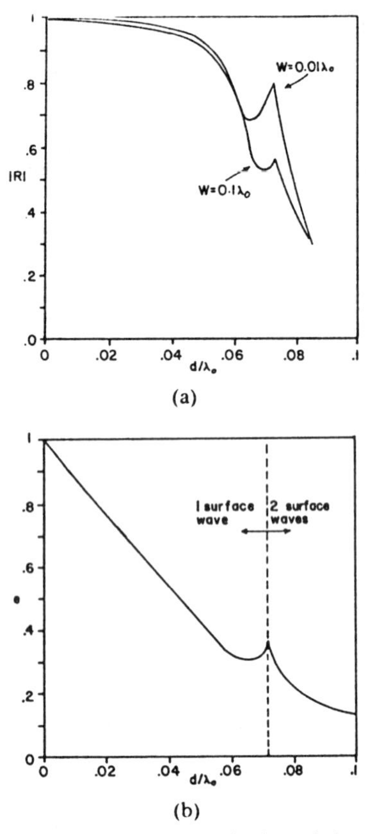

Fig. 5. (a) Reflection coefficient magnitude and (b) efficiency for an open-circuit microstrip line on an $\epsilon_r = 12.8$ substrate.

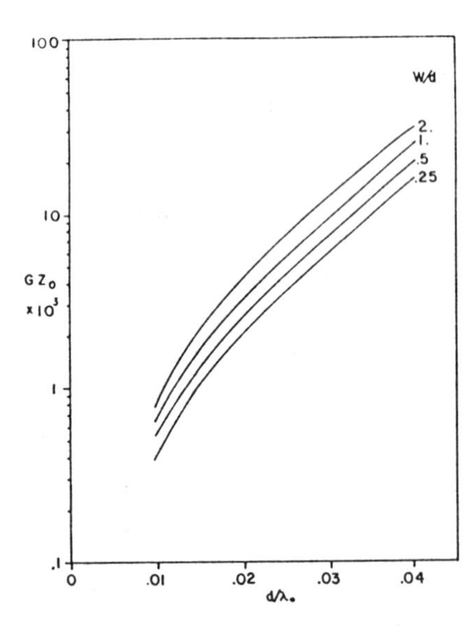

Fig. 6. Normalized end conductance for several common microstrip parameters on $\epsilon_r = 12.8$.

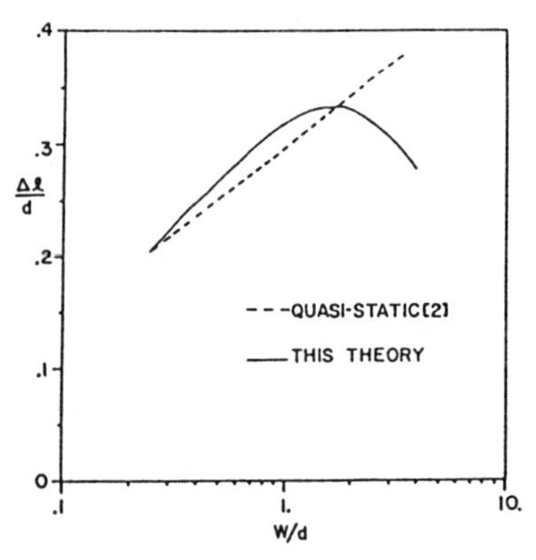

Fig. 7. Calculated length extension of an open-circuit microstrip line compared with quasi-state theory [2] on a substrate with $\epsilon_r = 12.8$, $d = 0.02\lambda_0$.

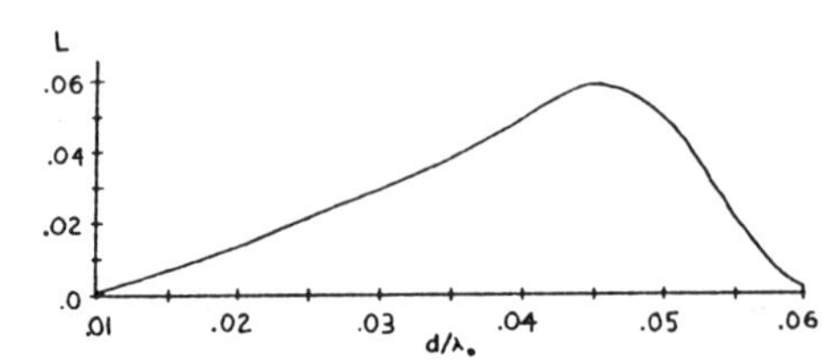

Fig. 8. Loss for a microstrip gap with $\epsilon_r = 12.8$, $G = 0.2d$, $w = 2.5d$.

lium Arsenide substrate. For lossless cases, the length extension has been computed by quasi-static analysis [1], [2], as well as other methods [3], [5]. In contrast to the end conductance and reflection coefficient magnitude, we have found the length extension to be much more sensitive to the number of expansion modes, current distribution across the microstrip line, and truncation of the integrations in (13)–(15), especially for thin substrates. Instead of a uniform current distribution with respect to y, we found that a current distributed to enforce the edge condition

$$I^{\text{inc}}, I^{\text{ref}} \propto \frac{1}{\sqrt{1-(2y/w)^2}} \tag{28}$$

gave better agreement with quasistatic results in the thin substrate limit. Fig. 7 presents calculated results for the length extension using this theory and the quasi-static result [2]. Good agreement occurs for narrow lines but not for lines greater than two substrate thicknesses. Based on this result and on results with substrates having smaller dielectric constants, we conclude that the simple transverse current distribution assumed in (28) will give a reasonable length extension result for widths less than an eighth of a wavelength in the dielectric. For precise length extension calculations, or for wider microstrip lines, a more complicated transverse current distribution is necessary. In this paper, our calculation of length extension is primarily to help validate the less sensitive conductance calculations.

For the gap, we have calculated total power loss due to the combination of surface-wave and space-wave radiation by computing

$$L = 1 - |R|^2 - |T|^2.$$

In Fig. 8, we plot this quantity versus substrate electrical thickness for a representative configuration on Gallium Arsenide. Loss increases, peaks, and then decreases as frequency increases. The decrease occurs due to the fact that a gap, intuitively considered as a series capacitance, looks like a series short at high frequencies and thus like less of a radiation-producing discontinuity. This loss data is insensitive to the distribution of current in the transverse direction. We have calculated circuit models for a gap using this method and found models which roughly agree with quasi-static models but which are somewhat sensitive to transverse current distribution. For very accurate calculation of gap and fringing capacitances, a higher degree of modal approximation is desirable. Investigations of this type are underway.

IV. Conclusion

A full-wave analysis has been presented for the problems of microstrip open-end and gap discontinuities. For the open end, the reflection coefficient, radiated power, and surface-wave power have been calculated and compared with previous calculations and measured data, when available. Plots of end conductance and length extension have been presented for a high dielectric substrate. Loss at a gap discontinuity has also been calculated. This type of analysis should aid in the design of microwave integrated circuits, particularly for higher frequencies and high dielectric constant substrates. Similar analysis can be used to characterize more complicated microstrip discontinuities.

References

[1] E. O. Hammerstad and F. Bekkadal, "Microstrip handbook," ELAB Rep. STF44A74169, University of Trondheim, 1975.

[2] K. C. Gupta, R. Garg, and I. J. Bahl, *Microstrip Lines and Slotlines*. Dedham, MA: Artech House, 1979.

[3] R. Jansen, "Hybrid mode analysis of end effects of planar microwave and millimeterwave transmission lines," *Proc. Inst. Elec. Eng.*, vol. 128, pt. H. pp. 77–86, Apr. 1978.

[4] L. Lewin, "Radiation from discontinuities in stripline," *Proc. Inst. Elec. Eng.*, vol. 107C, pp. 163–170, Feb. 1960.

[5] J. R. James and A. Henderson, "High frequency behavior of microstrip open-circuit terminations," *IEE J. Microwave Opt. Acoust.*, vol. 3, pp. 205–211, Sept. 1979.

[6] J. R. James, P. S. Hall, and C. Wood, *Microstrip Antenna Theory and Design*. London: Peter Peregrinus, 1981.

[7] M. Maeda, "An analysis of gaps in microstrip transmission lines," *IEEE Trans. Microwave Theory Tech.*, vol. MTT-20, pp. 390–396, June 1972.

[8] P. Benedek and P. Silvester, "Equivalent capacitances for microstrip gaps and steps," *IEEE Trans. Microwave Theory Tech.*, vol. MTT-20, pp. 729–733, Nov. 1972.

[9] Y. Rahmat-Samii, T. Itoh, and R. Mittra, "A spectral domain analysis for solving microstrip discontinuity problems," *IEEE Trans. Microwave Theory Tech.*, vol. MTT-24, pp. 372–378, Apr. 1984.

[10] A. Gopinath and C. Gupta, "Capacitance parameters of discontinuities in microstrip lines," *IEEE Trans. Microwave Theory Tech.*, vol. MTT-26, pp. 831–836, Oct. 1978.

[11] R. Jansen and N. Koster, "A unified CAD basis for the frequency dependent characterization of strip slot and coplanar MIC components," in *Proc. 11th Eur. Microwave Conf.* (Amsterdam), 1981, pp. 682–687.

[12] D. M. Pozar, "Input impedance and mutual coupling of rectangular microstrip antennas," *IEEE Trans. Antennas Propagat.*, vol. AP-30, pp. 1191–1196, Nov. 1982.

[13] D. M. Pozar, "Considerations for millimeter wave printed antennas," *IEEE Trans. Antennas Propagat.*, vol. AP-31, pp. 740–747, Sept. 1983.

[14] T. Itoh and R. Mittra, "Spectral-domain approach for calculating the dispersion characteristics of microstrip lines," *IEEE Trans. Microwave Theory Tech.*, vol. MTT-21, pp. 496–499, July 1973.

[15] C. P. Hartwig, D. Masse, and R. A. Pucel, "Frequency dependent behavior of microstrip," in *1968 G-MTT Int. Symp. Dig.*, pp. 11–116.

[16] I. J. Bahl and D. K. Trivedi, "A designer's guide to microstrip line," *Microwaves*, pp. 174–182, May 1977.

[17] P. B. Katehi and N. G. Alexopoulos, "On the modeling of electromagnetically coupled microstrip antennas—The printed strip dipole," *IEEE Trans. Microwave Theory Tech.*, this issue, pp. 1029–1035.

[18] P. B. Katehi and N. G. Alexopoulos, "On the effect of substrate thickness and permittivity on printed circuit dipole properties," *IEEE Trans. Antennas Propagat.*, vol. AP-31, pp. 34–39, Jan. 1983.

Paper 7.5

A Dynamic Model for Microstrip–Slotline Transition and Related Structures

HUNG-YU YANG, STUDENT MEMBER, IEEE, AND NICÓLAOS G. ALEXÓPOULOS, FELLOW, IEEE

Abstract —An analysis of microstrip to slotline transition is presented. The method of moments is applied to the coupled integral equations. In the formulation, the Green's function for the grounded dielectric substrate, which takes into account all the radiation, surface wave, and substrate effects, is used. Meanwhile, all the mutual coupling effects are included in the method of moments solution. Certain related structures, such as slotline and microstrip discontinuities, a slot fed by a microstrip line, and a printed strip dipole fed by a slotline, can also be solved with this analysis. The present approach may find applications to other related transitions in MIC design.

I. Introduction

QUASI-STATIC METHODS, equivalent waveguide models, and equivalent circuit models have been widely used in the modeling of microstrip or slot type discontinuities in microwave and millimeter-wave devices in the past [1]. Recently, a more rigorous approach, which takes into account all the physical effects including radiation and surface waves, has been applied to certain microstrip discontinuities [2]–[5]. In this scheme, the method of moments, which determines the current on the strip or electric field on the slot, is implemented in the solution of the Pocklington integral equation. The exact Green's function for a grounded dielectric substrate due to either an electric dipole or a magnetic dipole has been used, which includes all the physical effects. Based on this approach, a dynamic model for microstrip–slotline transition and its related structures such as a microstrip-fed slot and a slotline-fed printed dipole is proposed in this paper. The developed model with some modifications can be applied to other types of transitions in MIC and MMIC design.

In microstrip–slotline transition, a short-circuit slotline which is etched on one side of the substrate is crossed at a right angle by an open-circuit microstrip on the opposite side. This type of transition makes the two-level circuit design possible [1]. Some experimental work has been reported in [6], [7] and a transmission line circuit model was reported in [8]. In the present approach, the radiation and surface waves due to the cross-junction, the line discontinuities, and all the mutual coupling due to the dominant mode as well as higher order modes of each line are included in the method of moments solution. The

Manuscript received March 30, 1987; revised September 3, 1987. This work was supported in part by the U.S. Army Research Office under Contract DAAG 03-83-k-0090 and in part by TRW-Micro E56129R85S.

The authors are with the Electrical Engineering Department, University of California, Los Angeles, CA 90024.

IEEE Log Number 8717988.

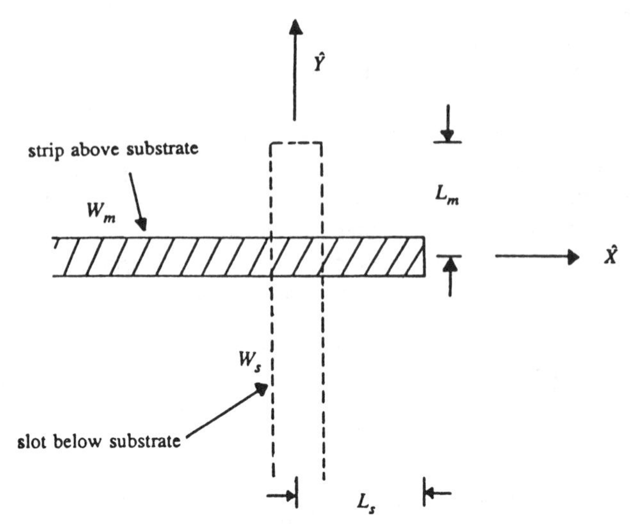

Fig. 1. Top view of microstrip–slotline transition.

VSWR and input impedance of the transition can be determined by the current distribution on the microstrip line in conjunction with transmission line theory. One of the main features of the present method is the modeling of two coupled half-infinite lines where expansion modes are composed of piecewise-sinusoidal modes and traveling-wave and standing-wave modes. In the formulation procedure, certain important problems in MIC, MMIC, or printed antenna design can also be solved. When only one of the lines is discussed, the present approach is simplified to the modeling of slotline or microstrip line discontinuities as reported in [2]–[5]. If one of the lines is of finite length, then the method presented herein can be applied to the modeling of microstrip line-fed slot or slotline-fed printed strip dipole.

II. Theory

A. Green's Function Formulation

The microstrip to slotline transition is shown in Fig. 1 and the cross section is shown in Fig. 2, where the lines are extended a certain distance beyond the cross-junction, so that their extension may act as a tuning stub. Due to the fact that the width of the lines is assumed to be much smaller than the wavelength, the transverse vector components (J_y and M_{mx}) on the lines are a second-order effect, and are neglected for simplicity. Therefore, only the $\hat{x}$-directed electric surface current J_x on the strip and the

Reprinted from *IEEE Trans. Microwave Theory Tech.*, vol. 36, no. 2, pp. 286–293, Feb. 1988.

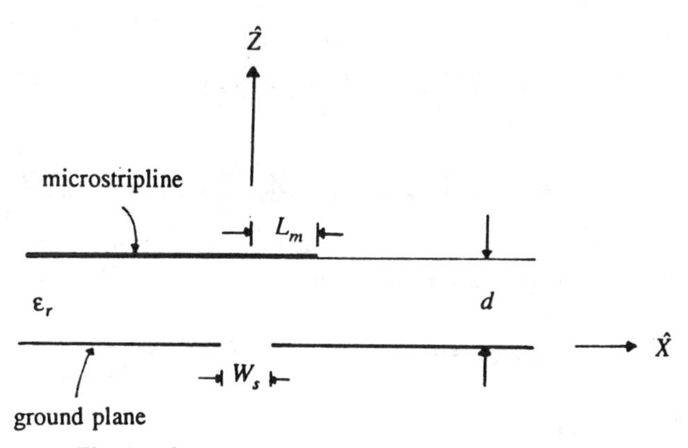

Fig. 2. Cross section of microstrip–slotline transition.

$\hat{y}$-directed magnetic surface current M_{my} ($\hat{x}$-directed electric field E_x) are considered. Under the above assumptions, the coupled integral equations can be formulated in terms of E_x and H_y. As a result, the electric field E_x at (x, y, d) due to the presence of both strip and slot is

$$E_x = \iint G_{xx} J_x \, ds_m + \iint G_{xy} M_{my} \, ds_s, \tag{1a}$$

and the difference of the magnetic field H_y at $(x, y, 0^+)$ and at $(x, y, 0^-)$ is

$$\Delta H_y = \iint G_{yx} J_x \, ds_m + \iint G_{yy} M_{my} \, ds_s, \tag{1b}$$

where G_{xx} and G_{yx} are the dyadic Green's function components due to an $\hat{x}$-directed infinitesimal electric dipole at $z = d$, and G_{xy} and G_{yy} are the dyadic Green's function components due to a $\hat{y}$-directed infinitesimal magnetic dipole at $z = 0$. J_x is the current on the microstrip s_m while M_{my} is the magnetic current (electric field) on the slotline s_s. The dyadic Green's function components G_{xx}, G_{xy}, G_{yx}, and G_{yy} can be obtained by solving the boundary value problem. The results when transformed into the spectral domain can be expressed as

$$G_{xx}(x, y|x_0, y_0) = \iint_{-\infty}^{\infty} \bar{G}_{xx}(\lambda_x, \lambda_y) e^{j\lambda_x(x-x_0)} \cdot e^{j\lambda_y(y-y_0)} \, d\lambda_x \, d\lambda_y \tag{2a}$$

$$G_{xy}(x, y|x_0, y_0) = \iint_{-\infty}^{\infty} \bar{G}_{xy}(\lambda_x, \lambda_y) e^{j\lambda_x(x-x_0)} \cdot e^{j\lambda_y(y-y_0)} \, d\lambda_x \, d\lambda_y \tag{2b}$$

$$G_{yx}(x, y|x_0, y_0) = -G_{xy}(x, y|x_0, y_0) \tag{2c}$$

and

$$G_{yy}(x, y|x_0, y_0) = \iint_{-\infty}^{\infty} \bar{G}_{yy}(\lambda_x, \lambda_y) e^{j\lambda_x(x-x_0)} \cdot e^{j\lambda_y(y-y_0)} \, d\lambda_x \, d\lambda_y \tag{2d}$$

where

$$\bar{G}_{xx}(\lambda_x, \lambda_y) = \frac{-jZ_0}{4\pi^2 k_0 \epsilon_r} \cdot \left[\frac{k_1^2 - \lambda_x^2}{D_e(\lambda)} + \frac{\lambda_x^2 q_1 (1-\epsilon_r) \sinh(q_1 d)}{D_e(\lambda) D_m(\lambda)} \right] \sinh(q_1 d) \tag{3a}$$

$$\bar{G}_{xy}(\lambda_x, \lambda_y) = \frac{-1}{4\pi^2} \cdot \left[\frac{q_1}{D_e(\lambda)} + \frac{\lambda_x^2 (\epsilon_r - 1) \sinh(q_1 d)}{D_e(\lambda) D_m(\lambda)} \right] \tag{3b}$$

$$\bar{G}_{yy}(\lambda_x, \lambda_y) = \frac{-j}{4\pi^2 Z_0 k_0} \cdot \left[(k_1^2 - \lambda_y^2) \cdot \frac{q_1 \cosh(q_1 d) + \epsilon_r q \sinh(q_1 d)}{q_1 D_m(\lambda)} + \frac{\lambda_y^2 q (1-\epsilon_r)}{D_e(\lambda) D_m(\lambda)} + \frac{(k_0^2 - \lambda_y^2)}{q} \right] \tag{3c}$$

$$\begin{aligned} D_e(\lambda) &= q \sinh(q_1 d) + q_1 \cosh(q_1 d) \\ D_m(\lambda) &= q_1 \sinh(q_1 d) + \epsilon_r q \cosh(q_1 d) \\ q &= \sqrt{\lambda^2 - k_0^2} \\ q_1 &= \sqrt{\lambda^2 - k_1^2} \\ \lambda &= \sqrt{\lambda_x^2 + \lambda_y^2} \\ k_0 &= \omega\sqrt{\mu_0 \epsilon_0} \\ k_1 &= \omega\sqrt{\mu_0 \epsilon_0 \epsilon_r} \\ Z_0 &= \sqrt{\mu_0/\epsilon_0}. \end{aligned} \tag{3d}$$

The integrands of (2a)–(2d) have poles whenever $D_e(\lambda)$ and/or $D_m(\lambda)$ become zero. The zeros of $D_e(\lambda)$ correspond to TE surface waves, while the zeros of $D_m(\lambda)$ correspond to TM surface waves [9]. The existence of these poles causes some problems in numerical integration. An efficient way of evaluating this type of integral is by a pole extraction technique, according to which the pole behavior is extracted out of the integrand, leaving a continuous function which may be integrated directly [10], [11]. The integrations in (2a)–(2d) may further be converted into polar coordinates such that a double infinite integration may be reduced to a finite and an infinite one.

B. The Choice of Expansion Modes

In the method of moments procedure, the J_x and M_{my} are expanded in terms of a set of known functions. The transverse dependence of J_x and M_{my} is assumed to be Maxwell's distribution, that is,

$$J_x = f(x) \cdot J_t(y)$$

and

$$M_{my} = g(y) \cdot M_t(x) \tag{4}$$

with

$$J_t(y) = \frac{1}{\pi W_m/2\sqrt{1-\left(\dfrac{y}{W_m/2}\right)^2}}$$

and

$$M_t(x) = \frac{1}{\pi W_s/2\sqrt{1-\left(\dfrac{x}{W_s/2}\right)^2}}$$

such that the edge condition is satisfied. For the transition under consideration, the modeling of two half-infinite lines is necessary, in which several mechanisms are possible. If subsectional expansion modes are used in the two finite lines with a δ gap source, then in order to characterize the cross-junction of this resonator, two sets of wave amplitudes on each line, corresponding to two different lengths of the parasitic line, are required. This scattering matrix formulation is found here to be very sensitive to error. A more reliable method is to simulate the physical situation where both the microstrip line and the slotline are terminated by a matched load. In this scheme, subsection expansion modes are used in both lines near the cross-junction region, while entire domain traveling waves are used to represent the transmitted wave in the parasitic line (slotline here) and the incident wave and the reflected wave on the feed line (microstrip line here). This method is similar to that proposed in [4], where only microstrip discontinuities are considered. Other choices of expansion modes are also possible, such as using subsectional basis functions in the feed line and subsectional modes and traveling-wave modes in the parasitic line, or the traveling-wave modes starting away from the cross-junction, such that the mutual coupling of traveling wave and PWS modes on different lines is negligible. These types of expansion modes have some advantages in the numerical analysis and will be discussed in the next section.

The traveling-wave mode used in the expansion function corresponds to the fundamental propagating mode of the microstrip line or the slotline. The microstrip line and slotline propagation constants k_m and k_s, respectively, can be obtained through the following characteristic equations [4] and [5]:

$$\int_{-\infty}^{\infty} \tilde{G}_{xx}(k_m, \lambda_y) F_y^2(\lambda_y)\, d\lambda_y = 0 \tag{5a}$$

$$\int_{-\infty}^{\infty} \tilde{G}_{yy}(\lambda_x, k_s) F_x^2(\lambda_x)\, d\lambda_x = 0 \tag{5b}$$

where

$$F_y(\lambda_y) = J_0(\lambda_y W_m/2) \tag{6a}$$

and

$$F_x(\lambda_x) = J_0(\lambda_x W_s/2). \tag{6b}$$

From the information of k_m and k_s, the unknown current distribution on the microstrip line can be expanded as

$$f(x) = I^{\text{inc}} + I^{\text{ref}} + \sum_{n=1}^{N} I_n f_n(x) \tag{7}$$

while the unknown electric field distribution in the slotline can be expanded as

$$g(y) = V^t + \sum_{n=1}^{M} E_n g_n(y) \tag{8}$$

where

$$I^{\text{inc}} = e^{-jk_m x} \tag{9a}$$

$$I^{\text{ref}} = -\Gamma e^{jk_m x} \tag{9b}$$

and

$$V^t = T e^{jk_s y}. \tag{9c}$$

Piecewise-sinusoidal (PWS) modes are used as subsectional expansion modes and are defined starting from the end of each line. These modes are defined as

$$f_n(x) = \frac{\sin k_e(h - |x + nh|)}{\sin k_e h} \quad \text{for } |x+nh| < h,\quad |y| < W_m/2 \tag{10a}$$

and

$$g_n(y) = \frac{\sin k_e(h - |y + nh|)}{\sin k_e h} \quad \text{for } |y+nh| < h,\quad |x| < W_s/2 \tag{10b}$$

where h is the half-length of the PWS mode. The choice of k_e can be quite arbitrary. Here k_e is chosen as $\sqrt{(\epsilon_r+1)/2}\, k_0$ such that it is close to its physical value and it facilitates the numerical integration when the asymptotic extraction technique is applied [12].

C. The Method of Moments and Matrix Formulation

The coupled integral equations can be obtained from (1a) and (1b) by forcing the boundary conditions that the E-field must be zero on the microstrip line and the H-field must be continuous across the slotline. When the expansion modes are substituted into (1a) and (1b),

$$\iint \left[I^{\text{inc}} + I^{\text{ref}} + \sum_{n=1}^{N} I_n f_n \right] J_t(y_0) G_{xx}(x, y|x_0, y_0)\, dx_0\, dy_0 + \iint \left[V^t + \sum_{n=1}^{M} E_n g_n \right] M_t(x_0) G_{xy}(x, y|x_0, y_0)\, dx_0\, dy_0 = 0 \tag{11}$$

is obtained for each (x, y) on the microstrip line and

$$\iint\left[I^{\text{inc}}+I^{\text{ref}}+\sum_{n=1}^{N} I_n f_n\right] J_t(y_0) G_{yx}(x, y|x_0, y_0)\, dx_0\, dy_0 + \iint\left[V^t+\sum_{n=1}^{M} E_n g_n\right] M_t(x_0) G_{yy}(x, y|x_0, y_0)\, dx_0\, dy_0 = 0 \quad (12)$$

is obtained for every (x, y) in the slotline.

In (11) and (12), there are $M+N+2$ unknowns. In Galerkin's procedure, $N+1$ PWS testing functions on the microstrip line are applied to (11), while $M+1$ PWS testing functions in the slotline are applied to (12), to obtain $M+N+2$ linear equations with $M+N+2$ unknowns which, when expressed in matrix form, are

$$\begin{bmatrix} [Z_{\text{self}}] & [Z_{\text{tself}}] & [T_{\text{meact}}] & [T_{\text{tmeact}}] \\ [T_{\text{emact}}] & [T_{\text{temact}}] & [Y_{\text{self}}] & [Y_{\text{tself}}] \end{bmatrix} \begin{bmatrix} [I] \\ -\Gamma \\ [E] \\ T \end{bmatrix} = \begin{bmatrix} [I_{\text{inc}}] \\ [V_{\text{tra}}] \end{bmatrix} \quad (13)$$

where $[Z_{\text{self}}]$ is an $(N+1)\times N$ matrix with matrix elements

$$Z_{\text{self}}^{nm} = \iint_{-\infty}^{\infty} \bar{G}_{xx}(\lambda_x, \lambda_y) F_y^2(\lambda_y) A^2(\lambda_x) \cdot \cos[\lambda_x(m-n)h]\, d\lambda_x\, d\lambda_y \quad (14)$$

$[Z_{\text{tself}}]$ is an $(N+1)\times 1$ column vector with elements

$$Z_{\text{tself}}^{n} = \iint_{-\infty}^{\infty} \bar{G}_{xx}(\lambda_x, \lambda_y) F_y^2(\lambda_y) P(\lambda_x) \cdot A(\lambda_x) e^{jnh\lambda_x}\, d\lambda_x\, d\lambda_y \quad (15)$$

$[Y_{\text{self}}]$ is an $(M+1)\times M$ matrix with matrix elements

$$Y_{\text{self}}^{mn} = \iint_{-\infty}^{\infty} \bar{G}_{yy}(\lambda_x, \lambda_y) F_x^2(\lambda_x) A^2(\lambda_y) \cdot \cos[\lambda_y(m-n)h]\, d\lambda_x\, d\lambda_y \quad (16)$$

$[Y_{\text{tself}}]$ is an $(M+1)\times 1$ column vector with elements

$$Y_{\text{tself}}^{m} = \iint_{-\infty}^{\infty} \bar{G}_{yy}(\lambda_x, \lambda_y) F_x^2(\lambda_x) Q(\lambda_y) \cdot A(\lambda_y) e^{jmh\lambda_y}\, d\lambda_x\, d\lambda_y \quad (17)$$

$[T_{\text{meact}}]$ is an $(N+1)\times M$ matrix with matrix elements

$$T_{\text{meact}}^{nm} = \iint_{-\infty}^{\infty} \bar{G}_{xy}(\lambda_x, \lambda_y) F_x(\lambda_x) F_y(\lambda_y) A(\lambda_x) A(\lambda_y) \cdot \cos[\lambda_x(L_m - nh)] \cos[\lambda_y(L_s - mh)]\, d\lambda_x\, d\lambda_y \quad (18)$$

$[T_{\text{emact}}]$ is an $(M+1)\times N$ matrix with matrix elements

$$T_{\text{emact}}^{mn} = -T_{\text{meact}}^{nm} \quad (19)$$

$[T_{\text{tmeact}}]$ is an $(N+1)\times 1$ column vector with elements

$$T_{\text{tmeact}}^{n} = \iint_{-\infty}^{\infty} \bar{G}_{xy}(\lambda_x, \lambda_y) F_x(\lambda_x) F_y(\lambda_y) Q(\lambda_y) A(\lambda_x) \cdot \cos[\lambda_x(L_m - nh)] e^{j\lambda_y L_s}\, d\lambda_x\, d\lambda_y \quad (20)$$

$[T_{\text{temact}}]$ is an $(M+1)\times 1$ column vector with elements

$$T_{\text{temact}}^{m} = \iint_{-\infty}^{\infty} \bar{G}_{yx}(\lambda_x, \lambda_y) F_x(\lambda_x) F_y(\lambda_y) P(\lambda_x) A(\lambda_y) \cdot \cos[\lambda_y(L_s - mh)] e^{j\lambda_x L_m}\, d\lambda_x\, d\lambda_y \quad (21)$$

$[V_{\text{tra}}]$ is an $(M+1)\times 1$ column vector with elements

$$V_{\text{tra}}^{m} = -\iint_{-\infty}^{\infty} \bar{G}_{yx}(\lambda_x, \lambda_y) F_x(\lambda_x) F_y(\lambda_y) R(\lambda_x) A(\lambda_y) \cdot \cos[\lambda_y(L_s - mh)] e^{j\lambda_x L_m}\, d\lambda_x\, d\lambda_y \quad (22)$$

$[I_{\text{inc}}]$ is an $(N+1)\times 1$ column vector with elements

$$I_{\text{inc}}^{n} = -\iint_{-\infty}^{\infty} \bar{G}_{xx}(\lambda_x, \lambda_y) F_y^2(\lambda_y) R(\lambda_x) \cdot A(\lambda_x) e^{jnh\lambda_x}\, d\lambda_x\, d\lambda_y \quad (23)$$

$$A(x) = 2k_e \frac{(\cos k_e h - \cos xh)}{(x^2 - k_e^2)} \quad (24)$$

$$P(\lambda_x) = \left[e^{(-j\lambda_x k_m/2k_e)} + j\right] \int_{-\infty}^{0} \sin k_m x e^{j\lambda_x x}\, dx \quad (25)$$

$$Q(\lambda_y) = \left[e^{(-j\lambda_y k_s/2k_e)} + j\right] \int_{-\infty}^{0} \sin k_s y e^{j\lambda_y y}\, dy \quad (26)$$

$$R(\lambda_x) = \left[e^{(-j\lambda_x k_m/2k_e)} - j\right] \int_{-\infty}^{0} \sin k_m x e^{j\lambda_x x}\, dx. \quad (27)$$

$[I]$ is an $(N)\times 1$ column vector with elements $I_1, I_2, \cdots, I_N$; $[E]$ is an $(M)\times 1$ column vector with elements $E_1, E_2, \cdots, E_M$; and L_s and L_m are the stub length for slotline and microstrip, respectively.

D. Some Aspects in Numerical Analysis

The numerical integrations in (14)–(23) are computed after transforming into polar coordinates. The convergence of the integrations depends mainly upon the Green's function. Since $\bar{G}_{xy}(\lambda_x, \lambda_y)$ or $\bar{G}_{yx}(\lambda_x, \lambda_y)$ decays exponentially, with the decay factor proportional to the thickness of the substrate, (18)–(22) are integrated directly. The integrands in (14)–(17) and (23) are slowly convergent. Therefore the asymptotic extraction technique [12] is applied, which requires additional computations of the self and mutual reactions of the PWS and traveling-wave modes in homogeneous medium with $\epsilon_{\text{eff}} = \sqrt{(\epsilon_r+1)/2}$.

Because of the method of numerical integration in (14)–(23), it is convenient to deal with real expansion modes; therefore the traveling-wave mode is usually decomposed into sine and cosine terms [4]. Also, since the current or electric field is zero at the end of the lines, the cosine expansion mode starts a quarter wavelength from the terminated end.

The formulation in the last section is quite flexible and can be easily modified to other types of mode expansion mechanisms. For example, the traveling-wave modes may start more than a wavelength away from the cross-junction, which modifies the Fourier transform of the traveling-wave mode in the above formulation. This type of expansion mode has the advantage that the mutual coupling between traveling-wave modes and PWS modes of the other line ($[T_{temact}]$ and $[T_{tmeact}]$) is negligible. Therefore, the computation effort in (13) can be reduced. Besides, the solutions are automatically convergent in the sense of the number of expansion modes. Another type of expansion mode is also possible where only PSW modes are used in the feed line. This has the advantage of providing some insight into the current distribution on the feed line and of avoiding the computations of the submatrices $[Y_{tself}]$ and $[T_{tmeact}]$ in (13). The above different mode expansion mechanisms may also be used to check the convergence and stability of the solution.

E. Related Problems

Each submatrix itself in (13) has a physical meaning, which contains the information of the self-reaction or mutual coupling of different expansion modes. With submatrices $[Z_{self}]$, $[Z_{tself}]$, and $[I_{inc}]$, the end effects of an open-circuit microstrip line can be determined, while with $[Y_{self}]$, $[Y_{tself}]$, and $[V_{inc}]$ (determined in a way similar to $[I_{inc}]$) the end effects of a short-circuit slotline can be determined. Also, if the parasitic line (either slotline or microstrip line) is finite where only PWS modes are used, the problems of the microstrip-line-fed slot or the slotline-fed printed strip dipole can be solved with this model.

III. Numerical Results

A. Microstrip–Slotline Transition

The results of microstrip–slotline transition are obtained based on the developed algorithm. The numerical analysis is performed on the IBM 3090 system. Typically, for each datum it takes about one minute and thirty seconds of computer time, in contrast to a half second, to obtain the propagation constant k_m, although a lot of effort has been made to reduce the computer cost. An example of a 50-Ω microstrip line to an 80-Ω slotline transition is given. The results of the VSWR and input impedance are shown in Figs. 3 and 4, respectively. The results of VSWR are first checked by interchanging the feed line and the parasitic line. The differences in $|\Gamma|$ are within 2 percent, which is consistent with the property of low-loss two-port networks. The obtained complex reflection coefficient is further checked by changing the number of modes and different mode expansion mechanisms as described in the last section. Two sets of input impedances with different numbers of expansion modes and base function size are shown in

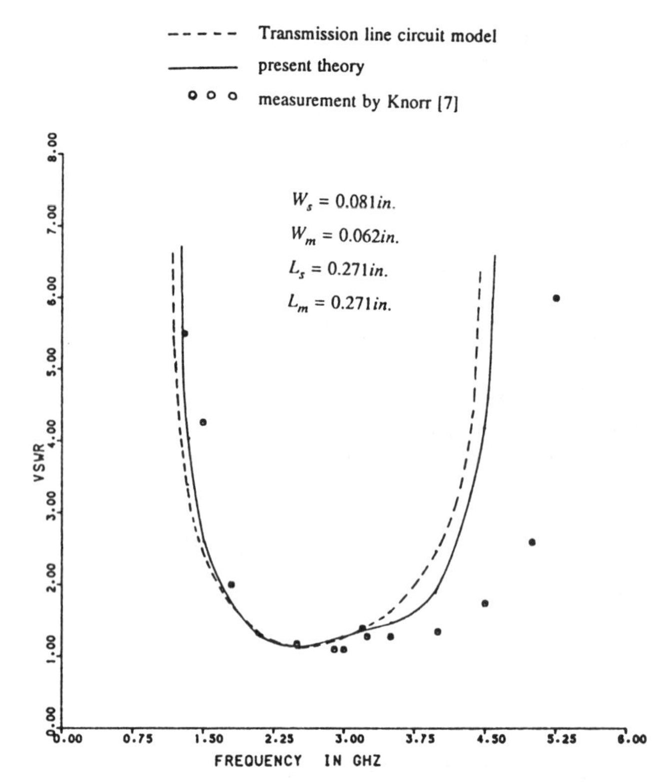

Fig. 3. VSWR versus frequency for microstrip–slotline transition. ϵ_r = 20 and d = 0.125 in.

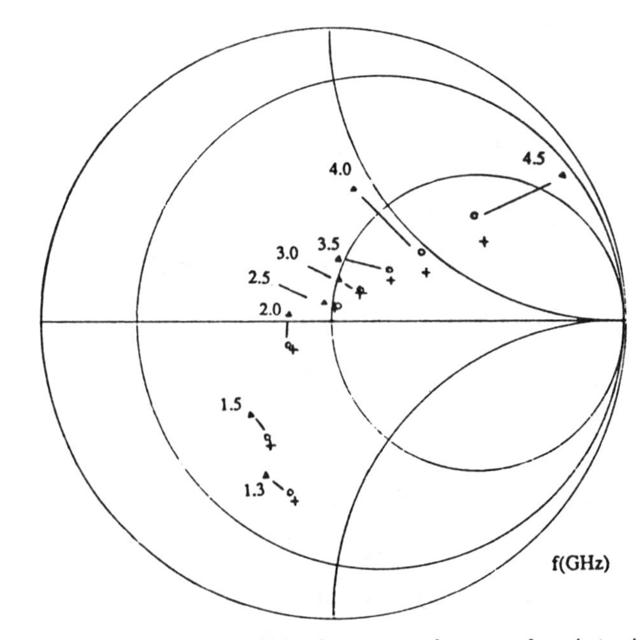

Fig. 4. Smith chart plot of impedance versus frequency for microstrip–slotline transition. The reference plane is on the center of the cross section.

Fig. 4 to illustrate a convergence test example. With the particular device parameters it is found that both the magnitude and the phase of the reflection coefficient converge very well (2 percent in $|\Gamma|$ and 5° in phase) for $d \leqslant 0.036\lambda_0$. However, for higher frequencies, the results are more unstable, and typically the results are 5–10 percent accurate in $|\Gamma|$ and 10–15° in phase before higher order modes turn on. This behavior may be due to two causes. First, when the radiated and surface waves are not weakly excited, the transmission line theory applied to the microstrip line or slotline is only an approximation, and the mode expansion approach to a certain extent involves brute force. Second, the transverse vector components (J_y and M_y), which are neglected in the present investigation, will become more important as the frequency gets higher. The VSWR's obtained by the transmission line circuit model [8] and the measurement [7] are also shown in Fig. 3 to provide a comparison. In the transmission line circuit model the stub length is assumed measured from the center of each line, and the propagation constants k_m and k_s and the excess length are obtained in the current analysis. It is seen from Figs. 3 and 4 that the present method agrees very well with the circuit model in the low-frequency range. The discrepancy for higher frequencies may be due to the higher order modes, surface waves, and radiation effects which are neglected in the circuit model. The measurements reported in [7] show wider bandwidth than that of either the circuit model or the present analysis. It is believed that the accuracy of the device parameters, the nonideal match load, and, especially, the coaxial to microstrip line transition will more or less affect the frequency-dependent results in the measurement. Besides, the material used in [7] is Custom HiK 707-20 ($\epsilon_r = 20$), which is usually very lossy especially for higher frequencies. The reasons mentioned above may explain the discrepancy between theory and experiment.

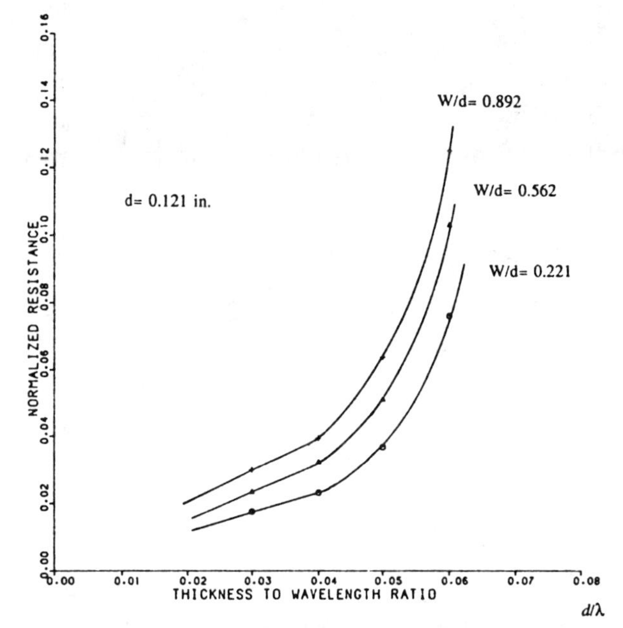

Fig. 5. Normalized end resistance of a shorted slot, $\epsilon_r = 12$.

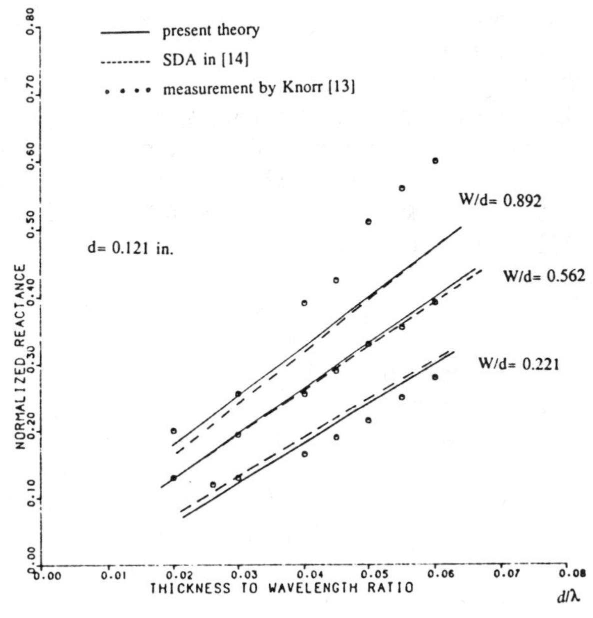

Fig. 6. Normalized end reactance of a shorted slot, $\epsilon_r = 12$.

B. Slotline Discontinuities

Since the results of open-end microstrip discontinuities have been reported with the present approach in [3] and [4], they are not repeated here. However, the results by this analysis have been checked (good agreement) with the results in either [3] or [4]. Experimental results of shorted end slotline discontinuities were reported in [13]. A spectral-domain approach (SDA) of this problem has been reported in [14], where closed coplanar waveguide is used and surface wave and radiation effects are not taken into account. Slotline discontinuities with full-wave analysis have been discussed in [5], but no direct result of the equivalent circuit is provided. The shorted end resistance and reactance of a slotline are shown in Figs. 5 and 6, respectively, as a function of various device parameters. The normalized reactance obtained here is also compared with the measurements reported in [13] and the SDA method reported in [14]. The comparison shows very good agreement with the SDA method. The discrepancy with measured results may be due to some difficulties in the measurements, as discussed in [14]. The resistance part of the equivalent circuit is due to radiation and surface waves. It is seen that the resistance increases with the increase of substrate thickness and slot width, which implies that the energy, in both surface waves and radiation, increases with substrate thickness and slot width. It is also observed that for the chosen substrate thickness ($d/\lambda_0 = 0.06$) the resistance and reactance are of the same order, which means that in this case radiated space waves and surface waves are strongly excited.

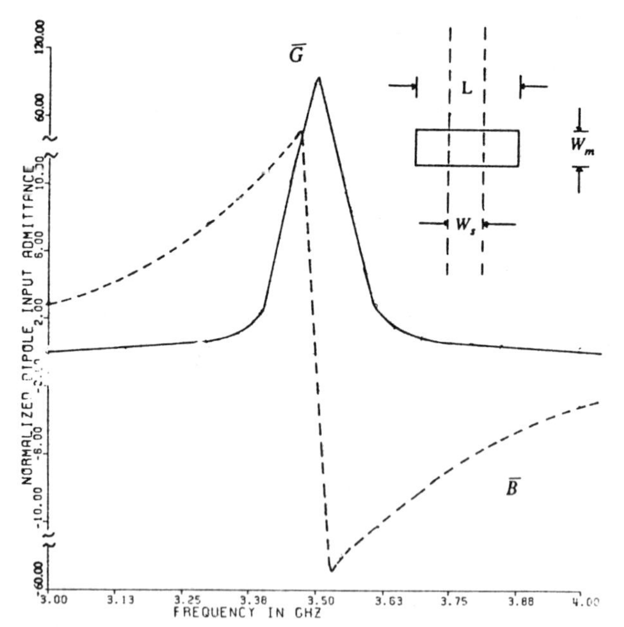

Fig. 7. Normalized equivalent admittance of a dipole fed by a slotline. $\epsilon_r = 12$, $d = 0.121$ in, $W_s = 0.562d$, $W_m = d$, and $L = 1.6$ cm.

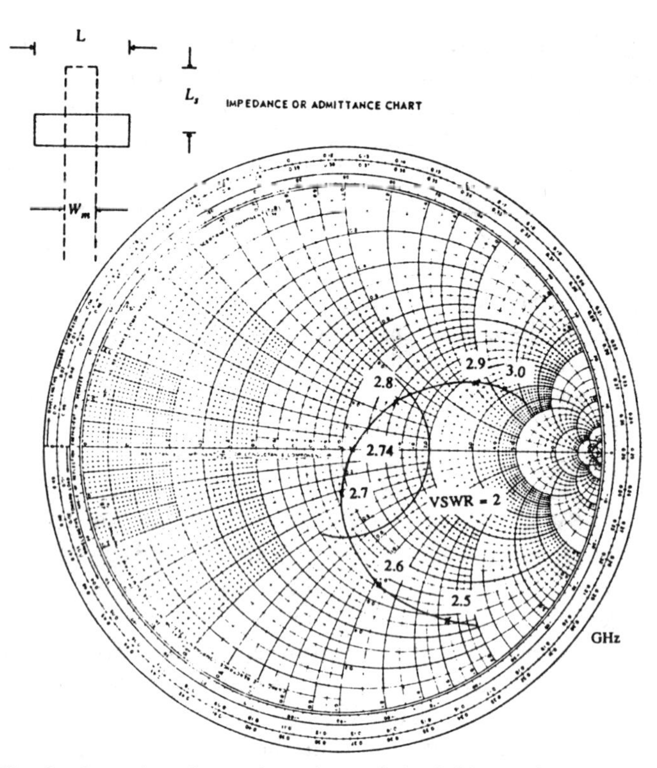

Fig. 8. Input impedance of a stub-tuned slot fed by a microstrip line. $\epsilon_r = 2.2$, $d = 0.2032$ cm, $W_m = 0.635$ cm, $W_s = 0.1016$ cm, $L = 4.0$ cm, and stub length = 0.22 cm.

C. A Slotline-Fed Printed Strip Dipole

In slotline circuit design, large-dielectric-constant materials are typically used to confine energy inside the material. Under this condition, the slotline is a possible candidate as a feed line of microwave or millimeter-wave printed antennas. An example of a slotline-fed printed strip dipole is shown in Fig. 7, where the input impedance of a strip dipole versus frequency is shown for $\epsilon_r = 12$ and $d = 0.121$ in. Here the input impedance does not include the tuning stub. In actual circuit design, a tuning stub is typically used to match the feed line. The results are obtained by using PWS modes on both the strip dipole and the finite but long slotline. Two δ-gap generators of the same magnitude are placed symmetrically near both ends with the dipole on the center position. Due to the symmetry, the input admittance seen at either side of the slotline is half of the input admittance of the printed strip dipole. It can be observed in Fig. 7 that resonance occurs at about 3.5 GHz. However, with a tuning stub, resonance occurs at about 3.35 GHz or at about 3.68 GHz where $\bar{G}$(normalized conductance) = 1. It can be seen that the bandwidth of this structure is quite small. To increase the bandwidth, the strip can be printed in low dielectric substrate while a large dielectric material is used on the other side of the ground to confine energy. One can see from Fig. 7 that when the strip dipole is off resonance, the susceptance is almost linear with frequency. This implies that strip dipole may be used as a wide-band capacitance in slotline coupler or filter design.

D. A Microstrip-Line-Fed Slot

The microstrip-line-fed slot can be used as either a radiating element or a circuit element in filter design [15]. The input impedance of a slot with a tuning stub is obtained from the information of the reflection coefficient in the microstrip line. The normalized input impedance of a slot fed by a microstrip as a function of frequency is plotted on the Smith chart in Fig. 8. It is seen that the bandwidth is mainly determined by the tuning stub since the resistance is quite insensitive to the frequency. Therefore to increase the bandwidth, the stub length should be chosen such that at resonant frequency the change of stub impedance with frequency is as small as possible. Another way of increasing bandwidth is to control the device parameters such that resonance occurs even without the tuning stub. Fig. 8 shows a typical example for this design where the bandwidth is about 6 percent with stub length $\approx 0.02\lambda_0$.

IV. Conclusions

A full-wave analysis is proposed in this paper to develop a generalized dynamic model for microstrip–slotline transition, microstrip-slot, and slotline-microstrip dipole, as well as slotline and microstrip discontinuities. The present approach may find some applications in other transitions in MIC design.

References

[1] K. G. Gupta, R. Garg, and I. J. Bahl, *Microstrip Lines and Slotlines*. Dedham, MA: Artech House, 1979.

[2] P. B. Katehi and N. G. Alexopoulos, "On the modeling of electromagnetic coupled microstrip antennas—The printed strip dipole," *IEEE Trans. Antennas Propagat.*, vol. AP-32, pp. 1179–1186, Nov. 1984.

[3] P. B. Katehi and N. G. Alexopoulos, "Frequency-dependent characteristics of microstrip discontinuities in millimeter wave integrated circuits," *IEEE Trans. Microwave Theory Tech.*, vol. MTT-33, pp. 1029–1035, Oct. 1985.

[4] R. W. Jackson and D. M. Pozar, "Full wave analysis of microstrip open-end and gap discontinuities," *IEEE Trans. Microwave Theory Tech.*, vol. MTT-33, pp. 1036–1042, Oct. 1985.

[5] R. W. Jackson, "Considerations in the use of coplanar waveguide for millimeter wave integrated circuits," *IEEE Trans. Microwave Theory Tech.*, vol. MTT-47, pp. 1450–1456, Dec. 1986.

[6] G. H. Robinson and J. L. Allen, "Slot line application to miniature ferrite devices," *IEEE Trans. Microwave Theory Tech.*, vol. MTT-17, pp. 1097–1101, Dec. 1969.

[7] J. B. Knorr, "Slot-line transitions," *IEEE Trans. Microwave Theory Tech.*, vol. MTT-22, pp. 548–554, May 1974.

[8] D. Chambers, S. B. Cohn, E. G. Cristol, and F. Young, "Microwave active network synthesis," Standford Resr. Inst., Semiannual Report, June 1970.

[9] N. G. Alexopoulos, D. R. Jackson, and P. B. Katehi, "Criteria for nearly omnidirectional radiation patterns for printed antennas," *IEEE Trans. Antennas Propagat.*, vol. AP-33, pp. 195–205, Feb. 1985.

[10] I. E. Rana and N. G. Alexopoulos, "Current distribution and input impedance of printed dipoles," *IEEE Trans. Antennas Propagat.*, vol. AP-29, pp. 99–106, Jan. 1981.

[11] I. E. Rana and N. G. Alexopoulos, "Correction to 'Current distribution and input impedance of printed dipoles', and 'Mutual impedance computation between printed dipoles'," *IEEE Trans. Antennas Propagat.*, vol. AP-30, p. 822, July 1982.

[12] D. M. Pozar, "Improved computational efficiency for the moment method solution of printed dipoles and patches," *Electromagnetics*, vol. 3, nos. 3–4, pp. 299–309, July–Dec. 1983.

[13] J. B. Knorr and J. Saenz, "End effect in shorted slot," *IEEE Trans. Microwave Theory Tech.*, vol. MTT-21, pp. 579–580, Sept. 1973.

[14] R. Jansen, "Hybrid mode analysis of end effects of planar microwave and millimeter wave transmission lines," *Proc. Inst. Elec. Eng.*, vol. 128, pt. H., pp. 77–86, Apr. 1981.

[15] E. A. Mariani and J. P. Agrios, "Slot-line filters and couplers," *IEEE Trans. Microwave Theory Tech.*, vol. MTT-21, pp. 1089–1095, Dec. 1970.

Part 8
Mode-Matching and Field-Matching Techniques

MODAL analysis is one of the oldest rigorous techniques for the solution of waveguide discontinuities. In a historical paper of 1944, Whinnery and Jamieson [8.1] presented data for the modeling of discontinuities in parallel-plane transmission lines. They applied the modal analysis method published three years earlier by Hahn [8.2] to compute the equivalent circuit parameters and at the same time give a physical picture of the phenomena involved. They showed, for instance, that the *local E waves* (below cutoff TM modes) and *local H waves* represent electric and magnetic energies, respectively, that may be represented by lumped reactive elements. This paper, in spite of the limited applicability of the results presented, contains all the features of the method, which is often referred to as the mode-matching technique.

A number of years later the modal analysis technique was presented first by Clarricoats and Slinn [8.3], and, immediately after, by Wexler (Paper 8.1) using more systematic and organized formulations to evaluate the scattering parameters of discontinuities. As a natural consequence of the modal expansion technique, the concept of the generalized scattering matrix was introduced by Clarricoats and Slinn. The term modal analysis was used by Wexler, who categorized boundary reduction- and boundary enlargement-type discontinuities, and considered diaphragms, offset waveguides, and multiple junctions as well. He observed the dependence of the convergence behavior of the mode ratio, expressing the feeling that an optimum ratio should exist.

This aspect was examined in greater detail by Masterman and Clarricoats (Paper 8.2). They developed an improved mode-matching formulation of abrupt junctions between parallel waveguides with arbitrary cross sections using an elegant matrix notation. A number of results are presented and discussed also for discontinuities in circular waveguides. They observed optimum convergence behavior when the highest spatial frequency was about the same on either side of the discontinuity.

The problem of the relative convergence phenomenon is more clearly stated and analyzed in several papers only the first of which could be included here for reasons of space. Paper 8.3 by Lee *et al.* discusses the relative convergence of the mode-matching solution of iris-type waveguide discontinuities and indicates how to perform efficient numerical computations. The same conclusions about the optimum modal ratio were stated by Mittra *et al.* [8.4]. More recently, a comparative study on various mode-matching formulations in connection with the planar waveguide model of the microstrip has been presented by Chu *et al.* (Paper 8.4).

A technique for the analysis of uniform waveguides with a cross section that can be divided into rectangular subregions is often referred to as the mode-matching technique. Strictly speaking, this name should be used to designate the above-mentioned method, where the EM fields in the discontinuity apertures are expanded in terms of *normal modes* of the waveguides. The method of subregions is better referred to as field-matching. It consists of solving the field equations in each subregion by the method of separation of variables, then imposing the continuity conditions at the boundaries between adjacent regions. A typical application of field-matching is the analysis of the shielded microstrip as developed by Kowalski and Pregla (Paper 8.5). An analysis technique involving both mode-matching and field-matching has been used by Strube and Arndt [8.5] for the analysis of the transition from waveguide to shielded image guide. The existence of backward waves and complex waves is pointed out and discussed in this paper. Complex modes have been shown by Omar and Schünemann (Paper 8.6) to play an important role in the analysis of finline discontinuities by mode-matching. This technique has recently received renewed attention from a number of researchers with the aim of developing more efficient formulations. Some examples are [8.6]–[8.8].

The analysis of junctions between different waveguides or transmission lines by mode-matching requires knowledge of their modal spectrum. This can be computed by other numerical techniques such as the spectral-domain approach. One such combined method of analysis was recently proposed by Hélard *et al.* [8.9] for the characterization of finline discontinuities.

Actually, mode-matching techniques are used also in conjunction with other methods, and the classification of such combined techniques may sometimes appear arbitrary. Other examples are some papers using the transverse resonance technique and others relevant to the planar-circuit approach. For completeness, the reader is referred also to Paper 9.2 by Peng and Oliner, Vahldieck and Bornemann [9.6], and Paper 11.3 by Wolff *et al.* which could just as well have been included here.

References

[8.1] J. R. Whinnery and H. W. Jamieson, "Equivalent circuits for discontinuities in transmission lines," *Proc. IRE*, vol. 32, pp. 98–116, Feb. 1944.

[8.2] W. C. Hahn, "A new method for the calculation of cavity resonators," *J. Appl. Phys.*, vol. 12, pp. 62–68, Jan. 1941.

[8.3] P. J. B. Clarricoats and K. R. Slinn, "Numerical solution of waveguide-discontinuity problems," *Proc. IEE*, vol. 114, pp. 878–886, July 1967.

[8.4] R. Mittra, T. Itoh, and T. S. Li, "Analytical and numerical studies of the relative convergence phenomenon arising in the solution of an integral equation by the moment method," *IEEE Trans. Microwave Theory Tech.*, vol. MTT-20, pp. 96–104, Feb. 1972.

[8.5] J. Strube and F. Arndt, "Rigorous hybrid-mode analysis of the transition from rectangular waveguide to shielded dielectric image guide," *IEEE Trans. Microwave Theory Tech.*, vol. MTT-33, pp. 391–401, May 1985.

[8.6] A. S. Omar and K. F. Schuenemann, "Transmission matrix representation of finline discontinuities," *IEEE Trans. Microwave Theory Tech.*, vol. MTT-33, pp. 765–769, Sept. 1985.

[8.7] R. R. Mansour and R. H. Macphie, "An improved transmission matrix formulation of cascaded discontinuities and its application to *E*-plane circuits," *IEEE Trans. Microwave Theory Tech.*, vol. MTT-34, pp. 1490–1498, Dec. 1986.

[8.8] F. Alessandri, G. Bartolucci, and R. Sorrentino, "Admittance-matrix formulation of waveguide discontinuity problems. Application to the design of branch-guide couplers," *IEEE Trans. Microwave Theory Tech.*, vol. MTT-36, pp. 394–403, Feb. 1988.

[8.9] M. Hélard, J. Citerne, O. Picon, and V. Fouad Hanna, "Theoretical and experimental investigations of finline discontinuities," *IEEE Trans. Microwave Theory Tech.*, vol. MTT-33, pp. 994–1003, Oct. 1985.

Solution of Waveguide Discontinuities by Modal Analysis

ALVIN WEXLER, MEMBER, IEEE

***Abstract*—A general method is presented for analysis of waveguide junctions and diaphragms by summing normal modes of propagation, giving solutions for the resulting scattered modes. Because interaction effects of dominant and higher-order modes between discontinuities are allowed, finite-length obstructions can be studied.**

Solutions are found without any prior assumption about the total fields existing at the discontinuities and, as a result, the formulation is applicable to a wide range of problems. The technique proves to be simple and is ideally suited to computers, involving mainly the solution of sets of simultaneous linear equations.

Thick and thin symmetrical bifurcations of a rectangular guide are studied. Forward-scattered mode amplitudes and input admittances are calculated, the computed admittance of the thin bifurcation is compared with well-known results, and transverse field patterns on both sides of the junction are plotted, thus showing the accuracy of the match.

The results of a finite-length bifurcation by a thick vane are presented for a range of lengths, the parameters of the equivalent *T* network being given in each case. For very short lengths, the problem corresponds to an inductive strip across the guide.

I. Introduction

VERY FEW waveguide discontinuities have been solved exactly, and these have been accomplished by integral transform techniques [1]. Other integral equation formulations, solved by quasi-static approximations, were reviewed by Lewin [2]. Although there is hope that some restrictions may eventually be alleviated if new ways of dealing with integral equations are found, the outlook is not particularly bright. Collin [3] presents examples illustrating the use of variational techniques. But the method as outlined requires much mathematical innovation when applied directly to particular problems. Other approaches employing static approximations and perturbational methods are very approximate and are usually unacceptable for the broad class of problems encountered in practice.

In the modal analysis method, the amplitudes of normal modes are chosen so as to satisfy boundary conditions at the discontinuity. Because the modal approach is direct and conforms closely to physical reality, it should have the widest application. The method gives excellent estimates to the aperture fields and scattered modes, and should be of particular value to multimode propagation studies, e.g., multimode techniques applied to aerial improvement [4]. Objections of slow convergence and involved numerical work are not particularly significant to digital computers, and so demand is increasing for a general and convenient formulation rather than for economy in computing effort. It is, therefore, the purpose of this paper to present a general method of normal mode summation, to renew interest in the technique that has never been fully exploited, and to argue the case for a complete change of emphasis in the solution of waveguide discontinuity problems.

Manuscript received September 17, 1966; revised May 8, 1967.

The author is with the Department of Electrical Engineering, University of Manitoba, Winnipeg, Canada.

II. Description of the Problem

Consider two uniform cylindrical waveguides having different cross sections and distributions of enclosed electrical properties. The junction formed by joining them end-to-end, with axial lines parallel, can be described as a function of two transverse coordinates u_1 and u_2. Boundary conditions, continuity of transverse fields through all apertures and zero tangential electric field at conducting obstacles, are satisfied by a suitable infinite series of modes appropriate to each side of the junction. If the modes of propagation in both guides and the scattering coefficients of succeeding discontinuities are known, the properties of the junction may be computed. The problem is to find how power is apportioned between the various scattered modes.

The transverse fields of each mode may be written as

$$\hat{e}_i(u_1, u_2, z) = a_i\bar{e}_i(u_1, u_2)\cdot e^{\pm \gamma_i z} \tag{1}$$

and

$$\hat{h}_i(u_1, u_2, z) = a_i\bar{h}_i(u_1, u_2)\cdot e^{\pm \gamma_i z}. \tag{2}$$

The sign of the exponent is fixed by the propagating direction. $\hat{e}_i$, $\hat{h}_i$, and γ_i are the transverse vector functions and propagation constant of the ith mode. If not known explicitly, they can often be derived numerically [5]. Factors a_i are the mode coefficients which, along with the reflection factor ρ of the incident mode, are to be determined. Modes are numbered in an arbitrary sequence, the variables i, j, k, m, n, and r being reserved for this purpose. In general, the fields must be described in a piecewise fashion, e.g., Fig. 1.

A waveguide cross section consists of a conducting boundary enclosing any distribution of magnetic, dielectric, and perfectly conducting regions. For our purposes, a waveguide boundary is defined as the perfectly conducting periphery containing all permeable regions only. Thus, for example, the cross-section boundary of a coaxial line is the inner wall of the outer conductor and the outer wall of the inner one.

Three classes of discontinuity, consisting of the following situations, must be considered: 1) the projection of the guide boundary nearer the klystron completely encompasses the guide boundary following the junction (Section

Reprinted from *IEEE Trans. Microwave Theory Tech.*, vol. MTT-15, no. 9, pp. 508–517, Sept. 1967.

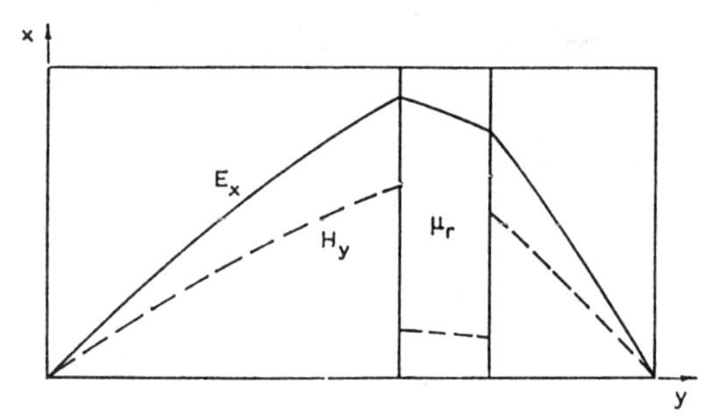

Fig. 1. A possible TE transverse-field configuration in a magnetic slab-loaded waveguide.

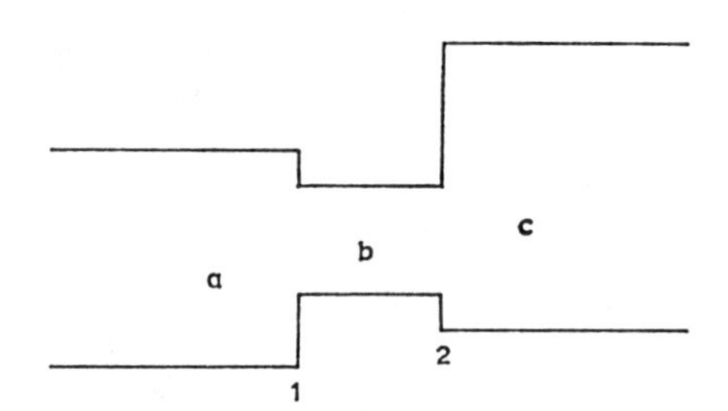

Fig. 2. Three waveguides with higher-order mode coupling between junctions.

III); 2) the guide nearer the klystron is contained within the projection of the guide boundary following the junction (Section IV); and, the remaining possibility, 3) neither guide boundary can be contained within the other (Section V). Classes 1) and 2) include coincident boundaries as a special case.

III. Boundary Reduction

Fig. 2 represents part of a general waveguide system consisting of three dissimilar guides a, b, and c. The junctions are numbered 1 and 2. Consider a mode $i=1$ emanating from a matched source in a and impinging on waveguide b at $z=0$. The coefficient of this mode is a_1, those for the back-scattered modes are $a_2, a_3, \cdots, a_i, \cdots$, and in an anisotropic guide ρa_1 as well. Taking $\bar{\mathscr{E}}$ to be the total transverse electric-field vector function within the aperture at the discontinuity, the field expanded in terms of modes just to the left of junction 1 is

$$\bar{\mathscr{E}} = (1 + \rho)a_1\bar{e}_{a1} + \sum_{i=2}^{\infty} a_i\bar{e}_{ai}. \tag{3}$$

Subscript a denotes quantities relative to the first waveguide. Similarly, the total magnetic field may be expressed by

$$\bar{\mathscr{H}} = (1 - \rho)a_1\bar{h}_{a1} - \sum_{i=2}^{\infty} a_i\bar{h}_{ai}. \tag{4}$$

In many cases, such as modes in the slab-loaded waveguide of Fig. 1 and hybrid modes in rod-loaded circular guides, unique wave admittances cannot be defined. Thus, for complete generality, the transverse electric and magnetic fields are expressed independently.

Refer again to Fig. 2. The aperture fields at $z=0$ will now be expressed in terms of modes in b. If waveguide b is matched, the transverse electric-field pattern of mode j is given by $b_j\bar{e}_{bj}$. However, each transmitted evanescent or propagating mode j reaching junction 2 partially reflects and scatters power into other modes k, some of which return to junction 1. Therefore, it is necessary to account for these returned waves, as well as for the positively directed ones, when summing modes.

Scattering coefficients are used to relate the amplitudes and phases of modes transmitted past junction 1 to those reflected from junction 2. Consider waveguide b to be excited from junction 1 by a single forward wave j whose coefficient is unity, i.e., $b_j=1$ at junction 1. Then, scattering coefficients s_{jk} are defined as equal to mode coefficients b_k, of the back-scattered waves k, transformed in amplitude and phase to junction 1 from 2. This is repeated for all j. Clearly, junction 2 must be solved before junction 1.

Multiplying b_j by s_{jk} gives the contribution of j to k as seen at junction 1. Because each forward-propagating mode j has an infinity of back-scattered modes k of the form $s_{jk}b_j\bar{e}_{bk}$ associated with it, the total transverse electric and magnetic fields just to the right of junction 1 result by summing over all j and k. This gives

$$\bar{\mathscr{E}} = \sum_{j=1}^{\infty} b_j(\bar{e}_{bj} + \sum_{k=1}^{\infty} s_{jk}\bar{e}_{bk}) \tag{5}$$

and

$$\bar{\mathscr{H}} = \sum_{j=1}^{\infty} b_j(\bar{h}_{bj} - \sum_{k=1}^{\infty} s_{jk}\bar{h}_{bk}). \tag{6}$$

Boundary conditions to be satisfied at the discontinuity are as follows: transverse electric and magnetic fields must be continuous across the aperture, and electric field tangential to the conducting obstacle must vanish. The single boundary condition is sufficient at a conducting surface.

Let m be any mode number in waveguide a. In all uniform guides with reflection symmetry and perfectly conducting walls, the following orthogonality relation [6] holds for nondegenerate modes:

$$\int_a \bar{e}_{ai} \times \bar{h}_{am} \cdot \bar{u}_z \, ds = 0 \tag{7}$$

when $i \neq m$. The surface integral extends over the entire cross section of the waveguide a. Degenerate modes should be orthogonalized by the Gram–Schmidt procedure [7].

Take the cross product of (3) with $\bar{h}_{am}$ and integrate over the cross section of waveguide a. Assuming orthogonality of modes and substituting (5) for the (as yet) unknown aperture field $\bar{\mathscr{E}}$, thus employing the continuity condition for transverse electric fields, we get

$$(1 + \rho)a_1 \int_a \bar{e}_{a1} \times \bar{h}_{a1} \cdot \bar{u}_z \, ds = \sum_{j=1}^{\infty} b_j\left(\int_b \bar{e}_{bj} \times \bar{h}_{a1} \cdot \bar{u}_z \, ds + \sum_{k=1}^{\infty} s_{jk} \int_b \bar{e}_{bk} \times \bar{h}_{a1} \cdot \bar{u}_z \, ds\right) \tag{8}$$

when $m=1$, and

$$a_m \int_a \bar{e}_{am} \times \bar{h}_{am} \cdot \bar{u}_z \, ds = \sum_{j=1}^{\infty} b_j \left(\int_b \bar{e}_{bj} \times \bar{h}_{am} \cdot \bar{u}_z \, ds + \sum_{k=1}^{\infty} s_{jk} \int_b \bar{e}_{bk} \times \bar{h}_{am} \cdot \bar{u}_z \, ds \right) \quad (9)$$

$$\frac{a_i}{a_1} = \frac{\sum_{j=1}^{N} \frac{b_j}{a_1} \left(\int_b \bar{e}_{bj} \times \bar{h}_{ai} \cdot \bar{u}_z \, ds + \sum_{k=1}^{N} s_{jk} \int_b \bar{e}_{bk} \times \bar{h}_{ai} \cdot \bar{u}_z \, ds \right)}{\int_a \bar{e}_{ai} \times \bar{h}_{ai} \cdot \bar{u}_z \, ds}, \quad (14)$$

when $m \neq 1$. Because $\bar{\mathscr{E}}$ exists over the aperture only and vanishes elsewhere, the integrals on the right-hand sides of (8) and (9) are taken over b. This completes the electric-field boundary conditions.

Now, take the cross product of (4) with $\bar{e}_{bn}$ and integrate over the cross section of waveguide b. Substituting (6) for the unknown aperture field $\bar{\mathscr{H}}$, and using the orthogonality relation

$$\int_b \bar{e}_{bn} \times \bar{h}_{bj} \cdot \bar{u}_z \, ds = 0 \quad (10)$$

for nondegenerate modes when $n \neq j$, we find that

$$(1-\rho)a_1 \int_b \bar{e}_{bn} \times \bar{h}_{a1} \cdot \bar{u}_z \, ds - \sum_{i=2}^{\infty} a_i \int_b \bar{e}_{bn} \times \bar{h}_{ai} \cdot \bar{u}_z \, ds = \left(b_n - \sum_{j=1}^{\infty} b_j s_{jn} \right) \int_b \bar{e}_{bn} \times \bar{h}_{bn} \cdot \bar{u}_z \, ds. \quad (11)$$

Continuity of transverse magnetic field was used in the derivation of (11).

Changing the index m to i in (9), substituting it into (11) so as to eliminate a_i, and rearranging, we get

$$\rho \int_b \bar{e}_{bn} \times \bar{h}_{a1} \cdot \bar{u}_z \, ds + \sum_{j=1}^{N} \frac{b_j}{a_1} \sum_{i=2}^{M} \frac{\int_b \bar{e}_{bj} \times \bar{h}_{ai} \cdot \bar{u}_z \, ds + \sum_{k=1}^{N} s_{jk} \int_b \bar{e}_{bk} \times \bar{h}_{ai} \cdot \bar{u}_z \, ds}{\int_a \bar{e}_{ai} \times \bar{h}_{ai} \cdot \bar{u}_z \, ds} \int_b \bar{e}_{bn} \times \bar{h}_{ai} \cdot \bar{u}_z \, ds$$

$$+ \left(\frac{b_n}{a_1} - \sum_{j=1}^{N} \frac{b_j}{a_1} s_{jn} \right) \int_b \bar{e}_{bn} \times \bar{h}_{bn} \cdot \bar{u}_z \, ds = \int_b \bar{e}_{bn} \times \bar{h}_{a1} \cdot \bar{u}_z \, ds; \quad (12)$$

and from (8),

$$\rho \int_a \bar{e}_{a1} \times \bar{h}_{a1} \cdot \bar{u}_z \, ds - \sum_{j=1}^{N} \frac{b_j}{a_1} \left(\int_b \bar{e}_{bj} \times \bar{h}_{a1} \cdot \bar{u}_z \, ds + \sum_{k=1}^{N} s_{jk} \int_b \bar{e}_{bk} \times \bar{h}_{a1} \cdot \bar{u}_z \, ds \right) = - \int_a \bar{e}_{a1} \times \bar{h}_{a1} \cdot \bar{u}_z \, ds. \quad (13)$$

For practical reasons, the infinite series were truncated at M and N which signify the number of modes in waveguides a and b, respectively. Equation (12) generates N linear equations where $n=1, 2, \cdots, N$ and (13) supplies one equation. There are $N+1$ unknowns (b_1/a_1, b_2/a_1, $\cdots$, b_n/a_1, $\cdots$, b_N/a_1 and ρ), and so the system of equations may be solved. For complicated problems (e.g., junctions between rectangular and circular guides or between guides slab-loaded differently, etc.), the integrations in (12) and (13) should be performed numerically [8].

By rewriting (9), the coefficients of the back-scattered modes in guide a may be found. Therefore,

where $i \neq 1$. Terms previously formed by the Gram–Schmidt orthogonalization procedure should now be decomposed into normal waveguide modes, thus completing the study of the junction.

By using (11) in place of (12), it is possible to solve for ρ, b_j/a_1, and a_i/a_1 all at the same time. This procedure has two disadvantages: 1) the computer store requirement approximately quadruples (assuming that M and N are about the same size); and 2) as the amount of computing is proportional to the cube of the number of unknowns, the work increases by eight times. It is certainly preferable to use (12) and then to find the a_i/a_1 through (14).

IV. Boundary Enlargement

This is the complement of the problem discussed in Section III. Many of the comments made previously are applicable here as well.

Call the first, and smaller, guide a and the larger one b. Equations (3) through (6) describe the fields at the junction as before.

The derivation of the simultaneous equations is almost identical to that of Section III. Briefly, cross-multiply (6) by $\bar{e}_{am}$ and integrate over the cross section of a. Express the aperture field by (4). Note that, as in (8) and (9), two cases occur: $m=1$ and $m \neq 1$. Also, cross-multiply (5) by $\bar{h}_{bn}$ and integrate over the cross section of b, substituting (3) for $\bar{\mathscr{E}}$. Assume orthogonality, and after some algebraic manipulation, the following equations result:

$$\rho \int_a \bar{e}_{a1} \times \bar{h}_{bn} \cdot \bar{u}_z \, ds - \sum_{j=1}^{N} \frac{b_j}{a_1} \sum_{i=2}^{M} \frac{\int_a \bar{e}_{ai} \times \bar{h}_{bj} \cdot \bar{u}_z \, ds - \sum_{k=1}^{N} s_{jk} \int_a \bar{e}_{ai} \times \bar{h}_{bk} \cdot \bar{u}_z \, ds}{\int_a \bar{e}_{ai} \times \bar{h}_{ai} \cdot \bar{u}_z \, ds} \int_a \bar{e}_{ai} \times \bar{h}_{bn} \cdot \bar{u}_z \, ds$$

$$- \left(\frac{b_n}{a_1} + \sum_{j=1}^{N} \frac{b_j}{a_1} s_{jn} \right) \int_b \bar{e}_{bn} \times \bar{h}_{bn} \cdot \bar{u}_z \, ds = - \int_a \bar{e}_{a1} \times \bar{h}_{bn} \cdot \bar{u}_z \, ds, \tag{15}$$

where $n = 1, 2, \cdots, N$;

$$\rho \int_a \bar{e}_{a1} \times \bar{h}_{a1} \cdot \bar{u}_z \, ds + \sum_{j=1}^{N} \frac{b_j}{a_1} \left(\int_a \bar{e}_{a1} \times \bar{h}_{bj} \cdot \bar{u}_z \, ds - \sum_{k=1}^{N} s_{jk} \int_a \bar{e}_{a1} \times \bar{h}_{bk} \cdot \bar{u}_z \, ds \right) = \int_a \bar{e}_{a1} \times \bar{h}_{a1} \cdot \bar{u}_z \, ds; \tag{16}$$

and

$$\frac{a_i}{a_1} = \frac{- \sum_{j=1}^{N} \frac{b_j}{a_1} \left(\int_a \bar{e}_{ai} \times \bar{h}_{bj} \cdot \bar{u}_z \, ds - \sum_{k=1}^{N} s_{jk} \int_a \bar{e}_{ai} \times \bar{h}_{bk} \cdot \bar{u}_z \, ds \right)}{\int_a \bar{e}_{ai} \times \bar{h}_{ai} \cdot \bar{u}_z \, ds}, \tag{17}$$

where $i \neq 1$. The system of $N+1$ linear equations, defined by (15) and (16), may be solved for $b_1/a_1, b_2/a_1, \cdots, b_N/a_1$ and ρ. The back-scattered modes a_i/a_1 can then be found from (17).

V. Systems of Connected Waveguides

Fig. 2 depicts part of a system in which interaction of dominant and higher-order modes between discontinuities occurs. As we have seen, if power flows from the left, it is necessary to know the scattering properties of junction 2 before solving the problem at junction 1. Similarly, before analyzing junction 2, the scattering properties of any discontinuity farther down the guide must be known. Ultimately, analysis must begin at a simple termination, such as a matched or single-mode guide or a short circuit which causes independent reflection of each mode incident upon it, regardless of the amplitude and phase of any other.

If the length l of a particular waveguide is small, many modes generated at one junction, figure in the field summation at the other. In other words, higher-order mode coupling occurs. Choose a finite number of modes in the waveguide consisting of the lowest-order modes likely to be set up at either junction.

As indicated earlier, the variable $i=1$ is not reserved for any particular mode but is allowed to represent any mode presumed incident on a junction. If r is the mode incident on say junction 1, then ρ_r is its reflection factor and a_i/a_r, with $i \neq r$, denotes the $M-1$ coefficients of other back-scattered modes. These are all found as previously described. Therefore, M scattering coefficients of junction 1, as defined at the next junction towards the klystron, are given by

$$\begin{aligned} S_{ri} &= \frac{a_i}{a_r} e^{-(\gamma_i + \gamma_r) l} \qquad & i \neq r, \\ &= \rho_r e^{-2\gamma_r l} \qquad & i = r \end{aligned} \tag{18}$$

thus including amplitude and phase change of both incident and scattered modes between junctions. Consider each mode $r = 1, 2, \cdots, M$, in turn, to be independently incident on the junction, and solve the resulting system of equations each time. In this way, all M^2 scattering coefficients are found.

Junctions can be represented by T, π, and transformer networks [9], [10]. To evaluate the equivalent circuit, three determinations of the input admittance y' as a function of a load in guide c are generally required.

A. Diaphragms and Offset Waveguides

Refer to Fig. 2. Clearly, as the length of waveguide b decreases, the coupling between modes generated at both junctions becomes more pronounced and we have an iris between two offset guides or, as a special case, simply two offset guides. In the study of diaphragms, it is necessary to match fields through the windows. However, it is incorrect to equate both sets of waveguide modes across the plane of the diaphragm, as this will not ensure zero transverse electric field at the conducting surfaces. By treating the diaphragm as a special case of a three-waveguide system, this problem does not arise. Furthermore, this method does not require any prior assumption as to the total field within the window. It is only necessary to know the form of the normal modes in the waveguide defined by the aperture shape.

B. Multiple-Guide Junctions

Fig. 3 typifies a class of multiple-guide junctions that this method can accommodate. A particular problem is the analysis of selective launching of modes into a large guide b by controlling the amplitude and phase of modes in a and a'. This problem is a significant one in the design of multimode aerials [4].

Briefly, the method is as follows. Think of the system to the left of the junction in Fig. 3 as one composite waveguide rather than as a number of guides. The modes in this composite system are defined in a special way. We define a set to conform to the interior of one of the constituent guides,

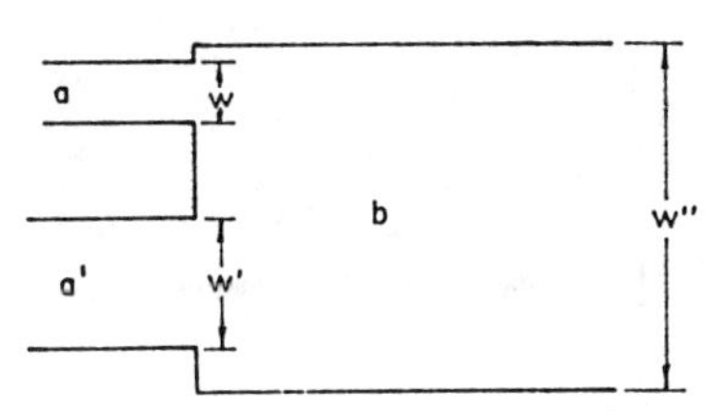

Fig. 3. A multiple-guide junction.

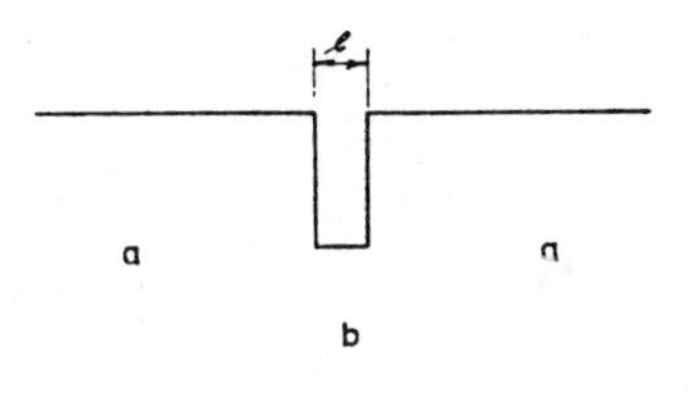

Fig. 4. A thick, symmetrical, inductive iris.

say a; elsewhere in the cross section these modes have zero field. Similarly, another set is defined to suit the other guide a', this time with no field over the first aperture and intervening region. These modes are denoted $i=1, 2, \cdots, M$, where M is the total number of modes assumed in both a and a'. It is this double set that forms the normal and complete mode system in the composite waveguide. Once modes are defined in this way, it is of no consequence to the matching procedure that they belong to electrically isolated guides.

Now, consider excitation only from a by mode ir. This mode will partially reflect into a, and other back-scattered modes will be generated, some traveling into a and others into a'. Mode coefficients are found by solving the set of simultaneous equations. The problem must then be repeated with excitation from a' and the resulting fields found by superposition.

The reverse situation, with a mode in the large guide impinging on the smaller guides, is handled similarly. The bifurcation, discussed in Section VI, is one problem of this type and the approach is described there.

C. Symmetry Considerations

Considerable simplification occurs when discontinuities exhibit certain symmetries and are either symmetrically or antisymmetrically excited. For example, the iris in Fig. 4 is symmetrical about a transverse plane. If both ports are excited symmetrically, an open circuit appears at the central plane; antisymmetrical excitation produces a short circuit. Under these conditions, only pure reflection occurs at the central plane, and so $s_{jk}=0$ when $j\neq k$. s_{jj} is given simply by

$$s_{jj} = \frac{1 - y'_{bj}}{1 + y'_{bj}}. \tag{19}$$

y'_{bj} is the normalized input admittance of the jth mode in b at the discontinuity, distance $l/2$ from the symmetry plane.

Two parameters are sufficient to specify the equivalent network of such discontinuities. For example, the upper-arm impedances of the equivalent T network are both given by $Z_{11}-Z_{12}$ and the common branch by Z_{12}. The computed input impedance, with symmetrical excitation, yields $Z_{11}+Z_{12}$, and that with antisymmetrical excitation gives $Z_{11}-Z_{12}$.

VI. Numerical Examples

The bifurcation of a rectangular waveguide by a thin vane is one of the few junction problems that have been solved rigorously. As a check on the theory just developed, it will be solved by the modal analysis method. Thick, semi-infinite and finite-length bifurcations will also be investigated. These examples serve to indicate the general approach and illustrate some practical difficulties.

A. H-Plane Bifurcation

Fig. 5 shows a rectangular waveguide loaded with a thick, perfectly conducting vane. y-coordinate dimensions are normalized with respect to the broad dimension w. Assuming excitation by an H_{01} mode, only symmetrical modes are generated at the discontinuity. Transverse field patterns of the two lowest-order modes in the bifurcated guide b are shown and are seen to be H_{01} and H_{03} modes deformed by the vane. Modes in a are the usual ones in an empty rectangular guide. Note that this is a boundary-reduction problem.

Expressions for the transverse fields in guide a are

$$\bar{e}_{ai} = \bar{u}_x \sin(p\pi y/w). \tag{20}$$

and

$$\bar{h}_{ai} = \bar{u}_y y_{ai} \sin(p\pi y/w). \tag{21}$$

The wave admittance of the ith mode is

$$y_{ai} = \sqrt{\frac{\varepsilon_0}{\mu_0}}\sqrt{1 - \left(\frac{p\lambda_0}{2w}\right)^2}. \tag{22}$$

Modes are numbered consecutively, i.e., $i=1, 2, \cdots, M$, and so

$$p = 2i - 1, \tag{23}$$

thus giving only symmetrical modes when substituted into (20) and (21).

In the left-hand region of waveguide b the transverse fields are

$$\bar{e}_{bj} = \bar{u}_x \sin\left(\frac{2q\pi y/w}{1 - t/w}\right) \tag{24}$$

and

$$\bar{h}_{bj} = \bar{u}_y y_{bj} \sin\left(\frac{2q\pi y/w}{1 - t/w}\right). \tag{25}$$

Equations (24) and (25) hold in the range

$$0 < y/w < 0.5(1 - t/w)$$

and are zero across the vane. Substitute $(1-y/w)$ for y/w when $0.5(1+t/w)<y/w<1$. The admittance of the jth mode is

$$y_{bj} = \sqrt{\frac{\varepsilon_0}{\mu_0}}\sqrt{1 - \left(\frac{q\lambda_0/w}{1 - t/w}\right)^2}. \tag{26}$$

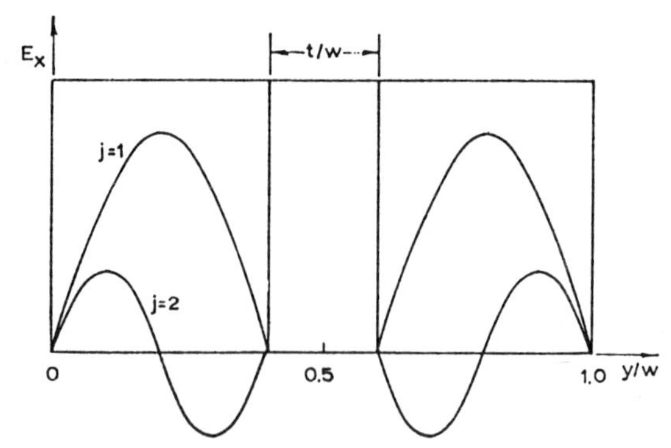

Fig. 5. Transverse fields of the two lowest-order TE modes in a centrally bifurcated rectangular guide.

Here, we have

$$q = j, \tag{27}$$

where $j = 1, 2, \cdots, N$. When calculating y_{ai} and y_{bj}, take the positive root when real and the negative imaginary root when imaginary.

If the vane were not central, one set of modes would have to be defined to conform to the interior of one of the partitioned guides with zero field elsewhere. Similarly, another set would have to be defined for the other section. This is similar to the multiple-guide input discussed in Section V. For the symmetrical problem, however, each mode has to be defined over both apertures. Otherwise, two columns of the matrix representing (12) and (13) will be identical, and the system will then be singular.

Referring to (12) and (13), it is seen that the following three integrals are required:

$$\int_a \bar{e}_{ai} \times \bar{h}_{ai} \cdot \bar{u}_z \, ds = 0.5wy_{ai} \tag{28}$$

$$\int_b \bar{e}_{bj} \times \bar{h}_{bj} \cdot \bar{u}_z \, ds = 0.5wy_{bj}(1 - t/w), \tag{29}$$

and

$$\int_b \bar{e}_{bj} \times \bar{h}_{ai} \cdot \bar{u}_z \, ds = 0.5wy_{ai}(1 - t/w) \cdot [\sin(f)/f - \sin(g)/g], \tag{30}$$

where

$$f = [0.5p(1 - t/w) - q]\pi \tag{31}$$

and

$$g = [0.5p(1 - t/w) + q]\pi. \tag{32}$$

The factor w is common to all terms in (12) and (13), and so it may be deleted. As all fields are uniform along x, the integrations above were performed only with respect to y. Having evaluated these integrals, all the analysis required is completed. This is a particularly simple problem, the s_{jk} terms vanishing due to the infinite length of the bifurcated guide.

Cases $t/w = 0$ and 0.2 were studied with the frequency parameter $w/\lambda_0 = 0.7$. Integrals (28) to (30) were computed and substituted into (12) and (13) as required. The same number of modes were employed in each guide, i.e., $M = N$. The resulting set of simultaneous linear equations was solved by standard Gaussian reduction and back-substitution techniques [11].

Forward- and back-scattered mode coefficients were computed and, from the reflection factor, the normalized input admittance

$$y' = \frac{1 - \rho}{1 + \rho} \tag{33}$$

was found. y' is tabulated in Table I as a function of different expansion sizes for the thick and thin bifurcation. Results of an exact analysis of the thin-vane case, which uses a transform method, are plotted by Marcuvitz [12]. He gives the resulting shift in the null point, and from it the normalized admittance was calculated to be $y' = -j2.416$. On this basis it is seen that the forty-mode expansion is less than 0.05 percent in error. Even with only ten modes the error is less than 1 percent. Also, corresponding computations for a vane with normalized thickness $t/w = 0.2$ are given. An exact solution is not known for this case, and so a comparison is not available. However, the rate of convergence is seen to be equally rapid.

In Table II the first five forward-scattered mode coefficients, normalized with respect to that of the incident mode $i = 1$, are presented for $t/w = 0$. It is clear that the apportionment of power between scattered modes can be closely calculated. For example, there is little difference in the computed b_j/a_1 values between the twenty- and forty-mode cases. Coefficients b_j/a_1 for lower-order modes are known to a higher accuracy than the remaining terms. This inaccuracy in the last few terms, due to an attempt to compensate for the missing modes, may be seen in the $N = 5$ case where $j = 5$ is the last mode of the finite series.

As a reasonable approximation, take $N = 40$ to furnish almost exact results in comparison to $N = 10$. On this basis, b_1/a_1, b_2/a_1, and b_3/a_1 (for $N = 10$) are known to within errors of 0.5, 2, and 4 percent, respectively. It is difficult to assess just how accurate the $N = 40$ values are, but they are probably of a very high order. Phase angles converge very rapidly. For this particular problem, all modes in a given solution have the same phase at the discontinuity, except for 180° phase reversals. Similar comments apply to the back-scattered coefficients a_i/a_1.

The junction fields are expressed by (3) through (6). Substituting (20), (21), (24), and (25) into them, and dividing by a_1, we obtain the following equations. In waveguide a

$$E_x = (1 + \rho)\sin(\pi y/w) + \sum_{i=2}^{M} \frac{a_i}{a_1}\sin(p\pi y/w) \tag{34}$$

$$H_y = (1 - \rho)y_{a1}\sin(\pi y/w) + \sum_{i=2}^{M} \frac{a_i}{a_1} y_{ai}\sin(p\pi y/w), \tag{35}$$

and in waveguide b within $0<y/w<0.5(1-t/w)$

$$E_x = \sum_{j=1}^{N} \frac{b_j}{a_1}\sin\left(\frac{2q\pi y/w}{1 - t/w}\right) \tag{36}$$

$$H_y = \sum_{j=1}^{N} \frac{b_j}{a_1} y_{bj}\sin\left(\frac{2q\pi y/w}{1 - t/w}\right). \tag{37}$$

When $0.5(1 + t/w)<y/w<1$, substitute $(1-y/w)$ for y/w in (36) and (37). These fields are plotted (by computer) in Figs. 6 to 9 for half the guide width only, the fields being symmetrical about $y/w=0.5$. Solid curves represent the field summation in guide a immediately preceding the junction, and the broken curves represent the fields just inside guide b.

First of all, consider the electric fields. The quality of the match is seen to improve as more modes are used. Notice that the electric field in b is zero at the vane. This is because each constituent mode of the Fourier series has zero electric field there. However, the summation of modes in guide a does not vanish there, although it is attempting to do so. In particular, over the thick vane, E_x in guide a oscillates about zero. The greater the number of modes used, the greater the frequency of oscillation, and the smaller their amplitudes, converging to zero in the limit.

The aperture electric fields are roughly what one might have expected. The resulting pattern is a compromise between the incident H_{01} mode and the requirement that the electric field should disappear at the vane; the maximum electric field does not occur at the center of the bifurcated region but slightly more to the center of the guide. As the wave proceeds down the guide, the higher-order modes attenuate very rapidly, and the pattern becomes substantially that of an undistorted half-sine wave in each region.

The magnetic field has to cope with a singularity at the edge of the vane. For this reason, H_y does not attain as good a match as does E_x for the same number of modes. Besides going to infinity as the corner is approached, an added difficulty is that H_y in guide b must vanish at the vane for the same reasons as E_x. This is attempted by rising to a high value near the vane, and then suddenly dropping sharply to zero.

In the limit, with increasingly large expansions, the oscillations disappear and H_y increases almost linearly with y/w, except in the vicinity of the edge where it goes to infinity. Directly in front of the vane, a magnetic field exists supported by surface currents.

In the preceding study, the number of modes employed in guides a and b were equal, i.e., $M=N$. For several vane thicknesses ($t/w=0$, 0.2, and 0.8), computations were made of input admittance y' versus N for a range of M/N values. It was found that if M/N was greater than unity, higher accuracy could be achieved. However, if M/N was too large, instabilities occurred and wrong answers resulted. It is felt that there may be a way of choosing an optimum ratio for a given discontinuity, but the matter has not yet been investigated. In a general way, it seems that the greater the discontinuity, the larger the optimum M/N ratio required.

B. Finite-Length Bifurcation

Having considered the semi-infinite bifurcation, the finite bifurcation of length l will now be studied. This is a symmetric problem, and so it will be treated as described in Section V.

The propagation constant of the jth mode in guide b is given by

$$\gamma_j = \frac{\pi}{w}\sqrt{\left(\frac{2q}{1 - t/w}\right)^2 - \left(2\frac{w}{\lambda_0}\right)^2}. \tag{38}$$

The normalized input admittance of the jth mode, distance $l/2$ from an open circuit at the central plane, is

$$y'_{bj} = \tanh(\gamma_j l/2) \tag{39}$$

and with a short circuit at the central plane,

$$y'_{bj} = \coth(\gamma_j l/2). \tag{40}$$

s_{jj} is then computed from (19). It could equally have been found by rewriting (18) to give

$$s_{jj} = \pm e^{-\gamma_j l}, \tag{41}$$

where the plus and minus signs correspond to symmetrical and antisymmetrical excitation, respectively.

Assuming an open circuit at the central plane, ρ was computed from (12) and (13) (using the relevant values of s_{jj}), and from it the normalized input impedance $Z_{11}+Z_{12}$ was found. With a short circuit at the central plane, the input impedance $Z_{11}-Z_{12}$ was found. Z_{12} was then easily calculated.

Table III presents the resulting equivalent T network values for different normalized lengths l/w. For long vanes, $l/w\geq 10$, $Z_{11}-Z_{12}$ is the inverse of y' computed for the semi-infinite case. As the length decreases, $Z_{11}-Z_{12}$ becomes smaller and Z_{12} increases. At about $l/w=0.1$, they are approximately equal. With further reduction in length, the bifurcation appears to be a zero-thickness inductive strip. Certain numerical difficulties arise when l/w is made exactly equal to zero. This is heralded by increasing disagreement between the Z_{12} values, computed for the $N=25$ and $N=50$ cases, as the length decreases. Finally, the results make no sense at all. Possibly, the simultaneous equations tend to become ill-conditioned in the limit. This point should be investigated more carefully.

TABLE I

SEMI-INFINITE BIFURCATIONS $w/\lambda_0=0.7$. NORMALIZED INPUT ADMITTANCE y'

N	$t/w=0$	$t/w=0.2$
5	$-j2.363$	$-j7.213$
10	$-j2.396$	$-j7.335$
20	$-j2.410$	$-j7.378$
40	$-j2.415$	$-j7.394$

TABLE II

SEMI-INFINITE BIFURCATION $t/w=0$, $w/\lambda_0=0.7$. FIRST FIVE FORWARD-SCATTERED MODE COEFFICIENTS b_j/a_1

j	$N=5$	$N=10$	$N=20$	$N=40$
1	$0.7956\underline{/67.06°}$	$0.7857\underline{/67.35°}$	$0.7824\underline{/67.46°}$	$0.7813\underline{/67.50°}$
2	$-0.1802\underline{/67.06°}$	$-0.1717\underline{/67.35°}$	$-0.1693\underline{/67.46°}$	$-0.1685\underline{/67.50°}$
3	$0.0975\underline{/67.06°}$	$0.0872\underline{/67.35°}$	$0.0847\underline{/67.46°}$	$0.0840\underline{/67.50°}$
4	$-0.0715\underline{/67.06°}$	$-0.0562\underline{/67.35°}$	$-0.0535\underline{/67.46°}$	$-0.0528\underline{/67.50°}$
5	$0.0789\underline{/67.06°}$	$0.0409\underline{/67.35°}$	$0.0379\underline{/67.46°}$	$0.0372\underline{/67.50°}$

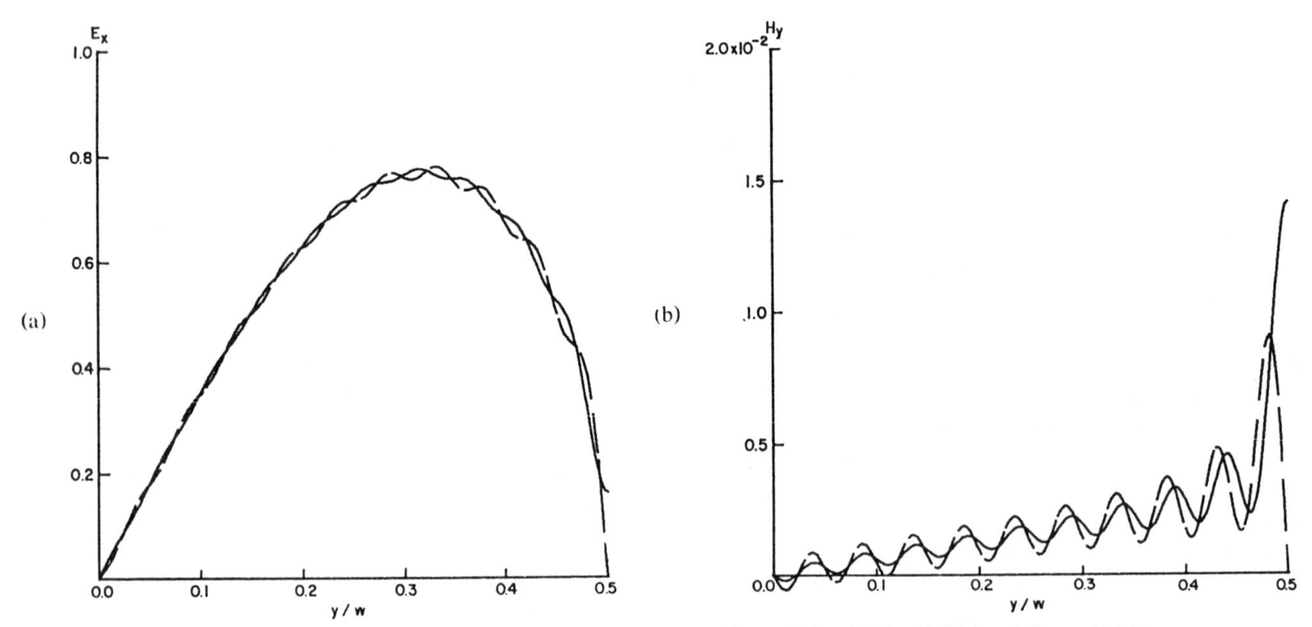

Fig. 6. H-plane bifurcation transverse fields: $t/w=0$, $N=20$. (a) Phase of $E_x=67.46°$. (b) Phase of $H_y=-22.54°$.

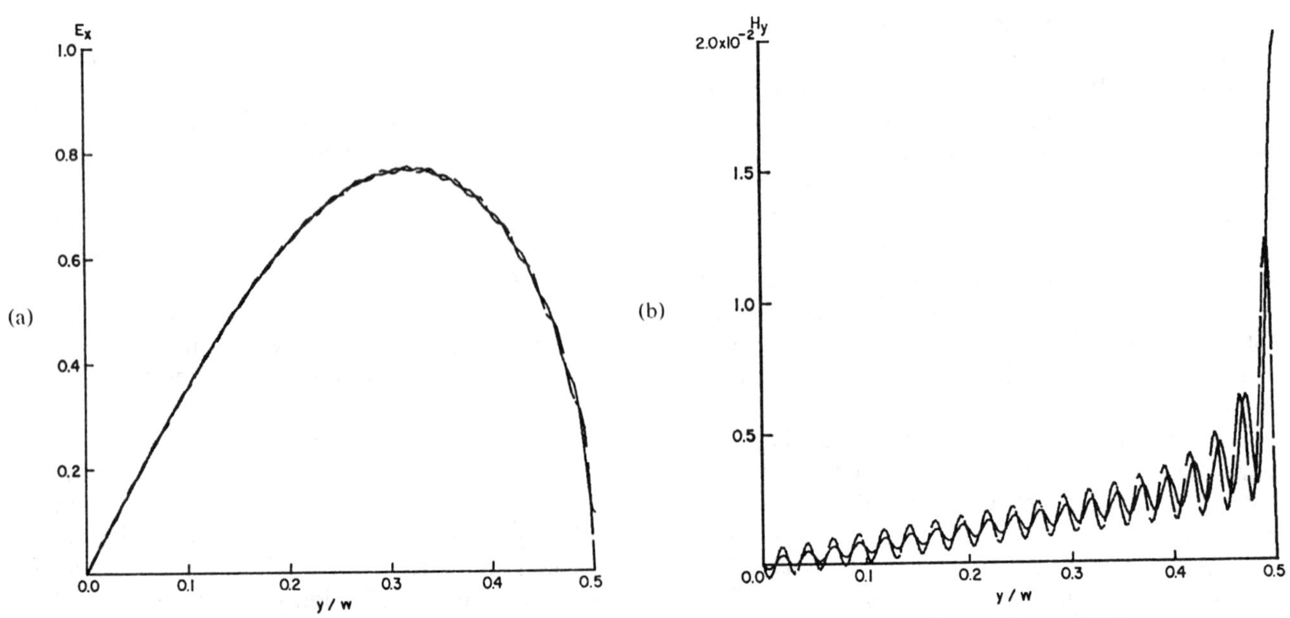

Fig. 7. H-plane bifurcation transverse fields: $t/w=0$, $N=40$. (a) Phase of $E_x=67.50°$. (b) Phase of $H_y=-22.50°$.

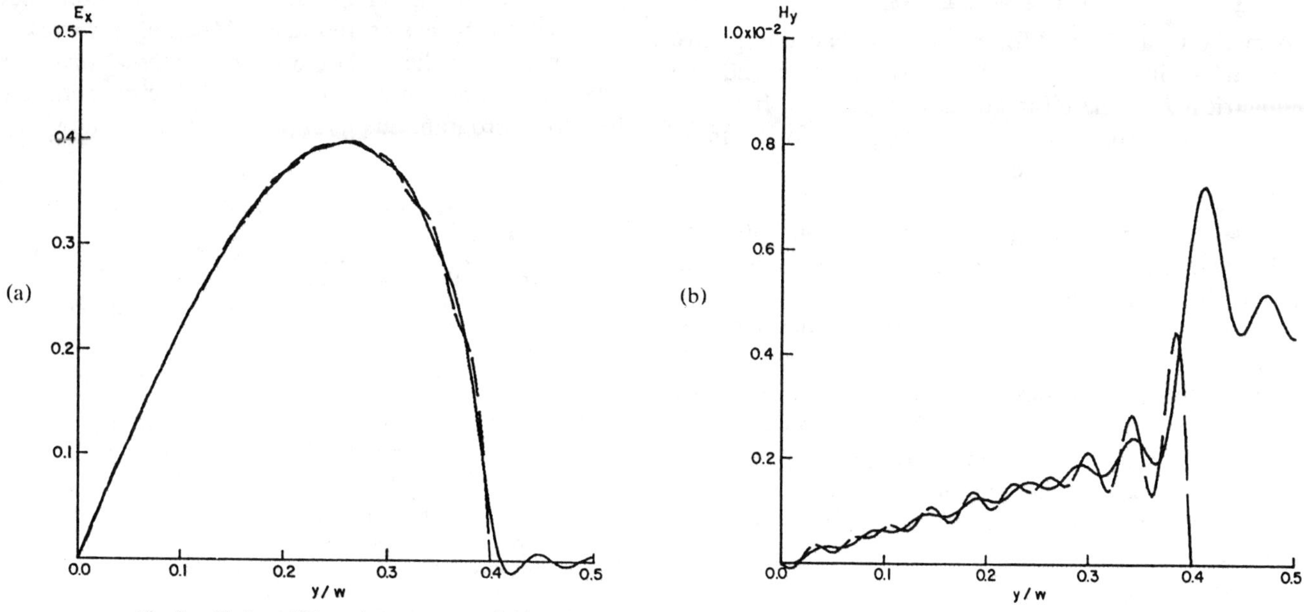

Fig. 8. H-plane bifurcation transverse fields: $t/w = 0.2$, $N = 20$. (a) Phase of $E_x = 82.28°$. (b) Phase of $H_y = -7.72°$.

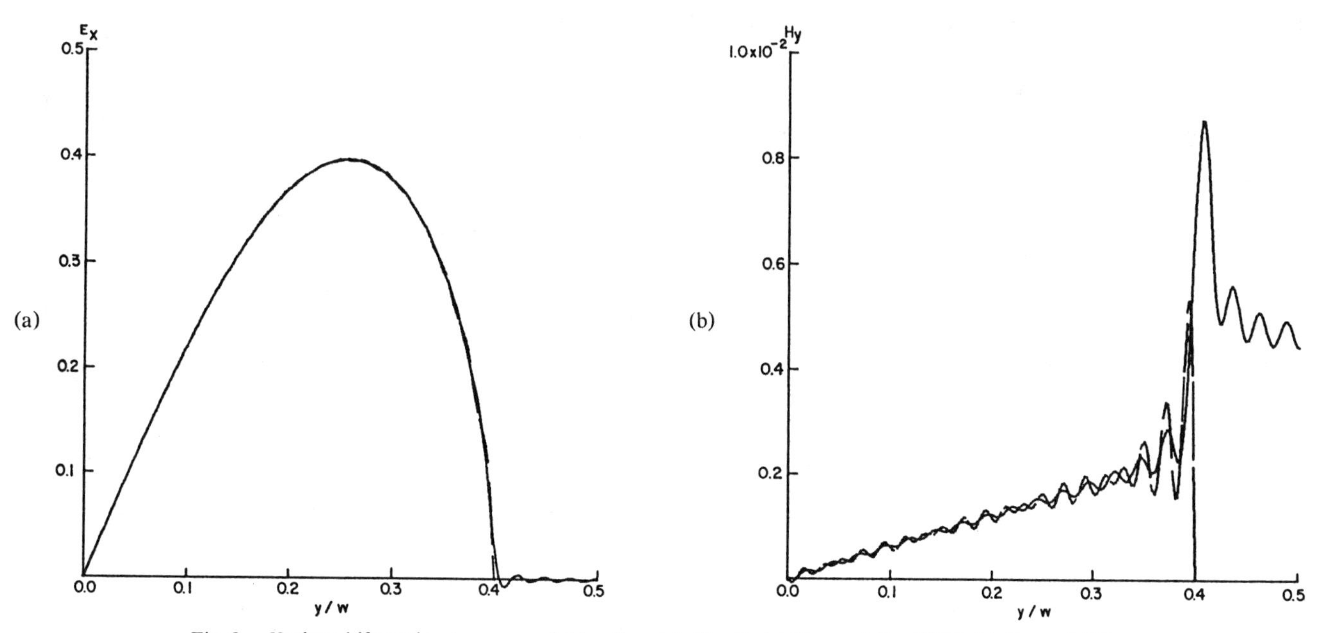

Fig. 9. H-plane bifurcation transverse fields: $t/w = 0.2$, $N = 40$. (a) Phase of $E_x = 82.30$. (b) $H_y = -7.70°$.

TABLE III

FINITE-LENGTH BIFURCATION $t/w = 0.2$, $w/\lambda_0 = 0.7$. SYMMETRICAL T EQUIVALENT NETWORK PARAMETERS

l/w	$N=25$		$N=50$	
	Z_{12}	$Z_{11}-Z_{12}$	Z_{12}	$Z_{11}-Z_{12}$
10	$j.0000$	$j.1354$	$j.0000$	$j.1352$
1	$j.0002$	$j.1352$	$j.0002$	$j.1350$
10^{-1}	$j.0769$	$j.0651$	$j.0767$	$j.0650$
10^{-2}	$j.1589$	$j.0089$	$j.1578$	$j.0089$
10^{-3}	$j.1801$	$j.0009$	$j.1766$	$j.0009$
10^{-4}	$j.1907$	$j.0001$	$j.1824$	$j.0001$
10^{-5}	$j.1986$	$j.0000$	$j.1859$	$j.0000$

VII. Conclusions

A method was derived for solving a wide class of waveguide discontinuities by modal analysis. Thick and thin symmetrical H-plane bifurcations were studied. Input admittance and scattered modes were computed for a number of expansion sizes. Accuracy improved with the number of modes used.

Transverse electric-field patterns, plotted on both sides of the junction, indicated a good match for twenty and forty modes. Magnetic field matches less well due to the singularity at the vane. Even so, with forty modes the actual pattern may be easily discerned. Because permeable media do not exhibit singularities, it is believed that convergence will be much more rapid for discontinuities formed by them.

More information is required on the effects of different numbers of modes on alternate sides of a junction and on the solution's behavior as the axial length of a discontinuity vanishes.

Acknowledgment

The author wishes to acknowledge the cooperation of the University of Manitoba Computer Centre. Particular appreciation is felt for the unstinting help of K. W. Schmidt who did the programming. Dr. J. W. Bandler contributed many helpful criticisms and ideas. Much of this work was developed while the author was with International Computers and Tabulators, Ltd., London. The cooperation of their Atlas programming group is gratefully acknowledged.

References

[1] R. E. Collin, *Field Theory of Guided Waves*. New York: McGraw-Hill, 1960, pp. 409–452.

[2] L. Lewin, "On the resolution of a class of waveguide discontinuity problems by the use of singular integral equations," *IRE Trans. Microwave Theory and Techniques*, vol. MTT-9, pp. 321–332, July 1961.

[3] R. E. Collin [1], pp. 314–367.

[4] S. W. Drabowitch, "Multimode antennas," *Microwave J.*, vol. 9, pp. 41–51, January 1966.

[5] J. B. Davies and C. A. Muilwyk, "Numerical solution of uniform hollow waveguides and boundaries of arbitrary shape," *Proc. IEE* (London), vol. 113, pp. 277–284, February 1966.

[6] R. E. Collin [1], pp. 229–232.

[7] P. M. Morse and H. Feshbach, *Methods of Theoretical Physics*. New York: McGraw-Hill, 1953, pp. 928–929.

[8] C. E. Fröberg, *Introduction to Numerical Analysis*. Reading, Mass.: Addison-Wesley, 1965, pp. 172–201, 221–225.

[9] N. Marcuvitz, *Waveguide Handbook*. New York: McGraw-Hill, 1951, pp. 117–126.

[10] E. L. Ginzton, *Microwave Measurements*. New York: McGraw-Hill, 1957, pp. 317–329.

[11] C. E. Fröberg [8], pp. 74–75.

[12] N. Marcuvitz [9], pp. 172–174.

Computer field-matching solution of waveguide transverse discontinuities

P. H. Masterman, Ph.D., B.Sc., and Prof. P. J. B. Clarricoats, D.Sc.(Eng.), Fel.I.E.E.E., F.Inst.P., C.Eng., F.I.E.E.

Indexing terms: *Waveguides, Computer applications, Electromagnetic fields and waves*

Abstract

A computational method for solving a wide range of transverse waveguide discontinuity problems is described. Results are obtained by the simultaneous solution of matrix equations, generated by Fourier analysis, which relate the complex amplitudes of orthogonal electric and magnetic field components. In some cases, the solution is found to be sensitive to the way in which infinite series of field functions are truncated, and it is shown how the optimum form of truncation can be determined for many configurations of practical importance. Several examples showing the application of the method are given, and comparison of results with those obtained by experiment, and by other analytical techniques, confirms its accuracy.

List of principal symbols

a, b, c, d = waveguide and/or iris dimensions
a_{im}, a'_{in} = coefficients of modes incident on discontinuity
a_{rm}, a'_{rn} = coefficients of modes travelling away from discontinuity
b' = waveguide or iris height
B = equivalent discontinuity susceptance
b_{ki} = aperture electric-field-function coefficient
c_{ki} = aperture magnetic-field-function coefficient
$\boldsymbol{e}_m, \boldsymbol{e}'_n$ = transverse electric-field components for waveguide modes
$\boldsymbol{e}_{ki}$ = transverse electric-field function for aperture k
$\boldsymbol{E}_T, \boldsymbol{E}'_T$ = transverse electric field in waveguide
$\boldsymbol{E}_{Tk}$ = transverse electric field in aperture k
h = height of waveguide
$\boldsymbol{h}_m, \boldsymbol{h}'_n$ = transverse magnetic-field components for waveguide modes
$\boldsymbol{h}_{kj}$ = transverse magnetic-field function for aperture k
$\boldsymbol{H}_T, \boldsymbol{H}'_T$ = transverse magnetic field in waveguide
$\boldsymbol{H}_{Tk}$ = transverse magnetic field in aperture k
$\hat{\imath}_x, \hat{\imath}_y$ = unit vectors in specified co-ordinate directions
p, p' = numbers of modes in waveguides
P = total number of coupling apertures
q, q_k, q_T = numbers of electric or magnetic aperture field functions
R = radius of iris aperture or smaller circular waveguide
R' = radius of larger circular waveguide
S, S' = waveguide cross-sectional surface
S_k = aperture surface
x, y, z = Cartesian co-ordinates
Y_L, Y_R, Y_0, Y'_0 = equivalent transmission-line characteristic admittances
$\bar{\gamma}_n, \bar{\gamma}_{1n}$ = complex propagation coefficients, normalised to that of free space
ϵ_{ij} = Kronecker symbol ($= 1, i = j, = 2, i \neq j$)
ϵ_0 = free-space permittivity
λ_0 = free-space wavelength
λ_g = guide wavelength
μ_0 = free-space permeability

1 Introduction

Determining the nature of the electromagnetic fields in the vicinity of a waveguide discontinuity is a classic microwave-engineering problem, for which many different methods of solution have been suggested during the past 30 years. In a few specific cases, an exact solution can be found, using, for example, the 'integral-transform' technique,[1] but, in general, this is not possible, and some degree of approximation must be made. Of the approximate techniques, the 'variational' and 'integral-equation' methods are applicable to a wide range of problems, and can produce sufficiently accurate results for most purposes. The former method is described by Collin,[2] while the work of Lewin[3] is well known in connection with the latter. Both methods are reviewed by Marcuvitz.[4] A serious drawback, particularly with the variational method, is that the formulation for a given structure may call for considerable mathematical ability on the part of the user. A further restriction is that these methods cannot easily be used in the case of a transverse discontinuity involving waveguides capable of supporting more than one propagating mode.

Many of the conventional techniques aim to produce a formula for a given type of discontinuity, from which the solution for a particular set of dimensions may be obtained by hand calculation. Fairly gross approximations must often be made to keep such formulas to manageable proportions. At one time, this approach was very valuable, but, now that access to high-speed digital computers is generally available, there is much to recommend an accurate technique involving lengthy but straightforward calculations, which can easily be applied to a wide variety of structures. Such a technique is described in this paper, its accuracy and limitations are investigated, and some applications are considered. In a later paper, the application of the method to discontinuities parallel to the waveguide axis will be considered.

The method has been known for some time in a restricted form; the original formulation being due to Whinnery and Jamieson[5] who considered discontinuities in parallel-plate waveguides.

The application to discontinuities in dielectric-loaded waveguides was first considered by Clarricoats.[6] Later, a computer formulation was developed by Clarricoats and Slinn[7,8] and these authors extended their computer method of solution to waveguide step discontinuities. A similar study was made independently by Wexler,[9] and recent applications include those of Cole[10] and Braekelmann.[11,12] Wexler's term 'modal analysis' will be used in this paper to refer to the restricted method. In modal analysis, the fields on either side of the discontinuity plane are expressed initially as infinite series of waveguide modes. The amplitude of each mode is then equal to a Fourier series of the mode amplitudes on the opposite side of the junction. For the modes on one side, this follows from the matching of transverse electric fields, whereas, for those on the other side, magnetic-field matching is considered. Two infinite sets of linear equations are thus generated, which, after truncation, are solved simultaneously to give the amplitude coefficients of the various modes. By taking into

Paper 6349 E, received 2nd September 1970
Dr. Masterman was formerly with the Department of Electrical Engineering, University of Leeds, Leeds LS2 9JT, and is now with the Signals Research & Development Establishment, Christchurch, Hants., England. Prof. Clarricoats is with the Department of Electrical & Electronic Engineering, Queen Mary College, University of London, Mile End Road, London E1, England

Reprinted with permission from *Proc. IEE,* vol. 118, no. 1, pp. 51–63, Jan. 1971.

account multiple reflections of modes between two adjacent step discontinuities, both Clarricoats and Slinn, and Wexler, were able to extend the method to deal with a thick iris, but found that, as the thickness decreased, progressively more modes had to be included in the calculation for an accurate solution. It was therefore found impossible to use the method to solve an infinitely thin iris.

A preliminary description of a method similar to modal analysis and applicable to thin irises has recently been given by the authors.[13] Instead of expressing the fields in the iris aperture in terms of waveguide modes, separate series of orthogonal functions are used for the transverse electric and magnetic components. The aperture magnetic-field coefficients are easily found from symmetry considerations, and field matching between the waveguide modes and aperture functions leads, as in modal analysis, to two sets of linear equations which can be solved simultaneously to give the required modal amplitudes. This, again, is a special case of the general computational method.

In this paper, the computational method is presented in a form applicable to transverse waveguide discontinuities consisting, in general, of abrupt junctions between two or more parallel waveguides with arbitrary cross-sections and filling media. Coupling between the guides is effected by means of any number of apertures in an infinitely thin conducting sheet lying in the discontinuity plane, which is normal to the waveguide axes.

2 General method of solution for transverse discontinuities

An example of the type of discontinuity described above is shown in Fig. 1. In the discontinuity plane, let the

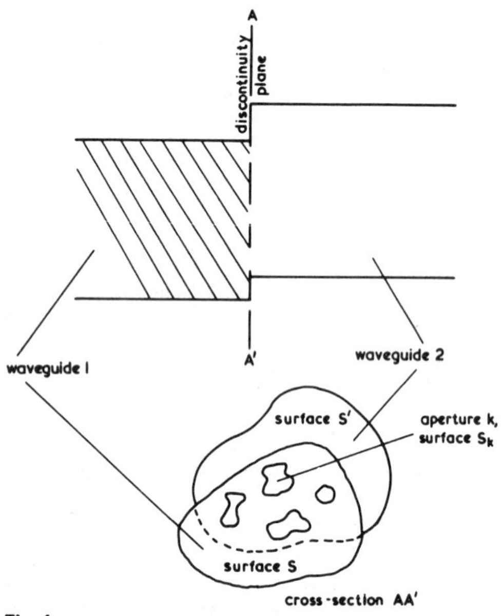

Fig. 1
General form of a waveguide discontinuity

transverse electric- and magnetic-field components of the mth mode in waveguide 1 be $\boldsymbol{e}_m$ and $\boldsymbol{h}_m$, respectively, and those of the nth mode in waveguide 2 be $\boldsymbol{e}'_n$ and $\boldsymbol{h}'_n$, respectively. The following orthogonality property applies to any closed waveguide structure, the symbols used here being those for waveguide 1:

$$\int_S \boldsymbol{e}_i \times \boldsymbol{h}_j \, . \, ds = 0 \quad \text{for } i \neq j \qquad (1)$$

The orthogonality condition for waveguide 2 is similar, but with S, $\boldsymbol{e}_i$ and $\boldsymbol{h}_j$ replaced by primed quantities. In general, there can be an incident field in waveguide 1 consisting of an infinitude of modes with coefficients a_{im}, and an incident field in waveguide 2 with mode coefficients a'_{in}. Usually, all such coefficients are zero, except for a_{i1} or a'_{i1}, but the general form may be required if a closely spaced pair of discontinuities is considered, where the field incident on one includes evanescent modes excited by the other. The coefficients are, in general, complex. The transverse fields in the discontinuity plane are then given by the following equations.

Waveguide 1:

$$\boldsymbol{E}_T = \sum_{m=1}^{\infty} (a_{im} + a_{rm})\boldsymbol{e}_m \qquad (2a)$$

$$\boldsymbol{H}_T = \sum_{m=1}^{\infty} (a_{im} - a_{rm})\boldsymbol{h}_m \qquad (2b)$$

Waveguide 2:

$$\boldsymbol{E}'_T = \sum_{n=1}^{\infty} (a'_{in} + a'_{rn}) + \boldsymbol{e}'_n a_n \qquad (3a)$$

$$\boldsymbol{H}'_T = \sum_{n=1}^{\infty} (a'_{rn} - a'_{in})\boldsymbol{h}'_n \qquad (3b)$$

The coefficients a_{rm} and a'_{rn} refer to modes travelling away from the junction in waveguides 1 and 2, respectively, and it is these that the method aims to evaluate.

The diaphragm separating the two waveguides is assumed to be infinitely thin and perfectly conducting. It follows from the latter assumption that the transverse electric field in the discontinuity plane vanishes everywhere except in the apertures, and also that the component of this field parallel to the edge of the aperture must vanish at that edge. In order to apply the method described in this Section, it is necessary to express the transverse electric field in each aperture as a series of vector functions which satisfy the above boundary condition. The transverse-electric-field components of the normal modes for a homogeneous waveguide, having the same cross-section as the aperture, form a convenient set for this purpose. Let $\boldsymbol{e}_{ki}$ be the ith such component for the kth aperture. The electric field in that aperture is then

$$\boldsymbol{E}_{Tk} = \sum_{i=1}^{\infty} b_{ki}\boldsymbol{e}_{ki} \qquad (4)$$

where b_{ki} is the complex amplitude coefficient of the ith component.

It is now necessary to represent the magnetic field in each aperture in a similar manner, and in such a way that the aperture magnetic- and electric-field functions have the following orthogonality property:

$$\int_{Sk} \boldsymbol{e}_{ki} \times \boldsymbol{h}_{kj} \, . \, ds = 0 \qquad \text{for } i \neq j \qquad (5)$$

This will be the case if the functions $\boldsymbol{h}_{kj}$ are derived from the same waveguide modes as the functions $\boldsymbol{e}_{ki}$

The transverse magnetic field in the kth aperture is then

$$\boldsymbol{H}_{Tk} = \sum_{i=1}^{\infty} c_{ki}\boldsymbol{h}_{ki} \qquad (6)$$

where c_{ki} is the complex amplitude coefficient of the ith component. By using functions derived from waveguide modes to represent the aperture magnetic field, the condition that the component of $\boldsymbol{H}_{Tk}$ normal to the aperture edge should vanish has been applied. In fact, this is not the case, since it can be shown that the magnetic field becomes infinite at an aperture edge. However, by taking a large number of functions to represent the aperture field, very high values of field immediately in front of the edge can be achieved, and, in the limit where the number of functions becomes infinite, the singularity is accurately represented.

It should be pointed out that, although the vector functions $\boldsymbol{e}_{ki}$ and $\boldsymbol{h}_{ki}$ have properties in common with waveguide modes, they do not form a 'mode' in the usual sense of the word. Whereas knowledge of the transverse electric field in a waveguide as a series of modes enables the magnetic field to be calculated using the wave admittances, a series representation for an aperture electric field as in eqn. 4 gives no information about the magnetic field in that aperture. In a waveguide, longitudinal electric field must vanish at the edge of the cross-section, but this does not hold for an aperture in an infinitely thin conducting sheet. A much wider variety of solutions to Maxwell's equations is therefore possible in the latter case.

The transverse electric and magnetic fields must be continuous across each aperture. Thus, for the kth aperture,

$$E_{Tk} = \sum_{m=1}^{\infty}(a_{im} + a_{rm})e_m = \sum_{i=1}^{\infty} b_{ki}e_{ki} = \sum_{n=1}^{\infty}(a'_{in} + a'_{rn})e'_n \quad . \quad . \quad . \quad . \quad (7)$$

$$H_{Tk} = \sum_{m=1}^{\infty}(a_{im} - a_{rm})h_m = \sum_{i=1}^{\infty} c_{ki}h_{ki} = \sum_{n=1}^{\infty}(a'_{rn} - a'_{in})h'_n \quad . \quad . \quad . \quad . \quad (8)$$

Note that eqns. 7 and 8 only hold over the aperture in question. On forming vector products with h_M, M being any positive integer, and integrating over aperture k, the second and third terms in eqn. 7 give rise to an infinite set of equations as follows:

$$\sum_{m=1}^{\infty}(a_{im} + a_{rm})\int_{S_k} e_m \times h_M . ds - \sum_{i=1}^{\infty} b_{ki}\int_{S_k} e_{ki} \times h_M . ds \qquad M = 1, 2, 3, \ldots \infty \quad . \quad . \quad . \quad . \quad (9)$$

The vanishing of the transverse electric field over the diaphragm implies that

$$\sum_{m=1}^{\infty}(a_{im} + a_{rm})\int_{S_0} e_m \times h_M . ds = 0 \quad . \quad . \quad . \quad . \quad (10)$$

for all M, where $S_0 = S - \sum_{k=1}^{P} S_k$, P being the total number of apertures. Equations of the same form as eqn. 9 having been added together, for all P apertures, eqn. 10 may be used to extend the left-hand-side integral over the entire waveguide cross-section S. Use of the orthogonality condition of eqn. 1 then enables all terms but one to be eliminated from the series, the result being

$$a_{iM} + a_{rM} = \frac{\sum_{k=1}^{P}\sum_{i=1}^{\infty} b_{ki}\int_{S_k} e_{ki} \times h_M . ds}{\int_S e_M \times h_M . ds} \qquad M = 1, 2, 3, \ldots, \infty \quad . \quad (11)$$

Starting with the third and fourth terms in eqn. 7, a similar process allows the mode coefficients for waveguide 2 to be expressed in terms of the aperture electric-field-function coefficients

$$a'_{iN} + a'_{rN} = \frac{\sum_{k=1}^{P}\sum_{i=1}^{\infty} b_{ki}\int_{S_k} e_{ki} \times h'_N . ds}{\int_{S'} e'_N \times h'_N . ds} \qquad N = 1, 2, 3, \ldots \infty \quad . \quad (12)$$

On vector premultiplying the second and third terms in eqn. 8 by e_{kI}, where I is any positive integer, and integrating over aperture k, a further set of equations results:

$$\sum_{m=1}^{\infty}(a_{im} - a_{rm})\int_{S_k} e_{kI} \times h_m . ds = \sum_{i=1}^{\infty} c_{ki}\int_{S_k} e_{kI} \times h_{ki} . ds \qquad I = 1, 2, 3, \ldots \infty \quad . \quad (13)$$

The orthogonality condition of eqn. 5 may be applied directly to this set of equations, which then becomes

$$c_{kI} = \frac{\sum_{m=1}^{\infty}(a_{im} - a_{rm})\int_{S_k} e_{kI} \times h_m . ds}{\int_{S_k} e_{kI} \times h_{kI} . ds} \qquad I = 1, 2, 3, \ldots \infty. \; k = 1, 2, 3, \ldots P \quad . \quad (14)$$

Similarly, starting with the third and fourth terms in eqn. 8,

$$c_{kI} = \frac{\sum_{n=1}^{\infty}(a'_{rn} - a'_{in})\int_{S_k} e_{kI} \times h'_n . ds}{\int_{S_k} e_{kI} . h_{kI} . ds} \qquad I = 1, 2, 3, \ldots \infty. \quad k = 1, 2, 3, \ldots P \quad . \quad (15)$$

To solve eqns. 11, 12, 14 and 15 for the unknown modal amplitudes a_{rm} and a'_{rn}, it is first necessary to truncate the infinite series involved, so that the equations can be conveniently processed by a computer. If the numbers of modes taken into account in waveguides 1 and 2 are p and p', respectively, and the number of functions taken into account in aperture k is q_k, then $m \leqslant p$, $M \leqslant p$, $n \leqslant p'$, $N \leqslant p'$ and, for aperture k, $i \leqslant q_k$ and $I \leqslant q_k$. It is then possible to write eqn. 11 as a matrix equation

$$\begin{bmatrix} a_{i1} \\ \cdot \\ \cdot \\ \cdot \\ a_{ip} \end{bmatrix} + \begin{bmatrix} a_{r1} \\ \cdot \\ \cdot \\ \cdot \\ a_{rp} \end{bmatrix} = \begin{bmatrix} R_1(1,1) \ldots R_1(1,q_1) & \vdots & & R_P(1,1) \ldots R_P(1,q_P) \\ \cdot \quad \cdot & \vdots & \ldots & \cdot \quad \cdot \\ R_1(p,1) \ldots R_1(p,q_1) & \vdots & & R_P(p,1) \ldots R_P(p,q_P) \end{bmatrix} \begin{bmatrix} b_{11} \\ \cdot \\ \cdot \\ b_{1q_1} \\ \cdots \\ \cdot \\ \cdot \\ \cdots \\ b_{P1} \\ \cdot \\ \cdot \\ b_{Pq_P} \end{bmatrix} \quad (16)$$

In this equation,

$$R_k(i, j) = \int_{S_k} e_{kj} \times h_i . ds \Big/ \int_S e_i \times h_i . ds \quad . \quad . \quad . \quad (17)$$

For convenience, eqn. 16 will be written as

$$a_i + a_r = Rb \quad . \quad . \quad . \quad . \quad . \quad . \quad . \quad . \quad . \quad . \quad (18)$$

If q_T is defined as $\sum_{k=1}^{P} q_k$, then R has dimensions $p \times q_T$ and b has q_T elements.

The following matrix equation results from eqn. 12:

$$a'_i + a'_r = R'b \quad . \quad . \quad . \quad . \quad . \quad . \quad . \quad . \quad . \quad . \quad (19)$$

The column vectors a'_i, a'_r have p' elements, and R' has dimensions $p' \times q_T$. The elements of R' are defined in a similar manner to those of R, but with h_i, e_i and S replaced by primed quantities.

The P sets of equations represented by eqn. 14 can be lumped together to give a single matrix equation

$$\begin{bmatrix} c_{11} \\ \cdot \\ \cdot \\ c_{1q_1} \\ \cdots \\ \cdot \\ \cdot \\ \cdots \\ c_{P1} \\ \cdot \\ \cdot \\ c_{Pq_P} \end{bmatrix} = \begin{bmatrix} S_1(1,1) \ldots S_1(1,p) \\ \cdot \quad \cdot \\ S_1(q_1,1) \ldots S_1(q_1,p) \\ \cdots \\ \cdot \quad \cdot \\ \cdots \\ S_P(1,1) \ldots S_P(1,p) \\ \cdot \quad \cdot \\ S_P(q_P,1) \ldots S_P(q_P,p) \end{bmatrix} \begin{bmatrix} a_{i1} - a_{r1} \\ \cdot \\ \cdot \\ \cdot \\ a_{ip} - a_{rp} \end{bmatrix} \quad (20)$$

In this equation,

$$S_k(i, j) = \int_{S_k} e_{ki} \times h_j . ds \Big/ \int_{S_k} e_{ki} \times h_{ki} . ds \quad . \quad . \quad (21)$$

Writing eqn. 20 in a more convenient form,

$$c = Sa_i - Sa_r \quad . \quad . \quad . \quad . \quad . \quad . \quad . \quad . \quad . \quad . \quad (22)$$

where c has q_T elements, and S has dimensions $q_T \times p$.

Similarly, from eqn. 15,

$$c = S'a'_r - S'a'_i \quad . \quad . \quad . \quad . \quad . \quad . \quad . \quad . \quad . \quad (23)$$

where elements of the $q_T \times p'$ matrix S' are defined as in eqn. 21, but with h_j replaced by h'_j.

By multiplying eqn. 18 through by $\boldsymbol{S}$, $\boldsymbol{a}_r$ can be eliminated from eqn. 22 to give

$$\boldsymbol{c} = 2\boldsymbol{S}\boldsymbol{a}_i - \boldsymbol{S}\boldsymbol{R}\boldsymbol{b} \quad (24)$$

Similarly, eliminating $\boldsymbol{a}_r'$ from eqns. 19 and 23,

$$\boldsymbol{c} = \boldsymbol{S}'\boldsymbol{R}'\boldsymbol{b} - 2\boldsymbol{S}'\boldsymbol{a}_i' \quad (25)$$

Equating eqns. 24 and 25,

$$(\boldsymbol{S}\boldsymbol{R} + \boldsymbol{S}'\boldsymbol{R}')\boldsymbol{b} = 2\boldsymbol{S}\boldsymbol{a}_i + 2\boldsymbol{S}'\boldsymbol{a}_i' \quad (26)$$

The column vectors $\boldsymbol{a}_i$ and $\boldsymbol{a}_i'$ are known, so that eqn. 26 represents a set of q_T simultaneous equations in q_T unknowns b_{ki}. Having solved these equations, the values for the elements of $\boldsymbol{b}$ can be substituted in eqns. 18 and 19 to give the required modal amplitude vectors $\boldsymbol{a}_r$ and $\boldsymbol{a}_r'$.

Fig. 2 shows the variation with q, the number of aperture

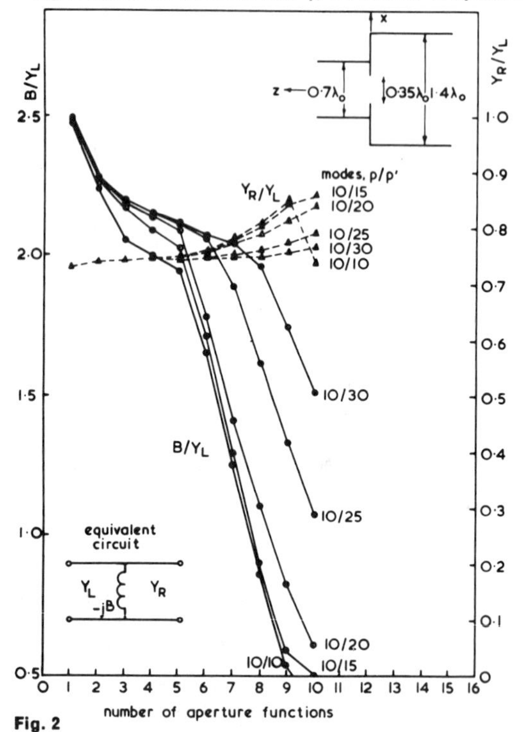

Fig. 2
Inductive-iris coupling between rectangular waveguides of different widths

functions, of the transmission-line equivalent-circuit parameters for an H plane junction between two rectangular waveguides with a thin single-aperture inductive iris in the discontinuity plane, for various values of p/p'. The equivalent circuit relates to the H_{10} mode, which, for the dimensions in question (see inset), is the only propagating mode in either guide, the H_{20} mode being unexcited because of symmetry about the plane $x = 0$. This example illustrates the way in which the choice of the ratio of modes and the aperture function affects the solution of discontinuity problems. It is clear that, for a given ratio p/p', even small changes in q can cause very large variations in the equivalent shunt susceptance, so that, for the method to be of practical value, some criteria for choosing the ratios must be determined. We shall return to this general problem after considering two simpler structures in which only one ratio is involved, namely a thin iris in a waveguide and a step discontinuity in which the smaller-guide cross-section is entirely contained within that of the larger at the discontinuity plane.

3 Special case of a thin iris in a waveguide

3.1 Modification of general method

A particular case of the general discontinuity described in Section 2, and one which has important applications in the design of matching devices for microwave circuitry, is that of a thin conducting iris in a waveguide, transverse to the direction of wave propagation. Here, again, there can be any number of apertures, but, in this case, the waveguides on either side of the discontinuity have identical properties, and there is no step between them. Thus, having put p equal to p', the matrices $\boldsymbol{R}$ and $\boldsymbol{R}'$ in the general analysis become identical, as do $\boldsymbol{S}$ and $\boldsymbol{S}'$. Eqn. 26 therefore becomes

$$\boldsymbol{S}\boldsymbol{R}\boldsymbol{b} = \boldsymbol{S}\boldsymbol{a}_i + \boldsymbol{S}\boldsymbol{a}_i' \quad (27)$$

In the more usual case, where incidence is from one side only, one of the terms on the right-hand side of eqn. 27 vanishes. The above result can then be alternatively derived by considering the symmetric nature of the discontinuity. Coefficients a_{rm} and a_{rm}' for the mth mode travelling away from the iris in guides 1 and 2, respectively, must now be identical, to ensure electric-field matching in the aperture(s). Substitution in eqn. 8 then shows the total transverse aperture magnetic field due to such modes to be zero, so that the aperture magnetic field $\boldsymbol{H}_{Tk}$ is simply that of the incident field alone. Using series representations for the magnetic fields in the various apertures, and matching these fields to the incident field, the following matrix equation can be obtained:

$$\boldsymbol{c} = \boldsymbol{S}\boldsymbol{a}_i \quad (28)$$

$\boldsymbol{a}_i$ is known, and so the elements of $\boldsymbol{c}$ can be found. When this result is substituted in eqn. 22, it is found that

$$\boldsymbol{S}\boldsymbol{a}_r = 0 \quad (29)$$

It might, at first, be thought that this set of equations could be solved to give the unknown coefficients a_{rm}. However, a unique solution is only possible if there are at least as many linearly independent equations as there are unknowns, and it is clear that, if this is the case, the solution will be $a_{rm} = 0$ for all m, which implies that no iris is present. To obtain the correct solution, the problem must be underspecified in eqn. 29, in which case $q_T < p$. Eqn. 18 can then be used to impose the condition for electric-field matching on the solution. If this equation is multiplied through by $\boldsymbol{S}$, eqn. 29 can be used to eliminate $\boldsymbol{a}_r$, the resulting equation being eqn. 27.

It is not sufficient merely to set $q_T < p$ to obtain optimum convergence; the ratios q_k/p are critical, and their values depend on the dimensions of the structure in question. To illustrate this point and to demonstrate the use of the method, a particular case will now be considered, namely that of a thin single-aperture inductive iris in a rectangular waveguide.

3.2 Thin inductive iris in rectangular waveguide

3.2.1 Formulation of problem

This case is shown in Fig. 3*a*. If the dimensions are chosen so that only the dominant H_{10} mode can propagate in the waveguide, the discontinuity can be represented by the simple equivalent circuit of Fig. 3*b*. The incident field is assumed to consist only of the dominant mode, and this can be taken to have unit amplitude without loss of generality. Thus

$$\left.\begin{aligned} a_{im} &= 1 \quad \text{for } m = 1 \\ a_{im} &= 0 \quad \text{for } m \neq 1 \end{aligned}\right\} \quad (30)$$

The susceptance of the iris can be deduced from the complex reflection coefficient a_{r1} from the expression

$$\frac{1 - a_{r1}}{1 + a_{r1}} = 1 + j\frac{B}{Y_0} \quad (31)$$

Both the discontinuity and the incident field are uniform in the $y$direction: it follows that modes excited by the discontinuity must also have this uniformity, and are therefore of type H_{n0}, where n can be any positive integer. Notice, however, that, if b is equal to c, only modes for which n is odd will be excited, because of the symmetry of the structure.

On substituting standard modal field expressions for $\boldsymbol{e}_n$ and $\boldsymbol{h}_n$, and suitable functions of a similar type for the aperture-field components $\boldsymbol{e}_{1m}$ and $\boldsymbol{h}_{1m}$, the form of the elements of matrix $\boldsymbol{R}$ becomes

$$R_{nm} = \frac{2}{a}\int_b^{a-c} \sin\frac{n\pi x}{a} \sin\frac{m\pi}{a-b-c}(x-b)dx \quad (32)$$

Using eqn. 21, the elements of $\boldsymbol{S}$ are found to be related to those of $\boldsymbol{R}$ through the equation

$$S_{mn} = j\frac{a}{a-b-c}\sqrt{\left(\frac{\epsilon_0}{\mu_0}\right)}\,\bar{\gamma}_n R_{nm} \quad . \quad . \quad . \quad . \quad . \quad (33)$$

where $\bar{\gamma}_n = \sqrt{\left\{\frac{n^2}{4(a/\lambda_0)^2} - 1\right\}}$

Evaluating the integral, eqn. 32 becomes

$$R_{nm} = \frac{2m}{\pi}\,\frac{a-b-c}{a}\,\frac{\sin n\pi b/a - (-1)^m \sin n\pi(a-c)/a}{m^2 - \left[\frac{a-b-c}{a}\right]^2 n^2}$$

$$\text{for } am \neq (a-b-c)n \quad . \quad (34a)$$

$$R_{nm} = \frac{a-b-c}{a}\cos\frac{n\pi b}{a} \quad \text{for} \quad am = (a-b-c)n \quad . \quad (34b)$$

The basic operation in obtaining results for this case, using a

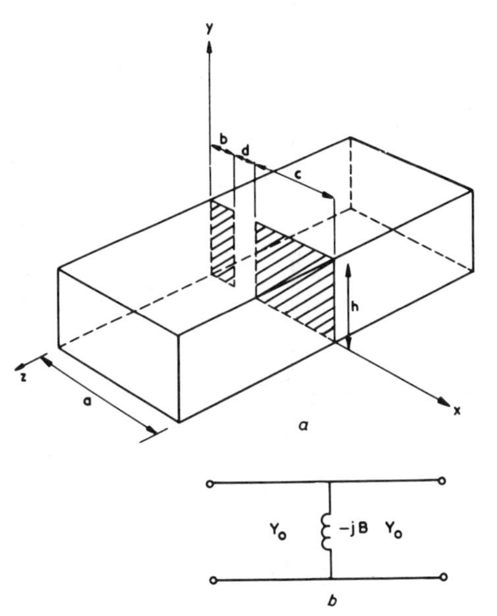

Fig. 3
Thin inductive iris in rectangular waveguide
a Waveguide
b Transimission-line equivalent circuit

digital computer, is the solution of the matrix equation obtained by substituting eqn. 30 in eqn. 27, i.e.

$$\boldsymbol{SRb} = \boldsymbol{S}_1 \quad . \quad . \quad . \quad . \quad . \quad . \quad . \quad . \quad . \quad . \quad . \quad (35)$$

where $\boldsymbol{S}_1$ is the first column of $\boldsymbol{S}$.

Note that the elements of $\boldsymbol{b}$ are complex, and those of $\boldsymbol{S}$ are imaginary if the suffix n in eqn. 33 refers to an evanescent waveguide mode. Since some computers cannot deal directly with complex numbers, it may be necessary to separate each equation in the set represented by eqn. 35 into its real and imaginary parts, yielding a set of $2q$ equations in $2q$ real unknowns. The simultaneous equations were solved, in practice, by the method of elimination and back substitution, although an iterative method could be used as an alternative, possibly with some saving of computing time. Using efficient subroutines for the matrix operations, run time for the case $p = 30$, $q = 15$ was about 16s on a KDF9 computer.

3.2.2 Computed results

As indicated above, the choice of q/p has a considerable effect on the computed results. The effect of varying this parameter was investigated for an inductive iris with an aperture width $a/2$ and waveguide width $a = 0{\cdot}8\lambda_0$ (see Fig. 3). In Fig. 4, results are given for three positions of the iris aperture; the symmetric case $b = c$, the case $b = 0$ in which the waveguide wall forms one edge of the aperture and an intermediate case $b = 0{\cdot}1a$. These cases are designated A, B and C, respectively. Twenty waveguide modes were included, and, in case A, these were such that only odd values of n were allowed in exprs. 32–34, the even modes not being excited because of symmetry.

In cases A and B, there is uniform convergence of the values

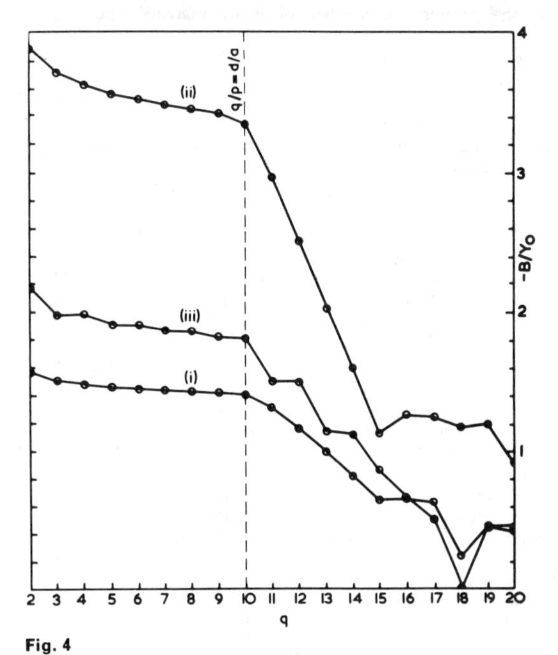

Fig. 4
Susceptance as a function of number of aperture components for thin inductive iris in rectangular waveguide
$a = 0{\cdot}8\lambda_0$
$d/a = 0{\cdot}5$
(i) Case A: $b/a = 0{\cdot}25$, $c/a = 0{\cdot}25$
(ii) Case B: $b/a = 0$, $c/a = 0{\cdot}5$
(iii) Case C: $b/a = 0{\cdot}1$, $c/a = 0{\cdot}4$

of normalised susceptance, calculated from eqn. 31 for q increasing from 2 to 9. Between $q = 9$ and $q = 10$, the slopes increase slightly, and, for $q > 10$, the value of $|B/Y_0|$ falls away rapidly, at first almost linearly, but with erratic behaviour as the situation $q = p$ is approached. In the previous Section, consideration of eqn. 29 showed that a_{r1}, and consequently B/Y_0, should be zero when p and q are equal, but this is not borne out by the results in Fig. 4. However, solutions obtained using fewer waveguide modes have behaved in the expected manner, and it seems likely that the irregular variations when q lies between 15 and 20 are the result of computational errors associated with the generation of very large numbers, and therefore have no real significance. The variation with q of the susceptance in case C follows the same general pattern as in the other two cases, but is not as regular. This is because additional aperture functions taken into account have alternate odd and even dependence. The odd functions are more strongly excited, and therefore have a greater effect on the calculated susceptance value. A feature that the three graphs have in common is the abrupt change in slope at the point $q/p = d/a$. Results have been obtained for other aperture widths, and, in every case, a distinct change of slope occurs at this point. The phenomenon is not restricted to the inductive iris, as will be shown in Section 4. The reason for the change of slope can be explained as follows. The aperture magnetic-field distribution is known exactly, being that of the incident waveguide mode, and the amplitudes of the aperture magnetic-field functions, which are derived from it by Fourier analysis (see eqn. 28), are therefore fixed and independent of the calculation process. Each of these field functions is matched across the aperture by a series of waveguide modes. If the highest-order aperture function has a field variation $\sin m\pi\,(x - b)/d$ across the aperture, the predominant term in the mode series will be that for which $n = ma/d$, since this has the same order of field variation across the aperture. If the series of modes is truncated before this term is reached (i.e. when $q/p > d/a$), magnetic-field matching for this function will therefore be poor, and an error will be introduced into the solution. Similarly, the highest-order electric-field function cannot be correctly matched to the waveguide field, and has been found to acquire an excessively large amplitude. The more q/p exceeds d/a, the greater the number of badly matched aperture functions, and the less accurate does the solution become. The field-matching argument also applies when $q/p < d/a$, but the effect of including excessively high-order waveguide modes is not as marked, because their amplitudes are determined in the calculation rather than initially, as was the case with the aperture magnetic-field functions. The excess modes are effectively ignored, having small amplitude coefficients in solution. The change of slope represents the transition between these two situations, and the solution corresponding to this point should be the most accurate, since field matching is good for all modes and aperture functions.

Since the accuracy of the solution is not markedly affected by an excess of waveguide modes, q/p should be chosen slightly less than d/a when d/a cannot be conveniently expressed as a ratio of integers.

When the optimum value of q/p for a particular configuration has been determined, it is important to investigate the convergence of the solution as q and p increase with their ratio constant, simultaneously comparing the results with an accepted value of known accuracy. Formulas derived by the 'equivalent-static' method are available, from which normalised susceptance values applicable in cases A and C can be found.[4] Results for such cases, obtained both from these formulas and by computation, are given in Table 1, the

Table 1

CONVERGENCE OF NORMALISED SUSCEPTANCE VALUES FOR THIN INDUCTIVE IRIS IN RECTANGULAR WAVEGUIDE

p	q	$b = c$	Error	$b = 0$	Error
			%		%
6	4	0·48382	1·13	0·83586	2·92
12	8	0·48060	0·45	0·85289	0·95
18	12	0·47958	0·24	0·85708	0·46
24	16	0·47911	0·14	0·85881	0·26
30	20	0·47885	0·09	0·85971	0·16
36	24	0·47869	0·05	0·86025	0·09
Extrapolated convergence value		0·47843		0·86106	
From 'wave-guide-hand-book' formula		0·4782		0·8485	

$a = 0{\cdot}8\lambda_0$ $d = 0{\cdot}667a$

dimensions having been chosen such that $a = 0{\cdot}8\lambda_0$, and $d = 0{\cdot}667a$. For the guidewidth in question, the 'equivalent-static' formulas are stated to be in error by about 1% or less. p and q were taken in the ratio 3 : 2, which corresponds to the condition for optimum field matching derived above. There is uniform convergence in both cases, and this has enabled the final values, obtained as p and q tend to infinity, to be estimated. The percentage error caused by truncating the series of modes and aperture functions could therefore be found, and this is given in the Table for each value of p. It is clear that p need not be very large to give an accurate solution: an error of 1% is generally acceptable, and this is not exceeded in either case when $p = 12$. In fact, the error when $p = 6$ in the symmetric case is only slightly greater than this. The accuracy for a given value of p is greater for $b = c$ than for $b = 0$, because only odd modes and aperture functions are considered, so that the order of approximation to the true fields is double that in the asymmetric case.

The agreement between the two methods of solution in the symmetric case (case A) is excellent, the discrepancy being less than 0·1%. For the asymmetric case (case C), the solutions agree to within 1·5%, a sufficiently small discrepancy to confirm the correctness of the computational method, bearing in mind the tolerance quoted for the standard value.

The form of the field distributions in the aperture and in the waveguide, as calculated from the complex amplitude coefficients generated by the program, has been examined in the case $q/p = d/a$. Of particular importance are the variations of the distributions as p and q increase, the degree of accuracy with which the aperture and waveguide fields match one another, and the degree of accuracy with which they match the theoretical distributions, when these are known.

Figs. 5a and b show, respectively, the electric- and magnetic-field distributions in the discontinuity plane, for an iris with an aperture of width $d = 0{\cdot}5a$, offset from the centre, whose dimensions correspond to case C in Fig. 4. The components of the aperture and waveguide electric fields are all excited

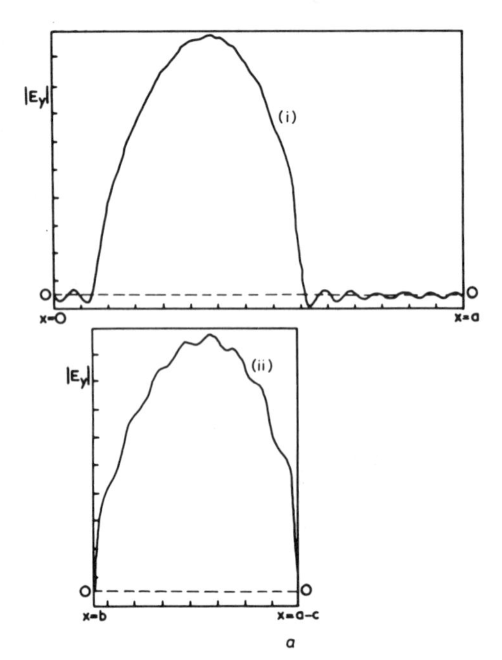

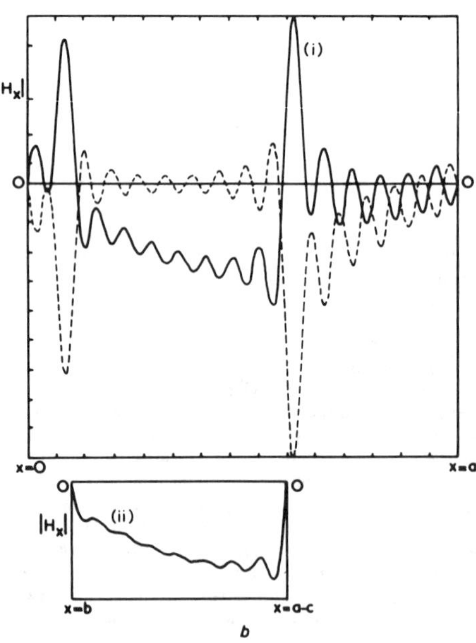

Fig. 5

Discontinuity-plane field distributions for thin inductive iris in rectangular waveguide

$a = 0{\cdot}8\lambda_0$ $b = 0{\cdot}1a$ $c = 0{\cdot}4a$
$p = 30$ $q = 15$
a Electric field
Phase relative to incident wave = 41·88°
(i) Waveguide field (ii) Aperture field
b Magnetic field
(i) waveguide field
—— real part
- - - imaginary part
(ii) Aperture field (real part)

with the same phase, so that the real and imaginary parts are identical except in amplitude. Theory predicts that the electric field should vanish over the iris, and that the imaginary part of the magnetic field should vanish in the aperture, because the aperture magnetic field is just that of the incident mode, and therefore purely real. It can be seen from Fig. 5 that, in both cases, the field oscillates about zero, the amplitude of oscillation increasing towards the aperture edges. Also, the magnetic field should theoretically become infinite at the edges. This does not happen when a finite number of modes is considered, but the field does become large at the edges compared with the field elsewhere. The aperture magnetic field, shown in Fig. 5*b*, is the approximation to the theoretical distribution (a truncated sinusoid) given by a Fourier series of the aperture functions, in this case containing 15 terms. The ripple is again most pronounced near the aperture edges. In the waveguide, the ripple in the magnetic field across the aperture has greater amplitude, because this field is related to the predetermined aperture field, which is only partially matched with the finite number of modes. The same effect can be seen in the electric field, except that a series of aperture functions is now used to match each waveguide mode, i.e. the reverse process to the magnetic field case; and so the ripple is larger in the aperture that in the waveguide. The effect on the field distributions of increasing p and q in the correct ratio has been investigated, and it is found that, whereas the frequency of the ripple increases, its amplitude decreases, except near the aperture edges, where there is evidence of Gibb's phenomenon. In the limit, as p and q tend to infinity, such overshoots will become spikes of zero width, which do not affect field matching as assessed by integrating the squared deviation of the two distributions. The waveguide magnetic field at the aperture edges is found to increase as p and q increase, and it seems likely that the theoretical infinite amplitude would be approached in the limit.

The extent to which the field distribution derived from the computed mode amplitudes matches that derived from the aperture-function amplitudes across the aperture has also been investigated quantitatively,[14] the approach being similar to that used by Cole[10] for a step discontinuity. The amount of discrepancy was calculated by deriving an expression for the squared modulus of the difference between the two fields at any point and integrating this over the aperture. It was expected that, if convergence were correct, any field-matching discrepancies would tend to zero as p and q were increased to infinity, their ratio being held constant. Results indicated that, for $q/p < d/a$, discrepancies did in fact tend to zero, although, for magnetic-field matching, this was only so if the iris edges were excluded from the integration range. It was, however, not possible to show this conclusively, because of the restrictions on the magnitude of p and q imposed by storage limitations. For $q/p > d/a$, on the other hand, results showed clearly that the electric-field matching grew progressively worse as p and q increased, emphasising the need for a correct choice of the ratio.

To conclude the treatment of the thin inductive iris, susceptance values of practical significance have been calculated and are given in Fig. 6. The curves show the variation of B/Y_0 with aperture displacement, for various aperture and waveguide widths. In every case, 30 modes were included, and $g/p = d/a$. The accuracy of the results, determined from convergence tests, is estimated to range from about 10% for $d = 0{\cdot}1a$ to less than 1% when $d = 0{\cdot}7a$. For the single-obstacle and symmetric cases, denoted by $2b/(a-d)$ equal to 0 and 1, respectively, the results could be compared with those of Marcuvitz,[4] and agreement was found to be satisfactory when the estimated accuracies of both methods of solution were taken into account.

3.3 Application to other iris configurations

Two other thin-iris configurations have been studied in some detail; a thin capacitive iris in rectangular waveguide and a thin circular iris in circular waveguide supporting the H_{11} mode. For the former, the method of analysis is similar to that for the inductive iris, but the expressions for the modes and aperture functions are slightly more involved, because an incident dominant mode now excites evanescent modes of types H_{1n} and E_{1n}, which have field variations in both x and y directions. In general, n can be any positive integer, but, when $c = d$, only modes for which n is even are excited. Since

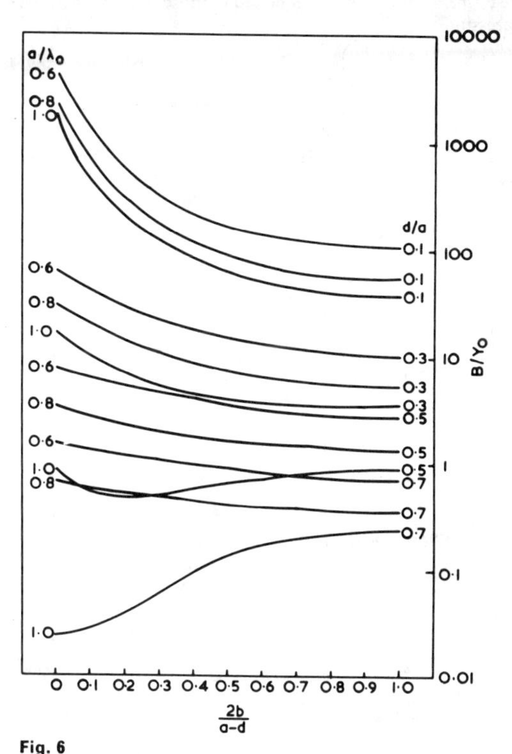

Fig. 6
Normalised susceptance of thin inductive iris as a function of aperture displacement

the incident electric field is polarised entirely in the y direction and no x component need be introduced to satisfy the iris boundary conditions, the E_{1n} and H_{1n} modes are excited with a relative amplitude and phase so that their x directed electric-field components cancel; together they form the LSE_{1n} mode. If the problem is formulated in terms of such LSE modes, rather than in individual H and E modes, there is a considerable saving of computer time and storage for a solution of given accuracy.

The transverse-field components in the plane of the iris for the LSE_{1n} mode can be written as follows (see inset to Fig. 8):

$$\boldsymbol{e}_n = \sin\frac{\pi x}{a}\cos\frac{n\pi y}{b}\,\hat{\boldsymbol{i}}_y \quad . \quad . \quad . \quad . \quad . \quad . \quad . \quad (36a)$$

$$\boldsymbol{h}_n = j\sqrt{\left(\frac{\epsilon_0}{\mu_0}\right)}\frac{1}{\bar{\gamma}_{1n}}\left\{\left(\frac{\lambda_0^2}{4a^2}-1\right)\sin\frac{\pi x}{a}\cos\frac{n\pi y}{b}\,\hat{\boldsymbol{i}}_x + \frac{n\lambda_0^2}{4ab}\cos\frac{\pi x}{a}\sin\frac{n\pi y}{b}\,\hat{\boldsymbol{i}}_y\right\} \quad (36b)$$

where $\bar{\gamma}_{1n} = \sqrt{\left\{\frac{\lambda_0^2}{4}\left(\frac{1}{a^2}+\frac{n^2}{b^2}\right)-1\right\}} = \sqrt{\left(\frac{\lambda_0^2 n^2}{4b^2}-\frac{\lambda_g^2}{\lambda_0^2}\right)}$

Notice that, if $n = 0$, the expressions reduce to those for the H_{10} mode; so the incident field can also be thought of as that of the lowest-order LSE mode. Suitable functions $\boldsymbol{e}_{1m}$ and $\boldsymbol{h}_{1m}$ for expanding the aperture fields are similarly expressed, except that b is replaced by b' and there is no multiplying term involving $\bar{\gamma}$ in the magnetic-field expression. The elements of matrix $\boldsymbol{R}$, defined by eqn. 17, become

$$R_{nm} = \frac{2n}{\pi}\left(\frac{b'}{b}\right)^2 \frac{\sin\frac{n\pi c}{b} + (-1)^{m+1}\sin\frac{n\pi}{b}(c+b')}{m^2 - n^2\left(\frac{b'}{b}\right)^2}$$

$$bm \neq b'n \quad . \quad (37a)$$

$$R_{nm} = \frac{b'}{b} \cos \frac{m\pi c}{b'} \qquad bm = b'n \quad . \quad . \quad . \quad . \quad (37b)$$

The elements of matrix $\boldsymbol{S}$ are related to those of $\boldsymbol{R}$ through the expression

$$S_{mn} = j\sqrt{\left(\frac{\epsilon_0}{\mu_0}\right)} \frac{1}{\gamma_{1n}} \frac{b}{b'} \frac{\epsilon_{m0}}{\epsilon_{n0}} R_{nm} \quad . \quad . \quad . \quad . \quad . \quad (38)$$

Notice the similarity between the above expressions and those for the inductive iris, eqns. 33 and 34.

Fig. 7 shows the dependence of the computed susceptance

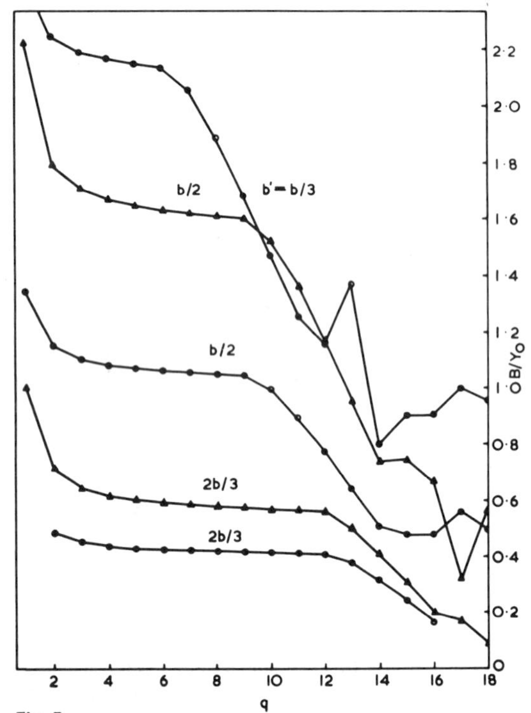

Fig. 7

Normalised susceptance for thin capacitive iris as a function of q, the number of aperture components

○ $c = d$, $b = 0{\cdot}6245\lambda_g$
△ $c = 0$ $b = 0{\cdot}4\lambda_g$

on the number of aperture functions q when 18 waveguide modes are included. Various values of b'/b are considered, and results are presented for both the symmetric case $c = d$ and the single-obstacle case $c = 0$. The behaviour of the solution is clearly very similar to that for the inductive iris, distinct changes of slope being shown where $q/p = b'/b$. The explanation of Section 3.2.2 applies equally to this case, so that this ratio should give greatest accuracy.

Table 2 shows the convergence of the solution for

Table 2

CONVERGENCE OF NORMALISED SUSCEPTANCE VALUES FOR A THIN CAPACITIVE IRIS IN A RECTANGULAR WAVEGUIDE

p	q	Case 1	Error	Case 2	Error
			%		%
6	4	0·4112	6·09	0·5889	6·83
12	8	0·3960	2·16	0·5641	2·32
18	12	0·3921	1·16	0·5578	1·18
24	16	0·3903	0·70	0·5551	0·69
30	20	0·3894	0·46	0·5536	0·42
36	24	0·3888	0·31	0·5527	0·25
Extrapolated convergence value		0·3876		0·5513	
From 'waveguide-handbook' formula		0·4031		0·5697	

Case 1: $c = d$, $b = 0{\cdot}6245\lambda_g$, $b'/b = 0{\cdot}667$
Case 2: $c = 0$, $b = 0{\cdot}4\lambda_g$, $b'/b = 0{\cdot}667$

$q/p = b'/b = 2/3$ for both symmetric and single-obstacle cases. This is not as rapid as in the inductive-iris example, and there is less difference between the convergence rates for symmetric and single-obstacle cases. About 20 modes would need to be included for accuracy to within 1%. The estimated converged values differ from those of Marcuvitz[4] by about 4% in the case $c = d$ and $b = 0{\cdot}6245\lambda_g$, and by about 3·2% for $c = 0$ and $b = 0{\cdot}4\lambda_g$. Since the accuracy for Marcuvitz's results is given as less than 5%, the agreement is satisfactory.

Fig. 8 shows the normalised susceptance of a symmetric

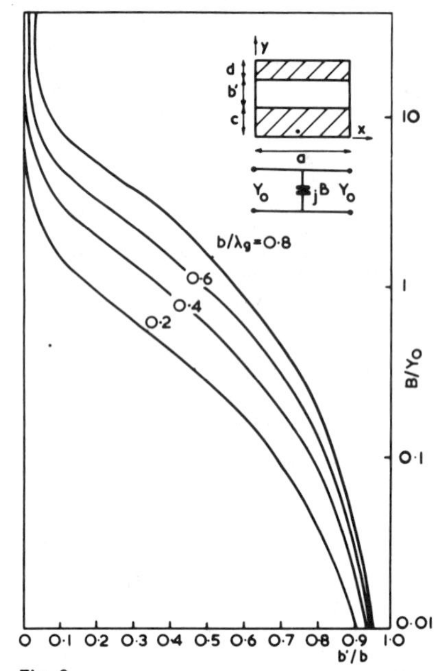

Fig. 8

Normalised susceptance of thin symmetric capacitive iris in rectangular waveguide

capacitive iris as a function of b'/b, with b/λ_g as a parameter. In the range $0{\cdot}1 < b'/b < 0{\cdot}7$, 40 waveguide modes were taken into account, sufficient to give solutions within 1% of the converged values.

A thin-iris problem with cylindrical geometry is depicted in Fig. 10 (inset). The H_{11} mode in circular waveguide is incident on a circular iris. Both H and E modes are excited at the discontinuity, in this case with unknown relative amplitudes. The susceptance in the transmission-line equivalent circuit for single-mode propagation can be either inductive or capacitive, depending on the dimensions of the structure, and a resonant condition is also possible under some circumstances. Because the structure has circular symmetry, the only modes excited at the iris are those with the same azimuthal field variation as the incident mode, and these are therefore of type H_{1n} and E_{1n}. Both H-type and E-type aperture functions are introduced, being derived, respectively, from H and E mode functions for a circular waveguide having the same diameter as the aperture. Expressions for such functions, and for the elements of matrices $\boldsymbol{R}$ and $\boldsymbol{S}$, are to be found in Reference 14. It is worth noting that matrix elements corresponding to the

Table 3

CONVERGENCE OF CIRCULAR-IRIS SOLUTION

p	q	B/Y_0	Error
			%
6	4	−4·111	1·91
12	8	−4·066	0·79
18	12	−4·051	0·42
24	16	−4·044	0·25
30	20	−4·040	0·15
36	24	−4·037	0·07

Estimated converged value = − 4·034
$R = 2R'/3$, $R' = 0{\cdot}3\lambda_0$

interaction of waveguide H modes and E-type aperture functions are zero.

The application to this problem of the technique for determining the optimum value of p/q by locating a change of slope is illustrated in Fig. 9, from which it is clear that

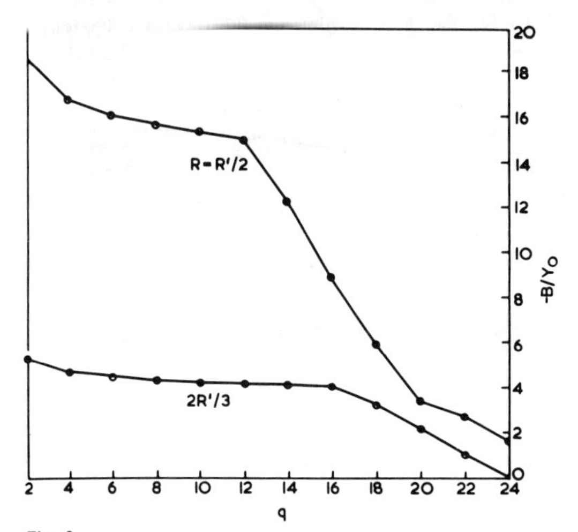

Fig. 9
Normalised susceptance for thin circular iris in circular waveguide supporting H_{11} mode as a function of the number of aperture components
$P = 24$ $R' = 0{\cdot}3\lambda_0$

$p/q = R'/R$ is the required relationship. Table 3 shows how the solution converges when $R'/R = 3/2$, $R' = 0{\cdot}3\lambda_0$ and $p/q = R'/R$. Convergence is rapid and compares favourably with that for the inductive iris in rectangular waveguide. For example, in both cases, only 12 modes are required to give better accuracy than within 1%. In both Fig. 9 and Table 3, only even values of p and q have been considered, to ensure that the correct ratio of modes to aperture functions holds independently for both E and H modes and functions.

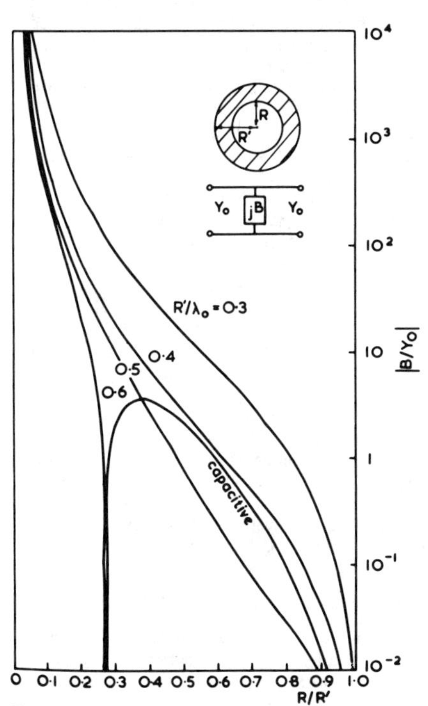

Fig. 10
Susceptance of thin circular iris in circular waveguide supporting H_{11} mode

In order to check the correctness of susceptance values computed for this case, a comparison was made with values calculated from a formula given by Marcuvitz.[4] This formula was derived by an integral-equation method, and is subject to the limitations that R should be small compared with R' and λ_0/π, and that the second mode should be well below cutoff. No estimates of accuracy are given. The results of the comparison are shown in Table 4. In general, the smaller

Table 4

COMPUTED SUSCEPTANCES FOR CIRCULAR IRIS COMPARED WITH 'WAVEGUIDE-HANDBOOK' VALUES

R/λ_0	R/R'	B/Y_0 (computed)	B/Y_0 (Reference 4)	Discrepancy
				%
0·3	0·1	−2632	−2780	5·6
0·3	0·3	−90·50	−94·13	4·0
0·3	0·5	−14·96	−13·16	12·0
0·4	0·1	−612·8	−653·9	6·7
0·4	0·3	−19·03	−22·14	16·4
0·4	0·5	−2·678	−3·095	15·5
0·5	0·1	−403·3	−439·4	9·0
0·5	0·3	−10·08	−14·88	47·6
0·5	0·5	−0·799	−2·080	160·1

R/R' and R'/λ_0 become, the better is the agreement between the two solutions. This is to be expected in view of the above limitations, and tends to confirm the correctness of the computational method.

Fig. 10 shows the normalised susceptance of a thin circular iris as a function of R/R', with R'/λ_0 as parameter. Forty waveguide modes were taken into account for $R/R' < 0{\cdot}7$, and the discrepancy compared with converged values is estimated to be less than 1% for R/R' greater than about 0·25. The susceptance is inductive except on the $R'/\lambda_0 = 0{\cdot}6$ curve when R/R' exceeds 0·273, for which value a resonant condition exists.

4 Waveguide step discontinuity

4.1 Modification of general method

In the general 2-waveguide junction of Fig. 1, consider now that the thin diaphragm in the discontinuity plane is removed, and that, in cross-section AA', surface S lies entirely within surface S'. The coupling aperture now has the same cross-section as waveguide 1, so that aperture functions $\boldsymbol{e}_{1m}$ and $\boldsymbol{h}_{1m}$ become identical to the field components $\boldsymbol{e}_m$ and $\boldsymbol{h}_m$ in waveguide 1. It follows from eqns. 2, 4 and 6 that

$$\left.\begin{aligned} a_{im} + a_{rm} &= b_{1m} \\ \text{and} \quad a_{im} - a_{rm} &= c_{1m} \end{aligned}\right\} \qquad (39)$$

for all m.

Having set $q = p$, it is clear from eqns. 17 and 21 that $\boldsymbol{R}$ and $\boldsymbol{S}$ will now be unit matrices. Substituting for $\boldsymbol{b}$ and $\boldsymbol{c}$ in eqn. 25 using eqn. 39, the equation becomes

$$\boldsymbol{a}_i - \boldsymbol{a}_r = \boldsymbol{S}'\boldsymbol{R}'\boldsymbol{a}_i + \boldsymbol{S}'\boldsymbol{R}'\boldsymbol{a}_r - 2\boldsymbol{S}'\boldsymbol{a}_i' \qquad (40)$$

The elements of the two matrices are now such that

$$R'_{ij} = \frac{\int_S \boldsymbol{e}_j \times \boldsymbol{h}'_i . ds}{\int_{S'} \boldsymbol{e}'_i \times \boldsymbol{h}'_i . ds} \quad \text{and} \quad S'_{ij} = \frac{\int_S \boldsymbol{e}_i \times \boldsymbol{h}'_j . ds}{\int_S \boldsymbol{e}_i \times \boldsymbol{h}_i . ds} \qquad (41)$$

In eqn. 40, $\boldsymbol{a}_i$ and $\boldsymbol{a}'_i$ are known, so that eqn. 40 represents a set of p linear equations in p complex unknowns a_{rm}, which may be solved as in the thin-iris case. Alternatively, a matrix equation involving the vector $\boldsymbol{a}'_r$ instead of $\boldsymbol{a}_r$ could be derived, but it has been found desirable to make p smaller than p', in which case eqn. 40 gives the least number of simultaneous equations to be solved.

In the usual situation where incidence is in the form of the dominant mode on one side of the junction only, eqn. 40 becomes, for incidence from waveguide 1,

$$(\boldsymbol{T}' + \boldsymbol{I})\boldsymbol{a}_r = \boldsymbol{\Delta} - \boldsymbol{T}_1 \qquad (42)$$

and, for incidence from waveguide 2,

$$(T' + I)a_r = 2S_1' \qquad (43)$$

In these expressions, Δ has p elements of the form: $\Delta_1 = 1$; $\Delta_i = 0$, $i \neq 0$. $T' = S'R'$. T_1 is the first column of T', I is a $p \times p$ unit matrix and S_1 is the first column of S'.

Having obtained the small-waveguide coefficients a_{rm}, those for the large waveguide may be obtained from eqn. 19, having substituted for b using eqn. 39. The version of eqn. 19 corresponding to eqn. 42 is

$$a_r' = R'a_r + R_1' \qquad (44)$$

where R_1' is the first column of R', and that corresponding to eqn. 43 is

$$a_r' = R'a_r - \Delta' \qquad (45)$$

Δ' is defined in the same way as Δ, but has p' elements.

As mentioned in Section 1, the use of the above approach to solve step problems is not new, although the formulation given here differs in some respects from that of other workers. For example, an improvement which has been made on the work of Clarricoats and Slinn[8] is the elimination of one unknown column vector to reduce the number of simultaneous equations to be solved. The advantage of this has also been recognised by Cole.[10] In Wexler's formulation,[9] matrix notation was not used.

In the particular case of a step discontinuity where incidence is in the form of a dominant-mode wave from one side, other modes being below cutoff, a variational approach based on that given by Collin and Brown[15] has been shown[14] to lead to the generation of a set of linear equations identical to those obtained by modal analysis, as represented by eqn. 42 or 43, except that the infinite series are not truncated. It should be pointed out, however, that, although the variational method leads to the same solution in this case and thus confirms the corectness of modal analysis, its application is much less straightforward. This is particularly so where multimode propagation is involved.

4.2 Choosing the ratio of the numbers of modes

By analogy with the thin iris, the choice of the mode ratio p/p' can be expected to affect convergence as p and p' increase, and a similar argument regarding correct matching of field components in the aperture can be applied. For junctions of waveguides with similar geometries, the ratios which will equate the spatial frequencies of the highest-order modes on either side of the discontinuity plane can easily be determined. For example, for H and Eplane steps in rectangular waveguide, the desired ratios are those of the guide-widths and heights, respectively. To confirm that such ratios also give optimum convergence, the equivalent normalised susceptance for a symmetric Eplane step discontinuity in which $b'/b = 0{\cdot}5$ has been calculated as a function of p, for various mode ratios p/p'. The results are shown in Fig. 11. In all cases, convergence is to very similar, if not identical, values. Any discrepancy there may be between the final values is less than $0{\cdot}3\%$. The choice of p/p' is therefore not nearly so critical as the choice of q/p for the thin iris, and it is clear how other workers succeeded in obtaining accurate results with arbitrarily chosen ratios. It can be seen, however, that, when $p/p' = 0{\cdot}5$, convergence is considerably faster than in the other cases. For example, an error of less than 1% can be achieved with only three modes in the small waveguide, whereas five modes are required in the case $p/p' = 0{\cdot}25$, and 12 when p and p' are equal. Since the value of p governs the number of simultaneous equations to be solved, it is clear that, for a solution of given accuracy, there will be a considerable saving of computer time and storage if p/p' is correctly chosen.

4.3 Computed results for step discontinuities

Several applications of the modal-analysis technique to step junctions involving single-mode propagation have been described by other authors,[8,9,11,12] so that attention in this Section will be confined to examples where more than one mode can propagate in one or both waveguides. The first of these concerns an Hplane junction of overmoded rectangular waveguides, which has also been solved by Felsen* using a ray-optical technique. Comparison has been made between his results and those derived by modal analysis

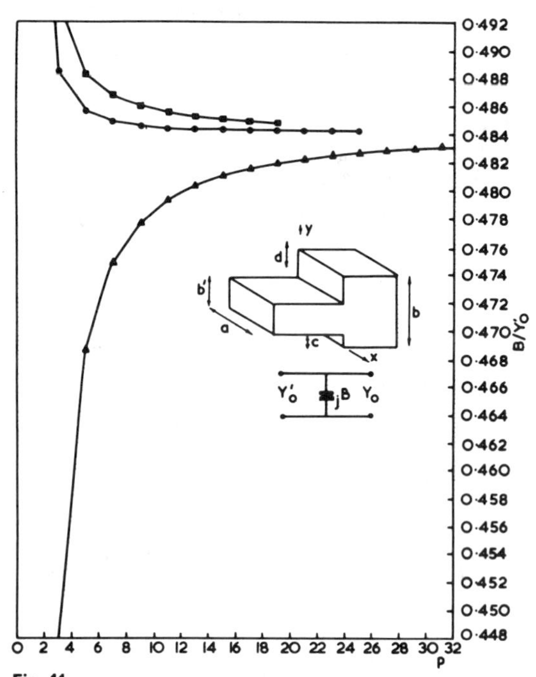

Fig. 11
Convergence of Eplane-step susceptance with increasing numbers of modes
$Y_0/Y_0' = 0{\cdot}5$ $b'/b = 0{\cdot}5$ $c = d$ $b = 0{\cdot}8\lambda_g$
△ $p' = p$
○ $p' = 2p$
□ $p' = 4p$

in both symmetric and asymmetric cases, and for incidence of the dominant mode from either waveguide. Agreement was found to be satisfactory in all cases. The results presented here (Figs. 12 and 13) are for a symmetric step in which the dominant mode is incident in the smaller guide, the guide widths being in the ratio 2 : 1. 24 modes were taken into account in the larger guide and 12 in the smaller. The amplitudes of H_{10}, H_{30} and H_{50} modes reflected from the junction are shown in Fig. 12 as functions of frequency, while Fig. 13 gives their phases relative to that of the incident H_{10} mode. Whereas Felsen's curves are smooth, those derived by modal analysis have several abrupt changes in slope, corresponding to higher-mode cutoff frequencies. It would seem that the ray-optic method will not reveal the fine structure of the curves, although it is recognised that manufacturing tolerances may eliminate this structure in practice. However, since the discrepancy between the solutions in the vicinity of such points can be as much as 90% in amplitude and 50° in phase, modal analysis is particularly valuable if an exact formulation is required.

Another multimode-step example is a circularly symmetric junction of circular waveguides, as shown in Fig. 14. It has an important application as a mode convertor in the 'Potter-horn' multimode antenna,[16] where an H_{11} mode incident in the small guide excites both H_{11} and E_{11} propagating modes in the large guide. The transverse-field patterns of these modes are included in Fig. 14. If other modes are to be evanescent, the dimensions must be chosen so that

$$0{\cdot}293 < R/\lambda_0 < 0{\cdot}610 \text{ and } 0{\cdot}610 < R'/\lambda_0 < 0{\cdot}847$$

The amplitude ratio with which the H_{11} and E_{11} modes are excited in the larger guide has been determined experimentally by Nagelberg and Shefer,[17] who express it in terms of a 'mode conversion-factor', being the ratio of the radial-electric-field amplitude of the E_{11} mode to that of the

* FELSEN, L.: Private communication

H_{11} mode, measured at the waveguide wall. In terms of decibels, this is written as

$$M = 20\log_{10}\left|\frac{a'_{r2}e'_1(E)}{a'_{r1}e'_2(H)}\right|_{r=R'} \quad (46)$$

a'_{r2} and a'_{r1} being the complex transmission coefficients of the E_{11} and H_{11} modes, respectively.

Computed values of M are compared with Nagelberg's experimental results in Fig. 15. 16 H and 16 E modes were taken into account in the larger waveguide, and the mode

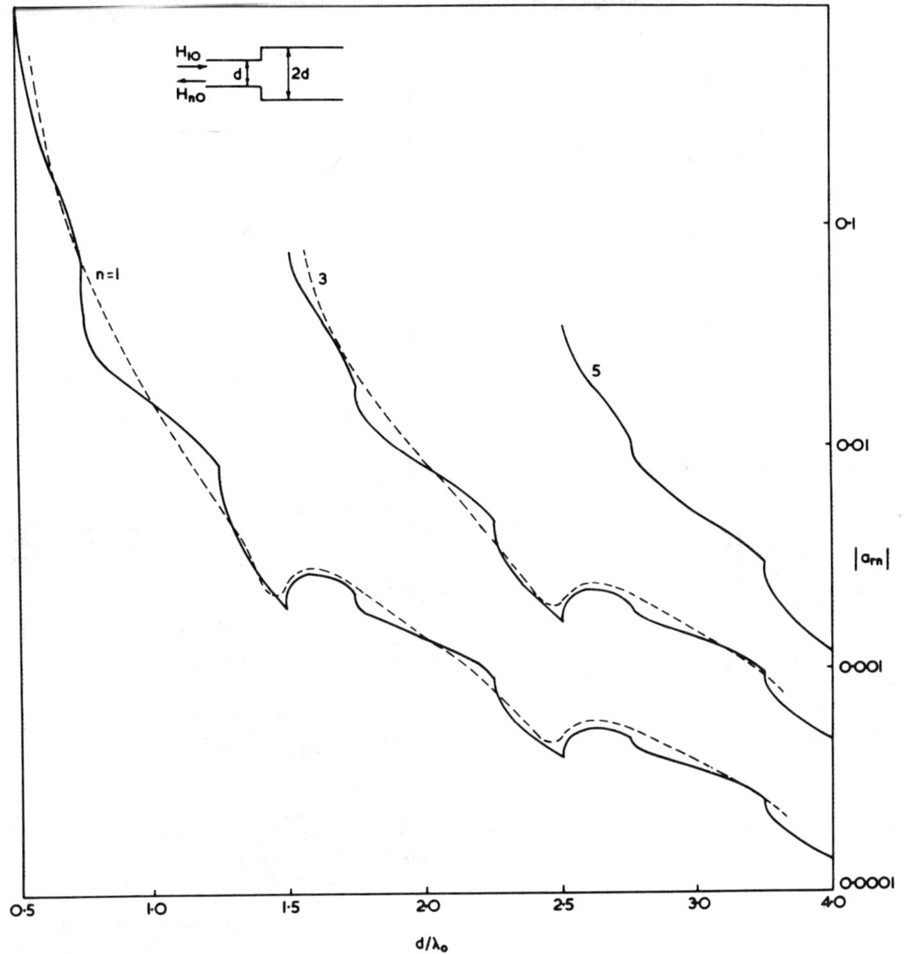

Fig. 12
Symmetric Hplane step in rectangular waveguide
Propagating-mode reflection coefficients (amplitude)
$p = 12 \quad p' = 24$
—— computational method
- - - ray-optical method (Felsen)

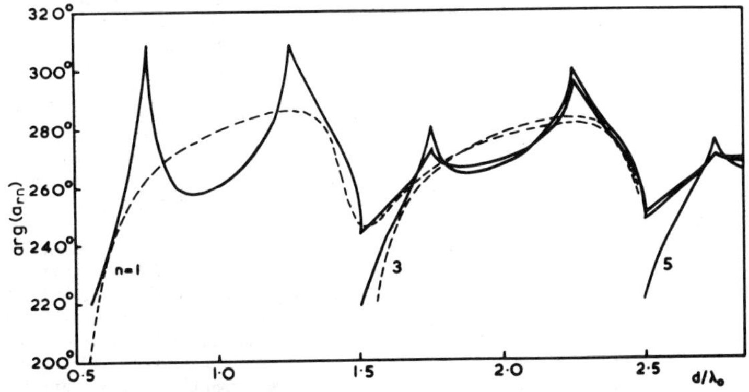

Fig. 13
Symmetric Hplane step in rectangular waveguide
Propagating-mode reflection coefficient (phase)
—— computational method
- - - ray-optical method (Felsen)

ratio was taken as $p'/p \simeq R'/R$. The agreement between theory and experiment is clearly good, especially for the smaller values of R/R'. When $2R = 2{\cdot}5$ in, there is a discrepancy of about 1 dB, but one would expect experimentally determined values to disagree with theory here, because the E_{11} mode is only very weakly excited, and any imperfections in the manufacture of the junction will have significant effects on its amplitude.

The breaks in the $2R = 2{\cdot}3$ in and $2R = 2{\cdot}5$ in curves correspond to E_{11} cut-off in the small waveguide.

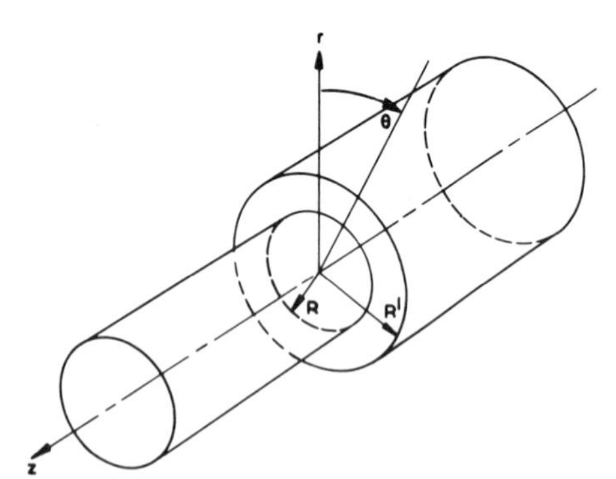

Fig. 14A *Junction of circular waveguide*

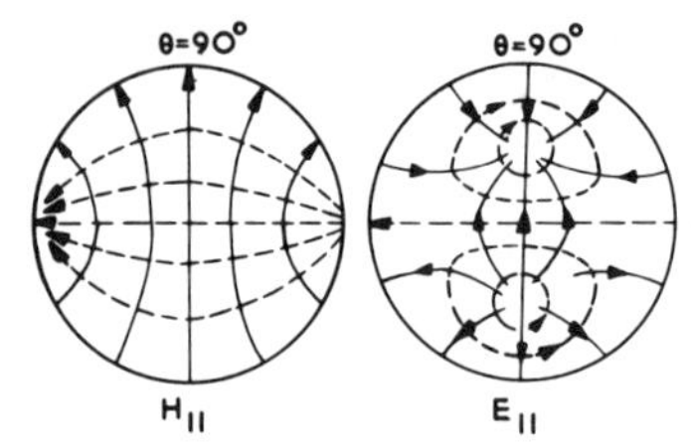

Fig. 14B *Transmitted-mode field patterns*
—— electric
--- magnetic

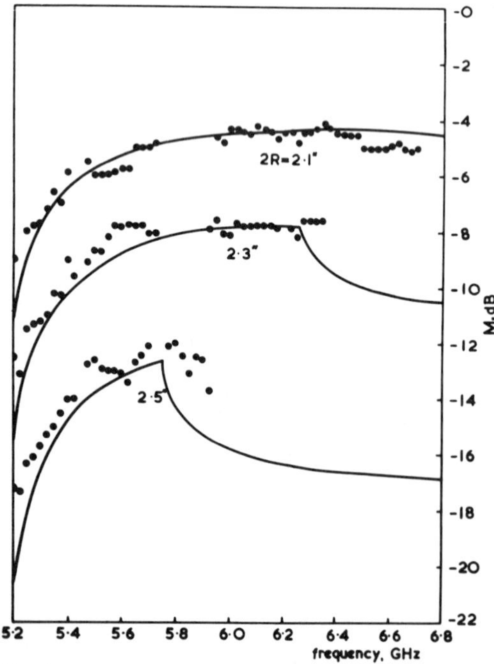

Fig. 15
Mode conversion at a circular waveguide junction
$2R' = 2{\cdot}8$ in $p' = 32$
● experimental points (Nagelberg and Shefer)

Values of voltage standing-wave ratio (v.s.w.r.) for the case $2R = 2{\cdot}5$ in are included in Fig. 16, with experimental points for comparison. There is good agreement in magnitude,

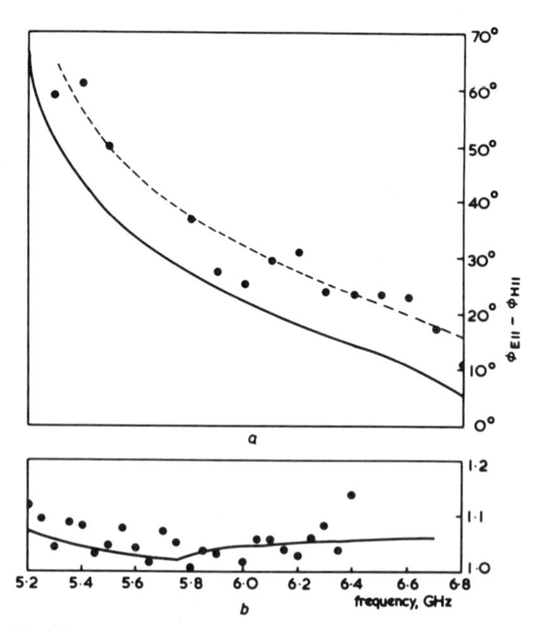

Fig. 16
Properties of circular waveguide junction
$2R' = 2{\cdot}8$ in $p' = 32$
a Relative excitation phase of H_{11} and E_{11} modes
$2R = 2{\cdot}1$ in
—— computational method
--- mean of experimental points
● experimental points (Nagelberg)
b Input v.s.w.r. (H_{11} mode)
$2R = 2{\cdot}5$ in
● experimental points (Nagelberg)

although the computed values show no sign of the oscillatory behaviour observed by Nagelberg.

Nagelberg† has also measured the relative phase with which the H_{11} and E_{11} modes are launched in the case $2R = 2{\cdot}1$ in. This is shown in Fig. 16, together with a graph of computed results. It can be seen that, on average, the experimental values differ from the computed ones by about 10°, a satisfactory figure in view of the scatter on the experimental points.

Subsequent to the completion of the present authors' study, Agarwal and Nagelberg[18] have published results obtained by modal analysis for this junction. Their findings are in agreement with those presented here.

4.4 Multiple and overlapping step discontinuities

The extension of modal analysis to discontinuity problems involving two or more step junctions (e.g. a thick iris) has been considered by previous authors.[8,9] However, solutions have not always been obtained in the most efficient way. Clarricoats and Slinn,[8] for instance, having generated a pair of matrix equations for each step by considering both electric- and magnetic-field matching, combined all the unknown amplitude coefficients for the three-waveguides in a single column vector, and solved for them all simultaneously. If this is done, so much storage is required that, with an average-sized computer, it is not always possible to take sufficient modes into account for satisfactory accuracy, particularly if the separation between steps is small. We have found it possible to combine the original matrix equations so as to eliminate the sets of unknown coefficients in all but one of the waveguides, and, having solved for this set, to obtain the other sets of unknowns by back substitution. In this way, the number of simultaneous equations to be solved initially is reduced, typically by a factor of four, and at least twice as many modes can be taken into account for a given store limit.

† NAGELBERG, E. R.: 'Design of dual-mode conical horns for communication antennas', Bell Telephone Laboratories, 1966 (unpublished)

Details of our work on one application of this technique have already been published.[19]

If the restriction imposed in this Section, that the cross-section of one waveguide should lie entirely within that of the other at the step, is removed, the coupling aperture is no longer equal to the cross-section of the smaller waveguide, and the general method of Section 2 must be used. If the two waveguides have identical cross-sections with fourfold symmetry, however, and there is no rotation of one guide relative to the other, the junction becomes equivalent to a thin iris in a continuous waveguide and may be solved as in Section 3. The choice of aperture functions for an overlapping step junction may not be straightforward if the aperture boundary has an irregular shape. For an overlapping circular waveguide junction, however, Knetsch[20, 21] has overcome this difficulty by introducing an intermediate circular waveguide whose cross-section encompasses both waveguides forming the junction. This guide is taken to have zero length, and separate series of orthogonal functions are used for electric and magnetic fields. The method of solution is then similar to that of Section 2, except that the extension of the surface of integration to allow use of orthogonality (eqns. 10 and 11) must now be performed for the intermediate region rather than for the waveguide cross-sections, and applies in the derivation of eqns. 14 and 15 instead of eqns. 11 and 12.

5 Discussion

We have shown in the preceding Sections that, by using the computational field-matching technique, discontinuity problems can be formulated in a simple way, and that, at least for steps and irises, very accurate results can be obtained. However, the accuracy is conditional on the correct choice of the ratios of modes and aperture functions, particularly for thin irises. When the junction involves only rectangular or circular geometry, the condition of spatial-frequency matching for the highest-order functions can be applied, and, for an iris, the choice is confirmed by the slope change in a graph of susceptance of dominant-mode reflection coefficient against mode ratio q/p. This principle can easily be extended to the combined step and iris of Fig. 2; spatial-frequency matching requirements are met when q, p and p' are in the ratios 1 : 2 : 4, so that the point $q = 5$ on the 10 : 20 curve should represent the most accurate solution. As would be expected, there is a change of slope at the point. If there is correct spatial-frequency matching between the aperture and one waveguide, even very large changes in the number of modes in the other waveguide do not affect the solution by more than 10%. This result parallels the non-critical mode ratio for a step discontinuity, where the aperture field is identical to that in the smaller waveguide, so that spatial-frequency matching on one side of the junction is always exact.

A problem which remains to be studied is that of the optimum choice of q/p for a thin iris involving a mixed geometry, for example a circular iris in a rectangular waveguide. Since the modes and aperture functions have quite different forms, the spatial-frequency-matching argument is inapplicable, and it is unlikely that any distinct change of slope would be seen on a curve of B/Y_0 against q/p. The results of Section 3.2.2 indicate that, for an inductive iris in rectangular waveguide, an accurate solution could be obtained with less-than-optimum values of q/p, although convergence would be slower. It therefore seems likely that good results for a mixed-geometry problem could be obtained with an arbitrary value of q/p, provided that it was small enough to avoid a divergent situation. A number of trials would reveal the value which gave the most rapid convergence.

6 Acknowledgments

The authors acknowledge many helpful discussions with C. D. Hannaford and some useful suggestions made by Prof. L. B. Felsen.

7 References

1 COLLIN, R. E.: 'Field theory of guided waves' (McGraw-Hill 1960), chap. 10
2 COLLIN, R. E.: 'Field theory of guided waves' (McGraw-Hill, 1960), chap. 8
3 LEWIN, L.: 'Advanced theory of waveguides' (Iliffe, 1951)
4 MARCUVITZ, N.: 'Waveguide handbook', (McGraw-Hill, 1951)
5 WHINNERY, J. R., and JAMIESON, H. W.: 'Equivalent circuits for discontinuities in transmission lines', *Proc. Inst. Radio. Engrs.*, 1944, **32**, pp. 98–114
6 CLARRICOATS, P. J. B.: 'Properties of waveguides containing ferrites with special reference to waveguides of circular cross-section'. Ph.D. thesis, Imperial College, London, 1958
7 CLARRICOATS, P. J. B., and SLINN, K. R.: 'Numerical method for the solution of waveguide-discontinuity problems', *Electron. Lett.*, 1966, **2**, pp. 226–227
8 CLARRICOATS, P. J. B., and SLINN, K. R.: 'Numerical solution of waveguide-discontinuity problems', *Proc. IEE*, 1967, **114**, (7), pp. 878–886
9 WEXLER, A.: 'Solution of waveguide discontinuities by modal analysis', *IEEE Trans.*, **MTT-15**, 1967, pp. 508–517
10 COLE, W. J., NAGELBERG, E. R. and NAGEL, C. M.: 'Iterative solution of waveguide discontinuity problems', *Bell Syst. Tech. J.*, 1967, **46**, pp. 649–672
11 BRAECKELMANN, W., and HENKE, H.: 'Reflection and transmission at an abrupt change in cross-section of a rectangular waveguide in the E and H plane', *Electron. Lett.*, 1969, **5**, pp. 332–333
12 BRAECKELMANN, W.: 'Bandstop filter for the H_{10} mode in a rectangular waveguide', *ibid.*, 1969, **5**, pp. 500–502
13 MASTERMAN, P. H., CLARRICOATS, P. J. B., and HANNAFORD, C. D.: 'Computer method of solving waveguide-iris problems', *ibid.*, 1969, **5**, pp. 23–25
14 MASTERMAN, P. H.: 'The numerical solution of waveguide discontinuity problems', Ph.D. thesis, Leeds University, 1969
15 COLLIN, R. E., and BROWN, J.: 'Calculation of the equivalent circuit of an axially unsymmetric waveguide junction', *Proc. IEE*, **1956**, **103C**, pp. 121–128
16 POTTER, P. D.: 'A new horn antenna with suppressed sidelobes and equal beamwidths', *Microwave J.*, 1963, **6**, pp. 71–78
17 NAGELBERG, E. R., and SHEFER, J.: 'Mode conversion in circular waveguides', *Bell Syst. Tech. J.*, 1965, pp. 1321–1338
18 AGARWAL, K. K., and NAGELBERG, E. R.: 'Phase characteristics of a circularly symmetric dual-mode transducer', *IEEE Trans.*, **1970**, **MTT-18**, pp. 69–71
19 HUCKLE, P. R., and MASTERMAN, P. H.: 'Analysis of a rectangular waveguide junction incorporating a row of rectangular posts', *Electron. Lett.*, 1969, **5**, pp. 559–561
20 KNETSCH, H. D.: 'Beitrag zur Theorie sprunghafter Querschnittsveranderungen von Hohlleitern', *Arch. Elekt. Ubertragung*, **22**, (12), 1968, pp. 591–600
21 KNETSCH, H. D.: 'Achsversetzungen von Rundhohlleitern mit beliebigen Radien', *ibid.*, 1969, **23**, pp. 23–32

Convergence of Numerical Solutions of Iris-Type Discontinuity Problems

SHUNG WU LEE, WILLIAM R. JONES, SENIOR MEMBER, IEEE, AND JAMES J. CAMPBELL

Abstract—**The convergence of numerical solutions of several iris-type discontinuity problems in waveguides and periodic structures is investigated. It is demonstrated that the numerical solution of a set of equations obtained from a mode-matching procedure (which corresponds to an integral equation formulation of the problem generally known as the moment method) may converge to an incorrect value if an improper ratio is chosen between the number of modal terms in the aperture and the number of terms retained in the kernel of the integral equation. Guidelines for efficient numerical computations are indicated.**

I. Introduction

THE development of high-speed computers has revolutionized the capability for solving electromagnetic boundary value problems. Many problems which were considered intractable only a few years ago now can be solved by numerical methods with the aid of computers. For example, in problems involving discontinuities in uniform waveguides (or periodic structures) the fields in different regions can be expanded in terms of discrete sets of modes with unknown coefficients, and the numerical problem involves the calculation of these amplitude coefficients from an appropriately truncated infinite system of linear equations derived from boundary and continuity conditions.

In the past few years, many papers have been published on the numerical solutions to various types of waveguide discontinuity problems, but few of them have seriously considered the convergence of the numerical solution with respect to the truncation of the infinite system of equations. Often, a numerical solution is regarded "satisfactory" because "it remains practically unchanged even though more and more modes are used in the computations," or because "energy is conserved." The purpose of this paper is to closely examine these somewhat vague criteria and establish some general guidelines for efficient numerical computation in a class of waveguide discontinuity problems.

The class of problems to be considered is iris-type discontinuity problems in uniform waveguides or periodic structures. Solutions to these problems are of great practical interest since they are used widely in the design of microwave circuits, phased arrays, diffraction gratings, and radomes. Except for one or two special cases, e.g., the half-iris in a parallel-plate waveguide (see Section III), exact solutions cannot be obtained. During World War II, scientists at the M.I.T. Radiation Laboratory developed some ingenious analytical techniques based on the Wiener–Hopf method, quasi-static approximations, and variational principles; however, the applicability of these techniques is generally restricted to simple two-dimensional problems or problems involving electrically small apertures [1]. Thus, for general discontinuity problems, computer-aided numerical methods provide a powerful, and for the most part, necessary means in order to obtain useful solutions.

In formulating these problems, there exists several different procedures, each of which may lead to a different system of matrix equations. The most popular procedure perhaps is the one based on an integral equation formulation, which is subsequently transformed into an infinite matrix equation in the Garlekin sense [2], [3] generally known as the *moment method.* In the present paper, we show through a simple example that the numerical solution to such a matrix equation exhibits a *relative convergence* phenomenon between the degrees of freedom in the kernel and that in the unknown (aperture field) of the integral equation. Unless their relative ratio is chosen properly, the numerical solution may not converge to the correct value as the degrees of freedom in the kernel and that in the unknown increase indefinitely.

The term relative convergence was first introduced by Mittra [4] in 1963 in solving a waveguide bifurcation problem through a particular mode-matching procedure different from the moment method. However, as discussed in [5], the relative convergence phenomenon in that problem was due to the choice of the particular mode-matching procedure, and it can be removed by using the more stable moment method. In the class of iris discontinuity problems, however, a relative convergence phenomenon can be clearly observed even when the moment method is used, as will be demonstrated in the following paragraphs.

In problems involving uniform waveguides or periodic structures, the moment method can be applied directly to the modal field representations without explicitly formulating an integral equation. This procedure is presented for a simple iris discontinuity problem in Section II. The relationship of the moment method derived in this fashion and the integral equation formulation is discussed in Section III. Numerical results and the discussion of their implications are given in Section IV. In Section V we consider some more general problems involving iris-type discontinuities,

Manuscript received September 3, 1970; revised January 11, 1971.
The authors are with the Ground Systems Group, Hughes Aircraft Company, Fullerton, Calif. 92634.

Reprinted from *IEEE Trans. Microwave Theory Tech.*, vol. MTT-19, no. 6, pp. 528–536, June 1971.

and, finally, in Section VI a summary of our results is presented.

II. Formulation of Iris Loaded Waveguide Problem by a Mode-Matching Procedure

The geometry of the waveguide discontinuity problem under consideration is shown in Fig. 1. For a TM mode incident from the left, the (magnetic) fields in the three separate regions can be expressed in terms of their respective normal modes

$$H_y(x,z) = \begin{cases} \sum_{n=1}^{\infty} (A_n^+ \exp(-\gamma_{na}z) + A_n^- \exp(\gamma_{na}z)) \sqrt{\frac{2}{\epsilon_n^1 a}} \cos\frac{(n-1)\pi}{a}x, & z<0 \quad (1) \\ \sum_{n=1}^{\infty} (B_n^+ \exp(-\gamma_{nb}z) + B_n^- \exp(\gamma_{nb}z)) \sqrt{\frac{2}{\epsilon_n^1 b}} \cos\frac{(n-1)\pi}{b}x, & 0<z<d \quad (2) \\ \sum_{n=1}^{\infty} C_n^+ \exp(-\gamma_{nc}(z-d)) \sqrt{\frac{2}{\epsilon_n^1 c}} \cos\frac{(n-1)\pi}{c}x, & z>d \quad (3) \end{cases}$$

where

$$\gamma_{ne} = \{[(n-1)\pi/e]^2 - k^2\}^{1/2}, \qquad e = a, b, c; \qquad k = 2\pi/\lambda$$

$$\epsilon_n^1 = \begin{cases} 2, & \text{if } n = 1 \\ 1, & \text{if } n \neq 1. \end{cases}$$

In the numerical computations, only the first P modes are retained in region A, N in region B, and Q in region C. In order to determine the unknown scattering coefficients $\{A_n^-, B_n^+, B_n^-, C_n^+\}$ in terms of incident field amplitudes $\{A_n^+\}$ for fixed values of P, N, and Q, the continuity of the tangential electric and magnetic fields at $z=0$ and $z=d$ is invoked.

First consider the junction at $z=0$. The continuity of H_y requires that

$$\sum_{n=1}^{P} (A_n^+ + A_n^-) \sqrt{\frac{2}{\epsilon_n^1 a}} \cos\frac{(n-1)\pi}{a}x = \sum_{n=1}^{N} (B_n^+ + B_n^-) \sqrt{\frac{2}{\epsilon_n^1 b}} \cos\frac{(n-1)\pi}{b}x, \qquad 0<x<b. \quad (4)$$

Multiplying both sides of (4) by $\sqrt{2/\epsilon_m^1 b}\cos(m-1)\pi x/b$ for $m=1, 2, \cdots, N$ and integrating from $x=0$ to b, leads to the matrix equation

$$T_{NP}^{(a)}(A_P^+ + A_P^-) = B_N^+ + B_N^- \quad (5)$$

where A_P^+, for example, is a vector with elements $\{A_n^+\}$ and the subscript P indicates its dimension. The $(N\times P)$ matrix $T_{NP}^{(a)}$ is defined by

$$T_{mn}^{(a)} = \sqrt{\frac{4}{ab\epsilon_m^1\epsilon_n^1}} \int_0^b \cos\frac{(m-1)\pi}{b}x \cdot \cos\frac{(n-1)\pi}{a}x\, dx. \quad (6)$$

Next, the continuity of E_x at $z=0$ gives

$$\sum_{n=1}^{P} (A_n^+ - A_n^-)\gamma_{na} \sqrt{\frac{2}{a\epsilon_n^1}} \cos\frac{(n-1)\pi}{a}x = \sum_{n=1}^{N} (B_n^+ - B_n^-)\gamma_{nb} \sqrt{\frac{2}{b\epsilon_n^1}} \cos\frac{(n-1)\pi}{b}x, \qquad 0<x<a \quad (7)$$

whereupon multiplying both sides of (7) by $\sqrt{2/a\epsilon_m^1} \cdot\cos(m-1)\pi x/a$ for $m=1, 2, 3, \cdots, P$ and integrating from $x=0$ to a, yields

$$\gamma_{ma}(A_m^+ - A_m^-) = \sum_{n=1}^{N} \gamma_{nb}(B_n^+ - B_n^-) \sqrt{\frac{4}{ab\epsilon_m^1\epsilon_n^1}} \cdot \int_0^a \cos\frac{(n-1)\pi}{b}x \cos\frac{(m-1)\pi}{a}x\, dx = \sum_{n=1}^{N} \gamma_{nb}(B_n^+ - B_n^-) \sqrt{\frac{4}{ab\epsilon_m^1\epsilon_n^1}} \cdot \int_0^b \cos\frac{(n-1)\pi}{b}x \cos\frac{(m-1)\pi}{a}x\, dx, \qquad m = 1, 2, 3, \cdots, P. \quad (8)$$

The second line in (8) follows from the boundary condition

$$E_x = 0, \qquad z = 0, \text{ and } b < x < a. \quad (9)$$

Equation (8) can be written in matrix form as

$$\Gamma_{PP}^{(a)}(A_P^+ - A_P^-) = \tilde{T}_{NP}^{(a)}\Gamma_{NN}^{(b)}(B_N^+ - B_N^-) \quad (10)$$

where $\Gamma_{PP}^{(a)}$, for example, is a diagonal matrix with elements $\{\gamma_{ma}\}$ and the "$\sim$" on $\tilde{T}_{NP}$ is the matrix transpose operator.

Application of the continuity and boundary conditions at the junction $z=d$ is accomplished by a similar mode-matching procedure and leads to the following systems of equations:

$$T_{NQ}^{(c)}C_Q^+ = F_{NN}^+B_N^+ + F_{NN}^-B_N^- \quad (11)$$

$$\Gamma_{QQ}^{(c)}C_Q^+ = \tilde{T}_{NQ}^{(c)}\Gamma_{NN}^{(b)}(F_{NN}^+B_N^+ - F_{NN}^-B_N^-) \quad (12)$$

which correspond, respectively, to (5) and (10). The matrices $T_{NQ}^{(c)}$, etc., are the same as $T_{NQ}^{(a)}$, etc., with a replaced by c. The two diagonal matrices (F_{NN}^+, F_{NN}^-) have elements $\exp(\pm\gamma_{nb}d)$ and account for the modal phase shifts corresponding to the distance d.

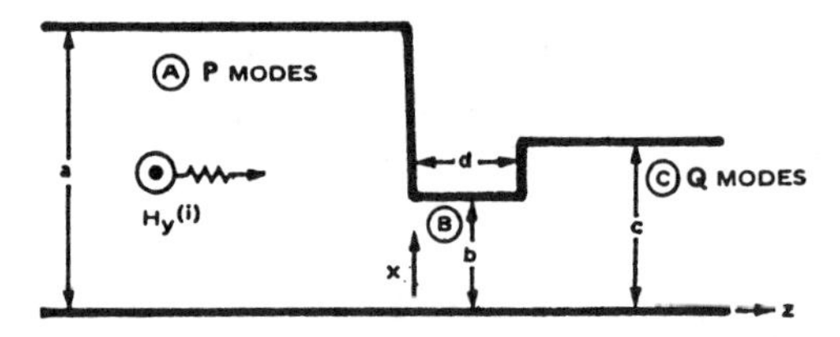

Fig. 1. Thick asymmetrical iris discontinuity problem.

The four matrix equations (5), (10), (11), and (12) provide a system of $(P+2N+Q)$ linear equations for the $(P+2N+Q)$ amplitude coefficients $(A_P^-, B_N^+, B_N^-, C_Q^+)$. Relative to this system of equations the following questions are posed.

1) What is the relation between the present matrix equations and an integral equation formulation of the problem?

2) What are the conditions on the numbers of modes N, P, and Q required to obtain a sequence of approximate solutions which converge to the solution of the problem, and how does the rate of convergence depend on (N, P, Q)?

3) How well is the power conserved and the edge condition satisfied?

The answers to the foregoing questions are considered in the following sections.

III. General Discussion of the Matrix Equations

In this section the first question posed in the previous section is considered and some general relations between the integers (N, P, Q) are described.

In order to clearly illustrate the problems at hand and avoid unnecessarily complicated algebraic manipulations, we will restrict our attention to the simple special case of the problem posed in Section II where $a=c$ and $d=0$ (cf. Fig. 1). In this case, the discontinuity becomes a capacitive iris in a uniform waveguide with the corresponding substitutions, $T_{NQ}^{(a)}=T_{NQ}^{(c)}=T_{NQ}$, $F_{NN}^+=F_{NN}^-=I_{NN}$, and $P=Q$; (5) and (10)–(12) assume the forms

$$(T_{NP}\Gamma_{PP}^{(a)^{-1}}\tilde{T}_{NP}\Gamma_{NN}^{(b)})(B_N^+ - B_N^-) = T_{NP}A_P^+ \quad (13)$$

$$A_P^- = A_P^+ - \Gamma_{PP}^{(a)^{-1}}\tilde{T}_{NP}\Gamma_{NN}^{(b)}(B_N^+ - B_N^-) \quad (14)$$

$$C_P^+ = A_P^+ - A_P^-. \quad (15)$$

Before presenting numerical data computed from these equations, the following observations are in order.

A. Integral Equation Formulation

As an alternative to the mode-matching procedure discussed in Section II, the iris waveguide problem can be formulated in terms of a Fredholm equation of the first kind.[1] Upon choosing the aperture electric field $\mathcal{E}(x) = -j\omega\epsilon_0 E_z(x, z=0)$ as the unknown, the integral equation takes the form

$$\int_0^b K(x, y)\mathcal{E}(y)\,dy = \sum_{m=1}^{\infty} A_m^+ \sqrt{\frac{2}{a\epsilon_m^1}} \cos\frac{(m-1)\pi}{a}x, \quad 0 < x < b \quad (16)$$

where the kernel $K(x, y)$ is given by

$$K(x, y) = \lim_{P\to\infty} \sum_{p=1}^{P} \frac{1}{\gamma_{pa}}\left(\frac{2}{a\epsilon_p^1}\right) \cdot \cos\frac{(p-1)\pi}{a}x \cos\frac{(p-1)\pi}{a}y. \quad (17)$$

This integral equation cannot be solved exactly.[2] However, numerical techniques may be employed to obtain an approximate solution. This may be accomplished as follows. Expand the aperture electric field $\mathcal{E}(y)$ in terms of N appropriate modes, e.g.,

$$\mathcal{E}(y) = \sum_{n=1}^{N} (B_n^+ - B_n^-)\gamma_{nb} \sqrt{\frac{2}{b\epsilon_n^1}} \cos\frac{(n-1)\pi}{b}y \quad (18)$$

where $\{B_n^+ - B_n^-\}$ are the unknown coefficients. Substituting (18) into (16) gives

$$\sum_{m=1}^{P} \left\{\left[\frac{1}{\gamma_{ma}}\sum_{n=1}^{N} T_{nm}\gamma_{nb}(B_n^+ - B_n^-)\right] - A_m^+\right\} \cdot \sqrt{\frac{2}{a\epsilon_m^1}} \cos\frac{(m-1)\pi}{a}x = 0, \quad 0 < x < b. \quad (19)$$

Following the moment method [2], [3], (8) is multiplied by $\sqrt{2/b\epsilon_l^1}\cos(l-1)\pi x/b$ for $l=1, 2, \cdots, N$ and integrated from $x=0$ to b. This procedure leads to a matrix equation, which is identical to (13), and establishes the correspondence between the integral equation and the mode-matching procedure used in Section II. For future reference, it is noted that in terms of the integral equation formulation, P is the number of terms used in computing the kernel in (17), and N is the number of terms used in approximating the aperture field in (18). In passing it should be recalled that the approximate version of (15) with a static kernel [i.e., $k\to 0$ in (16)] has been studied in great detail in [1]. For waveguides in which only the dominant mode can propagate and all higher modes are well below cutoff, a very good approximate solution can be obtained by direct analytical methods.

[1] For the general problem illustrated in Fig. 1, the formulation outlined in the text leads to a coupled system of integral equations.

[2] By taking the limit of letting the waveguide walls approach infinity in a proper manner, (15) becomes the integral equation for the well-known half-plane problem which can be solved exactly by the Wiener–Hopf technique [6].

B. Power Conservation

In many numerical studies of boundary value problems, the conservation of power is used as an indicator of the accuracy of the numerical solutions. In the present formulation, however, it is easy to show that the conservation of the (complex) power is automatically satisfied and is independent of the values of (P, N) [7]. Therefore, power conservation in this problem is only a check of algebra, computer programming, and roundoff error, and only in these ways is it a measure of the accuracy of the solution.

C. Case with $N=P$

A striking characteristic of (13)–(15) is that when $N=P$ (assuming T_{PP} nonsingular),

$$A_P^- = A_P^+ - \Gamma_{PP}^{(a)^{-1}}\tilde{T}_{PP}\Gamma_{PP}^{(b)}(T_{PP}\Gamma_{PP}^{(a)^{-1}} \cdot \tilde{T}_{PP}\Gamma_{PP}^{(b)})^{-1}T_{PP}A_P^+ \equiv 0 \quad (20)$$

and

$$C_P^+ = A_P^+. \quad (21)$$

In other words, if the *same* number of modes is used in the waveguide and the aperture, the reflection coefficient is identically *zero* (as if the iris were absent), and the result is clearly incorrect.

D. Case with $N>P$

For any two matrices U_{NP} and V_{PN} with $N>P$,

$$\det(U_{NP}V_{PN}) \equiv 0. \quad (22)$$

Therefore, when the number of the aperture modes N is greater than the number of waveguide modes P, the $(N\times N)$ matrix $(T_{NP}\Gamma_{PP}^{(a)^{-1}}\tilde{T}_{NP}\Gamma_{NN}^{(b)})$ is singular and the vector $B_N^+ - B_N^-$ cannot be obtained by inverting this matrix.

However, this is not to say that solutions for $B_N^+ - B_N^-$ do not exist. Consider (13) written in the form

$$T_{NP}\{\Gamma_{PP}^{(a)^{-1}}\tilde{T}_{NP}\Gamma_{NN}^{(b)}(B_N^+ - B_N^-) - A_P^+\} = 0 \quad (23a)$$

or

$$T_{NP}W_P = 0. \quad (23b)$$

The solutions W_P of (23b) depend on the rank of T_{NP}. If the rank of T_{NP} is equal to P, the unique solution of this equation is $W_P \equiv 0$ or

$$\Gamma_{PP}^{(a)^{-1}}\tilde{T}_{NP}\Gamma_{NN}^{(b)}(B_N^+ - B_N^-) - A_P^+ = 0. \quad (24)$$

If the system (24) has solutions either unique or not, they all lead to zero reflection coefficients ($A_P^- = 0$ for all P) via (14) and therefore an incorrect solution. It may be noted that the system (24) corresponds to setting the coefficients of $\cos(m-1/a)\pi x$, $m=1, 2, \cdots, P$ equal to zero in the sum in (19).

If the rank of T_{NP} is *less than* P, there exist nontrivial solutions W_P of (23b). In this case

$$\Gamma_{PP}^{(a)^{-1}}\tilde{T}_{NP}T_{NN}^{(b)}(B_N^+ - B_N^-) = A_P^+ + W_P \quad (25)$$

and a (nonunique) solution exists if and only if the rank of the augmented matrix $(\Gamma_{PP}^{(a)^{-1}}\tilde{T}_{NP}\Gamma_{NN}^{(b)} \vdots A_P^+ + W_P)$ is equal to the rank of $\Gamma_{PP}^{(a)^{-1}}\tilde{T}_{NP}\Gamma_{NN}^{(b)}$. If a solution exists, then all solutions give rise to the reflection coefficients $A^-_P = W_P$ via (14) and they are not unique since W_P is not unique.

The foregoing discussion reveals that in the numerical solution of the integral equation (16), the specification of more independent functions in the aperture field representation than the number of modes used in the approximate spectral representation of the kernal corresponds to an ill-posed problem. And as such the problem has at best a nonunique solution leading to identically zero or nonunique reflection coefficients.

E. Case with $N=1$ *and* $P>1$

If the aperture dimension (b/λ) is small, it is generally accepted [2] that the aperture field can be well approximated by one mode ($N=1$). Subject to this approximation, the transmission coefficient in (3) becomes

$$C_n^+ = \frac{t_{1n}\sum_{p=1}^{P} t_{1p}A_p^+}{\gamma_{na}\sum_{p=1}^{P}(t_{1p}^2/\gamma_{pa})}, \quad n = 1, 2, \cdots, P \quad (26a)$$

provided $a/\lambda \neq m+1$, $m=0, 1, 2, \cdots, P$, i.e., $\gamma_{ma} \neq 0$. If the propagation constant γ_{ma} for the mth mode in the waveguide is identically zero, (26a) assumes the form

$$C_n^+ = \begin{cases} 0, & \text{if } n \neq m \\ \sum_{p=1}^{P} t_{1p}A_p^+/t_{1m}, & \text{if } n = m. \end{cases} \quad (26b)$$

In this case the only mode with a nonzero transmission coefficient is the one at the onset of propagation (or cutoff), and consequently, no energy is transmitted. For the case $N>1$ it can be demonstrated numerically that the results given by (26b) no longer hold. Therefore, in general,

$$C_n^+\big|_{\gamma_{ma}=0} \neq 0, \quad n \neq m. \quad (27)$$

This indicates that under the condition $a/\lambda = m+1$, a one-mode approximation for the aperture field is not justified regardless how small (b/λ) is.[3]

[3] This explains why the grating lobe series method incorrectly predicts a transmission null at the grating lobe angle in some array problems. This fact was pointed out to the authors by Dr. E. C. DuFort.

IV. Numerical Results and Discussion

Based on the foregoing observations, numerical computations were confined to cases with $P>N$. For N close to P, the matrix of the linear system (13) became ill conditioned, and it was necessary to modify the method of solution of the system of equations to account for this difficulty.

As a first numerical example, consider the case

$$a = 0.4\lambda; \quad b = \frac{a}{2}; \quad A_n^+ = \begin{cases} 1, & n = 1 \\ 0, & n \neq 1 \end{cases}$$

which corresponds to an incident TEM wave in a waveguide with an iris equal to *half-width* of the guide. For this special case, the problem can be solved exactly by the Wiener–Hopf technique with a solution for the normalized input admittance [6]

$$Y = G + jB = \frac{1 + A_1^-}{1 - A_1^-} = 1 + j1.59.$$

The numerical solution obtained from (13) and (14) always gives $G=1$. Consequently, it is only necessary to study the convergence of B with respect to (P, N).[4]

In Fig. 2,[5] the susceptance B is plotted versus P for the fixed ratios $(N/P)=(2/3)$, $(1/2)$, and $(1/4)$. Note that B converges with respect to P to the correct value most rapidly when

$$N:P = b:a. \tag{28}$$

For $N=16$ and $P=32$, the value of B is 1.59, and for all practical purposes it can be regarded as the exact solution. However, it is interesting to point out that even though B is computed quite accurately, the coefficients of the higher order modes may not be very accurate. One test for the accuracy of the higher order modes is their decay rate. According to the edge condition, the asymptotic decay of the reflection coefficients should be

$$A_n \sim n^{-3/2}, \qquad n \to \infty. \tag{29}$$

In Fig. 3, we plot A_n versus n on a logarithmic scale. The slope of the set of points $\{A_n\}$ when computed accurately should be $(-3/2)$, i.e., parallel to the solid line. The coefficients $\{A_n\}$ computed from $N=16$ and $P=32$ satisfy this criterion quite well up to the 25th mode. Beyond that, the set of $\{A_n\}$ oscillates a little. Another test for the accuracy of the higher order modes is to examine the aperture field computed from them. Some criteria for the aperture field are 1) $E_x(x, 0)$ should tend to infinity as $1/\sqrt{|x-b|}$ for $x \to b$, the edge of the iris; 2) $E_x(x, 0)$ should be identically zero for $b<x<a$; 3) $H_y(x, z)$ should be continuous across $z=0$, $0<x<b$, and equal to the incident field there. As may be seen

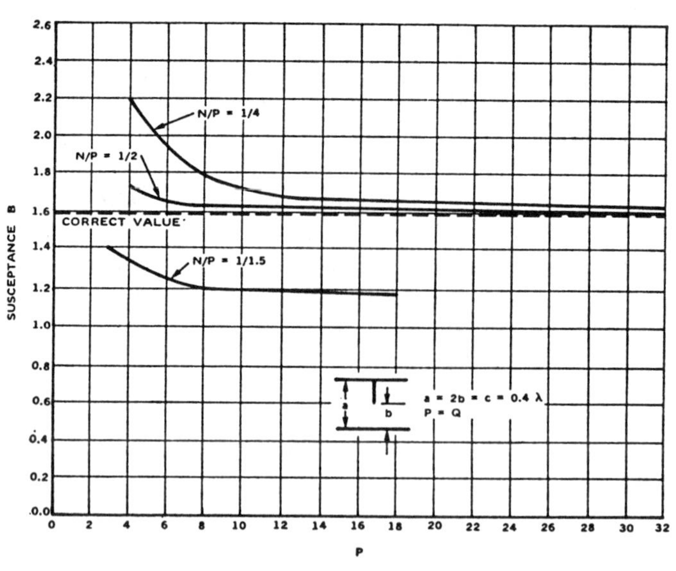

Fig. 2. TEM-mode susceptance of an iris discontinuity in a parallel-plate waveguide computed from (13)–(15) with fixed N/P.

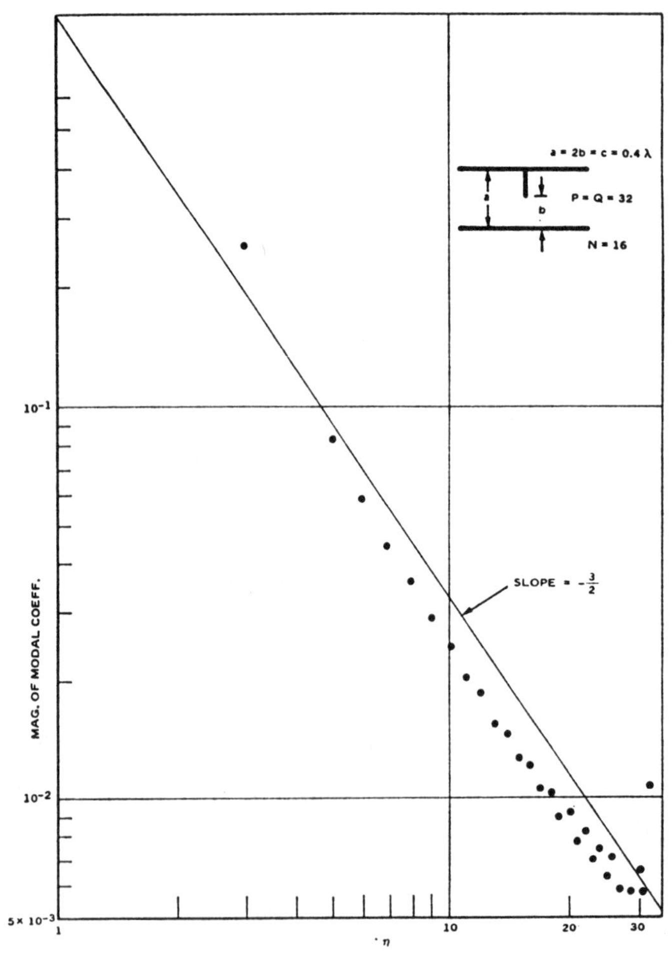

Fig. 3. Magnitude of the modal coefficients $\{A_n\}$ of the reflected field in an iris waveguide due to an incident TEM mode.

from Fig. 4, the aperture field for the case $N=16$ and $P=32$ satisfies the criteria for H_y very well, while for E_x the criteria are satisfied only approximately.

In Fig. 5, the numer N of aperture modes is fixed and the number P of waveguide modes is varied. Some

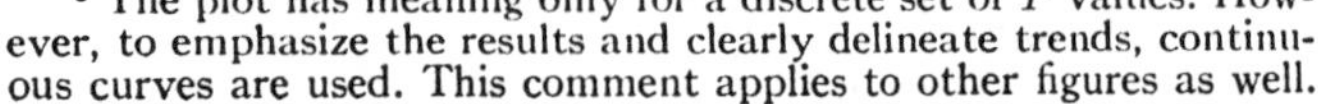

[4] Recall that $Q=P$ for the iris loaded waveguide.

[5] The plot has meaning only for a discrete set of P values. However, to emphasize the results and clearly delineate trends, continuous curves are used. This comment applies to other figures as well.

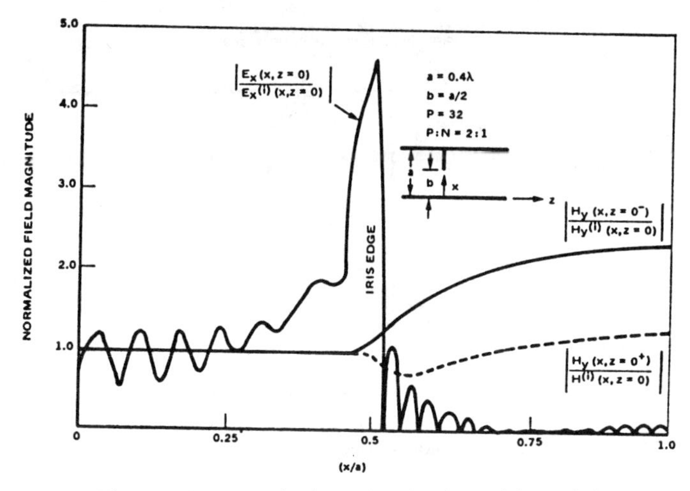

Fig. 4. Aperture field distribution in an iris loaded waveguide due to an incident TEM mode.

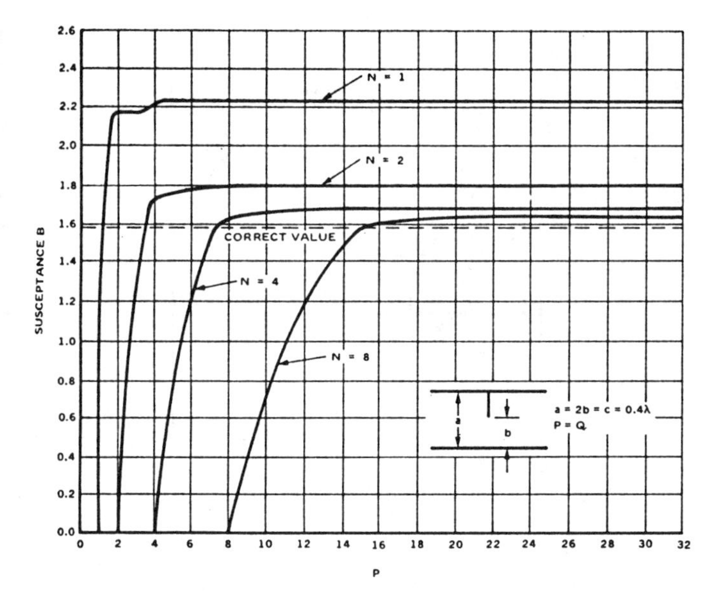

Fig. 5. TEM-mode susceptance of an iris discontinuity in a parallel-plate waveguide computed from (13)–(15) with fixed N.

observations are:

1) When $P=N$, B is identically zero as predicted from (20) and (21).

2) B increases with P. A striking fact is that, for a fixed value of N, there exists a critical P, say P_c, given by

$$P_c = \left[\frac{a}{b} N\right] \tag{30}$$

where $[x]$ means the largest integer less than or equal to x. For P beyond P_c, the value of B remains essentially a constant. From the integral equation point of view, this demonstrates that the accuracy of the kernel in (17) required to achieve a stable solution for B depends on a correlation between the approximate aperture distribution and the terms neglected in the truncated kernel, i.e., the spatial frequencies of the neglected terms must not be close to any of the spectra in the approximate aperture distribution. This may also be

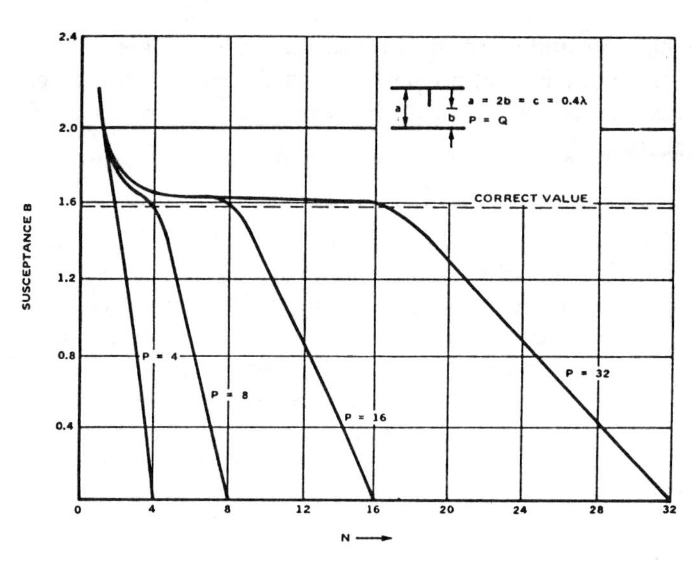

Fig. 6. TEM-mode susceptance of an iris discontinuity in a parallel-plate waveguide computed from equations (13)–(15) with fixed P.

explained as follows. The coupling between the terms of kernel $K(x, y)$ in (17) and the expansion functions in the unknown aperture distribution $\mathcal{E}(y)$ in (18) may be measured by the quantities

$$\begin{aligned} T_{np}^{(a)} &= \sqrt{\frac{4}{ab\epsilon_n{}^1\epsilon_p{}^1}} \cdot \int_0^b \cos\frac{(n-1)\pi}{b}x \cos\frac{(p-1)\pi}{a}x\,dx \\ &= (-1)^m\left[\sin\frac{(p-1)\pi b}{a}\right] \cdot \frac{\dfrac{(p-1)\pi}{a}}{\left[\dfrac{(n-1)\pi}{b}\right]^2 - \left[\dfrac{(p-1)\pi}{a}\right]^2}. \end{aligned} \tag{31}$$

For fixed n, say N, the strongest coupling or correlation occurs when $p=P_c$ as defined in (30). Beyond that $T_{np}^{(a)}$ attenuates as $0[(p)^{-1}]$. Therefore, the inclusion of more terms beyond P_c in the computation of the kernel in (17) should not significantly affect the solution for B. All terms less than P_c obviously must be included in the kernel since they correlate strongly with terms for which $n<N$ in the aperture distribution. Thus, *for efficient numerical computations one should not choose P much beyond P_c.*

3) For a given N, the solution for B, say B_N, obtained from the integral equation (16) with the *exact* kernel is always an upper bound of the correct susceptance, the difference (B_N-B) decreases as N increases. This well-known property follows from the positive definite character of the integral operator.

In the numerical solution of the integral equation (16) there is a tendency to fix P, contending that by choosing a large enough P the kernel $K(x, y)$ may be accurately computed, and then improve the solution for B by taking more and more aperture modes. In Fig. 6 it is demonstrated such an approach can lead to serious error. Once again, accurate results are obtained only when N/P satisfies (28).

V. General Iris-Type Discontinuity Problems

The problem discussed in the previous section is only one of a class of waveguide and periodic structure problems involving transverse iris-type discontinuities. For example, the problems shown in Fig. 7 also belong to this class. The dependence of relative convergence upon the number of modes used in the aperture and the number in the waveguide regions enters into *all* of these problems, except that in some problems the convergence of the numerical solutions may not depend so critically on the correct ratio as in those problems discussed in Section III. This point is illustrated in this section by considering some further examples.

In Fig. 8, the value of the admittance is presented for a TEM mode incident on an iris separating two waveguides of different cross-sectional dimensions. The computation is based on the system of equations derived in Section III. For the fixed values $P=(4/3)Q=16$, the correct result is obtained only if the number of modes used in the aperture $N=P(b/a)=8$. This curve should be compared with Fig. 6, where the waveguide possesses a symmetrical iris discontinuity. Note, for example, that for $N=10$, the computed value of B is about 18.7 percent off the correct value in Fig. 8, while there is about 21.4-percent error in Fig. 6. This indicates that as the guide dimensions a and c become farther apart, the dependence of the correct ratio for P, Q, and N is somewhat less critical. As a matter of fact, it can be shown that if $a \neq c$, the reflected field is no longer identically zero for the case $P=Q=N$.

As a second example, consider the problem of diffraction of a plane wave normally incident on an infinite array of conducting strips (cf. Fig. 9). Let $(2P+1)$ be the number of space harmonics used in the free-space region and N be that used in the aperture. By using the correct ratio between N and P, the solution converges quite rapidly. As shown in Fig. 9, the value of susceptance approaches $B=-0.97$[6] for $P \geq 5$. On the other hand, if N/P is not properly chosen the numerical results vary quite erratically. For brevity, these values are not given here.

All the problems discussed so far are two dimensional with the complete field derivable from a scalar quantity. Most practical problems, however, are of a three-dimensional nature, involving vector fields, and therefore it is important to investigate the effect of the relative convergence phenomenon in some of these.

Consider, for example, the problem of radiation from an infinite array of waveguides, as shown in Fig. 10.

[6] The value given by Marcuvitz [1] is about 0.88. Marcuvitz indicates that this value probably is in error by less than 10 percent.

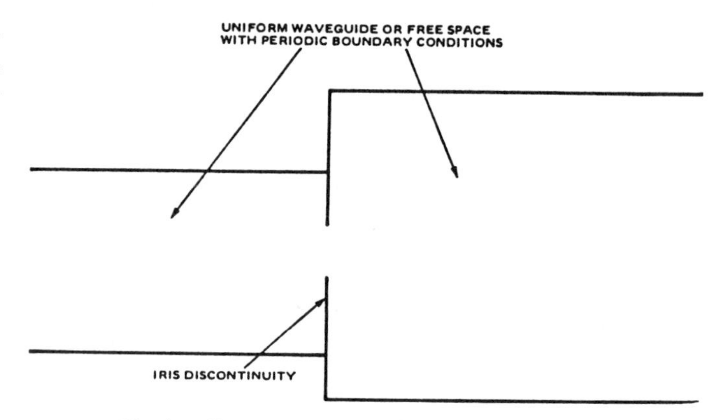

Fig. 7. General transverse iris loaded waveguide discontinuity problem.

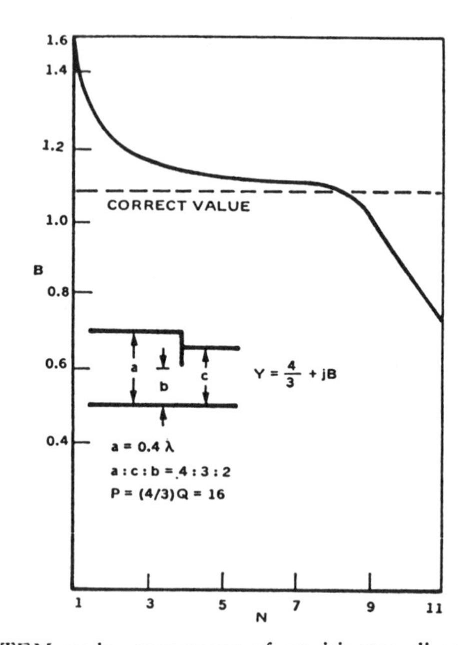

Fig. 8. TEM-mode susceptance of an iris-step discontinuity in a parallel-plate waveguide computed for fixed P and Q.

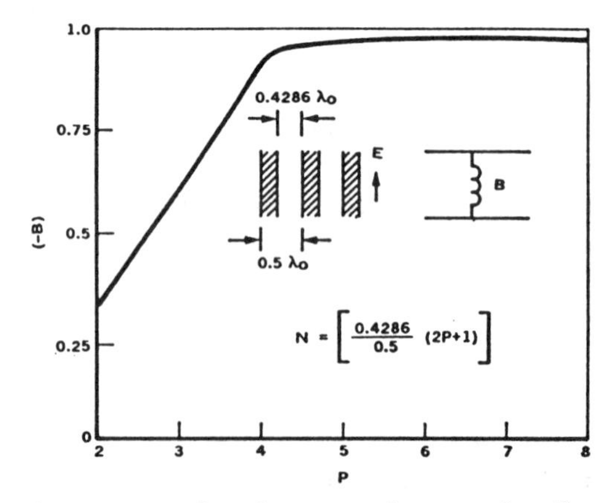

Fig. 9. Susceptance of a plane wave for a grating discontinuity computed for the correct ratio of N to P.

The waveguides are arranged in a periodic lattice, and each is loaded with irises at its opening. The field in the free-space region $z>0$ is expanded in terms of Bloch functions (space harmonics), and the z-directed vector

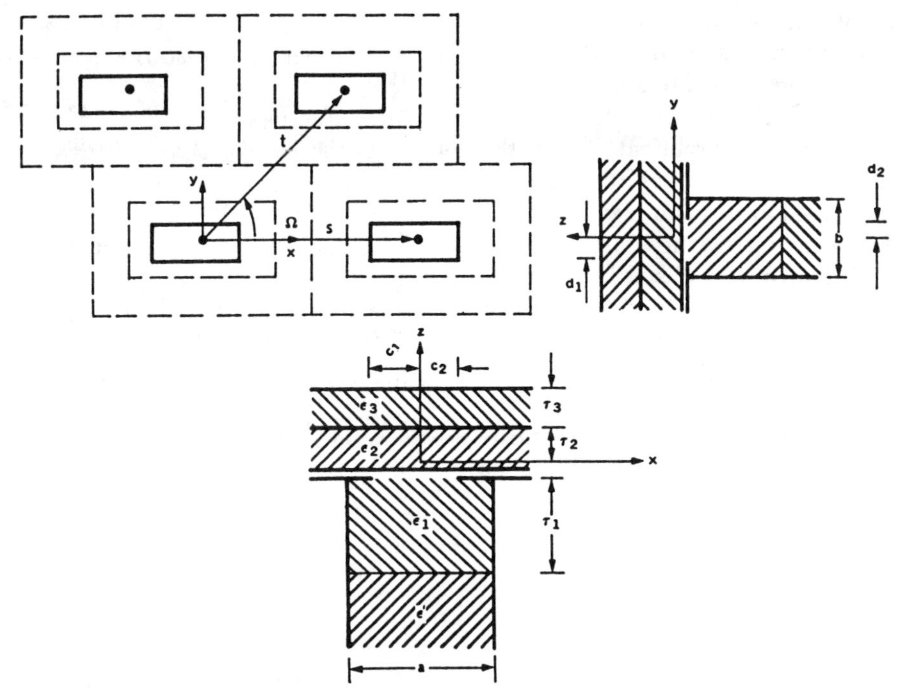

Fig. 10. Radiation from an infinite array of iris loaded rectangular waveguide with a dielectric radome.

TABLE I

REFLECTION COEFFICIENT FOR TE_{10} MODE IN A PHASED ARRAY[a]

	M						
	1	2	3[b]	4	5	6	7
R	0.9398 −55.56°	0.9400 −55.56°	0.9233 −59.85°	0.9340 −59.91°	0.8040 −83.96°	0.7930 −87.05°	0.5251 −132.04°

[a] Referring to Fig. 10, the parameters used for the computation are $k_0s=3.77$, $k_0t=2$, $\Omega=60°$; $k_0a=3.6$, $k_0b=1.2$, $k_0c=1.8$, $k_0d=1.2$ $\epsilon'=1$, $\epsilon_1=2$, $\epsilon_2=1$, $\epsilon_3=3$; $k_0\tau_1=3$, $k_0\tau_2=1$, $k_0\tau_3=1.5$; and the scan angle ($\theta_0=20°$, $\phi_0=30°$).

[b] Correct ratio.

potential assumes the form

$$\begin{matrix}\text{TE:}\\ \text{TM:}\end{matrix}\quad \psi=\sum_{p=-P}^{P}\sum_{q=-Q}^{Q}\begin{bmatrix}A_{pq}\\ B_{pq}\end{bmatrix}\exp(-ju_{p0}x)\cdot\exp(-jv_{pq}y)\exp(-j\gamma_{pq}z) \quad (32)$$

where $(u_{p0}, v_{pq}, \gamma_{pq})$ are the wavenumbers along (x, y, z) directions. There are a total of $2(2P+1)(2Q+1)$ space harmonics in the free-space region. The field in the waveguide region is expanded in terms of the normal modes

$$\psi=\begin{cases}\text{TE:}\ \displaystyle\sum_{m=0}^{\overline{M}}\sum_{n=0}^{\overline{N}}{}' C_{mn}\cos\frac{m\pi}{a}x\cos\frac{n\pi}{b}y\exp(-j\beta_{mn}z)\\ \text{TM:}\ \displaystyle\sum_{m=1}^{\overline{M}}\sum_{n=1}^{\overline{N}} D_{mn}\sin\frac{m\pi}{a}x\sin\frac{n\pi}{b}y\exp(-j\beta_{mn}z)\end{cases} \quad (33)$$

with a total of $(2\overline{M}\overline{N}+\overline{M}+\overline{N})$ modes. The field in the aperture is expressed in terms of modes similar to (33), except (a, b) is replaced by (c, d), with a total of $(2MN+M+N)$ modes. The mode-matching procedure used in the array problem is again based on the integral equation formulation, and it can be found elsewhere [8]. In Table I the reflection coefficient of an incident TE_{10} mode from the waveguide region is presented. The number of aperture modes is varied through M for fixed ($P=4$, $Q=2$, $\overline{M}=7$, $\overline{N}=2$, $N=2$). Since $c/a=0.5$, a good choice of M should be $[0.5\times 7]=3$. Note that the reflection coefficient varies quite significantly for $M>4$, and the relative convergence phenomenon is apparent in this three-dimensional phased array problem. However, it does appear that the solution is less critically dependent on the choice of the ratio of numbers of modes used in the numerical solution.

VI. CONCLUSION

The main results derived from the analyses and examples presented in this paper are twofold. 1) In the case where mode matching is employed to obtain analytical formulations of waveguide discontinuity problems for subsequent numerical analyses, that form of mode matching which is equivalent to an integral equation formulation of the problem should be employed. 2) The number of modes used in the mode-matching procedure which correspond to approximating the spectral representation of the kernel of the integral equation should be as large as practical. Once this number is fixed, the number of modal distributions selected for the aperture field approximation should be limited to that number which gives the maximum correlation with the modal distributions used in the truncated spectral representation of the kernel. Furthermore, it has been demonstrated that once the number of terms in the representation of the kernel has been fixed, the arbitrary inclusion of more and more modes in the aperture field approximation does not necessarily improve the calculation, but may even lead from good results to seriously degraded results.

References

[1] N. Marcuvitz, *Waveguide Handbook.* New York: McGraw-Hill, 1951.
[2] D. S. Jones, *The Theory of Electromagnetism.* New York: Macmillan, 1964, pp. 264–273.
[3] R. F. Harrington, *Field Computation by Moment Method.* New York: Macmillan, 1964.
[4] R. Mittra, "Relative convergence of the solution of a doubly infinite set of equations," *J. Res. Nat. Bur. Stand.*, 67D, pp. 245–254.
[5] S. W. Lee, W. R. Jones, and J. J. Campbell, "Convergence of numerical solution of iris-discontinuity problems," Hughes Aircraft Co., Fullerton, Calif., FR 70-14-594, Aug. 1970.
[6] L. A. Weinstein, *The Theory of Diffraction and the Factorization Method, Generalized Wiener—Hopf Technique.* Boulder, Colo.: Golem, 1969.
[7] N. Amitay and V. Galindo, "On energy conservation and the method of moments in scattering problems," *IEEE Trans. Antennas Propagat.*, vol. AP-17, Nov. 1969, pp. 747–751.
[8] S. W. Lee and W. R. Jones, "On the suppression of radiation nulls and broad-band impedance matching of rectangular waveguide phased arrays," *IEEE Trans. Antennas Propagat.*, vol. AP-19, Jan. 1971, pp. 41–51.

Comparative Study of Mode-Matching Formulations for Microstrip Discontinuity Problems

TAK SUM CHU, TATSUO ITOH, FELLOW, IEEE, AND YI-CHI SHIH, MEMBER, IEEE

Abstract —**Several matrix formulations for the microstrip step-discontinuity problem are compared. Although they are theoretically identical, one of them has an advantage in numerical labor, relative, and absolute convergence. Results of this method are checked with other published data and with those independently obtained by the modified residue calculus technique.**

I. Introduction

A STEP DISCONTINUITY is frequently encountered in microstrip line circuits and, hence, its analysis is important for circuit design. There are several approaches available. When the microstrip is enclosed in a waveguide-like case, it is possible to calculate the fundamental and higher order modes in both sides of the step and to impose the continuity conditions of the tangential field in the cross section of the shield case at the step location. This process leads to a system of mode-matching equations. When an open microstrip line circuit is dealt with, the higher order modes can become radiation modes which must be included in the mode-matching procedure.

In many applications, the so-called waveguide model has been found useful for calculation of the scattering at the microstrip discontinuity [1]–[3]. In this paper, we assume that the waveguide model is an acceptable technique for the calculation of the step-continuity problem. In addition, radiation and surface waves are not considered. The motivation for the present work is somewhat different from those published. Although a number of numerical data are presented in the literature [1]–[3], details of the numerical process are not clear. The objective of the present paper is not to duplicate the numerical data already available, but to place some foundation on how these data should be calculated. We present several alternative formulations. Although these formulations are theoretically identical, it is pointed out that the numerical labor and accuracy depend on the choice of formulation and some are better than others.

Manuscript received February 4, 1985; revised May 14, 1985. This work was supported in part by the U.S. Army Research Office under Contract DAAG 29-84-k-0076.

T. S. Chu and T. Itoh are with the Department of Electrical and Computer Engineering, University of Texas, Austin, TX 78712.

Y-C. Shih is with Hughes Aircraft Company, Microwave Products Division, Torrance, CA 90509.

Fig. 1. The waveguide model of the open microstrip line.

The best formulation can be decided based on the matrix size, relative and absolute convergence, and other numerical considerations. It turns out that the best formulation is the one we often choose without clear reasoning. The data for a microstrip step discontinuity are compared with available data. They are also compared with the modified residue calculus technique, which serves as an independent check of the numerical accuracy.

Before starting the formulation, let us briefly review the waveguide model. In this technique, the uniform microstrip line of width w_0 on the substrate of height h and relative dielectric constant ϵ_r is replaced with an equivalent parallel-plate waveguide with magnetic side walls (Fig. 1). The substrate height is kept identical. However, the effective width w_{eff} and the effective dielectric constant ϵ_{eff} are used to define the effective waveguide in such a way that the effect of the fringing field of the microstrip is taken into account. Specifically, these effective values are related to the propagation constant β and the characteristic impedance Z_0 via

$$\epsilon_{eff} = (\beta / k_0)^2$$
$$Z_0 = \left(120\pi / \sqrt{\epsilon_{eff}}\right)(h / w_{eff}). \quad (1)$$

Note that β and Z_0 can be found from a standard analysis such as the spectral-domain method, from curve fitting, or an empirical formula, once the structural parameters of the microstrip line are given.

For the analysis of the step discontinuity, both sides of the step are replaced with their respective equivalent waveguides. Note that the heights of these two waveguides are identical. Hence, the problem remains a two-dimensional one as the field is uniform in the y- (vertical) direction. Also, note that the dominant mode in the equivalent waveguide is TEM. In the following sections, all structural parameters used, except h, are presumed to be the "effective" ones, unless otherwise stated.

Reprinted from *IEEE Trans. Microwave Theory Tech.*, vol. MTT-33, no. 10, pp. 1018–1023, Oct. 1985.

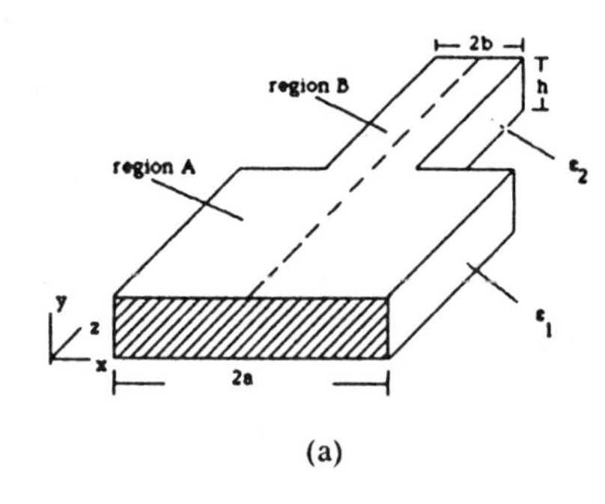

(a)

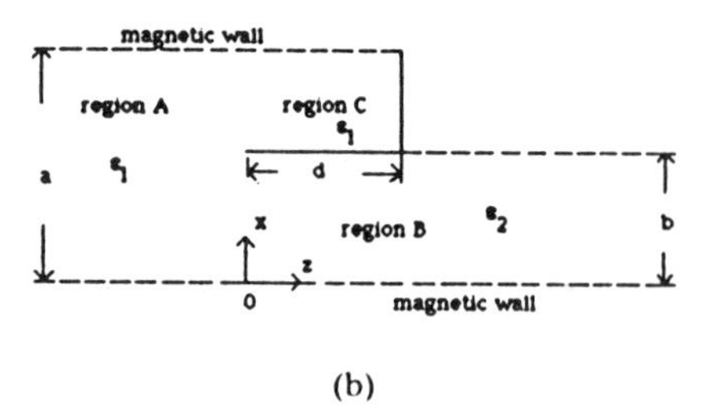

(b)

Fig. 2. (a) Waveguide model for symmetric microstrip step discontinuity. (b) Auxiliary geometry for the waveguide problem.

II. Formulation

The problem under study is the waveguide model for the microstrip step discontinuity shown in Fig. 2(a). The structure is assumed to be symmetrical, and the parallel-plate waveguide is idealized with magnetic side walls. For convenience of analysis, an auxiliary structure is introduced in Fig. 2(b). Only one half of the original structure is considered because of symmetry, and the transversal magnetic wall at the discontinuity is recessed to create a new region C. The original structure is recovered by letting $d = 0$.

The mode-matching procedure begins by expanding the tangential electric and magnetic fields at the junction in terms of the normal modes on both sides of the junction. For TE_{n0} $(n = 0, 1, \cdots)$ excitation, we write down the E_y continuity equation

$$\sum_{n=0}^{M-1} (A_n^+ + A_n^-)\phi_{an} = \begin{cases} \sum_{n=0}^{K-1} (B_n^+ + B_n^-)\phi_{bn}, & 0 \le x \le 0 \\ \sum_{n=0}^{L-1} C_n \phi_{cn}(1+\rho_n), & b \le x \le a \end{cases} \tag{2a}$$

and a corresponding one for H_x

$$\sum_{n=0}^{M-1} (A_n^+ - A_n^-) Y_{an}\phi_{an}$$

$$= \sum_{n=0}^{K-1} (B_n^+ - B_n^-) Y_{bn}\phi_{bn}, \qquad 0 \le x \le b$$

$$= \sum_{n=0}^{L-1} C_n Y_{cn}\phi_{cn}(1-\rho_n), \qquad b \le x \le a \tag{2b}$$

where

$$\phi_{an} = \sqrt{(\epsilon_{n0}/a)}\cos(k_{an}x), \qquad k_{an} = (n\pi/a)$$
$$\phi_{bn} = \sqrt{(\epsilon_{n0}/b)}\cos(k_{bn}x), \qquad k_{bn} = (n\pi/b)$$
$$\phi_{cn} = \sqrt{(\epsilon_{n0}/c)}\cos(k_{cn}(a-x)), \qquad k_{cn} = (n\pi/c)$$
$$\epsilon_{n0} = 1, \quad n = 0$$
$$2, \quad n \ne 0$$

and

$$Y_{an} = \sqrt{\left(\epsilon_1 k_0^2 - (n\pi/a)^2\right)} = \beta_n$$
$$Y_{bn} = \sqrt{\left(\epsilon_2 k_0^2 - (n\pi/b)^2\right)} = \gamma_n$$
$$Y_{cn} = \sqrt{\left(\epsilon_1 k_0^2 - (n\pi/c)^2\right)} = \bar{\beta}_n$$
$$\rho_n = \exp(-2j\bar{\beta}_n d).$$

In (2), ϕ_{an}, ϕ_{bn}, and ϕ_{cn} are normal modes in Regions A, B, and C, with propagation constants β_n, γ_n, and $\bar{\beta}_n$, respectively. A_n^+ and B_n^- are the given incident field coefficients from Regions A and B (usually only one A_n^+ or B_n^- is considered at a time), while A_n^-, B_n^+, and C_n are the unknown excited field coefficients in regions A, B, and C, respectively. ρ_n is the reflection from the magnetic wall in Region C.

From modal orthogonality, we obtain the linear simultaneous equations for the unknown modal coefficients

$$A_m^+ + A_m^- = \sum_{n=0}^{K-1} H_{mn}(B_n^+ + B_n^-) + \sum_{n=0}^{L-1} \bar{H}_{mn} C_n (1+\rho_n)$$

$$Y_{am}(A_m^+ - A_m^-) = \sum_{n=0}^{K-1} H_{mn} Y_{bn}(B_n^+ - B_n^-) + \sum_{n=0}^{L-1} H_{mn} Y_{cn} C_n (1-\rho_n),$$

$$m = 0, 1, 2, \cdots, M-1 \tag{3a}$$

$$\sum_{n=0}^{M-1} H_{nm}(A_n^+ + A_n^-) = B_m^+ + B_m^-$$

$$\sum_{n=0}^{M-1} H_{nm} Y_{an}(A_n^+ - A_n^-) = Y_{bm}(B_m^+ - B_m^-),$$

$$m = 0, 1, 2, \cdots, K-1 \tag{3b}$$

$$\sum_{n=0}^{M-1} \bar{H}_{nm}(A_n^+ + A_n^-) = C_m(1+\rho_m)$$

$$\sum_{n=0}^{M-1} \bar{H}_{nm} Y_{an}(A_n^+ - A_n^-) = C_m Y_{cm}(1-\rho_m),$$

$$m = 0, 1, 2, \cdots, L-1 \tag{3c}$$

where

$$H_{mn} = \int_0^b \phi_{am}\phi_{bn}\,dx = \left(\frac{\epsilon_{m0}\epsilon_{n0}}{ab}\right)^{1/2}\left(\frac{1}{2}\right)\left(\frac{2k_{am}(-1)^n \sin(k_{am}b)}{k_{am}^2 - k_{bn}^2}\right)$$

$$\bar{H}_{mn} = \int_b^a \phi_{am}\phi_{cn}\,dx = \left(\frac{\epsilon_{m0}\epsilon_{n0}}{ac}\right)^{1/2}\left(\frac{1}{2}\right)\left(\frac{2k_{am}(-1)^{n+1} \sin(k_{am}b)}{k_{am}^2 - k_{cn}^2}\right).$$

To condense the above equations, we define the following matrices:

$$Y_i = \begin{bmatrix} Y_{i0} & 0 & & \cdots & \\ & Y_{i1} & & & \\ 0 & & \cdot & & \\ & & Y_{i2} & & \\ \vdots & & & & \\ & & & \ddots & \\ & & & & Y_{in} \end{bmatrix},$$

$$i = a, b, c$$
$$n = M-1, K-1, L-1,$$

$$R = \begin{bmatrix} 1+\rho_0 & 0 & \cdots & \\ 0 & 1+\rho_1 & & \\ \vdots & & & \\ & & \ddots & \\ & & & 1+\rho_n \end{bmatrix}$$

$$R' = \begin{bmatrix} 1-\rho_0 & & 0 & \cdots \\ & 1-\rho_1 & & \\ 0 & & & \\ \vdots & & & \\ & & \ddots & \\ & & & 1-\rho_n \end{bmatrix}$$

$$n = L-1$$

$$Y_d = \left[\begin{array}{c|c} Y_b & 0 \\ \hline 0 & Y_c \end{array}\right] \quad \overline{R}' = \left[\begin{array}{c|c} I & 0 \\ \hline 0 & R' \end{array}\right] \quad \overline{R} = \left[\begin{array}{c|c} I & 0 \\ \hline 0 & R \end{array}\right]$$

$$G = [H \mid \overline{H}]$$

where I is the identity matrix, H is a matrix of size $M \times K$ with generic element H_{mn} as defined above, while $\overline{H}$ is a matrix of size $M \times L$ with generic element $\overline{H}_{mn}$.

Then, the mode-matching equations can be written in the following matrix form:

$$\underset{\sim}{a}^+ + \underset{\sim}{a}^- = G\overline{R}\underset{\sim}{d}^+ + G\underset{\sim}{d}^- \tag{4a}$$

$$Y_a(\underset{\sim}{a}^+ - \underset{\sim}{a}^-) = GY_d\overline{R}'\underset{\sim}{d}^+ - GY_d\underset{\sim}{d}^- \tag{4b}$$

$$G^T(\underset{\sim}{a}^+ + \underset{\sim}{a}^-) = \overline{R}\underset{\sim}{d}^+ + \underset{\sim}{d}^- \tag{4c}$$

$$G^TY_a(\underset{\sim}{a}^+ - \underset{\sim}{a}^-) = Y_d\overline{R}'\underset{\sim}{d}^+ - Y_d\underset{\sim}{d}^- \tag{4d}$$

where superscript T denotes transpose operation, and

$$\underset{\sim}{a}^+ = \begin{bmatrix} A_0^+ \\ A_1^+ \\ A_2^+ \\ \vdots \\ A_{M-1}^+ \end{bmatrix} \qquad \underset{\sim}{d}^- = \begin{bmatrix} B_0^- \\ B_1^- \\ \vdots \\ B_{K-1}^- \\ \hline 0 \\ \vdots \\ 0 \end{bmatrix}$$

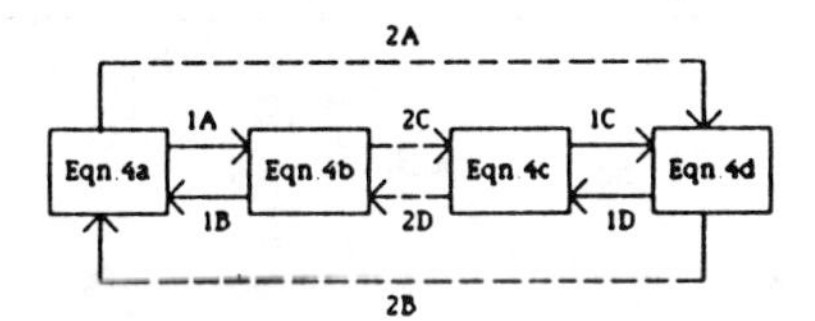

Fig. 3. Classification of formulations.

$$\underset{\sim}{d}^+ = \begin{bmatrix} B_0^+ \\ B_1^+ \\ \vdots \\ B_{K-1}^+ \\ \hline C_0 \\ C_1 \\ \vdots \\ C_{L-1} \end{bmatrix} \qquad \underset{\sim}{a}^- = \begin{bmatrix} A_0^- \\ A_1^- \\ A_2^- \\ \vdots \\ A_{M-1}^- \end{bmatrix}$$

$\underset{\sim}{a}^+$ and $\underset{\sim}{d}^-$ are column vectors of the excitation terms and $\underset{\sim}{a}^-$ and $\underset{\sim}{d}^+$ are column vectors of the unknown modal coefficients. All matrices are of size $(M \times M)$; this requires that $K + L = M$.

When $M \to \infty$, we can prove that $G^{-1} \equiv G^T$. Therefore, (4a) and (4b) are equivalent to (4c) and (4d). Two independent vector equations are required to solve for two unknown vectors. Hence, for four pairs of equations ((4a) and (4b), (4b) and (4c), (4c) and (4d), and (4d) and (4a)), substituting one equation into the other in the same pair, we have eight ways to solve for $\underset{\sim}{a}^+$ and $\underset{\sim}{d}^-$. They are defined graphically in Fig. 3. The approaches indicated by a solid arrow are classified as the formulations of the first kind and those indicated by a dashed arrow are of the second kind. Although the eight ways of solution are theoretically equivalent, their numerical behaviors are somewhat different, especially when the magnetic wall is introduced at the upper half of the junction ($d = 0$, $\rho_n = 1$).

For general cases $d \neq 0$, all the formulations require a matrix inversion of size $(M \times M)$. For our limiting case of $d = 0$, special modifications must be made for some cases. Specifically, 1D and 2B need to invert a $(M+L)\times(M+L)$ matrix and 2C (Appendix A) needs to invert a smaller $(K \times K)$ matrix. Hence, 2C is most attractive to us because of its potential of numerical efficiency. In the next section, we will examine the various approaches in terms of the numerical stability and convergence.

III. Numerical Results

To study the numerical behavior of the various formulations, we have chosen the structural parameters as: $a = 100$, $b = 26.1$ (in mils), $\epsilon_1 = 2.2$, $\epsilon_2 = 2.1$. The dominant mode (TEM) reflection and transmission coefficients at the junction are calculated by varying the matrix size for different K/M ratios.

Since formulations 1D and 2B have an apparent disadvantage in numerical calculations, they are not considered here. After extensive studies, we have found that 1A, 1B, and 1C are numerically identical. Similarly, 2A and 2D are

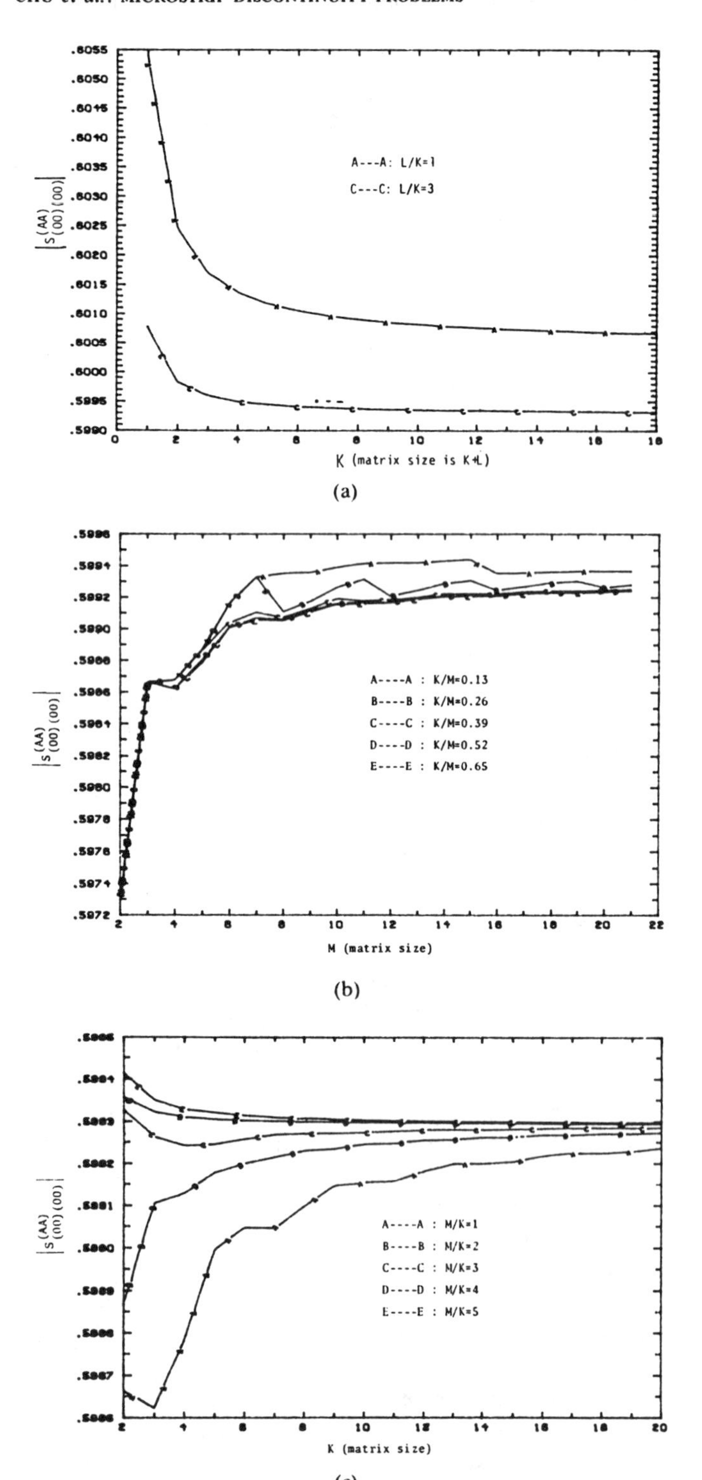

Fig. 4. Convergence study for various formulations: (a) formulation 1A; (b) formulation 2A; (c) formulation 2C.

numerically identical. Therefore, only three sets of data, corresponding to 1A, 2A, and 2C, are given.

In each formulation, the indices L, K, and M are involved. The numerical results are affected by the ratios among these indices. This is called the relative convergence phenomenon, and it has been thoroughly discussed in the literature [4], [5]. It is well known that the best approximation to the true solution is obtained for $L/K = c/b = 73.6/26.1$ or $M/K = a/b = 100/26.1$ (refer to as the right

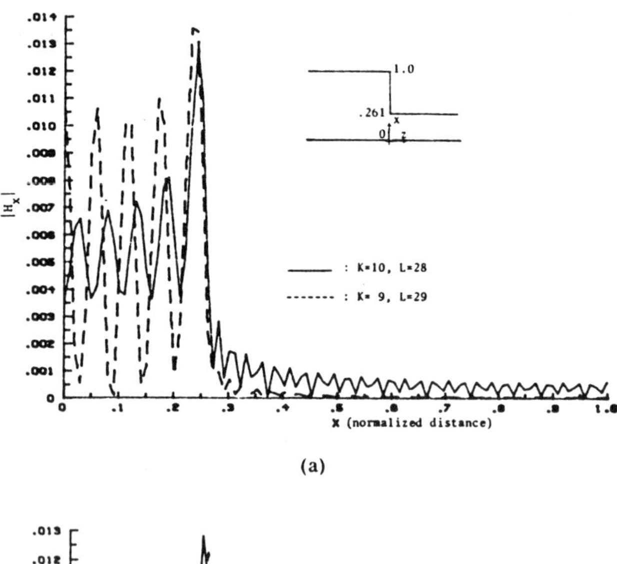

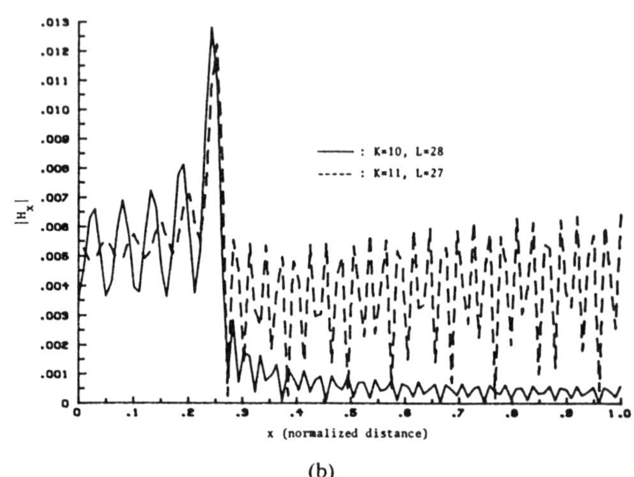

Fig. 5. Relative convergence problem of formulation 1A demonstrated by field plots.

ratio). It is observed that 2A and 2C suffer very little from the relative convergence problem (Fig. 4(b) and (c)). The problem is more serious in 1A, as can be seen in Fig. 4(a). With a ratio of one, the dominant mode reflection coefficient converges to an incorrect value (curve A in Fig. 4(a)). Curve C is calculated using a ratio of three, which is close to the right ratio.

The relative convergence effect can be more clearly observed from the plot of H_x at the junction. The resultant field calculated by 1A is shown in Fig. 5. In Fig. 5(a), we plot the fields calculated using a ratio of $L/K = 28/10$ (very close to the right ratio) and a ratio of 29/9 (higher than the right ratio). In Fig. 5(b), we compare the field calculated using a ratio of 27/11 (lower than the right ratio) and a ratio of 28/10. It is interesting to note that, with a higher ratio, the calculated field behaves better on the magnetic wall discontinuity than on the aperture, while the opposite is true for field calculated using a lower ratio. This might seem reasonable because, for a higher ratio, we use more modes on the magnetic wall than on the aperture and vice versa for a lower ratio. Fig. 6 shows the resultant fields calculated by 2C. Different ratios have no noticeable effect. The fields calculated by 2A behave similar to those calculated by 2C.

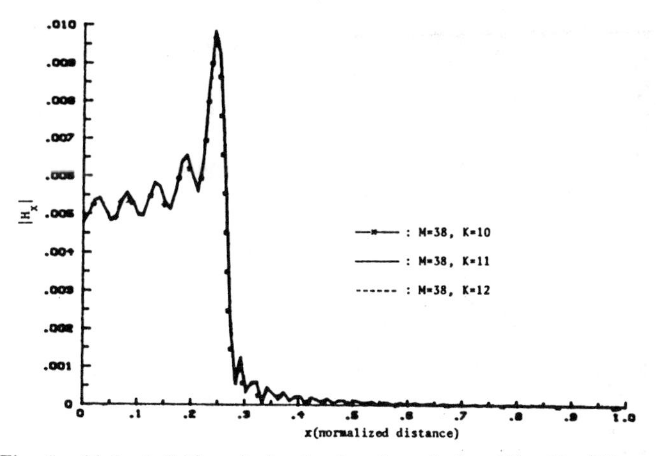

Fig. 6. Plot of fields calculated using formulation 2C with different M/K ratios.

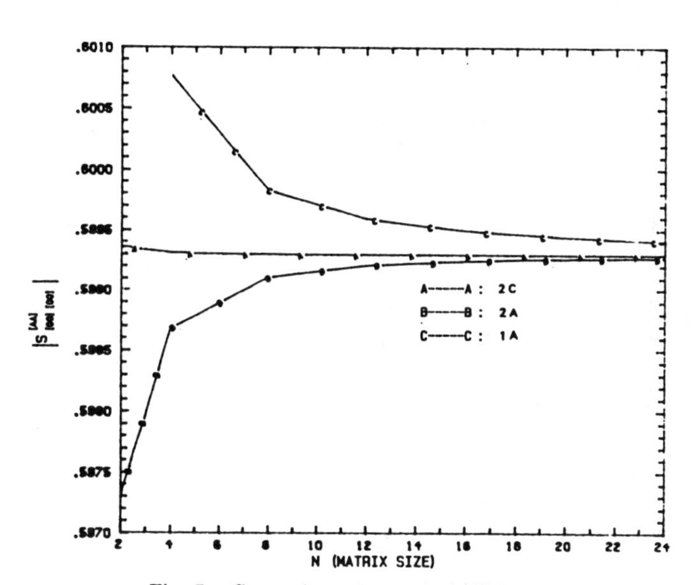

Fig. 7. Comparison of numerical efficiency.

A comparative study on the numerical efficiency for different approaches has also been done. In this case, $M/K = 4$, which is close to a/b, is chosen. The results of the dominant mode reflection coefficient $S[00][00]$ are evaluated as a function of the matrix size required and shown in Fig. 7. In addition, a comparison of how well the fields of the two sides match at the junction is done between formulations 2A and 2C. In both calculations, M is set to 40, K to 10. The result is shown in Fig. 8. The fields calculated by 2C match as well as, if not better than, those calculated by 2A. Keep in mind that we have to invert a matrix of size 40 in 2A, compared to a matrix of size 10 in 2C. It is now obvious that 2C has definite advantages over other approaches. This formulation is to be used for further studies.

Let us refer back to (2) at this point. In many attempts, E_y in the region $b < x < a$ is not used as $H_x = 0$ there. This is identical to 2C and, hence, is our preferred choice.

To check the validity of our calculations, we have calculated the frequency response of a microstrip step discontinuity using the same parameters given by Kompa [2]. The results of the dominant mode reflection and transmission coefficients $S[00][00]$ and $S[00][00]$ are shown in Fig. 9. They are in good agreement with Kompa's results. The small discrepancy is due to the different formulas used for obtaining the effective width and effective dielectric constant of the waveguide model. Furthermore, we have checked the results with those independently obtained by the modified residue calculus technique [6]. The results of $S[00][00]$ are shown in Table I and Fig. 10 for comparison. The calculations are performed using Kompa's parameters.

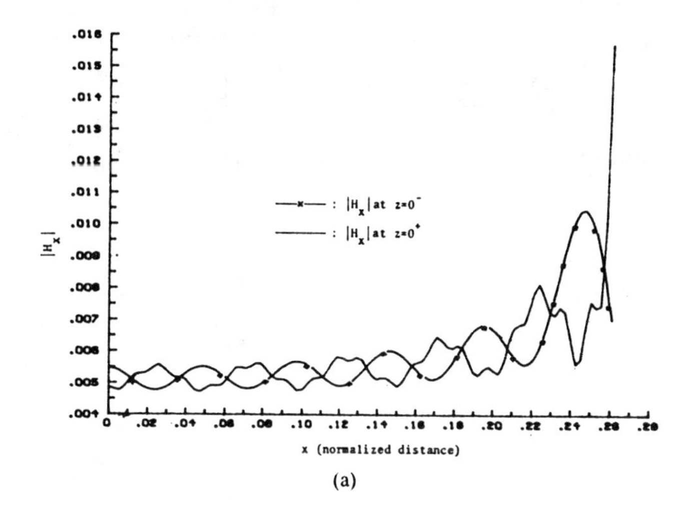

(a)

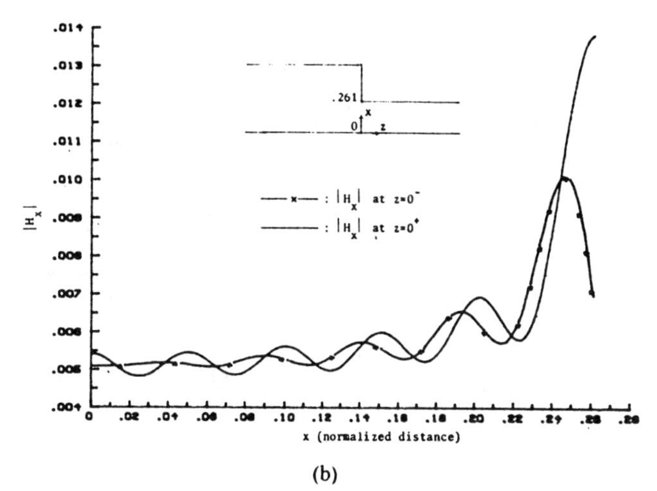

(b)

Fig. 8. Comparison of field plot between formulation 2A and 2C. (a) 2A, (b) 2C.

IV. Conclusion

The mode-matching method has been applied to analyze the microstrip step-discontinuity problem based on the waveguide model. A comparison has been made among the various mode-matching solutions based on the matrix size and relative and absolute convergence. Although they are theoretically identical, one of them proves to be most suitable for numerical calculations. The results by this method are in good agreement with other published data and with those independently obtained by the modified residue calculus technique.

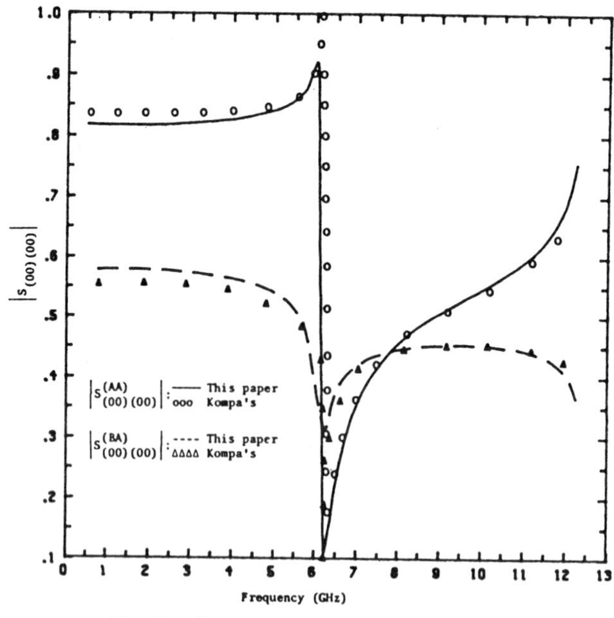

Fig. 9. Comparison with Kompa's results.

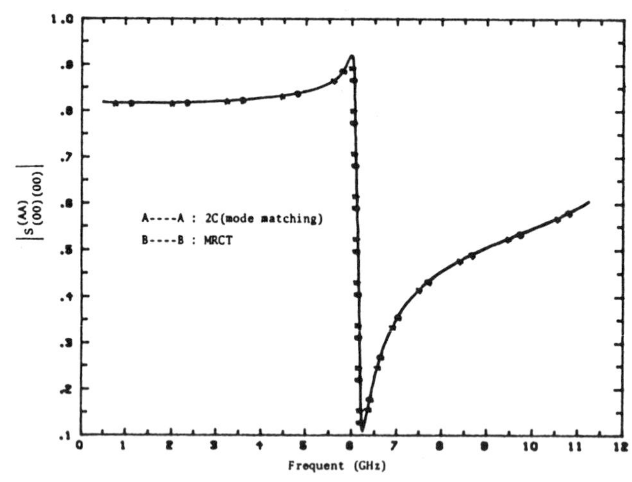

Fig. 10. Comparison with results of modified residue calculus tech nique.

TABLE I
COMPARISON OF THE RESULTS BY MODE-MATCHING METHOD AND BY MRCT*

	Mode Matching	MRCT
$S^{(AA)}_{(00)(00)}$	$0.1837 - j0.02291$	$0.1837 - j0.02297$
$S^{(BA)}_{(00)(00)}$	$-0.7856 + j0.2227$	$-0.7855 + j0.2233$

*Calculations are performed using Kompa's parameters at 2

REFERENCES

[1] I. Wolff, G. Kompa, and R. Mehran, "Calculation method for microstrip discontinuities and *T*-junctions," *Electron. Lett.*, vol. 8, pp. 177–179, Apr. 1972.

[2] G. Kompa, "*S*-matrix computation of microstrip discontinuities with a planar waveguide model," *Arch. Elec. Ubertragung.*, vol. 30, pp. 58–64, 1975.

[3] W. Menzel and I. Wolff, "A method for calculating the frequency-dependent properties of microstrip discontinuities," *IEEE Trans. Microwave Theory Tech.*, vol. MTT-25, pp. 107–112, Feb. 1977.

[4] S. W. Lee, W. R. Jones, and J. J. Campbell, "Convergence of numerical solutions of iris-type discontinuity problems," *IEEE Trans. Microwave Theory Tech.*, vol. MTT-19, pp. 528–536, June 1971.

[5] R. Mittra, T. Itoh, and T. S. Li, "Analytical and numerical studies of the relative convergence phenomenon arising in the solution of an integral equation by the moment method," *IEEE Trans. Microwave Theory Tech.*, vol. MTT-20, pp. 96–104, Feb. 1972.

[6] T. S. Chu and T. Itoh, "Analysis of microstrip step discontinuity by the modified residue calculus technique," submitted to *IEEE Trans. Microwave Theory Tech.*, (Special Issue on Numerical Technique).

Dispersion Characteristics of Shielded Microstrips with Finite Thickness

by GÜNTER KOWALSKI and REINHOLD PREGLA *

Communication from the Institut für Hochfrequenztechnik der Technischen Universität Braunschweig

DK 621.372.821

The dispersion characteristics of a shielded microstrip and the propagation constants of the even higher order modes are analyzed by dividing the cross-section into two parts with five subregions each. A complete set of modes is assumed in each subregion satisfying all the boundary conditions on the metalic surfaces and the fields are matched at the interfaces of the subregions. The calculation holds for finite or infinitely thin strips and for a rectangular screen around the microstrip, which is wide enough to approximate the characteristics of the open structure.

Dispersionscharakteristik der geschirmten Mikrostrip-Leitung mit endlicher Streifenleiterdicke

Die Dispersionscharakteristik einer geschirmten Mikrostrip-Leitung und der Ausbreitungskoeffizient der geraden höheren Wellentypen wird berechnet. Dazu wird die Querschnittsebene in zwei Teile mit je fünf Unterbereichen geteilt. In jedem Unterbereich wird ein vollständiger Satz von Wellentypen angesetzt, der die Randbedingungen auf allen leitenden Oberflächen erfüllt, und die Felder werden an den Grenzflächen der Unterbereiche angepaßt. Die Rechnung ist für endlich und unendlich dünnen Streifen anwendbar und für einen rechteckigen Schirm um die Streifenleitung, der weit genug entfernt ist, um das Verhalten der offenen Streifenleitung anzunähern.

1. Introduction

The dispersion characteristics of a microstrip were recently investigated by several authors [1]—[8]. In most of these papers infinitely thin strips placed symmetrically in an inhomogeneously filled waveguide were studied. The cross-section was divided into two homogeneous regions separated by the strip and the dielectric surfaces.

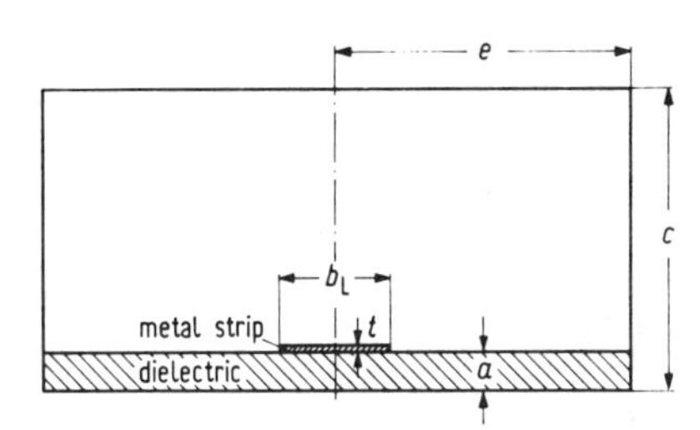

Fig. 1. Cross-section of the dielectrically supported shielded strip line.

This paper gives an analysis, which holds for finite or infinitely thin strips placed symmetrically in a waveguide (see Fig. 1). The method follows a treatment given by E. KÜHN and W.-D. SCHWARTZ [9]. The cross-section is divided into several subregions. In each region the electromagnetic field is represented by a set of functions with unknown coefficients. Each of these functions independently satisfies the Maxwell's equations and all boundary conditions on the metalic strip and the outer guide. The latter condition requires to divide the cross-section into two parts with five subregions each.

To match the tangential field components at the interfaces of the subregions, the components are expanded into Fourier series. This leads to an infinite set of linear equations for the coefficients of the subregion-functions. Only few terms of each equation are necessary to obtain accurate results. A non trivial solution requires the determinant of the set to be zero.

2. Formulation of the Problem

As long as we are only interested in modes of even order with respect to the electric field, we can replace the plane of symmetry by a magnetic wall. The half cross-section is divided into five subregions (see Fig. 2). The field components $\vec{E}$ and $\vec{H}$ of the subregions are derived from the Hertzian potentials according to

$$\vec{E}_\nu = -\nabla \times (u_z \Psi_\nu^{(h)} e^{-jk_z z}) + \frac{1}{j\omega\varepsilon_\nu} \nabla \times \nabla \times (u_z \Psi_\nu^{(e)} e^{-jk_z z}), \quad (1a)$$

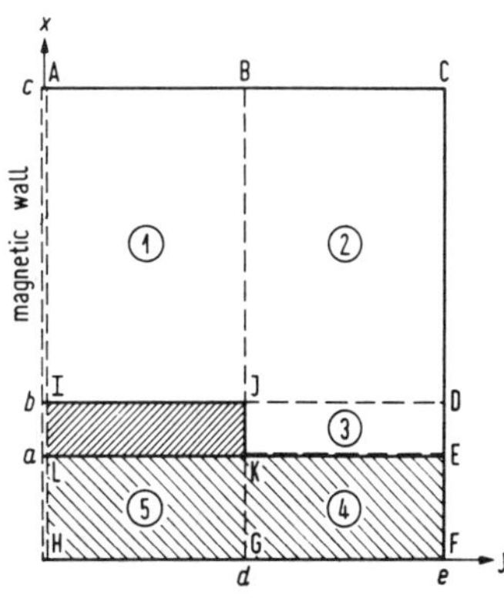

Fig. 2. Subdivision of the strip line cross-sectional regions.

* Dipl.-Ing. G. KOWALSKI, Dozent Dr. R. PREGLA, im Institut für Hochfrequenztechnik der Technischen Universität, D-33 Braunschweig, Postfach 7050.

Reprinted with permission from *Arch. Elek. Ubertragung.*, vol. 25, pp. 193–196, Apr. 1971.

$$\vec{H}_\nu = \frac{1}{j\omega\mu} \nabla \times \nabla \times (u_z \Psi_\nu^{(h)} e^{-jk_z z}) + + \nabla \times (u_z \Psi_\nu^{(e)} e^{-jk_z z}), \quad \nu = 1, \ldots, 5. \tag{1b}$$

Here $\Psi_\nu^{(e)}$ and $\Psi_\nu^{(h)}$ are the Hertzian potentials of the TM and the TE modes respectively in the region ν, u_z is the unit vector and k_z the propagation constant in the z-direction. The dielectric permittivity is $\varepsilon_\nu = \varepsilon_0$ in regions 1, 2, 3 and $\varepsilon_\nu = \varepsilon_r \varepsilon_0$ in regions 4, 5. To satisfy all boundary conditions at the metalic surfaces and at the magnetic wall the following functions are used:

$$\Psi_1^{(h)} = \sum_{n=1}^{N+1} A_n \cos \alpha_n^{(1)} (x - c) \sin \beta_n^{(1)} y\,, \tag{2a}$$

$$\Psi_1^{(e)} = \sum_{n=1}^{N} B_n \sin \alpha_{n+1}^{(1)} (x - c) \cos \beta_{n+1}^{(1)} y\,, \tag{2b}$$

$$\Psi_2^{(h)} = \sum_{n=1}^{N+1} \bar{C}_n \cos \alpha_n^{(1)} (x - c) \cos \beta_n^{(1)} (y - e) + + \tilde{C}_n \cos \alpha_n^{(2)} (x - c) \cos \beta_n^{(2)} (y - e)\,, \tag{2c}$$

$$\Psi_2^{(e)} = \sum_{n=1}^{N} \bar{D}_n \sin \alpha_{n+1}^{(1)} (x - c) \sin \beta_{n+1}^{(1)} (y - e) + + \tilde{D}_n \sin \alpha_{n+1}^{(2)} (x - c) \sin \beta_{n+1}^{(2)} (y - e)\,, \tag{2d}$$

$$\Psi_3^{(h)} = \sum_{n=1}^{N+1} E_n \cos \alpha_n^{(2)} (x - a) \cos \beta_n^{(2)} (y - e) + + F_n \sin \alpha_n^{(2)} (x - a) \cos \beta_n^{(2)} (y - e)\,, \tag{2e}$$

$$\Psi_3^{(e)} = \sum_{n=1}^{N} G_n \sin \alpha_{n+1}^{(2)} (x - a) \sin \beta_{n+1}^{(2)} (y - e) + + H_n \cos \alpha_{n+1}^{(2)} (x - a) \sin \beta_{n+1}^{(2)} (y - e) \tag{2f}$$

and

$$\alpha_n^{(1)} = \frac{(n-1)\pi}{c - b}\,, \tag{3a}$$

$$\beta_n^{(1)} = (k_0^2 - k_z^2 - \alpha_n^{(1)2})^{1/2}\,, \tag{3b}$$

$$\alpha_n^{(2)} = (k_0^2 - k_z^2 - \beta_n^{(2)2})^{1/2}\,, \tag{3c}$$

$$\beta_n^{(2)} = \frac{(n-1)\pi}{e - d}\,, \tag{3d}$$

$$\bar{\alpha}_n^{(1)} = \frac{(n-1)\pi}{a}\,, \tag{3e}$$

$$\bar{\beta}_n^{(1)} = (k_0^2 - k_z^2 - \bar{\alpha}_n^{(1)2})^{1/2}\,, \tag{3f}$$

$$\bar{\alpha}_n^{(2)} = (\varepsilon_r k_0^2 - k_z^2 - \beta_n^{(2)2})^{1/2}\,. \tag{3g}$$

k_0 is the free space propagation constant. Only the Hertzian potentials of the regions 1, 2, 3 are given. Those of the regions 4, 5 can easily be obtained in a similar way. The corresponding sets of coefficients are $\bar{Q}_n$, $\tilde{Q}_n$, $\bar{R}_n$, $\tilde{R}_n$, S_n, T_n, and the propagation constants in x- and y-direction respectively are $\bar{\alpha}_n^{(1)}$, $\bar{\alpha}_n^{(2)}$, $\bar{\beta}_n^{(1)}$, $\beta_n^{(2)}$. The coefficients $A_n, \ldots, T_n$ are as yet unknown. For an exact computation the series should be infinite. They are limited to N terms for TM-modes and $N+1$ terms for TE-modes. It must be noted, that the electric field components of the Hertzian potential in region 2 satisfy the boundary condition on the outer guide rectangle $\overline{\mathrm{BCD}}$ as well as at the point J for any coefficients $\bar{C}_n, \ldots, \tilde{D}_n$. The set with the coefficients $\bar{C}_n$ and $\bar{D}_n$ represents electric field vectors, which are normal on the contour $\overrightarrow{\mathrm{BCDJ}}$, and the set with the coefficients $\tilde{C}_n$ and $\tilde{D}_n$ represents those, which are normal on the contour $\overrightarrow{\mathrm{JBCD}}$. The corresponding is true in region 4.

Matching the tangential field components across the interfaces between the regions and expanding with a suitable set of orthogonal functions according to

$$\begin{matrix}\sin\\ \cos\end{matrix}\,(\alpha_n^{(1)} x); \quad b \leqq x \leqq c\,, \quad y = d\,, \tag{4a}$$

$$\begin{matrix}\sin\\ \cos\end{matrix}\,(\beta_n^{(2)} y); \quad x = a, b\,, \quad d \leqq y \leqq e\,, \tag{4b}$$

$$\begin{matrix}\sin\\ \cos\end{matrix}\,(\bar{\alpha}_n^{(1)} x); \quad 0 \leqq x \leqq a\,, \quad y = d\,, \tag{4c}$$

a homogeneous set of equations for the unknown coefficients $A_n, \ldots, T_n$ is obtained. Since the orthogonal functions are the same functions used in the series of the Hertzian potentials, many of the submatrices in the set of equations become diagonal matrices. Hence it is possible to eliminate certain sets of coefficients without inverting any matrix. Eliminating all coefficients except the $\tilde{C}_n$, $\tilde{D}_n$, $\tilde{Q}_n$, $\tilde{R}_n$ a matrix with $(4N+2) \times (4N+2)$ elements is obtained. It is possible to eliminate the $\tilde{C}_n$ and $\tilde{D}_n$ as well and to reduce the system to the size of $(2N+1) \times (2N+1)$ elements without inverting any matrix. This was not done, however, because the formulas would become quite involved while on the other hand the size of the matrix is already small enough to be handled by numerical computation.

Some special manipulations are necessary in order that the equations hold for infinitely thin strip. The poles at $t = 0$ were eliminated first by substracting half of the equations from the rest and secondly by multiplying with the terms, which become infinite at $t = 0$. The limiting case of an infinitely thin strip leads to a set of equations, which can as well be obtained by omitting the subregion 3 and matching the field components of region 2 to those of region 4.

The eigenvalues of k_z for a specified k_0 were found numerically by bisection.

3. Numerical Results

The numerical computations have been carried out with the computer ICL 1900 of the Computing Center of the Technical University Braunschweig. Typical computation times were as follows. One multiplication takes about 14 µs. The access time for a two dimensional array-element is 50 µs. The matrix is built up and the value of its determinant is computed once within 0.7 s for $N = 5$ and 4.0 s for $N = 10$. The computing time increases with N^3 for larger values of N.

In all our computations we determined the effective relative dielectric constant, which is defined as $\varepsilon_{eff} = (\lambda_0/\lambda_s)^2$, where λ_s is the wavelength of propagation along the strip and λ_0 is the free space wavelength. We investigated the effective dielectric constant as a function of the normalized frequency $a/\lambda_0 = fa/c_0$, where f is the actual frequency and c_0 is the free space velocity.

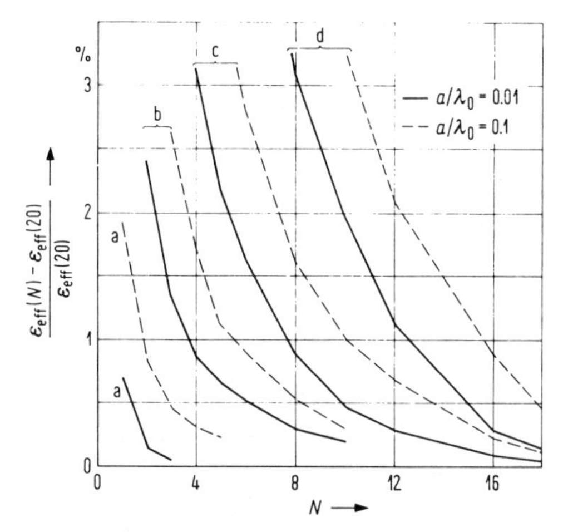

Fig. 3. Convergence of the effective dielectric constant versus the number of terms considered in the series;
a) $\varepsilon_r = 9.0$, $b_L/a = 2$, $c/a = 4$, $e/a = 3.5$, $t/a = 0$,
b) $\varepsilon_r = 9.7$, $b_L/a = 1$, $c/a = 5$, $e/a = 5$, $t/a = 0$,
c) $\varepsilon_r = 9.7$, $b_L/a = 1$, $c/a = 10$, $e/a = 10$, $t/a = 0$,
d) $\varepsilon_r = 9.7$, $b_L/a = 1$, $c/a = 20$, $e/a = 20$, $t/a = 0$.

At first the convergence was investigated. Fig. 3 shows as a function of N the deviation of the effective dielectric constant from the value computed for $N = 20$. The system converges more slowly for higher frequencies. If the size of the outer shield grows, more terms are needed to obtain sufficient accuracy. For the most widely spaced shield (see the contour denoted with d) in Fig. 3) the limiting value for large N is not yet achieved. It should be mentioned, that for the specific arrangement (denoted with a in Fig. 3), which was previously investigated by other authors [1], [5], [6], only $N = 2$ meaning two terms for TM-modes and three terms for TE-modes are needed resulting in a matrix of size 10×10 and in an accuracy of 1%.

Figs. 4 and 5 show the variation of the effective dielectric constant with the size of the shield. The influence of the shield decreases for higher frequencies. Note, that for $c/a = e/a = 8.0$ the characteristics of an open strip are closely approximated and that $N = 10$ gives sufficient accuracy. The results in Fig. 6 were computed for these parameters. Considering Alumina as dielectric the relative permittivity was chosen at $\varepsilon_r = 9.7$. The width of the strip was varied so, that the static characteristic impedance is between 25 Ω and 100 Ω for an infinitely thin strip. The effective dielectric constant decreases with the thickness of the strip. This effect is considerable in particular at low frequencies. But for practical application in printed circuit tech-

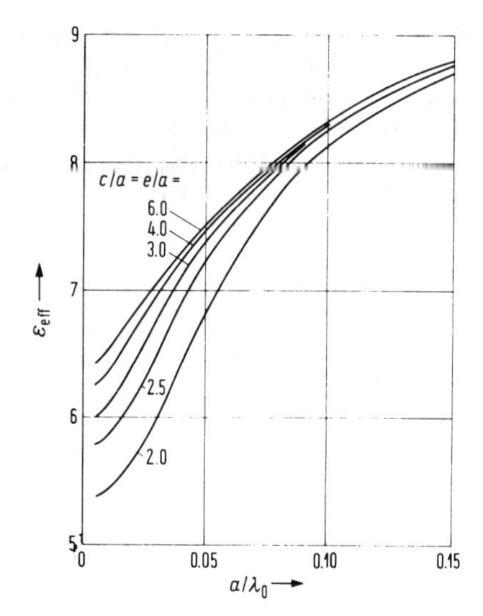

Fig. 4. Effective dielectric constant for various sizes of the outer shield; $\varepsilon_r = 9.7$, $b_L/a = 1$, $t/a = 0$.

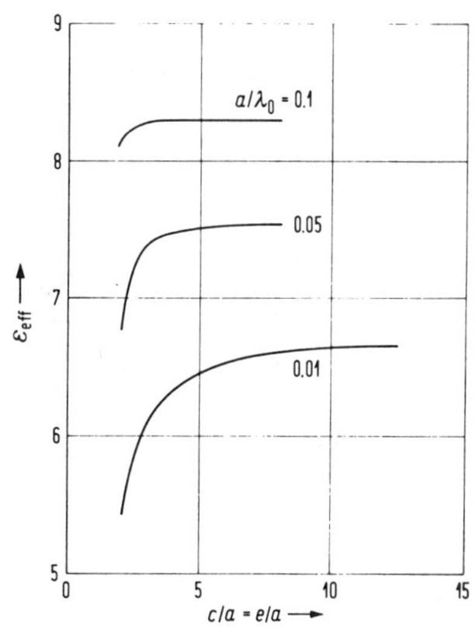

Fig. 5. Effective dielectric constant versus the size of the outer shield; $\varepsilon_r = 9.7$, $b_L/a = 1$, $t/a = 0$.

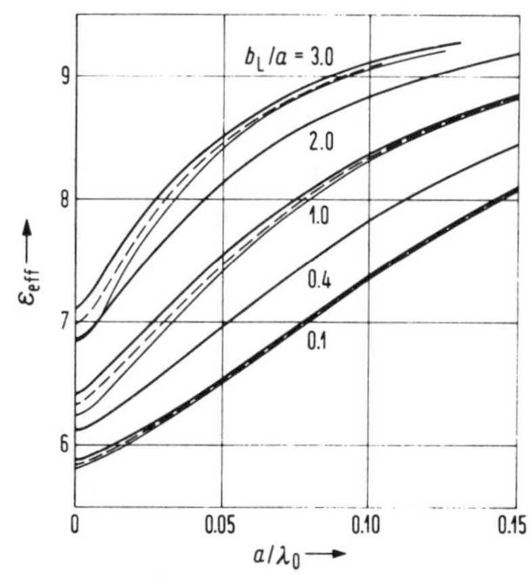

Fig. 6. Effective dielectric constant for a spaciously shielded strip line; $\varepsilon_r = 9.7$, $c/a = e/a = 8$;
——— $t/b_L = 0$,
– – – $t/b_L = 0.05$,
——— $t/b_L = 0.1$.

niques strip thickness is so small compared to the thickness of the dielectric, that its influence may be neglected in most cases. It increases, however, for narrow strips. As an example the low frequency value of ε_{eff} changes in the most unfavourable case about 1% if the strip thickness is 1% of the dielectric thickness ($b_L/a = 0.1$, $t/b_L = 0.1$, $t/a = 0.01$).

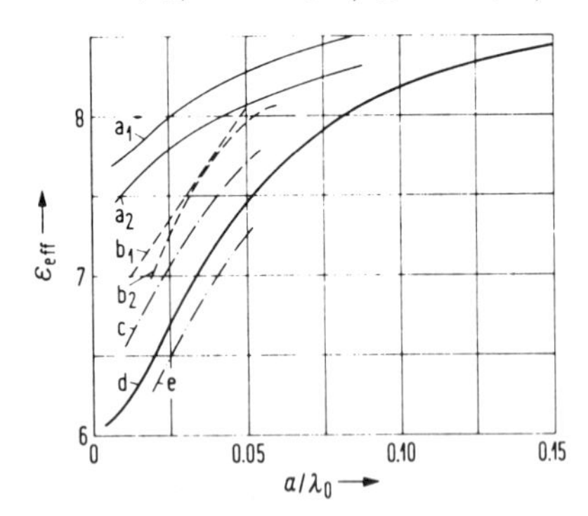

Fig. 7. Comparison with other results; $\varepsilon_r = 9$, $b_L/a = 2$, $c/a = 4$, $e/a = 3.5$, $t/a = 0$;

a₁: 20 terms } Hornsby and Gopinath [5] (Fourier
a₂: 10 terms } analysis),
b₁: coarse mesh points } Hornsby and Gopinath [6]
b₂: fine mesh points } (Finite-difference methods)
c: Mittra and Itoh [1],
d: this method,
e: Fromm [8].

To verify the results the two different arrangements were investigated, which were used by Mittra and Itoh [1] and also by Hornsby and Gopinath [5], [6] and a fourth method based on the concept, that the upper regions 2, 3, 4 are considered as one region, whose orthogonal functions are computed by the Ritz-method [8]. It must be mentioned, that the values from Mittra and Itoh are available only from a graphical representation and therefore may be somewhat in error. The same is valid for the set of parameters given in Fig. 8, where the first two higher modes are investigated. Our results do not agree with Mittra and Itoh for these higher order modes. We believe, however, that our results are more accurate, because the higher order modes are related to the LSE and LSM modes, which can be easily calculated [10] and which are given in the same figure. The higher order modes change to these modes, if the strip vanishes. Furthermore the cut-off frequencies in this and other cases, which were computed, agree with the results given by R. Pregla and W. Schlosser [11].

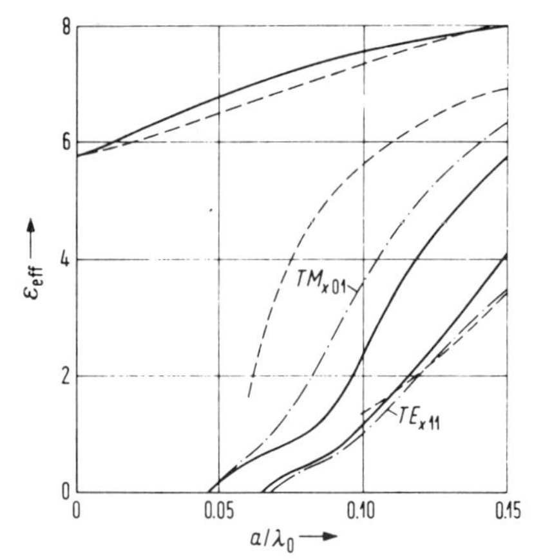

Fig. 8. Higher order modes in a shielded strip line; $\varepsilon_r = 8.875$, $b_L/a = 1$, $c/a = 10$, $e/a = 5$, $t/a = 0$;
— — — Mittra and Itoh [1],
——— this method,
—·—·— without strip.

Acknowledgment

Thanks are due to R. Mittra and T. Itoh, who made some of their unpublished work available to us and to E. Kühn from the Institut für Hochfrequenztechnik of the Technical University Braunschweig, who helped with discussions and suggestions.

(Received October 20th, 1970.)

References

[1] Mittra, R. and Itoh, T., A new technique for the analysis of the dispersion characteristics of microstrip-lines. To be published.

[2] Zysman, G. I. and Varon, D., Wave propagation in microstrip transmission lines. 1969 G-MTT Symposium Digest, pp. 3–9.

[3] Tai Tsun Wu, Theory of the microstrip. J. appl. Phys. **28** [1957], 299–302.

[4] Delonge, P., On Wu's theory of microstrip. Electron. Letters **6** [1970], 541–542.

[5] Hornsby, J. S. and Gopinath, A., Fourier analysis of a dielectric-loaded waveguide with a microstrip line. Electron. Letters **5** [1969], 265–267.

[6] Hornsby, J. S. and Gopinath, A., Numerical analysis of a dielectric-loaded waveguide with a microstrip line — Finite difference methods. Transact. Inst. Elect. Electron. Engrs. MTT-**17** [1969], 684–690.

[7] Veszely, G., On the coupled-mode analysis of the shielded strip-line. Proc. 4th Colloquium of Microwave Communication, Budapest, April 21–24, 1970, Vol. III, pp. ET-29/1–ET-29/5.

[8] Fromm, H. H., Bestimmung der Dispersionscharakteristik der niedrigsten und der ersten höheren Wellenformen auf der Mikrostrip-Leitung. Diplomarbeit am Institut für Hochfrequenztechnik der Technischen Universität Braunschweig, 1969.

[9] Kühn, E. and Schwartz, W.-D., Band-rejection filters for the TE_{01} mode in circular waveguide. AEÜ **25** [1971], to be published.

[10] Unger, H.-G., Elektromagnetische Wellen I. F. Vieweg, Braunschweig 1967, pp. 101–107.

[11] Pregla, R. and Schlosser, W., Waveguide modes in dielectrically supported strip-lines. AEÜ **22** [1968], 379–386.

The Effect of Complex Modes at Finline Discontinuities

ABBAS SAYED OMAR, MEMBER, IEEE, AND KLAUS F. SCHÜNEMANN, SENIOR MEMBER, IEEE

***Abstract*—The effect of ignoring complex modes on the solution of finline discontinuity problems is investigated. It is shown that the modal energy distribution at both sides of the discontinuity may be greatly affected by overlooking complex modes, even if they are not strongly excited. It is also shown that disregarding only one mode of a pair of complex modes, while taking the other into account, results in a contradiction to the principle of complex power continuity across the discontinuity plane. Comparison to measured data is also given to justify the validity of the numerical results.**

I. INTRODUCTION

THE ANALYSIS of discontinuities between planar transmission lines, in particular microstrip lines and finlines, has received increasing interest [1]–[7]. A proper modeling of such discontinuities is of fundamental importance for any successful printed-circuit design. Only finline discontinuity problems will be considered here. Extending the discussion to other planar structures is, however, straightforward.

Two rigorous approaches have been reported for the analysis of finline discontinuity problems. The first one depends on the transverse resonance concept (e.g., [1]–[3]). A determination of high-order finline modes is not needed for this approach. The problem is completely formulated in terms of homogeneously filled rectangular (or parallel-plate) waveguide modes in conjunction with a proper modeling of the tangential field in the metallization plane. This method inherently shows some disadvantages: The effect of the discontinuity is only available with respect to the dominant mode. No information concerning higher order modes can be obtained. A complex discontinuity, which is composed of a number of cascaded simple discontinuities (e.g. steps in the slot width), has then to be analyzed "as a whole." The properties of the individual simple discontinuities cannot, in general, be used to construct an accurate solution of the complex discontinuity due to the lack of information about high-order modes. The "as a whole" analysis of complex discontinuities may need a large number of basis functions to properly model the tangential field in the plane of the fins. This leads to dealing with oversized matrices, which greatly degrades the numerical efficiency of the method.

Manuscript received March 13, 1986; revised July 24,1986. This work was supported in part by the Deutsche Forschungsgemeinschaft (DFG).

The authors are with the Technische Universität Hamburg-Harburg, Arbeitsbereich Hochfrequenztechnik, Postfach 90 14 03, D-2100 Hamburg 90, West Germany.

IEEE Log Number 8610828.

The second approach depends on the modal expansion concept (e.g. [4]–[6]). It is an application of the method of moments, in which both the basis and the testing functions are the electromagnetic fields of the normal modes of propagation at both sides of the discontinuity. Another choice of basis and testing functions has been suggested in [7] following the method which has been presented in [8]. In the authors' opinion, modal fields are in general the best choice for basis and testing functions, because they individually satisfy the same equations (Maxwell's equations) and boundary conditions for the expanded field. Continuity of complex power across the discontinuity plane and hence the unitarity of the scattering matrix are also guaranteed in the modal expansion method, as has been shown in [5]. Generalized scattering and/or transmission matrices, which contain all information about the dominant as well as the higher order modes, are obtained for simple discontinuities (e.g. steps). Complex discontinuities can be analyzed by processing the generalized scattering or transmission matrices characterizing the individual simple steps. The main problem in this method, then, is the accurate determination of an approximately complete set of finline modes.

As has already been shown [9], the singular integral equation (SIE) technique is very efficient for determining such a set. It has also been shown that complex modes can be supported by finlines, so that ignoring these modes in constructing an approximately complete set of finline modes may lead to erroneous solutions.

The possibility of complex modes in a circular waveguide containing a coaxial dielectric rod were first predicted in [10]. It has been shown there that the appearance of a backward-wave mode in a certain frequency band is associated with the appearance of complex modes in a lower frequency band. It has also been shown that complex modes can occur under certain conditions even if there is no frequency range in which backward-wave modes can propagate. More theoretical and experimental investigations on complex modes in dielectric-loaded circular waveguides have been reported, e.g. in [11]–[14].

Complex modes in a shielded rectangular dielectric image guide, which can be considered as a rectangular waveguide with a rectangular dielectric insert, have been reported in [15] and [16]. We have recently shown that complex modes can exist in finlines [9], [17].

This paper addresses the effect of overlooking complex modes on the solution of finline discontinuity problems.

Reprinted from *IEEE Trans. Microwave Theory Tech.*, vol. MTT-34, no. 12, pp. 1508–1514, Dec. 1986.

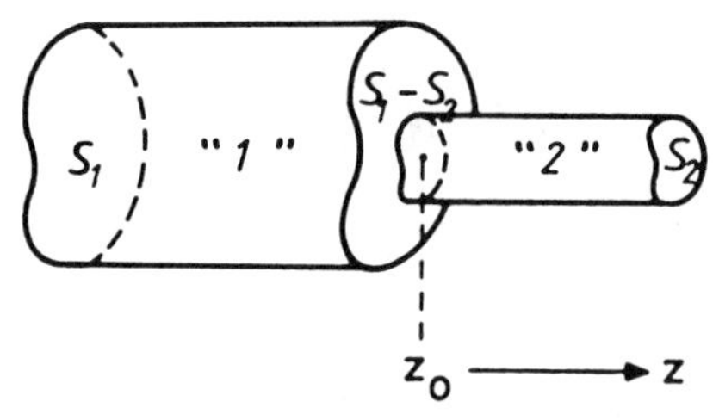

Fig. 1. A boundary-reduction discontinuity between two general waveguides.

II. Basic Formulation

The formulation of the modal expansion method is well documented in the literature and can be found in, e.g., [18] and [19] for homogeneously filled waveguides and in [5] for inhomogeneously filled waveguides. The describing equations can be equally obtained as a result of minimizing certain error quantities, as will be shown in the following.

Consider the boundary reduction discontinuity which is shown in Fig. 1. Waveguides 1 and 2 are assumed to be quite general, except for the restriction that both have discrete modal spectra; this simplifies the discussion to some extent, because all field expansions include in this case only summations and no integrations. Expanding the transverse fields of guides 1 and 2 at the discontinuity plane $z = z_0$ with respect to N-dimensional and M-dimensional sets, respectively, of their modes results in

$$E^{(1)} = \sum_{n=1}^{N} V_n^{(1)} e_n^{(1)}$$

$$H^{(1)} = \sum_{n=1}^{N} I_n^{(1)} h_n^{(1)} \tag{1}$$

$$E^{(2)} = \sum_{m=1}^{M} V_m^{(2)} e_m^{(2)}$$

$$H^{(2)} = \sum_{m=1}^{M}{}' I_m^{(2)} h_m^{(2)} \tag{2}$$

where $e_n^{(1)}(h_n^{(1)})$ and $e_m^{(2)}(h_m^{(2)})$ are the transverse electric (magnetic)-field vectors of the nth mode in guide 1 and the mth mode in guide 2, respectively, and $V_n^{(1)}$, $I_n^{(1)}$, $V_m^{(2)}$, and $I_m^{(2)}$ are expansion coefficients. Looking from side 1, the errors $R_e^{(1)}$ and $R_h^{(1)}$ in the electric and magnetic fields, respectively, are given by

$$R_e^{(1)} = \begin{cases} E^{(1)} - E^{(2)} & \text{on } (S_2) \\ E^{(1)} & \text{on } (S_1 - S_2) \end{cases}$$

$$R_h^{(1)} = \begin{cases} H^{(1)} - H^{(2)} & \text{on } (S_2) \\ H^{(1)} - J_s & \text{on } (S_1 - S_2) \end{cases} \tag{3}$$

where S_1 and S_2 are the cross sections of guides 1 and 2, respectively, and J_s is the induced surface current at $(S_1 - S_2)$. Two error quantities $F_e^{(1)}$ and $F_h^{(1)}$ can now be defined as

$$F_e^{(1)} = \int_{S_1} \left(R_e^{(1)} \times H^{(1)} \right) \cdot ds$$

$$F_h^{(1)} = \int_{S_1} \left(E^{(1)} \times R_h^{(1)} \right) \cdot ds. \tag{4}$$

The error quantity $F_h^{(1)}$ contains the unknown surface current J_s, and hence will not be used, in order not to increase the number of unknowns. The error quantity $F_e^{(1)}$ can be regarded as a functional of $H^{(1)}$ which must be minimized. This means that the expansion coefficients $I_n^{(1)}$ of $H^{(1)}$ must be adjusted in order to minimize $F_e^{(1)}$. Substituting (1) and (3) into (4) and carrying out the necessary integrations, one obtains

$$F_e^{(1)} = \sum_{n=1}^{N} I_n^{(1)} \left(P_n V_n^{(1)} - \sum_{m=1}^{M} A_{nm} V_m^{(2)} \right) \tag{5}$$

where

$$\int_{S_1} \left(e_n^{(1)} \times h_m^{(1)} \right) \cdot ds = P_n \delta_{nm} \tag{6}$$

$$\int_{S_2} \left(e_m^{(2)} \times h_n^{(1)} \right) \cdot ds = A_{nm} \tag{7}$$

and δ_{nm} is the Kronecker delta. Performing the usual Fourier expansion procedure by differentiating $F_e^{(1)}$ with respect to $I_n^{(1)}$ $(n = 1, 2, \cdots, N)$ and equating the result to zero, one obtains

$$[\Lambda_p] V^{(1)} = [A] V^{(2)} \tag{8}$$

where $[\Lambda_p]$ is an $(N \times N)$ diagonal matrix with elements P_n; $[A]$ is an $(N \times M)$ matrix with elements A_{nm}; and $V^{(1)}$ $(V^{(2)})$ is an $N(M)$-dimensional column vector with elements $V_n^{(1)}(V_m^{(2)})$.

Looking from side 2, the errors in the electric and magnetic fields are given by

$$R_e^{(2)} = E^{(2)} - E^{(1)}$$

$$R_h^{(2)} = H^{(2)} - H^{(1)}. \tag{9}$$

Again, two error quantities can be defined, namely

$$F_e^{(2)} = \int_{S_2} \left(R_e^{(2)} \times H^{(2)} \right) \cdot ds$$

$$F_h^{(2)} = \int_{S_2} \left(E^{(2)} \times R_h^{(2)} \right) \cdot ds. \tag{10}$$

The error quantities $F_e^{(2)}$ and $F_h^{(2)}$ can be regarded as functionals of $H^{(2)}$ and $E^{(2)}$, respectively, which must be minimized. Due to the completeness properties of the normal modes, it can be proven that in the limiting case when N, M tend to infinity, minimizing $F_e^{(2)}$ with respect to the expansion coefficients $I_m^{(2)}$ of $H^{(2)}$ results in a matrix equation, which is equivalent to (8). Hence, $F_h^{(2)}$ must be minimized with respect to the expansion coefficients $V_m^{(2)}$ of $E^{(2)}$. $F_h^{(2)}$ is readily proved to be given by

$$F_h^{(2)} = \sum_{m=1}^{M} V_m^{(2)} \left(Q_m I_m^{(2)} - \sum_{n=1}^{N} A_{nm} I_n^{(1)} \right) \quad (11)$$

where

$$\int_{S_2} \left(e_n^{(2)} \times h_m^{(2)} \right) \cdot ds = Q_m \delta_{mn}. \quad (12)$$

Differentiating $F_h^{(2)}$ with respect to $V_m^{(2)}$ $(m = 1, 2, \cdots, M)$ and equating the result to zero, one obtains

$$[\Lambda_Q] I^{(2)} = [A]' I^{(1)} \quad (13)$$

where $[\Lambda_Q]$ is an $(M \times M)$ diagonal matrix with elements Q_m, and $I^{(1)}(I^{(2)})$ is an $N(M)$-dimensional column vector with elements $I_n^{(1)}(I_m^{(2)})$. Equations (8) and (13) are the characteristic equations of the modal expansion method. They are equivalent to, e.g., [5, eq. (3)], which has been derived using another approach.

From the above discussion, it is easily seen that the expansion coefficients of the fields at both sides of the discontinuity are so adjusted that the error quantities $F_e^{(1)}$ and $F_h^{(2)}$ are minimized. This can be viewed as "a similarity balance" process, in which the fields at both sides of the discontinuity are "similar" with respect to minimum error quantities $F_e^{(1)}$ and $F_h^{(2)}$. The mode coupling coefficient A_{nm} defined by (7) represents a measure of the degree of similarity between the nth mode of guide 1 and the mth mode of guide 2. Two different modes in either guide 1 or guide 2 are then completely "dissimilar" due to the orthogonality relations (6) and (12).

III. Effect of Ignoring Modes at Either Side of the Discontinuity

According to the similarity balance concept discussed above, the nth mode excited in guide 1 (which will be called mode (a)) is balanced by exciting a similar field in guide 2. This similar field is, in general, composed of a superposition of all the M modes of guide 2, the magnitude of each depending on its degree of similarity to mode (a). In particular, the magnitude of a mode with a high degree of similarity will dominate the magnitudes of the other, less similar modes. This similarity balance holds for each of the N modes of guide 1.

Let us assume now that the mth mode of guide 2 (which will be called mode (b)) has the largest degree of similarity to mode (a). Omitting mode (b) from the M modes of guide 2 can only be compensated by increasing the magnitudes of modes that are less similar to mode (a), in order to restore the similarity balance. This will disturb the modal distributions (and hence the stored energy) at both sides of the discontinuity. It is important to note that this disturbance does not necessarily depend on how strongly mode (a) is excited, because balancing a weakly excited mode in guide 1 may require strongly excited modes in guide 2 which have a very weak degree of similarity to that mode. Omitting, however, both mode (a) and mode (b) will have a much smaller effect on the modal distributions, in particular, if both are just weakly excited.

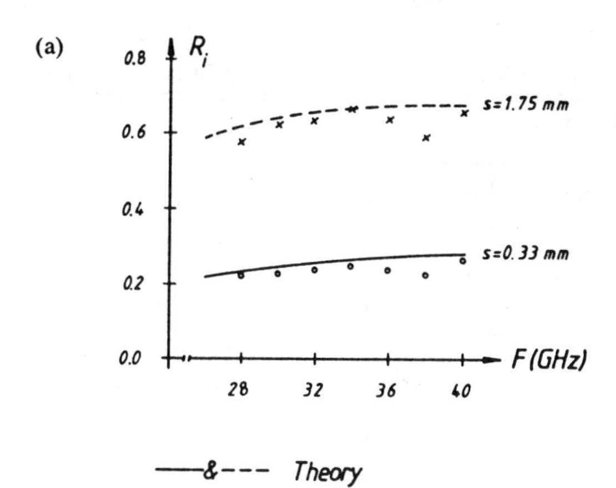

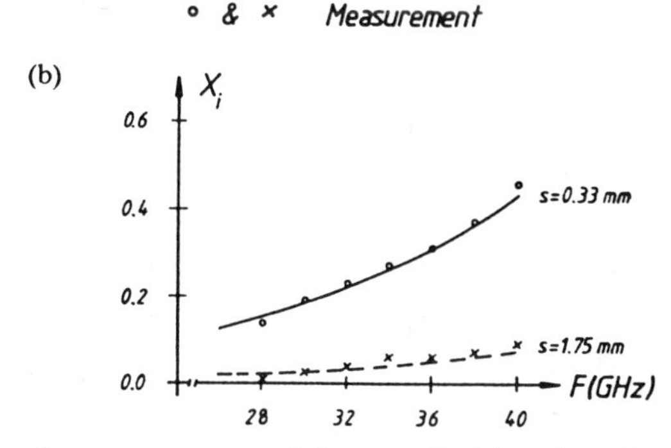

Fig. 2. Frequency response of the normalized input impedance of the waveguide–finline junction shown in Fig. 3. (a) Normalized input resistance. (b) Normalized input reactance.

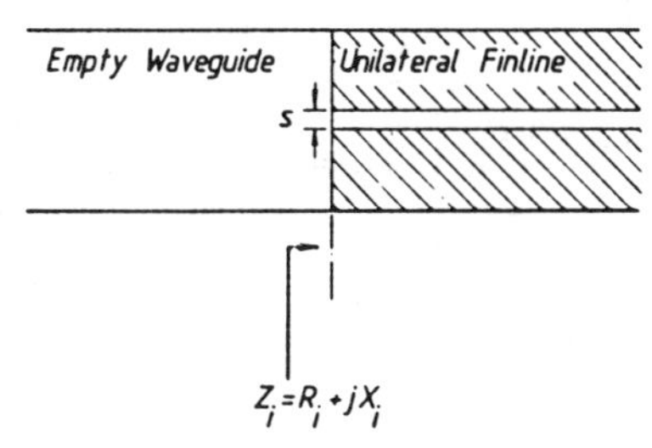

Fig. 3. A waveguide–finline discontinuity. Parameters: WR-28 housing; substrate thickness = 0.254 mm; substrate dielectric constant = 2.22.

IV. Effect of Overlooking Complex Modes on Finline Discontinuities

As has already been shown in [9], finline modes change their nature as any of the finline parameters changes. A pair comprising an inductive and a capacitive evanescent mode may become a complex pair, and vice versa, as, e.g., the slot width changes. If we would analyze a discontinuity using only usual (noncomplex) modes, it can happen that a pair (or more) of modes at one side of the discontinuity is noncomplex, while the corresponding pair at the other side, which has the largest degree of similarity, is a complex one. Both the modal distributions and the stored energy at both sides of the discontinuity would then be greatly affected, even if the former pair is not strongly excited. The situation is much more favorable if both pairs are complex, so that both would be ignored in the matching process.

Another interesting effect is that of disregarding only one mode of a pair of complex modes, while taking the

TABLE I
MODAL ENERGY DISTRIBUTION OF THE DISCONTINUITY SHOWN IN FIG. 4 WITH AND WITHOUT TAKING COMPLEX MODES INTO ACCOUNT

(a) at finline 1

mode order	1	2	3	4	5	6	7	8	9	10
ß 1/(mm)	0.4853	-j0.6209	-j1.1519	-j1.6490	-j1.6513	-j1.7001	-j1.7043	-j1.8729	-j1.8770	-j2.0855
energy with C.M.	+0.2554	+0.0046	+0.1409	+0.0001	+0.0043	-0.1749	+0.0976	+0.0009	-0.0009	-0.0632
energy without	-0.0003	+0.0044	+0.1204	+0.0001	+0.0033	-0.2510	+0.1377	+0.0009	-0.0008	-0.0911

mode order	11	12	13	14	15	16	17	18	19	20
ß 1/(mm)	-j2.1089	-j2.1094	-j2.4184	-j2.4409	-j2.5746	-j2.6949	-j2.7514	-j3.0220	-j3.1225	-j3.1641
energy with C.M.	+0.0610	+0.0192	+0.0005	-0.0007	+0.0065	-0.0279	+0.0081	+0.0724	+0.0030	-0.0012
energy without	+0.0000	+0.0000	+0.0006	-0.0011	+0.0037	-0.0458	+0.0061	+0.0264	+0.0005	-0.0392

(b) at finline 2

mode order	1	2	3	4	5	6	7	8	9	10
ß 1/(mm)	0.7024	-j0.6037	-j0.7271	-j1.5945	-j1.6488	-j1.6772	-j1.7427	0.0073 -j1.8699	-0.0073 -j1.8699	-j1.8886
energy with C.M.	-0.8219	+0.0630	+0.2718	+0.0381	+0.0000	+0.0054	+0.0375	-0.0008	-0.0008	-0.0016
energy without	-0.6629	+0.0879	+0.3971	+0.1023	+0.0000	+0.0198	+0.1278	-------	-------	-0.0000

mode order	11	12	13	14	15	16	17	18	19	20
ß 1/(mm)	-j1.9680	-j2.3999	-j2.4095	-j2.4667	-j2.4745	-j2.5318	-j2.6976	-j3.0679	-j3.1139	-j3.2065
energy with C.M.	+0.0006	+0.0164	-0.0174	-0.0035	+0.0181	+0.0002	+0.0175	+0.0035	-0.0167	-0.0153
energy without	+0.0221	+0.0058	-0.0132	-0.0053	+0.0022	+0.0002	+0.0089	+0.0032	-0.0058	-0.0027

Operating frequency = 30 GHz.

other into account. As has been shown in [9], one mode of a pair of complex modes propagates in the same direction in which it is attenuated (let it be called mode (c)), while the other propagates opposite to the direction in which it is damped (let it be called mode (d)). Each carries, by itself, neither active nor reactive power. Let us now assume that both modes have been excited in guide 2. Mode (c) (mode (d)) represents for guide 1 an energy loss (gain) mechanism, which occurs in guide 2. Guide 1 does not "know" that each of these modes carries, by itself, no power. If mode (d) (mode (c)) is disregarded while mode (c) (mode (d)) is retained in the matching process, the amplitudes of the different modes in guide 1 are adjusted to account for the energy loss (gain), which occurs in guide 2. In other words, guide 2 appears lossy (active) if it is looked at from guide 1. The complex power, which is calculated in guide 1, takes this energy loss (gain) into account. On the other hand, the complex power, which is calculated in guide 2, does not "feel" any energy loss (gain) due to the absence of the power carried by mode (c) (mode (d)). This consequently results in a discontinuity in the complex power across the junction. This effect cannot be compared to the truncation effect (i.e., matching finite number of modes at both sides of the discontinuity), because continuity of complex power is independent of the number of modes which enter the matching process, as has been shown in [5].

V. Numerical Results

In order to check the accuracy of the numerical results, comparison to measured data in *Ka*-band is demonstrated in Fig. 2, which shows the frequency response of the resistive (Fig. 2(a)) and reactive (Fig. 2(b)) parts of the normalized input impedance of the waveguide–finline junction shown in Fig. 3. For the two indicated cases, convergence of the numerical results has been achieved by using only five modes at each side of the discontinuity. For both slot widths ($s = 1.75$ mm, $s = 0.33$ mm), the order of the first complex modes is larger than five; hence, complex modes have negligible influence on the numerical results for these two cases. The reader can refer to [5] for further information about the convergence problem.

Table I shows the stored energy distributions at both sides of the finline discontinuity shown in Fig. 4, computed with and without taking complex modes into consideration. Due to the large slot width ratio ($s_1/s_2 = 35$), convergence had to be achieved by using 20 modes at both sides of the discontinuity. The incident field is the dominant mode of finline 1 carrying unit power. The dominant mode of finline 2 shows a standing wave pattern within the distance l between the discontinuity and the open circuit. It stores capacitive energy because l is slightly larger than half a guide wavelength for this mode ($l = 5.0$ mm, $\lambda_{g2} =$ 8.945 mm). If complex modes are omitted, the stored energy is calculated with an error of 19.34 percent. The total energy, on the other hand, which is stored in both the dominant mode and all higher order modes turns out to amount to -0.4059 taking complex modes into account while it is $+0.1272$ if these modes are omitted. The corresponding normalized input impedance at the discontinuity plane is $-j0.1039$ with and $+j0.0001$ without complex modes. The error is larger than 100 percent because the effect of overlooking complex modes has changed the capacitive nature of the structure between the discontinuity and the open circuit into an inductive one. It should be pointed out that this severe error is due to overlooking only one pair of complex modes (namely the eighth and ninth modes of finline 2), which are only weakly excited. The error would be much more disastrous if many complex mode pairs existed on one or both sides of the discontinuity.

The frequency response of the normalized input reactance of the same discontinuity (seen at the discontinuity plane) is plotted in Fig. 5, computed with and without complex modes. The irregularity of the curve representing the computation without complex modes is due to the fact

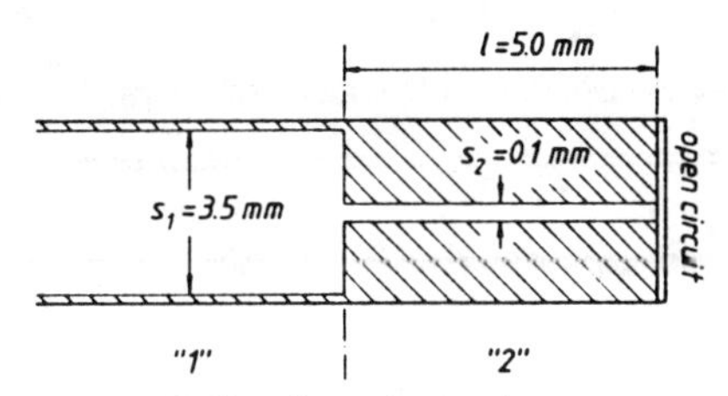

Fig. 4. A unilateral finline discontinuity. Parameters: as in Fig. 3.

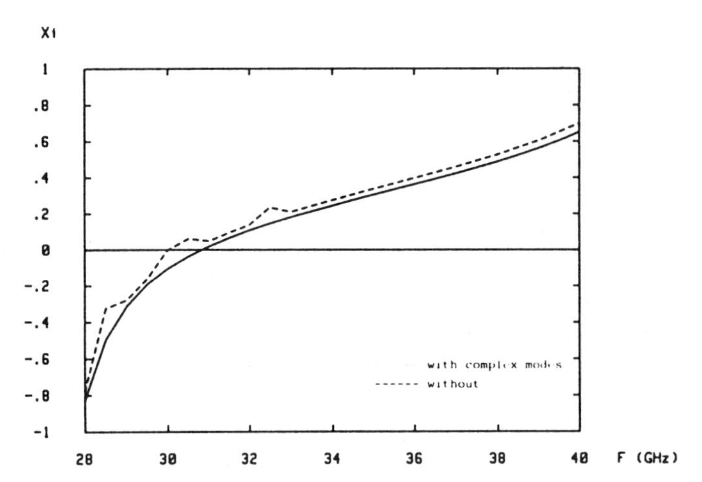

Fig. 5. Frequency response of the normalized input reactance of the finline discontinuity shown in Fig. 4.

that some of the modes change their nature (from complex to noncomplex and vice versa) as the frequency changes. A pair of complex modes which is not taken into account in a certain frequency range may become a noncomplex pair in another frequency range. The reader can refer to [16] for a deeper understanding of this phenomenon. Apart from this irregularity, a deviation of about 2.6 percent in the resonance frequency (at which $X_i = 0$) can be observed. This is a crucial error for any narrow-band application. The deviation in the upper half of the *Ka*-band is at least 8 percent, while the convergence of the results has been achieved within only 1 percent.

Finally, in order to demonstrate the effect of disregarding only one mode of a pair of complex modes while taking the other into account, the structure shown in Fig. 4 (with the distance l being shortened to $l = 0.5$ mm) has been analyzed. The complex powers carried by the first eight modes at both sides of the discontinuity once by disregarding the ninth mode and once by disregarding the eighth mode of finline 2 are tabulated in Tables II and III, respectively. Although the structure has been assumed lossless and passive, the complex power carried by the dominant mode of finline 1 has an active part, which is negative (positive) in Table II (III) in order to account for the disappearance of the ninth (eighth) mode of finline 2. The dominant mode of finline 2 stores capacitive energy due to its standing wave nature between the discontinuity and the nearby open circuit. The complex power carried individually by the eighth and ninth modes of finline 2 vanishes as expected. The total complex powers carried by

TABLE II
COMPLEX POWER DISTRIBUTION OF THE DISCONTINUITY SHOWN IN FIG. 4 WITH THE NINTH MODE OF FINLINE 2 DISREGARDED

(a) finline 1

mode order	1	2	3	4	5	6	7	8	Total
ß 1/(mm)	0.4853	-j0.6209	-j1.1519	-j1.6490	-j1.6513	-j1.7001	-j1.7043	-j1.8729	
complex power	- 0.0230 -j0.5002	+j0.0134	+j0.3870	+j0.0002	+j0.0112	-j0.4652	+j0.2661	+j0.0191	- 0.0230 -j0.2684

(b) finline 2

mode order	1	2	3	4	5	6	7	8	Total
ß 1/(mm)	0.7024	-j0.6037	-j0.7271	-j1.5945	-j1.6488	-j1.6772	-j1.7427	+ 0.0073 -j1.8699	
complex power	-j0.8674	+j0.0909	+j0.4834	+j0.2273	+j0.0000	+j0.0311	+j0.3401	+j0.0000	+ 0.0000 +j0.3054

Distance l shortened to 0.5 mm.

TABLE III
COMPLEX POWER DISTRIBUTION OF THE DISCONTINUITY SHOWN IN FIG. 4 WITH THE EIGHTH MODE OF FINLINE 2 DISREGARDED

(a) finline 1

mode order	1	2	3	4	5	6	7	8	Total
ß 1/(mm)	0.4853	-j0.6209	-j1.1519	-j1.6490	-j1.6513	-j1.7001	-j1.7043	-j1.8729	
complex power	+ 0.0239 -j0.5195	+j0.0139	+j0.4019	+j0.0002	+j0.0116	-j0.4831	+j0.2764	+j0.0198	+ 0.0239 -j0.2788

(b) finline 2

mode order	1	2	3	4	5	6	7	9	Total
ß 1/(mm)	0.7024	-j0.6037	-j0.7271	-j1.5945	-j1.6488	-j1.6772	-j1.7427	- 0.0073 -j1.8699	
complex power	-j0.9008	+j0.0944	+j0.5020	+j0.2360	+j0.0000	+j0.0323	+j0.3532	+j0.0000	+ 0.0000 +j0.3171

Distance l shortened to 0.5 mm.

the modes of finlines 1 and 2 are not equal, which contradicts the principle of complex power continuity.

ACKNOWLEDGMENT

The authors thank J. K. Piotrowski for supplying the measured data and C. Bühst for preparing the manuscript.

REFERENCES

[1] J. B. Knorr, "Equivalent reactance of a shorting septum in a finline: Theory and experiment," *IEEE Trans. Microwave Theory Tech.*, vol. MTT-29, pp. 1196–1202, 1981.
[2] R. Sorrentino and T. Itoh, "Transverse resonance analysis of finline discontinuities," *IEEE Trans. Microwave Theory Tech.*, vol. MTT-32, pp. 1633–1638, 1984.
[3] J. B. Knorr and J. C. Deal, "Scattering coefficients of an inductive strip in a finline: Theory and experiment," *IEEE Trans. Microwave Theory Tech.*, vol. MTT-33, pp. 1011–1017, 1985.
[4] H. El-Hennawy and K. Schünemann, "Hybrid finline matching structures," *IEEE Trans. Microwave Theory Tech.*, vol. MTT-30, pp. 2132–2138, 1982.
[5] A. S. Omar and K. Schünemann, "Transmission matrix representation of finline discontinuities," *IEEE Trans. Microwave Theory Tech.*, vol. MTT-33, pp. 765–770, 1985.
[6] M. Helard, J. Citerne, O. Picon, and V. F. Hanna, "Theoretical and experimental investigation of finline discontinuities," *IEEE Trans. Microwave Theory Tech.*, vol. MTT-33, pp. 994–1003, 1985.
[7] K. J. Webb and R. Mittra, "Solution of the finline step-discontinuity problem using the generalized variational technique," *IEEE Trans. Microwave Theory Tech.*, vol. MTT-33, pp. 1004–1010, 1985.
[8] H. Auda and R. F. Harrington, "A moment solution for waveguide junction problems," *IEEE Trans. Microwave Theory Tech.*, vol. MTT-31, pp. 515–520, 1983.
[9] A. S. Omar and K. Schünemann, "Formulation of singular integral equation technique for planar transmission lines," *IEEE Trans. Microwave Theory Tech.*, vol. MTT-33, pp. 1313–1322, 1985.

[10] P. J. B. Clarricoats and B. C. Taylor, "Evanescent and propagating modes of dielectric-loaded circular waveguide," *Proc. Inst. Elec. Eng.*, vol. 111, pp. 1951–1956, 1964.

[11] P. J. B. Clarricoats and K. R. Slinn, "Complex modes of propagation in dielectric loaded circular waveguide," *Electron Lett.*, vol. 1, pp. 145–146, 1965.

[12] S. B. Rayevskiy, "Some properties of complex waves in a double-layer, circular, shielded waveguide," *Radio Eng. Electron Phys.*, vol. 21, pp. 36–39, 1976.

[13] V. A. Kalmyk, S. B. Rayevskiy, and V. P. Ygryumov, "An experimental verification of existence of complex waves in a two-layer, circular, shielded waveguide," *Radio Eng. Electron Phys.*, vol. 23, pp. 17–19, 1978.

[14] H. Katzier and J. K. Lange, "Grundlegende Eigenschaften komplexer Wellen am Beispiel der geschirmten kreiszylindrischen dielektrischen Leitung," *Arch. Elek. Übertragung.*, vol. AEÜ-37, pp. 1–5, 1983.

[15] U. Crombach, "Complex waves on shielded lossless rectangular dielectric image guide," *Electron Lett.*, vol. 19, pp. 557–558, 1983.

[16] J. Strube and F. Arndt, "Rigorous hybrid-mode analysis of the transition from rectangular waveguide to shielded dielectric image guide," *IEEE Trans. Microwave Theory Tech.*, vol. MTT-33, pp. 391–401, 1985.

[17] A. S. Omar and K. Schünemann, "New type of evanescent modes in finlines," in *Proc. 15th EuMC* (Paris), 1985, pp. 317–322.

[18] R. Safavi-Naini and R. H. McPhie, "On solving waveguide junction scattering problems by the conservation of complex power technique," *IEEE Trans. Microwave Theory Tech.*, vol. MTT-29, pp. 337–343, 1981.

[19] R. de Smedt and B. Denturck, "Scattering matrix of junctions between rectangular waveguides," *Proc. Inst. Elec. Eng.*, vol. 130, Pt. H, pp. 183–190, 1983.

Part 9
Transverse Resonance Techniques

THE concept of transverse resonance is a rather old one. It originated as an application of the microwave circuit formalism in the direction perpendicular to the actual power flow in a cylindrical waveguide. The method was originally adopted for the analysis of certain types of waveguide configurations, such as dielectric loaded or ridged rectangular guides, that are derived from conventional waveguides modified by the addition of some discontinuities or obstacles placed across a transverse direction. Using elementary transmission-line theory, the characteristic equation of the waveguide can immediately be written as the resonance condition of the transverse equivalent network, provided the characterization of the discontinuity is available in the form of an equivalent lumped element. A historical paper by Cohn (Paper 9.1), concerning the analysis of ridge waveguides, is reprinted here as representative of this type of approach. This has also been extensively used in the characterization of leaky wave antennas obtained from slitted waveguides (see, for instance, the classic work by Goldstone and Oliner [9.1]).

When the characterization of the discontinuity is available with good approximation, this form of the transverse resonance technique (TRT) is of invaluable utility in providing the propagation properties of the guiding structure with minimum computational effort. It has, however, some obvious limitations. First and foremost, higher order mode interactions are neglected between longitudinal discontinuities and waveguide walls. Second, the method relies on the characterization of the longitudinal discontinuity, which is generally available only for given incident field distributions such as the principal TEM mode of the parallel-plate guide or the dominant TE_{10} mode of the rectangular guide. Third, the method does not permit the evaluation of the complete modal spectrum of the waveguide. Finally, the method does not allow the computation of the local field distribution close to the discontinuity, which may be important in some applications. To remove such limitations it is necessary to use a rigorous representation of the discontinuities.

The way in which the transverse resonance concept can be generalized into a rigorous analysis technique is illustrated in a very illuminating paper by Peng and Oliner (Paper 9.2), where the modal spectrum of open dielectric waveguides is evaluated. The TRT is applied first to determine the mode function and propagation wavenumber of a layered dielectric structure. A generalized TRT is then applied to determine the guiding characteristics of a dielectric strip waveguide, considered as the cascade connection of two dielectric steps in the transverse direction.

The generalized TRT is usefully adopted in the analysis of planar structures. The basic idea in the transverse resonance approach is that the boundary value problem can more easily be formulated as if the propagation would occur in the transverse instead of longitudinal direction, adopting a set of modes differing from the usual TE and TM modes with respect to the longitudinal axis. In the transverse direction, in fact, layered structures are seen as the cascade of homogeneous regions where TE and TM modes are uncoupled. In another historical paper [9.2], Cohn converted the analysis of the slotline into that of a capacitive iris in a rectangular waveguide formed by enclosing a slotline section $\lambda_g/4$ between two pairs of conducting walls. This idea is a useful one to reduce the complexity of a problem and was actually followed later in a number of analyses of printed transmission lines. In practice, it leads to a special form of modal analysis applied in the transverse direction of a transmission line. It differs from the conventional TRT because the transverse discontinuity is not represented by a lumped element, but is analyzed by a rigorous mode-matching procedure. The equivalent lumped element can now be replaced by a generalized network with a (theoretically) infinite number of ports. Cohn's method was limited by some approximations on the slot field distribution. A rigorous TRT following the same guidelines was years later applied to the analysis of coplanar waveguides on semiconductor substrates. The essence of the method is briefly outlined in Paper 9.3, while the results of the analysis have been reported in [9.3]. TE and TM waves on the various layers of the planar structure are coupled only at the plane of the metallization. In this form, the TRT can be simply viewed as a mode-matching procedure applied in the transverse direction. A quite similar technique, transverse modal analysis, was formulated by Yee [9.4].

The idea of viewing the cross section of a shielded microstrip or a finline structure as the cascade of a number of transmission lines in the transverse direction was adopted by Arndt and Paul [9.5] and applied recently to finlines by Vahldieck and Bornemann [9.6]. In these papers, the field expansions are made in terms of TE and TM modes with respect to the axial direction but propagating in the transverse direction. An efficient unified analysis for isotropic and anisotropic planar lines was developed recently by Mansour and MacPhie (Paper 9.4).

The latest extension of the concepts outlined is the transverse resonance analysis of discontinuities, which was first developed for finline discontinuities (Paper 9.5) by Sorrentino and Itoh. To perform an analysis of the structure in the transverse direction it was necessary to enclose the discontinuity between conducting walls so as to create a resonant cavity. The analysis problem can thus be converted into that of discontinuity in a rectangular waveguide. The parameters of the discontinuity are then easily computed via the resonances of the cavity. One of the advantages of this technique is that no complex modes are involved in the analysis, which is simply performed in terms of TE and TM modes. The characterization of the discontinuity, however, is made only in terms of the dominant mode. The method has also been applied to other types of discontinuities [9.7].

REFERENCES

[9.1] L. O. Goldstone and A. A. Oliner, "Leaky-wave antennas I: Rectangular waveguides," *IRE Trans. Antennas Propagat.*, vol. AP-7, pp. 307–319, Oct. 1959.

[9.2] S. B. Cohn, "Slot line on a dielectric substrate," *IEEE Trans. Microwave Theory Tech.*, vol. MTT-17, pp. 768–778, Oct. 1969.

[9.3] R. Sorrentino, G. Leuzzi, and A. Silbermann, "Characteristics of metal-insulator-semiconductor waveguides for monolithic microwave circuits," *IEEE Trans. Microwave Theory Tech.*, vol. MTT-32, pp. 410–416, Apr. 1984.

[9.4] H.-Y. Yee, "Transverse modal analysis of printed circuit transmission lines," *IEEE Trans. Microwave Theory Tech.*, vol. MTT-33, pp. 808–816, Sept. 1985.

[9.5] F. Arndt and G. U. Paul, "The reflection definition of the characteristic impedance of microstrips," *IEEE Trans. Microwave Theory Tech.*, vol. MTT-27, pp. 724–731, Aug. 1979.

[9.6] R. Vahldieck and J. Bornemann, "A modified mode-matching technique and its application to a class of quasi-planar transmission lines," *IEEE Trans. Microwave Theory Tech.*, vol. MTT-33, pp. 916–926, Oct. 1985.

[9.7] T. Uwano, R. Sorrentino, and T. Itoh, "Characterization of stripline crossing by transverse resonance analysis," *IEEE Trans. Microwave Theory Tech.*, vol. MTT-35, pp. 1369–1376, Dec. 1987.

Properties of Ridge Wave Guide*

SEYMOUR B. COHN†, MEMBER, I.R.E.

Summary—**Equations and curves giving cutoff frequency and impedance are presented for rectangular wave guide having a rectangular ridge projecting inward from one or both sides. It is shown that ridge wave guide has a lower cutoff frequency and impedance and greater higher-mode separation than a plain rectangular wave guide of the same width and height. The cutoff frequency equation is fairly accurate for any practical cross section. The impedance equation is strictly accurate only for an extremely thin cross section. Values found by the use of this equation have, however, been found to check experimental values very closely. A number of uses for this type of wave guide are suggested.**

I. Applications

THE CROSS-SECTIONAL shape of ridge wave guide is shown in Fig. 1. This type of wave guide is briefly described in a text by Ramo and Whinnery,[1] where a simple method of calculating the cutoff frequency is given. That method is used in this paper.

The lowered cutoff frequency, lowered impedance, and wide bandwidth free from high-mode interference obtainable with ridge wave guide make it useful in many ways. A few uses are listed below:

(a) It is useful as transmission wave guide, where a wide frequency range must be covered, and where only the fundamental mode can be tolerated. It will be shown that a frequency range of four to one or more can be easily obtained between the cutoff frequencies of the TE_{10} and TE_{20} modes, and six to one or more between those of the TE_{10} and TE_{30} modes. The attenuation is several times as great as that for ordinary wave guide, but is still much less than for ordinary coaxial cable. The reduced cutoff frequency of ridge wave guide also permits a compact cross section.

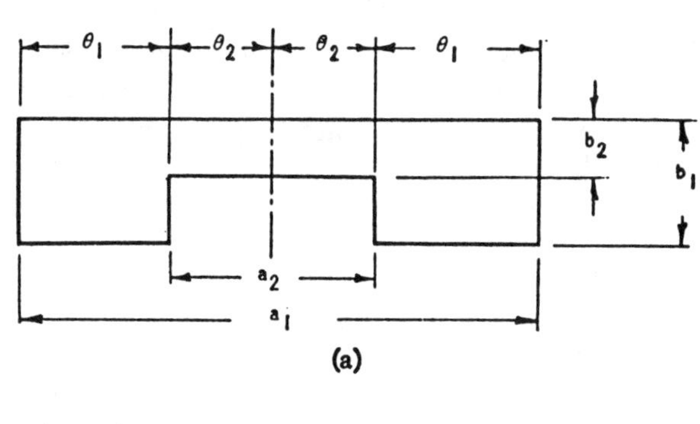

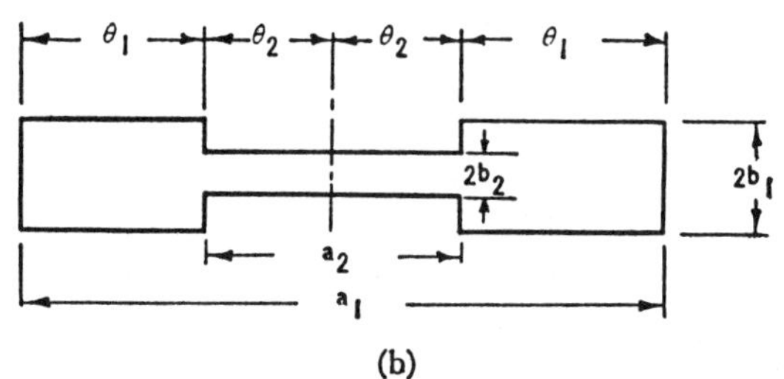

Fig. 1—Parameters for single-ridge (a) and double-ridge (b) wave-guide cross-sections.

* Decimal classification: R118.2. Original manuscript received by the Institute, May 9, 1946.

† Cruft Laboratory, Harvard University, Cambridge, Mass. The work reported in this paper was done at the Radio Research Laboratory under contract with the Office of Scientific Research and Development, National Defense Research Committee, Division 15.

[1] S. Ramo and J. R. Whinnery, "Fields and Waves in Modern Radio," John Wiley and Sons, New York, N. Y.; 1944.

Reprinted from *Proc. I.R.E.*, vol. 35, pp. 783–788, Aug. 1947.

(b) Ridge wave guide has been used successfully as matching or transition elements in wave-guide to coaxial junctions. In one type of junction, a quarter-wavelength section of ridge wave guide serves as a matching transformer from the impedance of the guide ("toll-ticket" wave guide, $2\frac{3}{4}\times\frac{3}{8}$-inch cross section) to the 50-ohm coaxial cable. In another junction, a tapered length of ridge wave guide gives a gradual match from standard $3\times1\frac{1}{2}$-inch rectangular wave guide to a 50-ohm coaxial line.[2]

(c) Various forms of ridge wave guide are useful also as filter elements, cavity elements, cavity terminations, etc. Wherever an element of line is needed having reduced cutoff frequency, reduced impedance, or wide mode separation, ridge wave guide provides a simple solution.

(d) The attenuation formula for ridge guide (8) shows that the attenuation may be made very high by making a_1 and $Z_{0\infty}$ as small as possible. If the guide, or just the ridges, are made of steel instead of copper, the attenuation may be made about 1000 times greater than that for ordinary copper wave guide without ridges. H. C. Early of the Radio Research Laboratory has made use of a length of such wave guide tapered to standard $3\times1\frac{1}{2}$-inch wave guide in the design of a broadband matched load.[3,4] The total length of the load and taper is only four feet.

(e) Another application, due to Early, is in a wide-band wattmeter,[3] in which a wave guide having nearly constant impedance over a wide band is required.

II. Design Data

The design equations use the notation of Fig. 1. a_1, a_2, b_1, and b_2 are inside dimensions in centimeters. θ_1 and θ_2 are the electrical phase lengths in terms of the cutoff wavelength in free space

$$\left(\text{e.g., } \theta_2 = \frac{a_2/2}{\lambda_c'} \times 360\right)$$

where λ_c' is the wavelength in free space at the ridge-guide cutoff frequency.

The cutoff of the TE_{10} mode occurs when the lowest root of the following equation is satisfied:

$$\frac{b_1}{b_2} = \frac{\cot\theta_1 - \dfrac{B_c}{Y_{01}}}{\tan\theta_2}. \qquad (1)^5$$

B_c is the equivalent susceptance introduced by the discontinuities in the cross-section, as explained in Appendix I.[6]

[2] S. B. Cohn, "Design of simple broad-band wave guide-to-coaxial line junction," to be published in Proc. I.R.E.

[3] H. C. Early, "A wide-band wattmeter for wave guide," Proc. I.R.E., vol. 34, pp. 803–807; October, 1946.

[4] H. C. Early, "A wide-band directional coupler for wave guide," Proc. I.R.E., vol. 34, pp. 883–887; November, 1946.

[5] (1), (3), (6), and (8) are derived in the appendix.

[6] B_c may be calculated from the curves in a paper by J. R. Whinnery and H. W. Jamieson, "Equivalent circuits for discontinuities in transmission lines," Proc. I.R.E., vol. 32, pp. 98–116; February, 1944.

Equation (1) is accurate if proximity effects are taken fully into account in calculating B_c. In the curves of this paper, proximity effects are neglected, but the results are highly accurate so long as $(a_1-a_2/2)>b_1$.

In terms of θ_1 and θ_2, λ_c' is given by

$$\lambda_c' = \left(\frac{90°}{\theta_1+\theta_2}\right)\lambda_c \qquad (2)$$

where $\lambda_c=2a_1$ is the cutoff wavelength of the guide without the ridge, and where θ_1 and θ_2 are values satisfying (1).

The TE_{10}-mode cutoff wavelength is plotted in Figs. 2 and 3 for a wide variety of ridge shapes in guide having cross-section ratios of $b_1/a_1=0.136$ ("toll-ticket" wave guide, $2\frac{3}{4}\times\frac{3}{8}$ inch) and 0.500, respectively. The ordinate $\lambda_c'/2a_1=\lambda_c'/\lambda_c=f_c/f_c'$ is the ratio of cutoff wavelength with the ridge to that without the ridge. The abscissa a_2/a_1 is the ratio of ridge width to guide width. Each solid curve corresponds to a constant value of b_2/b_1. As an example, if a particular ridge wave guide has $b_1/a_1=0.5$, $a_2/a_1=0.4$, and $b_2/b_1=0.1$, then from Fig. 3, $\lambda_c'/\lambda_c=f_c/f_c'=2.6$. If the cutoff frequency without the ridge is 2600 megacycles, the cutoff frequency with the ridge will be 1000 megacycles.

On comparing Fig. 2 and Fig. 3, it will be seen that there is not a great deal of difference between the corresponding constant b_2/b_1 curves. The only reason there is any difference is the size of the discontinuity susceptance term, B_c/Y_{01}, which is small for $b_1/a_1=0.136$, and fairly large for $b_1/a_1=0.5$. If b_1/a_1 has a value different from 0.136, or 0.5, Figs. 2 and 3 may still be used with little error. Fig. 2 should be used for values of b_1/a_1 between zero and about one-third, and Fig. 3 should be used for values of b_1/a_1 in the vicinity of 0.5.

The characteristic impedance at infinite frequency for the TE_{10} mode is given by

$$Z_{0\infty} = \frac{120\pi^2 b_2}{\lambda_c'\left\{\sin\theta_2 + \dfrac{b_2}{b_1}\cos\theta_2\tan\dfrac{\theta_1}{2}\right\}}. \qquad (3)^4$$

If $Z_{0\infty}$ and the cutoff frequency f_c' are known, the characteristic impedance at any frequency f is obtained by multiplying $Z_{0\infty}$ by the right-hand side of (4).

$$\frac{Z_0}{Z_{0\infty}} = \frac{\lambda_g}{\lambda} = \frac{1}{\sqrt{1-\left(\dfrac{f_c'}{f}\right)^2}}. \qquad (4)$$

The guide wavelength is also obtained by multiplying the space wavelength at the same frequency by the right-hand side of (4).

Equation (4) is plotted in Fig. 4.

Constant $Z_{0\infty}$ curves are plotted in Figs. 2 and 3 as dashed lines. In the example cited above, the impedance of a guide having $b_1/a_1=0.5$, $a_2/a_1=0.4$, $b_2/b_1=0.1$, and $\lambda_c'/\lambda_c=2.6$ would be 47 ohms at infinite frequency. At one and one-half times the cutoff frequency, the impedance is multiplied by the factor 1.34, found from Fig. 4, which gives $Z_0=47\times 1.34=63$ ohms.

Equation (3) does not take the discontinuity susceptance fully into account, and consequently it is truly accurate only if b_1/a_1 is small. In addition to this, it has

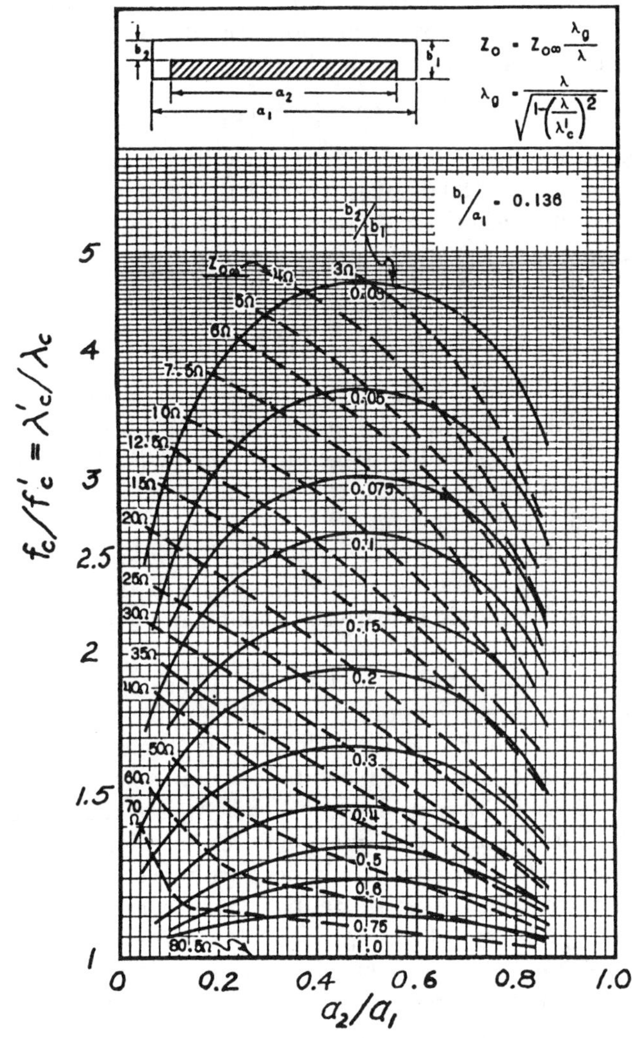

Fig. 2—Characteristic impedance and cutoff wavelength of ridge wave guide ($b_1/a_1=0.136$).

the same restrictions as (1). Experiments have given excellent checks of Fig. 2 ($b_1/a_1=0.136$), while for $b_1/a_1=0.5$ the impedance for the above example was found experimentally to be about 35 to 40 ohms. To obtain a 50-ohm impedance, b_2/b_1 has to be increased from 0.1 to about 0.133 (see Part III, below). But even for $b_1/a_1=0.5$ (3) is a useful approximation, and gives a good starting point in design work.

If b_1/a_1 is not equal to 0.136 or 0.5, $Z_{0\infty}$ may still be determined very closely from Figs. 2 and 3. For values of b_1/a_1 between about zero and one-third, multiply values of $Z_{0\infty}$ on Fig. 2 by the scale factor

$$\frac{b_1/a_1}{0.136}.$$

For values of b_1/a_1 between about one-third and two-thirds, multiply values on Fig. 3 by

$$\frac{b_1/a_1}{0.5}.$$

For example, if $b_1/a_1=0.2$, $b_2/b_1=0.3$, and $a_2/a_1=0.5$, Fig. 2 gives $Z_0=28$ ohms for $b_1/a_1=0.136$. Therefore,

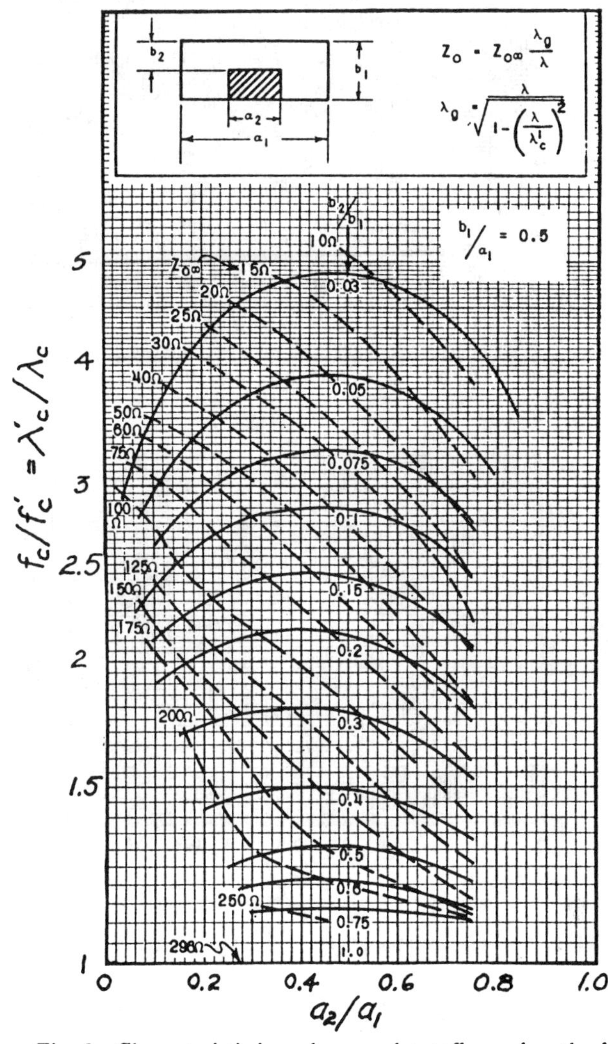

Fig. 3—Characteristic impedance and cutoff wavelength of ridge wave guide ($b_1/a_1=0.5$).

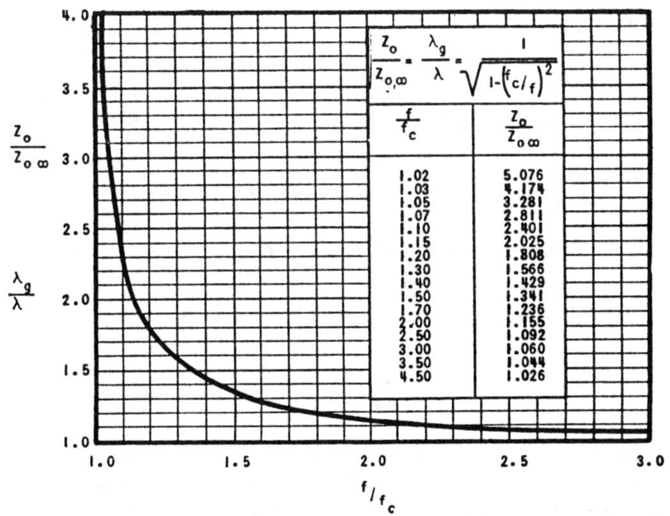

f/f_c	$Z_0/Z_{0\infty}$
1.02	5.076
1.03	4.174
1.05	3.281
1.07	2.811
1.10	2.401
1.15	2.025
1.20	1.808
1.30	1.566
1.40	1.429
1.50	1.341
1.70	1.236
2.00	1.155
2.50	1.092
3.00	1.060
3.50	1.044
4.50	1.026

Fig. 4—Function relating characteristic impedance and guide wavelength to frequency.

for $b_1/a_1=0.2$, $Z_{0\infty}=28\times0.2/0.136=41.1$ ohms. The characteristic impedance was checked experimentally for a cross section having $b_1/a_1=\frac{1}{4}$, and was found to be very close to the value calculated by the foregoing method.

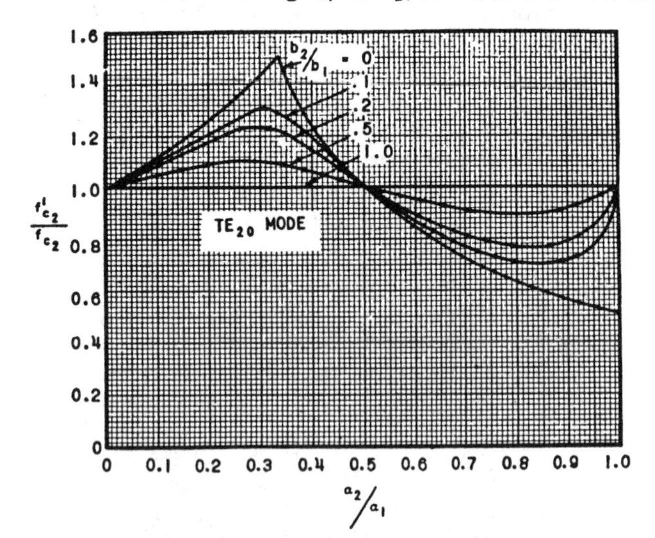

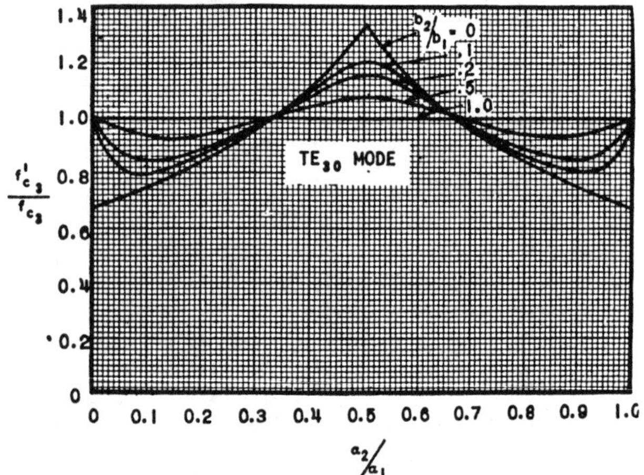

Fig. 5—Cutoff-frequency ratios for the TE_{20} and TE_{30} modes in ridge wave guide.

The higher roots of (1) give the cutoff frequencies of all the odd TE_{mo} modes. The TE_{30} mode is of considerable interest, since it is usually the lowest mode that can cause trouble in a transmission system having a symmetrical cross section in both E and H directions at every point including the ends. For $0\leqq\theta_1\leqq90$ degrees, choose the root of θ_2 between 180 and 270 degrees. For 90 degrees$\leqq\theta_1\leqq180$ degrees, choose the root of θ_2 between 90 and 180 degrees. For 180 degrees$\leqq\theta_1\leqq270$ derees, choose the root of θ_2 between 0 and 90 degrees. Once a pair of values θ_1 and θ_2 have been determined, the ratio f_{c3}'/f_{c3} can be determined from the relation

$$\frac{f_{c3}'}{f_{c3}}=\frac{\theta_1+\theta_2}{270°} \tag{5}$$

where f_{c3}' and f_{c3} are the cutoff frequencies for the TE_{30} mode with and without the ridge, respectively. f_{c3}'/f_{c3} is plotted in Fig. 5 as a function of a_2/a_1 for several values of b_2/b_1, with B_c/Y_{01} neglected. Note that when a_2/a_1 is one-half, f_{c3}'/f_{c3} is a maximum, and the greatest separation of the TE_{10} and TE_{30} cutoff frequencies is obtained. It is easily shown that, for $a_2/a_1=\frac{1}{2}$, f_{c3}'/f_{c3} increases as b_2/b_1 decreases, and in the limit approaches 4/3.

The even TE_{mo}-mode cutoffs are given by solutions of the following equation in which the discontinuity susceptance term has been neglected:

$$\theta_2=\tan^{-1}(-n\tan\theta_1) \tag{6}$$[5]

where $n=b_1/b_2$. For the TE_{20} mode, the θ_2 root lies between 90 and 180 degrees for $0\leqq\theta_1\leqq90$ degrees, and the θ_2 root between 0 and 90 degrees for 90 degrees$\leqq\theta_1\leqq180$ degrees. The cutoff frequency is given by

$$\frac{f_{c2}'}{f_{c2}}=\frac{\theta_1+\theta_2}{180°}\,. \tag{7}$$

This is plotted in Fig. 5 as a function of a_2/a_1 for several values of b_2/b_1. The maximum value of f_{c2}'/f_{c2} occurs between $a_2/a_1=\frac{1}{4}$ and $\frac{1}{3}$, depending upon b_2/b_1. As b_2/b_1 is made vanishingly small, the maximum value of f_{c2}'/f_{c2} approaches 3/2 at $a_1/a_2=\frac{1}{3}$.

Figs. 2, 3, and 5 show that when a wide frequency band free from TE_{20} and TE_{30} modes is desired, the ridge width should be between about $\frac{1}{3}$ and $\frac{1}{2}$ of the total guide width.

The formulas and curves for a single ridge in a wave guide are directly applicable to the double-ridge cross section shown in Fig. 1(b). In this case, the total height of the guide is $2b_1$ and the total spacing is $2b_2$. Thus, if the width is $2\frac{3}{4}$ inches and the height is $\frac{3}{4}$ inch, then $b_1/a_1=0.136$, and the cutoff curves in Fig. 1 apply exactly. The characteristic-impedance curves apply also, but their values must be doubled. Hence, for a double-ridge guide in which $b_1/a_1=0.136$, $a_2/a_1=0.35$, and $b_2/b_1=0.2$, the relative cutoff wavelength is $\lambda_c'/\lambda_c=1.9$, and the infinite-frequency characteristic impedance is $Z_{0\infty}=2\times26=52$ ohms, by Fig. 2.

The attenuation constant in decibels per meter for copper single-ridge wave guide may be calculated fairly closely from the following approximate formula:

$$\alpha=6.01(10)^{-7}k\sqrt{f}\left\{\frac{\dfrac{1}{a_1}+\dfrac{2}{b_1}\left(\dfrac{f_c'}{f}\right)^2}{\sqrt{1-\left(\dfrac{f_c'}{f}\right)^2}}\right\}\frac{60\pi^2\left(\dfrac{b_1}{a_1}\right)}{Z_{0\infty}}$$

decibels per meter (8)

where a_1 and b_1 are in centimeters, and f is in cycles per second. k is a correction constant a little larger than unity, which takes account of the more crowded current distribution in ridge wave guide than in plain wave guide. If a_2/a_1 is larger than about $\frac{1}{3}$, this term is probably not greater than 1.5.

For double-ridge wave guide, b_1 should be replaced by the total guide height, $2b_1$. If any metal other than copper is used, α is proportional to $\sqrt{\mu/\sigma}$.

III. Experimental Verification

A three-foot length of ridge wave guide having the cross-sectional dimensions shown in Fig. 6(a) has been tested. For this symmetrical cross section, the design

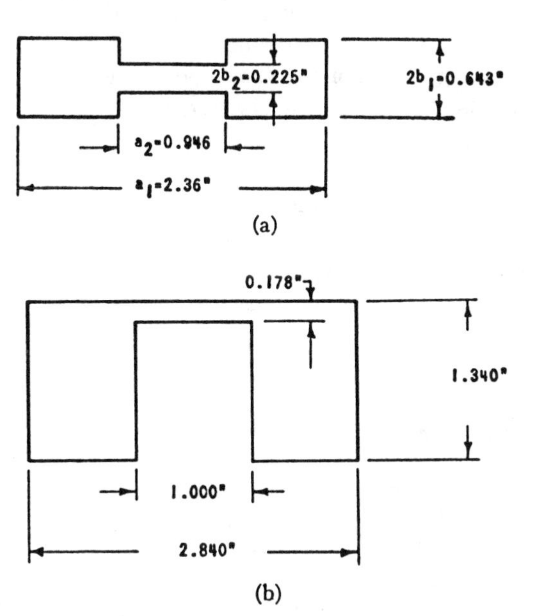

Fig. 6—Two experimental cross sections discussed in the text.

method of II is applicable. The parameters are $b_1/a_1 = 0.136$, $b_2/b_1 = 0.35$, and $a_2/a_1 = 0.40$. Without the ridges, the cutoff wavelength would be $2 \times 2.36 \times 2.54 = 12.0$ centimeters, and the cutoff frequency would be $30{,}000/12.0 = 2500$ megacycles. Fig. 2 gives $f_c/f_c' = 1.50$ and $Z_{0\infty}/2 = 37$ ohms. Therefore, $f_c' = 1670$ megacycles and $Z_{0\infty} = 74$ ohms. Fig. 5 gives approximately $f_{c2}'/f_{c2} = 1.10$ and $f_{c3}'/f_{c3} = 1.06$. Hence,

$$f_{c2}' = 2 \times 2500 \times 1.10 = 5500 \text{ megacycles, and}$$

$$f_{c3}' = 3 \times 2500 \times 1.06 = 7950 \text{ megacycles.}$$

The calculated and measured cutoff frequencies are tabulated below.

Mode Cutoffs

MODE	CALCULATED	MEASURED
TE_{10}	1670 megacycles	1675 megacycles
TE_{20}	5500 megacycles	5200 megacycles
TE_{30}	7950 megacycles	7900 megacycles.

Ridge wave guide has been used for elements in wide-band junctions between wave guide and coaxial line. In one type of junction designed for "toll-ticket" wave guide ($a_1 = 2.75$ inches, $b_1 = 0.375$ inch, $b_1/a_1 = 0.136$) a quarter-wavelength section of ridge wave guide is used as a matching transformer between the 103-ohm guide and the 50-ohm line. The experimental results checked the ridge wave guide calculated impedance within a few per cent. In another type of junction for $3 \times 1\frac{1}{2}$-inch wave guide, a tapered length of ridge guide is used to match the 50-ohm coaxial line. In this case, the impedance calculated for the ridge guide proved less accurate, because the approximations were less valid with this higher ratio of b_1 to a_1. The impedance in this case proved, however, to be about 25 per cent lower than the calculated value, and hence the impedance curves in Fig. 3, though not very accurate, serve as a valuable guide in the preliminary design of a piece of equipment. In a double-ridge type of junction for $3 \times 1\frac{1}{2}$-inch wave guide, the impedance curves checked very well. In this case the ratio of b_1 to a_1 was approximately 0.25.

A cross section in $3 \times 1\frac{1}{2}$-inch wave guide that has been found experimentally to have $Z_{0\infty}$ approximately equal to 50 ohms is shown in Fig. 6(b). Fig. 3 gives a value of 65 ohms for $b_1/a_1 = 0.5$, $b_2/b_1 = 0.133$, and $a_2/a_1 = 0.352$. For the above cross section, the impedance must be scaled by the factor 0.472/0.500, since b_1/a_1 is not quite 0.5. Therefore, the calculated impedance is 61.5 ohms, which is 23 per cent greater than the approximate measured impedance.

The paper now under preparation on wave-guide to coaxial-line junctions will give further details.[2]

APPENDIX

I. The Cutoff Equation

In the cross section of Fig. 1(a), the electromagnetic field at the cutoff frequency may be considered as the resultant of a wave traveling from side to side without any longitudinal propagation. As pointed out by S. Ramo and J. R. Whinnery,[1] such a cross section may be treated at cutoff by assuming it to be an infinitely wide, composite, parallel-strip transmission line short-circuited at two points. The TE_{10}-mode cutoff occurs at the frequency at which this strip transmission line has its lowest-order resonance. All the other TE_{m0} cutoffs occur at the corresponding m-order resonance frequencies. For m odd, the resonance must be of a type giving an infinite impedance at the center of the cross section. For m even, this impedance must be zero. A resonance condition may therefore be set up by setting the input admittance of half the cross section equal to zero or infinity (Fig. 7). The discontinuity susceptance B_c at the change in height must be included in the calculation.

If one examines the equivalent circuit, it is seen that it is a composite, dissipationless, passive line matched at both ends, and it is, therefore, matched at every point within. Hence, the sum of the admittances across $x-x$ must equal zero, and the following relation results:

$$-Y_{01} \cot \theta_1 + B_c + Y_{02} \tan \theta_2 = 0$$

$$\frac{Y_{02}}{Y_{01}} = \frac{Z_{01}}{Z_{02}} = \frac{\cot \theta_1 - \frac{B_c}{Y_{01}}}{\tan \theta_2}.$$

But in a strip transmission line, the characteristic impedance is proportional to the height. Therefore,

$$\frac{b_1}{b_2} = \frac{\cot \theta_1 - \frac{B_c}{Y_{01}}}{\tan \theta_2} \tag{9}$$

which is the cutoff condition for the odd TE_{mo} modes (1).

For the even modes, the equivalent circuit is shown in Fig. 7(c).

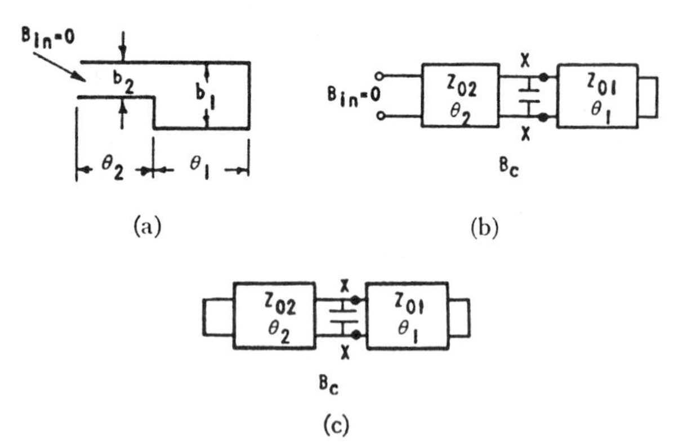

Fig. 7—Development of the equivalent circuit for ridge wave guide.

In this case, $-Y_{01}\cot\theta_1+B_c-Y_{02}\cot\theta_2=0$, and hence

$$\cot\theta_1+\frac{b_1}{b_2}\cot\theta_2=\frac{B_c}{Y_{01}}\cdot \tag{10}$$

Equation (6) follows readily from this.

The discontinuity-susceptance term B_c/Y_{01} is obtainable from a paper by J. R. Whinnery and H. W., Jamieson.[6]

II. Impedance Equation

In deriving the impedance equation for ridge wave guide, it will be assumed that b_1/a_1 is small, so that the discontinuity susceptance at the edges of the ridge may be neglected.

If the TE_{10} mode alone is set up in the wave guide, the E field distribution is the same at all frequencies, including $f=f_c$ and $f=\infty$. The E field can be calculated easily at the cutoff frequency by the approach used in deriving the cutoff equation. At $f=\infty$, the wave impedance is that of free space.[7] Hence, if the E field is known, the H field is given by $H=E/120\pi$. Both E and H are completely transverse at $f=\infty$, and the current on the top or bottom of the wave guide is completely longitudinal. The current per unit width is equal to the H field intensity at the surface of the conductor.

The guide impedance at infinite frequency will now be defined as the *ratio of voltage across the center of the guide to the total longitudinal current on the top face*

$$Z_{0\infty}=\frac{V_0}{I}=\frac{b_2E_0}{2\int_0^{a_1/2}i\,dx}=\frac{120\pi b_2E_0}{2\int_0^{a_1/2}E\,dx}\cdot \tag{11}$$

The impedance at any other frequency is related to $Z_{0\infty}$ by the expression

$$Z_0=\frac{Z_{0\infty}}{\sqrt{1-\left(\frac{f_c'}{f}\right)^2}} \tag{12}$$

where f_c' is the cutoff frequency of the ridge guide.

Since the guide has been assumed thin, the voltage across the step will be continuous. The voltage distribution in the right half of the cross section will therefore be the same as that along the shorted composite transmission line shown in Fig. 8.

Since the input impedance is infinite at the open end, the voltage across the guide is a maximum at that point. Transmission-line theory shows that the voltage distribution over the θ_2 range is given by $V=V_0\cos\theta$ from $\theta=0$ to $\theta=\theta_2$. Over the θ_1 range it is given by

$$V=V_1\sin\frac{(\theta_1+\theta_2-\theta)}{\sin\theta_1}=V_0\frac{\cos\theta_2}{\sin\theta_1}\sin(\theta_1+\theta_2-\theta),$$

from $\theta=\theta_2$ to $\theta=\theta_1+\theta_2$.

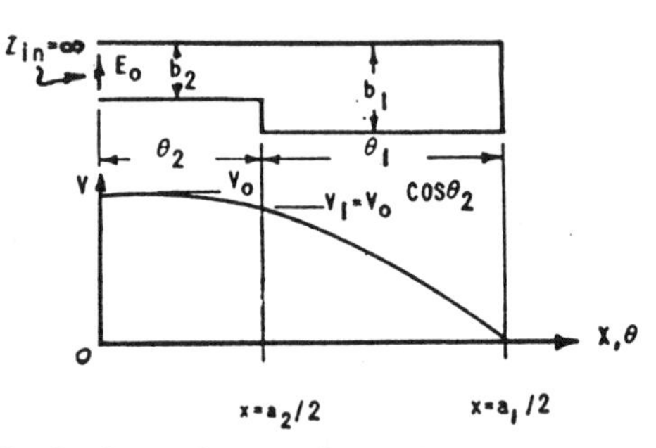

Fig. 8—Approximate voltage distribution across half of the cross section.

The E field is equal to V/b. Therefore,

$$E=E_0\cos\theta,\qquad 0\leqq\theta\leqq\theta_2 \tag{13}$$

$$E=\frac{b_2}{b_1}E_0\frac{\cos\theta_2}{\sin\theta_1}\sin(\theta_1+\theta_2-\theta),\qquad \theta_2\leqq\theta\leqq(\theta_1+\theta_2).$$

The integral in (11) may be evaluated as follows:

$$\int_0^{a_1/2}E\,dx=\frac{\lambda_c'}{2\pi}\int_0^{(\theta_1+\theta_2)}E\,d\theta$$

$$=E_0\frac{\lambda_c'}{2\pi}\left\{\int_0^{\theta_2}\cos\theta\,d\theta+\frac{b_2}{b_1}\frac{\cos\theta_2}{\sin\theta_1}\int_{\theta_2}^{\theta_1+\theta_2}\sin(\theta_1+\theta_2-\theta)\,d\theta\right\}$$

$$=E_0\frac{\lambda_c'}{2\pi}\left\{\left|\sin\theta\right|_0^{\theta_2}+\frac{b_2}{b_1}\frac{\cos\theta_2}{\sin\theta_1}\left|\cos(\theta_1+\theta_2-\theta)\right|_{\theta_2}^{\theta_1+\theta_2}\right\}$$

$$=E_0\frac{\lambda_c'}{2\pi}\left\{\sin\theta_2+\frac{b_2}{b_1}\frac{\cos\theta_2}{\sin\theta_1}(1-\cos\theta_1)\right\}$$

$$=E_0\frac{\lambda_c'}{2\pi}\left\{\sin\theta_2+\frac{b_2}{b_1}\cos\theta_2\tan\frac{\theta_1}{2}\right\}.$$

Substituting this relation in (11) gives

$$Z_{0\infty}=\frac{120\pi^2b_2}{\lambda_c'\left\{\sin\theta_2+\frac{b_2}{b_1}\cos\theta_2\tan\frac{\theta_1}{2}\right\}}\cdot \tag{14}$$

[7] For background on this derivation, consult J. C. Slater, "Micro-wave-Transmission," McGraw-Hill Book Co. New York, N. Y., 1942; chap. 4.

Guidance and Leakage Properties of a Class of Open Dielectric Waveguides: Part I—Mathematical Formulations

SONG-TSUEN PENG, MEMBER, IEEE, AND ARTHUR A. OLINER, FELLOW, IEEE

Invited Paper

Abstract—A class of open dielectric waveguides is discussed which is of direct importance to the areas of integrated optics and millimeter-wave integrated circuits. An accurate analysis of the properties of these waveguides reveals that interesting new physical phenomena, such as leakage and sharp cancellation or resonance effects, may occur under appropriate circumstances. The resulting leaky modes form a new class of such modes since the leakage, in the form of an exiting surface wave, has a polarization opposite to that which dominates in the bound portion of the leaky mode. These new effects are caused by TE–TM mode coupling, which was neglected in earlier approximate treatments. Part I presents the mathematical formulation based on a rigorous mode-matching procedure.

I. Introduction

A. General Remarks

OPEN dielectric waveguides have become increasingly important in the past few years, particularly in connection with the areas of integrated optics and millimeter-wave integrated circuits [1–3]. Optical fiber waveguides of circular cross section are, of course, central to the rapidly expanding area of fiber optics, but we shall not consider that class of structures because it has been rather exhaustively treated elsewhere. Furthermore, we restrict our concern here to those waveguides which are naturally suited for use in an integrated circuit context. One feature common to most such waveguides is the presence of a dielectric strip of rectangular cross section in conjunction with a uniform dielectric layered structure, so that the electromagnetic energy can be confined to the vicinity of the strip and be guided by it. For this class of waveguides a suitable generic name could be "dielectric strip waveguide."

The propagation characteristics of these open dielectric waveguides constitute a rich variety of phenomena, including the leakage of guided energy and leakage-related resonance effects under appropriate circumstances. The present authors were the first to predict these physical effects and to present an approximate theory describing them [4]; recent measurements [5] have confirmed their existence on a specific waveguiding structure. With respect to the leakage, it is not generally known that most modes on most of these waveguides can be *leaky*, instead of being purely bound, as is customarily assumed. On a lesser note, the hybrid guided modes on these waveguides possess six field components, and not five, as many believe. The basic reason for this incomplete understanding is that most of the published theoretical propagation characteristics have been obtained from approximate analyses [2], [6]–[11] that neglect those features which lead to the aforementioned effects. A more accurate analysis [12] is also incomplete in the sense that it furnishes the six field components but neglects any mention of the important leakage feature. It is intended in this two-part paper to offer a rigorous mathematical foundation and a clear physical picture for the explanation of these new phenomena.

This Part I of a two-part paper contains a *mathematical formulation* based on a rigorous mode-matching procedure that automatically takes into account all the features mentioned above. The key point that is neglected in the customary approximate treatments is the *coupling* produced between TE and TM waves at geometrical discontinuities. The *new physical effects* that result when the TE–TM coupling is taken correctly into account are described in Part II, together with various *numerical results* for typical waveguides which illustrate these effects quantitatively.

The new physical effects, namely, the presence of *leakage* and the appearance of sharp *resonance*, or cancellation, *effects*, are discussed in detail in Section III of Part II. The leakage, when it is present, occurs in the form of a surface wave which propagates away from the waveguide at some angle to it. The leakage effect can sometimes be used to advantage in the design of novel devices [13]. On the other hand, when this waveguide is part of an optical or millimeter-wave integrated circuit, such leakage can cause cross-talk between neighboring portions of the circuit, and deteriorate system performance. For these reasons, it is important to know in any specific case whether or not the waveguide will leak; this question is treated in detail in Section III of Part II.

The leakage to which we refer changes the guided mode from being purely bound to a *leaky mode*. There is also a point of fundamental interest here, since these leaky modes constitute *a new class* of leaky modes, in that the leakage

Manuscript received March 24, 1981; revised May 18, 1981. This work was supported in part by the Joint Services Electronics Program under Contract F49620-80-C-0077 and in part by the U.S. Army CORADCOM under Contract DAAK-80-79-C-0798.

The authors are with the Department of Electrical Engineering and Computer Science and the Microwave Research Institute, Polytechnic Institute of New York, Brooklyn, NY 11201.

Reprinted from *IEEE Trans. Microwave Theory Tech.*, vol. MTT-29, no. 9, pp. 843–854, Sept. 1981.

portion of the leaky mode has a *polarization opposite* to that which dominates in the bound portion of the leaky mode. For example, if the electric field of the guided mode is predominantly *vertically* polarized, the leaking surface-wave portion will have its electric field *horizontally* polarized. This novel feature distinguishes these leaky modes from the usual types of leaky mode.

The modes which we discuss here are all above cutoff, in contrast to certain high-loss below-cutoff leaky-wave solutions appearing in some treatments of optical fibers (open dielectric waveguides) of circular cross section. Furthermore, we do not consider the mechanisms which give rise to certain above-cutoff "tunneling" modes which occur on such optical fibers. The class of leaky modes we treat here do not occur in open dielectric waveguides of circular cross section, and they involve a coupling mechanism which is not applicable there.

Not all dielectric strip waveguides permit leakage, and, on those which can leak, some guided modes do leak and some do not. Such questions are treated in detail in Section III of Part II, which also presents physical explanations for the leakage effect, and for the resonance, or cancellation, effect, where sharp nulls in the leakage occur for specific values of strip width. In Section IV of Part II, many numerical results are given which illustrate the new physical effects. Included are measured results which verify the theoretical predictions in a specific case. The Introduction (Section I) of Part II provides guidance for further details.

B. The Approach and the Mathematical Procedures

In Part I, the aim is to provide a *rigorous* mathematical foundation for the analysis of this class of open dielectric waveguides. Stress is placed on network representations to establish physical pictures of the wave processes and to yield insight; in addition, a systematic microwave network approach is employed. This *building-block approach* first breaks the cross-sectional geometry into constituent parts, analyzes each part rigorously, and then combines them into a transverse resonance analysis of the complete waveguide.

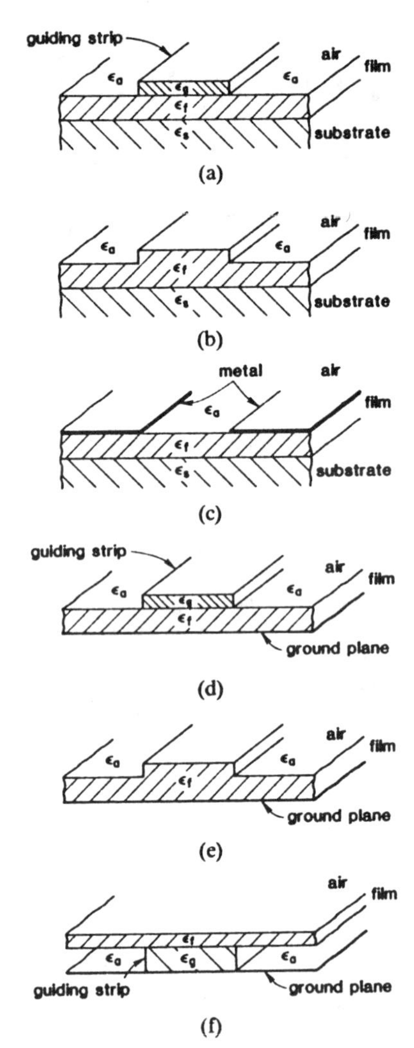

Fig. 1. Typical open dielectric waveguides for integrated optics and millimeter-wave integrated circuits. $\epsilon_f, \epsilon_g \geq \epsilon_s \geq \epsilon_a$. (a) Optical dielectric strip waveguide. (b) Optical rib waveguide. (c) Optical slot waveguide. (d) Millimeter-wave dielectric strip waveguide. (e) Millimeter-wave ridge waveguide. (f) Millimeter-wave inverted strip waveguide.

Some typical open dielectric waveguides are shown in Fig. 1. The waveguides in Figs. 1(a), (b), and (c) employ dielectric substrates and are intended for application to integrated optics, whereas those in Figs. 1(d), (e), and (f) are placed on metallic ground planes for use in millimeter-wave integrated circuits. It may be noted that when the dielectric constants of the strip and the film are the same, the waveguide is customarily termed a "rib waveguide" in optics (Fig. 1(b)), and a "dielectric ridge waveguide" in the millimeter-wave context (Fig. 1(e)).

It is customary in this class of waveguides to view the cross sections as consisting of a central, or *inside*, region sandwiched between two identical *outside* regions. This decomposition is to be viewed in the horizontal direction, where the inside region consists, in Fig. 1(a), for example, of the strip placed on the film on the substrate, and the outside regions have only the film on the substrate. Except for the optical slot waveguide (Fig. 1(c)), the presence of a dielectric strip in all the other structures makes the net, or effective, dielectric constant of the inside region higher than that of the outside regions. In the case of the optical slot guide, the metallic plates in the outside regions behave like an overdense plasma that possesses a predominantly negative-real dielectric constant in the optical frequency range. Thus the effective dielectric constant of the central region of the optical slot guide is higher than that of the outside regions, as in all the other structures in Fig. 1. Therefore, the electromagnetic energy is confined mostly to the inside region, with its higher effective dielectric constant. Further remarks about these waveguides are made in Section III-A of Part II.

Our approach here follows that of microwave network theory; the cross section of the waveguide is first viewed in terms of *constituent parts*, or building blocks, then each constituent is analyzed separately in its own simpler context, and finally all parts are put together to comprise the final structure of interest. In that way, the simpler parts are handled quickly, and the more difficult portions are paid the special attention required of them, and in a less cluttered context. Approximations, if they need to be made, can then be more systematically treated. When the parts are finally put together, the last step is easily handled.

The guiding of waves along the axis of these waveguides is customarily viewed in terms of surface waves which bounce back and forth inside the central region at an angle to the side walls, undergoing total reflection at each bounce; in the outside regions, the electromagnetic fields are transversely evanescent. For the structures under consideration, the waveguide side walls appear in the form of a step discontinuity, either in dielectric constant or in thickness, between two uniform dielectric-layered structures. The building blocks of the cross section are thus seen to be the inside and outside regions, which are simply portions of uniform dielectric-layered structures which support planar surface waves, and dielectric step discontinuities, or junctions, at the planes where the inside and outside regions meet.

In our building-block treatment, we first treat rigorously the separate uniform planar regions. The modes on such planar dielectric structures are well known, but we review them here both because we need them for the waveguiding problem and to explain our notation and procedure in a simpler context. The complete modal spectrum, comprising the surface waves and the non-surface waves, is discussed in Section II. In our analysis of these open structures, we follow the customary procedure of *discretizing* the continuous spectrum [12], [14], [15] by placing perfectly-conducting walls above and below (if necessary) the planar dielectric waveguide, thus replacing the open region with a partially-dielectric-filled parallel-plate waveguide, which supports, in addition to the surface waves [16], an infinite number of higher non-surface-wave modes, some propagating and the remainder nonpropagating.

The next constituent of the waveguide cross section to be analyzed is the *dielectric step junction*, or discontinuity, corresponding to the side of the dielectric strip waveguide. The rigorous mode-matching analysis is presented in Section III. Since guidance along the waveguide can be viewed in terms of surface waves bouncing back and forth at an angle to the sides of the waveguide and encountering "total reflection" at each bounce, the dielectric step junction problem involves the scattering by the step of a surface wave *obliquely* incident on it.

When the surface wave is incident normally on the step junction, the boundary-value problem is two-dimensional, and an incident TE surface wave produces reflected and transmitted waves of TE polarization only. For oblique incidence, on the other hand, the boundary-value problem becomes three-dimensional, and TE–TM coupling is produced at the step discontinuity. That means that an incident wave of a given polarization now *also* produces reflected and transmitted waves of the *opposite* polarization. This scattering problem is therefore of interest in its own right.

The rigorous analysis begins in Section III-A with a coordinate transformation which translates a TE or a TM surface wave propagating at an angle to the step discontinuity into an LSE or an LSM mode (or alternatively an *H*-type or an *E*-type mode) propagating *normally* to the step discontinuity, so that the step can be viewed as a *transverse* discontinuity. The mode-matching procedure for the boundary-value problem, which is the heart of the method, is described in Section III-B. The amplitudes of the scattered waves are determined by four infinite systems of equations, corresponding to the satisfaction of the boundary conditions. These equations, which are phrased in matrix form, must of course be truncated in practice to permit numerical results to be obtained.

The basic mode-matching procedure described in Section III-B is well known, and has been employed by others in determining the propagation characteristics of some waveguides. By utilizing certain matrix identities, however, an *alternative* matrix formulation is obtained. Both formulations yield numerically identical results when the matrices are of infinite order, but not when they are truncated. However, the alternative formulation always satisfies the conservation of power, and it is therefore useful for the development of equivalent networks for the dielectric step junction. The first formulation in truncated form does not satisfy the power conservation condition, so that its deviation from it can be used as a measure of the numerical accuracy of the scattering results.

It is then shown in Section III-C that one can readily develop an *input-admittance* formulation in terms of the matrix quantities involved in the mode-matching process. An equivalent network resulting from this formulation is employed in Section IV as a constituent of the general transverse equivalent network for the cross section of dielectric strip waveguides.

The analysis for the dielectric strip waveguides proceeds in Section IV by employing the building-block approach fundamental to microwave network theory. Since the cross section of this class of waveguides is seen to consist of two dielectric step junctions of the type just discussed connected by a length of uniform waveguide, the equivalent network for the step junction is employed in a rigorous overall transverse equivalent network from which one derives the dispersion relation for the waveguide propagation characteristics. In particular, the input admittance formulation mentioned earlier is used to obtain a rigorous *generalized transverse-resonance relation* for the determination of the waveguide properties. This transverse-resonance relation is expressed in terms of the admittance matrices looking both ways from a reference plane located at the step discontinuity. It is important to note that the procedure employed in deriving the transverse-resonance relation is general, and is independent of the detailed nature of the waveguiding structure; even the relation itself is general if one recognizes that the actual input admittance matrices will become altered when one changes from one waveguide type to another. The relation derived is in fact a generalization of the scalar transverse-resonance relation, valid when only a single mode is involved, to the matrix form for a multimode situation in which all mode-coupling effects are accounted for rigorously.

We conclude these introductory remarks by *summarizing* some of the principal features of our analyses. We employ the classic building-block approach of microwave network theory by viewing the cross sections of the dielectric strip

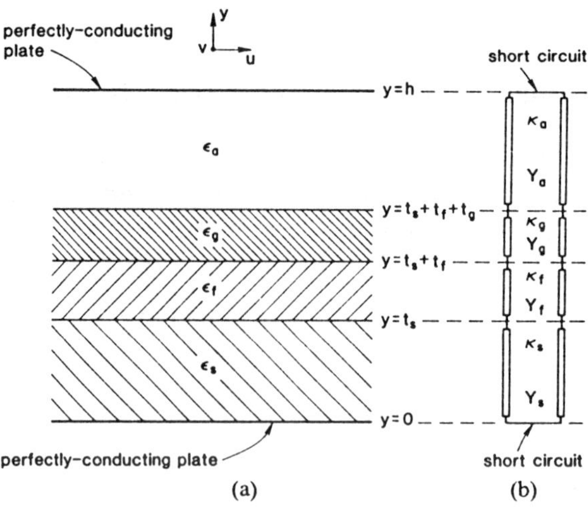

Fig. 2. Multilayer planar dielectric structure and its equivalent network representation. The physical dimensions shown are not in proportion; the perfectly-conducting plates are actually far removed from the central layers.

waveguides in terms of their constituent parts, namely, portions of planar dielectric waveguides and dielectric step junctions where they meet. The planar waveguides are well known, but the step junctions offer a major new challenge. In connection with the step junction, the oblique incidence angle transforms the scalar two-dimensional boundary-value problem that arises for normal incidence into a vector three-dimensional one. A mathematical consequence is TE–TM mode coupling at the discontinuity, and the physical consequence is that an incident wave of one polarization creates reflected and transmitted waves of the opposite polarization. The waveguide problem, which combines the constituent parts, is solved in terms of a generalized transverse-resonance relation, which represents a powerful and general approach, yielding accurate results or permitting in a systematic way various degrees of approximation. The mode-matching general procedure is phrased in admittance terms useful to those familiar with microwave networks, and equivalent networks are developed which summarize the wave processes pictorially and furnish insight into them. The TE–TM mode coupling at the step junctions corresponding to the sides of the waveguide results mathematically in complex eigenvalues under appropriate circumstances, and physically in several new effects, principally leakage and resonance, which are discussed in detail in Part II, together with many numerical results which illustrate the new physical effects quantitatively.

II. Surface Waves and Non-Surface Waves on Planar Dielectric Structures

As discussed in Section I-B, we adopt the building-block approach of microwave networks, and we recognize that the uniform regions present in the cross sections of the class of waveguides shown in Fig. 1 correspond to portions of planar dielectric layers. We therefore need to know the normalized mode functions of both the surface-wave modes and the non-surface-wave modes on these planar layered structures. These modes are well known, but they are reviewed and summarized here because they are needed later. The only new material in this section is relationship (9) involving mode functions of different types.

On these open dielectric structures, the non-surface-wave modes comprise a continuous spectrum. It is customary, however, in this class of problems [12], [14], [15] to *discretize* the continuous spectrum by placing perfectly-conducting walls above and below the planar dielectric waveguides constituting the uniform regions. The open region is thus replaced by a parallel-plate waveguide that is partially dielectric filled, which supports, in addition to a finite (small) number of surface waves, an infinite number of discrete higher modes, some of which are propagating but the remainder of which are below cutoff. These perfectly-conducting bounding walls are placed *far* above and below the guiding dielectric region so as to negligibly influence the properties of the surface waves. For the millimeter-wave structures (Figs. 1(d), (e), and (f)), which employ a ground plane, only an upper bounding plane is needed, of course.

It should also be appreciated that although the presence of these bounding planes may alter the fields far from the step discontinuity, as compared with the truly open environment, the essential physics of the scattering process is not affected. Furthermore, for the waveguide applications of concern here, the higher modes (and in fact the continuous spectrum in a truly open context) will all be below cutoff, so that all of the higher mode power excited at the step discontinuity, in the waveguide application, is completely stored and none of it is radiated.

Consistent with the method described in the preceding for discretizing the higher mode spectrum, a uniform multilayer planar dielectric structure enclosed by an oversize parallel-plate waveguide is shown in Fig. 2(a). Dimensions t_s and h are not in proportion, of course, since the perfectly-conducting plates are located far away. The guiding structure consists of four different dielectric media. For convenience, the media are designated as: air(ϵ_a), guiding strip (ϵ_g), film (ϵ_f), and substrate (ϵ_s). Such a structure is sufficiently general as a basis for the analysis of most

TABLE I
FIELD COMPONENTS OF A BASIC MODE

TE Mode	TM Mode
$E'_v = -\phi'(y)\exp(-jk'_u u)$	$H''_v = \phi''(y)\exp(-jk''_u u)$
$H'_y = \frac{k'_u}{\omega\mu_0}\phi'(y)\exp(-jk'_u u)$	$E''_y = \frac{k''_u}{\omega\epsilon_0\epsilon(y)}\phi''(y)\exp(-jk''_u u)$
$H'_u = \frac{1}{j\omega\mu_0}\frac{d}{dy}\phi'(y)\exp(-jk'_u u)$	$E''_u = \frac{1}{j\omega\epsilon_0\epsilon(y)}\frac{d}{dy}\phi''(y)\exp(-jk''_u u)$

practical dielectric-waveguide problems. For example, the outside region of a waveguide corresponds to the special case for which the thickness of the guiding strip vanishes, e.g., $t_g = 0$. For millimeter-wave applications, the lower perfectly-conducting plate may be regarded as a ground plane and, if no dielectric substrate is present, the thickness of the substrate becomes zero.

The basic modes, comprised of both surface waves and non-surface waves, of the partially-filled parallel-plate waveguide are well known in the literature [16]; some important properties of the basic modes that are relevant to the ensuing analysis are listed here.

For the rectangular coordinate system (u, y, v) indicated in Fig. 2(a), we assume that the basic modes are invariant along the v direction and propagate along the u direction. In this two-dimensional boundary-value problem, the structure supports independent TE and TM basic modes. The field components of the TE and TM modes are given in Table I. Here, we use a single prime to denote quantities for TE modes and a double prime for TM modes. It is also noted that an unprimed quantity will stand for either TE or TM modes. Such notation will be followed throughout this paper. In Table I, ϕ is the transverse mode function and k_u is the longitudinal propagation wavenumber of the basic mode. Evidently, a mode is completely determined by these two quantities.

Presented in the following are a summary of the well-known transverse-resonance technique applied to the structure of Fig. 2(a) and a brief listing of some relationships among the mode functions. Included, however, is a new result involving a mixture of TE and TM modes.

A. Transverse-Resonance Technique

Both the mode function ϕ and the propagation wavenumber k_u may be determined by the transverse-resonance technique. A transverse equivalent network for the dielectric-layer structure in Fig. 2(a) is shown in Fig. 2(b). The transmission-line parameters are known to be related to the longitudinal propagation wavenumber k_u by

$$Y_m = \begin{cases} Y'_m = \kappa'_m/\omega\mu_0, & \text{for the TE mode} \\ Y''_m = \omega\epsilon_0\epsilon_m/\kappa'', & \text{for the TM mode} \end{cases} \tag{1}$$

and

$$\kappa_m = \left(k_0^2\epsilon_m - k_u^2\right)^{1/2} \tag{2}$$

for $m = a, g, f$, or s, designating different media, and where Y_m is the characteristic admittance and κ_m is the propagation wavenumber of the transmission line representing the mth medium. It is noted that for lossless structures κ_m can be either purely real or purely imaginary. Therefore, the sign of the square root in (2) must be properly chosen such that the radiation condition is satisfied in each medium separately. The condition for resonance of the transmission-line system can be conveniently written in the general case as

$$Y_g\frac{Y_g\tan\kappa_g t_g - Y_a\cot\kappa_a t_a}{Y_g + Y_a\cot\kappa_a t_a\tan\kappa_g t_g} + Y_f\frac{Y_f\tan\kappa_f t_f - Y_s\cot\kappa_s t_s}{Y_f + Y_s\cot\kappa_s t_s\tan\kappa_f t_f} = 0 \tag{3}$$

which determines the surface-wave (or non-surface-wave) propagation wavenumber k_u via (1) and (2). Conventionally, such a surface-wave characteristic is expressed in terms of the normalized quantity

$$n_{\text{eff}} = k_u/k_0 \tag{4a}$$

or

$$\epsilon_{\text{eff}} = n_{\text{eff}}^2 = (k_u/k_0)^2. \tag{4b}$$

Here, n_{eff} is known as the effective index of refraction and ϵ_{eff} is the effective dielectric constant. Equation (3) is commonly known as the transverse-resonance relation or the dispersion relation of the waveguide. For a given set of structure parameters and a given operating frequency, the modal propagation constant is thus determined and so are the transmission-line parameters in Fig. 2(b). For the characteristic admittance defined by (1), the mode function is then determined by the transmission-line voltage for the TE mode and the transmission-line current for the TM mode.

B. Relationships Among Transverse Mode Functions

The transverse mode functions of a partially-filled parallel-plate waveguide are governed by the Sturm-Liouville eigenvalue problem

$$\left[\frac{d}{dy}p(y)\frac{d}{dy} + q(y)\right]\phi_n(y) = \kappa_n^2 w(y)\phi_n(y) \tag{5}$$

subject to the boundary conditions

$$\phi_n(0) = \phi_n(h) = 0, \quad \text{for TE modes} \tag{6a}$$

$$\dot{\phi}_n(0) = \dot{\phi}_n(h) = 0, \quad \text{for TM modes} \tag{6b}$$

where $\dot{\phi}_n(y)$ denotes the derivative of $\phi_n(y)$ with respect to y, and p, q, and w are known functions defined by

$$p(y) = w(y) = \begin{cases} 1, & \text{for TE modes} \\ 1/\epsilon(y), & \text{for TM modes} \end{cases} \tag{7a}$$

$$q(y) = \begin{cases} k_0^2\epsilon(y), & \text{for TE modes} \\ k_0^2, & \text{for TM modes.} \end{cases} \tag{7b}$$

Such an eigenvalue problem will yield an infinite set of eigenvalues, or longitudinal propagation constants of the modes, and a corresponding set of eigenfunctions, or transverse mode functions. The explicit solutions of the mode functions can be written down by inspection from the transmission-line system in Fig. 2(b) and are therefore omitted. Instead, we summarize here the general properties of the eigenvalues and eigenfunctions.

We assume that the dielectric materials forming the waveguide in Fig. 2(a) are lossless. The Sturm–Liouville eigenvalue problem defined by (5)–(7) is Hermitian, because of the perfectly-conducting bounding plates at $y=0$ and h. Therefore, all eigenvalues κ_n^2 are real and all eigenfunctions (mode functions) can be chosen to be real. Furthermore, the mode functions of the same type (TE or TM) are mutually orthogonal. With proper normalization, they can be chosen to satisfy the orthonormality relation

$$\langle \phi_m(y)|w(y)|\phi_n(y)\rangle = \int_0^h \phi_m(y)w(y)\phi_n(y)\,dy=\delta_{ij} \quad (8)$$

for every m and n. Here, δ_{ij} stands for the Kronecker delta. On the other hand, a relationship between the mode functions of different types may be obtained, by manipulating (5)–(7) for both TE and TM modes, as

$$(k'_{um})^2\left\langle \phi'_m(y)\left|\frac{1}{\epsilon(y)}\right|\dot{\phi}''_n(y)\right\rangle + (k''_{un})^2\left\langle \phi''_n(y)\left|\frac{1}{\epsilon(y)}\right|\dot{\phi}'_m(y)\right\rangle=0. \quad (9)$$

Such a relation has not previously appeared in the literature, possibly because it has not been needed in the past. For the general case of surface-wave scattering by a step discontinuity at an oblique incidence angle, however, this particular relation will ensure that the TE and TM modes are mutually orthogonal in power, even though the mode functions of one set may not be orthogonal to those of the other set. A proof of this new relation is presented elsewhere [17].

III. Scattering of a Surface Wave Obliquely Incident on a Dielectric Step Discontinuity

It was explained earlier that a dielectric step junction, or discontinuity, corresponds to the side of a dielectric strip waveguide. Following the building-block approach outlined in Section I-B, the step discontinuity is analyzed separately as a constituent in the waveguide cross section. Furthermore, since the waveguiding process is viewed in terms of surface waves bouncing back and forth between the waveguide sides at an angle to the sides, the constituent step-discontinuity problem must require that the surface wave be incident *obliquely* on the step structure. Most treatments in the literature of the scattering of surface waves by a step discontinuity involve normally-incident waves; for this reason, the oblique-incidence case is of interest in its own right.

When a surface wave is incident normally on a dielectric step discontinuity, the boundary-value problem is two-dimensional, and all higher modes excited at the discontinuity possess the same polarization as the incident mode. When the surface wave is incident at an oblique angle, however, the resulting three-dimensional boundary-value problem requires the coupling of TE and TM modes at the discontinuity, as is shown later. A rigorous phasing of the oblique-incidence case has recently appeared in the literature [15], but we also present an input admittance formulation and an equivalent network. Both the equivalent network and the input admittance form are valuable when the step discontinuity is employed as a constituent of more complex structures. It should be added that the new physical effects that emerge when the surface wave is incident obliquely can be exhibited when *only one* surface-wave mode *of each type* is included [18]; it is necessary to include many modes only if accurate numerical results are desired.

In order to simplify the analysis, the step discontinuity is to be treated as a *transverse* discontinuity even though the surface wave is incident at an angle. The TE or TM surface wave which is obliquely incident is thus to be subject to a coordinate transformation which establishes a transmission line formulation for a mode normally incident on the step structure. The surface-wave modes are then no longer TE or TM modes, with three field components, but they become modes with five field components, which have been characterized in the literature as LSE or LSM modes, or alternatively as H-type or E-type modes.

A. LSE (or $H^{(y)}$-Type) Modes and LSM (or $E^{(y)}$-Type) Modes

A dielectric step discontinuity with a surface wave incident on it at an oblique angle is depicted in Fig. 3. The structure is characterized by the xyz coordinate system, with the discontinuity located on the $x=0$ plane. Suppose the incident surface wave is a TE mode with respect to its direction of propagation, denoted by u. The electric-field vector has only one component, in the v direction, perpendicular to u and on the xz plane. Since the transverse mode function of a surface wave is independent of its direction of propagation, the y axis remains unchanged for any angle of incidence. In the uyv coordinate system, each TE or TM mode has only three field components whose spatial variations are well known. In solving the boundary-value problem of surface-wave scattering by a step discontinuity, it is necessary to deal with the two coordinate systems: x, y, and z, to be called the structure coordinate system, and u, y, and v, to be called the eigencoordinate system; evidently, they are mutually related by a coordinate rotation about the y axis. The transformation of the electromagnetic fields of a surface wave from the eigencoordinate system to the structure coordinate system results in an increase in the number of field components from three to five.

When the two coordinate systems are rotated with respect to each other by angle θ about the y axis, they are mutually related by

$$u=x\cos\theta+z\sin\theta \quad (10)$$

$$v=-x\sin\theta+z\cos\theta \quad (11)$$

and the wavenumbers of the field components in the x, y, z

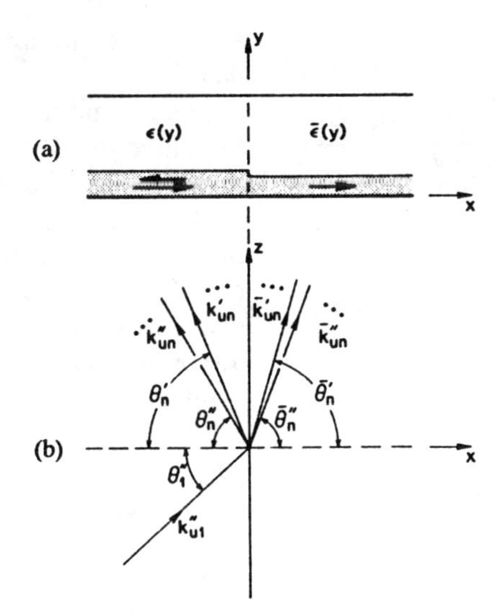

Fig. 3. Scattering of a surface wave by a dielectric step discontinuity. (a) Side view. (b) Top view.

TABLE II
FIELD COMPONENTS OF A SURFACE WAVE MODE IN THE STRUCTURE COORDINATE SYSTEM

LSE or $H^{(y)}$-Type Mode	LSM or $E^{(y)}$-Type Mode
$E'_x = \sin\theta' \exp(-jk'_x x)\phi'(y)$	$E''_x = \dfrac{\cos\theta''}{j\omega\epsilon_0\epsilon(y)}\exp(-jk''_x x)\dfrac{d}{dy}\phi''(y)$
$E'_y = 0$	$E''_y = \dfrac{k''_u}{\omega\epsilon_0\epsilon(y)}\exp(-jk''_x x)\phi''(y)$
$E'_z = -\cos\theta' \exp(-jk'_x x)\phi'(y)$	$E''_z = \dfrac{\sin\theta''}{j\omega\epsilon_0\epsilon(y)}\exp(-jk''_x x)\dfrac{d}{dy}\phi''(y)$
$H'_x = \dfrac{\cos\theta'}{j\omega\mu}\exp(-jk'_x x)\dfrac{d}{dy}\phi'(y)$	$H''_x = -\sin\theta'' \exp(-jk''_x x)\phi''(y)$
$H'_y = \dfrac{k'_u}{\omega\mu_0}\exp(-jk'_x x)\phi'(y)$	$H''_y = 0$
$H'_z = \dfrac{\sin\theta'}{j\omega\mu}\exp(-jk'_x x)\dfrac{d}{dy}\phi'(y)$	$H''_z = \cos\theta'' \exp(-jk''_x x)\phi''(y)$

system become

$$k_x = k_u \cos\theta \tag{12}$$

$$k_z = k_u \sin\theta \tag{13}$$

where k_x and k_y are the projections of the surface-wave propagation vector in the x and y directions. The field components that result from this simple coordinate transformation are listed in Table II.

The modes whose components are listed in Table II have five field components instead of the three possessed by TE and TM modes. These modes are no longer TE or TM, but they are characterizable by the absence of an electric or a magnetic field component in the y direction. Such modes are known in the literature as LSE or LSM modes [19], or as H-type or E-type modes with respect to the y direction [20], if $E_y = 0$ or $H_y = 0$, respectively. Thus, since the surface-wave mode in the first column of Table II possesses a y component of H but not of E, it may be designated an $H^{(y)}$-type mode or an LSE mode; similarly, the mode in the second column is an $E^{(y)}$-type mode or an LSM mode. We recall that these five-component modes enter here because we wish the transmission-line formulation to correspond to normal incidence on the step discontinuity. The physical waves are still TE or TM waves incident on the step junction at an angle.

The x variations of the field components in Table II represent the propagation of surface waves (and non-surface waves) in the forward (transmission-line) direction. In the scattering process, reflection of these modes takes place, resulting in propagation in the backward direction with the spatial variation $\exp(+jk_x x)$. Thus, the general field solution of a rotated mode must consist of both forward and backward propagating waves in the x direction. Such a general modal solution for each field component in the structure coordinate system is listed in Table III. Here, $V(x)$ and $I(x)$, with a single prime for an LSE or $H^{(y)}$-type mode and a double prime for an LSM or $E^{(y)}$-type mode, can be interpreted as the voltage and current satisfying the transmission-line equations

$$\frac{d}{dx}V(x) = -jk_x Z I(x) \tag{14a}$$

$$\frac{d}{dx}I(x) = -jk_x Y V(x) \tag{14b}$$

where k_x is the propagation wavenumber and $Z(=1/Y)$ is the characteristic impedance in the x direction. Further-

TABLE III
GENERAL MODAL SOLUTIONS FOR ALL THE FIELD COMPONENTS

LSE or $H^{(y)}$-Type Mode	LSM or $E^{(y)}$-Type Mode
$E'_x = -\frac{\omega\mu_0}{k'_u}\sin\theta' I'(x)\phi'(y)$	$E''_x = \frac{1}{\omega\epsilon_0} I''(x)\frac{1}{\epsilon(y)}\frac{d}{dy}\phi''(y)$
$E'_y = 0$	$E''_y = jV''(x)\phi''(y)\frac{1}{\epsilon(y)}$
$E'_z = V'(x)\phi'(y)$	$E''_z = \frac{1}{k''_u}\sin\theta'' V''(x)\frac{1}{\epsilon(y)}\frac{d}{dy}\phi''(y)$
$H'_x = j\frac{1}{\omega\mu_0}V'(x)\frac{d}{dy}\phi'(y)$	$H''_x = -j\frac{\omega\epsilon_0}{k''_u}\sin\theta'' V''(x)\phi''(y)$
$H'_y = -I'(x)\phi'(y)$	$H''_y = 0$
$H'_z = j\frac{1}{k'_u}\sin\theta' I'(x)\frac{d}{dy}\phi'(y)$	$H''_z = jI''(x)\phi''(y)$

more, k_x, together with the propagation wavenumber along the step discontinuity in the z direction, is related to the propagation wavenumber k_u of the basic TE or TM mode by

$$k_x^2 + k_z^2 = k_u^2. \tag{15}$$

The characteristic impedance is defined by [20]

$$Z = \begin{cases} \dfrac{\omega\mu_0 k'_x}{(k'_u)^2} = \dfrac{\omega\mu_0 \cos\theta'}{k'_u}, & \text{for LSE or } H^{(y)}\text{-type modes} \\ \dfrac{(k''_u)^2}{\omega\epsilon_0 k''_x} = \dfrac{k''_u}{\omega\epsilon_0\cos\theta''}, & \text{for LSM or } E^{(y)}\text{-type modes.} \end{cases} \tag{16a}$$

Invoking (4b), we obtain the alternative form

$$Z = \begin{cases} k'_x/\omega\epsilon_0\epsilon'_{\text{eff}}, & \text{for LSE or } H^{(y)}\text{-type modes} \\ \omega\mu_0\epsilon''_{\text{eff}}/k''_x, & \text{for LSM or } E^{(y)}\text{-type modes.} \end{cases} \tag{16b}$$

It is noted that for the scattering problem under consideration here, k_z is a known constant and is related to the parameters of the incident surface wave by

$$k_z = k_u \sin\theta \tag{17}$$

where θ is the incidence angle, as depicted in Fig. 3(b). Thus, the transmission-line parameters, k_x and Z, can readily be determined from (15) and (16), and the general modal solution for each field component in Table III can readily be written down, based on the solutions of the transmission-line equations in (14). Therefore, we assume from now on that, for each mode determined in Section III-A, the general modal solution for every field component in the structure coordinate system is known, as given in Table III.

B. Boundary-Value Problem for the Step Discontinuity

The uniform multilayer structure shown in Fig. 2 can support an infinite number of modes. Let k'_{un} and k''_{un} be the propagation wavenumbers of the nth TE and TM modes, respectively. For each mode, the general solutions for all the field components in terms of the structure coordinate system are given in Table III, with k'_u and k''_u replaced by k'_{un} and k''_{un}, respectively. We formulate here the boundary-value problem of surface-wave scattering by a step discontinuity for the general case of oblique incidence.

The two uniform multilayer regions on the two sides of the discontinuity in Fig. 3(a) are characterized by the distributions of dielectric constant, $\epsilon(y)$ and $\bar{\epsilon}(y)$, respectively. As an illustration, let us consider the case for which the TM fundamental mode is incident from the left at an oblique angle θ''_1. As will be shown, all the modes in the two constituent regions will generally be excited at the step discontinuity, some propagating and some decaying away from the discontinuity. The boundary conditions at the step discontinuity require that the total tangential field components be continuous across the step discontinuity, and a necessary condition for the continuity of the tangential field components is that every mode in the two constituent regions must have the same propagation wavenumber, k_z, in the direction along the step discontinuity. From (17), we then have the Snell's law for the various modes at a step discontinuity:

$$k_z = k'_{un}\sin\theta'_n = k''_{un}\sin\theta''_n = \bar{k}'_{un}\sin\bar{\theta}'_n = \bar{k}''_{un}\sin\bar{\theta}''_n \tag{18a}$$

or

$$\begin{aligned} n''_{\text{eff}\,1}\sin\theta''_1 &= n'_{\text{eff}\,n}\sin\theta'_n = n''_{\text{eff}\,n}\sin\theta''_n \\ &= \bar{n}'_{\text{eff}\,n}\sin\bar{\theta}'_n = \bar{n}''_{\text{eff}\,n}\sin\bar{\theta}''_n \end{aligned} \tag{18b}$$

which determines the angles of reflection and transmission for every TE or TM mode, as indicated in Fig. 3(b). With the knowledge of k_z, the propagation wavenumber in the x direction can be determined from (15) by replacing k_u by k_{un} for the nth mode (either TE or TM) to yield

$$k_{xn} = \left[k_{un}^2 - k_z^2\right]^{1/2}. \tag{19}$$

For the TM surface-wave incidence shown in Fig. 3(b), the last equation can be written conveniently in terms of the effective dielectric constants as

$$k''_{xn} = k_0\left[\epsilon''_{\text{eff}\,n} - \epsilon''_{\text{eff}\,1}\sin^2\theta''_1\right]^{1/2} \tag{20}$$

where θ''_1 is the given incidence angle. After k_{xn} is determined, the characteristic impedance of the mode is then specified by (16), with k_x replaced by k_{xn} to become

$$Z_n = \frac{1}{Y_n} = \begin{cases} \dfrac{k'_{xn}}{\omega\epsilon_0\epsilon'_{\text{eff}\,n}}, & \text{for LSE or } H^{(y)}\text{-type modes} \\ \dfrac{\omega\mu_0\epsilon''_{\text{eff}\,n}}{k''_{xn}}, & \text{for LSM or } E^{(y)}\text{-type modes.} \end{cases} \tag{21}$$

Thus, with respect to the x direction, the transmission-line parameters for every mode are determined, and the general modal solutions for all the field components in the structure coordinate system are considered completely determined, as described in the preceding subsection.

Referring to Fig. 3, we observe that the tangential components of the fields at the step discontinuity consist of the y and z components, and we shall therefore consider only those components explicitly. As stated earlier, the general field solution in each constituent region may be expressed in terms of the superposition of the complete set of mode functions. For the tangential field components for the $\epsilon(y)$ region ($x<0$), we have

$$E_y(x,y)=j\sum_{n=1}^{\infty} V_n''(x)\phi_n''(y)\frac{1}{\epsilon(y)} \tag{22a}$$

$$E_z(x,y)=-\sum_{n=1}^{\infty} V_n'(x)\phi'_n(y)-\sum_{n=1}^{\infty} V_n''(x)\psi_n''(y) \tag{22b}$$

$$H_y(x,y)=-\sum_{n=1}^{\infty} I_n'(x)\phi_n'(y) \tag{22c}$$

$$H_z(x,y)=-j\sum_{n=1}^{\infty} I_n'(x)\psi_n'(y)-j\sum_{n=1}^{\infty} I_n''(x)\phi_n''(y) \tag{22d}$$

where we employ the simplifying notation

$$\psi_n'(y)=\frac{1}{k'_{un}}\sin\theta'\frac{d}{dy}\phi_n'(y) \tag{23a}$$

$$\psi_n''(y)=\frac{1}{k''_{un}}\sin\theta''\frac{1}{\epsilon(y)}\frac{d}{dy}\phi_n''(y). \tag{23b}$$

It is noted that the z dependence $\exp(-jk_z z)$ has been suppressed in (22) for clarity. A similar set with an overbar may also be written for the $\bar{\epsilon}(y)$ region ($x>0$), but is omitted here for simplicity. At the step discontinuity at $x=0$, the tangential field components must be continuous. From (22) we obtain

$$\sum_{n=1}^{\infty} V_n''(0)\phi_n''(y)\frac{1}{\epsilon(y)}=\sum_{n=1}^{\infty} \bar{V}_n(0)\bar{\phi}_n''(y)\frac{1}{\bar{\epsilon}(y)} \tag{24a}$$

$$\sum_{n=1}^{\infty} V_n'(0)\phi_n'(y)+\sum_{n=1}^{\infty} V_n''(0)\psi_n''(y)=\sum_{n=1}^{\infty} \bar{V}_n'(0)\bar{\phi}_n'(y)+\sum_{n=1}^{\infty} \bar{V}_n''(0)\bar{\phi}_n''(y) \tag{24b}$$

$$\sum_{n=1}^{\infty} I_n'(0)\phi_n'(y)=\sum_{n=1}^{\infty} \bar{I}_n'(0)\bar{\phi}_n'(y) \tag{24c}$$

$$\sum_{n=1}^{\infty} I_n'(0)\psi_n'(y)+\sum_{n=1}^{\infty} I_n''(0)\phi_n''(y)=\sum_{n=1}^{\infty} \bar{I}_n'(0)\bar{\psi}_n'(y)+\sum_{n=1}^{\infty} \bar{I}_n''(0)\bar{\phi}_n''(y). \tag{24d}$$

These four equations hold for any y at $x=0$ within the enclosure. Scalar-multiplying these equations with either ϕ_m' or ϕ_m'' and making use of the orthogonality relation (8), we then obtain

$$V''=P''\bar{V}'' \tag{25a}$$

$$V'+R''V''=Q'\bar{V}'+S''\bar{V}'' \tag{25b}$$

$$I'=P'\bar{I}' \tag{25c}$$

$$R'I'+I''=S'\bar{I}'+Q''\bar{I}'' \tag{25d}$$

where V' and I' are column vectors with the transmission-line voltage and current of the nth TE mode, $V_n'(0)$ and $I_n'(0)$, at the nth positions; similar definitions hold for V'' and I'' for TM modes and also for those vectors with a superbar. The P's, Q's, R's, and S's are matrices characterizing the coupling of modes at the step discontinuity; their general elements are defined by the scalar products or overlap integrals of mode functions on the two sides of the discontinuity as

$$P'_{mn}=Q'_{mn}=\langle\phi'_m|\bar{\phi}_n'\rangle \tag{26a}$$

$$P''_{mn}=\left\langle\phi_m''\middle|\frac{1}{\bar{\epsilon}(y)}\middle|\bar{\phi}_n''\right\rangle \tag{26b}$$

$$Q''_{mn}=\left\langle\phi_m''\middle|\frac{1}{\epsilon(y)}\middle|\bar{\phi}_n''\right\rangle \tag{26c}$$

$$R'_{mn}=\left\langle\phi_m''\middle|\frac{1}{\epsilon(y)}\middle|\psi_n'\right\rangle \tag{26d}$$

$$R''_{mn}=\langle\phi_m'|\psi_n''\rangle \tag{26e}$$

$$S'_{mn}=\left\langle\phi_m''\middle|\frac{1}{\epsilon(y)}\middle|\bar{\psi}_n'\right\rangle \tag{26f}$$

$$S''_{mn}=\langle\phi_m'|\bar{\psi}_n''\rangle \tag{26g}$$

for any $m,n=1,2,3,\cdots$. It is evident from either (25) or (26) that the matrices P's and Q's are responsible for the coupling among modes of the same polarization, whereas R's and S's are responsible for the cross-coupling among modes of opposite polarization.

For a given incident surface wave, the amplitudes of the scattered modes are determined by the four infinite systems of equations in (25). In practice, these infinite systems of equations must be truncated for an approximate analysis, and we shall do this in connection with the waveguide problem in Section IV.

A set of modal relations alternative to that in (25) can be derived. On use of certain matrix identities, matrices $\boldsymbol{P}$ and $\boldsymbol{S}$ are eliminated from (25), and a new set obtained which is equivalent to that in (25) if all matrices are retained to infinite order. This new set is the following:

$$(Q'')^T V''=\bar{V}'' \tag{27a}$$

$$V'+(R')^T V''=Q'\left[\bar{V}'+(\bar{R}')^T\bar{V}''\right] \tag{27b}$$

$$(Q')^T I'=\bar{I}' \tag{27c}$$

$$R'I'+I''=Q''[\bar{R}'\bar{I}'+\bar{I}'']. \tag{27d}$$

When truncations are made in order to obtain numerical values for the scattering parameters, the two different formulations in (25) and (27) no longer yield identical results, and their convergence properties are also not identical. What is more important, however, is that the alternative formulation in (27) always satisfies the condition of power conservation across the junction, regardless of the number of modes retained after a truncation. A derivation of the alternative formulation in (27) and a proof that it always satisfies the condition of power conservation are presented elsewhere [17].

The set (27) that always preserves the law of conservation of power flow forms the basis for the development of equivalent networks for the dielectric step discontinuity, whereas the other set (25) is useful for numerical analyses of the problem. In fact, because the results obtained after truncation will not satisfy the conservation of power flow, we can use the deviation from conservation as a measure of the numerical accuracy obtained. The alternative formulation in (27) is therefore used as the basis for an equivalent network representation for the dielectric step discontinuity and for an input admittance formulation, presented under Section III-C, which is then employed in the development in Section IV of a generalized transverse-resonance relation for the dielectric strip waveguides.

C. *Input Admittance Matrix for the Step Discontinuity*

In many practical situations, it is necessary to determine only the reflection of a surface wave by a step discontinuity. Moreover, once the reflected mode amplitudes are determined, it is straightforward to determine the transmitted mode amplitudes. For the scattering of a surface wave incident from the left in Fig. 3(a), it is sufficient to have available an input admittance characterization that takes into account the effects of the step discontinuity and the semi-infinite uniform waveguide to the right. We derive now such an input admittance matrix for the step discontinuity.

Each of the higher mode transmission lines to the right of the step discontinuity can be simply represented by its characteristic admittance, and the voltage–current relation at each terminal is then

$$\bar{I}'_n = \bar{Y}'_n \bar{V}'_n \tag{28a}$$

$$\bar{I}''_n = \bar{Y}''_n \bar{V}''_n \tag{28b}$$

for every mode index n. In matrix form, the last two relations may be written as

$$\bar{I}' = \bar{Y}' \bar{V}' \tag{28c}$$

$$\bar{I}'' = \bar{Y}'' \bar{V}'' \tag{28d}$$

where $\bar{V}'$ and $\bar{I}'$ are voltage and current vectors for LSE (or $H^{(y)}$-type) modes in the outside region, with $\bar{V}'_n$ and $\bar{I}'_n$ as their nth positions, respectively, and similarly for $\bar{V}''$ and $\bar{I}''$ for LSM (or $E^{(y)}$-type) modes. $\bar{Y}'$ and $\bar{Y}''$ are the diagonal admittance matrices of the LSE and LSM modes, respectively, for the $\bar{\epsilon}(y)$ region. Substituting the last two equations into (27a) and (27b) and then eliminating $\bar{I}'$ and $\bar{I}''$ by invoking (27c) and (27d), we finally obtain

$$I' = Y_{11} V' + Y_{12} V'' \tag{29a}$$

$$I'' = Y_{21} V' + Y_{22} V'' \tag{29b}$$

where Y_{ij}, for $i, j = 1$ and 2, is an input admittance matrix that is related to the characteristic admittances, $\bar{Y}'_n$ and $\bar{Y}''_n$, and the mode-coupling matrices of the step discontinuity by

$$Y_{11} = Q' \bar{Y}' (Q')^T \tag{30a}$$

$$Y_{12} = Y_{11}\left[R^T - Q' \bar{R}^T (Q'')^T\right] \tag{30b}$$

$$Y_{21} = -\left[R - Q'' \bar{R} (Q')^T\right] Y_{11} \tag{30c}$$

and

$$Y_{22} = Q'' \bar{Y}'' (Q'')^T - Y_{21}\left[R^T - Q' \bar{R} (Q'')^T\right] \tag{30d}$$

where superscript T signifies "transpose." In (29), Y_{11} and Y_{22} are responsible for the coupling of modes of the same polarization and Y_{12} and Y_{21} represent the cross-coupling between modes of opposite polarization. These input admittance matrices can be computed in the straightforward manner described in the preceding, and require only the knowledge of the modal characteristics of the two constituent uniform planar waveguides. Expressions (30) are seen to be simple in form and afford with relative ease a systematic and effective analysis of step discontinuity problems.

IV. Guidance of Waves by Dielectric Strip Waveguides

In Section III we developed a mode-matching formalism for a dielectric step discontinuity between two uniform planar dielectric waveguides. The cross section of a dielectric strip waveguide may be viewed as consisting of two or more step discontinuities connected by a length of uniform waveguide, and the equivalent network for a step discontinuity may therefore be utilized in the analysis of waveguide characteristics. In this section, we follow such a building-block approach to the dielectric strip waveguide problem and we formulate the waveguide problem rigorously in the form of a *generalized transverse-resonance relation*.

A dielectric strip waveguide is shown in Fig. 4(a). As a wave is being guided along the strip, the process may be viewed in terms of multiple reflections which take place at the two step discontinuities forming the waveguide side walls. The basic modes of each constituent region of the waveguide are presented in Section III-A. If we employ the concept of input admittance and apply the equivalent network for the step discontinuities implied by relations (29), we obtain the transverse equivalent network shown in Fig. 4(b) for the analysis of transverse resonance in the lateral direction of the waveguide. The guiding characteristics of the waveguide are completely determined by a single parameter, i.e., the longitudinal propagation wavenumber k_z. As shown in the preceding section, all the parameters of the network in Fig. 4(b) are implicit functions of k_z, and the resonance condition of the network determines the allowable values of k_z for a given waveguide structure.

In practice, most dielectric strip waveguides are symmetric in geometry, such as the one shown in Fig. 4(a). Therefore, the transverse equivalent network in Fig. 4(b) is also symmetric with respect to the center plane. Such a network may be analyzed in terms of the two simpler networks obtained from open-circuit and short-circuit bisections, as shown in Fig. 5 for symmetric and antisymmetric distributions of voltage, or electric field, in the original waveguide. For simplicity, we shall deal only with symmetric structures in this paper; the generalization for asymmetric structures is almost trivial and is omitted.

Referring to Fig. 5, the relationship between the voltages and currents at the junctions at $x=0$ may be expressed in

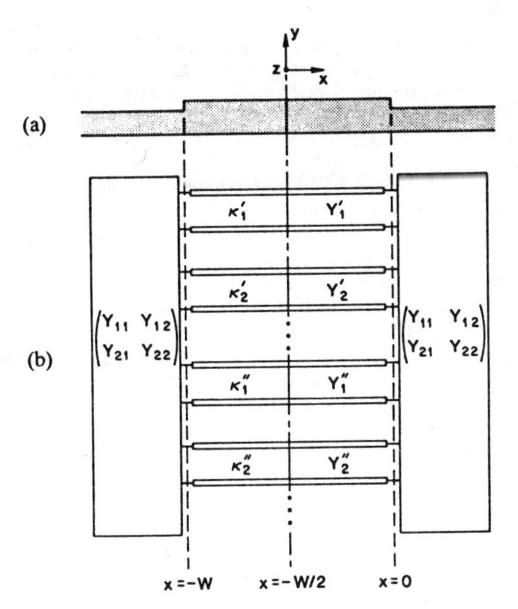

Fig. 4. Equivalent network for the transverse resonance in the lateral direction of a dielectric strip waveguide. (a) Dielectric strip waveguide. (b) Transverse equivalent network.

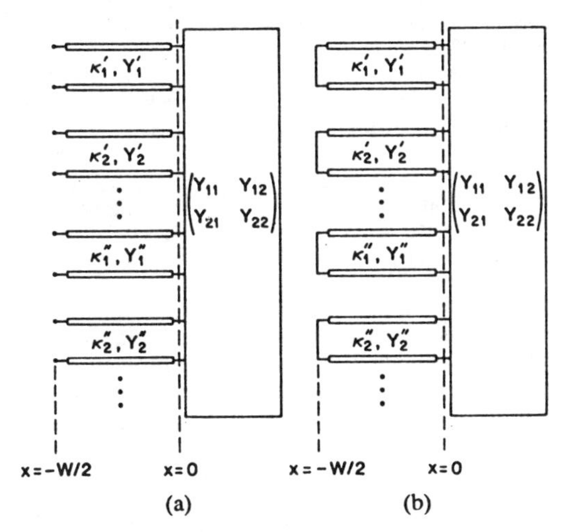

Fig. 5. Bisected transverse equivalent networks for the waveguide in Fig. 4. (a) Symmetric or open-circuit bisection. (b) Antisymmetric or short-circuit bisection.

terms of the transmission-line characteristics as

$$I' = -\overleftarrow{Y}'V' = -D(\overleftarrow{Y}_n')V' \tag{31a}$$

$$I'' = -\overleftarrow{Y}''V'' = -D(\overleftarrow{Y}_n'')V'' \tag{31b}$$

where $D(Y_n)$ stands for a diagonal matrix with Y_n as its element at the nth diagonal position, and therefore $\overleftarrow{Y}'$ and $\overleftarrow{Y}''$ are diagonal matrices with the diagonal elements defined by the input admittances of the transmission-line sections $\overleftarrow{Y}_n'$ and $\overleftarrow{Y}_n''$ for LSE and LSM modes, respectively. The negative signs in (31) reflect the fact that the currents in the transmission lines flow in the positive x direction, while the admittances are defined with respect to the opposite direction. Explicitly, for the two bisected networks, we have

$$\overleftarrow{Y}_n' = jY_n' \tan \kappa_n' W/2 \tag{32a}$$

$$\overleftarrow{Y}_n'' = jY_n'' \tan \kappa_n'' W/2 \tag{32b}$$

for the open-circuit bisection, and

$$\overleftarrow{Y}_n' = -jY_n' \cot \kappa_n' W/2 \tag{33a}$$

$$\overleftarrow{Y}_n'' = -jY_n'' \cot \kappa_n'' W/2 \tag{33b}$$

for the short-circuit bisection. More succinctly, the two equations in (31) may be combined to become

$$I = -\overleftarrow{Y}V \tag{34}$$

where I is a super column vector with the column vectors I' and I'' as its elements, and similarly for V. $\overleftarrow{Y}$ is a super matrix defined by

$$\overleftarrow{Y} = \begin{pmatrix} \overleftarrow{Y}' & 0 \\ 0 & \overleftarrow{Y}'' \end{pmatrix} \tag{35}$$

where the elements at the nth positions in the diagonal matrices $\overleftarrow{Y}'$ and $\overleftarrow{Y}''$ are given by either (32) or (33). For simplicity, $\overleftarrow{Y}$ will be called the admittance matrix looking to the left.

On the other hand, the relationship between the voltages and currents may also be expressed in terms of the input-admittance matrices for the step discontinuity as defined in (29). More succinctly, (29) may be written as

$$I = \overrightarrow{Y}V \tag{36}$$

where $\overrightarrow{Y}$ will be called the admittance matrix looking to the right, and the super matrix is defined by

$$\overrightarrow{Y} = \begin{pmatrix} Y_{11} & Y_{12} \\ Y_{21} & Y_{22} \end{pmatrix}. \tag{37}$$

Evidently, (34) and (36) are two different equations relating the same set of voltages to the same set of currents. Equating (34) and (36), we obtain

$$(\overleftarrow{Y} + \overrightarrow{Y})V = 0 \tag{38}$$

which is a system of linear homogeneous equations to determine the modal voltages at $x=0$. When the solution is obtained, the modal currents I at $x=0$ can be simply determined from either (34) or (36). With the terminal voltages and currents now known, the voltages and currents everywhere in the transmission-line system can be determined by standard techniques and the electromagnetic fields everywhere within the dielectric waveguide are completely specified via Table III.

For the linear homogeneous system of equation (38), the condition for the existence of a nontrivial solution is that the determinant of the coefficient matrix vanishes, namely

$$\det(\overleftarrow{Y} + \overrightarrow{Y}) = 0. \tag{39}$$

In other words, this is a condition under which nonzero voltages may exist in the absence of any source excitation in the network of Fig. 5(a) or (b); alternatively, the networks are in resonance. Again, the admittances in (38) are all functions of the longitudinal propagation wavenumber k_z. Therefore, (39) is the equation to determine the allowable values of k_z, and it will be simply referred to as the *generalized transverse-resonance relation* or *dispersion relation* for the dielectric waveguide.

For a single-mode case, $\overleftarrow{Y}$ and $\overrightarrow{Y}$ are scalar admittances, and (39) becomes the familiar transverse-resonance relation that states that, for a network system to be in resonance, the sum of the admittances (or impedances) looking into the two opposite directions at any point within the network system must vanish. Clearly, (39) is a generalization of the scalar transverse-resonance relation for a single-mode case to the matrix one for a multimode case, in order to account for the effect of mode coupling. Some important virtues of transverse-resonance relation (39) are 1) it is exact, 2) it is simple in form, 3) it is easily adaptable to more complex waveguide structures, and 4) it is an effective tool for a systematic numerical analysis. Furthermore, being an exact transverse-resonance relation, it can be used as a basis for developing approximation techniques that will exhibit the effect of mode coupling, and also to identify new physical phenomena that may take place in the waveguide structures.

REFERENCES

[1] a) H. Kogelnik, "Theory of dielectric wave guides," in *Integrated Optics*, T. Tamir, Ed. New York: Springer Verlag, 1975, ch. 2.
b) H. G. Unger, "Planar optical waveguides and fibers," Oxford, England: Oxford Univ. Press, 1978, ch. 3.

[2] W. V. McLevige, T. Itoh, and R. Mittra, "New waveguide structures for millimeter-wave and optical integrated circuits," *IEEE Trans. Microwave Theory Tech.*, vol. MTT-23, pp. 788–794, Oct. 1975.

[3] H. Jacobs and M. M. Chrepta, "Semiconductor dielectric waveguides for millimeter-wave functional circuits," in *Dig. IEEE Int. Microwave Symp.*, pp. 28–29, 1973.

[4] S. T. Peng and A. A. Oliner, "Leakage and resonance effects on strip waveguides for integrated optics," *Trans. Electronics and Commun. Eng. Jap.* (Special Issue on Integrated Optics and Optical Fiber Communications), vol. E61, pp. 151–154, Mar. 1978.

[5] K. Ogusu and I. Tanaka, "Optical strip waveguide: An experiment," *Appl. Opt.*, vol. 19, pp. 3322–3325, Oct. 1, 1980.

[6] E. A. J. Marcatili, "Dielectric rectangular waveguide and directional coupler for integrated optics," *Bell Syst. Tech J.*, vol. 48, pp. 2071–2102, Sept. 1969.

[7] R. M. Knox and P. P. Toulios, "Integrated circuits for the millimeter wave through optical frequency range," in *Proc. Symposium on Submillimeter Waves*. Brooklyn, NY: Polytechnic Press, Apr. 1970, pp. 497–516.

[8] H. Furuta, H. Noda, and A. Ihaya, "Novel optical waveguide for integrated optics," *Appl. Opt.*, vol. 13, pp. 322–326, Feb. 1974.

[9] N. Uchida, "Optical waveguide loaded with high refractive-index strip film," *Appl. Opt.*, vol. 15, pp. 179–182, Jan. 1976.

[10] V. Ramaswamy, "Strip-loaded film waveguides," *Bell Syst. Tech. J.*, vol. 53, pp. 697–704, Apr. 1974.

[11] T. Itoh, "Inverted strip dielectric waveguide for millimeter-wave integrated circuits," *IEEE Trans. Microwave Theory Tech.*, vol. MTT-24, pp. 821–827, Nov. 1976.

[12] R. Mittra, Y. L. Hou, and V. Jamnejad, "Analysis of open dielectric waveguides using mode-matching technique and variational methods," *IEEE Trans. Microwave Theory Tech.*, vol. MTT-28, pp. 36–43, Jan. 1980.

[13] E. W. Hu, S. T. Peng, and A. A. Oliner, "A novel leaky-wave strip waveguide directional coupler," in *Topical Meet. Integrated and Guided Wave Optics*, Paper No. WD 2 (Salt Lake City, UT) Jan. 1978.

[14] K. Solbach and I. Wolff, "The electromagnetic fields and the phase constants of dielectric image lines," *IEEE Trans. Microwave Theory Tech.*, vol. MTT-26, pp. 266–274, Apr. 1978.

[15] K. Ogusu, S. Kawakami and S. Nishida, "Optical strip waveguide: An analysis," *Appl. Opt.*, vol. 18, pp. 908–914, Mar. 15, 1979. *Ibid.*, Correction, vol. 18, p. 3725, Nov. 1979.

[16] W. Schlosser and H. G. Unger, "Partially filled waveguides and surface waveguides of rectangular cross section," in *Advances in Microwaves*, L. Young, Ed., New York: Academic Press, 1966, vol. 1, pp. 319–387.

[17] S. T. Peng and A. A. Oliner, "Mathematical formulations for the guidance and leakage properties of dielectric strip waveguides," forthcoming U.S. Army CORADCOM Report, Fort Monmouth, N.J.

[18] J. P. Hsu, S. T. Peng, and A. A. Oliner, "Scattering by dielectric step discontinuities for obliquely incident surface waves," in *Dig. URSI Meet.*, p. 46 (College Park, MD), May 1978.

[19] R. E. Collin, *Field Theory of Guided Waves*. New York: McGraw-Hill, 1960, ch. 6.

[20] H. M. Altshuler and L. O. Goldstone, "On network representations of certain obstacles in waveguide regions," *IRE Trans. Microwave Theory Tech.*, vol. MTT-7, pp. 213–221, Apr. 1959.

FULL-WAVE ANALYSIS OF INTEGRATED TRANSMISSION LINES ON LAYERED LOSSY MEDIA

Indexing terms: Integrated circuits, Transmission lines

Cohn's method of analysis of the slot line on a dielectric substrate is extended to the case of integrated lines on a lossy inhomogeneous substrate. The slow-wave characteristics of the metal-insulator-semiconductor coplanar waveguide for MMIC applications are calculated and found to be in good agreement with the experiments.

Introduction: The development of monolithic microwave integrated circuits (MMIC) is stimulating a great interest in the analysis of various configurations of printed transmission lines on semiconducting substrates. A particularly interesting feature consists of the possibility of slow-wave propagation which would permit a substantial size reduction of distributed passive components. Slow-wave propagation has been recognised both theoretically and experimentally in metal-insulator-semiconductor (MIS) and Schottky-barrier microstrip lines[1-3] and coplanar waveguides.[4] Analyses of these structures, however, have been generally performed under some rough approximations, a full-wave analysis having been presented only for the case of shielded MIS microstrip lines,[5] using the generalised spectral domain approach and assuming zero thickness of the metallisation. A different approach to the study of planar transmission lines for MMICs is presented here. Modifying the method developed by Cohn[6] for the analysis of the slot line on a dielectric substrate, a full-wave analysis of planar transmission lines can be developed which accounts for inhomogeneous and lossy substrates as well as for finite thickness of the strip and ground conductors. The method of analysis is outlined for the case of the coplanar waveguide which appears to have some advantages over the microstrip configuration;[4] with simple modifications, however, it can be applied to other printed transmission lines such as slot line, fin line etc.

Method of analysis: A MIS coplanar waveguide (CPW) such as the one tested by Hasegawa *et al.*[4] is sketched in Fig. 1*a*.

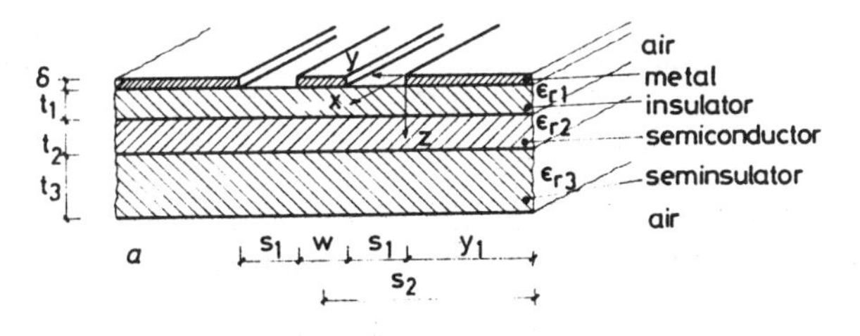

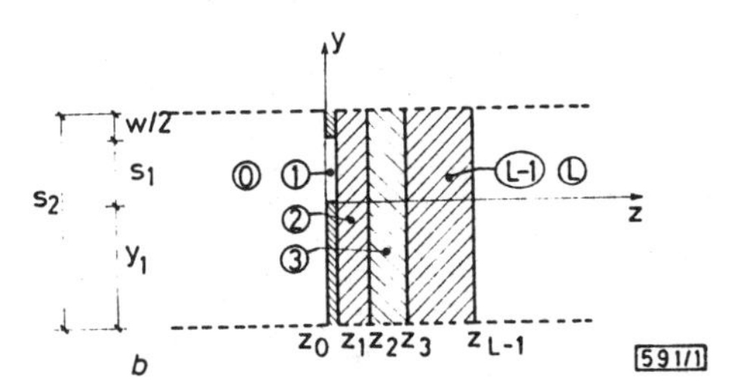

Fig. 1
a MIS coplanar waveguide
b Cross-section of reduced structure

This type of structure can be considered as a CPW fabricated on a substrate consisting of $L - 2$ isotropic and, generally, lossy layers having different permittivities. Contrary to Cohn's analysis of the slot line,[6] the present configuration cannot be converted into a rectangular waveguide problem by assuming the existence of a purely standing wave in the x direction; this is because losses prevent the existence of standing waves. It is possible, however, to convert the configuration of Fig. 1*a* into a parallel-plate waveguide (PPW) problem by inserting two longitudinal electric or magnetic walls perpendicular to the substrate, sufficiently apart from the slots, so that their effect on the field distribution, mainly concentrated in the proximity of the slots, can be neglected.[6] If the CPW is symmetrical, as we shall suppose for simplicity, the analysis of even (odd) modes can be performed by inserting a longitudinal magnetic (electric) wall at the centre of the strip conductor. This allows only one half of the structure to be analysed.

Fig. 1*b* shows the cross-section of the reduced structure, which appears as a PPW with magnetic plates, loaded with a metallic iris of finite thickness and with $L - 2$ lossy slabs. Because of its finite thickness, the iris itself may be regarded as a PPW section with electric plates. The structure can be analysed through a mode-matching procedure. Assuming an x-dependence as $\exp(-\gamma x)$, each of the four transverse components of the EM field inside the ith ($i = 0, 1, \ldots, L$) section of the structure can be expressed in the form

$$\phi(x, y, z) = \sum_n \psi_n(z) e^{-\gamma x} \begin{Bmatrix} \sin \dfrac{n\pi y}{s} \\ \cos \dfrac{n\pi y}{s} \end{Bmatrix} \tag{1}$$

where s is the spacing between the plates; sin or cos functions are used depending on the EM field component and on the nature (electric or magnetic) of the plates. Each term of the above series corresponds to four EM waves: two TE and two TM waves travelling in the $+z$ and $-z$ directions; the $\psi_n(z)$ expansion coefficients can be expressed in terms of the amplitudes of these waves:

$$\psi_n(z) = (a_n^+ A_n^+ + b_n^+ B_n^+) e^{-\beta_n z} + (a_n^- A_n^- + b_n^- B_n^-) e^{\beta_n z} \tag{2}$$

where

$$\beta_n^2 = -\gamma^2 + \left(\frac{n\pi}{s}\right)^2 - \omega^2 \mu \varepsilon$$

μ, ε being the permeability and complex permittivity of the ith section; the As and Bs are the unknown amplitudes of the TE and TM waves, while the as and bs are known quantities that can be easily obtained from Maxwell's equations, and depend on the EM field component represented by ϕ.

It should be observed that the number of terms required for a good approximation of the actual field distribution is different for the narrower section 1 and for the wider sections 0, 2, 3, ..., L, the latter being much higher, typically s_2/s_1 (see Fig. 1) times the former one. In Cohn's analysis of the slot line, for instance, only one mode, corresponding to a constant field distribution, is considered in the slot. While such an approximation is generally not allowed in the present case, because of the coupling between the slots, nevertheless only a few modes in section 1 are sufficient in practical cases.

Let N and M be the number of terms of the field expansions used in the narrower ($i = 1$) and wider ($i \neq 1$) sections, respec-

Reprinted with permission from *Electron. Lett.*, vol. 18, no. 14, pp. 607–609, July 8, 1982.

tively. Because of eqn. 2, each term depends on four-unknown constants; we have therefore $4 \times (N + M \times L)$ unknowns. The boundary conditions at $z = z_0, z_1, \ldots, z_{L-1}$ and $z = \pm\infty$ lead to a homogeneous system of equations in these unknowns. It is essential, for computational reasons, to manipulate the equations in order to reduce their number. It can be shown that the system of equations can be reduced to a system in the only unknowns representing the wave amplitudes in section 1; in other words the number of equations is reduced to $4 \times N$, which is a reasonable number for most practical applications. This results in a relatively short computing time for calculating the propagation constants of the structure.

Results: The method has been first tested on slot lines on dielectric substrates. An example is shown in Fig. 2, where the computed wavelength ratio λ_g/λ_0 against frequency is compared with the theoretical and experimental results quoted in Reference 7. The analysed slot line was realised on a 1·6 mm-thick dielectric substrate with $\varepsilon_r = 16{\cdot}3$; the slot was 1·57 mm wide and the metallisation 0·02 mm thick. As is seen, the computed results fit the experiment better than those obtained by the original Cohn's method; the differences are due to the inclusion of the finite thickness of the metallisation and to the higher number of terms ($N = 2$ instead of $N = 1$) used in the field expansion in the slot.

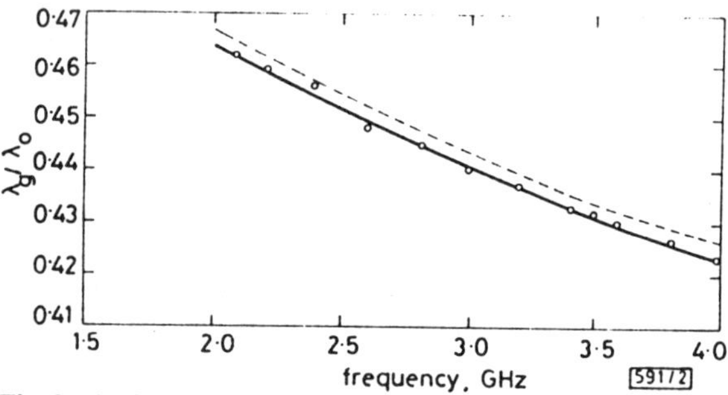

Fig. 2 *Slot-line wavelength ratio*

○ experiment[7]
- - - - Cohn's theory[6]
——— this theory

The three-layer MIS coplanar structure experimentally tested in References 4 and 8 has been then analysed. In this case, because of the strong coupling between the slots, a higher number of terms have to be used in the field expansion in section 1. The computed behaviour against frequency of the slow-wave factor λ_0/λ_g and the attenuation α using $N = 4$ terms are shown in Fig. 3 and are in agreement with the experimental ones.

We have furthermore investigated the characteristic of the propagation as a function of the semiconductor's resistivity ρ. Fig. 4 shows the computed behaviour of λ_0/λ_g and α against ρ at a frequency of 1 GHz. This Figure shows the existence of an optimum value of the resistivity ($1{\cdot}5 \times 10^{-2}$ Ω mm) at which the attenuation attains a minimum, while the slow-wave factor has a maximum. This behaviour is consistent with that predicted in Reference 2 for MIS microstrip structures. For the optimum value of ρ we have computed again λ_0/λ_g and α against frequency (dashed curves in Fig. 3). Better performance for slow-wave propagation is obtained in the whole frequency range considered. It can also be noted that the experimental points are somewhat intermediate between the curves for $\rho = 5{\cdot}5 \times 10^{-2}$ and $\rho = 1{\cdot}5 \times 10^{-2}$ Ω mm.

In conclusion the method of analysis presented allows the propagation characteristics of printed transmission lines for MMIC applications to be fully predictable. Some disagreement with experimental results may be ascribed to the critical values of the structural parameters and to the possible arising of nonlinear effects due to the presence of the semiconductor layer.

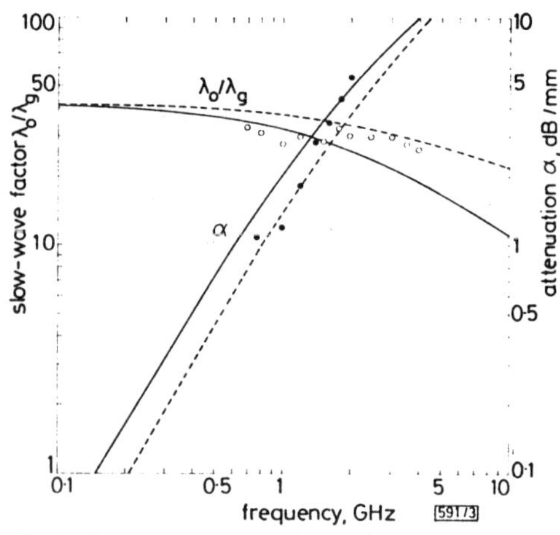

Fig. 3 *Slow-wave factor and attenuation against frequency of MIS CPW in Fig. 1*

$w = 100$ μm, $s_1 = 450$ μm, $\delta = 1$ μm, $t_1 = 0{\cdot}4$ μm, $t_2 = 3$ μm, $t_3 = 1$ mm, $\varepsilon_{r1} = 8{\cdot}5$, $\varepsilon_{r2} = 13{\cdot}1$, $\varepsilon_{r3} = 13{\cdot}1$

Resistivity of semiconductor:
——— $\rho = 5{\cdot}5 \times 10^{-2}$ Ω mm
- - - - $\rho = 1{\cdot}5 \times 10^{-2}$ Ω mm
●, ○ experiment[4,8]

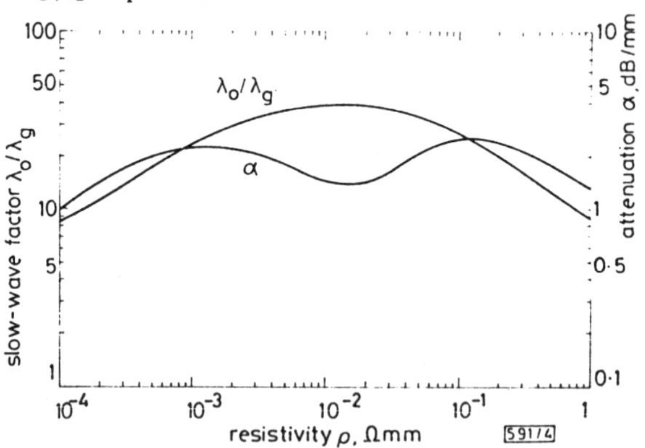

Fig. 4 *Slow-wave factor and attenuation against resistivity of semiconductor*

Acknowledgment: This work has been partially supported by the CNR, Italy.

R. SORRENTINO
G. LEUZZI

Istituto di Elettronica
Università di Roma
00184 Rome, Italy

2nd June 1982

References

1 HASEGAWA, H., FURUKAWA, M., and YARAI, H.: 'Properties of microstrip line on Si-SiO system', *IEEE Trans.*, 1971, **MTT-19,** pp 869–881

2 HUGHES, G. W., and WHITE, R. M.: 'Microwave properties of nonlinear MIS and Schottky-barrier microstrip', *ibid.*, 1975, **ED-22,** pp. 945–956

3 JÄGER, D., and RABUS, W.: 'Bias-dependent phase delay of Schottky contact microstrip line', *Electron. Lett.*, 1973, **9,** pp. 201–203

4 HASEGAWA, H., and OKIZAKI, H.: 'MIS and Schottky slow-wave coplanar striplines on GaAs substrates', *ibid.*, 1977, **13,** pp. 663–664

5 KENNIS, P., and FAUCON, L.: 'Rigorous analysis of planar MIS transmission lines', *ibid.*, 1981, **17,** pp. 454–456

6 COHN, S. B.: 'Slot line on a dielectric substrate', *IEEE Trans.*, 1969, **MTT-17,** pp. 768–778

7 MARIANI, E. A., HEINZMAN, C. P., AGRIOS, J. P., and COHN, S. B.: 'Slot line characteristics', *ibid.*, 1969, **MTT-17,** 1091–1096

8 SEKI, S., and HASEGAWA, H.: Cross-tie slow-wave coplanar waveguide on semi-insulating GaAs substrate', *Electron. Lett.*, 1981. **17,** pp. 940–941

9 LAMPARIELLO, P., and SORRENTINO, R.: 'The ZEPLS program for solving characteristic equations of electromagnetic structures', *IEEE Trans.*, 1975, **MTT-23,** pp. 457–458

A Unified Hybrid-Mode Analysis for Planar Transmission Lines with Multilayer Isotropic/Anisotropic Substrates

RAAFAT R. MANSOUR, MEMBER, IEEE, AND ROBERT H. MACPHIE, SENIOR MEMBER, IEEE

Abstract —**A unified hybrid-mode analysis is presented for determining the propagation characteristics of multiconductor, multilayer planar transmission lines. The analysis employs the conservation of complex power technique, and the emphasis is on numerical efficiency and simplicity. Numerical results, for finline and microstrip configurations, aim at the clarification of the effects of the metalization thickness, dielectric anisotropy, and substrate mounting grooves.**

I. Introduction

DURING THE PAST several years, a variety of techniques have been published for the characterization of planar transmission lines, techniques wherein there has been always a compromise between accuracy and numerical efficiency. In many cases, simplified approximations were introduced to achieve a good numerical efficiency at the expense of the accuracy. For example, most published techniques do not take into account the effect of the substrate mounting groove. Elsewhere numerous results are reported for idealized structures with zero metallization thickness.

With the increasing complexity of microwave and millimeter-wave circuit design, attention has been directed in recent years to generalized approaches that can treat a variety of planar transmission lines with more complicated configurations. Thus, in developing a new technique the "generalization" has become another challenging factor to be considered in addition to the accuracy and numerical efficiency.

Among the published rigorous techniques for analyzing planar transmission lines are the spectral-domain technique and the singular integral equation technique [1], [2]. Although these techniques have a very good numerical efficiency, they do not include the effects of metallization thickness and substrate mounting grooves. The metallization thickness has been taken into account by Beyer [3] and Vahldieck [4] using the mode-matching technique, and also by Kitazawa and Mittra [5] using the network analytical method. The effect of mounting grooves has been considered as well in [3] and [4], but these methods were presented to treat only finline structures. The effect of the metallization thickness on the propagation characteristics of multiconductor planar transmission lines has been approximately taken into account by Saad and Schunemann [6]. However, the approximation involved in this method is only valid for structures with large slot widths.

Very recently, an approach based on the mode-matching technique has been presented by Vahldieck and Bornemann in [7] to calculate the propagation constant, and extended by Bornemann and Arndt in [8] to calculate the characteristic impedance. However, as will be shown in Section II, this approach has a very poor numerical efficiency. Moreover, in view of the study given in [9] on the convergence of the modal analysis numerical solution, the mode-matching formulation used in this approach suffers from serious convergence problems and may fail to provide accurate results for structures with relatively small metallization thickness.

In this contribution, we present the details of a hybrid-mode approach for evaluating the propagation constants of the dominant and the higher order modes and the characteristic impedance for planar transmission lines with multiconductor and multilayer isotropic/anisotropic substrates. Besides the versatility and flexibility of this approach in treating complicated structures, it is numerically efficient and it includes the effects of the metallization thickness and substrate mounting grooves.

II. Formulation of the Problem

A generalized planar guiding structure is shown in Fig. 1. It consists of an arbitrary number of metallic strips deposited on various interfaces of a multilayer isotropic/anisotropic dielectric substrate. The hybrid nature of the electromagnetic field in this structure can be attributed to the coupling between LSE and LSM modes. By considering the propagation to take place along the transverse direction, the discontinuities at the various vertical planes serve to couple the LSE and LSM modes, and the hybrid modes are formed as a result of the repeated reflections of LSE and LSM modes from the short-circuited ends and the discontinuities.

Manuscript received April 2, 1987; revised August 17, 1987. This work was supported in part by the Natural Sciences and Engineering Research Council, Ottawa, Canada, under Grant A-2176.

R. R. Mansour is with COM DEV Ltd., Cambridge, Ontario, Canada NIR 7H6.

R. H. MacPhie is with the Department of Electrical Engineering, University of Waterloo, Waterloo, Ontario, Canada N2L 3G1.

IEEE Log Number 8717395.

Reprinted from *IEEE Trans. Microwave Theory Tech.*, vol. MTT-35, no. 12, pp. 1382–1391, Dec. 1987.

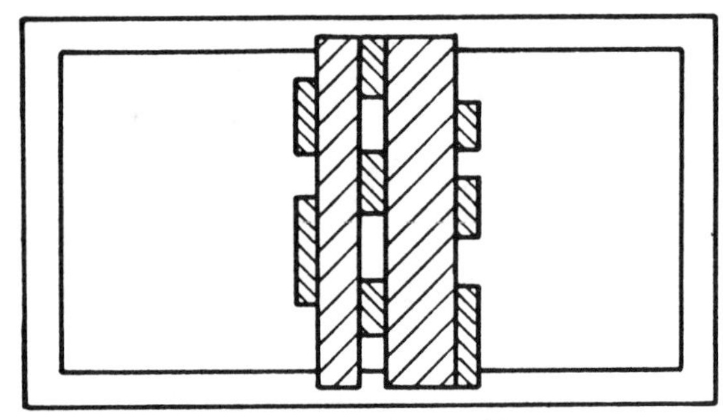

Fig. 1. A generalized planar guiding structure.

The first step in the analysis uses the conservation of complex power technique (CCPT) to treat the problem of scattering at the N-furcated waveguide discontinuity shown in Fig. 2 for LSE and LSM excitation. The LSE and LSM modes may be derived from magnetic- and electric-type Hertzian potential functions [10] having single components directed normal to the xz plane. The appropriate solutions for $\vec{\Pi}^h_{in}$ and $\vec{\Pi}^e_{in}$ in the ith guide are given by

$$\vec{\Pi}^h_{in} = \frac{1}{j\omega\mu K_{c,in}} \cos\frac{n\pi}{L_i}x\, e^{-j\alpha^h_{in}y}\, e^{-j\beta z} a_y \tag{1}$$

$$\vec{\Pi}^e_{in} = \frac{1}{j\alpha_{in} K_{c,in}} \sin\frac{n\pi}{L_i}x\, e^{-j\alpha^e_{in}y}\, e^{-j\beta z} \vec{a}_y. \tag{2}$$

The propagation constant along the z direction is represented by the exponential factor $e^{-j\beta z}$ and assumed to be the same in all guides and for both LSE and LSM modes. The transverse components of the fields may then given by

$$\vec{e}^{\,h}_{in} = \frac{1}{K_{c,in}}\left[-j\beta\cos\frac{n\pi}{L_i}x\,\vec{a}_x + \frac{n\pi}{L_i}\sin\frac{n\pi}{L_i}x\,\vec{a}_z\right] e^{-j\alpha^h_{in}y} e^{-j\beta z} \tag{3a}$$

$$\vec{h}^{h}_{in} = \frac{Y^h_{in}}{K_{c,in}}\left[+\frac{n\pi}{L_i}\sin\frac{n\pi}{L_i}x\,\vec{a}_x + j\beta\cos\frac{n\pi}{L_i}x\,\vec{a}_z\right] e^{-j\alpha^h_{in}y} e^{-j\beta z} \tag{3b}$$

$$\vec{e}^{\,e}_{in} = \frac{1}{K_{c,in}}\left[-\frac{n\pi}{L_i}\cos\frac{n\pi}{L_i}x\,\vec{a}_x + j\beta\sin\frac{n\pi}{L_i}x\,\vec{a}_z\right] e^{-j\alpha^e_{in}y} e^{-j\beta z} \tag{3c}$$

$$\vec{h}^{e}_{in} = \frac{Y^e_{in}}{K_{c,in}}\left[+j\beta\sin\frac{n\pi}{L_i}x\,\vec{a}_x + \frac{n\pi}{L_i}\cos\frac{n\pi}{L_i}x\,\vec{a}_z\right] e^{-j\alpha^e_{in}y} e^{-j\beta z} \tag{3d}$$

where

$$\alpha^h_{in} = \alpha^e_{in} = \left[\omega^2\mu\epsilon_i - \left(\frac{n\pi}{L_i}\right)^2 - \beta^2\right]^{1/2}$$

$$K_{c,in} = \left[\left(\frac{n\pi}{L_i}\right)^2 + \beta^2\right]^{1/2},$$

$n \quad = 0,1,2,\cdots$ for LSE

$n \quad = 1,2,3,\cdots$ for LSM

$$Y^h_{in} = \frac{\alpha^h_{in}}{\omega\mu} \qquad Y^e_{in} = \frac{\omega\epsilon}{\alpha^e_{in}} \qquad \text{(modal admittances).}$$

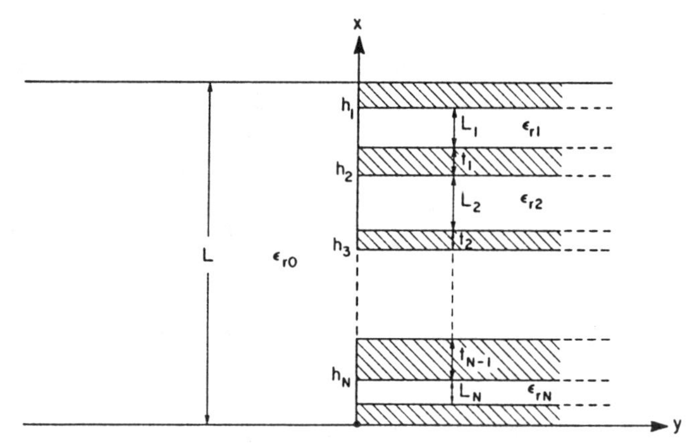

Fig. 2. An N-furcated waveguide junction.

Let the transverse fields in the ith guide at $y = 0$, $z = 0$ be represented[1] by an infinite sum of LSE-mode electric fields $\vec{e}^{\,h}_{in}$ and LSM-mode electric fields $\vec{e}^{\,e}_{in}$:

$$\vec{E}_i(x) = \sum_{n=0,1,2,\cdots} A^h_{in}\vec{e}^{\,h}_{in}(x) + \sum_{n=1,2,3,\cdots} A^e_{in}\vec{e}^{\,e}_{in}(x) \tag{4}$$

where A^h_{in} and A^e_{in} are the nth LSE- and LSM-mode amplitudes respectively in the ith guide, $i = 0,1,2,\ldots, N$. Because of the coupling between the LSE and LSM modes, the three basic matrices $\boldsymbol{H}$, $\boldsymbol{P}_A$, and $\boldsymbol{P}_B$ defined in the CCPT formulation [11] can be written

$$\begin{bmatrix} \underline{A}^h_0 \\ \underline{A}^e_0 \end{bmatrix} = \begin{bmatrix} \boldsymbol{H}^{hh}_1 & \boldsymbol{H}^{he}_1 & \boldsymbol{H}^{hh}_2 & \cdots & \boldsymbol{H}^{hh}_N & \boldsymbol{H}^{he}_N \\ \boldsymbol{H}^{eh}_1 & \boldsymbol{H}^{ee}_1 & \boldsymbol{H}^{eh}_2 & \cdots & \boldsymbol{H}^{eh}_N & \boldsymbol{H}^{ee}_N \end{bmatrix} \begin{bmatrix} \underline{A}^h_1 \\ \underline{A}^e_1 \\ \underline{A}^h_2 \\ - \\ - \\ \underline{A}^h_N \\ \underline{A}^e_N \end{bmatrix} \tag{5}$$

$$\boldsymbol{H} = \begin{bmatrix} \boldsymbol{H}^{hh}_1 & \boldsymbol{H}^{he}_1 & \boldsymbol{H}^{hh}_2 & \cdots & \boldsymbol{H}^{hh}_N & \boldsymbol{H}^{he}_N \\ \boldsymbol{H}^{eh}_1 & \boldsymbol{H}^{ee}_1 & \boldsymbol{H}^{eh}_2 & \cdots & \boldsymbol{H}^{eh}_N & \boldsymbol{H}^{ee}_N \end{bmatrix} \tag{6}$$

$$\boldsymbol{P}_A = \begin{bmatrix} \boldsymbol{P}^h_A & \boldsymbol{0} \\ \boldsymbol{0} & \boldsymbol{P}^e_A \end{bmatrix} \tag{7a}$$

$$\boldsymbol{P}_B = \begin{bmatrix} \boldsymbol{P}^h_{B1} & \boldsymbol{0} & \boldsymbol{0} & -- & \boldsymbol{0} & \boldsymbol{0} \\ \boldsymbol{0} & \boldsymbol{P}^e_{B1} & \boldsymbol{0} & -- & \boldsymbol{0} & \boldsymbol{0} \\ \boldsymbol{0} & \boldsymbol{0} & \boldsymbol{P}^h_{B2} & -- & \boldsymbol{0} & \boldsymbol{0} \\ - & - & - & -- & - & - \\ - & - & - & -- & - & - \\ \boldsymbol{0} & \boldsymbol{0} & \boldsymbol{0} & -- & \boldsymbol{P}^h_{BN} & \boldsymbol{0} \\ \boldsymbol{0} & \boldsymbol{0} & \boldsymbol{0} & -- & \boldsymbol{0} & \boldsymbol{P}^e_{BN} \end{bmatrix}. \tag{7b}$$

where h and e refer to LSE and LSM modes, respectively. $\boldsymbol{H}$ is the E-field mode-matching matrix which represents the coupling between the modes in the large guide and

[1] From now on we will use the following notations: † denotes Hermitian transpose; * denotes complex conjugate. All column matrices are denoted with an underbar, and all matrices are in boldface.

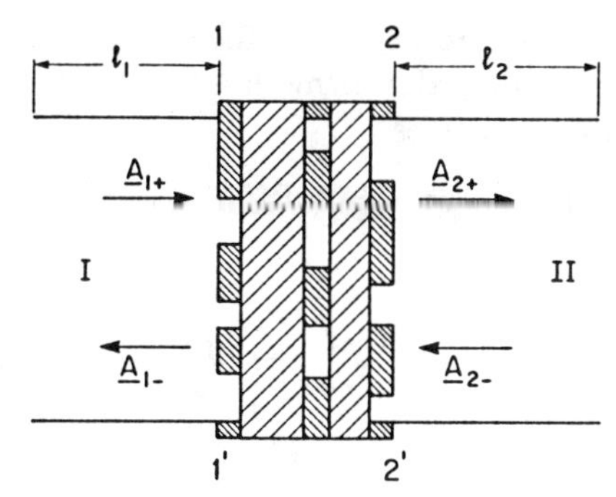

Fig. 3. A generalized discontinuity.

those of the N-furcated guide. P_A^h, P_A^e, P_{Bi}^h, and P_{Bi}^e for $i = 1, 2, 3, \ldots, N$, are diagonal matrices with diagonal elements representing the powers carried by unit amplitude LSE and LSM modes in the various guides. If $(P\,\text{LSE} + P\,\text{LSM})$ modes are retained in the large guide whereas $(Q\,\text{LSE} + Q\,\text{LSM})$ modes are retained in the N-furcated guide, the size of the matrices $\boldsymbol{H}$, $\boldsymbol{P_A}$, and $\boldsymbol{P_B}$ are respectively $(2P \times 2Q)$, $(2P \times 2P)$, and $(2Q \times 2Q)$.

The cross-coupling between LSE and LSM modes is readily understood from (5). However, it is interesting to note that if we let $\beta = 0$, all the elements of the matrices $\boldsymbol{H}_i^{eh}$ and $\boldsymbol{H}_i^{he}$ for $i = 1, 2, \ldots, N$, would vanish. In this case the LSE and LSM modes become TE and TM modes, respectively, with a direction of propagation normal to the xz plane. As has been seen in [11], these modes are not coupled by scattering at N-furcated parallel-plate waveguide junctions.

In view of [9] and after some manipulations, the transmission matrix of the N-furcated discontinuity can be determined in terms of the three basic matrices $\boldsymbol{H}$, $\boldsymbol{P_A}$, and $\boldsymbol{P_B}$ derived above for LSE and LSM excitation. The transmission matrix $\boldsymbol{T}^T$ of the overall discontinuity shown in Fig. 3 can then be easily obtained by simple matrix multiplications.

The next step is to apply the transverse resonance technique to find the eigenvalue equation for the propagation constant β. Let the parameters of the overall transmission matrix $\boldsymbol{T}^T$ be written as

$$\underline{A}_{1+} = \boldsymbol{T}_{11}^T\,\underline{A}_{2+} + \boldsymbol{T}_{12}^T\,\underline{A}_{2-} \tag{8a}$$

$$\underline{A}_{1-} = \boldsymbol{T}_{21}^T\,\underline{A}_{2+} + \boldsymbol{T}_{22}^T\,\underline{A}_{2-} \tag{8b}$$

where $\underline{A}_{i+}$ ($\underline{A}_{i-}$), $i = 1, 2$, are the forward (backward) LSE and LSM mode vectors at planes 11′, 22′, as indicated in Fig. 3. These mode vectors are also related as follows:

$$\underline{A}_{1+} = -\boldsymbol{L}_1^-\boldsymbol{L}_1^-\underline{A}_{1-} \qquad \underline{A}_{2-} = -\boldsymbol{L}_2^-\boldsymbol{L}_2^-\underline{A}_{2+} \tag{9a}$$

where

$$\boldsymbol{L}_1^- = \begin{bmatrix} \boldsymbol{L}_1^{-h} & \boldsymbol{0} \\ \boldsymbol{0} & \boldsymbol{L}_1^{-e} \end{bmatrix} \qquad \boldsymbol{L}_2^- = \begin{bmatrix} \boldsymbol{L}_2^{-h} & \boldsymbol{0} \\ \boldsymbol{0} & \boldsymbol{L}_2^{-e} \end{bmatrix}. \tag{9b}$$

$\boldsymbol{L}_1^{-h}$, $\boldsymbol{L}_1^{-e}$, $\boldsymbol{L}_2^{-h}$, and $\boldsymbol{L}_2^{-e}$ are diagonal matrices with diagonal elements given by

$$L_{1,nn}^{-h} = L_{1,nn}^{-e} = e^{-j[\omega^2\mu\epsilon - (n\pi/L)^2 - \beta^2]^{1/2} l_1} \tag{10a}$$

$$L_{2,nn}^{-h} = L_{2,nn}^{-e} = e^{-j[\omega^2\mu\epsilon - (n\pi/L)^2 - \beta^2]^{1/2} l_2} \tag{10b}$$

$$n = 0, 1, 2, \ldots, Q \text{ for LSE modes}$$
$$n = 1, 2, 3, \ldots, Q \text{ for LSM modes.}$$

In (10), l_1 and l_2 are, respectively, the lengths of the waveguide sections I and II shown in Fig. 3. Manipulating (8) and (9) gives

$$\begin{bmatrix} \boldsymbol{I} & -\left[\boldsymbol{T}_{11}^T - \boldsymbol{T}_{12}^T\boldsymbol{L}_2^-\boldsymbol{L}_2^-\right] \\ \boldsymbol{I} & \boldsymbol{L}_1^-\boldsymbol{L}_1^-\left[\boldsymbol{T}_{21}^T - \boldsymbol{T}_{22}^T\boldsymbol{L}_2^-\boldsymbol{L}_2^-\right] \end{bmatrix} \begin{bmatrix} \underline{A}_{1+} \\ \underline{A}_{2+} \end{bmatrix} = \boldsymbol{0}. \tag{11}$$

For a nontrivial solution the determinant should vanish:

$$\operatorname{Det}\left[\boldsymbol{L}_1^-\boldsymbol{L}_1^-\left[\boldsymbol{T}_{21}^T - \boldsymbol{T}_{22}^T\boldsymbol{L}_2^-\boldsymbol{L}_2^-\right] + \left[\boldsymbol{T}_{11}^T - \boldsymbol{T}_{12}^T\boldsymbol{L}_2^-\boldsymbol{L}_2^-\right]\right] = 0. \tag{12}$$

The solution for the propagation constant β may then be determined by seeking the zeros of the above determinantal equation.

The size of the eigenvalue matrix given in (12) is $(2Q \times 2Q)$, where $2Q$ is the number of modes (LSE + LSM) used in the N-furcated section, which is, in most practical configurations, much less than the number of modes $2P$ originally retained in the large guide to satisfy the edge condition. Since in the waveguide sections I and II indicated in Fig. 3, all the higher order modes are usually below cutoff, the elements associated with these modes in the matrices $\boldsymbol{L}_1^-$ and $\boldsymbol{L}_2^-$ are very small. Thus only the first few modes are needed in the application of the transverse resonance technique and a size of $(2Q \times 2Q)$ for the eigenvalue matrix is quite enough to provide accurate results for most practical applications.

In order to compare the numerical efficiency of the formulation presented in this paper to that reported in [7] and [8], let us consider as an example a typical unilateral finline structure with $d/b = 0.2$, where d is the slot width and b is the waveguide height. Let us also assume that $P = 25$ and $Q = 5$, i.e., (25 LSE modes + 25 LSM modes) are retained in the large guide, and (5 LSE modes + 5 LSM modes) are retained in the slot region. To avoid the relative convergence problem, it is noted that these numbers of modes are chosen such that $P/Q = b/d$. Thus, the size of the eigenvalue matrix derived in our formulation is only (10×10). On the other hand, as stated in [7], with 25 summation terms in the large waveguide (25 TE modes + 25 TM modes) the formulation handles transmission matrices of size (100×100) and leads to an eigenvalue equation of matrix size (50×50).

The considerable reduction of the matrix size we have achieved here is attributed to the use of the improved transmission matrix formulation for cascaded discontinuities, which has been introduced in [9]. This formulation does *not require the use of an equal number of modes*, which makes it possible to impose the edge condition and consequently guarantees accurate evaluation of the transmission parameters. Moreover it makes it possible to derive an eigenvalue matrix of size $(2Q \times 2Q)$ rather than $(2P \times 2P)$.

III. The Characteristic Impedance

With the eigenvalue β calculated, solving the eigenvalue equation for the eigenvector gives the field amplitudes in the various regions which are needed in the calculation of the characteristic impedance. There has been a considerable interest over the past several years in the theoretical investigation of the impedance. Most published results, however, are for structures with zero metallization thickness and no mounting grooves. In this contribution the effects of these parameters on the characteristic impedance based on the voltage–power definition will be investigated.

$$Z = \frac{VV^*}{2P} \qquad P = \sum_i P_i \tag{13}$$

where V is the slot voltage, which is obtained by line integrating the electric field over the width of the appropriate slot; the summation is on all guides involved and P_i is the power flow in guide i:

$$P_i = \frac{1}{2} \operatorname{Re} \int \vec{E}_i \times \vec{H}_i^* \cdot \vec{a}_z \, ds. \tag{14}$$

$\vec{E}_i$ and $\vec{H}_i$ are, respectively, the transverse components (with respect to z) of the electric and magnetic fields in the ith guide:

$$\vec{E}_i = \sum_{n=0,1,2,\cdots} \vec{\phi}^h_{in} + \sum_{n=1,2,3,\cdots} \vec{\phi}^e_{in} \tag{15}$$

$$\vec{H}_i = \sum_{n=0,1,2,\cdots} \vec{\psi}^h_{in} + \sum_{n=1,2,3,\cdots} \vec{\psi}^e_{in} \tag{16}$$

where

$$\vec{\phi}^h_{in} = \frac{-j\beta}{K_{c,in}} \cos \frac{n\pi}{L_i} x \, Q^h_{in} \, \vec{a}_x \tag{17a}$$

$$\vec{\psi}^h_{in} = \frac{Y^h_{in}}{K_{c,in}} \frac{n\pi}{L_i} \sin \frac{n\pi}{L_i} x \, W^h_{in} \, \vec{a}_x + \frac{K_{c,in}}{j\omega\mu} \cos \frac{n\pi}{L_i} x \, Q^h_{in} \, \vec{a}_y \tag{17b}$$

$$\vec{\phi}^e_{in} = \frac{-1}{K_{c,in}} \frac{n\pi}{L_i} \cos \frac{n\pi}{L_i} x \, Q^e_{in} \, \vec{a}_x + \frac{K_{c,in}}{j\alpha^e_{in}} \sin \frac{n\pi}{L_i} x \, W^e_{in} \, \vec{a}_y \tag{17c}$$

$$\vec{\psi}^e_{in} = \frac{jY^e_{in}\beta}{K_{c,in}} \sin \frac{n\pi}{L_i} x \, W^e_{in} \, \vec{a}_x. \tag{17d}$$

Q^h_{in}, W^h_{in}, Q^e_{in}, and W^e_{in} are scalar functions given by

$$Q^h_{in} = A^h_{in+} e^{-j\alpha^h_{in} y} + A^h_{in-} e^{+j\alpha^h_{in} y} \tag{18a}$$

$$W^h_{in} = A^h_{in+} e^{-j\alpha^h_{in} y} - A^h_{in-} e^{+j\alpha^h_{in} y} \tag{18b}$$

$$Q^e_{in} = A^e_{in+} e^{-j\alpha^e_{in} y} + A^e_{in-} e^{+j\alpha^e_{in} y} \tag{18c}$$

$$W^e_{in} = A^e_{in+} e^{-j\alpha^e_{in} y} - A^e_{in-} e^{+j\alpha^e_{in} y}. \tag{18d}$$

A_{in+}, A_{in-} denote, respectively, the amplitude of the incident and the reflected nth mode in guide i. Substituting (15)–(18) into (14) gives

$$P_i = P_i^{hh} + P_i^{ee} + P_i^{eh} \tag{19}$$

where

$$P_i^{hh} = \sum_{n=0,1,2,\cdots} \frac{\beta L_i}{4\omega\mu} \Gamma_n \int_0^{l_i} Q^h_{in} Q^{*h}_{in} \, dy \tag{20}$$

$$P_i^{ee} = \sum_{n=1,2,\cdots} \frac{\omega\epsilon\beta L_i}{4\alpha^e_{in}\alpha^{*e}_{in}} \int_0^{l_i} W^e_{in} W^{*e}_{in} \, dy \tag{21}$$

$$P_i^{eh} = \sum_{n=1,2,\cdots} \frac{-jL_i}{4\omega\mu} \frac{n\pi}{L_i} \int_0^{l_i} Q^e_{in} Q^{*h}_{in} \, dy + \frac{j\alpha^{*h}_{in} L_i}{4\omega\mu\alpha^e_{in}} \frac{n\pi}{L_i} \int_0^{l_i} W^e_{in} W^{*h}_{in} \, dy. \tag{22}$$

In (20) $\Gamma_n = 2$ for $n = 0$ and $\Gamma_n = 1$ for $n = 1, 2, \cdots$.

Originally, it was assumed that $2P$ modes were retained in the large guide and $2Q$ modes were retained in the N-furcated guide. The solution of the eigenvalue equation will provide, however, only information about the amplitudes of $2Q$ modes in all guides. Although a matrix size of $(2Q \times 2Q)$ for the eigenvalue equation is usually enough to achieve good results for the propagation constant β, in calculating the power flow, $2Q$ modes may not be sufficient to accurately represent the total power propagating in the large guide.

Indeed, investigation of the convergence of the calculated results showed that accurate calculation of the power flow requires the use of more modes in the large guide. The coefficients of those additional modes can be deduced *without increasing the size of the eigenvalue matrix*. This additional information is obtained from the E-field mode-matching matrix H, which has a size of $(2P \times 2Q)$ and has been derived at the beginning of the formulation.

IV. Planar Transmission Lines on Anisotropic Substrates

The problem of analyzing planar transmission lines on anisotropic substrates has been studied by many investigators using quasi-static as well as hybrid-mode analyses. Recently, a comprehensive review entitled "Integrated-Circuit Structures on Anisotropic Substrates" has been published by Alexopoulos [12]. This paper discusses in detail the different techniques used in analyzing these structures and contains an exhaustive list of references to the existing literature. It is noted, however, that the great majority of these references are based on quasi-static methods and in many cases only idealized structures are considered. In this section we will extend our formulation to take the effect of the dielectric anisotropy into account in addition to the effects of the metallization thickness and substrate mounting grooves.

For lossless anisotropic substrates, the permittivity tensor ϵ may be written in a diagonalized form [10]. Moreover, most of the substrates widely used in microwave integrated

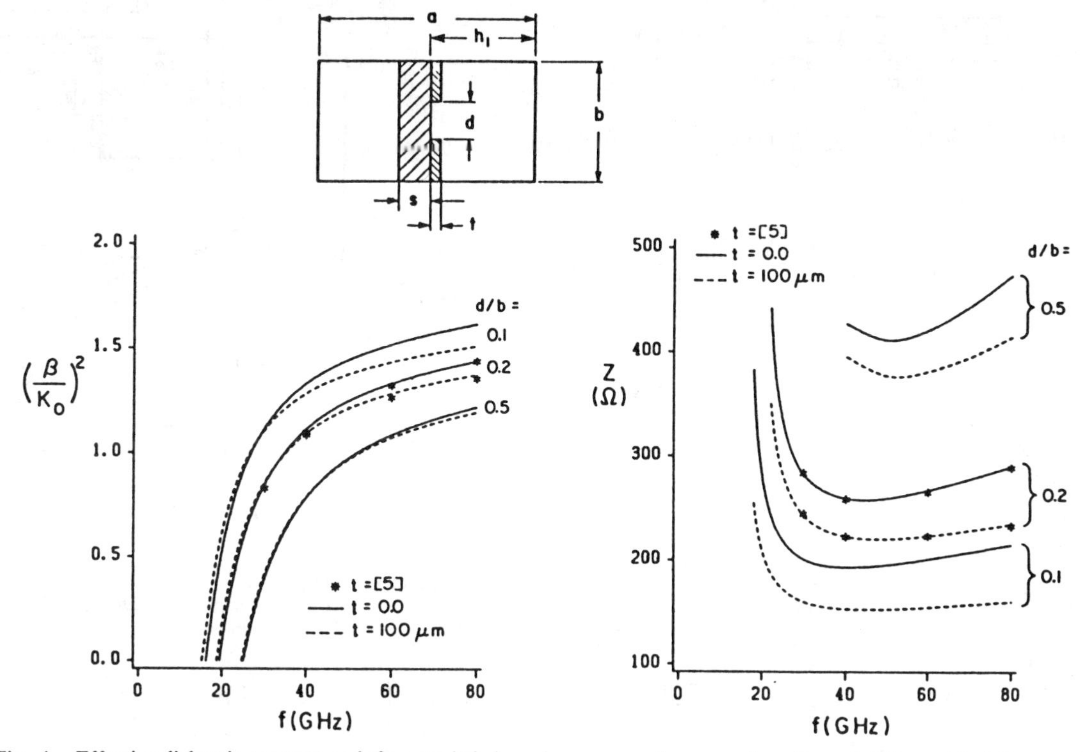

Fig. 4. Effective dielectric constant and characteristic impedance versus frequency in unilateral finlines with different values of slot width; $a = 2b = 4.7752$ mm, $S = 0.127$ mm, $h_1 = 2.3876$ mm, $\epsilon_r = 3.8$.

circuit design such as sapphire and Epsilam 10 are uniaxial crystals for which $\epsilon_{xx} = \epsilon_{zz}$. The permittivity tensor ϵ for this type of crystal can then be written as

$$\epsilon = \begin{bmatrix} \epsilon_2 & 0 & 0 \\ 0 & \epsilon_1 & 0 \\ 0 & 0 & \epsilon_2 \end{bmatrix}. \tag{23}$$

With the optic axis parallel to the y axis, the electric field of LSE modes will lie in the plane where no anisotropic effect is present. The solution in this case is basically the same as in the case of an isotropic medium with a dielectric permittivity ϵ_2. However, for LSM modes the situation is different because there is a component of the electric field in the y direction. In view of [10], the solution for LSM modes may be derived from

$$\vec{\prod}_n^e = \Psi_n^e(x) e^{-j\alpha_n^e y} e^{-j\beta z} \vec{a}_y$$

$$\vec{h}_n^e = j\omega\epsilon_2 \nabla \times \vec{\prod}_n^e$$

$$\vec{e}_n^e = \omega^2 \mu \epsilon_2 \vec{\prod}_n^e + \nabla \nabla \cdot \vec{\prod}_n^e \tag{24}$$

where Ψ_n^e is the solution of

$$\frac{\partial^2}{\partial x^2} \Psi_n^e(x) + \left[\omega^2 \mu \epsilon_1 - \frac{\epsilon_1}{\epsilon_2} (\alpha_n^e)^2 - \beta^2 \right] \Psi_n^e(x) = 0. \tag{25}$$

It can be readily shown that the transverse components of the fields in this case have the same form as those given in (3). However, α_n^h, α_n^e, Y_n^h, and Y_n^e are modified, becoming

$$\alpha_n^h = \left[\omega^2 \mu \epsilon_2 - \left(\frac{n\pi}{L} \right)^2 - \beta^2 \right]^{1/2} \tag{26}$$

$$\alpha_n^e = \left[\frac{\epsilon_2}{\epsilon_1} \right]^{1/2} \left[\omega^2 \mu \epsilon_1 - \left(\frac{n\pi}{L} \right)^2 - \beta^2 \right]^{1/2} \tag{27}$$

$$Y_n^h = \frac{\alpha_n^h}{\omega\mu} \qquad Y_n^e = \frac{\omega\epsilon_2}{\alpha_n^e}. \tag{28}$$

Thus, only a few minor changes are to be made in order to take the dielectric anisotropy into account. This illustrates the versatility of this approach and indicates how flexible and simple it is in dealing with complicated structures.

V. Numerical Results and Discussion

In Fig. 4 we show the effective dielectric constant $\epsilon_{\text{eff}} = (\beta/k_0)^2$ and the characteristic impedance Z versus frequency for unilateral finlines with different slot widths and for metallizations of $t = 0$ and $t = 100$ μm. Our results are in good agreement with those obtained by Kitazawa and Mittra [5], and our analysis can be used even for structures with infinitely thin fins. The results indicate that the metallization thickness has a significant effect on the characteristic impedance. Its effect, however, on the effective dielectric constant depends strongly on the slot width and is more pronounced for small slot widths.

To demonstrate the effect of the substrate mounting groove, Fig. 5 shows the normalized propagation constant of the dominant and the first higher order odd mode for groove depths of $e = 0$ and $e = 0.5$ mm. It is observed that

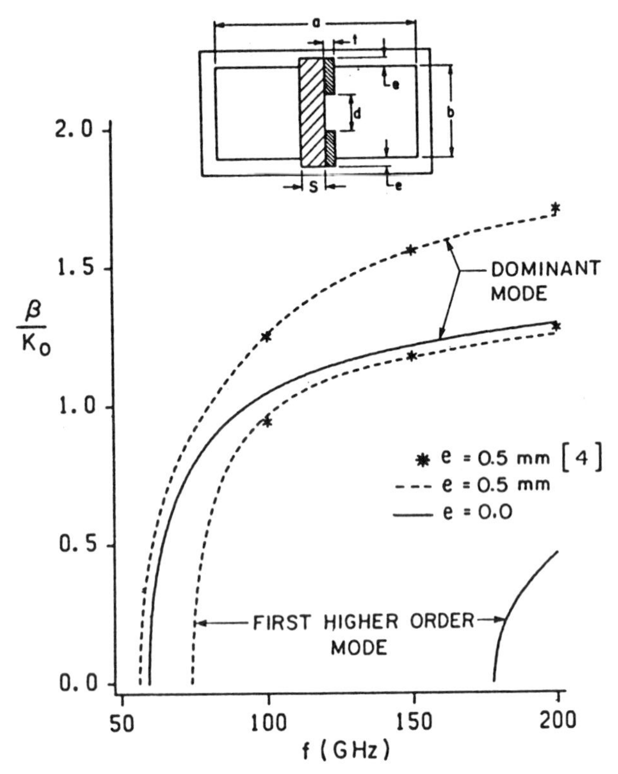

Fig. 5. Propagation characteristics of the dominant mode and first higher odd mode in a unilateral finline for $e = 0$ and $e = 0.5$ mm; $a = 2b = 1.65$ mm, $S = 0.11$ mm, $d = 0.3$ mm, $\epsilon_r = 3.75$, $t = 5$ μm.

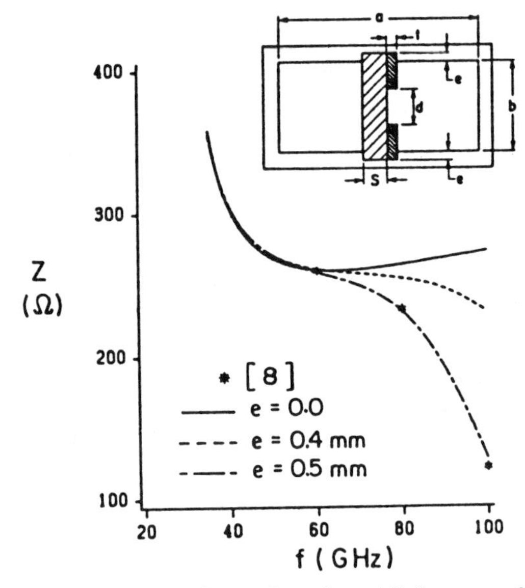

Fig. 6. Characteristic impedance of a unilateral finline versus frequency for different values of groove depth; $a = 2b = 3.1$ mm, $S = 0.22$ mm, $d = 0.4$ mm, $\epsilon_r = 3.75$, $t = 5$ μm.

neglecting the mounting groove leads to a higher cutoff frequency and a lower value for the propagation constant. This in fact is expected and is attributed to the part of the dielectric slab neglected in the mounting groove in the "ideal" case of $e = 0$. However, the effect of the groove on the dominant mode is significant when the first higher order mode starts to propagate. In addition, unlike the dominant mode's cutoff frequency, that of the first higher order mode decreases significantly when the mounting groove is used, leading to a very large reduction in the single-mode bandwidth.

Fig. 6 illustrates the effect of the mounting groove on the characteristic impedance. Our results confirm those recently published by Bornemann and Arndt [8]; the impedance follows the behavior of the dominant mode and starts to deviate from the ideal case ($e = 0$) only when the first higher order mode starts to propagate. In Fig. 7 we also show the effect of the metallization thickness on the single-mode bandwidth. It is observed that increasing the metallization thickness leads to a slight increase in the single-mode bandwidth and again the effect is more pronounced for smaller slot widths.

To demonstrate the fast convergence of the proposed analysis, Table I shows numerical results obtained for the propagation constant of the dominant and the first higher order modes and the characteristic impedance for a unilateral finline structure using different matrix sizes for the eignevalue equation. A matrix size of (6×6) is quite enough to provide convergent results within 0.5 percent.

It should be noted that the results given in Table I were obtained using (25 LSE modes+25 LSM modes) in the large guide; the number of modes retained in the slot was varied from (2 LSE modes+2 LSM modes) to (5 LSE modes+5 LSM modes). Although these results were obtained with $P/Q > b/d$, as has been shown in [9], in dealing with cascaded discontinuities the effect of the relative convergence problem becomes noticeable when $P/Q < R$ (in this case $R = b/d$), and as long as $P/Q \geqslant R$ good results can be achieved. It should be also mentioned that using a large number of modes in the large guide does not require a considerable computation effort since the sizes of the transmission matrices are determined by the number of modes retained in the slot [9]. The extra computation effort is involved in evaluating the matrix multiplication $[H^{\dagger}P_A^{\dagger}H]$.

In Fig. 8 we show the frequency dependence of the effect of the metallization thickness on the dominant and the first higher odd mode and characteristic impedance in bilateral finlines. The results show that unilateral and bilateral finlines have identical behavior as far as the effect of the metallization thickness is concerned. A comparison is also given in this figure between our results and those published by Schmidt and Itoh [13] using the spectral-domain technique, and good agreement is observed for both the propagation constant and the characteristic impedance.

Fig. 9 illustrates the effect of the metallization thickness on the normalized propagation constant and characteristic impedance of the basic even and odd modes in coplanar lines. The metallization thickness has a negligible effect on the propagation characteristic of the odd mode over most of the operating range. Its effect, however, on the propagation of the even mode is significant over the whole range. Increasing the metallization thickness lowers the characteristic impedance in both cases.

Fig. 10(a) shows the effect of the mounting groove on the propagation characteristic of the dominant and first higher odd modes in coplanar lines where, again, behavior

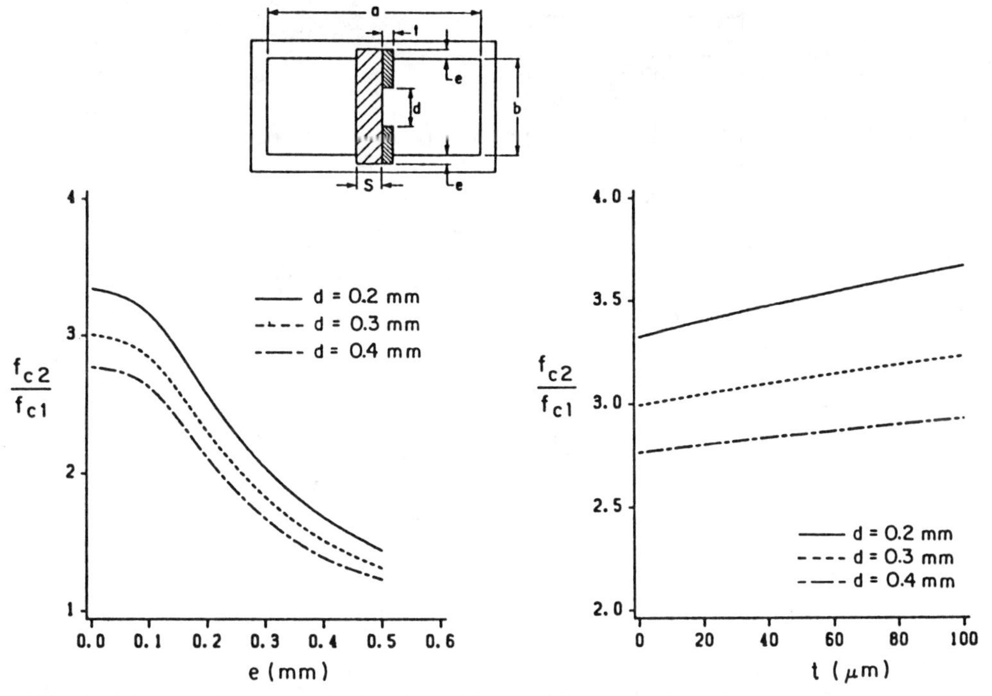

Fig. 7. Effects of the mounting groove depth and metallization thickness on the ratio f_{c2}/f_{c1}; $a=2b=1.65$ mm, $S=0.11$ mm, $d=0.3$ mm, $\epsilon_r=3.75$. (a) $t=5$ μm. (b) $e=0$.

TABLE I
THE NORMALIZED PROPAGATION CONSTANT AND CHARACTERISTIC IMPEDANCE OF A UNILATERAL FINLINE COMPUTED USING DIFFERENT MATRIX SIZES FOR THE EIGENVALUE EQUATION

Dominant Mode, f = 60 GHz

Matrix Size	β/K_o	Impedance (Ω)
(4x4)	1.1855	257.03
(6x6)	1.1866	258.75
(8x8)	1.1866	260.10
(10x10)	1.1854	260.83

First Higher Order Mode, f = 100 GHz

Matrix Size	β/K_o
(4x4)	0.36959
(6x6)	0.36911
(8x8)	0.36879
(10x10)	0.36851

$a=2b=3.1$ mm, $d=0.4$ mm, $S=0.22$ mm, $t=5$ μm, $\epsilon_r=3.75$.

similar to the unilateral finline is observed. The dispersion characteristics of coplanar lines with two dielectric layers are given in Fig. 10(b). A comparison between (a) and (b) in Fig. 10 shows that adding another dielectric layer may not enhance or compensate for the effect of the mounting groove.

In Fig. 11 the effect of the metallization thickness on the first four propagating modes in suspended microstrip lines is investigated. A noticeable effect is only observed on the propagation characteristic of the dominant mode, and our results agree well with those reported in [14] and [15]. Finally, we consider in Fig. 12 the effect of the metallization thickness on the normalized propagation constant in a coplanar line on sapphire substrate ($\epsilon_1=11.6, \epsilon_2=9.4$). We also compare our results with those given in [16] for unshielded coplanar lines. Our results are obtained for the shielded structure shown in Fig. 12 with electric walls at $\pm b/2$ for the even mode and with magnetic walls at $\pm b/2$ for the odd mode. A good agreement is observed between the results for the dominant even mode. The reason, however, for the discrepancy between the results of the odd mode at low frequencies is that for this mode the fields are not tightly bound to the slots, and the waveguide walls may not be far enough from the slots at low frequencies to simulate the open structure.

VI. CONCLUSIONS

The numerical results presented in this paper and the comparisons given with the other published data confirm the validity of the proposed analysis and show its simplicity in treating different planar structures with complicated configurations. With a reasonably small matrix size, the achievable accuracy of the solution exceeds most engineering requirements. The analysis can be extended to characterize planar transmission lines on semiconductor substrates. This also can be easily achieved since the dielectric layers are treated separately in the formulation. Thus, this approach promises to be useful as well in the design of monolithic microwave integrated circuits (MMIC's).

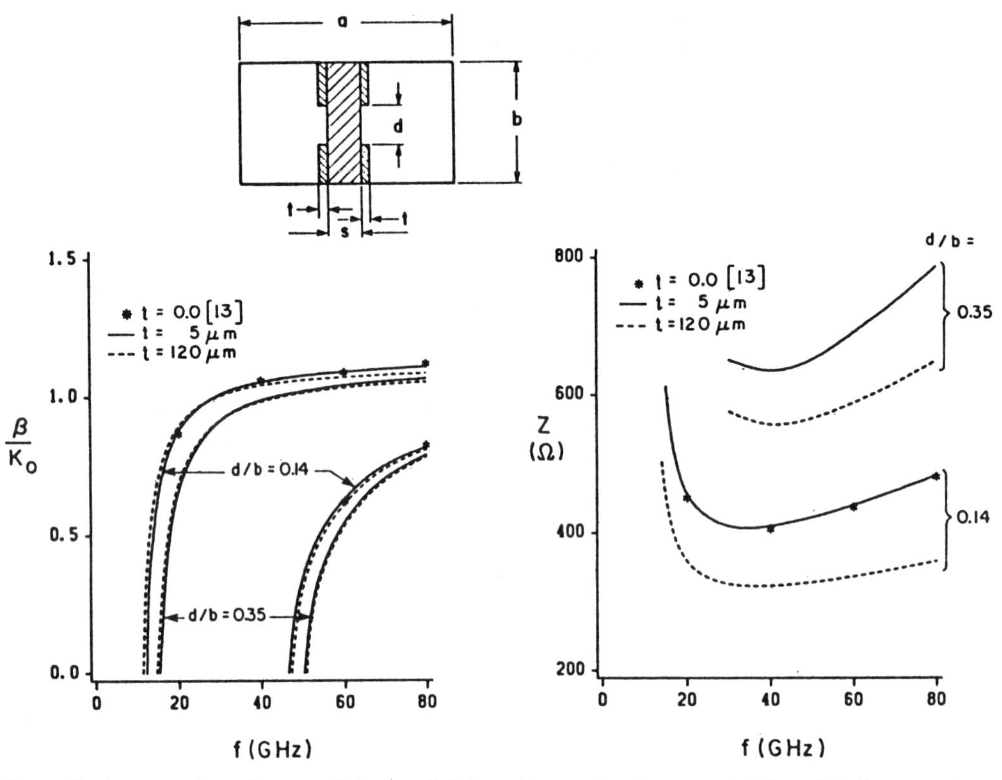

Fig. 8. Normalized propagation constant and characteristic impedance versus frequency in bilateral finlines; $a = 2b = 7.112$ mm, $S = 0.125$ mm, $\epsilon_r = 3$.

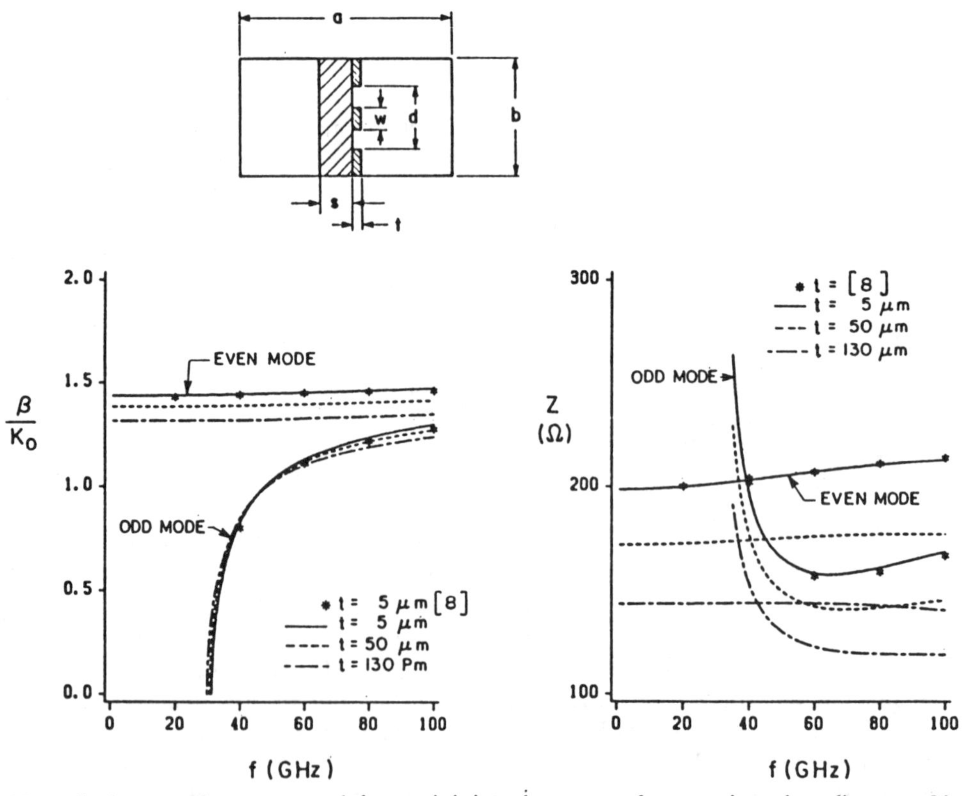

Fig. 9. Normalized propagation constant and characteristic impedance versus frequency in coplanar lines; $a = 2b = 3.1$ mm, $S = 0.22$ mm, $d = 0.6$ mm, $w = 0.2$ mm, $\epsilon_r = 3.75$.

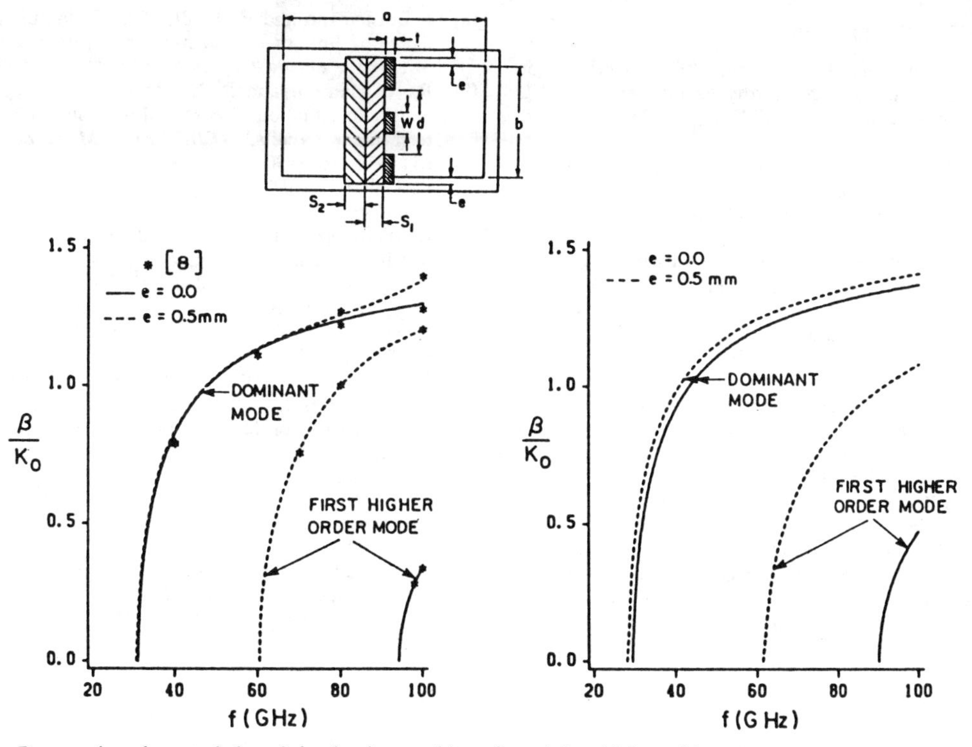

Fig. 10. Propagation characteristics of the dominant odd mode and first higher odd mode in a coplanar line for $e = 0$ and $e = 0.5$ mm; $a = 2b = 3.1$ mm, $S_1 = 0.22$ mm, $S_2 = 0.254$ mm, $d = 0.6$ mm, $w = 0.2$ mm, $t = 5$ μm. (a) $\epsilon_{r1} = 3.75$, $\epsilon_{r2} = 1.0$. (b) $\epsilon_{r1} = 3.75$, $\epsilon_{r2} = 2.2$.

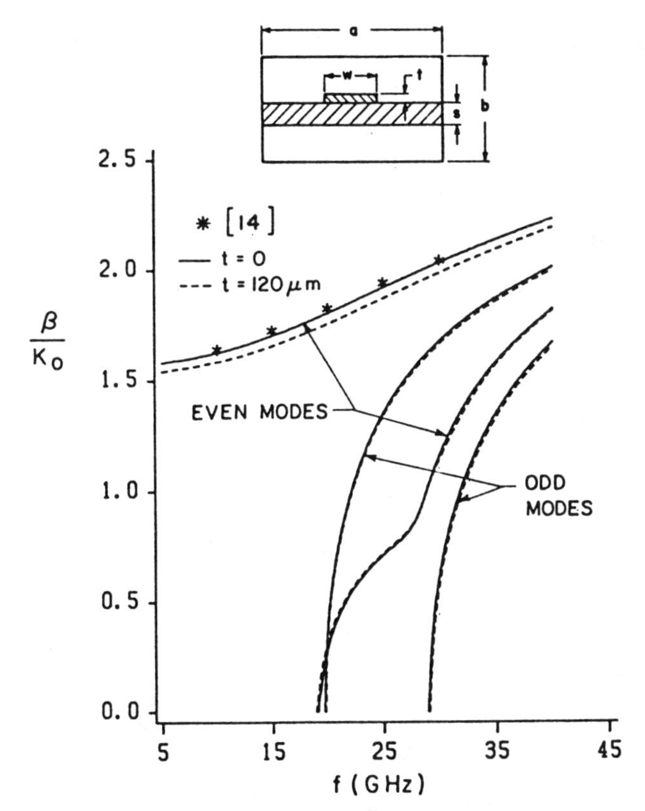

Fig. 11. Normalized propagation constant as a function of frequency for the suspended microstrip line; $a = 2b = 7.112$ mm, $S = 0.635$ mm, $w = 1.0$ mm, $\epsilon_r = 9.6$.

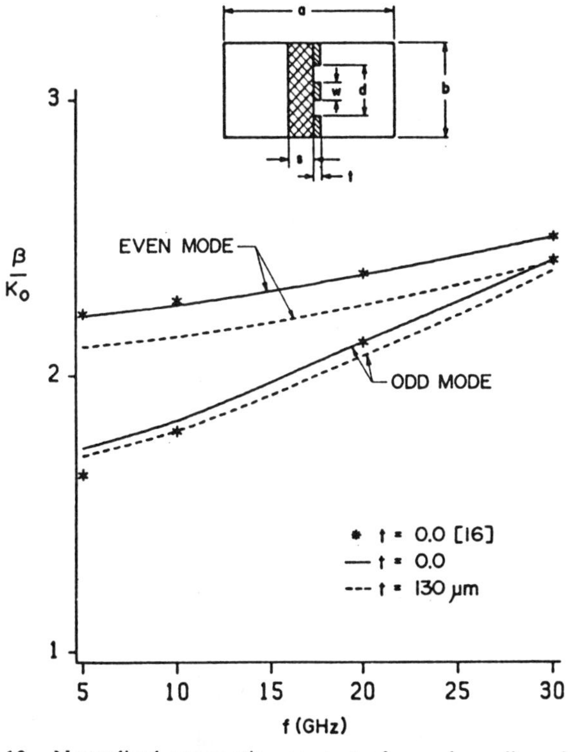

Fig. 12. Normalized propagation constant of a coplanar line of sapphire substrate; $a = 2b = 20$ mm, $S = 1.0$ mm, $d = 2.5$ mm, $w = 0.5$ mm, even mode (magnetic wall symmetry), odd mode (electric wall symmetry).

References

[1] L. P. Schmidt, T. Itoh, and H. Hofmann, "Characteristics of unilateral finline structure with arbitrarily located slots," *IEEE Trans. Microwave Theory Tech.*, vol. MTT-29, pp. 352–355, Apr. 1981.

[2] A. S. Omar and S. Schunemann, "Formulation of the singular integral equation technique for planar transmission lines," *IEEE Trans. Microwave Theory Tech.*, vol. MTT-33, pp. 1313–1322, Dec. 1985.

[3] A. Beyer, "Analysis of the characteristics of an earthed fin line," *IEEE Trans. Microwave Theory Tech.*, vol. MTT-29, pp. 676–680, July 1981.

[4] R. Vahldieck, "Accurate hybrid-mode analysis of various finline configurations including multilayered dielectrics, finite metallization thickness and substrate holding grooves," *IEEE Trans. Microwave Theory Tech.*, vol. MTT-32, pp. 1454–1460, Nov. 1984.

[5] T. Kitazawa and R. Mittra, "Analysis of finlines with finite metallization thickness," *IEEE Trans. Microwave Theory Tech.*, vol. MTT-32, pp. 1484–1487, Nov. 1984.

[6] A. M. K. Saad and K. Schunemann, "Efficient eigenmode analysis for planar transmission lines," *IEEE Trans. Microwave Theory Tech.*, vol. MTT-30, pp. 2125–2131, Dec. 1982.

[7] R. Vahldieck and J. Bornemann, "A modified mode matching technique and its application to a class of quasi-planar transmission lines," *IEEE Trans. Microwave Theory Tech.*, vol. MTT-33, pp. 916–926, Oct. 1985.

[8] J. Bornemann and F. Arndt, "Calculating the characteristic impedance of finlines by transverse resonance method," *IEEE Trans. Microwave Theory Tech.*, vol. MTT-34, pp. 85–92, Jan. 1986.

[9] R. R. Mansour and R. H. MacPhie, "An improved transmission matrix formulation of cascaded discontinuities and its application to *E*-plane circuits," *IEEE Trans. Microwave Theory Tech.*, vol. MTT-34, pp. 1490–1498, Dec. 1986.

[10] R. Collin, *Field Theory of Guided Waves.* New York: McGraw-Hill, 1960.

[11] R. R. Mansour and R. H. MacPhie, "Scattering at an *N*-furcated parallel-plate waveguide junction," *IEEE Trans. Microwave Theory Tech.*, vol. MTT-33, pp. 830–835, Sept. 1985.

[12] N. G. Alexopoulos, "Integrated-circuit structures on anisotropic substrates," *IEEE Trans. Microwave Theory Tech.*, vol. MTT-33, pp. 847–881, Oct. 1985.

[13] L. P. Schmidt and T. Itoh, "Spectral domain analysis of dominant and higher order modes in fin-lines," *IEEE Trans. Microwave Theory Tech.*, vol. MTT-28, pp. 981–985, Sept. 1980.

[14] H. Hofmann, "Dispersion of planar waveguides for millimeter-wave applications," *Arch. Elek. Übertragung.*, Bd. 31, I, pp. 40–44, 1977.

[15] J. Bornemann, "Rigorous field theory analysis of quasiplanar waveguides," *Proc. Inst. Elec. Eng.*, vol. 132 pt. H. no. 1, pp. 1–6, Feb. 1985.

[16] T. Kitazawa and Y. Hayashi, "Coupled slots on an anisotropic sapphire substrate," *IEEE Trans. Microwave Theory Tech.*, vol. MTT-29, pp. 1035–1040, Oct. 1981.

Transverse Resonance Analysis of Finline Discontinuities

ROBERTO SORRENTINO, MEMBER, IEEE, AND TATSUO ITOH, FELLOW, IEEE

Abstract —**A method of analysis is proposed for characterizing finline discontinuities. Two conducting or magnetic planes are inserted at some distances away from the discontinuity so as to obtain a closed resonant structure. A transverse resonance technique is then used to compute the resonant frequencies and, from these, the equivalent circuit parameters of the discontinuity. In the particular case when the discontinuity is removed, the method can be used to characterize uniform finlines.**

I. Introduction

FINLINES are now recognized as a suitable technology of millimeter-wave integrated circuits. While much theoretical work has been done concerning the analysis and characterization of uniform finline structures [1], [2], a relatively small number of analyses of finline discontinuities have been developed [3], [4].

This paper presents a new method of analysis and characterization of both uniform finlines and finline discontinuities. The method consists of computing the resonant frequencies of a resonator obtained by inserting two conducting or magnetic planes apart from the discontinuity; using a transverse resonance technique, the electromagnetic (EM) fields are expanded in terms of longitudinal section magnetic (LSM) and electric (LSE) modes of the rectangular waveguide. With respect to other approaches based on the field expansion in terms of finline modes [3], [4], the present one has the advantage of a substantial reduction of computer time. In this paper, this new method is applied to the simple step discontinuity as well as to the cascade of step discontinuities.

II. Characterization of the Discontinuity

The characterization of a finline discontinuity is obtained with the resonant frequencies of resonators which are obtained by introducing two shorting planes at some distances away from the discontinuity. The resultant structure is shown in Fig. 1 along with the dimensions and the coordinate system.

As long as the frequency is such that only dominant modes can propagate in the two finline sections and the higher order modes excited at the discontinuity have negligible amplitudes at the shorting planes, the discontinuity can be modeled as an equivalent two-port network, as shown in Fig. 2.

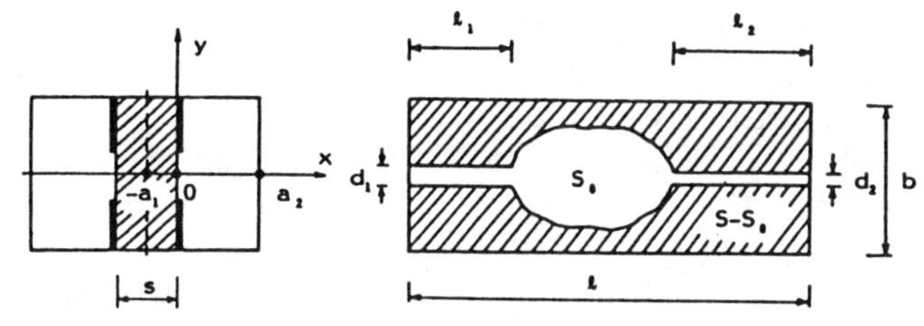

Fig. 1. Transverse and longitudinal cross sections of a finline discontinuity in a shorted cavity.

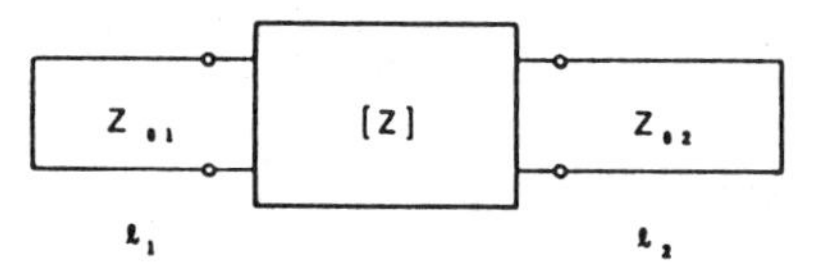

Fig. 2. Equivalent circuit of Fig. 1.

The resonance condition in terms of the impedance parameters of the discontinuity is

$$(Z_{11}+Z_1)(Z_{22}+Z_2)-Z_{12}^2=0 \tag{1}$$

where

$$Z_i = jZ_{oi}\tan(\beta_i l_i), \qquad i=1,2.$$

Z_{oi} is the characteristic impedance of the ith finline, and β_i is the corresponding phase constant. Alternatively, (1) can also be formulated in terms of the scattering parameters of the discontinuity.

If the same resonant frequency ω_r rad/s is obtained for three different pairs of l_1, l_2, (1) allows the evaluation of the three impedance parameters of the discontinuity at ω_r.

In the absence of the discontinuity, $\beta_1=\beta_2=\beta$, (1) then reduces to

$$\beta(\omega_r)=n\pi/l$$

with $l=l_1+l_2$. Thus, the length l corresponding to the resonant frequency ω_r yields the phase constant of a uniform finline at ω_r.

With simple modifications, the above procedure can be applied to other types of finline discontinuity problems, such as end effects in open- or short-circuited finline sections. In such cases, the equivalent circuit will consist of a line section terminated at one end with an unknown reactance. Its value can be computed by way of the resonant frequencies of a resonator obtained by short-circuit-

Manuscript received April, 26 1984; revised July 31, 1984. This work was supported in part by the U.S. Army Research Office under Contract DAAG 29-81-K-0053, and in part by the Joint Services Electronics Program.

R. Sorrentino is with the Department of Electronics, University of Rome La Sapienza, Via Eudossiana 18, 00184 Rome, Italy.
T. Itoh is with the Department of Electrical Engineering, University of Texas at Austin, Austin, TX 78712.

Reprinted from *IEEE Trans. Microwave Theory Tech.*, vol. MTT-32, no. 12, pp. 1633–1638, Dec. 1984.

ing the waveguide at some distance away from the line termination.

III. Computation of the Resonant Frequencies

The method for computing the resonant frequencies of the structure will be illustrated in the case of bilateral finline, shown in Fig. 1. The metallic fins are assumed to be infinitely thin, although the method can easily be modified to account the finite thickness of metallization.

Because of symmetry, a longitudinal magnetic plane can be inserted at the symmetric plane $x = -a_1$, so that only the region $x \geqslant -a_1$ has to be analyzed. The extension to nonsymmetrical structures, such as unilateral finlines, is straightforward and will not be considered here.

The EM field in the dielectric region (region 1: $-a_1 \leqslant x \leqslant 0$) and in the air region (region 2: $0 \leqslant x \leqslant a_2$) can be expanded in terms of TE and TM modes of a rectangular waveguide with inner dimensions l and b. We obtain the following expressions for the transverse E- and H-field components in the two regions:

Dielectric Region: $-a_1 \leqslant x \leqslant 0$

$$\begin{aligned} \boldsymbol{E}_{t1} &= \sum_{mn} A'_{mn} \cos k'_{mn}(x+a_1)\hat{\boldsymbol{x}} \times \nabla_t \psi_{mn} \\ &\quad + \frac{1}{j\omega\epsilon_0\epsilon_r} \sum_{mn} B'_{mn} k'_{mn} \cos k'_{mn}(x+a_1) \nabla_t \varphi_{mn} \\ \boldsymbol{H}_{t1} &= \frac{-1}{j\omega\mu_0} \sum_{mn} A'_{mn} k'_{mn} \sin k'_{mn}(x+a_1) \nabla_t \psi_{mn} \\ &\quad + \sum_{mn} B'_{mn} \sin k'_{mn}(x+a_1) \nabla_t \varphi_{mn} \times \hat{\boldsymbol{x}}. \end{aligned} \tag{2}$$

Air Region: $0 \leqslant x \leqslant a_2$

$$\begin{aligned} \boldsymbol{E}_{t2} &= \sum_{mn} A_{mn} \sin k_{mn}(x-a_2)\hat{\boldsymbol{x}} \times \nabla_t \psi_{mn} \\ &\quad - \frac{1}{j\omega\epsilon_0} \sum_{mn} B_{mn} k_{mn} \sin k_{mn}(x-a_2) \nabla_t \varphi_{mn} \\ \boldsymbol{H}_{t2} &= \frac{1}{j\omega\mu_0} \sum_{mn} A_{mn} k_{mn} \cos k_{mn}(x-a_2) \nabla_t \psi_{mn} \\ &\quad + \sum_{mn} B_{mn} \cos k_{mn}(x-a_2) \nabla_t \varphi_{mn} \times \hat{\boldsymbol{x}} \end{aligned} \tag{3}$$

where

$$\begin{aligned} \psi_{mn} &= P_{mn} \cos\frac{m\pi z}{l} \cos\frac{n\pi y}{b} \\ \varphi_{mn} &= P_{mn} \sin\frac{m\pi z}{l} \sin\frac{n\pi y}{b} \\ P_{mn} &= \sqrt{\frac{\delta_m \delta_n}{lb}} \frac{1}{\gamma_{mn}} \qquad \delta_i = \begin{cases} 1, & i=0 \\ 2, & i \neq 0 \end{cases} \\ \gamma_{mn}^2 &= \left(\frac{m\pi}{l}\right)^2 + \left(\frac{n\pi}{b}\right)^2 \\ k_{mn}^2 &= k_0^2 - \gamma_{mn}^2 \qquad k'^2_{mn} = k_0^2 \epsilon_r - \gamma_{mn}^2 \\ k_0^2 &= \omega^2 \mu_0 \epsilon_0 \end{aligned} \tag{4}$$

where ψ_{mn} and φ_{mn} are the TE and TM scalar potentials, and m and n are integers with starting values of 0 or 1, depending on whether the TE or TM mode is being considered. P_{mn} are determined from the normalization conditions for ψ_{mn}, φ_{mn}

$$\int_S |\nabla_t \psi_{mn}|^2 \, dS = 1$$

$$\int_S |\nabla_t \varphi_{mn}|^2 \, dS = 1.$$

Equations (2)–(4) already satisfy the boundary conditions at $x = -a_1$ and $x = a_2$. The boundary conditions at $x = 0$ are

$$\boldsymbol{E}_{t1} = \boldsymbol{E}_{t2} = \begin{cases} \boldsymbol{E}_{t0}, & \text{on } S_0 \\ 0, & \text{on } S - S_0 \end{cases} \tag{5}$$

$$\boldsymbol{H}_{t1} = \boldsymbol{H}_{t2} = \boldsymbol{H}_{t0}, \qquad \text{on } S_0 \tag{6}$$

where $\boldsymbol{E}_{to}$ and $\boldsymbol{H}_{to}$ are unknown functions of z, y. These functions are expanded in terms of a set of orthonormal vector functions $\boldsymbol{e}_\nu$, or $\boldsymbol{h}_\mu$ defined over aperture region S_o (see Appendix)

$$\boldsymbol{E}_{to} = \sum_\nu V_\nu \boldsymbol{e}_\nu \tag{7}$$

$$\boldsymbol{H}_{to} = \sum_\mu I_\mu \boldsymbol{h}_\mu. \tag{8}$$

Inserting (2), (3), (7), and (8) into (5) and (6), and making use of the orthogonal properties of $\psi_{mn}, \varphi_{mn}, \boldsymbol{e}_\nu$, and $\boldsymbol{h}_\mu$, we obtain a homogeneous system of equations in terms of unknown coefficients V_ν

$$\sum_\nu V_\nu \Bigg[\xi_{mn\nu} \zeta_{mn\mu} \left(k'_{mn} \tan k'_{mn} a_1 - k_{mn} \operatorname{cotan} k_{mn} a_2 \right) + \chi_{mn\nu} \theta_{mn\mu} k_0^2 \left(\epsilon_r \frac{\tan k'_{mn} a_1}{k'_{mn}} - \frac{\operatorname{cotan} k_{mn} a_2}{k_{mn}} \right) \Bigg] = 0, \qquad \mu = 1, 2 \cdots \tag{9}$$

where

$$\begin{aligned} \xi_{mn\nu} &= \int_{S_0} \hat{\boldsymbol{x}} \times \nabla_t \psi_{mn} \cdot \boldsymbol{e}_\nu \, dS \qquad \chi_{mn\nu} = \int_{S_0} \nabla_t \varphi_{mn} \cdot \boldsymbol{e}_\nu \, dS \\ \zeta_{mn\mu} &= \int_{S_0} \nabla_t \psi_{mn} \cdot \boldsymbol{h}_\mu \, dS \qquad \theta_{mn\mu} = \int_{S_0} \nabla_t \varphi_{mn} \times \hat{\boldsymbol{x}} \cdot \boldsymbol{h}_\mu \, dS. \end{aligned} \tag{10}$$

The condition for nontrival solutions determines the characteristic equation of the given structure. This equation may be regarded as a real function of ω, l_1, and l_2 equated to zero

$$f(\omega, l_1, l_2) = 0. \tag{11}$$

For a given value of $\omega = \omega_r$, (11) can be solved to evaluate three different pairs of l_1 and l_2 yielding the same resonant frequency ω_r. These values of l_1 and l_2 can be used for computing the discontinuity parameters discussed in the previous section.

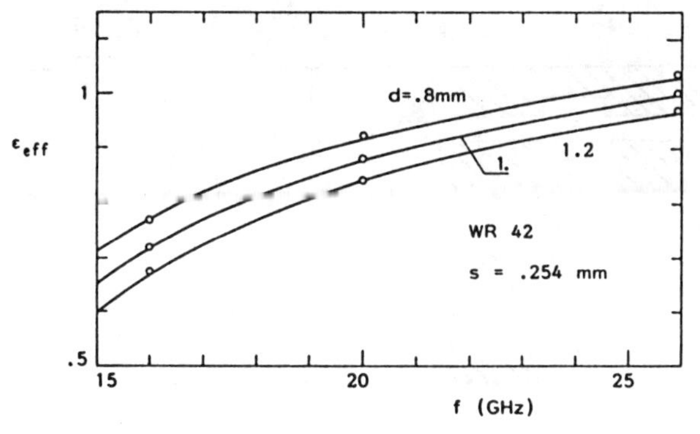

Fig. 3. Effective permittivity of a uniform finline. ∘ Spectral-domain method.

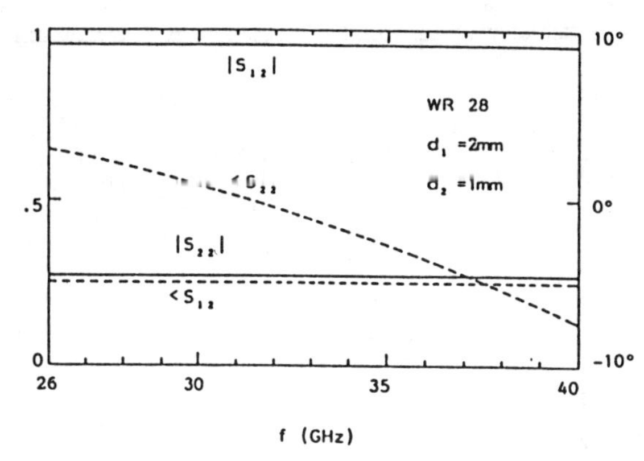

Fig. 5. Scattering parameters of the step discontinuity of Fig. 4.

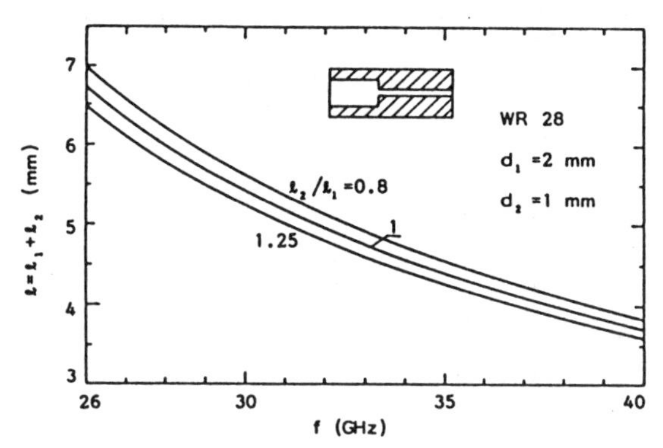

Fig. 4. Resonant frequency of a step discontinuity in a shorted cavity.

IV. Computed Results

According to the above-described technique, the EM fields in the air region, in the dielectric region, and in the aperture region of the resonator are expressed in terms of the series expansions (2), (3), (7), and (8). In the numerical computations, only a finite number of terms of each series can be retained. In order to obtain a proper convergent behavior of the solution, the number of terms in adjacent regions was chosen in such a way that the highest spatial frequencies of the EM field were about the same in the two regions [5], [6]. The method was first tested for computing the propagation characteristics of a uniform finline. In the absence of the discontinuity, the vector basis functions $\boldsymbol{e}_\nu$ and $\boldsymbol{h}_\mu$ on the aperature region (see Appendix) simply reduce to the transverse components of the normal modes of a rectangular waveguide with inner dimensions l and b.

The computed frequency behavior of the effective permittivity

$$\epsilon_{\text{eff}} = (\beta/\beta_0)^2$$

for different gap widths is shown in Fig. 3. Increasing the number of basis functions from 1 to 10, only small differences (less than 1 percent) have been obtained. The time required for computing one resonant frequency using four basis functions was typically 0.3 s on a Univac 1100 computer. The comparison with the results obtained using the spectral-domain approach is quite satisfactory.

In the presence of a step discontinuity, the vector basis functions $\boldsymbol{e}_\nu$ and $\boldsymbol{h}_\mu$ required to represent the EM field over the aperture, have a more complicated spatial distribution, and were evaluated as shown in Appendix. This required some additional computer time.

Fig. 4 shows the resonant frequency of the finline resonator containing a step discontinuity as a function of the total length $l = l_1 + l_2$, with the ratio l_2/l_1 as a parameter. Utilizing these data, the scattering parameters of the discontinuity have been computed using the procedure outlined in Section II, and are shown in Fig. 5. The computed scattering parameters of a unilateral finline discontinuity are compared in Fig. 6 with those computed by Schmidt [5] using the mode-matching procedure.

Although the procedure described above applies to a more complicated discontinuity structure, a certain simplification can be introduced if the discontinuity is longitudinally symmetric, such as the cascaded step discontinuities shown in Fig. 7. For instance, because of the symmetry, the analysis of the structure in Fig. 7(a) is reduced to the two equivalent structures containing a single step terminated by either a magnetic wall or an electric wall, as shown in Fig. 8. The equivalent circuits of the original and the two reduced structures are also shown there.

With obvious modifications of expressions (4) for ψ_{mn} and φ_{mn}, and of the basis functions $\boldsymbol{e}_\nu$ and $\boldsymbol{h}_\mu$ (see Appendix), the field analysis procedure described in Section III can be applied to the case of magnetic walls to obtain Z_{11}, Z_{22}, and Z_{21} by way of the resonant frequencies.

Fig. 9 shows the computed results at 26 GHz for the capacitive strips. The normalized reactance parameters of the equivalent T-network are shown as a function of the fin gap d_2 and the distance h. As expected, the capacitance associated with the shunt branch X_{12} increases with both h and the ratio d_1/d_2. On the contrary, the series branches have an inductive reactance whose value is much less sensitive to variations with respect to d_1/d_2. It can be shown that increasing h or d_1/d_2 results in an increase in the magnitudes of the reflection coefficient s_{11}. The phase of s_{11} varies almost linearly with h.

The dual case of inductive notches is shown in Fig. 10, where the normalized admittance parameters of the equivalent π-network are shown as functions of h and d_2/d_1. In this case, the inductance associated with the series branch

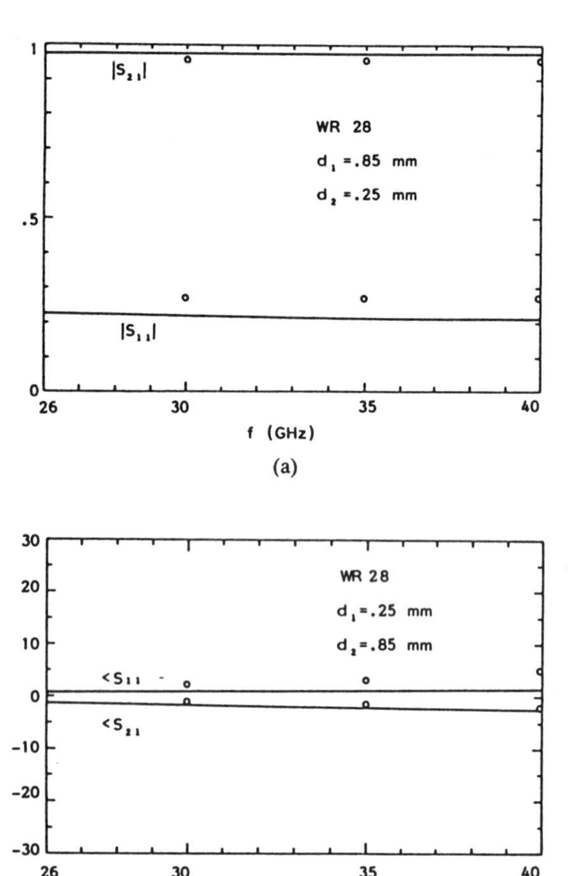

Fig. 6. Scattering parameters of a unilateral finline step discontinuity. Schmidt [7].

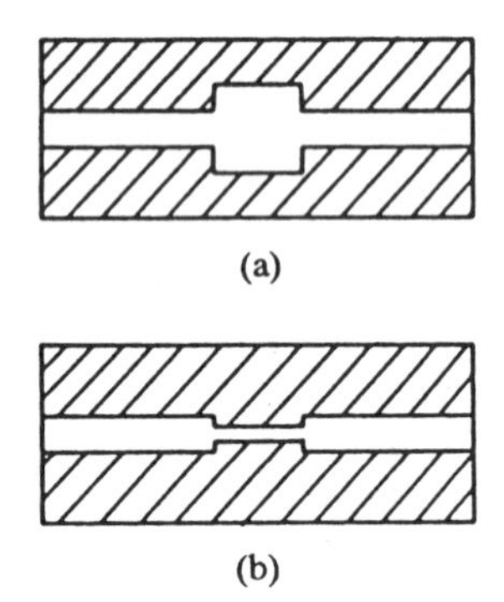

Fig. 7. Cascaded step discontinuities: (a) inductive notch and (b) capacitive strip.

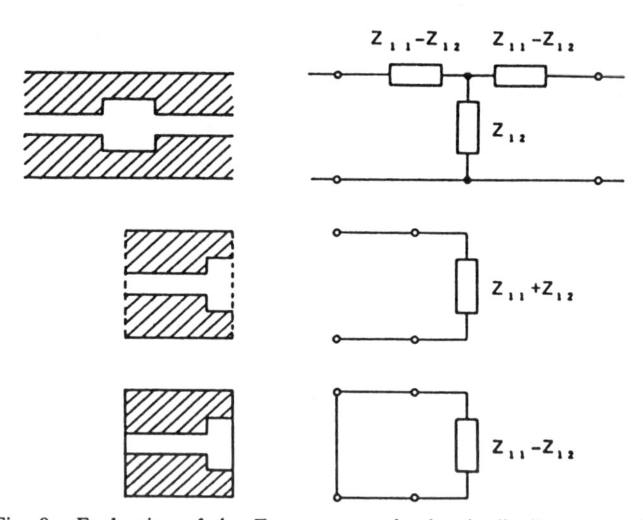

Fig. 8. Evaluation of the Z-parameters of a longitudinally symmetric discontinuity.

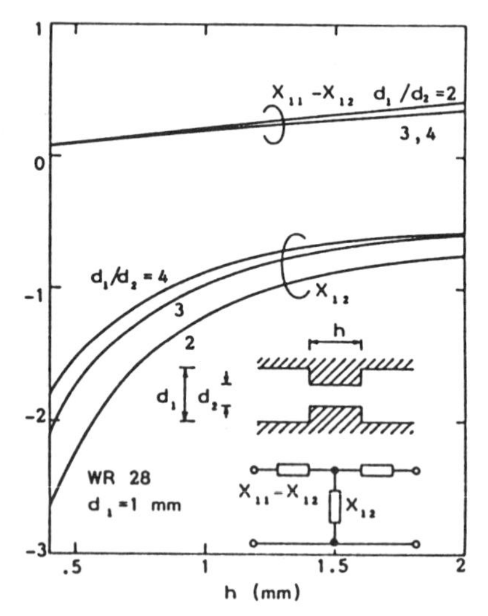

Fig. 9. Normalized reactance parameters of capacitive strips.

increases with h and d_2/d_1, while the capacitance of the shunt branches increases only slightly as a function of these parameters.

V. Conclusions

A new method of analysis has been proposed for the characterization of uniform finlines and finline discontinuities. The method is based on the computation of the resonant frequencies of a resonator obtained by short- (or open-) circuiting a finline section containing the discontinuity. The analysis procedure consists of a field expansion in terms of LSM and LSE modes of the rectangular waveguide. These expressions are matched with the field distribution in the plane of the fins. With respect to other approaches based on the field expansion in terms of finline modes, this procedure reduces computer time. The results are in good agreement with the numerical values obtained with other techniques.

Appendix

The two sets of orthonormalized vector functions $\boldsymbol{e}_\nu$, $\boldsymbol{h}_\mu$ used in (7) and (8) for expanding the EM field at $x = 0$ in the aperture region are derived in this Appendix in the case of a step discontinuity between two finline sections of different slot widths. Because of symmetry considerations, a longitudinal electric plane can be placed at $y = 0$ (see Fig. 1), so reducing the longitudinal section to that of Fig. 11.

The aperture region $S_0 \equiv (S_1 \cup S_2)$ may be viewed as the cross section of a waveguide having a stepped cross section. We can therefore expand the EM-field components $\boldsymbol{E}_{t0}$, $\boldsymbol{H}_{t0}$ lying in the yz plane in terms of the TE and TM scalar

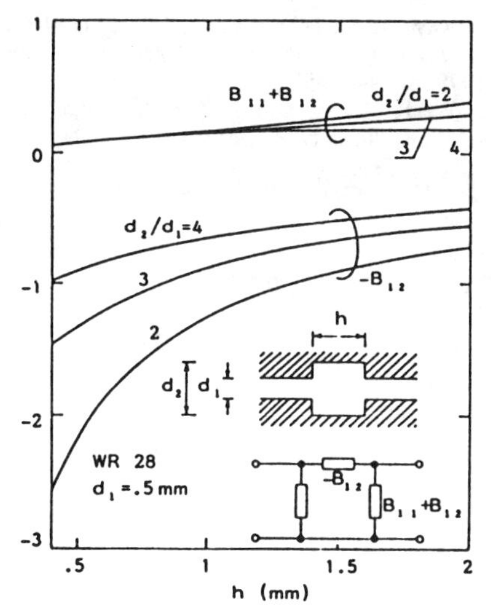

Fig. 10. Normalized admittance parameters of inductive notches.

potentials

$$E_{t0}=\sum_n V_n\hat{x}\times\nabla_t\psi_n+\sum_\nu V_\nu\nabla_t\varphi_\nu \quad \text{(A1)}$$

$$H_{t0}=\sum_n I_n\nabla_t\psi_n+\sum_\nu I_\nu\nabla_t\varphi_\nu\times\hat{x}. \quad \text{(A2)}$$

ψ_n and φ_ν represent the transverse potentials for TE and TM modes, respectively, satisfying the eigenvalue equations

$$\nabla_t^2\psi_n+k_{cn}^2\psi_n=0 \quad \text{(A3)}$$

$$\nabla_t^2\varphi_\nu+k_{c\nu}^2\varphi_\nu=0 \quad \text{(A4)}$$

in S_0 together with proper boundary conditions.

For the sake of brevity, only the solution of (A3) will be illustrated. Moreover, in order to simplify the notation, the index n will be dropped.

In order to solve (A3), the function ψ can be expressed as follows:

$$\psi=\begin{cases}\psi_1=\sum_r A_r\psi_r^{(1)}, & \text{in } S_1\\ \psi_2=\sum_s B_s\psi_s^{(2)}, & \text{in } S_2\end{cases} \quad \text{(A5)}$$

where

$$\psi_r^{(1)}=\cos k_{1r}(z+l_1)\cos\frac{r\pi y}{d_1/2} \quad \text{(A6)}$$

$$\psi_s^{(2)}=\cos k_{2S}(z-l_2)\cos\frac{s\pi y}{d_2/2} \quad \text{(A7)}$$

$$k_{ir}^2=k_c^2-\left(\frac{r\pi}{d_i/2}\right)^2,\qquad i=1,2. \quad \text{(A8)}$$

Expressions (A5)–(A8) are such that (A3) is satisfied together with the boundary conditions at $z=-l_1,l_2$ and $y=0,d_1/2,d_2/2$. The boundary conditions at $z=0$

$$\psi_1=\psi_2,\qquad 0\leqslant y\leqslant d_2/2 \quad \text{(A9)}$$

$$\frac{\partial\psi_1}{\partial z}=\begin{cases}\dfrac{\partial\psi_2}{\partial z}, & 0\leqslant y\leqslant d_2/2\\ 0, & d_2/2\leqslant y\leqslant d_1/2\end{cases} \quad \text{(A10)}$$

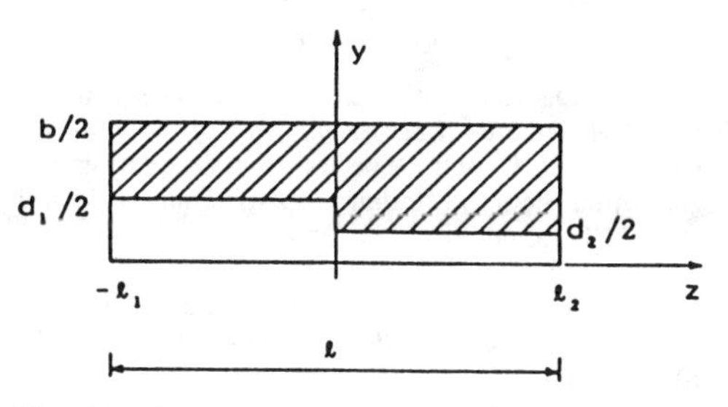

Fig. 11. Reduced geometry of the step discontinuity.

through the orthogonal properties of cosine functions, lead to a homogeneous system of equations in the expansion coefficients A_r, B_s

$$\sum_r A_r f_{rs}\cos k_{1r}l_1-\frac{d_2}{2\delta_s}B_s\cos k_{2s}l_z=0,\qquad s=0,1,2\cdots \quad \text{(A11)}$$

$$\frac{d_1}{2\delta_r}A_r k_{1r}\sin k_{1r}l_1+\sum_s B_s f_{rs}k_{2s}\sin k_{2s}l_2=0,\qquad r=0,1,2\cdots \quad \text{(A12)}$$

where

$$\delta_r=\begin{cases}1, & r=0\\ 2, & r\neq 0\end{cases}$$

$$f_{rs}=\int_0^{d_2/2}\cos\frac{r\pi y}{d_1/2}\cos\frac{s\pi y}{d_2/2}\,dy.$$

The condition for nontrivial solutions of (A11)–(A12) constitutes the characteristic equation from which the eigenvalues k_c^2 can be computed. For each k_c^2, the expansion coefficients A_r, B_s are determined using (A11)–(A12) and imposing the normalization condition

$$\int_{S_0}|\nabla_t\psi|^2\,dS=1.$$

Finally, it can be easily demonstrated that the ψ_n's so obtained satisfy the orthogonality condition

$$\int_{S_0}\nabla_t\psi_n\cdot\nabla_t\psi_m\,dS=0,\qquad n\neq m$$

even if, for numerical reasons, the series in (A5) will be truncated to a finite number of terms.

A similar procedure can be applied to the evaluation of the φ_ν's. The right-hand side of (A1) and (A2) finally provide the required expansions in terms of orthonormal vector functions.

If the resonator is terminated at $z=-l_1,l_2$ by magnetic walls, (A6) and (A7) are modified corresponding, in order to satisfy the open-circuit boundary conditions. Moreover, the eigenfunction φ_0, corresponding to the eigenvalue $k_c^2=0$, must also be included in expansions (A1) and (A2). This eigenfunction corresponds to the TEM mode of the stepped waveguide with mixed conducting and magnetic boundaries.

References

[1] H. Hofmann, "Dispersion of planar waveguides for millimeter-wave application," *Arch. Elek. Übertragung.*, vol. 31, pp. 40–44, Jan. 1977.
[2] L.-P. Schmidt and T. Itoh, "Spectral domain analysis of dominant and higher order modes in fin-lines," *IEEE Trans. Microwave Theory Tech.* vol. MTT-28, pp. 981–985, Sept. 1980.
[3] A. Beyer, "Calculation of discontinuities in grounded finlines taking into account the metallization thickness and the influence of the mount-slits," in *Proc. of the 12th European Microwave Conf.* (Helsinki, Finland), 1982, pp. 681–686.
[4] H. El Hennaway and K. Schunemann, "Analysis of fin-line discontinuities," in *Proc. of the 9th European Microwave Conf.* (Brighton, England), 1979, pp. 448–452.
[5] S. W. Lee, W. R. Jones, and J. J. Campbell, "Convergence of numerical solutions of iris-type discontinuity problems," *IEEE Trans. Microwave Theory Tech.*, vol. MTT-19, pp. 528–536, June 1971.
[6] Y. C. Shih and K. G. Gray, "Convergence of numerical solutions of step-type waveguide discontinuity problems by modal analysis," in *IEEE MTT-S Int. Symp. Dig.* (Boston, MA), 1983, pp. 233–235.
[7] L.-P. Schmidt, private communication.

Part 10
Spectral-Domain Approach

THE spectral-domain approach (SDA) or spectral-domain method is the most popular numerical method for the analysis of printed circuit configurations such as striplines, microstrips, coplanar lines, finlines, etc. It consists essentially of an integral equation formulation solved in the Fourier transform domain by Galerkin's method. Because of the possibility of incorporating into the basis functions the proper field behavior at the edges, the SDA has excellent numerical efficiency, but is restricted to regularly shaped structures. Like the method of lines and the transverse resonance technique, the SDA takes advantage of the planar configuration of the structures in order to reduce its dimensional complexity. A brief description of the method is given in Part 1 in the survey paper by Itoh (Paper 1.3).

The literature on the SDA is extensive and only a very small fraction can even be mentioned here. The reader is referred to Paper 10.1 by Jansen which is an extensive survey of the SDA furnished with a vast bibliography. Besides the mathematical formulation, the various aspects of the method are discussed in connection with the different types of problems at hand. A historical perspective of the progress of the SDA is also included.

The use of the Fourier transform in the area of microwave circuits was first introduced by Yamashita and Mittra [10.1]. In that paper Poisson's equation was solved in the Fourier domain to compute the microstrip capacitance through a variational expression. The SDA, however, was introduced under this name by Itoh and Mittra, in a historical paper reprinted here (Paper 10.2). The extension to planar microstrip resonators was made soon after by Itoh using a double Fourier transform [10.2].

One of the basic points in the application of the SDA is the evaluation of the dyadic Green's function for the structure. This problem has been substantially simplified by the generalized immittance approach developed by Itoh (Paper 10.3). Using a suitable coordinate transformation, TE and TM waves are separated in such a way that Green's function in the Fourier domain is derived by simple transmission-line theory. Thanks to this approach, the spectral-domain analysis of generalized printed transmission lines with conductors placed on different layers requires much less analytical effort and has therefore become quite feasible.

The possibility of extending the SDA to the computation of the continuous spectrum of open slot-like transmission lines has been shown by Citerne and Zieniutycz (Paper 10.4). The application of Floquet's theorem in conjunction with the SDA has recently been proposed by Glandorf and Wolff for characterizing periodically nonuniform microstrip lines [10.3].

So far the SDA has been applied to the solution of eigenvalue problems, either in the computation of the propagation characteristics of uniform transmission lines or in the computation of the resonances of printed resonators. An interesting extension of the SDA to excitation problems has recently been developed by Zhang and Itoh (Paper 10.5). By a proper formulation, the SDA is extended to the analysis of the scattering from E-plane strip discontinuities.

References

[10.1] E. Yamashita and R. Mittra, "Variational method for the analysis of microstrip lines," *IEEE Trans. Microwave Theory Tech.*, vol. MTT-16, pp. 251–256, Apr. 1968.

[10.2] T. Itoh, "Analysis of microstrip resonators," *IEEE Trans. Microwave Theory Tech.*, vol. MTT-22, pp. 946–952, Nov. 1974.

[10.3] F. J. Glandorf and I. Wolff, "A spectral-domain analysis of periodically nonuniform microstrip lines," *IEEE Trans. Microwave Theory Tech.*, vol. MTT-35, pp. 336–343, Mar. 1987.

The Spectral-Domain Approach for Microwave Integrated Circuits

ROLF H. JANSEN, SENIOR MEMBER, IEEE

(Invited Paper)

Abstract—A survey is given of the so-called spectral-domain approach, an analytical and numerical technique particularly suited for the solution of boundary-value problems in microwave and millimeter-wave integrated circuits. The mathematical formulation of the analytical part of this approach is described in a generalized notation for two- and three-dimensional strip- and slot-type fields. In a similar way, the numerical part of the technique is treated, keeping always in touch with the mathematical and physical background, as well as with the respective microwave applications. A discussion of different specific aspects of the approach is presented and outlines the peculiarities of shielded-, covered-, and open-type problems, followed by a brief review of the progress achieved in the last decade (1975–1984). The survey closes with considerations on numerical efficiency, demonstrating that spectral-domain computations can by speeded up remarkably by analytical preprocessing. The presented material is based on ten years of active involvement by the author in the field and reveals a variety of contributions by West German researchers previously not known to the international microwave community.

I. Introduction

GENERALLY SPEAKING, the term spectral-domain approach (SDA) refers to the application of integral transforms, such as the Fourier and Hankel transforms, to the solution of boundary-value and initial-value problems. As becomes obvious from the overview book and associated bibliography by Sneddon [1], this approach has been applied to mechanical and electromagnetic problems for at least a century. It provides an elegant tool for the reduction of the partial differential equations of mathematical physics to ordinary ones, which in many cases are amenable to further analytical processing. During the last 15 years, the spectral-domain approach has received considerably more interest together with the growing importance of printed circuits for very high frequencies, namely conventional and monolithic microwave and millimeter-wave integrated circuits (MIC's) fabricated by planar photolithographic technology. The actual and potential range of application of this technique implies hybrid thin- and thick-film circuits, monolithic MIC's on gallium arsenide, planar resonators and antennas, as well as multiconductor multilayer interconnections in high-speed computers. These circuits and components typically operate at frequencies between 0.1 and 100 GHz, and the main intention of using the approach has been the derivation of accurate, particularly frequency-dependent, design information.

Manuscript received February 5, 1985; revised May 31, 1985.

The author is with the University of Duisburg, Department of Electrical Engineering, FB9/ATE, Bismarckstrasse 81, D-4100 Duisburg 1, West Germany.

Already by 1957, Wu [2] had considered it an "obvious thing to do" to apply a Fourier transform in the analysis of microstrip lines. From the end of the 1960's on, several authors began to implement more and more of those steps which are characteristic for what today is denoted the spectral-domain approach for MIC's. Yamashita and Mittra [3], for example, solved Poisson's equation in the transform domain and computed microstrip line capacitance from a variational expression under application of parseval's theorem. Denlinger [4] in the United States and Schmitt and Sarges [5] in West Germany both derived an approximate solution to the microstrip dispersion problem in terms of the transformed strip current density. Itoh and Mittra [6], on the other hand, applied a spectral-domain approach in essentially the form it is still used today to the computation of slotline dispersion characteristics. Two years later, the same authors explicitly used the notation "spectral domain approach" for the specific technique (Galerkin's method in the transform domain) employed in one of their microstrip contributions [7]. Recent analyses still follow the basic outlines of this technique and the notation has been adopted by the microwave community.

In the initial research phase, a variety of fundamental applications and modifications of the spectral-domain approach and related methods had been reported within a few years. Coupled microstrip dispersion and characteristic impedances were computed by Kowalski and Pregla [8] and by Krage and Haddad [9]. Also, guided higher order modes in open microstrip lines were treated by Van de Capelle and Luypaert [10]. Itoh and Mittra [11] extended the spectral-domain formalism to shielded microstrip lines while Jansen [12] treated the same problem making use of a least-square criterion instead of Galerkin's method in the final step of the solution. As a first application to microstrip discontinuity problems, Rahmat-Samii *et al.* performed a quite general static spectral-domain analysis [13]. The first full-wave analyses of hybrid-mode microstrip resonator problems were reported by Itoh [14] and by Jansen [15], [16] in 1974, including rectangular, disk, ring, and concentric coupled shapes. Along the guidelines having emerged in this way, the spectral-domain approach has been used extensively for the characterization and analysis of elementary structures frequently appearing in MIC's. These structures can be classified as conducting thin patterns in one or more interfaces of a multilayer stratified dielectric medium. Therefore, the associated electromag-

Reprinted from *IEEE Trans. Microwave Theory Tech.*, vol. MTT-33, no. 10, pp. 1043–1056, Oct. 1985.

netic boundary-value problems lend themselves ideally to an SDA treatment. The partial differential equations considered are mainly the wave equation or, where small dimensions compared to wavelength prevail, the Laplace and the Poisson equation. Specific problems frequently tackled by the spectral domain approach are:

1) the static or frequency-dependent characterization of printed microwave transmission lines (a two-dimensional electromagnetic field problem).
2) the static or frequency-dependent analysis of problems concerning strip and slot transmission-line discontinuities, junctions and resonators, respectively and patch antennas (three-dimensional electromagnetic fields).

The contribution given here outlines the basic features of the analytical formulation of the spectral-domain approach as it applies to the above-mentioned problems. It is shown how for printed planar structures of arbitrary connected and disconnected shape embedded in a multilayer dielectric medium a single closed-form integral equation emerges from the application of the analytical steps of the SDA. As a result of explicit construction of that portion of the solution which depends on the vertical coordinate, this integral equation comes out reduced by one dimension compared to the original partial differential equation. From the beginning of the analysis, a considerable reduction in complexity is achieved and reduces the expense for the subsequent numerical part of the approach. This provides one of the important arguments for the superiority of the spectral-domain approach compared to other techniques.

In a discussion of the numerical procedure usually employed to solve the derived integral equation, the peculiarities of eigenvalue-type and deterministic MIC problems are treated briefly. There are arguments to prefer a Galerkin solution with certain symmetry properties for the former, while the latter do not generally result in a symmetric, respectively, Hermitian system of equations. From the obtained solutions, most of the quantities required in the characterization and analysis of MIC's can be obtained directly in the transform domain. Only one of the methods recently applied to MIC's shares several of the advantages of the SDA: the differential-difference approach (DDA), also called the method of lines [17]. A comparison with this, therefore, deserves a brief discussion.

After presenting these general features, the different aspects of the spectral-domain approach are outlined which have to be considered for shielded structures, laterally open structures, and configurations which are completely open electromagnetically. A specific implementation of the approach recently developed for the systematic frequency-dependent analysis of discontinuities and junctions in MIC's is described. It is discussed further as to how the radiation condition can be incorporated into the SDA formulation by proper choice of the integration path prevailing for the basic integral equation. To round out the

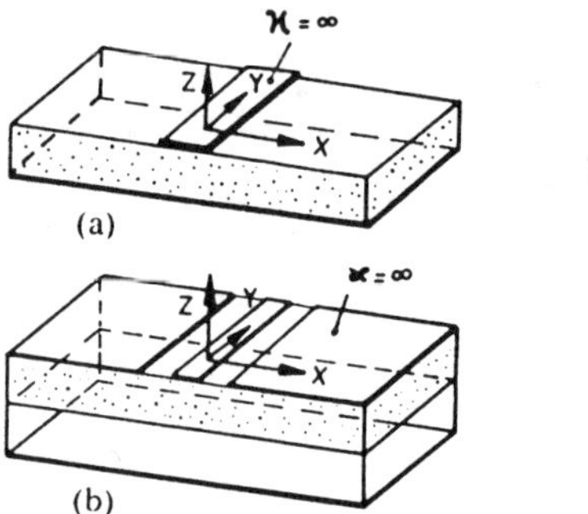

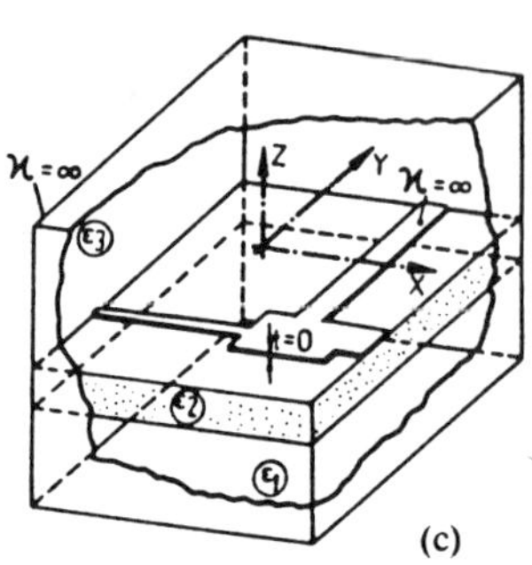

Fig. 1. (a) Microstrip line and (b) coplanar waveguide as examples for MIC transmission lines and (c) generalized MIC structure together with coordinate system used in the discussion.

picture given of the spectral-domain approach, the more important results achieved by its application are summarized in a subsequent section, and the state-of-the-art is described.

The last section of the paper is a discussion of the advantages and disadvantages of the spectral-domain approach. Emphasis is placed on the hybrid character of this technique which requires (and allows!) a certain amount of analytical preprocessing to achieve high efficiency. It is shown, further, how most of the disadvantages of the technique can be removed and to what degree, typically, a specific class of problems can be speeded up.

Remarks on the numerical problems associated with the development of user-oriented SDA packages are made and critical parts of these are illuminated. Finally, the main characteristics of the spectral-domain approach are summarized in a brief conclusion.

II. Mathematical Formulation

Some elements of the analytical steps necessary to apply the spectral-domain approach to specific problems, particularly the characterization of MIC transmission lines, have already been described in overview books [18], [19]. The treatment given here generalizes the formulation as far as possible and emphasizes those features which the different classes of MIC problems all have in common. For a visualization of the physical construction of the configurations to be considered, Fig. 1 shows (a) an open microstrip line, (b) a coplanar waveguide suspended above the ground plane of a circuit environment, (c) and a quite general shielded structure. The latter serves for the following discussion and could as well be laterally open or completely open. It provides an idealized view of the basic construction of MIC's indicating that the passive portions of these consist mainly of a thin conductor metallization in one or more interfaces of a double or multilayer dielectric medium; for an overview, see [20].

In agreement with common microwave practice, the formulation is in terms of time-harmonic electromagnetic fields, namely, a time dependence of $\exp(j\omega t)$. Vector quantities, like the electric field $\underline{E}$, are written by single underlining, matrices by double underlining. The involved conductors are assumed to have ideal conductivity κ and negligible thickness t. This is a very realistic assumption in hybrid MIC's, where strip and slot widths are usually large compared to conductor thickness. In monolithic MIC's on gallium-arsenide, this is not valid with the same generality. Here, the assumption is mainly a matter of convenience and simplification of the treatment. The consideration of finite thickness in the SDA formulation can be achieved by treating the thick metallization as a separate layer, see for example [21]–[23]. In addition, it is convenient in most cases to assume lossless dielectric media since this allows a ral number arithmetic for the SDA algorithm, except for cases where radiation or surface-wave excitation is involved. Loss parameters are usually introduced by perturbation methods subsequent to a numerical solution neglecting loss. This also applies to the evaluation of conductor loss, which can be taken into account if the asymptotic behavior of the field derived for zero metallization thickness is appropriately modified [24]–[27]. In each of the layers $i=1,2\ldots L$ of a general configuration like that of Fig. 1, the electromagnetic field is best formulated in terms of scalar LSM and LSE wave potentials [28], [29]. This is equivalent to the use of vector wave potentials having only one component in the z-direction, i.e., perpendicular to the stratified circuit medium. It allows a completely decoupled and, therefore, particularly simple analytical treatment of all classes of MIC problems [30]–[36] independent of the number of dielectric layers involved. This specific choice naturally leads to what Itoh [35] has named the spectral-domain inmittance approach. Just recently, Omar and Schünemann [37] have shown that only coupling of the LSE and the LSM contributions to the electromagnetic field occurs and is required in satisfying the edge condition with the last step of the solution. The scalar LSM and LSE wave potentials are denoted f and g here. They are subject to the homogeneous Helmholtz equation

$$(\Delta+k_i^2)f_i=(\Delta+k_i^2)g_i=0 \tag{1}$$

in each of the layers $i=1,2\ldots L$ of the circuit medium, with k_i denoting the wave number associated with the ith layer. It should be stressed that the homogeneous form of (1) applies even in the case of excitation problems. With the spectral-domain approach, sources are introduced in a natural way as impressed current densities or electric fields only into the interfaces between the layers [38]–[41]. Instead of considering the space-domain form of the Helmholtz equation (1) directly, its spectral-domain equivalent is used. Without loss of generality, the scalar potentials may be written in the form of inverse two-dimensional Fourier transforms, for example,

$$f_i(x,y,z)=\int_{C_x}\int_{C_y}\tilde{f}_i(k_x,k_y,z)\cdot\exp\left(-j(k_xx+k_yy)\right)dx\,dy. \tag{2}$$

For configurations of circular symmetry, the use of Hankel transforms is an adequate choice [1], [42]–[46]. In transmission-line problems, (2) is reduced by one dimension since these problems can be formulated in terms of the cross-sectional field alone postulating a longitudinal dependency of the form $\exp(-jk_yy)$. Alternatively, for these cases, the spectral wave potential $\tilde{f}_i$ may be viewed as factorized and containing a y-dependent factor in the form of a Dirac distribution, so that the integration in (2) reduces to one dimension. In the high majority of SDA contributions published, the integration paths C_x and C_y have been chosen to coincide with the respective coordinate axes. The spectral variables k_x and k_y may be interpreted each as a measure of spatial oscillation of the described field which is useful for later convergence considerations. This is immediately obvious if (2) implies a finite Fourier transform [11] and the spectral wave numbers k_x and/or k_y form an infinite numerable sequence. In that case, $\tilde{f}_i(k_{xm},k_{yn},z)$ describes the Fourier series coefficients of $f_i(x,y,z)$ and these values of k_{xm},k_{yn} are chosen in such a way that the boundary conditions on a lateral shielding parallel to the z-axis are satisfied. Further generalizing, one may consider such coefficients as being associated with any two suitably chosen complete orthogonal sets of solutions of the Helmholtz equation which are TM and TE with respect to the z-direction and satisfy the lateral boundary conditions [30], [31]. At the same time, this reveals how a suitable finite integral transform can be constructed for a given cross section of the shielding. In addition, this generalization makes clear that the mathematical formulation can be discussed completely independent of the special cross section or even the existence of a lateral shielding. The Helmholtz operator of (1), if applied in the transform domain, always appears as

$$\tilde{\Delta}+k_i^2=\frac{\partial^2}{\partial z^2}+k_i^2-k_x^2-k_y^2=\frac{\partial^2}{\partial z^2}+k_{zi}^2 \tag{3}$$

which is an ordinary differential operator. Due to the simple layered planar construction of MIC's, the transformed wave potentials $\tilde{f}_i$ and $\tilde{g}_i$ can be determined analytically and adopt the general form

$$\tilde{f}_i(k_x,k_y,z)=a_i(k_x,k_y)\cdot\cos\left(k_{zi}(z-z_i)\right)+b_i(k_x,k_y)\cdot\sin\left(k_{zi}(z-z_i)\right),\qquad i=1\cdots L. \tag{4}$$

The functions a_i and b_i are spectral distributions weighting the elementary plane-wave constituents with respect to the z-axis. The parameter z_i is arbitrary and is, for example, introduced to allow convenient satisfaction of the boundary conditions at the conducting ground plane and cover shielding usually existing in MIC's. With the relations

$$j\omega\epsilon_i\tilde{E}_{zi}=k_\rho^2\tilde{f}_i,\; j\omega\mu_i\tilde{H}_{zi}=k_\rho^2\tilde{g}_i,\; k_\rho^2=k_x^2+k_y^2 \tag{5}$$

it becomes clear that homogeneous boundary conditions prevail for $\tilde{f}_i$ and $\tilde{g}_i$ identical to those for the transformed field components $\tilde{E}_{zi}$ and $\tilde{H}_{zi}$. Further, since the transform defined by (2) affects only the x and y coordinates, all conditions specified for the spatial electromagnetic field in planes of constant values of z can be directly transferred into the spectral domain. Therefore, in a configuration involving the layers $i=1,2\ldots L$, the potential $\tilde{f}_i$ at the ground and top plane of the multilayer medium is

$$\tilde{f}_i(k_x,k_y,z)=a_i(k_x,k_y)\cos(k_{zi}(z-z_i)), \qquad i=1,L \tag{6}$$

if z_1 and z_L describe the positions of the ground and cover shielding. In complete analogy, $\tilde{g}_i$ is proportional to the respective sine function for $i=1$ and $i=L$ as a consequence of the vanishing of $H_{zi}(x,y,z)$ at $z=z_1$ and $z=z_L$. For vertically open structures like antennas and open microstrip, the potentials $\tilde{f}_L$ and $\tilde{g}_L$ both have an exponential z-dependence. The complete transformed electromagnetic field in all of the layers $i=1,2\ldots L$ is derived by application of the spectral-domain equivalent $\tilde{\nabla}$ of the ∇ operator, namely, by

$$\underline{\tilde{E}}_i=\frac{1}{j\omega\epsilon_i}\tilde{\nabla}\times\tilde{\nabla}\times(\tilde{f}_i\underline{u}_z)-\tilde{\nabla}\times(\tilde{g}_i\underline{u}_z)$$

$$\underline{\tilde{H}}_i=\frac{1}{j\omega\mu_i}\tilde{\nabla}\times\tilde{\nabla}\times(\tilde{g}_i\underline{u}_z)+\tilde{\nabla}\times(\tilde{f}_i\underline{u}_z). \tag{7}$$

These relations have the same form as their spatial counterparts [28], [29] and result by substituting the algebraic multiplicators jk_x and jk_y for the respective partial differential operators. Together with the foregoing discussion they show that, in a circuit medium of L layers, the total electromagnetic field can be described by $4(L-1)$ independent spectral LSM and LSE distributions $a_i(k_x,k_y)$, $b_i(k_x,k_y)$. For the dielectric interfaces between the different layers, exactly the same number of continuity conditions can be formulated in the spectral domain, i.e.,

$$\tilde{E}_{xi}-\tilde{E}_{xi+1}=0 \qquad \tilde{E}_{yi}-\tilde{E}_{yi+1}=0$$

$$\tilde{H}_{xi}-\tilde{H}_{xi+1}=-\tilde{J}_{yi} \qquad \tilde{H}_{yi}-\tilde{H}_{yi+1}=+\tilde{J}_{xi} \tag{8}$$

for $i=1\ldots L-1$ and z fixed at its interface value for each subscript i. They mirror the continuity of the electric field $\underline{E}_t$ tangential to the interfaces independent of the presence of a thin metallization. At the same time, they describe the discontinuity of the tangential magnetic field at such a metallization in terms of a surface current density $\underline{J}_t$. In interfaces which do not contain conductors, $\underline{J}_t$ is defined to be zero, enforcing the magnetic field continuity there. By analytical processing of the relations (8), all the unknown distributions a_i and b_i can be eliminated or expressed by the spectral-domain current density components $\tilde{J}_{xi}$ and $\tilde{J}_{yi}$. The latter may exist in only one of the interfaces or in several of these. In this stage of the analysis, the only boundary conditions which remain to be satisfied are those of the electric field $\underline{E}_t$ tangential to the conductor metallization and of the surface current density $\underline{J}_t$ to vanish in complementary regions. How the spectral-domain relations resulting from the analytical evaluation of (8) have to be arranged for further processing, therefore, depends on whether the metallized interfaces can be characterized as strip-type ($i=ist$) or slot-type ($i=isl$), respectively. In the general case, an algebraic spectral-domain equation results linking the mixed-type vectors

$$(\cdots\tilde{E}_{xist},\tilde{E}_{yist}\cdots\tilde{J}_{xisl},\tilde{J}_{yisl}\cdots)^T$$

and

$$(\cdots\tilde{J}_{xist},\tilde{J}_{yist}\cdots\tilde{E}_{xisl},\tilde{E}_{yisl}\cdots)^T \tag{9}$$

by a spectral immittance matrix; see for example [22], [35], [47]. T denotes transposition and is used only for convenience of writing. The lower one of the two vectors shown is put onto the right side of the described algebraic relation because its elements are better suited for expansion into known functions. These elements are typically confined to a small portion of the affected interfaces. For a similar reason, the upper one of the vectors is arranged onto the left side of the spectral-domain equation. By this, it is described by the other vector and needs to satisfy boundary conditions on only a small portion of its region of existence. The spectral algebraic relation between the vectors (9) is equivalent to a single integral equation which results by application of the transform inherent in (2). Since the whole discussion has been performed without recourse to particular shapes of the metallization pattern, this is true for arbitrary planar connected and disconnected conductors. From the procedure outlined, the kernel of the integral equation is available in analytical form. Further, this integral equation comes out reduced by one dimension compared to the original field problem associated with (1). It is one-dimensional for transmission-line problems and two-dimensional for discontinuities, resonators, and so on. As (4) shows, the z-dependency of the field is described in analytical form. In most cases of practical interest, the MIC configurations analyzed are either strip-type or slot-type exclusively, with only one layer of metallization. Under this presumption, the relation between the vectors (9) reduces to the simpler form

$$(\tilde{E}_{xist},\tilde{E}_{yist})^T=\underline{\underline{\tilde{Z}}}(p)\cdot(\tilde{J}_{xist},\tilde{J}_{yist})^T$$

or

$$(\tilde{J}_{xisl},\tilde{J}_{yisl})^T=\underline{\underline{\tilde{Y}}}(p)\cdot(\tilde{E}_{xisl},\tilde{E}_{yisl})^T. \tag{10}$$

The quantity p has been introduced to remind one of the fact that the elements of $\underline{\underline{\tilde{Z}}}$ and $\underline{\underline{\tilde{Y}}}$ depend in a known, analytical form on physical parameters defining the considered MIC problem (for example, vertical geometry, shielding dimensions, or operating frequency). Obviously, the spectral immittance matrix $\underline{\underline{\tilde{Y}}}$ is the inverse of $\underline{\underline{\tilde{Z}}}$ if both equations in (10) refer to the same problem and dielectric interface. For this reason, we may also write in the space domain

$$\underline{E}_t=L_\infty(p)\cdot\underline{J}_t \quad \text{or} \quad \underline{J}_t=L_\infty^{-1}(p)\cdot\underline{E}_t \tag{11}$$

where $\underline{E}_t=(E_x,E_y)^T$ and $\underline{J}_t=(J_x,J_y)^T$ are specialized to denote electric field and current density in the plane of the circuit metallization. The integral operators L_∞ and L_∞^{-1} are linear with respect to the vectors they operate on and both have a dyadic kernel defined by the spectral-domain Green's immittance matrices $\underline{\underline{\tilde{Z}}}$ and $\underline{\underline{\tilde{Y}}}$. The constituents of this kernel can be determined in a particularly elegant way by the so-called spectral-domain immittance approach [32], [35]. It has been shown by the author for shielded configurations that by suitable choice of an orthonormal vectorial function basis in (2), the kernel can be described by a single infinite scalar set of wave impedances $\tilde{Z}_n$ in conjunction with the elements of the function basis [30], [31]. In that case, the discrete impedance elements $\tilde{Z}_n$ are equal to the modal input wave impedances of the cylindrical shielding as seen in the plane of metallization. Independent of these details, there is always a duality between the strip-type and the slot-type formulation as visible in (10) and (11). For the former, the tangential electric field $\underline{E}_t$ has to vanish on the strip metallization F_{st}, while for the latter, the surface current density $\underline{J}_t$ does not exist in slot regions F_{sl}, i.e., outside of the metallization. Splitting up the right-hand quantities of (11) into an excited term (subscript $t$ex) and a source or impressed term (subscript $t$im) further yields

$$\underline{E}_t = \underline{0} = L_\infty(p)\cdot(\underline{J}_{t\text{ex}} + \underline{J}_{t\text{im}}) \text{ for } x, y \in F_{st}$$

$$\underline{J}_t = \underline{0} = L_\infty^{-1}(p)\cdot(\underline{E}_{t\text{ex}} + \underline{E}_{t\text{im}}) \text{ for } x, y \in F_{sl}. \quad (12)$$

These final integral equations are written here as in the formulation of a scattering problem [38]–[41]. They define, at the same time, the electric field $\underline{E}_t$ and current density $\underline{J}_t$ in the complementary regions F_{st}^{-1} and F_{sl}^{-1}, i.e., outside of strips and on the metallization around slots. If sources $\underline{J}_{t\text{im}}$, $\underline{E}_{t\text{im}}$ are note present, like in transmission-line and resonator problems, the equations in (12) each define a so-called nonstandard eigenvalue problem [48]–[50]. This notation applies since, without sources, (12) can be solved only for specific values of the parameter p (the eigenvalue), which is contained in the integral equation kernels in nonlinear, usually transcendental form. Which physical quantity is chosen in a problem as the parameter p is to some degree arbitrary. In transmission-line problems, the usual choice is $p = \beta$, i.e., the propagation constant, or $p = \beta^2$, the square of it. Resonator problems are conveniently treated in terms of $p = \omega_0$, the resonance frequency, or $p = l_0$, a dimension of the resonator.

III. Numerical Solution

The standard procedure applied in most computer solutions of the integral equations (12) today is Galerkin's method in the spectral domain, particularly for the eigenvalue problem. This is a preferable choice resulting from the self-adjointness of the involved integral operators and following the argumentation by Harrington [51]. The stationarity of such solutions has been discussed in an early contribution by the author in comparison to a least-squares alternative [31]. It has recently been shown by Lindell in a general context for the eigenvalue parameter p with respect to the trial field [48]–[50] which is $J_{t\text{ex}}$ for strip problems and $E_{t\text{ex}}$ for slot-type configurations. To simplify the discussion, restriction to strip-type problems is allowed without loss in generality. The numerical procedure is best understood if the equations prevailing in the spectral and the space domain are considered in parallel, i.e., writing briefly

$$\underline{E}_t = L_\infty(p)\cdot(\underline{J}_{t\text{ex}} + \underline{J}_{t\text{im}}) \quad \tilde{\underline{E}}_t = \tilde{\underline{\underline{Z}}}(p)\cdot(\tilde{\underline{J}}_{t\text{ex}} + \tilde{\underline{J}}_{t\text{im}}). \quad (13)$$

In the space domain, the physical vectors $\underline{E}_t$ and $\underline{J}_t = \underline{J}_{t\text{ex}} + \underline{J}_{t\text{im}}$ are different from zero in the complementary regions F_{st}^{-1} and F_{st}. The unknown surface current density $\underline{J}_{t\text{ex}}$ is expanded into a suitable, preferably complete set of expansion functions defined on F_{st} and vanishing outside. By this, continuity of the magnetic field outside the metallization is achieved at the same time. The expansion of $\underline{J}_{t\text{ex}}$ is actually performed in the original, spatial domain since this provides the best physical insight for a good choice. It depends on the specific problem under investigation whether the functions $\underline{J}_{tk}$ chosen should be easily transformable into the spectral domain or not. For the application of Galerkin's method, the set of testing functions necessary to enforce the vanishing of $\underline{E}_t$ on the conductor region F_{st} is the same as the expansion used, say $\underline{J}_{tj}$. The scalar product employed in the testing process is commonly defined by integration over F_{st} without a specific weighting factor and expressed here using parentheses (,). Making use of the linearity of the integral operator involved, the standard process of testing [51] finally yields

$$\sum_k \alpha_k(\underline{J}_{tj}, L_\infty(p)\cdot\underline{J}_{tk}) = -(\underline{J}_{tj}, L_\infty(p)\cdot\underline{J}_{t\text{im}})$$

or

$$\sum_k \alpha_k(\tilde{\underline{J}}_{tj}, \tilde{\underline{\underline{Z}}}(p)\cdot\tilde{\underline{J}}_{tk}) = -(\tilde{\underline{J}}_{tj}, \tilde{\underline{\underline{Z}}}(p)\cdot\tilde{\underline{J}}_{t\text{im}}). \quad (14)$$

The second alternative and mathematically identical equation applies as a consequence of Parseval's theorem [1], which also serves for a unique definition of the associated scalar product (,) in the spectral domain. In eigenvalue problems, the right-hand side of (14) vanishes and a nontrivial solution $\underline{J}_{t\text{ex}}$ exists only if the determinant of the respective linear system of equations is zero. This provides the nonlinear eigenvalue equation for the unknown parameter p and, subsequent to an iterative evaluation of p, the associated surface current distribution $\underline{J}_t = \underline{J}_{t\text{ex}}$. The electric field outside of the metallization is found from the application of (13). For MIC excitation problems, (14) is deterministic and can be solved in a single step for a prescribed value of the parameter p. The main difficulty in such cases is a realistic formulation of the excitation term $\underline{J}_{t\text{im}}$ such that it well describes the physical situation. Also, the introduction of such a source term may complicate the satisfaction of boundary conditions in its spatial vicinity as compared to an equivalent eigenvalue formulation [36]. As long as the source chosen has finite support in the $x-y$ plane, the field region is finite and the expansion functions are chosen properly, Galerkin's method can still be applied, for example, if the source is a slit voltage generator or a strip current sheet [38], [39]. However, if the field is excited by a transmission-line mode coming in from infinity and a reflected wave is involved, Galerkin's method cannot be applied any more since the associated scalar products become unbounded. In that case, which is a good description of practical MIC excitation problems, another version of moment methods has to be employed [40], [41] enforcing existence of the scalar products. Expansion and test functions have to be different then with the consequence that the final system of equations (14) is not symmetric or Hermitian any longer.

It should have become obvious from the discussion that interpreting the numerical procedure as "Galerkin's method in the transform domain" is too restrictive not only because of the last-mentioned details. Actually, it does not make a mathematical difference which one of the equations (14) is considered if the presumptions necessary for the application of Parseval's theorem are satisfied. As a rule of thumb, in laterally open problems, evaluation of the scalar products in (14) is alleviated if the spectral quantities are used. In these cases, expansion functions should be selected with explicitly available analytical transforms. On the other hand, for shielded configurations, it may have advantages to perform the scalar product operation in the spatial domain, particularly if a suitable orthogonal set of functions can be constructed for the description of the electromagnetic field [30], [31]. Also, it is generally easier to construct complete sets of expansion functions for conductors of complex shape in the space domain. So, the major advantage of the so-called spectral-domain approach is that it allows one to shift between the space and the transform domains in essentially all steps of the processing. The same applies to the computation of MIC design quantities from solutions obtained by the approach. Quantities like quality factors, dielectric and magnetic loss, conductor loss, and power transported in the cross section of transmission lines can with advantage be computed in either of the domains depending on the shielding situation and the specific problem. The evaluation of such design data involves volume or surface integration over the products

$$\underline{E}_i \cdot \underline{E}_i^* \quad \underline{H}_i \cdot \underline{H}_i^* \text{ and } \underline{E}_i \times \underline{H}_i^* \tag{15}$$

where the asterisk denotes complex conjugate. Integration over the vertical z-coordinate is always performed analytically due to the simple trigonometric dependencies associated with the layered MIC structure. Along the other coordinates, Parseval's theorem again allows a choice. Care has to be taken in conductor loss computations for metallizations of zero thickness. Because of the order of the edge singularity of the field for conductors of vanishing thickness [52], the square of the magnetic field tangential to the metallization is not integrable. However, this can be repaired to achieve a good approximation of conductor loss by proper modification of the asymptotic behavior of the transform of the magnetic field [24], [25]. The idea behind this is that, except for the immediate vicinity of the conductor edges, the field does not change noticeably if a small, finite thickness is introduced. The modification may also be performed in the spatial domain if a strip-type problem prevails for which the original current density distribution is available in closed form. Also, longitudinal strip current or transverse slot voltage may be evaluated in the space domain for the same reason. These quantities are often used in the calculation of characteristic impedances of strip and slot transmission lines [8], [9], [24], [34].

Similar advantageous properties as those described for the spectral-domain approach are shared to some degree by one of the methods recently applied to MIC's. This is named the differential-difference approach (DDA) here and is also called the method of lines [17], [53]–[55]. The fundamental similarity to the SDA formulation consists in the fact that it reduces the original Helmholtz equation (1) to a system of ordinary differential equations which can be solved explicitly. In contrast to the spectral-domain approach, the reduction in complexity and presumption for further analytical processing is achieved by discretization of the Helmholtz operator, writing, for example,

$$\left(\frac{\partial^2}{\partial z^2} + k_i^2\right) f_i^{m,n} + \frac{f_i^{m-1,n} - 2f_i^{m,n} + f_i^{m+1,n}}{h_x^2} + \frac{f_i^{m,n-1} - 2f_i^{m,n} + f_i^{m,n+1}}{h_y^2} = 0. \tag{16}$$

This implies a two-dimensional finite-difference representation of the field for each plane of constant coordinate z, i.e., a mesh of points m, n with $m = 1 \dots M, n = 1 \dots N$. It describes a band-structured system of coupled ordinary differential equations with a total number of $2MN$ unknowns for two scalar wave potentials. The system can be decoupled, i.e., brought into diagonal form, leading to the same number of discrete transformed potentials, say $\bar{f}_i^{m,n}$ and $\bar{g}_i^{m,n}$. For these, the z-dependency in the layered MIC structure can be described in analytical form including the boundary conditions at the ground and top planes. For the associated discretized tangential electric field and current density in the plane of metallization, the boundary conditions are formulated pointwise. This cannot be performed in terms of the transformed quantities $\bar{f}_i^{m,n}, \bar{g}_i^{m,n}$ and, therefore, requires a back-transformation into the original domain. As in the spectral-domain approach, the last step in the DDA procedure is the solution of a determinantal equation depending on one of the physical parameters p of the problem or the solution of a deterministic linear system of equations for prescribed values of p.

The method has been applied only to shielded structures so far, which is a consequence of the spatial discretization that makes it difficult to extend it to open regions. Several interesting similarities between the SDA and the DDA become plausible if one recalls that the application of finite integral transforms means a discretization in the spectral domain. Extension to open problems with the SDA is straightforward since fields of finite and infinite spatial support both have contributions over the infinite spectral domain. One of the advantages of the method of lines is that it exhibits a comparatively low numerical expense for the generation of each of the elements of the final matrix equation. In addition, it can in a very flexible way be used for the analysis of different conductor shapes and does not require a choice of expansion functions. On the other hand, application of the DDA to three-dimensional field problems results in very high matrix orders. If, for example, in a strip-type problem, a rectangular shielding with ground plane F and conductor area F_{st} is assumed, the order of the final DDA matrix is approximately

$$Q = 2M \cdot N \cdot \frac{F_{st}}{F}. \tag{17}$$

With growing complexity of the conductor shape, the spatial resolution $M \cdot N$ has to be increased and the number of floating-point operations in the differential-difference approach is proportional to Q^3. About the same resolution is achieved by an SDA treatment using $2MN$ Fourier coefficients, which, however, determines only the linear number of summations necessary to construct a matrix element. The order of the final SDA matrix is not directly related to $M \cdot N$, but mainly a question of the intelligent choice or systematic generation of expansion functions. It can be made extremely low, which makes the SDA a preferable technique for the repeated generation of MIC design information. From this point of view, it is an advantage that it allows the choice of expansion functions. Furthermore, the spectral-domain approach is specifically suited to analytical preprocessing and speedup measures as will be shown in the last section of this paper.

IV. Specific Aspects

In the analysis of MIC configurations by the spectral-domain approach three classes of structures have to be distinguished: shielded, covered, and open types. These are shown in Fig. 2 for the cross section of a single microstrip line. The shielded-type has been used extensively by contributors to the SDA in transmission-line and resonator problems. It presents a good description of real-life MIC structures only if radiation and surface-wave excitation from an adequate open structure are negligible. This applies, for example, for the technically used fundamental modes of printed strip and slot transmission lines under normal operating conditions [56]–[58] and to high-Q resonators with properly chosen, not too large, shielding dimensions [14]–[16], [30], [31]. However, practical MIC shielding cases are usually large in dimensions compared to the enclosed circuit elements, with the exception of finlines and related millimeter-wave components [59], [60]. Therefore, care has to be taken in the interpretation of data derived from a shielded-type analysis if these shall be used for MIC design purposes. With respect to this point of view, the use of the covered, i.e., laterally open-type of analysis seems to present a better choice for the characterization of MIC structures in general. A cover shielding can always be specified in the design of a practical circuit and, therefore, taken into account properly. The assumption of lateral openness is believed to provide the most realistic one if design quantities have to be computed for general applicability in the CAD of MIC's or as a basis for the generation of mathematical models. Nevertheless, nearly all of the SDA contributions to the analysis of covered-type configurations neglect energy leakage into the lateral direction. They are equivalent, therefore, to shielded-type SDA

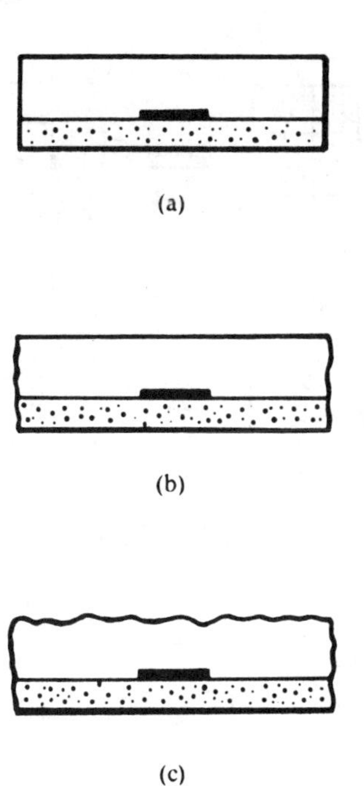

Fig. 2. Microstrip cross sections representing three different classes of MIC structures, (a) shielded, (b) covered, and (c) open type.

formulations with side walls removed left and right to infinity. Only recently, lateral energy leakage has been included in covered MIC analyses [40], [41], [61] and is considered a prerequisite to the description of dynamic coupling mechanisms. The open-type analysis of MIC structures, such as the one shown in Fig. 2(c), is applicable to nonradiating transmission-line modes, but mainly reserved to problems where radiation into free space is of dominant interest.

Using the shielded-type formulation, a systematic spectral-domain technique for the hybrid-mode characterization of MIC discontinuities and junctions has been developed by the author a few years ago [36], [62]. It is represented pictorially in the flow diagram of Fig. 3 to show how the SDA can be applied to derive design information in a very general way. The technique avoids the necessity of specifying sources and has, meanwhile, been applied successfully to a variety of strip- and slot-type problems [63]–[67]. It mainly refers to operating conditions where energy leakage into the volume field is not noticeable, but can, however, be extended to be valid without that restriction. The main idea behind the technique shown in Fig. 3 is a generalization of the Weissfloch or tangent method [28] in conjunction with a three-dimensional SDA resonator analysis. This generalization can be performed and becomes practicable here because the total electromagnetic field and current density is available from the analysis which would not be accessible or practicable in an equivalent measurement situation. On top of the left column of Fig. 3, the physical n-port investigated is shown in a shielding box with the field volume subdivided into short-circuited transmission-line sections (stubs) of length l_i and the n-port junction. The circuit representation using scattering parameters is depicted on the right-hand side with the respective reference planes RP_i, $i = 1 \ldots n$. For

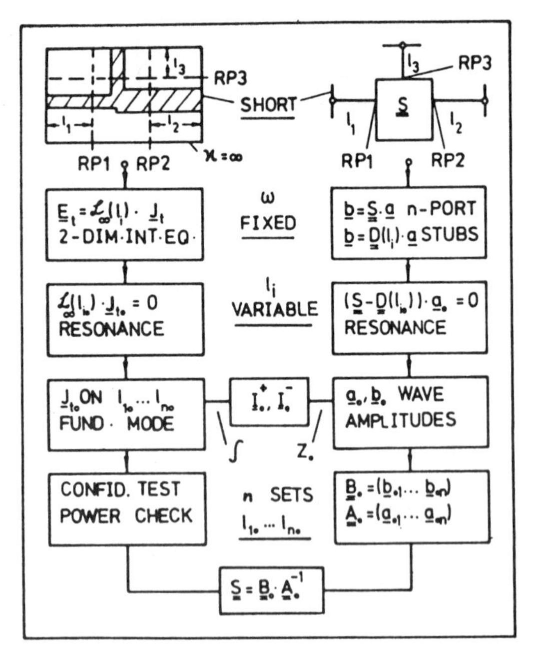

Fig. 3. A specific implementation of the SDA used for the frequency-dependent characterization of discontinuities in strip- and slot-type MIC's.

fixed operating frequency ω, the configuration is analyzed in terms of successively interchanged resonant lengths l_{i0} exactly n times. By introduction of precomputed strip current density distributions into the expansion functions used to describe $\underline{J}_t$, all the boundary conditions except those in the n-port region itself can be satisfied *a priori.* The resonant lengths and the stub current density amplitudes result from the numerical description of the n successive hypothetical resonator experiments. They are processed to obtain the complex amplitudes of the longitudinal strip currents or electric fields of the stubs. Then, using a power-related definition of characteristic stub impedance, the complex wave amplitudes associated with each of the n experiments are computed and assembled into matrices $\underline{\underline{A}}_0$ and $\underline{\underline{B}}_0$. The scattering matrix of the investigated n-port results from this easily. As a confidence test for the validity of the results it is checked in parallel that the power balance for the lossless n-port is satisfied to a good approximation. The technique has the advantage of providing phase information which is stationary with respect to the current density and field distribution, respectively [36]. It has its limitations if radiation mechanisms in MIC's are involved to a noticeable degree.

To understand leakage mechanisms in MIC's, the mathematical structure of the spectral immittance matrices of (10) has to be considered. Independent of the degree of openness and the number of dielectric layers prevailing in a specific problem, the spectral impedances can always be written in the form [31], [34], [36], [38], [41]

$$\tilde{\underline{\underline{Z}}}(p)=\begin{bmatrix} k_x^2 Z_{FE}+k_y^2 Z_{FH} & k_x k_y (Z_{FE}-Z_{FH}) \\ k_x k_y (Z_{FE}-Z_{FH}) & k_y^2 Z_{FE}+k_x^2 Z_{FH} \end{bmatrix} \tag{18}$$

with

$$Z_{FE}=Z_{FE}(k_\rho^2,p) \quad Z_{FH}=Z_{FH}(k_\rho^2,p).$$

The admittance matrix for slot-type problems follows from inversion of (18) and of the impedance elements Z_{FE}, Z_{FH}. It has exactly the same structure. Due to this duality, it is again sufficient to discuss the strip-type case for simplicity. The quantities Z_{FE} and Z_{FH} are the total LSM and LSE modal input wave impedances as seen into the medium below and above the plane of metallization. Thinking in terms of a transverse resonance approach [28], [29], therefore, makes clear that $1/Z_{FE}$ and $1/Z_{FH}$ have the properties of radial wave eigenfunctions in the layered circuit medium ($1/Y_{FE}$ and $1/Y_{FH}$ for slot-type problems). So, the elements Z_{FE} and Z_{FH} have poles for those values $k_\rho=k_{\rho p}$ of the radial wavenumber which are propagation constants of the LSM and LSE modes in the inhomogeneous parallel-plate medium between the ground and top planes if the conductor metallization is not present and they represent surface waves for open-type structures [36], [41]. The maximum discrete value of $k_{\rho p}^2$ corresponds to the dominant LSM_0 radial wave in the circuit medium which is the main cause for dynamic coupling in MIC's since this has zero cutoff frequency. With

$$k_{xp}=k'_{xp}+jk''_{xp}=\pm\left(k_{\rho p}^2-k_y^2\right)^{1/2} \tag{19}$$

the associated poles in the complex k_x-plane are all off the real k'_x-axis as long as k_y^2 is larger than the value of $k_{\rho p}^2$ of the LSM_0 mode. Physically, this means, for example, that MIC transmission-line modes with propagation constants k_y larger than that of the LSM_0 mode are nonradiating. This has already been discussed by Pregla in an early contribution [56]; however, his analysis has not been extended into the radiation region.

Higher order modes on covered and open MIC transmission lines do not generally exist in nonradiating form. Also, the respective MIC problems involving three-dimensional fields are always affected by energy leakage [40], [41] even if this may not be of practical concern at low frequencies. The SDA formulation can be extended in application to these cases by writing the scalar products of the final equations (14) in the form

$$\left(\underline{\tilde{J}}_{tj},\tilde{\underline{\underline{Z}}}(p)\underline{\tilde{J}}_{tk}\right)_x=\int_{C_x}\tilde{\underline{\underline{Z}}}(k_x,p)\underline{\tilde{J}}_{tk}(k_x)\cdot\underline{\tilde{J}}_{tj}(-k_x)\,dk_x \tag{20}$$

which is a consequence of Parseval's theorem [56] and by proper choice of the integration path C_x. The encountered immittance elements are meromorphic functions with respect to k_x in the covered case. The evaluation of (20) is achieved by residue calculus techniques [68]. In three-dimensional problems, in addition, integration in the proper related sheet of the complex k_y plane is involved. The simple principle of extension into the radiation region is a further fundamental advantage of the spectral-domain approach and allows rigorous treatment of complex MIC problems.

For the general rules of evaluating SDA integrals of the type (20) and a discussion of the physical background, a

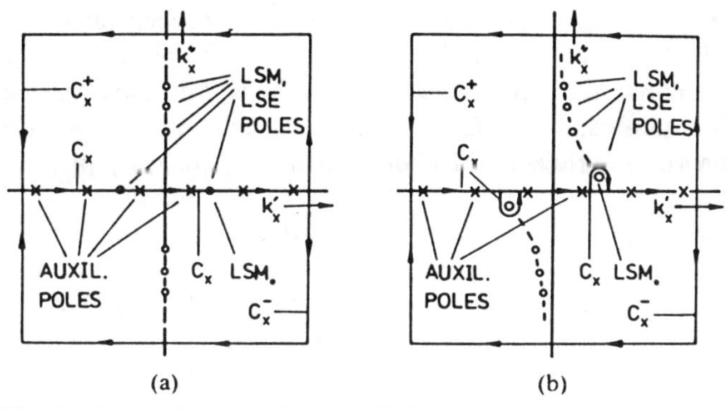

Fig. 4. Integration paths C_x in the SDA treatment of covered strip-type MIC problems, energy leakage (a) neglected and (b) taken into account, respectively.

covered transmission-line configuration such as that of Fig. 2(b) is considered. By introduction of the factor

$$1=\frac{\exp(jck_xw)}{2\cos(ck_xw)}+\frac{\exp(-jck_xw)}{2\cos(ck_xw)}, \qquad c>1 \quad (21)$$

the integrals in (20) can be split up into a sum of two contributions for each of which the integration path C_x can be closed in the complex k_x-plane at infinity [39], [41]. This is shown in Fig. 4(a) for an analysis which does not include energy leakage, and in Fig. 4(b) where this mechanism is properly accounted for. In both cases, the zeros of the auxiliary cosine function in (21) introduce additional, nonphysical poles onto the real k_x'-axis. According to the relation (19) and neglecting material loss, the LSM and LSE wave poles are located on the axes or not, depending on whether the leakage mechanism is incorporated into the solution (square of propagation constant k_y^2 complex) or not (k_y^2 a real number). The quantity w is a suitable normalization width. The positive real constant c is to some degree arbitrary and can be utilized for numerical check purposes and in convergence considerations. Integrating along the real k_x'-axis across the single poles artificially introduced by (21), Cauchy principal values are taken [68].

If, in Fig. 4(a), a nonradiating mode would be considered, i.e., with a propagation constant larger than that of the LSM radial wave, the LSM_0 pole would be located on the imaginary k_x''-axis. In that case, the dominant contributions to the integrals in (20) would come from the discrete, regular set of auxiliary poles, say k_{xm}, on the real axis. With the constant c in the factor (21) chosen sufficiently large, the set of k_{xm} becomes very dense and the problem could be described in terms of this set alone. In the limit of $c \to \infty$, this is nothing else than numerical integration along the real axis of the k_x-plane. However, describing an MIC problem under radiation conditions, as actually prevailing with the pole locations of Fig. 4(a), this becomes more complicated. Numerical integration along the k_x'-axis and across the LSM_0-pole (Cauchy principle value) now means introducing a discrete standing plane contribution k_{xp} into the electric-field distribution [61]. This is equivalent to the presence of a lateral shielding far away from the MIC configuration, reflecting the radiated LSM_0-field. The same type of result is achieved if one applies a transverse resonance approach to covered structures under operating conditions in the radiation region [57], [58].

The leaky character of higher order MIC transmission-line modes, discontinuities, and junctions of the covered-type are correctly described by the integration path shown in Fig. 4(b). This can be concluded from an investigation of the migration paths of the LSM and LSE wave poles in a slightly lossy dielectric medium [41], [61]. Those physical poles which in the lossless case of Fig. 4(a) are located on the real axis (here only the LSM_0-pole) are just below the k_x'-axis for a small dielectric tangent different from zero. They migrate across the real axis of the k_x-plane if an additional radiation mechanism is involved. Therefore, the original integration path C_x (the k_x'-axis) for nonradiating situations has to be distorted in the way indicated in Fig. 4(b). Otherwise, solutions would not pass over continuously into the radiation region of operation. As Pregla has already pointed out in his early study [56], the transition from one state of a solution to another, i.e., when the LSM_0-pole appears at the origin, does not present problems since the associated residues vanish then. The same arguments and integration path discussed here are valid also for the evaluation of integrals (scalar products) with respect to the k_y-variable in the SDA solution of three-dimensional field problems. However, depending on which integration is performed first, additional branch cuts have to be regarded either in the complex k_y- or k_x-plane. A broad and thorough treatment of the spectral-domain approach for a variety of representative leaky MIC problems has been elaborated in a recent dissertation by Boukamp [41].

V. Progress: 1975–1984

To round out the view given so far for the spectral-domain approach, a review is given of the improvements of the technique and the more important results achieved by its application during the last ten years. Emphasis is placed on frequency-dependent solutions since these become more and more important with the development of practical MIC's in the millimeter-wave region. Many of the contributions mentioned do not use the SDA in its pure form but deviate from it in the one step or another of the analysis. The reader may get an impression, therefore, that a high degree of flexibility is inherent in the details of the SDA formulation. The discussion is subdivided according to three different groups of MIC structures, namely transmission lines, resonators and antennas, and, finally, discontinuities.

Considering printed microwave and millimeter-wave transmission lines first, there has been a clear tendency since 1975 to treat this class of problems in a generalized way, allowing additional dielectric layers and more complicated strip and slot configurations [24], [26], [32], [34], [35], [59], [60], [69]–[78]. The inclusion of characteristic impedance data becomes standard in computer analysis programs, and also dielectric and conductor losses are considered frequently [24], [26], [34], [59], [60], [69], [72]–[78], [79]–[80]. This makes visible the beginning orientation of the SDA towards direct application in the design of MIC's. Together with this trend, computer time and storage requirements for the analysis programs become a more important point of view [80]. However, from an appraisal of methods applied to the microstrip dispersion problem published by Kuester and Chang in 1979 [81], it can be concluded that the majority of respective computer packages at that time still involved some problems. There was significant progress, therefore, when not only the number of applications of the SDA increased further, but in addition some conceptual simplifications, modifications, and basic numerical considerations were reported. As elegant and simple concepts, for example, the transfer matrix approach [32] and the very similar spectral-domain immittance concept [35] have been presented. Also, El-Sherbiny [82], [83] provided interesting aspects to the mathematical and physical background of the SDA and applied a modified Wiener–Hopf technique in the final step of solution. Some of the rules to be regarded in order to obtain stable, accurate solutions and a unified treatment of shielded, covered, and open strip and slot structures have been reported by the author [34]. The introduction of finite conductor thickness into the SDA formalism is mainly the result of Kitazawa's work [21]–[23], [84]. Its effect on coplanar waveguide properties, for example, is an increase of guided wavelength and a decrease of characteristic impedance, respectively.

With growing experience in the use of the SDA, application of the technique shifted to more involved transmission-line problems. Coupled strip–slot structures have been studied by various authors with regard to coupler design and an extension of the range of characteristic impedances achievable in microstrip [22], [35], [47], [71], [77]. Also, an increasing portion of SDA work on transmission-line structures with anisotropic media has been reported. Borburgh [85], [86] seems to have been the first to apply the technique to microstrip on a magnetized ferrite substrate and related analyses followed [87], [88]. A variety of authors have treated printed transmission lines in single- and double-layered anisotropic dielectric media [89]–[92]. Only recently, slow-wave MIS coplanar waveguide has been studied with respect to MMIC application [93]. Beyond this, the computation of the stopband properties of several periodic structures by the spectral-domain approach has been reported [88], [94]–[96]. The last-mentioned reference also contains some numerical results on Podell-type microstrip couplers. As a further example for inhomogeneous structures, an analysis of tapered MIC transmission lines combining the SDA for uniform lines with coupled-mode theory has been presented [97]. Finally, a very efficient hybrid-mode spectral-domain approach for conductor arrays has been used by Jansen and Wiemer [98] in the design of MIC interdigital couplers and lumped elements on small computers.

Results achieved for resonators and antennas are considered together here since a large class of planar antennas makes use of resonating open patch elements. The information given on patch antennas, however, is by far incomplete, as the emphasis is placed on MIC's in this paper. The first full-wave analyses of resonators concentrated on shielded structures and gave quite accurate results for the resonance frequencies and current density distributions of the open case if Q-factors were high and interaction with the volume field in the chosen shielding box low [30], [31]. Taking this into account, Jansen studied microstrip resonators of canonical and complicated shapes, with the latter described by a polygonal contour in terms of high-order finite-element polynomials for the current density [99].

Resonator shapes for which numerical results have been generated are rectangle, circular disk and ring, concentric coupled disk-ring and double-ring structure, stretched hexagon, and regular octagon. This work has recently been supplemented by Knorr [100] who analyzed a shielded short-circuited slotline resonator and by Sharma and Bhal [101], [102] who provided shielded-type results for the triangular shape and interacting rectangular microstrip structures.

Already by 1978, Pregla [43] had investigated open resonating microstrip rings including radiation using a Hankel transform and formulating the problem in terms of complex eigenfrequency. With increasing interest in microstrip antennas, further related analyses of circular shapes were performed in the years following [44]–[46], [103]–[105]. Itoh and Menzel presented a full-wave SDA treatment of open rectangular microstrip patches in 1981 [106] with clear emphasis on antenna applications. There is also direct antenna design work, for example, contributions by Bailey and Deshpande [107], [108] and by Newman *et al.* [109], which performs only part of the computational steps in the spectral domain. Numerical integration along the real axes in the spectral domain is the dominant choice in these papers; however, singularities near the integration path may cause problems (see, for example, Newman's remarks [109]). The first results for covered MIC geometries including the excitation of LSM and LSE waves in the layered circuit medium have been provided by Boukamp and Jansen [40], [41], [61]. The main intention of this research work was to study the mathematical and physical background and prepare the way for an extension of the SDA in application to dynamic MIC coupling problems. One of the practical results achieved in this context is, for example, that lateral leakage in MIC's can be minimized if the circuit cover height is chosen slightly lower than a value which would correspond to the onset of the first LSE mode.

The same covered-type spectral-domain approach has also been used to study the simplest case of a leaky microstrip discontinuity problem, i.e., the open end, with an excitation formulation [40], [41]. The motivation was again to provide a basis for the analysis of more complicated geometries. Interestingly, the numerical results in comparison with a resonator formulation indicate that there is a noticeable coupling effect between the open ends of half-wavelength resonators, such as those used, for example, in coupled line filters. The explanation is that, on the alumina substrate investigated, the distance between the respective open microstrip ends is small compared to the wavelength of the involved LSM_0-mode. Beyond this very elementary but rigorous example, MIC discontinuities have been computed using the SDA only for the static limit [13] and by the frequency-dependent shielded-type SDA implementation outlined in the foregoing section [36], [62]–[67], [110]. Due to the three-dimensional electromagnetic fields and relatively complicated geometries involved, this sparsity of results prevails for other methods to an even larger extent. Systematic and quite extensive design data have been published for open-circuited microstrip and suspended substrate lines, as well as short-circuited slots [36] and for the symmetrical and asymmetrical gap in microstrip and suspended substrate lines [63], [65]. Also, the inductive strip discontinuity in unilateral finlines, which is the related slot-type structure, has been treated [64]. Very recent work by Koster and Jansen provided a variety of microstrip impedance step data for use in MIC design [67], [110].

VI. Efficiency Considerations

The spectral-domain approach is a hybrid technique in the sense that it requires (and allows!) a certain amount of analytical preprocessing in order to achieve high computational efficiency for a specific problem or class of problems. One of its main disadvantages is the relatively high numerical expense which has to be spent to evaluate the coefficients of the final system of equations (14). These are improper integrals or infinite series with only moderate rate of convergence. The order of the final system, on the other hand, can be held extremely small compared to other techniques. This is achieved, for example, by regarding several criteria in the choice of the expansion functions [80]. Briefly summarizing, the set of expansion functions as a whole should be twice continuously differentiable in the interior of the region on which it is defined, so that it is in the domain of the original Helmholtz operator (1). Mathematical arguments and numerical experience indicate that this avoids the existence of spurious, nonphysical solutions [80]. Further, expansion functions in MIC problems should satisfy the edge condition, i.e., have the correct order of singularity at the boundary of the conductor metallization. This is a prerequisite to obtaining accurate solutions with a low number of terms or, equivalently, with a low order of the final system of equations. The set of functions used should be complete in order to allow convergence checks and investigation. It should be chosen with all the physical insight that is available for the specific problems, from static considerations, from idealizations such as the planar magnetic-wall waveguide model [19] and so on. The main rule is not to leave work to the computer for the evaluation of what is known in advance of the physical solution or can be obtained easier. This also implies the precomputation of expansion functions by a transmission-line SDA portion (two-dimensional fields) in computer programs for the SDA solution of three-dimensional field problems [36]. Finally, the use of static together with stationary precomputed information can provide a means to generate vector expansion functions based on the continuity equation. The analytical and programming expense required on the side of the investigator may be considerable, which mirrors the hybrid character of the SDA. However, in the way outlined, very efficient CAD tools can be developed by its application.

To come to a quantification of the numerical expense associated with the SDA, the number of point operations which have to be performed in the solution of typical MIC problems is estimated. Also, the possibilities of reducing this figure shall be discussed. Let us assume a not too elementary MIC transmission-line case, in parallel, a resonator problem formulated in Cartesian coordinates. The final system of SDA equations (14) is dense and has to be generated repeatedly in the iterative localization of the zeros of its determinant as a function of the eigenvalue parameter p. Even with an intelligently chosen start value of p, this has to be done about 10 times. Under the assumption of a reasonable choice of expansion functions, the number of point operations necessary to obtain the numerical value of the final SDA determinant is usually small compared to the expense investigated for its generation. This is a consequence of the fact that the number of summations required to compute a single coefficient (integral or series) of (14) is typically much larger than the order of the system matrix. The latter may be

$$Q = 2 \cdots 10 \text{ and } Q = 4 \cdots 100 \tag{22}$$

for the transmission-line and resonator problem, respectively.

For example, $Q = 4$ could apply to a simple, rectangular half-wavelength microstrip resonator [14], [15], [36]. The number of summations or discretization points to evaluate each single scalar product may be $100 \cdots 500$ for the two-dimensional and $100^2 \cdots 500^2$ for the three-dimensional case. In particular situations, this may be even higher [34] depending on the spectral distribution of the involved fields. The complexity of the immittance functions encountered depends only on the number of dielectric layers considered and may be characterized by a figure of at least $10 \cdots 100$ point operations. On the whole, this amounts to a total count of point operations of about

$$TC = 2 \cdot 10^4 \cdots 25 \cdot 10^6 \text{ and } TC = 8 \cdot 10^6 \cdots 125 \cdot 10^{10} \tag{23}$$

for the two cases considered (symmetric SDA matrices). This looks quite high, particularly for the very right-hand side. However, one has to keep in mind that the spatial resolution assumed there is equivalent to a mesh of $25 \cdot 10^4$ points in the plane of the MIC metallization. As a rough estimate, the matrix order in a respective DDA treatment would be $Q = 5000$, the matrix itself dense, and had to be processed repeatedly about 10 times.

A reduction of numerical expense in SDA solutions is achieved first by an optimization of the expansion functions. This is performed according to the outlined criteria and with some experience from a preliminary, crude version. It can be done with a relatively small amount of reprogramming and produces a typical speedup factor of

$5 \cdots 10$ for nonelementary two- and three-dimensional problems. Also, about a factor of 10 may be gained by choosing an excitation formulation instead of solving an eigenvalue problem which applies, however, only to the three-dimensional case. The estimated speedup results since the source formulation avoids repeated generation of the final system (14). An additional reduction in computer time can be obtained by splitting off asymptotic spectral contributions from the coefficients of (14) and integrating or summing up these by analytical techniques (a factor of 10). In eigenvalue problems, it is advisable to substitute the spectral immittances by accurate one-dimensional interpolants [36] and optimize CP-time at the cost of storage requirements [31] for that part of the computation which does not depend on the eigenvalue p (a factor of 10). SDA computer programs developed for regular industrial use in MIC design justify even more expense in analytical preprocessing. For these, the normal mode of application is the repeated solution of the same problem for several different operating frequencies. Therefore, a high speedup factor compared to the first solution can be achieved if this is employed to provide for the subsequent ones a very compact low-order set of expansion functions (tested by the author for the transmission-line problem described in [98]). Thus, average CP-time is further reduced by about a factor of 5. By a combination of such analytical measures, the total count of point operations may be brought down to

$$TC = 2 \cdot 10^3 \cdots 5 \cdot 10^5 \text{ and } TC = 8 \cdot 10^5 \cdots 25 \cdot 10^7 \quad (24)$$

which is hardly achievable by other techniques. However, great care has to be taken in properly designing the employed integration algorithms, i.e., choosing a correct spectral representation. This particularly effects cases where tight coupling is involved. For loose and multiple coupled situations, a sufficiently stable matrix inversion algorithm has to be chosen.

VII. Conclusion

The spectral-domain approach allows an elegant and closed-form integral equation formulation for a broad class of MIC problems which is reduced by one dimension compared to the original field problem. It results in a particularly low-order linear system of equations and provides design-relevant parameters in both the spectral and the space domain. In so far as it may require a considerable amount of analytical preprocessing to achieve highest efficiency, it is a hybrid method. A preference of the SDA for MIC problems is to some extent confirmed by the fact that the majority of rigorous frequency-dependent MIC design information has been generated using this technique. The survey presented here further confirms this preference; however, the advice should be deduced from the discussion not to apply the SDA in a crude and schematic way.

References

[1] I. N. Sneddon, *The Use of Integral Transforms.* New York: McGraw-Hill, 1972.

[2] Tai Tsun Wu, "Theory of the microstrip," *J. Appl. Phys.*, vol. 28, pp. 299–302, 1975.

[3] E. Yamashita and R. Mittra, "Variational method for the analysis of microstrip lines," *IEEE Trans. Microwave Theory Tech.*, vol. MTT-16, pp. 251–256, 1968.

[4] E. J. Denlinger, "A frequency dependent solution for microstrip transmission lins," *IEEE Trans. Microwave Theory Tech.*, vol. MTT-19, pp. 30–39, 1971.

[5] H. J. Schmitt and K. H. Sarges, "Wave propagation in microstrip," *Nachrichtentech. Z.*, vol. 5, pp. 260–264, 1971.

[6] T. Itoh and R. Mittra, "Dispersion characteristics of slot lines," *Electron. Lett.*, vol. 7, pp. 364–365, 1971.

[7] T. Itoh and R. Mittra, "Spectral-domain approach for calculating the dispersion characteristics of microstrip lines," *IEEE Trans. Microwave Theory Tech.*, vol. MTT-21, pp. 496–499, 1973.

[8] G. Kowalski and R. Pregla, "Dispersion characteristics of single and coupled microstrips," *Arch. Elek. Übertragung*, vol. 26, pp. 276–280, 1972.

[9] M. K. Krage and G. I. Haddad, "Frequency-dependent characteristics of microstrip transmission lines," *IEEE Trans. Microwave Theory Tech.*, vol. MTT-20, pp. 678–688, 1972.

[10] A. R. Van de Capelle and P. J. Luypaert, "Fundamental- and higher-order modes in open microstrip lines," *Electron. Lett.*, vol. 9, pp. 345–346, 1973.

[11] T. Itoh and R. Mittra, "A technique for computing dispersion characteristics of shielded microstrip lines," *IEEE Trans. Microwave Theory Tech.*, vol. MTT-22, pp. 896–898, 1974.

[12] R. H. Jansen, "A modified least-squares boundary residual (LSBR) method and its application to the problem of shielded microstrip dispersion," *Arch. Elek. Übertragung*, vol. 28, pp. 275–277, 1974.

[13] Y. Rahmat-Samii *et al.*, "A spectral-domain analysis for solving microstrip discontinuity problems," *IEEE Trans. Microwave Theory Tech.*, vol. MTT-22, pp. 372–378, 1974.

[14] T. Itoh, "Analysis of microstrip resonators," *IEEE Trans. Microwave Theory Tech.*, vol. MTT-22, pp. 946–951, 1974.

[15] R. H. Jansen, "Shielded rectangular microstrip disc resonators," *Electron. Lett.*, vol. 10, pp. 299–300, 1974.

[16] R. H. Jansen, "Computer analysis of edge-coupled planar structures," *Electron. Lett.*, vol. 10, pp. 520–522, 1974.

[17] U. Schulz, "The method of lines—A new technique for the analysis of planar microwave structures" (in German), Ph.D. dissertation, Univ. of Hagen, West Germany, 1980.

[18] R. Mittra, Ed., *Computer Techniques for Electromagnetics* (Int. Series of Monographs in El. Eng.). New York: Pergamon Press, 1973.

[19] K. C. Gupta *et al.*, *Microstrip Lines and Slotlines.* Dedham, MA: Artech House, 1979.

[20] K. C. Gupta and A. Singh, *Microwave Integrated Circuits.* New Delhi: Wiley Eastern Private Ltd., 1974.

[21] T. Kitazawa *et al.*, "A coplanar waveguide with thick metal-coating," *IEEE Trans. Microwave Theory Tech.*, vol. MTT-24, pp. 604–608, 1976.

[22] R. H. Jansen, "Microstrip lines with partially removed ground metallization-theory and applications," *Arch. Elek. Übertragung*, vol. 32, pp. 485–492, 1978.

[23] T. Kitazawa *et al.*, "Analysis of the dispersion characteristics of slot line with thick metal coating," *IEEE Trans. Microwave Theory Tech.*, vol. MTT-28, pp. 387–392, 1980.

[24] R. H. Jansen, "Fast accurate hybrid mode computation of non symmetrical coupled microstrip characteristics," in *Proc. 7th Eur. Microwave Conf.*, 1977, pp. 135–139.

[25] W. Schumacher, "Current distribution in the ground plane of covered microstrip and the effect on conductor losses" (in German), *Arch. Elek. Übertragung*, vol. 33, pp. 207–212, 1979.

[26] D.Mirshekar-Syakhal and J. B. Davies, "Accurate solution of microstrip and coplanar structures for dispersion and for dielectric and conductor loss," *IEEE Trans. Microwave Theory Tech.*, vol. MTT-23, pp. 694–699, 1979.

[27] R. Pregla, "Determination of conductor losses in planar waveguide structures," *IEEE Trans. Microwave Theory Tech.*, vol. MTT-28, pp. 433–434, 1980.

[28] R. E. Collin, *Field Theory of Guided Waves.* New York: McGraw-Hill, 1960.

[29] R. R. Harrington, *Time-Harmonic Electromagnetic Fields.* New York: McGraw-Hill, 1961.

[30] R. H. Jansen, "Computer analysis of shielded microstrip structures" (in German), *Arch. Elek. Übertragung*, vol. 29, pp. 241–247, 1975.

[31] R. H. Jansen, "Numerical computation of the resonance frequencies and current density distributions of arbitrarily shaped microstrip structures" (in German), Ph.D. dissertation, *Univ.* Aachen (RWTH), West Germany, 1975.

[32] J. B. Davies and D. Mirshekar-Syakhal, "Spectral domain solution of arbitrary coplanar transmission line with multilayer substrate," *IEEE Trans. Microwave Theory Tech.*, vol. MTT-25, pp. 143–146, 1977.

[33] L. J. van der Pauw, "The radiation of electromagnetic power by microstrip configurations," *IEEE Trans. Microwave Theory Tech.*, vol. MTT-25, pp. 719–725, 1977.

[34] R. H. Jansen, "Unified user-oriented computation of shielded, covered and open planar microwave and millimeterwave transmission-line characteristics," *IEE J. Microwaves, Opt., Acoust.*, vol. MOA-3, pp. 14–22, 1979.

[35] T. Itoh, "Spectral domain immitance approach for dispersion characteristics of generalized printed transmission lines," *IEEE Trans. Microwave Theory Tech.*, vol. MTT-28, pp. 733–736, 1980.

[36] R. H. Jansen, "Hybrid mode analysis of end effects of planar microwave and millimeter-wave transmission lines, *Proc. Inst. Elec. Eng.*, pt. H, vol. 128, pp. 77–86, 1981.

[37] A. S. Omar and K. Schünemann, "Space domain decoupling of LSE and LSM fields in generalized planar guiding structures," in *IEEE MTT Symp. Dig.*, 1984, pp. 59–61.

[38] L. J. Van der Pauw, "The radiation and propagation of electromagnetic power by a microstrip transmission line," *Philips Res. Rep.*, vol. 31, pp. 35–70, 1976.

[39] R. H. Jansen, unpublished results and computer program—SFPMIC—, "Source formulation approach to planar microwave integrated circuits," Univ. of Duisburg, West Germany, 1979/80.

[40] J. Boukamp and R. H. Jansen, "The high-frequency behaviour of microstrip open ends in microwave integrated circuits including energy leakage," in *Proc. 14th Eur. Microwave Conf.*, 1984, pp. 142–147.

[41] J. Boukamp, "Spectral domain techniques for the analysis of radiation effects in microwave integrated circuits" (in German), Ph.D. dissertation, Univ. of Aachen (RWTH), West Germany, 1984.

[42] T. Itoh and R. Mittra, "Analysis of a microstrip disk resonator," *Arch. Elek. Übertragung*, vol. 27, pp. 456–458, 1973.

[43] S. G. Pintzos and R. Pregla, "A simple method for computing the resonant frequencies of microstrip ring resonators," *IEEE Trans. Microwave Theory Tech.*, vol. MTT-26, pp. 809–813, 1978.

[44] W. C. Chew and J. A. Kong, "Resonance of the axial-symmetric modes in microstrip disk resonators," *J. Math. Phys.*, vol. 21, pp. 582–591, 1980.

[45] W. C. Chew and J. A. Kong, "Resonance of nonaxial symmetric modes in circular microstrip disk antenna," *J. Math. Phys.*, vol. 21, pp. 2590–2598, 1980.

[46] K. Araki and T. Itoh, "Hankel transform domain analysis of open circular microstrip radiating structures," *IEEE Trans. Antennas Propagat.*, vol. AP-29, pp. 84–89, 1981.

[47] D. Mirshekar-Syakhal and J. B. Davies, "Accurate analysis of coupled strip-finline structure for phase constant, characteristic impedance, dielectric and conductor losses," *IEEE Trans. Microwave Theory Tech.*, vol. MTT-30, pp. 906–910, 1982.

[48] I. V. Lindell, "Variational methods for nonstandard eigenvalue problems in waveguide and resonator analysis," *IEEE Trans. Microwave Theory Tech.*, vol. MTT-30, pp. 1194–1204, 1982.

[49] G. J. Gabriel and I. V. Lindell, "Comments on Variational methods for nonstandard eigenvalue problems in waveguide and resonator analysis," *IEEE Trans. Microwave Theory Tech.*, vol. MTT-31, pp. 786–789, 1983.

[50] G. J. Gabriel and I. V. Lindell, "Further comments on Variational methods for nonstandard eigenvalue problems in waveguide and resonator analysis," *IEEE Trans. Microwave Theory Tech.*, vol. MTT-32, pp. 474–476, 1984.

[51] R. F. Harrington, *Field Computation by Moment Methods.* New York: Macmillan, 1968.

[52] R. Mittra and S. W. Lee, *Analytical Techniques in the Theory of Guided Waves.* New York: Macmillan, 1971.

[53] S. B. Worm, "Analysis of planar microwave structures of arbitrary shape" (in German), Ph.D. dissertation, Univ. of Hagen, West Germany, 1983.

[54] H. Diestel and S. B. Worm, "Analysis of hybrid field problems by the method of lines with nonequidistant discretization," *IEEE Trans. Microwave Theory Tech.*, vol. MTT-32, pp. 633–638, 1984.

[55] S. B. Worm and R. Pregla, "Hybrid-mode analysis of arbitrarily shaped planar microwave structures by the method of lines," *IEEE Trans. Microwave Theory Tech.*, vol. MTT-32, pp. 191–196, 1984.

[56] G. Kowalski and R. Pregla, "Dispersion characteristics of single and coupled microstrips with double-layer substrates," *Arch. Elek. Übertragung*, vol. 27, pp. 125–130, 1973.

[57] H. Ermert, "Guided modes and radiation characteristics of covered microstrip lines," *Arch. Elek. Übertragung*, vol. 30, pp. 65–70, 1976.

[58] H. Ermert, "Guiding characteristics and radiation characteristics of planar waveguides," in *Proc. 8th Eur. Microwave Conf.*, 1978, pp. 94–98.

[59] H. Hofmann, "Dispersion of planar waveguides for millimeter-wave application," *Arch. Elek. Übertragung*, vol. 31, pp. 40–44, 1977.

[60] L. P. Schmidt and T. Itoh, "Spectral domain analysis of dominant and higher order modes in fin lines," *IEEE Trans. Microwave Theory Tech.*, vol. MTT-28, pp. 981–985, 1980.

[61] J. Boukamp and R. H. Jansen, "Spectral domain investigation of surface wave excitation and radiation by microstrip lines and microstrip disk resonators," *Proc. 13th Eur. Microwave Conf.*, 1983, pp. 721–726.

[62] R. H. Jansen, "A new unified numerical approach to the frequency dependent characterization of strip, slot and coplanar MIC components for CAD purposes," Univ. of Duisburg, West Germany, Res. Rep. FB9/ATE, pp. 29–33, 1980/1981.

[63] R. H. Jansen and N. H. L. Koster, "A unified CAD basis for the frequency dependent characterization of strip, slot and coplanar MIC components," in *Proc. 11th Eur. Microwave Conf.*, 1981, pp. 682–687.

[64] N. H. L. Koster and R. H. Jansen, "Some new results on the equivalent circuit parameters of the inductive strip discontinuity in unilateral fin lines," *Arch. Elek. Übertragung*, vol. 35, pp. 497–499, 1981.

[65] N. H. L. Koster and R. H. Jansen, "The equivalent circuit of the asymmetrical series gap in microstrip and suspended substrate lines," *IEEE Trans. Microwave Theory Tech.*, vol. MTT-30, pp. 1273–1279, 1982.

[66] R. H. Jansen and N. H. L. Koster, "New aspects concerning the definition of microstrip characteristic impedance as a function of frequency," in *IEEE MTT-Symp. Dig.*, 1982, pp. 305–307.

[67] N. H. L. Koster and R. H. Jansen, "The microstrip discontinuity: A revised description," *IEEE Trans. Microwave Theory Tech.*, vol. MTT-33, 1985.

[68] G. B. Arfken, *Mathematical Methods for Physicists.* New York: Academic Press, 1970.

[69] J. B. Knorr and K. D. Kuchler, "Analysis of coupled slots and coplanar strips on dielectric substrate," *IEEE Trans. Microwave Theory Tech.*, vol. MTT-23, pp. 541–548, 1975.

[70] N. Samardzija and T. Itoh, "Double-layered slot line for millimeter wave integrated circuits," *IEEE Trans. Microwave Theory Tech.*, vol. MTT-24, pp. 827–831, 1976.

[71 T. Itoh, "Generalized spectral domain method for multiconductor printed lines and its application to tunable suspended microstrips," *IEEE Trans. Microwave Theory Tech.*, vol. MTT-26, pp. 983–987, 1978.

[72] J. B. Knorr and P. M. Shayda, "Millimeter-wave fin line characteristics," *IEEE Trans. Microwave Theory Tech.*, vol. MTT-28, pp. 737–743, 1980.

[73] L. P. Schmidt *et al.*, "Characteristics of unilateral fin-line structures with arbitrarily located slots," *IEEE Trans. Microwave Theory Tech.*, vol. MTT-29, pp. 352–355, 1981.

[74] D. Mirshekar-Syakhal and J. B. Davies, "An accurate unified solution to various fin-line structures, of phase constant, characteristic impedance, and attenuation," *IEEE Trans. Microwave Theory Tech.*, vol. MTT-30, pp. 1854–1861, 1982.

[75] D. Mirshekar-Syakhal, "An accurate determination of dielectric loss effect in MMIC's including microstrip and coupled microstrip lines," *IEEE Trans. Microwave Theory Tech.*, vol. MTT-31, pp. 950–954, 1983.

[76] A. K. Sharma and W. J. R. Hoefer, "Propagation in coupled unilateral and bilateral finlines," *IEEE Trans. Microwave Theory Tech.*, vol. MTT-31, pp. 489–502, 1983.

[77] W. Schumacher, "Computation of guided wavelengths of planar transmission lines in two- and three-layered media using a moment method" (in German), Ph.D. dissertation, Univ. of Karlsruhe, West Germany, 1979.

[78] W. Schuhmacher, "Hybrid modes and field distributions in unsymmetrical planar transmission lines" (in German), *Arch. Elek. Übertragung*, vol. 34, pp. 445–453, 1980.

[79] J. B. Knorr and A. Tufekcioglu, "Spectral domain calculation of microstrip characteristic impedance," *IEEE Trans. Microwave Theory Tech.*, vol. MTT-23, pp. 725–728, 1975.

[80] R. H. Jansen, "High speed computation of single and coupled microstrip parameters including dispersion, high-order modes, loss and finite strip thickness," *IEEE Trans. Microwave Theory Tech.*, vol. MTT-26, pp. 75–82, 1978.

[81] E. F. Kuester and D. C. Chang, "An appraisal of methods, for computation of the dispersion characteristics of open microstrip," *IEEE Trans. Microwave Theory Tech.*, vol. MTT-27, pp. 691–694, 1979.

[82] A. M. A. El-Sherbiny, "Exact analysis of shielded microstrip lines and bilateral fin lines," *IEEE Trans. Microwave Theory Tech.*, vol. MTT-29, pp. 669–675, 1981.

[83] A. M. A. El-Sherbiny, "Millimeter-wave performance of shielded slot-linss," *IEEE Trans. Microwave Theory Tech.*, vol. MTT-30, pp. 750–756, 1982.

[84] T. Kitazawa and R. Mittra, "Analysis of finline with finite metallization thickness," *IEEE Trans. Microwave Theory Tech.*, vol. MTT-32, pp. 1484–1487, 1984.

[85] J. Borburgh, "Theoretical investigation of the dispersion and field distribution of guided modes on a microstrip line with gyrotropic substrate" (in German), Ph.D. dissertation, Univ. of Erlangen, West Germany, 1976.

[86] J. Borburgh, "The behaviour of guided modes on the ferrite-filled microstrip line with the magnetization perpendicular to the ground plane," *Arch. Elek. Übertragung*, vol. 31, pp. 73–77, 1977.

[87] Y. Hayashi and R. Mittra, "An analytical investigation of finlines with magnetized ferrite substrate," *IEEE Trans. Microwave Theory Tech.*, vol. MTT-31, pp. 495–498, 1983.

[88] C. Surawatpunya *et al.*, "Bragg interaction of electromagnetic waves in a ferrite slab periodically loaded with metal strips," *IEEE Trans. Microwave Theory Tech.*, vol. MTT-32, pp. 689–695, 1984.

[89] A. M. A. El-Sherbiny, "Hybrid mode analysis of microstrip lines on anisotropic substrates," *IEEE Trans. Microwave Theory Tech.*, vol. MTT-29, pp. 1261–1266, 1981.

[90] H. Lee and V. K. Tripathi, "Spectral domain analysis of frequency dependent propagation characteristics of planar structures on uniaxial medium," *IEEE Trans. Microwave Theory Tech.*, vol. MTT-30, pp. 1188–1193, 1982.

[91] T. Kitazawa and Y. Hayashi, "Propagation characteristics of striplines with multilayered anisotropic media," *IEEE Trans. Microwave Theory Tech.*, vol. MTT-31, pp. 429–433, 1983.

[92] M. Horno and R. Marques, "Coupled microstrips on double anisotropic layers," *IEEE Trans. Microwave Theory Tech.*, vol. MTT-32, pp. 467–470, 1984.

[93] Y. Fukuoka *et al.*, "Analysis of slow-wave coplanar waveguide for monolithic integrated circuits," *IEEE Trans. Microwave Theory Tech.*, vol. MTT-31, pp. 567–573, 1983.

[94] K. Ogusu, "Propagation properties of a planar dielectric waveguide with periodic metallic strips," *IEEE Trans. Microwave Theory Tech.*, vol. MTT-29, pp. 16–21, 1981.

[95] T. Kitazawa and R. Mittra, "An investigation of striplines and finlines with periodic stubs," *IEEE Trans. Microwave Theory Tech.*, vol. MTT-32, pp. 684–688, 1984.

[96] F. J. Glandorf, "Numerical solution of the electromagnetic eigenvalue problem of periodically inhomogenous microstrip lines" (in German), Ph. dissertation, Univ. of Duisburg, West Germany, 1982.

[97] D. Mirshekar-Syakhal and J. B. Davies, "Accurate analysis of tapered planar transmission lines for microwave integrated circuits," *IEEE Trans. Microwave Theory Tech.*, vol. MTT-29, pp. 123–128, 1981.

[98] R. H. Jansen and L. Wiemer, "Multiconductor hybrid-mode approach for the design of MIC couplers and lumped elements including loss, dispersion and parasitics," in *Proc. 14th Eur. Microwave Conf.*, 1984, pp. 430–435.

[99] R. H. Jansen, "High-order finite element polynomials in the computer analysis of arbitrarily shaped microstrip resonators," *Arch. Elek. Übertragung*, vol. 30, pp. 71–79, 1976.

[100] J. B. Knorr, "Equivalent reactance of a shorting septum in a fin line: Theory and experiment," *IEEE Trans. Microwave Theory Tech.*, vol. MTT-29, pp. 1196–1202, 1981.

[101] A. K. Sharma and B. Bhat, "Analysis of triangular microstrip resonators," *IEEE Trans. Microwave Theory Tech.*, vol. MTT-30, pp. 2029–2031, 1982.

[102] A. K. Sharma and B. Bhat, "Spectral domain analysis of interacting microstrip resonant structures," *IEEE Trans. Microwave Theory Tech.*, vol. MTT – 31, pp. 681–685, 1983.

[103] K. Kawano and H. Tomimuro, "Spectral domain analysis of an open slot ring resonator," *IEEE Trans. Microwave Theory Tech.*, vol. MTT-30, pp. 1184–1187, 1982.

[104] K. Araki *et al.*, "A study on circular disk resonators on a ferrite substrate," *IEEE Trans. Microwave Theory Tech.*, vol. MTT-30, pp. 147–154, 1982.

[105] S. M. Ali *et al.*, "Vector Hankel transform analysis of annular-ring microstrip antenna," *IEEE Trans. Antennas Propagat.*, vol. AP-30, pp. 637–644, 1982.

[106] T. Itoh and W. Menzel, "A full-wave analysis method for open microstrip structures," *IEEE Trans. Antennas Propagat.*, vol. AP-29, pp. 63–68, 1981.

[107] M. C. Bailey and M. D. Deshpande, "Integral equation formulation of microstrip antennas," *IEEE Trans. Antennas Propagat.*, vol. AP-30, pp. 651–656, 1982.

[108] M. D. Deshpande and M. C. Bailey, "Input impedance of microstrip antennas," *IEEE Trans. Antennas Propagat.*, vol. AP-30, pp. 645–650, 1982.

[109] E. H. Newman *et al.*, "Mutual impedance computation between microstrip antennas," *IEEE Trans. Microwave Theory Tech.*, vol. MTT-31, pp. 941–945, 1983.

[110] N. H. L. Koster, "Frequency-dependent characterization of discontinuities in planar microwave transmission lines" (in German), Ph.D. dissertation, Univ. of Duisburg, West Germany, 1984.

Spectral-Domain Approach for Calculating the Dispersion Characteristics of Microstrip Lines

TATSUO ITOH AND RAJ MITTRA

Abstract—**The boundary value problem associated with the open microstrip line structure is formulated in terms of a rigorous, hybird-mode representation. The resulting equations are subsequently transformed, via the application of Galerkin's method in the spectral domain, to yield a characteristic equation for the dispersion properties of the open microstrip line.**

Numerical results are included for several different structural parameters. These are compared with other available data and with some experimental measurements.

INTRODUCTION

Because microwave integrated circuits are being used at higher frequencies, it is often necessary to predict the dispersion characteristics of microstrip lines and similar configurations. However, only very recently has the hybrid-mode analysis been applied for rigorous formulation of the dispersion problem for both the shileded [1]–[3] and open versions [4] of the microstrip lines.

The method followed by Denlinger [4] for analyzing the open microstrip line is critically dependent on the forms of the distribution one assumes, for the two current components on the center strip of the line, in the process of solving for the unknown amplitude of these distributions. In this short paper, a new method is presented for circumventing the preceding difficulty and systematically solving for the current components to the desired degree of accuracy. The method is basically a modification of Galerkin's approach adapted for application in the Fourier transform, or spectral domain. One of the advantages of this approach is that it is numerically more efficient than the conventional methods that work directly in the space domain. This is due primarily to the fact that the process of Fourier transformation of the coupled integral equations in the space domain yields a pair of algebraic equations in the transform domain that are relatively easier to handle. Another important advantage is that the Green's function takes a much simpler form in the transform domain, as compared to the space domain where no convenient form of the Green's function is known to exist. Finally, the method itself is quite general, and hence, is applicable to a number of other structures, e.g., the slot line [5].

FORMULATION OF THE PROBLEM

Fig. 1 shows the cross section of the open microstrip line. The structure is assumed to be uniform and infinite in both x and z directions. The infinitely thin strip and the ground plane are perfect conductors. It is also assumed that the substrate material is lossless and its relative permittivity and permeability are ϵ_r and μ_r, respectively.

It is well known that the hybrid-field components can be expressed in terms of a superposition of the TE and TM fields, which are in turn derivable from the scalar potentials $\psi^{(e)}$ and $\psi^{(h)}$. For instance

$$E_{zi} = j\frac{k_i^2 - \beta^2}{\beta}\psi_i^{(e)}(x, y)e^{-j\beta z}$$

$$H_{zi} = j\frac{k_i^2 - \beta^2}{\beta}\psi_i^{(h)}(x, y)e^{-j\beta z} \tag{1}$$

$$k_1 = \omega\sqrt{\epsilon_1\mu_1} = \omega\sqrt{\epsilon_r\mu_r\epsilon_0\mu_0}$$

$$k_2 = \omega\sqrt{\epsilon_2\mu_2} = \omega\sqrt{\epsilon_0\mu_0} \tag{2}$$

where β is the unknown propagation constant and ω is the operating frequency. The superscripts (e) and (h) are associated with the TM and TE types of fields, respectively. The subscripts $i=1, 2$ serve to designate the regions 1 (substrate) or 2 (air). All the other field components are also easily derivable.

Manuscript received November 29, 1972; revised January 25, 1973. This work was supported in part by the United States Army Research Grant DA-ADO-D-31-71-G77 and in part by NSF Grants GK 33735 and GK 36854.

The authors are with the Department of Electrical Engineering, University of Illinois, Urbana, Ill. 61801.

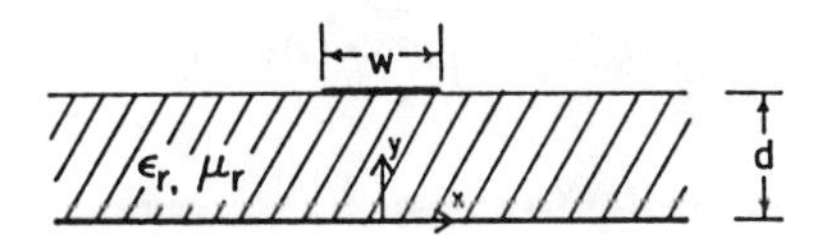

Fig. 1. Microstrip line.

As a first step, we define the Fourier transforms of the scalar potentials as

$$\tilde{\psi}_i^{(p)}(\alpha, y) = \int_{-\infty}^{\infty}\psi_i^{(p)}(x, y)e^{+j\alpha x}\,dx, \qquad i = 1, 2 \quad p = e \text{ or } h \tag{3}$$

and apply the continuity conditions to the field components in the Fourier transform domain. When this is done, the transforms of scalar potentials at the interface $y=d$ are expressed in terms of the transform of unknown current components on the strip $\tilde{J}_x(\alpha)$ and $\tilde{J}_z(\alpha)$. For instance,

$$\tilde{\psi}_2^{(e)}(\alpha, d) = \left\{\frac{1}{\det}\left[F_1 b_{22} + \frac{\alpha\beta}{k_1^2 - \beta^2}b_{12}\right]\tilde{J}_x(\alpha) + \frac{1}{\det}b_{12}\tilde{J}_z(\alpha)\right\}e^{-\gamma_2(y-d)} \tag{4a}$$

$$\tilde{\psi}_2^{(h)}(\alpha, d) = \left\{\frac{1}{\det}\left[F_1 b_{21} + \frac{\alpha\beta}{k_1^2 - \beta^2}b_{11}\right]\tilde{J}_x(\alpha) + \frac{1}{\det}b_{11}\tilde{J}_z(\alpha)\right\}e^{-\gamma_2(y-d)} \tag{4b}$$

where $\gamma_i = \alpha^3 + \beta^2 - k_i^2$, and

$$b_{11} = -b_{22} = j\alpha\left(\frac{k_2^2 - \beta^2}{k_1^2 - \beta^2} - 1\right) \tag{5a}$$

$$b_{12} = \frac{\omega\mu_0\gamma_1}{\beta}\left[\frac{\gamma_2}{\gamma_1} + \mu_r\frac{k_2^2 - \beta^2}{k_1^2 - \beta^2}\tanh\gamma_1 d\right] \tag{5b}$$

$$b_{21} = \frac{\omega\epsilon_0\gamma_1}{\beta}\left[\frac{\gamma_2}{\gamma_1} + \epsilon_r\frac{k_2^2 - \beta^2}{k_1^2 - \beta^2}\coth\gamma_1 d\right] \tag{5c}$$

$$\det = b_{11}b_{22} - b_{12}b_{21} \tag{5d}$$

$$F_1 = \frac{\omega\mu_0\mu_r\gamma_1}{j(k_1^2 - \beta^2)}\tanh\gamma_1 d. \tag{5e}$$

Note that b_{11}, b_{12}, etc., and F_1 are functions of the propagation constant β which is as yet unknown.

Up to this stage, the formulation presented herein is basically the same as that found in Denlinger [4]. The essential difference in the present method is in the application of the two final boundary conditions on the strip, which requires

$$E_{z2}(x, d) = 0, \qquad |x| < w/2 \tag{6a}$$

$$\frac{d}{dy}H_{z2}(x, d) = 0, \qquad |x| < w/2. \tag{6b}$$

Rather than applying (6) in the space domain (as in Denlinger [4]), we impose this condition in the Fourier transform domain instead.

As a first step we let

$$E_{z2}(x, d) = j\frac{k_2^2 - \beta^2}{\beta}u(x), \qquad |x| > w/2 \tag{7a}$$

$$\frac{d}{dy}H_{z2}(x, d) = j\frac{k_2^2 - \beta^2}{\beta}v(x), \qquad |x| > w/2. \tag{7b}$$

where u and v are unknowns. Taking the Fourier transform of E_{z2}, and $(d/dy)H_{z2}$, $|x| < \infty$ given by (6) and (7) and using the expressions given by (1) and (4) on the left-hand sides of the transform of (6) plus (7), we finally obtain the following coupled equations for the two current components

$$G_{11}(\alpha, \beta)\tilde{J}_x(\alpha) + G_{12}(\alpha, \beta)\tilde{J}_z(\alpha) = \tilde{U}_1(\alpha) + \tilde{U}_2(\alpha) \tag{8a}$$

$$G_{21}(\alpha, \beta)\tilde{J}_x(\alpha) + G_{22}(\alpha, \beta)\tilde{J}_z(\alpha) = \tilde{V}_1(\alpha) + \tilde{V}_2(\alpha) \tag{8b}$$

Reprinted from *IEEE Trans. Microwave Theory Tech.*, vol. MTT-21, pp. 496–499, July 1973.

where

$$\tilde{U}_1(\alpha) = \int_{-\infty}^{-w/2} u(x)e^{j\alpha x}\,dx$$

$$\tilde{U}_2(\alpha) = \int_{w/2}^{\infty} u(x)e^{j\alpha x}\,dx$$

$$\tilde{V}_1(\alpha) = \int_{-\infty}^{w/2} v(x)e^{j\alpha x}\,dx$$

$$\tilde{V}_2(\alpha) = \int_{w/2}^{\infty} v(x)e^{j\alpha x}\,dx$$

and

$$G_{11} = \frac{1}{\det}\left[F_1 b_{22} + \frac{\alpha\beta}{k_1^2 - \beta^2} b_{12}\right]$$

$$G_{12} = \frac{b_{12}}{\det}$$

$$G_{21} = \frac{\gamma_2}{\det}\left[F_1 b_{21} + \frac{\alpha\beta}{k_1^2 - \beta^2} b_{11}\right]$$

$$G_{22} = \frac{\gamma_2 b_{11}}{\det}.$$

Note that (8) is a set of two algebraic equations, in contrast to the coupled integral equations employed by Denlinger in the space-domain analysis. As alluded to earlier, this is the principal advantage of the present method of formulation.

Method of Solution

In this section we present an efficient method for solving the coupled equations (8). The method is essentially Galerkin's procedure applied in the Fourier transform domain. It is first noted that the two equations in (8) actually contain six unknowns. However, by using certain properties of these functions, we can eliminate four of the unknowns, viz., $\tilde{U}_1$, $\tilde{U}_2$, $\tilde{V}_1$, and $\tilde{V}_2$, from these equations.

To this end, let us first expand the unknown current components $\tilde{J}_x$ and $\tilde{J}_z$ in terms of known basis functions $\tilde{J}_{xn}$ and $\tilde{J}_{zn}$ as follows:

$$\tilde{J}_x(\alpha) = \sum_{n=1}^{M} c_n \tilde{J}_{xn}(\alpha) \tag{9a}$$

$$\tilde{J}_z(\alpha) = \sum_{n=1}^{N} d_n \tilde{J}_{zn}(\alpha). \tag{9b}$$

The basis functions $\tilde{J}_{xn}$ and $\tilde{J}_{zn}$ must be chosen such that their inverse Fourier transforms are nonzero only on the strip $|x| < w/2$. After substituting (9) into (8) we take the inner products with the basis functions $\tilde{J}_{zm}$ and $\tilde{J}_{xm}$ for different values of m. This yields the matrix equation

$$\sum_{n=1}^{M} K_{mn}^{(1,1)} c_n + \sum_{n=1}^{N} K_{mn}^{(1,2)} d_n = 0, \qquad m = 1, 2, \cdots, N \tag{10a}$$

$$\sum_{n=1}^{M} K_{mn}^{(2,1)} c_n + \sum_{n=1}^{N} K_{mn}^{(2,2)} d_n = 0, \qquad m = 1, 2, \cdots, M \tag{10b}$$

where

$$K_{mn}^{(1,1)} = \int_{-\infty}^{\infty} \tilde{J}_{zm}(\alpha) G_{11}(\alpha, \beta) \tilde{J}_{xn}(\alpha)\,d\alpha \tag{11a}$$

$$K_{mn}^{(1,2)} = \int_{-\infty}^{\infty} \tilde{J}_{zm}(\alpha) G_{12}(\alpha, \beta) \tilde{J}_{zn}(\alpha)\,d\alpha \tag{11b}$$

$$K_{mn}^{(2,1)} = \int_{-\infty}^{\infty} \tilde{J}_{xm}(\alpha) G_{21}(\alpha, \beta) \tilde{J}_{xn}(\alpha)\,d\alpha \tag{11c}$$

$$K_{mn}^{(2,2)} = \int_{-\infty}^{\infty} \tilde{J}_{xm}(\alpha) G_{22}(\alpha, \beta) \tilde{J}_{zn}(\alpha)\,d\alpha. \tag{11d}$$

One can verify via an application of Parseval's theorem that the right-hand sides of (8) are indeed eliminated by this procedure. Using this theorem we can show, for instance, that

$$\int_{-\infty}^{\infty} \tilde{J}_{zm}(\alpha)[\tilde{U}_1(\alpha) + \tilde{U}_2(\alpha)]\,d\alpha = \frac{1}{2\pi}\int_{-\infty}^{\infty} J_{zm}(x)\left[\frac{\beta}{j(k_2^2 - \beta^2)} E_{z2}(x, d)\right] dx = 0.$$

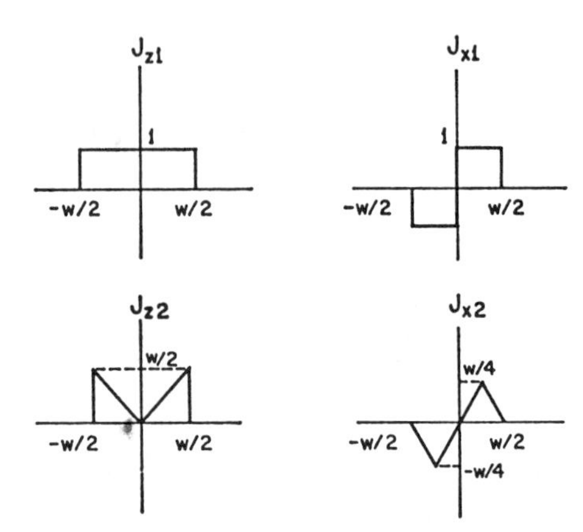

Fig. 2. Basis functions for J_x and J_z.

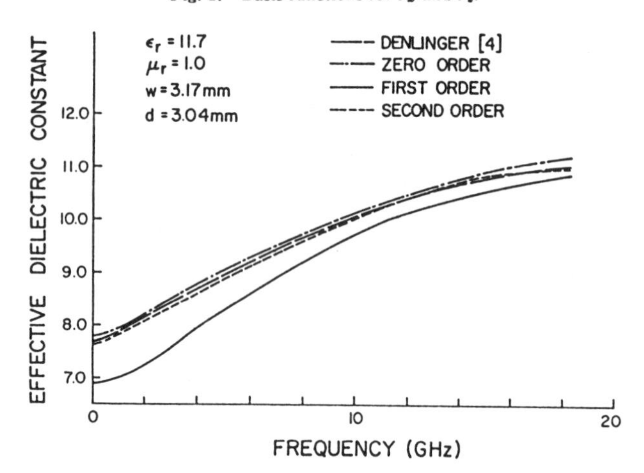

Fig. 3. Effective dielectric constant versus frequency.

The preceding relation is true since the current $J_{zm}(x)$, the inverse transform of $\tilde{J}_{zm}(\alpha)$, and $E_{z2}(x, d)$ are nonzero in the complementary regions of x.

The next step is to solve the simultaneous equations (10) for the propagation constant β, by setting the determinant of this set of equations equal to zero and seeking the root of the resulting equation. The propagation constant β is calculated for each frequency ω to obtain the dispersion relation for the microstrip line structure of Fig. 1.

As pointed out earlier, the numerical results obtained by Denlinger are critically dependent on the choice of the assumed forms of current components on the strip, because in his method the integral equations are solved for the amplitudes of the current components with assumed distribution. However, in the present method the solution can be systematically improved by increasing the number of basis functions and solving a larger size matrix.

Numerical Procedure and Results

The choice of the basis functions is rather arbitrary as long as they satisfy the required conditions that they are zero in the appropriate range and possess certain symmetry properties. For the dominant mode, it is easily seen that J_z is even-symmetric with respect to the y axis while J_x is odd-symmetric. We will first show how the solution of (10) improves with the increasing size of the matrix associated with (10).

Let us choose the set of functions J_{z1}, J_{z2}, J_{x1}, and J_{x2} as shown in Fig. 2. J_{xn} and J_{zn} for $n \geq 3$ can be defined in a similar manner. The Fourier transforms of these functions are easily obtained. These four functions are used in (11) to compute the matrix elements $K_{mn}^{(1,1)}$, $K_{mn}^{(1,2)}$, $K_{mn}^{(2,1)}$, and $K_{mn}^{(2,2)}$ for a given frequency. A dispersion relation has been calculated for three choices of matrix size, i.e.: 1) $N=1$, $M=0$; 2) $N=M=1$; and 3) $N=M=2$. In the first case, only the axial component J_{z1} of the strip current is retained, and this case may be called the zero-order approximation. The second case (the first-order approximation) uses J_{z1} and J_{x1}, while the third case is the second-order approximation with J_{z1}, J_{z2}, J_{x1}, and J_{x2} retained.

Fig. 3 shows the effective dielectric constant computed by the present method with different order of approximation. The effective dielectric constant is defined by

$$\epsilon_{\rm eff} = \left(\frac{\lambda}{\lambda_g}\right)^2 = \left(\frac{\beta}{k}\right)^2$$

where λ_g is the guide wavelength. In Fig. 3 the results computed by Denlinger [4] are also given for comparison. It is clear that both the zero- and the second-order approximations agree quite well with Denlinger's results. Some test calculations for the third-order approximation show that the results fall between the zero- and the second-order curves.

The following explanation may be offered as to why the first-order approximation does not give good results. An examination of J_{z1} and J_{z2} shows that they are good approximations for the z component of the strip current in the dc limit. Recall that in this approximation the effective dielectric constant is computed from the knowledge of the line capacitance only. However, the assumed form of J_{x1} is far from the actual distribution of the x-directed current component on the strip, because its true value actually goes to zero smoothly as one approaches the edge and the center of the strip. It is evident that on the basis of this criterion J_{x2} represents a much better approximation for the x component of the strip current and hence its inclusion results in better accuracy for the dispersion curves.

Fig. 4 shows the relative guide wavelength for several values of ϵ_r. The dispersion curves for the closed microstrip line are also included for comparison [2]. The closed microstrip line is placed in a rectangular shield case with a side dimension of 12.7 mm. The experimental results found in [2] are also reproduced. In Fig. 4 only the results for the zero-order approximation are plotted to retain the clarity of the figure. The second-order results fall between the zero-order results and the dispersion curves for the closed microstrip line. For the reasons given earlier, the first-order solution again yields poor approximation, giving values larger than the closed microstrip results. It should also be mentioned that the results for the zero-frequency limit agree well with the quasi-TEM solution, except in the case of the first-order approximation.

Finally, it is important to quote typical computation times for this method. The computation time on the CDC G-20 computer (approximately seven to ten times slower than the IBM 360/75) is about 30 s for the zero-order approximation, 120 s for the first-order, and 500 s for the second-order approximation. These times are typical for one point on the curve when the matrix elements given by (11) are accurate to three digits or better.

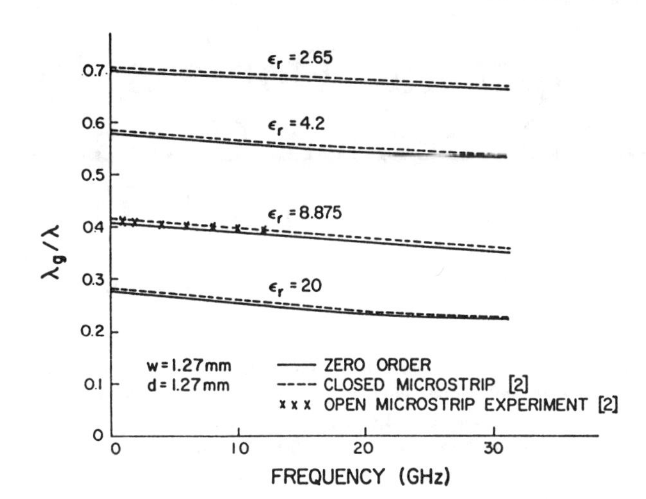

Fig. 4. Normalized guide wavelength versus frequency.

Conclusions

A numerical method has been presented for obtaining the dispersion relation of the open microstrip lines. The method is based upon application of Galerkin's procedure in the spectral domain. The accuracy of the numerical results obtained by the present method can be improved in a systematic manner by increasing the size of the matrix associated with the characteristic equation. Numerical results reported in this paper have been compared with other available data and experimental results.

References

[1] J. S. Hornsby and A. Gopinath, "Numerical analysis of a dielectric-loaded waveguide with a microstrip line—Finite-difference methods," *IEEE Trans. Microwave Theory Tech.*, vol. MTT-17, pp. 684–690, Sept. 1969.

[2] R. Mittra and T. Itoh, "A new technique for the analysis of the dispersion characteristics of microstrip lines," *IEEE Trans. Microwave Theory Tech.*, vol. MTT-19, pp. 47–56, Jan. 1971.

[3] G. Kowalski and R. Pregla, "Dispersion characteristics of shielded microstrips with finite thickness," *Archiv für Elektronik und Übertragungstechnik*, vol. 25, pp. 193–196, Apr. 1971.

[4] E. J. Denlinger, "A frequency dependent solution for microstrip transmission lines," *IEEE Trans. Microwave Theory Tech.*, vol. MTT-19, pp. 30–39, Jan. 1971.

[5] T. Itoh and R. Mittra, "Dispersion characteristics of slot lines," *Electron Lett.*, vol. 7, pp. 364–365, July 1971.

Paper 10.3

Spectral Domain Immitance Approach for Dispersion Characteristics of Generalized Printed Transmission Lines

TATSUO ITOH, SENIOR MEMBER, IEEE

***Abstract*—A simple method for formulating the dyadic Green's functions in the spectral domain is presented for generalized printed transmission lines which contain several dielectric layers and conductors appearing at several dielectric interfaces. The method is based on the transverse equivalent transmission line for a spectral wave and on a simple coordinate transformation. This formulation process is so simple that often it is accomplished almost by inspection of the physical cross-sectional structure of the transmission line. The method is applied to a new versatile transmission line, a microstrip-slot line, and some numerical results are presented.**

I. Introduction

A FEW YEARS AGO, a method called the spectral-domain technique was developed for efficient numerical analyses for various planar transmission lines and successfully applied to a number of structures [1], [2]. One difficulty in applying this technique is that a lengthy derivation process is required in the formulation stage, especially for the more complicated structures such as the one recently proposed by Aikawa [3], [4] in which more than one conductor are located at different dielectric interfaces. This paper presents a simple method for deriving the dyadic Green's functions (immittance functions) which is based on the transverse equivalent circuit concept as applied in the spectral domain in conjunction with a simple coordinate transformation rule. This technique is quite versatile and the formulation of the Green's function may be done almost by inspection in many structures. It is noted that symmetry in the structure is not required and that the analysis can be extended to finite circuit elements, such as the disk resonator.

In what follows, we first illustrate the formulation process for the microstrip line and subsequently extend it to a more general microstrip-slot structure. Numerical results for the microstrip-slot structure are also presented.

II. Illustration of the Formulation Process

To illustrate the formulation process, we will use a simple shielded microstrip line shown in Fig. 1. In conventional space-domain analysis [5], this structure may be analyzed by first formulating the following coupled homogeneous integral equations and then solving for the unknown propagation constant β:

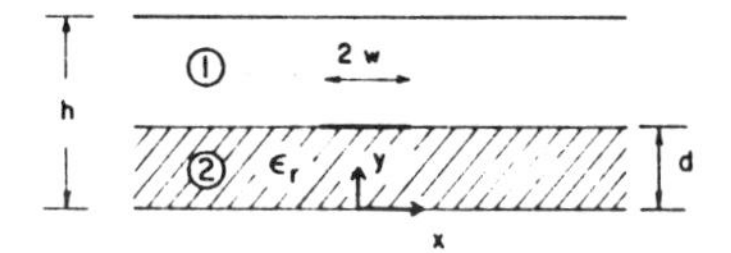

Fig. 1. Cross section of a microstrip line.

$$\int [Z_{zz}(x-x',d)J_z(x') + Z_{zx}(x-x',d)J_x(x')] \, dx' = 0 \tag{1}$$

$$\int [Z_{xz}(x-x',d)J_z(x') + Z_{xx}(x-x',d)J_x(x')] \, dx = 0 \tag{2}$$

where J_x and J_z are unknown current components on the strip and the Green's functions (impedance functions) Z_{zz}, etc., are functions of unknown β as well. The integration is over the strip, and (1) and (2) are valid on the strip. The left-hand sides of these equations give E_z and E_x components on the strip and, hence, are required to be zero to satisfy the boundary condition at the perfectly conducting strip. These equations may be solved provided that Z_{zz}, etc., are given. However, for the inhomogeneous structures, these quantities are not available in closed forms.

In the spectral domain formulation, we use Fourier transforms of (1) and (2) and deal with algebraic equations

$$\tilde{Z}_{zz}(\alpha,d)\tilde{J}_z(\alpha,d) + Z_{zx}(\alpha,d)\tilde{J}_x(\alpha,d) = \tilde{E}_z(\alpha,d) \tag{3}$$

$$\tilde{Z}_{xz}(\alpha,d)\tilde{J}_z(\alpha,d) + Z_{xx}(\alpha,d)\tilde{J}_x(\alpha,d) = \tilde{E}_x(\alpha,d) \tag{4}$$

instead of the convolution-type coupled integral equations (1) and (2). In (3) and (4), quantities with $\sim$ are Fourier transforms of corresponding quantities without $\sim$. The Fourier transform is defined as

$$\tilde{\phi}(\alpha) = \int_{-\infty}^{\infty} \phi(x) e^{j\alpha x} \, dx. \tag{5}$$

Notice that the right-hand sides of (3) and (4) are no longer zero because they are the Fourier transforms of E_z and E_x on the substrate surface which are obviously nonzero except on the strip. Hence, algebraic equations (3) and (4) contain four unknowns $\tilde{J}_z$, $\tilde{J}_x$, $\tilde{E}_z$, and $\tilde{E}_x$. However, $\tilde{E}_z$ and $\tilde{E}_x$ will be eliminated later in the solution process based on the Galerkin's procedure.

Manuscript received October 26, 1979; revised February 2, 1980. This work was supported in part by U.S. Army Research Office under Grant DAA29-78-G-0145.

The author is with the Department of Electrical Engineering, University of Texas, Austin, TX 78712.

Reprinted from *IEEE Trans. Microwave Theory Tech.*, vol. MTT-28, no. 7, pp. 733–736, July 1980.

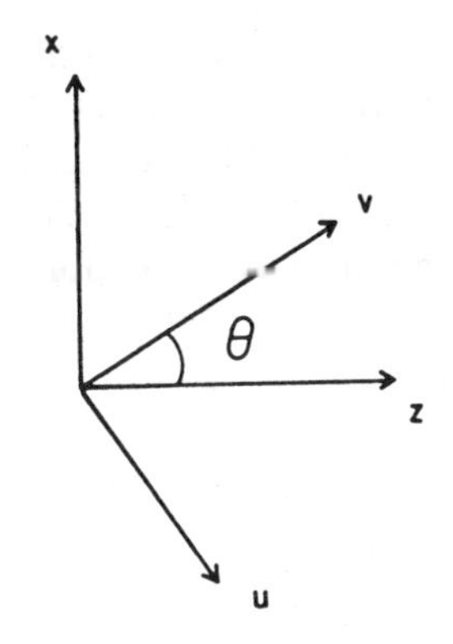

Fig. 2. Coordinate transformation.

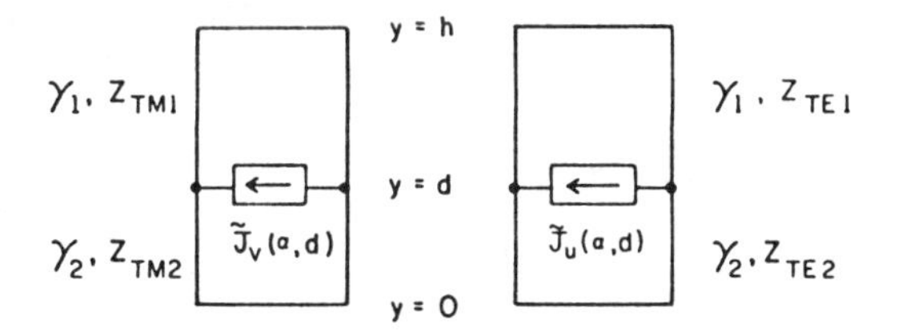

Fig. 3. Equivalent transmission lines for the microstrip line.

The closed forms of Green's impedance functions $\tilde{Z}_{zz}$, etc., can be derived by first writing the Fourier transforms of field components in each region in terms of superposition of TM-to-y and TE-to-y expressions by way of Maxwell's equations.

$$\begin{aligned}\tilde{E}_y(\alpha,y) &= A^e\cosh\gamma_1 y, & 0<y<d\\ &= B^e\cosh\gamma_2(h-y), & d<y<h \qquad (6)\end{aligned}$$

$$\begin{aligned}\tilde{H}_y(\alpha,y) &= A^h\sinh\gamma_1 y, & 0<y<d\\ &= B^h\sinh\gamma_2(h-y), & d<y<h \qquad (7)\end{aligned}$$

$$\gamma_1=\sqrt{\alpha^2+\beta^2-\epsilon_r k^2} \qquad \gamma_2=\sqrt{\alpha^2+\beta^2-k^2}\,. \qquad (8)$$

Next, we match tangential (x and z) components at the interface and apply appropriate boundary conditions at the strip [1], [2]. By eliminating A^e, B^e, A^h, and B^h from these conditions, we obtain expressions for Green's impedance functions $\tilde{Z}_{zz}$, etc.

In the new formulation process we will make use of equivalent transmission lines in the y direction. To this end, we recognize that from

$$E_y(x,y)e^{-j\beta z}=\frac{1}{2\pi}\int_{-\infty}^{\infty}\tilde{E}_y(\alpha,y)e^{-j(\alpha x+\beta z)}d\alpha \qquad (9)$$

all the field components are a superposition of inhomogeneous (in y) waves propagating in the direction of θ from the z axis where $\theta=\cos^{-1}(\beta/\xi)$, $\xi=\sqrt{\alpha^2+\beta^2}$. For each θ, waves may be decomposed into TM-to-y ($\tilde{E}_y,\tilde{E}_v,\tilde{H}_u$), and TE-to-$y$ ($\tilde{H}_y,\tilde{E}_u,\tilde{H}_v$) where the coordinates v and u are as shown in Fig. 2 and related with (x,z) via

$$\begin{aligned}u&=z\sin\theta-x\cos\theta\\ v&=z\cos\theta+x\sin\theta. \qquad (10)\end{aligned}$$

We recognize that $\tilde{J}_v$ current creates only the TM fields and $\tilde{J}_u$ the TE fields. Hence, we can draw equivalent circuits for the TM and TE fields as in Fig. 3. The characteristic admittances in each region are

$$Y_{TMi}=\frac{\tilde{H}_u}{\tilde{E}_v}=\frac{j\omega\epsilon_0\epsilon_i}{\gamma_i}, \qquad i=1,2 \qquad (11)$$

$$Y_{TEi}=-\frac{\tilde{H}_v}{\tilde{E}_u}=\frac{\gamma_i}{j\omega\mu}, \qquad i=1,2 \qquad (12)$$

where $\gamma_i=\sqrt{\alpha^2+\beta^2-\epsilon_i k^2}$ is the propagation constant in the y direction in the ith region. All the boundary conditions for the TE and TM waves are incorporated in the equivalent circuits. For instance, the ground planes at $y=0$ and h are represented by short circuits at respective places. The electric fields $\tilde{E}_v$ and $\tilde{E}_u$ are continuous at $y=d$ and are related to the currents via

$$\tilde{E}_v(\alpha,d)=\tilde{Z}^e(\alpha,d)\tilde{J}_v(\alpha,d) \qquad (13)$$

$$\tilde{E}_u(\alpha,d)=\tilde{Z}^h(\alpha,d)\tilde{J}_u(\alpha,d). \qquad (14)$$

$\tilde{Z}^e$ and $\tilde{Z}^h$ are the input impedances looking into the equivalent circuits at $y=d$ and are given by

$$\tilde{Z}^e(\alpha,d)=\frac{1}{Y_1^e+Y_2^e} \qquad (15)$$

$$\tilde{Z}^h(\alpha,d)=\frac{1}{Y_1^h+Y_2^h} \qquad (16)$$

where Y_1^e and Y_2^e are input admittances looking down and up at $y=d$ in the TM equivalent circuit and Y_1^h and Y_2^h are those in the TE circuit:

$$Y_1^e=Y_{TM1}\coth\gamma_1(h-d) \qquad Y_2^e=Y_{TM2}\coth\gamma_2 d \qquad (17)$$

$$Y_1^h=Y_{TE1}\coth\gamma_1(h-d) \qquad Y_2^h=Y_{TE2}\coth\gamma_2 d. \qquad (18)$$

The final step consists of the mapping from the (u,v) to (x,z) a coordinate system for the spectral wave corresponding to each θ given by α and β. Because of the coordinate transform (10), E_x and E_z are linear combinations of E_u and E_v. Similarly, J_x and J_z are superpositions of J_u and J_v. When these relations are used, the impedance matrix elements in (3) and (4) are found to be

$$\tilde{Z}_{zz}(\alpha,d)=N_z^2\tilde{Z}^e(\alpha,d)+N_x^2\tilde{Z}^h(\alpha,d) \qquad (19)$$

$$\tilde{Z}_{zx}(\alpha,h)=\tilde{Z}_{xz}(\alpha,d)=N_xN_z\left[-\tilde{Z}^e(\alpha,d)+\tilde{Z}^h(\alpha,d)\right] \qquad (20)$$

$$\tilde{Z}_{xx}(\alpha,d)=N_x^2\tilde{Z}^e(\alpha,d)+N_z^2\tilde{Z}^h(\alpha,d) \qquad (21)$$

where

$$N_x=\frac{\alpha}{\sqrt{\alpha^2+\beta^2}}=\sin\theta \qquad N_z=\frac{\beta}{\sqrt{\alpha^2+\beta^2}}=\cos\theta. \qquad (22)$$

Notice that $\tilde{Z}^e$ and $\tilde{Z}^h$ are functions of $\alpha^2+\beta^2$ and the ratio of α to β enters only through N_x and N_z.

It is easily shown that (19)–(21) are identical to those previously derived by means of boundary value problems imposed on the field expression [1], [2].

III. Extension to the Microstrip-Slot Structure

The method presented in the previous section may be extended to more complicated structures such as the microstrip-slot line structure in Fig. 4. This structure is

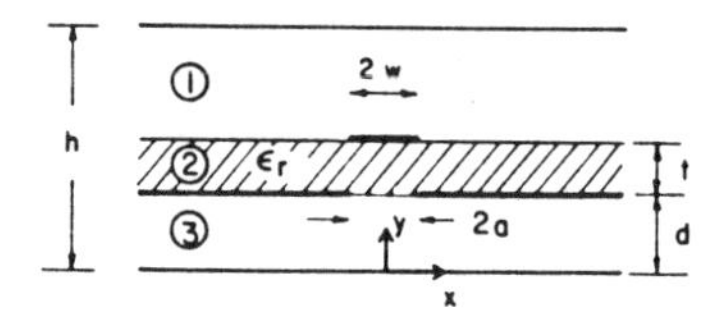

Fig. 4. Cross section of a microstrip-slot line.

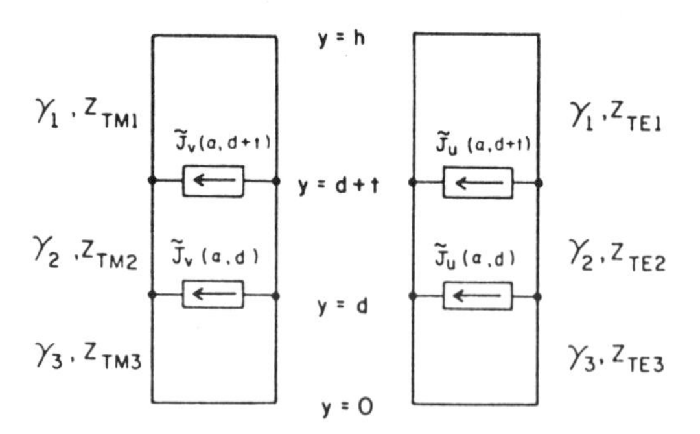

Fig. 5. Equivalent transmission lines for the microstrip-slot line.

believed to be useful in many microwave integrated circuit designs because there is an additional degree of freedom in design due to the existence of the slot [3], [4], [6]. The characteristic impedance and the propagation constant may be altered from those in the microstrip line by changing the slot width in the new structure.

By comparing the new structure with the microstrip line in Fig. 1, we may draw equivalent circuits in Fig. 5. From Fig. 5, we get

$$\tilde{E}_v(\alpha,d+t)=\tilde{Z}_{11}^e\tilde{J}_v(\alpha,d+t)+\tilde{Z}_{12}^e\tilde{J}_v(\alpha,d) \quad (23a)$$

$$\tilde{E}_u(\alpha,d+t)=\tilde{Z}_{11}\tilde{J}_u(\alpha,d+t)+\tilde{Z}_{12}^h\tilde{J}_u(\alpha,d) \quad (23b)$$

$$\tilde{E}_v(\alpha,d)=\tilde{Z}_{21}^e\tilde{J}_v(\alpha,d+t)+\tilde{Z}_{22}^e\tilde{J}_v(\alpha,d) \quad (24a)$$

$$\tilde{E}_u(\alpha,d)=\tilde{Z}_{21}^h\tilde{J}_u(\alpha,d+t)+\tilde{Z}_{22}\tilde{J}_u(\alpha,d) \quad (24b)$$

where $\tilde{Z}_{11}^e$ is the driving point input impedance at $y=d+t$ and $\tilde{Z}_{12}$ is the transfer impedance which expresses the contribution of the source at $y=d$ to the field at $y=d+t$. Other quantities may be similarly defined. Specifically

$$Z_{11}^e=\frac{1}{Y_1^e+Y_{2L}^e} \quad (25)$$

$$Y_1^e=Y_{\mathrm{TM}1}\coth\gamma_1(h-d-t) \quad (26)$$

$$Y_{2L}^e=Y_{\mathrm{TM}2}\frac{Y_{\mathrm{TM}2}+Y_3^e\coth\gamma_2 t}{Y_3^e+Y_{\mathrm{TM}2}\coth\gamma_2 t} \quad (27)$$

where

$$Y_3^e=Y_{\mathrm{TM}3}\coth\gamma_3 d. \quad (28)$$

It is readily seen that Y_3^e and Y_{2L}^e are input impedances looking down at $y=d$ and $d+t$, respectively, while Y_1^e is the one looking upward at $y=d+t$. On the other hand

$$\tilde{Z}_{12}^e=\frac{1}{Y_3^e+Y_{2u}^e}\cdot\frac{Y_{\mathrm{TM}2}/\sinh\gamma_2 t}{Y_1^e+Y_{\mathrm{TM}2}\coth\gamma_2 t}. \quad (29)$$

Here

$$Y_{2u}^e=Y_{\mathrm{TM}2}\frac{Y_{\mathrm{TM}2}+Y_1^e\coth\gamma_2 t}{Y_1^e+Y_{\mathrm{TM}2}\coth\gamma_2 t}$$

is the input admittance looking upward at $y=d$. We recognize that Z_{12}^e is the transfer impedance from Port 2 to Port 1 in the TM equivalent circuit. All other impedance coefficients in (23) and (24) may be similarly derived.

Impedance-matrix elements may be derived by the coordinate transform identical to the one used in the microstrip case. Some of the results are

$$\tilde{Z}_{zz}^{11}=N_z^2\tilde{Z}_{11}^e+N_x^2\tilde{Z}_{11}^h \quad (30)$$

$$\tilde{Z}_{zx}^{11}=N_zN_x(-\tilde{Z}_{11}^e+\tilde{Z}_{11}^h) \quad (31)$$

$$\tilde{Z}_{zz}^{12}=N_z^2\tilde{Z}_{12}^e+N_x^2\tilde{Z}_{12}^h. \quad (32)$$

The subscripts, say zx, indicate the direction of the field (E_z) caused by that of the contributing current (J_x). The superscripts, say 12, signify the relation between the interface where the field is observed (1) and the one where the current is present (2).

IV. Some Features of the Method

The method presented here is useful in solving many printed line problems. We will summarize the procedure for the formulation. 1) When the structure is given, we first draw TM and TE equivalent circuits. Each layer of dielectric medium is represented by different transmission lines and whenever conductors are present at particular interfaces, we place current sources at the junctions between transmission lines. At the ground planes, these transmission lines are shorted. 2) We derive driving point and transfer impedances from the equivalent circuits. 3) They are subsequently combined according to the sub- and superscript conventions described in the previous section, and we obtain the necessary impedance matrix elements.

The method has certain attractive features:

1) When the structures are modified, such changes are easily accommodated. For instance, when our structure has sidewalls, at say $x=\pm L$, to completely enclose the printed lines, all the procedures remain unchanged provided the discrete Fourier transform is used

$$\tilde{\phi}(\alpha)=\int_{-L}^{L}\phi(x)e^{j\alpha x}\,dx, \qquad \alpha=\frac{n\pi}{2L}. \quad (33)$$

On the other hand, when the top wall is removed, we only replace the shorted transmission line for the top-most layer with a semi-infinitely long one extending to $y\to+\infty$.

2) The formulation is independent of the number of strips and their relative location at each interface. Information on these parameters is used in the Galerkin's procedure to solve equations such as (3) and (4).

3) For some structures such as fin lines [8], it is more advantageous to use admittance matrix which provides the current on the fins due to the slot field. The formulation in this case almost parallels the present one. Instead of the current sources, we need to use voltage sources in the equivalent circuits.

4) It is easily shown that the method is applicable to finite structures such as microstrip resonators and antennas. Instead of (5), we need to use double Fourier transforms in x and z directions so that only the y dependence remains to allow the use of equivalent circuit concept.

5) Certain physical information is readily extracted. For instance, it is clear that denominators of typical impedance matrix elements give the transverse resonance equation when equated to zero. This implies that for certain spectral waves determined by α and β, surface wave poles may be encountered. How strongly the surface wave is excited, or if it is excited at all, is determined by the structure.

V. Numerical Example

Although the intention of this paper is to show the formulation process, the additional steps required to obtain numerical results are discussed for the sake of completeness. We computed dispersion characteristics of the microstrip-slot line with sidewalls at $x = \pm L$ by the present formulation followed by a Galerkin's procedure repeatedly used in the spectral-domain method.

In the previous section, the problem is formulated by using the impedance matrix with elements Z_{pq}^{ij}, $(i,j=1,2$ and $p,q=x,z)$ and we presumed that the current components on the conductors are unknown. It is more advantageous in numerical calculation if we choose the current components on the strip $\tilde{J}_z(\alpha, d+t)$ and $\tilde{J}_x(\alpha, d+t)$ and the aperture fields in the slot $\tilde{E}_z(\alpha, d)$ and $\tilde{E}_x(\alpha, d)$ for unknowns in the Galerkin's procedure. This is because the aperture field in the slot can be more accurately approximated than the current on the conductor at $y=d$ [4], [7]. To this end, we rearrange the impedance matrix equation to the one in which the above four unknown quantities are on the left-hand side. This modification can be readily accomplished. In the Galerkin's method, these unknowns are expressed in terms of known basis functions. Finally, we obtain homogeneous linear simultaneous equations as the right-hand side becomes identically zero by the inner product process [1], [2]. By equating the determinant to zero, we find the eigenvalue β.

There are two types of modes in the structure. One of them is a perturbed microstrip mode and another is a perturbed slot mode. For the perturbed microstrip quasi-TEM mode, we have computed dispersion relations by choosing only one basis function each for four unknowns. They are chosen such that appropriate edge conditions are satisfied at the edges of strip and slot. For instance, we can choose as the basis functions the Fourier transforms of

$$J_z(x,d+t) = \frac{1}{\sqrt{w^2 - x^2}} \qquad J_x(x,d+t) = x\sqrt{w^2 - x^2}$$

$$E_z(x,d) = \sqrt{a^2 - x^2} \qquad E_x(x,d) = \frac{x}{\sqrt{a^2 - x^2}}.$$

It is readily seen that Fourier transforms of these functions are analytically given in terms of Bessel functions. Fig. 6 shows some numerical examples of dispersion characteristics. The present results for a small slot width are compared with those of a shielded microstrip line [1]. It is clear that as the frequency increases, the presence of nonzero slot width becomes more significant. It is also seen that, as the slot width increases, the guide wavelength becomes larger because the effect for the free space below the slot is more pronounced. This suggests that the guide wavelength is adjustable by two means, one by changing the strip width and another by varying the slot width.

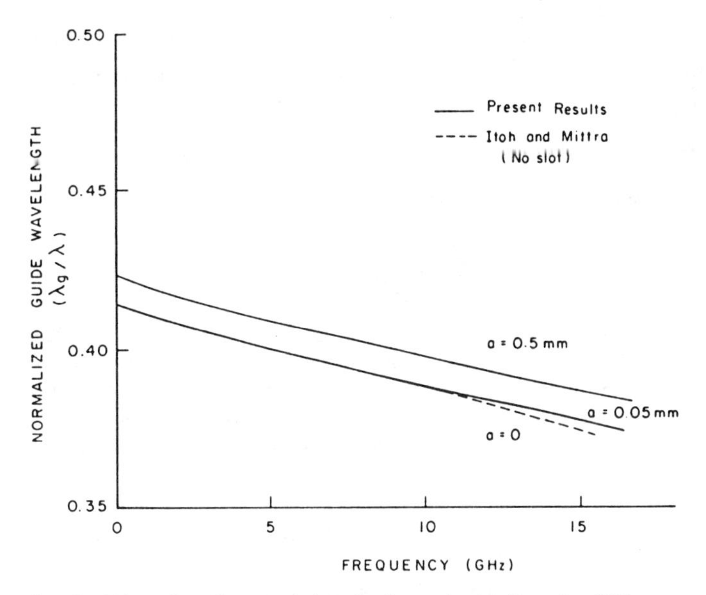

Fig. 6. Dispersion characteristics of microstrip-slot lines $L=6.35$ mm, $d=11.43$ mm, $t=1.27$ mm, $h=24.13$ mm, $w=0.635$ mm, $\epsilon_r=8.875$.

VI. Conclusions

We presented a simple method for formulating the eigenvalue problems for dispersion characteristics of general printed transmission lines. The method is intended to save considerable analytical labor for these types of problems. In addition, the method provides certain unique features. The method is applied to the problem of microstrip-slot line believed useful in microwave- and millimeter-wave integrated circuits. Numerical results are also presented.

References

[1] T. Itoh and R. Mittra, "A technique for computing dispersion characteristics of shielded microstrip lines," *IEEE Trans. Microwave Theory Tech.*, vol. MTT-22, pp. 896–898, Oct. 1974.

[2] T. Itoh, "Analysis of microstrip resonators," *IEEE Trans. Microwave Theory Tech.*, vol. MTT-22, pp. 946–952, Nov. 1974.

[3] M. Aikawa, "Microstrip line directional coupler with tight coupling and high directivity," *Electron. Commun. Jap.*, vol. J60-B, pp. 253–259, Apr. 1977.

[4] H. Ogawa and M. Aikawa, "Analysis of coupled microstrip-slot lines," *Electron. Commun. Jap.*, vol. J62-B, pp. 396–403, Apr. 1979.

[5] G. I. Zysman and D. Varon, "Wave propagation in microstrip transmission lines," presented at the Int. Microwave Symp. (Dallas, TX, May 1969, paper MAM-I-1.

[6] T. Itoh and A. S. Hebert, "A generalized spectral domain analysis for coupled suspended microstrip lines with tuning septums," *IEEE Trans. Microwave Theory Tech.*, vol. MTT-26, pp. 820–826, Oct. 1978.

[7] J. B. Davies and D. Mirshekar-Syahkal, "Spectral domain solution of arbitrary coupled transmission lines with multilayer substrate," *IEEE Trans. Microwave Theory Tech.*, vol. MTT-25, pp. 143–146, Feb. 1977.

[8] P. J. Meier, "Two new integrated circuit media with special advantages of millimeter wavelengths, presented at the 1972 IEEE G-MTT Int. Microwave Symp. (Arlington Heights, IL, May 1972).

Spectral-Domain Approach for Continuous Spectrum of Slot-Like Transmission Lines

J. CITERNE AND W. ZIENIUTYCZ

Abstract —**For the first time, the continuous spectrum part of slot-like transmission lines is described using the spectral-domain approach which has been successfully applied to the discrete part. Reliability of the approach is checked by numerical calculation of the surface current distribution across the slot plane in a simple illustrative example.**

I. INTRODUCTION

It is well known that surface waves do not form a complete set for open waveguides since the radiated field cannot be described by these modes alone [1]. Knowledge of the complete spectrum is required in order to analyze rigorously open discontinuities in which radiation cannot be neglected [2]. The discrete spectrum of slot-like lines (SL lines) has been successfully analyzed by the spectral-domain approach (SDA) [3]. It is the purpose of this paper to show that this technique also gives good results for the continuous spectrum.

II. FORMULATION

The SL lines under analysis (Fig. 1) consist of a combination of slots in an infinite conducting plane with a number of lossless dielectric layers superimposed on both sides. The $e^{-j\beta z}$ dependence and $e^{j\omega t}$ time variation are omitted in the analysis.

Using the Fourier transform in each region i ($i = O \cdots N$), the spectral densities of the axial field components of a continuous mode can be written as a combination of spectral plane waves

$$\begin{Bmatrix} \tilde{E}_{zi}(\alpha, y) \\ \tilde{H}_{zi}(\alpha, y) \end{Bmatrix} = \begin{Bmatrix} A_i(\alpha) \\ A_i'(\alpha) \end{Bmatrix} e^{-j\gamma_i y} + \begin{Bmatrix} B_i(\alpha) \\ B_i'(\alpha) \end{Bmatrix} e^{j\gamma_i y} \tag{1}$$

where

$$\gamma_i^2 = \rho_i^2 - \alpha^2 \quad \rho_i^2 = k_i^2 - \beta^2 \quad k_i^2 = \frac{\omega^2}{c^2}\epsilon_i.$$

The phase constant β of the forward-traveling wave may be either real ($0 \leqslant \beta \leqslant k_0$) for propagating modes or imaginary ($-j\infty < \beta \leqslant j0$) for evanescent modes. The whole continuous spectrum is then obtained by summing the modal fields (1) over the above-mentioned ranges of β.

In both the discrete and continuous spectra, the condition to be imposed at infinity is that modal fields are bounded [1]. This condition uniquely defines the modal fields in regions O and N as the inverse Fourier transform of (1) requiring for every constituent plane wave

$$\operatorname{Im}\gamma_O = \operatorname{Im}\gamma_N \geqslant 0. \tag{2}$$

Manuscript received October 17, 1984; revised April 15, 1985.

J. Citerne is with Laboratoire "Structures Rayonnantes," U.A. au C.N.R.S. 834, Institut National des Sciences Appliquées (I.N.S.A.), 35043 Rennes Cedex, France.

W. Zieniutycz is with the Telecommunication Institute, Technical University of Gdansk, Majakowskiego, 11-12, 80-952, Gdansk, Poland.

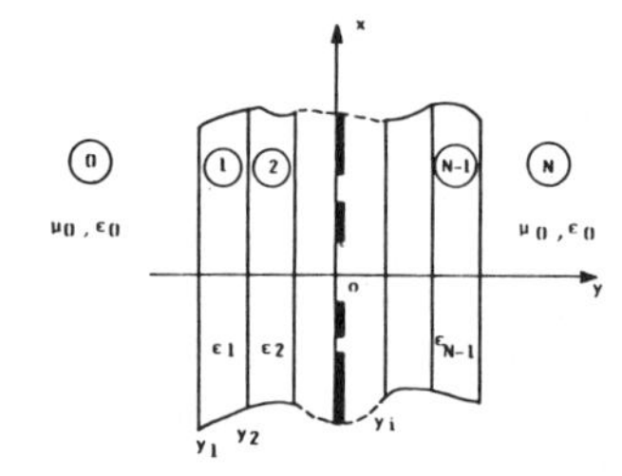

Fig. 1. Cross section of slot-like lines ("symmetric" configuration).

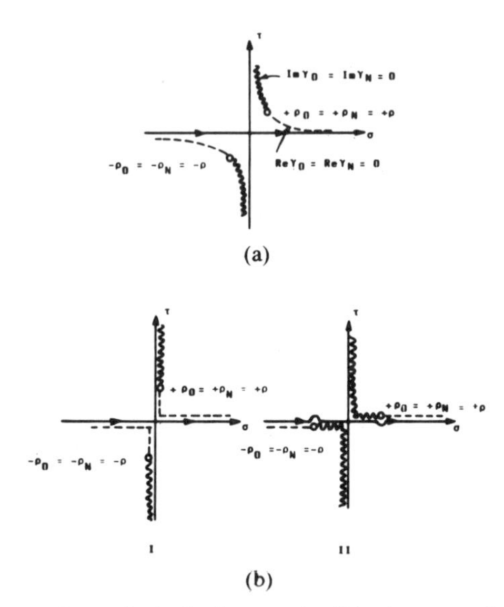

Fig. 2. Integration paths in the "proper sheet" in the α complex plane. (a) Lossy regions O and N. (b) Lossless regions O and N. I: Discrete spectrum case ($\rho^2 < 0$). II: Continuous spectrum case ($\rho^2 > 0$).

By assuming regions O and N to be lossy, the wiggly lines in Fig. 2(a) define the branch cuts of the double-valued functions $\gamma_O = \gamma_N$ in the complex $\alpha = \sigma + i\tau$ plane. The integration path from $\alpha = -\infty$ to $+\infty$ in Fig. 2(a) thus can be chosen as the real axis in the "proper" sheet of the α-plane where (2) is satisfied. Now, removing the losses in regions O and N, Fig. 2(b) describes integration paths for both the discrete spectrum with $\rho_O^2 = \rho_N^2 = \rho^2 \leqslant 0$ and the continuous spectrum with $\rho_O^2 = \rho_N^2 = \rho^2 > 0$. It can be noted in Fig. 2(b) that the discrete spectrum can use only the invisible range ($\gamma_O = \gamma_N = j\gamma$, $\gamma > 0$) of the plane-wave representation (1) for every real value of α lying on the integration path ($0 \leqslant |\alpha| < \infty$). So, in the whole spectral domain, we can write $B_O(\alpha) = B_O'(\alpha) = 0$ and $A_N(\alpha) = A_N'(\alpha) = 0$ in accordance with [3]. As for the plane-wave representation of the continuous spectrum, both invisible ($\gamma_O = \gamma_N = j\gamma$, $\gamma > 0$) and visible ($\gamma_O = \gamma_N = \gamma$, $\gamma > 0$) ranges must be used. Invisible and visible ranges correspond to parts $\rho < |\alpha| < \infty$ and $0 \leqslant |\alpha| < \rho$ of the spectral domain, respectively. The former is the evanescent part of the continuous spectrum for which we still have $B_O(\alpha) = B_O'(\alpha) = 0$ and $A_N(\alpha) = A_N'(\alpha) = 0$ and is related to the near-zone field. The latter is the propagating type and is responsible to the far-zone field; it provides the infinite complex power flow of a

Reprinted from *IEEE Trans. Microwave Theory Tech.*, vol. MTT-33, no. 9, pp. 817–818, Sept. 1985.

continuous mode [4] that can be written as

$$P=\delta(\gamma-\gamma')\Big\{\frac{\omega\epsilon_0\beta}{4\rho^2}\int_0^\rho\Big(|A_O(\alpha)|^2+|B_O(\alpha)|^2+|A_N(\alpha)|^2+|B_N(\alpha)|^2\Big)\,d\alpha+\frac{\omega\mu_0\beta^*}{4\rho^2}\int_0^\rho\Big(|A'_O(\alpha)|^2+|B'_O(\alpha)|^2+|A'_N(\alpha)|^2+|B'_N(\alpha)|^2\Big)\,d\alpha\Big\}. \quad (3)$$

From (3), it can be seen that a power separation arises between constituent spectral plane waves of TE and TM types as well as between spectral plane waves of a given type radiating in either the y or $-y$ direction. Therefore, continuous field solutions have to be constructed from four partial scattered fields corresponding to the illumination of the SL lines by TE and/or TM incident spectral plane waves denoted $A'_O(\alpha)e^{-j\gamma y}$ and/or $A_O(\alpha)e^{-j\gamma y}$, respectively, in region O, and TE and/or TM incident spectral plane waves denoted $B'_N(\alpha)e^{j\gamma y}$ and/or $B_N(\alpha)e^{j\gamma y}$, respectively, in region N.

These incident waves with arbitrary amplitudes and phases are created by filamentary sources at infinity. Selecting, for instance, the TE incident spectral wave $A'_O(\alpha)e^{-j\gamma y}$, we must write $A_O(\alpha)=0$ and $B_N(\alpha)=B'_N(\alpha)=0$ in the visible range of (1). Let us notice that the invisible range of the spectral domain does not exist in "symmetric" multilayered waveguides ($\epsilon_O=\epsilon_N$) with homogeneous boundaries at the interface $y=0$ [4]. On the contrary, this range is used in an "asymmetric" configuration ($\epsilon_O\neq\epsilon_N$) [5]. For each partial field, both homogeneous and inhomogeneous boundary conditions at interfaces y_i can be written in a general matrix notation [6]. This leads to pairs of functional equations relating the spectral densities of the tangential electric field to those of the surface current at the slot plane $y=0$. They are written as

$$\left[G^{\text{vis}}(\alpha,\beta)\right]\cdot\begin{bmatrix}\tilde{E}_x(\alpha,0)\\ \tilde{E}_z(\alpha,0)\end{bmatrix}=\begin{bmatrix}\tilde{J}_x(\alpha,0)\\ \tilde{J}_z(\alpha,0)\end{bmatrix}+\begin{bmatrix}\Delta_1(\alpha)\\ \Delta_2(\alpha)\end{bmatrix} \quad (4a)$$

for the visible (vis) range of the spectral domain, and as

$$\left[G^{\text{inv}}(\alpha,\beta)\right]\cdot\begin{bmatrix}\tilde{E}_x(\alpha,0)\\ \tilde{E}_z(\alpha,0)\end{bmatrix}=\begin{bmatrix}\tilde{J}_x(\alpha,0)\\ \tilde{J}_z(\alpha,0)\end{bmatrix} \quad (4b)$$

for the invisible (inv) range of the spectral domain. Quantities $\Delta_1(\alpha)$ and $\Delta_2(\alpha)$ in (4a), which are functions of the amplitude of the selected incident TE or TM spectral plane wave, represent sources at infinity. Obviously, no sources appear in (4b). Then, (4) can be solved in the spectral domain by using the Galerkin procedure as in [3]. Here, a set of inhomogeneous linear equations (deterministic problem) is obtained, whose solution gives the spectral densities of the partial hybrid field under consideration for each permissible value of the phase constant β.

III. Numerical Results

To verify the reliability of the method, the single-slot configuration without dielectric layers has been examined. Such an SL line supports continuous waves only. The partial field that corresponds to the TE spectral wave $A'(\alpha)e^{-j\gamma y}$ incident in the y direction becomes purely TE so that $E_z=H_z=0$ anywhere. We have in the invisible range $B'_O(\alpha)=A'_1(\alpha)=0$, while in the visible one, $B'_1(\alpha)=0$ and $A'_O(\alpha)=1$; such a source normalization in

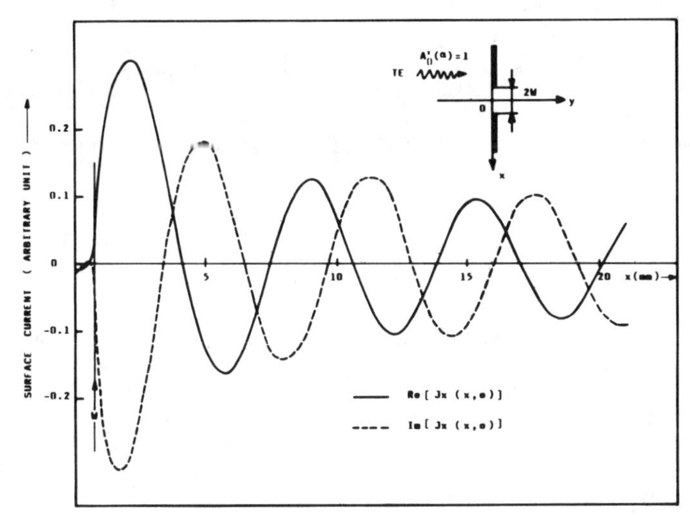

Fig. 3. Current density distribution $J_x(x,0)$ in the slot plane corresponding to a given partial field. $N_b=2$, $F=3$ GHz, $\beta=-j1$ rd/mm, $2W=1.66$ mm.

the spectral domain has to be distinguished from the field normalization given by (3).

The E_x electric-field component across the slot aperture ($|x|<W$) for all partial hybrid fields was expanded as

$$E_x(x,0)=a_0\Big/\sqrt{1-\left(\frac{x}{W}\right)^2}+\sum_{n=1}^{N_b}a_n\cos\frac{n\pi}{W}x. \quad (5)$$

Equations (4) have been solved numerically to yield the spectral partial field densities $\tilde{E}_x(\alpha,0)$ and $\tilde{J}_x(\alpha,0)$. The electric field $E_x(x,0)$ always satisfies the inhomogeneous boundary conditions at the slot plane $y=0$ because of the choice of expanding functions (5). To confirm the validity of the method, the surface current $J_x(x,0)$ must verify the prescribed inhomogeneous boundary conditions at the slot plane

$$J_x(x,0)=0,\quad |x|<W$$
$$J_x(x,0)\neq 0,\quad |x|\geq W. \quad (6)$$

Fig. 3 shows real and imaginary parts of the surface current at the slot plane $y=0$. Inasmuch as the current in the aperture is found insignificant in comparison with that on the conducting half plane, conditions (6) are satisfied.

IV. Conclusion

The spectral-domain approach for the continuous spectrum of slot-like lines is presented. Numerical results obtained for one of the four partial fields in a single-slot without dielectric layers confirm the reliability of the analysis. The method can be easily extended for microstrip-like transmission lines. Further results will be presented in the near future.

References

[1] V. V. Shevchenko, *Continuous Transitions in Open Waveguides*. Boulder, CO: Golem Press, 1971.

[2] T. E. Rozzi, "Rigorous analysis of the step discontinuity in a planar dielectric waveguide," *IEEE Trans. Microwave Theory Tech.*, vol. MTT-26, pp. 738–746, 1978.

[3] T. Itoh and R. Mittra, "Dispersion characteristics of slot-lines," *Electron. Lett.*, vol. 7, pp. 364–365, 1971.

[4] D. Marcuse, "Radiation losses of tapered dielectric slab-waveguides," *Bell Syst. Tech., J.*, vol. 49, pp. 273–290, 1970.

[5] D. Marcuse, *Theory of Dielectric Optical Waveguides*. New York: Academic Press, 1974, p. 19.

[6] J. B. Davies and D. Mirshekar-Shaykal, "Spectral domain solution and arbitrary coplanar transmission line with multilayered substrate," *IEEE Trans. Microwave Theory Tech.*, vol. MTT-25, pp. 143–146, 1977.

Spectral-Domain Analysis of Scattering from E-Plane Circuit Elements

QIU ZHANG AND TATSUO ITOH, FELLOW, IEEE

Abstract **—The spectral-domain method is used for analysis of the scattering characteristics of E-plane circuit components such as nontouching E-plane fins. The method deals with inhomogeneous algebraic equations instead of integral equations. It provides a number of attractive features. Numerical results have been compared with those reported in the literature for special cases. Several data items useful for E-plane configurations are included.**

I. INTRODUCTION

RECENTLY, finlines [1], [2] and other E-plane structures [3]–[5] have found wide applications in millimeter-wave integrated circuits. A number of passive, active, and nonreciprocal components have been developed with the E-plane technique. One of the key elements for passive E-plane components is the E-plane strip. A comprehensive design process of E-plane bandpass filters has been reported [6]. The analysis of the E-plane fin connecting the top and bottom walls is relatively straightforward [6], because the problem is a two-dimensional one. On the other hand, no extensive and accurate characterizations of nontouched E-plane fins seem to exist. A method based on a variational technique has been introduced for a special case where there is no dielectric substrate inserted in a waveguide [7]. The method in [7] is useful for a narrow strip, because only one current component along the E-plane direction is used and the assumed current distribution is constant in the axial (Z) direction.

This paper introduces a new analytical technique to characterize the scattering phenomena of a number of planar E-plane obstacles. For instance, it can handle a wide nontouching E-plane fin on a dielectric substrate. Unlike the method based on the variational technique, scattering coefficients of the dominant, as well as higher order, modes can be derived. The incident mode can be either dominant or a higher order.

The method in this paper is an extension of the spectral-domain method commonly used for characterizations of eigenmodes in a printed transmission line. It is extended to the excitation problem and hence provides a set of algebraic equations corresponding to coupled integral equations that would be derived in the space domain. Compared to the integral equation method, the new

Manuscript received June 12, 1986; revised September 6, 1986. This work was supported in part by the U.S. Army Research Office under Contract DAAG-29-84-K-0076.

The authors are with the Department of Electrical and Computer Engineering, University of Texas, Austin, TX 78712.

IEEE Log Number 8611630.

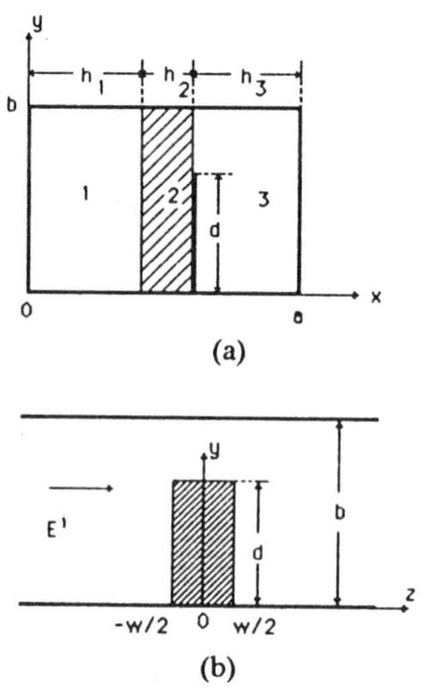

Fig. 1. E-plane strip in a rectangular waveguide. (a) End view. (b) Side view.

technique has a number of advantages. For instance, algebraic equations are easier to handle numerically. Also, the spectral-domain Green's functions have simple closed-form expressions. Compared to the variational method [7], the present method is not only more versatile but is also attractive from a computational point of view. In the new method, it is necessary to calculate the eigenvalue of only the particular scattered mode of interest. The variational method requires evaluations of all eigenvalues. Furthermore, the method in [7] assumes that only the TE modes are scattered. By the nature of the formulation, the present method contains both TE and TM modes in its formulation.

The calculated results for a special case ($\epsilon_1 = \epsilon_2 = \epsilon_3 = 1$) are compared with experimental and computed data in [7] to check the accuracy. Several curves of normalized input admittance and equivalent circuit element values are presented for a number of different parameters of the structure.

II. FORMULATION

In this paper, only the unilateral E-plane fin is treated. An application to a bilateral configuration is straightforward.

With reference to Fig. 1, the strip is assumed to be perfectly conducting and infinitesimally thin. For a given incident field E^i, the field scattered by the strip can be expressed by modal expansion. For instance, the E_y com-

Reprinted from *IEEE Trans. Microwave Theory Tech.*, vol. MTT-35, no. 2, pp. 138–150, Feb. 1987.

ponent is

$$E_y^S(x,y,z) = E_{y\,\mathrm{LSE}}^S + E_{y\,\mathrm{LSM}}^S$$

$$= \begin{cases} \sum_{m=1}^{\infty}\sum_{n=0}^{\infty} C_{mn}^{-}\phi_m(x)\cos(\alpha_n y)e^{j\beta_{mn}(z+W/2)}, & z<-W/2 \\ \sum_{m=1}^{\infty}\sum_{n=0}^{\infty} \phi_m(x)\cos(\alpha_n y)\left(E_{mn}^{-}e^{j\beta_{mn}z}+E_{mn}^{+}e^{-j\beta_{mn}z}\right), & |z|<W/2 \\ \sum_{m=1}^{\infty}\sum_{n=0}^{\infty} C_{mn}^{+}\phi_m(x)\cos(\alpha_n y)e^{-j\beta_{mn}(z-W/2)}, & z>W/2 \end{cases}$$

$$+\begin{cases} \sum_{m=1}^{\infty}\sum_{n=1}^{\infty} D_{mn}^{-}\psi_m(x)\sin(\alpha_n y)e^{j\beta'_{mn}(z+W/2)}, & z<-W/2 \\ \sum_{m=1}^{\infty}\sum_{n=1}^{\infty} \psi_m(x)\sin(\alpha_n y)\left(F_{mn}^{-}e^{j\beta'_{mn}z}+F_{mn}^{+}e^{-j\beta'_{mn}z}\right), & |z|<W/2 \\ \sum_{m=1}^{\infty}\sum_{n=1}^{\infty} D_{mn}^{+}\psi_m(x)\sin(\alpha_n y)e^{-j\beta'_{mn}(z-w/2)}, & z>W/2 \end{cases} \tag{1}$$

where

$$\phi_m(x) = \begin{cases} C_m\sinh(\gamma_{1m}x), & 0<x<h_1 \\ A_m\sinh[\gamma_{2m}(x-h_1)]+B_m\cosh[\gamma_{2m}(x-h_1)], & h_1<x<h_1+h_2 \\ \sinh[\gamma_{3m}(a-x)], & h_1+h_2<x<a \end{cases} \tag{2}$$

$$\psi_m(x) = \begin{cases} C'_m\cosh(\gamma'_{1m}x), & 0<x<h_1 \\ A'_m\sinh[\gamma'_{2m}(x-h_1)]+B'_m\cosh[\gamma'_{2m}(x-h_1)], & h_1<x<h_1+h_2 \\ \cosh[\gamma'_{3m}(a-x)], & h_1+h_2<x<a \end{cases} \tag{3}$$

$$\alpha_n = \frac{n\pi}{b} \tag{4}$$

$$\alpha_n^2+\beta_{mn}^2 = k_i^2+\gamma_{im}^2 \tag{5a}$$

$$\alpha_n^2+\beta_{mn}'^2 = k_i^2+\gamma_{im}'^2 \tag{5b}$$

$$k_i = \omega\sqrt{\epsilon_0\mu_0\epsilon_i}, \qquad i=1,2,3. \tag{6}$$

Here, ϵ_i is the relative dielectric constant; γ_{im} and γ'_{im} are the mth eigenmodes of the LSE and LSM modes in the partially filled waveguide in region i, and they can be obtained by solving the eigenvalue equations [8]. Similar equations can be written for E_z. On the other hand, the scattered fields E_y^S and E_z^S can be expressed as follows, if the induced current components $J_y(y,z)$ and $J_z(y,z)$ are provided:

$$E_y^S(x,y,z) = \int_0^b dy' \int_{-\infty}^{\infty} dz' \left[Gyy(x,y-y',z-z')J_y(y',z') + G_{yz}(x,y-y',z-z')J_z(y',z')\right] \tag{7a}$$

$$E_z^S(x,y,z) = \int_0^b dy' \int_{-\infty}^{\infty} dz' \left[G_{zy}(x,y-y',z-z')J_y(y',z') + G_{zz}(x,y-y',z-z')J_z(y',z')\right]. \tag{7b}$$

One way to find J_y and J_z is by applying the integral equations which require the total tangential electric-field components to be zero on the strip

$$E_y^i(x,y,z)+E_y^S(x,y,z)=0 \tag{8a}$$

$$(x=h_1+h_2,\ y,z \text{ on strip}).$$

$$E_z^i(x,y,z)+E_z^S(x,y,z)=0 \tag{8b}$$

The integral equations are

$$E_y^i(h_1+h_2,y,z)+\int_0^b dy'\int_{-W/2}^{W/2} dz'\left[G_{yy}(h_1+h_2,y-y',z-z')J_y(y',z')+G_{yz}(h_1+h_2,y-y',z-z')J_z(y',z')\right]=0 \qquad \left(0<y<b,|z|<\frac{W}{2}\right) \tag{9a}$$

$$E_z^i(h_1+h_2,y,z)+\int_0^b dy'\int_{-W/2}^{W/2} dz'\left[G_{zy}(h_1+h_2,y-y',z-z')J_y(y',z)+G_{zz}(h_1+h_2,y-y',z-z')J_z(y',z')\right]=0 \qquad \left(0<y<b,|z|<\frac{W}{2}\right). \tag{9b}$$

If those equations are solved, J_y and J_z can be obtained. These J_y and J_z are then substituted into (7), so that E_y^S and E_z^S are available everywhere. If the scattered field coefficient of a particular mode is needed, the E_y^S and E_z^S can be used in (1). Each coefficient may be found from orthogonality of the expansion functions.

Although the bove formulation is correct, we do not follow such an approach. Instead, we adopt a corresponding procedure in the Fourier-transformed domain. There are two reasons for using this new technique.

1) In the Fourier-transformed (spectral) domain, we deal with coupled algebraic equations instead of the coupled integral equations (9a) and (9b).

2) Derivations of the Green's functions in the space domain are very complicated. In the spectral domain, the Fourier-transformed Green's functions are given in closed form.

Let us introduce the Fourier transform as

$$\tilde{F}(\alpha_n,\beta)=\int_{-b}^{b} e^{j\alpha_n y}\,dy\int_{-\infty}^{\infty} f(y,z)e^{j\beta z}\,dz$$

$$\alpha_n=\frac{n\pi}{b}. \tag{10}$$

The Fourier transform of (1) at $x=h_1+h_2$ is given by

$$\begin{aligned}\tilde{E}_y^S(\alpha_n,\beta)=\tilde{E}_{y\,\mathrm{LSE}}^S+\tilde{E}_{y\,\mathrm{LSM}}^S=&\sum_{m=1}^{\infty} b\delta_n\sinh(\gamma_{3m}h_3)\left\{C_{mn}^-\left[\frac{e^{-jW\beta/2}}{j(\beta+\beta_{mn})}-\pi\delta(\beta+\beta_{mn})e^{j\beta_{mn}W/2}\right]\right.\\&+\left[E_{mn}^-\frac{\sin\frac{(\beta+\beta_{mn})W}{2}}{(\beta+\beta_{mn})}+E_{mn}^+\frac{\sin\frac{(\beta-\beta_{mn})W}{2}}{(\beta-\beta_{mn})}\right]+\left.C_{mn}^+\left[\pi\delta(\beta-\beta_{mn})e^{j\beta_{mn}W/2}-\frac{e^{j\beta W/2}}{j(\beta-\beta_{mn})}\right]\right\}\\&+\sum_{m=1}^{\infty}k_n jb\cosh(\gamma_{3m}'h_3)\left\{D_{mn}^-\left[\frac{e^{-j\beta W/2}}{j(\beta+\beta_{mn}')}-\pi\delta(\beta+\beta_{mn}')e^{j\beta_{mn}'W/2}\right]\right.\\&+\left[F_{mn}^-\frac{\sin\frac{(\beta+\beta_{mn}')W}{2}}{\beta+\beta_{mn}'}+F_{mn}^+\frac{\sin\frac{(\beta-\beta_{mn}')W}{2}}{\beta-\beta_{mn}'}\right]\\&+\left.D_{mn}^+\left[\pi\delta(\beta-\beta_{mn}')e^{j\beta_{mn}'W/2}-\frac{e^{j\beta W/2}}{j(\beta-\beta_{mn}')}\right]\right\}\qquad n=0,1,2,\cdots\end{aligned} \tag{11}$$

where

$$\delta_n=\begin{cases}2 & n=0\\1 & n\neq 0\end{cases} \tag{12a}$$

$$k_n=\begin{cases}1 & n\neq 0\\0 & n=0.\end{cases} \tag{12b}$$

The Fourier transform of (9) at $x=h_1+h_2$ is given by

$$\tilde{E}_y^i(\alpha_n,\beta)+\tilde{E}_y^S(\alpha_n,\beta)=\tilde{E}_y^t(\alpha_n,\beta) \tag{13a}$$

$$\tilde{E}_z^i(\alpha_n,\beta)+\tilde{E}_z^S(\alpha_n,\beta)=\tilde{E}_z^t(\alpha_n,\beta) \tag{13b}$$

where

$$\tilde{E}_y^S(\alpha_n,\beta)=\tilde{G}_{yy}(\alpha_n,\beta)\tilde{J}_y(\alpha_n,\beta)+\tilde{G}_{yz}(\alpha_n,\beta)\tilde{J}_z(\alpha_n,\beta) \tag{14a}$$

$$\tilde{E}_z^S(\alpha_n,\beta)=\tilde{G}_{zy}(\alpha_n,\beta)\tilde{J}_y(\alpha_n,\beta)+\tilde{G}_{zz}(\alpha_n,\beta)\tilde{J}_z(\alpha_n,\beta) \tag{14b}$$

are the scattered electric fields in the spectral domain. $\tilde{G}_{yy}$, $\tilde{G}_{yz}$, and $\tilde{G}_{zz}$ can be obtained by the immittance approach [9] (see Appendix). Notice that the right sides of (13) are not zero. This is because the application of the Fourier transform requires the use of fields not only on the strip but also the one outside. Hence, (13) contains four unknowns, $\tilde{J}_y$, $\tilde{J}_z$, $\tilde{E}_y$, and $\tilde{E}_z$. In the process of solution by Galerkin's method, $\tilde{E}_y$ and $\tilde{E}_z$ are eliminated. To this end, let us expand $\tilde{J}_y$ and $\tilde{J}_z$ as follows:

$$\tilde{J}_y(\alpha_n,\beta) = \sum_{i=1}^{I} a_i \tilde{J}_{yi}(\alpha_n,\beta) \tag{15a}$$

$$\tilde{J}_z(\alpha_n,\beta) = \sum_{j=1}^{J} b_j \tilde{J}_{zj}(\alpha_n,\beta). \tag{15b}$$

We substitute (15) into (13) and take the inner product of the resultant equations with basis functions. Use of the Parseval's relation eliminates $\tilde{E}_y^t$ and $\tilde{E}_z^t$. The inner products of $\tilde{E}_y^t$ and $\tilde{J}_{yi}$ and of $\tilde{E}_z^t$ and $\tilde{J}_{zj}$ are zero, because $\tilde{J}_{yi}$ and $\tilde{J}_{zj}$ are Fourier transforms of the functions nonzero on the strip while $\tilde{E}_y^t$ and $\tilde{E}_z^t$ are transforms of the functions nonzero outside the strip.

The results are

$$\sum_{i=1}^{I} K_{pi}^{yy} a_i + \sum_{j=1}^{J} K_{pj}^{yz} b_j = -S_{yp}, \qquad p=1,2,\cdots I \tag{16a}$$

$$\sum_{i=1}^{I} K_{qi}^{zy} a_i + \sum_{j=1}^{J} K_{qj}^{zz} b_j = -S_{zq}, \qquad q=1,2,\cdots J \tag{16b}$$

where

$$K_{pi}^{yy} = \sum_{n=-\infty}^{\infty} \int_{-\infty}^{\infty} \tilde{J}_{yp}^*(\alpha_n,\beta)\tilde{G}_{yy}(\alpha_n,\beta)\tilde{J}_{yi}(\alpha_n,\beta)\,d\beta \tag{17a}$$

$$K_{pj}^{yz} = K_{qi}^{zy} = \sum_{n=-\infty}^{\infty} \int_{-\infty}^{\infty} \tilde{J}_{yp}^*(\alpha_n,\beta)\tilde{G}_{yz}(\alpha_n,\beta)\tilde{J}_{zj}(\alpha_n,\beta)\,d\beta \tag{17b}$$

$$K_{qj}^{zz} = \sum_{n=-\infty}^{\infty} \int_{-\infty}^{\infty} \tilde{J}_{zq}^*(\alpha_n,\beta)\tilde{G}_{zz}(\alpha_n,\beta)\tilde{J}_{zj}(\alpha_n,\beta)\,d\beta \tag{17c}$$

$$S_{py} = \sum_{n=-\infty}^{\infty} \int_{-\infty}^{\infty} \tilde{J}_{yp}^*(\alpha_n,\beta)\tilde{E}_y^i(\alpha_n,\beta)\,d\beta \tag{18a}$$

$$S_{qz} = \sum_{n=-\infty}^{\infty} \int_{-\infty}^{\infty} \tilde{J}_{zq}^*(\alpha_n,\beta)\tilde{E}_z^i(\alpha_n,\beta)\,d\beta \tag{18b}$$

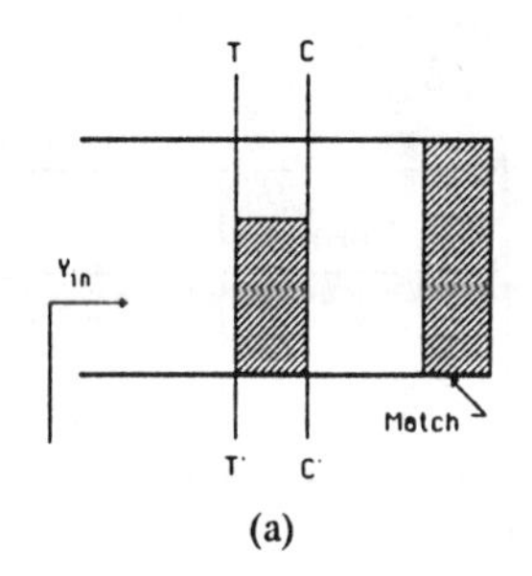

(a)

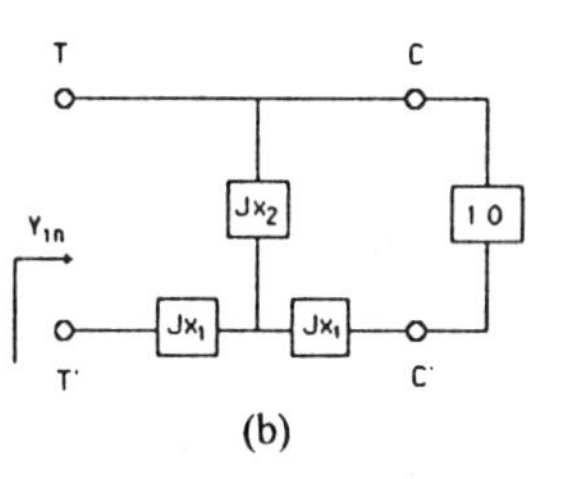

(b)

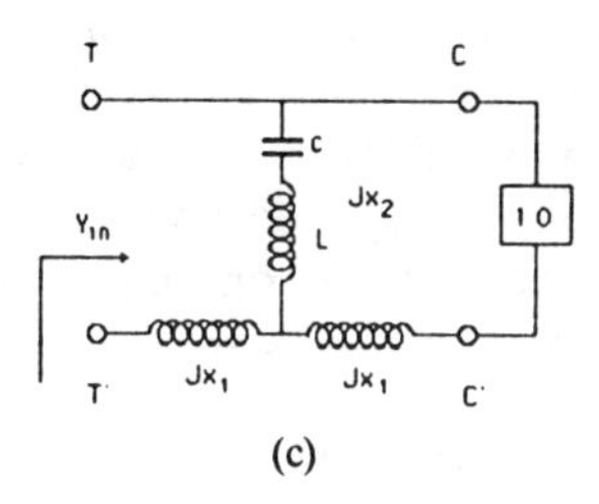

(c)

Fig. 2. (a) An E-plane strip circuit with a matched termination. (b) Two-port equivalent circuit. (c) Equivalent T-network for a narrow E-plane strip.

The scattering coefficients $C_{mn}^{\pm}$ and $D_{mn}^{\pm}$ can now be obtained. Let us express the left-hand side of (11) with (14a). Since $\tilde{G}_{yy}$ and $\tilde{G}_{yz}$ are given in closed forms and $\tilde{J}_y$ and $\tilde{J}_z$ are now known, the left-hand side of (11) is completely known. Furthermore, the left-hand side contains poles at $\beta = \pm\beta_{mn}$ and $\beta = \pm\beta'_{mn}$, since they are zeros of the denominators of $\tilde{G}_{yy}$ and $\tilde{G}_{yz}$. These values provide the eigenvalues of the LSE and LSM modes, γ_{im} and γ'_{im}. The right-hand side of (11) contains LSE poles at $-\beta_{mn}$ in the C_{mn}^- term and at $+\beta_{mn}$ in the C_{mn}^+ term. It contains LSM poles at $-\beta'_{mn}$ in the D_{mn}^- term and at $+\beta'_{mn}$ in the D_{mn}^+ term. Therefore, $C_{mn}^{\pm}$ and $D_{mn}^{\pm}$ can be obtained by residue calculus.

$$C_{mn}^{\pm} = \pm \lim_{\beta\to\pm\beta_{mn}} \left[(\beta\mp\beta_{mn})\frac{\tilde{G}_{yy}(\alpha_n,\beta)\tilde{J}_y(\alpha_n,\beta)+\tilde{G}_{yz}(\alpha_n,\beta)\tilde{J}_z(\alpha_n,\beta)}{jb\delta_n \sinh(\gamma_{3m}h_3)} e^{\mp jW\beta/2}\right]$$

$$\left(n=0,1,2,\cdots,\ m=1,2,3\cdots\ \delta_n=\begin{cases}2 & n=0\\ 1 & n\neq 0\end{cases}\right) \tag{19a}$$

$$D_{mn}^{\pm} = \mp \lim_{\beta\to\pm\beta'_{mn}} \left[(\beta\mp\beta'_{mn})\frac{\tilde{G}_{yy}(\alpha_n,\beta)\tilde{J}_y(\alpha_n,\beta)+\tilde{G}_{yz}(\alpha_n,\beta)\tilde{J}_z(\alpha_n,\beta)}{b\cosh(\gamma'_{3m}h_3)} e^{\mp jW\beta/2}\right]$$

$$(n=1,2,3\cdots,\ m=1,2,3\cdots). \tag{19b}$$

where * indicates the complex conjugate.

For a given incident field, we solve (16) and find a_i and b_j. Hence, $\tilde{J}_y$ and $\tilde{J}_z$ are now given.

This process corresponds to the use of the orthogonality relationship to find the modal coefficients in the space domain.

TABLE I
CONFIRMATION OF POWER CONSERVATION

d(mm)	R		T		$\|R\|^2 + \|T\|^2$
0.5	-.02008	-.02598 j	.79080	-.61119 j	1.00000
1.0	-.14712	-.13549 j	.66376	-.72071 j	1.00000
1.5	-.70508	-.10766 j	.10580	-.69288 j	1.00000
2.0	-.78485	.61825 j	.02602	.03304 j	1.00000
2.5	-.53859	.77455 j	.27229	.18934 j	1.00000
3.0	-.45482	.79016 j	.35606	.20495 j	1.00000
3.5	-.38694	.79227 j	.42394	.20705 j	1.00000

$h_1 = h_3 = 3.43$ mm; $h_2 = 0.254$ mm; $\epsilon_1 = \epsilon_3 = 1$; $b = 3.56$ mm; $w = 1.0$ m; $f = 35$ GHz; $\epsilon_2 = 2.2$.

It should be noted that in the case of E-plane fins connecting the top and bottom wall, the above equations are simplified. Since there is no field variations in the y-direction and only TE_{n0} modes are scattered for a TE_{10} excitation, we have only the one equation (13a). All the Fourier-transformed quantities are functions of β only.

III. NUMERICAL RESULTS

To compare the present method with the experimental and computed data in the previous publication [7], the special case $\epsilon_1 = \epsilon_2 = \epsilon_3 = 1$ is considered first. We assume that a dominant-mode incident electric field $E_{y10} = -j\beta_{10}\phi_1(x)e^{-\gamma\beta_{10}z}$, where $\phi_1(x)$ is defined in (2), comes from the left of the waveguide. The E-plane fin shown in Fig. 2(a) may be represented by an equivalent T-network, as shown in Fig. 2(b). When the waveguide is terminated with a matched load, the normalized input admittance may be represented by

$$Y_{in} = G_{in} + jB_{in} = \frac{1-R}{1+R} \tag{20}$$

where R is the reflection coefficient for the dominant mode, and can be determined in the present method by means of (19). In this study, the current basis functions are chosen as

$$J_{yi}(y,z) = \frac{\cos\left[(k-1)\left(\frac{\pi z}{W} + \frac{\pi}{2}\right)\right]}{\sqrt{1-\left(\frac{2z}{W}\right)^2}} \cdot \frac{\sin\left[(2l-1)\left(\frac{\pi y}{2d} + \frac{\pi}{2}\right)\right]}{\sqrt{1-\left(\frac{y}{d}\right)^2}} \tag{21a}$$

$$J_{zj}(y,z) = \frac{\sin\left[k\left(\frac{\pi z}{W} + \frac{\pi}{2}\right)\right]}{\sqrt{1-\left(\frac{2z}{W}\right)^2}} \frac{\cos\left[(2l-1)\left(\frac{\pi y}{2d} + \frac{\pi}{2}\right)\right]}{\sqrt{1-\left(\frac{y}{d}\right)^2}} \tag{21b}$$

where k and l are integers, and i and j are given by a combination of k and l. The equivalent circuit for a nontouching fin can be expressed as in Fig. 2(c) if the strip is not too wide. We can determine the normalized reactances X_1 and X_2 after Y_{in} in (20) is obtained. The expressions of X_1 and X_2 are given by

$$X_2 = \pm\left[\left(\frac{B_{in}}{B_{in}^2 + G_{in}^2 - G_{in}}\right)^2 + \frac{G_{in} - 1}{B_{in}^2 + G_{in}^2 - G_{in}}\right]^{1/2} \tag{22a}$$

$$X_1 = \frac{B_{in}}{G_{in} - (B_{in}^2 + G_{in}^2)} \mp X_2. \tag{22b}$$

The sign in (22a) and (22b) can be determined by the Foster reactance theorem in which $dx/dw > 0$ has to be satisfied for a lossless element. We can also determine the values of C and L in the equivalent circuit Fig. 2(c), once X_2 is found, under the assumption that the variations of C and L with frequency are small. The capacitance and inductance can be obtained by solving the coupled equations

$$\omega L - \frac{1}{C\omega} = X_2 \tag{23a}$$

$$L + \frac{1}{C\omega^2} = \frac{dX_2}{d\omega}. \tag{23b}$$

The numerical result has been checked by the power conservation law in which the equation $|R|^2 + |T|^2 = 1$ has to be satisfied. R and T are the reflection and transmission coefficients for the dominant mode, and can be determined from C_{10}^- and C_{10}^+ which are given by (19). The calculated results for different parameters are given in Table I. It shows that $|R|^2 + |T|^2$ is essentially 1. The convergence test with different numbers of basis functions has been performed. Although the accuracy can be increased with a larger number of basis functions, computation efforts also increase.

A comparison of the normalized susceptance versus frequency between the numerical results obtained by the present approach and those given in [7] is shown in Fig. 3(a) and (b). It is found that the numerical results agree

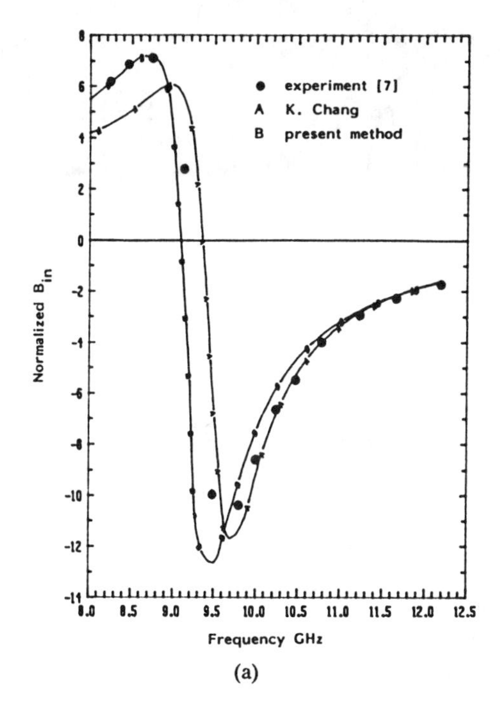

(a)

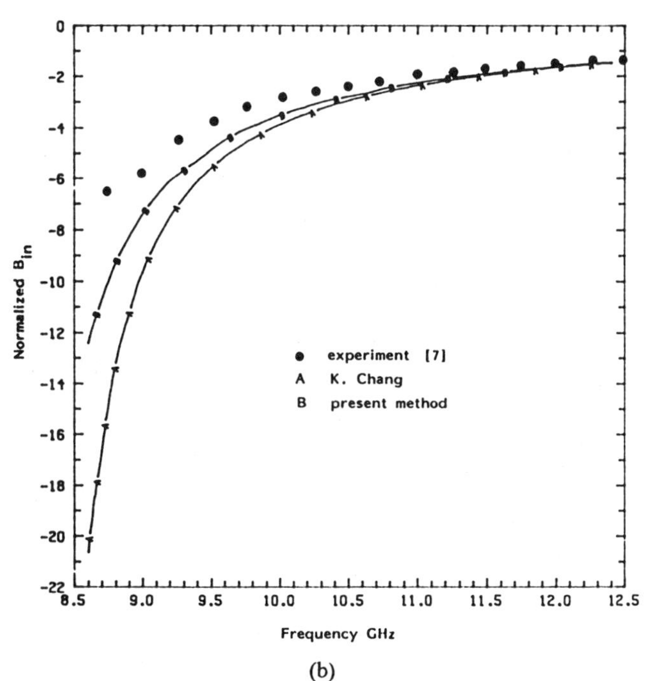

(b)

Fig. 3. Normalized susceptance of an E-plane strip in an X-band waveguide terminated with a matched load for $\epsilon_1 = \epsilon_2 = \epsilon_3 = 1$; $a = 22.86$ mm and $b = 10.16$ mm. (a) $h_1 + h_2 = 12.57$ mm, $d = 7.37$ mm, $w = 3.38$ mm. (b) $h_1 + h_2 = 11.43$ mm, $d = 9.19$ mm, $w = 1.7$ mm.

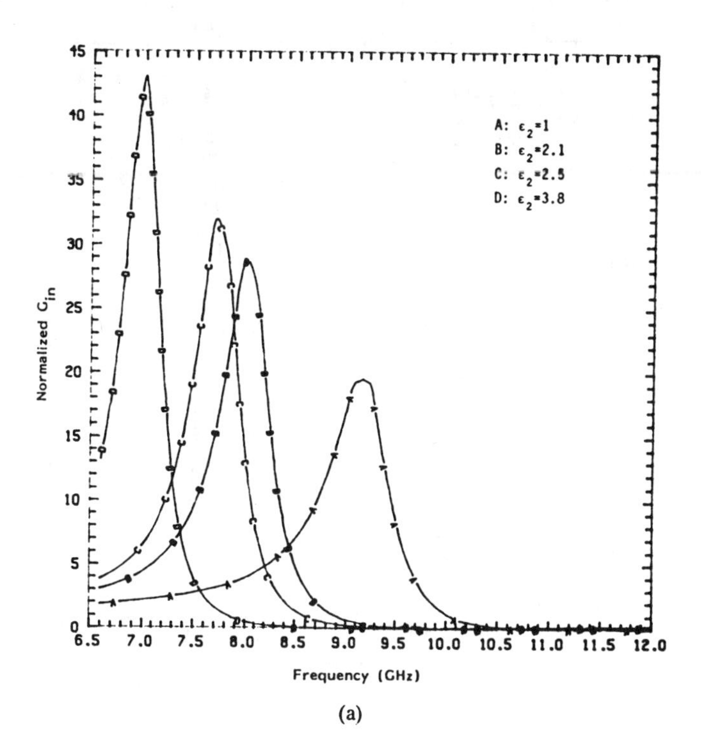

(a)

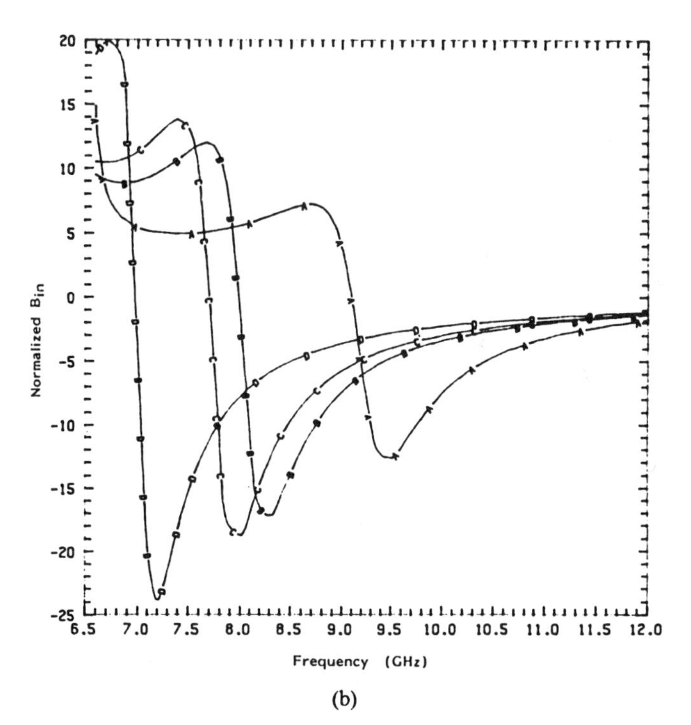

(b)

Fig. 4. Normalized admittance of an E-plane strip in an X-band waveguide terminated with a matched load versus frequency for different values of dielectric constant of the substrate. $\epsilon_1 = \epsilon_3 = 1$, $h_1 = 11.57$ mm, $h_2 = 1$ mm, $d = 7.37$ mm, and $w = 3.38$ mm. (a) Normalized conductance versus frequency. (b) Normalized susceptance versus frequency.

well with the experimental data and Chang's data. It is believed that the present method is more accurate and can be improved systematically with the use of more basis functions. Fig. 4(a) and (b) show the variations in normalized admittance versus frequency with different values of dielectric constant of the substrate in region 2. As the dielectric constant of the substrate increases, the resonant frequency at which B_{in} becomes zero decreases. This phenomenon happens because as the dielectric constant of the substrate increases, the wavelengths corresponding to each mode become shorter. The resonant frequency and the characteristics of input admittance of the considered structure can be controlled by the dielectric constant of the substrate.

Figs. 5 and 6 show the variations of normalized admittance with the height and width of the strip at different frequencies for an E-plane fin inserted in a Ka-band rectangular waveguide. Figs. 7 and 8 show the values of X_1 and X_2 versus the height d and the width w of the strip in an X-band rectangular waveguide. We note that for a

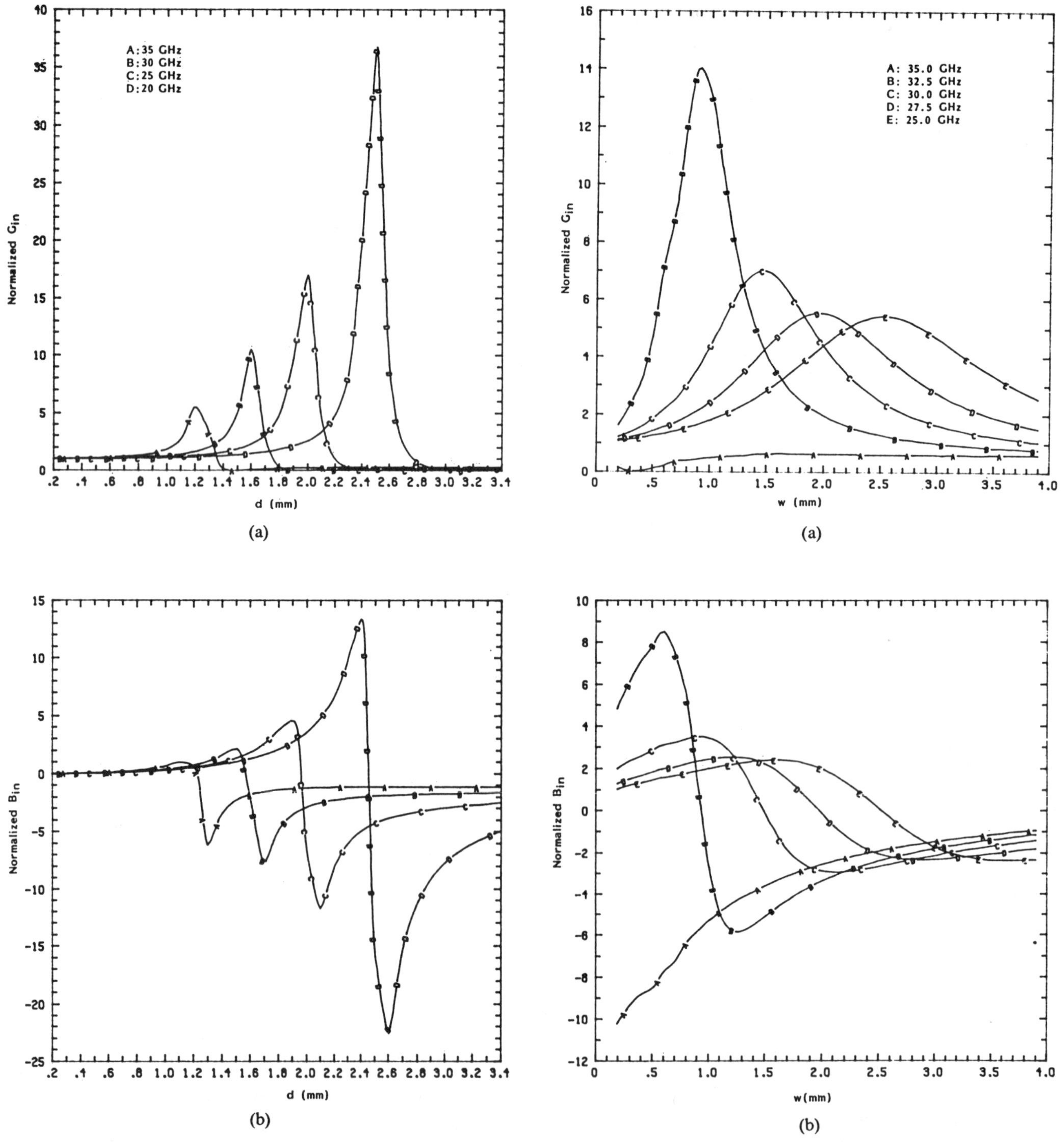

Fig. 5. Normalized admittance of an E-plane strip in a Ka-band waveguide terminated with a matched load versus strip height for different values of frequency. $\epsilon_1 = \epsilon_3 = 1$, $\epsilon_2 = 2.2$, $h_1 = 3.43$ mm, $h_2 = 0.254$ mm, $a = 7.11$ mm, $b = 3.56$ mm, $w = 1$ mm. (a) Normalized conductance versus strip height. (b) Normalized susceptance versus stri height.

Fig. 6. Normalized admittance of an E-plane strip in a Ka-band waveguide terminated with a matched load versus strip width for different values of frequency. $\epsilon_1 = \epsilon_3 = 1$, $\epsilon_2 = 2.2$, $h_1 = 3.43$ mm, $h_2 = 0.254$ mm, and $d = 1.78$ mm. (a) Normalized conductance versus strip width. (b) Normalized susceptance versus strip width.

given frequency, X_1 is not sensitive to d whereas X_2 increases with d. When w increases, X_1 increases. On the other hand, X_2 decreases with w for a higher frequency and increases in a certain region of w for lower frequency (curve A in Fig. 8(b)). Figs. 9 and 10 are the data corresponding to those in Figs. 7 and 8 except that the frequencies are in the Ka-band. Figs. 11 and 12 show X_1 and X_2 versus frequency with different values of height and width for the Ka-band waveguide. In Fig. 12, it is seen that for a narrow strip, X_1 varies slowly as the frequency is increased, while X_2 varies faster for a narrower strip than for a wider one. Fig. 13 shows the variation of normalized capacitance C and normalized inductance L with d. We note that there are two regions. In one of them (approximately corresponding to $d > w$, in this calculation $w = 1.0$ mm) C increases and L decreases as the frequency in-

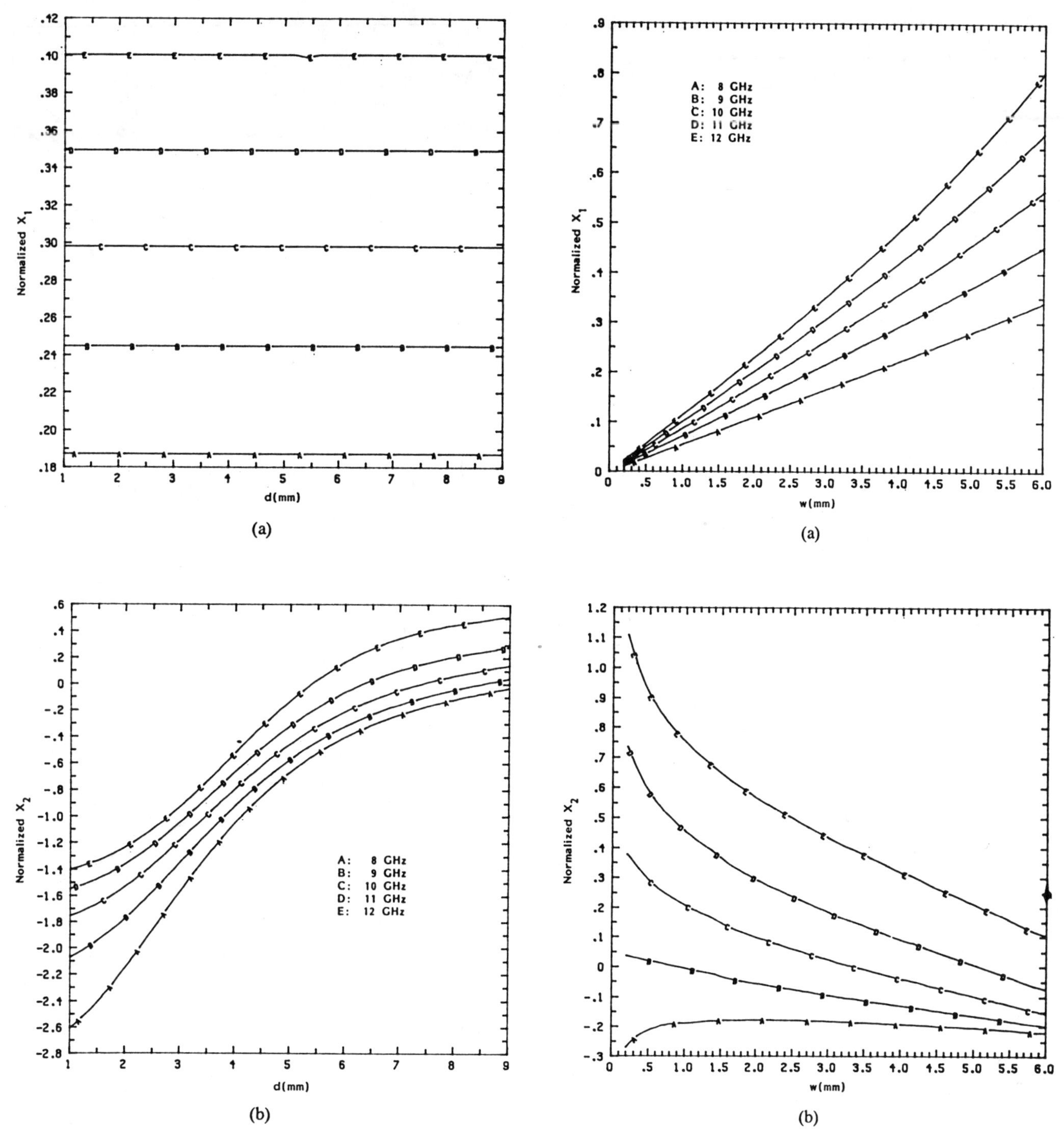

Fig. 7. Normalized reactance of equivalent circuit of an E-plane strip in an X-band waveguide versus strip height for different values of frequency. $\epsilon_1 = \epsilon_3 = 1$, $\epsilon_2 = 2.1$, $h_1 = 11.57$ mm, $h_2 = 1$ mm, and $w = 3.38$ mm. (a) Normalized X_1 versus strip height. (b) Normalized X_2 versus strip height.

Fig. 8. Normalized reactance of equivalent circuit of an E-plane strip in an X-band waveguide versus strip width for different values of frequency. $\epsilon_1 = \epsilon_3 = 1$, $\epsilon_2 = 2.1$, $h_1 = 11.57$ mm, $h_2 = 1$ mm, and $d = 7.37$ mm. (a) Normalized X_1 versus strip width. (b) Normalized X_2 versus strip width.

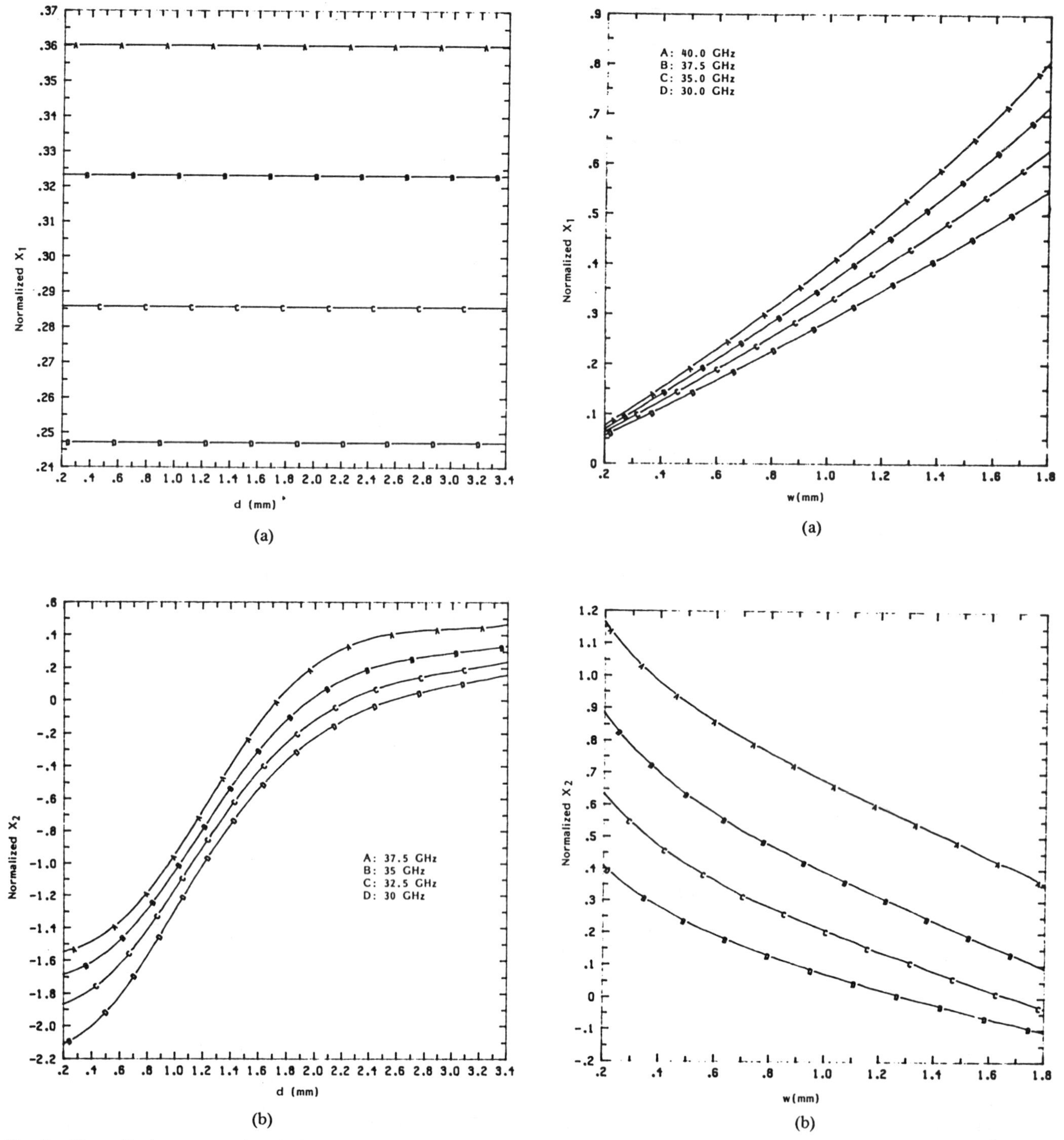

Fig. 9. Normalized reactance of equivalent circuit of an E-plane strip in a Ka-band waveguide versus strip height for different values of frequency. $\epsilon_1 = \epsilon_3 = 1$, $\epsilon_2 = 2.2$, $h_1 = 3.43$ mm, $h_2 = 0.254$ m, and $w = 1$ mm. (a) Normalized X_1 versus strip height. (b) Normalized X_2 versus strip height.

Fig. 10. Normalized reactance of equivalent circuit of an E-plane strip in a Ka-band waveguide versus strip width for different values of frequency. $\epsilon_1 = \epsilon_3 = 1$, $\epsilon_2 = 2.2$, $h_1 = 3.43$ mm, $h_2 = 0.254$ mm, and $d = 2.45$. (a) Normalized X_1 versus strip width. (b) Normalized X_2 versus strip width.

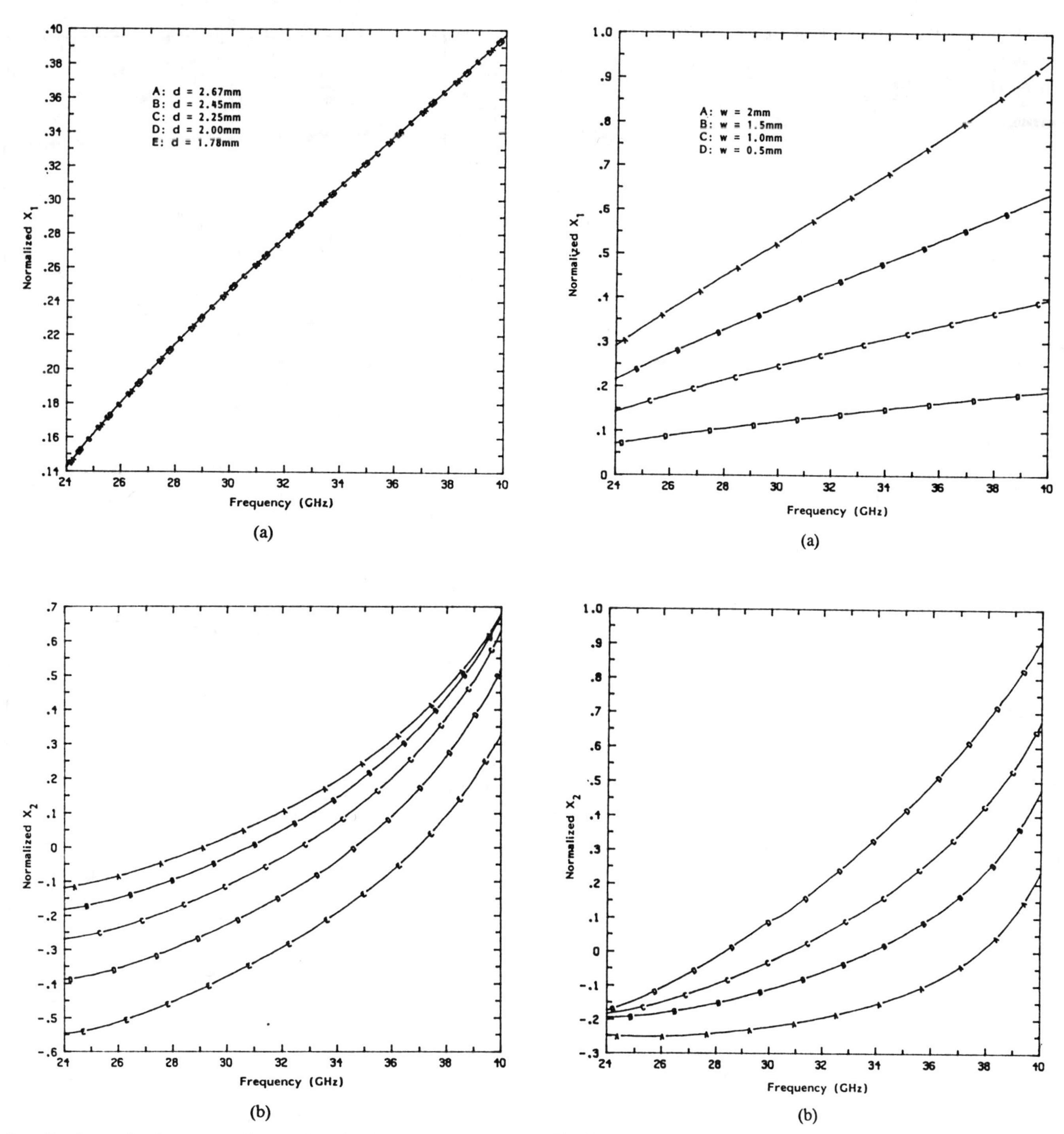

Fig. 11. Normalized reactance of equivalent circuit of an E-plane strip in a Ka-band waveguide versus frequency for different strip heights. $\epsilon_1 = \epsilon_3 = 1$, $\phi e_2 = 2.2$, $h_1 = 3.43$ mm, $h_2 = 0.254$ mm, and $w = 1$ mm. (a) Normalized X_1 versus frequency. (b) Normalized X_2 versus frequency.

Fig. 12. Normalized reactances of equivalent circuit of an E-plane strip in a Ka-band waveguide versus frequency for different strip widths. $\epsilon_1 = \epsilon_3 = 1$, $\epsilon_2 = 2.2$, $h_1 = 3.43$ mm, $h_2 = 0.254$ mm, and $d = 2.45$ mm. (a) Normalized X_1 versus frequency. (b) Normalized X_2 versus frequency.

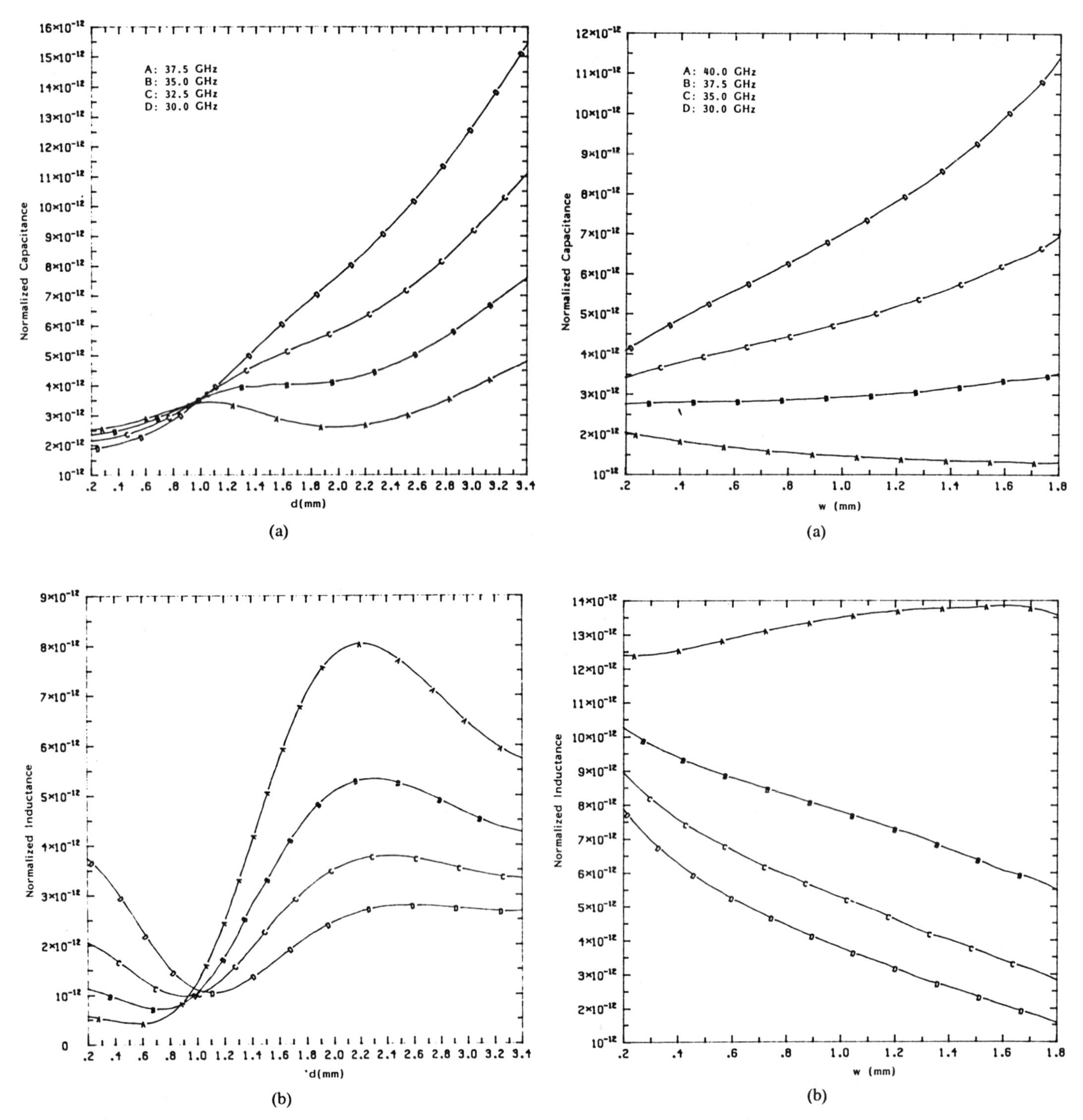

Fig. 13. Normalized equivalent circuit element values C and L of a narrow E-plane strip as a function of strip height. $\epsilon_1 = \epsilon_3 = 1$, $\epsilon_2 = 2.2$, $h_1 = 3.43$ mm, $h_2 = 0.254$ mm, and $w = 1$ mm. (a) Normalized C versus strip height. (b) Normalized L versus strip height.

Fig. 14. Normalized equivalent circuit element values C and L of a narrow E-plane strip as a function of strip width. $\epsilon_1 = \epsilon_3 = 1$, $\epsilon_2 = 2.2$, $h_1 = 3.43$ mm, $h_2 = 0.254$ mm, and $d = 2.45$ mm. (a) Normalized C versus strip width. (b) Normalized L versus strip width.

creases. In the other region, C decreases and L increases as frequency becomes higher. It is conjectured that this phenomenon is related to the field distributions and to the use of the equivalent circuit chosen here. Fig. 14 shows the variation of C and L with w for different values of frequency.

IV. Conclusions

A new analytical technique has been developed here which can be used for characterizations of the scattering phenomena of planar E-plane obstacles. The numerical results for a special case agree well with experimental and published data. The curves of normalized input admittance and equivalent circuit element values are presented for a number of different parameters of this structure. This technique is believed to be useful in the design of microwave filters and other planar circuit components.

Appendix
Closed-Form Expression of Green's Functions in the Spectral Domain

According to [9], the TM-to-X (LSM) and TE-to-X (LSE) equivalent transmission lines for the E-plane strip described in Fig. 1 can be drawn in Fig. 15. Here,

$$\gamma_i=\sqrt{\alpha_n^2+\beta^2-\epsilon_i k^2} \tag{A1}$$

$$Z_{\mathrm{TM}i}=\frac{\gamma_i}{j\omega\epsilon_0\epsilon_i} \tag{A2a}$$

$$Z_{\mathrm{TE}i}=-\frac{j\omega\mu}{\gamma_i} \qquad i=1,2,3, \quad n=0,1,2,3\cdots. \tag{A2b}$$

The driving point input impedance Z^e for the TM equivalent circuit is given by

$$Z^e=\frac{1}{y_2^e+y_{2\gamma}^e} \tag{A3a}$$

where

$$y_{2\gamma}^e=y_{\mathrm{TM}3}\cdot\coth\gamma_3 h_3 \tag{A3b}$$

$$y_2^e=y_{\mathrm{TM}2}\frac{y_{\mathrm{TM}2}+y_1^e\coth\gamma_2 h_2}{y_1^e+y_{\mathrm{TM}2}\coth\gamma_2 h_2} \tag{A3c}$$

$$y_1^e=y_{\mathrm{TM}1}\coth\gamma_1 h_1. \tag{A3d}$$

Notice that y_1^e and y_2^e are input admittances looking left at $x=h_1$ and $x=h_1+h_2$, respectively, while $y_{2\gamma}^e$ is the one looking right at $x=h_1+h_2$. Similar equations can be written for the TE equivalent circuit

$$Z^h=\frac{1}{y_2^h+y_{2\gamma}^h} \tag{A4a}$$

$$y_{2\gamma}^h=y_{\mathrm{TE}3}\coth\gamma_3 h_3 \tag{A4b}$$

$$y_1^h=y_{\mathrm{TE}1}\coth\gamma_1 h_1 \tag{A4c}$$

$$y_2^h=y_{\mathrm{TE}2}\frac{y_{\mathrm{TE}2}+y_1^h\coth\gamma_2 h_2}{y_1^h+y_{\mathrm{TE}2}\coth\gamma_2 h_2}. \tag{A4d}$$

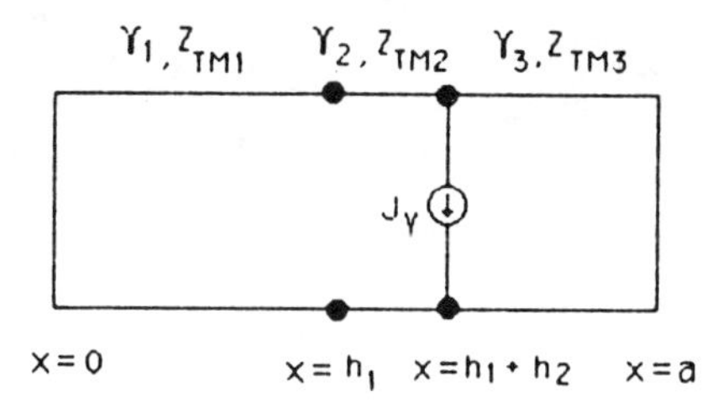

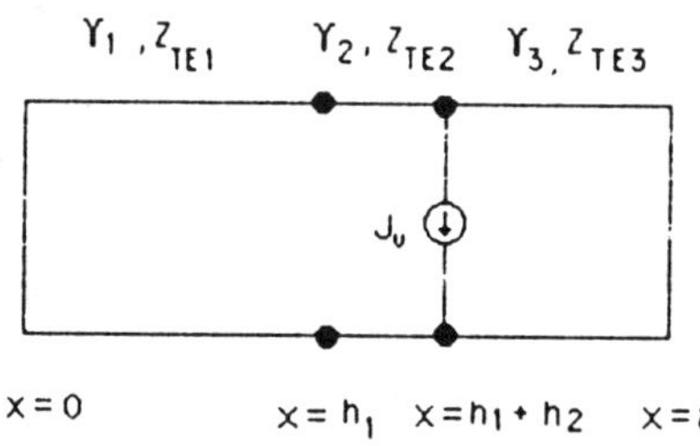

Fig. 15. The TM and TE equivalent transmission lines for the E-plant strip.

Finally, the Green's functions in the spectral domain can be represented by Z^h and Z^e as follows [8]:

$$\tilde{G}_{yy}=Z^eN_z^2+Z^hN_y^2 \tag{A5a}$$

$$\tilde{G}_{yz}=\tilde{G}_{zy}=(Z^e-Z^h)N_zN_y \tag{A5b}$$

$$\tilde{G}_{zz}=Z^hN_z^2+Z^eN_y^2 \tag{A5c}$$

where

$$N_z=\frac{\beta}{\sqrt{\alpha_n^2+\beta^2}} \tag{A6a}$$

$$N_y=\frac{\alpha_n}{\sqrt{\alpha_n^2+\beta^2}}. \tag{A6b}$$

Note that the denominators of Z^e and Z^h are the eigenvalue equations of the LSE and LSM modes.

References

[1] P. J. Meier, "Two new integrated circuit media with special advantages at millimeter wavelengths," in *IEEE-GMTT Int. Microwave Symp. Dig.*, pp. 221–223, 1972.
[2] P. J. Meier, "Integrated fin-line millimeter components," *IEEE Trans. Microwave Theory Tech.*, vol. MTT-22, pp. 1209–1216, Dec. 1974.
[3] Y. Konishi, K. Uenakada, and N. Hoshino, "The design of planar circuit mounted in waveguide and the application to low-noise 12-GHz converter," in *IEEE S-MTT Int. Microwave Symp. Dig.*, pp. 168–170, 1974.
[4] Y. Konishi and K. Uenakada, "The design of a bandpass filter with inductive strip-planar circuit mounted in waveguide," *IEEE Trans. Microwave Theory Tech.*, vol. MTT-22, pp. 869–873, Oct. 1974.
[5] Y. Konishi, "Planar circuit mounted in waveguide used as a down-converter," *IEEE Trans. Microwave Theory and Tech.*, vol. MTT-26, pp. 716–719, Oct. 1978.
[6] Y. C. Shih, T. Itoh, and L. Bui, "Computer-aided design of millimeter-wave E-plane filter," *IEEE Trans. Microwave Theory Tech.*, vol. MTT-31, pp. 135–141, Feb. 1983.
[7] K. Chang and P. J. Khan, "Equivalent circuit of a narrow axial strip in waveguide," *IEEE Trans. Microwave Theory Tech.*, vol. MTT-24, pp. 611–615, Sept. 1976.
[8] R. Collin, *Field Theory of Guided Waves.* New York: McGraw-Hill, 1960, pp. 224–232.
[9] T. Itoh, "Spectral domain immittance approach for dispersion characteristics of generalized printed transmission lines," *IEEE Trans. Microwave Theory Tech.*, vol. MTT-28, pp. 733–736, July 1980.

Part 11
Planar-Circuit Approach

THE planar-circuit approach (PCA) is not really a numerical method, but rather a mathematical formalism to model microstrip circuits so as to overcome the limitations inherent to the conventional transmission-line model. The PCA provides descriptions of microstrip components and discontinuities that are accurate enough for most practical applications, still permitting a relatively simple analytical treatment. This is in contrast with three-dimensional full wave analyses which require a high degree of mathematical and numerical processing.

The so-called planar waveguide model of microstrip circuits which was developed in Germany by Wolff's group is considered here under the general name of the PCA since it somewhat paralleled and complemented the development of the PCA.

The planar-circuit approach to the analysis and design of microstrip circuits and its theoretical basis are reviewed by Sorrentino in Paper 11.1. Planar-circuit analyses to obtain terminal descriptions of microwave planar elements are discussed and numerical methods in the context of the PCA, such as segmentation and desegmentation, are briefly described along with effective models to account for fringing field effects and radiation loss.

Research on the theory of planar circuits started in Italy and Japan independently, in the late sixties. Ridella and his coworkers developed planar-circuit models in terms of the impedance or admittance matrix formulation. A very comprehensive theory of two-dimensional networks was developed by Civalleri and Ridella [11.1]. The concept of planar circuit, however, was introduced in Japan by Okoshi and his group and was presented in the western literature in a now classic paper, which is reprinted here as Paper 11.2. In this paper a method is developed for the analysis of arbitrarily shaped planar circuits. The wave equation is converted into a boundary integral equation which is then solved by discretization. The method can be considered a boundary-element method in two dimensions, as discussed in Part 5. For circuits of simple geometrical shapes, the analytical solution in terms of Green's function is presented. The limitation of this method for microstrip circuits is that no account is taken of effects occurring at the metal edges.

At the same time, the planar waveguide model for calculating microstrip discontinuities was proposed by Wolff *et al.* (Paper 11.3). The microstrip structure is replaced by an equivalent waveguide with magnetic sidewalls, characterized in terms of an effective width and an effective permittivity. Numerical methods such as the mode-matching method can then be easily applied to the *effective* structure (see, for instance, Paper 8.4). Though rather arbitrary, this model is more accurate than a mere transmission-line model based only on the dominant mode. Further improvements of this model were obtained by adopting a frequency-dependent effective width in addition to the frequency-dependent effective permittivity [11.2], [11.3].

The concept of an effective magnetic wall model to account for fringe field effects was then extended to planar circuits by D'Inzeo *et al.* (Paper 11.4). The resonant mode expansion developed in this paper allows fringe field effects to be taken into account by using different effective models for different resonant modes. A reasonably high accuracy of the theoretical results was achieved up to fairly high frequencies. In a subsequent paper by Sorrentino and Pileri (Paper 11.5) a properly defined surface admittance at the planar-circuit boundary allowed radiation loss to be included in the model.

Specific numerical methods have been developed in the context of the PCA to extend the applicability of analytical approaches such as the Green's function method or the resonant mode expansion technique to geometries more complicated than simple rectangles or circles. The segmentation method, developed by Okoshi *et al.* (Paper 11.6), consists of subdividing the structure into elementary shapes. It has been complemented by the desegmentation method developed by Sharma and Gupta (Paper 11.7). These methods constitute an effective basis for the computer-aided design of microstrip circuits. A number of microstrip components have been designed accordingly. See, for instance, [11.4].

It may be worth mentioning that planar-circuit models can be incorporated into synthesis techniques to get very accurate and efficient designs, which are particularly needed in MMICs. One of very few such attempts is described in [11.5].

References

[11.1] P. P. Civalleri and S. Ridella, "Impedance and admittance matrices of distributed three-layer *N*-ports," *IEEE Trans. Circuit Theory*, vol. CT-17, pp. 392–398, Aug. 1970.

[11.2] I. Wolff and N. Knoppik, "Rectangular and circular microstrip disk capacitors and resonators," *IEEE Trans. Microwave Theory Tech.*, vol. MTT-22, pp. 857–864, Oct. 1974.

[11.3] G. Kompa, "*S*-matrix computation of microstrip discontinuities with a planar waveguide model," *Arch. Elek. Ubertragung.*, vol. 30, pp. 58–64, Feb. 1976.

[11.4] R. Chadha and K. C. Gupta, "Compensation of discontinuities in planar transmission lines," *IEEE Trans. Microwave Theory Tech.*, vol. MTT-30, pp. 2151–2156, Dec. 1982.

[11.5] M. Salerno and R. Sorrentino, "Planim: A new concept in the design of MIC filters," *Electron. Lett.*, vol. 22, no. 20, pp. 1054–1056, Sept. 1986.

Planar Circuits, Waveguide Models, and Segmentation Method

ROBERTO SORRENTINO, SENIOR MEMBER, IEEE

(Invited Paper)

Abstract—The planar-circuit approach to the analysis and design of microwave integrated circuits (MIC's), with specific reference to microstrip circuits, is reviewed. The planar approach overcomes the limitations inherent to the more conventional transmission-line approach. As the operating frequency is increased and/or low-impedence levels are required, in fact, the transverse dimensions of the circuit elements become comparable with the wavelength and/or the longitudinal dimensions. In such cases, one-dimensional analyses give inaccurate or even erroneous results.

The analysis of planar elements is formulated in terms of an N-port circuit and results in a generalized impedance-matrix description. Analysis techniques for simple geometries, such as the resonant mode expansion, and for more complicated planar configurations, such as the segmentation method, are discussed along with planar models for accounting for fringing fields effects and radiation loss.

I. Introduction

As MICROWAVE TECHNOLOGY evolves toward the use of higher frequencies and more sophisticated circuits and components, a considerable theoretical effort is required in order to improve the characterization and modeling of microwave structures. This is the basis for reliable computer-aided design (CAD) techniques.

In the setup of CAD techniques, one has to compromise between accuracy and simplicity. Exact analyses are often impractical because of the exceedingly high computer time required. From this viewpoint, the planar-circuit approach is a very powerful technique, which has been basically developed for the analysis of microstrip circuits, but can be extended to other microwave circuit configurations, such as reduced-height waveguide, stripline, suspended microstrip, etc.

Though the planar circuit is an approximate model of microstrip components, it constitutes a substantial improvement over conventional transmission-line models, providing accurate descriptions of their performances. On the other hand, planar-circuit models are simple enough to keep computer analyses reasonably inexpensive.

It is the scope of this paper to review the theoretical basis of the planar-circuit approach and to stress its suitability to the characterization, modeling, and design of two-dimensional microwave structures, with specific reference to microstrip circuits. This paper is not intended to provide details on planar-circuit analysis and design, which can be found in the referenced papers and overview books [1]–[3], but to illustrate the main features of the planar approach in contrast with the more conventional transmission-line approach.

The concept and definition of planar circuits are introduced in the next section, and the advantages of such an approach are briefly described. Starting from Maxwell's equations, the theoretical bases for the analysis of planar microwave components in terms of a two-dimensional circuit model are assessed in Section III. The terminal description of planar circuit is derived in Section IV; this is the basis for a brief discussion on the filtering properties and lumped-element equivalent circuits of planar elements. Once the terminal description of a single planar element has been obtained, the techniques mentioned in Section V, such as the segmentation method, can be applied to the analysis of more complicated planar configurations. The techniques for modeling a microstrip component such as a planar circuit, so as to account for effects of fringe fields and radiation loss, are discussed in Section VI. Finally, in order to describe the effects of planarity in microstrip circuits, a simple stub structure is taken as an example and its behavior illustrated in some detail in Section VII.

II. The Planar Circuit

The concept of a planar circuit was introduced by Okoshi and Miyoshi [4] as an approach to the analysis of microwave integrated circuits (MIC's). Depending on the number of dimensions which are comparable with the operating wavelength, conventional circuit elements can be classified into three categories: zero-dimensional (lumped), one-dimensional (uniform transmission lines), and three-dimensional (waveguides). The fourth category is represented by two-dimensional or planar circuits (Fig. 1). A planar circuit is defined as an electrical circuit having two dimensions comparable with the wavelength, while the third dimension is a negligible fraction of the wavelength. Strictly speaking, a distinction should be made between a microwave planar element and a planar circuit, the latter being the mathematical model, phrased in terms of voltage and current, of the former; in some instances throughout this paper, however, the two terms can be used indistinctly.

As will be shown in the next section, a two-dimensional circuit theory can be developed for planar components by extending to the two-dimensional case the concepts of

Manuscript received February 5, 1985; revised June 3, 1985.

The author is with the Department of Electronics, University of Rome La Sapienza, Rome, Italy.

Reprinted from *IEEE Trans. Microwave Theory Tech.*, vol. MTT-33, no. 10, pp. 1057–1066, Oct. 1985.

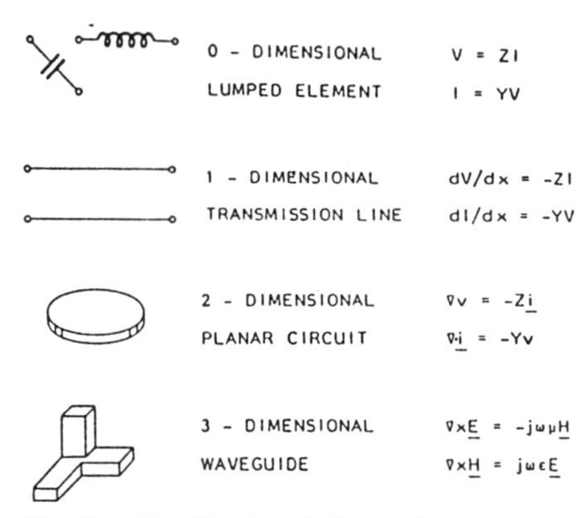

Fig. 1. Classification of electrical components.

voltage and current usually defined in transmission-line theory.

The planar approach can be used to characterize a number of MIC components, basically in stripline or microstrip configuration, which typically have one dimension, the substrate thickness, much smaller than the operating wavelength. Our attention will be focused on microstrip circuits, which presently play a major role in the area of MIC's.

With reference to a microstrip component, it should be observed that it can be only approximately considered as a planar circuit, as the electromagnetic (EM) field is not entirely confined to the substrate region but, particularly near the edges of the metallization, extends into air outside the dielectric substrate. In other words, the presence of stray fields makes the planar-circuit concept not rigorously applicable to microstrip components. Nonetheless, as discussed in Section VI, provided suitable modifications in terms of effective parameters are made, planar models provide accurate enough characterizations of microstrip circuits and components.

The planar-circuit model is intermediate between transmission-line and full-wave three-dimensional models. In some respects, it combines advantages of both approaches. On the one hand, with respect to the usual transmission-line description of microstrip circuits, the planar description is far more accurate, while, on the other hand, it is much more simple and computationally affordable than a full-wave description.

The advantages associated with the planar-circuit approach can be summarized as follows.

1) The planar-circuit approach provides accurate descriptions of microstrip components and discontinuities. As the operating frequency is increased and low-impedance values are required, the performance of microstrip circuits designed on a transmission-line basis deteriorates because of unwanted reactances associated with discontinuities. The EM field cannot any longer be assumed to have a uniform distribution in the transverse direction so that a planar approach is required to obtain accurate characterizations of the circuit performances.

2) New classes of components can be analyzed and designed using the planar-circuit approach. The wider degree of freedom of planar elements can be used to obtain specific performances and to overcome the limitation inherent to the one-dimensional approach. Several new components have been designed which utilize the planar concept, such as 3-dB hybrid circuits [5], bias filter elements [6], coupled-mode filters [7], in-phase 3-dB power dividers [8], etc.; circular polarization in microstrip antennas is obtained exciting two degenerate orthogonal modes in a planar structure [9].

3) Planar circuits are simpler to analyze than three-dimensional circuits. Although a three-dimensional full-wave analysis is the only rigorous approach to characterize microstrip circuits and components, it is too laborious and computer time consuming for most practical purposes. Planar-circuit analyses, on the contrary, require reasonably short computer times, while providing descriptions which are generally accurate enough for the needs of the microstrip circuit designer.

III. Planar-Circuit Analysis

The basic equations for the analysis of N-port planar circuits are derived in this section. The case of magnetic wall boundaries is considered, as is usually assumed for representing a microstrip or stripline component. A terminal description in terms of an impedance matrix can be derived for this type of planar circuit. The case of electrically conducting boundaries, which is representative for reduced-height waveguide, can be treated in a similar manner and described in terms of an admittance matrix [11].

Fig. 2 shows a schematic of a N-port planar element. The EM field is confined by two parallel perfectly conducting plates (top and bottom) bounded by the contour $\mathscr{C}$ and, laterally, by a cylindrical magnetic-wall surface. The excitation of the EM field inside this structure may take place either through some apertures produced at the lateral wall to couple the planar element to the external circuit (edge-fed microstrip) or by some internal current sources J_z. The latter case is normally encountered only in antenna applications, while the former is the only one usually considered in MIC applications. Both cases, however, can be formally treated in the same way; it can be easily demonstrated, in fact, that the coupling aperture produced in the magnetic wall is equivalent to an electric-current density flowing on the aperture surface.

Because of planarity (thus $\partial/\partial z = 0$) and open-circuit boundary conditions, Maxwell's equations reduce to

$$\nabla_t E_z = -j\omega\mu \hat{z} \times \boldsymbol{H}_t \tag{1}$$

$$\nabla_t \times \boldsymbol{H}_t = (j\omega\epsilon E_z + J_z)\hat{z} \tag{2}$$

where ∇_t is the two-dimensional nabla operator, $\hat{z}$ is the unit vector normal to the plane of the circuit, μ and ϵ are the permeability and permittivity of the filling substrate material. The E-field has only the z-component, while the H-field lies in the xy plane.

A two-dimensional form of telegraphists' equations can be obtained from (1) and (2) defining at each point $\boldsymbol{r}$ of the planar circuit a voltage v and a surface current density $\boldsymbol{J}_s$

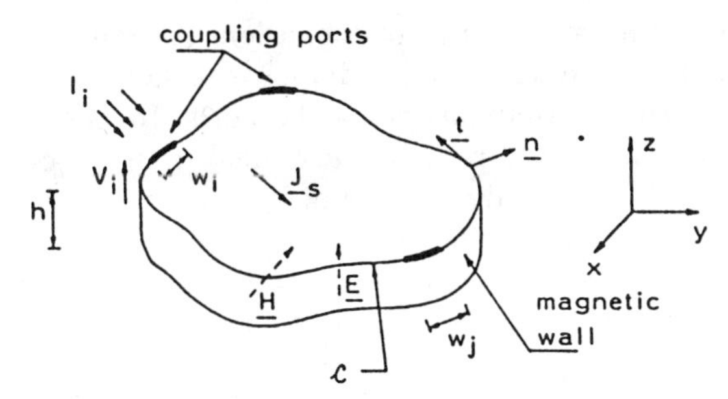

Fig. 2. Geometry of the planar circuit.

flowing on the top conductor as

$$v(r) = -hE_z \qquad V \tag{3}$$

$$J_s(r) = -\hat{z} \times H_t \qquad A/m \tag{4}$$

where h is the substrate thickness.

Note that the Poynting vector is given by

$$P = \frac{1}{2} E \times H^* = \frac{1}{2h} v J_s^* \qquad W/m^2$$

so that the quantity

$$P = \frac{1}{2} v J_s^* \qquad W/m$$

represents the linear power-density vector flowing on the planar circuit.

Inserting (3) and (4) into (1) and (2), we get

$$\nabla_t v = -j\omega\mu h J_s \tag{5}$$

$$\nabla_t \cdot J_s = -j\frac{\omega\epsilon}{h} v + J_z, \tag{6}$$

These equations represent a two-dimensional form of inhomogeneous telegraphists' equations, involving the voltage v and surface current density on the top metallic plate. The voltage wave equation is obtained taking the divergence of (5) and substituting into (6)

$$\nabla_t^2 v + k^2 v = -j\omega\mu h J_z \tag{7}$$

where

$$k^2 = \omega^2 \mu\epsilon.$$

The boundary condition associated with (7) is

$$\frac{\partial v}{\partial n} = \begin{cases} -j\omega\mu h J_s \cdot n, & \text{on } w_i,\ i = 1, 2, \cdots N \\ 0, & \text{elsewhere on } \mathscr{C} \end{cases} \tag{8}$$

w_i being the ith port of the planar circuit and n the outward directed normal to the periphery $\mathscr{C}$.

It can be demonstrated easily that the inhomogeneous boundary condition (8) can be replaced by a homogeneous boundary condition along the whole periphery $\mathscr{C}$, provided an additional equivalent current density

$$J_z = -J_s \cdot n \delta(r - r') \tag{9}$$

is assumed in (7), r' being the source location at w_i on the periphery $\mathscr{C}$.

A formal solution of (7) and (8) for the voltage v can be obtained using either the resonant-mode expansion technique [10], [11] or the Green's function approach [4]. The two approaches are substantially equivalent, since the Green's function is normally not known in closed form but is itself expressed in terms of a resonant mode expansion. In some cases, however, as discussed in Section VII, the Green's function approach can provide more compact analytical expressions leading to more efficient computer analyses. On the other hand, however, the resonant-mode technique provides a deeper physical insight, as it lends itself to a physical interpretation of the filtering properties of planar elements and is the basis for the modeling in terms of equivalent circuits.

The resonant-mode technique for planar structures with magnetic-wall boundaries can be obtained by suitably modifying the theory on field expansion in resonant cavities [12]. Let $\phi_\nu (\nu = 1, 2, \cdots)$ be the orthonormalized eigenfunctions of the following eigenvalue problem:

$$\nabla_t^2 \phi_\nu + k_\nu^2 \phi_\nu = 0, \qquad \text{in } S$$

$$\frac{\partial \phi_\nu}{\partial n} = 0, \qquad \text{on } \mathscr{C} \tag{10}$$

where S is the planar region bounded by the contour $\mathscr{C}$. The lowest eigenvalue of (10) is $k_0^2 = 0$, corresponding to the electrostatic mode.

Once the set of eigenfunctions ϕ_ν is known, the solution of (7) can be expressed as

$$v(r) = \sum_{\nu=0}^{\infty} A_\nu \phi_\nu \tag{11}$$

where

$$A_\nu = \frac{-j\omega\mu h}{k^2 - k_\nu^2} \int_S \phi_\nu J_z \, dS. \tag{12}$$

When the planar element is edge fed and no volume current sources are present inside it, because of (9), the integral over the planar surface S reduces to the integral over the ports at the periphery $\mathscr{C}$, so that (12) reduces to

$$A_\nu = \frac{-j\omega\mu h}{k^2 - k_\nu^2} \sum_{i=1}^{N} \int_{w_i} J_s \cdot (-n) \, dl. \tag{13}$$

The solution of (7) for a unit current density pulse δ located at r' gives the Green's function in terms of the set of eigenfunctions

$$G(r, r') = j\omega\mu h \sum_{\nu=0}^{\infty} \frac{\phi_\nu(r)\phi_\nu(r')}{k_\nu^2 - k^2} \tag{14}.$$

Using (14), the expression (11) for v can be replaced by

$$v(r) = \int_S G(r, r') J_z(r') \, dS \tag{15}$$

or, when (9) applies

$$v(r) = \sum_{i=1}^{N} \int_{w_i} G(r, r')(-n \cdot J_s) \, dl. \tag{16}$$

It is worth noting that the Green's function is a frequency-dependent function, while eigenfunctions ϕ_ν are not. As a consequence, the frequency dependence of v is not apparent in (15) while it results in the form of a partial fraction expansion in (11) and (12).

IV. Terminal Description

From now on we shall assume the planar element to be edge fed, as in usual microstrip circuit applications.

Voltage and surface currents on the planar circuit are excited by a linear current density $-\boldsymbol{n}\cdot\boldsymbol{J}_s$ injected through the various ports. At the ith port, these quantities can be expressed by their Fourier expansions

$$v = \sum_{m=0}^{\infty} V_i^{(m)}\sqrt{\delta_m}\cos\frac{m\pi l}{w_i} \tag{17}$$

$$-\boldsymbol{n}\cdot\boldsymbol{J}_s = \sum_{n=0}^{\infty} J_i^{(n)}\sqrt{\delta_n}\cos\frac{n\pi l}{w_i} \tag{18}$$

where l is the coordinate along the ith port ($0 \leqslant l \leqslant w_i$), w_i the port width, δ_m the Neumann delta ($\delta_m = 1$ for $m = 0$, $\delta_m = 2$ for $m \neq 0$). If the port is terminated by a transmission line having the same width w_i, $V_i^{(n)}$ and $J_i^{(n)}$ represent voltage and longitudinal current density amplitudes of the mode of nth order, $n = 0$ being the dominant TEM mode. The voltage v and the current density entering the ith port, $-\boldsymbol{n}\cdot\boldsymbol{J}_s$, can be thus represented by their Fourier expansion coefficients

$$V_i^{(m)} = \frac{\sqrt{\delta_m}}{w_i}\int_{w_i} v\cos\frac{m\pi l}{w_i}\,dl \tag{19}$$

$$J_i^{(n)} = \frac{\sqrt{\delta_n}}{w_i}\int_{w_i}(-\boldsymbol{n}\cdot\boldsymbol{J}_s)\cos\frac{n\pi l}{w_i}\,dl. \tag{20}$$

The above definitions are such that the complex power entering the circuit through the ith port is given by

$$P_i = \frac{1}{2}\int_{w_i}\boldsymbol{E}\times\boldsymbol{H}^*\cdot(-\boldsymbol{n})\,dl = \frac{1}{2}\sum_{m=0}^{\infty} V_i^{(m)} I_i^{(m)*} \tag{21}$$

where we have defined

$$I_i^{(m)} = w_i J_i^{(m)} \tag{22}$$

as the current entering the ith port associated with the mth order mode.

Inserting (11) and (13) into (19) and using (18) and (22), the relationship between voltage and currents is obtained in terms of the generalized impedance matrix of the planar circuit

$$V_i^{(m)} = \sum_{n=0}^{\infty}\sum_{j=1}^{N} Z_{ij}^{(mn)} I_j^{(n)} \tag{23}$$

where

$$Z_{ij}^{(mn)} = \frac{j\omega\mu h\sqrt{\delta_m\delta_n}}{w_i w_j}\sum_{\nu=0}^{\infty}\frac{g_{\nu i}^{(m)} g_{\nu j}^{(n)}}{k_\nu^2 - k^2} \tag{24}$$

$$g_{\nu i}^{(m)} = \int_{w_i}\phi_\nu\cos\frac{m\pi l}{w_i}\,dl \tag{25}$$

$Z_{ij}^{(mn)}$ gives the mth order voltage at the ith port due to a unit nth order current injected at the jth port, all other currents being zero. With this procedure, each physical port corresponds to an infinite number of electrical ports, relative to the spatial harmonics of voltage and current. In practice, only a few terms of expansions (17) and (18) are required to represent with good approximation the voltage and current distributions along the ports. In most cases, the width w_i of the port is much smaller than both the wavelength and the dimensions of the circuit, so that only the 0th order terms need to be retained in (17) and (18).

A formally more compact expression for the Z-elements is obtained using the Green's function

$$Z_{ij}^{(mn)} = \frac{\sqrt{\delta_m\delta_n}}{w_i w_j}\int_{w_i} dl\int_{w_j} G(\boldsymbol{r},\boldsymbol{r}')\cos\frac{m\pi l}{w_i}\cos\frac{n\pi l'}{w_j}\,dl'. \tag{26}$$

Expression (24), or (26), forms the basis for the description of the microwave planar element as an electrical circuit. Descriptions in terms of a generalized scattering matrix can be easily obtained through known formulas.

It is worth noting that the frequency dependence of the Z-parameters is not apparent in (26), as it is incorporated in the Green's function. Since coefficients (25) are frequency independent, on the contrary, the partial fraction expansion (24) explicitly shows the frequency dependence of the impedance parameters. Expression (24) is therefore useful to interpret the filtering properties of planar circuits and lends itself to the evaluation of equivalent-lumped circuits. These aspects are briefly discussed herein in the case of two-port ($N = 2$) planar elements.

As mentioned above, a notable simplification can be made when, as in many practical circuits, the ports are very narrow with respect to both the wavelength and the dimensions of the planar element. In such cases, the contribution of higher order modes ($n \geqslant 1$) at the ports can be neglected, so that voltages and current densities are assumed to be constant along the ports. For two-port elements, the Z-matrix reduces to a 2×2 matrix relating the voltages and currents of the dominant (0th order) modes at the two ports. Omitting for simplicity the indexes $m = n = 0$, (24) and (25) become

$$Z_{ij} = \frac{j\omega h}{w_i w_j \epsilon}\sum_{\nu=0}^{\infty}\frac{g_{\nu i} g_{\nu j}}{\omega_\nu^2 - \omega^2} \tag{27}$$

$$g_{\nu i} = \int_{w_i}\phi_\nu\,dl \tag{28}$$

where we have put $k_\nu^2 = \omega_\nu^2\mu\epsilon$ and $k^2 = \omega^2\mu\epsilon$. The above expressions can be used to obtain a general equivalent-lumped circuit in the form of a series connection, through ideal transformers, of anti-resonant LC cells, each cell corresponding to a resonant mode of the structure. If the planar element is symmetrical, then $|g_{\nu 1}| = |g_{\nu 2}|$, and the equivalent circuit can be put in the form of a symmetrical lattice network without any use of transformers [13].

For practical applications, only a finite number of cells are to be included in the equivalent circuit; such a number depends on the frequency range of interest and on the approximation required. In a low-frequency approximation, only the first two resonant modes can be taken into account, i.e., the static mode resonating at zero frequency and the first higher mode. When the latter is an odd mode, it can be easily shown that the equivalent circuit has the same structure as a third-order elliptic filter [14]. On this basis, low-pass filters with elliptic function responses have been designed by cascading microstrip rectangular elements [15].

The physical nature of the transmission zeros occurring in planar circuits can be easily interpreted on the basis of (27) [10]. It is well known that transmission zeros ($s_{21}=0$) in two-port networks occur in two cases: a) for $Z_{21}=0$ and b) for Z_{21} finite and Z_{11} or Z_{22} infinite. For planar elements, the latter case can occur at some specific resonant frequencies of the structure. This type of transmission zero has been called a modal transmission zero and occurs when a resonant mode 'ν' can be excited from one port ($g_{\nu 1} \neq 0$), but is uncoupled to the other port ($g_{\nu 2}=0$). The former type of transmission zero ($Z_{21}=0$), on the contrary, is due to the destructive interaction between resonant modes. While the frequency location of a modal transmission zero is determined only by the shape and dimensions of the planar element, the frequency location of an interaction transmission zero can be controlled by varying the position of the ports [6].

The same distinction between transmission zeros applies to waveguide circuits and has been used to improve selectivity in cylindrical filters [16]. A liquid-crystal field mapping technique for MIC's [17], [18] can been used to give perceptible evidence to these results [19].

V. Segmentation of Planar Elements

Numerical techniques [1], [4], [21], [22], [47], should be used for the analysis of planar elements with completely arbitrary shapes. Eigenfunctions ϕ_ν and Green's function G, in fact, are known for a limited number of simple geometrical shapes, such as rectangles, circles, circular sectors, rings, and annular sectors, etc. A number of Green's functions for such and other simple shapes are listed in [2] and [20].

The analysis of planar elements, however, can be easily extended to those geometries which result from the connection of elementary shapes. This is illustrated in a simple example. The structure of Fig. 3 can be decomposed into the cascade of two subelements (Fig. 3(b)) for which the impedance matrices $[Z_a]$ and $[Z_b]$ can be computed as discussed in the previous section. By grouping together voltages and currents at the connected ports, $[Z_a]$ and $[Z_b]$ can be put in the form

$$[Z_a]=\begin{bmatrix}[Z_{a11}] & [Z_{a12}]\\ [Z_{a21}] & [Z_{a22}]\end{bmatrix} \quad [Z_b]=\begin{bmatrix}[Z_{b11}] & [Z_{b12}]\\ [Z_{b21}] & [Z_{b22}]\end{bmatrix}. \tag{29}$$

Using the conditions imposed by the cascade connection, i.e.,

$$[V_{a2}]=[V_{b1}] \quad [I_{a2}]=-[I_{b1}] \tag{30}$$

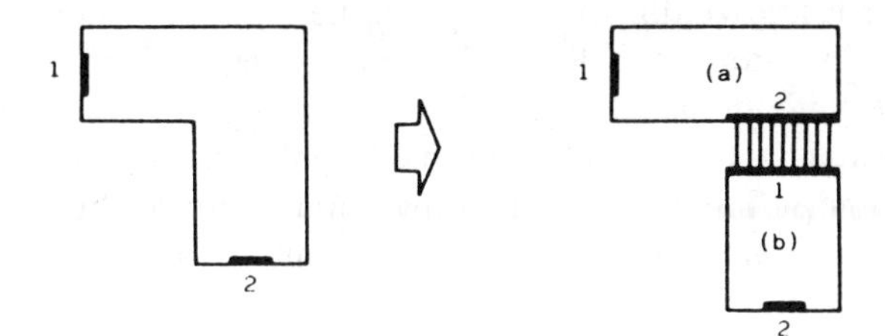

Fig. 3. Segmentation of planar elements.

the Z-matrix of the resulting network is expressed by

$$[Z]=\begin{bmatrix}[Z_{11}] & [Z_{12}]\\ [Z_{21}] & [Z_{22}]\end{bmatrix}$$

$$\begin{aligned}
[Z_{11}]&=[Z_{a11}]-[Z_{a12}][Y_{ab}][Z_{a21}]\\
[Z_{12}]&=[Z_{a12}][Y_{ab}][Z_{b12}]\\
[Z_{21}]&=[Z_{b21}][Y_{ab}][Z_{a21}]\\
[Z_{22}]&=[Z_{b22}]-[Z_{b21}][Y_{ab}][Z_{b12}]
\end{aligned} \tag{31}$$

where

$$[Y_{ab}]=([Z_{a22}]+[Z_{b11}])^{-1}. \tag{32}$$

Note that, according to the technique described in the previous section, voltage and current distributions at the interconnection between the planar elements are expressed in terms of their Fourier expansions, so that the same physical port corresponds to different (infinite, in theory) electrical ports. In practice, voltage and currents are approximated by truncated Fourier expansions, corresponding to a finite number of electrical ports.

An alternative technique is the segmentation method, in which, on the contrary, the interconnection is discretized into a finite number of ports. Voltage and current are assumed to be constant along each port, which thus corresponds to one electrical port. Voltage and current distributions along the interconnection are so approximated by stepped functions. The segmentation method has been originally formulated in terms of scattering matrix [23], but can be also implemented with higher computational efficiency in terms of impedance matrices [24].

The analysis of complicated geometries can be further extended by the desegmentation method [25], [26]. This technique can be applied to those geometries which, after addition of a simple element, can be analyzed by either the elementary or segmentation methods. In other words, the desegmentation method applies to geometries resulting from the subtraction of a simple shape from another geometry which can be analyzed by segmentation.

VI. Planar Models of Microstrip Circuits

As previously mentioned, a microstrip circuit cannot be considered as a planar circuit but only in an approximate way. Substantial discrepancies may arise due to the fields at the edges of the metallization. Nevertheless, an equivalent planar circuit can be used to model the microstrip circuit. This section will discuss how to define such an equivalent planar model.

As is known, the dynamic properties of a uniform microstripline can be calculated using an equivalent planar waveguide model [27], [28]. This is a waveguide with lateral magnetic walls, having the same height h as the substrate thickness. The width w_e and the permittivity of the filling dielectric are determined by the conditions that both the phase velocity and the characteristic impedance be the same as for the microstripline. As the dominant mode of the planar waveguide is a TEM mode, the equality of the phase velocities imposes that the filling dielectric has the same effective permittivity ϵ_e of the quasi-TEM mode of the microstripline, while the condition on the characteristic impedance imposes that

$$Z_0 = \eta_o h / \left(w_e \sqrt{\epsilon_e} \right)$$

where Z_0 is the characteristic impedance defined for the microstripline, $\eta_0 = \sqrt{\mu_0 / \epsilon_0}$ is the free-space impedance. A frequency dependence of ϵ_e and w_e can be introduced to account for dispersion on the phase velocity and characteristic impedance [28].

The planar waveguide model has been found to provide a good approximation of the cutoff frequencies of higher order modes of the microstripline. The usefulness of this model relies on the reduction of an open structure into a closed one; it therefore substantially simplifies the calculation of microstrip discontinuities. In some sense, the planar waveguide is a two-dimensional model, as it takes into account the transverse variations of the EM field. This model has found a number of applications from the analysis of the frequency-dependent properties of microstrip discontinuities [29], [30] to the design of stepped microstrip components [31], microstrip power dividers [32], and computer-aided design of microstrip filters [33].

It seems reasonable to extend the planar waveguide model to the case of two-dimensional microstrip circuits using a planar circuit with effective dimensions and an effective permittivity. The case of two-dimensional elements, however, is more complicated, as the EM field is allowed to vary along two directions. To illustrate this point with an example, let us consider the case of a two-port circular microstrip. Fig. 4(a) shows the experimental frequency behavior of the scattering parameter $|s_{21}|$. The theoretical behavior of Fig. 4(b) has been obtained by applying the mode expansion technique and by suitably choosing the effective parameters of the circular microstrip so as to optimize the agreement with Fig. 4(a).

A good agreement between theory and experiment is observed up to ~12 GHz. Above this frequency, the theory appears to be inconsistent with the experiment. Such an inconsistency can be explained by the following argument.

The effective permittivity is used to account for the electric-field lines being more or less confined to the substrate material and therefore depends on the electric-field distribution along the edge of the planar element. Considering the EM field as the superposition of the resonant modes of the structure, it is evident that a different effective permittivity should be ascribed to resonant modes having a different field distribution along the periphery of the circuit.

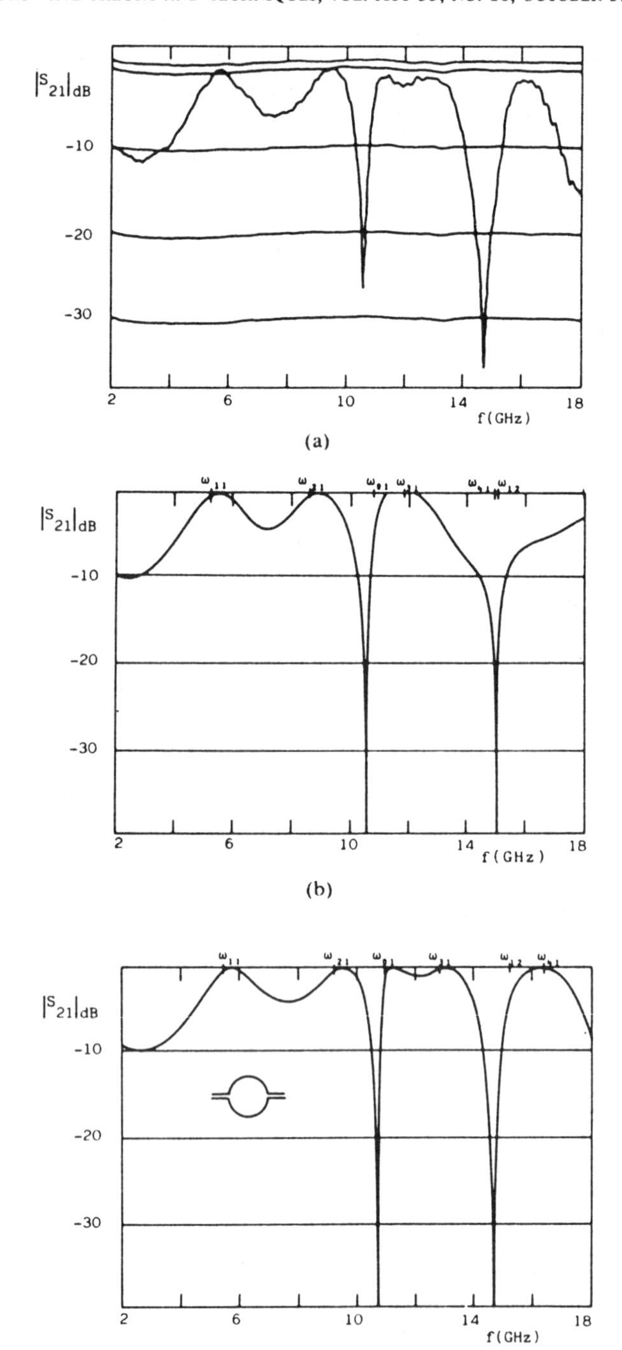

Fig. 4. Scattering parameter $|s_{21}|$ of a circular microstrip. (a) Experiment (after [10]). (b) Theoretical analysis using a single effective model. (c) Theoretical analysis using a different effective model for each resonant mode (after [10]).

Wolff and Knoppik have developed a theory [34] for computing the resonant frequencies of circular and rectangular microstrip resonators using a planar model. The theory can be extended to ring resonators [35]–[37]. This model is characterized by effective dimensions and effective permittivities which depend on the resonant mode; it can be used in conjunction with the resonant-mode technique to get the results of Fig. 4(c) [7]. The agreement with the experiment of Fig. 4(a) is quite good up to 18 GHz. The comparison between Fig. 4(b) and (c) shows that each resonant frequency undergoes a different shift; in particular, the resonant frequencies ω_{14} and ω_{21} are interchanged.

This example demonstrates that an accurate characterization of a two-dimensional microstrip circuit can be achieved through an effective planar-circuit model in conjunction with the resonant-mode technique, provided a suitable effective permittivity is used for each resonant mode. The effective permittivity accounts for the reactive energy associated with the fringing field of the corresponding mode and may produce modal inversions. Since, as previously demonstrated, the response of a planar element is tightly related to the sequence of resonant modes, the alteration of that sequence produces strong alterations of the frequency behavior of the circuit.

An additional source of discrepancy may be due to radiation loss. Differently from dielectric and conductor losses, which are associated with the field distribution inside the microstrip element, radiation loss depends on the field distribution along the periphery of the circuit. Conductor and dielectric losses do not alter substantially the circuit performance, but usually only introduce some degradations. On the contrary, radiation loss may produce effects substantially different from those predicted on the basis of a lossless theory.

A typical example is represented by nonsymmetrical structures. For a lossless reciprocal two-port network, $|s_{11}| = |s_{22}|$ at any frequency; this equality is no longer valid in the presence of losses. It has been experimentally observed in nonsymmetrical rectangular microstrips, in fact, that at some particular frequencies $|s_{11}| \cong 0$ while $|s_{22}| \cong 1$ [41]. This phenomenon occurs when a resonant mode, which strongly radiates, can be excited from the first port, but is uncoupled with the second one.

An exact theory for calculating radiation loss from microstrip structures has not been developed. The resonant-mode technique applied to planar circuits, however, can be extended to include radiation loss in an approximate way. This technique, in fact, has been extensively applied to the analysis of microstrip antennas [9], [38]–[40]. Basically, one has to account for radiation by assuming that the tangential magnetic field at the periphery of the circuit is different from zero, so that the field expansion coefficients must be modified accordingly. The general formulation presented in [41] is of impractical use; neglecting the coupling between resonant modes, which arises from the inhomogeneous boundary conditions, a simplified theory can be derived which reduces the problem to the evaluation of the complex power radiated by each unperturbed resonant mode. In spite of the approximations involved, which include neglecting surface waves, this theory was shown to accurately predict in a quantitative way the frequency behavior of the scattering parameters of rectangular and circular microstrip structures.

VII. Planar Analysis of Stub Structure

The limitations of the transmission-line (one-dimensional) approach to the analysis and design of MIC's are illustrated and discussed in this section, through the simple but typical and significant example of a stub structure.

The stub is used to provide a zero-impedance level, thus a transmission zero, at the frequency corresponding to a quarter of wavelength. Low-characteristic impedances of the stub are required for broad-band applications, so that, in microstrip circuits, this function is more conveniently realized as the parallel of two stubs. We are therefore led to the consideration of the double-stub structure of Fig. 5. From a planar point of view, this is a rectangular element $b \times l$ symmetrically connected to a main line of width w. With reference to the discussion of Section VI on fringe field effects, a different effective model should be used for each resonant mode of the structure. In order to simplify the discussion, which is aimed to point out two-dimensional effects arising even in a simplified model independently of fringe field effects, this structure will be characterized by a unique effective model, i.e., by effective dimensions w_e, b_e, l_e and effective dielectric constant ϵ_e. (This is a planar waveguide model, which is strictly valid as long as $l_e > b_e$ and higher order modes excited at the connection with the main line are rapidly decaying toward the open ends of the stub structure.)

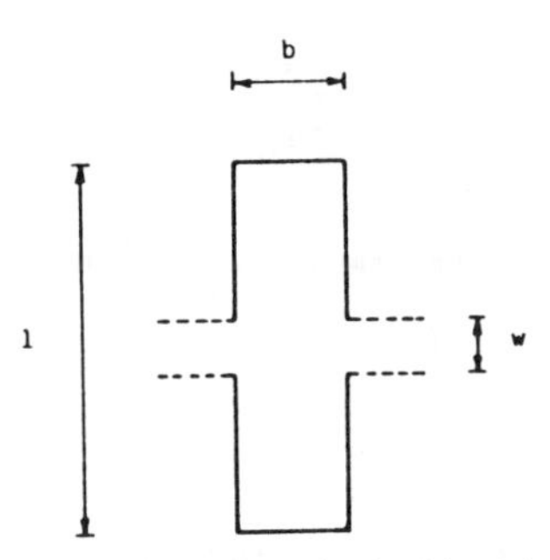

Fig. 5. Geometry of a double stub.

Depending on the dimensions and the frequency range, the rectangular structure behaves as: a) a shunt capacitor $C = \epsilon_e l_e b_e / h$ in the limit of very low frequencies, so that $w_e, b_e, l_e \ll \lambda$; b) a shunt stub of transmission line with characteristic impedance $Z_0/2 = (1/2)h\eta_0/(b_e\sqrt{\epsilon_e})$ and length $l_e/2$ as long as $w_e, b_e \ll \lambda$ and $w_e \ll l_e$; c) a planar circuit in all other cases.

The general case c), which includes a) and b) as special cases, can be treated as in Section III by evaluating the generalized impedance matrix, which accounts also for reflected and transmitted higher order modes on the main line. If these modes are evanescent and the line is long enough at both ends so that the discontinuity represented by the stub does not interact with other possible discontinuities, higher order modes have a negligible effect. In such a case, the Z-matrix computation can be reduced to the only terms relative to the dominant modes ($m = n = 0$ in (24)).

Using the resonant-mode technique, the Z-matrix is expressed in the form of a double series over indexes r and s corresponding to the resonances along l and b, respectively. More specifically, one obtains

$$Z_{11} = Z_{22} = j\omega c Z_0 \sum_{r=0}^{\infty} \sum_{s=0}^{\infty} \frac{g_{rs}^2}{\omega_{rs}^2 - \omega^2}$$

$$Z_{12} = Z_{21} = j\omega c Z_0 \sum_{r=0}^{\infty} \sum_{s=0}^{\infty} \frac{(-1)^s g_{rs}^2}{\omega_{rs}^2 - \omega^2} \tag{33}$$

with

$$c = 1/\sqrt{\mu_0 \epsilon_0 \epsilon_e}$$

$$Z_0 = \frac{h}{b_e} \eta_0 / \sqrt{\epsilon_e}$$

$$\omega_{rs} = c\sqrt{(r\pi/l_e)^2 + (s\pi/b_e)^2}$$

$$g_{rs} = \sqrt{\frac{\delta_r \delta_s}{l_e}} \operatorname{sinc}\left(\frac{r\pi w_e}{2l_e}\right) \cos\frac{r\pi}{2}. \tag{34}$$

In the numerical computation of the Z-parameters, it would be convenient to evaluate analytically the series in (33). Actually, it could be shown that either the series over r or that over s can be expressed in a closed form. This can be done by regarding the rectangular structure as a section of planar waveguide with its longitudinal axis directed along either the length l or the width b, respectively. In both cases, the Green's function is obtained as a single series involving the modes of the planar waveguide. It is found, in particular, that (33)–(34) can be replaced by

$$Z_{11} = Z_{22} = j \sum_{s=0}^{\infty} X_s \quad Z_{12} = Z_{21} = j \sum_{s=0}^{\infty} (-1)^s X_s \tag{35}$$

with

$$X_s = -Z_0 \cdot \frac{\beta_0}{\beta_s} \delta_s \left[\frac{1}{2} \operatorname{cotan}\left(\frac{\beta_s l_e}{2}\right) \operatorname{sinc}^2\left(\frac{\beta_s w_e}{2}\right) + \frac{\beta_s w_e - \sin(\beta_s w_e)}{(\beta_s w_e)^2} \right] \tag{36}$$

$$\beta_s^2 = \omega^2 \mu \epsilon_0 \epsilon_e - (s\pi/b_e)^2.$$

Trigonometric functions in (36) reduce to corresponding hyperbolic functions when the frequency and the index s are such that $\beta_s^2 < 0$. This corresponds to the sth order mode being below cutoff. Clearly, (35) and (36) are computationally much more efficient than (33) and (34). Alternative expressions can be obtained by evaluating analytically the series over s in (33) [42], [43]. In such a case, the rectangular structure is viewed as a longitudinally symmetric cascade of two step discontinuities.

Expressions (35) and (36) permit one to point out the differences between the planar approach and the transmission-line approach; they reduce to the usual expressions for the parallel of two shunt open stubs: a) retaining only the 0th order terms in (35) and b) in the limit for $w_e/\lambda \to 0$

$$\lim X_0 = \frac{Z_0}{2} \operatorname{cotan}\left(\frac{\beta_0 l_e}{2}\right).$$

It is worth noting that discrepancies between planar and transmission-line models arise not only because of the excitation of higher order modes ($s > 0$) in the stub structure, but also because of the finite width w_e of the ports. Even if the stub has a high characteristic impedance, and higher order modes can therefore be neglected, in fact, the finite width of the ports produces both a shift of the zero impedance frequency f_0 (because of the additive term appearing in (36)), and a different impedance slope (because of the coefficient of the cotangent).

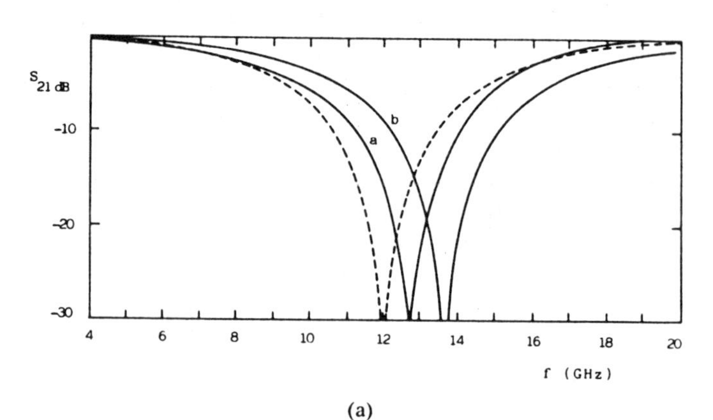

(a)

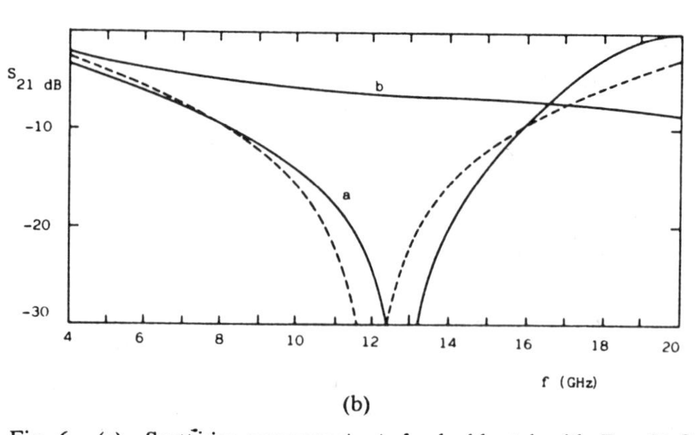

(b)

Fig. 6. (a) Scattering parameter $|s_{21}|$ of a double stub with $Z_0 = 90\ \Omega$, $f_0 = 12$ GHz: ideal response (dashed line); a planar analysis neglecting higher mode and b with higher modes. (b) Same as Fig. 6(a), but with $Z_0 = 30\ \Omega$.

These effects are illustrated in Fig. 6(a) and (b). Fig. 6(a) shows the frequency behavior of the scattering parameter $|s_{21}|$ of a double stub with $Z_0 = 90\ \Omega$ inserted on a 50-Ω line. The length of the stub has been chosen so that $(l_e - w_e) = \lambda/2$ at the frequency $f_0 = 12$ GHz. The dotted curve represents the response obtained by an ideal transmission-line model. Curve a has been computed including in (35) only the 0th order terms. A notable shift of the transmission zero frequency is observed because of the finite width of the 50-Ω line. The inclusion of higher order terms in (35), curve (b), gives rise to a further shift of the frequency f_0, which is about 13.7 GHz instead of 12 GHz.

These effects become even more marked if the stub impedance is reduced. Because of the excitation of higher order modes, the transmission zero may eventually disappear, as shown in Fig. 6(b), where the double stub impedance has been chosen as 30 Ω.

The difficulties of designing microstrip stubs with low-characteristic impedances have suggested the use of alternative structures, such as radial line stubs [44]–[46]. Linear stubs, however, can still be used, provided a planar approach is used in the design. This is demonstrated in Fig. 7, where the response of the planar structure designed is compared with that of an ideal transmission-line single stub of 15 Ω. Although the rectangular structure exhibits a somewhat more selective behavior, nevertheless the response appears to be satisfactory for practical applications. This simple example shows the wider design possibilities of the planar approach, which permits one to overcome the limitations inherent to the one-dimensional approach.

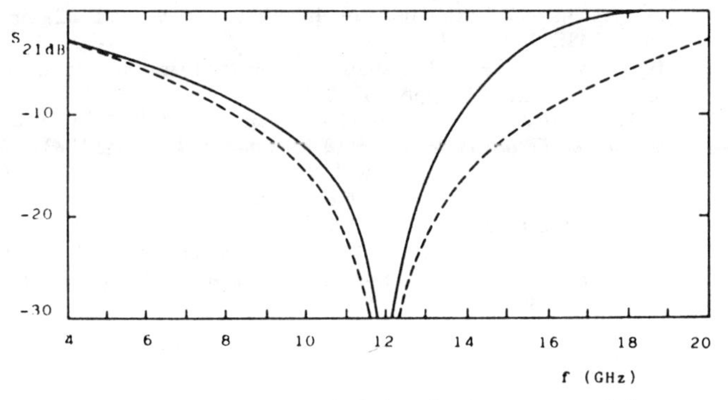

Fig. 7. Behavior of a planar stub designed to have a transmission zero at $f_0 = 12$ GHz, compared with the ideal response of a 15-Ω stub (----.)

VIII. Conclusion

An attempt has been made to review the planar-circuit concept and its theoretical basis for analysis and design of microwave planar components.

It has been stressed that the planar approach to MIC's is an approximate technique and therefore cannot be expected to provide extremely accurate results in all actual problems: to this scope, hybrid-mode full-wave techniques should be used. In conjunction with the planar models discussed in Section VI, however, accurate characterizations are obtained in most practical cases, including radiation loss, in such a way as to overcome the limitations inherent to the conventional transmission-line approach in the analysis and design of MIC's.

Acknowledgment

Since 1976, most of the work the author has done in the area of MIC's has been promoted and encouraged by Prof. F. Giannini. Prof. M. Salerno is gratefully acknowledged for helpful discussions.

References

[1] K. C. Gupta *et al.*, *Microstriplines and Slotlines*, Dedham, MA: Artech House, 1979.

[2] K. C. Gupta *et al.*, *Computer-Aided Design of Microwave Circuits*, Dedham, MA: Artech House, 1981.

[3] T. C. Edwards, *Foundations for Microstrip Circuit Design*, New York: Wiley, 1981.

[4] T. Okoshi and T. Miyoshi, "The planar circuit—An approach to microwave integrated circuitry," *IEEE Trans. Microwave Theory Tech.*, vol. MTT-20, pp. 245–252, 1972.

[5] T. Okoshi, T. Takeuchi, and J. P. Hsu, "Planar 3-dB hybrid circuit," *Electronics Comm. Japan*, vol. 58-B, pp. 80–90, 1975.

[6] G. D'Inzeo, F. Giannini, and R. Sorrentino, "Design of circular planar networks for bias filter elements in microwave integrated circuits," *Alta Freq.*, vol. 48, pp. 251e–257e, July 1979.

[7] T. Miyoshi and T. Okoshi, "Analysis of microwave planar circuits," *Electronics Comm. Japan*, vol. 55-B, pp. 24–31, 1972.

[8] K. C. Gupta, R. Chadha, and P. C. Sharma, "Two-dimensional analysis for stripline microstrip circuits," *IEEE MTT-S Int. Microwave Symp. Dig.* (Los Angeles), 1981, pp. 504–506.

[9] K. R. Carver and J. W. Mink, "Microstrip antenna technology," *IEEE Trans. Antennas Propagat.*, vol. AP-29, pp. 2–24, Jan. 1981.

[10] G. D'Inzeo, F. Giannini, C. M. Sodi, and R. Sorrentino, "Method of analysis and filtering properties of microwave planar networks," *IEEE Trans. Microwave Theory Tech.*, vol. MTT-26, pp. 462–471, July 1978.

[11] P. P. Civalleri and S. Ridella, "Impedance and admittance matrices of distributed three-layer *N*-ports," *IEEE Trans. Circuit Theory*, vol. CT-17, pp. 392–398, Aug. 1970.

[12] K. Kurokawa, *An Introduction to the Theory of Microwave Circuits*. New York: Academic Press, 1967, ch. 4.

[13] G. D'Inzeo, F. Giannini, and R. Sorrentino, "Wide-band equivalent circuits of microwave planar networks," *IEEE Trans. Microwave Theory Tech.*, vol. MTT-24, pp. 1107–1113, Oct. 1980.

[14] G. D'Inzeo, F. Giannini, and R. Sorrentino, "Novel integrated low-pass filter," *Electron. Lett.*, vol. 15, pp. 258–260, Apr. 1979.

[15] F. Giannini, M. Salerno, and R. Sorrentino, "Design of low-pass elliptic filters by means of cascaded microstrip rectangular elements," *IEEE Trans. Microwave Theory Tech.*, vol. MTT-30, pp. 1348–1353, Sept. 1982.

[16] D. E. Kreinheder and T. D. Lingren, "Improved selectivity in cylindrical TE_{011} filters by TE_{211}/TE_{311} mode control," *IEEE Trans. Microwave Theory Tech.*, vol. MTT-30, pp. 1383–1387, Sept. 1982.

[17] F. Giannini, P. Maltese, and R. Sorrentino, "Liquid crystal technique for field detection in microwave integrated circuitry," *Alta Freq.*, vol. 46, pp. 80e–88E, Apr. 1977.

[18] F. Giannini, P. Maltese, and R. Sorrentino, "Liquid crystal improved technique for thermal field measurements," *Appl. Opt.*, vol. 18, no. 17, pp. 3048–3052, Sept. 1979.

[19] G. D'Inzeo, F. Giannini, P. Maltese, and R. Sorrentino, "On the double nature of transmission zeros in microstrip structures," *Proc. IEEE*, vol. 66, pp. 800–802, July 1978.

[20] K. C. Gupta, "Two-dimensional analysis of microstrip circuits and antennae," *J. Inst. Electronics Telecomm. Engrs.*, vol. 28, pp. 346-364, July 1982.

[21] P. Silvester, "Finite element analysis of planar microwave circuits," *IEEE Trans. Microwave Theory Tech.*, vol. MTT-21, pp. 104–108, Feb. 1973.

[22] G. D'Inzeo, F. Giannini, and R. Sorrentino, "Theoretical and experimental analysis of non-uniform microstrip lines in the frequency range 2–18 GHz," in *Proc. 6th Euro. Microwave Conf.*, 1976, pp. 627–631.

[23] T. Okoshi, Y. Uehara, and T. Takeuchi, "The segmentation method —An approach to the analysis of microwave planar circuits," *IEEE Trans. Microwave Theory Tech.*, vol. MTT-24, pp. 662–668, Oct. 1976.

[24] R. Chadha and K. C. Gupta, "Segmentation method using impedance matrices for analysis of planar microwave circuits," *IEEE Trans. Microwave Theory Tech.*, vol. MTT-29, pp. 71–74, Jan. 1981.

[25] P. C. Sharma and K. C. Gupta, "Desegmentation method for analysis of two-dimensional microwave circuits," *IEEE Trans. Microwave Theory Tech.*, vol. MTT-29, pp. 1094–1097, 1981.

[26] P. C. Sharma and K. C. Gupta, "An alternative procedure for implementing the desegmentation method," *IEEE Trans. Microwave Theory Tech.*, vol. MTT-32, pp. 1–4, Jan. 1984.

[27] I. Wolff, G. Kompa, and R. Mehran, "Calculation method for microstrip discontinuities and T-junctions," *Electron. Lett.*, vol. 8, pp. 177–179, Apr. 1972.

[28] G. Kompa and R. Mehran, "Planar waveguide model for calculating microstrip components," *Electron. Lett.*, vol. 11, pp. 459–460, Sept. 1975.

[29] G. Kompa, "*S*-matrix computation of microstrip discontinuities with a planar waveguide model," *Arch. Elek. Übertragung.*, vol. 30, pp. 58–64, Feb. 1976.

[30] W. Menzel and I. Wolff, "A method for calculating the frequency-dependent properties of microstrip discontinuities," *IEEE Trans. Microwave Theory Tech.*, vol. MTT-25, pp. 107–112, Feb. 1977.

[31] G. Kompa, "Design of stepped microstrip components," *Radio Electron. Eng.*, vol. 48, pp. 53–63, Jan./Feb. 1978.

[32] W. Menzel, "Design of microstrip power dividers with simple geometry," *Electron. Lett.*, vol. 12, no. 24, pp. 639–640, Nov. 1976.

[33] R. Mehran, "Computer-aided design of microstrip filters considering dispersion, loss and discontinuity effects," *IEEE Trans. Microwave Theory Tech.*, vol. MTT-27, pp. 239–245, Mar. 1978.

[34] I. Wolff and N. Knoppik, "Rectangular and circular microstrip disk capacitors and resonators," *IEEE Trans. Microwave Theory Tech.*, vol. MTT-22, pp. 857–864, Oct. 1974.

[35] G. D'Inzeo, F. Giannini, R. Sorrentino, and J. Vrba, "Microwave planar networks: The annular structure," *Electron. Lett.*, vol. 14, no. 16, pp. 526–528, Aug. 1978.

[36] J. Vrba, "Dynamic permittivity of microstrip ring resonator," *Electron. Lett.*, vol. 15, no. 16, pp. 504–505, Aug. 1979.

[37] I. Wolff and V. K. Tripathi, "The microstrip open-ring resonator," *IEEE Trans. Microwave Theory Tech.*, vol. MTT-32, pp. 102–107, Jan. 1984.

[38] K. R. Carver, "A modal expansion theory for the microstrip antenna," in *AP-S Int. Symp. Dig.*, vol. I, June 1979, pp. 101–104.

[39] A. G. Derneryd and A. G. Lind, "Cavity model of the rectangular microstrip antenna," *IEEE Trans. Antennas Propagat.*, vol. AP-27, pp. 12-1/12-11, Oct. 1979.

[40] Y. T. Lo, D. Solomon, and W. F. Richards, "Theory and experiment on microstrip antennas," *IEEE Trans. Antennas Propagat.*, vol. AP-27, pp. 137–145, Mar. 1979.

[41] R. Sorrentino and S. Pileri, "Method of analysis of planar networks including radiation loss," *IEEE Trans. Microwave Theory Tech.*, vol. MTT-29, pp. 942–948, Sept. 1981.

[42] B. Bianco and S. Ridella, "Nonconventional transmission zeros in distributed rectangular structures," *IEEE Trans. Microwave Theory Tech.*, vol. MTT-20, pp. 297–303, May 1972.

[43] B. Bianco, M. Granara, and S. Ridella, "Filtering properties of two-dimensional lines' discontinuities," *Alta Freq.*, vol. 42, pp. 140E–148E, June 1973.

[44] J. P. Vinding, "Radial line stubs as elements in strip line circuits," in *NEREM Rec.*, 1967, pp. 108–109.

[45] A. H. Atwater, "Microstrip reactive circuit elements," *IEEE Trans. Microwave Theory Tech.*, vol. MTT-31, pp. 488–491, June 1983.

[46] F. Giannini, R. Sorrentino, and J. Vrba, "Planar circuit analysis of microstrip radial stub," *IEEE Trans. Microwave Theory Tech.*, vol. MTT-32, pp. 1652–1655, Dec. 1984.

[47] E. Tonye and H. Baudrand, "Multimode *S*-parameters of planar multiport junctions by boundary element method," *Electron. Lett.*, vol. 20, no. 19, pp. 799–802, Sept. 1984.

The Planar Circuit—An Approach to Microwave Integrated Circuitry

TAKANORI OKOSHI, MEMBER, IEEE, AND TANROKU MIYOSHI, STUDENT MEMBER, IEEE

***Abstract*—Three principal categories have been known in electrical circuitry so far. They are the lumped-constant (0-dimensional) circuit, distributed-constant (1-dimensional) circuit, and waveguide (3-dimensional) circuit. The planar circuit to be discussed in general in this paper is a circuit category that should be positioned as a 2-dimensional circuit. It is defined as an "electrical circuit having dimensions comparable to the wavelength in two directions, but much less thickness in one direction."**

The main subject of this paper is the computer analysis of an arbitrarily shaped, triplate planar circuit. It is shown that a computer analysis based upon a contour-integral solution of the wave equation offers an accurate and efficient tool in the design of the planar circuit. Results of some computer calculations are described.

It is also shown that the circuit parameters can be derived directly from Green's function of the wave equation when the shape of the circuit is relatively simple. Examples of this sort of analysis are also shown for comparison with the computer analysis.

I. Introduction

THREE PRINCIPAL categories have been known in electrical circuitry so far. They are the lumped-constant (0-dimensional) circuit, distributed-constant (1-dimensional) circuit, and waveguide (3-dimensional) circuit. The planar circuit to be discussed in general in this paper is a circuit category that should be positioned as a 2-dimensional circuit. It is defined as an "electrical circuit having dimensions comparable to the wavelength in two directions, but much less thickness in one direction."

Then three types of the planar circuit are possible. They are the triplate type, the open type, and the cavity type, as shown in Fig. 1. However, in this paper mainly the triplate-type planar circuit will be dealt with to avoid confusion.

There are three reasons that the planar circuit should be investigated in general at present [1], [2].

1) The planar circuit has wider freedom in the circuit design than the stripline circuit does. In other words, the former includes the latter as a special case. Therefore, if the design technique for the planar circuit is established in future, it will offer an exact and efficient tool in the design of microwave integrated circuits.

2) The planar circuit can offer a lower impedance level than the stripline circuit does. The recently developed microwave semiconductor devices, such as Gunn, IMPATT, or Schottky-barrier diodes, usually require a low-impedance circuitry.

Manuscript received March 10, 1971; revised September 7, 1971.
The authors are with the Department of Electronic Engineering, University of Tokyo, Tokyo, Japan.

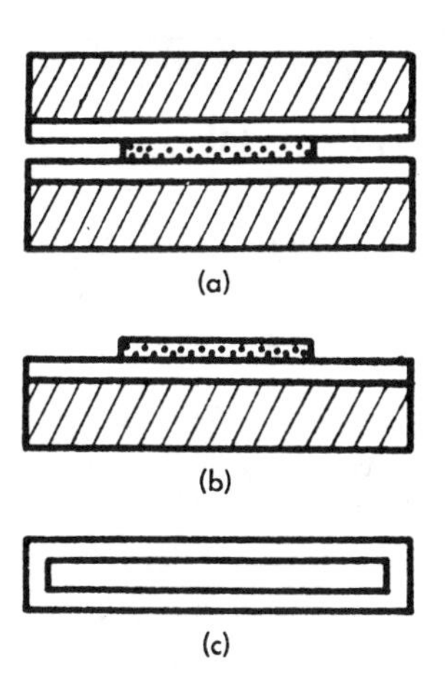

Fig. 1. Three types of the planar circuit. (a) Triplate type. (b) Open type. (c) Cavity type.

3) The planar circuit is easier to analyze and design than the waveguide circuit. By virtue of the recent progress in the computer, the analysis of an arbitrarily shaped planar circuit is within our reach if we rely on the computer.

We should note that the planar circuit is not an entirely new concept. A special case of this circuitry, the disk-shaped resonator, has been used in the stripline circulator or even as a filter [3]–[5]. The so-called "radial line" is also a special case of the planar circuit. However, to the authors' knowledge, general treatment of the planar circuit, or, in other words, the analysis of an arbitrarily shaped planar circuit, has never been presented.

The main subject of this paper is the analysis of an arbitrarily shaped, triplate planar circuit. The term "analysis" denotes here the determination of the circuit parameters of the equivalent multiport as shown in Fig. 2.

II. Basic Equations

A symmetrically excited, triplate planar circuit as shown in Fig. 2(a) will be considered throughout this paper. The model to be considered is as follows.

An arbitarily shaped, thin conductor plate is sandwiched between two ground conductors, with a spacing d from each of them. The circuit is assumed to be excited symmetrically with respect to the upper and lower ground conductors. There are several coupling ports, and their widths are denoted by W_i, W_j, $\cdots$. The rest of the periphery is assumed to be open circuited. The

Reprinted from *IEEE Trans. Microwave Theory Tech.*, vol. MTT-20, no. 4, pp. 245–252, Apr. 1972.

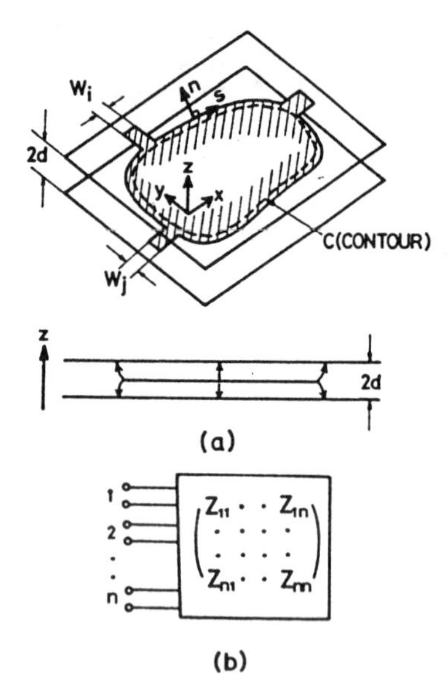

Fig. 2. (a) An arbitrarily shaped planar circuit. (b) Its equivalent multiport circuit.

xy coordinates and the z axis, respectively, are set parallel and perpendicular to the conductors.

When the spacing d is much smaller than the wavelength and the spacing material is homogeneous and isotropic, it is deduced directly from Maxwell's equation that a two-dimensional Helmholtz equation dominates the electromagnetic field in the planar circuit:

$$(\nabla_T^2 + k^2)V = 0, \qquad k^2 = \omega^2\epsilon\mu, \quad \nabla_T^2 = \frac{\partial^2}{\partial x^2} + \frac{\partial^2}{\partial y^2} \tag{1}$$

where V denotes the RF voltage of the center conductor with respect to the ground conductors; ω, ϵ, and μ are the angular frequency, permittivity, and permeability of the spacing material, respectively.[1] The network characteristics can be determined by solving (1) under given boundary conditions.

At most of the periphery where the coupling ports are absent, no current flows at the edge of the center conductor in the direction normal to the edge, because the circuit is excited symmetrically with respect to the upper and lower ground conductors.[2] Hence, the following boundary condition must hold:

$$\partial V/\partial n = 0 \tag{2}$$

where n is normal.

[1] In most of the discussions in this paper the circuit is assumed to be lossless. When we consider a small circuit dissipation, k is given, approximately, as

$$k = k' - jk'', \qquad k' \gg k'' \tag{F1}$$

where

$$k' = \omega\sqrt{\epsilon\mu}$$

$$k'' = \omega\sqrt{\epsilon\mu}\,(\tan\delta + r/d)/2 \tag{F2}$$

δ is loss angle of the spacing material, and r is skin depth.

[2] This is equivalent to assuming that the periphery is a perfect magnetic wall. Actually, however, a fringing field [see Fig. 2(a)] is always present. A simple but reasonable correction for it is to extend the periphery outwards by $2d(\log_e 2)/\pi$ to simulate the *static* fringing capacitance.

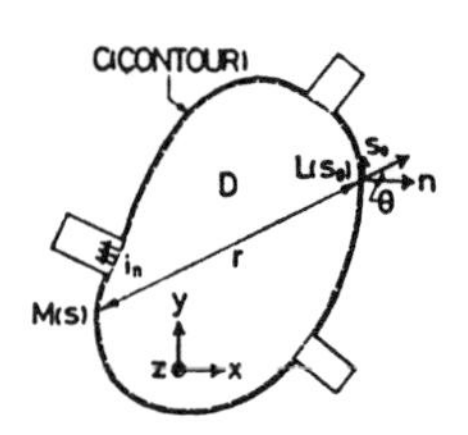

Fig. 3. Symbols used in the integral equation.

At a coupling port, (2) is no longer valid. Let the width of the port and the surface current density normal to the periphery C be denoted by W and i_n, respectively. If an admittance Y is connected to this port,

$$Y \doteqdot \frac{2\int_W i_n ds}{\int_W V ds/W} = \frac{-2jW\int_W \left(\frac{\partial V}{\partial n}\right) ds}{\omega\mu d\int_W V ds} \tag{3}$$

holds. The factor 2 expresses the fact that the current flows on both the upper and lower surfaces of the center conductor.

III. Computer Analysis

A. Integral Equation

The main feature of the planar circuit, as compared with waveguide circuit, is that we can analyze an arbitrarily shaped planar circuit within a reasonable computer time.

We consider an arbitrarily shaped, triplate planar circuit with several coupling ports, as shown in Fig. 3. Solving the wave equation over the entire area inside the contour C will require a long computer time. However, when we are concerned only with the RF voltage along the periphery, such a computation is not necessary. Using Weber's solution for cylindrical waves [6], the potential at a point upon the periphery is found to satisfy the following equation (refer to the Appendix for the detail of the derivation):

$$2jV(s) = \oint_c \{k\cos\theta H_1^{(2)}(kr)V(s_0) - j\omega\mu d\, i_n(s_0)H_0^{(2)}(kr)\}ds_0. \tag{4}$$

In this equation $H_0^{(2)}$ and $H_1^{(2)}$ are the zeroth-order and first-order Hankel functions of the second kind, respectively, i_n denotes the current density flowing outwards along the periphery, s and s_0 denote the distance along contour C. The variable r denotes distance between points M and L represented by s and s_0, respectively, and θ denotes the angle made by the straight line from point M to point L and the normal at point L, as shown in Fig. 3. When i_n is given, (4) is a second-kind Fredholm equation in terms of the RF voltage.

B. Circuit Parameters of an Equivalent N-Port [7]

For numerical calculation we divide the periphery into N incremental sections numbered as 1, 2, $\cdots$, N,

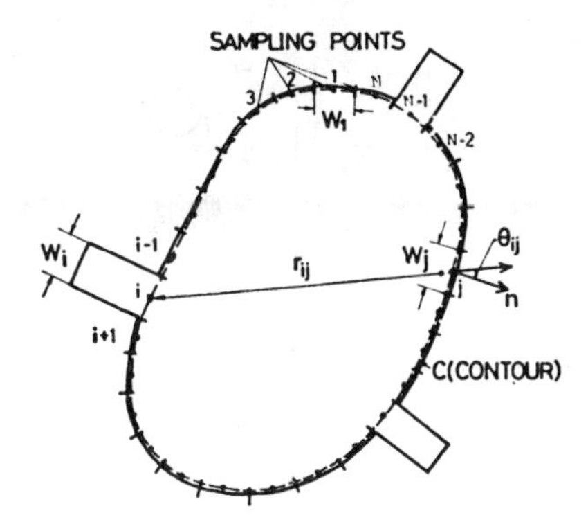

Fig. 4. Symbols used in the computer analysis.

having widths $W_1, W_2, \cdots, W_N$, respectively, as illustrated in Fig. 4. Coupling ports are assumed to occupy each one of those sections. Further, we set N sampling points at the center of each section.

When we assume that the magnetic and electric field intensities are constant over each width of those sections, the above integral equation results in a matrix equation:

$$2jV_i = \sum_{j=1}^{N} \{kV_jG_{ij} + j\omega\mu dI_jF_{ij}\} \tag{5}$$

where

$$F_{ij} = \begin{cases} \dfrac{1}{W_j}\displaystyle\int_{W_j} H_0^{(2)}(kr)ds, & (i \neq j) \\ 1 - \dfrac{2j}{\pi}\left(\log\dfrac{kW_i}{4} - 1 + \gamma\right), & (i = j) \end{cases}$$

$$G_{ij} = \begin{cases} \displaystyle\int_{W_j} \cos\theta H_1^{(2)}(kr)ds, & (i \neq j) \\ 0, & (i = j) \end{cases} \tag{6}$$

$\gamma = 0.5772 \cdots$ is Euler's constant, and $I_j = -i_n W_j$ represents the total current flowing into the jth port. The formulas for F_{ii} and G_{ii} in (6) have been derived assuming that the ith section is straight.

We can temporarily consider that all the N sections upon the periphery are coupling ports and that the planar circuit is represented by an N-port equivalent circuit. Then, from the above relations, the impedance matrix of the equivalent N-port circuit is obtained as

$$Z = U^{-1}H \tag{7}$$

where U and H denote N-by-N matrices determined by the shape of the circuit, whose components are given as

$$\begin{cases} u_{ij} = -kG_{ij}, & (i \neq j) \\ u_{ii} = 2j \end{cases} \qquad h_{ij} = j\frac{\omega\mu d}{2}F_{ij} \tag{8}$$

and U^{-1} denotes the inverse matrix to U.

In practice, most of the N ports described above are open circuited. When external admittances are connected to several of them and the rest of the ports are left open circuited, the reduced impedance matrix can be derived without difficulty.

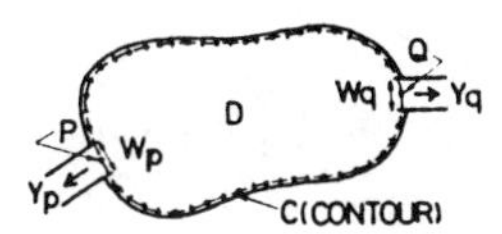

Fig. 5. Center conductor of a two-port planar circuit.

C. Transfer Parameters of a Two-Port Circuit [8]

In the case of a two-port circuit, the transfer parameters A, B, C, and D of the equivalent two-port can be given more simply as follows.

Suppose P and Q denote the driving terminal and load terminal, respectively, as shown in Fig. 5. Admittances Y_p and Y_q are connected to those terminals:

$$Y_p = 2i_n(P)W_p/V_p$$
$$Y_q = 2i_n(Q)W_q/V_q. \tag{9}$$

Then Y_p has a negative conductance component. Equation (5) can be applied to all the N sampling points. Thus the RF voltage at each point can be given by the following matrix equation:

$$[U + Y_pV + Y_qW]\begin{bmatrix} V_1 \\ \cdot \\ V_N \end{bmatrix} = 0 \tag{10}$$

where V and W are again matrices determined by the shape of the circuit:

$$V = \begin{bmatrix} 0 & \cdot & v_{1p} & \cdot & 0 \\ \cdot & \cdot & \cdot & \cdot & \cdot \\ \cdot & \cdot & \cdot & \cdot & \cdot \\ 0 & \cdot & v_{Np} & \cdot & 0 \end{bmatrix}, \qquad v_{ip} = h_{ip}$$

$$W = \begin{bmatrix} 0 & \cdot & w_{1q} & \cdot & 0 \\ \cdot & \cdot & \cdot & \cdot & \cdot \\ \cdot & \cdot & \cdot & \cdot & \cdot \\ 0 & \cdot & w_{Nq} & \cdot & 0 \end{bmatrix}, \qquad w_{iq} = h_{iq}.$$

(column p in V, column q in W)

In order that a steady field exists in the circuit, from the nontrivial condition,

$$\det[U + Y_pV + Y_qW] = 0 \tag{11}$$

must hold. This equation directly gives a bilinear relation between $-Y_p$, the driving point admittance, and Y_q, the load admittance, as

$$-Y_p = \frac{C' + D'Y_q}{A' + B'Y_q} \tag{12}$$

where A', B', C', and D' are given as the following determinants:

$$A' = \begin{vmatrix} u_{11} & \cdot & \overset{p}{v_{1p}} & \cdot & u_{1N} \\ \cdot & \cdot & \cdot & \cdot & \cdot \\ \cdot & \cdot & \cdot & \cdot & \cdot \\ u_{N1} & \cdot & v_{Np} & \cdot & u_{NN} \end{vmatrix}$$

$$B' = \begin{vmatrix} u_{11} & \cdot & \overset{p}{v_{1p}} & \cdot & \overset{q}{w_{1q}} & \cdot & u_{1N} \\ \cdot & \cdot & \cdot & \cdot & \cdot & \cdot & \cdot \\ \cdot & \cdot & \cdot & \cdot & \cdot & \cdot & \cdot \\ u_{N1} & \cdot & v_{Np} & \cdot & w_{Nq} & \cdot & u_{NN} \end{vmatrix}$$

$$C' = \det [U]$$

$$D' = \begin{vmatrix} u_{11} & \cdot & \overset{q}{w_{1q}} & \cdot & u_{1N} \\ \cdot & \cdot & \cdot & \cdot & \cdot \\ \cdot & \cdot & \cdot & \cdot & \cdot \\ u_{N1} & \cdot & w_{Nq} & \cdot & u_{NN} \end{vmatrix}.$$

Equation (12) shows that A', B', C', and D' are quantities proportional to the so-called transfer parameters A, B, C, and D of the equivalent two-port circuit. In order that the reciprocity condition ($\sqrt{AD-BC}=1$) holds, we should divide A', B', C', and D' by $\sqrt{A'D'-B'C'}$ to get A, B, C, and D, respectively, as

$$\begin{pmatrix} A & B \\ C & D \end{pmatrix} = \frac{1}{\sqrt{A'D'-B'C'}} \begin{pmatrix} A' & B' \\ C' & D' \end{pmatrix}. \tag{13}$$

When the circuit is a one-port circuit, the input admittance is given simply as (C'/A').

When the circuit has no coupling port and no circuit loss, $C'=0$ gives the proper frequency; that is, the resonant frequency of the circuit. In this situation the planar circuit is the Babinet dual of a metal wall TE-mode waveguide at its cutoff frequency.

D. Examples of Computer Analysis

In computing G_{ij} and F_{ij}, the integrals in (6) can be subdivided into as many subsections as necessary to assure the desired accuracy. However, in the following calculations the simplest approximation is used:

$$G_{ij} = \cos \theta_{ij} H_1^{(2)}(kr_{ij}) W_j \tag{14}$$

$$F_{ij} = H_0^{(2)}(kr_{ij}). \tag{15}$$

1) One-Port Disk-Shaped Circuit: As an example of the computer analysis, the input admittance of a one-port disk-shaped circuit with $\epsilon_r=2.62$, $a=1.841$ [m], $d=0.628$ [m] was computed first. (These values are not realistic ones; $a=1.841$ [m] is used so that the fundamental resonant mode is given by $k=1$ [m^{-1}].) The result is shown in Fig. 6. This figure shows the variation of the input admittance, given by (C'/A'), around the fundamental resonant frequency $f_0=1.841/2\pi a\sqrt{\epsilon\mu}$ where 1.841 is the first root of $J_1'(x)$. The parameter N denotes the number of the sampling points along the periphery.

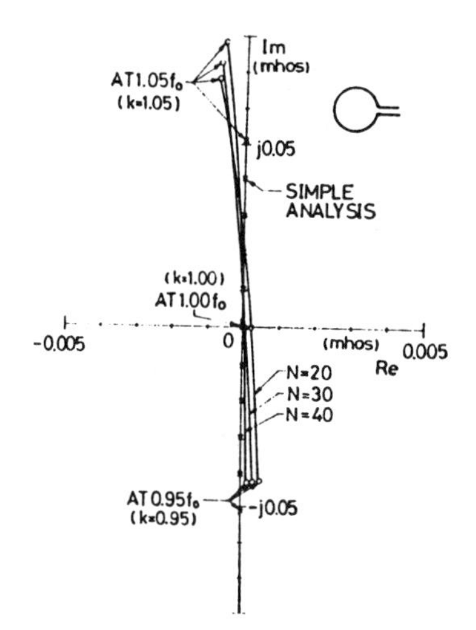

Fig. 6. Input admittance of an one-port disk-shaped circuit.

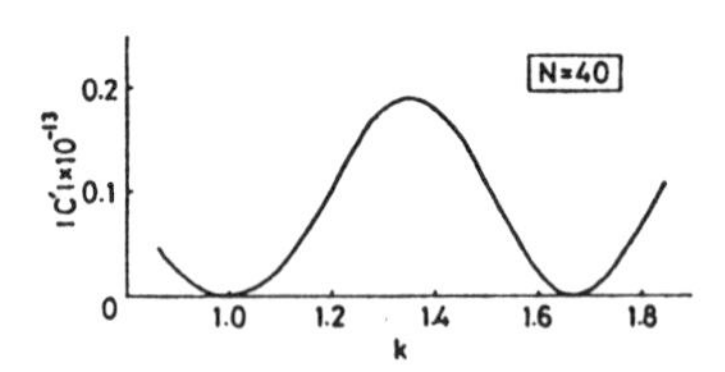

Fig. 7. The variation of $|C'|$ as a functions of k of a disk-shaped circuit.

TABLE I

COMPUTED FIRST EIGENVALUE k CORRESPONDING TO DIPOLE MODE OF A DISK-SHAPED CIRCUIT FOR VARIOUS N

Number of Sections N	Eigenvalue k
20	1.00013
30	1.00008
40	1.00007

As N increases, the real frequency locus approaches the values obtained by the simple theories as described in Section IV, shown as the small crosses along the ordinate in Fig. 6. Note that the abscissa is expanded by a factor of ten to exaggerate the computation error.

The values of k giving $C'=\det [U]=0$ corresponds to the resonant frequencies of the circuit. From the simple analyses to be described in Section IV, they should satisfy $J_m'(ka)=0$. For $a=1.841$ [m], k should then be 1.000 [m^{-1}], 1.659 [m^{-1}], and so forth. This fact gives a good check of the computation accuracy.

Since C' is complex due to the computation error and $C'=0$ is never realized for real k, we define k which gives the minimum of $|C'|$ as the eigenvalue. The variation of $|C'|$ is shown as a function of k in Fig. 7, which shows the first ($k=1.00$) and the second ($k=1.66$) minima. The former corresponds to the fundamental dipole mode (the first root of $J_1'(ka)=0$) and the latter to the quadrupole mode (the first root of $J_2'(ka)=0$). Table I shows the former k obtained for various N. As N increases k tends toward unity.

2) Two-Port Disk-Shaped Circuit: Next the transfer parameters A, B, C, and D of a disk-shaped circuit

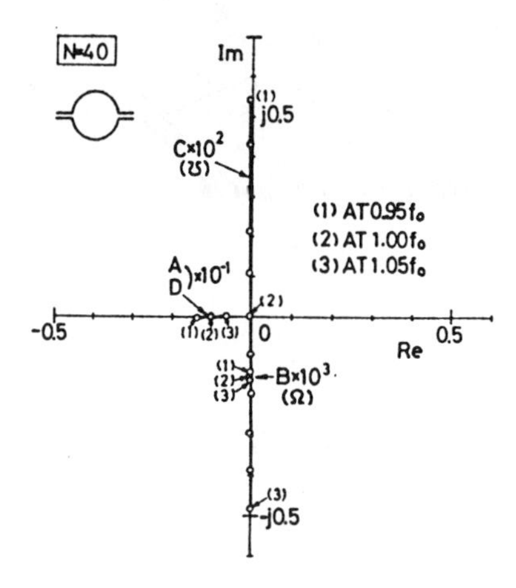

Fig. 8. Transfer parameters of a two-port disk-shaped resonator.

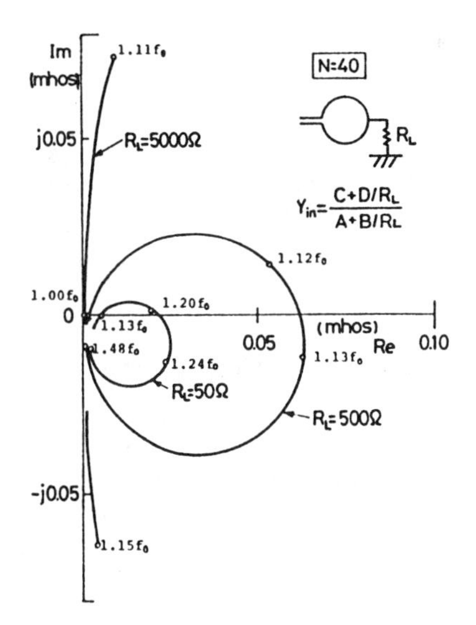

Fig. 9. Input admittance of a two-port disk-shaped resonator loaded by various load resistances R_L.

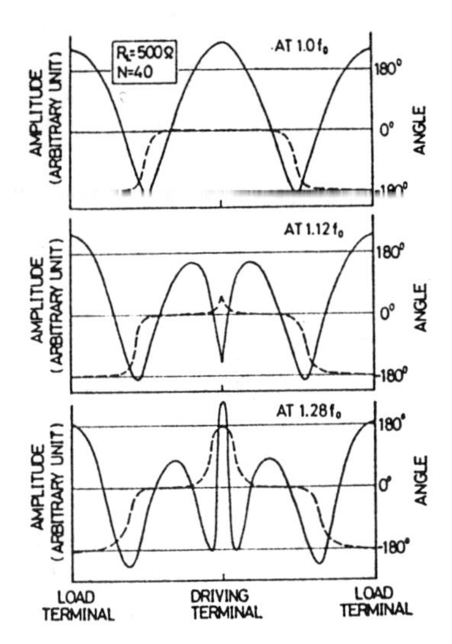

Fig. 10. The RF voltage distribution, magnitude (solid curve), and phase (broken curve), along the periphery of a disk-shaped circuit for $R_L=500\ \Omega$ and $N=40$, for various frequencies.

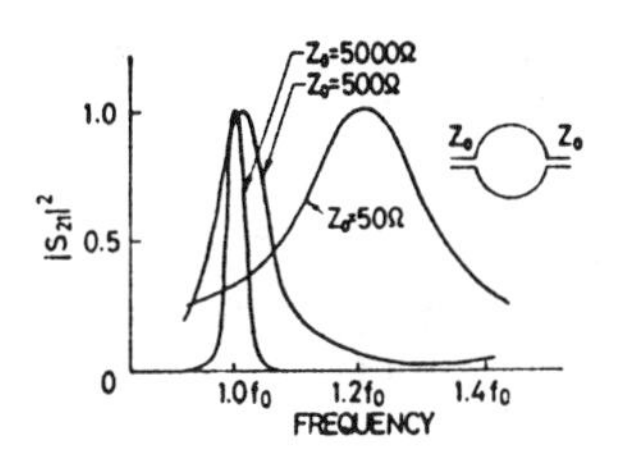

Fig. 11. The power transmission of a disk-shaped circuit calculated numerically for various characteristic impedances.

having two-ports on its opposite sides were computed, and the result is shown in Fig. 8. In this figure the abscissa gives the real part and the ordinate the imaginary part of the transfer parameters obtained in the case $N=40$. Parameters A and D are equal to each other, as the circuit is symmetrical, and they take -1.0 at the resonant frequency.

By using the obtained transfer parameters and the relation $Y_{in}=(C+D/R_L)/(A+B/R_L)$, the input admittance of the disk-shaped circuit loaded by a pure resistance R_L was computed as shown in Fig. 9. The curves show the computer calculations for load resistances of 50, 500, and 5000 Ω. The loci in this figure cover a frequency range from the dipole mode of resonance to the quadrupole mode of resonance. In between these two low-admittance, parallel resonant points we find the frequency where the input admittance is very high, that is, a series resonant point, on the right-hand sides of loci. Note that such frequencies can never be found except by computer analysis.

Fig. 10 shows the RF voltage distribution along the periphery for $R_L=500\ \Omega$ and $N=40$, for various frequencies. In this figure, both ends and the center of the abscissa correspond to the load terminal and the driving terminal, respectively. The solid and broken curves show the magnitude (arbitrary scale) and phase of the RF voltage along the periphery, respectively. It is found that as the frequency increases, the distribution of the RF voltage changes from a dipole mode to a quadrupole mode. At the frequency of $1.12f_0$, the RF voltage at the input port is minimized; this corresponds to the series resonance of the circuit.

Fig. 11 shows the power transmission calculated numerically by using the relation $S_{21}=2Z_0/(AZ_0+B+CZ_0^2+DZ_0)$ for the case when the characteristic impedances Z_0 of the input and output lines are equal to each other and are pure resistance. It is found that both the frequency giving the maximum power transmission and the transmission bandwidth increase as the line impedance Z_0 is lowered.

3) Arbitrarily Shaped Circuit: As an example of more irregularly shaped circuit, a planar circuit, as shown in Fig. 12, was studied. Fig. 13 shows the frequency loci of the input admittance for $R_L=500\ \Omega$ and $N=32$. In this figure f_0 denotes the resonant frequency of the quadrupole mode of a regular square circuit having dimensions $2a$ by $2a$. In this figure parallel resonances are found at $0.54f_0$ and $0.86f_0$, and a series resonance at $0.64f_0$.

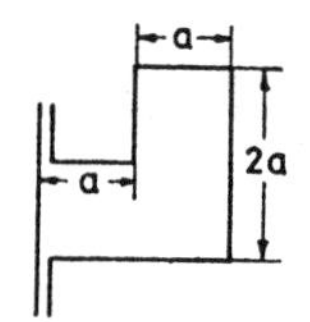

Fig. 12. Center conductor of an irregularly shaped circuit.

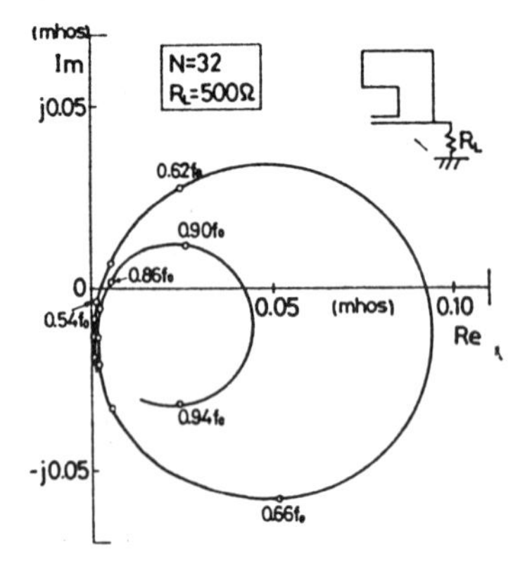

Fig. 13. The frequency locus of the input admittance for $R_L = 500\ \Omega$ and $N = 30$.

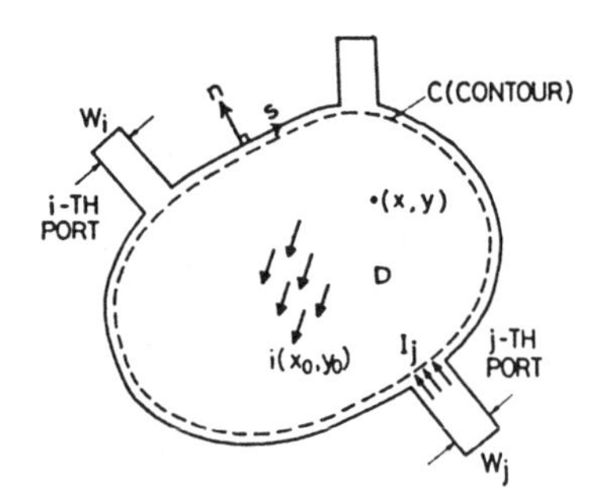

Fig. 14. Symbols used in the Green's function analysis.

IV. Analysis Based Upon Green's Function

When the shape of the circuit is relatively simple (a disk, for example) and we can get the Green's function of the wave equation analytically, the equivalent circuit parameters can be derived directly from the Green's function as follows.

We introduce the Green's function G of the second kind, having a dimension of impedance which satisfies

$$V(x, y) = \iint_D G(x, y \mid x_0, y_0) i(x_0, y_0) dx_0 dy_0 \tag{16}$$

inside the contour C shown in Fig. 14, and an open boundary condition

$$\partial G / \partial n = 0 \tag{17}$$

along C. In (16), $i(x_0, y_0)$ denotes an assumed (fictitious) RF current density injected normally into the circuit (see Fig. 14).

In a real planar circuit, current is injected from the periphery where a coupling port is present. Hence the RF voltage at a point upon the periphery is given by a

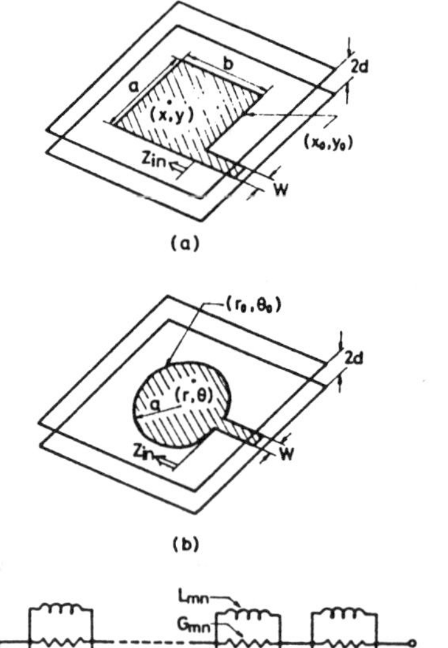

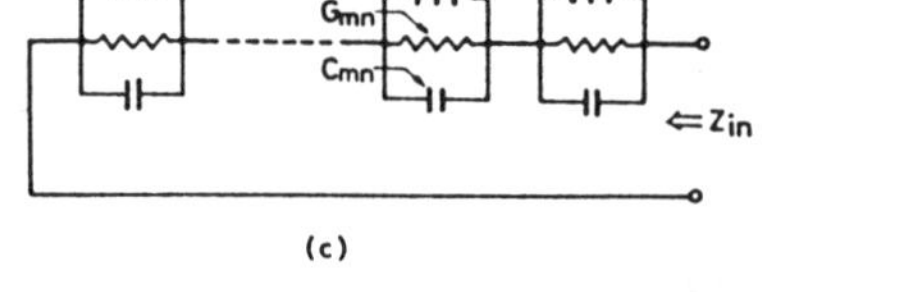

(c)

Fig. 15. (a) One-port square resonator. (b) One-port disk-shaped resonator. (c) The equivalent circuit describing the input admittance of one-port resonator.

line integral

$$V(s) = -\oint_c G(s \mid s_0) i_n(s_0) ds_0 \tag{18}$$

where s and s_0 are used to denote distance along C, and i_n is the line current density normal to C at coupling ports. Since i_n is present only at coupling ports, the RF voltage at the ith port is given approximately as

$$V_i \doteqdot \sum_j I_j \frac{1}{2W_i W_j} \int_{W_i} \int_{W_j} G(s \mid s_0) ds_0 ds \tag{19}$$

where $I_j = -2\int_{W_j} i_n(s_0) ds_0$ represents the current flowing into the jth port on both the upper and lower surfaces. Hence, the elements of the impedance matrix of the equivalent N-port circuit are

$$z_{ij} = \frac{1}{2W_i W_j} \int_{W_i} \int_{W_j} G(s \mid s_0) ds_0 ds. \tag{20}$$

As examples of the analysis based upon Green's function, the input impedances of one port disk and square circuits as shown in Fig. 15 are calculated.

A. Square Circuit

For a square circuit pattern [see Fig. 15(a)] having $a \times b$, the Green's function is given as [9]

$$G(x, y \mid x_0, y_0) = j\omega\mu d \frac{4}{ab} \sum_n \sum_m \frac{\cos(k_x x_0) \cos(k_y y_0)}{k_x^2 + k_y^2 - k^2} \cdot \cos(k_x x) \cos(k_y y) \tag{21}$$

where $k_x = m\pi/a$ and $k_y = n\pi/b$.

We compute the input impedance [Z_{in} shown in Fig. 15(a)] of a one-port square circuit having the

TABLE II

THE PROPER FUNCTION AND EQUIVALENT CIRCUIT PARAMETERS OF THE TRIPLATE-TYPE, SQUARE, AND DISK CIRCUITS

Planar Resonator	Square Resonator [Fig. 15(a)]	Disk Resonator [Fig. 15(b)]
Proper function	$\cos(k_x x)\cos(k_y y)$	$J_m(k_{mn}r)\cos(m\theta)$
Resonant frequency f_{mn}	$\frac{\sqrt{(m/a)^2+(n/b)^2}}{2\sqrt{\epsilon\mu}}$	$\frac{k_{mn}}{2\pi\sqrt{\epsilon\mu}}$
C_{mn}	$\frac{\epsilon ab}{2d}\frac{1}{F}$	$\epsilon\frac{\pi a^2}{d}\{1-m^2/(ak_{mn})^2\}\frac{1}{F}$
L_{mn}	$\frac{2\mu d}{ab\{(m\pi/a)^2+(n\pi/b)^2\}}F$	$\frac{\mu d}{\pi}\frac{1}{(ak_{mn})^2-m^2}F$
G_{mn}	$2\pi f_{mn}C_{mn}/Q_0$	$2\pi f_{mn}C_{mn}/Q_0$
F	$\left(\frac{\sin(k_x W)}{k_x W}\right)^2$	$\left(\frac{\sin(mW/a)}{mW/a}\right)^2$
Q_0	$Q_0^{-1}=Q_d^{-1}+Q_c^{-1}$; $Q_d=1/\tan\delta$ (δ is the loss angle of the dielectrics); $Q_c=d/r$ (r is the skin depth of the conductor)	

coupling port at one of the corners as shown in Fig. 15(a). Equations (20) and (21) directly give

$$Z_{\text{in}} = \sum_n \sum_m \frac{j\omega\mu d(\sin(k_x W)/k_x W)^2}{ab(k_x^2 + k_y^2 - k^2)}. \tag{22}$$

When we use (F1) and (F2) to consider the circuit loss, we obtain, after some computations,

$$Z_{\text{in}} = \sum_n \sum_m \frac{1}{\left(j\omega C_{mn} - j\dfrac{1}{\omega L_{mn}} + G_{mn}\right)} \tag{23}$$

where C_{mn}, L_{mn}, and G_{mn} are the equivalent circuit parameters corresponding to each mode, and are tabulated in Table II.[3] Equation (23) suggests that the equivalent circuit describing the input impedance is given [see Fig. 15(c)] as a series connection of many parallel resonating circuits representing each resonance.

B. Disk Circuit

The disk circuit is shown in Fig. 15(b). The Green's function is given as

$$G(r, \theta \mid r_0, \theta_0) = \sum_n \sum_m \frac{2j\omega\mu d J_m(k_{mn}a)J_m(k_{mn}r)\cos(m(\theta-\theta_0))}{(k_{mn}^2 - k^2)a^2(1 - m^2/a^2k_{mn}^2)J_m^2(k_{mn}a)} \tag{24}$$

where k_{mn} satisfies

$$\frac{\partial}{\partial r}J_m(k_{mn}r)\Big|_{r=a} = 0 \quad (n\text{th root}). \tag{25}$$

[3] When we are concerned only with the circuit performance near a single resonant frequency, we can also derive the equivalent circuit parameters from the resonant frequency, stored energy, and the unloaded Q factor. The parameters thus obtained agree with those shown in Table II, except for the factor F describing the effect of the width of the terminal. This sort of analysis is fairly common in microwave circuit analyses. For example, one of the reviewers of this paper called the authors' attention to [10].

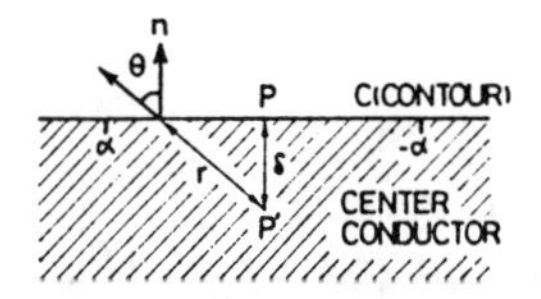

Fig. 16. Symbols used in the derivation of (4).

The equivalent circuit parameters of a one-port disk-shaped circuit can be computed by using (20), and are tabulated in Table II.

C. Multiport Disk and Square Circuits

The Green's function analysis can be applied to the circuit of this sort, which is useful in practical integrated circuitry as filters or hybrids. However, those examples are omitted for space limitations and will be reported elsewhere.

V. CONCLUSION

What is emphasized is that we can analyze an arbitrarily shaped planar circuit within a reasonable computer time. The *design* of a planar circuit, based upon the high-speed computer analysis and the trial-and-error principle, will also be possible within several years.[4]

Among possible applications of the planar circuit, the applications in Gunn and IMPATT oscillators seem to be promising. Since they are oscillation devices having relatively low impedances, the oscillator performance can be improved by using the planar circuit instead of the stripline circuitry.

APPENDIX

DERIVATION OF EQUATION (4)

The RF voltage at a piont P' *inside* the periphery satisfies the following Weber's solution for cylindrical waves [6],

$$4jV(P') = \oint_c \left\{H_0^{(2)}(kr)\frac{\partial V(Q)}{\partial n} - V(Q)\frac{\partial H_0^{(2)}(kr)}{\partial n}\right\} ds. \tag{A1}$$

To obtain the RF voltage of the point P just *upon* the periphery, a little algebra is required. We first define a point P' just inside the point P as shown in Fig. 16, where we assume that $\delta \ll \alpha \ll k^{-1}$. By using the following approximations of the Hankel function near the origin

$$H_0^{(2)}(kr) \doteqdot -\frac{2j}{\pi}\log\frac{k\sqrt{s^2+\delta^2}}{2}$$

$$\frac{\partial H_0^{(2)}(kr)}{\partial n} \doteqdot -\frac{2j}{\pi}\frac{\delta}{s^2+\delta^2}$$

[4] For example, the time required to obtain the entire data in Fig. 8 is about 100 s using a typical Japanese high-speed computer HITAC-5020E. The improvement in the speed by a factor of (1/10) may be needed for the design.

we can rewrite

$$4jV(P') = \int_{-\alpha}^{\alpha}\left\{-\frac{2j}{\pi}\log\frac{k\sqrt{s^2+\delta^2}}{2}\frac{\partial V}{\partial n}+\frac{2j}{\pi}\frac{\delta}{s^2+\delta^2}V\right\}ds + \int_{\Gamma}\left\{H_0^{(2)}(kr)\frac{\partial V}{\partial n}-V\frac{\partial H_0^{(2)}(kr)}{\partial n}\right\}ds \tag{A2}$$

where Γ denotes the contour excluding the section $-\alpha\sim+\alpha$. If V and $\partial V/\partial n$ vary slowly in the minute section between $-\alpha$ and α, the integrals in (A2) become

$$-\frac{2j}{\pi}\int_{-\alpha}^{\alpha}\log\frac{k\sqrt{s^2+\delta^2}}{2}\frac{\partial V}{\partial n}ds = -\frac{4j}{\pi}\frac{\partial V}{\partial n}\left\{\alpha\left(\log\frac{k\sqrt{\alpha^2+\delta^2}}{2}-1\right)-\frac{k\delta}{2}\left(\operatorname{arc\,cosec}\frac{\sqrt{\alpha^2+\delta^2}}{\delta}-\frac{\pi}{2}\right)\right\} \tag{A3}$$

$$\frac{2j}{\pi}\int_{-\alpha}^{\alpha}\frac{s}{s^2+\delta^2}V\,ds = \frac{2j}{\pi}V\tan^{-1}\frac{\alpha}{\delta}. \tag{A4}$$

When P' approaches P, and hence δ tends to zero, (A2), (A3), (A4) give

$$4jV(P') = -\frac{4j}{\pi}\frac{\partial V}{\partial n}\left\{\alpha\left(\log\frac{k\alpha}{2}-1\right)\right\}+2jV(P) + \int_{\Gamma}\left\{H_0^{(2)}(kr)\frac{\partial V}{\partial n}-V\frac{\partial H_0^{(2)}(kr)}{\partial n}\right\}ds. \tag{A5}$$

Next, as α tends to zero, the first term in the right-hand side of (A5) vanishes, and hence

$$2jV(P) = \oint_c\left\{H_0^{(2)}(kr)\frac{\partial V}{\partial n}-V\frac{\partial H_0^{(2)}(kr)}{\partial n}\right\}ds. \tag{A6}$$

This equation and the relations

$$\frac{\partial V}{\partial n} = -j\omega\mu d\, i_n$$

$$\frac{\partial H_0^{(2)}(kr)}{\partial n} = -k\cos\theta H_1^{(2)}(kr)$$

give (4) in the text.

Acknowledgment

The authors wish to thank M. Hashimoto of Osaka University, Osaka, Japan, for stimulating discussions.

References

[1] T. Okoshi, "The planar circuit," in *Rec. of Professional Groups, IECEJ*, Paper SSD68-37/CT68-47, Feb. 17, 1969.
[2] ——, "The planar circuit," *J. IECEJ*, vol. 52, no. 11, pp. 1430–1433, Nov. 1969.
[3] S. Mao, S. Jones, and G. D. Vendelin, "Millimeter-wave integrated circuits," *IEEE Trans. Microwave Theory Tech. (Special Issue on Microwave Integrated Circuits)*, vol. MTT-16, pp. 455–461, July 1968.
[4] Y. Tajima and I. Kuru, "An integrated Gunn oscillator," in *Rec. of Professional Groups, IECEJ*, Paper MW70-9, June 26, 1970.
[5] H. Bosma, "On stripline Y-circulation at UHF," *IEEE Trans. Microwave Theory Tech. (1963 Symposium Issue)*, vol. MTT-12, pp. 61–72, Jan. 1964.
[6] J. A. Stratton, *Electromagnetic Theory*. New York: McGraw-Hill, 1941, p. 460.
[7] T. Okoshi and T. Miyoshi, "The planar circuit—An approach to microwave IC," in *Proc. 1971 European Microwave Conf.*, Paper C4, Aug. 1971.
[8] —— "The planar circuit—A novel approach to microwave circuitry," in *Proc. Kyoto Int. Conf. on Circuit and System Theory*, Paper B-5-1, Sept. 1970.
[9] P. M. Morse and H. Feshbach, *Method of Theoretical Physics*, pt. II. New York: McGraw-Hill, 1953, p. 1360.
[10] S. B. Cohn, P. M. Sherk, J. K. Shimizu, and E. M. T. Jones, "Final report on strip transmission lines and components," Stanford Res. Institute, Contract DA36-0393SC-63232, Final Rep., pp. 79–162.

CALCULATION METHOD FOR MICROSTRIP DISCONTINUITIES AND T JUNCTIONS

Indexing terms: Stripline components, Waveguide junctions, Modelling

A method for calculating microstrip discontinuities and T junctions is described, and a waveguide model for the microstrip line is defined. With the help of this model and the use of an orthogonal series expansion, a solution of the above problems is found. Numerical results for the scattering matrices of both the discontinuities and the junctions are given.

Calculation methods for stripline discontinuities and junctions have been described by Oliner,[1] Altschuler and Oliner,[2] Franco and Oliner[3] and Campell,[4] using Babinet's principle and the well known solutions for equivalent problems given by Marcuvitz.[5] Leighton and Milnes[6] published a method for calculating T junctions in microstrip techniques, expanding the theory given by Altschuler and Oliner[2] to the case of the microstrip problem.

As is known, there are well tested methods for calculating discontinuities and junctions in the waveguide techniques. These methods use orthogonal-series expansions of the fields in the waveguide (e.g. see References 7–11). A necessary condition for applying orthogonal-series-expansion methods is that a complete set of field solutions for the problem considered must be known. A further condition should be that

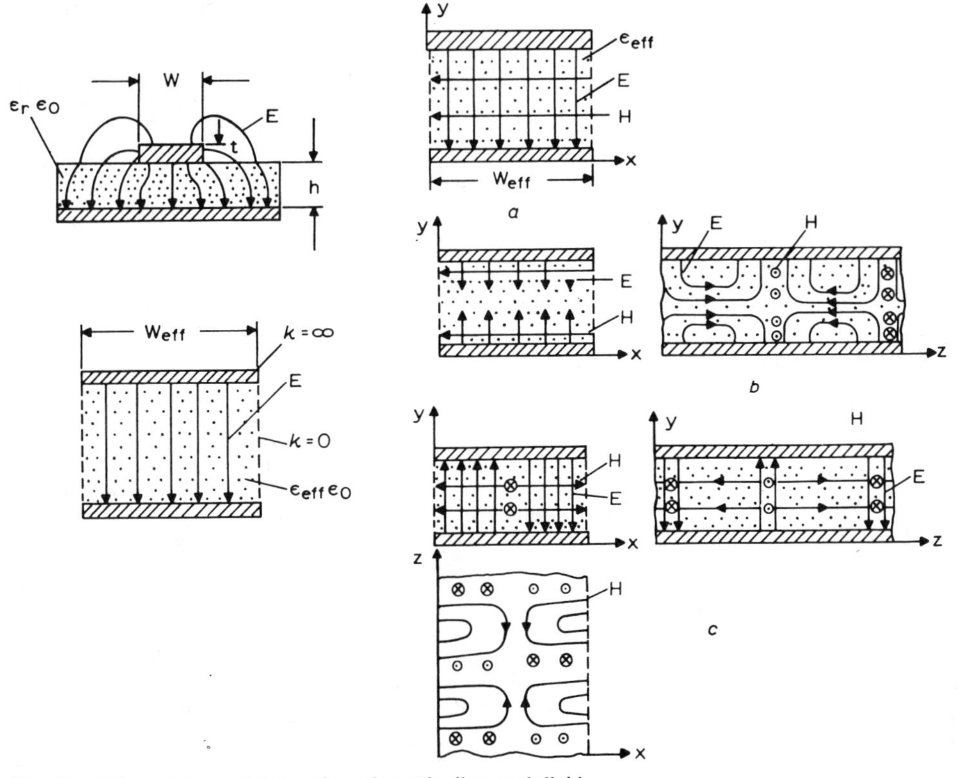

Fig. 1 *Waveguide model for the microstrip line and field distribution of the lowest-order modes of the model*

$W_{eff} = f(w, h, t)$
$\epsilon_{eff} = g(w, h, t, \epsilon_r)$
a TEM mode
b E_{10} mode
c H_{10} mode

Reprinted with permission from *Electron. Lett.*, vol. 8, no. 7, pp. 177–179, Apr. 6, 1972.

the solutions are orthogonal. As is well known, no complete set of solutions for the field problem of the microstrip line has yet been published.

For this reason, we took the waveguide model given by Wheeler[12, 13] for the lowest-order mode, which, to a first approximation, is a TEM mode. Wheeler showed that the behaviour of the TEM mode on the microstrip line can be described by a parallel-plate waveguide of width W_{eff} and relative permittivity ε_{eff}:

$$\left.\begin{aligned} W_{eff} &= \frac{h}{Z_w}\sqrt{\left(\frac{\mu_0}{\varepsilon_{eff}\,\varepsilon_0}\right)} \\ \varepsilon_{eff} &= \left(\frac{\lambda_0}{\lambda_g}\right)^2 \end{aligned}\right\} \quad . \; . \; . \; . \; . \; . \quad (1)$$

where Z_w is the characteristic impedance defined by Wheeler[12] and λ_g is the wavelength on the microstrip line. Leighton and

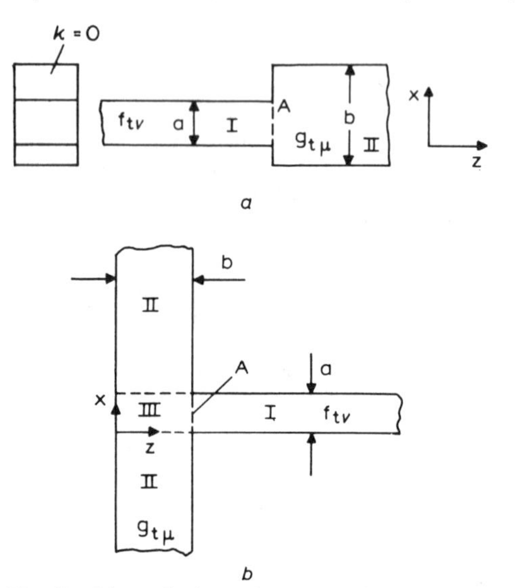

Fig. 2 *Discontinuity and T junction in microstrip technique*
a Discontinuity
b *T* junction

Milnes[6] also use this waveguide model, but they apply Babinet's principle to make use of a solved problem for an equivalent T junction.[5] In contrast to all solutions of the above problem that have been published, we make direct use of the waveguide model of the microstrip line described above to compute the energy stored in the discontinuities and T junctions. This is done by making the assumption that the higher-order modes of the above waveguide model describe, to a first approximation, the physical fields on the microstrip line. This assumption may, at first sight, look very arbitrary, but for solving an eigenvalue problem, an arbitrary, infinite and complete set of solutions, which satisfy the boundary conditions, can be taken. Then, if we take the waveguide model for the microstrip line which is exact only for the

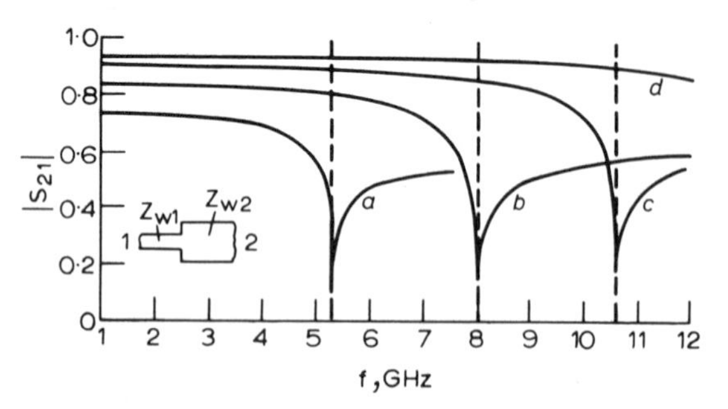

Fig. 3 *Numerical results for the transmission coefficients of a symmetric microstrip discontinuity on a polyguide substrate material (ε_r = 2·33, h = 1·5 mm)*
Change of the characteristic impedance from Z_{w1} = 50 Ω to (*a*) Z_{w2} = 10 Ω, (*b*) Z_{w2} = 15 Ω, (*c*) Z_{w2} = 20 Ω and (*d*) Z_{w2} = 25 Ω

TEM mode, and take into account the higher-order modes of this model, we shall obtain better results than those obtained by methods which do not consider any higher-order modes on the microstrip line. Fig. 1 shows the waveguide model and the lowest-order modes on this waveguide.

Fig. 2 shows the discontinuity and the T junction which has been calculated. Using the waveguide model described, we can find an orthogonal and complete set of field solutions $f_{t\nu}$ in the left-hand guide with effective width a (Fig. 2a) and a complete set $g_{t\mu}$ in the waveguide with effective width b. To compute the scattering matrix of the discontinuity, the transversal magnetic-field strength of the left-hand part of the structure is developed into a series expansion of the functions $g_{t\mu}$, and the transversal electric-field strength of the right-hand part is developed into a series expansion of the functions $f_{t\nu}$, as follows:

$$\left.\begin{aligned} \boldsymbol{H}_{tI} &= \sum_{\mu=1}^{\infty} \frac{B_\mu}{Z_\mu}(\boldsymbol{e}_z \times \boldsymbol{g}_{t\mu}) \\ \boldsymbol{E}_{tII} &= \sum_{\nu=1}^{\infty} A_\nu \boldsymbol{f}_{t\nu} \end{aligned}\right\} \quad . \; . \; . \; . \; . \; . \; . \; . \quad (2)$$

The coefficients A_ν and B_μ can be computed so that the boundary conditions in the discontinuity are met. From the amplitudes of the lowest-order TEM mode, the scattering matrix of the discontinuity can be derived. Fig. 3 shows the calculated transmission coefficients for a symmetric discontinuity with the characteristic impedances Z_{w1} = 50 Ω and (*a*) Z_{w2} = 10 Ω, (*b*) Z_{w2} = 15Ω, (*c*) Z_{w2} = 20 Ω and (*d*) Z_{w2} = 25 Ω. The curves are computed for microstrip lines on a polyguide substrate material with a relative permittivity ε_r = 2·33 and a height h = 0·625 mm. As can

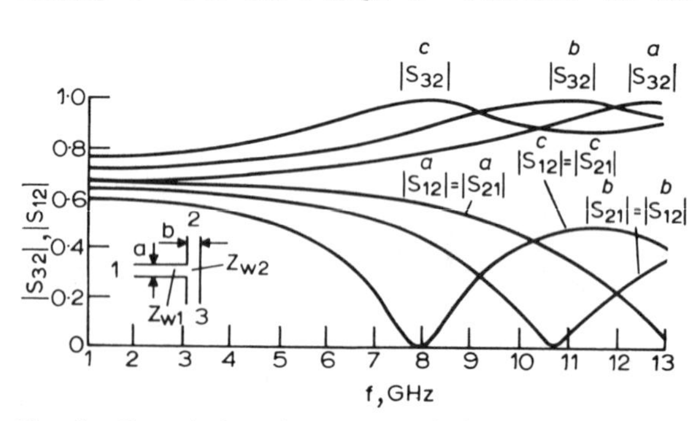

Fig. 4 *Numerical results for transmission coefficients of a microstrip T junction on polyguide substrate material (ε_r = 2·33, h = 1·5 mm)*
Z_{w1} = 50 Ω
a Z_{w2} = 50 Ω
b Z_{w2} = 40 Ω
c Z_{w2} = 30 Ω

be seen from Fig. 3, the scattering matrix depends strongly on the frequency for those substrate materials, especially if the characteristic impedance Z_{w2} is small. As calculations for aluminium substrate material show, the frequency dependence is negligible at frequencies up to 20 GHz.

If the T junction (Fig. 2*b*) is calculated in the same way, it must be noticed that the field solutions $g_{t\mu}$ of region II do not satisfy the boundary conditions in region III, as has been shown by Lewin[10] for a waveguide problem. Therefore the series expansion which has been used to calculate the discontinuity must be substituted by an integral representation, as shown in References 10 and 11. This means that, for the H_x component of the magnetic-field strength in regions II and III, a relationship

$$H_x = \int_{-\infty}^{+\infty} f(\alpha z)\frac{B}{\beta z_w}\sqrt{\left(\frac{2}{bh}\right)}\sin(\alpha z)\,dz \quad . \; . \quad (3)$$

must be used. The function $f(\alpha z)$ can be chosen so that the boundary conditions between region I and region III are satisfied and the tangential magnetic-field strength vanishes on the magnetic walls of region II ($z = 0$, $z = b$). From the amplitudes of the TEM mode in both microstrip lines, the scattering matrix of the T junction can be evaluated. Fig. 4 shows the calculated transmission coefficients of the T junction as functions of the frequency. For the T junction also, the frequency dependence is great if substrate material with a small relative permittivity is used (Fig. 4, $\varepsilon_r = 2{\cdot}33$). The frequency dependence once more is small for aluminium substrate material.

Measurements show that the agreement between our theory and experiments is good. For polyguide substrate material, a large influence of radiation losses on the scattering matrix can be measured. For aluminium substrate material, this influence is small.[6] The radiation loss can be calculated by a first approximation (e.g. see Reference 15), and taken into account in our theory. More detailed results of the theory and the experimental work will be published shortly.[14]

I. WOLFF
G. KOMPA
R. MEHRAN

Institut für Hochfrequenztechnik
Technische Hochschule Aachen
W. Germany

10th March 1972

References

1 OLINER, A. A.: 'Equivalent circuits for discontinuities in balanced strip transmission line', *IRE Trans.*, 1955, **MTT–3,** pp. 134–143

2 ALTSCHULER, H. M., and OLINER, A. A.: 'Discontinuities in the center conductor of symmetric strip transmission line', *ibid.*, 1960, **MTT–8,** pp. 328–338

3 FRANCO, A. G., and OLINER, A. A.: 'Symmetric strip transmission line tee junction', *ibid.*, 1962, **MTT–10,** pp. 118–124

4 CAMPBELL, J. J.: 'Application of the solution of certain boundary value problems to the symmetrical four-port junction and specially truncated bends in parallel plate waveguides and balanced strip transmission lines', *IEEE Trans.*, 1968, **MTT–16,** pp. 165–176

5 MARCUVITZ, N.: 'Waveguide handbook' (MIT Radiation Laboratory Series, 1950)

6 LEIGHTON, W. M., and MILNES, G.: 'Junction reactance and dimensional tolerance effects on X-Band 3 dB directional couplers', *IEEE Trans.*, 1971, **MTT–19,** pp. 818–824

7 MASTERMAN, P. H., and CLARRICOATS, P. J. B.: 'Computer field-matching solution of waveguide transverse discontinuities', *Proc. IEE*, 1971, **118,** (1), pp. 51–63

8 KNETSCH, H. D.: 'Beitrag zur Theorie spunghafter Querschnittsveränderungen von Hohlleitern', *Arch. Elek. Übertrag.*, 1968, **22,** pp. 591–600

9 SHARP, E. D.: 'An exact calculation for a T-junction of rectangular waveguide having arbitrary cross-section', *IEEE Trans.*, 1967, **MTT–15,** pp. 109–116

10 LEWIN, L.: 'On the inadequacy of discrete mode-matching techniques in some waveguide discontintuity problems', *ibid.*, 1970, **MTT–18,** pp. 364–369

11 BRÄCKELMANN, W.: 'Hohlleiterverbindungen für Rechteckhohlleiter', *Nachrichtentech. Z.*, 1970, **23,** pp. 2–7

12 WHEELER, H. A.: 'Transmission-line properties of parallel wide strips by a conformal mapping approximation', *IEEE Trans.*, 1964, **MTT–12,** pp. 280–289

13 WHEELER, H. A.: 'Transmission-line properties of parallel wide strips separated by a dielectric sheet', *ibid.*, 1965, **MTT–13,** pp. 172–185

14 WOLFF, I., KOMPA, G., and MEHRAN, R.: 'Streifenleitungsdiskontinuitäten und-Verzweigungen', *Nachrichtentech. Z.* (to be published)

15 LEWIN, L.: 'Radiation from discontinuities in strip-line', *Proc. IEE*, 1960, **107C,** pp. 163–170

Paper 11.4

Method of Analysis and Filtering Properties of Microwave Planar Networks

GUGLIELMO D'INZEO, FRANCO GIANNINI, CESARE M. SODI, AND ROBERTO SORRENTINO, MEMBER, IEEE

Abstract—A method of analysis of planar microwave structures, based on a field expansion in term of resonant modes, is presented. A first advantage of the method consists in the possibility of taking into account fringe effects by introducing, for each resonant mode, an equivalent model of the structure. Moreover, the electromagnetic interpretation of the filtering properties of two-port networks, particularly of the transmission zeros, whose nature has been the subject of several discussions, is easily obtained. The existence of two types of transmission zeros, modal and interaction zeros is pointed out. The first ones are due to the structure's resonances, while the second ones are due to the interaction between resonant modes. Several experiments performed on circular and rectangular microstrips in the frequency range 2–18 GHz have shown a good agreement with the theory.

I. Introduction

After the study of the transmission properties of microstrip lines, the great diffusion of microwave integrated circuits has led to the analysis of general planar circuits. To this purpose, analytical methods, applied to structures of simple geometry [1]–[3], and numerical methods, apt to the study of more complex geometries [4]–[6], have been developed. In both cases a magnetic wall model has been adopted for the structure because of the formidable boundary value problems. In such a way, however, one not only neglects the dispersion properties of the circuit, which are due to fringe effects, but often obtains erroneous results [7].

To overcome this difficulty, in the case of step discontinuities, i.e., of structures with separable geometry in rectangular coordinates, Menzel and Wolff [8] have recently proposed a method of analysis based on the correction of the magnetic wall model by means of frequency dependent effective parameters. However, it must be observed that effective parameters depend not only on the frequency, but also on the field distribution inside the structure. It is sufficient to instance the disk resonators for which Wolff and Knoppik [9] have shown a frequency dependent equivalent model to exist for each resonant mode, in such a way that a unique equivalent model for the structure cannot be defined. This fact strongly limits the applicability of all the analyses of microstrip structures presented until now. Considerable attention has been devoted to nonuniform lines, i.e., lines with continuously or not continuously varying cross sections. The existence of transmission zeros has been stressed both theoretically and experimentally. In the particular case of a double step discontinuity, the physical nature of such zeros has been discussed for a long time [2], [10]–[13] and they have been ascribed to the excitation of higher order modes of propagation in the line section between the two discontinuities. As will be shown below, such an interpretation, in our opinion, is not correct, also because transmission zeros are present in generic nonuniform lines where the EM field cannot propagate as exp $(-j\beta z)$.

In this paper an analysis of planar circuits based on the theory of resonant cavities is presented. Three important advantages are so obtained. The first consists in the possibility of introducing frequency dependent effective parameters for each resonant mode of the structure in such a way as to obtain an accurate characterization of its frequency behavior. The second is an electromagnetic interpretation of the network's filtering properties, particularly of the transmission zeros, is easily obtained and the above mentioned problems are clarified. Finally, the present method leads to the evaluation of the impedance matrix of the network in the form of a partial fraction expansion with the advantages pointed out by Silvester [6].

The analysis is limited to the important case of two-port networks, since the extension to the general case is

Manuscript received May 16, 1977; revised December 7, 1977. This work has been supported in part by the Consiglio Nazionale delle Ricerche (C. N. R.), Italy.

The authors are with the Istituto di Elettronica, Università di Roma, Rome, Italy.

Reprinted from *IEEE Trans. Microwave Theory Tech.*, vol. MTT-26, no. 7, pp. 462–471, July 1978.

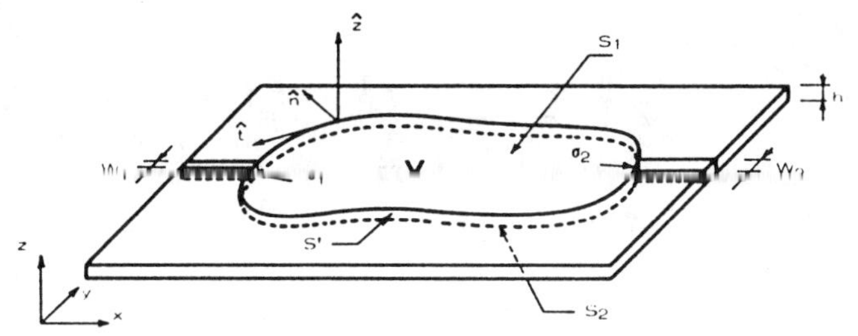
Fig. 1. The planar two-port circuit.

straightforward. The general filtering properties are discussed and criteria for locating transmission zeros are given. Several experimental results for circular and rectangular structures in the frequency range 2–18 GHz show a good agreement with the theoretical ones, obtained using the effective parameters proposed in [9]. Structures with nonseparable geometries could also be studied with the same technique through a numerical method (e.g., a finite element method).

II. Formulation of Field Problem

Fig. 1 shows a microstrip two-port circuit. The main difficulty in the study of such a structure is due to the fact that it is an open one, i.e., the EM field extends to infinity. The central section may be considered as an open resonator; the EM field is mainly concentrated in the cylindrical volume V bounded by the two conducting surfaces S_1 and S_2 and, laterally, by the cylindrical surface S'. It may be expressed as a function of the tangential magnetic field $\boldsymbol{H}_\tau$ on S' in terms of the modes of the cavity V. Following a procedure analogous to that of Kurokawa [14], one obtains

$$\boldsymbol{E}=\Sigma_a e_a \boldsymbol{E}_a+\Sigma_\alpha e_\alpha \boldsymbol{E}_\alpha \tag{1}$$

$$\boldsymbol{H}=\Sigma_a h_a \boldsymbol{H}_a+\Sigma_\alpha h_\alpha \boldsymbol{H}_\alpha \tag{2}$$

where $\boldsymbol{E}_a$ and $\boldsymbol{E}_\alpha$ are the orthonormalized eigenvectors of the following eigenvalue problem:

$$\nabla\times\nabla\times \boldsymbol{E}-\nabla\nabla\cdot\boldsymbol{E}-k^2\boldsymbol{E}=0, \quad \text{inside } V \tag{3a}$$

$$\boldsymbol{n}\times\boldsymbol{E}=0 \quad \nabla\cdot\boldsymbol{E}=0, \quad \text{on } S_1, S_2 \tag{3b}$$

$$\boldsymbol{n}\cdot\boldsymbol{E}=0 \quad \boldsymbol{n}\times\nabla\times\boldsymbol{E}=0, \quad \text{on } S' \tag{3c}$$

with the further conditions:

$$\nabla\cdot\boldsymbol{E}_a=0 \quad \nabla\times\boldsymbol{E}_a\neq 0, \quad \text{inside } V \tag{4a}$$

$$\nabla\times\boldsymbol{E}_\alpha=0, \quad \text{inside } V. \tag{4b}$$

Similarly, $\boldsymbol{H}_a$ and $\boldsymbol{H}_\alpha$ are the orthonormalized eigenvectors of

$$\nabla\times\nabla\times \boldsymbol{H}-\nabla\nabla\cdot\boldsymbol{H}-k^2\boldsymbol{H}=0, \quad \text{inside } V \tag{5a}$$

$$\boldsymbol{n}\cdot\boldsymbol{H}=0 \quad \boldsymbol{n}\times\nabla\times\boldsymbol{H}=0, \quad \text{on } S_1, S_2 \tag{5b}$$

$$\boldsymbol{n}\times\boldsymbol{H}=0 \quad \nabla\cdot\boldsymbol{H}=0, \quad \text{on } S' \tag{5c}$$

with the conditions

$$\nabla\cdot\boldsymbol{H}_a=0 \quad \nabla\times\boldsymbol{H}_a\neq 0, \quad \text{inside } V \tag{6a}$$

$$\nabla\times\boldsymbol{H}_\alpha=0, \quad \text{inside } V. \tag{6b}$$

It is possible to demonstrate that the eigenvalues of (3)–(4) coincide with those of (5)–(6) and that

$$\nabla\times\boldsymbol{H}_a=k_a\boldsymbol{E}_a$$
$$\nabla\times\boldsymbol{E}_a=k_a\boldsymbol{H}_a. \tag{7}$$

The coefficient of the expansions (1) and (2) may be calculated imposing that the EM field satisfies Maxwell's equations. One obtains

$$e_a=\frac{j\omega\mu}{k_a^2-\omega^2\mu\epsilon}\int_{S'}\boldsymbol{n}\times\boldsymbol{H}_\tau\cdot\boldsymbol{E}_a\,dS$$

$$h_a=\frac{-k_a}{j\omega\mu}e_a$$

$$e_\alpha=\frac{1}{j\omega\mu}\int_{S'}\boldsymbol{n}\times\boldsymbol{H}_\tau\cdot\boldsymbol{E}_\alpha\,dS$$

$$h_\alpha=0. \tag{8}$$

Once the set of eigenvalue of (3) and (5) is known, the evaluation of the EM field inside V depends on the knowledge of the tangential magnetic field $\boldsymbol{H}_\tau$ on S'. In a first approximation we may assume that $\boldsymbol{H}_\tau$ is different from zero only at the connections σ_i between the cavity and the lines where it has a TEM distribution[1]. Thus it is constant. However, $\boldsymbol{H}_\tau$ is not exactly zero on the remainder of S'; fringe effects can be taken into account by ascribing to the structure effective dimensions and an effective permittivity, according to the widely adopted magnetic wall model of microstrip structures. We shall come back to this point later.

Because of the above simplifying hypotheses, the EM field in the cavity is determined as a function of the magnetic field $\boldsymbol{H}_{\tau1}=H_1\boldsymbol{t}$ and $\boldsymbol{H}_{\tau2}=H_2\boldsymbol{t}$ at the outputs, which is independent of z. The structure may, therefore, be considered as a two-dimensional one. It is easily seen that, imposing the condition $\partial/\partial_z=0$ on (1)–(8), the $\boldsymbol{E}_a$'s have only the z component, while the $\boldsymbol{E}_\alpha$'s do not exist, with the exception of only the mode $\boldsymbol{E}_0$ having zero divergence. After simple manipulations, the EM field in the cavity may be expressed as follows:

$$\boldsymbol{E}=\hat{z}\Sigma_a e_a E_a+\hat{z}\,e_0 V^{-1/2} \tag{9}$$

$$\boldsymbol{H}=\frac{1}{j\omega\mu}\Sigma_a e_a \hat{z}\times\nabla_t E_a \tag{10}$$

where

$$e_a=\frac{j\omega\mu}{k_a^2-\omega^2\mu\epsilon}\left(\sqrt{\sigma_1}\,P_{a_1}H_1+\sqrt{\sigma_2}\,P_{a_2}H_2\right) \tag{11a}$$

$$P_{ai}=\sigma_i^{-1/2}\int_{\sigma_i}E_a\,dS, \qquad i=1,2 \tag{11b}$$

$$e_0=\frac{V^{-1/2}}{j\omega\epsilon}(\sigma_1 H_1+\sigma_2 H_2) \tag{11c}$$

where $\hat{z}$ is the unit vector of the z axis, V is the volume of

[1] Higher order modes on the uniform lines may be neglected with good approximation if the uniform sections are long enough and their widths are much smaller than the cavity's dimension [1]. In any case, when necessary, higher modes can be taken into account with a rather more complicate algebra.

the cavity, and σ_1 and σ_2 are the surfaces of the outputs of the cavity, i.e., the portions of S' where $\boldsymbol{H}_\tau$ is different from zero. $\boldsymbol{E}_0=\hat{z}V^{-1/2}$ is the mode having zero curl and zero divergence, belonging to the $\boldsymbol{E}_\alpha$'s. Since (11c) can be obtained from (11a) and (11b) by putting $k_a^2=0$ and $E_a=E_0=V^{-1/2}$, later on this mode will be included among the E_a's.

The eigenfunctions E_a have to satisfy the two-dimensional eigenvalue equation deriving from (3)

$$\nabla_t^2 E + k^2 E = 0 \tag{12}$$

together with the boundary condition

$$\frac{\partial E}{\partial n}=0 \tag{12'}$$

which derives from the second of (3c); the other boundary conditions are automatically satisfied.

One can note that the a modes are, in this case, TM with respect to the z direction; the o mode, on the contrary, corresponds to the electrostatic field problem.

Once (12) is solved for a particular geometry, the EM field inside the cavity is fully determined through (9)–(11) as a function of the magnetic fields supported by the uniform lines. Nevertheless, a terminal description of the structure as a two-port network is generally preferable. This can be obtained by evaluating the impedance matrix, relative, of course, to the dominant TEM modes of the lines. The amplitude of the electric field E_i on the ith line is obtained by projecting the field (9), calculated at σ_i, on the abstract vector space of the modes of the line and retaining the TEM component [15], i.e.,

$$E_i=\frac{1}{\sigma_i}\int_{\sigma_i}\hat{z}\cdot\boldsymbol{E}\,dS, \qquad i=1,2.$$

Through (9) and (11) E_i can be expressed as a function of H_1, H_2

$$\begin{aligned} E_1 &= H_1 j\omega\mu\Sigma_a\frac{P_{a1}^2}{k_a^2-\omega^2\mu\epsilon} + H_2 j\omega\mu\Sigma_a\frac{P_{a1}P_{a2}}{k_a^2-\omega^2\mu\epsilon}\cdot[\sigma_2/\sigma_1]^{1/2}\\ E_2 &= H_1 j\omega\mu\Sigma_a\frac{P_{a2}P_{a1}}{k_a^2-\omega^2\mu\epsilon}\cdot[\sigma_1/\sigma_2]^{1/2} + H_2 j\omega\mu\Sigma_a\frac{P_{a2}^2}{k_a^2-\omega^2\mu\epsilon}. \end{aligned} \tag{13}$$

If one defines equivalent voltages and currents in such a way as to normalize to unity the characteristic impedances of the lines, i.e.,

$$\begin{aligned} V_i &= E_i\left[\sigma_i\sqrt{\epsilon/\mu}\,\right]^{1/2}\\ I_i &= H_i\left[\sigma_i\sqrt{\mu/\epsilon}\,\right]^{1/2}, \qquad i=1,2 \end{aligned} \tag{14}$$

from (13) and (14) the following expression of the $[Z]$ matrix is easily obtained

$$[Z]=\Sigma_a[Z_a] \tag{15}$$

with

$$[Z_a]=\frac{j\omega c}{\omega_a^2-\omega^2}\begin{bmatrix} P_{a1}^2 & P_{a1}P_{a2}\\ P_{a2}P_{a1} & P_{a2}^2\end{bmatrix} \tag{15'}$$

μ and ϵ are the substrate's permeability and permittivity, respectively, and

$$\omega_a=ck_a$$

are the resonant frequencies of the cavity. If there are ν_a linearly independent eigenfunctions corresponding to the same eigenvalue k_a^2

$$E_a^{(1)}, E_a^{(2)}, \cdots E_a^{(\nu_a)}$$

which, without loss of generality, may be supposed to be ortogonal, (15') should be replaced by

$$[Z_a]=\frac{j\omega c}{\omega_a^2-\omega^2}\sum_{\nu=1}^{\nu_a}\begin{bmatrix} P_{a1}^{(\nu)^2} & P_{a1}^{(\nu)}P_{a2}^{(\nu)}\\ P_{a1}^{(\nu)}P_{a2}^{(\nu)} & P_{a2}^{(\nu)^2}\end{bmatrix},$$

$$c=1/\sqrt{\mu\epsilon} \tag{15''}$$

while, in (15), the summation over a should include only distinct ω_a's. $[Z]$ is a purely imaginary matrix since the structure has been supposed without losses. If the network is symmetrical

$$P_{a2}=\epsilon_a P_{a1} \tag{16}$$

where $\epsilon_a=1$ for even modes and $\epsilon_a=-1$ for odd modes. The impedance parameters may be written

$$\begin{aligned} Z_{11}&=Z_{22}=Z_{ev}+Z_{od}\\ Z_{12}&=Z_{21}=Z_{ev}-Z_{od} \end{aligned} \tag{17}$$

where

$$\begin{aligned} Z_{ev}&=j\omega c\Sigma\frac{P_{ev}^2}{\omega_{ev}^2-\omega^2}\\ Z_{od}&=j\omega c\Sigma\frac{P_{od}^2}{\omega_{od}^2-\omega^2} \end{aligned} \tag{17'}$$

ev being the index of the even modes, od of the odd modes.

The calculation of the $[Z]$ matrix requires the evaluation of the eigenfunctions and eigenvalues E_a, k_a^2 and then of the P_{a1}. This can be done analytically if the structure has a separable geometry; if the geometry is not separable, a numerical method could be adopted.

III. General Filtering Properties

The formulation given in the previous section has led to a complete characterization of the microwave network in terms of its impedance matrix. In order to discuss the filtering properties of the structure, a description in terms of the scattering parameters is preferable since the impedance matrix elements are not quantities easily measurable at microwave frequencies; moreover the scattering matrix provides a more appropriate physical description of the structure behavior. In terms of the impedance parameters the scattering parameters are given by

$$s_{11}=[(Z_{11}-1)(Z_{22}+1)-Z_{12}^2]/D$$
$$s_{22}=[(Z_{11}+1)(Z_{22}-1)-Z_{12}^2]/D$$
$$s_{12}=s_{21}=2Z_{12}/D \tag{18}$$

where

$$D=(Z_{11}+1)(Z_{22}+1)-Z_{12}^2. \tag{18'}$$

Let us start examining the structure's behavior at the resonant frequency ω_p of one of the modes. It is convenient to write the Z parameters as follows:

$$Z_{11}=j\omega c\frac{Q_{11}}{\omega_p^2-\omega^2}+\hat{Z}_{11}$$
$$Z_{22}=j\omega c\frac{Q_{22}}{\omega_p^2-\omega^2}+\hat{Z}_{22}$$
$$Z_{12}=j\omega c\frac{Q_{12}}{\omega_p^2-\omega^2}+\hat{Z}_{12} \tag{19}$$

where $\hat{Z}_{ij}$ remains finite for $\omega\to\omega_p$. Let us distinguish two cases.

1) $Q_{11}Q_{22}=Q_{12}^2$. This equality is always verified for nondegenerate modes. We further distinguish two subcases.

a) $Q_{11}Q_{22}=Q_{12}^2=0$. Since the case $Q_{11}=Q_{22}=Q_{12}=0$ may be excluded,[2] the structure has to be nonsymmetrical ($Q_{11}\neq Q_{22}$). From (18) and (19) one immediately obtains

$$s_{12}(\omega_p)=0.$$

According to whether Q_{11} or Q_{22} is different from zero, $s_{11}=1$, or $s_{22}=1$, respectively. The resonant frequency ω_p, therefore, corresponds to a transmission zero. This can be easily explained from an electromagnetic point of view. Suppose $Q_{22}=0$: this means that the p mode is uncoupled to the second port. When an EM field is incident to the first port at the frequency $\omega=\omega_p$, the EM field inside the cavity would become infinite (see (11a)) unless the total (incident plus reflected) magnetic field at the first port is zero; this implies $s_{11}=1$, $s_{12}=0$.

We may, therefore, conclude that in nonsymmetrical structures a transmission zero takes place at the resonant frequency of one mode which is uncoupled to one of the ports. Later on, transmission zeros taking place at resonant frequencies ω_a will be referred to as modal zeros.

b) $Q_{11}Q_{22}=Q_{12}^2\neq 0$. In this case, when $\omega\to\omega_p$ the scattering parameters do not generally assume significant values. Nevertheless, it is worth considering the case of a symmetrical structure ($Q_{11}=Q_{22}=\pm Q_{12}$). For $\omega\to\omega_p$ one obtains from (18) and (19)

$$s_{11}=s_{22}=\frac{\hat{Z}_{11}-\epsilon_p\hat{Z}_{12}}{1+\hat{Z}_{11}-\epsilon_p\hat{Z}_{12}} \qquad s_{12}=s_{21}=\frac{\epsilon_p}{1+\hat{Z}_{11}-\epsilon_p\hat{Z}_{12}}.$$

According to whether p is an even ($\epsilon_p=1$) or an odd ($\epsilon_p=-1$) mode the quantity $\hat{Z}_{11}-\epsilon_p\hat{Z}_{12}$ is equal to $2Z_{\text{od}}$ or to $2Z_{\text{ev}}$ (see (17')), respectively. These quantities are often negligible with respect to unity, so that $s_{11}=s_{22}\cong 0$, $s_{12}\cong\epsilon_p$. In other words, in a symmetrical nondegenerate structure modal transmission zeros do not take place; on the contrary, the ω_a's give generally place to approximate reflection zeros. It is worth specifying that the existence of a reflection zero at or near ω_p depends on the widths of the ports; in some cases, in fact, the reflection zero takes place only if the ports are small enough. Typical examples will be shown below. For the sake of brevity we omit to demonstrate the above statements which, on the other hand, can be easily proved.

2) $Q_{11}Q_{22}\neq Q_{12}^2$. This case can be verified only for a degenerate mode. It is easily seen that at the frequency $\omega=\omega_p$ $s_{11}=s_{22}=1$, $s_{12}=0$. This is another case of modal transmission zero, which is due to a degenerate mode of the cavity, or rather to the superposition of degenerate modes.

Having examined the structure's behavior at the resonant frequencies of the cavity, let us now consider the cases when a transmission zero takes place.

From (18) it follows that for s_{12} to be zero there are only two cases: a) $|D|=\infty$. This condition holds only if $\omega=\omega_a$, and therefore is that of a modal transmission zero. b) $Z_{12}=0$. Since

$$Z_{12}=j\omega c\Sigma_a\frac{P_{a1}P_{a2}}{\omega_a^2-\omega^2} \tag{20}$$

it can be easily inferred that between two consecutive resonant frequencies ω_p, ω_q such that[3]

$$\operatorname{sgn}(P_{p1}P_{p2})=\operatorname{sgn}(P_{q1}P_{q2}) \tag{21}$$

there is necessarily a frequency $\omega_z\in(\omega_p,\omega_q)$ such that

$$Z_{12}(\omega_z)=0,\ s_{12}(\omega_z)=0.$$

In case of mode degeneracy, (21) should be replaced by

$$\operatorname{sgn}\left(\sum_{\nu=1}^{\nu_p} I_{p1}^{(\nu)}I_{p2}^{(\nu)}\right)=\operatorname{sgn}\left(\sum_{\nu=1}^{\nu_q} I_{q1}^{(\nu)}I_{q2}^{(\nu)}\right). \tag{21'}$$

In order to find a physical interpretation of this type of transmission zero, suppose the cavity is excited by a field incident to the first port at a frequency located between ω_p and ω_q; if (21) is satisfied, the p and q modes will give place to opposite contributions to the field at the output. In other words, they interact destructively at the second port. At the frequency $\omega=\omega_z$, whose location between ω_p and ω_q depends also on the contribution of all the other modes, there is a totally destructive interaction in such a way that no power can be transferred towards the output. This type of transmission zero will be called interaction zero.

For symmetrical nondegenerate structures, because of (16), (21) becomes

[2]In that case, in fact, the p mode cannot be excited in the structure and therefore can be excluded from any consideration.

[3]If $P_{q1}P_{q2}=0$ (i.e., there is a modal transmission zero at ω_q), in (21) the successive mode must be considered, say, the r mode, such that $P_{r1}P_{r2}\neq 0$.

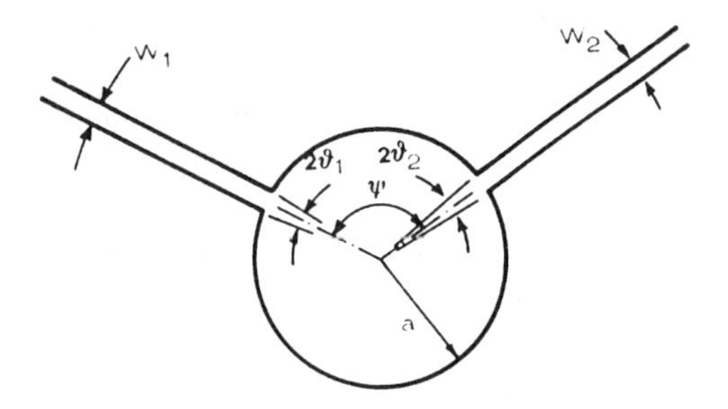

Fig. 2. The circular microstrip.

$$\epsilon_p = \epsilon_q$$

i.e., if two consecutive resonant modes are both even or odd, an interaction transmission zero is located between their resonant frequencies.

IV. The Circular Microstrip

The first case of the two-port network we have considered is the circularly shaped microstrip line shown in Fig. 2. The orthonormalized eigenfunctions of (12) are, in this case,

$$E_{mn} = C_{mn} J_m(k_{mn} r) \begin{Bmatrix} \cos m\phi \\ \sin m\phi \end{Bmatrix}, \qquad \begin{matrix} m = 0, 1, 2 \cdots \\ n = 1, 2, 3 \cdots \end{matrix} \tag{22}$$

where

$$C_{mn} = \frac{1}{\sqrt{V}\, J_m(\xi'_{mn})} \left[\frac{\delta_m}{1 - \left(\frac{m}{\xi'_{mn}} \right)^2} \right]^{1/2}$$

$$V = \pi a^2 h$$

$$k_{mn} = \xi'_{mn}/a = \omega_{mn}/c$$

and

$$\delta_m = \begin{cases} 1, & \text{for } m = 0 \\ 2, & \text{for } m \neq 0 \end{cases}$$

is the Neumann factor; ξ'_{mn} is the nth root of the equation

$$\frac{d}{dx} J_m(x) = 0$$

and h is the substrate's thickness. Besides the set of eigenfunctions (22), the eigenfunction corresponding to $k^2 = 0$ must be considered

$$E_{00} = 1/\sqrt{V}.$$

Such an eigenfunction can be obtained from (22) by putting conventionally

$$\left. \frac{m}{\xi'_{mn}} \right|_{\substack{n=0 \\ m=0}} = 0.$$

Equation (22) shows the existence of a pair of degenerate modes for any $m \neq 0$. We shall restrict our attention to the important case of symmetrical structures ($w_1 = w_2 = w$; $\theta_1 = \theta_2 = \theta$). Following the procedure described in the previous section, one obtains for the $[Z]$ matrix elements[4]

[4]As a consequence of the hypothesis that the widths of the ports are much smaller than the cavity's radius, the arcs θ_1 and θ_2 may be confused with the corresponding chord.

$$Z_{11} = Z_{22} = j\omega c \frac{4\theta^2}{\pi w} \sum_{m=0}^{\infty} \sum_{n=0}^{\infty} \frac{A_{mn}^2}{\omega_{mn}^2 - \omega^2}$$

$$Z_{21} = Z_{12} = j\omega c \frac{4\theta^2}{\pi w} \sum_{m=0}^{\infty} \sum_{n=0}^{\infty} \frac{A_{mn}^2}{\omega_{mn}^2 - \omega^2} \cos m\psi \tag{23}$$

where

$$A_{mn}^2 = \left(\frac{\sin m\theta}{m\theta} \right)^2 \frac{\delta_m}{1 - (m/\xi'_{mn})^2}. \tag{23'}$$

Let us consider the structure's behavior at the resonant frequencies of the cavity. The condition of case 2) of the previous section becomes

$$\cos^2 m\psi \neq 1. \tag{24}$$

Therefore, for generic values of ψ and for $m \neq 0$, each resonant frequency corresponds to a modal transmission zero. On the contrary, if $m\psi = s\pi$ ($s = 0, 1, 2, \cdots$) case 2b) is verified, i.e., for θ small enough, a reflection zero takes place near such resonant frequencies. This happens for all the modes of a doubly symmetrical structure ($\psi = \pi$) and, in general, for the $(0, n)$ modes.

Besides modal transmission zeros, interaction zeros take place between consecutive resonant frequencies $\omega_{m_1 n_1}$ and $\omega_{m_2 n_2}$ such that[5]

$$\operatorname{sgn}(\cos m_1\psi) = \operatorname{sgn}(\cos m_2\psi). \tag{25}$$

In the doubly symmetrical case ($\psi = \pi$) (31) becomes[6]

$$(-1)^{m_1} = (-1)^{m_2} \tag{25'}$$

i.e., an interaction zero is located between the resonant frequencies of two consecutive modes having both an even, or an odd, azimutal dependence.

From (25) it follows that transmission zeros can be located between any pair of resonant frequencies by varying the angle ψ between the two uniform lines.

Fig. 3 shows the theoretical behavior of the scattering parameter $|s_{12}|$ of a doubly symmetrical circular microstrip versus the frequency in the range 2–18 GHz. (Expressions (23) have been evaluated taking into account the first 62 modes). This curve has been obtained completely neglecting fringe effects, i.e., ascribing to the EM model the physical dimension of the structure and assuming for ϵ the permittivity of the substrate (alumina, $\epsilon = 10\,\epsilon_0$). The locations of the resonant frequencies ω_{mn} are also indicated in the figure. The structure presents two transmission zeros, which are due to the interaction between the pair of modes (2,1)–(0,1) and (1,2)–(5,1), accordingly to (25'). (The last resonant frequency is not indicated in the figure, because it is out of scale). Reflection zeros are located near each resonant frequency, with the exception of the modes (4,1) and (1,2) which are very close together; for the assumed port widths, corresponding to

[5]If $\cos m_2\psi = 0$, (thus a modal transmission zero takes place at $\omega_{m_2 n_2}$) the successive mode must be considered in (25). See footnote 3.

[6]It may be noted that in this case the structure behaves as a nondegenerate one, since for each k_{mn}^2 only one of the two degenerate modes can be excited, both from the input or from the output.

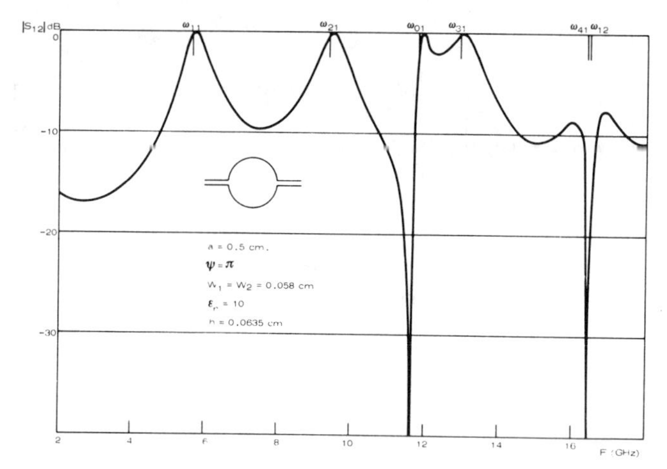

Fig. 3. Transmission coefficient $|s_{12}|$ versus the frequency for a circular microstrip (magnetic wall model without effective parameters).

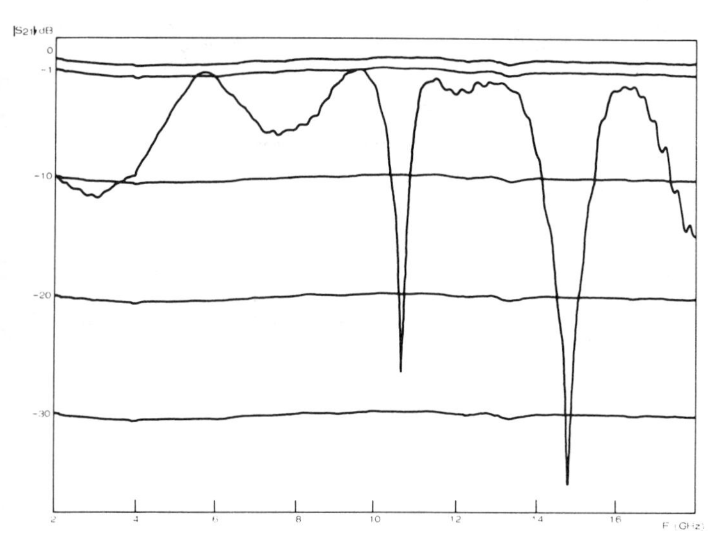

Fig. 4. Experimental behavior of the transmission coefficient $|s_{12}|$ versus the frequency for the same structure as in Fig. 3. Substrate material alumina.

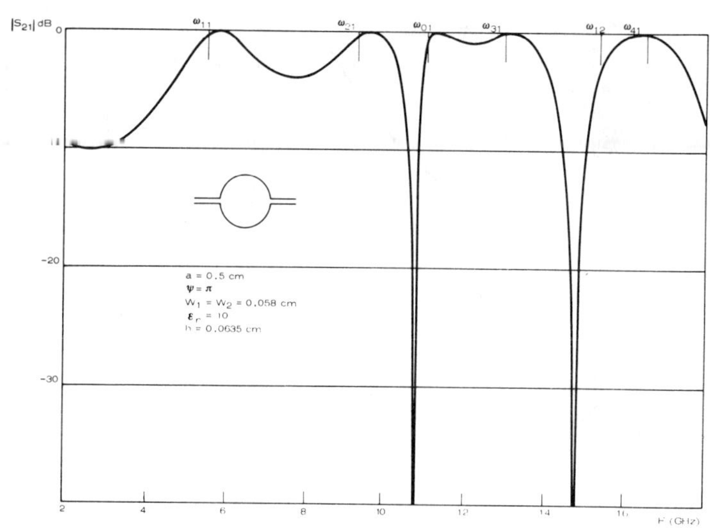

Fig. 5. Transmission coefficient $|s_{12}|$ versus the frequency for the same structure as in Fig. 3 (magnetic wall model according to the present theory).

50-Ω lines, reflection zeros do not take place at these frequencies.

Fig. 4 shows the experimental behavior of $|s_{12}|$ for the same structure as in Fig. 3. The comparison between these diagrams shows a good qualitative agreement between theory and experiment up to ~12 GHz. The disagreement consists on the one hand in a slight shifting of the reflection and transmission zero frequencies and, on the other hand, in lower values of the theoretical $|s_{12}|$. The latter fact can be easily explained, since the EM coupling between the uniform lines and the cavity is, in reality, stronger than the theory predicts, because the fringing field of the lines has been neglected. With regard to the frequency shifting of the two diagrams, it must be observed that the experimental resonant frequencies are different from the theoretical ones, as previously pointed out. Over ~12 GHz, the theory yields to inacceptable errors; the two diagrams do not agree even from a qualitative point of view. As has been previously noted [7], the experimental resonant frequencies may differ from the theoretical ones in such a way that the sequence of modes is different in the two cases. Since interaction zeros depend on such a sequence, their location might be strongly altered.

The above considerations indicate that account must be taken of fringe effects both of the lines and of the cavity. This can be done by ascribing to the lines effective widths and effective permittivities accordingly, for instance, to Wheeler [16] or to Schneider [17], [18]. With regard to the circular resonator, fringe effects depend on the resonant mode and can be taken into account through an equivalent model for each mode, accordingly to Wolff and Knoppik [9]. Expressions (23) should be therefore modified by introducing an effective port width w_{eff}, an effective frequency dependent permittivity of the lines $\epsilon_{\text{eff}}(f)$, an effective cavity radius r_{eff} and, finally, an effective dynamic permittivity of the cavity $\epsilon_{\text{dyn},mn}(f)$, which also depends on the resonant mode.

The results obtained in this way are shown in Fig. 5 and agree very well with the experiments in Fig. 4. In particular, one can note that the resonant frequencies of the modes (4,1) and (1,2) are now interchanged: the second transmission zero is therefore due to the interaction between the (3,1) and (1,2) modes. This phenomenon is analogous to the modal inversion in circular waveguides [19] and could be experimentally verified by means of a field mapping technique [20]. The residual differences between the theoretical and experimental magnitudes of $|s_{12}|$ are essentially due to losses, particularly to radiation losses, which have been completely neglected.

Fig. 6(a) and (b) shows the theoretical behavior of $|s_{12}|$ versus the frequency for a circular microstrip with $\psi = \pi/2$. This structure presents a modal transmission zero at each resonant frequency of one mode having an odd azimutal dependence (see (24)), i.e., in the frequency range considered, of the modes (1,1), (3,1), and (1,2). For $\psi = \pi/2$, (25) shows that interaction zeros take place between the resonant frequencies of modes $(4m_1, n_1) - (4m_2, n_2)$ or $(4m_1+2, n_1) - (4m_2+2, n_2)$; in the present case

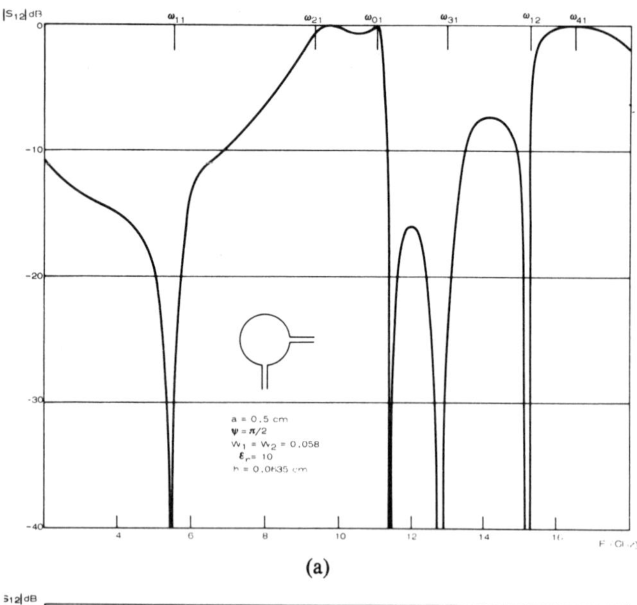

(a)

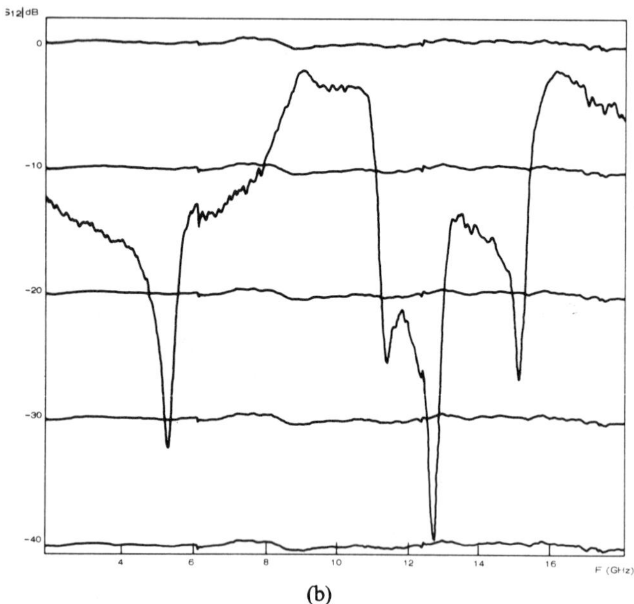

(b)

Fig. 6. Transmission coefficient $|s_{12}|$ versus the frequency, for a circular microstrip. (a) Theory. (b) Experiment.

there is only one interaction zero between the modes (0, 1) and (4, 1).

It is easy to verify that, for $\psi<\pi/2$, an interaction zero takes place between the (0,0) and (1,1) modes. For $\psi\rightarrow\pi/2$ the interaction zero tends to the modal zero due to the (1,1) mode. Fig. 7 shows the theoretical frequency location of such transmission zero as a function of the angle ψ for a given cavity's radius. As can be seen, a good agreement with the experiments is obtained. This figure shows the possibility of locating a transmission zero at a given frequency by suitably positioning the output port of the network.

V. The Rectangular Microstrip

Another type of two-port planar network consists of a rectangular central section connected with two uniform lines (Fig. 8). This structure may be considered also as a

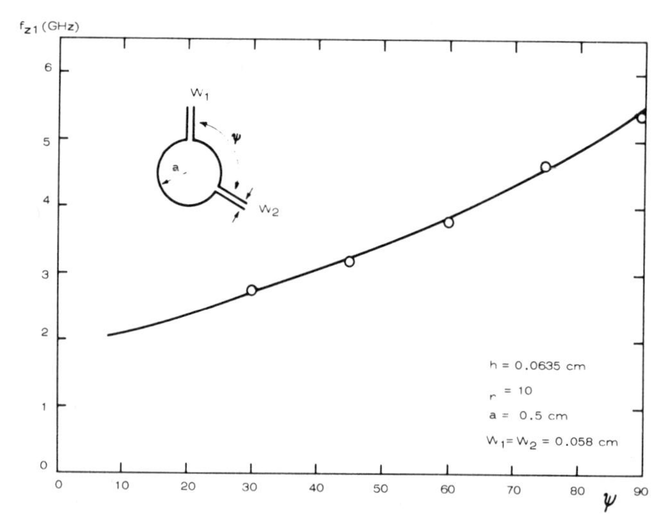

Fig. 7. Frequency location of the first transmission zero presented by a circular microstrip as a function of the angle between the two ports.

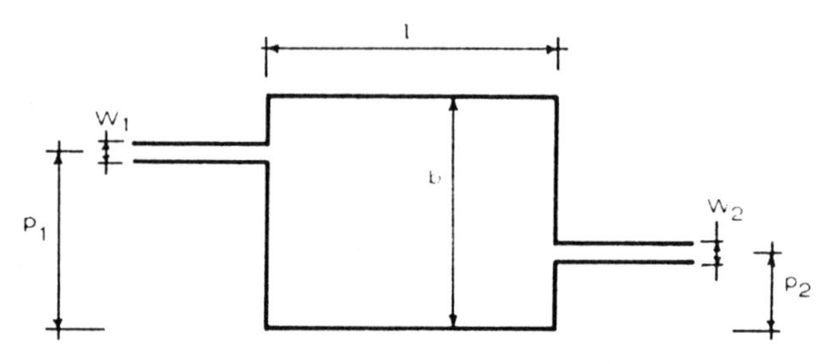

Fig. 8. The rectangular microstrip.

double step discontinuity; on the other hand, for $l \ll b$ it becomes a stub structure.

Simple calculations yield to the following expressions of the Z parameters of the structure[7]

$$Z_{11}=\frac{j\omega c w_1}{b l}\sum_{m=0}^{\infty}\sum_{n=0}^{\infty}\frac{\delta_m\delta_n f_{n_1}^2}{\omega_{mn}^2-\omega^2}$$

$$Z_{22}=\frac{j\omega c w_2}{b l}\sum_{m=0}^{\infty}\sum_{n=0}^{\infty}\frac{\delta_m\delta_n f_{n2}^2}{\omega_{mn}^2-w^2}$$

$$Z_{21}=Z_{12}=\frac{j\omega c\sqrt{w_1 w_2}}{b l}\sum_{m=0}^{\infty}\sum_{n=0}^{\infty}(-1)^m\frac{\delta_m\delta_n f_{n1} f_{n2}}{\omega_{mn}^2-\omega^2} \tag{26}$$

where

$$\omega_{mn}=c\pi\sqrt{(m/l)^2+(n/b)^2}$$

$$f_{ni}=\begin{cases}\cos\dfrac{n\pi p_i}{b}\,\dfrac{\sin\dfrac{n\pi w_i}{2b}}{\dfrac{n\pi w_i}{2b}}, & \text{for } n\neq 0\\ 1, & \text{for } n=0\end{cases}.$$

[7]The series over m could be evaluated analytically and equivalent expressions to those in [1] would be obtained; nevertheless, they are not suitable for our purpose, since it is necessary to introduce effective parameters for each resonant mode, i.e., for each term of the series (26), in the same way as has been done in the case of the circular microstrip.

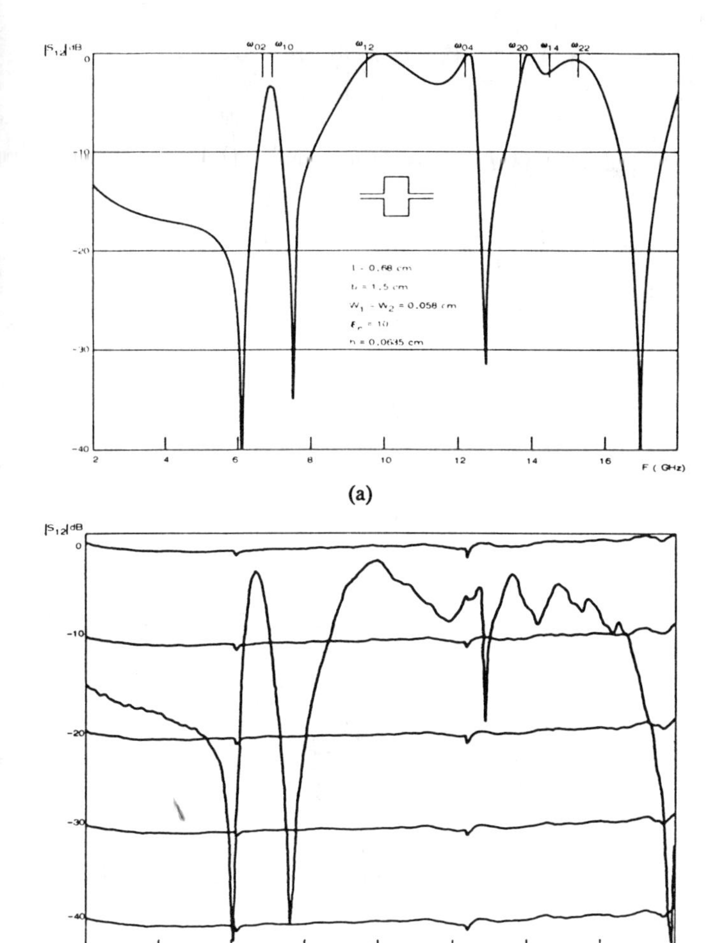

Fig. 9. Transmission coefficient $|s_{12}|$ versus the frequency for a symmetrical rectangular microstrip. (a) Theory. (b) Experiment.

For symmetrical structures ($w_1 = w_2$, $f_{n1}^2 = f_{n2}^2$), as previously stated, transmission zeros are only of the interaction type; in particular, if $p_1 = p_2$ condition (21) becomes

$$(-1)^{m_1} = (-1)^{m_2}$$

i.e., transmission zeros are located between the resonant frequencies of consecutive modes having both an even or an odd dependence with respect to the x direction.

Nonsymmetrical structures can also present modal transmission zeros, which are located at the resonant frequencies ω_{mn} such that $f_{n1} = 0$ and $f_{n2} \neq 0$, or vice versa.

Fig. 9(a) and (b) shows the theoretical and experimental behaviors of $|s_{12}|$ of a symmetrical rectangular microstrip versus the frequency in the range 2–18 GHz. Theoretical results have been obtained by adopting for the resonator the model suggested by Wolff and Knoppik [9]; the first 400 modes have been retained in (26). The structure presents four transmission zeros due to the interaction of the following pairs of modes: (0,0)–(0,2), (1,0)–(1,2), (0,4)–(2,0), (2,2)–(0,6). It is important to note that the first transmission zero is located close to the resonant frequency of the (0,2) mode. Since this frequency is nothing but the cutoff frequency of the $TE_{20}^{(x)}$ mode of propagation in the wider microstrip section, that transmission zero has been ascribed to the excitation of the $TE_{20}^{(x)}$ mode [10]–[12]. Such an interpretation is, in our opinion, inacceptable. As has been previously pointed out, given the symmetry of the structure, a resonant mode cannot give place to a transmission zero, but on the contrary it can give place to a reflection zero: the (0,2) mode can produce a transmission zero only through the interaction with another resonant mode, which is even with respect to the x direction (i.e., a $(2m,n)$ mode). For a different $1/b$ ratio, in fact, this mode does not give place to any transmission zero, as will be shown later. One may moreover observe that the (0,2) and (1,0) modes, which are rather close together, do not give place to any reflection zero. This is a typical case when two resonant modes interact together in such a way that no reflection zero takes place. However, one could verify that if the port widths would be very small (about 5μm) two reflection zeros would take place at these resonant frequencies.

The experimental results in Fig. 9(b) agree fairly well with the theoretical ones until ~12.5 GHz; at higher frequencies there is a little shifting between the theoretical and experimental resonant frequencies. Moreover, the effect of the losses becomes appreciable over ~15 GHz. A more accurate and complete model of the structure would therefore require a better evaluation of the resonant frequencies and, at the same time, the introduction of the losses.

In order to confirm what is stated above with regard to the (0,2) mode, another structure has been made with proper dimensions in such a way that this mode is located between two odd modes. Fig. 10(a) and (b) shows the corresponding theoretical and experimental results. As the theory predicts, in this case, the (0,2) mode does not give place to any transmission zero; the three transmission zeros are due to the interaction between the modes (4,0) and (2,2), (3,2) and (5,0), and (4,2) and (6,0). As can be noted the theory agrees very well with the experiment in the whole frequency range 2–18 GHz.

Fig. 11(a) and (b) shows the theoretical and experimental results for a nonsymmetrical rectangular microstrip. In this structure odd modes with respect to the y axis can be excited at the second port, but not at the first one. As a consequence, all the $(m,2n+1)$ modes give place to modal transmission zeros. It is worth observing that the first zero is located at the resonant frequency of the (0,1) mode, corresponding to the cutoff frequency of the $TE_{10}^{(x)}$ mode of propagation in the central section. This accounts for previous observations [10]–[13] with regard to nonsymmetrical step discontinuities; in this case, the statement that transmission zeros are due to higher order modes is correct, since it is a modal and not an interaction zero. The structure in Fig. 11 presents also three transmission zeros due to the interaction between the modes (2,0)–(0,2), (1,2)–(3,0), and (2,2)–(4,0). In spite of the complexity of the frequency behavior of the structure, the theoretical results may be considered highly satisfactory.

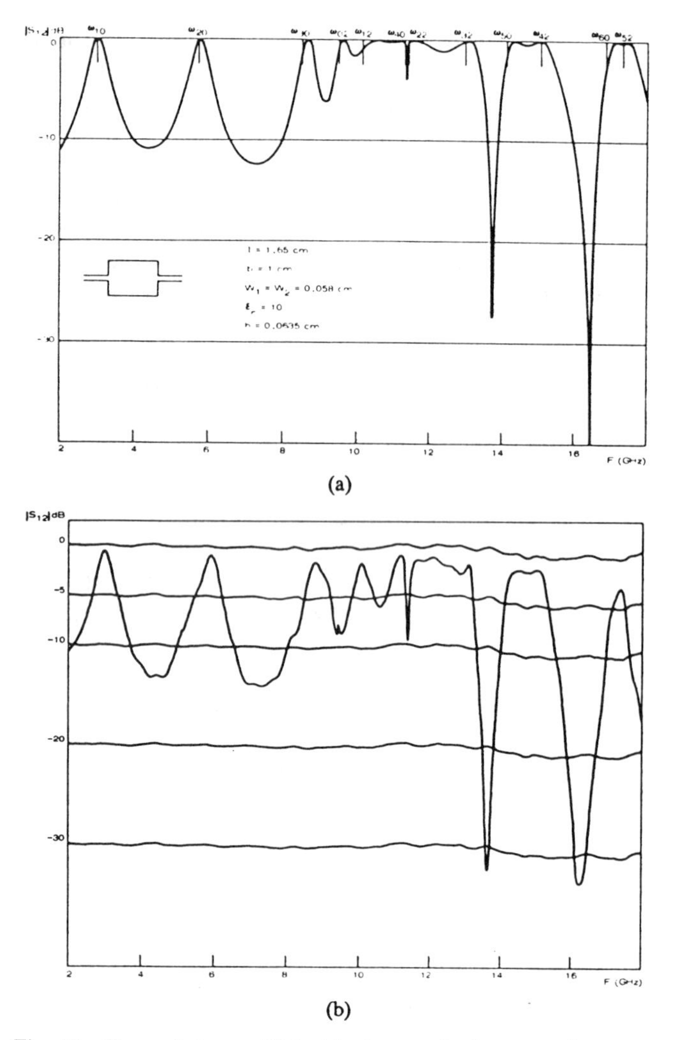

(a)

(b)

Fig. 10. Transmission coefficient $|s_{12}|$ versus the frequency for a symmetrical rectangular microstrip. (a) Theory. (b) Experiment.

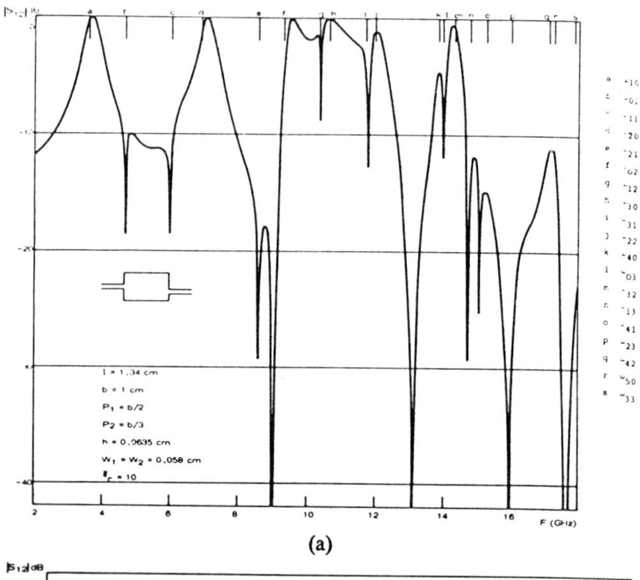

(a)

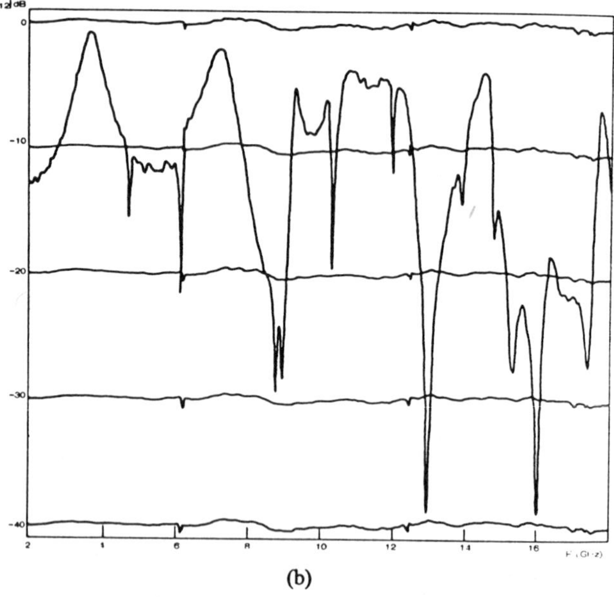

(b)

Fig. 11. Transmission coefficient $|s_{12}|$ versus the frequency for a nonsymmetrical rectangular microstrip. (a) Theory. (b) Experiment.

VI. Conclusions

A method of analysis of planar microwave structures is presented, which is based on a field expansion in terms of resonant modes. The case of two-port networks is considered, but the extension to N-port circuits is straightforward. The general filtering properties are discussed and, in particular, the physical nature of transmission zeros, which has been the subject of several discussions, is clarified. The existence of two types of transmission zeros, modal and interaction zeros, is pointed out. The first ones are due to the structure's resonances, while the second ones are due to the interaction between resonant modes. The latter are the only ones present in symmetrical nondegenerate structures.

Fringe effects are accounted for in a simple way, namely by introducing in the magnetic wall model effective parameters for each resonant mode, while other methods presented until now are limited by the impossibility of taking into account fringe effects in an adequate way. It is also shown that these effects should not be neglected, since they can produce a modal inversion and, consequently, a strong alteration of the structure's frequency behavior.

Several experimental results performed on circular and rectangular structures are compared with the theoretical ones deduced by assuming the effective parameters suggested in [9]. A good agreement is obtained, particularly for the circular microstrips; for rectangular microstrips a better characterization of the alterations due to fringing fields would be desirable. To that purpose it is necessary to adopt a full-wave analysis for calculating the resonant frequencies and the coupling between the lines and the resonant modes. This can be done on the basis of some methods presented in the literature [21]–[23].

Finally, it is worth pointing out that the method could be also adopted for planar circuits with any geometry, for instance through a finite element analysis.

Acknowledgment

The authors are indebted to the Servizio Microonde of the Elettronica S.p.a. for aid in the measurements; they also wish to thank B. De Santis for drawing the diagrams.

References

[1] B. Bianco and S. Ridella, "Nonconventional transmission zeros in distributed rectangular structures," *IEEE Trans. Microwave Theory Tech.*, vol. MTT-20, pp. 297–303, May 1972.

[2] G. Kompa, "S-Matrix computation of microstrip discontinuities with a planar waveguide model," *Arch. Elek. Übertragung*, vol. 30, pp. 58–64, Feb. 1976.

[3] G. Kompa and R. Mehran, "Planar waveguide model for calculating microstrip components," *Electron. Lett.*, vol. 11, pp. 459–460, Sept. 1975.

[4] T. Okoshi, Y. Uehara, and T. Takeuchi, "The segmentation method—An approach to the analysis of microwave planar circuits," *IEEE Trans. Microwave Theory Tech.*, vol. MTT-24, pp. 662–668, Oct. 1976.

[5] T. Okoshi and T. Miyoshi, "The planar circuit—An approach to microwave integrated circuitry," *IEEE Trans. Microwave Theory Tech.*, vol. MTT-20, pp. 245–252, Apr. 1972.

[6] P. Silvester, "Finite-element analysis of planar microwave networks," *IEEE Trans. Microwave Theory Tech.*, vol. MTT-21, pp. 104–108, Feb. 1973.

[7] G. D'Inzeo, F. Giannini, and R. Sorrentino, "Theoretical and experimental analysis of non-uniform microstrip lines in the frequency range 2–18 GHz," in *Proc. of 6th European Microwave Con.* (Rome, Italy) pp. 627–631, 1976.

[8] W. Menzel and I. Wolff, "A method for calculating the frequency-dependent properties of microstrip discontinuities," *IEEE Trans. Microwave Theory Tech.*, vol. MTT-25, pp. 107–112, Feb. 1977.

[9] I. Wolff and N. Knoppik, "Rectangular and circular microstrip disk capacitors and resonators," *IEEE Trans. Microwave Theory Tech.*, vol. MTT-22, pp. 857–864, Oct. 1974.

[10] G. Kompa, "Excitation and propagation of higher modes in microstrip discontinuities," presented at *Proc. of 3th European Microwave Conf.*, Brussels, 1973.

[11] I. Wolff, G. Kompa, and R. Mehran, "Calculation methods for microstrip discontinuities and T-junctions," *Electron. Lett.*, vol. 8, pp. 177–179, Apr. 1972.

[12] B. Bianco, M. Granara, and S. Ridella, "Comments on the existence of transmission zeros in microstrip discontinuities," *Alta Frequenza*, vol. XLI, pp. 533E–534E, Nov. 1972.

[13] B. Bianco, M. Granara, and S. Ridella, "Filtering properties of two-dimensional lines' discontinuities," *Alta Frequenza*, vol. XLII, pp. 140E–148E, July 1973.

[14] K. Kurokawa, *An Introduction to the Theory of Microwave Circuits.* New York: Academic Press, 1969, ch. 4.

[15] P. M. Morse and H. Feshbach, *Methods of Theoretical-Physics*, vol. I. New York: McGraw-Hill, 1953, ch. 6, pp. 716–719.

[16] H. A. Wheeler, "Transmission-line properties of parallel wide strips separated by a dielectric sheet," *IEEE Trans. Microwave Theory Tech.*, vol. MTT-13, pp. 172–185, Mar. 1965.

[17] M. V. Schneider, "Microstrip lines for microwave integrated circuits," *Bell Syst. Tech. J.*, vol. 48, pp. 1421–1444, May 1969.

[18] ——, "Microstrip dispersion," *Proc. Inst. Elect. Electron. Engrs.*, vol. 60, pp. 144–146, 1972.

[19] G. N. Tsandoulas and W. J. Ince, "Modal inversion in circular waveguides—Part I: Theory and phenomenology," *IEEE Trans. Microwave Theory Tech.*, vol. MTT-19, pp. 386–392, Apr. 1971.

[20] F. Giannini, P. Maltese, and R. Sorrentino, "Liquid crystal technique for field detection in microwave integrated circuitry," *Alta Frequenza*, vol. XLVI, pp. 80E–88E, Apr. 1977.

[21] S. Akhtarzad and P. B. Johns, "Three-dimensional transmission-line matrix computer analysis of microstrip resonators," *IEEE Trans. Microwave Theory Tech.*, vol. MTT-23, pp. 990–997, Dec. 1975.

[22] R. Jansen, "High-order finite element polinomials in the computer analysis of arbitrarily shaped microstrip resonators," *Arch. Elek. Übertragung*, vol. 30, pp. 71–79, Feb. 1976.

[23] T. Itoh, "Analysis of microstrip resonators," *IEEE Trans. Microwave Theory Tech.*, vol. MTT-22, pp. 946–951, Nov. 1974.

Paper 11.5

Method of Analysis of Planar Networks Including Radiation Loss

ROBERTO SORRENTINO, MEMBER, IEEE, AND STEFANO PILERI, MEMBER, IEEE

Abstract—**A general approach to the analysis of microwave planar structures, specifically intended to account for radiation loss is presented. By expanding the internal electromagnetic (EM) field in terms of resonant modes, also the external field is obtained in the form of a series, each term of which corresponds to the field radiated by a resonant mode excited in the structure. Neglecting the effect of the thin dielectric substrate on the power radiated and the coupling between the modes, occurring at the lateral surface of the structure, a simplified formulation is obtained which is shown to be in very good agreement with experiments, performed in the frequency range 2–12.4 GHz, which could not be explained on the basis of a lossless model.**

I. Introduction

THE PROBLEM of evaluating radiation losses by microstrip configurations has been approached in the literature mainly in the case of microstrip discontinuities (particularly open ends or open-ended resonators) under a quasi-TEM approximation [1]–[7]. The results obtained for single microstrip discontinuities, however, can hardly be used for predicting radiation losses in many circuit applications. When several discontinuities are present, in fact, coupling between them should be taken into account in order to evaluate the resultant radiation loss.

A more general approach to the above problem is presented in this paper. The method is an extension of the resonant mode expansion technique formerly developed for characterizing lossless planar structures [8]. A similar technique has been also used in theoretical investigations on microstrip antennas [9]–[13] for which single resonant mode approximation [14]–[16] may cause notable errors [12]. These microstrip antenna theories have a limited frequency range of applicability due to the fact that fringe field effects are not accounted for in an adequate way.

A formal solution for the electromagnetic (EM) field inside and outside a planar network is derived in Section II. This is obtained by expanding the internal field in terms of the eigenfunctions of the structure with lateral magnetic walls and expressing the external field in terms of the equivalent magnetic currents flowing over the lateral walls; enforcing the boundary conditions at these walls results in an infinite set of equations in the EM field expansion coefficients. The analysis shows that coupling between resonants modes arising from the inhomogeneous boundary conditions at the walls of the structure can be accounted for by introducing a mutual wall admittance properly defined for each pair of resonant modes, as well as a selfwall admittance for each mode. Computation of the EM field, however, should require the evaluation of the Green's function of the space outside the planar structure, which is not known.

An approximate solution of the general problem is then derived in Section II. On the one hand, in evaluating the external field, the Green's function of a half-space bounded by a perfectly conducting plane is used. Secondly, coupling between resonant modes is neglected so that only the complex wall admittance of each mode has to be calculated and a simple expression for the EM field expansion coefficients is obtained. The real part of the wall admittance may be calculated, by applying Poynting's theorem, in terms of the far field radiated by the mode. In this way an expression is obtained, which is identical to that derived by Richards and Lo [11], [12] on the basis of fairly heuristical considerations. On the contrary, the imaginary part of the wall admittance of the mode is evaluated through an effective dimension and an effective permittivity peculiar of the resonant mode, in much the same way as done by D'Inzeo *et al.* [8]. It is worth specifying that the dependence of the wall susceptance on the resonant mode cannot, in general, be neglected when wide-band simulations of the frequency behavior of the planar circuit are required. As pointed out in [8], [17], in fact, this phenomenon is responsible in particular for modal inversion occurring in microstrip resonant structures, which cannot be predicted by simply ascribing to the structure a unique effective permittivity. The approximate solution therefore combines the results by Richards and Lo, with regard to the effect of real power loss and those by D'Inzeo *et al.*, with regard to the effects of reactive power (fringing fields). A specific example is finally discussed in Section IV, showing that experimental results performed in the frequency range 2–12.4 GHz, which, at particular frequencies, were in manifest disagreement with the previous lossless theory [8], are well matched by the present one.

II. General Approach to the Field Problem

Fig. 1 shows the geometry of a generic N-port planar network. Since we are concerned only with the evaluation of the effects of radiation loss, the metallic surfaces on the top and ground planes will be supposed to be perfectly

Manuscript received December 8, 1980; revised April 3, 1981. This work was supported in part by the Consiglio Nazionale delle Ricerche (C.N.R.), Italy.

The authors are with Istituto di Elettronica, Università di Roma, 00184 Rome, Italy.

Reprinted from *IEEE Trans. Microwave Theory Tech.*, vol. MTT-29, no. 9, pp. 942–948, Sept. 1981.

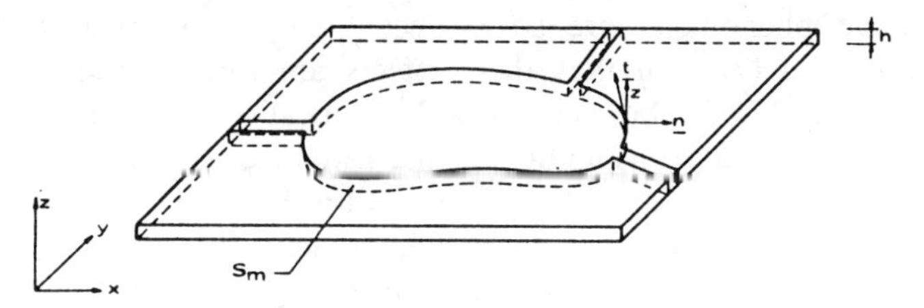

Fig. 1. The planar network.

conducting. Under the hypothesis of planarity ($\partial/\partial z=0$ for all field components), the EM field inside the volume V of the structure can be expressed as [8]

$$E=\hat{z}\Sigma_a e_a E_a \tag{1}$$

$$H=\frac{1}{j\omega\mu}\Sigma_a e_a \hat{z}\times\nabla_t E_a \tag{2}$$

where the E_a's are the orthonormalized eigenfunctions of the following eigenvalue problem:

$$\nabla_t^2 E_a + k_a^2 E_a = 0, \qquad \text{inside } V$$

$$\boldsymbol{n}\cdot\nabla_t E_a = 0, \qquad \text{on } S_m \tag{3}$$

$\boldsymbol{n}$ being the outward directed normal on the lateral surface S_m of V. The expansion coefficients of (1), (2) are expressed in terms of the tangential magnetic field component on S_m

$$e_a=\frac{j\omega\mu}{k_a^2-\omega^2\mu\epsilon}\int_{S_m}\boldsymbol{n}\times\boldsymbol{H}\cdot\boldsymbol{E}_a\,dS \tag{4}$$

where $\boldsymbol{E}_a=\hat{z}E_a$; μ, ϵ are the substrate permeability and permittivity, respectively. When fringe fields and radiation are ignored, only those portions of S_m, corresponding to the ports of the structure, contribute to the above integral, since $\boldsymbol{n}\times\boldsymbol{H}$ is assumed to be zero elsewhere on S_m. In such a case one may write

$$e_a=\frac{j\omega\mu}{k_a^2-\omega^2\mu\epsilon}f_a \tag{5}$$

where, as shown in [8], the quantity f_a can be calculated fairly easily in terms of the magnetic fields supported by the connecting lines.

In the actual case, a small tangential H-field component exists on the remaining portion Γ of S_m, which not only can produce a certain amount of power radiated, but also affects both the internal field distribution and the resonant frequencies of the structure. Instead of (5), the expression for the expansion coefficients is

$$e_a=\frac{j\omega\mu}{k_a^2-\omega^2\mu\epsilon}\left(\int_\Gamma \boldsymbol{n}\times\boldsymbol{H}\cdot\boldsymbol{E}_a\,dS+f_a\right). \tag{6}$$

The problem is, of course, to determine the tangential component of H over Γ. This can be done in the following way.

The field outside the planar structure can be evaluated, via the application of the theorem of equivalence, in terms of equivalent magnetic and electric currents flowing over the lateral walls of the structure [18]. In other words, the E- and H-fields tangential components over these walls allow the calculation of the EM field outside the structure. Let $G(\boldsymbol{r},\boldsymbol{r}')$ be the Green's function for the space outside V, the following expression for the external magnetic field is obtained:

$$\boldsymbol{H}^e=\int_{S_m}\Big[j\omega\epsilon(\boldsymbol{n}'\times\boldsymbol{E}')G(\boldsymbol{r},\boldsymbol{r}')-\frac{1}{j\omega\mu}(\boldsymbol{n}'\times\boldsymbol{E}'\cdot\nabla')\cdot\nabla'G(\boldsymbol{r},\boldsymbol{r}')+(\boldsymbol{n}'\times\boldsymbol{H}')\times\nabla'G(\boldsymbol{r},\boldsymbol{r}')\Big]\,dS' \tag{7}$$

where primes indicate dependence on source point $\boldsymbol{r}'$. In deriving (7), the presence of the connecting lines has been ignored; in particular, in the hypothesis that their widths are much smaller than the structure's dimension, the external field is evaluated as if the equivalent currents would extend over the entire lateral surface S_m. The same approximation will be used in (6), thus approximating the integral over Γ with the integral over S_m

$$e_a=\frac{j\omega\mu}{k_a^2-\omega^2\mu\epsilon}\left(\int_{S_m}\boldsymbol{n}\times\boldsymbol{H}^e\cdot\boldsymbol{E}_a\,dS+f_a\right) \tag{8}$$

where the condition $\boldsymbol{n}\times\boldsymbol{H}=\boldsymbol{n}\times\boldsymbol{H}^e$ on S_m has been used. In principle, the external magnetic field $\boldsymbol{H}^e$ may be obtained from (7) enforcing the condition that its tangential component over S_m be equal to $\boldsymbol{n}\times\boldsymbol{H}$. Through (1) and (8), in fact, $\boldsymbol{n}'\times\boldsymbol{E}'$ on the right-hand side of (7) can be expressed in terms of $\boldsymbol{n}'\times\boldsymbol{H}'$ over S_m, as to reduce the integrand to a function of $\boldsymbol{n}'\times\boldsymbol{H}'$ only. Imposing that the tangential component of the left-hand side of (7) be equal to $\boldsymbol{n}\times\boldsymbol{H}$ over S_m results in an integral equation in $\boldsymbol{n}\times\boldsymbol{H}$. Such an equation, however, may be very difficult to solve. A notable simplification is obtained under the hypothesis usually adopted that $\boldsymbol{n}\times\boldsymbol{H}$ over S_m is small enough as to neglect its contribution to the integral in (7). In such a way, as shown below, $\boldsymbol{H}^e$ can be expressed as a series whose coefficients are proportional to the e_a's, so that the continuity condition for $\boldsymbol{n}\times\boldsymbol{H}$ results in an infinite set of linear equations in the expansion coefficients.

Inserting (1) into (7), neglecting $\boldsymbol{n}'\times\boldsymbol{H}'$, and using the index b instead of a, one obtains

$$\boldsymbol{H}^e=\Sigma_b\frac{e_b}{j\omega\mu}\mathcal{H}_b \tag{9}$$

with

$$\mathcal{H}_b=\int_{S_m}E_b'\left[k^2\boldsymbol{t}'G(\boldsymbol{r},\boldsymbol{r}')+(\boldsymbol{t}'\cdot\nabla')\nabla'G(\boldsymbol{r},\boldsymbol{r}')\right]dS' \tag{9a}$$

where $\boldsymbol{t}=\hat{z}\times\boldsymbol{n}$, as shown in Fig. 1. Similarly, the external electric field is expressed as

$$\boldsymbol{E}^e=\Sigma_b e_b\mathcal{E}_b \tag{10}$$

with

$$\mathcal{E}_b=\int_{S_m}(\boldsymbol{n}'\times\boldsymbol{E}_b')\times\nabla'G(\boldsymbol{r},\boldsymbol{r}')\,dS'. \tag{10a}$$

Expressions (9)–(10) show that the external EM field consists of the superposition of an infinite number of field distributions; in other words, each internal field distribution $e_b\boldsymbol{E}_b$, corresponding to an eigenfunction of (3) produces an external electric field $e_b\mathcal{E}_b$ and an external magnetic field $(e_b/j\omega\mu)\mathcal{H}_b$.

Substitution of (9) into (8) yields to an infinite set of equations in the expansion coefficients

$$\left(k_a^2-\omega^2\mu\epsilon\right)e_a+j\omega\mu\Sigma_b e_b I_{ab}=j\omega\mu f_a, \qquad a=1,2,\cdots \tag{11}$$

where

$$I_{ab}=\frac{1}{j\omega\mu}\int_{S_m}\boldsymbol{E}_a\times\mathcal{H}_b\cdot\boldsymbol{n}\,dS. \tag{11a}$$

A more explicit expression for I_{ab} is obtained by inserting (9a) into (11a):

$$I_{ab}=\frac{1}{j\omega\mu}\int_{S_m}\int_{S_m}E_aE_b'\left[k^2\boldsymbol{t}\cdot\boldsymbol{t}'G(\boldsymbol{r},\boldsymbol{r}')\right.$$
$$\left.-(\boldsymbol{t}'\cdot\nabla')(\boldsymbol{t}\cdot\nabla)G(\boldsymbol{r},\boldsymbol{r}')\right]dS\,dS' \tag{12}$$

clearly showing that $I_{ba}=I_{ab}$, as must be in order that the planar network is reciprocal. The coefficient I_{ab} characterizes the interaction between the H-field produced by the bth mode and the E-field of the ath mode; it corresponds to the flux of mutual complex power resulting from the interaction of the two modes, which is given by

$$P_{ab}=\frac{1}{2}\int_{S_m}(e_a\boldsymbol{E}_a)\times\left(\frac{e_b}{j\omega\mu}\mathcal{H}_b\right)^*\cdot\boldsymbol{n}\,dS=\frac{e_ae_b^*}{2}I^*_{ab} \tag{13}$$

having accounted for the fact that $\boldsymbol{E}_a$ is a real quantity. The presence of the interaction term I_{ab} is such that, if one wishes to interpret the result expressed by (11) in terms of a wall admittance of the surface S_m, both an admittance for each resonant mode and a mutual admittance for pairs of resonant modes should be adopted. Equation (11), in fact, is equivalent to assume in (8)

$$\int_{S_m}\boldsymbol{n}\times\boldsymbol{H}\cdot\boldsymbol{E}_a\,dS=-\Sigma_b e_b Y_{ab}\int_{S_m}\boldsymbol{E}_a\cdot\boldsymbol{E}_b\,dS \tag{14}$$

where

$$Y_{ab}=\frac{1}{j\omega\mu}\left(\int_{S_m}\boldsymbol{E}_a\times\mathcal{H}_b\cdot\boldsymbol{n}\,dS\right)\Big/\left(\int_{S_m}\boldsymbol{E}_a\cdot\boldsymbol{E}_b\,dS\right)$$
$$=I_{ab}\Big/\left(\int_{S_m}\boldsymbol{E}_a\cdot\boldsymbol{E}_b\,dS\right) \tag{15}$$

may be interpreted as the mutual wall admittance between the ath and bth modes.

III. Simplified Theory

Since the Green's function required for evaluating the I_{ab} coefficients (12) is not known, an approximate solution of the problem stated in the previous section can be obtained by neglecting the presence of the thin dielectric substrate and the metallic patch, thus using the free space Green's function. (The presence of ground plane can be simply taken into account through a factor of 2.) Surface wave phenomena are so neglected, as it has been recognized [7], [10] that their effects are relatively small. System (11) can be then solved by truncation, so obtaining an approximate evaluation of the EM field as well as the other related quantities. The procedure for calculating the scattering parameters of the planar network will not be given here since it has been described in [8]. Calculation of the I_{ab} coefficients, however, remains quite onerous, since it involves a four-dimensional integration. In many cases this can be avoided by introducing a further simplifying hypothesis, as discussed below.

If radiation loss is not too high, at frequencies close to ω_a one may assume that only the ath mode is strongly excited inside the structure, so that (11) immediately gives

$$e_a=\frac{j\omega\mu f_a}{k_a^2-\omega^2\mu\epsilon+j\omega\mu I_{aa}}. \tag{16}$$

Although this expression is valid only in the proximity of ω_a, nevertheless its frequency behavior is such that e_a becomes smaller and smaller as ω departs from ω_a; if the resonant frequencies are sufficiently apart, it seems therefore reasonable to adopt (16) even at frequencies far from ω_a.[1]

As shown by (13), the quantity I_{aa} in the denominator of (16) is related to the complex power flowing out from the volume V; its real part corresponds to power radiated by the ath mode, while the imaginary part corresponds to an EM energy storage in the space outside V, which yields to a shift of the resonant frequency of the ath mode. Via application of the Poynting's theorem to the volume between S_m and a sphere of large radius, the real part of I_{aa}, or, equivalently, because of (15), the wall conductance of the ath mode, can be calculated in terms of the far EM field due to the ath mode. On the other hand, the shift of the resonant frequency due to the imaginary part of I_{aa}, i.e., to the wall susceptance of the ath mode, can be accounted for by ascribing to the structure an effective dimension and effective permittivity depending on the resonant mode, in much the same way as in [8]. It can be shown that the imaginary term in the denominator of (16), resulting from the real part of I_{aa}, is identical to that previously obtained by Richards and Lo [11] and substantially equivalent to that by Chew and Kong [13]. The difference between their result and the present one is in the way of evaluating the imaginary part of I_{aa}, which is due to

[1]Although, in general, this approximation might not be used for structures with very close resonant frequencies, nevertheless expression (16) may still be adopted with good approximation in many cases of this type, because of the following argument. Let us firstly observe that assuming the validity of (16) is equivalent to neglect the cross interaction between the resonant modes (thus $I_{ab}=0$ for $a\neq b$ in (11)). In a first order approximation, the magnetic field distribution $\mathcal{H}_b$ (9a) may be assumed, over S_m, to be simply proportional to $\boldsymbol{n}\times\boldsymbol{E}_b$, as in [13]. If the eigenvectors $\boldsymbol{E}_a$ and $\boldsymbol{E}_b$ are orthogonal also over S_m, then (13) shows that $I_{ab}=0$, i.e., there is no coupling between them. This is precisely the case of the (0,1) and (1,0) resonant modes of nearly square patches used as radiators for circular polarization [12].

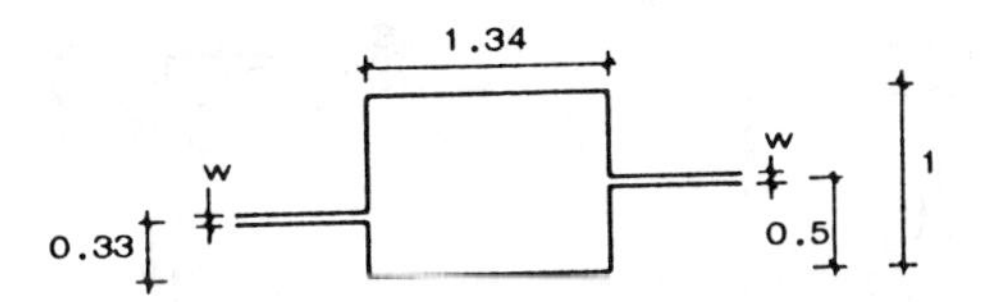

RESONANT FREQUENCIES

a ≡ $\omega_{1,0}$	f ≡ $\omega_{0,2}$
b ≡ $\omega_{0,1}$	g ≡ $\omega_{1,2}$
c ≡ $\omega_{1,1}$	h ≡ $\omega_{3,0}$
d ≡ $\omega_{2,0}$	i ≡ $\omega_{3,1}$
e ≡ $\omega_{2,1}$	l ≡ $\omega_{2,2}$

(a)

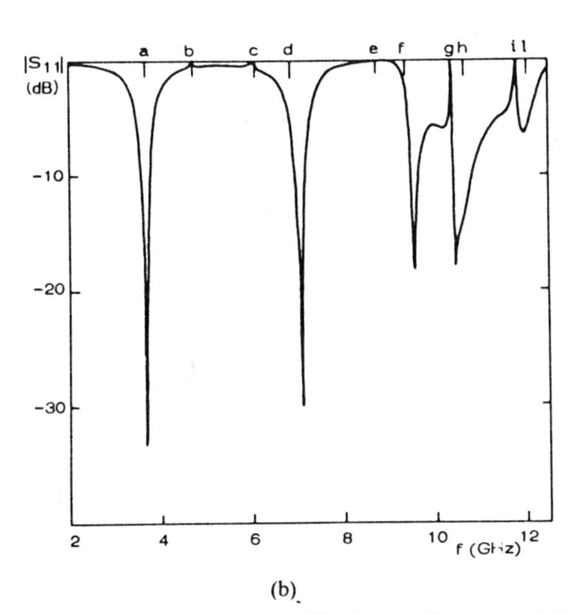

(b)

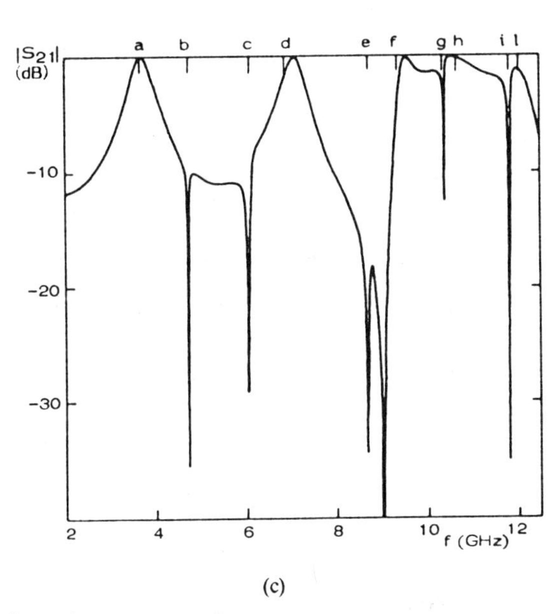

(c)

Fig. 2. (a) Geometry of the rectangular microstrip structure made on alumina substrate ($\epsilon_r = 10$, $h = 0.0635$ cm, $w = 0.06$ cm). Dimensions are in centimeters. (b), (c) Scattering parameters versus the frequency, evaluated according to the lossless theory [8].

the presence of fringing fields. As shown by D'Inzeo *et al.* [8], particularly in the case of circular microstrips, these fields produce strong alterations of the frequency behavior of planar networks, which, in wide-frequency ranges, cannot be accounted for by simply ascribing to the structure effective dimensions and a unique effective permittivity as in [11], [12]. The present theory therefore synthesizes the results by Richards and Lo and by D'Inzeo *et al.*

IV. Results

The simplified theory presented in the previous section, which neglects the mutual wall admittances between resonant modes, has been used to characterize a number of microstrip structures, made on alumina substrate ($\epsilon_r = 10$, $h = 0.635$ mm). A typical example of the results obtained is discussed here. The geometry of the structure, a nonsymmetrical rectangular one, is shown in Fig. 2(a). Fig. 2(b), (c), shows the frequency behavior up to X band of the scattering parameters calculated according to the lossless theory [8]. The resonant frequencies of the structure are indicated in the same diagrams. Resonant modes (m, n) which are odd with respect to the y-direction (n odd) can be excited from the first port, but are uncoupled to the second one: according to [8], so-called modal transmission zeros take place at their resonant frequencies. This is the case of all the transmission zeros presented by the structure, with the exception of the fourth and fifth ones, which are due to destructive interactions between resonant modes. While in the lossless case $|s_{11}| = |s_{22}|$, the experimental results quoted in Fig. 2(d), (e), (f) show notable differences between these two quantities: some sharp valleys in the reflection coefficient at the first port are totally absent in the reflection coefficient at the second port. Moreover, one can note that these valleys occur at frequencies where also valleys in the transmission coefficient s_{21} take place: this demonstrates that notable amounts of power are lost at

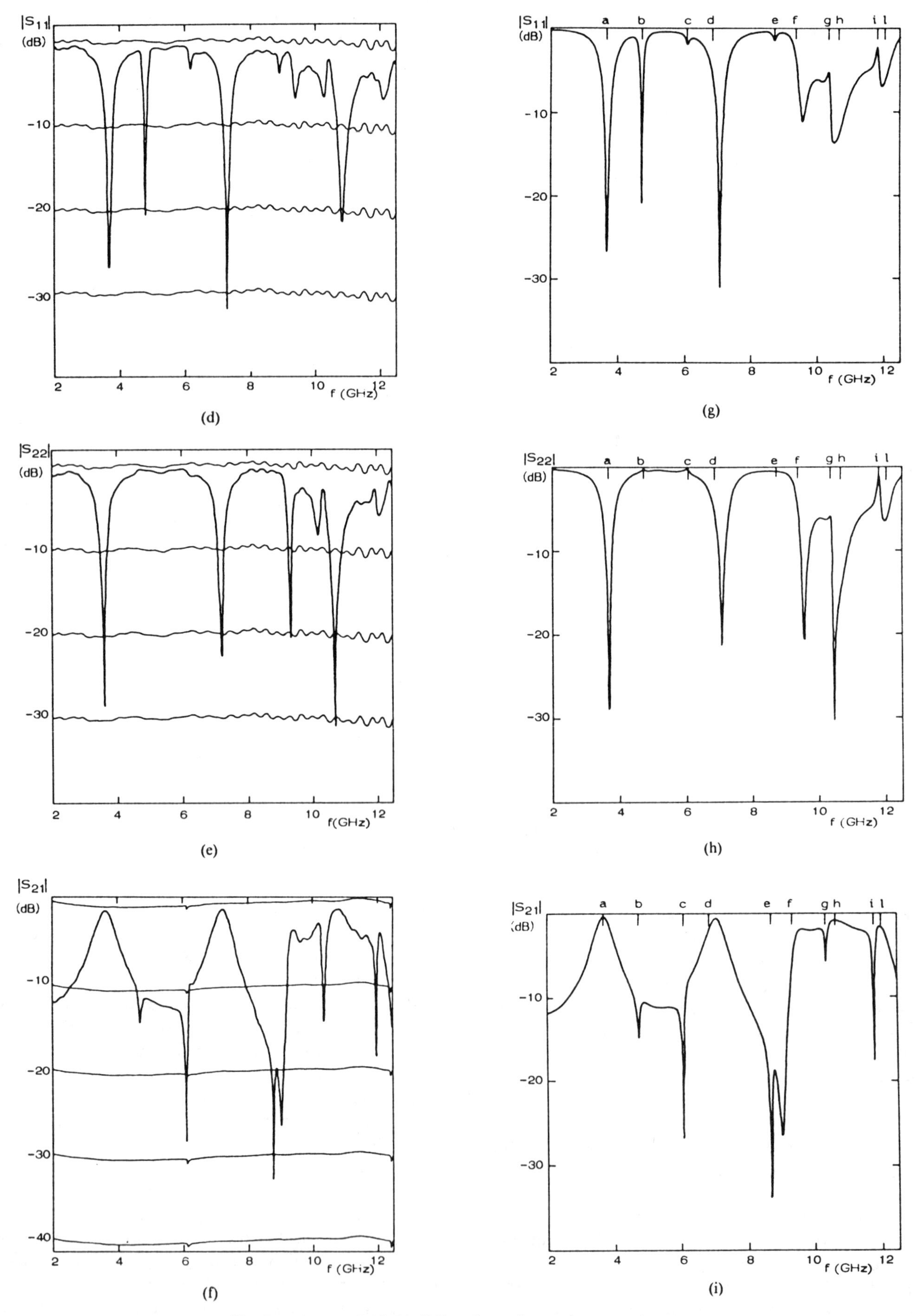

Fig. 2. (*Continued*) (d), (e), (f) Experimental scattering parameters versus the frequency. (g), (h), (i) Scattering parameters versus the frequency, evaluated according to the present theory.

these frequencies, when the structure is fed at the first port, so that the lossless theory yields to incorrect results for $|s_{11}|$.

The results obtained through the present theory are shown in Fig. 2(g), (h), (i). Besides obtaining a better agreement with the experiments for $|s_{22}|$ and $|s_{21}|$, the theory is capable of fully predicting the frequency behavior of $|s_{11}|$. The filtering properties of the structure can be easily explained from a physical point of view in terms of the resonant mode expansion of the EM field. The first sharp valley in s_{11}, for instance, is due to the excitation of the TM_{10} resonant mode, whose field distribution is independent of the y-direction; because of the symmetry of the field distribution, at this frequency the structure behavior is about the same when it is fed by the first port or by the second one and power is transmitted between them ($s_{11} \sim s_{22} \sim 0$, $|s_{21}| \sim 1$). It is worth noting that the valleys in $|s_{11}|$ and $|s_{22}|$ due to the excitation of the TM_{10} mode are not exactly coincident with its resonant frequency. This is due to the finite impedance of the connecting lines [19].

Let us now consider the second valley in s_{11}, which occurs at the resonant frequency of the TM_{01} mode, which is odd with respect to the y-direction. As previously noted, this is the frequency of a modal transmission zero. However, the field distribution is completely different if the structure is excited from the first port or from the second one: in the first case it is practically coincident with the field distribution of the TM_{01} mode, while, in the second case, it results from the superposition of several mode distributions [20]. While in both cases power is not transmitted from one to the other port ($s_{21} = s_{12} \sim 0$), when the structure is fed from the first port the TM_{01} mode is strongly excited and almost all the power is radiated ($s_{11} \sim 0$); if the structure is fed from the second port, this mode cannot be excited and the input power is reflected ($|s_{22}| \sim 1$).

A similar behavior is found at the frequencies of other modal transmission zeros; however, the power radiated is different for different resonant modes. It can be seen, for example, that the TM_{11} mode produces only a small valley in s_{11} at its resonant frequency. It is finally worth observing that the amount of radiated power does not strictly depend on the frequency, but mainly on the resonant mode excited in the structure; in the present case the maximum amount of radiated power occurs at the frequency of about 4.7 GHz, corresponding to the resonant frequency of the TM_{01} mode.

V. Conclusions

A general approach to the analysis of microwave planar networks including radiation loss has been presented. By expanding the internal field in terms of resonant modes, also the external field is obtained in the form of a series expansion, each term of which corresponds to the field radiated by a resonant mode. The resonant modes are coupled together by the inhomogeneous boundary conditions at the lateral surface of the planar structure. Computation of the EM field, however, should require the evaluation of the Green's function of the space outside the planar structure. It is then shown that, approximating the Green's function with that of a half-space bounded by a conducting plane and neglecting coupling between resonant modes, a simplified theory is obtained which synthesizes the results by Richards and Lo [11] and by D'Inzeo *et al.* [8]. Experiments on microstrip structures, which could not be explained on the basis of a lossless model, are shown to be in excellent agreement with the simplified theory.

Acknowledgment

Part of this work has been done by S. Pileri together with M. Bazzani and L. Monaco as their graduation thesis. The authors wish also to thank Prof. F. Giannini for helpful discussions and the Servizio Microonde of Elettronica S.p.a. for aid in the measurements.

References

[1] L. Lewin, "Radiation from discontinuities in strip-line," *Proc. Inst. Elec. Eng.*, vol. 107 C, pp. 163–170, 1960.

[2] ——, "Spurious radiation from microstrip," *Proc. Inst. Elec. Eng.*, vol. 125, pp. 663–642, July 1978.

[3] H. Sobol, "Radiation conductance of open-circuit microstrip," *IEEE Trans. Microwave Theory Tech.*, vol. MTT-19, pp. 885–887, Nov. 1971.

[4] G. Kompa, "Approximate calculation of radiation from open-ended wide microstrip lines," *Electron. Lett.*, vol. 12, pp. 222–224, Apr. 29, 1976.

[5] C. Wood, P. S. Hall, and J. R. James, "Radiation conductance of open-circuit low dielectric constant microstrip," *Electron. Lett.*, vol. 14, pp. 121–123, Feb. 16, 1978.

[6] F. Belohoubek and F. Denlinger, "Loss considerations for microstrip resonators," *IEEE Trans. Microwave Theory Tech.*, vol. MTT-23, pp. 522–526, June 1975.

[7] L. J. Van der Pauw, "The radiation of electromagnetic power by microstrip configurations," *IEEE Trans. Microwave Theory Tech.*, vol. MTT-25, pp. 719–725, Sept. 1977.

[8] G. D'Inzeo, F. Giannini, C. M. Sodi, and R. Sorrentino, "Method of analysis and filtering properties of microwave planar networks," *IEEE Trans. Microwave Theory Tech.*, vol. MTT-26, pp. 462–471, July 1978.

[9] K. R. Carver, "A modal expansion theory for the microstrip antenna," in *IEEE AP-S Int. Symp. Dig.*, vol. I, pp. 101–104, June 1979.

[10] Y. T. Lo, D. Solomon, and W. F. Richards, "Theory and experiment on microstrip antennas," *IEEE Trans. Antennas Propagat.*, vol. AP-27, pp. 137–145, Mar. 1979.

[11] W. F. Richards, and Y. T. Lo, "An improved theory for microstrip antennas and applications," in *IEEE-AP Int. Symp. Dig.*, (Seattle, WA), pp. 113–116, June 1979.

[12] Y. T. Lo, W. F. Richards, and D. D. Harrison, "An improved theory for microstrip antennas and applications—Part I," Interim Rep. RADC-TR-79-111, May 1979.

[13] W. C. Chew and J. A. Kong, "Radiation characteristics of a circular microstrip antenna," *J. Appl. Phys.*, vol. 51, pp. 3907–3915, July 1980.

[14] A. G. Derneryd, "A theoretical investigation of the rectangular microstrip antenna element," *IEEE Trans. Antennas Propagat.*, vol. AP-26, pp. 532–535, July 1978.

[15] ——, "Analysis of the microstrip disk antenna element," *IEEE Trans. Antenna Propagat.*, vol. AP-27, pp. 660–664, Sept. 1979.

[16] A. G. Derneryd and A. G. Lind, "Cavity model of the rectangular microstrip antenna," *IEEE Trans. Antenna Propagat.*, vol. AP-27, pp. 12-1/12-11, Oct. 1979.

[17] G. D'Inzeo, F. Giannini, and R. Sorrentino, "Theoretical and experimental analysis of non-uniform microstrip lines in the frequency range 2–10 GHz," in *Proc. 6th European Microwave*

Conf., (Rome, Italy), pp. 627–631, 1976.
[18] S. A. Schelkunoff, *Electromagnetic Waves.* Princeton, NJ: Van Nostrand, 1943, ch. 6.
[19] G. D'Inzeo, F. Giannini, and R. Sorrentino, "Wide-band equivalent circuits of microwave planar networks," *IEEE Trans. Microwave Theory Tech.*, vol. MTT-28, pp. 1107–1113, Oct. 1980.
[20] G. D'Inzeo, F. Giannini, P. Maltese, and R. Sorrentino, "On the double nature of transmission zeros in microstrip structures," *Proc. IEEE*, vol. 66, pp. 800–802, July 1978.

The Segmentation Method—An Approach to the Analysis of Microwave Planar Circuits

T. OKOSHI, MEMBER, IEEE,
Y. UEHARA, AND T. TAKEUCHI

***Abstract*—In many practical planar circuitries, the circuit pattern can be divided into several segments which themselves have simpler shapes such as rectangles. The segmentation method proposed in this short paper is a method in which the characteristics of a planar circuit are computed by combining those of the segmented elements. It features a relatively short computer time required. The principle and computer algorithm are described. Finally, as an example, the application of the proposed method to the trial-and-error optimum design of a ladder-type 3-dB hybrid is described.**

I. Introduction

A planar circuit is a 2-dimensional circuit that should be classified between the distributed-constant (1-dimensional) circuit and waveguide (3-dimensional) circuit. It is defined as an electrical circuit having dimensions comparable to the wavelength in two directions, but much less thickness in one direction [1].

Several methods have been known for the analysis of planar circuits. When the circuit pattern is as simple as square, rectangular, circular, or annular, the impedance matrix can be obtained in a series-expansion form from the Green's function of the wave equation [1]. When the circuit pattern is more arbitrary, the numerical analysis based upon the contour-integral representation of the wave equation is most efficient [1]. Other numerical approaches applicable to arbitrary circuit patterns are the variational method, relaxation method, and the finite-element method [2].

When the circuit pattern is entirely arbitrary, we must rely upon one of those numerical analyses. However, in actual planar circuitry, an entirely arbitrary circuit pattern is not very common; in many cases the pattern consists of several "segments," which themselves have simpler shapes such as rectangles.

The "segmentation method" proposed in this short paper is a method in which the characteristics of a planar circuit are computed by combining those of the segmented elements. It features relatively short computer time required in the analysis.

In this short paper, the principle and computer algorithm of the segmentation method are described first in Section II. In Section III, to demonstrate the feasibility of the proposed method, its application to the trial-and-error optimum design of a ladder-type 3-dB hybrid is described as an example. The obtained optimum design is verified by the experiment in Section IV.

II. Segmentation Method

A. Basic Concepts

A planar circuit consisting of two ground conductors and a circuit-conductor plate sandwiched by them as shown in Fig. 1 is called a triplate-type planar circuit. In the following only the triplate-type circuit is considered.

The basic equation governing the electromagnetic field in a

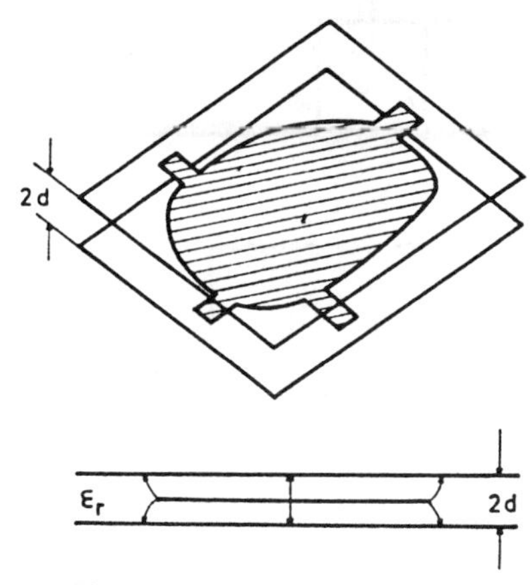

Fig. 1. A triplate-type planar circuit.

triplate-type planar circuit is the 2-dimensional wave equation

$$(\nabla_T^2 + k^2)V = 0 \tag{1}$$

where

$$\nabla_T^2 = \frac{\partial^2}{\partial x^2} + \frac{\partial^2}{\partial y^2}, \qquad V = E_z d, \qquad k^2 = \omega^2 \varepsilon \mu$$

ε and μ denote the permittivity and permeability of the spacing material, ω the angular frequency, d the spacing between conductors, and E_z denotes the electric-field strength normal to the conductors.

To analyze such a circuit means to solve the wave equation under given boundary conditions, and determine the circuit parameters of its equivalent n-port network as functions of the frequency. When the circuit pattern is simple (for instance, rectangular), the problem may be solved analytically [1]. The segmentation method is a method of the analysis suited to somewhat more complex patterns, as shown in Fig. 2(a). We divide such a circuit pattern into simpler segments, as shown in Fig. 2(b), compute the characteristics of each rectangular segments from their Green's functions [1], and finally obtain the characteristics of the entire circuit by synthesizing the characteristics of the segments.

B. Interface Network

To facilitate the synthesis of the segments in a unified manner, a new circuit called the "interface network" is introduced. As shown in Fig. 2(c), all connections between segments are made through this interface network. The interface network is nothing but a directly connecting network; it only connects directly those couples of ports facing each other, as shown in Fig. 2(c). However, a unified formulation of the synthesis is only possible with the aid of this fictitious network.

In the following, we consider a two-port circuit for the simplicity of the formulation. Fig. 2(c) may also be expressed as Fig. 2(d). We define, with respect to ports $1p,2p,1,2,\cdots,n$, incident and reflected waves of the interface network as

$$a_{1p},a_{2p},a_1,a_2,\cdots,a_n$$

and

$$b_{1p},a_{2p},b_1,b_2,\cdots,b_n.$$

In numbering ports, we use (numerals + p) for those of the

Manuscript received December 24, 1975; revised April 2, 1976.
The authors are with the Department of Electronic Engineering, University of Tokyo, Bunkyo-ku, Tokyo 113, Japan.

Reprinted from *IEEE Trans. Microwave Theory Tech.*, vol. MTT-24, pp. 662–668, Oct. 1976.

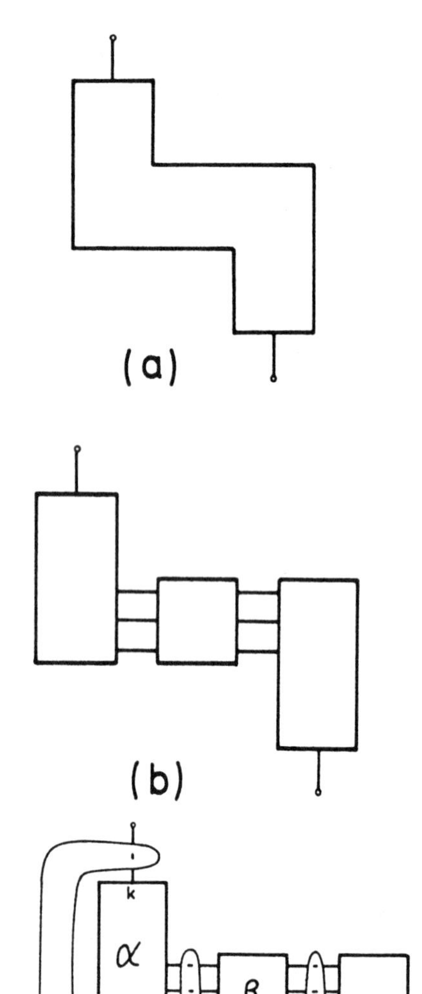

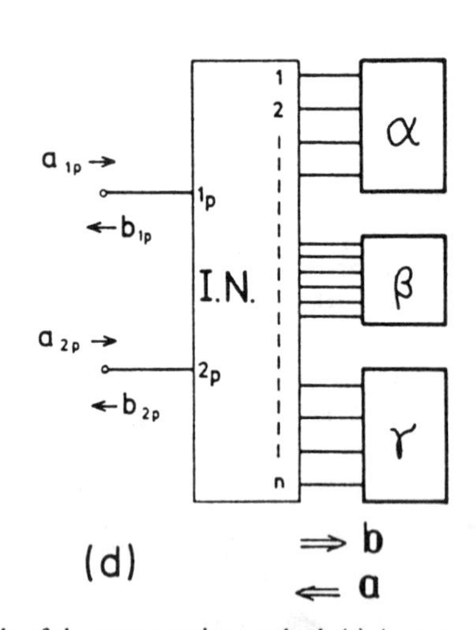

Fig. 2. Principle of the segmentation method. (a) An example of a circuit pattern to which the segmentation method is suitable. (b) Dividing into segments. (c) Introducing the interface network (I.N.). (d) A modified representation of Fig. 2(c).

interface network connected to external ports, and simple numerals for those connected to segmentary circuits.

The aforementioned incident and reflected waves are related in terms of the scattering matrix S_N of the interface network as

$$\begin{bmatrix} b_{1p} \\ b_{2p} \\ b_1 \\ b_2 \\ \vdots \\ b_n \end{bmatrix} = [S_N] \begin{bmatrix} a_{1p} \\ a_{2p} \\ a_1 \\ a_2 \\ \vdots \\ a_n \end{bmatrix} \tag{2}$$

where elements of $[S_N]$ are given by

$$[S_N]_{ij} = \begin{cases} 1, & \text{if ports } i \text{ and } j \text{ are connected} \\ 0, & \text{if ports } i \text{ and } j \text{ are not connected} \end{cases} \tag{3}$$

where $i,j = 1p,2p,1,2,\cdots,n$. When $i = j$, $[S_N]_{ij} = 0$.

C. Computation of S-Matrix

If ports $1p$ and $2p$ are connected directly in the interface network to ports k and l, respectively, S_N may be expressed as

$$[S_N] = \left[\begin{array}{cc|c} 0 & 0 & S_1^{\rightarrow} \\ 0 & 0 & S_2^{\rightarrow} \\ \hline S_1^{\downarrow} & S_2^{\downarrow} & S_n \end{array}\right] \tag{4}$$

where $S_1^{\rightarrow}$ and $S_2^{\rightarrow}$ are row vectors, and $S_1^{\downarrow}$ and $S_2^{\downarrow}$ are column vectors defined by

$$S_1^{\rightarrow} = (0,\cdots,0,\underset{\hat{k}}{1},0,\cdots,0)$$

$$S_2^{\rightarrow} = (0,\cdots,0,\underset{\hat{l}}{1},0,\cdots,0) \tag{4a}$$

$$S_1^{\downarrow} = [S_1^{\rightarrow}]^t$$

$$S_2^{\downarrow} = [S_2^{\rightarrow}]^t \tag{4b}$$

where the superscript t denotes transposition. The submatrix S_n in (4) expresses connections between segments.

The S-matrix of the entire circuit should be a (2×2) matrix giving b_{1p} and b_{2p} in terms of only a_{1p} and a_{2p}. However, we should note that, because $1p$ and $2p$ are connected to k and l, respectively,

$$b_{1p} = a_k \qquad b_{2p} = a_l \tag{5}$$

hold. Therefore, in the following we compute a_k and a_l in terms of a_{1p} and a_{2p}.

From (2), taking out n rows from the bottom, we obtain

$$\begin{bmatrix} b_1 \\ b_2 \\ \vdots \\ b_n \end{bmatrix} = [S_n] \begin{bmatrix} a_1 \\ a_2 \\ \vdots \\ a_n \end{bmatrix} + S_1^{\downarrow} a_{1p} + S_2^{\downarrow} a_{2p}. \tag{6}$$

On the other hand, we may also write

$$\begin{bmatrix} a_1 \\ a_2 \\ \vdots \\ a_n \end{bmatrix} = [S_c] \begin{bmatrix} b_1 \\ b_2 \\ \vdots \\ b_n \end{bmatrix} \tag{7}$$

where S_c is the composite S-matrix of the segmentary circuits expressed as

$$[S_c] = \begin{bmatrix} S_\alpha & & 0 \\ & S_\beta & \\ 0 & & S_\gamma \end{bmatrix} \tag{8}$$

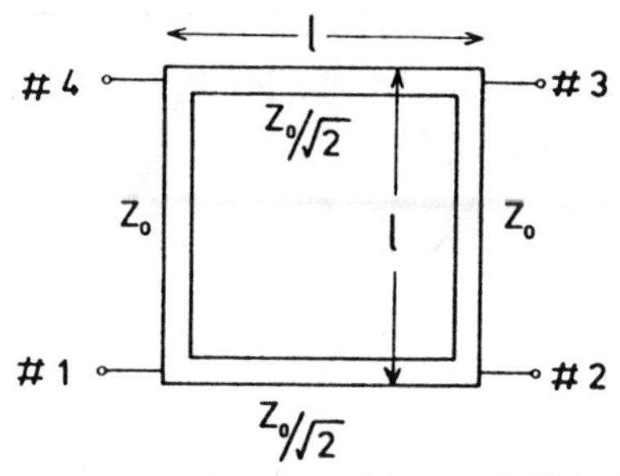

Fig. 3. Basic configuration of a ladder-type 3-dB hybrid circuit.

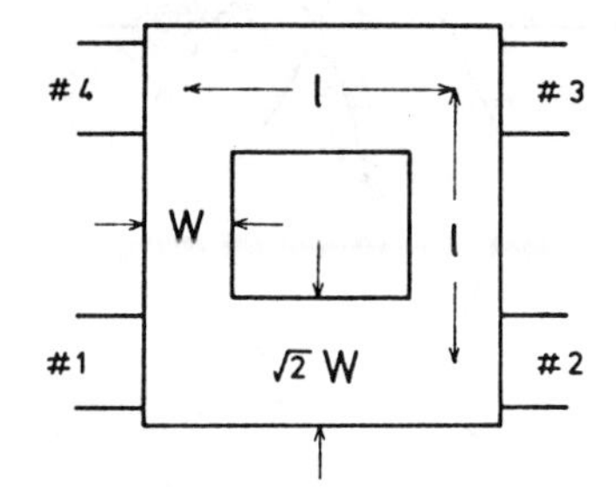

Fig. 4. A ladder-type planar circuit which requires planar-circuit approach.

S_α, S_β, and S_γ denoting S-matrices of the segments. When the segments are rectangular these matrices are obtainable analytically [1, eq. (20)].

Eliminating the b-vector from (6) and (7), we obtain

$$\begin{bmatrix} a_1 \\ a_2 \\ \vdots \\ a_n \end{bmatrix} = TS_1^{\downarrow} a_{1p} + TS_2^{\downarrow} a_{2p} \tag{9}$$

where

$$T = [E - S_c S_n]^{-1} [S_c] \tag{9a}$$

E denoting an $(n \times n)$ unit matrix. From (4b)

$$TS_1^{\downarrow} = \text{the } k\text{th column of } T$$
$$TS_2^{\downarrow} = \text{the } l\text{th column of } T. \tag{10}$$

Putting (5) into (9), we finally obtain

$$b_{1p} = a_k = T_{kk} a_{1p} + T_{kl} a_{2p}$$
$$b_{2p} = a_l = T_{lk} a_{1p} + T_{ll} a_{2p}. \tag{11}$$

Equation (11) shows that the S-matrix of the entire circuit is a (2×2) submatrix of T; it consists of four elements of T at crosspoints between the kth, lth columns, and the kth, lth rows. Thus the circuit characteristics may be derived from S_n and S_c.

In the aforementioned derivation, the entire circuit has been assumed as a 2-port circuit. The characteristics of a multiport circuit may be obtained in a similar manner.

III. Optimum Design of a Ladder-Type 3-dB Hybrid—An Example of Application

To demonstrate the usefulness of the segmentation method in design problems, its application to the optimum design of a ladder-type 3-dB hybrid will be described as an example.

A. Computer Analysis

We consider a ladder-type 3-dB hybrid as shown in Fig. 3. At a frequency for which $l = \lambda_g/4$, where λ_g denotes the wavelength in the stripline, this circuit shows hybrid characteristics; when power is incident at port 1, it will appear at ports 2 and 3 half-and-half, and no power appears at port 4. Similar relations hold for inputs at other ports.

However, in the following two cases, the stripline width becomes comparable to the quarter wavelength as shown in Fig. 4. The first case is when the required impedance level of the hybrid is relatively low as in a mixer circuit employing Schottky-barrier diodes which have low matching impedance. The second case is when the frequency is relatively high and the length of the arms are shortened. In either case, we have to deal with the circuit as a planar circuit.

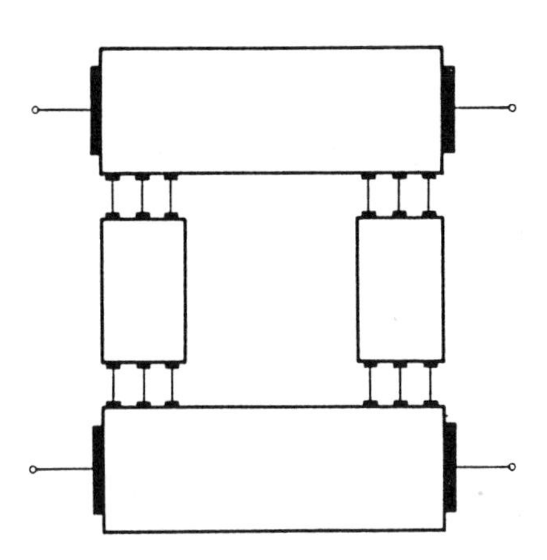

Fig. 5. Segmentation of the ladder-type 3-dB hybrid.

The segmentation method is most conveniently used in analyzing such a planar hybrid circuit. In the following analyses, we assume that the circuit is of triplate type, with dielectric-spacing material having ε_r (relative dielectric constant) = 2.53 and d = 1.52 mm, in accordance with the material used in the experiment.

We first divide the circuit into four segments as shown in Fig. 5. Next the S-matrices of each segment are computed by transforming the Z-matrices given in [1] into S-matrix form. Finally, we synthesize them by the method previously described to obtain the S-matrix of the entire 4-port.

The method of taking into account the width of external terminals (see Fig. 5) might require some comment. When the external terminals are assumed to have certain widths, components of the impedance matrix are computed by averaging the Green's function of the wave equation over the widths of the corresponding terminals [1]. When the terminal shifts, the averaging range should be moved accordingly.

B. Starting Circuit Pattern

As a starting circuit pattern, we take one as shown in Fig. 5 and assume that l = 10.8 mm for each arm. According to the distributed-constant model (Fig. 4), this length and ε_r = 2.53 lead to f_0 (center frequency) = 4.4 GHz. Further, if we assume Z_0 = 50 Ω, we obtain W = 3.6 mm and $\sqrt{2}\,W$ = 5.1 mm. Comparison of l and W suggests that the planar-circuit model must be employed in exact analysis and design.

The result of the segmentation-method analysis of the aforementioned circuit pattern is shown in Fig. 6(b). These characteristics show increase of the center frequencies (approximately 20 percent above f_0) as well as asymmetries. Obviously, this starting circuit pattern requires improvement.

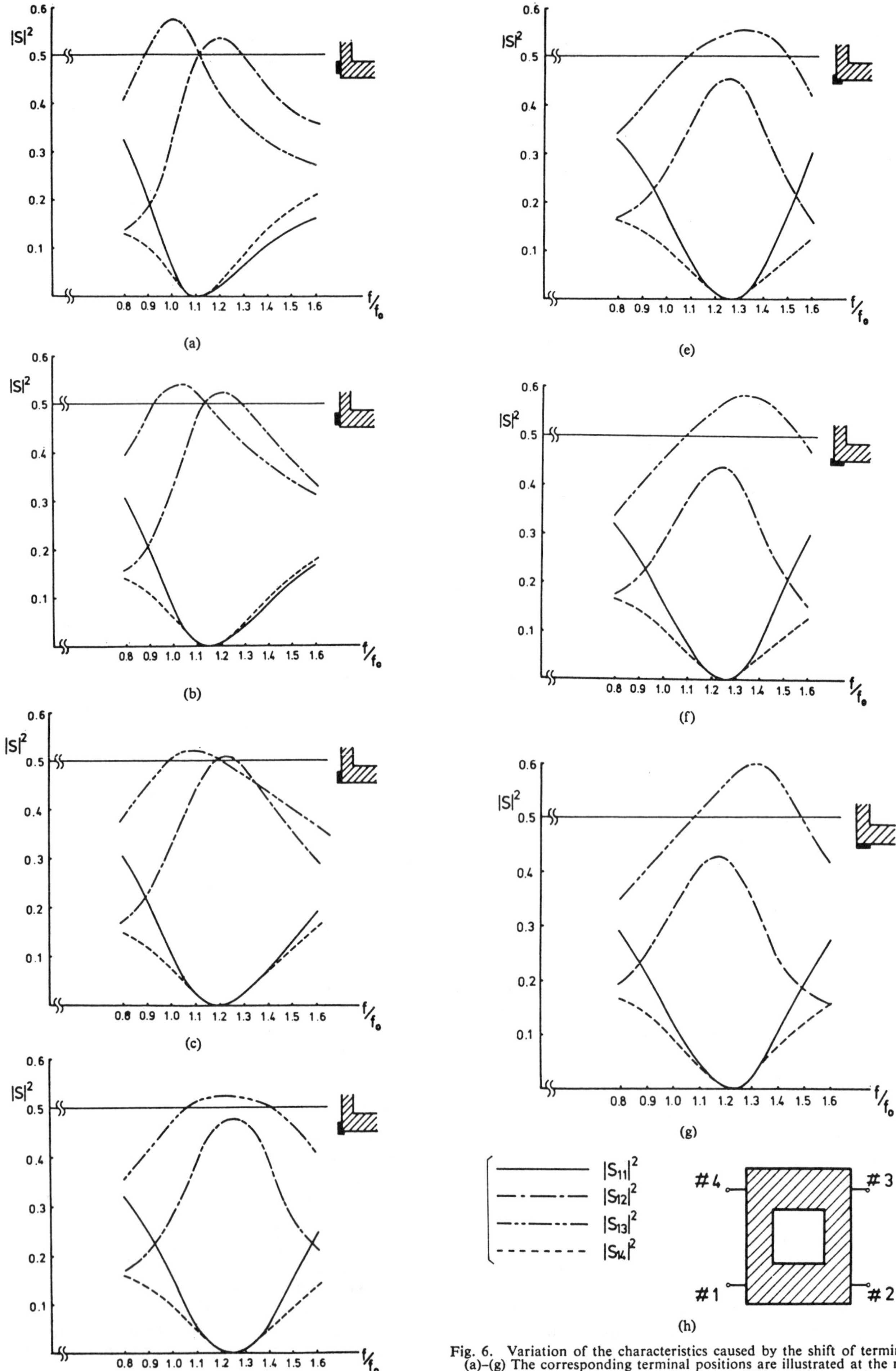

Fig. 6. Variation of the characteristics caused by the shift of terminals. (a)–(g) The corresponding terminal positions are illustrated at the right top. (h) The curve designations. Symbol f_0 denotes the center frequency based upon the distributed-constant theory.

C. Figure of Merit

Starting at the aforementioned circuit pattern we look for better ones. As the figure of merit of a circuit pattern we use the "bandwidth" defined as follows: at any frequency within the "band," incident power at port 1 is divided and appears at ports 2 and 3 with percentages less than 55 percent and greater than 45 percent, and at ports 1 and 4 with percentages less than 5 percent. The circuit pattern is modified successively upon trial-and-error basis so that the "bandwidth" be maximized and the symmetry of the frequency response be optimized.

In modifying the circuit pattern, symmetries of the circuit pattern with respect to two axes remain. Hence, the circuit characteristics may be determined by four parameters: $|S_{11}|^2$, $|S_{12}|^2$, $|S_{13}|^2$, and $|S_{14}|^2$.

D. Variation of Characteristics by Modification of Circuit Pattern

To collect basic information needed in the optimization, we first investigate the variation of characteristics caused by modification of design parameters.

First, effects of the position of terminals were investigated. The computed characteristics are shown in Fig. 6(a)–(g). This figure shows that when terminals move from the ends of the $\sqrt{2}\,W$-width arms toward those of the W-width arms, the S-parameters vary as shown in Table I(a). The optimum pattern is found somewhere between Figs. 6(c) and Fig. 6(d). The obtained terminal position is shown in Fig. 7, for which the center frequencies of $|S_{11}|^2$, $|S_{12}|^2$, and $|S_{14}|^2$ show the best coincidence. Hereafter, we consider only the terminal positions shown in Fig. 7.

Next, for further improvement of the characteristics, variations caused by shortening the $\sqrt{2}\,W$-width arms were investigated. It was found that the variations as shown in Table I(b) took place. (The detailed data are omitted to save space.)

Finally, the variations caused by widening the $\sqrt{2}\,W$-width arms were investigated. It was found that the variations as shown in Table I(c) took place.

If we disregard a common shift of the center frequencies, the variations of the length and/or width of the W-width arms may be replaced by those of the $\sqrt{2}\,W$-width arms.

E. Optimum Circuit Pattern

Based upon the aforementioned tendencies obtained by the high-speed computer analyses using the segmentation method, we performed the final optimization as follows.

1) Determine the optimum position of ports so that center frequencies of $|S_{11}|^2$, $|S_{12}|^2$, and $|S_{14}|^2$ coincide with each other (Fig. 7).

2) Determine the optimum width of the $\sqrt{2}\,W$-width arms so that the peak value of $|S_{12}|^2$ is optimized. (At this stage, the center frequency of $|S_{13}|^2$ is somewhat deviated.)

3) Determine the length of the $\sqrt{2}\,W$-width arms so that the center frequency of $|S_{13}|^2$ coincides with the others.

The optimum circuit pattern obtained finally and the corresponding circuit characteristics are shown in Fig. 8.

IV. Comparison with Experiment

Experiment was performed using the optimum circuit pattern obtained theoretically to confirm the validity of the analysis and design procedures.

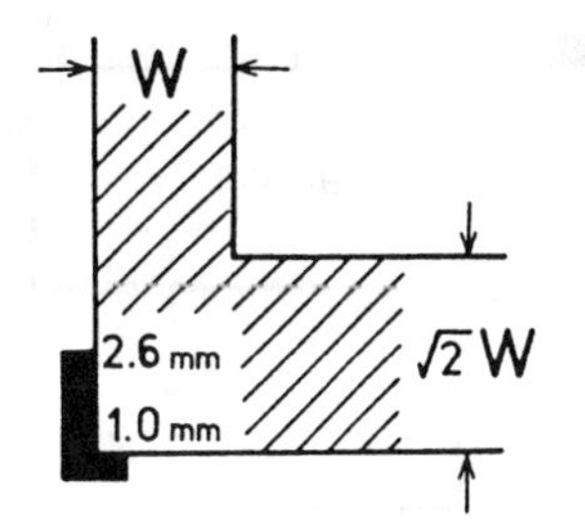

Fig. 7. The optimum terminal position.

TABLE I
VARIATIONS OF S-PARAMETERS CAUSED BY MODIFICATION OF THE CIRCUIT PATTERN

	$\lvert S_{11}\rvert^2$	$\lvert S_{12}\rvert^2$	$\lvert S_{13}\rvert^2$	$\lvert S_{14}\rvert^2$
Center frequency (p:peak, v:valley)	↗ v	× p	↗↗ p	↗ v
Peak or valley value	—	↘	↘↗	—

(a) Effects of the shift of terminal positions. (For their shift from ends the √2W-width arms toward those of the W-width arms.)

	$\lvert S_{11}\rvert^2$	$\lvert S_{12}\rvert^2$	$\lvert S_{13}\rvert^2$	$\lvert S_{14}\rvert^2$
Center frequency (p:peak, v:valley)	× v	× p	↗↗ p	× v
Peak or valley value	—	×	↘↗	—

(b) Effects of shortening the √2W-width arms.

	$\lvert S_{11}\rvert^2$	$\lvert S_{12}\rvert^2$	$\lvert S_{13}\rvert^2$	$\lvert S_{14}\rvert^2$
Center frequency (p:peak, v:valley)	× v	× p	↗ p	× v
Peak or valley value	—	↗↗	↘↘	—

(c) Effects of widening the √2W-width arms.

↗ : Increases gradually.
↗↗ : Increases abruptly.
↘↘ : Decreases abruptly.
↘↗ : Once decreases and then increases.
× : No remarkable variation.
— : These "valley" values are always almost zero.

A center conductor having dimensions shown in Fig. 9 was sandwiched by two ground conductors, with spacers of Rexolite 1422 (d = 1.52 mm, ε_r = 2.53 ± 0.03). The center conductor was produced by a photoetching technique. In the analysis described in Section III, the open-circuit boundary was assumed along the edges of the circuit. Actually, however, the electromagnetic field extends outward due to the edge effect. One method to take this effect into account in the analysis is to shift all the circuit boundaries outward by $0.442d$, in this case by 0.67 mm [3]. The circuit dimensions shown in Fig. 9 were chosen so that the corrected pattern agrees with Fig. 8.[1]

[1] However, due to error in the photoetching process, the measured arm lengths in the completed pattern (11.1 mm for all arms) were longer than the design value by 2 and 3 percent for the $\sqrt{2}W$-width and the W-width arms, respectively. The computed characteristics shown in Fig. 10 are based upon the actual circuit dimensions.

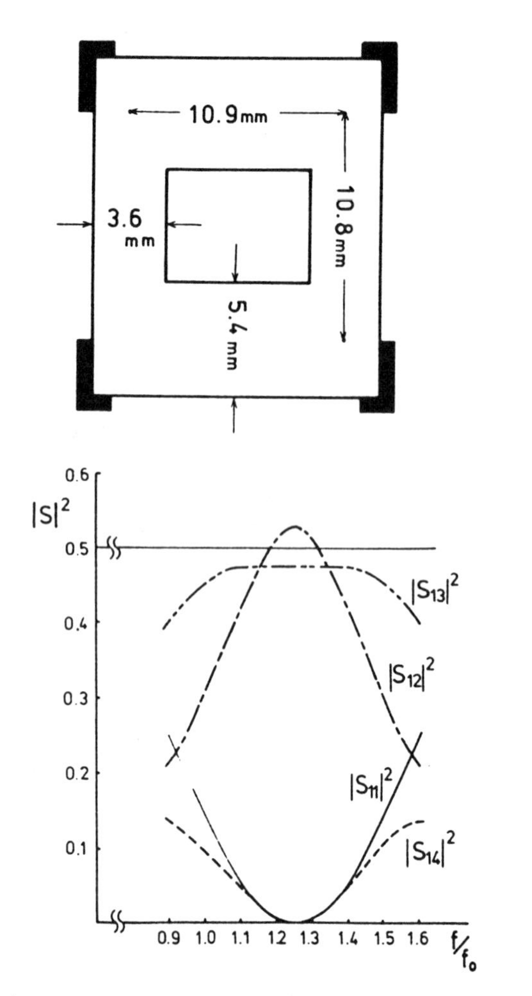

Fig. 8. The optimum circuit pattern and the corresponding characteristics.

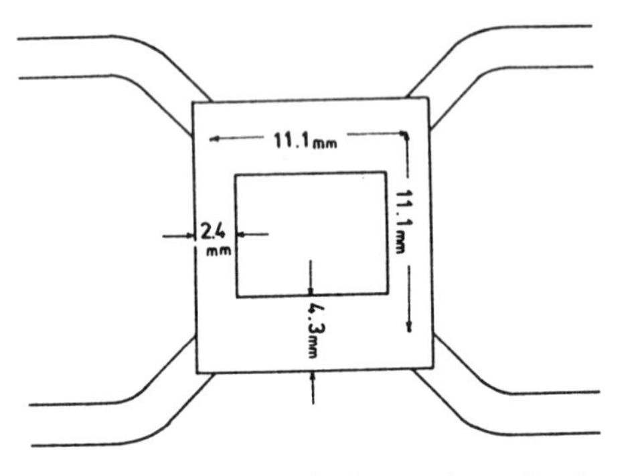

Fig. 9. The circuit pattern used in the experiment (see footnote).

The computed and measured circuit characteristics are shown in Fig. 10 by solid curves and dots, respectively. They show good agreement with each other, except for a certain loss (approximately -0.7 dB) found in the measured value of $|S_{13}|^2$.

V. Computer Time

A high-speed analysis technique is a premise for the trial-and-error optimum design of a microwave integrated circuit.

In the present case, the required computer time for analyzing the circuit pattern shown in Fig. 4 for a single frequency was about 2.5 s when 10×10 modes were considered in each segment. (The computer used was a HITAC-8800.) However, by decreasing the number of considered modes to a reasonable

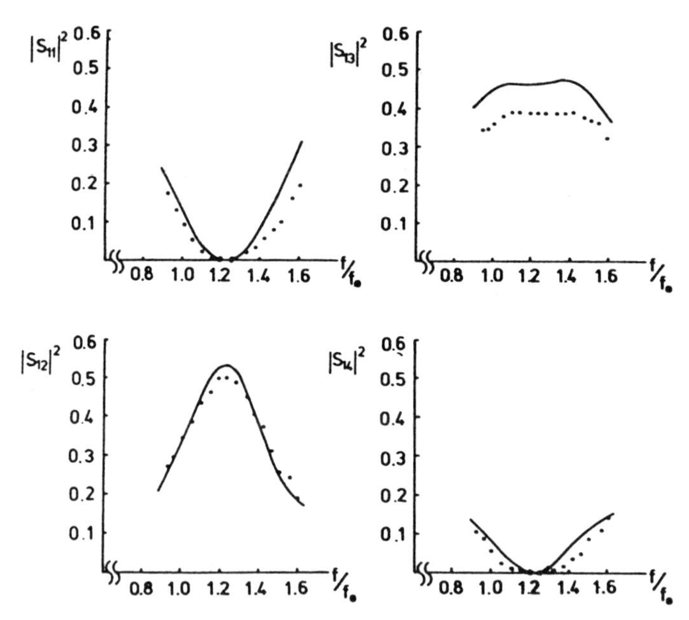

Fig. 10. Comparison of the computed and measured circuit characteristics. (Solid curves: computed; dots: measured.)

one (for example, 8×3), the required time will easily be reduced below 0.7 s. On the other hand, when we used the contour-integral method [1], the estimated time required for the corresponding accuracy was about 5 s.

Generally speaking, comparison of the previous two methods is rather difficult. However, in many cases reduction of the computer time by almost an order of magnitude was possible by the segmentation method without great loss of accuracy as compared with the contour-integral method.

VI. Conclusion

The principle and computation procedure of an efficient method for analyzing a planar circuit, the segmentation method, was presented. To show the usefulness of the proposed method, its application to the trial-and-error optimum design of a hybrid circuit was demonstrated.

The application of the same principle to the short-boundary planar circuit, which has mainly been dealt with by contour-integral analysis [4], will be another interesting task in near future.

Acknowledgment

The authors wish to thank Prof. Y. Okabe of University of Tokyo for his helpful advice, and Prof. J. P. Hsu of Kanagawa University and his colleagues for their assistance in preparing the circuit used in the experiment.

References

[1] T. Okoshi and T. Miyoshi, "The planar circuit—An approach to microwave integrated circuitry," *IEEE Trans. Microwave Theory Tech.*, vol. MTT-20, pp. 245–252, Apr. 1972.

[2] P. Silvester, "Finite element analysis of planar microwave networks," *IEEE Trans. Microwave Theory Tech.*, vol. MTT-21, pp. 104–108, Feb. 1973.

[3] N. Marcuvitz, *Waveguide Handbook.* New York: McGraw-Hill, 1951, p. 160.

[4] T. Okoshi and S. Kitazawa, "Computer analysis of short-boundary planar circuits," *IEEE Trans. Microwave Theory Tech.*, vol. MTT-23, pp. 299–306, Mar. 1975.

Desegmentation Method for Analysis of Two-Dimensional Microwave Circuits

P. C. SHARMA, STUDENT MEMBER, IEEE, AND KULDIP C. GUPTA, SENIOR MEMBER, IEEE

***Abstract*—A new method for the analysis of two-dimensional planar circuits called the "desegmentation" method is proposed. This method is applicable to configurations which can be converted into regular shapes (for which Green's functions are known) by adding one or more regular shaped segments to them. Two examples of planar circuits, chosen such that the results could be verified by the previously known segmentation technique, illustrate the validity of the method.**

I. INTRODUCTION

TWO-DIMENSIONAL planar circuit elements have been proposed [1]–[3] for use in microwave integrated circuits (MIC's) in stripline and microstrip line configurations. The analysis of this type of circuit involves determination of the circuit parameters of its equivalent n-port network as a function of frequency by solving the two-dimensional wave equation subject to the boundary conditions of a magnetic wall around the periphery. When the circuit pattern is of regular shape (i.e., for which Green's function is known as, for example, rectangular, circular, equilateral triangular, etc.), the Green's function technique can be used to solve the problem analytically [1]. Also if the pattern can be divided into segments having regular shapes the segmentation methods [2], [3] can be used for analysis. Sometimes a planar circuit can be extended to a regular shape by adding another segment (or segments) of regular shape to it. Two examples of this kind are shown in Fig. 1. The trapezoidal planar circuit of Fig. 1(a) can be converted into a triangular circuit of Fig. 1(b) by the addition of a triangular segment while the rectangular circuit with a slot (Fig. 1(c)) can be converted into a complete rectangle by the addition of another rectangle as illustrated in Fig. 1(d). In general, in the desegmentation method being proposed here, a regular pattern β is added to a nonregular pattern α (for which Green's function is not available) such that the resulting combination of α- and β-segments is also a regular pattern γ. The characteristics of β- and γ-segments can be computed using Green's function method. The characteristics of the α-circuit can be calculated by the method of "desegmentation" described in this paper.

The desegmentation method can be formulated to evaluate either the S-matrix or Z-matrix for the α-network, in terms of those for the β- and the γ-elements. However, for the segmentation method, it has been shown that the use of Z-matrices is computationally more efficient because Green's function for planar elements yields the Z-matrices directly. Also when this technique is used for analyzing microstrip antennas [4], one is interested in evaluating the equivalent magnetic current or the voltage along the periphery of the antenna. The voltage along the periphery can be calculated by dividing the periphery into a large number of ports and finding the voltage at these ports from the Z-matrix of the multiport network thus formed. Thus the Z-matrix approach is convenient for the antenna analysis also. Consequently, only the Z-matrix formulation is discussed in this paper.

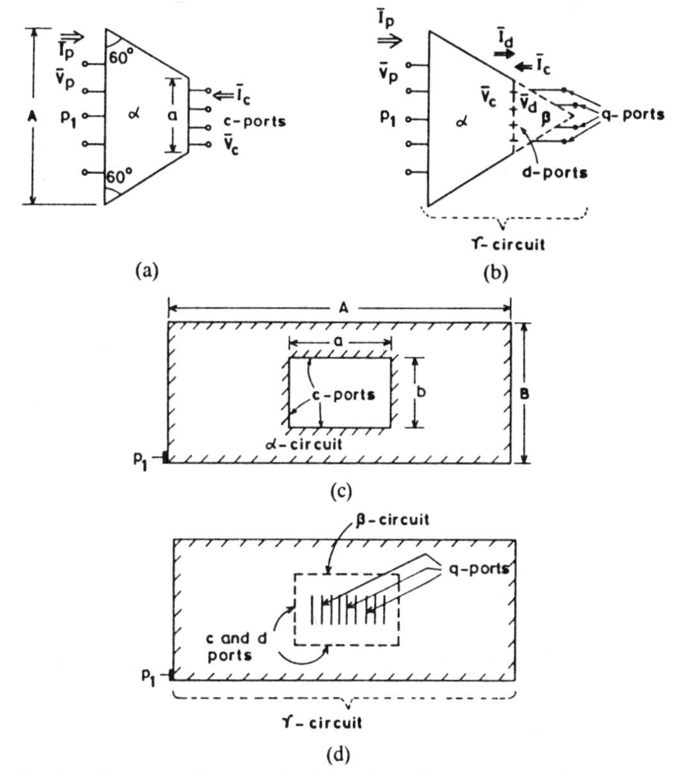

Fig. 1. Examples of planar circuits analyzed by desegmentation method. (a) A trapezoidal planar circuit. (b) Desegmentation method applied to (a). (c) Rectangular ring planar circuit. (d) Desegmentation method applied to (c).

II. THEORETICAL FORMULATION

Consider the circuit shown in Fig. 1(a) and its extension by desegmentation as shown in Fig. 1(b). As in the segmentation method, the continuous interconnection between

Manuscript received January 14, 1981; revised April 10, 1981. Part of this paper has been presented at the June 1981 IEEE MTT-S International Microwave Symposium.

The authors are with the Department of Electrical Engineering, Indian Institute of Technology, Kanpur-208 016, India.

Reprinted from *IEEE Trans. Microwave Theory Tech.*, vol. MTT-29, no. 10, pp. 1094–1098, Oct. 1981.

α- and β-segments is replaced by a discrete number of interconnected ports, named c-ports on α-segment and d-ports on β-segment. Their number $(=C=D)$ depends on the variation of the field along the interconnection and is found by an iterative process in numerical computations. Ports p and q are the external (unconnected) ports of α- and β-segments, respectively.

We define I_p, I_c, I_d, and I_q as the currents, and V_p, V_c, V_d, and V_q as the voltages, at the ports p, c, d, and q, respectively, where $p=1,2,\cdots,P$; $c=1,2,\cdots,C$; $d=1,2,\cdots,D$; and $q=1,2,\cdots,Q$. Since c-ports are connected to respective d-ports, we have $C=D$ and

$$V_c=V_d \qquad I_c=-I_d. \tag{1}$$

The Z-matrices for α, β- and γ-segments, namely $\tilde{Z}_\alpha$, $\tilde{Z}_\beta$, and $\tilde{Z}_\gamma$, respectively, can be partitioned into submatrices corresponding to the external (unconnected) and connected ports as follows:

$$\tilde{Z}_\alpha=\begin{bmatrix}\tilde{Z}_{pp\alpha} & \tilde{Z}_{pc}\\ \tilde{Z}_{cp} & \tilde{Z}_{cc}\end{bmatrix} \tag{2}$$

$$\tilde{Z}_\beta=\begin{bmatrix}\tilde{Z}_{dd} & \tilde{Z}_{dq}\\ \tilde{Z}_{qd} & \tilde{Z}_{qq\beta}\end{bmatrix} \tag{3}$$

$$\tilde{Z}_\gamma=\begin{bmatrix}\tilde{Z}_{pp\gamma} & \tilde{Z}_{pq}\\ \tilde{Z}_{qp} & \tilde{Z}_{qq\gamma}\end{bmatrix}. \tag{4}$$

The third subscript with submatrices in (2)–(4) is used to distinguish the submatrices of the same order in $\tilde{Z}_\alpha$, $\tilde{Z}_\beta$, and $\tilde{Z}_\gamma$. If $\tilde{Z}_\alpha$ and $\tilde{Z}_\beta$ are known, $\tilde{Z}_\gamma$ can be computed using the segmentation method [3]. $\tilde{Z}_\gamma$ thus obtained, by using (1), (2), and (3), is given by

$$\tilde{Z}_\gamma=\begin{bmatrix}\tilde{Z}_{pp\alpha}-\tilde{Z}_{pc}\tilde{Z}'_{dp} & \tilde{Z}_{pc}\tilde{Z}'_{dq}\\ \tilde{Z}_{qd}\tilde{Z}'_{dp} & \tilde{Z}_{qq\beta}-\tilde{Z}_{qd}\tilde{Z}'_{dq}\end{bmatrix} \tag{5}$$

where

$$\tilde{Z}'_{dp}=[\tilde{Z}_{cc}+\tilde{Z}_{dd}]^{-1}\tilde{Z}_{cp}$$

$$\tilde{Z}'_{dq}=[\tilde{Z}_{cc}+\tilde{Z}_{dd}]^{-1}\tilde{Z}_{dq}.$$

In the desegmentation method we express $\tilde{Z}_\alpha$ in terms of $\tilde{Z}_\beta$ and $\tilde{Z}_\gamma$ using (4) and (5). This is equivalent to obtaining $\tilde{Z}_\alpha$ in terms of $\tilde{Z}_\beta$ and $\tilde{Z}_\gamma$ employing the interconnection relations given in (1).

If the currents at ports p and q are considered as known excitations and other variables, namely, the voltages at p-, c-, d-, and q-ports and the currents at c- and d-ports are determined, the impedance matrix $\tilde{Z}_\alpha$ can be evaluated. In this formulation the unknowns, P voltages at p-ports, Q voltages at q-ports, C voltages and C currents at c-ports, and D voltages and D currents at d-ports, are added to $(P+Q+4D)$. The number of equations available is as follows: $(D+Q)$ equations from the definition of $\tilde{Z}_\beta$, $(P+Q)$ equations from the definition of $\tilde{Z}_\gamma$, and $2D$ equations from (1). These add up to only $(P+2Q+3D)$ equations. In order that $\tilde{Z}_\alpha$ obtained be unique, it is necessary and sufficient that

$$(P+2Q+3D)\geqslant(P+Q+4D) \tag{6}$$

which implies $Q\geqslant D$. Therefore, in the desegmentation method the number of q-ports $(=Q)$ on the periphery common to β- and γ-segments should be at least equal to the number of interconnected ports $(=C=D)$. Since β- and γ-segments are regular shapes, the submatrices of $\tilde{Z}_\beta$ and $\tilde{Z}_\gamma$ can be computed using the corresponding Green's functions. Therefore, the left-hand side of (5) is known. Thus for $Q\geqslant D$, $\tilde{Z}_\alpha$ can be computed from (5) as follows. Comparing submatrices of (4) and (5), we have

$$\begin{aligned}\left[\tilde{Z}_{qq\gamma}-\tilde{Z}_{qq\beta}\right]&=-\tilde{Z}_{qd}\tilde{Z}'_{dq}\\ &=-\tilde{Z}_{qd}\left[\tilde{Z}_{cc}+\tilde{Z}_{dd}\right]^{-1}\tilde{Z}_{dq}.\end{aligned} \tag{7}$$

This equation can be solved for $\tilde{Z}_{cc}$ to give

$$\tilde{Z}_{cc}=-\tilde{Z}_{dd}-\tilde{Z}_1 \tag{8}$$

where

$$\tilde{Z}_1=\tilde{Z}_{dq}\tilde{Z}^t_{dq}\left[\tilde{Z}^t_{qd}\left[\tilde{Z}_{qq\gamma}-\tilde{Z}_{qq\beta}\right]\tilde{Z}^t_{dq}\right]^{-1}\tilde{Z}^t_{qd}\tilde{Z}_{qd} \tag{8a}$$

and the superscript t indicates transpose of a matrix. Employing (5) and (8), other submatrices of $\tilde{Z}_\gamma$ can be expressed as

$$\tilde{Z}_{pq}=-\tilde{Z}_{pc}\tilde{Z}_1^{-1}\tilde{Z}_{dq} \tag{9}$$

$$\tilde{Z}_{qp}=-\tilde{Z}_{qd}\tilde{Z}_1^{-1}\tilde{Z}_{cp} \tag{10}$$

$$\tilde{Z}_{pp\gamma}=\tilde{Z}_{pp\alpha}+\tilde{Z}_{pc}\tilde{Z}_1^{-1}\tilde{Z}_{cp}. \tag{11}$$

Equations (9)–(11) can be rearranged to give submatrices of $\tilde{Z}_\alpha$ as

$$\tilde{Z}_{pc}=-\tilde{Z}_{pq}\tilde{Z}^t_{dq}\left[\tilde{Z}_{dq}\tilde{Z}^t_{dq}\right]^{-1}\tilde{Z}_1 \tag{12}$$

$$\tilde{Z}_{cp}=-\tilde{Z}_1\left[\tilde{Z}^t_{qd}\tilde{Z}_{qd}\right]^{-1}\tilde{Z}^t_{qd}\tilde{Z}_{qp} \tag{13}$$

$$\tilde{Z}_{pp\alpha}=\tilde{Z}_{pp\gamma}-\tilde{Z}_{pc}\tilde{Z}_1^{-1}\tilde{Z}_{cp}. \tag{14}$$

The Z-matrix of α-segment, as partitioned in (2), is thus given by (8), (12), (13), and (14). The procedure for computing $\tilde{Z}_\alpha$ from (8) and (12) to (14) is, therefore, to compute $\tilde{Z}_1$, $\tilde{Z}_{cc}$ from (8), $\tilde{Z}_{pc}$ and $\tilde{Z}_{cp}$ from (12) and (13), and finally $\tilde{Z}_{pp\alpha}$ from (14). As mentioned earlier, the expressions for $\tilde{Z}_\alpha$ obtained above hold good for $Q\geqslant D$. The computations get simplified if D can be made equal to Q. In planar circuits this condition can be met, since the number of interconnected ports can always be made greater than the minimum needed for the convergence of the results or Q can be increased by adding additional ports on the segment β. In this case, when $D=Q$, we have

$$\tilde{Z}_1=\tilde{Z}_{dq}\left[\tilde{Z}_{qq\gamma}-\tilde{Z}_{qq\beta}\right]^{-1}\tilde{Z}_{qd} \tag{8b}$$

and $\tilde{Z}_\alpha$ can be expressed as

$$\tilde{Z}_\alpha=\begin{bmatrix}\tilde{Z}_{pp\gamma}-\tilde{Z}_{pq}\tilde{Z}'_{qp} & -\tilde{Z}_{pq}\tilde{Z}'_{qd}\\ -\tilde{Z}_{dq}\tilde{Z}'_{qp} & -\tilde{Z}_{dd}-\tilde{Z}_{dq}\tilde{Z}'_{qd}\end{bmatrix} \tag{15}$$

where

$$\tilde{Z}'_{qp} = \left[\tilde{Z}_{qq\gamma} - \tilde{Z}_{qq\beta}\right]^{-1}\tilde{Z}_{qp}$$

$$\tilde{Z}'_{qd} = \left[\tilde{Z}_{qq\gamma} - \tilde{Z}_{qq\beta}\right]^{-1}\tilde{Z}_{qd}.$$

It can be seen from (15) that it is not necessary to follow any sequence, as is required in the general case, for computing the submatrices of $\tilde{Z}_\alpha$ when $D=Q$.

The $\tilde{Z}_\alpha$, obtained in both the cases discussed above, is the Z-matrix of the α-segment to which several c-ports have been added at the interface between α- and β-segments. Some of these c-ports may be the original c-ports, of the α-segment, for which the Z-parameters are required, and others are the ports added for computations in the desegmentation method outlined above. The Z-matrix pertaining to the original ports only, of α circuit, can be written by discarding the rows and columns corresponding to the c-ports added during computations of $\tilde{Z}_\alpha$ from (15).

One of the methods, of selecting the number of Q-ports, which has been used successfully in several cases is as follows. Consider a special case with $P=1$. Starting with an assumed value of $Q=1$, evaluate $\tilde{Z}_{pp\alpha}$ by using a part of (15) which may be rewritten as

$$\tilde{Z}_{pp\alpha} = \tilde{Z}_{pp\gamma} - \tilde{Z}_{pq}\left[\tilde{Z}_{qq\gamma} - \tilde{Z}_{qq\beta}\right]^{-1}\tilde{Z}_{qp}. \tag{16}$$

It may be noted that for computing $\tilde{Z}_{pp\alpha}$ we do not need to evaluate $\tilde{Z}_{dd}$, $\tilde{Z}_{dq}$, and $\tilde{Z}_{qd}$. Only the evaluations of $\tilde{Z}_\gamma$ and $\tilde{Z}_{qq\beta}$ are required which do not involve the Z-parameters corresponding to c- and d-ports. The value of the number Q is increased iteratively until the value of $\tilde{Z}_{pp\alpha}$, calculated from (16), converges. This gives the value of Q, that is, the minimum number of q-ports needed for computations. The number of c- and d-ports should be at least equal to this value. In the number of examples that have been studied, it is found that this is also the sufficient number of c-ports required for convergence of $\tilde{Z}_\alpha$.

It may be noted that, if $\tilde{Z}_{qq\gamma} = \tilde{Z}_{qq\beta}$, both (8a) and (8b) become indeterminate and the desegmentation method cannot be used. However, such a situation is rare in the case of two-dimensional planar circuits.

III. Examples and Discussion

For illustrating the validity and applications of the desegmentation method, two examples of the analysis of planar circuits are discussed in this section. The circuits chosen are such that the results could be verified by the previously used segmentation method also.

A. Example 1

Consider the trapezoidal planar circuit configuration shown in Fig. 1(a). It is desired to evaluate the input impedance, for this one port circuit, at port p_1. This impedance has been evaluated using both the desegmentation method proposed in this paper and the segmentation method known earlier [3].

For employing the desegmentation method, an equilateral triangle β is added to the trapezoid (α-segment) so that the combination of α and β is also an equilateral triangle γ as shown in Fig. 1(b). The Z-matrices for β- and γ-segments are computed using Green's function [6]. The number of the q-ports is decided by following the procedure outlined in the previous section and is found to be 4. The width of the q-ports is chosen small enough so that the field variation, over a port width, can be assumed to be negligible. As there is no specified port of the initial α-circuit corresponding to the c-ports, (16) yields the value of the input impedance needed. Fig. 2 shows the variation of this input impedance with frequency for the case when $A=3.1711$ cm, $a=0.25$ A, $\epsilon_r=2.55$, and the thickness of the dielectric substrate $d=1.6$ mm.

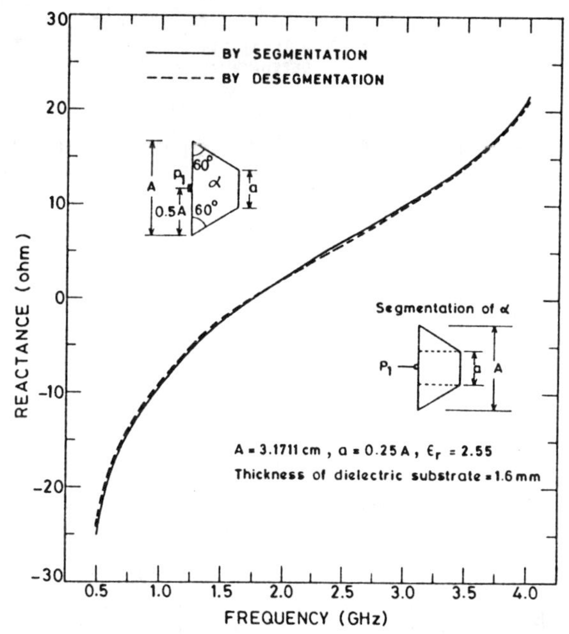

Fig. 2. Variations of input impedance of trapezoidal circuit as calculated by the segmentation and the desegmentation methods.

For the analysis using the segmentation method, the trapezoidal circuit is divided in three segments, i.e., two 30°–60° right triangles and one rectangle as illustrated in Fig. 2 (inset). The Z-matrices for these segments are computed using Green's functions [1], [6]. $\tilde{Z}_{pp\alpha}$ is then computed using the segmentation method for Z-matrices [3]. It is found that the number of interconnecting ports (between the rectangle and two 30°–60° triangles) required for convergence of $\tilde{Z}_{pp\alpha}$ is 16. The results obtained by this method are also plotted in Fig. 2 and agree very well with those obtained by the desegmentation method.

B. Example 2

Another example considered is a rectangular ring type planar circuit shown in Fig. 1(c). In this case also it is required to find out the input impedance at the port p_1 of the α-circuit. In this case β-segment is a rectangular element of length "a" and width "b" which when added to α-pattern of Fig. 1(c) results in a complete rectangular planar circuit γ as shown in Fig. 1(d). The characteristics of the small rectangle β and the outer (filled) rectangle γ, of dimensions $A \times B$, are computed using Green's function [1]. In this case the total periphery of the β-segment is common

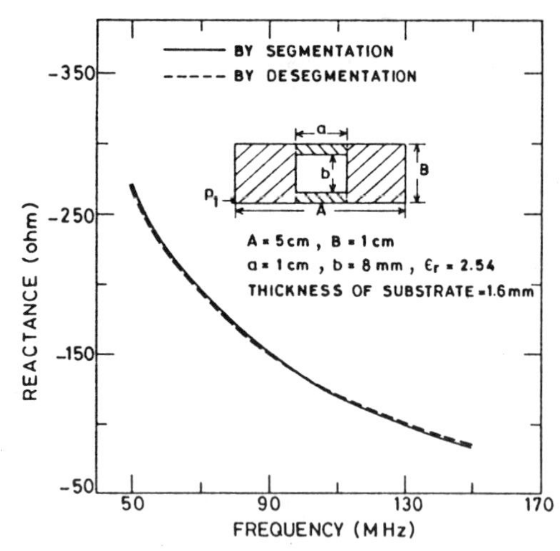

Fig. 3. Variation of input impedance of rectangular ring circuit, as calculated by segmentation and desegmentation methods.

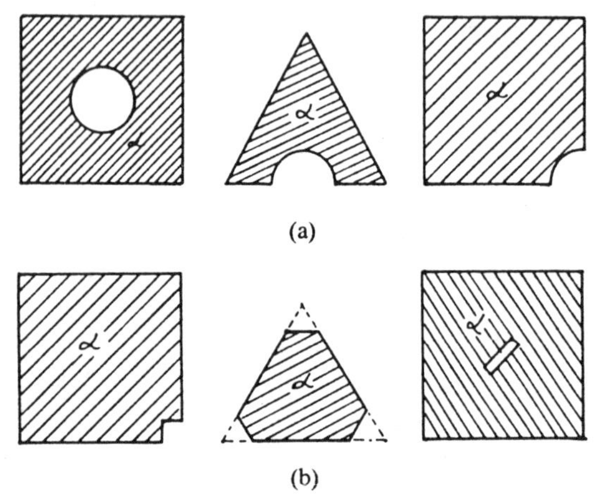

Fig. 4. (a) Examples of circuit patterns which can be analyzed only by desegmentation. (b) Examples of circuits for which desegmentation is more efficient.

with the inner periphery of the α-segment. The ports c and d are located along this interconnection. Thus there is no part of the periphery, of the β-segment, left for locating the q-ports. Since Q (the number of q-ports) cannot be made zero, these ports are located inside the β-segment (i.e., inside the small rectangle). These are the fictitious ports, with port voltages measured between the port locations and the ground plane. As in the case of the ports on periphery, in this case also the voltages are averaged over each port width. The port current is considered to flow in the direction normal to the plane of the paper. The formulation presented in the previous section holds good for such a case also. The minimum number of q-ports is decided by following the same procedure as used in Example 1 and is found to be 9. These q-ports are indicated by vertical bars in Fig. 1(d). The widths of these ports have also been selected iteratively for fast convergence of numerical computations. The values of the input impedance obtained are plotted in Fig. 3, as a function of frequency, for the case when $A=5$ cm, $B=1$ cm, $a=1$ cm, $b=8$ mm, $\epsilon_r=2.54$, and thickness of substrate is 1.6 mm.

This circuit is analyzed using the segmentation method also. The α-circuit is divided into four segments as shown in Fig. 3 (inset). In this example, the number of interconnected ports required, between the various segments, is found to be 4. The results agree very well with those obtained from the desegmentation method as shown in Fig. 3.

The two examples discussed above illustrate the validity and the applications of the desegmentation method proposed in this paper. Of course, the examples chosen are such that the segmentation method is possible, so that comparisons can be made. There are several situations where the segmentation is not possible and the desegmentation method can be used. Some examples of this type are shown in Fig. 4(a). Also there could be situations where the desegmentation method is more efficient. This is likely to happen when the size of the β-segment (needed to convert the α-segment into a regular shape γ) is small compared to the α- and the γ-segments. Some examples of this type are shown in Fig. 4(b).

IV. Concluding Remarks

The desegmentation method is also applicable to transmission line circuits and lumped element circuits. This method can be considered a generalization of the "de-embedding" problem discussed in [7] for eliminating the effects of connectors, etc., from the measured data. In this case, the total system can be treated as the γ-segment whose characteristics are measured. The embedding network (connectors etc.) becomes the β-segment (characterized by previous calibration) and the device or the network under test is the α-segment whose characteristics are to be found.

The desegmentation method, as presented in this paper, extends the applicability of the Green's function technique for analysis of two-dimensional planar circuits. It is expected to find applications in the analysis of microstrip and stripline circuits and also for studying two-dimensional microstrip antenna configurations.

Acknowledgment

Discussions with R. Chadha and G. Kumar are thankfully acknowledged.

References

[1] T. Okoshi and T. Miyoshi, "The planar circuit—an approach to microwave integrated circuitry," *IEEE Trans. Microwave Theory Tech.*, vol. MTT-20, pp. 245–252, Apr. 1972.

[2] T. Okoshi, Y. Uehara, and T. Takeuchi, "The segmentation method—an approach to the analysis of microwave planar circuits," *IEEE Trans. Microwave Theory Tech.*, vol. MTT-24, pp. 662–668, Oct. 1976.

[3] R. Chadha and K. C. Gupta, "Segmentation method using impedance matrices for analysis of planar microwave circuits," *IEEE Trans. Microwave Theory Tech.*, vol. MTT-29, pp. 71–74, Jan. 1981.

[4] P. C. Sharma and K. C. Gupta, "Segmentation and desegmentation techniques for analysis of two-dimensional microstrip antennas," to be presented at *IEEE Int. Conf. APS*, June 1981.

[5] V. A. Monaco and P. Tiberio, "Automatic scattering matrix computation of microwave circuits," *Alta. Freq.*, vol. 39, pp. 59–64, Feb. 1970.

[6] R. Chadha and K. C. Gupta, "Green's functions for triangular segments in planar microwave circuits," *IEEE Trans. Microwave Theory Tech.*, vol. MTT-28, pp. 1139–1143, Oct. 1980.

[7] R. F. Bour and P. Penfield, Jr., "De-embedding and unterminating," *IEEE Trans. Microwave Theory Tech.*, vol. MTT-22, pp. 282–288, Mar. 1974.

Author Index

Subject Index

(Number denotes first page of article.)

W

Y

Editor's Biography

Roberto Sorrentino (M'77–SM'84) received the "Laurea" degree in electronic engineering from "La Sapienza" University of Rome, Rome, Italy.

In 1971 he joined the Department of Electronics of the same university and became an Assistant Professor of Microwaves in 1974. He was also an Associate Professor at the University of Catania, Catania, Italy (1975–1976), and the University of Ancona, Ancona, Italy (1976–1977). From 1977 to 1986 he was with "La Sapienza" University of Rome as an Associate Professor teaching solid-state electronics (1977–1981) and microwave measurements (1981–1986). In 1983 and 1986 he was appointed a Research Fellow at the University of Texas at Austin. Since 1986 he has been a Professor at Tor Vergata University of Rome.

His research activities have been concerned with electromagnetic wave propagation in anisotropic media, interaction of electromagnetic fields with biological tissues, and mainly with the analysis and design of microwave and millimeter wave integrated circuits, including the development of numerical techniques for solving electromagnetic field problems. In particular, he has contributed to the planar-circuit approach for the analysis of microstrip circuits and to numerical techniques for the modeling of discontinuities in planar and quasi-planar configurations.

Dr. Sorrentino was Chairman of the Central and South Italy Section (1984–1987) of the Institute of Electrical and Electronics Engineers (IEEE). He founded the local MTT/AP Joint Chapter and served as Chapter Chairman from 1984 to 1987.

Dr. Sorrentino is a member of the editorial boards of the *IEEE Transactions on Microwave Theory and Techniques*, the *International Journal on Numerical Modelling*, and the *Journal of Electromagnetic Waves and Applications*. He is also a member of the Microwave Field Theory Committee of the IEEE Microwave Theory and Techniques Society and a member of the Management Committee of the European Microwave Conference. He is a Senior Member of the IEEE.